Jean-Pierre Camilleri · Colin L. Berry
Jean-Noël Fiessinger · Jean Bariéty (Eds.)

DISEASES

of the

ARTERIAL WALL

With 333 Figures

Springer-Verlag
London Berlin Heidelberg New York
Paris Tokyo

Jean-Pierre Camilleri
Professeur, Service d'Anatomie Pathologique et Unité de Recherche sur la Pathologie
Rénale et Vasculaire, INSERM U28, Paris, France

Colin L. Berry
Professor, Department of Morbid Anatomy, Institute of Pathology, The London Hospital,
Whitechapel, London E1 1BB, UK

Jean-Noël Fiessinger
Professeur, Service de Médecine Interne et de Pathologie Vasculaire, Hôpital Broussais,
Paris, France

Jean Bariéty
Professeur, Service de Néphrologie et Unité de Recherche sur la Pathologie Rénale et
Vasculaire, INSERM U28, Hôpital Broussais, Paris, France

ISBN-13:978-1-4471-1466-6   e-ISBN-13:978-1-4471-1464-2
DOI: 10.1007/978-1-4471-1464-2

British Library Cataloguing in Publication Data
Diseases of the arterial wall.
   1. Man. Arteries. Walls. Diseases   I. Camilleri, Jean-Pierre. *1936–*   616.1'3
ISBN-13:978-1-4471-1466-6

Library of Congress Cataloging-in-Publication Data
Diseases of the arterial wall/Jean-Pierre Camilleri . . . [et al.] (eds.).
      p.   cm.
   Translated from the French.
   Includes bibliographies and index
   ISBN-13:978-1-4471-1466-6
   1. Arteriosclerosis.   2. Arteries—Diseases.   3. Arteries—Pathophysiology.   I. Camilleri, Jean-Pierre, 1936–
      [DNLM: 1. Arteries.   2. Vascular Diseases.   WG 510 D611]
RC692.D56   1988
616.1'36—dc19
DNLM/DLC
for Library of Congress   88–24887
                                                                      CIP

Originally published in 1987 as *Les Maladies de la Paroi Arterielle* by Flammarion, 4 Rue Casimir-Delavigne, 75006 Paris
French Edition © Flammarion 1987
English Translation © Springer-Verlag Berlin Heidelberg 1989
Softcover reprint of the hardcover 1st edition 1989

2128/3196/543210 (printed on acid-free paper)

Preface to the French Edition

La pathologie vasculaire a été très marquée par les progrès biologiques de ces vingt dernières années. Le système artèriel est maintenant considéré comme un organe à part entière. Modelée au cours de l'organogenèse par les facteurs hémodynamiques, le paroi artérielle maintient une structure hautement organisée et des propriétés mécaniques qui dépendent directement des conditions de pression et de débit. La monocouche endothéliale développe une surface de plusieurs centaines de m² à l'interface sang-tissu; elle est à la fois un organe endocrine complexe synthétisant de nombreuses protéines qui participent à l'hémostase, une surface thromborésistante et hémocompatible, une barrière de perméabilité contrôlant les échanges sang-tissus. Les cellules musculaires lisses constituent un tissu multifonctionnel, contractile, assurant la synthèse des composants structuraux responsables des propriétés mécaniques de la paroi artérielle, la transmission de la force contractile, et une étonnante activité réparatrice en réponse aux agressions. Tout ceci est soumis à un ensemble complexe de communications cellulaires qui font de l'endothélium un véritable système récepteur pour la paroi vasculaire.

Parallèlement, ou à la suite de ces progrès, l'angéiologie s'est progressivement affirmée comme une spécialité clinique. Débordant le cadre de la chirurgie vasculaire, elle intègre les concepts physiopathologiques au diagnostic et au traitement des maladies artérielles. De cet effort d'intégration est né cet ouvrage, cherchant à concilier les connaissances fondamentales es plus récentes et la démarche clinique. L'objectif est de réunir l'ensemble des connaissances récemment acquises sur la physiologie et le physiopathologie de la paroi artérielle, dans la mesure où elles interviennent dans la définition d'une stratégic diagnostique et thérapeutique. Les acrosyndromes ne sont pas individualisés, ceci pour éviter qu'à travers la pathologie de la microcirculation cer ouvrage n'englobe l'ensemble des maladies systémiques: ils sont abordés au chapitre consacré aux vascularites. La plupart des maladies artérielles sont en revanche envisagées. L'athérosclérose est considérée comme modèle de pathologie des communications cellulaires avec les particularités épidémiologiques, hémodynamiques, pharmacologiques propres à chacune de ses localisations. L'angiopathie diabétique est envisagée à part du fait des interrelations complexes entre les perturbations biologiques et la micro-angiopathie. Les artériopathies inflammatoires, malgré les incertitudes pathogéniques qui subsistent, sont abordées du point de vue anatomoclinique. Ceci permet d'individualiser largement l'atteinte des grosses artères (maladie de Horton, Buerger, Takaysau) du cadre plus vaste des vascularites. Les maladies héréditaires du tissu conjonctig ont fréquemment une expression vasculaire: elles sont traitées dans un souci exhaustif mais aussi pragmatique dans les critères diagnostiques. Les aspects anatomocliniques des dysplasies fibromusculaires, des anévrysmes, de l'hypertension artérielle, de la pathologie traumatique et péri-opértoire des artères, des angiodysplasies et des tumeurs vasculaires sont envisagés dans les autres chapitres. Dans chacun de ces domaines, l'analyse des relations structure-fonction nous paraît essentielle pour comprendre les phénomènes pathologiques et pour établir la stratégie diagnostique et thérapeutique.

En concrétisant notre projet nous avons pris conscience du risque de déséquilibre entre une biologie en progrès incessant et une démarche clinique pragmatique et parfois empirique souvent envisagée de façon trop fragmentaire. Nous souhaitons que l'ambition du projet et son originalité nous permettent de bénéficier de l'indulgence de nos lecteurs. D'ores et déjà, nous remercions vivement tous les auteurs francophones et anglophones qui, en acceptant de participer à la rédaction, ont pris le pari de l'angéiologie.

Jean-Pierre Camilleri
Colin L. Berry
Jean-Noël Fiessinger
Jean Bariéty

REMERCIEMENTS

A tous ceux et toutes celles qui ont contribué, par leur compétence, à la réalisation de cet ouvrage.

Nous remercions tout particulièrement Maurice Wolfelsperger qui a réalisé la majeure partie de l'iconographie et nous a aidés à sa mise en pages, de même que Sylviane Robin et Brigitte Jarrin qui ont assuré avec dévouement et compétence les tâches de secrétariat.

Preface to the English Edition

The emphasis on the development of understanding of the arterial tree as an organ system, its modelling by physical forces, the complexity of interactions among its components and the specialist properties of the component cells and their products are emphasized in the Preface to the French edition of this book. These factors and the progress made in clinical diagnostic techniques have allowed progressively better evaluation of structure/function relationships in the system, and assessment of their disturbance in disease.

For these reasons this volume was conceived as a text helpful to those who work at the clinical/laboratory interface so essential for the proper management of vascular disease. The large segment of the population affected by these diseases requires considered care based on a thorough knowledge of the background of the natural history of the various processes affecting the wall and of the way in which investigation is organized.

The generosity of non-English-speaking colleagues in permitting changes in their translated texts has been unstinting. It is hoped that the English text has not altered expressions of view significantly; we have tried only to facilitate expression of the authors' meaning.

London
February 1989

Colin Berry

Contents

II PATHOPHYSIOLOGY OF EXPERIMENTAL AND HUMAN ATHEROSCLEROSIS

Contents

23 Diabetic Retinopathy

24 Diabetic Nephropathy

V INFLAMMATORY DISEASES OF THE ARTERIAL WALL

Contents   xix

28 The Vasculitis Syndromes in Aorta and Large Arteries

Pathological Aspects

Principles of Diagnosis and Therapeutic Approach

Contributors

F. Alhenc-Gelas
Chargé de Recherche, Unité de Recherche sur la Pathologie Vasculaire et L'Endocrinologie
Rénale, INSERM U36, Paris, France

J. P. Assal
Professeur, Policlinique Universitaire et Unité d'Angiologie, Hôpital Cantonal
Universitaire, Genève, Switzerland

J. Bariéty
Professeur, Service de Néphrologie et Unité de Recherche sur la Pathologie Rénale et
Vasculaire, INSERM U28, Hôpital Broussais, Paris, France

C. L. Berry
Professor, Department of Morbid Anatomy, Institute of Pathology, The London Hospital,
Whitechapel, London E1 1BB, UK

P. Birembaut
Professeur, Laboratoire d'Histologie Pol-Bouin, CHU de Reims, France

J. Bonnet
Chef de Clinique Assistant, Hôpital Cardiologique de Bordeaux, Bordeaux, France

H. Bouissou
Professeur, Service d'Anatomie Pathologique, CHU Rangueil, Chemin du Vallon, 31054
Toulouse Cédex, France

Monique Breton
Chargée de Recherche, Laboratoire de Biochimie, Faculté de Médecine Saint-Antoine, 27
rue Chaligny, 75571 Paris Cédex 12, France

P. Bruneval
Assistant, Service d'Anatomie Pathologique, Hôpital Toulouse-Rangueil, Toulouse,
France

G. Burnstock
Professor, Department of Anatomy and Embryology and Centre for Neuroscience,
University College London, Gower Street, London WC1E 6BT, UK

J.-P. Camilleri
Professeur, Service d'Anatomie Pathologique et Unité de Recherche sur la Pathologie
Rénale et Vasculaire, INSERM U28, Paris, France

L. Capron
Chef du Service de Rééducation Vasculaire, Hôpital Broussais, 75674 Paris, Cédex 14,
France

G. Chatelier
Chef de Clinique Assistant, Institut National de la Santé et de la Recherche Médicale,
Unité 36, 17 rue du Fer-à-Moulin, 75005, Paris, France

G. Chomette
Professeur, Service d'Anatomie Pathologique, Hôpital de la Pitié-Salpétrière, Paris, France

J. M. Cormier
Professeur, Service de Chirurgie Vasculaire, Hôpital Saint-Joseph, Paris, France

P. Corvol
Professeur, Unité de Recherche sur la Pathologie Vasculaire et l'Endocrinologie Rénale,
INSERM U36, Paris, France

C. F. Degos
Professeur, Service de Neurologie, Hôpital Saint-Joseph, Paris, France

M. Depairon
Assistant, Clinique Dermatologique, Université De Liège, Liège, Belgium

K. K. Dhital
Research Director, Department of Anatomy and Embryology and Centre for
Neuroscience, University College London, Gower Street, London WC1E 6BT, UK

J. Diebold
Professeur, Service d'Anatomie Pathologique, Hôtel-Dieu, Paris, France

L. O. Drouet
Chef de Travaux Assistant, Paroi Vasculaire et Atherogenese, INSERM U150, Hôpital
St. Louis, 1 av. Claude Vellefaux, F-75010, Paris, France

J. N. Fiessinger
Professeur, Service de Médecine Interne et de Pathologie Vasculaire, Hôpital Broussais,
Paris, France

M. Fievez MD
Institut de Morphologie Pathologique, Gerpinnes, Loverval, Belgium

G. Gabbiani
Professor, Department of Pathology, University of Geneva, 1211 Geneva 4, Switzerland

A. Garner
Professor, Institute of Ophthalmology, London, UK

D. P. Giddens
Professor, Department of Mechanical Engineering, Georgia Institute of Technology,
Atlanta, Georgia 30332, USA

S. Glagov
Professor, Department of Pathology, The University of Chicago, Chicago, Illinois 60637,
USA

I. Hüttner
Professor, Department of Pathology, McGill University, Lyman Duff Medical Science
Building, 3775 University Street, Montreal, Quebec, Canada, H3A 2B4

C. Jacquot
Assistant, Service de Néphrologie, Hôpital Broussais, Paris, France

M. Julian
Chargé de Recherche, Service d'Anatomie Pathologique, CHU Rangueil, Chemin du
Vallon, 31054 Toulouse Cédex, France

M. D. Kazatchkine
Professeur, Unité d'Immunopathologie and INSERM U28, Hôpital Broussais, Paris,
France

O. Kocher
Assistant, Department of Pathology, University of Geneva, 1211 Geneva 4, Switzerland

P. Lagneau
Chef du Service De Chirurgin Vasculaire, Hôpital Saint-Joseph, Paris, France

C. M. Lapière
Professeur, Clinique Dermatologique, Université de Liège, Liège, Belgium

Y. J. Legrand
Directeur de Recherche, Paroi Vasculaire et Atherogenese, INSERM U150, Hôpital
St. Louis, 1 av. Claude Vellefaux, F-75010, Paris, France

J. Levenson
Maître de Recherche, Centre de Diagnostic et U28 (INSERM), 96 rue Didot, Hôpital
Broussais, 75674, Cédex 14, France

J. Ménard
Professeur, Unité de Recherche sur la Pathologie Vasculaire et l'Endocrinologie Rénale,
INSERM U36, Hôpital

J.-B. Michel
Chargé de Recherche, Institut National de la Santé et de la Recherche Médicale, Unité
36, 17 rue Fer-à-Moulin, 75005, Paris, France

R. P. Michel
Professor, Department of Pathology, McGill University, Montreal, PQ, Canada

S. Moore
Professor, Department of Pathology, McGill University, 3775 University Street, Montreal,
Quebec, Canada, H3A 2B4

A. Mossaz
Chef de Clinique Assistant, Policlinique Universitaire et Unité d'Angiologie, Hôpital
Cantonal Universitaire, Genève, Switzerland

J. P. Muh
Professeur, Laboratoire de Biochimie, UER de Médecine, Université de Tours-BP 3223,
37032 Tours Cédex, France

U. E. Nydegger
Professor, Central Laboratory of Haematology, University of Bern, Inselspital, Bern,
Switzerland

J. Picard
Professeur, Laboratoire de Biochimie, Faculté de Médecine Saint-Antoine, 27 rue
Chaligny, 75571, Paris Cédex 12, France

M. T. Pieraggi
Chargé de Recherche, Service d'Anatomie Pathologique, CHU Rangueil, Chemin du
Vallon, 31054 Toulouse Cédex, France

J. L. Richard
Directeur de Recherche, Unité de Recherche d'Epidemiologie Cardiovasculaire, U258
INSERM, Hôpital Broussais, 96 rue Didot 75674, Paris, Cédex 14, France

L. Robert
Directeur de Recherche, Laboratoire de Biochimie du Tissu Conjonctif, CNRS no. 40, Crétail, France

M. Safar
Professeur, Centre de Diagnostic et U28 (INSERM), 96 rue Didot, Hôpital Broussais, 75674, Paris, Cédex 14, France

J. L. Salzmann
Assistant, Service d'Anatomie Pathologique, INSERM U28, Hôpital Broussais, Paris, France

P. Seifert MD
Unité d'Immunopathologie et INSERM U28, Hôpital Broussais, Paris, France

A. Simon
Professeur, Centre de Diagnostic et U28 (INSERM), 96 rue Didot, Hôpital Broussais, 75674, Paris, Cédex 14, France

P. N. Vuong
Chef du Département d'Anatomie Pathologique, Hôpital Saint-Michel, Paris, France

T. N. Wight
Professor, Department of Pathology, School of Medicine, University of Washington, Seattle, Washington 98195, USA

C. K. Zarins
Professor, Department of Surgery, The University of Chicago, Chicago, Illinois 60637, USA

Section I

Cellular and Molecular Biology of the Normal Arterial Wall

Chapter 1

Endothelial and Smooth-Muscle Cells

I. Hüttner, O. Kocher and G. Gabbiani

Structure of the Normal Arterial Wall

Normal muscular and elastic arteries consist of three morphologically distinct layers [20,325] which will be considered in more detail in Chap. 4. The intima consists of a narrow region bounded on the luminal side by a single continuous layer of endothelial cells and peripherally by a fenestrated sheet of elastic fibres, the internal elastic lamina. Between these boundaries are smooth-muscle cells and various components of extracellular connective-tissue matrix. With increasing age in man, intimal smooth-muscle cells and extracellular matrix components accumulate slowly and generally at a uniform rate, except in certain areas where nodular accumulations called "cushions" develop (see Chap 3). Why so few smooth-muscle cells are initially present in the intima, and why the number of these cells normally increases with age, are important questions that remain to be answered.

The media, or middle layer of the muscular artery, consists entirely of diagonally oriented smooth-muscle cells, surrounded by variable amounts of collagen, small elastic fibres and proteoglycans (mucopolysaccharides). These smooth-muscle cells form helices through the vessel wall and in most instances are coupled to one another by gap junctions (nexuses). It is important to note that no cells equivalent to skin fibroblasts are present in the media of the mammalian arteries and that the morphology of the media, in contrast to that of the intima, does not generally alter with age. The media of the elastic artery is characterized by units in which several layers of smooth-muscle cells are bounded on each side by fenestrated sheets of elastic tissue.

The adventitia, or outermost layer of the artery, consists principally of recognizable fibroblasts intermixed with smooth-muscle cells loosely arranged between bundles of collagen and surrounded by proteoglycans. It is usually separated from the media by a discontinuous sheet of elastic tissue, the external elastic lamina.

These three morphologically distinct layers are recognizable also in various segments of the microcirculation, although elastic laminae disappear at the level of arterioles, and smooth-muscle cells are replaced by pericytes in capillaries [198,199,231,298,299,300].

In the following pages we will describe briefly the morphology and function of the two principal cellular components of the arterial wall, the endothelial cell and the smooth-muscle cell.

The Endothelial Cell

Endothelial cells line the insides of arteries, veins, capillaries and lymphatics as a monolayer. These cells have important roles in physiological haemostasis, in the permeability of blood vessels and in the mediation of their response to a variety of physiological and pathological stimuli [144,191,232]. Abnormalities in the structure and function of endothelial cells may play a significant role in diseases of blood vessel walls, particularly thrombosis and atherosclerosis [180,308,331,373,380].

4

Biological Behaviour and Morphology of Endothelial Cells

In culture, endothelial cells are different from most other cells. Growth of endothelial cells, unlike that of fibroblast or smooth-muscle cells, is characterized by the formation of a highly ordered monolayer [393]. This monolayer adopts a morphological appearance and differentiated properties similar to those of the vascular endothelium in vivo. The closely apposed and non-dividing cells of the monolayer have a distinctive membrane asymmetry; they have a non-thrombogenic luminal surface and can no longer internalize bound ligands such as low-density lipoprotein, while fibronectin disappears from the luminal surface and concomitantly accumulates close to the basal surface [258,393]. Once the cells have formed this highly ordered structure, the only agents shown to be able to stimulate growth are those that disrupt the continuity of the monolayer. This is true both in vitro and in vivo [330] and suggests that contact inhibition of growth may be particularly important for endothelial cells when compared with other cells. There is evidence for the appearance of a unique cell-surface protein, CSP-60, in quiescent endothelial cells [393], although there is no evidence that CSP-60 controls growth.

The biological behaviour of the endothelial monolayer is reflected by the extremely low basal rate of replication in normal adult arterial endothelium in vivo [327,328]. It has been demonstrated, however, that there are focal areas of endothelium with increased replication, and that the overall replication of aortic endothelium is increased in response to hypertension [82,328] or endotoxaemia [293,294]; we will return to the significance of this phenomenon in Chap. 11 of this book.

Normal quiescent arterial endothelium consists of a layer of fairly uniform flattened spindle-shaped cells oriented in vivo with their long axis in the direction of blood flow (Fig. 1.1). Endothelial cells are invariably continuous in arteries in contrast with the fenestrated endothelial cells occurring in some visceral capillaries [64,231]. Some studies suggest that the arrangement of endothelial cells in arteries is under the influence of the extracellular matrix. However, blood flow is also clearly a factor since endothelial cells will return to a position parallel to flow when a segment of a vessel is rotated 90° [103]. In fact, flow alone appears to be all that is necessary to induce the normal orientation and elongated shape of endothelial cells suggesting the orien-

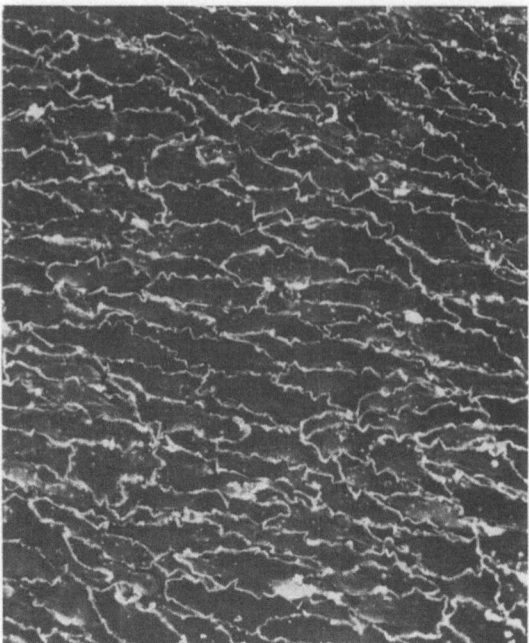

Fig. 1.1. Endothelium from normal rat aorta viewed "en face" in the scanning electron microscope. The flat, polygonal endothelial cells, aligned along the long axis of the vessel, are outlined by silver nitrate. (From Hüttner and Gabbiani [181], with the permission of the publisher) × 600

tation of the extracellular matrix may result indirectly from the direction of the blood flow.

The cytoskeleton plays an important role in maintaining the shape of endothelial cells. When endothelial cells are grown in tissue culture they lose their elongated spindle shape giving a cobble-stone appearance to the monolayer. Concomitant with the change in cell shape is a dramatic change in the arrangement and distribution of the different components of the cytoskeleton [305]. While the intermediate filament network remains relatively unchanged when endothelial cells are grown in culture [111,305], the microtubule network, which does not appear to focus on a single microtubule-organizing centre in the cytoplasm of endothelial cells in vivo, becomes highly concentrated in the perinuclear area, and individual microtubules run radially from this region to the cell periphery in vitro [152,305].

The most important cytoskeletal difference between in vivo and in vitro endothelial cells is seen in the distribution of microfilaments. While normal endothelial cells in vivo contain a small number of microfilament bundles (stress fibres) running the full length of endothelial cells oriented with blood flow, in cultured endothelial cells there are a large number of stress fibres running both circumferentially along the cell boundary and across its width [305]. These changes in the cytoskeleton may be due either to the lack of proper subendothelial matrix components or the lack of blood flow or both under culture conditions.

Interactions between adjacent endothelial cells and their attachment to the basal lamina are of crucial importance for the proper functioning of vascular endothelium as a continuous monolayer. There is now evidence that microfilaments play a significant role in these interactions. In normal in vivo endothelial cells, cytoplasmic actin microfilaments are present in three well-defined structural organizations: (a) a peripheral network or web, most prominent in endothelial cell flaps adjacent to intercellular clefts; (b) subplasmalemmal microfilament condensations that occur in connection with the abluminal plasma membrane and also in connection with the lateral endothelial plasma membrane at the level of tight junctions; and (c) stress fibres that are connected to subplasmalemmal microfilament condensations exclusively at the abluminal aspect of endothelial cells [184]. The presence of subplasmalemmal microfilament condensations at the lateral endothelial cell membrane in relation to tight junctions underlines the role of microfilaments in maintaining tight junction structure between adjacent endothelial cells [238,289,340]. However, subplasmalemmal microfilament condensations at the abluminal aspect of endothelial cells are also related to attachment of endothelial cells to the basal lamina.

These cytoplasmic structures are associated with a coincident distribution of densely stained and closely apposed extracellular matrix deposits in the subendothelial connective tissue compartment [184]. Such deposits acquire a fibrous appearance and become easily recognizable in connection with formation of stress fibres in endothelial cells [184; see this topic also in Chap. 11]. Using immunofluorescence and immunoelectron microscopy, similar extracellular matrix deposits have been shown in other in vitro and in vivo systems to consist of fibronectin: the close association between cytoplasmic stress fibres and extracellular fibronectin containing fibres has been named fibronexus [186,353,356].

Stress fibres are bundles of actin filaments containing myosin and alpha-actinin that are found in many cultured cells, but they are present only in vascular endothelial cells in living organisms; their function is probably related to isometric contraction and adhesion [41,45,415]. They are relatively rare in normal vascular endothelium in vivo [125], but are more prominent in endothelial cells of regions exposed to high velocity flow, such as the left ventricle, aortic valve and aorta [410,415]. Even in one selected segment, the aorta, they occur mainly in endothelial cells located immediately below intercostal artery branches [125]. However, stress fibres in vascular endothelial cells in vivo become numerous during hypertension [122,124,410,411] and they develop early and become common in all aortic endothelial cells during the regeneration that follows experimental endothelial denudation [125,184; see also Chap. 11 of this book]. This implies that stress fibres play a role in increased cellular adhesion to the subendothelial matrix during increased haemodynamic stress (as in branching regions of normal arteries and during hypertension in general) or possibly when the composition of subendothelial matrix is altered, say during endothelial regeneration in vivo or in culture conditions.

The total surface density of membrane domains covered by subplasmalemmal microfilament condensations does not change in regeneration when compared with normal endothelium suggesting that endothelial cells are normally equipped with structures to which stress fibres can be connected when formed under stressful conditions [184]. Subplasmalemmal microfilament condensations at the abluminal endothelial plasma membrane are similar to the peripheral dense bands in smooth-muscle cells implicated in force transmission [129]. It is interesting to note that membrane domains specialized for stress fibre attachment are different from the rest of the endothelial plasma membrane; they are resistant to formation of filipin–sterol complexes suggesting that they are probably too rigid to be easily deformed [184]. Furthermore, intracytoplasmic actin and actin-associated proteins, particularly vinculin and alpha–actinin, as well as concomitant extracellular fibronectin, have been localized at cell to matrix attachment sites in various in vitro and in vivo systems [72,73,134,135,136,186, 213,354]. Recently an integral membrane protein associated with sites of microfilament–membrane attachment has been identified [304]. These data suggest complex but specific interactions between cytoplasmic stress fibres and extracellular fibronectin fibres across the abluminal endothelial plasma membrane at the site of attachment of endothelial cells to the basal lamina.

While specific domains of the abluminal endothelial plasma membrane are important in maintaining firm attachment of the endothelial lining to the inside of arteries, membrane specializations at the lateral endothelial cell membrane are essential to establish interactions between adjacent endothelial cells and to maintain proper functioning of the vascular endothelium as a continuous monolayer.

Adjacent endothelial cells in arteries are interconnected by both tight junctions and gap junctions [176,177,183,345,346] (Figs. 1.2, 1.3 and 1.4). The function of these two types of cell junction is entirely different. Whereas tight junctions form local permeability seals between cells and are intimately related to the barrier function of the vascular endothelium, gap junctions are low-resistance pathways for intercellular communication; they are implicated in cell-to-cell transfer of ions (ionic or electronic coupling) and in cell-to-cell transfer of cellular metabolites (metabolic coupling) [114,142,279,297,369]. The

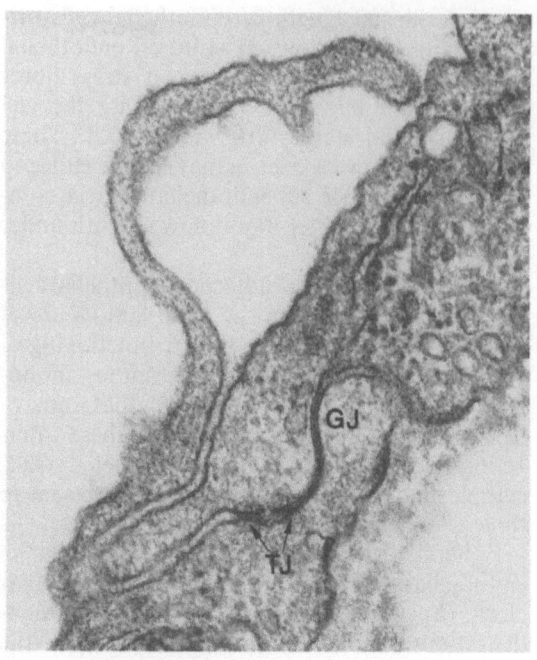

Fig. 1.3. Adjacent endothelial cells are interconnected both by tight junctions (*TJ*) and gap junctions (*GJ*). (From Hüttner and Gabbiani [181] with the permission of the publisher) × 112 000

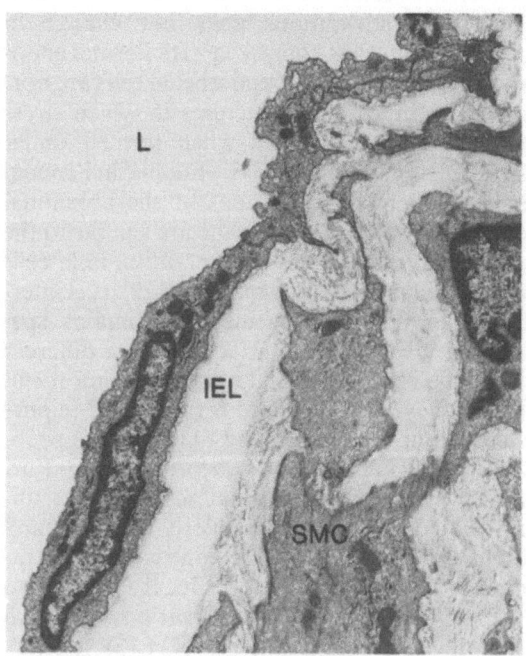

Fig. 1.2. Endothelium from normal rat aorta as seen on thin section electron microscopy. The endothelial cells forming a continuous monolayer line the inside of the vessel. *L*, lumen: *IEL*, internal elastic lamina; *SMC*, medial smooth-muscle cell. Tissues for all thin-section electron micrographs were treated with uranyl actetate en bloc and stained in section with uranyl acetate and lead citrate. (From Hüttner and Gabbiani [181] with the permission of the publisher) × 15 000

presence of gap junctions in arterial endothelium provides a structural basis for a synchronized function between adjacent endothelial cells [176].

In the region of a tight junction, the plasma membranes of two adjacent epithelial cells appear to fuse, providing a region of intimate contact that completely encircles each cell [101,369] and contributes to maintaining asymmetric polarity of individual epithelial cells [95]. The two-dimensional structural organization of the membranes in the region of the tight junction can be deduced from freeze-fracture studies [27,28,288,357]. In freeze-fracture replicas the tight junction is characterized by a network of ridges on the P face and complementary grooves on the E face of epithelial plasma membranes (Fig. 1.4).

According to classical views, the tight junction ridges are composed of two rows of tightly packed intramembrane particles (integral membrane proteins), with one row contributed by each of the adjacent plasma membranes. These rows of particles make head-to-head contact in the form of a modified zipper, holding the two membranes so close together that the inter-

201,289]. However, little information is available regarding the biochemical composition of the tight junction.

Very recently a high-molecular-weight (225 000) polypeptide associated with tight junctions has been identified in a variety of epithelia and termed "ZO-1". It is a potentially ubiquitous component of all mammalian tight junctions [374]. The lipidic nature of tight junction strands is consistent with experimental data indicating that tight junctions can rapidly adjust their organization in response to specific and non-specific stimuli. Rapid massive assembly of tight junctions has been observed in the absence of protein synthesis in vitro [200,376]. In addition, proliferation of cell junctions has been observed in various experimental conditions in vivo [182,184]. The dynamic structure of tight junctions is also reflected by Ca^{2+}-dependent disassembly and reassembly of tight junctional components [244,262] as well as charge-related alterations of junctional permeability in various in vitro and in vivo systems [264,266]. We have already noted the role of cytoskeleton in maintaining tight junctional structure [238,289,340]. This role is also supported by data on the regulation of epithelial tight junction permeability by cyclic adenosine monophosphate (cAMP).

In spite of their dynamic structure, there is good correlation between the structural complexity of tight junctions and junctional permeability [63,89,114,320,351]. Tight junctions vary from one tissue to another in the number of sealing strands they possess. By measuring transepithelial resistance, tissues in which the tight junction networks incorporate only one or two sealing strands (as in the proximal tubule of the kidney) offer little resistance to the passage of ions and may be defined as "leaky" epithelia. However, junctions incorporating six or more sealing strands (as in the distal tubule of the kidney or in the urinary bladder) possess a high electrical resistance and allow the formation of steep concentration gradients across the epithelium. These latter may be defined as "tight" epithelia [63,89,369]. Thus, the organization of the sealing strands into a network makes it possible for the tightness of the junction to be varied according to the physiological needs of the tissue.

In vascular endothelium, the complexity of tight junctions varies also from one vascular segment to another. This correlates well with the permeability characteristics of the endothelial cell layer in these vascular segments. The presence of single-stranded and widely discontinuous

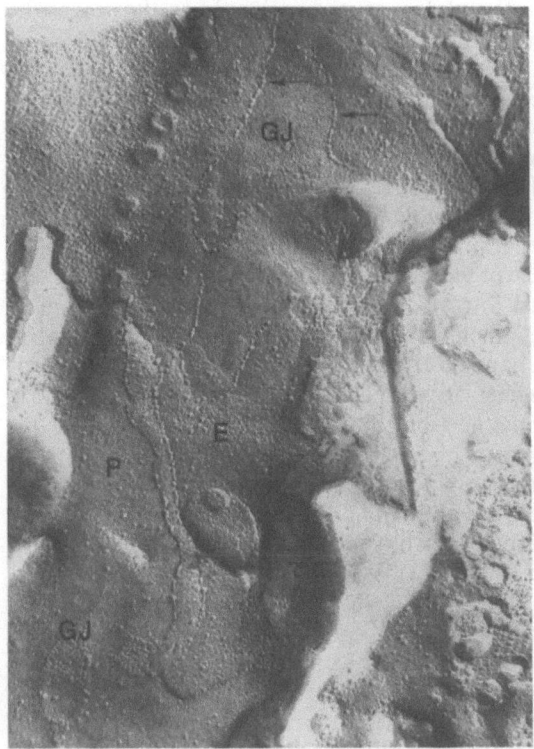

Fig. 1.4. Junctional complex in the lateral plasma membrane of adjacent endothelial cells from normal rat aorta visualized by freeze-fracture electron microscopy. The complex consists of a double-stranded tight junction associated with multiple gap junctions. The fracture plane exposes the E face (*E*) of the lateral membrane of one endothelial cell, then breaks in the lateral membrane of the other endothelial cell along the line of a junctional strand exposing the P face (*P*). On the E face, tight junctional strands (*arrows*) are seen as grooves marked along the midline by rows of intramembrane particles which partly surround gap junctions (*GJ*), seen here as a regular lattice pattern of pits. On the P face, tight junctional strands are seen as ridges marked by loosely packed rows of particles while gap junctions (*GJ*) appear as densely packed disc-shaped arrays of particles. (From Hüttner et al. [182] with the permission of the publisher) × 79 000

cellular space is obliterated. As a result, intramembrane particles form sealing strands or fibrils that physically block the passage of molecules across the epithelium [369]. Variations of this model propose that tight junctions are composed of single fibrils [397] or of two mutually offset fibrils [33] rather than two superimposed fibrils. A recent hypothesis suggests that tight junctional strands or fibrils correspond to intramembranous cylindrical inverted micelles, while integral membrane proteins could play a role in the maintenance of the structure, in particular through association with cytoskeletal components at the cytoplasmic surface [200, tight junctions in the endothelium of venous capillaries and of post-capillary venules correlates with leakiness of the endothelium particularly evidenced by preferential opening of interendothelial routes in these vascular segments for protein-rich exudate and cells during inflammation [233,234,345,351]. This contrasts with the multistranded and continuous belts of tight junctions present in the endothelium of the arterial ends of capillaries and of arterioles [345]. An extreme example of "tight endothelium" is that found in cerebral capillaries; the presence of complex multistranded tight junctions in cerebral capillary endothelium, in combination with the absence of leaky post-capillary venules in the cerebral circulation, is a major structural component of the blood–brain barrier to macromolecules [26,102,265].

The complexity of tight junctions in the endothelium of large arteries such as the aorta lies between "leaky" venous-type and "tight" capillary-type tight junctions, being characterized by mostly continuous single- or double-stranded junctional belts [177,178,182,183,184,346]. However, there is considerable heterogeneity from cell to cell in tight junctional structure even within the same aortic segment; while most adjacent endothelial cells are interconnected by continuous zonular tight junctions, discontinuous macular tight junctions also occur focally in normal aortic endothelium [182,183]. Thus, large-vessel endothelium with its "tight" and "leaky" regions seems to mirror the structure of the microvasculature with its juxtaposed "tight" and "leaky" segments.

In the region of a gap junction (also called a communicative junction or nexus), thin sections reveal that adjacent cell membranes are very closely apposed; the intercellular space appears to be narrowed from its normal width of about 25 nm to a width of about 2–3 nm. Heavy-metal tracers such as colloidal lanthanum permeate the narrow intercellular space, and on tangential section outline a hexagonal array of cylindrical structures into which the stain has not penetrated. The stain also fills a central hole in these cylinders, suggesting that it is the opening of a narrow channel [176]. In freeze-fracture replicas, the adjacent plasma membranes in the region of a gap junction contain a disc-shaped array of closely-spaced intramembrane particles, and it can be shown that the particles of the two arrays are aligned within the intercellular space [183,369]. Thus the gap-junction particles appear to form intercellular pipes or channels that bridge both the adjacent membranes and the intercellular space, thereby allowing the exchange of molecules between cells [369].

Biochemical studies on isolated gap junctions suggested that they contained one major protein, connexin, which has a molecular weight of 18 000 and consists of two chains of amino acid units [150]. Electron microscopic and x-ray crystallographic techniques furnished evidence furthermore that the gap junction particles are made up of six dumbell-shaped subunits that aggregate to form a cylindrical structure about 7 nm in diameter with a sixfold symmetry and a central channel [47,237]. These integral membrane proteins, organized in the form of tubes, provide channels of communication between the cytoplasm of adjacent cells. On the basis of more recent biochemical studies it is now generally agreed that the constituent protein of the gap junction is the polypeptide with molecular weight 27 000 [16,315]. Each channel is thought to be a dodecamer of this protein, a hexamer forming a hemichannel in each of the joined membranes.

In electrically excitable tissues, gap junctions serve as electrotonic synapses which allow the electrical activity of one cell to be transmitted to an adjacent cell without the need for mediation by a neurotransmitter, or messenger substance [278,279]. In other tissues the main role of gap junctions is probably metabolic coupling [123,142,176,228]. Molecules with a molecular weight of up to 1200 and with an upper size limit of 12 to 14 Å can readily pass from one cell to another through cell-to-cell channels of the gap junctions [352,392]. Ions, most sugars, amino acids, nucleotides, vitamins and "messenger" molecules such as steroid hormones and cAMP fall into this range. Cells connected by gap junctions can therefore draw from intercellular pools of such substances. Gap junction differentiation is hormonally regulated [81]; rapid induction of gap junction formation in vitro by chemical agents has also been reported [203]. The permeability of junctional channels is sensitive to the cytoplasmic free Ca^{2+} concentration [228,306,315].

In vascular endothelium, the role of gap junctions between adjacent endothelial cells is largely unknown; their main function is probably related to metabolic cooperation as in other non-muscle tissues. A possible role for intercellular communication via gap junctions is growth control in vascular endothelium; this may complement the proven humoral interactions which occur via specific growth factors. This is evidenced by the initial loss followed by neoformation of gap junctions during endothelial regeneration [184,365] that may indicate a relationship between presence of gap junctions and contact inhibition in vascular endothelium [332,333]. The presence of abundant gap junctions between arterial and venous (but not capillary) endothelial cells [176,183,345,346] suggests that gap junctions may have a role in transmitting some signal to underlying smooth-muscle cells. Coupled endothelial cells in arteries may act as a unit and this interaction, along with that of myoendothelial contacts, may help explain the ability of an intravascular hormone to affect vascular tone without gaining direct access to all cells of the vascular wall [77,84, 121,176,221,298,301,343]. The potential role of endothelium as a receptor system for smooth-muscle cells, whereby the smooth muscle responds to secondary signals generated in the endothelium and transferred intracellularly via gap junctions, was recently supported by identification of gap junctions at myoendothelial contacts in vivo [366,377], and the demonstration of rapid bidirectional transfer of nucleotides between cocultured endothelial and smooth-muscle cells in vitro [77,78]. Immunoreactivity between the major gap-junction polypeptides (molecular weight approximately 27 000) prepared from different tissues is also consistent with the possibility of gap-junction-mediated intercellular communication between heterologous cells such as endothelial and smooth-muscle cells.

The cytoplasm of individual endothelial cells, in addition to usual intracellular organelles such as endoplasmic reticulum, Golgi apparatus and mitochondria located mostly in the paranuclear region, contains Weibel–Palade bodies, unique to endothelial cells (Fig. 1.5). These structures are membrane-bound rod-shaped vesicles of 0.1×2–$3\,\mu m$ size that contain regularly spaced tubular components aligned parallel to the longitudinal axis. Recent immunoelectron microscopic investigation suggests that the Weibel–Palade bodies of endothelial cells are storage and/or processing organelles for factor VIII-related antigen/von Willebrand protein [400,407]. Von Willebrand protein is a large glycoprotein synthesized by megakaryocytes [261; see also Chap. 8]. Various processing steps in the biosynthesis of von Willebrand protein have now been identified [189,190,398,399].

A prominent feature of endothelial cells is that a relatively large fraction of the body of each endothelial cell is occupied by plasmalemmal vesicles of $\sim 70\,nm$ outer diameter (range: ~ 50 to $90\,nm$), limited by a usual unit membrane

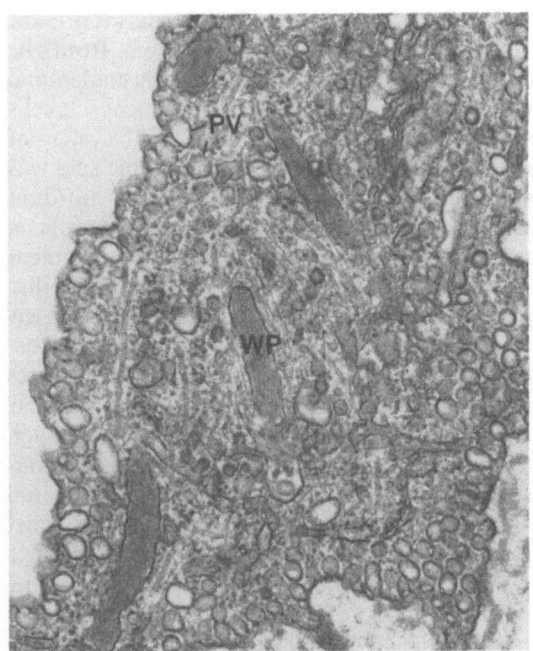

Fig. 1.5. The cytoplasm of individual endothelial cells is characterized by a large population of plasmalemmal vesicles (*PV*) and the presence of Weibel–Palade bodies (*WP*). × 76 000

(Fig. 1.5). The vesicles appear either as isolated single units in the cytoplasm, or as chains of fused vesicles, or as vesicles opened on either the blood or the tissue front of the endothelial cells [31,276,277,344]. In the latter case, the limiting membrane of the vesicles is in continuity with the plasmalemma, a detail which explains their name. On the basis of studies with macromolecular tracers it has been suggested that plasmalemmal vesicles ferry small amounts of material between blood and pericapillary space. This picture implies that the vesicles go through a cycle. First of all, they are temporarily attached to the cell surface, where they appear as invaginations of the plasmalemma. In this state the vesicular content equilibrates with the extracellular fluid. Subsequently, vesicles pinch off from the surface and move through the cytoplasm to the opposite cell front where the vesicular membrane fuses with the plasmalemma and the content is discharged to the cell exterior. This process of "vesicular transport" is assumed to be bidirectional, explaining, in particular, macromolecular transport in capillaries [32, 277,349]. Recent observations, however, necessitate a revised view of endothelial vesicles.

Electron microscopic studies of frog mesenteric capillaries with tannic acid indicated that endothelial plasmalemmal vesicles may represent a system of branching invaginations from the cell surface similar to those in plasmalemmal vesicles found in pericytes or smooth-muscle cells and in a manner reminiscent of the T-system of striated muscle fibres [36]. When tannic acid was added to an already fixed tissue, this mordant reached apparently free vesicles in the cytoplasm under conditions in which vesicular movement was excluded and in which the impermeability of cell membrane was preserved. These observations imply that most of the endothelial vesicles are permanent or semipermanent structures which may be labelled by macromolecular tracers through diffusion rather than through transport [36,180]. Three-dimensional reconstructions of plasmalemmal vesicles from serial sections of both frog mesenteric and rat heart capillaries further substantiated these findings by showing that less than 1%, if any, of the plasmalemmal vesicles are free and isolated in the endothelial cytoplasm [37,116]. This view of the plasmalemmal vesicles as parts of cell-surface invaginations implies that a search for alternative functions to macromolecular transport for these vesicles is necessary [75]. We will discuss the potential but questionable role of endothelial vesicles in macromolecular transport later in this chapter.

There is general agreement on the function of the coated vesicles, found in a significantly lower density than the plasmalemmal vesicles, in the endothelial cytoplasm. With the exception of the endothelial cells of hepatic sinusoids, the density of coated vesicles in vascular endothelial cells is lower, by at least three orders of magnitude, than that of the usual plasmalemmal vesicles [277]. Coated vesicles are slightly larger in diameter than regular plasmalemmal vesicles and are provided with a "coat", reminiscent in three dimensions of a geodetic cage. The coat is associated with the cytoplasmic aspect of the vesicular membrane and is formed by the polymerization of a single large-molecular-weight protein, clathrin [219,282], which is known to bind actin and alpha-actinin. In vitro studies on a number of cell types indicate that coated vesicles are directly involved in the binding and the subsequent endocytosis of macromolecules such as low-density lipoproteins and growth factors by high-affinity receptors [2,148,245]. Studies of high-affinity receptor binding and internalization of ferritin-labelled ligands indicate association of the ligand with coated regions of the cell surface that invaginate (coated pits) and pinch off to form coated vesicles.

Functional Properties of Vascular Endothelium

The functional properties of the vascular endothelium as a continuous cell layer fall into three major fields [380]: (a) the normal endothelium is metabolically active and highly versatile. Endothelial cells synthesize prostacyclin (PGI_2) and an activator of plasminogen, they produce factor VIII antigen and von Willebrand factor, they contribute several components to the subendothelial connective tissue and they contain receptors to a variety of vasoactive agents [341]; (b) the normal endothelial cell layer provides a thromboresistant surface that prevents platelet or leucocyte adherence and activation of intrinsic coagulation systems [313,373]; (c) normal endothelium forms a barrier to the passage of blood constituents into the artery wall; this is essential to maintain the normal microenvironment and growth pattern of underlying smooth-muscle cells in large arteries [308].

Vascular Endothelium as a Metabolically Active, Secretory Tissue

The vascular endothelium is the most extensive simple endothelium in the body. This tissue has a surface area of several hundred square metres and has a total mass comparable to other vital organs of the body. The organization of endothelium as a broad monolayer amplifies its impact as a metabolically active, secretory tissue. Various blood-borne substrates are efficiently delivered to ectoenzymes present on the luminal endothelial surface, and the resulting metabolites can diffuse into surrounding tissues or be carried by the circulation to distant sites. Substances synthesized and secreted by endothelial cells similarly can have local or systemic effects. In addition, all materials entering or leaving the blood-vessel wall are potentially susceptible to biochemical alteration by the endothelium [232,341].

Our current understanding of vascular endothelium as a metabolically active and highly versatile tissue is related to recent advances in endothelial culture techniques including the development of long-term monolayer cultures of both large vessel [144,191] and capillary endothelial cells [105,419]. Culturing endothelial cells on microcarrier beads [44] as well as culturing cloned endothelial cells [187] promise to provide further information on interactions between endothelial cells and circulating ligands and/or cells.

The synthesis of PGI_2 is an example of the biochemical versatility of endothelial cells. PGI_2 is the most potent naturally occurring inhibitor of platelet aggregation known. The presence of thromboxane A_2 (TXA_2) in the platelets with strong pro-aggregating and vessel constricting properties, and of PGI_2 in the endothelial cells with opposite properties on both counts provides a powerful regulatory system; by synthesizing PGI_2, the vascular endothelium tends to counteract any unnecessary or excessive platelet activation [251,253]. Endothelial cells also produce von Willebrand factor (factor VIII-VWF), a cofactor involved in the adherence of platelets to subendothelial connective tissues [189,190].

Synthesis by cultured endothelial cells of thrombospondin [256,291], a major platelet alpha-granule glycoprotein, and synthesis of proteins structurally and immunologically related to platelet membrane glycoproteins has also been demonstrated recently [269]. It has been suggested that secretion of these proteins (also present in circulating blood platelets) by appropriately stimulated endothelial cells into local microenvironments in vivo may support platelet interaction and/or interaction of endothelial cells with the underlying extracellular matrix. In addition, endothelial cells exhibit an ecto-ADPase activity, which can convert pro-aggregating adenosine diphosphate released from platelets to adenosine, a potent platelet inhibitor, and also contribute to the regulation of local vascular tone [283,284]. Endothelial cells also synthesize plasminogen activator(s) and inhibitor(s) and thus may be an important source of fibrinolytic activity.

Endothelial cells actively interiorize and/or degrade catecholamines, serotonin, bradykinin and other vasoactive amines which are systemic regulators of vascular tone and also mediators of platelet aggregation [341,380]. Angiotensin-converting enzyme localized in vascular endothelium is another example of how endothelial cells participate in the regulation of vascular tone and blood pressure [196,314].

Endothelial cells are capable of synthesizing most of the extracellular matrix components of the arterial wall, including basement membrane, type IV collagen, type III collagen, elastin, glycosaminoglycans and fibronectin, all essential for the anchorage of the endothelial monolayer in the face of constant haemodynamic stress [19,210,230,255]. The synthesis of specific matrix components by the endothelial cells is temporally ordered and may play a significant role in the differentiation of endothelial cells during angiogenesis [105,106,151].

Individual endothelial cells contain specific receptors for neurotransmitters, as well as for polypeptide and steroid hormones, further illustrating that endothelium is a hormonally responsive tissue [4,38,301,341]. Bovine aortic endothelial cells have been demonstrated to metabolize acetylated low-density lipoprotein (AC-LDL), but not normal LDL, at an accelerated rate compared to other cell types [372]. This macrophage-like property subsequently was demonstrated and used as a useful metabolic marker in capillary endothelial cells also [396]. The integrity of endothelium is essential also in interpreting responses of the vascular smooth muscle to alpha-adrenoceptor agonists [4] and to variations in blood flow [249]. These and other examples of secretory function by the endothelium effecting the underlying smooth-muscle cell layers in the arterial wall are illustrated in more detail later in this chapter, and other endocrine functions related to vascular endothelium are discussed in Chap. 8.

Finally, recent data indicate that endothelial cells may be both a source of and target for interleukin-1 [94,226], and that endothelial cells may provide iron for their own destruction in various forms of oxygen-derived free radical injury [100]. These novel endothelial autocrine mechanisms may play a role particularly in the pathogenesis of vascular injury in haematogenous infections and in immune vasculitides (see Chap. 26), as the vascular endothelium, located strategically at the interface between the blood and tissue, is the first to encounter blood-borne injurious stimuli.

Vascular Endothelium as a Thromboresistant Surface

Vascular endothelium as a cell layer provides a thromboresistant blood-compatible surface. This surface does not promote blood clotting or the adherence of circulating platelets, leucocytes or red blood cells. Several properties of endothelium appear to be essential to this vital function. Conceptually, they are in two broad categories: first, the physicochemical characteristics of the endothelial cell surface; and, second, active metabolic effects on platelet function and blood clotting. Studies with endothelial cultures show a definite organizational pattern in the cellular matrix, with localization beneath, but never on top of the confluent monolayer [19]

Failure to maintain this structural polarity in vivo conceivably could result in loss of endothelial attachment and thus in promotion of thrombosis.

The continuity of the endothelial cell layer is essential for the normal non-thrombogenicity of endothelium. The negatively charged endothelial surface coat and alpha$_2$-macroglobulin, a protease inhibitor, located at the luminal surface of endothelial cells, have been implicated in thromboresistance. More important, as discussed above, endothelial cells are able to secrete quantities of PGI$_2$ sufficient to inhibit thrombosis [251,253] and they have surface ADPase activity [283,284]. It has been shown that soluble mediators, such as interleukin-1 (produced during inflammatory immune responses), induce endothelial biosynthesis and surface expression of a tissue factor-like pro-coagulant activity [18] as well as changes in the pericellular matrix of endothelial cells [255]. Thus, vascular endothelium appears to participate in both anti-coagulant and pro-coagulant reactions, probably partly through the presence of specific binding sites that modulate coagulant activity on the cell surface [267]. While anti-coagulant mechanisms predominate on quiescent endothelial cells, induction of endothelial dysfunction by perturbants can shift this balance towards promotion of localized clot formation [18,70, 130,229,313].

Although blood flow in vivo is likely to produce a high dilution ratio, therefore the significance of these properties for thrombosis on the surface of arteries remains uncertain (and for now we can relate thrombosis only to actual denudation of the vessel surface), there is growing evidence that certain injuries and stimuli result in changes in the haemostatic potential of endothelial cells that may lead to occlusion of the microcirculation [313]. A more detailed view on the thromboresistance of vascular endothelium is presented in Chap. 8.

Vascular Endothelium as a Selective Barrier

The ultimate purpose of the circulatory system is to allow continuous exchange of substances between blood and tissues. Because of its strategic location, the endothelial cell layer in most areas assumes the role of a barrier regulating this exchange. In arteries, the endothelial cell layer acts as a selectively permeable barrier and is essential to regulate the influx of plasma macromolecules into the arterial wall and particularly

to maintain the normal microenvironment and growth pattern of underlying smooth-muscle cells [308].

Structurally, a continuous endothelial cell layer provides a number of potential routes to the subendothelial space, via channels along intercellular tight junctions, via plasmalemmal vesicles and/or transendothelial channels and in the lipid phase of the endothelial plasma membrane.

Intercellular Tight Junctions. Open or leaky tight junctions are well established as pathways of water-soluble macromolecules across continuous endothelium at certain segments of the microvascular bed, notably the venous ends of the capillaries and/or in the post-capillary venules under both physiological and pathological conditions [231,233,235,351]. The permeability characteristics of tight junctions in arterial capillaries [205,246,277,295,350], and in arteries [104,141,177,178,326,371], however, are a matter of some controversy. Nevertheless, it appears that the tight junctions of arterial endothelium are normally impermeable or have low permeability to macromolecules in most regions. They are, however, relatively labile structures that may widen under the influence of haemodynamic factors such as high blood pressure and possibly of vasoactive agents [179,263,380].

The leakiness of the endothelium in the venous capillaries and post-capillary venules to macromolecules indicates the presence of single-stranded and discontinuous tight junctions while the relative impermeability of the endothelium in the arterial end of capillaries and of arterioles indicates the presence of multistranded and continuous belts of tight junctions between adjacent endothelial cells in these vascular segments [345,351]. Structurally discontinuous, mechanically weak tight junctions between adjacent endothelial cells in the venous ends of the microcirculation may provide sites for a small number of leaks allowing the low-level non-selective passage of macromolecules across the endothelial cell layer under normal conditions [35]. They are the obvious sites of endothelial cell separation which results in a large increase in protein efflux and elevated lymph-to-plasma total protein ratio in inflammation [153]. This view of microvascular permeability is supported by recent physiological data [153,155,270,271] and is consistent with the notion that all trans-capillary solute transport occurs by passive processes in accordance with the original pore theory [36,280].

The pore theory of capillary permeability, based primarily on theoretical physiological considerations, was defined first by Pappenheimer [280] in quantitative terms to describe and predict the transport of hydrophilic solutes across the microvascular membrane. It was further advanced by Grotte [159] who described a small-pore/large-pore system, with perhaps one large pore for every 30 000–40 000 small pores, to explain the restricted passage of macromolecules across the microvascular membrane [218,239]. This concept of the microvascular membrane can be depicted by a brick wall model where the cement between adjacent bricks represents the small-pore system, and an occasional missing brick represents the large-pore system [55].

On the basis of current morphological knowledge it appears that the small-pore system structurally corresponds to hydrophilic channels located along tight junctions in interendothelial clefts throughout the entire microcirculation; there, channels along endothelial tight junctions, in a way similar to those along tight junctions of other epithelia [89], allow the passage of ions and small water-soluble substances (probably those < 10 Å in diameter) across the endothelial cell layer [277]. The large-pore system, however, corresponds structurally to infrequent interendothelial openings located at structurally discontinuous open junctions in the venous end of the microcirculation; these openings allow a low level of non-selective passage of macromolecules across the endothelial cell layer.

The diffusion pathway is labyrinthine at cell junctions [263] and there is probably a fixed charged matrix between the endothelial cells, therefore flow of the type occurring in chromatographic columns is likely in these hydrophilic pathways [35]. The permeability of these hydrophilic pathways, accordingly, may be influenced by various factors. For example, omission of protein from the perfusion solution leads to increase in transcapillary filtration in peripheral capillaries [7,76,218]; the passage of negatively charged macromolecules is retarded through capillaries [76,247], in a way similar to that found in glomerular basement membrane [9, 53,204,296,311,312,391]; and polycations effect not only endocytosis by redistributing charged sites of cell membranes [79,217,286, 307,367, 383,409] but also integrity and permeability of tight junctions [264]. Thus the view expressed by Chambers and Zweifach [55] and earlier by Starling [218] is probably correct, that in the normal physiological state, the microvascular membrane is dominated by passive processes through the intercellular cement.

Pathophysiological situations, particularly inflammation, open up interendothelial avenues (thus dramatically increasing the number of "large pores"), primarily in venous endothelium where the junctions are mechanically weak [265,351]. Such openings, although transient and very focal, may result in significant protein extravasation [271]. In models of osmotic blood–brain barrier opening, theoretical considerations indicate that opening of a very small (0.001) percentage of endothelial tight junctions is sufficient for marked tracer extravasation into the neuropil (Dorovinidzis et al. unpublished data).

In more recent years, on the basis of these data, the concept of microvascular membrane with exclusively static porosity characteristics has been replaced in physiological terms by a model of microvascular membrane that considers, in addition to the static pathway, the variable mediator-stimulated pathway subjected to direct physiological regulation [153,154]. The clearance of protein via the mediator-stimulated pathway is far in excess of protein clearance via the non-mediator-stimulated pathway and it corresponds evidently to transient endothelial cell separation in the venous end of microcirculation [153,154,173,375]. A role for contractile proteins in mediator-related opening of interendothelial pathways in post-capillary venules is supported by morphological observations [235] as well as by the presence of these contractile proteins [12,96,248] and of histamine receptors in venular endothelium [167] and by association between receptor stimulation, changes of cytosolic Ca^{2+} and/or of actin filaments and endothelial cell separation, as demonstrated in cultured endothelial cell systems [207,310,339,342,416].

In large arteries, such as the aorta, the complexity of endothelial tight junctions lies generally between "leaky" venous-type and "tight" capillary-type tight junctions, being characterized by mostly continuous single- or double-stranded junctional belts [177,178,182,183,184, 346]. There is, however, considerable heterogeneity in tight junctional structure from cell to cell even within the same aortic segment, with discontinuous tight junctions also occurring focally in normal aortic endothelium [182,183]. Cell junction heterogeneity correlates with areas of different permeability in this cell layer as evidenced by preferential labelling of some interendothelial clefts following in vivo injection of ultrastructural tracers both under normal and under pathophysiological conditions [177,178, 179]. It correlates also with observations made on aortic endothelium following the addition of ruthenium red after fixation of the endothelial cell layer. This cationic extracellular tracer was observed to penetrate the entire length of intercellular spaces between some endothelial cells but it was stopped by tight junctional elements between other endothelial cells (Martinez-Palomo, unpublished data). Preferential penetration of some interendothelial clefts by silver salts has also been observed in aortic endothelium [236,420]. Thus, large-vessel endothelium with its "tight" and "leaky" regions seems to mirror the situation in the microcirculation with its juxtaposed "tight" and "leaky" segments, where the mechanically weak "leaky" regions react primarily to various haemodynamic and chemical stimuli [179].

Plasmalemmal Vesicles and/or Transendothelial Channels. The large population of plasmalemmal vesicles in continuous endothelium is its most conspicuous anatomical feature [31,403]. It has been suggested, on the basis of studies with fine structural tracers, that these vesicles and/or transendothelial channels represent the sole avenue for exchange of water-soluble macromolecules across capillary endothelium under physiological conditions [277,349,350]. It has further been suggested that vesicular transport across the microvascular endothelium occurs in the perfused rat heart [25] and in frog mesenteric capillaries [65, 384], and vesicular transport has also been implicated as a mechanism responsible for increased permeability of arterial endothelium in various pathophysiological situations.

Recent investigations have identified a number of distinct properties of the membrane of endothelial plasmalemmal vesicles, including the paucity or absence of anionic sites detected by cationized ferritin binding, suggesting that these structures may favour the penetration of anionic molecules (such as the majority of plasma proteins) for diffusion or vesicular transport [348], the high concentration of lectin receptors [347] and of binding sites for albumin [140] as well as the presence of a specific antigen, detected by its monoclonal antibody [317] only on vesicles and not on the adjacent plasmalemma proper. A characteristic striped surface structure has been also observed on the cytoplasmic aspect of plasmalemmal vesicles, distinct from that of coated

pits and plasmalemma proper [287]. Quick-freeze, deep-etch studies [11] have identified furthermore an internal structure in the diaphragm of plasmalemmal vesicles, similar to that of fenestral diaphragms [64]. All these data seem to rule out derivation of plasmalemmal vesicles by random invaginations from the plasmalemma and suggest their potential role in macromolecular transport. Vesicular transport of various plasma constituents such as low-density lipoproteins [390], glycosylated albumin [413], albumin and fibrinogen [14,418], transferrin [192] and insulin [209] has been suggested in various in vitro and in vivo systems.

Although these data certainly implicate plasmalemmal vesicles in receptor-mediated transfer of ligands across vascular endothelium and are consistent with observations on the uptake at the luminal surface and discharge at the tissue front of cationized ferritin by vesicles, the role of plasmalemmal vesicles and/or transendothelial channels in passive, non-selective permeation of macromolecules across the endothelial cell layer is questionable. Studies using a rapid freezing–substitution method of fixation suggest that the number of plasmalemmal vesicles and the fusion of vesicles to form transendothelial channels may be related to a slow influence of chemical fixatives [240,241,302,401,402]. An extreme example illustrating how chemical fixatives may affect the plasma membrane is the extensive vesiculation that occurs, including formation of plasmalemmal vesicles and transendothelial channels in endothelial cells fixed with glutaraldehyde, in the presence of the membrane detergent dimethylsulphoxide. The electron microscopic studies of frog mesenteric capillaries with tannic acid as a mordant described on pp. 9–10 imply that most of the endothelial vesicles are permanent or semi-permanent structures which may be labelled by macromolecular tracers through diffusion rather than through transport [34,35,36,37,116]. In this context, extensive labelling of endothelial vesicles by ultrastructural tracer molecules reported in various pathophysiological situations may imply that these tracers diffused into the vesicles from the subendothelial space after they had traversed focally opened endothelial tight junctions [180].

Lipid Phase of the Endothelial Plasma Membrane. Diffusion in the lipid phase of the endothelial cell membrane has recently been proposed as a mechanism of lipid transport across continuous endothelium [334]. According to this model of transport, the relatively water-insoluble products of lipolysis (diglycerides, monoglycerides and free fatty acids) would preferentially dissolve in the outer layer of the endothelial plasma membrane. Having entered this lipid phase, they could rapidly diffuse along the cell surface to reach the abluminal side, where binding to extracellular matrix or transfer to other cells might occur [334]. Lipoprotein lipase, an enzyme which hydrolyses the triglyceride component of plasma chylomicrons, appears to be bound to the glycoprotein coat of the luminal endothelial surface [93], thus this hypothetical scheme indicates the potential for active involvement of endothelium in the transport of complex lipid into the arterial wall. Another mechanism whereby lipoproteins may cross the endothelial cell layer is receptor-mediated uptake and discharge [390] by the endothelial cells as we have noted above. In this context, receptor density on the endothelial cell surface may influence lipid deposition in the arterial wall under various pathophysiological conditions [80]. The fate of plasma lipids and other macromolecules which have reached the subendothelial space may also be influenced by endothelium-associated connective tissue elements. For example, certain glycosaminoglycans can selectively bind low-density lipoproteins [188], thus changes of intimal glycosaminoglycans synthesized by the endothelial cells may be an important factor contributing to intimal lipid accumulation [250] (see also Chap. 13).

In a unique fashion, endothelium also may influence its own permeability characteristics through the local secretion or degradation of vasoactive mediators. The endothelial lining of large blood vessels can release prostaglandins, degrade bradykinin, generate angiotensin II and synthesize histamine. Each of these hormones has been implicated in the control of vascular permeability. Aortic endothelium in particular has an active "histamine-forming capacity" (the histidine decarboxylase system), which appears to be responsive to haemodynamic (wall shear) stresses. This mechanically coupled enzymatic mechanism for histamine generation may explain the focal increases in aortic permeability often observed in areas of increased haemodynamic stress [83]. Thus, the endothelial cell layer in arteries is a highly dynamic permeability barrier; abnormal permeation and subsequent retention of plasma constituents in the arterial wall may result from metabolic dysfunction as well as structural modification of the endothelial lining.

The Smooth-Muscle Cell

Smooth-muscle cells in the arterial wall form a multifunctional component with active contractility, capacity for synthesis of strain-bearing wall structures and a considerable proliferative activity in repair situations after injury. The smooth-muscle cell is the principal cell involved in the pathogenesis of the lesions of atherosclerosis (see Chap. 11).

Biological Behaviour of Smooth-Muscle Cells

Smooth muscle consists of long, tapering cells that have a single nucleus and contain both thick (myosin) and thin (actin) filaments aligned with the long axis of the cell. However, these filaments are not arranged in the strictly ordered pattern found in skeletal and cardiac muscle and do not appear to form well-defined myofibrils [363].

Contraction in striated (skeletal and cardiac) muscle cells is produced by the sliding of actin filaments against myosin filaments. The head regions of myosin molecules, which project from myosin filaments, undergo an ATP-driven cycle in which they attach to adjacent actin filaments, undergo a conformational change that pulls one filament against the other and then detach [185,368]. This cycle is facilitated by accessory muscle proteins that hold the actin and myosin filaments in parallel overlapping arrays with the correct orientation and spacing for sliding to occur. Alpha-actinin and desmin, found in the region of the Z disc, are believed to help determine the arrangement of filaments in each sarcomere and to tie adjacent sarcomeres together, respectively [224,318,408]. Two other accessory proteins, troponin and tropomyosin, allow the contraction of muscle to be regulated by Ca^{2+} [260,378].

Actin and myosin are also found in smooth muscle where they produce contraction in fundamentally the same way as in striated muscle. In smooth-muscle cells, however, the two types of filaments are less highly ordered than in skeletal muscle, and their movement is dependent on a Ca^{2+}-regulated phosporylation of myosin [1,163,321]. Both the actin and the myosin of smooth muscle are of a kind special to that tissue. The actin differs slightly in amino acid sequence from the actin of skeletal or cardiac muscle, although the differences have no known functional significance. But although smooth-muscle myosin closely resembles that of skeletal muscle, it differs functionally in two very important ways: firstly the level of its ATPase activity, even under optimal conditions, is tenfold lower than that of myosin of skeletal muscle and is subject to a more direct Ca^{2+} regulation; secondly, the myosin of smooth muscle, in common with the myosins of non-muscle cells, is able to interact with actin filaments and thereby cause contraction only when its light chains are phosphorylated; when the myosin light chains are dephosphorlyated, the myosin cannot interact with actin and the muscle relaxes.

Both the phosphorylation and the dephosphorylation of smooth-muscle myosin light chains are carried out by specific enzymes. Smooth-muscle myosin ATPase is Ca^{2+}-dependent because the phosphorylating enzyme, myosin light-chain kinase, is regulated by Ca^{2+} levels. The effect of Ca^{2+} is mediated by a Ca^{2+}-binding protein, calmodulin, that is structurally and functionally very similar to troponin C. Calmodulin has been identified in all animal and plant cells that have been examined. It has been highly conserved during evolution and appears to be a ubiquitous intracellular Ca^{2+} receptor, playing a part in the majority of the Ca^{2+}-regulated processes that have been studied in eukaryotic cells. Calmodulin is a single polypeptide chain of 148 amino acid residues whose sequence is related to that of troponin C, suggesting that the latter is a specialized form of calmodulin. Like tropinin C, calmodulin has four high-affinity Ca^{2+}-binding sites and undergoes a large conformational change when it binds Ca^{2+}.

Among the increasing numbers of cellular proteins known to be regulated by calmodulin in a Ca^{2+}-dependent manner are some forms of cyclic nucleotide phosphodiesterase and adenylate cyclase, as well as membrane-bound Ca^{2+}-ATPases, phosphorylase kinase, and the myosin light-chain kinase of both muscle and non-muscle cells [62,211,243]. The complex of calmodulin with Ca^{2+} activates myosin light-chain kinase, which otherwise has little tendency to phosphorylate myosin. Several other factors also affect the activity of the myosin light-chain kinase, for example, the level of cyclic AMP in the cell, which is regulated in turn by hormones that influence smooth-muscle contraction. Smooth muscle, therefore, is triggered to contract by an influx of Ca^{2+}, as is skeletal muscle, but the mechanism of triggering is different. It is activated not by impulses from voluntary nerves but by nerves of the autonomic system or by

hormones, as is also the case in cardiac muscle. The enzymatic steps that follow Ca^{2+} binding to the soluble protein calmodulin act relatively slowly to activate smooth-muscle contraction. Rapid activation is not required, since the much slower myosin cross-bridge cycle in smooth-muscle cells only allows them to contract slowly.

While calmodulin is an ubiquitous protein that mediates Ca^{2+}-dependent processes in many cells, troponin C is best considered a specialized form of calmodulin that has evolved in skeletal and cardiac muscle cells to bind to thin filaments directly and thereby provide a more rapid Ca^{2+} regulation of contraction.

Smooth-muscle cells should not be viewed as slow, poorly constructed versions of skeletal-muscle cells that are adequate merely because fewer demands are made on them. In fact, they are specially designed to be able to maintain tension for prolonged periods while hydrolysing five- to tenfold less ATP than would be required by a skeletal muscle cell that performed the same task. The slow myosin cross-bridge cycle in smooth muscle prevents it from contracting quickly, but it enables the muscle to maintain a constant tension with much greater efficiency. This is particularly important in vascular smooth muscle regulating and maintaining the tone of blood vessels.

Contractile and Cytoskeletal Proteins as Biochemical Markers of Differentiation in Smooth Muscle

Blood vessels arise from mesenchyme as a budding network of small endothelial-lined channels (see Chap. 3). These channels become surrounded by locally derived, irregularly shaped mesenchymal cells lacking smooth-muscle-cell-specific features. With time, cell layers delimited by elastic lamellae become more identifiable in large arteries, and the cells take on a more characteristic ultrastructural appearance. At parturition, most arteries appear to have developed their adult number of layers of smooth muscle [414]. After this stage, the number of layers in the wall does not increase, and the medial thickening is due to the production of connective tissue [139]. The signals controlling this differentiation of smooth-muscle cells are not known. The fact that vessels begin as endothelial channels suggests that endothelium triggers differentiation of smooth-muscle cell from the surrounding embryonic milieu [329].

Different cell types contain distinct species of contractile and cytoskeletal proteins which serve also as excellent biochemical markers of differentiation. Contractile proteins in muscle cells include specific isotypes characteristic of the differentiational states. Although actin is a very conserved protein, six isoforms have been identified in mammals: α-cardiac muscle, α-skeletal muscle, α- and γ-smooth muscle and β- and γ-cytoplasmic, these last isoforms being present in all eukaryotic cells. Vascular smooth-muscle cells contain α- and γ-smooth muscle, and β- and γ-cytoplasmic. Bidimensional gel electrophoresis allows α, β and γ isoforms to be distinguished [387]. In the normal adult aorta, smooth-muscle cells of the media show a predominance of α-actin, some β-actin, and very little γ-actin, whereas in the foetus and newborn animal the β-isoform predominates [215,216]. Cultured vascular smooth-muscle cells also undergo differential expression of isoactins in relation to their growth state with predominance of smooth-muscle specific α-isoactin following growth arrest [8,274]. Mammalian non-muscle cells, including fibroblasts, contain only β- and γ-isoform. The predominance of α-actin in adult vascular smooth-muscle cells contrasts with the predominance of γ-actin in adult non-vascular smooth-muscle cells [126].

Recently, the use of a monoclonal antibody that recognizes exclusively α-smooth-muscle actin has allowed identification of a population of smooth-muscle cells negative for α-actin in rat aortic media. This cell population was shown to represent about half of the medial smooth-muscle cells in 5-day-old rats but only 8% in 6-week-old animals, indicating that the presence of α-actin correlates with the state of differentiation in vascular smooth muscle [358]. In cardiac muscle, myosin can be found in three isoenzymatic forms, designated V1, V2 and V3. To date, a similar resolution of myosin variants has not been shown in smooth muscle, although several studies suggest they do exist. Heterogeneity of myosin antigenic expression has been demonstrated in smooth-muscle cells of elastic arteries versus other blood vessels [220].

Predominance of Vimentin Filaments in Vascular Smooth-Muscle Cells

In addition to contractile proteins, the cytoplasm contains structural proteins (intermediate-sized, 10 nm, filaments) which are also different in different cell types and change with differentiation [224]. Thus epithelial cells contain intermediate filament proteins related to epidermal prekeratin and mesenchymal cells contain filaments of the vimentin type [15]. In differentiated smooth-muscle cells of digestive, respiratory and urogenital tracts, intermediate filaments of the desmin type are predominant [110,126,224]. In contrast with non-vascular smooth-muscle cells, however, the major intermediate filament protein present in vascular smooth-muscle cells is vimentin [126,216,319]. Smooth-muscle cells of the arterial wall are also heterogeneous as far as their contents of vimentin and desmin are concerned. Of adult rat aortic medial smooth-muscle cells, 51% contain vimentin alone, 48% contain both vimentin and desmin and 1% contain desmin alone [215]. More desmin-positive cells are present in the aorta towards the iliac arteries than at the arch [273,319]. Vimentin is found in 87% of the medial cells in the foetal aorta, 13% contain both vimentin and desmin and none contain desmin alone. With development, there is a gradual decrease in the percentage of cells containing vimentin alone and an increase in the percentage of cells in which both vimentin and desmin co-exist, until the adult values are reached 12 weeks after birth [216].

These data indicate that adult mammalian vascular smooth-muscle cells are heterogeneous and profoundly different from non-vascular smooth-muscle cells in their contractile and intermediate filament composition [126,216, 319]. The abundance of vimentin in vascular smooth-muscle cells suggests a close relationship to other mesenchymal cells such as fibroblasts [109,126]. It has been shown that vimentin filaments appear in non-mesenchymal cells (epithelial and myogenic cells included) when these are grown in culture [112]. These observations indicate that abundance of vimentin may be related to the known proliferative and motile potential of vascular smooth-muscle cells as observed in blood-vessel differentiation and pathological situations such as atherosclerotic plaque formation [216; see Chap. 11.]

There is strong evidence from cell culture studies for the reversibility of the differentiated vascular smooth-muscle phenotype [143,329]. When the media of adult arteries is enzyme-dispersed into single smooth-muscle cells and seeded into primary culture, the cells are morphologically and functionally similar to those of the intact vessel, that is, they either spontaneously contract or can be induced to contract by electrical, mechanical or chemical stimuli, and they synthesize only small amounts of extracellular matrix material. However, if these cells are seeded below a critical cell density, after about 1 week the cells undergo a spontaneous change in phenotype [57]. Functionally, they have the cacapity to contract and gain the capacity to divide, and do so logarithmically in response to serum mitogens [59]. In addition, they synthesize four to five times the amount of extracellular matrix molecules as the contractile state cells. Morphologically, the cells lose their thick, myosin-containing filaments and greatly increase the amount of organelles involved with synthesis, such as rough endoplasmic reticulum and free ribosomes, and thus resemble immature smooth-muscle cells [58]. These two distinct phenotypes have been termed "contractile" and "synthetic" [56,381]. Cells having features of both contractile and synthetic cells also exist [329].

In vivo, the reversibility of the differentiated vascular smooth-muscle phenotype is evidenced both in experimental intimal thickening and in lesions of human atherosclerosis [67,146; see Chap. 11]. The normal human arterial media is largely comprised of "contractile" smooth-muscle cells, that is, cells filled with contractile filaments and characterized by expression of the cytoskeletal proteins unique to the differentiated vascular smooth-muscle cells. In contrast, most of these features are absent in diffuse intimal thickenings adjacent to atherosclerotic plaques in which an increased proliferative activity of smooth-muscle cells has been suggested [329]. Instead, these cells have a predominant rough endoplasmic reticulum implying a "synthetic phenotype", and show a typical switch in actin expression with a predominance of the β-form and a noticeable amount of γ-form [329].

All these data suggest the possibility that the commitment of vascular smooth-muscle cells to replicate requires a switch from the contractile phenotype to a less differentiated, synthetic phenotype. An alternative possibility is that distinctive smooth-muscle cell subtypes exist even in the uninjured arterial media [329]. This latter hypothesis would suggest that commitment to proliferation depends on recruitment or activation of a subset of smooth-muscle cells that may represent a foetal rest, similar to the stem cells seen in adult striated muscle.

Interestingly, as noted above, vessels subjected to atherosclerosis contain a subset of smooth-muscle cells lacking desmin, one of the proteins characteristic of smooth-muscle differentiation [215]. If a few widely scattered immature "foetal" smooth-muscle cells with a propensity for proliferation were stimulated to migrate into the intima by mitogens following endothelial injury,

the cells could be released from controls on proliferation imposed by neighbouring "adult" cells. If these cells also retained the ability of foetal smooth-muscle cells to produce mitogen [336], we would also be able to explain the sustained proliferation that continues after the initial stimulation of platelet release during trauma [329]. Selection among a smooth-muscle population containing such a proliferogenic sub-population could account for both the monoclonal phenotype of chronic human atherosclerotic lesions and the suggestion that monoclonality arises gradually as the human lesion evolves [225,285; see also Chap. 11].

Whether "foetal" or "blast" cells are present or not in the adult arterial wall, this potential functional heterogeneity is not apparent from ultrastructural studies [139,216,272]. The vascular smooth muscle of adult mammals is composed of a morphologically homogeneous cell type, which shows similar, if not identical, ultrastructural features to those of non-vascular smooth-muscle cells [127,363].

Functional Morphology of Vascular Smooth Muscle

In the following pages we will review briefly the ultrastructure of the vascular smooth-muscle cell within the context of the basic ultrastructural features of smooth muscle, and whenever possible, we will relate it to general aspects of cell physiology.

Contraction and Force Transmission

In transverse sections three types of filaments are readily apparent in vascular smooth muscle: thick (myosin), thin (actin) and intermediate (10 nm) filaments (Figs. 1.6 and 1.7). The filamentous organization of actin and myosin, as well as the length–tension curve of these muscles, are compatible with a sliding filament mechanism of contraction [363]. The function of the ubiquitous intermediate filament is not well understood.

Myosin (Thick) Filaments. Although myosin was biochemically demonstrated in smooth muscle as early as 1959, the in situ form of myosin in smooth muscle has been controversial until recently [5,87,321,363] due to poor preservation of thick filaments as well as their propensity to artefact formation. The myosin filaments of mammalian smooth muscle are

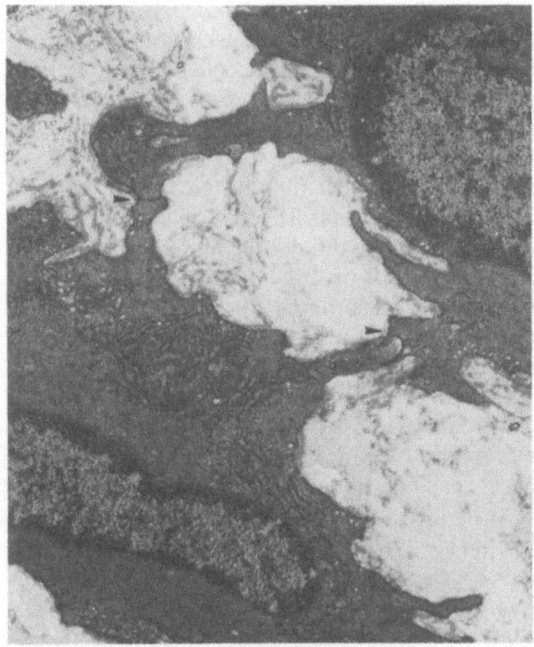

Fig. 1.6. Smooth-muscle cells from the media of normal rat carotid artery as seen on thin-section electron microscopy. Note gap junctions, sites of electrotonic and metabolic coupling, between profiles of adjacent smooth-muscle cells (*arrowheads*). × 24 000

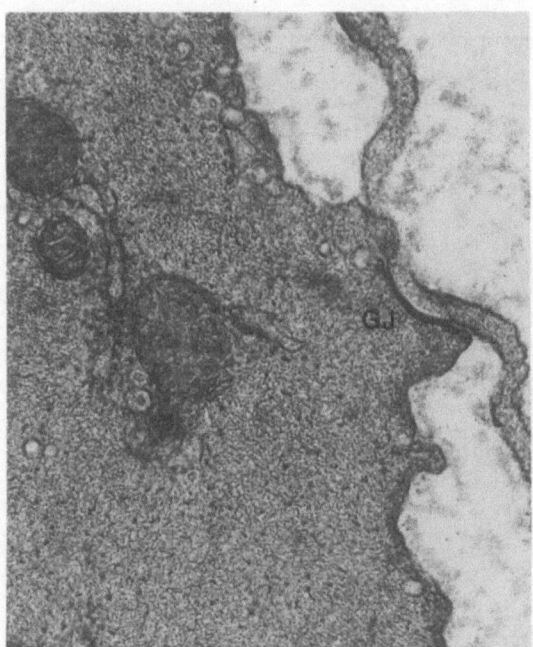

Fig. 1.7. High-power detail of smooth-muscle cell cut transversally visualizes thick and thin as well as intermediate filaments in the cytoplasm. *GJ*, gap junction, × 99 000

approximately 15–19 nm in diameter, and their length is 2.2 µm ± 0.14 SD [5]. Although the lateral or longitudinal distribution of thick filaments in smooth muscle is not as regular as in striated muscles, some longitudinal order was observed by following thick filaments through serial sections in which they were found to occur in groups of three to five. This longitudinal ordering of neighbouring groups of filaments could be considered to represent mini-sarcomeres [5,363].

The maximal force developed by smooth muscle is equal to or greater than that developed by striated muscle, in spite of the approximately fivefold lower concentration of myosin in smooth muscle [259]. The maximal force developed by muscle is proportional to the sarcomere length and to the number of cross-bridges acting in parallel. The greater length of the myosin filaments, 2.2 µm compared to 1.5–1.6 µm in vertebrate striated muscle, and the larger thin-to-thick filament ratio in smooth muscle could result in the simultaneous engagement of a larger number of cross-bridges in parallel during tension development in smooth than in striated muscle [5]. This would at least partially compensate for the lower concentration of myosin. The arrangement of the smooth-muscle cells within a strip, either in parallel or in series, will also contribute to the effective force per cross-sectional area. Greater forces for an intact strip than for isolated cells have been calculated and the greater forces have been attributed to lateral cell-to-cell attachments resulting in a parallel arrangement. Also, in estimating forces per cross-sectional area of a vascular strip, it is necessary to bear in mind that cells may be laterally dispersed relative to each other along a strip, and the forces can be transmitted to the ends of the preparation by connective tissue [97]. This argument can also be applied to myosin filaments per cross-sectional area of a single smooth-muscle cell. A smaller number of myosin filaments in a given fibre can develop equal or greater force than a greater number of myosin filaments, if they are laterally dispersed along the length of the fibre [363].

Actin (Thin) Filaments. Filamentous actin (F-actin) is 6–8 nm in diameter and is found in all muscle cells as well as many other cell types. In the best organized examples of rabbit portal anterior mesenteric vein, myosin filaments are surrounded by an orbit of thin filaments with a thin-to-thick ratio of approximately 15 : 1 [87,363]. This is in agreement with the ratio of

actin to myosin determined by biochemical isolation procedures [164,259]. In many preparations thin filaments are very regularly
arranged and are grouped into bundles. The
regulatory protein tropomyosin lies in the groove
of the actin filament and the ratio of tropomyosin
to actin is similar to that in striated muscle.
However, vertebrate smooth muscles do not
contain troponin, and the regulatory system
differs from skeletal muscle. Myosin is found not
only in lower concentration in smooth muscle
as compared to striated muscle, but it is also
characterized by a tenfold lower level of ATP-
ase activity. Its interaction with actin, furthermore, is dependent on a Ca^{2+}-regulated phosphorylation. All these features indicate a slow
cross-bridge cycle in smooth muscle that prevents it from contracting quickly, but it enables
the smooth muscle to maintain a constant
tension with much greater efficiency [1,163].

Intermediate (10 nm) Filaments. The intermediate filaments are the third type of filamentous structure found in smooth muscle. They
measure around 10 nm in diameter and have
a clear-cut cross-sectional profile. Intermediate
filaments are well preserved with conventional
fixatives, and are less affected by extraction or
by swelling of the muscle cells than are thin and
thick filaments [363]. The intermediate filaments
are ubiquitous components of a variety of contractile and non-contractile cells, and although
morphologically identical they are distinguishable biochemically and immunologically
in different cell types [224]. In differentiated
smooth-muscle cells of visceral organs the intermediate filaments of the desmin type are predominant [3,224]. In contrast, the major
intermediate filament protein present in vascular
smooth-muscle cells is vimentin [126], and
smooth-muscle cells of the arterial wall are also
heterogeneous as far as their contents of vimentin and desmin are concerned [215,216].

Structurally, intermediate filaments in
smooth-muscle cells are often localized in the
proximity of dense bodies, around which they
may form rosettes; other intermediate filaments
are gathered in small bundles or are lined
beneath dense bands (also called surface patches)
near the cell surface. Although the intermediate
filaments do not seem to penetrate into the dense
bodies, there is probably a mechanical connection between the two structures. It has been
suggested that this third class of filaments forms
a network, called the cytoskeleton, which gives
support to the myofilaments [363].

Dense Bodies and Dense Bands Dense bodies
are electron-dense structures scattered in the sarcoplasm of smooth-muscle cells. They are elongated and lie parallel to the myofilaments. The
material of the dense bodies appears similar to
the material forming the dense bands which are
attached to the cell membrane and in certain
vascular smooth-muscle cells some dense bodies
are in continuity with dense bands. Dense bodies
and dense bands probably correspond to Z-lines
of striated muscle fibres. In both structures the
presence of α-actinin has been demonstrated by
immunocytochemistry [323].

The distribution of myofilaments and the
occurrence of dense bands probably accounts for
the remarkable amount of shortening a smooth-
muscle cell can undergo; they may also account
for the wide range of changes in the shape of the
cell profile [127]. The cytoplasmic side of the
muscle-cell membrane is heavily encrusted with
electron-dense material forming the so-called
membrane-associated dense bodies, dense
plaques, dense patches [363] or dense bands
[127,129] (see Fig. 1.8). In transversely sectioned
muscle cells the dense bands are segments of the
cell profile interposed between groups of caveo-

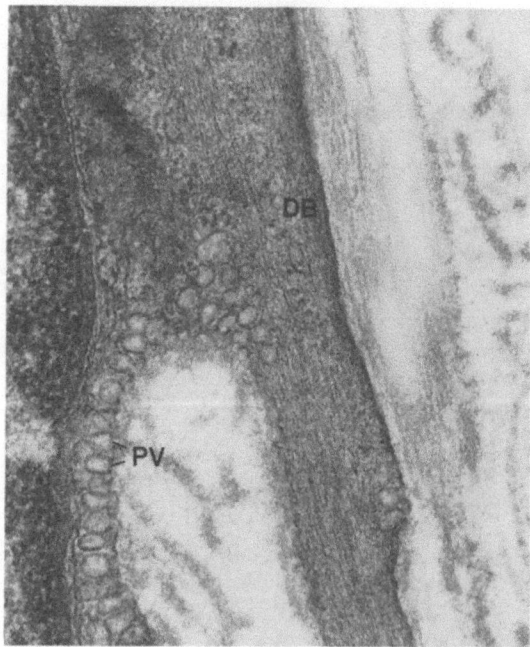

Fig. 1.8. Detail of smooth-muscle cell cut longitudinally
demonstrates alternating segments of the surface membrane
occupied by the surface caveolae or plasmalemmal vesicles
(*PV*) and by dense bands (*DB*). Dense bands represent sites of
force transmission between cytoplasmic contractile filaments
and stroma. × 99 000

lae or plasmalemmal vesicles. They lie approximately parallel to the long axis of the cell, and are found over the entire cell surface. Dense bands occupy on the whole between 30% and 50% of the cell profile in the middle portions of the smooth-muscle cell, and towards the end of the cell the percentage of the cell perimeter occupied by dense bands may reach 100%. In many vascular muscle cells, dense bands are much thicker than in visceral-muscle cells and protrude into the sarcoplasm for up to 1 μm. Characteristically, in small arteries and in arterioles, dense bands are markedly larger and more numerous on the adventitial side of the smooth-muscle cells than on the luminal side; this asymmetrical distribution has been interpreted as essential to an efficient mechanism for constriction of the lumen [128].

Force transmission from the contractile apparatus to the cell membrane in smooth-muscle cells occurs mainly via the insertion of bundles of actin filaments into the dense bands [129]. Remarkably, the insertion sites of actin filaments in smooth muscles are not limited to the ends of the cell (as in cardiac and skeletal muscles) but are scattered over its entire length. The pull produced by the myofilaments is thus distributed all over the "lateral" walls of the cell. The intermediate filaments that are anchored to the dense bands probably also contribute to the transmission of force. As we have mentioned above, the dense bands (and the dense bodies) of smooth-muscle cells contain α-actinin [322], a major component of Z-bands of striated muscles. Alpha-actinin is an actin-associated protein of molecular weight 100 000; it seems to act as a cross-linker and a spacer between actin filaments [193]. A number of proteins are known to accumulate at sites of actin–membrane association where cells adhere to substrates [408]. Vinculin, intracellular protein with molecular weight 130 000 [134] and the vinculin-binding protein talin [42,43], are found at focal contacts or adhesion plaques of non-muscle cells and also at the dense bands (but not at the dense bodies) of smooth-muscle cells [136]. These proteins localized at the cytoplasmic side of the cell membrane have been implicated in the linkage of actin filaments to the cell membrane although the structural nature of the linkage is not clear [13]; a direct interaction between α-actinin and vinculin [74] and the role of an 82 000 molecular weight protein [13] have been recently suggested.

Another actin-binding protein that is particularly plentiful in smooth muscle is filamin, a 250 000 molecular weight protein [406]. Recent immunocytochemical studies on the contractile and cytoskeletal proteins have distinguished two domains in the smooth-muscle cell: (a) actomyosin domains, made up of continuous longitudinal arrays of actin and myosin filaments, and (b) longitudinal, fibrillar, intermediate filament domains, free of myosin but containing actin and α-actinin-rich dense bodies. Filamin was found to be localized specifically in the latter intermediate filament–actin domains, but was excluded from the core of the dense bodies. Filamin was also localized close to the cell border at the inner surface of the plasmalemma-associated dense bands. In isolated cells, the surface filamin label showed a rib-like distribution similar to that displayed by vinculin [359].

On the basis of these studies it was suggested that two functionally distinct systems exist in smooth-muscle cells: an actomyosin system required for contraction, and an intermediate filament–actin system, with associated gelation proteins, that is responsible, at least in part, for the slow relaxation and tone peculiar to smooth muscle [359]. According to this hypothesis there is a subset of actin filaments that has a non-contractile function. These filaments being co-distributed with the intermediate filaments and apparently originating from the dense bodies contained within the intermediate filament arrays, would serve some kind of cytoskeletal role. It is conceivable for example that the intermediate filament–dense body system may be fixed, transitorily, at any length, through the cross-linking by filamin of interdigitating arrays of actin filaments originating from the dense bodies [22]. In this way filamin could serve a role in stress maintenance and thus act as an important modulator of the mechanical response. Alternatively, or additionally, filamin may act as a coupler between the contractile and cytoskeletal and membrane domains [359].

Force transmission occurs across cell membranes and from cell membranes to stroma in smooth muscle. Although it seems clear that the pull generated by the myofilaments is transmitted to the dense bands scattered all over the smooth-muscle cells, it is not fully understood how filaments are anchored to the membrane and how their pull is transmitted across the membrane and to the extracellular stroma. However, intracytoplasmic actin and actin-associated proteins, particularly α-actinin and vinculin, as well as concomitant extracellular fibronectin have been localized at cell-to-matrix attachment sites in various non-muscle systems both in vitro and in vivo [45,72,73,134,135,136,186,213,354].

Recent studies using monoclonal antibody technology suggest that the fibronectin receptor is a transmembrane glycoprotein complex [30,166, 214,304] that has an extracellular binding site for fibronectin [175] in addition to a domain that interacts with the vinculin-binding protein, talin [174]. The transmembrane linkage of talin with the extracellular matrix via the fibronectin receptor provides one mechanism by which cytoplasmic components can become functionally coupled to extracellular information [13].

These data suggest complex but specific interactions between cytoplasmic actin filaments and extracellular fibronectin fibres across the plasma membrane at cell-to-matrix attachment sites. The close association between bundles of cytoplasmic actin filaments (stress fibres) and bundles of extracellular fibronectin microfibrils in transformed fibroblasts and in myofibroblasts of granulation tissue in vivo has been named the fibronexus, an organelle specialized for increased cell-to-matrix adhesion [186,353,356]. As we have noted above, endothelial cells also develop abundant actin stress fibres and adhesive structures structurally identical to fibronexus in situations requiring their increased attachment to the basal lamina, such as in hypertension and during regeneration [184; see also Chap 11].

Subplasmalemmal microfilament condensations at endothelial cell stress fibres are similar if not identical to the dense bands of smooth-muscle cells [184]. Membrane domains specialized for stress fibre attachment in endothelial cells [184] as well as membrane domains occupied by dense bands in smooth-muscle cells [254, 337,338] are resistant to formation of filipin–sterol complexes suggesting that they are probably too rigid to be easily deformed and are different from the rest of the plasma membrane [184,370]. These data are consistent with the hypothesis that in smooth-muscle cells a subset of actin filaments, similar to stress fibres of non-muscle cells, possesses a non-contractile cytoskeletal function [359]. These actin filament bundles are linked to specific membrane domains at dense bands by actin associated proteins and connected across the cell membrane to extracellular fibronectin. Fibronectin, a 200 000 molecular weight glycoprotein, has been shown to mediate adhesion of the cell membrane to the stroma in various in vitro and in vivo systems [186]. Fibronectin has an affinity for collagen, glycosaminoglycans and also for the surface of many cell types; it also has a tendency towards self-association, which can result in the formation of fibrils. The evidence obtained from

smooth muscle in situ for co-extensive fibronectin matrix fibrils with actin microfilaments is not as good as in granulation tissue myofibroblasts [355,356], but it nonetheless strongly suggests that certain sites of the smooth-muscle plasma membrane, marked by a net of electron-dense material, are anchored intracellularly to actin filament bundles and extracellularly to fibronectin fibrils and collagen fibres [129].

The basal lamina is present over the entire smooth-muscle cell surface, except at the level of the gap junctions and of other intimate appositions between smooth-muscle cells. It is usually more conspicuous where it lies over a dense band. Microfibrils, 11 nm in diameter, are embedded in the basal lamina and extend into the adjacent extracellular structures. Extracellular microfibrils, located over dense bands and coaxial with cytoplasmic microfilament bundles, extend among the nearest collagen fibrils and probably represent fibronectin. Other 11-nm microfibrils mix with those distributed around elastic fibres [129]. Recently a glycoprotein of molecular weight 350 000, called fibrillin, has been isolated from the medium of fibroblast cell cultures and suggested to represent a principal structural component of microfibrils [316]. Laminin, a cross-shaped protein of molecular weight 1 000 000 [99,382], has been identified as one of the major cell-binding components of basal laminae [213] and in this respect appears to be functionally similar to fibronectin [417]. It is thought that laminin anchors the cells to basal laminae by forming bridges to collagen IV [212]. In addition to known interaction patterns for laminin and fibronectin, recent studies also indicate potential direct interactions between cell membrane and collagen IV at the level of the basal laminae [6].

The major elements of the stroma in smooth muscle are collagen fibrils and elastic fibres arranged in lamellae. Collagen concentrations in mammalian smooth muscles are invariably greater than in striated muscles. The high collagen content of smooth muscle supports the idea that their stroma serves a mechanical role in transmittal of force and could perhaps be regarded as constituting intramuscular micro-tendons [129]. Most of the collagen content is accounted for by the collagen fibrils. Some of the collagen, however, is part of the basal lamina and is non-fibrillar. Type III collagen is the predominant collagen extractable from normal vascular smooth muscle. Elastic fibres usually branch extensively and are joined to one another forming continuous laminae, fenestrated sheets or small elastic plates. The elastic material is composed of amorphous bands surrounded by microfibrils 11 nm in diameter. Both components have been chemically characterized in cultures of vascular smooth muscles [257].

As we have noted above, microfibrils are often interposed between elastic fibres and basal laminae of smooth-muscle cells. It is generally accepted that passive mechanical properties of smooth muscles (especially vascular smooth muscles) are due to a large extent to the extracellular material; in view of the extensive attachments of the cells to the stroma, the latter should account for a major part of the series elastic component [129].

Direct force transmission also occurs from cell to cell. In addition to force transmission from smooth-muscle cells to surrounding stroma, sites of direct adherence between smooth-muscle cells probably allow transmission of pull from one cell to another. The sites of direct adherence, called adherens junctions or intermediate junctions, are symmetrical structures formed by two dense bands that match each other in adjacent smooth-muscle cells [129]. These intermediate junctions are invariably straight and remain straight in shortened muscles, presumably being rather rigid. The intercellular space at the level of junctions measures 40–60 nm, and a single, ill-defined layer of electron-dense material, which is in continuity with the basal laminae of the two cells, runs along the middle of it. These adherens junctions are different from desmosomes, in which the intercellular space is wider and has a multilayered appearance resulting probably from transmembrane linkers. Moreover, desmosomes are connected to intermediate filaments, such as tonofilaments in epithelial cells, whereas intermediate junctions are mainly associated with actin microfilaments [129].

Recently a protein of molecular weight 135 000 has been identified at adherens junctions using a monoclonal antibody raised against integral membrane proteins of chick cardiac intercalated discs and subsequently characterized as adherens junction-specific cell adhesion molecule (A-CAM), a Ca^{2+}-dependent membrane glycoprotein involved in intercellular adhesion [394,395]. It is also known that while vinculin is ubiquitously present in all adherens junctions, the protein with molecular weight 135 000 is associated only with intercellular contacts and talin with attachments to non-cellular matrices exclusively [137]. These data also suggest a possible heterogeneity of the smooth-muscle cell membranes at sites of dense bands facing the connective tissue stroma as compared to dense bands facing each other at intermediate junctions.

In summary, smooth muscle is a tissue of outstanding performance in terms of force generated, amount of shortening and adaptability to functional demands. The high cellularity of smooth muscle points to major roles for the junctions between cells and between cell and stroma. Numerous bundles of actin filaments are associated with the cell membrane. The latter appears linked to the membrane of other cells at the intermediate junctions, and to elements of the stroma such as elastic fibres and collagen fibrils [129].

Excitation–Contraction Coupling

Muscle cells, like neurons, can be excited chemically, electrically and mechanically to produce an action potential which is transmitted along their plasma membrane. Unlike neurons, however, they have a contractile mechanism which is activated by the action potential and they possess specialized membrane structures for the excitation–contraction coupling. Skeletal muscle does not normally contract in the absence of nervous stimulation as it lacks structural and functional connections between individual muscle fibres and is generally under voluntary control. Cardiac muscle has cross-striations as does fast-contracting skeletal muscle, but is functionally syncytial in character and contracts rhythmically in the absence of external innervation owing to the presence in the myocardium of pacemaker cells that discharge spontaneously and of gap junctions (nexuses) that provide low-resistance pathways for the spread of excitation from one cell to another. Slow-contracting smooth muscle lacks cross-striations and is functionally syncytial in most cases.

In general, smooth muscle can be divided into single-unit (visceral) smooth muscle and multi-unit smooth muscle. Single-unit smooth muscle occurs in large sheets, it is functionally syncytial in character owing to the presence of low-resistance gap junctions between individual cells and contains pacemakers that discharge irregularly. Single-unit smooth muscle is found in most hollow viscera and essentially this type of smooth muscle is present in the walls of blood vessels. Multi-unit smooth muscle is made up of individual units without interconnecting gap junctions and pacemaker activity. It is found in structures such as the iris of the eye, in which fine, graded contractions occur. It is not under voluntary control, but it has many functional similarities to skeletal muscle. There are no reports of vascular smooth muscles with similar electrical properties [194].

Single-unit smooth muscle is characterized morphologically by the presence of gap junctions (also called communicative junctions or nexuses) between individual smooth-muscle cells, providing a structural basis for the functionally syncytial character of this type of smooth muscle. Gap junctions are specialized regions of the plasma membrane where proteins are specifically oriented to construct a direct hydrophilic pathway providing continuity between the cytoplasm of contracting cells [170,227] (see above). In vascular smooth muscle, well-defined gap junctions are evident morphologically in pulmonary artery [86,117] and in large arteries of the systemic circulation (Figs. 1.6 and 1.7), but only close membrane contacts are seen on thin-section electron microscopy of electronically coupled arterioles. Electronic coupling has been shown also to develop between smooth-muscle cells in vitro [58]; synchronous contractions have been shown to correlate with appearance of gap junctions both between enzyme-dispersed smooth-muscle cells in primary culture [46,242] and between smooth-muscle cells in continuous cell lines [208,324].

Single-unit smooth muscle is unique in that unlike other types of muscle, it contracts when stretched in the absence of any extrinsic innervation, a feature particularly important in vascular smooth muscle. Single-unit activity in small blood vessels has important functional implications even if the size of the unit (i.e. the population of synchronized cells) is much smaller than in the larger vessels with propagating smooth muscle (e.g. portal vein). For example, the active response to stretch typical of single-unit smooth muscle provides a possible basis for myogenic autoregulation of blood flow [195].

Another important functional property of smooth muscle (both in single-unit and multi-unit types) is its sensitivity to chemical agents released from nerves locally or brought to it in the circulation. In mammals, visceral muscle usually has a dual nerve supply, from the two divisions of the autonomic nervous system. The function of the nerve supply is not to initiate activity in the muscle but rather to modify it. Stimulation of one division of the autonomic nervous system usually increases smooth-muscle activity, whereas stimulation of the other decreases it. However, in some organs, noradrenergic stimulation increases and cholinergic

stimulation decreases smooth-muscle activity; in others the reverse is true.

Although vascular smooth muscle is an electrically coupled system in general (corresponding to single-unit type visceral smooth muscle), there are electrophysiological differences between the smooth muscle in various vascular segments [389]. Action potentials are not normally generated by smooth-muscle cells in the main distributing arteries (such as the main pulmonary artery); this type of smooth muscle has been called tonic vascular smooth muscle. Other vessels, most characteristically the portal mesenteric and other veins, commonly exhibit spontaneous depolarization (pacemaker potentials) and action potential; this type of smooth muscle has been called phasic vascular smooth muscle [197]. Another electrophysiological classification distinguishes between non-spike-generating vascular smooth muscle activated by graded depolarization (main pulmonary artery) and spike-generating vascular smooth muscle showing action potentials that synchronize contraction in a small or larger population of cells (portal-mesenteric vein) [194].

Physiological control factors and vasoactive changes that influence the peripheral circulation can do so by affecting the membrane potential, the pacemaker waves and the spike discharge in vascular smooth muscle [194]. Stimulation of portal-mesenteric veins by norepinephrine or epinephrine through alpha-adrenoceptors is characterized by depolarization and an increase in the frequency of action potentials. The smooth muscle of main pulmonary artery is depolarized in a dose-dependent manner by norepinephrine but no action potentials normally occur. Thus a given vasoactive agent can act differently on different vascular smooth muscle; the mechanism of action on a spike-generating cell may not be the same as in a gradedly responding cell [194].

Another outstanding characteristic of vascular smooth muscle is the striking individuality of the contractile activity of the various vascular segments even in the same type of vascular smooth muscle. Many differences in reactivity of vascular smooth muscle can be explained in terms of interference with its innervation, the degree of which varies considerably, or in terms of modulating influences which profoundly affect the smooth muscle [389]. These influences will be reviewed in detail in Chap. 7. In spite of the striking heterogeneity of contractile reactivity in various vascular segments, the basic events in the contractile process of all vascular and non-vascular smooth muscle are essentially identical.

In skeletal and cardiac muscle the time from initial depolarization to initiation of contraction is less than 10 ms. In contrast, smooth muscle starts to contract about 200 ms after the start of the spike with the peak of contraction being reached as long as 500 ms after the spike. Thus, the excitation–contraction coupling in smooth muscle is a very slow process compared to that in skeletal and cardiac muscle. Contraction is initiated in smooth muscle, as in striated muscle, by a rise in cytoplasmic Ca^{2+}. The rise in cytoplasmic Ca^{2+} is generally triggered by some signal through the surface membrane: an action potential and/or graded depolarization (electromechanical coupling) or by pharmacomechanical coupling, a trigger mechanism that is independent of, although it may occur in conjunction with, changes in surface membrane potential [194,360]. It is likely, although not directly demonstrated in the case of graded depolarization, that each of these mechanisms of excitation–contraction coupling can release Ca^{2+} from an intracellular store in smooth muscle [194,360]. Recently, the recycling of intracellular stores of Ca^{2+} in depolarized vascular smooth muscle in calcium-free solutions has been demonstrated, indicating that these intracellular stores of Ca^{2+} are sufficient to activate a maximal contraction and that pharmacomechanical coupling can release intracellular Ca^{2+}, and operate independently of Ca^{2+} influx from the extracellular space [21].

The sarcoplasmic reticulum (SR) is the organelle primarily responsible for the physiological regulation of cytoplasmic Ca^{2+} in smoothmuscle cells as it is in skeletal muscle. Although the volume of SR is relatively small in smooth muscle as compared to skeletal muscle, it has been shown that an SR that occupies no more than 2% of the cell volume can store enough Ca^{2+} to activate contraction [21]. Furthermore, the real discrepancy between the volume of intracellular Ca^{2+} stores in the SR of mammalian smooth muscle and that of striated muscle is not as great as indicated by the differences in total SR volume (e.g. 1%–2.5% in mammalian phasic smooth muscle versus 9% in frog twitch muscle). Ca^{2+} storage in SR of frog striated muscle is largely confined to the terminal cisternae [364] which constitute only about 3%–4% of the cell volume. It is generally accepted that activation of vertebrate skeletal (and mammalian cardiac) muscle releases Ca^{2+} from the terminal cisternae of the SR. During relaxation

Ca^{2+} is pumped back into the SR.

The SR is an intracellular membrane system of tubules and, in some species, flattened cisternae. It occupies 1.5% to 7.5% of the total smooth-muscle cell volume, being most extensive in the large elastic arteries (e.g. rabbit aorta and main pulmonary artery) and in oestrogen-treated or pregnant uterus [133,363]. Surface couplings are structural specializations where a 12–20 nm space separating the cytoplasmic leaflets of the plasma membrane and the adjacent junctional SR is spanned by periodic bridging structures. The Ca^{2+} concentration in the SR is approximately 30–50 mmol/kg dry weight [21]. This is less than the concentration of Ca^{2+} in the terminal cisternae of striated muscle that contain calsequestrin, but very significantly higher than the Ca^{2+} concentration in the longitudinal reticulum in striated muscle that is largely devoid of calcium-binding proteins. Therefore, it is likely that in smooth (as in striated) muscle, a significant fraction of the Ca^{2+} within the lumen of the SR is in bound form, possibly associated with calcium-binding proteins [360].

Accumulation of Ca^{2+} by the SR has been demonstrated by direct electron optical methods in smooth-muscle preparations [360]. Physiological experiments demonstrating the ATP dependence of such Ca^{2+} uptake also identified the SR, rather than the plasma membrane, as the site of active Ca^{2+} accumulation in smooth muscle [98]. Electron probe analysis of the SR in smooth muscles indicates that excitatory stimuli release Ca^{2+} mostly from the junctional SR, although release can also take place from the more deeply located, central elements of SR that are not connected by bridging structures to the surface membrane [360]. The mechanism of electrical triggering of Ca^{2+} release from the "junctional SR" of smooth-muscle cells (as also from the terminal cisternae of striated muscle) is a major unresolved problem of muscle physiology [71,360].

Mitochondria have a relatively low affinity but very large capacity for accumulating Ca^{2+}. Massive loading of isolated as well as in situ mitochondria with Ca^{2+} occurs when mitochondria are exposed to high concentrations of these cations in the presence of an energy source and precipitating anion [362]. However, this massive, energy-supported mitochondrial Ca^{2+} transport (resulting usually in calcium phosphate granules in the mitochondrial matrix) occurs only in severely damaged smooth-muscle cells. Vascular calcification may be initiated from foci of such calcified mitochondria [361]. The normal mitochondrial Ca^{2+} concentration is in the range compatible with the suggestion that the free Ca^{2+} in the mitochondrial matrix regulates some of the mitochondrial enzymes. Thus mitochondria do not regulate cytoplasmic Ca^{2+} but may themselves be metabolically regulated by small changes in the level of free Ca^{2+} in the mitochondrial matrix space [162].

The surface membrane of smooth-muscle cells can be partitioned, rather arbitrarily, into three structural domains: (a) the surface caveolae, also called surface vesicles or plasmalemmal vesicles, (b) the regions forming surface couplings and connected to the junctional SR by bridging structures and (c) the "non-specialized" surface membrane. The latter is partially occupied by dense bands specialized for actin filament attachment and force transmission in smooth muscle (see Fig. 1.8). The surface caveolae are invaginations of the surface membrane that communicate with the extracellular space. It is questionable whether surface caveolae can detach or are capable of pinocytotic activity, in contrast to the coated pits or coated vesicles that contain low-density lipoprotein receptors and can be internalized [147]. Surface caveolae make up a significant part (about 25%–50%) of the membrane area; their significance in smooth-muscle cells however is unknown. It has been proposed that they may serve as stretch receptors or they may be involved in ionic transport and cell volume regulation.

The bidirectional transport of Ca^{2+} across the plasma membrane and the role of membrane-bound Ca^{2+} itself are the two major functional aspects of the participation of the surface membrane in cellular Ca^{2+} metabolism. There is evidence showing that excitatory agents can increase Ca^{2+} influx across the surface membrane which normally has a low permeability to Ca^{2+} and other divalent cations. An active Ca^{2+} efflux system removes Ca^{2+} from the smooth muscle, and maintains normal, steady-state cellular Ca^{2+}, in the face of periods of Ca^{2+} influx. The major efflux pathway is probably via the calmodulin-stimulated Ca^{2+} transport ATPase of smooth-muscle surface membranes [85].

The contribution of Ca^{2+} influx to activation of contraction, as distinguished from its role as a charge carrier in the action potential, is uncertain both under physiological and experimental conditions. Experimental data suggest that Ca^{2+} accumulation by the SR plays the dominant role in relaxation. The findings, together with observations showing that sufficient Ca^{2+} is stored in the SR for activating contraction [21], raise the possibility that transmembrane Ca^{2+} fluxes make only a very minor contribution to contractile activation during physiological smooth-muscle contraction. If this is indeed the case, then the major role of the calmodulin-stimulated calcium-ATPase would be the removal of the small amounts of Ca^{2+} gained through action potentials or graded depolarization that while insufficient to activate contraction, could essentially overload the SR. This conjecture however remains to be verified [360], as does the suggestion that filling of the Ca^{2+} stores (sarcoplasmic reticulum) can occur directly from the extracellular space through channels in "bridging structures", rather than indirectly (from the cytoplasm) through the plasma membrane [48]. The role of membrane-bound Ca^{2+} itself similarly has not been established in excitation–contraction of smooth muscle.

Electron probe microanalysis has shown that the amount of Ca^{2+} bound to the surface membrane is very much lower than that contained in the lumen of the SR [21], and experimental data suggest that although the minute amount of Ca^{2+} bound to the surface membrane may be physiologically significant, it is not necessary for either excitation–contraction coupling or for activation of smooth muscle in normal conditions [360].

An increase in peripheral vascular resistance is the fundamental cause of essential hypertension and in models of genetic hypertension a generalized defect in membrane calcium regulation, that is genetically transmitted rather than acquired as a consequence of hypertension, has been postulated [223,290,386]. A vascular smooth-muscle cell characteristic that has been consistently associated with increased reactivity and hypertension in the spontaneously hypertensive rat is increased activity of the electrogenic Na^+–K^+ pump (membrane transport ATPase) that contributes to membrane potential both at rest and during excitation [168,169]. The exact mechanism of increased smooth-muscle cell reactivity in this hypertensive model however is not known. Although increased activity of the Na^+–K^+–ATPase pump may be sufficient to explain the increased smooth-muscle reactivity, it is also possible that the altered Na^+ influx and extrusion may effect intracellular Ca^{2+}-release steps important in excitation–contraction coupling [169]. Furthermore, vascular reactivity is the most complex expression of the total composition of the vascular smooth-muscle cell itself, as well as of the elements with which it is organized into blood vessels. Vascular hyper-responsiveness in hypertension is probably a process dominated by hyper-reactivity of smooth-muscle cells in its early phase which merges imperceptibly into a state dominated by geometric redesign of the vascular wall and new equilibrium in later phases [113,275].

Synthetic Activities

By incorporating labelled amino acids into collagen and into elastic fibre proteins, it has been demonstrated that in the artery wall the smooth-muscle cells are the principal cells that form connective tissue both in the media and in the intima of developing and mature vessels [138,139,309]. Synthesis of collagen by arterial smooth-muscle cells has also been shown in culture [39] (see Chaps. 2, 5 and 12).

The mature elastic fibre is composed of two morphologically distinct structures representing two distinctly different proteins. The principal component of the elastic fibre, the protein elastin, has an amorphous appearance. This component is characterized by its paucity of polar amino acids and by its unique cross-linking amino acids (desmosine and isodesmosine) derived from quaternary linkages of lysine residues. The second component of the elastic fibre is a microfibril that is approximately 11 nm in diameter. This microfibril is a glycoprotein that is rich in polar amino acids, contains between 30 and 80 residues of cystine per 1000 amino acids and is highly disulphide cross-linked (see Chap. 2). It has been demonstrated that arterial smooth-muscle cells are capable of forming both of the elastic fibre proteins in culture [268]. It has been shown also that arterial smooth-muscle cells secrete lysyl oxidase [149], an enzyme playing a pivotal role in the formation of cross-linkages that stabilize the fibrous structures of both collagen and elastin, thus providing anchoring points important to the tensile and/or elastic properties exhibited by these proteins [202].

Proteoglycans are complex macromolecules that consist of carbohydrate polymers (glycosaminoglycans) covalently bound to a protein core [165]. Correlating with this structural model, recent observations suggest that proteoglycans exist as extended bottlebrush structures within the extracellular matrix of blood vessels [60]. These macromolecules occupy large hydrodynamic domains within the extracellular matrix of blood vessels and serve as important structural links between the fibrous and cellular components. Since these macromolecules are extremely hydrophilic because of their anionic charge, they contribute to the turgor pressure and viscoelasticity of arteries and veins. Studies using cultured arterial smooth-muscle cells of primates have demonstrated that 60%–80% of the glycosaminoglycan formed by the smooth-muscle cells is dermatan sulphate, approximately 10% is a mixture of chondroitin sulphates A and C and less than 5% is hyaluronic acid. The ability of primate arterial smooth-muscle cells to form such large quantites of dermatan sulphate in cultures is of interest because of the potential importance of the interactions of dermatan sulphate with low-density lipoproteins [188] and the potential ability of these glycosaminoglycans to bind and hold low-density lipoproteins during atherosclerotic lesion formation in vivo (see Chap. 13).

At the ultrastructural level connective tissue synthesis correlates with the presence of rough endoplasmic reticulum, Golgi complexes and free ribosomes in the cytoplasm of smooth-muscle cells. An extensive synthetic apparatus in mammalian uterine smooth-muscle cells appears in parallel with progressive increases in uterine collagen during pregnancy and a similar effect can be elicited under oestrogen stimulation [58]. In aortic smooth muscle, rough endoplasmic reticulum and Golgi complexes are prominent during development and in various pathophysiological situations when synthesis of the extracellular matrix is greatly increased as compared to that occurring in the aortic media of the normal adult [58].

In addition to their capacity to form enormous amounts of connective tissue matrix [40], vascular smooth-muscle cells can synthesize several prostaglandin derivates, including prostaglandin I_2 (prostacyclin) and prostaglandin E [252]. Vascular smooth-muscle cells contain receptors for low-density lipoprotein (LDL) [54], and for growth factors, including platelet-derived growth factor (PDGF) [24,412]; they can accumulate lipid and can take on the appearance of foam cells. Lysosomes play an important role in catabolic metabolism within cells and have been shown to be important in cellular regulatory processes [29]. Lysosomes have been identified in vascular smooth-muscle cells both in situ and in culture conditions [303]. Changes in smooth-muscle cell lysosomes have been observed in models for vascular disease. These changes include, for example, intralysosomal accumulation of lipid in atherosclerosis, stimulation of lysosomal activities in hypertension and decreased lysosomal function in diabetes mellitus [107]. These and other examples of synthetic and metabolic activities as well as endocrine functions (e.g. renin synthesis) will be illustrated in Chap. 9.

We have discussed the significance of two different phenotypic states of smooth muscle, the so-called "contractile" and "synthetic" states. It has been suggested that these phenotypic states may represent different states of activity and suceptibility to mitogens and other blood constituents [59]. There have also been observations that smooth-muscle cells in certain situations can secrete a growth factor that appears to be a form of PDGF. Secretion of PDGF by embryonic or appropriately stimulated smooth-muscle cells may be responsible for autocrine or paracrine stimulation [404] and for the altered growth characteristics of these cells in atherosclerotic lesion formation as discussed in more detail in Chap. 11 of this book.

Interactions Between Endothelial and Smooth-Muscle Cells

Communication between cells is an essential process in embryological development and is important for the maintenance of normal tissue physiology. It is also necessary for a number of pathophysiological responses. In vascular tissue, structural and metabolic interactions occur between endothelium and smooth-muscle cells, the normal resident cells of arterial intima and media. Additional cell interactions are established during the development of atherosclerotic lesions when the adhesion of circulating leucocytes to endothelium is followed by mononuclear cell infiltration into the vessel wall. The monocytes differentiate to become tissue macrophages, and thus become a major component of the intima which can interact with endothelial and smooth-muscle cells [77]. Cell interactions during the development of atherosclerotic lesions will be the subject of Chap. 11. Here we discuss briefly examples of interactions only between the endothelial and smooth-muscle cells, the two major cellular components of normal artery wall.

Communication between endothelial and smooth-muscle cells may involve humoral mechanisms in which soluble mediators pass between these cells in one or both directions or there may be direct cell-to-cell contact between these cells where regions of the respective plasma membranes fuse to form gap junctional channels [77].

Communication Between Endothelial and Smooth-Muscle Cells Mediated Via Humoral Mechanisms

Two of the best examples of humoral communication between endothelial and vascular smooth-muscle cells are: (a) the regulation of smooth-muscle proliferation by endothelial-derived growth factors and growth inhibitors and (b) the regulation of smooth-muscle vasoactivity by endothelial-derived relaxing factor.

Endothelial-Cell-Derived Growth Factors (EDGF)

Endothelial-cell-derived growth factors (EDGF) are mitogens. As a significant proportion of these factors are PDGF-like [91,92], they are also called cellular PDGF (PDGF$_c$) [77]. Using a V-sis cDNA probe, transcriptional expression of PDGF$_c$ has been demonstrated in cultured human and bovine endothelial cells [10,69,405] and low levels of expression were also detected in human and bovine endothelium scraped from the lumen of human umbilical vein and bovine aorta respectively [10]. Studies on the regulation of PDGF$_c$ production by cultured endothelial cells suggested furthermore that injurious agents may stimulate PDGF$_c$ production in vivo (108) and that PDGF$_c$ message levels may be associated with endothelial cell proliferation during events such as neovascularization or repair of vascular injury. In addition to the role of a stimulated or altered endothelium in smooth-muscle cell proliferation, the normally quiescent and non-growing vascular smooth muscle also can synthesize and release PDGF$_c$, thus promoting its own proliferation in an autocrine fashion when it undergoes hyperplasia [336,404]. This situation may apply to normally developing media and pathological intimal proliferation [77].

Endothelial-Mediated Inhibition of Vascular Smooth-Muscle Cell Proliferation

Substantial evidence that endothelial cells produce inhibitors of smooth-muscle proliferation has accumulated from in vitro studies. Endothelial cells synthesize and secrete sulphated mucopolysaccharides and glycosaminoglycans, including heparan sulphates [23]. It has been demonstrated that heparin suppresses intimal smooth-muscle cell proliferation after endothelial injury in vivo [66,160], and inhibits serum-stimulated smooth-muscle cell proliferation in vitro [171], and it has also been shown that endothelial-cell-derived heparin inhibits smooth-muscle cell growth [49,50,51, 52,206]. It was suggested furthermore that a platelet endoglycosidase liberates directly active heparin-like residues from the endothelial cell surface [51]. Subsequently smooth-muscle cells were also found to secrete an anti-proliferative heparan sulphate and a lysosomal endoglycosidase which was identical to the platelet enzyme, suggesting autocrine inhibition of proliferation [115]. These data suggest that a tendency towards mitogen-mediated intimal smooth-muscle cell proliferation in arteries may be continuously suppressed by heparin-like inhibitors, the secretion of which is regulated by autocrine and exocrine mechanisms. If the balance of mitogen to heparin is tipped in favour of the mitogen by compromised endothelial cell function or increased mitogen availability, the equilibrium may tend towards proliferation [77].

Endothelial-Cell-Derived Relaxing Factor (EDRF) and Smooth-Muscle Cyclic Guanosine Monophosphate (cGMP)

Both convective transvascular fluid motion towards the lymph drainage and passive diffusion in extracellular fluid favour redistribution of endothelial products in the intimal and medial interstitium of blood vessel [77]. Several important discoveries concerning vasoactive agents produced by endothelial cells and acting specifically upon medial smooth-muscle cells have been made since the mid-1970s using intact blood vessels. It was discovered that endothelium is a major source of PGI_2 [252], a potent stimulator of cAMP formation which results in vasodilatation [172]. However, PGI_2 is also synthesized and released by smooth-muscle cells themselves [88] and thus regulation of vasoactivity by PGI_2 may not be related strictly to endothelial–vascular smooth muscle interaction.

Recently, the existence of an endothelially derived agent apparently directed specifically at vascular smooth-muscle cells was identified. It was discovered that in isolated blood-vessel preparations, endothelial cells are essential for smooth-muscle cell relaxation induced by acetylcholine [118,120]. Several studies have suggested that vasodilatation induced by a number of compounds (serotonin [119], histamine [388], bradykinin [61]) is dependent upon stimulation of the release of an EDRF. The humoral nature

of EDRF was established in part by using isolated vessels with intact or denuded endothelium [156,157,158]. The synthesis and release of EDRF from pure cultures of endothelial cells and its confirmation in vitro as a humoral agent has also been demonstrated [68,131,132,281].

Although the chemical structure and cellular mechanisms involved in EDRF metabolism are still unclear, its action upon vascular smooth muscle involves the stimulation of cyclic guanosine monophosphate (cGMP) formation and is linked to the phosphorylation of several cellular proteins [90,292]. In vitro EDRF has an extremely short half-life (approximately 6 s) although recent reports indicate that the half-life of EDRF in the vessel wall may be increased by a factor of almost ten over that measured in vitro [145,161]. It does not require contact between endothelial cells and smooth-muscle cells for its activity [132]. Recent co-culture studies have confirmed the specific increase of only smooth muscle cGMP following appropriate stimulation. By separating endothelial and smooth-muscle cells 2 to 3 mm, a distance which could not be traversed by active EDRF, an increase in cGMP levels in the smooth-muscle cells was prevented [77]. Thus, the production and release of EDRF by endothelial cells is an excellent example of the unidirectional effect of a short-lived humoral agent [77].

Communication Between Endothelial and Smooth-Muscle Cells Mediated Via Direct Cell-to-Cell Contacts

These cell-to-cell contacts are based on the presence of myoendothelial junctions in arteries and arterioles. As has been particularly well documented for the microcirculation [298,299] and to a lesser extent for larger vessels [366,379], endothelium and smooth-muscle cells frequently extend processes to form heterocellular zones of contact, the myoendothelial junctions. In arteries, these cytoplasmic extensions pass through the fenestrations of the internal elastic lamina and they were first recognized as early as 1921 [379]. It is only recently, however, that the gap junctions have been demonstrated in myoendothelial bridges [366]. In fact, the majority of myoendothelial junctions are probably gap junctions. These are sites of coupling where intra-membrane protein channels allow communication between the aqueous cytoplasm of both endothelial cells and smooth-muscle cells [228,385] (see above). Small aggregates of gap junctional channels may be sufficient for communication, explaining the relative rarity of structurally well-defined gap junctions at sites of myoendothelial contacts (Hüttner; unpublished observations).

In contrast, structural [176,177,183,221,222, 265,343,345,346] and functional [78,221,222] studies have demonstrated abundant and regular arrays of communicating gap junctions between arterial and venous (but not capillary) endothelial cells in vivo, in isolated confluent sheets of endothelial cells in primary culture and in multipassaged cultures. Gap junctions clearly represent a potential open channel for the equilibration of intracellular metabolites throughout the endothelial cell monolayer although there may be some regionality in the extent of communication.

Recent investigations using a method of heterocellular contact co-culture and microcarrier techniques have demonstrated rapid, bidirectional transfer of nucleotide between endothelial and smooth-muscle cells [78]. It was concluded from these studies that the attachment of microcarrier-bound endothelial cells to monolayers of vascular smooth-muscle cells resulted in the rapid formation of gap junctions between the two cell types and that intercellular transport of small molecules such as uridine phosphates occurred in both directions [78]. Nevertheless the role of gap junctions between endothelial cells and in myoendothelial junctions is largely unknown in vivo. Gap junctional communication in vivo may play a role in growth control complementing the proven humoral interactions which occur via specific growth factors between cells of the arterial tissue.

A suggested role of endothelium as a receptor system for smooth-muscle cells is particularly intriguing [176,298,301]. In this scheme, coupled endothelial cells in arteries may act as a unit and this interaction, along with that of myoendothelial junctions, may help explain the ability of an intravascular hormone to affect vascular tone without gaining access to all cells of the vascular well. Vascular endothelium thus would function as a receptor system for smooth-muscle cells in which the smooth muscle responds to secondary signals generated in the endothelium and transferred intracellularly via gap junctions [78,176,222,298,299]. In support of this idea are studies which have demonstrated aortic contractility in response to the application of norepinephrine-coated glass beads to the endothelium [17] and of angiotensin–protein complexes which could not reach medial smooth-muscle cells by the time of contraction [301].

Coordination of vasodilation among resistance vessels is necessary for optimal distribution of blood flow in various physiological and pathophysiological situations. Direct interactions between the endothelial cell layer and the coupled system of medial smooth-muscle cells is also a likely explanation for propagated vasodilation as observed in intact arteriole preparations [335].

References

1. Adelstein, RS, Eisenberg E (1980) Regulation and kinetics of the actin-myosin-ATP interaction. Ann Rev Biochem 49:921–956
2. Anderson RGW, Brown MS, Goldstein JL (1977) Role of the coated endocytic vesicle in the uptake of receptor-bound low density lipoprotein in human fibroblasts. Cell 10:351–364
3. Anderton BH (1981) Intermediate filaments: a family of homologous structures. J Muscle Res Cell Motil 2:141–166
4. Angus JA, Cocks TM, Satoh K (1986) The alpha-adrenoceptors on endothelial cells. Fed Proc 45:2355–2359
5. Ashton FT, Somlyo AV, Somlyo AP (1975) The contractile apparatus of vascular smooth muscle: intermediate high voltage stereo electron microscopy. J Mol Biol 98:17–29
6. Aumailley M, Timpl R (1986) Attachment of cells to basement membrane collagen type IV. J Cell Biol 103:1569–1575
7. Baldwin AL, Winlove CP (1984) Effect of perfusate composition on binding of ruthenium red and gold colloid to glycoalyx of rabbit aortic endothelium. J Histochem Cytochem 32:259–266
8. Barja F, Coughlin C, Belin D et al. (1986) Actin isoform synthesis and mRNA levels in quescent and proliferating rat aortic smooth muscle cells in vivo and in vitro. Lab Invest 55:226–233.
9. Barnes JL, Radnik RA. Gilchrist EP et al. (1984) Size and charge selective permeability defects induced in glomerular basement membrane by a polycation. Kidney Int 25:11–19
10. Barrett TB, Gajdusek CM, Schwartz SM et al. (1984) Expression of the *sis* gene by endothelial cells in culture and in vivo. Proc Natl Acad Sci USA 81:6772–6774
11. Bearer EL, Orci L, Sors P (1985) Endothelial fenestral diaphragms: a quick-freeze, deep-etch study. J Cell Biol 100:418–428
12. Becker CG, Nachman RL (1973) Contractile proteins of endothelial cells, platelets and smooth muscle. Am J Pathol 71:1–22
13. Beckerle MC (1986) Identification of a new protein localized at sites of cell substrate adhesion. J Cell Biol 103:1679–1687
14. Bendayan M (1980) Use of the protein A-gold technique for the morphological study of vascular permeability. J Histochem Cytochem, 28:1251–1254
15. Bennett GS, Fellini SA, Croop JM et al. (1978) Differ-

ences among 100-Å filament subunits from different cell types. Proc Natl Acad Sci USA 75:4364–4368

16. Bennett MV, Spray DC (1985) Gap junctions. Cold Spring Harbor Laboratory, Cold Spring Harbor, New York

17. Bevan JA, Duckles SP (1975) Evidence for alpha-adrenergic receptors on intimal endothelium. Blood Vessels 12:307–319

18. Bevilacqua MP, Pober JS, Wheeler ME et al. (1985) Interleukin-1 activation of vascular endothelium. Effects on procoagulant activity and leukocyte adhesion. Am J Pathol 121:393–403

19. Birdwell CR, Gospodarowicz D, Nicolson GL (1978) Identification, localization, and role of fibronectin in cultured bovine endothelial cells. Proc Natl Acad Sci USA 75:3273–3277

20. Bloom W, Fawcett DW (1986) A textbook of histology, 11th edn. Saunders, Philadelphia, pp 367–405.

21. Bond M, Kitazawa T, Somlyo AP et al. (1984) Release and recycling of calcium by the sarcoplasmic reticulum in guinea pig portal vein smooth muscle. J Physiol (Lond) 355:677–695

22. Bond M, Somlyo AV (1982) Dense bodies and actin polarity in vertebrate smooth muscle. J Cell Biol 95:403–413

23. Bounassisi V (1973) Sulfated mucopolysaccharide synthesis and secretion in endothelial cell cultures. Exp Cell Res 76:363–368

24. Bowen-Pope DF, Seifert RA, Ross R (1985) The platelet-derived growth factor receptor. In: Boynton AL, Leffert HL (eds) Control of animal cell proliferation: recent advances, vol 1. Academic Press, New York, pp 281–312

25. Boyles J, L'Hernault N, Laks H et al. (1981) Evidence for vesicular shuttle in heart capillaries. J Cell Biol 91:418 (abstract)

26. Bradbury MWB (1985) The blood–brain barrier. Transport across the cerebral endothelium. Circ Res 57:213–222

27. Branton D, Bullivant S, Gilula NB et al. (1975) Freeze-etching nomenclature. Science 190:54–56

28. Bretcher MS, Raff MC (1975) Mammalian plasma membranes. Nature 258:43–49

29. Brown MS, Ho Y, Goldstein J (1980) The cholesteryl ester cycle in macrophage foam cells. Continued hydrolysis and reesterification of cytoplasmic cholesteryl esters. J Biol Chem 255:9344–9352

30. Brown PJ, Juliano RL (1986) Expression and function of a putative cell surface receptor for fibronectin in hamster and human cell lines. J Cell Biol 103:1595–1603

31. Bruns RR, Palade GE (1968) Studies on blood capillaries. I. General organization of muscle capillaries. J Cell Biol 37:244–276

32. Bruns RR, Palade GE (1968) Studies on blood capillaries. II. Transport of ferritin molecules across the wall of muscle capillaries. J Cell Biol 37:277–299

33. Bullivant S (1981) Possible relationship between tight junction structure and function. In: Leader JP (ed) Epithelial ion and water transport. Raven Press, New York, pp 265–275

34. Bundgaard M (1980) Transport pathways in capillaries in search of pores. Ann Rev Physiol 42:325–336

35. Bundgaard M, Frokjaer-Jensen J (1982) Functional aspects of the ultrastructure of terminal blood vessels: a quantitative study on consecutive segments of the frog mesenteric microvasculature. Microvasc Res 23:1–30

36. Bundgaard M, Frokjaer-Jensen J, Crone C (1979) Endothelial plasmalemmal vesicles as elements in a system of branching invaginations from the cell surface. Proc Natl Acad Sci USA 76:6439-6442

37. Bundgaard M, Hayman P, Crone C (1983) The three dimensional organization of plasmalemmal vesicular profiles in the endothelium of rat heart capillaries. Microvasc Res 25:358–369

38. Buonassisi V, Venter JC (1976) Hormone and neurotransmitter receptors in an established vascular endothelial cells line. Proc Natl Acad Sci USA 73:1612–1616

39. Burke JM, Ross R (1977) Collagen synthesis by arterial smooth muscle cells during proliferation and quiescence in culture. Exp Cell Res 107:387–395

40. Burke JM, Ross R (1979) Synthesis of connective tissue macromolecules by smooth muscle. Int Rev Connect Tissue Res 8:119–157

41. Burridge K (1981) Are stress fibers contractile? Nature 294:691–692

42. Burridge K, Connell L (1983) A new protein of adhesion plaques and ruffling membranes. J Cell Biol 97:359–367

43. Burridge K, Mangeat P (1984) An interaction between vinculin and talin. Nature 308:744–745

44. Busch C, Cancilla PA, De Bault et al. (1982) Use of endothelium cultured on microcarriers as a model for the microcirculation. Lab Invest 47:498–504

45. Byers HR, Fujiwara K (1982) Stress fibers in cells in situ: immunofluoresence visualization with antiactin, antimyosin and anti-alpha-actinin. J Cell Biol 93:804–811

46. Campbell GR, Chamley JH, Burnstock G (1984) Development of smooth muscle cells in tissue culture. J Anat 117:295–312

47. Caspar D, Goodenough D, Makowski L et al. (1977) Gap junction structures. I. Correlated electron microscopy and x-ray diffraction. J Cell Biol 74:605–628

48. Casteels R, Droogmans G (1981) Exchange characteristics of the noradrenaline-sensitive calcium store in vascular smooth muscle cells of rabbit ear artery. J. Physiol (Lond) 317:263–279

49. Castellot JJ, Addonizio ML, Rosenberg RD et al. (1981) Cultured endothelial cells produce a heparin-like inhibitor of smooth muscle cell growth. J Cell Biol 90:372–379

50. Castellot JJ, Choay J, Lormeau J-C et al. (1986) Structural determinants of the capacity of heparin to inhibit the proliferation of vascular smooth muscle cells. II. Evidence for a pentasaccharide sequence that contains a 3-0-sulfate group. J Cell Biol 102:1979–1984

51. Castellot JJ, Favreau LV, Karnovsky MJ et al. (1982) Inhibition of vascular smooth muscle cell growth by endothelial cell-derived heparin: possible role of a platelet endoglycosidase. J Biol Chem 257:11256–11260

52. Castellot JJ, Rosenberg RD, Karnovsky MJ: Endothelium, heparin and the regulation of vascular smooth muscle cell growth. In: Jaffe E (ed) Biology of endothelial cells. Martinus Nijhoff, Boston, Massachusetts, pp 118–128

53. Caulfield JP, Farquhar MG (1986) Distribution of anionic sites in glomerular basement membrane, their possible role in filtration and attachment. Proc Natl. Acad Sci USA 73:1646–1650

54. Chait A, Ross R, Alberts JJ et al. (1980) Platelet-derived growth factor stimulates activity of low density lipoprotein receptors. Proc Natl Acad Sci USA 77:4084–4088

55. Chambers RW, Zweifach BW (1947) Intercellular cement and capillary permeability. Physiol Rev 27:436–463

56. Chamley-Campbell JH, Campbell GR (1981) What controls smooth muscle phenotype? Atherosclerosis 40:347–357

57. Chamley-Campbell JH, Campbell GR, McConnell JD et al. (1977) Comparison of vascular smooth muscle cells from adult human, monkey and rabbit in primary culture and in subculture. Cell Tissue Res 177:503–522

58. Chamley-Campbell JH, Campbell GR, Ross R (1979) The smooth muscle cell in culture. Physiol Rev 59:1–61

59. Chamley-Campbell JH, Campbell GR, Ross R (1981) Phenotype-dependent response of cultured aortic smooth muscle to serum mitogens. J. Cell Biol 89:379–383

60. Chen K, Wight TN (1984) Proteoglycans in arterial smooth muscle cell cultures: an ultrastructural histochemical analysis. J Histochem Cytochem 32:347–357

61. Cherry PD, Furchgott RF, Zawadzki JV et al. (1982) Role of endothelial cells in relaxation of isolated arteries by bradykinin. Proc Natl Acad Sci USA 79:2106–2210

62. Cheung WY (1982) Calmodulin. Sci Am 246:48–56

63. Claude P, Goodenough DA (1973) Fracture faces of zonulae occludentes from "tight" and "leaky" epithelia. J Cell Biol 58:390–400

64. Clementi F, Palade GE (1969) Intestinal capillaries. I. Permeability to peroxidase and ferritin. J Cell Biol 41:33–58

65. Clough G, Michel CC (1981) The role of vesicles in the transport of ferritin through frog endothelium. J Physiol (Lond) 315:127–142

66. Clowes AW, Karnovsky MJ (1977) Suppression by heparin of smooth muscle cell proliferation in injured arteries. Nature 265:625–626

67. Clowes AW, Reidy MA, Clowes MM (1983) Mechanisms of stenosis after arterial injury. Lab Invest 49:208–215

68. Cocks TM, Angus JA, Campbell JH et al. (1985) Release and properties of endothelium-derived relaxing factor (EDRF) from endothelial cells in culture. J Cell Physiol 123:310–320

69. Collins T, Korman AJ, Wake CT et al. (1984) Immune interferon activates multiple class II major histocompatibility complex genes and the associated invariant chain gene in human endothelial cells and dermal fibroblasts. Proc Natl Acad Sci USA 81:4917–4921

70. Colucci M, Balconi G, Lorezet R et al. (1983) Cultured human endothelial cells generate tissue factor in response to endotoxin. J Clin Invest 71:1893–1896

71. Costello B, Chadwick C, Saito A et al. (1986) Characterization of the junctional face membrane from terminal cisternae of sarcoplasmic reticulum. J Cell Biol 103:741–753

72. Couchman JR, Badley RA, Rees DA (1983) Redistribution of microfilament-associated proteins during the formation of focal contacts and adhesions in chick fibroblasts. J Muscle Res Cell Motil 4:647–661

73. Couchman JR, Hook M, Rees DA, Timpl R (1983) Adhesion, growth and matrix production by fibroblasts on laminin subunits. J Cell Biol 96:177–183

74. Craig SW (1985) Alpha-actinin, an f-actin cross-linking protein, interacts directly with vinculin. J Cell Biol 101:136 (abstract)

75. Crone CY (1986) Modulation of solute permeability in microvascular endothelium. Fed Proc 45:77–83

76. Curry FE, Michel CC (1980) A fiber matrix model of capillary permeability. Microvasc Res 20:96–99

77. Davies PF (1986) Biology of disease: vascular cell interactions with special reference to the pathogenesis of atherosclerosis. Lab Invest 55:5–24

78. Davies PF, Ganz P, Diehl PS (1985) Methods in laboratory investigation: reversible microcarrier-mediated junctional communication between endothelial and smooth muscle cell monolayers. An in vitro model of vascular cell interactions. Lab Invest 53:710–718

79. Davies PF, Rennke HG, Cotran RS (1981) Influence of molecular charge upon the endocytosis and intracellular fate of peroxidase activity in cultured arterial endothelium. J Cell Sci 49:69–86

80. Day AS, Alavi M, Moore S (1985) Influx of (^{3}H, ^{14}C) cholesterol-labelled lipoprotein into re-endothelialized and de-endothelialized areas of ballooned aortas in normal-fed and cholesterol-fed rabbits. Atherosclerosis 55:313–330

81. Decker RS (1976) Hormonal regulation of gap junction differentiation. J Cell Biol 69:669–685

82. De Chastonay C, Gabbiani G, Elemer G et al. (1983) Remodeling of the rat aortic endothelial layer during experimental hypertension. Changes in replication rate, cell density, and surface morphology. Lab Invest 48:45–52

83. De Forrest JM, Hollis TM (1978) Shear stress and aortic histamine synthesis. Am J Physiol 234:H701–H705

84. De Mey JG, Vanhoutte PM (1982) Heterogenous behavior of the canine arterial and venous wall. Importance of the endothelium. Circ Res 51:439–447

85. De Schutter G, Wuytack F, Verbist J, Casteels R (1984) Tissue levels and purification by affinity chromatography of the calmodulin-stimulated Ca^{2+}-transport ATP-ase in pig antrum smooth muscle. Biochim Biophys Acta 773:1–10

86. Devine CE (1978) Vascular smooth muscle morphology and ultrastructure. In: Kaley G, Altura BM (eds) Micro-circulation, vol II. University Park Press, Baltimore, pp 4–39

87. Devine CE, Somlyo AP (1971) Thick filaments in vascular smooth muscle. J Cell Biol 49:636–649

88. De Witt DL, Day JS, Sonnenburg WK et al. (1983) Concentrations of prostaglandin endoperoxide synthase and prostaglandin I_2 synthase in the endothelium and smooth muscle of bovin aorta. J Clin Invest 72:1882–1888

89. Diamond JM (1978) Channels in epithelial cell membranes and junctions. Fed Proc 37:2639–2644

90. Diamond J, Chu EB (1983) Possible role for cyclic GMP in endothelium-dependent relaxation of rabbit aorta by acetylcholine. Comparison with nitroglycerin. Res Commun Chem Pathol Pharmacol 41:369–381

91. Di Corletto PE, Bowen-Pope DF (1983) Cultured endothelial cells produce a platelet-derived growth factor-like protein. Proc Natl Acad Sci USA 80:1919–1923

92. Di Corletto PE, Gajdusek CM, Schwartz SM et al. (1983) Biochemical properties of the endothelium-derived growth factor: comparison to other growth factors. J Cell Physiol 114:339–345

93. Dicorleto PE, Zilversmit DB (1975) Lipoprotein lipase activity in bovine aorta. Proc Soc Exp Biol Med 148:1101–1105

94. Dinarello CA (1985) An update on human Interleukin-1: from molecular biology to clinical relevance. J Clin Immunol 5:287–297

95. Dragsten PR, Blumenthal R, Handler JS (1981) Membrane asymmetry in epithelia: is the tight junction a barrier to diffusion in the plasma membrane? Nature 294:718–722

96. Drenckhahn D (1983) Cell motility and cytoplasmic filaments in vascular endothelium. Prog Appl Microcirc 1:53–70

97. Driska SP, Murphy RA (1978) An estimate of cellular force generation in an arterial smooth muscle with a high actin:myosin ratio. Blood Vessels 15:26–32

98. Endo M, Yagi S, Iino M (1982) Tension-pCa relation and sarcoplasmic reticulum responses in chemically skinned smooth muscle fibers. Fed Proc 41:2245–2250

99. Engel J, Odermatt E, Engel A et al. (1981) Shapes, domain organizations, and flexibility of laminin and fibronectin, two multifunctional proteins of the extracellular matrix. J Mol Biol 150:97–120

100. Fantome JE, Ward PA (1982) Role of oxygen-derived free radicals and metabolites in leukocyte dependent inflammatory reactions. Am J Pathol 107:405–410

101. Farquhar MG, Palade G (1963) Junctional complexes in various epithelia. J Cell Biol 17:375–412

102. Fenstermacher JD, Rapoport SI (1984) Blood–brain barrier. In: Renkin EM, Michel CC (eds) Handbook of physiology, section 2: the cardiovascular system, vol IV: microcirculation, part 2. American Physiological Society, Bethesda, pp 969–1000

103. Flaherty JT, Pierre JE, Ferrans VJ et al. (1972) Endothelial nuclear patterns in the canine arterial tree with particular reference to hemodynamic events. Circ Res 30:23–33

104. Florey, Lord, Sheppard BL (1970) The permeability of arterial endothelium to horseradish peroxidase. Proc R Soc Lond (Biol) 174:435–443

105. Folkman J, Haudenschild CC (1980) Angiogenesis in vitro. Nature 288:551–556

106. Form DM, Pratt BM, Madri JA (1986) Endothelial cell proliferation during angiogenesis. In vitro modulation by basement membrane components. Lab Invest 55:521–530

107. Fowler S, Wolinsky H (1980) Lysosomes in vascular smooth cells. In: Bohr DF, Somlyo AP, Sparks HV (eds) Handbook of physiology, section 2: the cardiovascular system, vol II: vascular smooth muscle. American Physiological Society, Bethesda, pp 133–160

108. Fox PL, Di Corleto PE (1984) Regulation of production of a platelet-derived growth factor-like protein by cultured bovine aortic endothelial cells. J Cell Physiol 121:298–308

109. Frank ED, Warren L (1981) Aortic smooth muscle cells contain vimentin instead of desmin. Proc Natl Acad Sci USA 78:3020–3024

110. Franke WW, Schmid E, Freudenstein C et al. (1980) Intermediate-sized filaments of the prekeratin type in myoepithelial cells. J Cell Biol 84:633–654

111. Franke WW, Schmid E, Osborn M et al. (1979) Intermediate-sixed filaments of human endothelial cells. J Cell Biol 81:570–580

112. Franke WW, Schmid E, Winter S et al. (1979) Widespread occurrence of intermediate-sized filaments of the vimentin-type in cultured cells from diverse vertebrates. Exp Cell Res 123:25–46

113. Friedman SM (1983) Vascular reactivity. In: Genest J, Kuchel O, Hamet P et al. (eds) Hypertension. Physiopathology and treatment, 2nd edn. McGraw-Hill, New York, pp 457–473

114. Friend DS, Gilula NB (1972) Variations in tight and gap junctions in mammalian tissues. J Cell Biol 53:758–776

115. Fritze LMS, Reilly CF, Rosenberg RD (1985) An antiproliferative heparan sulfate species produced by postconfluent smooth muscle cells. J Cell Biol 100:1041–1049

116. Frokjaer-Jensen J (1980) Three-dimensional organization of plasmalemmal vesicles in endothelial cells. An analysis by serial sectioning of frog mesenteric capillaries. J Ultrastruct Res 73:9–20

117. Fry GN, Devine CE, Burnstock G (1977) Freeze–fracture studies of nexuses between smooth muscle cells. J Cell Biol 72:26–34

118. Furchgott RF (1983) Role of endothelium in responses of vascular smooth muscle. Circ Res 53:557–572

119. Furchgott RF (1984) Role of endothelium in responses of vascular smooth muscle to drugs. Annu Rev Pharmacol Toxicol 24:175–197

120. Furchgott RF, Zawadzki JV (1980) The obligatory role of endothelial cells in the relaxation of arterial smooth muscle by acetylcholine. Nature 288:373–376

121. Furchgott RF, Zawadzki JV, Cherry PD (1981) Role of the endothelium in the vasodilator response to acetylcholine. In: Vanhoutte PM, Leusen I (eds) Vasodilatation. Raven Press, New York, pp 49–66

122. Gabbiani G, Badonnel M-C, Rona G (1975) Cytoplasmic contractile apparatus in aortic endothelial cells of hypertensive rats. Lab Invest 32:227–234

123. Gabbiani G, Chaponnier C, Hüttner I (1978) Cytoplasmic filaments and gap junctions in epithelial cells and myofibroblasts during wound healing. J Cell Biol 96:561–568

124. Gabbiani G, Elemer G, Guelpa C et al. (1979) Morphologic and functional changes of the aortic intima during experimental hypertension. Am J Pathol 96:399–422

125. Gabbiani G, Gabbiani F, Lombardi D et al. (1983) Organization of actin cytoskeleton in normal and regenerating arterial endothelial cells. Proc Natl Acad Sci USA 80:2361–2364

126. Gabbiani G, Schmid E, Winter S et al. (1981) Vascular smooth muscle cells differ from other smooth muscle cells: predominance of vimentin filaments and a specific alpha-type actin. Proc Natl Acad Sci USA 78:298–302

127. Gabella G (1981) Structure of smooth muscle. In: Bulbring E, Brading Af, Jones AW et al. (eds) Smooth muscle: an assessment of current knowledge. Edward Arnold, London, pp 1–46

128. Gabella G (1983) Asymmetric distribution of dense bands in muscle cells of mammalian arterioles. J Ultrastruct Res 84:24–33

129. Gabella G (1984) Structural apparatus for force transmission in smooth muscles. Physiol Rev 64:455–477

130. Gajdusek C, Carbon S, Ross R et al. (1986) Activation of coagulation releases endothelial cell mitogens. J Cell Biol 103:419–428

131. Ganz P, Davies PF, Leopold JA et al. (1985) Endothelial-vascular smooth muscle interactions in culture. Circulation [Suppl III] 72:265A

132. Ganz P, Davies PF, Leopold JA et al. (1986) Short and long-term interactions of endothelium and vascular smooth muscle in co-culture: effects of cyclic GMP production. Proc Natl Acad Sci USA 83:3552–3556

133. Garfield RE, Somlyo AP (1985) Structure of smooth muscle. In: Grover AK, Daniel EE (eds) Calcium and smooth muscle contractility. Humana Press, Clifton, New Jersey, pp 1–36

134. Geiger B (1979) A 130K protein from chicken gizzard.

Its localization at the termini of microfilaments bundles in cultured chicken cells. Cell 18:193–205

135. Geiger B, Avnur Z, Kreis TE et al. (1984) The dynamic of cytoskeletal organization in areas of cell contact. In: Shay JW (ed) Cell and muscle motility. Plenum Press, New York, pp 195–234

136. Geiger B, Tokuyasu KT, Dutton AH et al. (1980) Vinculin, an intracellular protein localized at specialized sites where microfilament bundles terminate at cell membranes. Proc Natl Acad Sci USA 77:4127–4131

137. Geiger B, Volk T, Volberg T (1985) Molecular heterogeneity of adherens junctions. J Cell Biol 101:1523–1531

138. Gerrity RG, Adams EP, Cliff WJ (1975) The aortic tunica of the developing rat. II. Incorporation by medial cells of ^{3}H-proline into collagen and elastin: autoradiographic and chemical studies. Lab Invest 32:585–600

139. Gerrity RG, Cliff WJ (1975) The aortic tunica media of the developing rat. I. Quantitative sterologic and biochemical analysis. Lab Invest 32:585–600

140. Ghitescu L, Fixman A, Simionescu M et al. (1986) Specific binding sites for albumin restricted to plasmalemmal vesicles of continuous capillary endothelium: receptor-mediated transcytosis. J Cell Biol 102:1304–1311

141. Giacomelli F, Wiener J (1974) Regional variation in the permeability of rat thoracic aorta. Am J Pathol 75:513—528

142. Gilula NB, Reeves OR, Steinbach A (1972) Metabolic coupling, ionic coupling, and cell contacts. Nature 235:262–265

143. Gimbrone MA Jr, Cotran RS (1975) Human vascular smooth muscle in culture. Growth and ultastructure. Lab Invest 33:16–27

144. Gimbrone MA Jr, Cotran RS, Folkman J (1974) Human vascular endothelial cells in culture: Growth and DNA synthesis. J Cell Biol 60:673–684

145. Ginsburg R, Zera PH (1984) Endothelial relaxant factor in the human epicardial coronary artery. Circulation [Suppl II]. 70:122A

146. Goldberg ID, Stemerman MB, Ransil BJ et al. (1980) In vivo aortic muscle cell growth kinetics. Differences between thoracic and abdominal segments after intimal injury in the rabbit. Circ Res 47:182–189

147. Goldstein JL, Anderson RGW, Brown MS (1979) Coated pits, coated vesicles and receptor-mediated endocytosis. Nature 279:679–685

148. Goldstein JL, Brown MS (1977) The low density lipoprotein pathway and its relation to atherosclerosis. Annu Rev Biochem 46:897–930

149. Gonnerman WA, Ferrara R, Franzblau C (1981) Measurement of medium lysil oxidase activity in aorta smooth muscle cells. Effects of multiple medium changes and inhibition of protein synthesis. Biochemistry 20:3864–3867

150. Goodenough DA (1974) Bulk isolation of mouse hepatocyte gap junctions. Characterization of the principal protein, connexin. J Cell Biol 61:557–563

151. Gospodarowicz D, Ill CR (1980) Extracellular matrix and the control of proliferation of vascular endothelial cells. J Clin Invest 65:1351–1364

152. Gotlieb AI, McBurnie ML, Subrahmanyan L et al. (1981) Distribution of microtubule organizing centers in migrating sheets of endothelial cells. J Cell Biol 91:589–594

153. Grega GJ (1986) Role of the endothelial cell in the regulation of microvascular permeability to molecules. Introductory remarks. Fed Proc 45:75–76

154. Grega GJ, Adamski SW, Dobbins DE (1986) Physiological and pharmacological evidence for the regulation of permeability. Fed Proc 45:96–100

155. Grega GJ, Svensjö E, Haddy FJ (1981) Macromolecular permeability of the microvascular membrane: physiological and pharmacological regulation. Microcirculation 1:325–341

156. Griffith TM, Edwards DH, Collins P et al. (1985) Endothelial derived relaxant factor. JR Coll Physicians Lond 19:74–79

157. Griffith TM, Edwards DH, Lewis MJ et al. (1984) The nature of endothelium-derived vascular relaxant factor. Nature 308:645–647

158. Griffith TM, Henderson AH, Hughes-Edwards D et al. (1984) Isolated perfused rabbit coronary artery and aortic strip preparations: the role of endothelium-derived relaxant factor. J Physiol 351:13–24

159. Grotte G (1956) Passage of dextran molecules across the blood–lymph barrier. Acta Clin Scand [Suppl] 211:1–84

160. Guyton JR, Rosenberg RD, Clowes AW et al. (1980) Inhibition of rat arterial smooth muscle cell proliferation by heparin. In vivo studies with anticoagulant and nonanticoagulant heparin. Circ Res 46:625–634

161. Habib JB, Wells SL, Williams CL et al. (1984) Atherosclerosis impairs endothelium-dependent arterial relaxation. Circulation [Suppl II] 70:123 (abstract)

162. Hansford RG (1985) Relation between mitochondrial calcium transport and control of energy metabolism. Rev Physiol Biochem Pharmacol 102:1–72

163. Hartshorne DJ (1982) Phosphorylation of myosin and the regulation of smooth-muscle actomyosin. In: Dowben RM, Shay JW (eds) Cell and muscle motility II. Plenum Press, New York, pp 188–220

164. Hartshorne DJ, Gorecka A (1980) Biochemistry of the contractile proteins of smooth muscle. In: Bohr DF, Somlyo AP, Sparks HV (eds) Handbook of physiology, section 2: the cardiovascular system, vol II: vascular smooth muscle. American Physiological Society, Bethesda, pp 93–120

165. Hascall VC, Hascall GK (1981) Proteoglycans. In: Hay ED (ed) Cell biology of extracellular matrix. Plenum Press, New York, pp 39–63

166. Hasegawa T, Hasegawa E, Chen W-T et al. (1985) Characterisation of a membrane-associated glycoprotein complex implicated in cell adhesion to fibronectin. J Cell Biochem 28:307–318

167. Heltianu C, Simionescu M, Simioniescu N (1982) Histamine receptors of the microvascular endothelium revealed in situ with a histamine–ferritin conjugate: characterstic high-affinity binding sites in venules. J Cell Biol 93:357–364

168. Hermsmeyer K (1984) Altered arterial muscle ion transport mechanism in the spontaneously hypertensive rat. J Cardiovasc Pharmacol 6:S10–S15

169. Hermsmeyer K, Harder D (1986) Membrane ATPase mechanism of K^+-return relaxation in arterial muscles of stroke-prone SHR and WKY. Am J Physiol 250:C557–C562

170. Herzberg EL, Lawrence RS, Gilula NB (1981) Gap junctional communication. Annu Rev Physiol 43:479–491

171. Hoover RL, Rosenberg R, Haering W et al. (1980) Inhibition of rat arterial smooth muscle cell proliferation by heparin. II. In vitro studies. Circ Res 47:578–583

172. Hopkins NK, Gorman RR (1981) Regulation of endothelial cell cyclic nucleotide metabolism by prostacyclin. J Clin Invest 67:540–546

173. Horan KL, Adamski SW, Ayele W et al. (1986) Evidence that prolonged histamine suffusions produce transient increases in vascular permeability subsequent to the formation of venular macromolecular leakage sites. Proof of the Majno–Palade hypothesis. Am J Pathol 123:570–576

174. Horwitz A, Duggan K, Buck C et al. (1986) Interaction of the plasma membrane fibronectin receptor with talin: a transmembrane linkage. Nature 320:531–533

175. Horwitz A, Duggan K, Greggs R et al. (1985) The cell-substrate attachment (CSAT) antigen has properties of a receptor for laminin and fibronectin. J Cell Biol 101:2134–2144

176. Hüttner I, Boutet M, More RH (1973) Gap junctions in arterial endothelium. J Cell Biol 57:247–252

177. Hüttner I, Boutet M, More RH (1973) Studies on protein passage through arterial endothelium. I. Structural correlates of permeability in rat arterial endothelium. Lab Invest 28:672–677

178. Hüttner I, Boutet M, More RH (1973) Studies on protein passage though arterial endothelium. II. Regional differences in permeability to fine structural protein tracers in arterial endothelium of normotensive rat. Lab Invest. 28:678–685

179. Hüttner I, Boutet M, Rona G et al. (1973) Studies on protein passage through arterial endothelium. III. Effect of blood pressure levels on the passage of fine structural protein tracers through rat arterial endothelium. Lab Invest 29:536–546

180. Hüttner I, Gabbiani G (1982) Vascular endothelium: recent advances and unanswered questions. Lab Invest 47:409–411 (editorial)

181. Hüttner I, Gabbiani G (1983) Vascular endothelium in hypertension. In: Genest J, Kuchel O, Hamel P et al. (eds) Hypertension: pathophysiology and treatment. McGraw Hill, New York, pp 473–488

182. Hüttner I, Mocostabella P, De Chastonay C, Gabbiani G (1982) Volume, surface and junctions of rat aortic endothelium during experimental hypertension. A morphometric and freeze–fracture study. Lab Invest 46:489–504

183. Hüttner I, Peters HY (1978) Heterogeneity of cell junctions in rat aortic endothelium: a freeze–fracture study. J Ultrastruct Res 64:303–308

184. Hüttner I, Walker C, Gabbiani G (1985) Aortic endothelial cell during regeneration. Remodeling of cell junctions, stress fibers, and stress fiber–membrane attachment domains. Lab Invest 53:287–302

185. Huxley HE (1969) The mechanism of muscular contraction. Science 164:1356–1366

186. Hynes RO, Yamada KM (1982) Fibronectins: multifunctional modular glycoproteins. J Cell Biol 95:369–377

187. Isomura T, Dvorak Am, Garcia RI et al. (1986) Inbred guinea pig aortic endothelial cell clones. Model for studying the vascular endothelium under totally isologous conditions. Lab Invest 55:703–716

188. Iverius PH (1972) The interaction between human plasma lipoproteins and connective tissue glycosaminoglycans. J Biol Chem 247:2607–2613

189. Jaffe EA (1982) Synthesis of factor VIII by endothelial cells. Ann NY Acad Sci 401:163–169

190. Jaffe EA, Hoyer LW, Nachman RL (1974) Synthesis of von Willebrand factor by cultured human endothelial cells. Proc Natl Acad Sci USA 71:1906–1909

191. Jaffe EA, Nachman RL, Becker CC et al. Culture of human endothelial cells derived from umbilical veins: Identification by morphologic and immunologic criteria. J Clin Invest 52:2745–2756

192. Jefferies WA, Brandon MR, Hunt SV et al. (1984) Transferrin receptor on endothelium of brain capillaries. Nature 312:162–163

193. Jockusch BM, Isenberg G (1981) Interaction of alpha-actinin and vinculin with actin: opposite effects on filament network formation. Proc Natl Acad Sci USA 78:3005–3009

194. Johansson B, Somlyo AP (1980) Electrophysiology and excitation–contraction coupling. In: Bohr DF, Somlyo AP, Sparks HV (eds) Handbook of physiology, section 2: the cardiovascular system, vol II: vascular smooth muscle. American Physiological Society, Bethesda, pp 301–323

195. Johnson PC (1978) Myogenic tone in resistance vessels. In: Vanhoutte P, Leusen I (eds) Mechanism of vasodilatation. Karger, Basel, pp 73–78

196. Johnson AR, Erdos EG (1977) Metabolism of vasoactive peptides by human endothelial cells in culture. Angiotensin I converting enzyme (kininase II) angiotensinase. J Clin Invest 59:684–695

197. Jones AW (1981) Vascular smooth muscle and alterations during hypertension. In: Bülbring E, Brading AF, Jones AW et al. (eds) Smooth muscle: an assessment of current knowledge, Edward Arnold, London pp 397–429

198. Joyce NC, Haire MF, Palade GE (1985) Contractile proteins in pericytes. I. Immunoperoxidase localization of tropomyosin. J Cell Biol 100:1379–1386

199. Joyce NC, Haire MF, Palade GE (1985) Contractile proteins in pericytes. II. Immunocytochemical evidence for the presence of two isomyosins in graded concentrations. J Cell Biol 100:1387–1395

200. Kachar B, Pinto Da Silva P (1981) Rapid massive assembly of tight junction strands. Science. 213:541–544

201. Kachar B, Reese TS (1982) Evidence for the lipidic nature of tight junction strands. Nature 296:464–466

202. Kagan HM, Vaccaro CA, Bronson RE et al. (1986) Ultrastructural immunolocalization of lysil oxidase in vascular connective tissue. J Cell Biol 103:1121–1128

203. Kannon MS, Daniel EE (1978) Formation of gap junctions by treatment in vitro with potassium conductance blockers. J Cell Biol 78:338–348

204. Kanwar YS (1984) Biology of disease. Biophysiology of glomerular filtration and proteinuria. Lab Invest 51:7–21

205. Karnovsky MJ (1967) The ultrastructural basis of capillary permeability studied with peroxidase as a tracer. J Cell Biol 35:213–236

206. Karnovsky MJ (1981) Endothelial–vascular smooth muscle cell interactions. Am J Pathol 100:200–206

207. Killackey JJF, Johnston MG, Movat HZ (1986) Increased permeability of microcarrier cultured endothelial monolayers in response to histamine and thrombin. Am J Pathol 122:50–61

208. Kimes BW, Brandt BL (1976) Characterization of two putative smooth muscle cell lines from rat thoracic aorta. Exp Cell Res 98:349–366

209. King GL, Johnson SM (1985) Receptor-mediated transport of insulin across endothelial cells. Science 227:1583–1586

210. Kinsella MG, Wight TN (1986) Modulation of sulfated proteoglycan synthesis by bovine aortic endothelial cells during migration. J Cell Biol 102:679–687

211. Klee CB, Crouch TH, Richman PG (1980) Calmodulin. Ann Rev Biochem 49:489–515
212. Kleinman HK, Cannon FB, Laurie GW et al. (1985) Biological activities of laminin. J Cell Biochem 27:317–325
213. Kleinman HK, Klebe RJ, Martin GR (1981) Role of collagenous matrices in the adhesion and growth of cells. J Cell Biol 88:473–485
214. Knudsen K, Horwitz A, Buck C (1985) A monoclonal antibody identifies a glycoprotein complex involved in cell-substratum adhesion. Exp Cell Res 157:218–226
215. Kocher O, Skalli O, Bloom WS et al. (1984) Cytoskeleton of rat aortic smooth muscle cells. Normal conditions and experimental intimal thickening. Lab Invest 50:645–562
216. Kocher O, Skalli O, Cerutti D et al. (1985) Cytoskeletal features of rat aortic cells during development. An electron microscopic, immunohistochemical, and biochemical study. Circ Res 56:829–838
217. Kontos HA (1985) Oxygen radicals in cerebral vascular injury. Circ Res 57:508–516
218. Landis EM, Pappenheimer JR (1963) Exchange of substances through the capillary walls. In: Hamilton WF, Dow P (eds) Handbook of physiology, section 2: circulation, vol II. American Physiological Society, Washington DC pp 961–1034
219. Larkin JM, Donzell WC, Anderson RGW (1986) Potassium-dependent assembly of coated pits: new coated pits form as planar clathrin lattices. J Cell Biol 103:2619–2627
220. Larson DM, Fujiwara K, Alexander RW et al. (1984) Heterogeneity of myosin antigenic expression in vascular smooth muscle in vivo. Lab Invest 50:401–407
221. Larson DM, Sheridan JD (1982) Intercellular junctions and transfer of small molecules in primary vascular endothelial cultures. J Cell Biol 92:183–191
222. Larson DM, Sheridan JD (1985) Junctional transfer in cultured vascular endothelium. I. Dye and nucleotide transfer. J Membr Biol 83:157–167
223. Lau K, Thomas D, Eby B (1986) The nature and role of disturbances in calcium metabolism in genetic hypertension. Fed Proc 45:2752–2757
224. Lazarides E (1980) Intermediate filaments as mechanical integrators of cellular space. Nature 283:249–256
225. Lee KT, Janakidevi K, Kroms M et al. (1985) Mosaicism in female hybrid hares heterozygous for glucose-6-phosphase dehydrogenase. VII. Evidence for selective advantage of one phenotype over the other in ditypic samples from aortas of hares fed cholesterol oxidation products. Exp Mol Pathol 42:71–77
226. Libby P, Ordovas JM, Auger KR et al. (1986) Endotoxin and tumor necrosis factor induce Interleukin-1 gene expression in adult human vascular endothelial cells. Am J Pathol 124:179–185
227. Loewenstein WR (1979) Junctional intercellular communications and the control of growth. Biochim Biophys Acta 560:1–65
228. Loewenstein WR (1981) Junctional intercellular communication: the cell-to-cell membrane channel. Physiol Rev 61:829–913
229. Lyberg T, Galdal K, Evensen S et al. (1983) Cellular cooperation in endothelial thromboplasmin synthesis. Br J Haematol 53:85–95
230. Madri JA, Stenn KS (1982) Aortic endothelial cell migration. I. Matrix requirements and composition. Am J Pathol 106:80–186
231. Majno G (1965) Ultrastructure of the vascular membrane. In: Hamilton WF, Dow P (ed) Handbook of physiology, section 2: circulation, vol III. American Physiological Society, Washington DC, pp 2293–2375
232. Majno G, Joris I (1978) Endothelium 1977. A review. In: Chandler AB, Eurenius K, McMillan GC et al. (eds) The thrombotic process in atherogenesis. Plenum Press, New York, pp 169–225, 481–526
233. Majno G, Palade GE (1961) Studies on inflammation. I. The effect of histamine and serotonin on vascular permeability: an electron microscopic study. J Biophys Biochem Cytol 11:571–605
234. Majno G, Palade GE, Schoefl GI (1961) Studies on inflammation. II. The site of action of histamine and serotonin along the vascular tree: a topographic study. J Biophys Biochem Cytol 11:607–626
235. Majno G, Shea SM, Leventhal M (1969) Endothelial contraction induced by histamin-type mediators: an electron microscopic study. J Cell Biol 42:647–672
236. Majno G, Underwood JM, Zand T et al. (1985) The significance of endothelial stomata and stigmata in the rat aorta. An electron microscopic study. Virchows Arch [A] 408:75–91
237. Makowski L, Caspar DLD, Phillips WC et al. (1977) Gap junction structures. II. Analysis of the x-ray diffraction data. J Cell Biol 74:629–645
238. Martinez-Palomo A, Meza I, Beaty G et al. (1980) Experimental modulation of occluding junctions in a cultured transporting epithelium. J Cell Biol 87:736–745
239. Mayerson HS (1963) The physiological importance of lymph. In: Hamilton WF, Dow P (eds) Handbook of physiology, section 2: circulation, vol II. American Physiological Society, Washington DC, pp 1035–1073
240. Mazzone RW, Kornblau SM, (1980) Pinocytic vesicles in the endothelium of rapidly frozen rabbit lung. Microvasc Res 21:193–211
241. McGuire PG, Twietmeyer TA (1983) Morphology of rapidly frozen aortic endothelial cells. Glutaraldehyde fixation increases the number of caveolae. Circ Res 53:424–429
242. McLean MJ, Sperelakis NY (1977) Electrophysiological recordings from spontaneously contracting reaggregates of cultured vascular smooth muscle cells from chick embyros. Exp Cell Res 104:309–318
243. Means AR, Dedman JR (1980) Calmodulin – an intracellular calcium receptor. Nature 285:73–77
244. Meldolesi J, Castiglioni G, Parma R et al. (1978) Ca^{++}-dependent disassembly and reassembly of occluding junctions in guinea pig pacreatic acinar cells. Effect of drugs. J Cell Biol 79:156–172
245. Mello RJ, Brown MS, Goldstein JL et al. (1980) LDL receptors in coated vesicles isolated from bovine adrenal cortex: binding sites unmasked by detergent treatment. Cell 20:829–837
246. Michel CC (1979) The investigation of capillary permeability in single vessels. Acta Physiol Scand [Suppl] 463:67–74
247. Michel CC (1984) Fluid movement through capillary walls. In: Renkin EM, Michel CC (eds) Handbook of physiology, section 2: the cardiovascular system, vol IV: microcirculation, part 1. American Physiological Society, Bethesda, pp 375–409
248. Miller FN, Sims DE (1986) Contractile elements in the regulation of macromolecular permeability. Fed Proc 45:84–88
249. Miller VM, Aarhus LL, Vanhoutte PM (1986) Modulation of endothelium-dependent responses by chronic

alterations of blood-flow. Am J Physiol 251:H520–H527

250. Minick CR, Stemerman MB, Insull W Jr (1977) Effect of regenerated endothelium on lipid accumulation in the arterial wall. Proc Natl Acad Sci USA 74:1724–1728

251. Moncada S, Gryglewski R, Bunting S et al. (1976) An enzyme isolated from arteries transforms prostaglandin endopeptides to an unstable substance that inhibits platelet aggregation. Nature 263:663–665

252. Moncada S, Herman AF, Higgs EA et al. (1977) Differential formation of prostacyclin (PGX or PGI$_2$) by layers of the arterial wall: an explanation for the anti-thrombotic properties of vascular endothelium. Thromb Res 11:323–344

253. Moncada S, Vane JR (1979) Arachidonic acid metabolites and the interactions between platelets and blood-vessel walls. N Engl J Med 300:1142–1147

254. Montesano R (1979) Inhomogenous distribution of filipin–sterol complexes in smooth muscle cell plasma membrane. Nature 280:328–329

255. Montesano R, Mossaz A, Ryser JE et al. (1984) Leukocyte interleukins induce cultured endothelial cells to produce a highly organized, glycosaminoglycan-rich pericellular matrix. J Cell Biol 99:1706–1715

256. Mosher DI, Doyle MJ, Jaffe EA (1982) Synthesis and secretion of thrombospondin by cultured human endothelial cells. J Cell Biol 93:343–348

257. Muir L, Bornstein WP, Ross R (1976) A presumptive subunit of elastic fiber microfilbrils secreted by arterial smooth-muscle cells in culture. Eur J Biochem 64:105–114

258. Muller WA, Gimbrone MA (1986) Plasmalemmal proteins of cultured vascular endothelial cells exhibit apical–basal polarity: analysis by surface-selective iodination. J Cell Biol 103:2389–2402

259. Murphy RA, Herlhy JT, Megerman J (1974) Force-generating capacity and contractile protein content of arterial smooth muscle. J Gen Physiol 64:691–705

260. Murray JM, Weber A (1974) The cooperative action of muscle proteins. Sci Am 230(2):59–71

261. Nachman R, Levine R, Jaffe EA (1977) Synthesis of Factor VIII antigen by cultured guinea pig megakaryocytes. J Clin Invest 60:914–921

262. Nagy Z, Goehlert UG, Wolfe LS et al. (1985) Ca^{2+} depletion-induced disconnection of tight junctions in isolated rat brain microvessels. Acta Neuropathol (Berl) 68:48–52

263. Nagy Z, Mathieson G, Hüttner I (1979) Blood–brain barrier opening to horseradish peroxidase in acute arterial hypertension. Acta Neuropathol (Berl) 48:45–53

264. Nagy Z, Peters H, Hüttner I (1983) Charge-related alterations of the cerebral endothelium. Lab Invest 49:662–671

265. Nagy Z, Peters H, Hüttner I (1984) Fracture faces of cell junctions in cerebral endothelium during normal and hyperosmotic conditions. Lab Invest 50:313–322

266. Nagy Z, Szabo M, Hüttner I (1985) Blood–brain barrier impairment by low pH buffer perfusion via the internal carotid artery in rat. Acta Neuropathol (Berl) 68:160–163

267. Nawroth P, Stern D (1985) An endothelial cell pre-coagulant pathway. J Cell Biochem 28:253–264

268. Narayanan S, Sandberg LB, Ross R et al. (1976) The smooth muscle cell. III. Elastin synthesis in arterial smooth muscle cell culture. J Cell Biol 68:411–419

269. Newman PJ, Kawai Y, Montgomery RR et al. (1986) Synthesis by cultured human umbilical vein endothelial cells of two proteins structurally and immunologically related to platelet membrane glycoproteins IIb and IIIa. J Cell Biol 103:81–86

270. Olesen S-P, Crone C (1984) Serotonin increases microvascular permeability in the brain. Int J Microcirc Clin Exp. 3:466 (abstract)

271. Olesen S-P, Crone C (1986) Substances that rapidly augment ionic conductance of endothelium in cerebral venules. Acta Physiol Scand 127:233–241

272. Olivetti G, Anversa P, Melissari M et al. (1980) Morphometric study of early postnatal development of the thoracic aorta in the rat. Circ Res 47:417–424

273. Osborn M, Caselitz J, Weber K (1981) Heterogeneity of intermediate filament expression in vascular smooth muscle: a gradient in desmin positive cells from the rat aortic arch to the level of the arterial iliaca communis. Differentiation 20:196–202

274. Owens GK, Loeb A, Gordon D et al. (1986) Expression of smooth muscle-specific alpha-isoactin in cultured vascular smooth muscle cells: relationship between growth and cytodifferentiation. J Cell Biol 102:343–352

275. Owens GK, Schwartz SM (1982) Alterations in vascular smooth muscle mass in the spontaneously hypertensive rat. Role of cellular hypertrophy, hyperploidy and hyperplasia. Circ Res 51:280–289

276. Palade GE, Bruns RR (1968) Structural modulations of plasmalemmal vesicles. J Cell Biol 37:663–649

277. Palade GE, Simionescu M, Simionescu N (1979) Structural aspects of the permeability of the microvascular endothelium. Acta Physiol Scand [Suppl] 463:11–32

278. Pappas GD (1973) Junctions between cells. Hosp Pract 8:39–46

279. Pappas GD, Asada Y, Bennett MVL (1971) Morphological correlates of increased coupling resistance at an electrotonic synapse. J Cell Biol 49:173–188

280. Pappenheimer JR (1953) Passage of molecules through capillary walls. Physiol Rev 33:387–423

281. Peach MJ, Loeb AL, Singer HA et al. (1984) Endothelium-derived vascular relaxing factor. Hypertension [Suppl 1] 7:I94–I100

282. Pearse BMF (1976) Clathrin: a unique protein associated with intracellular transfer of membrane by coated vesicles. Proc Natl Acad Sci USA 73:1255–1259

283. Pearson JD, Carleton JS, Gordon JL (1980) Metabolism of adenine nucleotides by ectoenzymes of vascular endothelial and smooth-muscle cells in culture. Biochemistry 190:421–429

284. Pearson JD, Gordon JL (1979) Vascular endothelial and smooth muscle cells in culture selectively release adenine nucleotides. Nature 281:384–386

285. Pearson TA, Dillman JM, Heptinstall RH (1983) The clonal characteristics of human aortic intima. Comparison with fatty streaks and normal media. Am J Pathol 113:33–40

286. Pelikan P, Gimbrone MA, Cotran RS (1979) Distribution and movement of anionic cell surface sites in cultured human vascular endothelial cells. Atherosclerosis 32:69–80

287. Peters K-R, Carley WW, Palade GE (1985) Endothelial plasmalemmal vesicles have a characteristic striped bipolar surface structure. J Cell Biol 101:2233–2238

288. Pinto Da Silva P (1972) Translational mobility of the membrane intercalated particles of human erythrocyte ghost. pH-dependent, reversible aggregation. J Cell Biol 53:777–787

289. Pinto Da Silva P, Kachar B (1982) On tight-junction structure. Cell 28:441–450

290. Postnov YV (1981) Alteration of cell membrane control over intracellular calcium in essential hypertension and in spontaneously hypertensive rats. In: Laragh, JH, Buckler FR, Seldin DW (eds) Frontiers in hypertension research. Springer-Verlag, Berlin Heidelberg New York, pp 91–93

291. Raugi GJ, Mumby SM, Abbot-Brown D et al. (1982) Thrombospondin: synthesis and secretion by cells in culture. J Cell Biol 95:351–354

292. Rapoport RM, Murad F (1983) Agonist-induced endothelium-dependent relaxation in rat thoracic aorta may be mediated through cGMP. Circ Res 52:352–357

293. Reidy MA (1985) Biology of disease. A reassessment of endothelial injury and arterial lesion formation. Lab Invest 53:513–520

294. Reidy MA, Schwartz SM (1983) Endothelial injury and regeneration. IV. Endotoxin – a non-denuding injury to aortic endothelium. Lab Invest 48:25–34

295. Renkin EM (1979) Relation of capillary morphology to transport of fluid and large molecules: a review. Acta Physiol Scand [Suppl] 463:81–91

296. Rennke HG, Cotran RS, Venkatachalam MA (1975) Role of molecular charge in glomerular permeability. Tracer studies with cationized ferritins. J Cell Biol 67:638–646

297. Revel JP, Yee AF, Hudspeth AJ (1971) Gap junctions between electrotonically coupled cells in tissue culture and in brown fat. Proc Natl Acad Sci USA 68:2924–2927

298. Rhodin JAG (1967) The ultrastructure of mammalian arterioles and precapillary sphincters. J Ultrastruct Res 18:181–223

299. Rhodin JAG (1968) Ultrastructure of mammalian venous capillaries, venules and small collecting veins. J Ultrastruct Res 25:452–500

300. Rhodin JAG (1980) Architecture of the vessel wall. In: Bohr DR, Somlyo AP, Sparks HV (eds) Handbook of physiology, section 2: the cardiovascular system, vol II, vascular smooth muscle. American Physiological Society, Bethesda pp 1–31

301. Richardson JB, Beaulnes A (1971) The cellular site of action of angiotensin. J Cell Biol 51:419–432

302. Robinson SM, Hoover RL, Karnovsky MJ (1984) Vesicles (caveolae) number is reduced in cultured endothelial cells prepared for electron microscopy by rapid-freezing. J Cell Biol 99:287 (abstract)

303. Robinson JM, Okada T, Castellot JJ Jr et al. (1986) Unusual lysosomes in aortic smooth muscle cells: presence in living and rapidly frozen cells. J Cell Biol 102:1615–1622

304. Rogalski AA, Singer SJ (1985) An integral glycoprotein associated with the membrane attachment sites of actin microfilaments. J Cell Biol 101:785–801

305. Rogers KA, Kalnins VI (1983) Comparison of cytoskeleton in aortic endothelial cells in situ and in vitro. Lab Invest 49:650–654

306. Rose B, Simpson I, Loewenstein WR (1977) Calcium ion produces graded changes in permeability of membrane channels in cell junction. Nature 267:625–627

307. Rosenblum WI (1986) Aspects of endothelial malfunction and function in cerebral microvessels. Lab Invest 55:252–268

308. Ross R (1986) The pathogenesis of atherosclerosis – an update. N Engl J Med 314:488–500

309. Ross R, Klebanoff SJ (1971) The smooth muscle cell. I. In vivo synthesis of connective tissue proteins. J Cell Biol 50:159–171

310. Rotrosen D Gallin JI (1986) Histamin type I receptor

occupancy increases endothelial cytosolic calcium, reduces F-actin, and promotes albumin diffusion across cultured endothelial monolayers. J Cell Biol 103:2379–2387

311. Ryan GB, Hein SJ, Karnovsky MJ (1976) Glomerular permeability to proteins. Effects of hemodynamic factors on the distribution of endogenous immunoglobulin G and exogenous catalase in the rat glomerulus. Lab Invest 34:415–427

312. Ryan GB, Karnovsky MJ (1976) Distribution of endogenous albumin in the rat glomerulus. Role of hemodynamic factors in glomerular barrier function. Kidney Int 9:36–45

313. Ryan US (1986) The endothelial surface and responses to injury. Fed Proc. 45:101–108

314. Ryan US, Ryan JW, Whitaker C et al. (1976) Localisation of angiotensin converting enzyme (kininase II). II. Immunocytochemistry and immunofluorescence. Tissue Cell 8:125–145

315. Saez JC, Spray DC, Nairn AC et al. (1986) cAMP increases junctional conductance and stimulates phosphorilation of the 27-kDa principal gap junction polypeptide. Proc Natl Acad Sci USA 83:2473–2477

316. Sakai LY, Keene, DR, Engvall E (1986) Fibrillin, a new 350-kD glycoprotein, is a component of extracellular microfibrils. J Cell Biol 103:2499–2509

317. Schlingemann RO, Dingjan GM, Emeis JJ et al. (1985) Monoclonal antibody PAL-E Specific for endothelium. Lab Invest 52:71–76

318. Schliwa M (1981) Proteins associated with cytoplasmic actin. Cell 25:587–590

319. Schmid E, Osborn M, Rungger-Brandle E et al. (1982) Distribution of vimentin and desmin filaments in smooth muscle tissue of mammalian and avian aorta. Exp Cell Res 137:329–340

320. Schneeberger EE, Lynch RD (1984) Tight junctions: their structure, composition and function. Circ Res 55:723–733

321. Schoenberg CF, Neehham DM (1976) A study of the mechanism of contraction in vertebrate smooth muscle. Biol Rev 51:53–104

322. Schollmeyer JE, Furcht LT, Goll DE et al. (1976) Localization of contractile proteins in smooth muscle cells and in normal and transformed fibroblasts. In: Goldman R, Pollard T, Rosenbaum J (eds) Cell Motility, vol A. Cold Spring Harbor Laboratory, Cold Spring Harbor, New York, pp 361–388

323. Schollmeyer JE, Goll DE, Robson RM et al. (1973) Localization of alpha-actinin and topomyosin in different muscles. J Cell Biol 59:306 (abstract)

324. Schubert D, Harris AJ, Devine CE et al. (1974) Characterization of a unique muscle cell line. J Cell Biol 61:398–413

325. Schwartz CJ, Werthessen NT, Wolf S (1980) Structure and function of the circulation, vol 1. Plenum Press, New York

326. Schwartz SM, Benditt EP (1972) Studies on aortic intima. I. Structure and permeability of rat thoracic aortic intima. Am J Pathol 66:241–264

327. Schwartz SM, Benditt EP (1976) Clustering of replicating cells in aortic endothelium. Proc Natl Acad Sci USA, 73:651–653

328. Schwartz SM, Benditt EP (1977) Aortic endothelial cell replication. I. Effects of age and hypertension in the rat. Circ Res 41:248–255

329. Schwartz SM, Campbell GR, Campbell JH (1986) Replication of smooth muscle cells in vascular disease. Circ Res 58:427–444

330. Schwartz SM, Gajdusek CM, Reidy MA et al. (1980) Maintenance of integrity in aortic endothelium. Fed Proc 39:2618–2625

331. Schwartz SM, Gajdusek CM, Selden SC III (1981) Vascular wall growth control: The role of the endothelium. Arteriosclerosis 1:107–126

332. Schwartz SM, Haudenschild CC, Eddy EM (1978) Endothelial regeneration I. Quantitative analysis of initial stages of endothelial regeneration in rat aortic intima. Lab Invest 38:568–580

333. Schwartz SM, Stemerman MB, Benditt EP (1975) The aortic intima. II. Repair of the aortic lining after mechanical denudation. Am J Pathol 81:15–42

334. Scow RO, Blanchette-Mackie EJ, Smith LC (1976) Role of capillary endothelium in the clearance of chylomicrons: a model for lipid transport from blood by lateral diffusion in cell membranes. Circ Res 39:149–162

335. Segall SS, Duling BR (1986) Flow control among microvessels coordinated by intercellular conduction. Science 234:868–870

336. Seifert RA, Schwartz SM, Bowen-Pope DF (1984) Developmentally regulated production of platelet-derived growth factor-like molecules. Nature 311:669–671

337. Severs NJ, Simons HL (1983) Failure of filipin to detect cholesterol-rich domains in smooth muscle plasma membrane. Nature 303:637–638

338. Severs NJ, Simons HL (1986) Caveolar bands and the effects of sterol-binding agents in vascular smooth muscle plasma membrane. Single and double labeling with filipin and tomatin in the aorta pulmonary artery, and vena cava. Lab Invest 55:295–307

339. Shasby DM, Lind SE, Shasby SS et al. (1985) Reversible oxidant-induced increases in albumin transfer across cultured endothelium: alterations in cell shape and calcium homeostasis. Blood 65:605–614

340. Shasby DM, Shasby SS, Sullivan JM et al. (1982) Role of endothelial cell cytoskeleton in control of endothelial permeability. Circ Res 51:657–661

341. Shepro D, D'Amore PA (1984) Physiology and biochemistry of the vascular wall endothelium. In: Renkin EM, Michel CC (eds) Handbook of physiology, section 2: the cardiovascular system, vol IV: microcirculation, part 1. American Physiological Society, Bethesda, pp 103–164

342. Shepro D, Hechtman HB (1985) Endothelial serotonin uptake and mediation of prostanoid secretion and stress fiber formation. Fed Proc 44:2616–2619

343. Sheridan JP, Larson DM (1982) Junctional communication in the peripheral vasculature. In: Pitts JD, Finbow ME: The functional integration of cells in animal tissues. Cambridge University Press, Cambridge pp 263–283

344. Simionescu M, Simionescu N, Palade GE (1974) Morphometric data on the endothelium of blood capillaries. J Cell Biol 60:128–152

345. Simionescu M, Simionescu N, Palade GE (1975) Segmental differentiations of cell junctions in the vascular endothelium: the microvasculature. J. Cell Biol 67:863–885

346. Simionescu M, Simionescu N, Palade GE (1976) Segmental differentiations of cell junctions in the vascular endothelium: arteries and veins. J Cell Biol 68:705–723

347. Simionescu M, Simionescu N, Palade GE (1982) Differential microdomains on the luminal surface of capillary endothelium. Distribution of lectin receptors. J Cell Biol 94:406–413

348. Simionescu M, Simionescu N, Santoro F et al. (1985) Differentiated microdomains of the luminal plasmalemma of murine muscle capillaries: segmental variations in young and old animals. J Cell Biol 100:1396–1407

349. Simionescu N (1983) Cellular aspects of transcapillary exchange. Physiol Rev 63:1536–1577

350. Simionescu N, Simionescu M, Palade GE (1975) Permeability of muscle capillaries to small hemipeptides. Evidence for the existence of patent transendothelial channels. J Cell Biol 64:586–607

351. Simionescu N, Simionescu M, Palade GE (1978) Open junctions in the endothelium of the postcapillary venules of the diaphragm. J Cell Biol 79:27–44

352. Simpson I, Rose B, Loewenstein WR (1977) Size limit of molecules permeating the junctional membrane channels. Science 195:294–296

353. Singer II (1979) The fibronexus: a transmembrane association of fibronectin-containing fibers and bundles of 5 nm microfilaments in hamster and human fibroblasts. Cell 16:675–685

354. Singer II (1982) Association of fibronectin and vinculin with focal contacts and stress fibers in stationary hamster fibroblasts. J Cell Biol 92:398—408

355. Singer II, Kazazis DM, Kawka DW (1985) Localization of the fibronexus at the surface of granulation tissue myofibroblasts using double-label immunogold electron microscopy on ultrathin frozen sections. Eur J Cell Biol 38:94–101

356. Singer II, Kawka DW, Kazazis DM et al. (1984) In vivo codistribution of fibronectin and actin fibers in granulation tissue: immunofluorescence and electron microscope studies of the fibronexus at the myofibroblast surface. J Cell Biol 98:2091–2106

357. Singer SJ, Nicholson GL (1972) The fluid mosaic model of the structure of cell membranes. Science 175:720–731

358. Skalli O, Ropraz P, Trzeciak A et al. (1986) A monoclonal antibody against alpha-smooth muscle actin: a new probe for smooth muscle differentiation. J Cell Biol 103:2787–2796

359. Small JV, Furst DO, De Mey J (1986) Localization of filamin in smooth muscle. J Cell Biol 102:210–220

360. Somlyo AP (1983) Excitation–contraction coupling and the ultrastructure of smooth muscle. Circ Res 57:497–507

361. Somlyo AP, Somlyo AV, Shuman H et al. (1978) Electron probe analysis of calcium compartments in cryo sections of smooth and striated muscles. Ann NY Acad Sci 307:523–544

362. Somlyo AP, Somlyo AV, Shuman H (1979) Electron probe analysis of vascular smooth muscle: composition of mitochondria, muscles and cytoplasm. J Cell Biol 81:316–335

363. Somlyo AV (1980) Ultrastructure of vascular smooth muscle. In: Bohr DF, Somlyo AP, Sparks HV (eds): Handbook of physiology, section 2: the cardiovascular system, vol II: vascular smooth muscle. American Physiological Society, Bethesda, pp 33–67

364. Somlyo AV, Gonzalez-Serratos H, Shuman H et al. (1981) Calcium release and ionic changes in the sarcoplasmic reticulum of tetanized muscle: an electron probe study. J Cell Biol 90:577–594

365. Spagnoli LG, Pietra GG, Villaschi S et al. Morphometric analysis of gap junctions in regenerating arterial endothelium. Lab Invest 46:139–148

366. Spagnoli LG, Villaschi S, Neri L et al. (1982) Gap junction in myo-endothelial bridges of rabbit carotid

arteries. Experientia 38:124–125

367. Sprague EA, Kelley JL, Suenram CA et al. (1985) Stimulation of albumin endocytosis by cationized ferritin in cultured aortic smooth muscle cells. Am J Pathol 121:443–443

368. Squire J (1981) The structural basis of muscular contraction. Plenum Press, New York

369. Staehelin LA, Hull BE (1978) Junctions between living cells. Sci Am 238:141–152

370. Steer CJ, Bisher M, Blumenthal R et al. (1984) Detection of membrane cholesterol by filipin in isolated rat liver coated vesicles is dependent upon removal of the clathrin coat. J Cell Biol 99:315–319

371. Stein O, Stein Y (1972) An electron microscopic study of transport of peroxidases in the endothelium of mouse aorta. Z Zellforsch Mikrosk Anat 133:211–222

372. Stein O, Stein Y (1980) Bovine aortic endothelial cells display macrophage-like properties toward acetylated (^{125}I)-labelled low density lipoprotein. Biochim Biophys Acta 620:631–635

373. Stemerman MB (1974) Vascular intimal components: Precursors of thrombosis. Prog Hemost Thromb 2:1–47

374. Stevenson BR, Siliciano JD, Mooseker MS et al. (1986) Identification of ZO-1: a high molecular weight polypeptide associated with the tight junction (zonula occludens) in a variety of epithelia. J Cell Biol 103:755–766

375. Svensjö E, Grega GJ (1986) Evidence for endothelial cell-mediated regulation of macromolecular permeability by postcapillary venules. Fed Proc 45:89–95

376. Talmon A, Cohen E, Bacher A et al. (1984) Separation of induction and expression of tight junction formation mediated by proteases. Biochim Biophys Act a 769:505–507

377. Taugner R, Kirchheim H, Forssmann WG (1984) Myoendothelial contacts in glomerular arterioles and in renal interlobular arteries of rat, mouse and *Tupaia belangeri*. Cell Tissue Res 235:319–325

378. Taylor KA, Amos LA (1981) A new model for the geometry of the binding of myosin crossbridges to muscle thin filaments. J Mol Biol 147:297–324

379. Thoma R (1921) Über die intima des arterien. Virchows Arch [A] 230:1–45

380. Thorgeirsson G, Robertson AL (1978) The vascular endothelium – pathobiologic significance. A review. Am J Pathol 93:803–848

381. Thyberg J, Nilsson J, Palmberg L et al. (1985) Adult human arterial smooth muscle cells in primary culture. Modulation from contractile to synthetic phenotype. Cell Tissue Res 239:69–74

382. Timpl R, Rhode H, Gehron-Robey P et al. (1979) Laminin – a glycoprotein from basement membrane. J Biol Chem 254:9933–9937

383. Trout JJ, Koenig H, Goldstone AD et al. (1986) Blood–brain barrier breakdown by cold injury. Polyamine signals mediate acute stimulation of endocytosis, vesicular transport and microvillus formation in rat cerebral capillaries. Lab Invest 55:622–631

384. Turner MR, Clough G, Michel CC (1983) The effects of cationized ferritin and native ferritin upon the filtration coefficient of single frog capillaries. Evidence that proteins in the endothelial cell coat influence permeability. Microcirc Res 25:205–222

385. Unwin PNT, Zampighi G (1980) Structure of the junction between communicating cells. Nature 283:545–549

386. Van Breemen C, Cauvin C, Johns A et al. (1986) Ca²⁺ regulation of vascular smoooth muscle. Fed Proc 45:2746–2751.

387. Vandekerckhove J, Weber K (1979) The complete amino acid sequence of actins from bovine aorta, bovine heart, bovine fast skeletal muscle and rabbit slow skeletal muscle. Differentiation 14:123–133

388. Van de Vorde J, Leuen I (1983) Role of endothelium in the vasodilator response of rat thoracic aorta to histamine. Eur J Pharmacol 87:113–120

389. Vanhoutte PM (1978) Heterogeneity in vascular smooth muscle. In: Kaley G, Altura BM (eds) Microcirculation vol II. University Park Press, Baltimore, pp 181–309

390. Vasile E, Simionescu M, Simionescu N (1983) Visualization of the binding, endocytosis, and transcytosis of low density lipoprotein in the arterial endothelium in situ. J Cell Biol 96:1677–1689

391. Venkatachalam MA, Rennke HG (1978) The structural and molecular basis of glomerular filtration. Circ Res 43:337–347

392. Verselis V, White RL, Spray DC et al. (1986) Gap junctional conductance and permeability are linearly related. Science 234:461–464

393. Vlodavsky I, Gospodarowicz D (1979) Structural and functional alterations in the surface of vascular endothelial cells associated with the formation of a confluent cell monolayer and with the withdrawal of fibroblast growth factor. J Supramol Struct 12:73–114

394. Volk T, Geiger B (1986) A-CAM: a 135-kD receptor of intercellular adherens junctions. I. Immunoelectron microscopic localization and biochemical studies. J Cell Biol 103:1441–1450

395. Volk T, Geiger B (1986) A-CAM: a 134-kD receptor of intercellular adherens junctions. II. Antibody-mediated modulation of junction formation. J Cell Biol 103:1451–1464

396. Voyta JC, Via DP, Butterfield CE et al. (1984) Identification and isolation of endothelial cells based on their increased uptake of acetylated low density lipoprotein. J Cell Biol 99:2034–2040

397. Wade JB, Karnovsky MJ (1974) The structure of the zonula occludens. A single fibril model based on freeze-fracture. J Cell Biol 60:168–180

398. Wagner DD, Marder VJ (1984) Biosynthesis of von Willebrand protein by human endothelial cells: processing steps and their intracellular localization. J Cell Biol 99:2123–2130

399. Wagner DD, Mayadas T, Marder VJ (1986) Initial glycosylation and acidic pH in the golgi apparatus are required for multimerization of von Willebrand factor. J Cell Biol 102:1320–1324

400. Wagner DD, Olmstead JB, Marder VJ (1982) Immunolocalization of von Willebrand protein in Weibel-Palade bodies of human endothelial cells. J Cell Biol 95:355–360

401. Wagner RC, Andrew SB (1985) Utrastructure of the vesicular system in rapidly frozen capillary endothelium of the rete mirabile. J Ultrastruct Res 90:172–182

402. Wagner RC, Robinson CS (1982) Tannic acid tracer analysis of permeability pathways in the capillaries of the rete mirabile: demonstration of the discreteness of endothelial vesicles. J Ultrastruct Res 81:37–46

403. Wagner RC, Robinson CS (1984) High-voltage electron microscopy of capillary endothelial vesicles. Microvasc Res 28:197–205

404. Walker LN, Bowen-Pope DF, Ross R et al. (1986) Production of platelet-derived growth factor-like molecules by cultured arterial smooth muscle cells accompanies proliferation after arterial injury. Proc Natl Acad Sci USA (1986) 83:7311–7315

405. Warren HB, Collins T, Davies PF (1986) C-sis RNA expressed by cholesterol loaded bovine aortic endothelial cells. Fed Proc 45:1073 (abstract)

406. Wang K, Ash JF, Singer SJ (1975) Filamin, a new high-molecular-weight protein found in smooth muscle and non-muscle cells. Proc Natl Acad Sci USA 72:4483–4486

407. Warhol MJ, Sweet JM (1984) The ultrastructural localization of von Willebrand factor in endothelial cells. Am J Pathol 117:310–315

408. Weeds A (1982) Actin-binding proteins – regulators of cell architecture and motility. Nature 296:811–816

409. Wei EP, Ellison MD, Kontos HA et al. (1986) O_2 radicals in arachidonate-induced increased blood–brain barrier permeability to proteins. Am J Physiol 251:H693–H699

410. White GE, Fujiwara K (1986) Expression and intracellular distribution of stress fibers in aortic endothelium. J Cell Biol 103:63–70

411. White GE, Gimbrone MA Jr, Fujiwara K (1983) Factors influencing the expression of stress fibers in vascular endothelial cells in situ. J Cell Biol 97:416–424

412. Williams LT, Tremble P, Antoniades HN (1982) Platelet-derived growth factor binds specifically to receptors on vascular smooth muscle cells and the binding becomes nondissociable. Proc Natl Acad Sci USA 79:5867–5870

413. Williams SK, Devemy JJ, Bittensky MW (1981) Micropinocytotic ingestion of glycosylated albumin by isolated microvessels: possible role in pathogenesis of diabetic microangiopathy. Proc Natl Acad Sci USA 78:2393–2397

414. Wolinsky H, Glagov S (1967) A lamellar unit of aortic medial structure and function in mammals. Circ Res 20:99–101

415. Wong AJ, Pollard TD, Herman IM (1983) Actin filament stress fibers in vascular endothelial cells in vivo. Science 29:867–869

416. Wysolmerski R, Lagunoff D (1985) The effect of ethchlorvynol on cultured endothelial cells. A model for the study of the mechanism of increased vascular permeability. Am J Pathol 119:505–512

417. Yamada KM (1983) Cell surface interactions with extracellular materials. Ann Rev Biochem 52:761–799

418. Yokota S (1983) Immunocytochemical evidence for transendothelial transport of albumin and fibrinogen in rat heart and diaphragm. Biomed Res 4:577–586

419. Zetter BR (1981) The endothelial cells of large and small blood vessels. Diabetes 30 [Suppl] 2:24–28

420. Zand T, Underwood JM, Nunnari JJ et al. (1982) Endothelium and "silver lines". An electron microscopic study. Virchows Arch [A] 395:133–144

Chapter 2

Extracellular Matrix of the Arterial Vessel Wall

L. Robert and P. Birembaut

The arterial vessel wall is characterized by a particularly abundant extracellular matrix, the composition of which varies with the size, the nature and the position in the arterial tree of the vessel. Nevertheless, the major components of the vessel wall extracellular matrices are collagens of various types and elastin. The cohesion between cells and these fibrous proteins is ensured by glycoproteins such as fibronectin and laminin and by proteoglycans.

In this chapter, a brief report of the biochemistry of these four classes of macromolecules will be followed by a morphological description.

Collagens

Biochemistry

Eleven types of collagen have been isolated from various extracellular matrices. The analysis and evaluation of the different types of vascular collagens have been extensively studied and all details will be found in recent reviews by Borel and Bellon [5] and Barnes [2].

In capillary walls, type IV collagen is found, along with type V. Analysis has shown that type IV collagen molecules contain some non-triple helical sequences, unlike types I, II and III (the "interstitial" collagens) which express a characteristic ultrastructural pattern with a major axial repeat period of 67 nm. Type IV collagen, like some other types, remains as procollagen. Non-helical parts of the molecule have not been cleaved and it has been suggested that this type of collagen forms a network called "chicken wire". In the meshes of this network, other macromolecules may be found, representing a large fraction of the lamina densa of basement membranes [54].

Large vessel walls, both adventitia and media, are particularly rich in interstitial collagens type I and III, and both are synthesized by adventitial fibroblasts and the muscle cells of the media. Both types I and III procollagens are first synthesized by the cell with C- and N-propeptides which are cleaved when the molecules leave the cell and join together, starting the process of fibrillogenesis (Fig. 2.1). An important step in biosynthesis is the hydroxylation of numerous prolyl and lysyl residues and the glycosylation of hydroxylysine with glucose and galactose. All collagen molecules are glycoproteins but their glucidic content varies considerably. For example, type IV collagen contains more glucids than type I or type III. Apart from these glycosydyl residues added to the hydroxylysine, C- and N-terminal propeptides and non-helical parts of various types of collagen molecules also contain glucidid residues, N-glycosydyl being linked to aspartic acid.

Types I and III collagens are mainly responsible for the tensile strength of vascular walls (see Chap. 3). Age-related and atherosclerotic intimal changes are characterized by abundant type I collagen produced by phenotypically modulated smooth-muscle cells invading the intima, although in the earlier phase of plaque formation, type III collagen and fibronectin may predominate [36] (see Chap. 12).

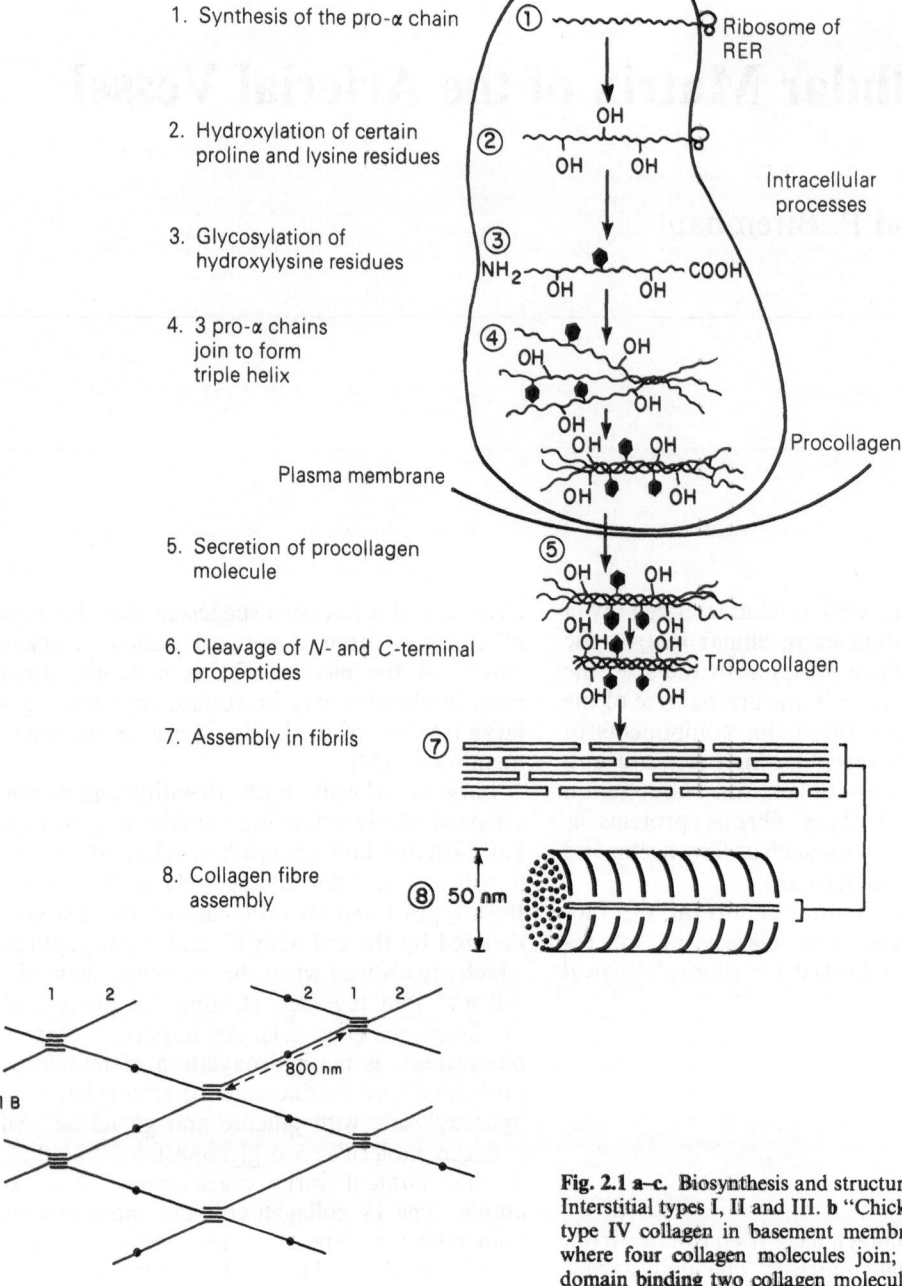

1. Synthesis of the pro-α chain

2. Hydroxylation of certain proline and lysine residues

3. Glycosylation of hydroxylysine residues

4. 3 pro-α chains join to form triple helix

5. Secretion of procollagen molecule

6. Cleavage of *N*- and *C*-terminal propeptides

7. Assembly in fibrils

8. Collagen fibre assembly

Fig. 2.1 a–c. Biosynthesis and structure of collagen fibres. **a** Interstitial types I, II and III. **b** "Chicken wire" structure of type IV collagen in basement membranes: *1*, 7-S domain where four collagen molecules join; *2*, non-triple helical domain binding two collagen molecules (from). **c** Type VI "intimal" collagen structure: filamentous form; non-triple helical domains are figured as ovals.

Interstitial collagens only possess a continuous triple helical part, making them more neutral proteinase resistant. The other types of collagen (and even type III collagen) possess non-triple helical parts in important amounts, rendering them sensitive to proteases.

Other types of collagen have also been described in the vascular wall. Type VI, the so-called "short chain collagen" or "intimal collagen", and type V collagen are also present. Type V collagen is found around smooth-muscle cells and along basement membranes. Endothelial cells of the intima principally synthesize type VI and type III collagens. In vitro endothelial cells synthesize type III collagen but little or no type I. The basement membranes surrounding smooth-muscle cells are rich in type IV collagen.

The singular amino acid composition of collagen should be noted. Every third residue in the helical chains is glycine, whereas proline and hydroxyproline together account for another 20% of the residues. Glycine–X–Y sequences make up the remainder where X and Y are most often prolyl and hydroxyprolyl respectively. Similar sequences can be found in proteins such as C19 and acetylcholinesterase. As noted by Gray and Miller [22], these molecules are distinguished from collagen by their functions: they do not tend to aggregate and are not involved in the extracellular matrix. Hydroxyproline is thus a convenient marker for collagen.

Three individual chains (α chains) are coiled around a central axis to form a right-handed helix in the molecule of tropocollagen (Fig. 2.1), which represents the basic unit of interstitial collagen fibres (types I, II and III). At both amino- and carboxy-termini of each α chain, there are non-helical extensions called telopeptides. At the time of secretion by the cell, part of these telopeptides is cleaved and the presence and amount of these cleaved fractions in the blood may be related to the biosynthetic activity of the cells (for example, serum N-propeptide levels of type III collagen are raised in fibrosis and in liver cirrhosis). The cleavage is not identical for the other types of collagen. As far as type I collagen is concerned the major part of the non-helicoidal extensions is cleaved, whereas type IV collagen keeps a non-negligible globular portion at one extremity.

These findings are significant in the arrangement and the aggregation of the tropocollagen molecules [6]. In the case of types I, II and III collagens, the tropocollagen molecules in the extracellular space aggregate to form fibrils then fibres, the diameter of which varies from one tissue to another. Fibres become stabilized by covalent intermolecular cross-links initiated by a specific copper-dependent lysyloxidase which converts lysyl and hydroxylysyl ε-amino groups to aldehydes. Intra- and intermolecular cross-links can form by the reaction of two aldehydic groups (aldol condensation) or by the formation of a Schiff base [1]. Bound tropocollagen molecules are aligned end-to-end and parallel with a two-dimensional quarter stagger as shown in Fig. 2.1. This characteristic stagger is responsible for the ultrastructural pattern, i.e. the major axial repeat period of 67 nm of interstitial collagen fibres. Other types of collagen do not express this periodicity and the arrangement of

the molecular structure is different. For example, the cohesion of the type IV collagen molecule network is partly related to disulphide bonds.

Specific enzymes (tissue or bacterial collagenases) may attack collagen fibres, cleaving the three helical chains of the native molecule. Other proteinases (cathepsins, elastases) degrade non-helical parts. Such enzymes play a major role in tissue degradation in the normal remodelling of connective tissues during growth, after injury and repair and in pathological states (inflammations, tumours).

Morphology

Types I and III tropocollagen molecules aggregate to form fibrils of variable diameters which then assemble as fibres. Other types of collagen, especially types IV and V, are involved in the formation of basement membranes.

Types I and III collagen fibres form large tortuous bundles in the media and in the adventitia of arteries. They are bi-refringent and acidophilic and vary in diameter from 1 to 20 μm. They are well stained with eosin and Masson's trichrome, and appear red or yellow (type I) and green (type III) when stained by Picrosirius red and observed in polarized light [31]. Picrosirius red is an acidic dye which binds to collagen with an orientation parallel to tropocollagen molecules, resulting in increased bi-refringence. Nevertheless, the specificity of the various types of collagen is debatable; the thickness of the sections interferes in the differential staining and immunohistochemistry provides more precise information on the nature and type of collagen fibres.

In response to acute or chronic distension, collagen fibres are found in excess in arterial walls where they are associated with elastic fibres [10,14].

"Reticular fibres" (reticulin) are considered to be related to type III collagen by some authors [32]. They form a delicate network of fibres in the intima, in the media around muscle cells and pericytes and in the adventitia, and are stained by alkaline solutions of silver salts. The current view is that the argyrophilia of reticulin fibres is due to a coating of proteoglycans and glycoproteins on thin collagen fibrils.

Electron microscopy reveals the characteristic "native" pattern of cross-banding with a major axial repeat period of 67 nm for interstitial types I and III collagen. Types IV and V collagen, present in the lamina densa of basement membranes, do not express this periodicity. Light

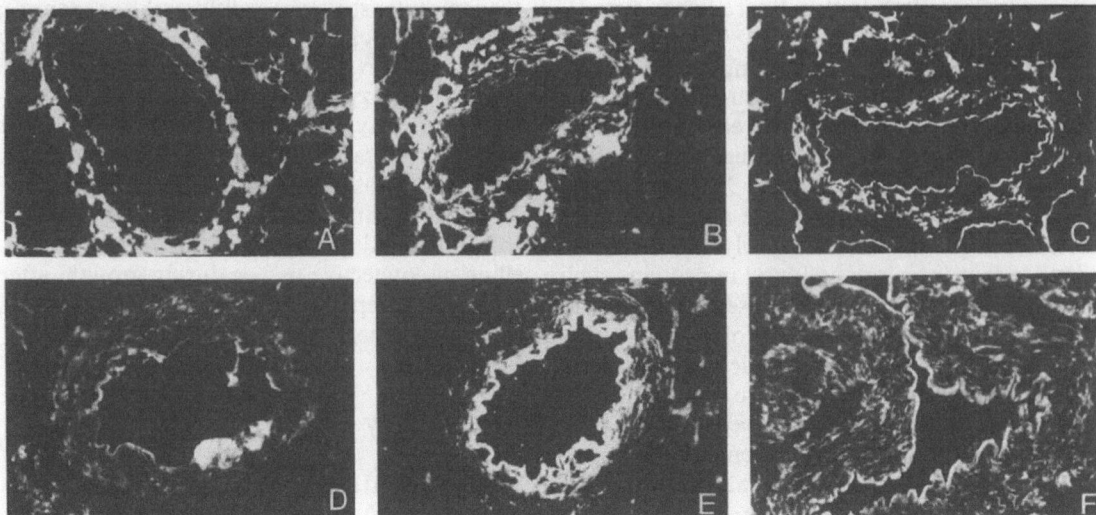

Fig. 2.2 a–f. Distribution of various types of collagen, of heparan sulphate proteoglycan and structural glycoproteins (laminin, fibronectin) in arterial walls (studied by immunofluorescence). (Courtesy of Prof. J. M. Foidart, University of Liège) × 250 **a** Type I collagen is present principally in the media and in the adventitia. **b** Type III collagen: its distribution is more extensive, especially in the intima. **c** Type IV collagen is found in basement membranes underlying the endothelium and around smooth-muscle cells. **d** Heparan sulphate proteoglycan: its distribution is analogous to that of type IV collagen. **e** Fibronectin is found in the intima, in basement membranes and in the adventitia. **f** Laminin is also seen exclusively in basement membranes.

microscopy reveals fibrils assembled as fibres which are packed in bundles. Type III collagen fibrils are generally thinner (30 nm diameter) than type I collagen fibrils (60 to 90 nm diameter). The distribution of these various collagen types is well detected by immunohistochemistry using specific antibodies raised against the different types of collagen (Fig. 2.2). The use of this technique confirms the presence of collagen types I and III in the whole vessel wall. Types IV and V collagen (basement membrane collagens) line the endothelium and surround smooth-muscle cells of the media [20].

Type VI collagen ("short chain collagen" or "intimal collagen") is not localized exclusively in the intima but is also found in the media, around smooth-muscle cells [30,55].

Elastin

Biochemistry

Elastin is a fibrous protein largely distributed in vascular vessel walls but less abundant than collagen. Elastin is not present in basement membranes and the resilience of capillary walls is certainly related to the particular molecular structure of type IV collagen, to proteoglycans and structural glycoproteins. Elastin is thus not the only component of extracellular matrix responsible for the elasticity of the connective tissue. Nevertheless, in some vessels, like "elastic arteries", submitted to rhythmic variations of their diameter due to pulsatile blood flow, elastin plays a major role.

Like collagen, about one-third of the residues of elastin are glycine. Elastin is also rich in proline, but hydroxyproline is not abundant (1%–1.5%). By contrast, elastin contains characteristic amino acids – desmosine, isodesmosine and lysino-norleucine (Fig. 2.3).

Tropoelastin is a soluble precursor of elastin, with a molecular weight of about 70 000, rich in lysine. A copper-dependent lysyl oxidase reacts with lysyl residues and both desmosine and isodesmosine are formed with a steric orientation on a microfibrillar scaffold. Without the formation of these cross-links, elastin would not be elastic. This elasticity is related to the hydrophobic nature of the molecule.

With ageing and in atherosclerosis, lipids (mainly esters of cholesterol and free fatty acids) are deposited upon elastic fibres and attract Ca^{2+} [8,29]. Ca^{2+} deposits promote lipid accumulation and vice versa. As a consequence, elastin loses its extensibility and its degradation by elastases is accelerated (Fig. 2.4). These elastases may reach the tissue from the blood. Poly-

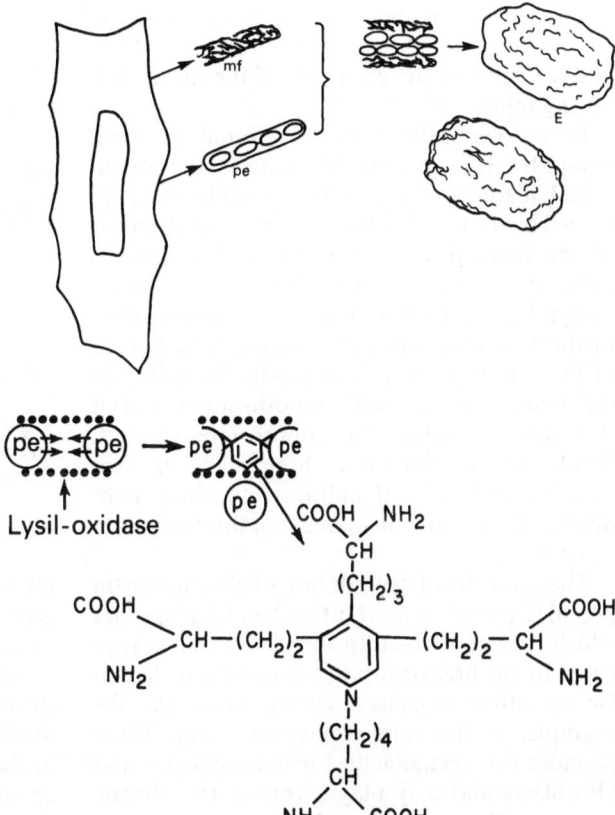

Fig. 2.3. Schematic representation of biosynthesis of elastic fibres: the smooth-muscle cells secrete microfibrils (*mf*) on which precursors of elastin (*pe*) or tropoelastin settle. After the action of lysyl oxidase, desmosine and isodesmosine are formed between two adjacent tropoelastin molecules. Mature elastic fibres then form, always associated with microfibrils (*E*).

morphonuclears contain one particularly active serine-protease elastase. Macrophages, which may invade the vessel wall, possess a metallo-proteinase and smooth-muscle cells also have a serine-proteinase on their plasma membrane. Fibroblasts in the adventitia also secrete a metal-lo-protease. The collective action of these enzymes results in progressive fragmentation of elastic fibres which begins in the intima and progresses to the media. Although neo-synthesis of elastin is possible as shown by the deposition of thin elastic fibrils in the neo-intima and in

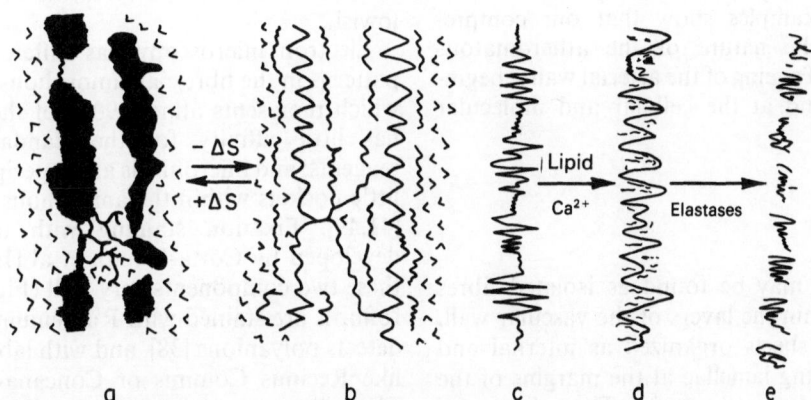

Fig. 2.4. Schematic representation of a contracted elastic fibre. **a** H_2O molecules (*V*) are excluded from hydrophobic dense regions (*black*). **b** After stretching, the contact between exposed solvent and hydrophobic amino acid residues becomes closer and the water has to "reorganize" around these residues (decrease of entropy, $-S$). The return to the original length will be spontaneously accompanied by an increase of entropy ($+S$). **c, d** With ageing and in atheroma, lipids and Ca^{2+} settle in the fibres, blocking this gain of entropy. **e** The degradation of fibres by elastases is increased.

atheromatous plaques, there is never complete reconstitution of the concentric elastic lamellae nor recovery of the elasticity of the artery following injury.

In venous walls, elastin is found in small amounts. This presumably reflects the mechanical loading of the wall, a conclusion supported by the fact that when a direct anastomosis of the femoral vein of a dog to the femoral artery is performed, a 1000-fold increase in the biosynthetic activity of elastin is found in vitro in the vein [39]. Thus the smooth-muscle cells of the venous walls clearly retain the ability to differentiate as "arterial" smooth-muscle cells if they are submitted to appropriate stresses. Similar observations have been made in vitro with "stretched" cell cultures. In these conditions, there is also an increase of the biosynthesis of elastin.

The products of degradation of elastin, elastin peptides, possess important biological properties which may contribute to the vicious circle triggered by the breakdown of arterial elastic lamellae or other organized elastic fibres (as for example, in the emphysematous lung). These peptides are chemotactic for macrophages and fibroblasts and may play a role in the fibrotic process. They are also antigenic and when injected along with Freund's adjuvant are able to initiate the formation of atheromatous plaques in the rabbit [28,40]. This is due to the fact that elastin peptides, like antibodies against whole elastin and immune complexes, react with the plasma membranc of smooth-muscle cells and modify ionic flux. The consequences are a massive calcium influx in the cell with a loss of contractility and an increase of both chemotactic motility and biosynthetic activity of these cells. These few examples show that our comprehension of the nature of the atheromatous process and of ageing of the arterial wall is beginning to expand at the cellular and molecular level.

Morphology

Elastic fibres may be found as isolated fibres dispersed within the layers of the vascular wall, or as elastic sheets organized as internal and external limiting lamellae at the margins of the media in middle-sized arteries. They also occur as a system of concentric lamellae in the media of large arteries like the aorta and pulmonary artery (elastic conducting arteries). In these vessels, the number of elastic lamellae is deter-

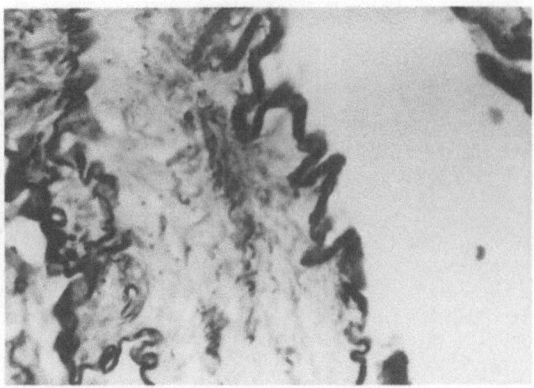

Fig. 2.5. Middle-sized artery, stained with orcein. Internal and external elastic laminae are seen with a scalloped pattern due to retraction of fibres after fixation. × 400

mined by mechanical forces and there is a gradual reduction of their number from the heart to the iliac bifurcation (see Chap. 3).

With ageing, there is a progressive lysis of the concentric lamellae. In contrast, thin elastic fibrils which connect them or which are found in the neo-intima of atheromatous plaques are apparently synthesized throughout life. An internal elastic lamina is present in nearly all arteries, appearing as a tortuous structure due to the retraction of the fibres after fixation (Fig. 2.5). With ageing, this lamina appears to duplicate. Recently, Godeau [23] has used a specific stain for the elastic network which allows a good evaluation of these fibres by morphometry and computerized image analysis. This technique has been applied for quantitative studies on dermal elastic fibres [24]. When examined by immunofluorescence with UV, elastic fibres are yellowish.

Electron microscopy has differentiated two patterns in the fibres: an amorphous component which represents almost 90% of the fibres and has little affinity for the standard staining reagents; and microfibrils at the periphery and in little pockets within the amorphous component [41,49]. Fixation staining with tannic acid, developed by Cotta-Pereira et al [12], discloses these two components very well (Fig. 2.6). Microfibrils are stained with Ruthenium red, which detects polyanions [38], and with labelled lectins like Recinus Commis or Concanavalin A [4]. Their diameter varies from 7 to 17 nm, but the average size is 10 to 15 nm. Cleary et al. [9] have shown that the diameter of microfibrils of the internal elastic lamina of aortas increases significantly throughout the fetal and post-natal

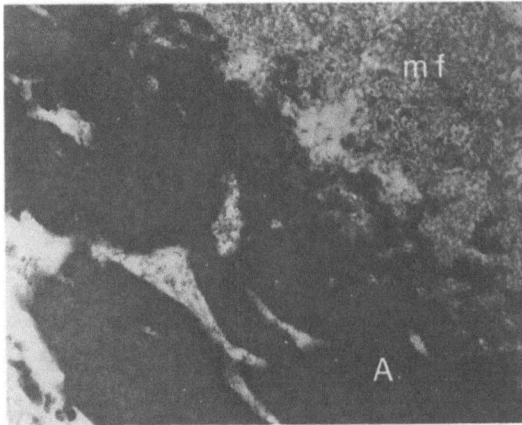

Fig. 2.6. Elastic internal lamina, stained by tannic acid. This fixation–staining method using tannic acid and electron microscopy reveals the two components of elastic fibres: the amorphous component (*A*), corresponding to elastin, and microfibrillar (*mf*) component formed principally by structural glycoproteins. These microfibrils here have a diameter of 12 nm and do not express periodicity. × 50 000

stages of development. At high resolution, in cross-section, elastin-associated microfibrils have been reported to consist of four to six filaments, each surrounded by an irregular coating of electron-dense material. These filaments, arranged around the periphery of the microfibril, leave an electron-lucent central zone, producing a "tubular" appearance [17]. This microfibril may in fact be formed by some glycoproteic components around a precursor of elastin. These elastin-associated microfibrils have been compared and sometimes related to oxytalan fibres, precursors of elastic fibres, and described by Elder [16] in invertebrate aortas. These microfibrils, like interstitial collagens, intervene in platelet–subendothelium interactions [4] (see Chap. 8). They can be associated with basal laminae, in particular those present around smooth-muscle cells, playing a role in maintaining the integrity of distended arterial walls. The amorphous component, well stained by tannic acid, can be digested by elastases [26,43,48]. Goldfischer et al. [25] have visualized the complex infrastructure of this amorphous part of elastic fibres using fixation and staining with a glutaraldehyde–ruthenium red mixture in phosphate buffer. They have described two interlacing but distinct structural components: (a) a framework of circumferentially orientated microfibrils which extend from the core of the elastic fibre into the surrounding matrix, where they appear to function as anchoring fibres, and (b)

a three-dimensional meshwork of filaments that permeate the fibre, which may represent the supramolecular organization of elastin.

Proteoglycans

Biochemistry

Proteoglycans, previously called acid mucopolysaccharides, have been studied extensively. First isolated from cartilage, where they represent the major part of the extracellular matrix, they have also been extracted from vessel walls. They are formed from numerous proteoglycan subunits, consisting of a core protein of variable length to which long polysaccharide chains called glycosaminoglycans (GAG) are linked covalently.

These various GAG are listed in Table 2.1. Among them, hyaluronic acid alone is not linked to a core protein. The proteoglycan subunits are non-covalently associated with hyaluronic acid long chains to form the macromolecular proteoglycan aggregate. Some core proteins possess a site of recognition for hyaluronic acid which binds to this molecule with a strong affinity. Heparan sulphate proteoglycan and sometimes chondroitin sulphates in small amounts, are found in basement membranes. In large vessels, chondroitin sulphate, dermatan sulphate and heparan sulphate proteoglycans have been isolated [3]. It must be noted that some proteoglycans, specially dermatan sulphate proteoglycans, form insoluble precipitates with LDL. This chemical reaction, described first by Gero and Bihari-Varga in Budapest and studied afterwards by numerous authors, is considered

Table 2.1 Glycosaminoglycans found in vascular proteoglycans

Designation	Abbreviation	Components
Hyaluronic acid	HA	Glucuronic acid, *N*-acetyl, D-glucosamine
Chondroitin-4-sulphate	C4S	Glucuronic acid, *N*-acetyl, D-galactosamine-4-sulphate
Chondroitin-6-sulphate	C6S	Glucuronic acid, *N*-acetyl, D-galactosamine-6-sulphate
Dermatan-sulphate	DS	L-iduronic acid, *N*-acetyl, D-galactosamine sulphate
Heparan-sulphate	HS	Glucuronic acid, L-iduronic sulphate, D-glucosamine, *N*-sulphate, *N*-acetyl

by some to be responsible for the extracellular deposition of lipid in atherogenesis, and an increase of proteoglycans in the vessel wall detected by histochemistry in the early stages of the formation of atheromatous plaques is thought to be pathogenic. Glycosaminoglycans have been extracted and are able to form complexes with lipoproteins. Such associations have also been extracted from atheromatous plaques [56] (see Chap. 12).

The physiological role of proteoglycans is to control molecular traffic and ionic balance in tissues. They form a meshwork which may impede the transfer of molecules from the circulating blood to the tissues, and as polyanions they intervene in ionic movements. The degradation of proteoglycans is accomplished by proteinases, which attack core proteins, and by endoglycosaminidases such as hyaluronidase, chondroitinase and heparitinase. This process diminishes the polymerization of these macromolecules and increases tissue permeability.

Morphology

Proteoglycans are one of the major components of the amorphous ground substance of connective tissues, surrounding cells, fibres and fibrils. They are polyanions. In cartilage, where they are abundant, they are stained by conventional basic dyes such as haematoxylin. In lower concentrations in vessel walls, special cationic dyes like alcian blue or colloidal iron must be used for their detection. The proteoglycans present in the intima are generally more alcianophilic than those found in the media. This affinity for alcian blue disappears after selective enzymatic treatment digests GAGs (chondroitinase ABC, hyaluronidase) [57]. Scott and Dorling [52,53] have described several methods which can be used to differentiate the GAG in the tissues, using alcian blue with variations of the pH or the PAS method with prolonged oxidation. Proteoglycans react with some cationic dyes, such as toluidine blue or crystal violet, to give a characteristic metachromasia.

Electron microscopy also characterizes proteoglycans as polyanions, being stained by ruthenium red [38] and digested by enzymatic treatments (chondroitinase ABC, hyaluronidase) (Fig. 2.7). Scott [51] has used specific dyes derived from alcian blue for their detection.

These macromolecules are frequently associated with collagen fibres in vessel walls, especially dermatan sulphate, and they are present at the

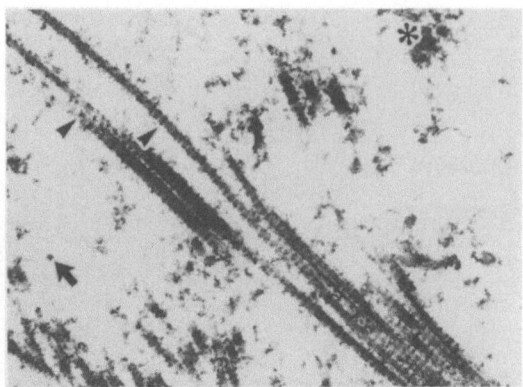

Fig. 2.7. Arterial adventitia stained with ruthenium red. Collagen fibres are seen with their major axial repeat period of 67 nm. The granules corresponding to proteoglycans (*arrowhead*) bind to collagen fibres. Other granules are isolated in the ground substance (*arrow*) or associated with microfibrils (*). × 60 000

surface of elastic fibres, in the interfibrillar ground substance (chondroitin sulphate) and in basement membranes (heparan sulphate proteoglycan) [15,57]. They appear as numerous, irregularly shaped, 20–50 nm diameter granules possessing 3–6 nm diameter filamentous projections. These granules are interconnected via their filamentous projections. In addition, there are smaller 10–20 nm diameter granules present within the basement membranes surrounding smooth-muscle cells and underlying endothelial cells. The collagen-associated granules are located at intervals of 70 nm and their distribution suggests close and organized relations between these macromolecules.

Recent studies have shown that these ruthenium-red-stained particles are present in increased concentration in the innermost part of the mixed fatty-gelatinous lesions and in a decreased concentration in the fibrous cap of atheromatous plaques [27].

Immunohistochemistry identifies the distribution of some of these proteoglycans more precisely, particularly heparan sulphate proteoglycan which is found in basement membranes, along with type IV collagen and laminin [11] (Fig. 2.2).

Structural Glycoproteins

Biochemistry

Evidence for specific glycoproteins in the extracellular matrix was first discovered in the cornea

Table 2.2 Structural glycoproteins (SG) of the vascular wall and/or synthesized by the vascular cells in culture

SG	Molecular weight (10^3)	Localization
Fibronectin	450	Subendothelium, media, adventitia
Laminin	1000	Basal lamina
Entactin*	158	Basal lamina
Nidogen	160	Basal lamina
SG 115 (mesonectin)	115–130	Media of aorta
SG 36	36	Media of aorta
SG 50	50	Media of aorta
SG 130	130	Media of aorta
SG 43	43	Media of aorta
Thrombospondin	3×160	Endothelial cells, smooth-muscle cells
SG bound to collagen	70	Endothelial cells
SG bound to albumin	43 (after reduction)	Endothelial cells, smooth-muscle cells
Von Willebrand factor	800 more than 1200	Endothelial cells

* May be identical to nidogen.

in 1963 [41,46]. Numerous other glycoproteins have since been isolated and characterized and Table 2.2 lists the principal structural glycoproteins found in basement membrane of capillaries and vessel walls.

Fibronectin (FN) is one of the most studied glycoproteins. Its composition and structure are shown in Fig. 2.8. One of the functions of FN, like other glycoproteins, is to establish the cohesion between the plasma membrane and the extracellular matrix [58]. Transformed malignant cells lose their ability to retain FN on their cell membranes and their migration through the tissues is thus facilitated. In basement membranes, laminin plays a major role in the anchoring of endothelial or epithelial cells to type IV collagen. The biological function and the localization of the other glycoproteins listed in Table 2.3 have not yet been elucidated. Some of these glycoproteins, like FN and laminin, are also present in the plasma in variable amounts. It seems likely that plasma FN may leak and penetrate into the tissues and associate with fibrous networks where it appears in excess in

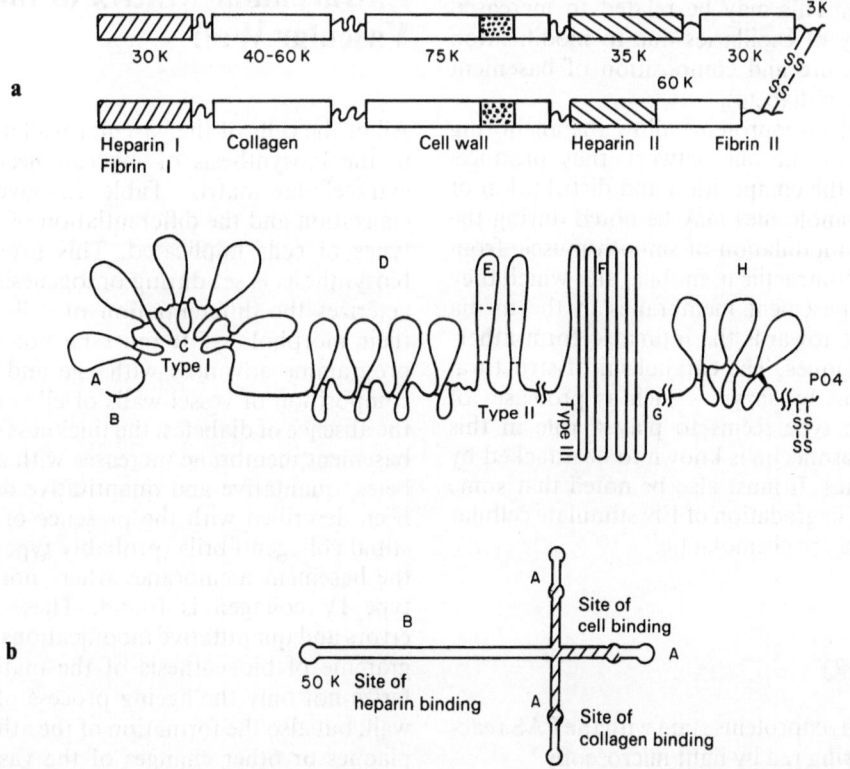

Fig. 2.8. Structural glycoproteins: **a** Schematic representation of the structure of fibronectin: the two peptide chains are bound by two disulphide bonds. On each chain, the various sites of fixation of different macromolecules and plasma membrane are indicated with the molecular weight. Through these domains, fibronectin acts as a "biological glue" and ensures the adhesion of cells (fibroblasts) to the fibrous network. **b** Laminin: the domains of fixation of heparin, collagen and plasma membrane are shown.

Table 2.3 Differentiation and programme of biosynthesis of extracellular matrix macromolecules of vascular wall

Type of cells	Extracellular matrix macromolecules			
	Collagens	Elastin	Proteo-glycans	Glyco-proteins
Endothelial cells	III, IV, VI	+	HS	Fibronectin, thrombo-spondin, Von Willebrand factor
Smooth-muscle cells	I, III, IV, V, VI	+	CS, DS, HS	Fibronectin, SG 30, 140
Fibroblasts	I, III	+	HA, CS	Fibronectin

pathological conditions such as diabetes [35,37] and Werner's disease, two diseases with an accelerated ageing of connective tissues [44]. Plasma FN levels increase exponentially with age and the slope of this semilogarithmic diagram declines in diseases like diabetes [35]. In such conditions, there is an increase of biosynthesis of FN by fibroblasts and of the tissue penetration of plasma FN. This may be related to increased permeability of capillaries due to modification of the structure and composition of basement membranes in diabetes.

Structural glycoproteins allow the anchoring of cells to the fibrous network they produce. Changes in the composition and distribution of these macromolecules may be noted during the phenotypic modulation of smooth muscle from sessile and contractile to mobile, afer which they shed their basement membranes in the media and migrate towards the intima to form atheromatous plaques. The catabolism of structural glycoproteins by enzymes such as proteases of the elastase type seems to play a role in this process. Fibronectin is known to be attacked by such enzymes. It must also be noted that some products of degradation of FN stimulate cellular division and are chemotactic.

Morphology

Structural glycoproteins stain with the PAS reaction, appearing red by light microscopy.

Electron microscopy reveals them as thin microfibrils, bound to collagen fibres and to elastin as seen previously. They are stained with ruthenium red and lectins like ricinus communis and concanavalin A. Some have been isolated,

purified and used to raise specific antibodies, permitting immunohistochemistry to be used for their detection. Fibronectin is present in large amounts in the vessel walls in the subendothelial layer in the media, around smooth-muscle cells and in the adventitia as thin filaments (Fig. 2.2) [7]. Some authors have considered that FN is one of the microfibrillar components bound to amorphous elastin [33]. Fibronectin has also been detected in the microvasculature in the pericyte–endothelial interstitium. The arrangement of these elements suggests a mechanical linkage between the two cells and such a linkage would allow contraction or relaxation of pericyte to affect vessel diameter [13]. Laminin is a basement membrane glycoprotein and is therefore found underlying endothelial cells and around smooth-muscle cells of the media, with a distribution similar to that of type IV collagen (Fig. 2.2) [19]. Nidogen and entactin are also observed within basement membranes.

General Arrangement of Extracellular Matrix of the Vascular Wall

All of the cells of the vascular wall are involved in the biosynthesis of macromolecules of the extracellular matrix. Table 2.3 gives the specialization and the differentiation of the various types of cells implicated. This programme of biosynthesis is set during ontogenesis and characterizes the differentiation of cells as well as their morphology. It must be noted that this programme advances with age and affects the composition of vessel walls of all sizes. Even in the absence of diabetes, the thickness of capillary basement membrane increases with age. In diabetes, qualitative and quantitative defects have been described with the presence of thin interstitial collagen fibrils (probably type III) within the basement membranes where normally only type IV collagen is found. These qualitative errors and quantitative modifications of the programme of biosynthesis of the matrix characterize not only the ageing process of the vessel wall, but also the formation of the atheromatous plaques or other changes of the vascular wall, such as the increase of sinuosity of capillaries with age and the formation of aneurysms.

From all these observations, some physiopathological and pharmacological notions may be derived. To understand the diseases of the vascular wall, the programme of biosynthesis of the extracellular matrix and its pathological modifications must be understood. To treat them, drugs able to correct deviations from this programme and the interactions between circulating blood cells and the vessel wall are necessary. Inhibitors of proteolytic enzymes such as elastases may play a major role in pharmacological approaches to vascular diseases in the future.

References

1. Bailey AJ, Robins SP (1973) Development and maturation of the crosslinks in the collagen fibers of the skin. In: Robert L, Robert B (eds) Frontiers of matrix biology. I. Aging of connective tissues – the skin. S Karger, Basel, pp 130–156
2. Barnes MJ (1985) Collagens in atherosclerosis. Coll Relat Res 5:65–97
3. Berenson GS, Radhakrishnamurthy SR, Srinivasan SR et al. (1984) Recent advances in molecular pathology: carbohydrate–protein macromolecules and arterial wall integrity: role in atherogenesis. Exp Mol Pathol 41:267–287
4. Birembaut P, Legrand YJ, Bariety J et al. (1982) Histochemical and ultrastructural characterization of subendothelial glycoprotein microfibrils interacting with platelets. J Histochem Cytochem 30:75–80
5. Borel JP, Bellon G (1985) Les collagènes vasculaires. Pathol Biol 33:16–22
6. Burgeson RE (1982) Genetic heterogeneity of collagens. J Invest Dermatol [suppl.] 79:25–30
7. Chemnitz J, Collatz-Christiansen B: (1983) Repair in arterial tissue: demonstration of fibronectin in the normal and healing rabbit thoracic aorta by the indirect immunoperoxidase technique. Virchows Arch [A] 399:307–316
8. Claire M, Jacotot B, Robert L (1976) Characterization of the lipids associated with macromolecules of the intercellular matrix of human aorta. Connect Tissue Res 4:61–71
9. Cleary EG, Fanning JC, Prosser IW (1981) Possible roles of microfibrils in elastogenesis. Connect Tissue Res 8:161–166
10. Cleary EG, Moont M (1976) Hypertension in weanling rabbits. In: Sandberg LB, Gray WR, Franzblau C (eds) Advances in experimental medicine and biology, vol 79. Plenum Press, New York, pp 477–490
11. Clowes AW, Clowes MM, Gown AM et al. (1984) Localization of protoheparan sulfate in rat aorta. Histochemistry 80:379–384
12. Cotta-Pereira G, Guerra Rodrigo F, David-Ferreira JF (1976) The use of tannic acid glutaraldehyde in the study of elastic and elastic-related fibers. Stain Technol 51:7–11
13. Courtoy PJ, Boyles J (1983) Fibronectin in the microrovasculature: localization in the pericyte endothelial interstitium. J Ultrastruct Res 83:258–273
14. Doscher W, Upmanas S, Mandell C et al. (1983) The response of elastin and collagen to acute hyperdistension of the rabbit abdominal aorta. Vasa 12:20–24
15. Eisenstein R, Kuetter K (1976) The ground substance of the arterial wall. Part 2. Electron microscopic studies. Atherosclerosis 27:37–46
16. Elder HY (1973) Distribution and functions of elastic fibers in the invertebrates. Biol Bull 144:44–63
17. Fanning JC, Yates NG, Cleary EG (1981) Elastin associated microfibrils in aorta: species differences in large mammals. Micron 12:339–348
18. Fauvel F, Grant ME, Legrand YJ et al. (1983) Interaction of blood platelets with a microfibrillar extract from adult bovine aorta: requirement for Von Willebrand factor. Proc Natl Acad Sci USA 80:551–554
19. Foidart JM, Bere EW, Yaar M et al. (1980) Distribution and electron microscopic localization of laminin, a noncollagenous basement membrane glycoprotein. Lab Invest 42:336–342
20. Furthmayr H, Von Der Mark K (1982) The use of antibodies to connective tissue proteins in studies on their localization in tissues. In: Furthmayr H (ed) Immunohistochemistry of the extracellular matrix, vol II. CRC Press, Boca Raton pp 89–117
21. Furthmayr H, Wiedermann H, Timpl R et al. (1983) Electron microscopical approach to a structural model of intima collagen. Biochem J 211:303–311
22. Gray S, Miller EJ (1983) What is collagen, what is not? Overview. Ultrastruct Pathol 4:365–377
23. Godeau G (1984) Selective staining technique for identification of human skin elastic fibers. Pathol Biol 32:215–216
24. Godeau G, Gonnord G, Wegrowski J et al. (1985) Effect of colchicine on atherosclerosis. II. Study of dermal elastic fibers by quantitative histochemistry, automated image analysis. Clin Physiol Biochem 3:234–239
25. Goldfischer S, Coltoff Schiller B, Schwartz E et al. (1983) The infrastructure of aortic elastic fibers. Tissue Cell 15:429–435
26. Hornebeck W, Legrand YJ (1980) Possible implication of two elastinolytic proteases isolated from tissue aorta and blood platelets in atherosclerosis. In: Robert AM, Robert L (eds) Frontiers of matrix biology: biology and pathology of elastic tissues. S Karger, Basel, pp 199–212
27. Huang W, Haust MD (1984) Proteoglycans in human atherosclerotic lesions – a pilot qualitative and quantitative study by ruthenium red. Histopathology 8:835–845
28. Jacob MP, Hornebeck W, Lafuma C et al. (1984) Ultrastructural and biochemical modifications of rabbit arteries induced by immunization with soluble elastin peptides. Exp Mol Pathol 41:171–190
29. Jacotot B, Beaumont JL, Monnier G et al. (1973) Role of elastic tissue in cholesterol deposition in the arterial wall. Nutr Metab 15:46–58
30. Jander R, Rauterberg J, Glanville RW (1983) Further characterization of the three polypeptide chains of bovine and human short chain collagen (intima collagen). Eur J Biochem 133:39–46
31. Junqueira LC, Bignolas G, Brentani RR (1979) Picrosirius red staining plus polarization microscopy: a specific method for collagen detection in tissue sections. Histochem J 11:447–455
32. Junqueira LC, Carneiro J (1983) Basic histology, 4th edn. Lange Medical, Los Altos, pp 89–119
33. Kraus JM (1983) Microfibrils in the aorta. Connect Tissue Res 11:153–167
34. Labat-Robert J, Leutenegger M, Llopis G et al. (1984) Plasma and tissue fibronectin in diabetes. Clin Physiol Biochem 2:39–48
35. Labat-Robert J, Robert L (1984) Modifications of fibronectin in age-related diseases: diabetes and cancer.

Arch Gerontol Geriatr 3:1–10

36. Labat-Robert J, Szendroi M, Godeau GJ et al. (1985) Comparative distribution patterns of type I and III collagens and fibronectin in human arteriosclerotic aorta. Pathol Biol 33:261–265

37. Leutenegger M, Birembaut P, Poynard JP et al. (1983) Distribution of fibronectin in diabetic skin. Pathol Biol 31:45–48

38. Luft JH (1971) Ruthenium red and violet. Part I. Chemistry purification methods use for electron microscopy and mechanism of action. Anat Rec 171:347–368

39. Moczar M, Allard R, Robert L et al. (1976) Biosynthesis of elastin and other matrix molecules in venous arterial prothesis. Pathol Biol 24:37–41

40. Robert AM, Grosgeat Y, Reverdy V et al. (1971) Lésions artérielles produites chez le lapin par immunisation avec l'élastine et les glycoprotéines de structur de l'aorte. Etudes biochimiques et morphologiques. Atherosclerosis 13:427–449

41. Robert B, Parlebas J, Robert L (1963) Etude immunochimique d'une glycoprotéine de la cornée, la kératoglycosaminoglycanne. CR Acad Sci 256:323–325

42. Robert B, Szigetti M, Derouette JC et al. (1971) Studies on the nature of the "microfibrillar" component of elastic fibers. Eur J Biochem 21:507–516

43. Robert L (1982) The intercellular matrix: maturation and aging of the vascular wall. In: Kalsner S (ed) The coronary artery. Billing and Sons, Guildford, pp 391–416

44. Robert L (1983) Mécanismes cellulaires et moléculaires du vieillissement. Masson, Paris (Collection biologie moléculaire 5)

45. Robert L, Chaudiere JC, Jacotot B (1984) Interaction between lipids and the intercellular matrix of the arterial wall: its role in the evolution of atherosclerotic lesions. In: Malinow MR, Blaton VH (eds) Regression of atherosclerotic lesions. Experimental studies and observations in humans. NATO ASI Series Life Sciences 79:145–173

46. Robert L, Dische Z (1963) Analysis of a sulfated sialofucoglycosaminogalactomannosidoglycan from

corneal stroma. Biochem Biophys Res Comm 10:209–214

47. Robert L, Robert AM (1980) Biochimie des tissus conjonctifs normaux et pathologiques. Colloque CNRS n°287

48. Robert L, Robert AM (1980) Elastin, elastase and arteriosclerosis. In: Robert AM, Robert L (eds) Frontiers of matrix biology: biology and pathology of elastic tissues. S Karger, Basel, pp 130–167

49. Ross R (1973) The elastic fiber: a review. J Histochem Cytochem 21:199–208

50. Santoro SA, Cunningham LW (1981) The interaction of platelets with collagen. In: Gordon LJ (ed) Platelets in biology and pathology, vol II. Elsevier, Amsterdam, pp 249–264

51. Scott JE (1980) The molecular biology of histochemical staining by cationic phthalocyanin dyes: the design of replacements for Alcian Blue. J Microsc 119:373–381

52. Scott JE, Dorling J (1965) Differential staining of acid glycosaminoglycans (mucopolysaccharides) by Alcian Blue in salt solutions. Histochemie 5:221–233

53. Scott JE, Dorling J (1969) Periodate oxidation of acid polysaccharides. III. A PAS method for chondroitinsulphates and other glycosaminoglycuronans. Histochemie 19:295–301

54. Timpl R, Wiedermann H, Van Delden V et al. (1981) A network model for the organization of type IV collagen molecules in basement membranes. Eur J Biochem 120:203–211

55. Von Der Mark H, Aumailley M, Wick G et al. (1984) Immunochemistry genuine size and tissue localization of collagen type VI. Eur J Biochem 142:493–502

56. Wegrowski J, Moczar M, Robert L (1985) Proteoglycans from porcine aorta. Comparative study for their interactions with lipoproteins. Biochem J 235:823–831

57. Wight TN, Ross R (1975) Proteoglycans in primate arteries. I. Ultrastructural localization and distribution in the intima. J Cell Biol 67:660–674

58. Yamada KM (1983) Cell surface interactions with extracellular materials. Ann Rev Biochem 52:761–799

Chapter 3

Organogenesis of the Arterial Wall

C. L. Berry

It is evident that the growth of the embryo beyond its earliest stages requires the establishment of a circulatory system, to supply nutrients to and remove waste from developing tissues. The subsequent differentiation and development of the organism eventually produces a specialized cardiovascular system with distinct variations in form in different parts. In between embryonic and adult life vascular systems will have developed, involuted, been obliterated, modified their form or changed their microarchitecture, all in response to stimuli that are, with few exceptions, not well understood.

It is essential that the first physiologically effective system of the embryo is the cardiovascular system. Pulsations may be seen in the heart at 32 d by ultrasonography in man and in their absence it may be presumed that proper development cannot continue. This functional demand on the system in turn requires that vessels, while functioning as effective conduits, may need to fulfil physiological roles and that when these change, as say, in the pulmonary artery at term, the changes necessary for the maintenance of the physiological milieu must not interfere with flow.

What are the functions that arteries perform? Arteries are not simple conduits but have distinct mechanical properties which make a contribution to the function of the cardiovascular system [52]. A central feature is their non-uniformity of mechanical functions, with particular regard to distensibility. Proximal vessels are compliant and become progressively less distensible as we approach the periphery. The effects of this elastic non-uniformity are that:

1. The oscillatory component of the cardiac work is reduced. The presence of a proximal elastic reservoir which can store energy allows the heart to work at a rate and stroke volume that is optimal for a given cardiac output. If the large arteries were a rigid system of pipes, changes in stroke volume would be difficult to accommodate and pulse rate would become the main variable in changing cardiac output, a less mechanically effective solution.

2. The decrease in distensibility of arteries that occurs as we move distally greatly reduces a problem that exists with distensible systems, that is, a tendency for oscillation to occur between various parts. This should not conjure up a vision of blood slopping to and fro in the arterial tree, but a tendency to this type of behaviour would exist which could damage arteries by the establishment of standing waves. These do occur in diseases of human arteries and may further damage the intima [15].

3. The pulse-wave disturbance is continuously amplified as it moves distally and travels in less distensible vessel walls. The form of the pulse wave is known to be critical in the function of some tissues and organs, for example the kidney, where hypertension may develop if the organ is enveloped in a rigid wrapping, and the lung, where in experiments using isolated perfused lungs, gas exchange has been shown to be more effective in the presence of pulsatile flow.

The role of the smooth muscle of the arteriolar wall in controlling blood pressure will not be considered in detail here. However, it is evident that a variable peripheral resistance adds to the control options available in the circulation; its loss in certain disease states is dramatically illustrated by the failure to adjust flow in the fainting attacks which occur in the elderly with rigid vessels.

How are these various properties conferred on the arterial wall? The wall of larger arteries acts as a two-phase material, like the GRP (glass-reinforced plastic) of which many small boats are made. These materials have composite properties which differ considerably from those of their various components, due to their inter-actions. In arteries which are being inflated, the vessel radius initially increases rapidly with small pressure increments. Subsequently the vessel becomes progressively less distensible and at pressures around twice the mean systolic level little further change in radius occurs for each pressure increment. Practically speaking, at pressures of less than 75 mmHg, the arterial wall behaves like a tube of pure elastin. As inflation occurs, more collagen fibres are recruited in the lamellar unit structure which distributes stress in the wall, and at pressure greater than 200 mmHg the wall is very stiff and its mechanical properties approximate to those of collagen. This changing mechanical behaviour is a critical arterial function and may be quantitatively varied as a response to altered haemodynamics.

Scleroproteins are not the only variables affecting functions. We shall see later how wall thickness affects them, but it is clear from consideration of these factors alone that vessels are complex functional units. At all times in organogenesis, changes in arteries must occur in a manner which does not compromise functional needs, themselves far more varied than the conduit role assigned to vessels by many biologists and physicians.

The structure of the wall in terms of cells, extracellular material and their organization is considered in the following section.

Structural Organization of Vessel Walls

In considering how the structure of vessels is determined, understanding some simple mechanical principles helps greatly in understanding the varied structures seen. These will be considered in detail later, but briefly, the flow in any tube depends on the tube's diameter, the pressure difference across the ends and the viscosity of the fluid flowing through it. If we assume for the purposes of this discussion that blood is *not* thicker than water, we can concentrate our attention on flow and pressure.

A large diameter pipe will permit a greater flow than a narrow one for the same pressure gradient. Raising the pressure at one end of a narrow pipe will permit a flow comparable to that in a wide pipe to be established.

As a simplification, we may say that in blood vessels, flow changes in response to metabolic demand. An increased flow may be achieved by a rise in pressure differential across the system or by vascular dilatation – assuming that the wall of the vessel is infinitely plastic, that is to say, that there are no constraints on its form. This is certainly not the case in reality and other aspects of vessel function require that a particular structure is maintained (see below), but these ideas help us to follow certain early changes in development.

For obvious reasons, very early vasculogenesis has largely been studied in the chick.

Early Development of the Circulation

The first signs of vessel formation are seen in the form of the so-called blood islands in the yolk sac. These clusters of mesodermal cells form both endothelium and blood cells. This process is repeated over a considerable area and the endothelial cells formed from the outer layer of the islands link up to form a plexus of small blood vessels [42]. New vessels form from this network by sprouting and migration, although it is not clear whether all the activities involved in new vessel formation occur in the embryo or foetus. Clearly, proliferation and migration occur but whether the latter is assisted by the action of protease production which may ease the extension of the network without producing obvious tissue damage [32] is not known. Ryan and Barnhill [47] have suggested that epithelial angiogenic factors in the overlying chorio-allantoic membrane may be important at this stage. Undoubtedly, flow subsequently affects vessel development as has been shown in chick experiments where occlusive lesions produced by thermal injury alter the vascular pattern and change the form of the vessels in which flow is increased.

Mesodermal cells subsequently invest vessels to produce what will be media and adventitia. Investment of this kind is evident by the time the aortic arches begin to form from the plexus on each side of the neural groove, by the 22-somite stage in man.

It is not the purpose of this account to describe the morphogenesis of the vascular system. This account only serves to illustrate that very early in development flow affects vessel formation and that specialized mesenchyma structures form in the vessel walls presumably in response to haemodynamic factors. However, in acardiac foetuses a muscular vessel may be present at the site of the thoracic aorta without evidence of lamellation (Berry, unpublished observations) and the "aorta" may be present in the absence of the heart and pulmonary arteries [2]. Some genetically determined template of the major components of the circulation must exist but their form can certainly be altered by local factors which may be exemplified in the branchial arteries and the iliac vessels (see below). How are these local influences expressed? We have seen that there are functional demands on vessels – are these demands met by the expression of specific genes? This seems improbable for the reasons outlined below.

Structural Organization of the Arterial Wall

Despite the degenerative problems that now beset our cardiovascular systems it is evident that the basic morphological plan of large arteries is a biologically sound one, as it has persisted for more than 350 million years (Fig. 3.1a–d). Other aspects of cardiovascular function are known to be similar in mammals and primitive fish, for example, the dogfish (*Scyliorhinus canicula*) has a heart which obeys Starling's law [16].

The lamellar unit structure of the aortic media, proposed by Wolinsky and Glagov [59] and

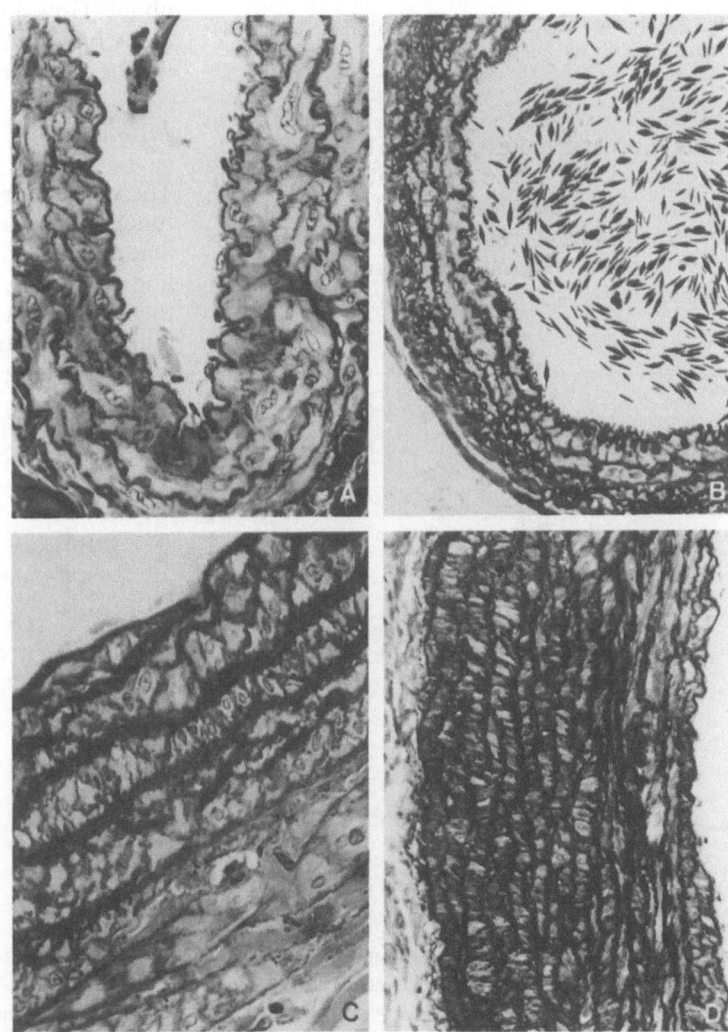

Fig. 3.1A Ventral aorta of the dogfish, *Scyliorhinus canicula*. Smooth-muscle cells and intermuscular elastin are seen. Methylene blue, × 225. **B** Aorta of toad (*Buffo buffo*). A lamellar structure is evident. Toluidine blue, × 150. **C** Aorta of snake (rat snake, *Elaphe schrencki*). Toluidine blue, × 225. **D** Aorta of marmosette (*Callithrix jaccus*). Toluidine blue, × 150

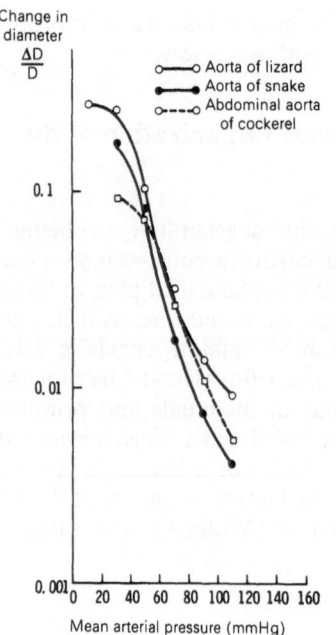

Fig. 3.2. Incremental strain values for morphologically similar vessels in different animals.

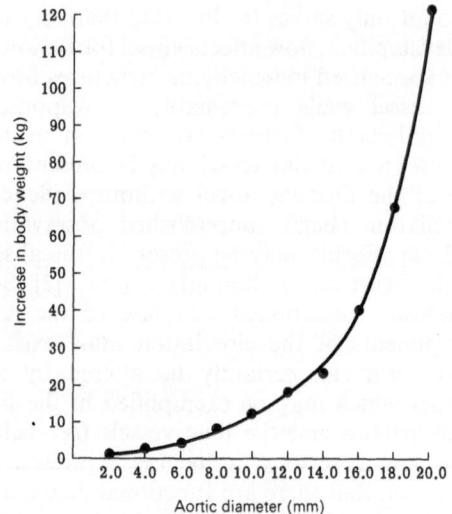

Fig. 3.4. Increase in aortic diameter with total body weight.

recently revised and extended as a concept [19], is present in all major vertebrate classes with relatively little modification, and this in turn is accompanied by a remarkable mechanical identity when morphologically comparable vessels are examined (Fig. 3.2 and 3.3). How is this brought

about? How does the structure of the wall relate to its function? The main segments of the wall will be considered in turn.

Intima

The intima contributes little to the mechanical function of arteries. One relevant point may be made, however, in terms of growth and remodelling in vessels. The component cells of endothelium have a morphological appearance that reflects their situation, being flattened hexagons with their long axis in the direction of flow in

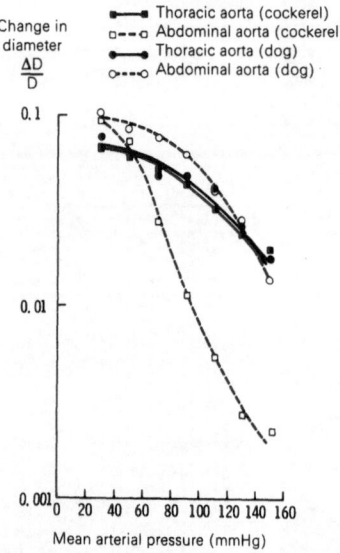

Fig. 3.3. Incremental strain values, indicating that the cockerel abdominal aorta (predominantly muscular) has different mechanical properties from the elastic vessels with which it is compared.

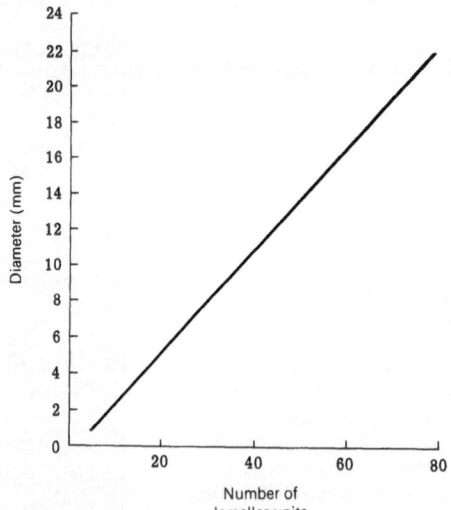

Fig. 3.5. Increase in the numbers of elastic lamellae with vessel diameter. (After Wolinsky and Glagov [59]).

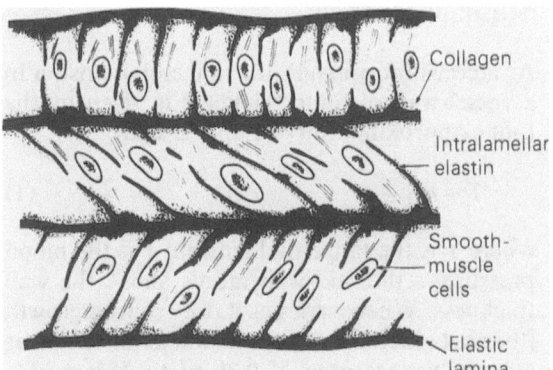

Fig. 3.6. Lamellar unit structure.

high pressure areas like the aorta but tall and cuboidal in some veins. Local proliferations of endothelium occur at sites of pressure drop [45,49] and this may be of significance in the production of so-called "cushions".

Media

In 1893 Thoma stated that "the growth in thickness of a vessel wall is proportional to the tension in the wall, which itself is determined by the diameter of the vessel lumen and by the blood pressure". As already indicated, this statement forms the basis of our understanding of the relationships between structure and function in arteries.

The mechanical properties of elastic arteries depend largely on the tunica media which is com-

posed of a number of roughly concentric lamellae, familiar to all pathologists. The lamellar units described by Wolinsky and Glagov [59] consist of an elastic lamina and the intervening smooth-muscle cells and scleroprotein. The number of these units varies with the diameter of the vessel, increasing progressively with increase in size and weight in different animals (see Figs. 3.4 and 3.5). Vessels with less than 29 concentric lamellae have no vasa vasora [60]. The human aorta is atypical in some respects, the abdominal part having fewer than the expected number of lamellar units with a correspondingly lighter wall tension per unit. This part of the aorta also has no vasa vasora, despite a wall thickness that suggests they should be present.

The medial structure is, if we speak anthropomorphically, designed to distribute tension uniformly throughout the wall. Various apects of the mechanical properties of vessels depend on this type of structure, which allows "two-phase" behaviour.

The distribution of scleroproteins and muscle cells is shown in Fig. 3.6. The "unit" consists of elastic lamellae, arranged as interconnecting sheets of fibres in flattened cylinders around the artery, separated by smooth-muscle cells arranged in helices of varying pitch. Scanning and transmission electron microscopy show the appearance of these structures and longitudinal views indicate the sheet-like structure of the elastic lamina (Figs. 3.7, 3.8 and 3.9). There is a separate network of collagen fibres, closely investing the elastic tissue but not directly con-

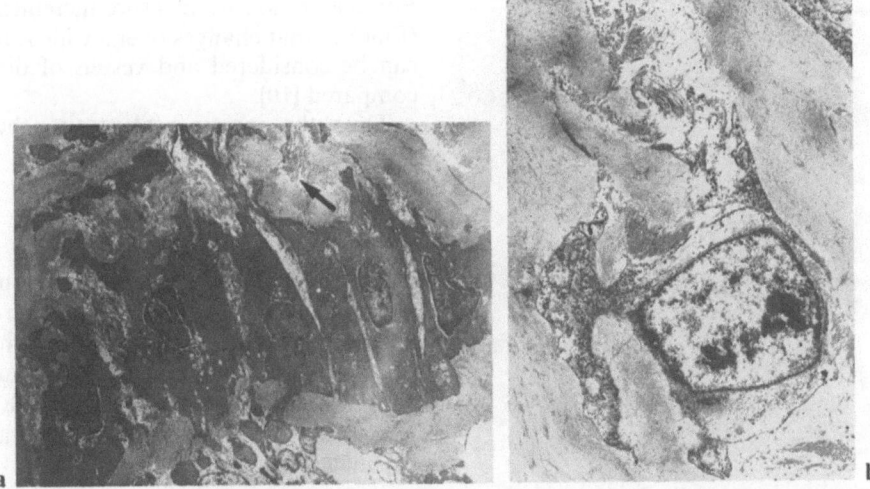

Fig. 3.7a. Low-power electron photomicrograph of aorta. Smooth-muscle cells run between laminae of elastic tissue. Collagen can be seen near the elastic sheets (*arrow*). × 2000. **b** The close approximation of collagen to elastin is clearly evident. × 3000

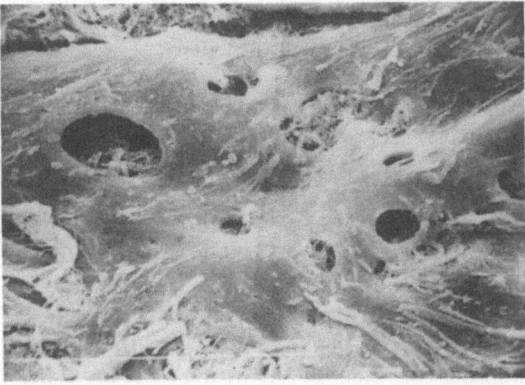

Fig. 3.8. The sheet-like nature of the elastic tissue is evident in this longitudinal view, obtained using scanning electron microscopy. × 2000

nected with it. The relationship between force/ unit area on the vessel wall and the number of lamellar units present is such that 1 unit is added for around 2000 dynes/cm increase in tangential tension.

The muscle probably contributes little to the mechanical properties of large arteries although its role as a "spacer" for scleroproteins may be important. Even if the highest value obtained for its elastic modulus are assumed to be correct, the effects of contraction would be small (see Berry et al. [10] for discussion).

What measurements help us to determine the factors determining the structure of the arterial wall? We need to consider a number of basic mechanical variables.

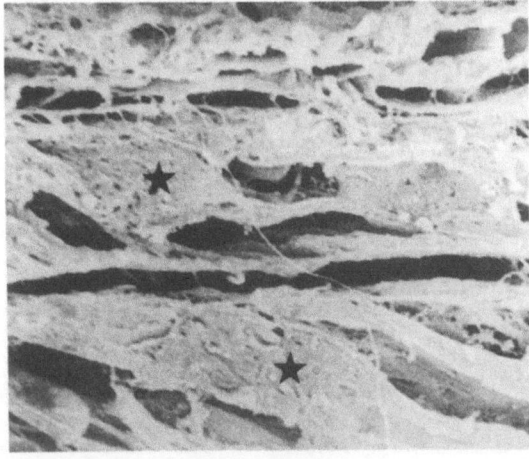

Fig. 3.9. Transverse view of elastic tissue using scanning electron miscoscopy. The elastic tissue has retracted, showing smooth-muscle cell bodies (*stars*). × 3000

Static Mechanics

At a constant pressure the tangential tension in a vessel wall of finite thickness is given by the Lamé approximation of Laplace's law:

$$T = P \frac{r}{h} \tag{1}$$

where T is the tangential tension, P is the blood pressure, r the mid-wall radius and h the wall thickness. When r rises as it does during growth [the aorta increases from a diameter of 12 mm at a body surface area of 0.30 m^2 to 25 mm at a body surface area of 1.70 m^2 [26]] T will rise. With age P also rises and h falls (see later).

These changes alter the stresses and strains on vessels. The use of these terms in common speech has meant that their precise meaning has often become blurred. *Stress* is the force exerted per unit area and was measured by all of us in school physics when verifying Hooke's law by hanging weights on a wire. For a given stress the wire increased its length by a given increment and the ratio of extension in length to the original length is the *strain*. Young's modulus is calculated by dividing stress by strain. The mechanical properties of a vessel will determine the change in radius (strain) for a given pressure increment (stress).

Young's modulus is not constant for arteries over a wide range of stresses, as it is for steel wire. In elastic vessels the ratio

$$\frac{\text{stress}}{\text{strain}}$$

increases with applied stress; that is, the vessel becomes functionally stiffer at higher pressures. For this reason we measure incremental strains (Einc) so that changes over a wide range of radii can be considered and vessels of different size compared [10].

Thus the incremental strain, the ratio of change in radius to a given radius

$$\frac{r}{r}$$

for one 10 mmHg increment in pressure will not necessarily be the same as for the 10 mmHg increment in pressure in materials which do not obey Hooke's law. The stress on an arterial wall is the tangential tension developed in it, and the elastic modulus of an artery can thus be calculated from

$$Y = 0.75 \times \frac{\text{stress}}{\text{strain}}$$

The figure 0.75 is a constant included because the mechanical properties of an artery are anisotropic, that is to say, not constant in all directions. However, in practice this factor is not important as vessels in situ elongate very little until pressures in excess of 300 mmHg are reached. It can be assumed that the vessel increases in diameter only as luminal pressure is increased.

In practice Einc is measured by a radiographic technique [10] and calculated from

$$Einc = \frac{0.75 Pr}{r} \frac{r}{h}$$

where r is radius, P the pressure at which measurements are made and h the wall thickness at that pressure.

Einc can therefore be regarded as a measure of structural stiffness and depends on the composition and arrangement of the components of the wall. At a given value of Einc the incremental strain which governs pulse-wave velocity and is a measure of the effectiveness of the vessel as an elastic reservoir, is determined by the wall thickness.

This makes it clear that Einc alone provides inadequate information to permit us to design a vessel. If one thinks of an analogy with a bridge, knowing the properties of a structure made of iron and concrete arranged in a network would tell us only that the materials are suitable for bridge building. To build the bridge, we need to know the loads that are expected in order to know how much material to use and, to desert our analogy, to see how thick the vessel wall should be. This would determine the functional stiffness required and would depend on both the structural stiffness and the relative wall thickness (the ratio of the wall thickness to the vessel diam-

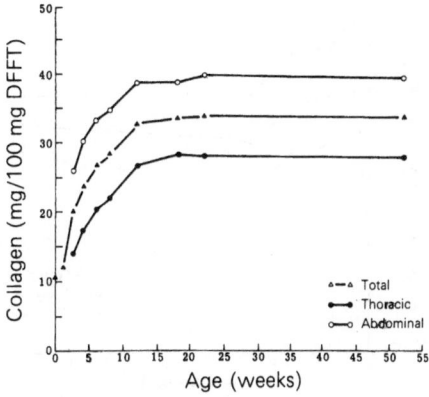

Fig. 3.11. Change in collagen content of rat aorta with age.

eter). This is measured as Ep, a pressure/strain elastic modulus (see Fig. 3.10).

The relative wall thickness varies considerably during development – there are obvious changes occurring rapidly in both the neonatal aorta and pulmonary artery. It is generally true that as pressures increase relative wall thickness falls and thus, by simple geometry, the functional stiffness goes down. However, the changes in structural stiffness work in the opposite direction as more collagen is recruited as the vessels dilate and this is generally the quantitatively more important effect.

Scleroproteins in the Aortic Wall

A number of quantitative determinations of the scleroprotein content of animal aortas have been made (see Looker and Berry [32] for review). However, sequential studies of scleroproteins and nucleic acids throughout development have not been made apart from our studies on the rat [32] and man [11].

In the rat we have documented a marked decline in cellularity of the walls during growth which is accompanied by an increase in collagen content in absolute terms and an increase to a steady 40% of the dry fat-free weight of maturity (20 weeks in the rat) for the abdominal aorta and 27% for the thoracic aorta (Fig. 3.11). These changes are accompanied by an increase in the absolute amount of elastin (Fig. 3.12) and a relative decline post-natally with an inverse relationship to collagen.

Thus the thoracic aorta contains significantly more elastin and less collagen than the abdominal vessel, a fact clearly related to its functional

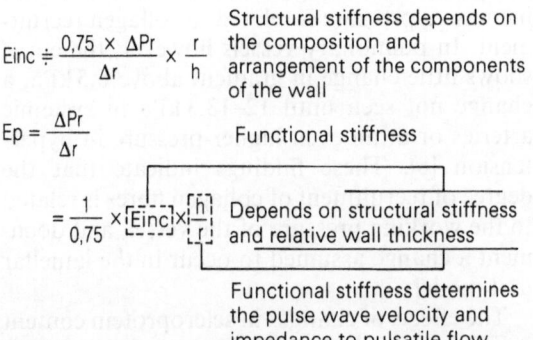

Fig. 3.10. Definition of terms used and their role in the assessment of arterial function.

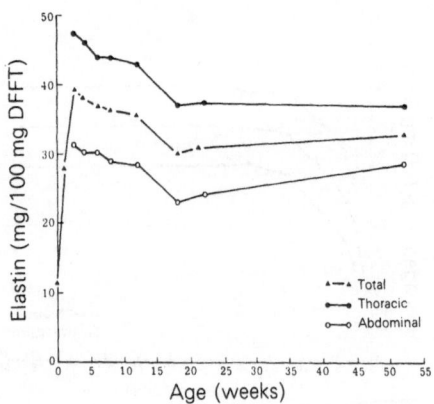

Fig. 3.12. Change in elastin content of the rat aorta.

role as the distensible proximal part of the circulation.

Similar changes are seen in man between 10 weeks gestation and 1 year of age. During development the water content of the vessel falls steadily, a change which itself modifies function. An important difference in the pattern of elastin and collagen accumulation was found, however. Although elastin appeared in the aorta at around 15 mm crown–rump length, the majority of this component was added to the aorta in the last 4 weeks of gestation and the first 8 weeks after birth (Fig. 3.13). Collagen showed a different pattern, with a steady increase during intra-uterine life reaching a plateau at term – expressed as the proportion of the dry fat-free weight. Maximum concentrations of RNA occur at around 25 weeks gestation indicating con-

siderable metabolic activity in the aorta at this time. These findings suggest that changes in the aorta at birth, in pressure, flow and dimensions, all of which act to increase tangential tension in the wall, are responsible for the change in proportions of elastic seen.

What functional changes accompany this biochemical maturation? We have measured internal radius, relative wall thickness and incremental elastic modulus in developing rats [10]. It became apparent that most of the differences in properties of the thoracic and abdominal aortas were not due to changes in chemical composition but rather to differences in wall thickness. Incremental elastic modulus is not significantly different for the two parts of the vessel but strain values do differ, as others have found [36,39]. Here is an example of how vessels may change their function by "geometrical" rather than biochemical variation.

Other demands on vessels may produce different responses. The stimulus to much of this work was the observation in 1969 [3] that the pulmonary artery in man could be rapidly remodelled in response to altered pressure, as for example following pulmonary artery banding in congenital heart disease (Fig. 3.14). This is a pathological situation but it is also clear that abrupt changes in flow and pressure normally occur at birth. Our own work [29] makes it clear that there are important differences between the major and minor circulations; in particular there is no amplification of pulse-wave disturbance towards the periphery in the pulmonary vessels, in contrast to what is found systemically. This finding agrees with studies in man showing a single 'reflecting" site at the periphery of the lung [37] rather than a number of sites established by the elastic non-uniformity of systemic arteries [53]. The pressure/radius curve for pulmonary arteries is not identical to that in the aorta where a reduction in the gradient of the curve with increasing pressure is related to collagen recruitment. In pulmonary vessels however, the curve shows little change in gradient above 6.5 kPa, a change not seen until 12–13.3 kPa in systemic arteries or at an even higher pressure in hypertension [8]. These findings indicate that the degree of recruitment of collagen fibres is related to the working pressure of the vessel, and document a change assumed to occur in the lamellar unit model.

The effects of changes in scleroprotein content on the mechanical properties of the wall can be explored further [28]. The relationships are complex and although Roach and Burton [43]

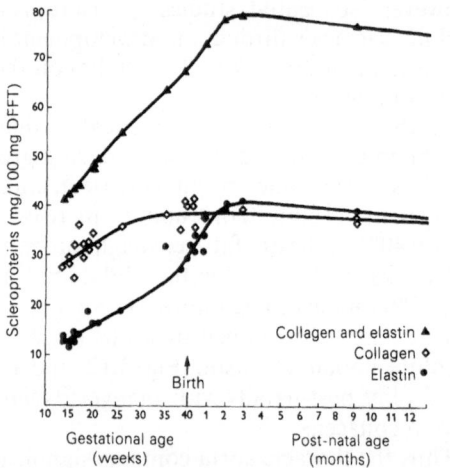

Fig. 3.13. Scleroprotein content of the human aorta in early life.

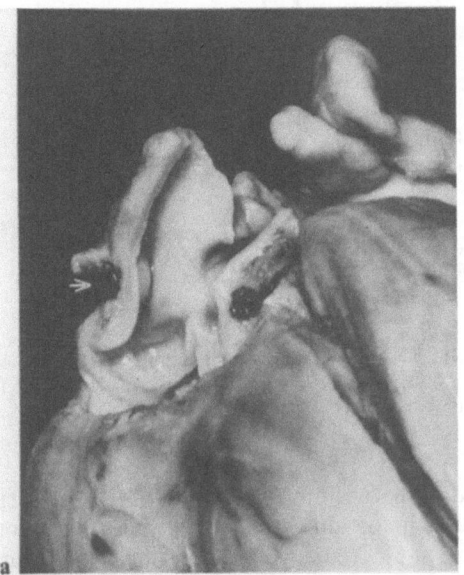
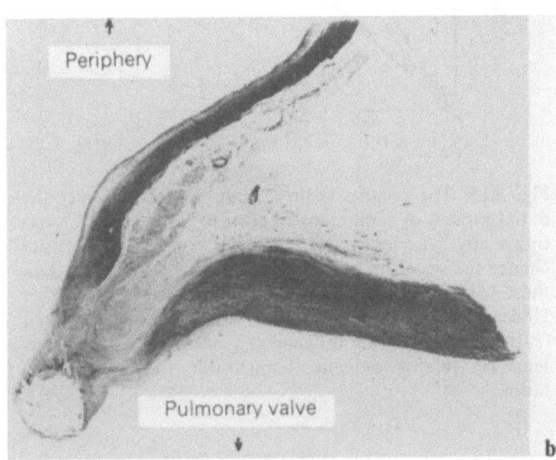

Fig. 3.14a A band (*arrowed*) is tied around the pulmonary artery to reduce blood flow to the lungs where there is a large left to right shunt. (Male infant, death 3 d post-operatively.) **b** The thickness of the pulmonary artery wall on the side nearest the heart (high pressure) is greater than on the distal side. Pulmonary artery from child 3 weeks post-operatively. Elastic van Gieson, × 20

and Apter et al. [1] have found that the elastic constants of the aorta measured at a low level of strain are related to its elastic content, relationships to collagen content were not clear. Cox has made a considerable contribution in this area [21,22,23].

We adopted a simple model of the aorta in which elastic and collagen fibres bear stress in parallel. We studied mechanical properties and scleroprotein content in the aortas of normal animals, hypertensive animals and animals treated with β-amino-proprionitrile in order to alter the mechanical properties of their collagen [9]. In normotensive animals Einc increases with age. Hypertensive animals have more elastin in the aorta than normotensive ones, an adaptive change [6] related to the increased distensibility of the vessels in the early stages of hypertension (see later). Analysis of our data showed that at a given degree of distension of the vessel the increase in the amount of collagen present in pathologically altered vessels is associated with a decrease in the proportion bearing stress. It is likely that for all normal vessels a given degree of distension is associated with a fixed absolute amount of collagen bearing stress. The changes that occur with age and with the development of pathological changes in vessels only "dilute" the stress-bearing component, they do not contribute to it. (Those interested in the mathematical arguments which support this assertion

should see Greenwald and Berry 1980 [28]). In other words "new" collagen in a vessel is analogous to a scar in the skin; it does not help the tissue to function properly.

Developmental Implications

What evidence have we that changes in these variables are important in the development and growth of vessels in man? Consider an artificial increase in wall tension in developing animals when hypertension is induced.

In experimental studies in growing animals we have shown [8] that the vessel walls of hypertensive animals become thickened, as many others have found, with the relative wall thickness (h/r) increasing rapidly, and falling rapidly in turn when hypertension was relieved. However, when blood pressure is raised between weeks 4 and 10 in developing rats (they are weaned at 3 weeks and sexually mature at about 12 weeks) the changes that occur in the wall are not entirely reversible and h/r is still abnormal at 1 year, i.e. 42 weeks after blood pressure has been restored to normal levels. The functional changes that accompanied these changes were equally striking. Incremental strain (the measurement of the functional stiffness of a vessel at any particular pressure) differed markedly in

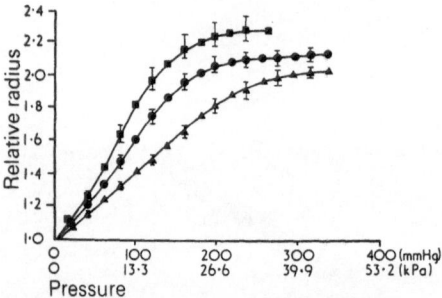

Fig. 3.15. The relative radius (change in radius over the initial radius) of normotensive (■), hypertensive (▲) and transiently hypertensive (●) animals. For a given pressure change the vessels of hypertensive animals distend less than those of normotensive animals, suggesting "stiffness". However, if the operating pressure of the vessel is considered (c. 100 mmHg in normotensives, 200 mmHg in hypertensives), relative radii are comparable, indicating adaptation.

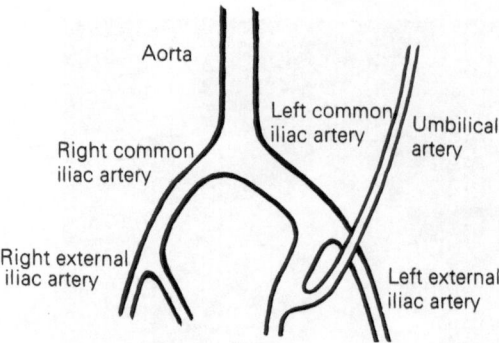

Fig. 3.16. The form of the circulation in the single umbilical artery syndrome. Here the assumption is made that a left-sided umbilical artery is present.

the hypertensive and normotensive animals.

At a pressure of 100 mmHg, hypertensive vessels were more distensible than normotensive ones, but at the effective systolic pressure of each group the aorta showed an incremental strain which was lower than normal but which tended towards normal values (see also Fig. 3.15) (findings subsequently confirmed by Cox [22]). This suggests an adaptive response in vessels, enabling them to maintain functional characteristics that preserve their ability to act as a functional unit in the circulation.

This adaptive response has morphological consequences which are complex. To illustrate this complexity, thickening of the wall of an artery, assuming that no change occurs in the physiochemical arrangements of the constituents of the wall, will result in a decrease in strain (functional stiffness) rather than the increase that many assume, and we have found that at a given pressure hypertensive vessels are more distensible than normal ones.

There is plenty of evidence that developing vessels adapt in man. The effects of pulmonary artery banding have been illustrated, and radical alteration in the umbilical vessels has been shown to occur when flow is changed, as in the single umbilical artery syndrome (SUA). The pattern of the iliac circulation in this syndrome is shown in Fig. 3.16. All of the placental flow, and thus the majority of the total blood flow, will pass through the common iliac vessel on the side from which the solitary umbilical artery arises. The majority of children with the syndrome are entirely normal, without evidence of growth

retardation or limb shortening [14], but their vessels can be shown to have abnormal compliance at 5–9 years, when this is measured using non-invasive techniques. This change is found only in the segment of vessel between the aortic bifurcation and the inguinal ligament (Fig. 3.17) [7]. The compliance of the arteries of the lower leg is normal.

As we have shown that the elastic properties of the components of the arterial wall do not change with age [10] the increased compliance of the vessels was predicted to be due to an alteration in the ratio of vessel radius to wall thickness [8]. This prediction was later fulfilled when we were able to examine sections of the arteries of

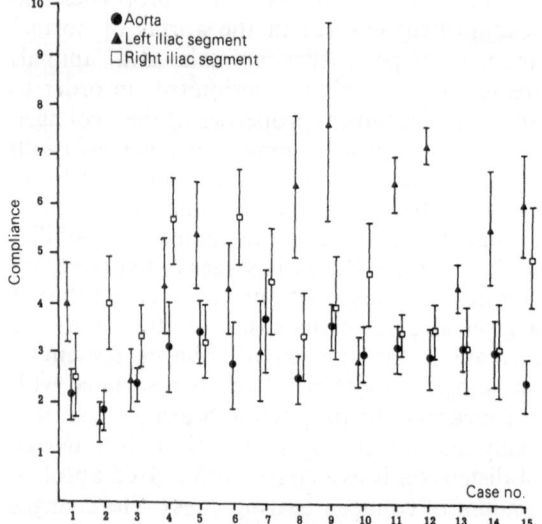

Fig. 3.17. Fifteen cases of the single umbilical artery syndrome had compliance values for their aorta and iliac vessels determined non-invasively. In each case, it can be seen that the compliance of one common iliac artery resembles that of the aorta while the other is much more compliant.

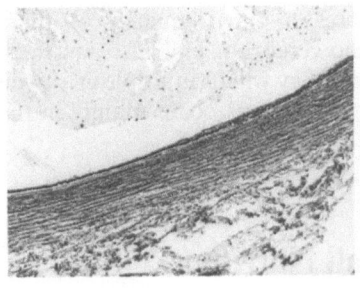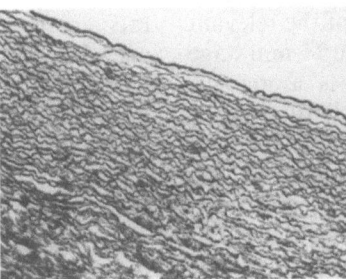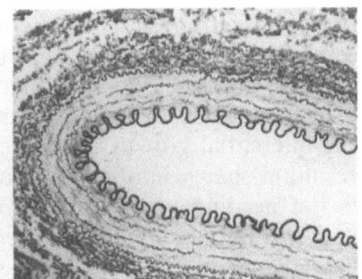

Fig. 3.18. Aorta of infant aged 28 d, with a single umbilical artery on the *left side*. Elastic van Gieson, × 100

infants with SUA who did not survive (see Fig. 3.18). The changes shown in wall thickness and form are apparent as early as the 150 mm stage, in foetuses we have been able to study by courtesy of Dr T. Shephard.

The radical change in morphology of the vessels, which in this situation is flow- rather than pressure-dependent, is remarkable and indicates that vessel-wall morphology is not entirely genetically determined, but adapts to local loads.

The mechanical principles involved in vessel-wall development certainly control the establishment of the elastic structure of arterial branchings and bifurcations [4]. There are only two patterns of medial elastic structure which relate to the tension acting on the vessel wall, and these are consistent in form at different sites. The patterns are determined late in foetal life and post-natally and are remodelled as vessel orifices migrate during growth. The two types are illustrated in Figs. 3.19 and 3.20. Type 1 occurs where the wall of the branch forms an angle of less than 100° with the parent vessel and shows a central raphe into which elastic lamellae appear to be inserted. Type 2 is found where angles are greater than 100° and the number of lamellar units are progressively reduced over a short distance into the branch. Mechanical details are described in Berry [4] but in general it is true to say that the type 1 structure is of considerable strength and represents the high-load aspect of the bifurcation where abrupt changes in wall tension occur as the larger vessel, with a bigger diameter, gives rise to the smaller. In type 2 the wall thickness changes more gradually as loads fall progressively.

Other aspects of the development of bifurcations are important in vascular organogenesis. Mechanical factors play a part in the pathogenesis of the medial defects which subsequently permit the development of berry aneurysms in the cerebral circulation. From Fig. 3.13 it is evident that in man as in other mammals, collagen is the "early" structural element that is laid down in vessels to provide a framework for vascular development, and elastin appears later

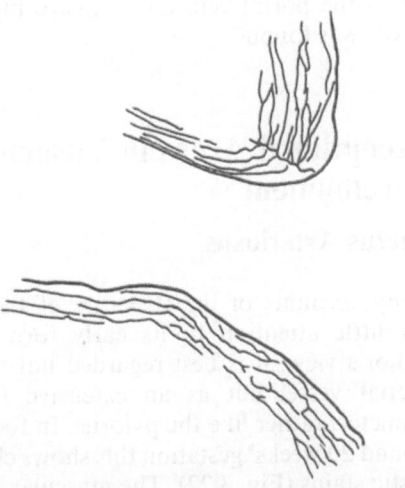

Fig. 3.19. Arterial bifurcation. Type I structure with a raphe at the junction and abrupt change in lamellar unit number opposite a type 2 change.

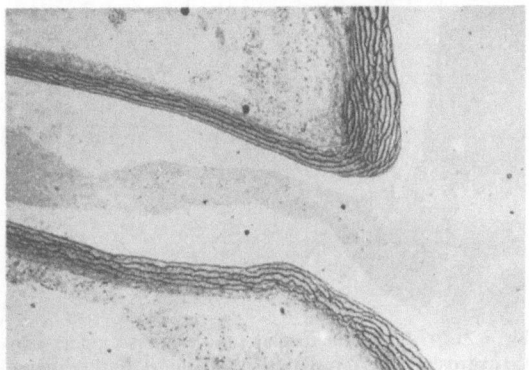

Fig. 3.20. The two types of structure found at the mouth of a vessel. Rat, Toluidine blue, × 120

in response to stress. The pattern of the relevant cerebral vessels is established by the 24-mm stage in man and subsequently there is a massive expansion of this part of the vasculature due to the interstitial growth of the hemispheres. The resultant change in orientation of major arteries is not accompanied by extensive remodelling, and so bifurcations are widened as the parenchyma expands. In 1930, Forbus suggested that the "self-supporting" zone of a vessel on the opposed inner aspect of the branches of a bifurcation may not develop an appropriate structure (Fig. 3.21) [27]. This "defect" is found at a high frequency in all primates [48], a group in which late expansion of the hemispheres occurs. The absence of media means that no cells are available to produce the appropriate load-bearing structures.

As would be expected from this explanation, these defects occur at a high frequency – a frequency which might be an embarrassment to those who propose that they are the cause of berry aneurysms. However, as Crawford [24] suggested, it is probable that they are only a contributing factor although an essential one, and that age and blood-pressure levels are major determinants in aneurysm formation.

Changes in load on the vessel wall may also influence the structure of the smaller cerebral vessels. Microaneurysms on cerebral arteries were suggested as the cause of cerebral haemorrhage by Charcot and Bouchard in 1868 [17], and shown to occur more frequently in hypertensives by Cole and Yates [20]. We have shown that the induction of hypertension in developing

rats greatly increases the frequency of these aneurysms in the cerebrum and also increases the likelihood that they will be multiple in a given individual [33]. The lesions occur mainly in the basal ganglia.

Arterial Wall Nutrition

It is probable that the intima of all large vessels (arteries and veins) is nourished directly from the lumen. The media of all arteries with more than 29 medial lamellar units have vasa vasorum, as demonstrated by Wolinsky and Glagov [59,60] in injection studies in 12 mammalian species. Aortas with vasa vasorum always had an avascular zone in the subintima which increased slightly in width with age but which was invariably around 29 lamellar units in extent.

The medial distribution of vasa vasorum in the aorta has been studied in man [18]. The adult form of vascularization of the media is achieved by 13 years of age; the addition of lamellar units to the outside of the vessel as growth occurs results in the gradual increase in the extent of the penetration of the vasa vasorum. There is evidence to support an essential nutritive role of these vessels in the larger vessels since Wilens et al. [58] produced medial necrosis in the dog by "stripping" the adventitia off the aorta.

Papadia and Setti [38] have described the lymphatic system of blood vessels. This consists mainly of perivascular networks with short vessels draining the outer layers of the media, but in the portal vein an extensive intramural plexus was found.

Exceptional Areas in Vascular Development

Ductus Arteriosus

Many accounts of the structure of the ductus pay little attention to its early form. In the author's view, it is best regarded not as a true arterial vessel but as an extensive muscular sphincter, rather like the pylorus. In foetuses of around 20 weeks' gestation this shows clearly on elastic stains (Fig. 3.22). The muscular development is organized and interrupts an elastic laminar structure precisely and in an identical way in all the specimens we have examined (Fig.

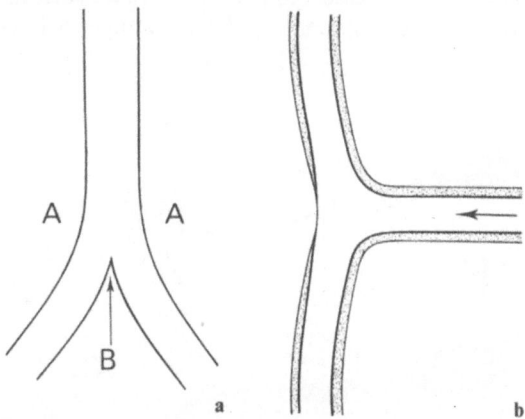

Fig. 3.21a. At "*B*", the media of the vessel is supported by the hydrostatic pressure of the blood within it. As the cerebral hemispheres expand, the angle between the blood vessels increases rapidly so that the position shown in **b** is ultimately arrived at and there is a medial deficiency.

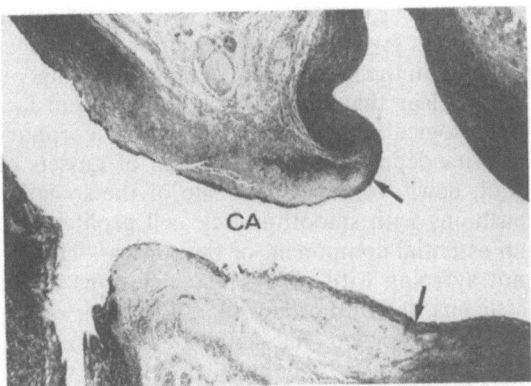

Fig. 3.22. Ductus arteriosus in a 250 mm foetus. A clear muscular interruption of elastic structures is seen. Elastic van Gieson, × 30

3.23). Analogies between changes in endothelium and intima in this duct and in elastic arteries should be made with extreme caution, and the apparent later development of elastic tissue in the "outer" (i.e. large vessel) ends of the ductus may be due to shortening. It has been suggested that abnormal elastic development in the ductus may cause it to remain patent because of failure of the normal contraction mechanism and Toda et al. [54] and others have suggested that the ductus should be regarded as a muscular artery. The development of a muscular structure in what "ought" to be an elastic vessel is attributed by Leonard et al. [34] to a failure of development of tension in the wall because of support supplied by the recurrent laryngeal nerve wrapping around the ductus arteriosus during growth and development. This loss of this support in abnormal development results in development of a high load-bearing (elastic) structure which may itself prevent closure.

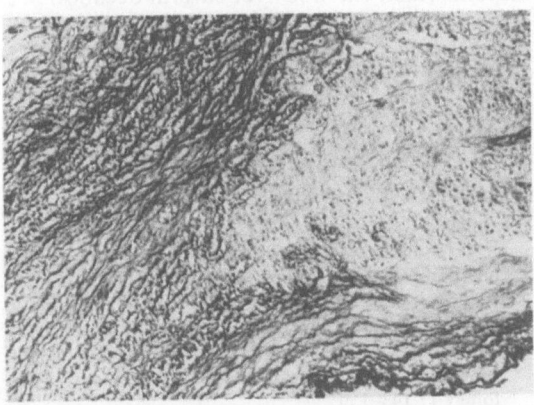

Fig. 3.23. Detail of the wall at the site of a muscular interruption. Elastic van Gieson, × 150

The cross-sectional area of the ductus remains unchanged in late foetal life while the wall thickness increases [13]. Although inferences about flow and pressure may be made from data on lumen size and wall thickness in elastic arteries, this is less clearly true in muscular vessels where contraction and shortening makes extrapolation problematical.

The ductus closes gradually by muscular contraction in the first day of life, rather less suddenly in man than in other animals [31], and the closure is clearly reversible initially. The aortic pressure will be lower than the pulmonary artery pressure at term and reversal of flow begins soon after delivery. The wall contracts under the influence of the rise in PO_2 from below 25 to above 50 mmHg, apparently without involvement of the extensive adrenergic nerve supply, which may be functionally effective in severe foetal hypoxia [31].

Dawes and his group, working with mature foetal lambs [25] have shown that the narrowing of the ductus produces a pressure gradient between the aorta and the pulmonary artery and that despite this, flow decreases. There is an increasing velocity of flow through the ductus with turbulence, giving rise to a murmur. It has been suggested that this acceleration of blood reduces lateral pressure on the walls via a true Venturi effect, facilitating the occlusive process. The author has been able to find no direct evidence for this. With progressive decrease in flow this murmur eventually decreases.

Coarctation

Coarctation of the aorta is a common defect with a well-defined association with other cardiac anomalies; Tawes et al. [50], in a large series, found that 75% of cases had an associated cardiac lesion. The associated abnormalities include ventricular septal defects and left ventricular inlet/outlet anomalies which materially affect prognosis in infants. Bicuspid aortic valves are present in around 25% of cases [50]. Coarctation is commoner in males and there is a tendency to familial aggregation; the heritability of the defect is calculated at 58% [12]. No distinct aetiological factors have been identified but Rudolph et al. [46] have suggested that altered flow in the aortic arch during development may radically alter the structure of the wall.

The classical systems of classification of the defect have largely been abandoned and the current view of the defect is that it represents one end of a spectrum extending from classical coarctation to arch atresia [5]. Altered flow in the developing great arteries and through the ductus may clearly affect the wall of the aorta by the mechanisms described above and this seems likely to be more important in pathogenesis now that older theories of extension of ductal tissues into the aortic wall have been shown to be invalid.

The terms infantile and adult are inappropriate to describe the lesion in modern practice and the distinction is less important with early surgical intervention the normal course. It may still, in practical terms, be useful to categorize coarctation by the presence or absence of a ductus arteriosus, because of the clinical effects. True coarctation does not occur distal to the ductal insertion into the aorta; in the vast majority of cases the site of obstruction is opposite the position of the ductus where a shelf-like projection intrudes into the lumen of the aorta from the postero-superior wall immediately beyond the origin of the left subclavian artery. In infants this change may be accompanied by varying degrees of narrowing of the arch between the subclavian artery and the ductus.

Histological examination reveals infolding of the media with considerable intimal thickening in which there is cell proliferation, glycosaminoglycan accumulation and, later, marked collagen and elastin production.

Coronary Vessels and Vascular Cushions

There is a considerable body of literature on the significance of what have usually been described as "endothelial cushions", mainly in the coronary arteries (see Robertson [44] for review of the early literature). Robertson concluded that the lesions, which could be found in other arteries, were not related to subsequent atherosclerosis but were a normal growth phenomenon. These studies however, and the subsequent careful work of the Velicans [55,56], have been ignored in recent years. In 1969, Vlodaver et al. published a report describing this type of morphological appearance in three different ethnic groups [57]. The authors suggested that early changes seen in the coronary vessels of the infants studied indicated the differential susceptibility of the various races to subsequent atherosclerosis. A description of fragmentation of the elastic lamina with "splitting" and "reduplication" followed by proliferation of smooth-

muscle cells, with subsequent degenerative change, is given. Although the paper describes changes in the next decade of life convincingly it is not clear that the early "lesions" are in fact pathological. The normal growth of coronary arteries depends on the insertion of gussets of wall, newly formed at the site of the so-called cushions with smooth-muscle cell proliferation an essential component of the process. Though not agreeing with all of the conclusions of the Velicans, the author considers that their observations support the concept that intimal cushions are growth points in developing vessels. The conclusions of Pesonen [40] and Pesonen et al. [41] on the differential thickening seen in infant coronary vessels in different parts of Finland do not seem to be justified. In this area, as in so many others, a thorough understanding of the mechanisms of normal growth and development is essential in interpreting apparently abnormal findings.

Conclusions

The walls of vessels develop under two influences, a genetically determined influence on the pattern of the circulation, and one which depends on the effects of haemodynamic load and determines the nature and arrangements of the components of the vessel wall. This latter set of influences permits the modification of the wall in a way which preserves optimal functions. These changes may have consequences for vascular health in later life [6] and the ability of the cells of the wall to divide and synthesize extracellular proteins certainly persists into the 5th decade [30]. Vascular development is a continuum which extends over several decades.

References

1. Apter JT, Rabinowitz M, Cumming MT (1966) Correlation of visco-elastic properties of large arteries with microscopic structure. Circ Res 19:104–121
2. Averback P, Wigglesworth FW (1978) Congenital absence of the heart: observation of human funicolopagous twinning with insertio funiculi furcata, fusion, forking, and interpositio velamentosa. Teratology 17:143–150
3. Berry CL (1969) Changes in the wall of the pulmonary artery after banding. J Pathol 99:29–32

4. Berry CL (1973) The establishment of the elastic structure of arterial bifurcation and branches, its relevance to medial defects of cerebral arteries. Atherosclerosis 18:117–127

5. Berry CL (1975) Congenital heart disease. In: Pomerance A, Davies MJ (eds) The pathology of the heart. Blackwell, Oxford

6. Berry CL (1978) Hypertension and arterial developments: long term considerations. Br Heart J 15:709–717

7. Berry CL, Gosling RG, Laogun AA et al. (1976) Anomalous iliac compliance in children with a single umbilical artery. Br Heart J 38:510–515

8. Berry CL, Greenwald SE (1976) The effects of hypertension on the static mechanical properties and scleroprotein content of the rat aorta. Cardiovasc Res 10:437

9. Berry CL, Greenwald SE, Menahem N (1981) Effect of Beta-aminopropionitrile on the static elastic properties and blood pressure of spontaneously hypertensive rats. Cardiovasc Res 15:373–381

10. Berry CL, Greenwald SE, Rivett JF (1975) Static mechanical properties of the developing and mature rat aorta. Cardiovasc Res 9:669–678

11. Berry CL, Looker T, Germain J (1972) Nucleic acid and scleroprotein content of the developing human aorta. J Pathol 108:265

12. Boon AR, Roberts DG (1976) A family study of coarctation of the aorta. J Med Genet 13:420–433

13. Bruins CLDC, Gittenberger-de Groot AL (1978) Flow and pressure: the mechanics of closure of the human ductus arteriosus: In: Van Mierop LHS, Oppenheimer-Dekker A, Bruins CLDC (eds) Embryology and teratology of the heart and the great arteries. Leiden University Press, pp 210–220 (Boerhaave series 19)

14. Bryan EM, Kohler HG (1974) The missing umbilical artery. I. Prospective study based on a maternity unit. Arch Dis Child 49:844

15. Busse R, Wetherer E, Bauer RD et al. (1975) The genesis of the pulse contours of the distal leg arteries in man. Pflugers Arch 360:63–79

16. Butler PJ, Short S, Taylor EW (1976) Factors affecting blood flow through the ventral aorta of the dogfish (Scyliorhinus canicula). Proceedings of the Physiological Society, Dec 1975. J Physiol 256:74P–76P

17. Charcot JM, Bouchard C (1868) Nouvelle recherches sur la pathogenese de l'hemorrhagie cerebrale. Arch Physiol 1:110

18. Clarke JA (1964) X-ray microscopic study of the vasa vasorum of the normal human aorta. Z Anat Entwicklungsgeschichte 124:261

19. Clark JM, Glagov C (1985) Transmural organisation of the arterial media. The lamellar unit revisited. Atherosclerosis 5:19–34

20. Cole PM, Yates PO (1967) The occurrence and significance of intracerebral microaneurysms. J Pathol Bacteriol 93:393

21. Cox RH (1977) Carotid artery mechanics and composition in renal and DOCA hypertension in the rat. Cardiovasc Med 2:761–766

22. Cox RJ (1978) Passive mechanics and connective tissue composition of canine arteries. Am J Physiol 234:H535–H541

23. Cox RH, Jones AW, Fischer GM (1974) Carotid artery mechanics, connective tissue and electrolyte changes in puppies. Am J Physiol 227:563–568

24. Crawford T (1959) Some observations on the pathogenesis and natural history of intracranial aneurysms. J Neurol Neurosurg Psychiatr 22:259

25. Dawes GS, Mott JC, Widdicombe JC (1955) The cardiac murmur from the patient ductus arteriosus in newborn lambs. J Physiol 128:344

26. Epstein ML, Goldberg SJ, Allen HD et al. (1975) Great vessel, cardiac chamber and wall growth patterns in normal children. Circulation 51:1124–1129

27. Forbus WD (1930) On the origin of miliary aneurysms of the superficial cerebral arteries. Bull J Hotz Hosp 47:239

28. Greenwald SE, Berry CL (1980) The effects of alterations in scleroprotein content on the static elastic properties of the arterial wall. In: Kovach AGB, Monos E, Rubanyi G (eds) Cardiovascular physiology. Heart, peripheral circulation and methodology. Adv Physiol Sci 8:203–212

29. Greenwald SE, Berry CL, Haworth, SG (1982) Changes in the distensibility of the intrapulmonary arteries in the normal newborn and growing pig. Cardiovasc Res 16:716–726

30. Henrichs KJ, Berry CL (1979) Morphometrische Untersuchungen un Arterien bei Hypertonie-im Tierexperiment und an menschlichem Untersuchungsgut. Verh Dtsch Ges Pathol 63:684

31. Heymann MA, Rudolph AM (1975) Control of the ductus arteriosus. Physiol Rev 55:62

32. Kuettner KE, Pauli BU (1982) Vascularity of cartilage. In: Hall BK (ed) Cartilage. Academic Press, New York, pp 281–312

33. Lee J, Berry CL (1978) Cerebral micro-aneurysm formation in the hypertensive rat. J Pathol 124:7–11

34. Leonard ME, Hutchins GM, More GW (1983) Role of the vagus nerve and its recurrent laryngeal branch in the development of the human ductus arteriosus. Am J Anat 167:313–327

35. Looker T, Berry CL (1972) The growth and development of the rat aorta. II. Changes in nucleic acid and scleroprotein content. J Anat 113:17

36. McDonald GA (1974) Blood flow in arteries, 2nd edn. Edward Arnold, London

37. Milnor WR, Bergel DH, Barganier JD (1966) Hydraulic power associated with pulmonary blood flow and its relation to heart rate. Circ Res 18:467–480

38. Papadia F, Setti GC (1972) The lymphatic drainage system of the great blood vessels in normal, pathologic and experimental conditions. Arteno Parmese 43:133–135

39. Patel DJ, Vaishnav RN (1972) The rheology of large blood vessels. In: Bergel DH (ed) Cardiovascular fluid dynamics, vol 2. Academic Press, London, pp 1–64

40. Pesonen E (1974) Intimal cushions in vascular development coronary wall thickening in children. Atherosclerosis 20:173–187

41. Pesonen E., Norio R, Sarna S (1975) Thickenings in the coronary arteries in infancy as an indication of genetic factors in coronary heart disease. Circulation 51:218–225

42. Reagen FP (1917) Experimental studies on the origin of vascular endothelium and of erythrocytes. Am J Anat 21:39

43. Roach MR, Burton AC (1959) The effect of age on the elasticity of human iliac arteries. Canad J Biochem Physiol 37:557–570

44. Robertson JH (1960) The significance of intimal thickening in the arteries of the newborn. Arch Dis Child 35:588–590

45. Rodbard S (1958) Physical factors in the progression of stenotic vascular lesions. Circulation 17:410

46. Rudolph AM, Heymann MA, Spitznas U (1972)

Haemodynamic considerations in the development of narrowing of the aorta. Am J Cardiol 30:514

47. Ryan TJ, Barnhill RL (1983) Physical factors in angiogeneis. In: Development of the vascular system. Ciba Foundation Symposium 100. Pitman, London, pp 80–94

48. Stehbens WE (1963) Cerebral aneurysms of animals other than man. J Pathol Bacteriol 86:161

49. Tawes RL, Aberdeen E, Berry CL (1968) The growth of an aortic anastomosis, an experimental study in piglets. J Pediatr Surg 64:1122–1132

50. Tawes R., Aberdeen E., Waterston DJ et al. (1969) Coarctation of the aorta in infants and children. A review of 333 operative cases including 179 Infants. Circulation [Suppl 1] 39:173

51. Tawes RL, Berry CL, Aberdeen E (1969) Congenital biscupid aortic valves associated with coarctation of the aorta in Children. Br Heart J 31:127–128

52. Taylor MG (1964) Wave travel in arteries and the design of the cardiovascular system. In: Pulsatile blood flow. McGraw-Hill, New York, pp 343–372

53. Taylor MG (1966) Wave transmission through an assembly of randomly branching elastic tubes. Biophys J 6:697–716

54. Toda T., Tsuda N, Takagi T et al. (1980) Ultrastructure of developing human ductus arteriosus. J Anat 131:25–37

55. Velican C, Velican D (1977) Studies on human coronary arteries. I. Branch pads or cushions. Acta Anat 99:377–385

56. Velican C, Velican D (1977) Histogenetic differences between parent and daughter vessels of human coronary arteries. Atherosclerosis 26:273–287

57. Vlodaver MD, Kahn HA, Neufeld HN (1969) The coronary arteries in early life in three different ethnic groups. Circulation 39:541–550

58. Wilens SL, Malcolm JA, Vazquez JM (1965) Experimental infarction (medial necrosis) of the dog's aorta. Am J Pathol 47:695

59. Wolinsky H, Glagov S (1967) A lammellar unit structure of aortic medial structure and function in mammals. Circ Res 20:99–111

60. Wolinsky H, Glagov S (1967) Nature of species differences in the medial distribution of aortic vasa vasorum in mammals. Circ Res 20:409–421

Chapter 4

Age-Related Morphological Changes of the Arterial Wall

H. Bouissou, M. T. Pieraggi and M. Julian

The arterial system, formed by the division of the aorta into a large number of distributing and irrigating arteries of decreasing diameter, has a uniform basic wall structure, consisting of the intima, the media and the adventitia.

Arteries are divided into two types according to the composition of their middle layer: the elastic arteries and the free and intraparenchymal muscular arteries. The ageing process of the arteries, which may begin more or less early, varies according to their composition and their position within the arterial system [1].

Vessels from 2–4-year-old children, showing no significant pathological change, are seen in Fig. 4.1.

"Elastic"-Type Arteries

These consist of the aorta and its branches (brachiocephalic trunk, prescalenic subclavian, common carotid, internal mammary arteries). The transitional arteries, elastic arteries which progressively become muscular (common iliac, extrascalenic subclavian, axillary, external carotid arteries), undergo the same changes as the aorta.

In the child and the young adult [18,22] (Fig. 4.2a), the intima is thin, made up of a fine layer of loose connective tissue and rich glycosaminoglycans (GAG's), containing few or no cells, and separates the endothelium from the first elastic layer of the media. During the early

years of life, this layer becomes thicker and more organized. Glycosaminoglycans, elastic fibrils and small collagen bundles accumulate progressively and the cell population increases at the same time (smooth-muscle cells, macrophages). The media is made up of parallel elastic layers, which at birth are around 35 in number in the aorta and increase to about 56 in the young adult. In these elastic layers are the smooth-muscle cells surrounded by GAGs and a small quantity of collagen. The adventitia consists of fibroblasts in a loose network of collagen surrounding the vasa vasorum.

During normal ageing [8,22], or "physiological arteriosclerosis", the aorta and its large branches become thick and sclerotic to varying degrees (Fig. 4.2b). The first sign of ageing, which occurs around the age of 15, is thickening of the intima, due to an accumulation of GAGs and lysis of one or several elastic layers. The corresponding layers of the media thus become part of the intima. This "intimalization" [22,23] (Fig. 4.3) varies in extent and depth along the vessel and continues throughout life. At the same time, the remaining elastic layers of the media become irregular and fragmented and are increasingly separated by accumulations of GAGs and collagen fibres. The artery becomes fibrous. Sometimes the lysis of elastic material is accompanied by proliferation of smooth-muscle cells in the internal part of the media, and the artery becomes progressively musculo-elastic. When there is a large accumulation of GAG, pseudo-cysts of mucinus material may appear in the damaged media. The artery becomes muco-

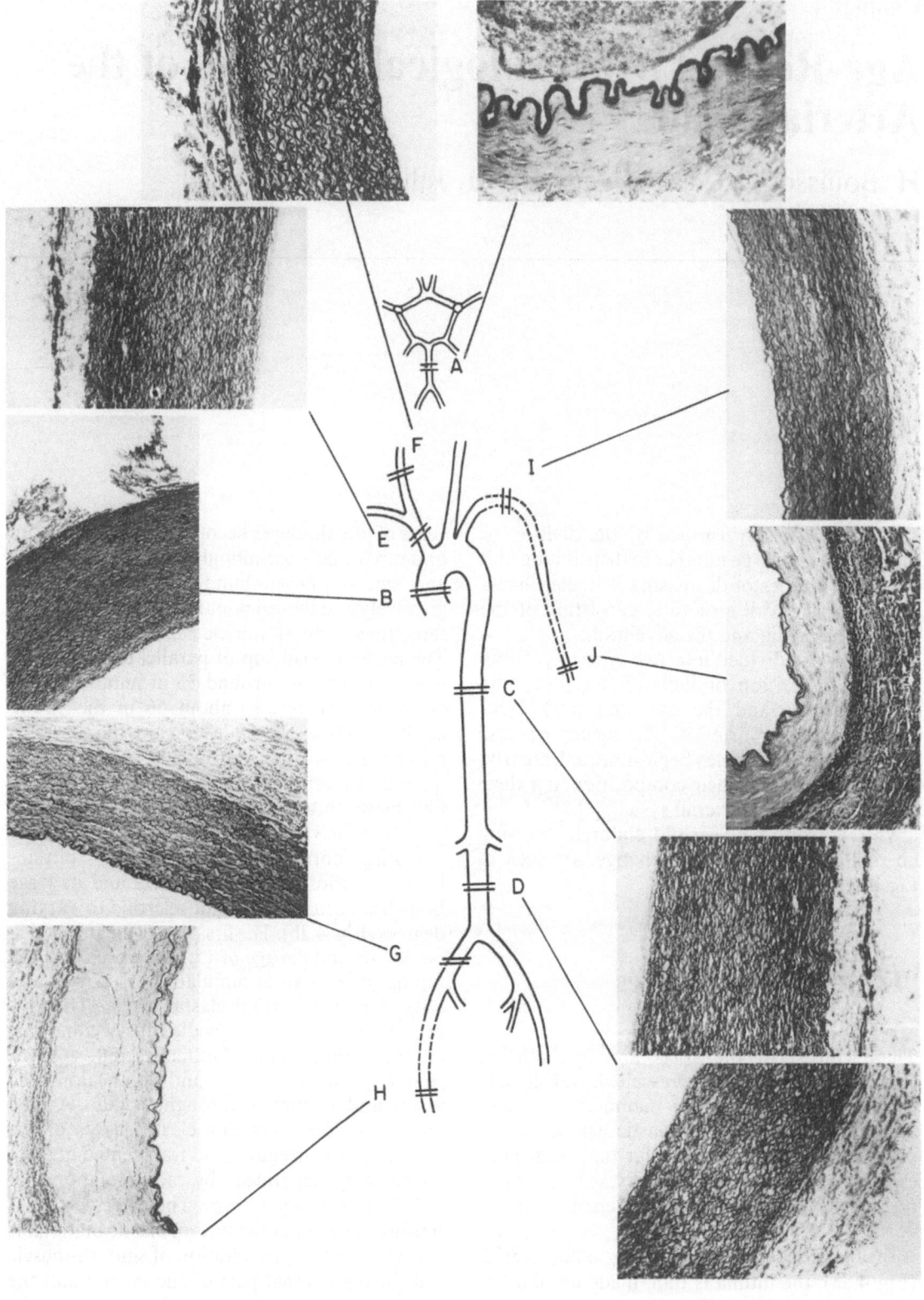

Fig. 4.1. Vessels from 2–4-year-old children showing no significant pathological change. All magnifications are objective magnifications. (We are indebted to CL Berry for providing the pictures.) **A** Basilar artery. Elastic van Gieson, × 240. **B** Ascending aorta (the adventitia has been removed). Elastic van Gieson, × 50. **C** Descending thoracic aorta at the level of the sixth rib. Elastic van Gieson, × 100. **D** Abdominal aorta. Elastic van Gieson, × 100. **E** Brachio-cephalic artery. Elastic van Gieson, × 50. **F** Common carotid artery. Elastic van Gieson, × 100. **G** Common iliac artery mid-way between the bifurcation of the external iliac artery and the internal iliac artery. Elastic van Gieson, × 150. **H** Popliteal artery. Elastic van Gieson, × 240. **I** Brachial artery at the level of the deck of the humerus. Elastic van Gieson, × 100. **J** Radial artery, Elastic van Gieson, × 200

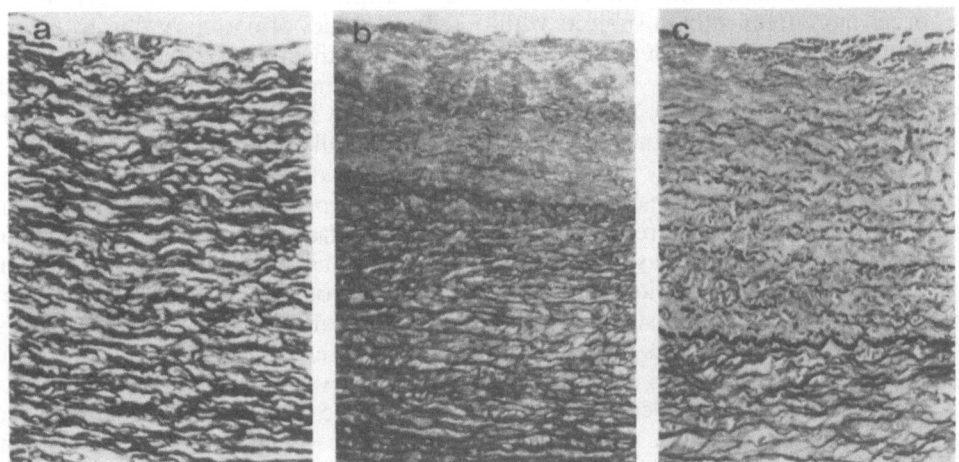

Fig. 4.2. a Normal aorta (18 years). Distinct intima and well-differentiated first elastic layer. In the media there is a dense elastic network with long parallel fibres. Verhoeff's stain, × 25. **b** Aorta showing physiological arteriosclerosis (30 years). Thickened, fibrous intima resulting from influx of proteoglycans and slight successive intimalization. Verhoeff's stain, × 10. **c** Aorta showing ateriosclerotic disease (30 years). Very thick intima formed by lysis of several medial elastic layers (intimalization), together with a major influx of proteoglycans. Within the present media, separated from the new intima by a well-defined elastic layer, there is fragmentation of the elastic layers and widening of the interlamellar spaces (accumulation of GAG). Verhoeff's stain × 25

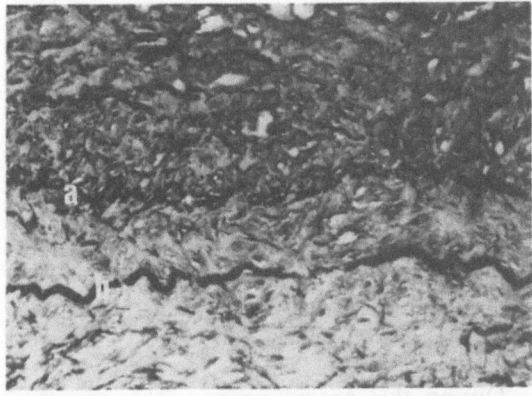

Fig. 4.3. Aorta (40 years). Intimalization process of the intimal–medial junction. Stippled appearance (**a**) of a disintegrating elastic layer (lysis). It is incorporated into the media with the interlamellar tissue which separates it from (**b**), the distinct elastic layer presently separating the intima and the media. Orcein, × 100

elastic. The culmination of this process has been described by Cellina [11] as "medianecrosis disseminata aortae". A grading system for the most severe degrees of this lesion has been proposed by Edwards et al. [10,12]. The relationship between these non-specific medial changes and aortic dissection is discussed in Chap. 31 [2,24]. Schlatmann and Becker [24], studying histological changes in the media of 100 normal aortas at various ages, showed four types of changes which could be observed in ageing aortas with increasing incidence: (a) pooling of mucoid material described as cystic medial necrosis; (b) elastin fragmentation; (c) fibrosis; (d) medianecrosis.

These changes result from physiological modifications in the components of the arterial wall [18]: the endothelium is more permeable, resulting in thickening of the intima and the interlamellar spaces nearest the lumen and the smooth-muscle cells are no longer able to maintain the elastic tissue and secrete collagens and GAGs, resulting in lysis of elastic layers, fibrosis and larger interlamellar spaces.

Focal medianecrosis, characterized by loss of nuclei, can be found in normal aortas. Extending to one- or two-thirds of the media, such luminar medial necrosis has been reported in cases of acute aortic dissection [14,15]. The significance of this change remains a matter of debate (see Chap. 31). It seems unlikely that these medial changes occur more frequently in subjects with hypertension when compared with age-matched normotensive subjects [16,24].

Between the ages of 15 and 30, fibrous elevations (Fig. 4.4) appear in elastic and some muscular arteries; in the coronary arteries in particular [29]. Their number increases with age. Of varying extent (a few millimetres in the coronary artery to 1.5 cm in the elastic arteries), they are oriented parallel to the blood flow and are whitish, opalescent, firm and slightly raised above the normal intima. In cross-section they are of more of less homogeneous fibrous consistency, with no yellow centre. Microscopically, they consist of parallel collagen fibres interspersed with a few elastic fibres [8]. There may be some foam cells, mainly in the abdominal aorta and at the entrance regions of the coronary arteries where these vessels age earliest and most severely.

We consider these elevations to be hypertrophic fibrous scars, which form upon already large and destructive gelatinous plaques formed during periods of marked deposition which were followed by the arrival of secretory collagen-forming smooth-muscle cells. Smith [27] also notes this origin. Successive microthrombi may have been integrated and become fibrous. They may also simply result from the deposition or successive depositions of blood lipids in the arterial wall; if these are not metabolized by the smooth-muscle cells, they cause hypertrophic fibrosis of the intimal–medial junction. We do not consider these fibrous plaques to be advanced atherosclerotic lesions, unlike Small and Shipley [26] Ross and Glomset [20] and Ross [21]. Together with Velican [29], who made several studies of coronary fibrous plaques, we distinguish them from atheromatous lesions, for marked intimal fibrosis rarely occurs at the site of secondary lipid overload, as we have observed experimentally [3,17].

Theoretically, ageing in elastic arteries should result in progressive sclerosis, substitution of destroyed elastic fibres and some accumulation of GAGs and lipids, with persistence or even hyperplasia of smooth-muscle cells. These changes, accompanied by dilatation, take place slowly and are observed in most autopsies of the elderly. However, some autopsies of subjects aged more than 45–50 years show intense, precocious arteriosclerosis, often associated with marked stage III atheroma. This "arteriosclerotic disease" [8] shows identical but more marked lesions than "physiological arteriosclerosis".

It is probable that the biological quality of the vessel and the environment of life in the wider sense of the term are not the same in the two cases. In arteriosclerotic disease, the marked and brutal penetration of GAGs and lipids in the arterial wall accounts for the extent of damage caused, the disorganization and the impossibility of adaptation or effective repair. Thus the stagnation and accumulation of GAGs, and in particular lipids, in an arterial wall which is incapable of metabolizing them or transporting them to the vasa vasorum account for the intensity of the lesions. More marked in certain portions of the arteries than in others (haemodynamics is certainly a factor in the preferential localizations of ageing effects), they will extend with the passage of time, and further lesions will appear at points formerly spared or only slightly involved. The elastic artery is prematurely worn, unequally and irregularly aged.

During this ageing process, the adventitia becomes more fibrous. In the vasa vasorum, few changes other than banal microangiopathy are noted with age.

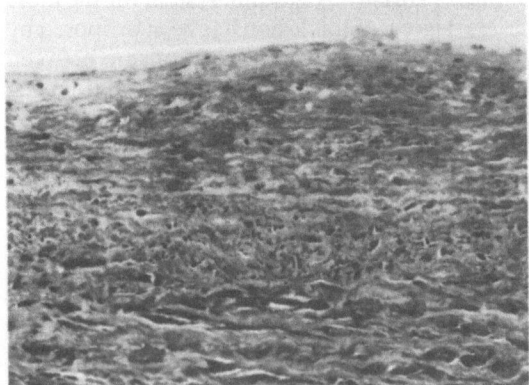

Fig. 4.4. Aorta (38 years). Fibrous plaque. The thickened, fibrous intima bulges into the lumen (hypertrophic scar). Masson's trichrome, × 25

"Muscular"-Type Arteries

The muscular arteries of a young subject are composed of: (a) an endothelium attached by a narrow layer of loose connective tissue to the internal elastic lamina (IEL), an undulating, thick and regular layer where splitting is rare and limited; (b) a media made up of an accumulation of smooth-muscle cells which are arranged in a pattern of helices. Each of these cells is surrounded by a small amount of connective tissue. The thickness of the media decreases from the artery to the arteriole, where it consists of a single layer of smooth-muscle cells; (c) a fibrous adventitia containing the vasa vasorum and separated from the media by an external elastic lamina less clearly characterized than the IEL and more often discontinuous.

These structures change with age (Figs. 4.5 and 4.6). The intima is thickened and becomes fibrous due to substances transported by the plasma and to migration of smooth-muscle cells from the media. This change in the inner lining of the artery sometimes results in concentric "onion-like" layers in arterioles, a type of "proliferative endarteritis" caused by the successive accumulation of cells and collagen and/or elastic fibres. These intimal changes, of whatever kind, injure the endothelial cells and cause thrombi which may be sequential. The latter become endothelialized and fibrous and further thicken the arterial wall.

The internal elastic lamina duplicates in the form of fibres and entangled fibrils. These elastic neofibres undergo constant transformation; they disappear over areas of varying extent, reform and become more or less evident. They have no physiological value. The media becomes fibrous and atrophic; the smooth-muscle cells disappear and are replaced by fibroblastically differentiated cells which elaborate collagens. In some arteries, there is hyalinization of the medial layer, which may be partial or circular, total or limited, in the latter case only affecting some of the myocyte layers. This "vitreous degeneration" (hyalinosis) follows plasma infiltration and results in a homogeneous, eosinophilic media which may become calcified (Monckeberg's medial sclerosis) or even ossified. It is frequent in certain internal organs such as the spleen, the liver and the kidney and is the "rust of life".

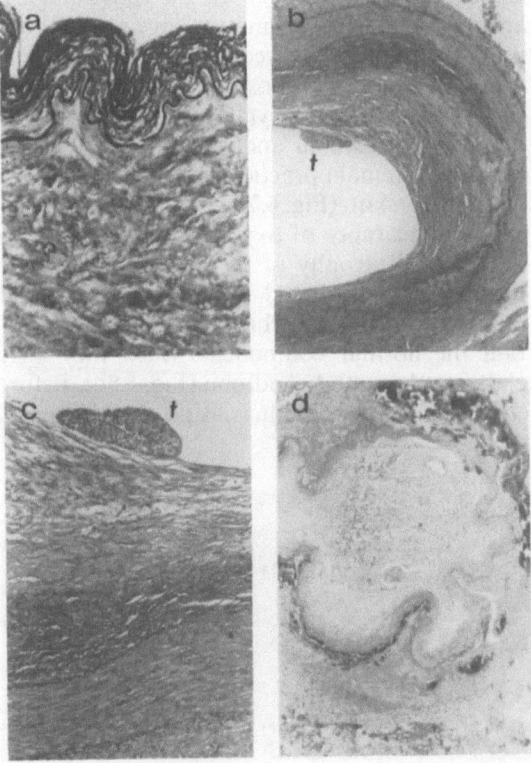

Fig. 4.5. Tibial artery. **a** 64 years. Discrete ageing: slight even intimal thickening with neogenesis of elastic fibres. Slightly fibrous media. Verhoeff's stain, × 25. **b** 71 years. Advanced arteriosclerosis: a thrombus (*t*) is seen on an intima which has become thickened by successive integrated fibrotic thrombi. The internal elastic lamina is dystrophic. The media is thin and fibrous. Verhoeff's stain, × 4. **c** Details of the same artery, showing adhering thrombi and modifications of the internal elastic lamina. Verhoeff's stain, × 10. **d** 65 years. Advanced complicated arteriosclerosis. Presence of an occlusive revascularized thrombus. The media and internal elastic lamina are calcified. Masson's stain, × 4

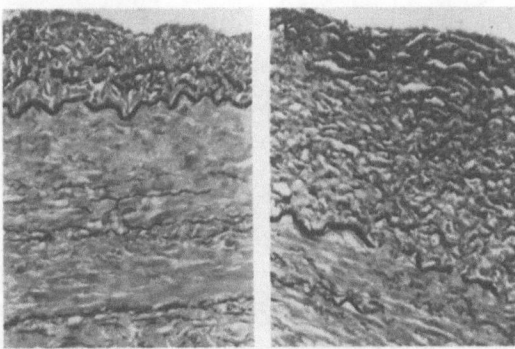

Fig. 4.6. Coronary artery. **a** 67 years. Common trunk of the left coronary. Very slight physiological arteriosclerosis. Discrete intimal thickening with transient stippled neoformation of elastic lamellae. Practically normal elastic layer (only one dehiscence). Verhoeff's stain, × 25. **b** 67 years. Common trunk of the left coronary. Pronounced arteriosclerosis, very marked thickening of the intima. The internal elastic lamina is largely destroyed. Fibrous media. Verhoeff's stain, × 25

Not all portions of the arteries age at the same time and in the same way [5]. The tibial artery ages more quickly than the radial artery, the coronary artery shows signs of wear from the age of 5 years [13], whereas ageing in the cerebral arteries is not marked until the age of about 60 [28]. In these arteries, the media becomes fibrous whereas the intima and the internal elastic lamina remain normal for a considerable time. In the coronary artery the significance of what have been described as endothelial cushions is discussed in Chap. 3. In the tibial artery there is also marked thickening of the intima, with fragmentation and splitting of the internal elastic lamina. The media of the radial artery becomes fibrous and its intima may remain unchanged for a long period.

Assessment of Ageing of the Arterial Wall

The aorta and the coronary artery age concurrently with the dermis [6] and morphological examination of a cutaneous biopsy specimen has been shown to give a reliable indication of the arteriosclerotic state (normal or pathological) of these two vessels.

In normal (type 0) skin (Fig. 4.7a), the superficial papillary dermis is formed of bundles of dense collagen fibres placed closely together, and

of elastic oxytalan fibres made up solely of structural glycoproteins (SGPs). By electron microscopy, they are of fibrillar appearance; by light microscopy they are fine and slender, rising vertically and perpendicularly towards the stratum germinativum. At their base, they are continuous with the network of elaunin fibres, formed of SGPs and elastin, which lie in the reticular dermis, parallel to the epidermal basal layer. In the deep reticular dermis, the collagen fibres form bundles and the elastic fibres, made up solely of elastin, occupy the interstitial spaces. These various collagen and elastic fibrillar components are secreted and provided by active fibroblasts, which, during maturity, maintain their condition by slow turnover (Fig. 4.8a). Type 0 skin may be observed at any age but is more typical of subjects under 30 years old.

Discretely aged skin (type I, Fig. 4.7b), which is normal after the age of 30 (physiological ageing), indicates premature ageing of connective tissue if observed before that age. The only difference in composition from type 0 skin is the growing rarity of oxytalan fibres; the collagen bundles, the elaunin and elastic fibres and the fibroblasts remain practically identical.

Type II skin (Fig. 4.7c) is characterized by the disappearance of oxytalan fibres, a varying degree of dystrophy of the elaunin and elastic fibres, thin, disjointed collagen bundles and quiescent fibroblasts which are unable to maintain the fibrillar component which they have elaborated in good condition (Fig. 4.8b). Before the age of 40, type II skin is a sign of pathological

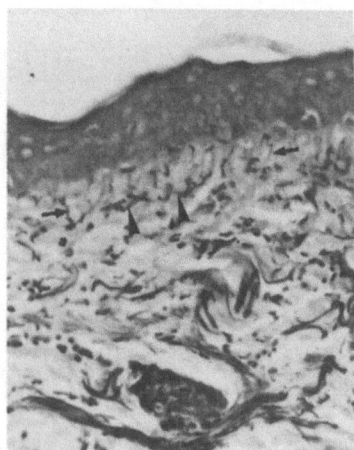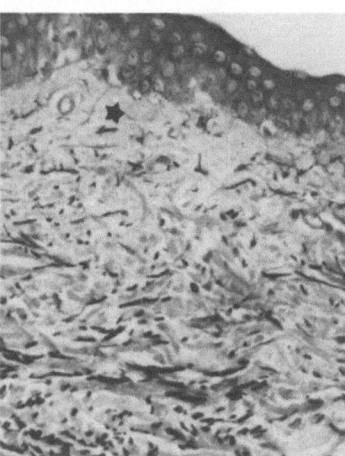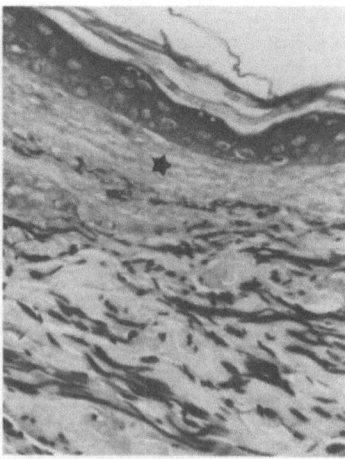

Fig. 4.7. Light microscopy. **a** Skin type 0. Papillary dermis showing numerous oxytalan fibres perpendicular to the basal layer. In the reticular dermis, thick elaunic fibres run parallel to the surface. In the deep dermis between collagen bundles there are abundant elastic lamellae. Verhoeff's stain, × 25. **b** Skin type I. Disappearance places of oxytalan fibres (*star*) in papillary dermis. Verhoeff's stain, × 25. **c** Skin type II. Disappearance of oxytalan fibres (*star*) in the papillary dermis. Dystrophy of elaunic and elastic fibres in reticular and deep dermis. Verhoeff's stain, × 25

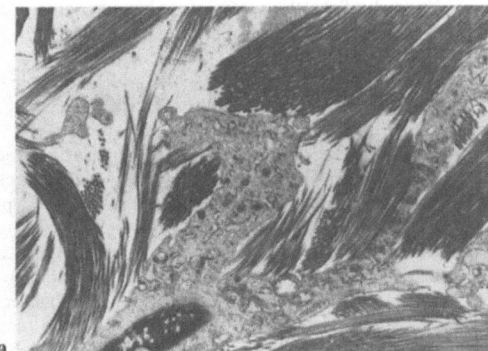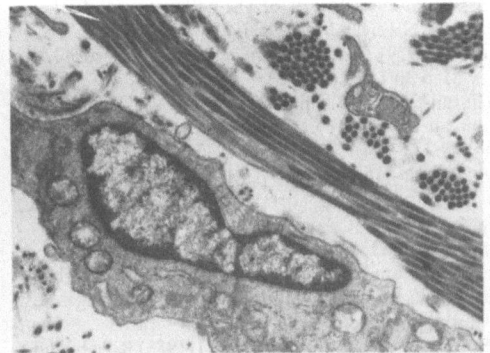

Fig. 4.8. Transmission electron microscopy. **a** Skin type 0. Active fibroblast with close connection with collagen bundles. Uranyl acetate – lead citrate, × 10 000. **b** Skin type II. Quiescent fibroblast without direct contact with collagen. Uranyl acetate – lead citrate, × 10 000

ageing and may indicate arteriosclerotic disease. After that age it is a normal finding. Numerous observations during autopsy show there is a close correlation between dermal and vascular (aortic and coronary) elastolysis [6,9]. Skin biopsy observations reveal:

1. Marked and accelerated ageing of connective tissue (type II) in the skin of subjects with known coronary heart disease aged under 40.
2. A high percentage of pathological findings in subjects aged 20 years with young parents themselves presenting with cardiovascular disease (myocardial infarction, severe arterial hypertension) [7] or with diabetes [19].

Microscopic examination of the dermis would thus seem to be an indirect, easy and reliable method of determining the arteriosclerotic state of subjects while still young, and detecting "arteriosclerotic potential" in those whose background shows them to be at risk, but who are as yet free of any arteriosclerotic symptom.

References

1. Berry CL (1973) Growth, development, and healing of large arteries. Ann R Coll Surg Engl 53:246–257
2. Bouissou H, Alzieu J, Pieraggi MTh, Julian M (1978) Les dissections aortiques aigues communicantes (essai pathogénique). Paroi artérielle/Arterial Wall 4:135–147
3. Bouissou H, De Graeve J, Julian M, Pieraggi MTh, Thiers JC (1980) Chronic lathyrism. Plasma, aorta and rat skin treated for five months with Beta-amino-propionitrile with or without high fat diet. Pathol Biol 28:645–650
4. Bouissou H, De Graeve J, Solera ML et al. (1982) Le cholesterol cutané du coronarien jeune et agé. Arch Mal Coeur 5:621–626
5. Bouissou H, Emery MCR, Sorbara R (1975) Age related changes of the middle cerebral artery and a comparison with the radial and coronary artery. Angiology 26:257–268
6. Bouissou H, Fabre MTh, Julian M, Rumeau JL, Huron R (1970) La biopsie cutanée permet-elle de connaitre l'état de la paroi vasculaire? Rev Europ Etudes Clin Biol 15:444–543
7. Bouissou H, Julian M, Pierragi MT, Aouidet A, Marie W (1984) Microangiopathie et atteinte conjonctive dermiques du sujet jeune avec ou sans antécédent direct cardiovasculaire ou diabétique. Nouv Presse Méd 13:2815–2816
8. Bouissou H, Pieraggi MTh, Julian M (1981) L'athérome (mise au point) Ann Pathol 1:5–20
9. Bouissou H, Pieraggi MTh, Julian M, Douste-Blazy L (1976) Simultaneous degradation of elastin in dermis and in aorta. In: Robert L (ed) Frontiers of matrix biology, vol 3. Karger, Basel, pp 242–255
10. Carlson RG, Lillehei CW, Edwards JE (1970) Cystic medial necrosis of the ascending aorta in relation to age and hypertension. Ann J Cardiol 25:411–415
11. Cellina M (1931) Medianecrosis disseminata aortae. Virch Arch [A] 280:65–86
12. Edwards JE (1961) An atlas of acquired diseases of the heart and great vessels, vol III. Saunders, Philadelphia, pp 1067–1120
13. Fabre MTh, Julian M, Bouissou H (1968) L'artère coronaire chez l'enfant et l'adulte jeune. Presse Med 76:1915–1918
14. Gore I (1952) Pathogenesis of dissecting aneurysms of the aorta. Arch Pathol 53:142–153
15. Gore I (1953) Dissecting aneurysms of the aorta in persons under forty years of age. Arch Pathol 55:1–3
16. Hirst AE, Johns VJ, Kime SW (1958) Dissecting aneurysms of the aorta; a review of 505 cases. Medicine 37:217–279
17. Julian M, Pieraggi-Fabre MTh, Bouissou H (1971) Lathyrisme experimental. La Paroi aortique au cours de l'intoxication chronique avec régime athérogène. Pathol Biol 19:1037–1044
18. Pierragi MTh, Bouissou H (1981) Aorte humaine: endothélium et athérome. Ann Pathol 1:271–279
19. Pieraggi MT, Bouissou H, Julian M, Aouidet A, Marie W (1984) Relations entre les lésions du derme et le "terrain artériscléreux" Nouv Presse Méd 13:2518–1519

20. Ross R, Glomset JA (1976) The pathogenesis of atherosclerosis. N Engl J Med 295:369–377, 420–425

21. Ross R (1979) The arterial wall and atherosclerosis. Ann Rev Med 30:1–5

22. Sendrail-Pesque M, Julian M, Durroux R, Fabre MTh, Rumeau JL (1969) Le vieillissement aortique. Pathol Biol 17:223–224

23. Schlatmann IJM, Becker AE (1977) Histologic changes in the normal aging aorta: implication for dissecting aneurysm. Am J Cardiol 39:13–20

24. Schlatmann IJM, Becker AE (1977) Pathogenesis of dissecting aneurysm of aorta. Comparative histopathologic study of significance of medial changes. Am J Cardiol 39:21–26

25. Schwartz SM, Reidy A, Hansson G (1983) Role of endothelial dysfunction in vivo in smooth muscle prolifer-

ation. In: Schettler G, Giotto AM (eds) Atherosclerosis VI. Springer-Verlag, Berlin Heidelberg New York, pp 350–357

26. Small DM, Shipley GG (1974) Physical chemical basis of lipid deposition in atherosclerosis. Science 185:222–229

27. Smith B (1983) Identification of the gelatinous lesions. In: Schettler G. Giotto AM (eds) Atherosclerosis VI. Springer-Verlag, Berlin Heidelberg New York, pp 170–173

28. Sorbara R, Bouissou H (1972) Arteriosclerose des artères cerebrales. Nouv Presse Méd 1:3193–3195

29. Velican D, Velican C (1982) Discrepancies between data on atherosclerotic involvement of human coronary arteries furnished by gross inspection and by light microscopy. Atherosclerosis 43:39–49

Chapter 5

Effects of "Ageing" on Arterial Cells in Vitro

Monique Breton and J. Picard

Introduction

Current theories relating to the genesis of athero-sclerosis assign an important role to smooth-muscle cells. The first stages of this disease consist of a thickening of the intima caused by the proliferation of smooth-muscle cells which have migrated from the media towards the intima [60,61]. According to certain inves-tigators, this stimulation of the growth of smooth-muscle cells, which are usually quiescent in normal arteries, could be explained either by a mutation conferring a selective advantage on a single cell which could then divide upon stimu-lation [2] or by injury to the intimal barrier causing the smooth-muscle cells to be exposed to circulating serum factors [60], and, in particu-lar, to platelet-derived growth factors which seems to be the principal mitogenic agent for smooth-muscle cells [6,62].

In man, the development of atherosclerosis seems to progress exponentially with age, and the theory developed by Martin et al. [45] is interesting because it takes into consideration the role of ageing in the pathogenesis of athero-sclerosis: it suggests the existence of an intimate relationship between the age of the cells and their potential for division. According to these investigators, the stem cells of the media divide to give rise to differentiated smooth-muscle cells which, under normal circumstances, exercise a negative control on cell division in their vicin-ity by secreting inhibitory substances called "chalones". With age, the capacity of the stem cells to divide slackens and the concentration of "chalones" diminishes, and this leads to an accumulation of smooth-muscle cells in fibrous plaques.

Recent progress in cell culture techniques makes it possible to grow endothelial cells and arterial smooth-muscle cells of very varied origin which continue to express properties of differ-entiated cells similar to those found in vivo. This methodology represents a powerful tool for both the study of the behaviour of one cell type inde-pendently of all of the other phenomena which interfere with investigations in vivo and the dem-onstration in vitro of regulatory pathways which cannot be studied in vivo. It has been shown, for example, that endothelial cells are capable of exercising a controlling influence on the growth of smooth-muscle cells: depending on the con-ditions of culture, endothelial cells synthesize and secrete factors which are capable of either inhibiting [10,11] or stimulating [21] the growth of smooth-muscle cells.

In the remainder of this chapter, the "ageing" of endothelial and smooth-muscle cells in vitro will be taken to mean the changes in the prop-erties of the cells after several subcultures.

Cell Culture Techniques

Smooth-Muscle Cells

Different methods can be employed to prepare cultures of the smooth-muscle cells of blood vessels, but in general the following techniques are used.

The artery is excised aseptically by the appro-priate surgical operation and this is followed by the removal of the adventitia, which is rich in fibroblasts, and, possibly, of the intima, the surface of which is lined with endothelial cells. The smooth-muscle cells can then be placed in culture after being disaggregated by a 0.075%–

0.2% solution of collagenase [24,49,77], a 0.1%–0.2% solution of collagenase containing 0.05% trypsin [52] or a 0.05%–0.10% solution of collagenase containing 0.025% to 0.050% elastase [12,44].

The technique currently most often used is that described by Ross [59]: a segment of the media or intima–media of the artery is cut into small pieces of 1–2 mm^2 cross-section which are then dispersed in sterile dishes or flasks containing just sufficient culture medium to cover them. When the explants have become attached to the walls sufficiently well, the medium may be changed and its volume may be increased. The proportion of the explants capable of releasing cells and the latent period necessary before the first cells migrate away from the explant seem to be inversely proportional to the age of the animal donor [3,16]. For example, in the case of porcine aorta [18], the results obtained with explants taken from 8–12-week-old animals indicated that the first cells appear 4–5 d after the explant has been placed in culture and that 70%–80% of the explants release cells after being cultured for 15 to 20 d. In contrast, when the animal donor is 18 weeks old, a small number of explants (less than 20%) release cells after 5 d of culture and only 40% of the explants are capable of growth after 25 d of incubation.

The secondary cultures are usually obtained by dissociation of the cells with a 0.25% solution of trypsin to which 0.025% EDTA may be added. The cells are rinsed beforehand with a medium containing neither calcium nor magnesium.

Endothelial Cells

The isolation of endothelial cells for culture may be carried out by treating an arterial segment maintained at 37°C with a solution containing 1–2 mg/ml of collagenase for 5–15 min [25,36,67] and then centrifuging the cell suspension at low speed after the addition of medium to stop the enzymatic reaction. The cells are then resuspended in fresh culture medium. Another method involves lightly swabbing the internal surface of the artery [26] and transferring the cells entrapped in the swab to a culture medium.

Secondary cultures can be obtained by dissociation of the cells with a 0.05% solution of trypsin containing 0.025% EDTA.

Usually the primary cultures are contaminated with smooth-muscle cells and it is necessary to clone the endothelial cells [23] in order to obtain a pure cell preparation. Furthermore, the endo-

thelial cells must be cultured on extracellular matrices or in the presence of growth factors if they are to preserve their differentiated characteristics [19,28].

Growth of Arterial Cells

Many investigators have confirmed in endothelial and smooth-muscle cells, the observations made by Hayflick [32] on fibroblasts: an intimate relationship seems to exist between the age of the cells and their potential for division.

However, the conditions of culture, that is the medium used, nature of the serum, presence or absence of particular growth factors, origin and plating density of the cells, vary from one investigator to another and all of these factors need to be borne in mind when comparing different results.

Smooth-Muscle Cells

The capacity of smooth-muscle cells to divide diminishes with the age of the animal donor. Bierman [3] has, in fact, observed that the number of doublings of smooth-muscle cells from the arteries of human subjects of less than 20 years of age is about twice as high as that seen with cells taken from individuals of more than 50 years of age.

Furthermore, a decrease in the time between cell divisions and an increase in cell density at confluence have been observed when cells "age" in vitro. Breton [7] has shown that the doubling time of smooth-muscle cells obtained from porcine aorta is about 34 h at the third subculture and about 20 h at the 11th subculture, whereas the cell density at confluence increases from 86 000 cells/cm^2 to 136 000 cells/cm^2. Similarly, Nilson et al. [49] have demonstrated an acceleration of the proliferation of smooth-muscle cells from rat aorta in culture with the age of the animal: the nuclei of cells maintained for 24 h in a medium containing serum and tritiated thymidine became radiolabelled to the extent of 80% and 55%, depending on whether the cells were obtained from 5-day-old or 8-month-old rats, respectively.

The acceleration of cell division observed in vitro when the "ageing" of smooth-muscle cells has attained a certain stage may be the result of a dedifferentiation of the cells or, on the contrary, of a characteristic property of such cells. Nonetheless, the experiments performed in vivo by Stemerman et al. [71] have shown that the response of aortic smooth-muscle cells to injury to the endothelium by a catheter is greater and lasts longer in the aortae of old rats than in those of young rats.

These results should be compared with observations made by Pietila and Nikkari [53] who have shown that aortic smooth-muscle cells taken from atheromatous lesions, while morphologically similar to smooth-muscle cells isolated from non-pathological aortae, do show accelerated growth when compared with control cells.

From these data from different investigators, it is presently impossible to know whether the acceleration of the proliferation of smooth-muscle cells as a function of "ageing" in vitro is due to a particular population of cells or all of the cells. However, a recent study [31] has shown large variability in the rate of growth of smooth-muscle cells. This study, which was performed by video cinematography on clones of smooth-muscle cells from rat aorta, showed that some cells can double in number in less than 10 h whereas others require about 30 h to do so. In contrast, clones of smooth-muscle cells obtained from aortae, the endothelium of which has been subjected to a mechanical lesion, proliferated more rapidly and the time interval between two generations varied widely: the proportion of mother cells capable of giving rise to two daughter cells in less than 14 h was about 55% in the case of cells obtained from undamaged aortae and 80% for cells derived from damaged aortae. The time interval required between two successive cycles of mitosis varied by a factor of 3.6 for the cells obtained from intact aortae but only by a factor of 1.7 in the case of cells derived from damaged aortae; the duration of mitosis was not significantly different for the two types of cell and was about 40 min.

Endothelial Cells

It has been reported in various publications that the doubling time of endothelial cells increases with the number of subcultures [48,67]. Thus, Schwartz [67] observed that bovine aortic endothelial cells divide within 3 to 4 d in the first subcultures but require up to 17 days to divide after 18 passages.

A study carried out by Mueller et al. [48] on endothelial cells obtained from foetal calf aorta led to the establishment of a relationship between the growth rate of the cells and the total number of generations passed through after the cells have been placed in culture. A comparison between the cells of the 55th generation and those of the 79th generation showed a slow but progressive decrease in the initial growth rate and a considerable fall in the cell density attained at confluence. At the 55th generation the doubling time of the cell population in the exponential phase of growth was less than 2 d, at the 70th generation it had increased to almost 3 d and at the 80th generation it was more than 2 weeks. This increase of the doubling time as a function of "ageing" in vitro seemed to be due to an increase in the number of cells incapable of undergoing division: after incorporation of tritiated thymidine, 90% of the total number of nuclei were labelled up to the 50th generation, at the 70th generation this proportion had fallen to between 70% and 75% and at the 80th generation only 10% to 15% of the cells were capable of incorporating radioactive thymidine.

Nonetheless, depending on the conditions of culture, the capacity of aortic endothelial cells to divide can extend beyond the 80th generation. Johnson and Longenecher [37] have observed that cells from the bovine adult can be cultured for 130 generations if they are grown on extracellular matrices synthesized by smooth-muscle cells and the best results obtained by Gospodarowicz et al. have shown that endothelial cells grown on extracellular matrices in the presence of fibroblast growth factor (FGF) can be maintained for 390 generations [29].

Karyotype analyses performed on a clone of endothelial cells established from foetal calf aortas have shown that the number of chromosomal aberrations increases with the "ageing" in vitro of these cells [48]. In fact, at the 43rd generation 12% of the endothelial cells showed chromosomal translocations and this proportion had risen to 20% by the 66th generation.

Growth Factors and the Proliferation of Arterial Cells

For a number of years many groups of researchers have been interested in the role played by certain growth factors in the proliferation of arterial cells in vitro. In fact, the proliferation of endothelial cells in the processes of repair of the internal membrane of the vessels and/or the proliferation of smooth-muscle cells during the development of atherosclerosis seem to be the decisive events in the pathogenesis of this disease, according to the most commonly accepted theories.

The growth factor derived from platelets (PDGF) has been identified as one of the mitogenic agents of connective tissue cells and, in particular, of aortic smooth-muscle cells [6,62]. It has been suggested that this factor may play a fundamental role in the development of atherosclerosis since, following a lesion to the endothelium, the platelets can adhere to the subendothelial collagen, aggregate and release the PDGF which is stored in the alpha granules [60]. PDGF is a glycoprotein which exists in two distinct forms with apparent molecular weights of 31 000 (PDGF I) and 28 000 (PDGF II). PDGF I appears to differ from PDGF II in having a higher carbohydrate content. The biological properties and, in particular, the mitogenic potential of the two forms of PDGF are similar.

Studies performed by Nilsson et al. [49] on the aortae of rats aged 5 days and 8 months have shown that the maximal stimulation of DNA synthesis by PDGF is similar despite the difference in age of the animal donors but that the response of smooth-muscle cells obtained from the older rats required higher concentrations of growth factor (25 ng/ml instead of 10 ng/ml). In contrast, the effect of PDGF on the multiplication of smooth-muscle cells from old animals always remains weaker than that observed for cells obtained from young animals, irrespective of the concentration of PDGF. It would appear, in fact, that the action of PDGF on the growth of smooth-muscle cells requires the presence of other factors in the plasma [84]. Studies carried out on the mitogenic factors of serum have indeed shown that serum contains two distinct classes of factors which act synergistically to induce the synthesis of DNA [56,83] and cell division [62]: so-called "competence" factors which confer competence to synthesize DNA on cells blocked in the G_0 phase and so-called "progression" factors which allow cells to progress through the G_1 phase and enter the S phase. The mode of action of these factors appears to be very different.

Experiments carried out using Balb/c 3T3 cells have shown that brief exposure of quiescent cells to "competence" factors is sufficient for them to enter the G_1 phase but that "progression" factors must be present continuously for the cells to progress through the G_1 phase and into the S phase [72]. Nonetheless, it appears to be necessary for the "competence" factors to occupy their binding sites throughout the G_1 phase for the S phase to be initiated [80]. PDGF and FGF would appear to be "competence" factors whereas insulin and insulin-like growth factors (IGF I and IGF II) would appear to be "progression" factors.

Recently, Clemmons [13] has shown that PDGF and somatomedin-C (Sm-C), also known as IGF I, have a synergistic action on DNA replication in the smooth-muscle cells of porcine aorta. The addition of PDGF alone to the culture medium led to the incorporation of tritiated thymidine in only 11% of the cells whereas the addition of PDGF and Sm-C brings about the labelling of 82% of the nuclei. The contradictory findings made by different authors in their attempts to discover whether or not the PDGF is capable of stimulating the growth of smooth-muscle cells in the absence of any other plasma factor could result from the fact that the smooth-muscle cells, in the absence of Sm-C from the culture medium, might be able to synthesize a somatomedin-like factor which could partially compensate for the absence of Sm-C and thus allow the replication of DNA to proceed [13].

Studies [7] have also shown that the presence of a "progression" factor on its own (insulin) is insufficient to stimulate the growth of the smooth-muscle cells of porcine aorta blocked in the G_0 phase but that the simultaneous addition of serum factors, including PDGF, and insulin brings about DNA replication.

The principal characteristics of the different growth factors which are capable of acting on aortic cells are summarized in Table 5.1.

Smooth-muscle cells and endothelial cells exhibit different sensitivities to growth factors. Thus, PDGF, which has a very marked effect on the growth of smooth-muscle cells, does not seem to have a significant mitogenic action on endothelial cells [33,47,74,76]. In contrast, another growth factor, called vascular endothelial cell proliferation factor (VEPF) which is isolated from platelets like PDGF, seems to have a specific action on the growth of endothelial cells [47].

Growth factors are not only mitogens but also factors which enable cells to maintain their functional and morphological characteristics [30,77,78].

Table 5.1. Characteristics of the different growth factors

Name	Abbreviations	Molecular weight	Origin	Action on the growth of		References
				Smooth-muscle cells	Endothelial cells	
Platelet-derived growth factors I and II	PDGF I PDGF II	31 000 28 000	Platelet	+	−	6,13,17, 33,49,62 76
Endothelial cell growth factor	ECGF	75 000 17 000–25 000	Bovine hypothalamus	ND	+	41,42
Fibroblast growth factor	FGF	13 600	Bovine brain	+	+	26,27,35
Epidermal growth factor	EGF	6 000	Mouse salivary glands	+	+	35,75
Insulin	—	6 000	Plasma	+	−	7,26,55, 74
Insulin-like growth factor	IGF	7 500	Human plasma	+	ND	13
Vascular endothelial cell proliferation factor	VEPF	20 000	Platelet	ND	+	47
Endothelial-derived growth factor	ECDGF (EDGF)	10 000–30 000	Culture medium of aortic endothelial cell	+	−	21,22

Metabolic Activities of Arterial Cells

The biochemical and morphological changes observed in the arteries during the development of atherosclerosis relate not only to cell proliferation (as pointed out in the previous section), but also to lipid metabolism and the biosynthesis of the components of connective tissue, i.e. collagen, elastin and proteoglycans.

Glycosaminoglycans and Proteoglycans

For some years several groups have concentrated their attention on the study of proteoglycans. These macromolecules are, in fact, capable of influencing the structural integrity, elasticity and permeability of the arterial wall and of affecting lipid metabolism, cell proliferation, haemostasis and thrombosis [10,81,85].

The proteoglycans synthesized by smooth-muscle cells in culture exhibit physicochemical properties which are scarcely different from those found for the proteoglycans isolated from tissue. Studies have shown [7,34] that the proteoglycans isolated from the media of porcine aorta and the proteoglycans synthesized and secreted into the culture medium by smooth-muscle cells derived from the same tissue present:

1. a similar distribution of the different glycanic chains, although the proteoglycans secreted into the medium have a higher proportion of hyaluronic acid than those isolated from tissue;

2. a similar proportion of monomers and proteoglycans capable of forming complexes;

3. a similar mean buoyant density for a considerable proportion of the proteoglycans.

The principal properties of the proteoglycans isolated from the media of porcine aorta or synthesized by the smooth-muscle cells derived from

Table 5.2. Properties of the proteoglycans isolated from the media of porcine aorta or secreted into the medium by the smooth-muscle cells derived from the same tissue

Properties	Tissue	Culture medium
Proportion of proteoglycans present in complex form (%)	20	19
Hydrodynamic size of subunits after dissociation of complexes (K_{AV} on Sepharose CL-2B column)	0.70	0.68
Hydrodynamic size of monomers		
1. Proportion of proteoglycans excluded on Sepharose CL-6B column (%)	0	4
2. K_{AV} of proteoglycans retarded on Sepharose CL-6B column	0.40	0.25
Buoyant density	Two populations of different buoyant density: 1. 1.42 g/ml 2. Superior to 1.60 g/ml	A single population of buoyant density superior to 1.60 g/ml
Glycosaminoglycan composition (%):		
1. Hyaluronic acid	5	23
2. Heparan sulphate	21	10
3. Chondroitin sulphate–dermatan sulphate	74	67

the same tissue are summarized in Table 5.2. It should be noted that the proteoglycans secreted into the medium in the form of monomers are characterized as being slightly larger than the monomers isolated from tissue since they possess a K_{av} of 0.25 and 0.40, respectively, after chromatography under associative conditions on a column of Sepharose CL–6B. This size difference may be explained by the fact that the proteoglycans which remain in the extracellular stroma for a relatively long time may be subjected to more extensive degradation than those in the culture medium. This hypothesis has been confirmed in the case of the proteoglycans of cartilage. Buckwalter and Rosenberg [8] have shown that the region of the protein chain rich in chondroitin sulphate shows variations in size which may result from proteolytic attack at this end of the protein chain. These variations in the length of the protein chain have not been observed in the proteoglycans synthesized by chondrocytes in culture which are isolated from either normal tissue [46] or a chondrosarcoma [18].

The smooth-muscle cells of porcine aorta maintain their phenotypic character with respect to the synthesis of proteoglycans up to the seventh or eighth passage (the cells are plated at a density of 12 000 cells/cm² and trypsinized every week). In contrast, the proteoglycans synthesized

from the 10th passage onwards showed structural differences [7]. As the cells "age" in vitro, so the extent of aggregation of the proteoglycans secreted into the medium increased: at the 10th passage, 39% of the proteoglycans were present in the form of complexes whereas at the third passage the proportion accounted for by complexes was about 19%. The size of the subunits which make up the proteoglycan complexes was similar, irrespective of the number of passages, but the size of 90% of the monomers decreased with "ageing" in vitro. This diminution in size of the monomers can be explained, in part, by a decrease in the size of their glycanic chains.

With regard to the variation in the composition of the glycanic chains [16], an increase of 55% in the proportions of hyaluronic acid and dermatan sulphate was observed when the smooth-muscle cells of porcine media were subjected to 12 passages: at the second passage, hyaluronic acid and dermatan sulphate accounted for 11% and 22%, respectively, of the total glycosaminoglycans, whereas at the 12th passage the proportions of hyaluronic acid and dermatan sulphate were 17% and 34%, respectively.

Certain physicochemical properties of the proteoglycans appear to develop during "ageing" of the arterial wall in vivo in a manner which is paralleled in the "ageing" of smooth-muscle cells in vitro; in particular, an increase in dermatan sulphate is observed in both situations [16, 38,39,86]. However, studies concerned with the influence of age on the structural properties of the proteoglycans of arterial tissue are very few. It is thus very difficult to know whether the structural variations observed in the smooth-muscle cells in culture are only the result of a dedifferentiation of the cells in vitro, or whether they correspond, in part, to changes which are characteristic of ageing in vivo. Studies carried out on cartilage [20,50,51,73] or on chondrocytes in culture [51] have shown that some structural properties change in the same manner whether the ageing be in vivo or in vitro. In particular, a decrease in the size of the monomers has been documented in both types of study. This could be attributed, in part, to a shortening of the chondroitin sulphate chains. Furthermore, Sweet et al. [73] have demonstrated a decrease in the number of chondroitin sulphate chains bound to the protein chain. It is thus possible, by analogy with the observations made on cartilage, that the diminution in the size of the monomers and of the carbohydrate chains of the proteoglycans is a characteristic of "ageing" in the arterial wall.

Fibrous Proteins: Collagen and Elastin

The arterial cells in culture maintain their capacity to synthesize fibrous proteins just as has been described for the proteoglycans. Arterial cells synthesize the soluble precursor of collagen, procollagen, and the major types of collagen identified in tissues, collagen types I and III [9,40,64]. The soluble precursor of elastin (tropoelastin) and the amino acids derived from lysine (desmosine and isodesmosine) which are unique to elastin in which they form cross-links, have also been identified in the culture medium [1,44,69].

The conditions of culture appear to play a fundamental role in the regulation of the synthesis of fibrous proteins. It has been shown that the addition of ascorbate to the culture medium of aortic smooth-muscle cells stimulates the synthesis of collagen fibres [1,15,69], whereas it inhibits [1,68] or does not change [66] the accumulation of insoluble elastin.

During the development of atheromatous plaques, an accumulation of collagen has been observed which, in some cases, seemed to be the result of increased proline hydroxylase activity

[65]. Variations in collagen metabolism have also been observed with smooth-muscle cells isolated from the normal aorta of the rabbit were compared with those obtained from atherosclerotic aorta [54]: the synthesis of total collagen by the cells originating in the atherosclerotic aorta was found to be greater than that found with normal cells. This stimulation of synthesis affected collagen types I and III to the same extent, whereas observations in vivo seemed to point to a change in the proportions of collagens type I and III in the atheromatous plaque: in the normal human aorta the quantitative ratio for collagen type I/collagen type III is about 3/7 whereas in atheromatous lesions the ratio is reversed and is found to be as high as 6.5/3.5 [43].

In most individuals, the elastic fibres are progressively degraded with age and this process of degradation seems to be considerably accelerated in diabetic and atherosclerotic subjects [58]. In contrast with the other macromolecules of connective tissue, the elastin of healthy tissue is characterized by a very low rate of turnover [63].

Experiments carried out recently by McMahon et al. were concerned with the synthesis of elastin by the smooth-muscle cells of rat aorta as a function of the age of the animal donor and the number of subcultures [44]. The insoluble elastin isolated from cultures of cells from 2–3-day-old or 6-week-old rats represents 20%–50% of the total protein after 21 d of culture (first passage), whereas in cultures derived from animals 13 months old and older, elastin is present only in trace amounts or may even be undetectable. Analytical data for the insoluble elastin synthesized by cells isolated from the aorta of rats aged 2–3 d or 6 weeks showed great similarity to those for the insoluble elastin isolated from the arterial tissue of young rats, i.e. in amino acid composition and proportions of desmosine and isodesmosine. However, if the cells were isolated from the oldest rats and maintained in culture for longer times (49 d) or were subjected to a second passage after trypsinization, only a small amount of elastin with a composition close to that of insoluble elastin could be isolated.

Low-Density Lipoproteins

Endothelial and smooth-muscle cells in culture are capable of binding specifically to low-density lipoproteins (LDL) and degrading them [14, 57,82]. In addition, endothelial cells are characterized by possessing specific receptors for modified LDL such as the acetyl-LDL [79], similar to those described on the surface of macrophages [70].

Working with smooth-muscle cells of human aorta taken from subjects of from 5 to 71 years of age, Bierman et al. [4] have shown that the binding and internalization of the LDL do not vary significantly with the age of the donor whereas their degradation is markedly diminished in older donors. The authors of this report compared their results with observations made in vivo which were concerned with the progressive accumulation with age of LDL and cholesterol esters in arterial smooth-muscle cells.

The studies carried out by Bierman and Schwartz, in which they used endothelial cells of bovine aorta derived from the same clone [5], have shown that the activity of the receptors for native and acetyl-LDL changes differently with the "ageing" of the cells in vitro: the binding and degradation of the native LDL increased in relation to the amount of intracellular protein as the cells progressed from the 16th to the 33rd generation, whereas the binding and degradation of the acetylated LDL did not vary significantly. If the increase in the amount of the intracellular protein observed by these investigators is taken into account (the amount of intracellular protein amounted to $342\,\mu g/10^6$ cells after 16 divisions and to $572\,\mu g/10^6$ cells after 33 divisions), the binding and degradation of the native LDL were increased 2.7- and 3.9-fold as the endothelial cells progressed from the 16th to the 33rd generation.

In the case of the endothelial cells of bovine aorta, the activity of the receptors for acetylated LDL was found to be distinctly higher than that of the receptors for native LDL [5,70], irrespective of the extent of "ageing" of the endothelial cells in vitro [5]. In the case of young cells (16 generations), the binding of the acetylated LDL was almost five times higher than that in the native LDL. When the cells had "aged" in vitro (33 generations), the difference between the binding of the two types of LDL to endothelial cells was less, although the binding of the acetylated LDL was still more than three times higher than that of the native LDL. Similarly, the degradation of the acetylated LDL was always greater than that of the native LDL although the difference in the degradation of the two types of LDL diminished with the number of passages: the degradation of the acetylated LDL was 24 times higher than that of the native LDL in young cells and 10 times higher in older cells [5].

Thus, the activity leading to the degradation of the LDL by arterial endothelial cells seems to increase with "ageing" in vitro.

Conclusion

The development of atherosclerosis in man advances exponentially with age. This relationship seems to be the result of considerable metabolic changes in arterial cells which originate from cells which will have undergone several cell divisions.

As we have related in this chapter, arterial cells in culture, when placed in a favourable environment, are capable of differentiating and maintaining most of their metabolic properties. Changes in these properties with increasing number of cell divisions show analogies with ageing in vivo. Thus, this model system provides a very interesting tool for observing the ageing of cells when they have been stimulated to divide in an accelerated fashion or for studying the cellular and molecular mechanisms which play a role in the regulation of the growth or metabolism of arterial cells. This type of study is inaccessible in vivo. It was in this way that the fundamental regulatory role which endothelial cells may play in the growth of smooth-muscle cells was recently demonstrated.

The study of arterial cells in culture can contribute to a better understanding of the changes which occur during ageing of the arterial wall and its response to immunological, chemical and mechanical injury.

References

1. Barone LM, Faris B, Chipman SD et al. (1985) Alteration of the extracellular matrix of smooth muscle cells by ascorbate treatment. Biochim Biophys Acta 840:245–254
2. Benditt EP, Benditt JM (1973) Evidence for a monoclonal origin of human atherosclerotic plaques. Proc Natl Acad Sci USA 70:1753–1756
3. Bierman EL (1978) The effect of donor age on the in vitro life span of cultured human arterial smooth muscle cells. In Vitro 14:951–955
4. Bierman EL, Albers JJ, Chait A (1979) Effect of donor age on the binding and degradation of low density lipoproteins by cultured human arterial smooth muscle cells. J Gerontol 34:483–488
5. Bierman EL, Schwartz SM (1984) Effect of clonal sen-

escence on low density lipoprotein-receptor activity of bovine arterial endothelial cells. In Vitro 20:809–814

6. Bowen-Pope DF, Ross R (1982) Platelet-derived growth factor. II. Specific binding to cultured cells. J Biol Chem 257:5161–5171

7. Breton M (1985) Protéoglycanes synthétisés par les cellules musculaires lisses d'aorte: étude sur le tissu et en culture cellulaire, effet de l'insuline. Thèse de Doctorat d'Etat ès Sciences Naturelles, Université Pierre et Marie Curie, Paris

8. Buckwalter JA, Rosenberg LC (1982) Electron microscopic studies of cartilage proteoglycans. Direct evidence for the variable length of the chondroitin sulfate-rich region of proteoglycan subunit core protein. J Biol Chem 257:9830–9839

9. Burke JM, Balian G, Ross R et al. (1977) Synthesis of types I and III procollagen and collagen by monkey aortic smooth muscle cells in vitro. Biochemistry 16:3243–3249

10. Castellot JJ, Addonizio ML, Rosenberg R et al. (1981) Cultured endothelial cells produce a heparinlike inhibitor of smooth muscle cell growth. J Cell Biol 90:372–379

11. Castellot JJ Jr, Favreau LV, Karnovsky MJ et al. (1982) Inhibition of vascular smooth muscle cell growth by endothelial cell-derived herparin. Possible role of a platelet endoglycosidase. J Biol Chem 257:11256–11268

12. Chamley JH, Campbell GT, McConnel JD et al. (1977) Comparison of vascular smooth muscle cells from adult human monkey and rabbit in primary culture and in subculture. Cell Tissue Res 177:503–522

13. Clemmons DR (1985) Exposure to platelet derived growth factor modulates the porcine aortic smooth muscle cell response to somatomedin C. Endocrinology 117:83

14. Coetzee GA, Stein O, Stein Y (1979) Uptake and degradation of low density lipoproteins (LDL) by confluent, contact-inhibited bovine and human endothelial cells exposed to physiological concentrations of LDL. Atherosclerosis 33:425–431

15. DeClerk YA, Jones PA (1980) The effect of ascorbic acid on the nature and production of collagen and elastin by rat smooth muscle cells. Biochem J 186:217–225

16. Deudon E (1978) Hétérogénéité métabolique des protéoglycanes de la paroi artérielle (étude sur le tissu entier et en culture de cellules). Thèse De Doctorat d'Etat ès Sciences Naturelles, Université Pierre et Marie Curie, Paris

17. Deuel TF, Huang JS, Profitt RT et al. (1981) Human platelet-derived growth factor: purification and resolution into two active protein fractions. J Biol Chem 256:8896–8899

18. Fellini SA, Kimura JH, Hascall VG (1981) Polydispersity of proteoglycans synthesized by chondrocytes from the Swarm rat chondrosarcoma. J Biol Chem 256:7883–7889

19. Freshney RI, Frame MC (1983) Culture of endothelial cells for the study of angiogenesis in vitro. In: Thilo-Korner DGS, Freshney RI (eds) First international endothelial cell symposium of the European Tissue Culture Society, Paris 1982. Karger, Basel, pp 13–28

20. Garg HB, Swann DA (1981) Age-related changes in the chemical composition of bovine articular cartilage. Biochem J 193:459–468

21. Gajdusek C, Di Corleto P, Ross R et al. (1980) An endothelial cell derived growth factor. J Cell Biol 85:467–472

22. Gajdusek CM, Schwartz SM (1982) Ability of endothelial cells to condition culture medium. J Cell Physiol 110:35–42

23. Gajdusek CM, Schwartz SM (1983) Technique for cloning bovine aortic endothelial cells. In Vitro 19:394–402

24. Gee AP, Minta JO (1984) Migration of porcine endothelial and smooth muscle cells in response to platelet-associated factors. Am J Hematol 17:29–38

25. Gimbrone MA, Cotran RS, Folkman S (1974) Human vascular endothelial cells in culture. Growth and DNA synthesis. J Cell Biol 60:673–684

26. Gospodarowicz D, Moran J, Braun D et al. (1976) Clonal growth of bovine vascular endothelial cells: fibroblast growth factor as a survival agent. Proc Natl Acad Sci USA 73:4120–4124

27. Gospodarowicz D, Bialecki H, Greenburg G (1978) Purification of the fibroblast growth factor activity from bovine brain. J Biol Chem 253:3736–3743

28. Gospodarowicz D, Ill CR (1980) The extracellular matrix and the control of proliferation of vascular endothelial cells. J Clin Invest 65:1351–1364

29. Gospodarowicz D, Vlodavsky I, Savion N (1980) The extracellular matrix and the control of proliferation of vascular endothelial and vascular smooth muscle cells. J Supramol Struct 13:339–400

30. Grotendorst GR, Seppä HEJ, Kleinman HK et al. (1981) Attachment of smooth muscle cells to collagen and their migration towards platelet derived growth factor. Proc Natl Acad Sci USA 78:3669–3672

31. Haudenschild CC, Grunwald J (1985) Proliferative heterogeneity of vascular smooth muscle cells and its alteration by injury. Exp Cell Res 157:364–370

32. Hayflick L (1965) The limited in vitro lifetime of human diploid cell strains. Ex Cell Res 37:614–636

33. Heldin CH, Westermark B, Wasteson A (1981) Specific receptors for platelet-derived growth factor on cells derived from connective tissue and glia. Proc Natl Acad Sci USA 78:3664–3668

34. Horn MC, Breton M, Deudon E et al. (1983) The structural characterization of proteoglycans of cultured aortic smooth muscle cells and arterial wall. Biochim Biophys Acta 755:95–105

35. Hoshi H, McKeehan WL (1984) Brain and liver cell-derived factors are required for growth of human endothelial cells in serum-free culture. Proc Natl Acad Sci USA 81:6413–6417

36. Jaffe EA, Nachman R, Becker C et al. (1973) Culture of human endothelial cells: identification by morphologic and immunologic criteria. J Clin Invest 52:2745–2756

37. Johnson LK, Longenecker JP (1982) Senescence of aortic endothelial cells in vitro: influence of culture conditions and preliminary characterization of the senescent phenotype. Mech Ageing Dev 18:1–18

38. Kaplan D, Meyer K (1960) Mucopolysaccharides of aorta at various ages. Proc Soc Exp Biol Med 105:78–81

39. Kumar V, Berenson GS, Ruiz H et al. (1967) Acid mucopolysaccharides of human aorta. Part 1. Variations with maturation. J Atherosclerosis Res 7:573–581

40. Layman DL, Titus J (1975) Synthesis of type I collagen by human smooth muscle cells in vitro. Lab Invest 33:103–107

41. Maciag T, Cerundolo J, Ilsley S et al. (1979) An endothelial cell growth factor from bovine hypothalamus: identification and partial characterization. Proc Natl Acad Sci USA 76:5674–5678

42. Maciag T, Hoover GA, Weinstein R (1982) High and low molecular weight forms of endothelial cell growth factor. J Biol Chem 257:5333–5336

43. McCullagh KG, Balian G (1975) Collagen characterization and cell transformation in human atherosclerosis. Nature 258:73–75

44. McMahon MP, Faris B, Wolfe BL et al. (1985) Aging effects on the elastin composition in the extracellular matrix of cultured rat aortic smooth muscle cells. In Vitro 21:674–680

45. Martin GM, Sprague CA (1973) Symposium on *in vitro* studies related to atherogenesis, life histories of hyperplastoid cell lines from aorta and skin. Exp Mol Pathol 18:125–141

46. Mitchell D, Hardingham T (1981) The effects of cycloheximide on the biosynthesis and secretion of proteoglycans by chondrocytes in culture. Biochem J 196:521–529

47. Miyazono K, Okabe T, Urabe A et al. (1985) A platelet factor that stimulates the proliferation of vascular endothelial cells. Biochem Biophys Res Commun 126:83–88

48. Mueller EM, Rosen EM, Levine EM (1980) Cellular senescence in a cloned strain of bovine fetal aortic endothelial cells. Science 207:889–891

49. Nilsson, J, Ksiazek T, Heldin CH et al. (1983) Demonstration of stimulatory effects of platelet-derived growth factor on cultivated rat arterial smooth muscle cells. Differences between cells from young and adult animals. Exp Cell Res 145:231–238

50. Oohira A, Nogami H (1980) Age-related changes in physical and chemical properties of proteoglycans synthesized by costal and matrix-induced cartilages in the rat. J Biol Chem 255:1346–1350

51. Pacifici M, Fellini SA, Holtzer H et al. (1981) Changes in the sulfated proteoglycans synthesized by "aging" chondrocytes. I. Dispersed cultured chondrocytes and *in vivo* cartilages. J Biol Chem 256:1029–1037

52. Pascal M, Sepulchre C, Chazan JB et al. (1983) Evidence for the inhibition of platelet-derived growth factor induced rat smooth muscle cells DNA synthesis by fenofibric acid at the C_0/G_1 cell cycle level. Life Sci 33:925–934

53. Pietilä K, Nikkari T (1980) Enhanced growth of smooth muscle cells from atherosclerotic rabbit aortas in culture. Atherosclerosis 36:241–248

54. Pietilä K, Nikkari T (1980) Enhanced synthesis of collagen and total protein by smooth muscle cells from atherosclerotic rabbit aortas in culture. Atherosclerosis 37:11–19

55. Pfeifle B, Ditschuneit H (1981) Effect of insulin on growth of cultured human aterial smooth muscle cells. Diabetologia 20:155–158

56. Pledger WJ, Stiles CD, Antoniades HN et al. (1977) Induction of DNA synthesis in Balb/c 3T3 cells by serum components: reevaluation of the commitment process. Proc Natl Acad Sci USA 74:4481–4485

57. Reckless JPD, Weinstein DB, Steinberg D (1978) Lipoprotein and cholesterol metabolism in rabbit arterial endothelial cells in culture. Biochim Biophys Acta 529:475–487

58. Robert L, Robert B, Robert AM (1970) Molecular biology of elastin as related to aging and atherosclerosis. Exp Gerontol 5:339–356

59. Ross R (1971) The smooth muscle cell. II. Growth of smooth muscle in culture and formation of elastic fibers. J Cell Biol 50:172–186

60. Ross R, Glomset JA (1973) Atherosclerosis and arterial smooth muscle cells. Science 180:1332–1339

61. Ross R, Glomset JA (1976) The pathogenesis of atherosclerosis. N Engl J Med 295:369–377

62. Ross R, Glomset J, Kariya B et al. (1974) A platelet-dependent serum factor that stimulates the proliferation of arterial smooth muscle cells *in vitro*. Proc Natl Acad Sci USA 71:1207–1210

63. Rucker RB, Tinker D (1977) Structure and metabolism of arterial elastin. Int Rev Exp Pathol 17:1–47

64. Sage H, Crouch E, Bornstein P (1979) Collagen synthesis by bovine aortic endothelial cells in culture. Biochemistry 18:5433–5442

65. St Clair RW (1976) Metabolism of the arterial wall and atherosclerosis. In: Paoletti R, Gotto AM (eds) Atherosclerosis reviews, vol I. Raven Press, New York, pp 61–117

66. Schor AM, Schor SL, Allent D (1984) The synthesis of subendothelial matrix by bovine aortic endothelial cells in culture. Tissue Cell 16:677–692

67. Schwartz S (1978) Selection and characterization of bovine aortic endothelial cells. In Vitro 14:966–980

68. Schwartz E, Bienkowski RS, Coltoff-Schiller B et al. (1982) Changes in the components of extracellular matrix and in growth properties of cultured aortic smooth muscle cells upon ascorbate feeding. J Cell Biol 92:462–470

69. Snider R, Faris B, Verbitski V et al. (1981) Elastin biosynthesis and cross-link formation in rabbit aortic smooth muscle cell cultures. Biochemistry 20:2614–2618

70. Stein O, Stein Y (1980) Bovine aortic endothelial cells display macrophage-like properties towards ^{125}I-acetylated low-density lipoprotein. Biochim Biophys Acta 620:631–635

71. Stemerman MB, Weinstein R, Rowe JW et al. (1982) Vascular smooth muscle cell growth kinetics *in vivo* in aged rats. Proc Natl Acad Sci USA 79:3863–3866

72. Stiles CD, Capone GT, Scher CD et al. (1979) Dual control of cell growth by somatomedins and platelet-derived growth factor. Proc Natl Acad Sci USA 76:1279–1283

73. Sweet MBE, Thonar EJ-MA, Marsh J (1979) Age related changes in proteoglycan structure. Proteoglycans of articular cartilage. Arch Biochem Biophys 198:439–448

74. Taggart H, Stout RW (1980) Control of DNA synthesis in cultured vascular endothelial and smooth muscle cells, response to serum, platelet-deficient serum, lipid-free serum, insulin and oestrogens. Atherosclerosis 37:549–557

75. Taylor J, Mitchell WM, Cohen S (1972) Epidermal growth factor physical and chemical properties. J Biol Chem 247:5928–5934

76. Thorgeirsson G, Robertson Jr AL (1978) Platelet factors and the human vascular wall: variation in growth response between endothelial and medial smooth muscle cells. Atherosclerosis 30:67–78

77. Thyberg J, Palmberg L, Wilsson J et al. (1983) Phenotype modulation in primary cultures of arterial smooth muscle cells. On the role of platelet-derived growth factor. Differentiation 25:156–167

78. Tseng SCG, Savion N, Stern R et al. (1982) Fibroblast growth factor modulates synthesis of collagen in cultured vascular endothelial cells. Eur J Biochem 122:355–360

79. Van Hinsbergh VWM, Havekes L, Emeis JJ et al. (1983) Metabolism of low-density lipoprotein (LDL) and acetylated LDL by subconfluent and confluent endothelial cells from human umbilical cord arteries and veins. Atherosclerosis 3:547–559

80. Van Obberghen-Schilling E, Perez-Rodriguez R, Poyssegur J (1982) Hirudin, a probe to analyze the growth-promoting activity of thrombin in fibroblasts: reevalu-

ation of the temporal action of competence factors. Biochem Biophys Res Commun 106:79–86

81. Vijayagopal P, Radhakrishnamurthy B, Srinivasan SR et al. (1980) Studies of biological properties of proteoglycans from bovine aorta. Lab Invest 42:190–196

82. Vlodavsky I, Fielding PE, Fielding CJ et al. (1978) Role of contact inhibition in the regulation of receptor-mediated uptake of low-density lipoprotein in cultured vascular endothelial cells. Proc Natl Acad Sci USA 75:356–360

83. Vogel A, Raines E, Kariya B et al. (1978) Coordinate control of 3T3 cell proliferation by platelet-derived growth factor and plasma composants. Proc Natl Acad Sci USA 75:2810–2814

84. Weinstein R, Stemerman MB, Maciag T (1981) Hormonal requirements for growth of arterial smooth muscle cells *in vitro*: an endocrine approach to atherosclerosis. Science 212:818–819

85. Wight TN (1980) Vessel proteoglycans and thrombogenesis. Prog Hemost Thromb 5:1–39

86. Yamamoto H, Kanaide H, Nakamura M (1983) Metabolism of glycosaminoglycans of cultured rat aortic smooth muscle cells altered during subculture. Br J Exp Pathol 64:156–165

Chapter 6

Structural Modifications of the Arterial Wall in Hypertension

J.-B. Michel, J. L. Salzmann and M. Safar

Introduction

Physiological and pathological observations of the vascular system in clinical and experimental models of hypertension, normotension and cardiac insufficiency, suggest that the level of blood pressure is related directly to both primary functions of the cardiovascular system: the peripheral blood flow and filtration pressure in capillaries necessary for oxygen delivery to tissues [8,15]. For example, in spontaneously hypertensive rats, these two variables are probably in the normal range [22] which is an indication of the cardiovascular system's ability to adopt its functional and structural parameters for survival. In human essential hypertension, cardiac output is in the normal range [3]. In experimental models, as in human beings, the homeostatic mechanisms of blood flow regulation are conserved. The adaptation of the cardiovascular system in hypertension probably is related to the minimal energy level necessary for the maintenance of a cardiovascular function compatible with survival.

Nevertheless, the functional adaptive process is associated with structural modifications of the cardiovascular system which have their own consequences. In hypertension, it is not only the heart which undergoes structural changes, but also the vascular system, the arterial wall and the venous system. The relationship between high blood pressure and structure of the vessel walls is probably the predominant factor in the elevation of blood pressure; the other is that the increase in stress and strain leads to structural modification of the arterial walls which participate in the maintenance of high blood pressure when the initial functional determinant of hypertension has disappeared [17].

Structural Modifications of the Arterial System as a Consequence of Hypertension

For the arterial wall, high blood pressure represents an increase in stress, i.e. an increase in force which strains the living tissue within the wall. Laplace's law states that the increase in stress and strain is proportional to the pressure and to the diameter of the arterial lumen and inversely proportional to the thickness of the arterial wall. Thus, the increase in stress is quickly associated with thickening of the arterial wall, an adaptive process which attempts to normalize the stress at the level of each contractile unit within the wall. The increase in thickness of the arterial wall affects all layers [10][1,32]. This increase in wall thickness is consistent in all the experimental models of hypertension and it is positively correlated to the level of blood pressure. The increase in thickness of the intima is predominantly due to the effect of hyperplasia of subendothelial cells; the number of endothelial cells does not change but there is a proliferation of subendothelial cells between endothelium and internal elastic lamina. This hyperplasia is associated with an increase in intimal content of extracellular matrix proteins, particularly collagens. The subendothelial cells are predominantly smooth-muscle cells and monocytes [4].

While endothelial cells do not proliferate, they probably increase their rates of degradation and production. They are also slightly hypertrophied, which modifies their shape so that they become more cuboidal and more prominent in the arterial lumen. These morphological modifications of the endothelial layer are associated with a functional increase in permeability. These changes suggest an increase in regeneration turnover of endothelium in hypertension (see Chap. 11).

The increase in arterial wall thickness seen in hypertension is due predominantly to the change in medial thickness. This increase in media thickness is related to an increase in both smooth-muscle mass and extracellular collagen content. Elastin content probably increases slightly in absolute value [1] but its relative value does not change [31,17]. The mechanism of increase in smooth-muscle mass probably depends on the type of arterial wall studied. In the wall of large elastic arteries such as the aorta, the increase is due predominantly to hypertrophy of smooth-muscle cells: each smooth-muscle cell increases in mass and length; and there is an increase in contractile and cytoskeletal protein biosynthesis and in biosynthesis and secretion of collagens into the extracellular space [25]. This functional hyperactivity is associated with an increase in the genomic mass, permitting the polyvalent increase in protein biosynthesis: the nuclei of smooth-muscle cells become diploid or polyploid [27].

The increase in relative and absolute content of extracellular matrix proteins in the arterial wall in hypertension depends partly on the model and the assay used. In the experimental renovascular model of hypertension in rats (one clip, two kidneys), the relative and absolute values of collagen content consistently increased without a marked change in the relative amount of elastin in the aortic wall. A similar observation has been made in genetically spontaneously hypertensive rats of the Okamoto strain. In contrast, in the Doca-salt model (an experimental model of hypertension induced in rats using deoxycorticosterone acetate, salt excess and unilateral nephrectomy), Berry and co-workers [1] reported a significant increase in the elastin content within the aortic wall. In the same experimental model Ooshima demonstrated an increase in the biosynthesis of collagens by smooth-muscle cells

in the aortic wall. It seems that the increase in biosynthesis can be modified during the hypertensive process by factors more complex than the direct increase in stress-strain relationship at a cellular level.

In small-calibre resistant arteries, hypertension is predominantly associated with a hyperplasia of smooth-muscle cells [12,13,14] comparable with the proliferation of cells observed in the subendothelial layer of the intima in large arteries. This hyperplasia is also associated with an increase in collagen biosynthesis [26] and in arterioles which do not possess an elastic lamina, can lead to the successive stratification of myofibroblastic cells, producing the classical "onion skin" chracteristic of the hypertensive microangiopathy. The microscopic lesion can be defined as a hyperplastic arteriosclerosis because hyalinosis is markedly associated with hyperplastic lesions in arterioles.

Hyalinosis is characterized by the extracellular accumulation of materials which are slightly granular in appearance under electron microscopy, predominantly glycoproteins of plasma origin. The increase in wall thickness associated with the decrease in luminal area of arterioles acts to increase the peripheral resistance and to decrease the flow of blood to the tissues. The presence of fibrin within the wall may accompany hyalinosis. This phenomenon is observed in severe hypertension and particularly in malignant hypertension. The presence of fibrin in the arteriolar wall indicates a problem with permeability, which is quickly reversed following a decrease in blood pressure. The question which is not resolved, is the nature of the signal between the increase in strain and the increase in protein biosynthesis in smooth-muscle cells. It is probable that activation of second messengers, such as phosphinositols, not only activates the functional mechanism of actin–myosin interaction but also interferes with the expression of genes encoding for growth factors. Recently it has been shown that activation of the phosphoinositol pathway increased the expression levels of oncogenes encoding for growth factors [18]. For example, in the cardiovascular system, myocardial hypertrophy is associated with an increase in expression of c-*myc* oncogene [28]. Analogous phenomena could be the basis of the development of the smooth-muscle cell hypertrophy and hyperplasia associated with hypertension.

Hypertension is also associated with a significant reduction of capillary density within the heart. This phenomenon is consistent in all the studied models of hypertension and may lead to reduced oxygen delivery to the myocardium. However, this decrease in capillary density, predominantly in the subendothelial layer of the left ventricular wall, is more related to myocardial hypertrophy, which leads to a relative decrease in the number of capillaries by mass unit, rather than to a true absolute decrease in the number of capillaries within the heart [21].

In summary, depending on its type, the arterial wall submitted to an increase in stress develops a hyperactivity of smooth-muscle cells which leads to hypertrophy in large arteries, to hyperplasia in resistant vessels and to hyalinosis in arterioles.

Structural Modifications of the Arterial Wall as a Result of High Blood Pressure

Whatever the mechanism initiating the increase the stress, hypertension is associated with an increase in thickness of arterial and arteriolar walls which leads to an increase in peripheral resistance and to a decrease in arterial compliance.

Moreover, hypertrophy and hyperplasia of smooth-muscle cells amplify the functional contractile response to vasoactive agents [23]. For a similar nervous or hormonal signal leading to vasoconstriction in a normotensive model, more effectors respond in hypertension, inducing generation of more force. In vivo, for example, an increased response to catecholamines has been demonstrated in hypertensive patients when compared with normotensive controls. This phenomenon can be related to structural changes of the arterial wall rather than to a specific increase in functional response to catecholamines.

In isolated perfused vascular systems, where flow and pressure can be controlled and adjusted, and where hormonal and nervous afferents have been suppressed, the increase in thickness of vascular wall and in collagen content are directly responsible for the increase in resistance of the system [24].

In isolated large arteries, such as carotids or aortas, of hypertensive animals, the compliance of the wall (the ratio between the variation in volume and the variation in pressure) is diminished for a determined pressure [1,16]. This decrease in compliance of vessel walls can be related to the increase in smooth-muscle mass in the normal range of pressures and to the increase in collagen content for high perfusion pressures. The role played by elastin in this phenomenon remains unclear (but see Chap. 4 for findings in young animals).

It seems apparent that at these high blood-pressure levels increased response to pressor agents, increased peripheral resistance, decreased arterial wall compliance and the structural changes of the arterial wall play a predominant role in the maintenance of hypertension independent of the initial mechanism.

Complications: Lacunes and Atherosclerosis

Hyperplasia of smooth-muscle cells and hyalinosis of arteriolar walls can be complicated by complete obliteration of the arteriolar lumen. These thromboses lead to infarctions which are generally small because of the low calibre of the thrombosed arteriole. In the brain, these small infarcts or lacunes can be asymptomatic or may give rise to symptoms which are related to the functional significance of the infarcted tissue. In the kidney, these lesions lead to localized nephron destruction which results in nephroangiosclerosis and the late appearance of renal insufficiency [20].

In large arteries, structural changes of the walls play an important role in the relationship between hypertension and development of atheroma. However, hypertension is only one predictive factor in the development of atheroma: others include high lipid intake, smoking and environmental stress. The biological events which relate atheroma to hypertension are not completely defined. Two mechanical factors seem to be important in this relationship: firstly the increased permeability of the endothelium to macromolecules; secondly the decreased filtration of the macromolecules through the thicker and collageneous intimal and medial layers of the arterial wall [29]. These two phenomena are related and lead to an increase in the accumulation of proteins in the intimal layer and to the formation of plaques.

Other phenomena may also play a role in atherogenesis associated with hypertension.

Damaged endothelium could activate platelet aggregation and release of growth factors. Activation of the endothelial layer could be related to the increase in blood velocity associated with sympathetic nervous system activation accompanying some hypertension.

Effect of Treatment

One of the questions posed by the development of structural changes in the arterial vessel wall concerns the reversibility of these phenomena by hypotensive treatments.

In the study by Wolinski on rat renovascular hypertension [32], aortic media hypertrophy was reversed by unclipping, but this result was obtained only in female rats, not in males. More recently, Berry and Greenwald have reported different results on this subject. In 1976, Berry [1] reported the reversal of aortic media hypertrophy in Doca-salt hypertension in rats, following the removal of the Doca pellet and prolonged normalization of blood pressure. More recently, the same authors [9] reported controversial results including the absence of regression of aortic media hypertrophy in spontaneously hypertensive rats of the Okamoto strain, but only one of the hypotensive drugs used was a diuretic, and the decrease in blood pressure was incomplete and of short duration. Carlier and Rorive [2] reported similar results. The absence of reversal of the aortic media mass in these studies could be due to the short duration of treatment and the absence of blood pressure normalization. Recently, the normalization of blood pressure, obtained within 1 month of treatment with converting enzyme inhibition in renovascular hypertensive rats [17], completely reversed the hypertrophy of smooth-muscle cells of the aortic wall but failed to reverse the increase in collagen content. However, Ooshima [25], in the Doca-salt hypertensive model in rats, has shown that the increase in collagen biosynthesis *was* inhibited by the anti-hypertensive treatment (reserpine). Thus, the persistent increase in collagen content within the aortic wall during treatment may be related to accumulation because of slow turnover of the protein, despite a decrease of biosynthesis. In a long-duration experiment using spontaneously hypertensive rats, 3 months of treatment with converting enzyme inhibitors decreased blood pressure and decreased collagen content within the aortic wall.

In summary, the decrease in blood pressure, and particularly its normalization over a long period of time, using drugs which are active for 24 hours, is associated with reversal of aortic media hypertrophy and a decrease in collagen biosynthesis by the smooth-muscle cells.

Similar events occur during treatment of hyperplasia of the arteriolar wall. The prolonged normalization of blood pressure is associated with a reversal of nephroangiosclerosis in renovascular hypertensive rats [20]. Similarly, mesenteric blood-pressure normalization maintained over a long period (23 weeks) is associated with reversal of the hyperplasia of smooth-muscle cells in resistance vessels [30].

Similarity Between the Ageing Process and Hypertension in the Arterial Wall

The ageing process leads to structural changes of the arterial wall which are similar to those observed in hypertension. Medial thickness increases with age [5], as does intimal thickness [11]. In the aortic wall, there is hypertrophy of the smooth-muscle cells, as in hypertension. The collagen content of the media increases with age as in hypertension, whereas the elastic tissue is degraded.

The data reported in Table 6.1 were obtained from the descending thoracic aorta of male Wistar rats after 1 year of ageing. They confirm

Table 6.1. Changes that occur in the descending thoracic aorta of male Wistar rats with age and in renovascular hypertension

	Male normotensive rats		Renovascular hypertension
Age	4 months ($n=12$)	16 months ($n=14$)	4 months ($n=12$)
Media thickness (μm)	97.5 ± 1.6	120 ± 15	122 ± 4
Collagen Density (%)	7.3 ± 13	15.8 ± 4.4	8.6 ± 3.6
Elastin density (%)	38.6 ± 13	13 ± 7	32 ± 8.3
Nuclear density (n/field)	38 ± 4	29 ± 4	35 ± 5

the results reported in the literature: the media thickness increases with age, at a rate of 20% per year; the collagen density increases twofold in 1 year, and the elastin density does not change with time. The nuclear density reflects the number of cells lost with age, and in relative values shows that the ageing process is more of a hypertrophic phenomenon than a hyperplastic phenomenon in the aortic media.

The effect of vasodilators observed in hypertension can also be shown during ageing in normotensive animals. Haudenschild [11] has shown that a hypotensive treatment (using two drugs, dihydralazine and a diuretic) decreased stress, not only in hypertensive animals but also in normotensive animals

In the normotensive ageing animals, hypotension can be associated with a decrease in intima and media thickness. Recently [19] in a long-term treatment using two types of vasodilators (dihydralazine and a converting enzyme inhibitor) in 1-year-old normotensive Wistar rats, we have shown that a decrease in wall stress is associated with a decrease in media thickness. However, the two different treatments interact with smooth-muscle function in different ways and thus probably do not have the same effect on the proteins of the extracellular matrix. Converting enzyme inhibitor, which decreases the activation of the phosphoinositol pathway, did not change the relative content of elastin, whereas dihydralazine, which probably directly or indirectly increases the cyclic GMP pathway, had a significant positive effect on the elastin content in the aortic wall. In summary, the ageing process seems to have a similar effect to hypertension on the structure of the wall in large arteries. These structural changes are time dependent and can be prevented by vasodilator treatment which significantly decreases the wall stress in the normotensive animals. Moreover, two drugs which do not have the same intracellular messenger pathway in smooth-muscle cells can have different effects on the biosynthesis of extracellular matrix proteins.

References

1. Berry, CL, Greenwald SE (1976) Effects of hypertension on the static, mechanical properties and chemical composition of the rat aorta. Cardiovasc Res 10:437–451
2. Carlier P, Rorive G (1985) Pathogenesis and reversibility of the aortic changes in experimental hypertension. Cardiovasc Pharmacol 7:S46–S51
3. Chau NPH, Safar ME, London GM, Weiss YA (1979) Essential hypertension: an approach to clinical data by the use of models. Hypertension 2:87–97
4. Chobanian AV, Prescott MF, Haudenschild C (1984) The effects of hypertension on the arterial wall. Exp Mol Pathol 41:153–169
5. Cliff WJ (1970) The aortic tunica media in aging rats. Exp Mol Pathol 13:172–189
6. Furanaya M (1982) Histometrical investigations of arteries in reference to arterial hypertension. Tokohu J Exp Med 75: 388–414
7. Folkow B (1982) Physiological aspect of primary hypertension. Physiol Rev 62:347–503
8. Guyton AC (1980) Circulatory physiology III: arterial pressure and hypertension. Saunders, Philadelphia, pp 293–306
9. Greenwald SE, Berry CL, Ramsey RE (1985) The static elastic properties and chemical composition of the rat aorta in spontaneously induced hypertension: the effect of an antihypertensive drug. Br J Exp Pathol 66:633–642
10. Haudenschild CC, Chobanian AV (1984) Blood pressure lowering diminishes age related changes in the rat aortic intima. Hypertension [Suppl I] 6:62–68
11. Haudenschild CC, Prescott MF, Chobanian AV (1980) Effects of hypertension and its reversal on aortic intima lesions of the rat. Hypertension 2:23–44
12. Lee RMKW (1985) Vascular changes at the prehypertensive phase in mesenteric arteries from spontaneously hypertensive rats. Blood Vessels 22:105–126
13. Lee RMKW, Forrest JB, Garfield RE, Daniel EE (1983) Ultrastructural changes of mesenteric arteries from spontaneously hypertensive rats: a morphologic study. Blood Vessels 20:72–91
14. Lee RMKW, Triggles CR (1986) Morphometric study of mesenteric arteries from genetically hypertensive Dahl strain rats. Blood Vessels 23:199–224
15. Levenson JA, Peronneau PP, Simon ACH, Safar ME (1981) Pulsed doppler: determination of diameter, blood flow velocity and volume flow of brachial artery in man. Cardiovasc Res 15:164–170
16. Levy BI, Benessiano J, Poitevin P, Lubin L, Safar ME (1985) Systemic arterial compliance in normotensive and hypertensive rats. J Cardiovasc Pharmacol [Supple II] 7:22–27
17. Levy BI, Michel JB, Salzmann JL et al. (1988) Effect of chronic inhibition of converting enzyme on mechanical and structural properties of arteries in rat renovascular hypertension. Circ Res (in press)
18. Marx JL (1987) Polyphosphonositide research undated. Science 235:274–276
19. Michel JB, Azizi M, Salzmann JL, Levy B, Menard J (1987) Effects of vasodilators on the structure of the aorta in normotensive aging rats. J Hypertension 5:5165–5168
20. Michel JB, Dussaule JC, Choudat L et al. (1985) Effects of antihypertensive treatment in one-clip two kidney hypertension in rats. Kidney Int 20:1011–1020
21. Michel JB, Salzmann JL, Ossondo NM, Bruneval L, Barres D, Camilleri JP (1986) Morphometric analysis of collagen network and plasma perfused capillary bed in the myocardium of rats during evolution of cardiac hypertrophy. Basic Res Cardiol 81:142–154
22. Mulvany MJ (1984) Pathophysiology of vascular smooth muscle in hypertension. J Hypertension [Suppl. III] 2:413–420
23. Mulvany MJ (1983) Do resistance vessel abnormalities contribute to the elevated blood pressure of spon-

taneously hypertensive rats? Blood Vessels 20:1–22

24. Mulvany MJ, Hansen PK, Palkjaer C (1978) Direct evidence that the greater contractility of resistance vessels in spontaneously hypertensive rats is associated with a narrow lumen, a thickened media and an increased number of smooth muscle cell layers. Circ Res 43:854–864

25. Ooshima A, Fuller GC, Cardinale G, Spector S, Udenfriend S (1974) Increased collagen synthesis in blood vessels of hypertensive rats and its reversal by antihypertensive agents. Proc Natl Acad Sci USA 74:3019–3023

26. Ooshima A, Fuller GC, Cardinale G, Spector S, Udenfriend S (1975) Collagen biosynthesis in blood vessels of brain and other tissues of the hypertensive rat. Science 190:898–900

27. Owens GK, Rabinovitch PS, Schwartz SM (1981) Smooth muscle cell hypertrophy versus hyperplasia in hypertension. Proc Natl Acad Sci USA 78:7759–7763

28. Starksen NF, Simpson PC, Bishopric N et al. (1986) Cardiac myocyte hypertrophy is associated with c-*myc* protooncogene expression. Proc Natl Acad Sci USA 83:8348–8350

29. Tedgui A, Chiron B, Curmi P, Juan L (1987) Effect of nicardipine and verapamil on in vitro albumin transport in rabbit thoracic aorta. Arteriosclerosis 7:80–87

30. Washaw DM, Root DT, Halpern W (1980) Effects of antihypertensive drug therapy on the morphology and mechanics of resistance arteries from spontaneously hypertensive rats. Blood Vessels 17:257–270

31. Wiener J, Loud AD, Giacomelli F, Anversa P (1977) Morphometric analysis of hypertension induced hypertrophy of rat thoracic aorta. Am J Pathol 88:619–634

32. Wolinski H (1970) Response of the rat aortic media to hypertension: morphological and chemical studies. Circ Res 26:507–522

33. Wolinski H (1971) Effects of hypertension and its reversal on the thoracic aorta of male and female rats. Circ Res 38: 622–637

Chapter 7

Adrenergic and Non-adrenergic Neural Control of the Arterial Wall

K. K. Dhital and G. Burnstock

Introduction

The availability of new and improved techniques in fluorescence histochemistry (particularly immunohistochemistry), electron microscopy, electrophysiology and pharmacology since the 1960s has led to a wealth of discoveries that have profoundly reshaped our understanding of the autonomic nervous system. There has also been a dramatic rise in the number of putative neurotransmitter substances, which show a marked overlap in their localization and function. These findings add new concepts and further complexity to autonomic neuroeffector mechanisms and demand a reappraisal of our definitions for subclassing these neurohumoral agents into neurotransmitters; neuromodulators and trophic factors.

This chapter reviews some of these recent discoveries in relation to the neurohumoral control of the vasculature, with special emphasis on non-classical neurotransmission, co-transmission and neuromodulation.

Structure of the Vascular Neuroeffector Junction

Unlike the classical arrangement of the neuroeffector junction in skeletal muscle and that in ganglia, the model of the autonomic vascular neuroeffector junction based on that proposed by Burnstock and Iwayama [77], is characterized by an extensive terminal branching of the autonomic nerve which forms a two-dimensional perivascular "autonomic ground plexus" confined to the adventitial–medial border (see Fig. 7.1) [58,224]. In this manner the whole muscle bundle and not merely a single smooth-muscle cell becomes the effector. A special feature of this extensive terminal apparatus is the presence of varicose regions which are free of Schwann-cell covering. The pearl-like varicosities (1–2 μm in diameter), of which there are a few hundred per millimetre in mammalian adrenergic terminals [78,183] are separated by narrow inter-varicose regions (0.1–0.3 μm).

Neurohumoral substances stored in vesicles or granules within the varicosities are released en passant during the conduction of nerve impulses and then interact with receptors on the post-

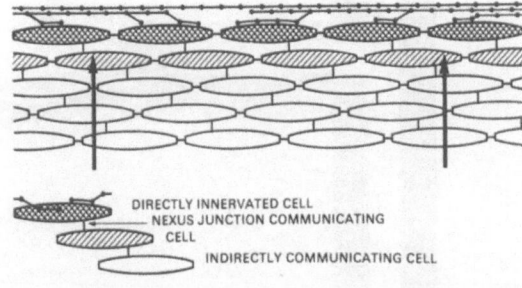

DIRECTLY INNERVATED CELL
NEXUS JUNCTION COMMUNICATING CELL
INDIRECTLY COMMUNICATING CELL

Fig. 7.1. Model of autonomic vascular neuroeffector junction where smooth-muscle cells are influenced by nerve-released (●-●-●) and circulating (↑) neurohumoral substances. Adapted from Burnstock [60] with kind permission of the authors and the publisher.

junctional membrane of the smooth-muscle cells of the outer coat of the tunica media. The eventual change in membrane potential leads to the electrotonic spread of activity via specialized regions of low-resistance pathways termed "nexuses" or "gap-junctions" which interconnect and couple smooth-muscle cells [80].

Although thickenings are sometimes shown in the pre-junctional varicosity membrane, post-junctional specializations as seen in skeletal neuromuscular junctions [256] and ganglionic synapses [391] are never seen in vascular neuro-effector junctions. The vascular nerve–muscle junctional cleft is a variable feature, ranging in width from 50–100 nm in some densely innervated small muscular arteries to as much as 2000 nm in certain large elastic arteries [60,80]. In contrast to arteries, it appears that the cleft width is independent of vessel diameter in veins (375). In densely innervated tissues like the vas deferens and the iris this distance can be as little as 20 nm.

Although the structure of the autonomic ground plexus and the wide variation in cleft width makes it difficult to calculate the percentage of active varicosities during a single nerve impulse, the asymmetrical geometry suggests that the vascular neuroeffector junction is open to both pre- and post-junctional modulatory influences by both nerve-mediated, locally produced and circulating neurohumoral substances.

Pattern of Perivascular Innervation

Classical silver impregnation and methylene blue staining techniques [1,205] were limited in their specificity and led only to basic descriptions of the presence of perivascular nerve fibres. It was not until the introduction of fluorescence histochemistry [163] that at least nerves containing biogenic amines could be recognized. More recent modifications of fluorescence histochemical techniques, improvements in electron microscopy and the use of immunohistochemical techniques have allowed further and more specific differentiation of perivascular nerve types. Our detailed knowledge of perivascular neuronal geometry, largely based on sympathetic adrenergic nerves, allows us to make certain generalizations about the pattern and distribution of vascular innervation. Such a generalized model is illustrated in Fig. 7.2.

The density of innervation is inversely related to the size of the vessel such that it is sparse in large elastic arteries and increases to reach a peak at the level of small arteries and large arterioles. Thickenings at sites of arterial branching known as "intimal cushions" are also densely innervated as are some pre capillary sphincters. Indeed the degree of innervation appears to be dependent on the amount of resistance that a particular type of vessel presents in a given vascular bed. It is therefore appropriate that the small pre

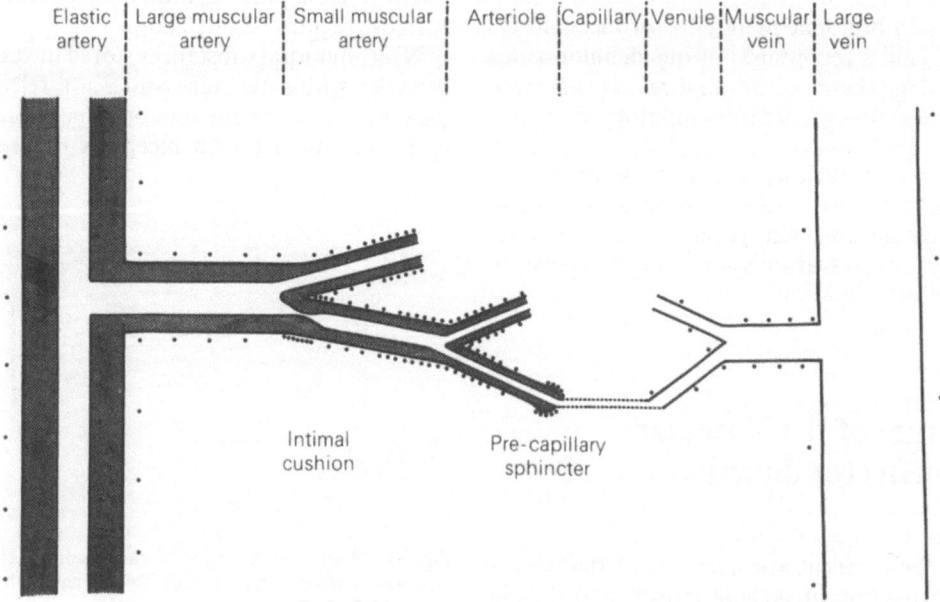

Fig. 7.2. Perivascular innervation density in different regions of the vascular system [60].

capillary vessels which can account for up to 80% of total peripheral resistance [323], and which represent the largest influence on local blood flow are the most densely innervated vascular structures. Similarly, pharmacological studies have shown that the neuronal noradrenaline uptake capacity and arterial diameter are also inversely related [403]. Nexuses are more frequent [35] and the mechanism of electrotonic spread of excitation is more prominent [36,75] in smaller arteries.

Although there is still much controversy about the innervation of capillaries, recent studies suggest that at least in some areas these vessels may be innervated. There is evidence for adrenergic and non-adrenergic innervation of feline hypothalamic capillaries based on ultrastructural and light microscopic studies respectively [365–367]. Acetylcholinesterase-positive axons have been shown to be in contact with cerebral capillaries in the turtle [237], bullfrog [410] and the lamprey [238]. There is also physiological evidence for serotoninergic innervation of brain microvessels [240,363,364]. Substance P-, cholecystokinin-, enkephalin- and neurotensin-containing nerve fibres have also been shown immunohistochemically to be closely associated with cerebral microvessels [90, 222,249]. The immunohistochemical findings of actomycin in both pericyte and endothelial cells of the capillary wall [345] and the presence of adrenoceptors [241,262] suggest the possibility of extrinsic neural regulation of capillary function with regard not only to microcirculatory haemodynamics but also to transcapillary transport processes. Although the width of the nerve–muscle junctional cleft in these capillaries may be less than the neuromuscular distance in the intestine where the smooth-muscle cells are considered to be influenced by transmitter release, morphological proximity alone cannot be considered to imply the presence of synapses and functional transmission. While there is increasing evidence for the innervation of capillaries, it must be noted that at the level of the microcirculation, circulating and locally produced agents probably play a greater role than do perivascular nerves in the control of circulatory mechanisms [83].

The venous system, which in man can contain up to 70% of the blood volume, serves as an adaptable reservoir under the influence of perivascular sympathetic nerves. Although veins in general are less innervated, their basic innervation pattern is similar to that described for arteries. Thus large veins are sparsely innervated while some medium-sized muscular veins receive a fairly rich innervation. Collecting venules and small veins are sparsely innervated, if at all [176]. Veins which are exposed to high hydrostatic pressure loads are characterized by thickening of the smooth-muscle media and dense sympathetic innervation. Veins located at or above the level of the heart, where capacitance effects are small, receive a sparse innervation. An exception to this generalization is the dense innervation of intracranial veins [140], a feature possibly related to the importance of cerebral blood volume in regulating intracranial pressure.

The perivascular nerve fibres which make up the neural plexus around the media are not to be mistaken for the larger perivascular nerves which run along the outer adventitia of the vessel to innervate both vascular and non-vascular structures more distally. The presence of ganglion cells within the vessel wall, as observed in feline aorta and coronary arteries [205], rabbit portal vein [79], canine cerebral arteries [45] and skeletal muscle arterioles [234,333], appears to be a relatively rare occurrence.

Development and Regeneration of Perivascular Innervation

Studies on the development of perivascular innervation have, until recently, been confined to the adrenergic system [22,104,130,223,290, 381,404]. While the generalized concept for the development of these nerve fibres still holds true, i.e. pre-natal and early post-natal increases in density of vascular innervation are followed by a decrease in later stages of life, the time course of this trend shows remarkable variation in different vascular beds and species. Old-age rats do not show a reduction in vascular adrenergic neuroeffector function [136], yet in old rabbits there is a reduction in both vascular smooth-muscle reactivity and density of adrenergic innervation [104,133,378]. Some vessels such as the canine femoral and brachial arteries, the rabbit aorta and human temporal arteries show a decline in the density of perivascular sympathetic innervation in early post-natal life [22,129,130, 389].

Adrenergic vasoconstrictor fibres in the dog hindlimb are not functional until 2 weeks after birth [43], while in the sheep, sympathetic control of blood pressure appears to be greater in the newborn than in the adult [454]. It is suggested that the increased activity of the autonomic nervous system and the subsequent establishment of sympathetic vasopressor tone is responsible for the increase in blood pressure during development [330].

A recent study on the development of perivascular peptidergic vasoactive intestinal polypeptide (VIP), substance P (SP) and calcitonin gene-related peptide (CGRP) in nerves in guinea pigs shows a decline in their density of innervation in early post-natal stages [122] and is consistent with the demonstration of decreased concentration of neuropeptide Y (NPY), SP and VIP from the age of <1 year to 16–41 years in human cerebral vessels [154]. These results suggest that peptides may have a trophic role in early development of perivascular innervation in different vessels [78].

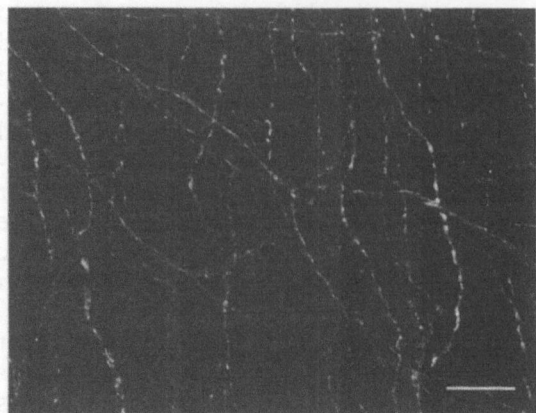

Fig. 7.3. Adrenergic nerves in a whole-mount stretch preparation of human mesenteric artery visualized by the glyoxylic acid method. Calibration bar = 50 μm.

There has been little experimental investigation of regeneration of perivascular nerves and what little there has been was limited to the study of perivascular adrenergic nerves after surgical crush lesions in the guinea pig carotid and mesenteric arteries [103]. A perivascular adrenergic plexus of normal distribution was re-established after 3–8 weeks in the mesenteric artery while similar regeneration took over 8 weeks in the renal artery; hyperinnervation was also observed in some vessels. Further studies on regeneration of perivascular nerves, particularly of non-adrenergic, non-cholinergic nerves, are very much warranted, especially in view of their implications for transplanted vessels in restorative surgery. Clearly there is much variation in the pattern and density of innervation of different blood vessels, reflecting not only changes in local physiological requirements in relation to development and the particular organ system they supply, but also responses to injury and species variation.

Adrenergic Nerves

Modifications of the Falck–Hillarp technique [5,117,164,177,282] and more recently the use of specific antibodies against enzymes of catecholamine synthesis [190] have led to a detailed knowledge of the neuronal geometry of adrenergic perivascular nerves. Adrenergic nerves (Fig. 7.3) are the most prolific nerve type and are the main neurogenic elements in the homeostatic regulation of vascular resistance.

Adrenergic nerves largely mediate vasoconstriction by the release of noradrenaline and its subsequent activation of adrenoceptors. Once

limited to α- and β-receptor types [4] they have since been subdivided into α_1/α_2 [269] and β_1-β_2 [268] on the basis of pharmacological studies. Post-junctional α_1- and α_2-receptors are responsible for the pressor effects on vascular smooth muscle while pre-junctional α_2-receptors are involved in the inhibition of noradrenaline release [318,319]. Adrenergic receptors have been demonstrated autoradiographically in arterial vessels of the rat and man [115,241,262].

While α-adrenoceptor-mediated pressor effects represent the main physiological role of the adrenergic system, vasodilatation upon activation of β-receptors has also been reported. Such vasodilatory responses to noradrenaline after α-receptor blockade have established the presence of β-adrenoceptors in cerebral, digital [372], pulmonary [44], coronary, facial [324,352, 453], skeletal [307] and hepatic circulations [201].

Recent studies on rat thoracic aorta [9], porcine and canine coronary, mesenteric and renal arteries [95] and rabbit and cat cerebral arteries [444] show that noradrenaline (NA) can cause vasodilation after α_1- and β-adrenoceptor blockade, by causing the release of endothelium-derived relaxing factor (EDRF) via activation of α_2-receptors. The chemical identity of EDRF remains unknown.

A third type of adrenergic receptor, namely the γ-receptor, has been proposed to account for the pressor responses to electrical stimulation after α-receptor blockade [225]. However, due to a lack of specific receptor antagonists, the high concentration of noradrenaline needed to activate them, the persistence of prazosin-resistance responses after depletion of NA by reserpine, and the compelling evidence in support of release of adenosine 5'-triphosphate (ATP) as a co-transmitter with NA during sympathetic nerve stimulation [66,73,392], the γ-receptor theory still remains controversial [393].

Perivascular adrenergic nerves originate from cell bodies located in pre- and paravertebral ganglia of the sympathetic nervous system. In addition to this peripheral source, cerebral blood vessels also receive adrenergic nerve fibres of central origin, i.e. from the locus ceruleus and related cell groups in the lower brain stem [117]. Electrical stimulation of the locus ceruleus in the anaesthetized rat has been shown recently to elicit a sharp rise in arterial blood pressure [131]. A similar arrangement for autoregulation has also been proposed for spinal vessels [127,128] and peripheral nerve vasa nervorum [20,123] whereby there may be local or segmental control of blood flow according to physiological requirements.

Loss of perivascular adrenergic innervation has been reported in alcoholism [288], amyloidosis [376], orthostatic hypotension (Shy–Drager syndrome) [25] and subarachnoid haemorrhage [118,148,171,286,408]. In diabetes both loss and increase in density of innervation have been shown depending on the vascular bed and duration of the disease [123,388]. There is an increase in the density of perivascular adrenergic innervation of cerebral vessels as well as a diminished pre-junctional modulation of NA release from perivascular nerves via P_2-purinoceptors in the spontaneously hypertensive rat [248,264, 275]. Changes in adrenergic receptors have also been demonstrated in autonomic failure [116].

Cholinergic Nerves

The distribution of perivascular cholinergic nerves (Fig. 7.4) in the cardiovascular system is less extensive than that of adrenergic nerves. Unlike their constrictor role in poikilotherms, post-ganglionic cholinergic nerves in mammals are largely vasodilatory. Except in some pulmonary arteries of certain species [113], some veins [199] and isolated human umbilical vessels [15,197], the presence of cholinergic constrictor nerves in vessels is rare in mammals.

Histochemical and physiological studies have shown that cholinergic vasomotor control of parasympathetic origin is present in the brain, heart, kidney, bladder, salivary gland, tongue and external genitalia [57,65] while sympathetic

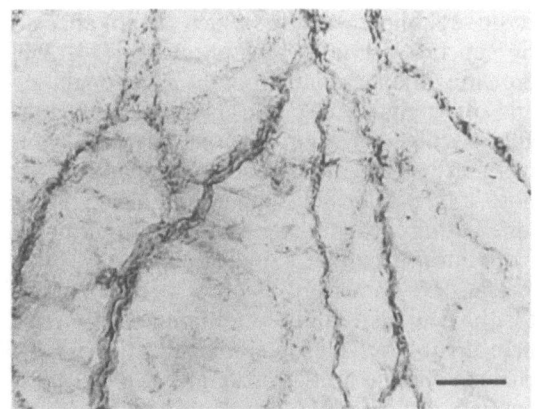

Fig. 7.4. Acetylcholinesterase-positive nerve fibres in a whole-mount stretch preparation of the guinea pig mesenteric artery. Calibration bar = 50 μm.

cholinergic vasodilator fibres [438], once thought to be limited to a functional role in skeletal muscle, are also found in skin [50], the uterus [377] and possibly in the renal vasculature [320] in certain species. In skeletal muscle, these vasodilatory sympathetic fibres are involved in reflex changes in muscle blood flow during behavioural reactions to emotional stress [2] and are not activated by postural changes or exercise [40,370]. However it has been suggested that in the renal vasculature these cholinergic sympathetic fibres may have a role in local reflexes [281] and be involved in the redistribution of renal blood flow.

Histochemical demonstration of cholinergic nerves has so far relied on the localization of acetylcholinesterase (AChE), the degradatory enzyme of acetylcholine [251,263]. Although cholinergic nerves are very rich in AChE, this method is not specific and has been demonstrated to stain adrenergic nerve fibres [76,160] as well as to be associated with substance P [93]. A small population of cholinesterase-positive nerve fibres may well be sensory substance P fibres in view of their depletion after capsaicim treatment [347]. Furthermore, there is anatomical evidence for the enclosure of both adrenergic and cholinergic nerve varicosities within the same Schwann cell sheath [26,78]. Consequently our knowledge of the origin of cholinergic nerves is limited when compared to that of the adrenergic system. Nevertheless it has been claimed that cholinergic nerves innervating blood vessels supplying skeletal muscle originate from sympathetic chain ganglia [55] and those of the uterine artery originate from the paracervical ganglia [33]. Intracranial perivascular AChE-positive cholinergic nerves which are not affected by superior cervical ganglionectomy [141,260] are thought to originate from the sphenopalatine and otic ganglia [215]. The sphenopalatine ganglion is also the source of perivascular cholinergic fibres of the mucous membrane of the nose and palate [87]. Cholinergic nerve fibres innervating coronary vessels appear to originate from intrinsic cardiac ganglia [344].

The development of a more specific method for the demonstration of autonomic cholinergic neurons [64] relying on the localization of the enzyme choline acetyltransferase (ChAT) [252] has so far proved largely inadequate. An improvement in this method involves immunohistochemical use of specific antibodies raised against ChAT [137,250,353] or choline storage vesicles [111]. This technique has been used successfully to demonstrate enteric cholinergic neurons in the guinea pig and mouse [179,180] and more recently in an ultrastructural study showing ChAT immunoreactivity in vascular endothelial cells and axons associated with cerebral vessels in the rat brain [349].

Non-adrenergic, Non-cholinergic Nerves

It was not until the early 1960s that the long-recognized residual responses to electrical stimulation after adrenergic and cholinergic blockade received extensive experimental scrutiny. Experiments mainly carried out on mammalian gut led to the firm estalishment of the presence of non-adrenergic non-cholinergic (NANC) neuro-effector mechanisms [57,59].

On the basis of experimental evidence for the cardiovascular actions of purines, serotonin and peptides, and their localization in autonomic nerves, this section will outline the current status of vascular non-adrenergic, non-cholinergic neurotransmission.

Purinergic Nerves

The potent vasoactive effects of purine nucleotides and nucleosides on the vasculature had been noted as early as 1929 [132]. The possibility that ATP might be the vasodilatory substance released upon antidromic stimulation of sensory nerves was proposed for the rabbit ear vasculature [231].

Investigations to identify the transmitter substance in NANC nerves supplying the gut and bladder led to the "purinergic nerve hypothesis" [59] whereby ATP or a related purine nucleotide was proposed as the NANC transmitter. Since then, both vasodilator and vasoconstrictor effects of purine compounds, mediated via purinoceptors, have been reported extensively for the vasculature [65].

Analogous to the subdivision of adrenergic, cholinergic and histaminergic receptors, purinoceptors have been classified into P_1 and P_2 [63] mainly on the basis of selective actions of agonists and antagonists. P_1-purinoceptors have an agonist potency order of adenosine greater than ATP, their activation causes changes in intracellular levels of adenosine $3':5'$-phosphate (cAMP) and they are completely antagonized by methylxanthines. P_2-purinoceptors have an agonist potency order of ATP greater than adenosine, their activation does not affect cAMP levels but does stimulate prostaglandin synthesis and they are not antagonized by methylxanthines, though they are partially blocked by quinidine, high concentrations of 2-substituted imidazoline compounds, $2'$-2-dipyridilisatogen, ararylazidoaminopropionyl-ATP (ANAPP3) [226] and by the stable analogue of ATP, α-β-methylene-ATP [252,322,392]. P_1-purinoceptors have been further subdivided into A_1/A_2 [440] or R_i/R_a [287]. Recently the P_2-receptors has also been classified into P_{2x} and P_{2y} subtypes [85].

The distribution of both P_1- and P_2-purinoceptors has been demonstrated in many vascular beds [81,402]. P_1-receptors, which are more widely distributed than P_2-receptors, mediate vasodilatation. These receptors are also located on perivascular nerve terminals where they are involved in pre-junctional inhibition of transmitter release [120,350,445]. P_2-purinoceptors are located exclusively post-junctionally and mediate both pressor and depressor responses [258], P_{2x}-receptors mediating vasoconstriction and P_{2y}-receptors mediating vasodilatation. Particularly in the microcirculation, purinergic vasodilatation appears to be mediated via the activation of P_{2y}-receptors located on the endothelial cells with the subsequent release of EDRF [121,174,257]

Nerve fibres containing ATP can be demonstrated by a fluorescence histochemical-method using quinacrine or mepacrine binding (Fig. 7.5). Quinacrine has been shown to bind strongly to ATP [114,343]. Where there is pharmacological evidence for purinergic nerves, quinacrine-positive nerve fibres are present [79].

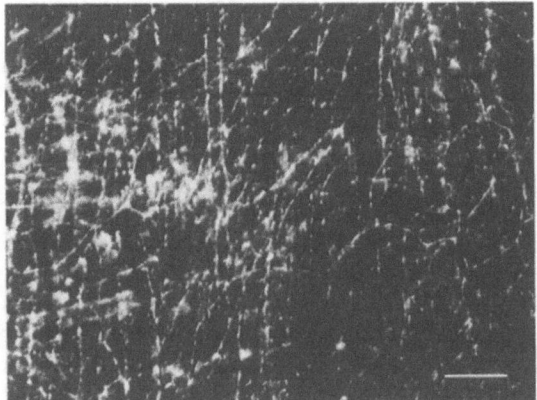

Fig. 7.5. Whole-mount stretch preparation of the rabbit portal vein showing quinacrine-positive nerve fibres. Calibration bar = 100 μm.

There is good evidence for ATP as the predominant transmitter in nerves supplying the rabbit portal vein [79] and for ATP as a co-transmitter with NA in perivascular sympathetic nerves [68,71–74,84,86]. Recent studies have revealed dense plexuses of neurons showing adenosine deaminase-like immunoreactivity in the basal hypothalamus of the rat suggesting that in these nerves adenosine may be a neurotransmitter [334,335].

The potent vasoactive action of purines in most vascular beds and the implications of purinergic nerves in coronary insufficiency and angina, transient cerebral ischaemia and stroke, migraine, skin inflammation, chronic ulcers and colitis [342] and hypertension [248] have sparked keen interest in the possible therapeutic values of purinergic drugs. The considerable evidence that has accumulated in support of the purinergic nerve hypothesis strongly warrants the development of specific and potent agonists and antagonists to both P_1- and P_2-purinoceptors. It is becoming increasingly clear that purines have a far wider role in neuroregulatory mechanisms that was originally envisaged [70,86].

Aminergic Nerves

Serotonin

Originally discovered as a vasoactive substance in blood serum [358], serotonin (5-hydroxytryptamine; 5-HT) is a leading candidate for the role of a non-adrenergic, non-cholinergic neurotransmitter in autonomic nerves. Autoradiographic [18,89,90], fluorescent histochemi-

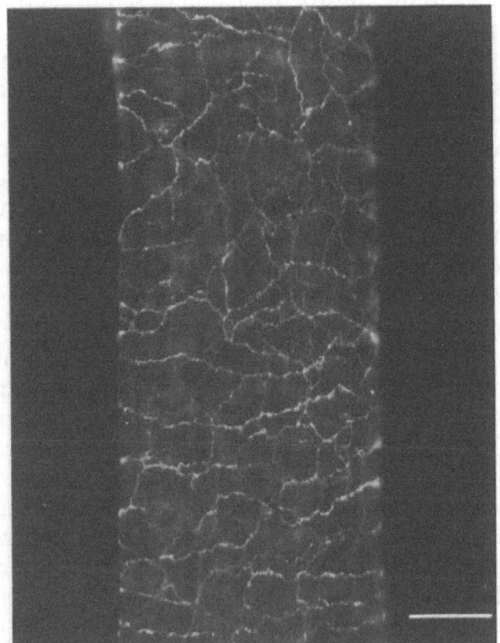

Fig. 7.6. Serotonin-like immunoreactive nerve fibres in a whole-mount stretch preparation of the guinea pig posterior cerebral artery. Calibration bar = 50 μm.

Table 7.1. Immunohistochemical and fluoresence histochemical demonstration of 5-HT-containing perivascular nerves

Location of vessel	Species	References
Brain	Man (foetus), monkey, cat, rabbit, lamprey, guinea pig, rat, gerbil	7, 32, 82, 106, 125, 126, 150, 165, 204, 238, 240, 336, 337, 383
Spinal cord	Monkey, cat, rat	127, 128, 267
Peripheral nerves	Rabbit	20
Gastrointestinal tract	Man (foetus), lamprey	203, 204, 235
Hand	Man	23

cal [23,125–128,235,238,337], biochemical [363] and more recently immunohistochemical (Fig. 7.6 and Table 7.1) and pharmacological studies provide strong support for the role of serotonin as a perivascular autonomic neurotransmitter.

Depending on the vascular bed and the concentration, serotonin can mediate both vasoconstriction and vasodilatation [346,442]. Serotonin induces vasoconstriction in most isolated large blood vessels, and it is particularly potent in the cerebral and coronary vasculatures. In some cerebral vessels serotonin is more potent than noradrenaline [416]. Venules are more sensitive to the constrictor action of serotonin than arterioles at the level of resistance vessels [14].

Serotonin-induced vasoconstriction may take place by activation of post-junctional vascular 5-HT receptors [443], by interaction with post-junctional α-adrenoceptors [21,38,356] or by enhancement of pressor responses to other vaso-constrictor agonists [443] and sympathetic nerve stimulation [326]. The existence of two serotonin receptor types was proposed by Gaddum and Picarelli [184]. D-receptors were suggested to mediate serotoninergic effects in nervous tissue and M-receptors to mediate the effects in muscular tissue. A further class of serotonin receptors (5-HT$_1$/5-HT$_2$) as later described in the brain [354]. Pharmacological studies provide indirect evidence that 5-HT$_2$ receptors mediate vaso-constriction [39,96]. Recently the development of high-affinity M-receptor antagonists have allowed the identification of three subtypes of this receptor [369]. One of these subtypes is implicated in the painful effects of serotonin. There is much inconsistency in the terminology and classification of 5-HT receptors. The availability of selective serotoninergic antagonists as well as agonists is proving to be of much help in redressing this situation.

Vasodilatory actions of serotonin, which are particularly marked during elevated sympathetic tone, are mainly mediated via inhibition of adrenergic neurotransmission. This is achieved either by pre-junctional inhibition via α_2-adrenoceptors [236] of noradrenaline release or by activation of vascular β-adrenoceptors [142]. Other inhibitory mechanisms of serotonin may involve the release of vasodilator substances such as vasoactive intestinal polypeptide [159], the stimulation of prostaglandin production [102] or the activation of endothelial-dependent relaxations [95,442]. Some endothelium-denuded vascular preparations, like the coronary vessels, show an enhanced pressor response to serotonin [97]. Vasodilatory action via activation of inhibitory serotonergic receptors is also a possibility [247].

Although serotonin has been shown to have a vasomotor action on various vascular beds, perivascular serotoninergic nerve fibres, with the exception of brain stem and spinal vessels [125–128,267], human digital and foetal mesenteric vessels [23,203,204] and peripheral nerve vasa nervorum [20], have only been located in cerebral vessels. They are strongly implicated in the pathogenesis of a wide range of cerebrovascular disorders involving cerebral vasospasm, such as subarachnoid haemorrhage, the prodromal phase of migraine and transient ischaemic

attacks. The central serotoninergic neuron system which supplies nerve fibres to some cerebral vessels has been shown recently to play an important role in vascular CO_2 reactivity [240]. Serotonin is also thought to be involved in coronary vasospasm [19]. Serotonin has been shown to induce a Ca^{2+}-dependent increase in blood–brain barrier permeability via activation of 5-HT$_2$-receptors located on the luminal surface of the endothelium in frog cerebral vessels [340]. The development of selective 5-HT antagonists is of great therapeutic value as evidenced by their ability to block 5-HT-induced pain in human forearm blisters and in the treatment of migraine and cluster headaches. In the diabetic rat, there is a reduction in the density of 5-HT-LI nerve fibres supplying the cerebral circulation [266].

Substantial evidence suggests that serotoninergic nerve fibres innervating cerebral blood vessels have a central origin in the raphe nuclei [89,90,150,328,383]. However, the immunocytochemical demonstration of serotonin in the superior cervical [214] and nodose ganglia [186], and the disappearance of perivascular serotoninergic nerve fibres after superior cervical ganglionectomy indicate that there is also a peripheral origin for these fibres [7,106,312]. These studies also suggest that the relative contribution of perivascular serotoninergic fibres from the two sources may vary in different species.

Some serotoninergic cell bodies and their dendrites have been observed in close contact with brain stem parenchymal vessels in the rat, cat and monkey. It is suggested that these dendrites might have a receptive role in monitoring blood temperature and mechanical changes of the vessel wall [127]. Such a sensory role for serotoninergic nerve fibres is consistent with the demonstration of 5-HT-L1 cell bodies in the trigeminal and spinal ganglia [246].

Dopamine

There is increasing evidence that dopamine is also involved in vascular neuroeffector mechanisms. Dopamine has been shown to mediate both pressor and depressor effects on cerebral blood vessels in vitro [169,315]. These studies, and the identification of dopamine-containing neurons, whose dendrites are closely associated with cerebral arterioles, venules and capillaries in the substantia nigra [165], strongly support a role for dopamine in the pathophysiology of cerebrovascular disorders. Dopamine-induced vasoconstriction is however not due to a direct action via stimulation of dopaminergic receptors but instead is mediated indirectly via activation of α_2-adrenoceptors and 5-HT receptors. Blockade of these latter receptors shows that dopamine causes vasodilatation via activation of specific dopamine D1-receptors [155,169,198,384,417].

Peptidergic Nerves

The presence of similar peptides throughout the animal kingdom, from man to primitive organisms such as protozoa and urochordates, shows a remarkable preservation of these chemical substances during evolution. From possibly being simple mediators of cell-to-cell communication, the role of peptides seems to have evolved to that of hormones, neurotransmitters, neuromodulators and tissue factors.

The rapid advances in increasingly specific immunohistochemical methods and techniques for the synthesis and characterization of peptides have revealed a widespread distribution of these substances in exocrine, endocrine and paracrine tissue. Notable is the demonstration of many peptidergic perivascular nerve types which strongly suggests that they have an important role in the control of cardiovascular function.

Substance P-, vasoactive intestinal polypeptide-, neuropeptide Y- and calcitonin gene-related peptide-containing perivascular nerves are the more prolific of the peptidergic nerve types and will be discussed in detail. Other peptides implicated in the neurogenic regulation of the vasculature will also be described.

Substance P (SP)

Substance P, the first "brain–gut" peptide originally described by Euler and Gaddum [162] and later characterized as an undecapeptide [91], is now widely accepted as a neurotransmitter in primary sensory neurons [227,228] and has been demonstrated immunohistochemically in perivascular nerves of many vascular beds (Table 7.2 and Fig. 7.7). It has been implicated in blood-flow regulation and vascular pain [327].

Perivascular SP-like immunoreactive (SP-L1) nerve fibres which are capsaicin sensitive [29,135,178] and unaffected by sympathectomy [149,339], most probably originate from the sensory trigeminal [284,455] and spinal ganglia [30,304]. Excision of the trigeminal ganglion in cats has been shown to reduce the vascular SP content in the ipsilateral intracranial and extracranial arteries by an average of 55% and 78% respectively [339]. Capsaicin-insensitive non-

Table 7.2. Immunohistochemical demonstration of SP-containing perivascular nerves

Location of vessel	Species	References
Brain	Man, pig, cat, guinea pig, rat, gerbil	8, 82, 90, 144–147, 149, 154, 178, 181, 239, 249, 284, 311, 429, 432, 436, 455
Peripheral nerves	Rabbit	20
Heart and associated vessels	Man, monkey, dog, cat, chicken, quail, frog, lizard, fish	28, 29, 122, 178, 181, 219, 359, 361, 362, 451
Respiratory tract	Man, cat, guinea pig, rat	178, 181, 293, 301, 355, 382, 405, 448
Gastrointestinal tract	Cat, guinea pig, rat, chicken, salamander	28, 29, 49, 52, 99, 100, 122, 178, 181, 244, 293
Liver	Rat, guinea pig	27–30, 178
Spleen	Guinea pig, rat	29, 178
Skeletal muscle	Guinea pig, rat	28, 29, 122, 178, 181
Kidney and urogenital tract	Man, pig, dog, cat, opossum, rabbit, guinea pig, rat, mouse	12, 29, 122, 166, 168, 178, 181, 313, 348, 390, 396, 450
Skin	Cat, guinea pig	112, 178, 227, 228, 293
Eye	Monkey, cat, rabbit, guinea pig	397, 413
Ear	Guinea pig	433
Dental pulp	Cat	189, 341
Endocrine gland	Rat	218

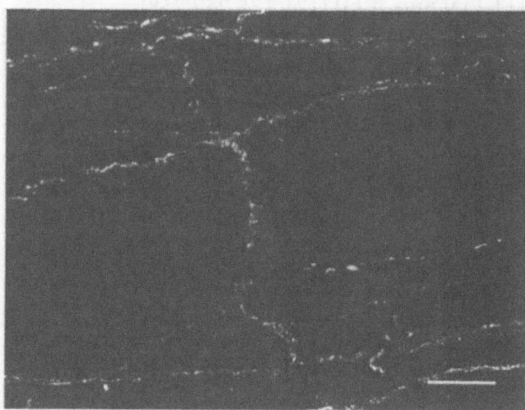

Fig. 7.7. SP-like immunoreactive nerve fibres in a whole-mount stretch preparation of the guinea pig uterine artery. Calibration bar = 50 μm. (Courtesy of Dr M. C. Mione)

perivascular SP-L1 fibres are found in the enteric nervous system [232,244], vertebral spinal cord and various regions of the brain [108,185], and in rat adrenal gland [54], salivary gland and lingual mucosa [108]. The presence of capsaicin-insensitive perivascular SP-L1 fibres supplying arterioles of the distal colon and rectum in rats is surprising [108].

The pattern of perivascular SP-L1 nerve fibres shows an opposite trend to that of those containing VIP and noradrenaline [181], such that larger arteries and veins have a greater density of innervation than smaller ones.

Substance P has potent vasodilatatory [56, 145,208] as well as vasoconstrictor [34,220] actions in many blood vessels. It is reported to be involved in reflex vasodilatation [62,188,278]. Receptor blockers to SP, namely (D-Pro$_7$, DPhe$_7$, Trp$_9$)SP and (D-Pro$_7$, D-Trp$_7$, 9)SP, have been shown to block SP-induced and neurogenic vasodilatation in the dental pulp, oral mucosa [374], rat hindlimb [279] and SP-induced relaxation of cerebral arteries [147].

However, chemical and ganglionic lesions and the lack of correlation between the density of innervation and the vascular responsiveness to SP tend to support the concept that perivascular SP-L1 fibres have a purely sensory function and do not mediate direct vasomotor influences. Studies in the rat show that the superior mesenteric artery which receives a dense SP-L1 perivascular plexus is unresponsive to exogenous SP, while the carotid artery which lacks SP innervation exhibits a dose-dependent relaxation [29]. The direct action of SP on vascular smooth-muscle receptors has also been challenged by studies showing SP-induced relaxation to be endothelium dependent [216,457]. This action is also shared by bradykinin [92], acetylcholine [174], 5-HT [95] and ATP [121,175,257,258].

These studies and the suggestion that SP fibres may serve to signal peripheral nerve damage [446], indicate that perivascular SP-L1 nerve fibres do not play a role in direct short-term autoregulation of blood flow but that they may have an important influence on the regulation of vascular tone in response to noxious stimuli or during pathological conditions of the vasculature such as ischaemia, claudication and aneurysms.

Vasoactive Intestinal Polypeptide (VIP)

VIP, a 28-amino acid residue peptide structurally related to secretin, glucagon and gastric inhibitory peptide, which was first isolated from the

Table 7.3. Immunohistochemical demonstration of VIP-containing perivascular nerves

Location of vessel	Species	References
Brain	Monkey, cow, pig, dog, cat, rabbit, guinea pig, rat, hamster, gerbil, mouse	8, 143, 146, 154, 156, 181, 192, 215, 261, 283, 310, 329, 429, 432
Peripheral nerve	Rabbit	20
Heart and associated vessels	Cat, guinea pig	122, 181, 293, 430
Respiratory tract	Man, dog, cat	119, 181, 293, 355, 420, 422, 424, 430
Gasrointestinal tract	Man, monkey, pig, dog, cat, guinea pig, rat, mouse, chicken, frog, fish	28, 53, 105, 122, 181, 230, 244, 270, 279, 293, 360, 379, 421, 430
Liver and gallbladder	Man, pig, cat, guinea pig, rat, mouse	27, 221, 243, 406
Kidney and urogenital tract	Man, monkey, pig, dog, cat, rabbit, guinea pig, rat, mouse, chicken, frog, fish	13, 107, 122, 168, 181, 206, 272, 273, 291, 293, 313, 325, 348, 355, 396, 399, 430, 439, 450, 452
Skeletal muscle	Cat	24, 122, 181, 242, 279, 293, 430
Skin	Cat	279, 293, 355
Eye	Cat, guinea pig, hamster	124, 413, 427
Ear	Cat, guinea pig	423, 433
Dental pulp	Man, cat	425
Endocrine gland	Cat, rat	218, 229, 428
Exocrine gland	Pig, cat, rat	41, 293, 294, 426, 447
Carotid body	Cat	292, 293, 355, 449
Paravertebral ganglia	Cat	293

porcine duodenum, is a potent vasodilator [380]. Originally considered a classic gut hormone, VIP-like immunoreactivity (VIP-L1) has since been shown to be widely distributed in the body and especially in association with the cardiovascular system (Table 7.3 and Fig. 7.8).

The pattern of perivascular VIP-L1 nerve fibres closely resembles that of the adrenergic and cholinergic systems [181,260,430]. VIP-L1 fibres are more abundant in the cerebral than in peripheral blood vessels [134,270,430] and for this reason most pharmacological and physio-

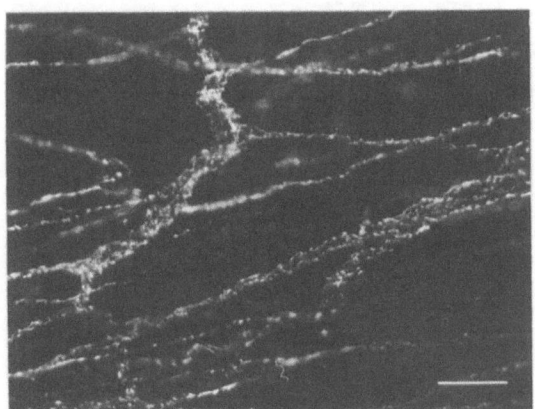

Fig. 7.8. VIP-like immunoreactive nerve fibres in a whole-mount stretch preparation of the guinea-pig uterine artery. Calibration bar = 50 μm. (Courtesy of Dr M. C. Mione)

logical studies on VIP-ergic responses have been carried out in the cerebral vasculature.

Microapplication or infusion of VIP induces a dose-dependent vasodilatation of all in vitro and in vivo vascular preparations so far studied [143,209,309,317,338,430], the peptide being effective in concentrations as low as 0.1 nM [139]. Small cerebral arteries are more responsive than large arteries to microapplication of VIP [317].

Physiological studies in primates have shown that responses to intracarotid infusions of VIP of cerebral blood flow and metabolism are greater after opening of the blood–brain barrier [317]. This indicates that mechanisms involving the activation of central neurons may be important in mediating the vasodilatory action of VIP. VIP-induced increase in firing of cortical neurons, stimulation of cortical glucose utilization, dilatation of cortical arteries and increase in cortical blood flow [316] are consistent with this suggestion.

Unlike the depressor actions of SP, ATP, acetylcholine (ACh) and bradykinin, VIP-induced responses are independent of an intact endothelium [216,275]. The failure of β-adrenoceptor, cholinergic and histamine H_2 blockers to antagonize responses to VIP [299] and the recent study showing a dose-dependent and reversible inhibition of these responses by an antibody to VIP [48] suggest that there are specific VIP receptors. Autoradiographic studies have recently localized VIP receptors in vascular smooth muscle in both human and guinea pig lung [31].

Several studies indicate that VIP fibres innervating the cerebral vasculature have a central origin. 6-Hydroxydopamine treatment and the removal of the pterygopalatine and superior cervical ganglia, both of which contain large numbers of VIP-positive cell bodies and fibres, did not affect the density of VIP-like immunoreactive fibres in cerebral vessels [143,261, 310]. Removal of the sphenopalatine ganglion however, leads to a great reduction in the content of VIP and choline acetyltransferase as shown in studies on the nasal mucosa of cats [297]. Added to this, the co-existence of VIP and AChE in cerebral perivascular nerves [215,261] strongly supports a central origin for VIP-positive fibres as is postulated for cholinergic fibres. The co-existence of these two nerve types is discussed in more detail later.

There is evidence that VIP-L1 nerve fibres innervating vessels in the urinary bladder originate from the pelvic ganglia [206,313], those innervating splenic vessels originate from the coeliac ganglion complex [304] and those innervating vessels of the lower limb originate from the lumbo-sacral sympathetic chain [291]. Perivascular VIP-containing fibres are not sensitive to capsaicin treatment [181,313].

There is a marked reduction of VIP-L1 perivascular nerve fibres in diabetic penile tissue in rat and man [107], and in the cerebral circulation of diabetic rats [266]. Conversely, elevated levels of VIP have been reported in perivascular nerves innervating the choroid in pre-diabetic and diabetic Chinese hamsters [124]. Furthermore, the ability of VIP to mimic many of the metabolic and cardiovascular symptoms of endotoxin shock [172] strongly implicates this peptide in vascular pathology. While we can confer an important role for VIP in the cardiovascular system, an understanding of its precise physiological function and any therapeutic advantages of manipulating the VIP-ergic system must await the development of specific antagonists.

Neuropeptide Y (NPY)

The recent isolation of the 36-amino acid peptide NPY from porcine brain [412] has added yet another substance to the growing list of putative non-adrenergic, non-cholinergic, peptidergic neurotransmitters. NPY, which is similar to pancreatic polypeptide (PP) and which shows a close sequence homology with avian PP, is present in a wide range of tissues as evidenced by immunohistochemical studies. Since the original studies showing NPY-induced vasoconstriction in the submandibular gland [300] and NPY-like immunoreactivity (NPY-L1) and its action in cerebral vasculature [151], the localization and vasomotor actions of NPY have been widely

Table 7.4. Immunohistochemical demonstration of NPY-containing perivascular nerves

Location of vessel	Species	References
Brain	Man, cat, rat, gerbil	6, 8, 151, 152, 154, 302, 386, 437
Heart and associated vessels	Man, pig, rabbit, guinea pig, rat, mouse	157, 207, 298, 302, 305, 395, 437
Respiratory tract	Man, cat, rabbit, guinea pig, rat, mouse	157, 298, 302, 305, 395, 435, 437
Gastrointestinal tract	Man, pig, cat, guinea pig, rat	157, 302, 407, 437
Spleen	Man, cat, rabbit, guinea pig	157, 304, 305, 437
Liver	Man, cat, rabbit, guinea pig	11, 157, 302, 437
Pancreas	Man, pig, cat, guinea pig, rat	298, 302, 407
Kidney and urogenital tract	Man, cat. rabbit, guinea pig, rat, mouse	3, 157, 298, 302, 305, 313, 325, 437
Skeletal muscle	Man, guinea pig, rat	157, 302, 305
Skin	Man	302
Eye	Guinea pig, rat	459
Ear	Cat, guinea pig, rat	435
Tongue	Guinea pig	157, 437
Endocrine gland	Rat	218, 386
Exocrine gland	Man, cat	298, 302
Carotid body	Man	302

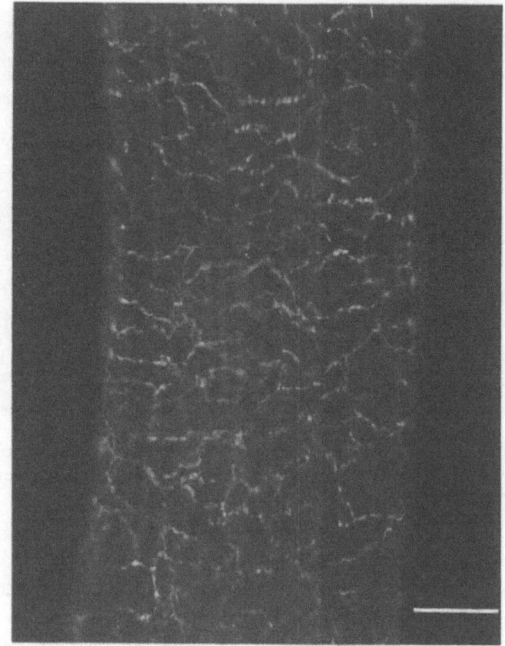

Fig. 7.9. NPY-like immunoreactive nerve fibres in a whole-mount stretch preparation of the rat anterior cerebral artery. Calibration bar = 50 μm.

reported in the cardiovascular system (Table 7.4 and Fig. 7.9).

The potent, calcium-dependent [110,151], pressor action of NPY, both in vivo and in vitro, is resistant to combined α- and β-adrenergic blockade as well as to the serotonin antagonist ketanserin [98,110,151,152]. These pressor responses are potentiated by reserpine and markedly reduced by the calcium channel blocker nifedipine [308].

Studies on isolated cerebral vessels and microapplication in feline pial vessels have shown NPY to be an equally if not a more potent vasoconstrictor than noradrenaline [151,153]. However, systemic administration of NPY in cats has recently shown it to be 100 times and 7 times less potent than angiotensin II and noradrenaline respectively in its pressor response [98], and to enhance pressor responses to phenylepinephrine and to sympathetic nerve stimulation [110]. These results may be explained by the greater sensitivity of pial arteries as compared with extracranial arteries to NPY [210] and by the dual post-junctional action on vascular smooth muscle where it enhances α-adrenoceptor-mediated pressor effects at low doses and induces vasoconstriction per se at high doses [110]. However, in the rat and rabbit basilar arteries where NA is not implicated in the pressor effects, NPY induces vasoconstriction at low doses suggesting that in some vessels NPY may represent the main transmitter substance mediating this response [138].

Recent physiological studies have demonstrated NPY-induced reduction of cerebral [10,419] and splenic [304] blood flow, strongly implicating this peptide in the maintenance of vasomotor tone in normal and pathological conditions.

The distribution of perivascular NPY-L1 nerve fibres is similar to that of adrenergic fibres [68,139] and there is strong evidence that NPY and NA co-exist in the autonomic innervation to many vessels [157,196,298,395,437,459]. In the heart, NPY is reported to inhibit sympathetic neurotransmission via a pre-synaptic mechanism on NA release, causing coronary vasoconstriction [170]. NPY is also implicated in the neuromodulation of the cardiac vagus [259]. There is also evidence for the co-existence of

NPY with VIP and dynorphin II in the guinea-pig uterine artery [325]. NPY-L1 perivascular nerve fibres originate from the ipsilateral peripheral sympathetic ganglia [6,8,152,304,313, 386]. Six weeks after bilateral superior cervical ganglionectomy in the rat, the NPY content of pial arteries has been reported to fall to 40% of control values [385]. The persistence of this population of NPY-L1 fibres in some cerebral vessels after bilateral superior cervical ganglionectomy suggests that they may have a central origin as has been suggested for the adrenergic system [6].

Calcitonin Gene-Related Peptide (CGRP)

CGRP, a recently discovered 37-amino residue peptide which has been immunohistochemically localized in both the central and peripheral nervous systems, is the main product generated from calcitonin gene expression in non-thyroidal tissue [16,373]. Perivascular CGRP-like immunoreactive (CGRP-L1) nerve fibres are distributed extensively in the cardiovascular system (Table 7.5 and Fig. 7.10). The vascular innervation pattern of CGRP clearly resembles that of SP, and the two peptides have been shown to co-exist in perivascular and non-perivascular tissues [194,277,306,411,436).

Systemic administration of CGRP results in a marked dose-dependent vasodilatation and a prolonged increase in heart rate in the rat, while intracerebroventricular administration results in vasoconstriction and increased plasma noradrenaline and adrenaline levels [167,170]. The

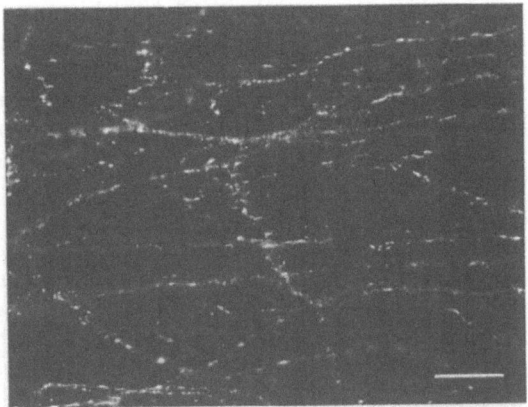

Fig. 7.10. CGRP-like immunoreactive nerve fibres in a whole-mount stretch preparation of the guinea pig mesenteric artery. Calibration bar = 50 μm.

mechanisms by which CGRP acts in the central nervous system to modulate sympathetic and adrenomedullary activity remains unclear. In vitro studies in the rabbit, cat and human cerebral arteries, and rabbit and cat mesenteric vessels have shown that CGRP-induced vasomotor responses like those of VIP are not endothelium dependent and are not mediated by activation of adrenergic, cholinergic or histaminergic receptors [211,212,216]. On the other hand, vasodilatory responses to CGRP are endothelium dependent in rat aortic rings [46].

CGRP is reported to be more abundant than SP and VIP in renal arteries and large veins [331], indicating that it might have a particularly important role in these vessels. Only 3 years after its isolation, CGRP is a leading candidate for a NANC, peptidergic neurotransmitter in the neuronal systems in both central and peripheral nervous systems subserving both motor and sensory functions. These studies and the recent evidence that circulating CGRP comes from perivascular nerves [465], strongly implicate CGRP in the regulation of blood flow and in the processing of sensory information during vascular injury. The potentiating action of CGRP on histamine- and bradykinin-induced oedema [47] and its central actions in inhibiting gastric acid secretion and in raising plasma noradrenaline levels [409], indicate that CGRP may also be involved in neurogenic inflammation and that it may also be an extracellular modulator.

The loss of perivascular CGRP-L1 nerve fibres in guinea pigs treated with capsaicin suggests that these nerves originate from sensory ganglia [194,306,415]. CGRP-L1 has been shown in the vagal and trigeminal ganglia, within the brain

Table 7.5. Immunohistochemical demonstration of CGRP-containing perivascular nerves

Location of vessel	Species	References
Brain	Man, cat, rabbit, guinea pig	8, 211, 436
Heart and associated vessels	Guinea pig, rat	122, 306, 301
Respiratory tract	Guinea pig	306
Gastrointestinal tract	Guinea pig, rat	94, 122, 376, 331
Kidney and urogenital tract	Guinea pig, rat	122, 331
Skeletal muscle	Guinea pig	122, 331
Eye	Monkey, cat, guinea pig, rat	414
Tongue	Rat	373, 415
Palate	Monkey, cat, rat	371
Dental pulp	Cat	189

stem and in co-existence with ChAT-L1 in motor neurons of the facial, hypoglossal and ambiguous nuclei [371,373,411]. No CGRP-LI was observed in the superior cervical ganglion in this species [194]. CGRP-LI is also present in enteric neurons [94,182,194] and there is evidence that some positively stained varicose fibres follow blood vessels in the submucosa. These intrinsic gut CGRP-L1 fibres are resistant to capsaicin treatment. Except for a small population of CGRP-L1 nerve fibres supplying lingual blood vessels [415], perivascular CGRP-L1 nerves in the rat are reported to be insensitive to capsaicin [331]. The origin of these fibres needs to be elucidated. It would appear that CGRP-L1 perivascular nerve fibres may have sympathetic as well as sensory, motor and enteric origins [195,415].

Bilateral superior cervical ganglionectomy in the rat has been shown to result, after 6 weeks, in a 70% increase in the CGRP content of iris and pial arteries, and a 34% increase in its concentration in the trigeminal ganglion [385]. The time course of this increase in CGRP content strongly implicates this substance in pain mediation in human sympathalgia.

Other Peptides

A number of other peptides besides those described above have also been demonstrated immunohistochemically in perivascular nerves, albeit with a more limited distribution, and shown to induce vasomotor responses (Table 7.6).

Gastrin-releasing peptide (GRP), which contains 27 amino acid residues, is the mammalian counterpart of bombesin (BOM) with which it shares a close sequence homology at the active carboxyl-terminal region. BOM elicits a rise in blood pressure in the anaesthetized dog, cat, rabbit, rat and chicken, while in the monkey it causes hypotension [161]. These responses are not affected by either adrenergic or cholinergic blockade. However, in physiological studies

Table 7.6. Immunohistochemical demonstration of peptides other than SP, VIP, NPY and CGRP in nerves associated with blood vessels

Peptide	Location of vessel	Species	References
Neurotensin (NT)	Brain	Rat	90
	Kidney	Man, monkey, pig, dog, opossum, cat, guinea pig, rat	168
	Heart and associated vessels	Man, monkey, dog, guinea pig, chicken, fish, frog, lizard	361,362
Somatostatin (SOM)	Kidney	Man, monkey, pig, dog, opossum, cat, guinea pig	168
Avian pancreatic polypeptide (APP)	Brain	Man, monkey, cow, pig, cat, rabbit, guinea pig, rat, mouse, chicken	152,431,432
	Salivary gland	Rat	296
	Ear	Guinea pig	433
Gastrin/cholecystokinin (G/CCK)	Brain	Monkey	222,329
	Gastrointestinal tract	Cat, guinea pig, rat	293,387
Enkephalin (ENK)	Brain	Cat	249,329
	Gastrointestinal tract	Cat, guinea pig, rat	293,387
Gastrin-releasing peptide (GRP)	Brain and spinal cord	Cat, guinea pig, rat, mouse	434
	Urogenital tract	Rabbit, guinea pig, mouse	396
	Ear	Guinea pig	433
Vasopressin-like peptide (VLP)	Brain	Rat	245
	Liver	Rat	213
	Kidney	Mouse	213
Peptide histidine Isoleucine (PHI)	Brain	Cat	156
Galanin (GAL)	Gastrointestinal tract	Pig, guinea pig, rat	158,321
Dynorphin (DYN)	Urogenital tract	Guinea pig	325

neither GRP nor BOM induced any vasoactive responses in cerebral vessels, and the distribution of GRP-L1 nerve fibres in this vascular bed was unaffected by superior cervical ganglionectomy [434]. Furthermore, the localization of GRP-L1 in spinal sensory ganglia suggests that GRP along with somatostatin (SOM) may subserve a sensory role in the vasculature [135, 434]. SOM causes vasoconstriction of isolated feline cerebral vessels [209] and increases arterial blood pressure when administered intracerebroventricularly in anaesthetized rats [418]. Similarly, demonstration of enkephalin (ENK)-L1 perivascular nerve fibres in the brain and gut [249,295,387] and dynorphin-B (DYN)-L1 perivascular nerve fibres in the urogenital system [325] indicates that opioid peptides may also play an important role in the modulation of vascular sensory information.

Recent studies in the guinea-pig iris have shown that DYN co-exists with both noradrenaline and SP, indicating that this peptide has a dual origin from both peripheral and sensory ganglia [193]. Systemic administration of DYN causes bradycardia and hypotension in anaesthetized rats [187]. ENK has been shown to induce vasodilatation of feline pial vessels. This response, which is blocked by naloxone, suggests that there are opioid receptors present in pial vessels [209].

Neurotensin (NT), a tridecapeptide originally isolated from the bovine hypothalamus [88,191], has been demonstrated in nerve fibres closely associated with microvessels in the heart and brain [90,168,361,362]. NT causes vasoconstriction of some isolated blood vessels [357] and vasodilatation when administered in vivo, although in dogs it induces systemic vasoconstriction [37]. NT was found to have no vasomotor action in isolated feline cerebral vessels [209].

Immunoreactive vasopressin-like peptide has recently been shown to be distributed in the sympathetic nervous system, including perivascular nerves of mammals. It appears to co-exist with noradrenaline and has a distribution similar to that of NPY [213]. Physiological studies show that while intracerebroventricular injection of arginine-vasopressin in anaesthetized rats causes a fall in mean arterial blood pressure, its administration in conscious rats and application to isolated feline cerebral vessels causes pressor responses [209,458]. Intrathecal administration into the lower thoracic region of anaesthetized rats also causes a rise in arterial blood pressure [368].

Peptide histidine isoleucine (PHI), a 27-amino acid peptide structurally related to VIP and contained with it in the precursor protein pre-pro-VIP, has recently been shown to be present in perivascular nerve fibres. It appears to co-exist with VIP in some cerebral perivascular nerves and induces dose-dependent depressor responses [156]. Its co-release with VIP into the venous effluent of the feline submandibular gland after parasympathetic nerve stimulation, has implicated this novel peptide in the non-cholinergic vasodilatation in this exocrine gland [303,338].

Cholecystokinin-L1 (CCK-L1) nerve fibres have also been shown to be intimately associated with microvessels in the gut [387] and brain [222]. In the cerebral vasculature, it is suggested that CCK-L1 neurons may have an important role in either modulating the permeability of the blood–brain barrier or in monitoring the chemical composition of the blood and that of the extracellular environment [314]. The recent discovery of galanin and its presence in nerves associated with the vasculature in the gut [158,321] adds further to the list of putative neurotransmitter substances related to the neural control of vascular tone.

The number of perivascular peptidergic nerve types and their varying degrees of vasomotor action support the speculation that not all of them are directly involved in the regulation of vascular tone. It is becoming increasingly clear that they may serve as neuromodulators or neurogenic elements involved in the processing of many sensory modalities. In the cerebral circulation they are strongly implicated in maintaining the integrity of the blood–brain barrier.

Neuromodulation and Co-transmission

Extensive comparative studies on the evolution of the autonomic nervous system [57] and the evidence for co-existence of biologically active substances in some invertebrate nerves [51,101] led Burnstock to question the validity of the principle that one nerve releases only one transmitter [62]. The possibility that some nerves store and release more than one transmitter has now received abundant experimental support which strongly indicates that the majority of nerve fibres utilize several neuroactive substances, some with neurotransmitter roles, others functioning as neuromodulators or trophic factors [66,70].

Neuromodulators are substances which influence neurotransmission either by action on receptors on the pre-junctional varicosity membrane to alter transmitter release or by action on post-junctional receptors to affect the degree and time course of transmitter action on the effector cell. Such neuromodulatory mechanisms are illustrated in Fig. 7.11.

The inhibition of transmitter release by NA, ACh and ATP by action on pre-junctional α_2-adrenoceptors, muscarinic and P_2-purinoreceptors respectively are all examples of auto-inhibition. The concept of "cross talk" between different nerve types, as seen in the inhibitory actions of ACh on responses to sympathetic nerve stimulation via pre-junctional muscarinic receptors [398] and in the NA-induced reduction of ACh release from cholinergic nerves [351], is supported by both pharmacological studies [401,441] and by the close anatomical apposition of adrenergic and cholinergic nerve varicosities which are often enclosed within the same Schwann cell sheath [78]. Recently studies have also shown that opioid receptors on adrenergic nerve terminals of rabbit-ear arteries may serve to modulate NA release by pre-synaptic action [173].

The demonstration of several vesicle types in ultrastructural examination of nerve profiles and recent pharmacological findings are consistent with the multitransmitter concept. However, further electron microscopical studies using specific cytochemical methods for labelling transmitters and related enzymes are needed before vesicle types can be identified as being characteristic of particular transmitter substances.

The use of tritium-labelled adenosine and NA has demonstrated the co-release of ATP and NA from perivascular sympathetic nerves of the rabbit aorta and portal vein [400,280; see also Fig. 7.12]. The co-transmission of ATP with NA has also been shown in the rat-tail artery [394], in the dog basilar artery [332] and in the rabbit-ear and mesenteric arteries [217,254,255,265,400]. ATP and NA release from guinea-pig portal vein and ATP-specific quinacrine binding in rat portal vein are both abolished by sympathectomy [79].

Studies on "axonal reflexes" involving the release of transmitter following antidromic impulses in collateral branches of primary sensory fibres indicate that SP and ATP may co-exist in these nerve fibres [62].

The co-existence of ACh and VIP in para-sympathetic nerves supplying the salivary gland,

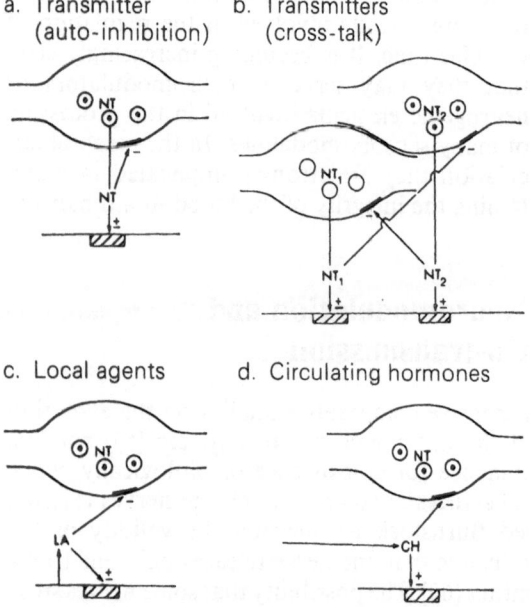

Fig. 7.11. Pre-junctional neuromodulatory mechanisms. **a** Autoinhibition by a single neurotransmitter (*NT*) on specific pre-junctional receptors. **b** Interaction ("cross-talk") between two neurotransmitters released from closely opposed varicosities. **c** Pre-junctional modulation by locally produced agents (*LA*) **d** Pre-junctional modulation by circulating hormones (*Ch*).

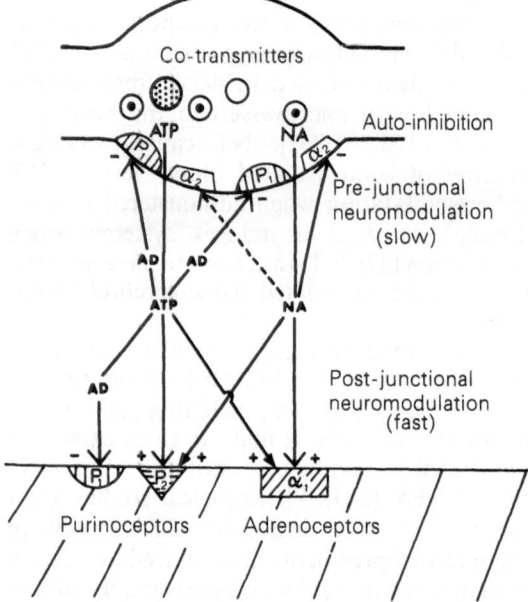

Fig. 7.12. Release, action and interactions of NA and ATP released from sympathetic perivascular nerves. The essential features are the synergistic actions of the co-transmitters, the mutual post-junctional enhancement of their actions and the subsequent inhibition of their release via pre-junctional receptors [82].

Table 7.7. Co-existence of transmitters in perivascular nerves

Classical Transmitter	Putative Transmitter							Vessel	References
	ATP	NPY	SP	CGRP	DYN	VIP	PHI		
NA		NPY						Many vessels	157,196,302,304,305,325, 395,435,437,459
NA		NPY			DYN	VIP		Uterine artery	325
NA	ATP							Basilar artery	332
								Pulmonary artery	254,255
								Mesenteric artery	265
								Portal vein	79,400
								Tail artery	394
								Ear artery	217
Ach						VIP		Cerebral arteries	215,261
								In submandibular gland	289
								In bladder ·	291
	ATP		SP					Cutaneous vessels	63
			SP	CGRP				Many vessels	194,277,306,436
						VIP	PHI	Cerebral arteries	156
								In submandibular gland	303

Evidence for co-existence in perivascular nerves is appearing rapidly in the literature. As such it should be noted that this table was valid only until April 1986

a preparation which has been studied extensively for conditions of release, receptor sites and transmitter action, illustrates one of the unique advantages of co-transmission [289]. At low-frequency stimulation, ACh is released from parasympathetic nerves causing salivary secretion and some vasodilatation. At high frequencies the nerves release VIP causing marked vasodilatation and indirect enhancement of ACh release via pre-junctional modulation.

There is now increasing evidence not only for the co-existence of established transmitters with several peptides, but also for peptide/peptide interactions and co-transmission in perivascular nerves (Table 7.7). For example the localization of CGRP-LI as well as dynorphin-LI in sensory and sympathetic perivascular nerves and ganglia [194,306,331,373] and the inhibition of SP release from primary sensory fibres by opiates and somatostatin [279], support the concept that many peptides serve as neuromodulators of various neurotransmitters in different types of perivascular nerves which are involved in regulating diverse aspects of cardiovascular function.

Concluding Comments

The shift in our understanding of the autonomic nervous system, from the concept of antagonistic adrenergic and cholinergic nerves to the present concept of a remarkably complex system involving the interaction of many substances and nerve types with each other as well as with central and ganglionic mechanisms, has been a rapid one. The multitransmitter concept, co-transmission and neuromodulation now have extensive experimental evidence in their favour and are widely recognized and accepted as fundamental characteristics of the autonomic nervous system. Development of highly selective and sensitive techniques has led to the demonstration of an ever increasing number of substances involved in the neurogenic regulation of the cardiovascular system and more effort must now be directed towards resolving their precise physiological roles in normal as well as experimental and human vascular pathology. The demonstration of tissue-specific regulation of RNA processing, as shown in the discovery of CGRP, strongly warrants further studies to resolve the genetic contribution as well as that of tissue factors, and their possible manipulation in regulation to the development, regeneration and trophic control of vascular neurogenic elements.

The presence of "sensory-motor" nerves and those containing trophic substances suggests that the role of perivascular nerves has expanded from that of merely regulating the calibre of the vessel wall. It is becoming increasingly clear that neuromodulation and co-transmission lead to a high degree of interaction between perivascular nerve types. The search for therapeutic agents for vascular disorders must take into account the importance of this neurohumoral environment, where each perivascular nerve type and the multiple vascular neuroeffector mechanisms play a vital part in the overall fine tuning of the vasculature.

References

1. Abráhám A (1981) Microscopic innervation of the heart and blood vessels in vertebrates including man. Pergamon Press, Oxford
2. Abrahams VC, Hilton SM (1958) Active muscle vasodilatation and its relation to the "fight and flight" reactions in the conscious animal. J Physiol (Lond) 140:16P
3. Adrian TE, Gu J, Allen JM et al. (1984) Neuropeptide Y in the male genital tract. Life Sci 35:2643–2648
4. Ahlqvist RP (1948) A study of the adrenotropic receptors. Am J Physiol 153:586–600
5. Ajelis V, Björklund B, Falk B et al. (1979) Application of the aluminium formaldehyde (ALFA) histofluorescence method for demonstration of peripheral stores of catecholamines and indoleamines in freeze-dried paraffin embedded tissue, cryostat sections and whole mounts. Histochemistry 65:1–15
6. Alafaci C, Cowen T, Crockard HA et al. (1985) The original distribution of noradrenergic and NPY-containing nerves in the cerebral blood vessels of the gerbil. J Cereb Blood Flow Metab 5 [Suppl 1]:S543–S544
7. Alafaci C, Cowen T, Crockard HA et al. (1986) Cerebral perivascular serotonergic fibres have a peripheral origin in the gerbil. Brain Res Bull 16:303–304
8. Alafaci C, Cowen T, Crockard HA et al. (1986) Perivascular nerve types supplying cerebral blood vessels of the gerbil. Acta Physiol Scand 127 (Suppl 552):9–12
9. Allan G, Brook CD, Cambridge D et al. (1983) Enhanced responsiveness of vascular smooth muscle to vasoconstrictor agents after removal of endothelial cells. Br J Pharmacol 79:334P
10. Allen JM, Schon F, Todd N et al. (1984) Presence of neuropeptide Y in human circle of Willis and its possible role in cerebral vasospasm. Lancet II:550–552
11. Allen JM, Gu J, Adrian TE et al. (1984) Neuropeptide Y in the guinea-pig biliary tract. Experientia 40:765–767
12. Alm P, Alumets J, Brodin E et al. (1978) Peptidergic (substance P) nerves in the genitourinary tract. Neuroscience 3:419–425
13. Alm P, Alumets R, Hakanson R et al. (1980) Origin and distribution of VIP (vasoactive intestinal polypeptide) nerves in the genito-urinary tract. Cell Tissue Res 205:337–347
14. Altura BM (1981) Pharmacology of venules: some current concepts and clinical potential. J Cardiovasc Pharmacol 3:1413–1428
15. Altura BM, Malaviy D, Reich CF et al. (1972) Effects of vasoactive agents on isolated human umbilical arteries and veins. Am J Physiol 222: 345–355
16. Amara SG, Jonas V, Rosenfeld MG et al. (1982) Alternative RNA processing in calcitonin gene expression generates mRNAs encoding different polypeptide products. Nature 298:240–244
17. Amenta F, Cavallotti C, Collier WL (1985) The adrenergic innervation of rat uterine tubes in old age: a fluorescence histochemical study. Arch Gerontol Geriatr 4:37–42
18. Amenta F, De Rossi M, Mione MC et al. (1985) Characterization of [^{3}H] 5-hydroxytryptamine uptake within rat cerebrovascular tree. Eur J Pharmacol 112:181–186
19. Angus JA, Brazenor RM, Le Duc MA (1982) Verapil-A selective antagonist of constrictor substance in dog coronary artery: implications for variant angina. Clin Exp Pharmacol Physiol [Suppl 6]:15–28
20. Appenzeller O, Dhital KK, Cowen T et al. (1984) The nerves to blood vessels supplying blood to nerves: the innervation of vasa nervorum. Brain Res 304:383–386
21. Apperley E, Humphrey PPA, Levy GP (1976) Receptors for 5-hydroxytryptamine and noradrenaline in rabbit isolated ear artery and aorta. Br J Pharmacol 58:211–221
22. Armati-Gulson P, Burnstock G (1983) The development of adrenergic innervation in some human foetal blood vessels. J Auton Nerv Syst 7(2):111–118
23. Arneklo-Nobin B, Owman C (1985) Adrenergic and serotoninergic mechanisms in human hand arteries and veins studied by fluorescence histochemistry and in vitro pharmacology. Blood Vessels 22:1–12
24. Azanza MJ, Garin P (1986) The autonomic innervation of the rat diaphragm. Gen Pharmacol 17(1):109–112
25. Bannister R, Crowe R, Burnstock G (1981) Adrenergic innervation in autonomic failure. Neurology 31:1501–1506
26. Barajas L, Wang P (1975) Demonstration of acetyl-cholinesterase in the adrenergic nerves of the renal glomerular arterioles. J Ultrastruct Res 53:244–253
27. Barja F, Mathison R (1982) Adrenergic and peptidergic (substance P and vasoactive intestinal polypeptide) innervation of the rat portal vein. Blood Vessels 19:263–272
28. Barja F, Huggel H, Mathison R (1982) Substance P nerve fibres in peripheral blood vessels. Neuroscience 7:S18
29. Barja F, Mathison R, Huggel H (1983) Substance P-containing nerve fibres in large peripheral blood vessels of the rat. Cell Tissue Res 229:411–422
30. Barja F, Mathison R (1984) Sensory innervation of the rat portal vein and the hepatic artery. J Auton Nerv Sys 10:117–125
31. Barnes PJ, Carstairs JR (1985) Autoradiographic localization of VIP receptors in guinea pig and human lung. Proc Br Pharmacol Soc 18–20 December, 128
32. Baumgarten HG (1972) Biogenic amines in the cyclostome and lower vertebrate brain. Prog Histochem Cytochem 4:1–90
33. Bell C (1974) Selective cholinergic denervation of the uterine artery in the guinea-pig. Experimentia 30:257–258
34. Bérubé A, Marceau F, Drouin JN et al. (1978) The rabbit mesenteric vein: a specific bioassay for substance P. Can J Physiol Pharmacol 56:603–609
35. Bevan JA, Su C (1973) Sympathetic mechanisms in blood vessels: nerve and muscle relationships. In: Elliot HW, George R, Okon R (eds) Annual review of pharmacology. 13:269–285
36. Bevan JA, Ljung B (1974) Longitudinal propagation of myogenic activity in rabbit arteries and in the rat portal vein. Acta Physiol Scand 90:703–751
37. Bissette G, Manberg JW, Nemeroff CB et al. (1978)

Neurotensin, a biologically active peptide. Life Sci 23(22):2173–2182

38. Black JL, French RJ, Mylecharane EJ (1981) Receptor mechanisms for 5-hydroxytryptamine in rabbit arteries. Br J Pharmacol 74:619–626

39. Blackshear JL, Orlandi C, Garnic JD et al. (1985) Differential large and small vessel responses to serotonin in the dog hindlimb in vivo: role of the 5HT$_2$ receptor. J Cardiovasc Pharmacol 7:42–49

40. Blair DA, Glover WE, Roddie IC (1961) Vasomotor responses in the human arm during leg exercise. Circ Res 9:264–74

41. Bloom SR, Bryant MG, Polak JM et al. (1979) Vasoactive intestinal peptide-like immunoreactivity in salivary glands of the rat. J Physiol (Lond) 289:23P

42. Bloom SR, Edwards AV (1980) Vasoactive intestinal peptide in relation to atropine resistant vasodilatation in the submaxillary gland of the cat. J Physiol (Lond) 300:41–53

43. Boatman DL, Shaffer RA, Dixon RL (1965) Function of vascular smooth muscle and its sympathetic innervation in the newborn dog. J Clin Invest 44:241–246

44. Boe J, Boe A-M, Simonsson B et al. (1980) In vitro affects on parasympathetic agonists and atropine on human segmental pulmonary arteries. Lung 157:65–70

45. Borodulya AV, Pletchkova EC (1976) Cholinergic innervation of vessels of the base of the brain. Acta Anat (Basel) 96:135–147

46. Brain SD, Williams TJ, Tippins JR et al. (1985) Calcitonin gene-related peptide is a potent vasodilator. Nature 313:54–56

47. Brain SD, Girgis S, MacIntyre I et al. (1985) Inflammatory oedema induced by synergism between calcitonin gene-related peptide (CGRP) and mediators of increased vascular permeability. Br J Pharmacol 86:855–860

48. Brayden JE, Bevan JA (1985) Inhibition of cerebral neurovasodilatation by VIP-specific antiserum. J Cereb Blood Flow Metab 5 [Suppl 1]:S509

49. Brodin E, Alumets J, Hakanson R et al. (1981) Immunoreactive substance P in the chicken gut: distribution, development and possible functional significance. Cell Tissue Res 216:455–469

50. Brody MJ, Shaffer RA (1970) Distribution of vasodilator nerves in the canine hindlimb. Am J Physiol 218:470–474

51. Brownstein MJ, Saavedra JM, Axelrod J (1974) Coexistence of several putative neurotransmitters in single identified neurons of aplysia. Proc Natl Acad Sci USA 71:4662–4665

52. Buchan AMJ, Polak JM, Pearse AGE (1980) Gut hormones in Salamandra salamandra. Cell Tissue Res 211:331–343

53. Buchan AMJ, Polak JM, Bryant MG et al. (1981) Vasoactive intestinal polypeptide (VIP)-like immunoreactivity in anuran intestine. Cell Tissue Res 216:413–422

54. Bucsics A, Saria A, Lembeck F (1961) Substance P in the adrenal gland: origin and species distribution. Neuropeptides 1:329

55. Bülbring E, Burn JH (1935) The sympathetic dilator fibres in the muscles of the cat and dog. J Physiol (Lond) 83:483–501

56. Burcher E, Atterhög J-H, Pernow B et al. (1977) Cardiovascular effects of substance P: effects on the heart and regional blood flow in the dog. In: Von Euler US, Pernow B (eds) Substance P. Raven Press, New York, pp 261–268

57. Burnstock, G (1969) Evolution of the autonomic innervation of visceral and cardiovascular systems in vertebrates. Pharmacol Rev 21:247–324

58. Burnstock G (1970) Structure of smooth muscle and its innervation. In: Bülbring E, Brading A, Jones A, Tomita T (eds) Smooth muscle. Edward Arnold, London, pp 1–69

59. Burnstock G (1972) Purinergic nerves. Pharmacol Rev 24:509–581

60. Burnstock G (1975) Innervation of vascular smooth muscle: histochemistry and electron microscopy. In: Physiological and pharmacological control of blood pressure. Clin Exp Pharmacol Physiol [Suppl 2]:7–20

61. Burnstock G (1976) Do some nerve cells release more than one transmitter? Neuroscience 1:239–248

62. Burnstock G (1977) Autonomic neuroeffector junctions – reflex vasodilatation of the skin. J Invest Dermatol 69:47–57

63. Burnstock G (1978) A basis for distinguishing two types of purinergic receptors. In: Straub RW, Bolis L (eds) Cell membrane receptors for drugs and hormones: a multidisciplinary approach. Raven Press, New York, pp 107–118

64. Burnstock G (1979) The ultrastructure of autonomic cholinergic nerves and junctions. Prog Brain Res 49:3–21

65. Burnstock G (1980) Cholinergic and purinergic regulation of blood vessels. In: Bohr DF, Somylo AP, Sparks HV (eds) Handbook of physiology, section 2: the cardiovascular system, vol II: vascular smooth muscle. American Physiological Society, Bethesda, pp 567–612

66. Burnstock G (1980) Neurotransmitters, cotransmitters, neuromodulators and trophic factors in the autonomic nervous system. In: Levi-Montalcini R (ed) Nerve cells, transmitters and behaviour. Elsevier, Amsterdam, pp 253–286

67. Burnstock G (1982) The co-transmitter hypothesis, with special reference to the storage and release of ATP with noradrenaline and acetylcholine. In: Cuello AC (ed) Cotransmission. Macmillan, London, pp 151–163

68. Burnstock G (1982) Neuropeptides as trophic factors. In: Bloom SR, Polak JM, Lindenlaub EFK (eds) Systemic role of regulatory peptides. Schattauer Verlag, Stuttgart, pp 423–441

69. Burnstock G (1983) Recent concepts of chemical communication between excitable cells. In: Osborne NN (ed) Dale's principle and communication between neurones. Pergamon Press, Oxford, pp 7–35

70. Burnstock G (1985) Purinergic mechanisms broaden their sphere of influence. TINS 8(1):5–6

71. Burnstock G (1985) Neurohumoral control of blood vessels; some future directions. J Cardiovasc Pharmacol 7 [Suppl 3]:S137–S146

72. Burnstock G (1985) Neurogenic control of cerebral circulation. Cephalagia [Suppl 2]:25–33

73. Burnstock G (1985) Nervous control of smooth muscle by transmitters, cotransmitters and modulators. Experimentia 41:869–874

74. Burnstock G (1986) Purines as cotransmitters in adrenergic and cholinergic neurones. In: Hökfelt T et al. (eds) Coexistence of neuronal messengers: a new principle in chemical transmission. (Progress in brain research 68). Elsevier, Amsterdam, pp 193–203

75. Burnstock G, Prosser CL (1960) Conduction in smooth muscles: comparative electrical properties. Br J Pharmacol 43:180–189

76. Burnstock G, Robinson PM (1967) Localization of

catecholamines and acetylcholinesterase in autonomic nerves. Circ Res 21 [Suppl 3]:43–55

77. Burnstock G, Iwayama T (1971) Fine structural identification of autonomic nerves and their relation to smooth muscle. Prog Brain Res 34:389–404

78. Burnstock G, Costa M (1975) Adrenergic neurons. Chapman and Hall, London

79. Burnstock G, Crowe R, Wong HK (1979) Comparative pharmacological and histochemical evidence for purinergic inhibitory innervation of the portal vein of the rabbit, but not guinea-pig. Br J Pharmacol 65:377–388

80. Burnstock G, Chamley JM, Campbell GR (1980) The innervation of arteries. In: Schwartz CJ, Wethessen NT, Wolf S (eds) Structure and function of the circulation, vol. 1. Plenum Press, New York, pp 729–767

81. Burnstock G, Brown CM (1981) An introduction to purinergic receptors. In: Burnstock G (ed) Purinergic receptors. Chapman and Hall, London, pp 1–45

82. Burnstock G, Griffith SG (1983) Neurohumoral control of the vasculature. In: Woolf N (eds) Biology and pathology of the vessel wall. Praeger, New York, pp 15–40

83. Burnstock G, Griffith SG (1983) Innervation of microvascular smooth muscle. Prog Appl Microcirc 3:19–39

84. Burnstock G, Griffith SG, Sneddon P (1984) Autonomic nerves in the precapillary vessel wall. J Cardiovasc Pharmacol 6:S344–S353

85. Burnstock G, Kennedy C (1985) Is there a basis for distinguishing two types of P_2-purinoceptors? Gen Pharmacol 16(5):433–440

86. Burnstock G, Kennedy C (1986) A dual function for ATP in the regulation of vascular tone: excitatory cotransmitter with noradrenaline from perivascular nerves and locally released inhibitory intravascular agent. Circ Res 58:319–330

87. Carpenter MB (1976) Human neuroanatomy, 7th edn. Williams and Wilkins, Baltimore, p 198

88. Carraway R, Leeman SE (1973) The isolation of a new hypotensive peptide, neurotensin from bovine hypothalami. J Biol Chem 248:6854–6861

89. Chan-Palay V (1976) Serotonin axons in the supra- and subependymal plexuses and in the leptomeninges; their roles in local alterations of cerebrospinal fluid and vasomotor activity. Brain Res 102:103–130

90. Chan-Palay V (1977) Innervation of cerebral blood vessels by norepinephrine, indoleamine, substance P and neurotensin fibers and the leptominengeal indoleamine axons: their roles in vasomotor activity and local alterations of brain blood composition. In: Owman C, Edvinsson L (eds) Neurogenic control of brain circulation. Pergamon Press, Oxford, pp 39–53

91. Chang MM, Leeman SE, Niall HD (1971) Amino acid sequence of substance P. Nature New Biol 232:86

92. Cherry PD, Furchgott RF, Zawadzki JV et al. (1982) Role of endothelial cells in relaxation of isolated arteries by bradykinin. Proc Natl Acad Sci USA 79:2106–10

93. Chubb IW, Hodgson AJ, White GH (1980) Acetylcholinesterase hydrolyses substance P. Neuroscience 5:2065–2072

94. Clague JR, Sternini C, Brecha NC (1985) Localization of calcitonon gene-related peptide-like immunoreactivity in neurons of the rat gastrointestinal tract. Neurosci Lett 56:63–68

95. Cocks TM, Angus JA (1983) Endothelium-dependent relaxation of coronary arteries by noradrenaline and serotonin. Nature 305:627–630

96. Cohen ML, Fuller RW, Wiley KS (1981) Evidence for 5-HT$_2$ receptors mediating contraction in vascular smooth muscle. J Pharmacol Exp Therap 218:421–425

97. Cohen RA, Shepherd JT, Vanhoutte PM (1983) Prejunctional and postjunctional actions of endogenous norepinephrine at the sympathetic neuroeffector junction in canine coronary arteries. Circ Res 52:16–25

98. Corder R, Lowry PJ, Ramage AG et al. (1986) Comparison of the haemodynamic actions of neuropeptide Y, angiotensin II and noradrenaline, in anaesthetized cats. Eur J Pharmacol 121:25–30

99. Costa M, Cuello AC, Furness JB et al. (1980) Distribution of enteric neurons showing immunoreactivity for substance P in the guinea-pig ileum. Neuroscience 5:323–331

100. Costa M, Furness JB, Llewellyn-Smith IJ et al. (1981) Prejections of substance P-containing neurons within guinea-pig small intestine. Neuroscience 6:411–424

101. Cottrell GA (1976) Does the giant cerebral neurone of Helix release two transmitters: ACh and serotonin? J Physiol (Lond) 259:44–45

102. Coughlin SR, Moskowitz MA, Antoniades HN et al. (1981) Serotonin receptor-mediated stimulation of bovine smooth muscle cell prostacyclin synthesis and its modulation by platelet-derived growth factor. Proc Natl Acad Sci USA 78:7134–7138

103. Cowen T, MacCormick DEM, Toff WD et al. (1982) The effect of surgical procedures on blood vessel innervation. A fluorescence histochemical study of degeneration and regrowth of perivascular adrenergic nerves. Blood Vessels 19:65–78

104. Cowen T, Haven AJ, Wen-Qin C et al. (1982) Development and ageing of perivascular adrenergic nerves in the rabbit. A quantitative fluorescence histochemical study using image analysis. J Auton Nerv Syst 5:317–336

105. Cowen T, Haven AJ, Burnstock G (1985) Pontamine sky blue: a counterstain for background autofluorescence in fluorescence and immunofluorescence histochemistry. Histochemistry 82:205–208

106. Cowen T, Alafaci C, Crockard HA et al. (1986) 5-HT-containing nerves to major cerebral arteries of the gerbil originate in the superior cervical ganglia. Brain Res 384:51–59

107. Crowe R, Lincoln J, Blacklay PF et al. (1983) Vasoactive intestinal polypeptide-like immunoreactive nerves in diabetic penis. A comparison between streptozotocin-treated rats and man. Diabetes 32:1075–1077

108. Cuello AC, Gamse R, Holzer P et al. (1981) Substance P-immunoreactive neurons following neonatal administration of capsaicin. Naunyn-Schmiedeberg's Arch Pharmacol 315:185–194

109. Cummings JP, Felten DL (1979) A raphe dendrite bundle in the rabbit medulla. J Comp Neurol 183:1–24

110. Dahlof C, Dahlof P, Lundberg JM (1985) Neuropeptide Y (NPY): enhancement of blood pressure increase upon α-adrenoceptor activation and direct pressor effects in pithed rats. Eur J Pharmacol 109:289–292

111. Dahlström A, Bööj S, Carlsson SS et al. (1981) Rapid accumulation and axonal transport of "cholinergic vesicles" in rat sciatic nerve, studied by immunohistochemistry. Acta Physiol Scand 111:217–219

112. Dalsgaard C-J, Jonsson C-E, Hökfelt T et al. (1983) Localization of SP-immunoreactive fibres in the human digital skin. Experientia 39:1018–1020

113. Daly I, De Burgh I, Hebb C (1966) Pulmonary and bronchial vascular systems. Edward Arnold, London

114. Da Prada M, Richards JG, Lorez HP (1978) Blood platelets and biogenic monoamines: biochemical, pharmacological and morphological studies. In: de Gaetano G, Garattini S (eds) Platelets: a multi-disciplinary approach. Raven Press, New York, pp 331–353

115. Dashwood M, Bagnall J (1982) An autoradiographic demonstration of prazosin binding to arterial vessels in the rat. Eur J Pharmacol 78:121–123

116. Davies B (1983) Adrenergic receptors in autonomic failure. In: Bannister R (ed) Autonomic failure. A textbook of clinical disorders of the autonomic nervous system. Oxford University Press, Oxford, pp 174–200

117. De La Torre JC, Surgeon JW (1976) A methodological approach to rapid and sensitive monoamine histofluorescence using a modified glyoxylic acid technique: the SPG method. Histochemistry 49:81–93

118. Delgado TJ, Brismar J, Svendgaard NA (1985) Subarachnoid haemorrhage in the rat: angiography and fluorescence microscopy of the major cerebral arteries. Stroke 16: 595–601

119. Dey RD, Shannon WA, Said SI (1981) Localization of VIP-immunoreactive nerves in airways and pulmonary vessels of dogs, cats and human subjects. Cell Tissue Res 220:231–238

120. DeMey J, Burnstock G, Vanhoutte PM (1979) Modulation of the evoked release of noradrenaline in canine saphenous vein via presynaptic receptors for adenosine but not ATP. Eur J Pharmacol 55:401–405

121. DeMey JG, Vanhoutte PM (1981) Role of the intima in cholinergic and purinergic relaxation of isolated canine femoral arteries. J Physiol (Lond) 316:347–355

122. Dhall U, Cowen T, Haven AJ et al. (1986) Perivascular noradrenergic and peptidergic nerves show different patterns of changes during development and ageing in the guinea-pig. J Auton Nerv Syst 16:109–126

123. Dhital KK, Lincoln J, Appenzeller O et al. (1986) Adrenergic innervation of vasa and nervi nervorum of optic, sciatic, vagus and sympathetic nerve trunks in normal and streptozotocin-diabetic rats. Brain Res 367:39–44

124. Diani AR, Peterson T, Sawada GA et al. (1985) Elevated level of vasoactive intestinal peptide in the eye and urinary bladder of diabetic and prediabetic clinical hamsters. Diabetologia 28:302–307

125. Di Carlo V (1977) Histochemical evidence for a serotonergic innervation of the microcirculation in the brain stem. In: Owman C, Edvinsson L (eds) Neurogenic control of the brain circulation. Pergamon Press, Oxford, pp 55–58

126. Di Carlo V (1981) Serotoninergic innervation of extrinsic brain stem blood vessels. Neurology 31:104

127. Di Carlo V (1984) Perivascular serotonergic neurons: somatodendritic contacts and axonic innervation of blood vessels. Neurosci Lett 51:295–302

128. Di Carlo V (1984) Segmental serotoninergic innervation of spinal cord arterial circulation. Neurosci Lett 49:225–231

129. Dolezel S, Gerova M, Gero J (1973) Sympathetic construction and monoaminergic innervation of large arteries. Folia Morphol (Praha) 21:364–366

130. Dolezel S, Gerova M, Gero J. (1974) Postnatal development of the sympathetic innervation in skeletal muscles of the dog. Physiol Bohemoslov 23:138–139

131. Drolet G, Gauthier P (1985) Peripheral and central mechanisms of the pressor response elicited by stimulation of the locus coeruleus in the rat. Can J Physiol Pharmacol 63:599–605

132. Drury AN, Szent-Györgyi A (1929) The physiological activity of adenine compounds with special reference to their action upon the mammalian heart. J Physiol (Lond) 68:213–237

133. Duckles SP (1983) Age-related changes in adrenergic neuronal function of rabbit vascular smooth muscle. Neurobiol Aging 4:151–156

134. Duckles SP. Said SI (1982) Vasoactive intestinal peptide as a neurotransmitter in the cerebral circulation. Eur J Pharmacol 78:371–374

135. Duckles SP, Buck SM (1982) Substance P in the cerebral vasculature: depletion by capsaicin suggests a sensory role. Brain Res 245:171–174

136. Duckles SP, Carter BJ, Williams CL (1985) Vascular adrenergic neuroeffector function does not decline in aged rats. Circ Res 56:109–116

137. Eckenstein F, Thoenen H (1982) Production of specific antisera and monoclonal antibodies to choline acetyltransferase: characterization and use for identification of cholinergic neurons. EMBO 1:363–368

138. Edvinsson L (1985) Characterization of the contractile effect of neuropeptide Y in feline cerebral arteries. Acta Physiol Scand 125:33–41

139. Edvinsson L (1985) Functional role of perivascular peptides in the control of cerebral circulation. TINS 8(3):126–131

140. Edvinsson L, Nielsen KC, Owman C et al. Sympathetic adrenergic influence on brain vessels as studied by changes in cerebral blood volume of mice. Eur Neurol 6:193–202

141. Edvinsson L, Nielsen KC, Owman C et al. (1972) Cholinergic mechanism in pial vessels. Histochemistry, electron microscopy and pharmacology. Z Zellforsch Mikrosk Anat 134:311–325

142. Edvinsson L, Hardebo JC, Owman C (1978) Pharmacological analysis of 5-hydroxytryptamine receptors in isolated intracranial and extracranial vessels of cat and man. Circ Res 42:143–151

143. Edvinsson L, Fahrenkrug J, Hako J et al. (1980) VIP (vasoactive intestinal polypeptide)-containing nerves of intracranial arteries in mammals. Cell Tissue Res 208:135–142

144. Edvinsson L, Uddman R (1981) Adrenergic, cholinergic and peptidergic nerve fibres in dura mater – involvement in headache? Cephalagia 1:175–179

145. Edvinsson L, McCulloch J, Uddman R (1981) Substance P: immunohistochemical localization and effect upon cat pial arteries in vitro and in situ. J Physiol (Lond) 318:251–258

146. Edvinsson L, McCulloch J, Uddman R (1982) Feline cerebral veins and arteries: composition of autonomic innervation and vasomotor responses. J Physiol (Lond) 325:161–173

147. Edvinsson L, Uddman R (1982) Immunohistochemical localization and dilatory effect of substance P on human cerebral vessels. Brain Res 232:466–471

148. Edvinsson L, Egund N, Owman CH et al. (1982) Reduced noradrenaline uptake and retention in cerebrovascular nerves associated with angiographically visible vasoconstriction following experimental subarachnoid hemorrhage in rabbits. Brain Res Bull 9:799–805

149. Edvinsson L, Rosendal-Hegesen S, Uddman R (1983) Substance P: localization, concentration and release in cerebral arteries, choroid plexus and dura mater. Cell Tissue Res 234:1–7

150. Edvinsson L, Degueurce A, Duverger D et al. (1983) Central serotonergic nerves project to the pial vessels of the brain. Nature 306:55–57

151. Edvinsson L, Emson P, McCulloch J et al. (1983) Neuropeptide Y: cerebrovascular innervation and vasomotor effects in the cat. Neurosci Lett 43:79–84

152. Edvinsson L, Emson P, McCulloch J et al. (1984) Neuropeptide Y: immunocytochemical localization to and effect upon feline pial arteries and veins in vitro and in situ. Acta Physiol Scand 122:155–163

153. Edvinsson L, Ekblad E, Hakanson R et al. (1984) NPY-potentiates the effect of various vasoconstrictor agents on rabbit blood vessels. Br J Pharmacol 83:519–525

154. Edvinsson L, Edman R, Ottoson A et al. (1985) Distribution, concentration and effects of neuropeptide-Y (NPY), substance P (SP) and vasoactive intestinal polypeptide (VIP) in human cerebral blood vessels. J Cereb Blood Flow Metab 5 [Suppl 1]:S545

155. Edvinsson L, McCulloch J, Sharkey J (1985) Vasomotor responses of cerebral arterioles in situ to putative dopamine receptor agonists. Br J Pharmacol 85:403–410

156. Edvinsson L, McCulloch J (1985) Distribution and vasomotor effects of peptide HI (PMI) in feline cerebral blood vessels in vitro and in situ. Regul Pept 10:345–356

157. Ekblad E, Edvinsson L, Wahlestedt C et al. (1984) Neuropeptide Y co-exists and co-operates with noradrenaline in perivascular nerve fibres. Regul Pept 8:225–235

158. Ekblad E, Rökaeus A, Hakanson R et al. (1985) Galanin nerve fibres in the rat gut: distribution, origin and projections. Neuroscience 16:355–363

159. Eklund S, Fahrenkrug J, Jodal M et al. (1980) Vasoactive intestinal polypeptide, 5-hydroxytryptamine and reflex hyperaemia in the small intestine of the cat. J Physiol (London) 302:549–557

160. Eränko O, Eränko L (1971) Loss of histochemically demonstrable catecholamines and acetylcholinesterase from sympathetic nerve fibres of the pineal body of the rat after chemical sympathectomy with 6-hydroxydopamine. Histochem J 3:357–363

161. Erspamer V, Melchiorri P, Sopranzi N (1972) The action of bombesin on the systemic arterial blood pressure of some experimental animals. Br J Pharmacol 45:442–450

162. Euler US von, Gaddum JH (1931) An unidentified depressor substance in certain tissue extracts. J Physiol (Lond) 72:74

163. Falck B, Hillarp NA, Thieme G et al. (1962) Fluorescence of catecholamines and related compounds condensed with formaldehyde. J Histochem Cytochem 10:348–354

164. Falck B, Owman Ch (1965) A detailed methodological description of the fluorescence method for the cellular demonstration of biogenic monoamines. Acta Univ Lund Sect 27:1–23

165. Felten DL, Crutcher KA (1979) Neuronal–vascular relationships in the raphe nuclei, locus coeruleus, and substantia nigra in primates. Am J Anat 155:467–482

166. Ferguson M, Bell C (1985) Substance P-immunoreactive nerves in the rat kidney. Neurosci Lett 60:183–188

167. Fisher LA, Kikkawa DO, Rivier JE et al. (1983) Stimulation of noradrenergic sympathetic outflow by calcitonin gene-related peptide. Nature 305:534–536

168. Forssman BB, Hock D, Metz J (1982) Peptidergic innervation of the kidney. Neurosci Lett [Suppl] 10:S183

169. Forster C, Drew GM, Hilditch A et al. (1983) Dopamine receptors in human basilar arteries. Eur J Pharmacol 87:227–235

170. Franco-Cereceda A, Dahlöf C, Lundberg JM (1985) Role of neuropeptide Y and calcitonin-gene related peptide (CGRP) in cardiac sympathetic and sensory control. Acta Physiol Scand [Suppl] 124:136

171. Fraser RAR, Stein BM, Barrett RE et al. (1970) Noradrenergic mediation of experimental cerebrovascular spasm. Stroke 1:356–362

172. Freund H, Ebeid AM, Fischer JE (1981) An increase in vasoactive intestinal peptide levels in canine endotoxin shock. Surg Gynecol Obstet 152:604–606

173. Fukuda M, Hosoki E, Ishida Y et al. (1985) Opioid receptor types on adrenergic nerve terminals of rabbit ear artery. Br J Pharmacol 86:539–545

174. Furchgott RF (1981) The requirement for endothelial cells in the relaxation of arteries by acetylcholine and some other vasodilators. Trends Pharmacol Sci 2:173–176

175. Furchgott RF (1983) Role of endothelium in responses of vascular smooth muscle. Circ Res 53:557–573

176. Furness JB (1973) Arrangement of blood vessels and their relation with adrenergic nerves in the rat mesentery. J Anat 115:346–364

177. Furness JB, Costa M, Wilson AJ (1977) Water-stable fluorophores produced by reaction with aldehyde solution, for the histochemical localization of catechol- and indole thylamine. Histochemistry 52:159–170

178. Furness JB, Papka RE, Della NG et al. (1982) Substance P-like immunoreactivity in nerves associated with the vascular system of guinea-pigs. Neuroscience 7:447–459

179. Furness JB, Costa M, Eckenstein F (1983) Neurones localized with antibodies against choline acetyltransferase in the enteric nervous system. Neurosci Lett 40:105–109

180. Furness JB, Costa M, Keast JR (1984) Choline acetyltransferase and peptide immunoreactivity of submucous neurons in the small intestine of the guinea-pig. Cell Tissue Res 237:329–336

181. Furness JB, Costa M, Papka RE et al. (1984) Neuropeptides contained in peripheral cardiovascular nerves. Clin Exp Hypertens [A] 6 (12):91–106

182. Furness JB, Costa M, Gibbins IL et al. (1985) Neurochemically similar myenteric and submucous neurons directly traced to the mucosa of the small intestine. Cell Tissue Res 241:155–163

183. Gabella G (1976) Structure of the autonomic nervous system. Chapman and Hall, London

184. Gaddum JH, Picarelli ZP (1957) Two kinds of tryptamine receptor. Br J Pharmacol 12:323–328

185. Gamse R, Leeman SE, Holzer P et al. (1981) Differential aspects of capsaicin on the content of somatostatin, substance P and neurotensin in the nervous system of the rat. Naunyn Schmiedebergs Arch Pharmacol 317:140

186. Gaudin-Chazel G, Portalier P, Barrit MC et al. (1982) Serotonin-like immunoreactivity in paraffin sections of the nodose ganglia of the cat. Neurosci Lett 33:169–172

187. Gautret B, Schmidt H (1985) Central and peripheral sites for cardiovascular actions of Dynorphin- (1–13) in rats. Eur J Pharmacol 111:263–266

188. Gazelius B, Olgart L (1980) Vasodilatation in the dental pulp produced by electrical stimulation of the inferior olveolar nerve in the cat. Acta Physiol Scand 108:181–186

189. Gazelius B, Edwall B, Lundberg J et al. (1985) Calcitonin gene-related peptide (CGRP), a potent vaso-

dilator related to sensory nerves in the cat. Acta Physiol Scand [Suppl] 124:134

190. Geffen LB, Livett DB, Rush RA (1969) Immunohistochemical localization of protein components of catecholamine storage vesicles. J Physiol (Lond) 204:593–605

191. Geller RG, Govier WC, Pisano JJ et al. (1970) The action of ranatensin, a new polypeptide from amphibian skin, on the blood pressure of experimental animals. Br J Pharmacol 40:605–616

192. Gibbins IL, Brayden JE, Bevan JA (1984) Perivascular nerves with immunoreactivity to vasoactive intestinal polypeptide in cephalic arteries of the cat: distribution, possible origins and functional implications. Neuroscience 13:1327–1346

193. Gibbins IL, Morris J-L, Furness JB et al. (1985) Neuropeptide Y and dynorphin in noradrenergic neurons and substance P in presumptive cholinergic neurons innervating the guinea-pig iris. Proc Aust Physiol Pharmacol Soc 16(2):91P

194. Gibbins IL, Furness JB, Costa M et al. (1985) Co-localization of calcitonin gene-related peptide-like immunoreactivity with substance P in cutaneous vascular and visceral sensory neurons of guinea-pigs. Neurosci Lett 57:125–130

195. Gibson SJ, Polak JM, Bloom SR et al. (1984) Calcitonin gene-related peptide immunoreactivity in the spinal cord of man and of eight other species. J Neurosci 4:3101–3111

196. Glover WE (1985) Increased sensitivity of rabbit ear artery to noradrenaline following perivascular nerve stimulation may be a response to neuropeptide Y released as cotransmitter. Clin Exp Pharmacol Physiol 12:227–230

197. Gokhale SD, Gulati OD, Kelkar LV et al. (1966) Effects of some drugs on human umbilical artery in vitro. Br J Pharmacol 27:332–346

198. Goldberg LI, Kohli JD (1981) Specific dopamine receptors in vascular smooth muscle. In: Vanhoutte PM, Leussen I (eds) Vasodilatation. Raven Press, New York, pp 131–140

199. Gräb W, Jenssen S, Rein N (1929) Die leberals Blutdepot. Klin Wochschr 8:1539–1548

200. Greenwald SE, Lever MJ, MacIntyre I et al. (1985) Human calcitonin gene-related peptide is a potent vasodilator in the pig coronary circulation. Proc Br Pharmacol Soc (Abstracts) 1985, 18–20

201. Greenway CV, Lawson AE (1966) The effects of adrenaline and noradrenaline on venous return and regional blood flow in the anaesthetised cat with special reference to intestinal blood flow. J Physiol (Lond) 186:579–595

202. Griffith SG, Lincoln J, Burnstock G (1982) Serotonin as a neurotransmitter in cerebral arteries. Brain Res 247:388–392

203. Griffith SG, Burnstock G (1983) Serotoninergic neurons in human fetal intestine: an immunohistochemical study. Gastroenterology 85:929–937

204. Griffith SG, Burnstock G (1983) Immunohistochemical demonstration of serotonin in nerves supplying human cerebral and mesenteric blood-vessels: some speculations about their involvement in vascular disorders. Lancet I:561–562

205. Grigor'eva TA (1962) The innervation of blood vessels. Pergamon Press, New York

206. Gu J, Huang WM, Blank M et al. (1983) Measurement of VIP and origin of its innervation in the rat urinary bladder. Regul Pept 6(3):305

207. Gu J, Polak JM, Allen JM, et al. (1984) High concentrations of a novel peptide, Neuropeptide Y, in the innervation of mouse and rat heart. J Histochem Cytochem 32:467–472

208. Hallberg D, Pernow B (1975) Effect of substance P on various vascular beds in the dog. Acta Physiol Scand 93:277–285

209. Hanko J, Hardebo JE, Owman Ch (1981) Effects of various neuropeptides on cerebral blood flow. J Cereb Blood Flow Metab 1:S346–S347

210. Hanko JH, Tornebrandt K, Hardebo JE et al. (1985) Neuropeptide Y induces and modulates vasoconstriction in intracranial and peripheral vessels of animals and man. J Cereb Blood Flow Metab 5 [Suppl 1]:S511

211. Hanko J, Hardebo JE, Kahrström J et al. (1985) Calcitonin gene-related peptide is present in mammalian cerebrovascular nerve fibres and dilates pial and peripheral arteries. Neurosci Lett 57:91–95

212. Hanko J, Sundler F, Hardebo JE et al. (1985) Dilatation of pial and peripheral arteries by calcitonin gene-related peptides. J Cereb Blood Flow Metab 5 [Suppl 1]:S507

213. Hanley MR, Benton HP, Lightman SL et al. (1984) A vasopressin-like peptide in the mammalian sympathetic nervous system. Nature 309:258–261

214. Häppölä O, Pä värinta H, Soinila S et al. (1986) Pre- and postnatal development of 5-hydroxytryptamine-immunoreactive cells in the superior cervical ganglion of the rat. J Auton Nerv Syst 15:21–31

215. Hara H, Hamill GS, Jacobowitz DM (1985) Origin of cholinergic nerves to the rat major cerebral arteries: coexistence with vasoactive intestinal polypeptide. Brain Res Bull 14:179–188

216. Hardebo JE, Hanko J, Kährström J et al. (1985) Endothelium-dependent relaxation in cerebral arteries. J Cereb Flood Flow Metab 5 [Suppl 1]:S533

217. Head RJ, Stitzel RE, Delaland IS et al (1977) Effect of chronic denervation on activities of monoamine-oxidase and catechol-O-methyl transferase and on contents of noradrenaline and adenosine-triphosphate in rabbit ear artery. Blood Vessels 14:229–239

218. Hedge GA, Huffman LF, Grunditz T et al. (1984) Immunocytochemical studies of the peptidergic innervation of the thyroid gland in the Brattleboro rat. Endocrinology 115:2071–2076

219. Helke CJ, O'Donohue TL, Jacobowitz DM (1980) Substance P as a baro- and chemoreceptor afferent neurotransmitter: immunocytochemical and neurochemical evidence in the rat. Peptides 1:1–9

220. Hellstrand P, Järhult J (1980) Effects of nine different gastrointestinal polypeptides on vascular smooth muscle in vitro. Acta Physiol Scand 110:89–94

221. Hellstrand P, Fahrenkrug J, Uddman R et al. (1985) Role of vasoactive intestinal polypeptide (VIP) in the neurogenic vasodilatation of the portal vein in the rabbit. Regul Pept 12:309–316

222. Hendry SHC, Jones EG, Beinfeld MC (1983) Cholecystokinin-immunoreactive neurons in rat and monkey cerebral cortex make symmetric synapses and have intimate associations with blood vessels. Proc Natl Acad Sci USA 80:2400–2404

223. Hill CE, Hirst GDS, Van Helden DF (1983) Development of sympathetic innervation to proximal and distal arteries of the rat mesentery. J Physiol (Lond) 338:129–147

224. Hillarp N-A (1946) Structure of the synapse and the peripheral innervation apparatus of the autonomic

nervous system. Acta Anat [Suppl 4]:1–153

225. Hirst GDS, Neild TO (1980) Evidence for two populations of excitatory receptors for noradrenaline on arteriolar smooth muscle. Nature 283:767–768

226. Hogaboom GK, O'Donnell JP, Fedan JS (1980) Purinergic receptors: photoaffinity analog of adenosine triphosphate is a specific adenosine triphosphate antagonist. Science 208:1273–1275

227. Hökfelt T, Kellerth J-O, Nilsson G et al. (1975) Experimental immunohistochemical studies on the localization and distribution of substance P in cat primary sensory neurons. Brain Res 100:235–252

228. Hökfelt T, Johansson O, Kellerth J-O (1977) Immunohistochemical distribution of substance P. In: Von Euler, US, Pernow B (eds) Substance P. Raven Press, New York, pp 117–145

229. Hökfelt T, Lundberg JM, Schultzberg M et al. (1981) Immunohistochemical evidence for a local VIP-ergic neuron system in the adrenal gland of the rat. Actas Physiol Scand 113:575–576

230. Holmgren S, Vaillant C, Dimaline R (1982) VIP-, substance P-, gastrin/CCK-, bombesin-, somatostatin- and glucagon-like immunoreactivities in the gut of the rainbow trout, *Salmo gairdneri*. Cell Tissue Res 223:141–153

231. Holton P (1959) The liberation of adenosine triphosphate on antidromic stimulation of sensory nerves. J Physiol (Lond) 145:494–504

232. Holzer P, Gamse R, Lembeck F (1980) Distribution of substance P in the rat gastrointestinal tract – lack of effect of capsaicin pretreatment. Eur J Pharmacol 61:303

233. Holzer P, Ducsics A, Lembeck F (1982) Distribution of capsaicin-sensitive nerve fibres containing immunoreactive substance P in cutaneous and visceral tissues of the rat. Neurosci Lett 31:253

234. Honig CR, Frierson JL (1976) Neurons intrinsic to arterioles initiate postcontraction vasodilation. Am J Physiol 230:493–507

235. Honma S (1970) Presence of monoaminergic neurons in the spinal cord and intestine of the Lamprey, *Lampetra japonica*. Arch Histol Jpn 32:383–393

236. Humphrey PPA, Feniuk W, Watts AD (1983) Prejunctional effects of 5-hydroxytryptamine on noradrenergic nerves in the cardiovascular system. Fed Proc 42:218–222

237. Iijima T (1977) A histochemical study of the innervation of cerebral blood vessels in the turtle. J Comp Neurol 176:307–314

238. Iijima T, Wasano T (1980) A histochemical and ultrastructural study of serotonin-containing nerves in cerebral blood vessels of the lamprey. Anat Rec 198:671–680

239. Itakura T, Okuno T, Nakakita K et al. (1984) A light and electron microscopic immunohistochemical study of vasoactive intestinal polypeptide- and substance P-containing nerve fibres along the cerebral blood vessels: Comparison with aminergic and cholinergic nerve fibres. J Cereb Blood Flow Metab 4:407–414

240. Itakura T, Yokote H, Kumura H et al. (1985) 5-Hydroxytryptamine innervation of vessels in the rat cerebral cortex. Immunohistochemical findings and hydrogen clearance study of CBF. J Neurosurg 62:42–47

241. Hakura T, Nakai K, Nakakita K et al. (1985) Autoradiographic demonstration of adrenergic receptors in cerebral blood vessels. J of Cereb Blood Flow Metab 5 [Suppl 1]:S495

242. Järhult J, Hellstrand P, Sundler F (1980) Immunohistochemical localization and vascular effects of vasoactive intestinal polypeptide in skeletal muscle of the cat. Cell Tissue Res 207:55–64

243. Järhult J, Fahrenkrug J, Hellstrand P et al. (1982) VIP (vasoactive intestinal polypeptide)-immunoreactive innervation of the portal vein. Cell Tissue Res 221:617–625

244. Jessen KR, Saffrey MJ, Van Noorden S et al. (1980) Immunohistochemical studies of the enteric nervous system in tissue culture and in situ: localization of vasoactive intestinal polypeptide (VIP), substance P and enkephalin immunoreactive nerves in the guinea-pig gut. Neuroscience 5:1717–1735

245. Jójárt I, Joó F, Siklós L, László FA (1984) Immunoelectronhistochemical evidence for innervation of brain microvessels bu vasopressin-immunoreactive neurons in the rat. Neurosci Lett 51:259–264

246. Kai-Kai MA, Ken P (1985) Localization of 5-hydrozytryptamine to neurons and endoneurial mast cells in rat sensory ganglia. J Neurocytol 14:63–78

247. Kalkman HO, Boddeke HWGM, Doods HN et al. (1983) Hypotensive activity of serotonin receptor agonists in rats is related to their affinity for 5-HT, receptors. Eur J Pharmacol 91:155–156

248. Kamilkawa Y, Cline JR, Su C (1980) Diminished purinergic modulation of the vascular adrenergic neutrotransmission in spontaneously hypertensive rats. Eur J Pharmacol 66:347–353

249. Kapadia SE, de Lanerolle NC (1984) Immunohistochemical and electron microscopic demonstration of vascular innervation in the mammalian brain stem. Brain Res 292:33–39

250. Karenkan K-S, Chao LP (1981) Localization of choline acetyltransferase at neuromuscular junctions. Muscle Nerve 4:91–93

251. Karnovsky MJ, Roots L (1964) A "direct coloring" thiocholine method for cholinesterase. J Histochem Cytochem 12:219–221

252. Kása P, Mann SP, Hebb C (1970) Localization of choline acetyltransferase. Nature 226:812–814

253. Kasakov L, Burnstock G (1983) The use of the slowly degradable analog, α–β -methylene ATP, to produce desensitisation of the P_2-purinoceptor: effect on nonadrenergic, noncholinergic responses of the guinea-pig urinary bladder. Eur J Pharmacol 86:291–294

254. Katsuragi T, Su C (1980) Purine release from vascular adrenergic nerves by high potassium and a calcium ionophore A-23187. J Pharmacol Exp Ther 215:685–690

255. Katsuragi T, SU C (1982) Augmentation by theophylline of (^{3}H) purine release from vascular adrenergic nerves: evidence for presynaptic autoinhibition. J Pharmacol Exp Ther 220:152–156

256. Katz B (1966) Nerve, muscle and synapse. McGraw-Hill, New York

257. Kennedy C, Burnstock G (1985) Evidence for two types of P_2-purinoceptors in the longitudinal muscle of the rabbit portal vein. Eur J Pharmacol 111:49–56

258. Kennedy C, Delbro D, Burnstock G (1985) P_2-Purinoceptors mediate both vasodilation (via the endothelium) and vasoconstriction of the isolated rat femoral artery. Eur J Pharamcol 107:161–168

259. Kilborn MJ, Potter EK, McCloskey DI (1985) Neuromodulation of the cardiac vagus: comparison of neuropeptide Y and related peptides. Regul Pept 12:155–161

260. Kobayashi S, Tsukahara S, Sugita K et al. (1981) Adre-

nergic and cholinergic innervation of rat cerebral arteries. Histochem 70:129–138

261. Kobayashi S, Kyoshima K, Olschowka JA et al. (1983) Vasoactive intestinal polypeptide immunoreactive and cholinergic nerves in the whole mount preparation of the major cerebral arteries of the rat. Histochemistry 79:377–381

262. Kobayashi H, Wada A, Izumi F et al. (1985) Alpha-adrenergic receptors in cerebral microvessels of normotensive and hypertensive rats. Circ Res 56:402–409

263. Koelle GB, Friedenwald JS (1949) A histochemical method for localizing cholinesterase activity. Proc Soc Exp Biol Med 70:617–622

264. Kubo T, Su C (1983) Effects of adenosine on [^{3}H] norepinephrine release from perfused mesenteric arteries of SHR and renal hypertensive rats. Eur J Pharmacol 87:349–352

265. Kügelen IV, Starke K (1985) Noradrenaline and adenosine triphosphate as co-transmitters of neurogenic vasoconstriction in rabbit mesenteric artery. J Physiol (Lond) 367:435–455

266. Lagnado MLJ, Crowe R, Lincoln J et al. (1988) A reduction of nerves containing vasoactive intestinal polypeptide and serotonin, but not neuropeptide Y and noradrenaline in cerebral blood vessel of the 8 week streptozotocin-induced diabetic rat. Blood Vessels (in press)

267. Lamotte CC, Johns DR, de Lanerolle NC (1982) Immunohistochemical evidence of indolamine neurons in Monkey spinal cord. J Comp Neurol 206:359–370

268. Lands AM, Arnold A, McAuliff JP et al. (1967) Differentiation of receptor systems activated by sympathomimetic amines. Nature 214:597–598

269. Langer SZ (1974) Presynaptic regulation of the release of catecholamines. Pharmacol Rev 23:1793–1800

270. Larsson L-I, Fahrenkrug J, Schaffalitzky de Muckadell O et al. (1976) Localization of vasoactive intestinal polypeptide (VIP) to central and peripheral neurons. Proc Natl Acad Sci USA 73:3197–3200

271. Larsson L-I, Edvinsson L, Fahrenkrug J et al. (1976) Immunohistochemical localization of a vasoactive polypeptide (VIP) in cerebrovascular nerves. Brain Res 113:400–404

272. Larsson L-I, Fahrenkrug J, Schaffalitzky de Muckadell O (1977) Occurrence of nerves containing vasoactive intestinal polypeptide immunoreactivity in the male genital tract. Life Sci 21:503–508

273. Larsson L-I, Fahrenkrug J, Schaffalitzky de Muckadell O (1977) Vasoactive intestinal polypeptide occurs in nerves of the female genitourinary tract. Science 197:1374–1375

274. Lechin F, Van der Dijs B, Lechin E et al. (1978) The dopaminergic and noradrenergic blockades: a new treatment for headache. Headache 18: 69–74

275. Lee TJ-F, Saito A (1984) Altered cerebral vessel innervation in the spontaneously hypertensive rat. Circ Res 55:393–403

276. Lee TJ-F, Saito A, Berezin I (1984) Vasoactive intestinal polypeptide-like substance: the potential transmitter for cerebral circulation. Science 224:898–901

277. Lee Y, Takami K, Kawai Y et al. (1985) Distribution of calcitonin gene-related peptide in the rat peripheral nervous system with reference to its coexistence with substance P. Neuroscience 15:1227–1237

278. Lembeck F, Holzer P (1979) Substance P as neurogenic mediator of antidromic vasocilation and neurogenic plasma extravasation. Naunyn Schmiedebergs Arch Pharmacol 310:175

279. Lembeck F, Donnerer J, Bartho L (1982) Inhibition of neurogenic vasodilatation and plasma extravasation by substance P antagonists, somatostatin and (D-METZ, PRO5) enkephalinamide. Eur J Pharmacol 85:171–176

280. Levitt B, Westfall DP (1982) Factors influencing the release of purines and norepinephrine in the rabbit portal vein. Blood Vessels 19:30–40

281. Liang CC (1975) A possible sympathetic cholinergic mechanism in the renal reflex elicited by stimulation of abdominal viscera in the dog. Clin Exp Pharmacol Physiol 2:103–117

282. Lindvall O, Björklund A (1974) The glyoxylic acid fluorescence histochemical method: a detailed account of the methodology for the visualization of central catecholamine neurons. Histochemistry 39:97–127

283. Lindvall M, Alumets J, Edvinsson L et al. (1978) Peptidergic (VIP) nerves in the mammalian choroid plexus. Neurosci Lett 9:77–82

284. Liu-Chen LY, Mayberg M, Moskowitz MA (1983) Immunohistochemical evidence for a substance P-containing trigemino-vascular pathway to pial arteries in cats. Brain Res 268:162–166

285. Liu-Chen LY, Gillespie SA, Norregaard TV et al. (1984) Cholecystokinin-8 (CCK8) immunoreactivity in cerebral arteries and pia arachnoid and the effect of unilateral trigeminal ganglionectomy. Fed Proc 43:304

286. Lobato RD, Marin J, Salaices M et al. (1980) Effect of experimental subarachnoid hemorrhage on the adrenergic innervation of cerebral arteries. J Neurosurg 53:477–479

287. Londos C, Cooper DMF, Wolff J (1980) Subclasses of external adenosine receptors. Proc Natl Acad Sci USA 77:2551–2554

288. Low PA, Walsh JC, Huang CY et al. (1975) The sympathetic nervous system in alcoholic neuropathy. A clinical and pathological study. Brain 98:357–364

289. Lundberg JM (1981) Evidence for coexistence of vasoactive intestinal polypeptide (VIP) and acetylcholine in neurons of cat exocrine glands: morphological, biochemical and functional studies. Acta Physiol Scand [Suppl 496]:1–57

290. Lundberg J, Ljung B, Stage D et al. (1976) Postnatal autogenic development of the adrenergic innervation pattern in rat portal vein: a histochemical study. Cell Tissue Res 172:15–27

291. Lundberg JM, Hökfelt T, Schultzberg M et al. (1979) Occurrence of vasoactive intestinal polypeptide (VIP)-like immunoreactivity in certain cholinergic neurons of the cat: evidence from combined immunohistochemistry and acetylcholinesterase staining. Neuroscience 4:1539–1559

292. Lundberg JM, Hökfelt T, Fahrenkrug J et al. (1979) Peptides in the cat carotid body (glomus caroticum): VIP-, enkephalin-, and substance P-like immunoreactivity. Acta Physiol Scand 107:279–281

293. Lundberg JM, Hökfelt T, Änggärd et al. (1980) Peripheral peptide neurons: distribution, axonal transport, and some aspects on possible function. In: Costa E, Trabucchi M (eds) Advances in biochemical psychopharmacology, vol 22. Neural peptides and neuronal communication. Raven Press, New York, pp 25–36

294. Lundberg JM, Änggärd A, Fahrenkrug J et al. (1980) Vasoactive intestinal polypeptide in cholinergic neurons of exocrine glands: functional significance of coexisting transmitters for vasodilation and secretion. Proc Natl Acad Sci USA 77:1651–1655

295. Lundberg JM, Hökfelt T, Änggärd A et al. (1980) Peripheral peptide neurons: distribution, axonal trans-

port, and some aspects on possible function. In: Costa E, Trabucchi M (eds) Advances in biochemical psychopharmacology, vol 22. Neural peptides and neuronal communication. Raven Press, New York, pp 25–36

296. Lundberg JM, Hökfelt T, Änggård A et al. (1980) Coexistence of an avian pancreatic polypeptide (APP) immunoreactive substance and catecholamines in some peripheral and central neurons. Acta Physiol Scand 110:107–109

297. Lundberg JM, Änggård A, Emson P et al. (1981) Vasoactive intestinal polypeptide and cholinergic mechanisms in cat nasal mucosa: studies on choline acetyltransferase and release of vasoactive intestinal polypeptide. Proc Natl Acad Sci USA 78:5255–5259

298. Lundberg JM, Terenius L, Hökfelt T et al. (1982) Neuropeptide Y (NPY)-like immunoreactivity in peripheral noradrenergic neurons and effects of NPY on sympathetic function. Acta Physiol Scand 116:477–480

299. Lundberg JM, Änggård A, Fahrenkrug J (1982) VIP as a mediator of hexamethonium-sensitive, atropine-resistant vasodilation in the cat tongue. Acta Physiol Scand 116:387–392

300. Lundberg JM, Tatemoto K (1982) Pancreatic polypeptide family (APP, BPP, NPY and NPY) in relation to sympathetic vasoconstriction resistant to α-adrenoceptor blockade. Acta Physiol Scand 116:393–402

301. Lundberg JM, Saria A (1982) Vagal substance P nerves involved in control of vascular permeability and smooth muscle tone in trachea and bronchi. Br J Pharmacol 77:441P

302. Lundberg JM, Terenius L, Hökfelt T et al. (1983) High levels of neuropeptide Y in peripheral noradrenergic neurons in various mammals including man. Neurosci Lett 42:167–172

303. Lundberg JM, Fahrenkrug J, Larsson O et al. (1984) Corelease of vasoactive intestinal polypeptide and peptide histidine isolencine in relation to atropine-resistant vasodilatation in cat submandibular salivary gland. Neurosci Lett 52:37–42

304. Lundberg JM, Änggård A, Pernow J et al. (1985) Neuropeptide Y-, substance P- and VIP-immunoreactive nerves in cat spleen in relation to autonomic vascular and volume control. Cell Tissue Res 239:9–18

305. Lundberg JM, Saria A, Franco-Cerceda A et al. (1985) Differential effects of reserpine and 6-hydroxydopamine on neuropeptide Y (NPY) and noradrenaline in peripheral neurons. Naunyn Schmiedebergs Arch Pharmacol 328:331–340

306. Lundberg JM, Franco-Cereceda A, Hua X et al. (1985) Coexistence of substance P and calcitonin gene-related peptide-like immunoreactivities in sensory nerves in relation to cardiovascular and bronchoconstrictor effects of capsaicin. Eur J Pharmacol 108:315–319

307. Lundvall J, Järhult J (1976) Beta-adrenergic dilator component of the sympathetic vascular response in skeletal muscle. Acta Physiol Scand 96:180–192

308. Mabe Y, Tatemoto K, Huidobro-Turo JP (1985) Neuropeptide Y- induced pressor responses: Activation of a non-adrenergic mechanism, potentiation by reserpine and blockade by nifedipine. Eur J Pharmacol 116:33–39

309. Malm L, Sundler F, Uddman R (1980) Effects of vasoactive intestinal polypeptide on resistance and capacitance vessels in the nasal mucosa. Acta Otolaryngol (Stockh) 90:304–308

310. Matsuyama T, Shiosaka S, Matsumoto M et al. (1983) Overall distribution of vasoactive intestinal poly-

peptide-containing nerves on the wall of cerebral arteries: an immunohistochemical study using whole-mounts. Neuroscience 10:89–96

311. Matsuyama T, Matsumoto M, Shiosaka S et al. (1984) dual innervation of substance P-containing neuron system in the wall of the cerebral arteries. Brain Res 322:144–147

312. Matsuyama T, Matsumoto M, Shiosaka S et al. (1984) Serotonergic and peptidergic innervation of the cerebral arteries: an immunohistochemical and electron microscopic study. Neurosci Lett [Suppl 17]:S118

313. Mattiason A, Ekblad E, Sundler F et al. (1985) Origin and distribution of neuropeptide Y-, vasoactive intestinal polypeptide- and substance P- containing nerve fibres in the urinary bladder of the rat. Cell Tissue Res 239:141–146

314. McCulloch J (1983) Peptides and the microregulation of blood flow in the brain. Nature 304:120

315. McCulloch J (1984) Role of dopamine in interactions among cerebral function, metabolism and blood flow. In: MacKenzie ET et al. (eds) Neurotransmitters and the cerebral circulation. (LERS Monograph, vol 2) Raven Press, New York, pp 137–155

316. McCulloch J (1984) Perivascular nerve fibres and the cerebral circulation. TINS 7(5):135–138

317. McCulloch J, Edvinsson L (1980) The effects of vasoactive intestinal polypeptide upon pial arteriolar calibre, cerebral blood flow, cerebral oxygen consumption and the electroencephalogram. Am J Physiol 238:H449–456

318. McGrath JC (1982) Evidence for more than one type of postjunctional α-adrenoceptor. Biochem Pharmacol 31:467–484

319. McGrath JC (1983) The variety of vascular α-adrenoceptors. TIPS 4:14–18

320. McKenna OC, Angelakos ET (1968) Acetylcholinesterase-containing nerve fibres in the canine kidney. Circ Res 23:645–651

321. Melander T, Hökfelt T, Rökaeus A et al. (1985) Distribution of galanin-like immunoreactivity in the gastro-intestinal tract of several mammalian species; Cell Tissue Res 239:253–270

322. Meldrum LA, Burnstock G (1983) Evidence that ATP acts as a cotransmitter with noradrenaline in sympathetic nerves supplying the guinea-pig vas deferens. Eur J Pharmacol 92:161–163

323. Mellander S, Johansson B (1968) Control of resistance, exchange and capacitance junctions in the peripheral circulation. Pharmacol Rev 20:117–196

324. Mellander S, Andersson P-O, Afzelius L-E et al. (1981) Neural beta-adrenergic dilation of the facial vein in man. Possible mechanism in emotional blushing. Acta Physiol Scand 114:393–399

325. Morris JL, Gibbins IL, Furness JB et al. (1985) Co-localization of neuropeptide Y, vasoactive intestinal polypeptide and dynorphin in non-noradrenergic axons of the guinea-pigs uterine artery. Neurosci Lett 62:31–37

326. Moritoki H, Su I (1981) Potentiating effects of 5-hydroxytryptamine and histamine on nerve stimulation induced contractions of the rabbit mesenteric artery. Arch Int Pharmacodyn Ther 252(2):186–195

327. Moskowitz MA (1984) The neurobiology of vascular head pain. Ann Neurol 16:157–168

328. Moskowitz MA, Liebmann JE, Reinhard JF et al. (1979) Raphe origin of serotonin-containing neurons within choroid plexus of the cat. Brain Res 169:590–594

329. Moskowitz MA, Norregaard TV, Liu-Chen L-Y et al.

(1984) VIP, CCK, and met-enkephalin in pia arachnoid and cerebral arteries after unilateral lesion of cat trigeminal ganglia. Abstr Soc Neurosci 13:576

330. Mott JC (1961) The stability of the cardiovascular system. In: Wolstenholme GEW, O'Connor M (eds): Somatic stability in the newly born. A Ciba Foundation symposium. Little Brown, Boston, pp 192–214

331. Mulderry PK, Ghatei MA, Rodrigo J et al. (1985) Calcitronin gene-related peptide in cardiovascular tissues of the rat. Neuroscience 14:947–954

332. Muramatsu I, Fujiwara M, Mivra A et al. (1981) Possible involvement of adenine nucleotides in sympathetic neuroeffector mechanisms of dog basilar artery. J Pharmacol Exp Ther 216:401–409

333. Myers HZ, Axhwnk EA, Honig CR (1975) Ganglion cells in arterioles of skeletal muscles role in sympathetic vasodilation. Am J Physiol 229:126–138

334. Nagy JI, LaBella LA, Buss M (1984) Immunohistochemistry of adenosine deaminase: implications for adenosine neurotransmission. Science 224:166–168

335. Nagy JI, Buss M, LaBella LA, Daddona PE (1984) Immunohistochemical localization of adenosine deaminase in primary afferent neurons of the rat. Neurosci Lett 48:133–138

336. Napoleone P, Sancesario G, Amenta F (1982) 5-Hydroxytryptophan uptake in the indoleaminergic nerve fibres within rat cerebrovascular tree. Neurosci Lett 28:57–60

337. Napoleone P, Sancesario G, Amenta F (1982) Indoleaminergic innervation of rat choroid plexus: a fluorescence histochemical study. Neurosci Lett 34:143–147

338. Nilsson SFE, Bill A (1985) Effects of facial nerve stimulation, VIP and PHI on ocular blood flow. Acta Physiol Scand 124 [Suppl 542]: 132

339. Norregaard TV, Moskowitz MA (1985) Substance P and the sensory innervation of intracranial and extracranial feline cephalic arteries. Implications for vascular pain mechanisms in man. Brain 108:517–533

340. Oleson S-P (1985) A calcium-dependent reversible permeability increase in microvessels in frog brain, induced by serotonin. J Physiol (Lond) 361:103–113

341. Olgart L, Hökfelt T, Nillson G et al. (1977) Localization of substance P-like immunoreactivity in nerves in the tooth pulp. Pain 4:153–159

342. Osias MB, Siegel NJ, Chaudry IH et al. (1977) Postischemic renal failure: accelerated recovery with adenosine triphosphate magnesium chloride infusion. Arch Surg 112:729–731

343. Olson L, Alund M, Norberg K-A (1976) Fluorescence microscopical demonstration of a population of gastrointestinal nerve fibres with a selective affinity for quinacrine. Cell Tissue Res 171:407–423

344. Osborne LW, Silva DG (1970) Histological, acetylcholinesterase and fluorescence histochemical studies on the atrial ganglia of the monkey heart. Exp Neurol 27:497–511

345. Owman C, Edvinsson L, Hardebo JE et al. (1977) Immunohistochemical demonstration of actin and myosin in brain capillaries. Acta Neurol Scand 56 [Suppl 64]:384–385

346. Page IH (1954) Serotonin (5-hydroxytryptamine) Physiol Rev 34:563–588

347. Papka RE, Furness JB, Della NG et al. (1981) Depletion by capsaicin of substance P-immunoreactivity and acetylcholinesterase activity from nerve fibres in the guinea pig heart. Neurosci Lett 27:47–54

348. Papka RE, Cotton JP, Traurig HH (1985) Comparative distribution of neuropeptide tyrosine-vasoactive intestinal polypeptide-, substance P- immunoreactive, acetylcholinesterase-positive and noradrenergic nerves in the reproductive tract of the female rat. Cell Tissue Res 242:475–490

349. Parnavelas JG, Kelly W, Burnstock G (1985) Ultrastructural localization of choline acetyltransferase in vascular endothelial cells in rat brain. Nature 316:724–725

350. Paton DM (1981) Presynaptic neuromodulation mediated by purinergic receptors. In: Burnstock G (ed) Purinergic receptors. Chapman and Hall, London, pp 199–219 (Receptors and recognition, series B, vol 12)

351. Paton WDM, Vizi ES (1969) The inhibitory action of noradrenaline and adrenaline on acetylcholine output by guinea-pig ileum longitudinal muscle strip. Br J Pharmacol 35:10–18

352. Pegram BL, Bevan RD, Bevan JA (1976) Facial vein of the rabbit: neurogenic vasodilation mediated by β-adrenergic receptors. Circ Res 39:854–860

353. Peng JH, Kimura H, McGreer PL et al. (1981) Anticholine acetyltransferase fragments antigen binding (Fab) for immunohistochemistry. Neurosci Lett 21:281–285

354. Peroutka SJ, Snyder SH (1979) Multiple serotonin receptors: differential binding of [^{3}H]5-hydroxytryptamine, [^{3}H]lysergic acid diethylamide and [^{3}H]spirperidol. Mol Pharmacol 16:687–699

355. Polak JM, Bloom SR (1980) Peripheral localization of regulatory peptides as a clue to their function. J Histochem Cytochem 28:918–924

356. Puroy RE, Hurlbutt DE, Rains LA (1981) Receptors for 5-hydroxytryptamine in rabbit isolated ear artery and aorta. Blood Vessels 18:16–27

357. Quirion R, Rioux F, St Pierre S et al. (1979) Increased sensitivity to neurotensin in fasted rats. Life Sci 25:1969–1973

358. Rapport MM, Green AA, Page I (1948) Serum vasoconstrictor (serotonin). IV. Isolation and characterization. J Biol Chem 176:1243–1251

359. Reinecke M, Weihe E, Forsmann WG (1980) Substance P- immunoreactive nerve fibres in the heart. Neurosci Letts 20:265–269

360. Reinecke M, Schlüter P, Yanaihara N et al. (1981) VIP immunoreactivity in enteric nerves and endocrine cells of the vertebrate gut. Peptides 2 (Suppl. 2):149–156

361. Reinecke M, Weihe E, Carraway RE et al. (1982) Localization of neurotensin immunoreactive nerve fibres in the guinea-pig heart: evidence derived by immunohistochemistry, radioimmunoassay and chromatography. Neuroscience 7:1785–1795

362. Reinecke M, Vallance C, Weihe E et al. (1982) Neuropeptide (substance P, neurotensin) immunoreactive nerve fibres in the heart of higher vertebrates. Neurosci Lett [Suppl] 10:S403

363. Reinhard JF, Koskowitz MA, Elspas S et al. (1978) Serotoninergic innervation of rat and bovine parenchymal brain blood vessels: biochemical and pharmacological studies. Neurosci Abstr 4:451

364. Reinhard JF, Liebmann JE, Schlosberg AJ et al. (1979) Serotonin neurons project to small blood vessels in the brain. Science 206:85–87

365. Rennels ML, Nelson E (1975) Capillary innervation in the mammalian central nervous system: an electron microscopic demonstration. Am J Anat 144:233–241

366. Rennels ML, Fujimoto K, Nelson E (1979) Capillary innervation in the hypothalamus: a fluorescence histochemical and electron microscopic study. Acta Neurol

Scand 60 [Suppl 72]:92–93

367. Rennels ML, Gregory TF, Fujimoto K (1983) Innervation of capillaries by local neurons in the cat hypothalamus: a light microscopic study with horseradish peroxidase. J Cereb Blood Flow Metab 3:535–542

368. Riphagen CL, Pittman Q (1985) Cardiovascular responses to intrathecal administration of arginine vasopression in rats. Regul Pept 10:293–298

369. Richardson BP, Engels G, Donatsch P et al. (1985) Identification of serotonin M-receptor subtypes and their specific blockade by a new class of drugs. Nature 316:126–131

370. Roddie IC, Shepherd JT, Whelan RF (1957) Contribution of constrictor and dilator nerves to skin vasodilatation during body heating. J Physiol (Lond) 136:489

371. Rodrigo J, Polak JM, Terenghi G et al. (1985) Calcitonin gene-related peptide (CGRP)-immunoreactive sensory and motor nerves of the mammalian palate. Histochemistry 82:67–74

372. Rose GA, Moulds RFW (1979) Pharmacological comparison of isolated human cerebral and digital arteries. Stroke 10:736–741

373. Rosenfeld MG, Mermod J-J, Amara SG et al. (1983) Production of a novel neuropeptide encoded by the calcitonin gene via tissue-specific RNA processing. Nature 304:129–135

374. Rossel S, Olgart L, Gazelius B (1981) Inhibition of antidromic and substance P-induced vasodilation by a substance P-antagonist. Acta Physiol Scand 111:381–382

375. Rowan RA, Bevan JA (1982) Distribution of adrenergic synaptic cleft width in vascular and non-vascular smooth muscle. In: Bevan JA (ed) Vascular neuroeffector mechanisms IV. Proceedings of the 4th international symposium on vascular neuroeffector mechanisms, Kyoto, Japan. Raven Press, New York

376. Rubenstein AE, Rudansky MC, Yahr MD (1983) Autonomic failure due to amyloid. In: Bannister R (ed) Autonomic failure. A textbook of clinical disorders of the autonomic nervous system. Oxford University Press, Oxford, pp 574–595

377. Ryan ML, Clark KE, Brody MJ (1974) Neurogenic and mechanical control of canine uterine vascular resistance. Am J Physiol 227:547–555

378. Saba H, Cowen T, Haven AJ et al. (1984) Reduction in noradrenergic perivascular nerve density in the left and right cerebral arteries of old rabbits. J Cereb Blood Flow Metab 4:284–289

379. Saffrey MJ, Polak JM, Burnstock G (1982) Distribution of vasoactive intestinal polypeptide-, substance P, enkephalin- and neurotensin-like immunoreactive nerves in the chicken gut during development. Neuroscience 7:279–293

380. Said SI, Mutt V (1970) Polypeptide with broad biological activity: isolation from small intestines. Science 169:1217–1218

381. Santer RM (1982) Fluorescence histochemical observations on the adrenergic innervation of the cardiovascular system in the aged rat. Brain Res Bull 9:667–672

382. Saria A, Martling C-R, Dalsgaard C-J et al. (1985) Evidence for substance P-immunoreactive spinal afferents that mediate bronchoconstriction. Acta Physiol Scand 125:407–414

383. Scatton B, Duverger D, L'Heureux R et al. (1985) Neurochemical studies on the nature of the serotoninergic innervation of the cerebral circulation of the cerebral circulation. J Cereb Blood Flow Metab 5 [Suppl 1]:S527

384. Schmidt M, Imbs J-L, Schwartz J (1981) The vascular dopamine receptor: a review. J Pharmacol 12:355–382

385. Schon F, Ghatei MA, Allen JM et al. (1985) The effect of sympathectomy on calcitonin gene related peptide levels in the rat trigeminovascular system. Brain Res 348:197–200

386. Schon F, Allen JM, Yeats JC et al. (1985) Neuropeptide Y innervation of the rodent pineal gland and cerebral blood vessels. Neurosci Lett 57:65–71

387. Schultzberg M, Hökfelt T, Nilsson G et al. (1980) Distribution of peptide- and catecholamine-containing neurons in the gastro-intestinal tract of rat and guinea-pig: immunohistochemical studies with antisera to substance P, vasoactive intestinal polypeptide, enkephalins, somatostatin, gastrin/cholecystokinin, neurotensin and dopamine β-hydroxylase, Neuroscience 5:689–744

388. Scott TM, Foote J, Galway G (1984) Changes in the catecholaminergic innervation of jejunal arteries following induction of diabetes with streptozotocin. J Anat 139:738

389. Shibata S, Hattori K, Sakurai I, Mori I et al. (1971) Adrenergic innervation and cocaine-induced potentiation of adrenergic responses of aortic strips from young and old rabbits. J Pharmacol Exp Ther 177:621–632

390. Sikri KL, Hoyes AD, Barber P et al. (1981) Substance P-like immunoreactivity in the intramural nerve plexuses of the guinea-pig ureter: a light and electron microscopical study. J Anat 133:425–442

391. Skok VI (1973) Physiology of autonomic ganglia. Igaku Shoin, Tokyo

392. Sneddon P, Burnstock G (1984) Inhibition of excitatory junction potentials in guinea-pig vas deferens by α, β-methylene-ATP: further evidence for ATP and noradrenaline as cotransmitters. Eur J Pharmacol 100:85–90

393. Sneddon P, Burnstock G (1984) Do we need γ-receptors? TIPS 5:264–265

394. Sneddon P, Burnstock G (1984) ATP as a co-transmitter in rat tail artery. Eur J Pharmacol 106:149–152

395. Sternini C, Brecha N (1985) Distribution and colocalization of neuropeptide Y- and tyrosine hydroxylase-like immunoreactivity in the guinea-pig heart. Cell Tissue Res 241:93–102

396. Stjernquist M, Hakanson R, Leander S et al. (1983) Immunohistochemical localization of substance P, vasoactive intestinal polypeptide and gastrin-releasing peptide in vas deferens and seminal vesicle, and the effect of these and eight other neuropeptides on resting tension and neurally evoked contractile activity. Regul Pept 7:67–86

397. Stone RA, Laties AM, Brecha NC (1982) Substance P-like immunoreactive nerves in the anterior segment of the rabbit, cat and monkey eye. Neuroscience 7:2459–2468

398. Story DF, Allen GS, Glover AB et al. (1975) Modulation of adrenergic transmission by acetylcholine. Clin Exp Pharmacol Physiol [Suppl 2]:27–33

399. Ström C, Lundberg J, Ahlman H et al. (1981) On the VIP-ergic innervation of the uterotubal junction. Acta Physiol Scand 11:213–215

400. Su C (1975) Neurogenic release of purine compounds in blood vessels. J Pharmacol Exp Ther 195:159–166

401. Su C (1978) Modes of vasoconstrictor and vasodilator neurotransmission. Blood Vessels 15:183–189

402. Su C (1981) Purinergic receptors in blood vessels. In:

Burnstock G (ed) Purinergic receptors. Chapman and Hall, London, pp 93–117 (Receptors and recognition, series B, vol 12)

403. Su C, Duckles SP, Florence V (1977) Uptake of ^{3}H-norepinephrine in rabbit mesenteric blood vessels. Blood Vessels 14:65–76

404. Su C, Bevan JA, Assali NS et al. (1977) Development of neuroeffector mechanisms in the carotid artery of the fetal lamb. Blood Vessels 14:12–24

405. Sundler F, Alumets J, Brodin E (1977) Perivascular substance P-immunoreactive nerves in tracheobronchial tissue. In: Von Euler US, Pernow B (eds) Substance P. Raven Press, New York, pp 271–273

406. Sundler F, Alumets J, Hakanson R et al. (1977) VIP innervation of the gallbladder. Gastroenterology 72:1375–1377

407. Sundler F, Moghimzadeh E, Hakanson R et al. (1983) Nerve fibres in the gut and pancreas of the rat displaying neuropeptide-Y immunoreactivity. Intrinsic and extrinsic origin. Cel Tissue Res 230:487–493

408. Svendgaard N-A, Edvinsson L, Olin T et al. (1977) On the pathophysiology of cerebral vasospasm: transmitter changes in perivascular sympathetic nerves, and increased pial artery sensitivity to norepinephrine and serotonin. In: Owman C, Edvinsson (eds) Neurogenic control of brain circulation. Proceedings of international symposium, Wenner-Gren. Pergamon Press, Oxford, pp 143–152

409. Taché Y, Gonion M, Lauffenberger M et al. (1984) Inhibition of gastric acid secretion by intrathecal injection of calcitonin gene-related peptide in rats. Life Sci 35:871–878

410. Tagawa T, Ando K, Wasano T et al. (1979) A histochemical study of the innervation of cerebral blood vessels in the bullfrog. J Comp Neurol 183:25–32

411. Takami K, Kawai Y, Shiosaka S et al. (1985) Immunohistochemical evidence for the coexistence of calcitonin gene-related peptide and choline acetyltransferase-like immunoreactivity in neurons of the rat hypoglossal facial and ambiguous nuclei. Brain Res 328:386–389

412. Tatemoto K, Carlquist M, Mutt V (1982) Neuropeptide Y – a novel brain peptide with structural similarities to peptide YY and pancreatic polypeptide. Nature 296:659–660

413. Terenghi G, Polak JM, Proberg L et al. (1982) Mapping, quantitative distribution and origin of substance P- and VIP-containing nerves in the uvea of guinea-pig eye. Histochemistry 75:399–417

414. Terenghi G, Polak JM, Ghatei MA et al. (1985) Distribution and origin of calcitonin gene-related peptide (CGRP) immunoreactivity in the sensory innervation of the mammalian eye. J Comp Neurol 233:506–516

415. Terenghi G, Polak JM, Rodrigo J et al. (1986) Calcitonin gene-related peptide-immunoreactive nerves in the tongue, epiglottis and pharynx of the rat: occurrence, distribution and origin. Brain Res 365:1–14

416. Toda N, Fujita Y (1973) Responsiveness of isolated cerebral and peripheral arteries to serotonin, norepinephrine and transmural electrical stimulation. Circ Res 33:98–104

417. Toda N (1976) Influence of dopamine and noradrenaline on isolated cerebral arteries of the dog. Br J Pharmacol 58:121

418. Tonoue T, Hata H, Ohnishi T et al. (1985) Somatostatin and GABA correlate with cervical autonomic nerve activity. Regul Pept 10:299–307

419. Tuor UI, Kelly P, Edvinsson L et al. (1985) Neuro-peptide Y and the regulation of local cerebral blood flow. A quantitative autoradiographic study. J Cereb Blood Flow Metab 5 [Suppl 1]:S513

420. Uddman R, Alumets J, Densert O et al. (1978) Occurrence and distribution of VIP nerves in the nasal mucosa and tracheobronchial wall. Acta Otolaryngol (Stockh) 86:443–448

421. Uddman R, Alumets J, Edvinsson L et al. (1978) Peptidergic (VIP) innervation of the oesophagus. Gastroenterology 75:5–8

422. Uddman R, Sundler F (1979) Vasoactive intestinal polypeptide nerves in human upper respiratory tract. Otol Rhinol Laryngol 41:221–226

423. Uddman R, Alumets J, Densert O et al. (1979) Innervation of the feline eustachian tube. Ann Otol Rhinol Laryngol 88:557–561

424. Uddman R, Malm L, Sundler F (1980) The origin of vasoactive intestinal polypeptide (VIP) nerves in the feline nasal mucosa. Acta Otolaryngol (Stockh) 89:152–156

425. Uddman R, Björlin G, Möller B et al. (1980) Occurrence of VIP nerves in mammalian dental pulps. Acta Odontol Scand 38:325–328

426. Uddman R, Fahrenkrug J, Malm L et al. (1980) Neuronal VIP in salivary glands: distribution and release. Acta Physiol Scand 110:31–38

427. Uddman R, Alumets J, Ehinger B et al. (1980) Vasoactive intestinal peptide nerves in ocular and orbital structures of the cat. Invest Ophthalmol Vis Sci 19:878–885

428. Uddman R, Alumets J, Hakanson R et al. (1980) Vasoactive intestinal peptide (VIP) occurs in nerves of the pineal gland. Experientia 36:1119–1120

429. Uddman R, Edvinsson L, Sundler F (1981) Perivascular substance P: occurrence and distribution in mammalian pial vessels. J Cereb Blood Flow Metab 1:227–232

430. Uddman R, Alumets J, Edvinsson L et al. (1981) VIP nerve fibres around peripheral blood vessels. Acta Physiol Scand 112:65–70

431. Uddman R, Edvinsson L, Hakanson R et al. (1982) Immunohistochemical demonstration of APP (avian pancreatic polypeptide) – immunoreactive nerve fibres around cerebral blood vessels. Brain Res Bull 9:715–718

432. Uddman R, Edvinsson L, Malm L (1982) Perivascular nerves in the feline carotid rete. Cell Tissue Res 226:301–308

433. Uddman R, Ninoyu O, Sundler F (1982) Adrenergic and peptidergic innervation of cochlear blood vessels. Arch Otol Rhinol Laryngol 236:7–14

434. Uddman R, Edvinsson L, Owman C et al. (1983) Nerve fibres containing gastrin-releasing peptide around pial vessels. J Cereb Blood Flow Metab 3:386–390

435. Uddman R, Sundler F, Emson P (1984) Occurrence and distribution of neuropeptide-Y-immunoreactive nerves in the respiratory tract and middle ear. Cell Tissue Res 237:321–327

436. Uddman R, Edvinsson L, Ekman R et al. (1985) Innervation of the feline cerebral vasculature by nerve fibres containing calcitonin gene-related peptide: trigeminal origin and co-existence with substance P. Neurosci Lett 62:131–136

437. Uddman R, Ekblad E, Edvinsson L et al. (1985) Neuropeptide Y-like immunoreactivity in perivascular nerve fibres of the guinea-pig. Regul Pept 10:243–257

438. Uvnäs B (1966) Cholinergic vasodilator nerves. Fed Proc 25:1613–1622

439. Vaalasti A, Linnoila I, Hervonen A (1980) Immu-

nohistochemical demonstration of VIP, [Met⁵]- and [Leu⁵]-enkephalin immunoreactive nerve fibres in the human prostate and vesicles. Histochemistry 66:89–98

440. Van Calker D, Muller M, Hamprecht B (1979) Adenosine regulates via two different types of receptors, the accumulation of cyclic AMP in cultured brain cells. J Neurochem 33:999–1005

441. Vanhoutte PM (1974) Inhibition of acetylcholine of adrenergic neurotransmission in vascular smooth muscle. Circ Res 34:317–326

442. Vanhoutte PM, Cohen RA, Van Nueten JM (1984) Serotonin and arterial vessels. J Cardiovasc Pharmacol 6:S421–S428

443. Van Neuten JM, Janssen PAJ, De Ridder W et al. (1982) Interaction between 5-hydroxytryptamine and other vasoconstrictor substances in the isolated femoral artery of the rabbit: effect of ketanserin (R41468). Eur J Pharmacol 77:281–287

444. Verrecchia C, Hamel E, Sercombe R et al. (1985) Constriction of cerebral arteries to noradrenaline is modulated by the endothelium. J Cereb Blood Flow Metab 5 [Suppl 1]:S549

445. Vizi ES (1979) Presynaptic modulation of neurochemical transmission. Prog Neurobiol 12:181–290

446. Wall PD, Fitzgerald M (1982) If substance P fails to fulfill the criteria as a neurotransmitter in somatosensory efferents, what might be its function? In: Ciba Found Symp 91:249–266

447. Wharton J, Polak JM, Bryant MG et al. (1979) Vasoactive intestinal polypeptide (VIP)-like immunoreactivity in salivary glands. Life Sci 25:273–280

448. Wharton J, Polak JM, Bloom SR et al. (1979) Substance P-like immunoreactive nerves in mammalian lung. Invest Cell Pathol 2:3

449. Wharton J, Polak JM, Pearse AGE et al. (1980) Enkephalin-, VIP- and substance P-like immunoreactivity in the carotid body. Nature 284:269–271

450. Wharton J, Polak JM, Probert L et al. (1981) Peptide containing nerves in the ureter of the guinea-pig and cat. Neuroscience 6:969–982

451. Wharton J, Polak JM, McGregor GP et al. (1981) The distribution of substance P-like immunoreactive nerves in the guinea-pig heart. Neuroscience 6:2193–2204

452. Willis E, Ottesen B, Wagner G et al. (1981) Vasoactive intestinal polypeptide (VIP) as a possible neurotransmitter involved in penile erection. Acta Physiol Scand 113:545–547

453. Winquist RJ, Bevan JA (1981) Relation location of alpha- and beta-adrenoceptors to site of release of sympathetic transmitter in rabbit facial vein. Circ Res 49:486–491

454. Woods JR, Dandavino A, Murayama K et al. (1977) Autonomic control of cardiovascular functions during neonatal development and in adult sheep. Circ Res 40:401–407

455. Yamamoto K, Matsuyama T, Shiosaka S et al. (1983) Overall distribution of substance P-containing nerves in the wall of cerebral arteries of the guinea pig and its origins. J Comp Neurol 215:421–426

456. Zaidi M, Bevis PJR, Girgis SI et al. (1985) Circulating CGRP comes from the perivascular nerves. Eur J Pharmacol 117:283–284

457. Zawadzki JV, Furchgott RF, Cherry P (1981) The obligatory role of endothelial cells in the relaxation of arterial smooth muscle by substance P. Fed Proc 40:689

458. Zerbe RL, Feverstein G (1985) Cardiovascular effects of centrally administered vasopressin in conscious and anaesthetised rats. Neuropeptides 6:471–484

459. Zhang SQ, Terenghi G, Unger WG et al. (1984) Changes in substance P- and neuropeptide Y-immunoreactive fibres in rat and guinea-pig irides following unilateral sympathectomy. Exp Eye Res 39:365–372

Chapter 8

The Arterial Wall and the Haemostatic Process

Y. J. Legrand and L. O. Drouet

Introduction

An important feature of the endothelium is the thromboresistance of its luminal surface, which protects haemostatic factors in the blood from being activated by the vessel itself. Any alteration in the endothelium transforms the vascular wall into a thrombogenic surface to which blood platelets will adhere, and triggers the multistep process of haemostatis in which both platelets

and plasma are involved. This phenomenon is beneficial because it leads to endothelial repair, but may have dramatic consequences if it also leads to thrombosis and provokes the vascular alterations associated with atherosclerosis. In this chapter, we will describe the properties of the endothelium which confer on it its thromboresistance, and we will present some properties of particular constituents which confer haemostatic and thrombogenic properties on the subendothelium (Fig. 8.1).

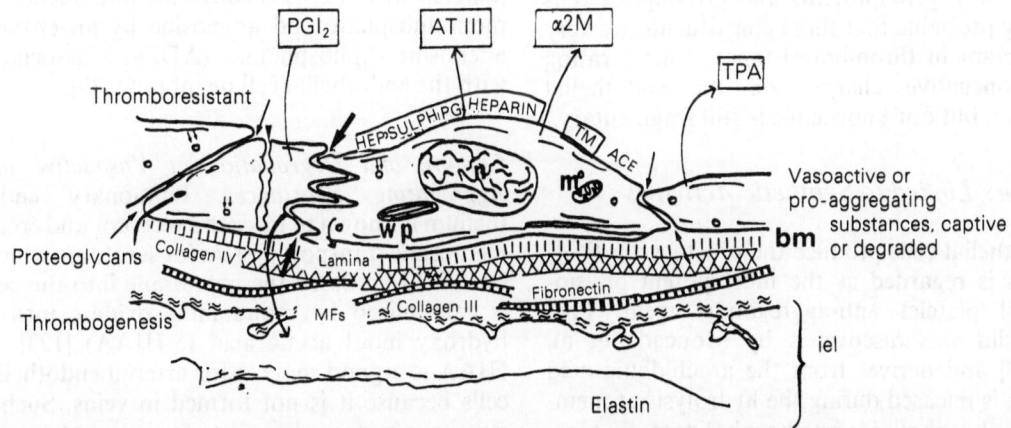

Fig. 8.1. Schematic representation of an arterial intima. An endothelial cell (*end*) is attached by the constituents of basement membrane (*bm*) to the collagenous matrix of the subendothelium which separates basement membrane from the internal elastic lamina (*iel*). Substances secreted into blood (*thin long arrows*) together with constituents of the cell coat [heparan sulphate proteoglycan (*HEP SULPH PG*), heparin] or "activities" expressed on the membrane [thrombomodulin (*TM*), angiotensin-converting enzyme (*ACE*)] contribute to the thromboresistance of the endothelium. All components of basement membrane and subendothelium are produced by the endothelium (*thick arrows*). Some of them, such as collagen type III and the microfibrils (*MFs*) associated with elastin in the internal elastic lamina, participate in the thrombogenicity of the subendothelium. Transendothelial migrations occur through interendothelial junctions (→) and transendothelial channels (→). An endocytosis vesicle (↓) and several cytoplasmic vesicles (↓↓) are also seen α2m, α2 macroglobulin; *AT III*, antithrombin III; *tP*, tissue plasminogen activator; *wp*, Weibel–Pallade bodies; *m*, mitochondria.

Thromboresistance of the Endothelium

The normal endothelium forms a haemocompatible surface which does not react with platelets and leucocytes and fails to activate the coagulation process and the complement and kinin systems. This thromboresistance is due to particular properties of the endothelial cells which influence the three main steps of the haemostatic process, primary haemostasis, plasma coagulation and fibrinolysis. The properties of the endothelial cells depend on three groups of factors: firstly, on the nature of their surfaces, secondly, on synthetic activities, and thirdly, on their ability to degrade, transform or inactivate certain substances which are thrombogenic.

Thromboresistance in Relation to Primary Haemostasis

This bears upon phenomena in which blood platelets are directly involved. Three groups of factors must be considered.

Factors Linked to the Endothelial Surface

The endothelial cell surface (glycocalix) is rich in sulphated proteoglycans (heparan sulphate proteoglycan), glycoproteins and glycolipids. It is highly probable that these constituents are very important in thromboresistance, concentrating electronegative charges on the endothelial surface, but our knowledge is still fragmentary.

Factors Linked to Synthetic Activities

Endothelial cells produce the prostacyclin PGI_2 which is regarded as the most potent physiological platelet anti-aggregating agent. Prostacyclin was discovered by Moncada et al. [95,96] and derives from the arachidonic acid which is released during the hydrolysis of membrane phospholipids by phospholipase A_2. Normally, its rate of synthesis is low, but it is enhanced when the cell is stimulated, for example by localized thrombin generation [30,129,130]. Arachidonic acid is then oxidized by cyclooxygenase into endoperoxides which are rapidly transformed into prostacyclin by the enzyme prostacyclin synthetase. Prostacyclin has two main properties: it is a potent vasodilator and it prevents the aggregation of platelets [56]. At high doses, it also inhibits their adhesion to subendothelium and to collagen [70].

It is not certain that prostacyclin contributes to endothelial thromboresistance under normal physiological conditions, even though a hypothetical balance between prostacyclin (vasodilator and anti-aggregating) and plateletderived thromboxane (vasoconstrictor and proaggregating) helps to explain the homeostasis of the endothelial–platelet system. In particular, diminution of the circulating PGI_2 has been shown in atherosclerosis [54], diabetes [34], toxaemia of pregnancy, pre-eclampsia [87] and so on. However, these examples are not a direct demonstration of the role of prostacyclin in thromboresistance; the half-life of PGI_2 is so short that it cannot be measured directly in the plasma, and it seems that the plasma levels of its stable derivative, 6-keto PGF_1, are often artefactual.

Inactivation of Thrombogenic Substances by the Endothelial Cells

The endothelial cells are able to degrade, pick up, transform and inactivate substances which directly participate in arterial thrombosis.

Degradation of Pro-aggregating Substances. Adenosine $5'$ diphosphate (ADP) released from platelets or red cells is converted into adenosine monophosphate and adenosine by an enzyme adenosine diphosphatase (ADPase) associated with the endothelial cell membrane [89].

Capture and Degradation of Vasoactive and Aggregating Substances. Pulmonary endothelium is known to possess a receptor and crossmembrane transport system for serotonin which, once it has crossed the membrane into the cell, is catabolized by monoamine oxidase into 5-hydroxy indol acetic acid (5-HIAA) [123]. 5-HIAA is a good marker for arterial endothelial cells because it is not formed in veins. Such a clearance system also exists for catecholamines, for which endothelial cells bear α- and β-adrenergic receptors on their surface [21], these receptors being associated with an active transport system across the membrane. For example, pulmonary endothelium [1,68] metabolizes norepinephrine. Other vasoactive substances are also degraded: histamine, adenosine, angiotensin [the angiotensin-converting enzyme is a good marker for endothelial cells [38]], bradykinin and so on.

Thromboresistance in Relation to Plasma Coagulation

This is mostly due to the properties of the endothelial cell surface. In this regard, proteoglycans such as heparan sulphate proteoglycan are important because of their electronegative charge. Engelberg [40] has recently shown the existence of anionic sites at heparan sulphate and heparin-rich microdomains of the luminal surface of the cells. Circulating heparin can exchange with heparan sulphate [40]: heparin can thus restore the normal endothelial electronegativity which has been depressed by cations such as protamine. The binding of heparin to endothelial cells [51] is important for two reasons: for its anti-coagulant properties and for its ability to preserve the anionic sites.

Anti-proteases are also present on the endothelial pericellular coat and are possibly associated with certain glycosaminoglycans. For example, anti-thrombin III, although mainly produced by the liver, may also originate from the endothelium [9]. Immunofluorescence has also revealed the presence of α_2 macroglobulin on the luminal face of endothelial cells [9]. The occurrence of these anti-proteases on the cell coat explains observations showing that thrombin can fix to endothelial cells [14,72] after which it becomes inactive. The thrombomodulin–protein C system was described by Esmon and Owen [41]: activated protein C is a potent inhibitor of coagulation by inactivating factors Va and VIIIa [122]. Protein C is activated by thrombin if this enzyme binds to an endothelial cell surface protein, thrombomodulin. It has been shown that when plasma is coagulated in vitro, the activation of protein C is at too low a level to inhibit factors Va and VIIIa and consequently prothrombin activation is limited [42,73]. In the presence of endothelial cells, the situation is different: for example, in the isolated rabbit heart protein C becomes activated if it is perfused together with thrombin, and conversely remains inactivated if it is perfused alone, in the absence of thrombin [42]. The mechanism of the system was understood when thrombomodulin was isolated from pulmonary endothelial cells and was found to be a protein of molecular weight 74 000. One molecule of thrombin binds to one molecule of thrombomodulin, and then activates protein C in the presence of calcium. Thrombin can thus initiate a negative feedback mechanism which regulates its own generation.

Thromboresistance in Relation to Fibrinolysis [39]

The endothelial cells synthesize two different plasminogen activators: one is urokinase-like, the other is the tissue plasminogen activator (tPA) which was isolated from human melanoma cells [27,108]. tPA is released from endothelial cells by certain stimuli such as stress and exercise, under pathological conditions such as venous occlusion [29] and electric shock and also by pharmacologically active agents such as thrombin, histamine, bradykinin, acetylcholine and dDAVP [92]. tPA is irregularly distributed: for example, more tPA is released at the level of the arms than in the legs [103]. This may explain observations showing that venous thrombosis occurs more frequently in the legs than in the arms, perhaps because a chronic stimulation of tPA release due to orthostatic pressure results in relative depletion in the legs.

Chemically, tPA is a serine protease of molecular weight 128 000, with two chains of amino acids. The first chain (A chain) has 278 amino acids, the second (the B chain, bearing the active sites) has 275 [105,128]. Its activity is very specific because it acts almost exclusively in the presence of fibrin and not of fibrinogen: tPA has a high affinity for fibrin with a Kd = 150 nM [77,109]. It finds a fibrin-binding site which is masked on fibrinogen and becomes accessible during the release of the fibrinopeptide [2]. tPA therefore acts only when the fibrin clot is formed, since both tPA and its substrate fibrinogen bind to the clot so that plasmin is generated. Another important point is that on plasmin, the binding site (lysin-binding site) and its active site are both occupied: if they were free, plasmin would be rapidly inactivated by circulating α_2-antiplasmin. In the complete fibrinolytic system, plasmin is very slowly inactivated by α_2-antiplasmin [28]. This results in a continuous renewal of its inactivated molecules by the constant generation of new plasmin molecules resulting from the activation of their precursor by tPA bound to fibrin. For a more detailed account, see Lijnen and Collen [90].

The thromboresistance of the endothelium is thus due to various activities of the endothelial cells. Those which oppose platelet activation are particularly important in arteries where the haemorheological conditions facilitate contact between platelets and the vessel wall. Those which impair or slow down coagulation and those which induce or facilitate fibrinolysis are more important in veins where static conditions facilitate the activation of the clotting factors. It is interesting to note that thromboresistance is due not only to passive conditions linked to the nature of the endothelial surface, but also to active factors due to the life of the cells themselves. Also noteworthy is the existence of a polarity both in the cell coat and in the secretion of the molecules involved, towards the lumen of the vessel. This is highly relevant to the situation of the cells at the interface between the tissue and the blood. The next section will deal with the interaction of blood with molecules which are, in contrast, secreted towards the parietal face of the endothelium.

The Thrombogenicity of the Subendothelium

When endothelial cells are injured or when the endothelium is destroyed, the haemocompatible surface which coats the wall disappears and blood platelets and plasma constituents involved in blood clotting come into contact with the subendothelium. This contains macromolecules which have been synthesized and secreted by the endothelial cells. At least two kinds of macromolecules react with platelets and initiate the process which leads to the formation of a haemostatic plug or of a thrombus: these are collagen and the microfibrils of the elastic lamina. The interactions between the vessel wall and the platelets are extremely complex and depend on three series of factors including:

1. The composition of the subendothelium: the expression of the endothelial synthetic reactivity varies all along the vascular tree, and, as a consequence, the subendothelium also varies in its constitution and thickness.

2. The haemodynamic conditions: these govern the forces which influence the displacement of the circulating cells and molecules. They depend on the physical characteristics of the flow: velocity, diameter of the vessel, parietal elasticity and contractility, on the number and properties of the cells (red cell deformability for example) and on the constitution and viscosity of the plasma.

3. The functional characteristics of the cellular and molecular elements which are activated on the exposed subendothelial surface. They cooperate: for example, plasma factors influence cell deposition (von Willebrand factor influences the adhesion of platelets, and fibrinogen the platelet–platelet cohesion); and platelets activate plasma coagulation by providing a pro-coagulant surface which facilitates the activation of factor X and prothrombin on their membrane [11,131,132].

Understanding of the mechanisms has been made possible by the use of ex vivo and in vitro methods; in vivo methods have also been used to follow the time course of the events which follow the lesion of the endothelium as the triggering step of platelet deposition.

Ex Vivo Study of the Thrombogenicity of the Subendothelium

The original technique for the de-endothelialization of an artery in rabbit was described by Baumgartner and Haudenschild in 1972 [6]. They use a balloon catheter to expose subendothelium in the aorta. In another method, de-endothelialization was induced by air flow which replaced blood flow for a short period in a rabbit carotid artery [47]. The arterial segment was then collected and everted on a plastic rod which was placed into an annular perfusion chamber integrated into a closed intravascular circuit. The interaction of platelets with the subendothelium was quantified either by morphometry to determine the percentage of surface covered by isolated, adhesive platelets and by thrombi [6] or by isotopic methods after labelling of the platelets with chromium-51 [35] or indium-111 [119]. This latter method was most frequently applied to rabbit to establish the haemorheological conditions and identify the thrombogenic components in the subendothelium. It has also been used with human arteries to define the role of plasma von Willebrand factor in the platelet/vessel wall interaction. These three aspects will now be discussed.

Haemorheological Aspect of Platelet/Vessel Wall Interactions

In Baumgartner's annular chamber, shear forces can be varied between 10/s and 10 000/s [125]. Below 650/s, the percentage of surface covered with adhering platelets depends directly on the wall shear rate; above 800/s, it becomes independent of the wall shear rate. In addition to the general flow conditions, transport phenomena of blood flow are important. Dosne et al. [35], showed that the deposition of ^{51}Cr-labelled platelets depended on the convective diffusion flux of the erythrocytes: when the haematocrit increases from 10% to 40% platelet deposition also increases, and this is explained by the effect of secondary fluxes created by the rotation of the erythrocytes in blood flow. In this regard, it is worth noting that when the diameter of the vessel is wide enough to permit laminar blood flow, there is a gradient in the haematocrit from the centre of the flow towards the periphery. The rotational motion of the erythrocytes is zero in the centre but high in the neighbourhood of the wall, where they facilitate platelet deposition. As far as the erythrocytes are concerned, not only is their number important, but also their deformability, which participates in the secondary diffusion fluxes due to their rotation. The viscosity of the plasma is also important.

Identification of the Thrombogenic Macromolecules

The thrombogenicity of collagen and its role in the initiation of haemostatis were known before the development of the ex vivo experiments. In those models, the treatment of aortic segments by commercial collagenase preparations was reported to diminish the reactivity of subendothelium [6]. Birembaut et al. [13], using a collagenase preparation highly specific for collagen, have shown that this macromolecule was not the only thrombogenic constituent of the subendothelium, and that the elastin-associated microfibrils were also able to provoke platelet adhesion and thrombus formation. This was the conclusion of a series of histochemical and ultrastructural observations in which de-endothelialized collagenase-treated rabbit aortas were submitted to blood flow in the conditions of Baumgartner and Haudenschild [6], then examined with an electron microscope after the microfibrils were stained with various appropriate histochemical markers. In spite of complete digestion of collagen fibres, platelets still continued to adhere to the microfibrils of the elastic lamina: that microfibrils and not elastin were involved, was shown by the fact that after combined treatment of the aortic segment with collagenase and chymotrypsin (which digested all proteins including microfibrils in the suben-

dothelium), no platelets were observed in contact with the remaining elastin.

Role of Plasma Factors in the Thrombogenicity of the Subendothelium

The experimental model of Baumgartner has also been used with rabbit and human arterial segments to demonstrate the crucial role of von Willebrand factor in the adhesion of platelets to subendothelium, which is particularly evident under high shear-rate conditions [37,125]. Below or at a shear rate of 800/s, the absence or presence of von Willebrand factor was not significant. At higher shear rates, the absence of von Willebrand factor results in defective adhesiveness: the higher the shear rate, the greater the defect in the adhesiveness. This indicates that von Willebrand factor is important in the microcirculation where shear forces are high. This observation has been confirmed by the inhibition of platelet deposition on the vessel wall by human homologous or rabbit heterologous antibodies directed against human von Willebrand factors [8].

At low shear rate, Turritto et al. [126] showed that von Willebrand factor is involved in the size and thickness of the thrombus, and that it has a role in the adhesion and spreading of platelets. Sakariasen et al. [115] and Bolhuis et al. [15] have studied the adhesion of platelets and binding of the von Willebrand factor to arterial subendothelium as a function of time and shown that the binding of von Willebrand factor preceded an increase in adhesion and strengthened the spreading of the platelets. These observations however were made with plasma von Willebrand factor and did not take into account the pre--existence of the factor in the subendothelium. Similar ex vivo systems were also used to establish the map of the structural domains of the factors which are implicated in each function: it was shown that the two functions of monoclonal antibodies [120], i.e. recognizing particular epitopes and inhibiting both the ristocetin-induced platelet agglutination and adhesion to the subendothelium, are borne by the same structural domain of the molecule.

Von Willebrand factor is in fact a mixture of multimers with a very wide range of molecular weights (1×10^6 to 20×10^6): the largest multimers are involved in the platelet–vessel wall interaction [57] and preferentially bind to collagen [118]. The discovery of variants of the Willebrand disease [57,114] in which the high polymers are absent has shown their importance in the biological activity of the von Willebrand protein.

In Vitro Study of the Vascular Thrombogenic Macromolecules

Macromolecules have been extracted from various tissues and organs, e.g. bovine skin or aorta and human placenta, and more rarely from cell culture media. Other rare systems are provided by tissues neosynthesized in vitro from cells in culture or by isolated aortic segments. For a long time, the basic methodology has been the measurement of the aggregation of platelets. Stirring platelet-rich plasma with an inducer provokes the aggregation which is measured by the progressive increment in the optical transmission of the medium, using an aggregometer coupled to a recorder [this is the classical method of Born and Cross [16]]. In fact, the aggregation of platelets is the result of a series of events, the first step of which is adhesion. Several techniques have been proposed for the evaluation of platelet adhesion to collagen, in conditions that ensure that the platelet–collagen interaction is blocked before activation starts. Some of these techniques are based upon the use of a surface coated with fibrillar collagen or with an extracellular matrix synthesized by endothelial cells in culture [100,116]. These surfaces can be introduced into perfusion chambers [100, 116] or represent the wells of titration plates [88]. In other systems, a rotating probe [22] coated either by the molecule or by an aortic segment is rotated in a [111]indium-labelled platelet suspension, or blood can be syringed through collagen-coated capillary tubing [101].

Other methods apply the principle of affinity chromatography using collagen sepharose columns [19] or gel filtration which separate the adhesive platelets from non-adhesive platelets [82]. In a last group of methods, separation is by transmembrane filtration. Following this procedure, adhesion has been quantified either by microscopic observations or by isotopic evaluation after labelling the platelets with chromium-51 [35], indium-111 [36,119] or ^{14}C serotonin [82]. This last procedure is interesting since it allows simultaneous evaluation of adhesion and secretion. In fact, microscopy and the isotopic evaluation procedure are complementary; the first is time consuming, but the second can be interpreted only if the conditions have been established so that the number of micro-aggregates be as small as possible: only mor-

phometry can assess this point. Scanning electron microscopy permits the morphological changes which occur to platelets after adhering to collagen to be followed: shape change, pseudopodia production and the formation of microaggregates are seen.

Some of these methods have been applied on a quantitative and on a qualitative basis to identify the structural determinants of the thrombogenic macromolecules which are implicated in their reaction with platelets. The main data from the literature will be presented below.

Structural Determinants of Collagen and Microfibrils Involved in the Thrombogenicity of Subendothelium

We have already stated that collagen and microfibrils are the two main subendothelial macromolecules which activate platelets; however, other glycoproteins may well have similar effects.

Interaction of Blood Platelets with Collagen

In the early 1960s, microscopic observations of damaged vessels [18] and torn rabbit mesenteries [74] were the first showing adherence of platelets to the fibres of collagen. This adherence was confirmed in 1964 by Hughes and Lapiere [60] who circulated platelet-rich plasma on a collagen-coated glass surface. They described an immediate and irreversible adhesion of the platelets to collagen, followed by the spreading of platelets along the fibres and progressive building of aggregates. This opened the way to a great number of studies on what is still now regarded as a triggering step in haemostasis, thrombosis and atherosclerosis. Most of them concerned type I collagen, in which the importance of the structural determinants linked to its molecular organization (quaternary and tertiary structure) has been established.

Importance of the Quaternary and Tertiary Structures. It is established that collagen fibres result from a particular arrangement (quaternary structure) of triple helical molecules (tertiary structure). The preservation of this structure is an absolute prerequisite for the induction of platelet adhesion, activation and aggregation. Monomeric collagen does not aggregate platelets [63,98] but when it is heated to between 33°C and 37°C its ability to interact with platelets is restored: the reactivity of collagen depends on the size of the fibres rather than on their molecular organization. The reader will know (see Chap. 3) that in native collagen there is a staggering of 67 nm between two adjacent molecules, which corresponds to 1/4.4 of their length; Muggli has shown that this native arrangement is not essential and that different molecular organizations (segment-long spacing, fibre-long spacing, amorphous collagen) are also reactive [99]. Size and diameter seem to be the main two factors: Kronick and Jimenez [76] calculated that the threshold of the platelet aggregation was reached with striated fibrillar elements with a diameter of 40 nm and length of 900 nm. In view of the organization of native collagen, this length corresponds to that of approximately four molecules with the 67 nm staggering. In fact, collagen must present a regular recurrence of structural elements and it is this which gives it its reactivity with platelet membranes; collagen also has an affinity for platelets which is high enough to trigger adhesion and the following step of activation. Santoro and Cuningham [117] proposed the idea that the aggregation of platelets by collagen results from multiple and simultaneous interactions between platelet membrane receptors and sites which are indefinitely repeated along the rigid structure formed by collagen fibres.

The preservation of the collagen quaternary structure requires the preservation of its tertiary structure. If the tertiary structure is damaged by enzymatic cleavage [23,66] or by denaturation [66], collagen ceases to aggregate platelets. Fauvel et al. [43] have shown that the $\alpha_1(I)$ and the $\alpha_2(I)$ chains isolated from denatured type I collagen and re-associated as monomeric [α_1(I)]3 and (α_2)3 molecules did not provoke platelet adhesion; however, when [α_1(I)]3 trimers were re-associated as fibrils, their re-

activity was restored. Similar observations have recently been made by Barnes [5] on reconstituted (α_2)3 fibrils.

Collagen Primary Structure and Platelet Adhesion. The reactivity of collagen with platelets seems linked to particular domains of its amino acid sequences which can be identified after cleavage by cyanogen bromide and isolation of the resulting peptide. In the α_1 chains of type I collagen, Fauvel et al. [43] showed the reactivity of the *C*-terminal CB6 peptide (214 amino acids between residues 838 and 1055) and Barnes [5] that of the CB8 peptide between residues 139 and 418. In type III collagen an active sequence is located in the central region of the chains between residues 403 and 551 at the level of the CB4 peptide [44]. Chemical and enzymic cleavage of this peptide led to the identification of an octapeptide (position 74–81 in the CB4 peptide) which seems to be involved in platelet activation. Its sequence is lys–pro–gly–glu–pro–gly–pro–lys: this peptide inhibits platelet aggregation [71,83], release of the granule content [71], expression of the fibrinogen receptor [106] and exposure of the pro-coagulant activity associated with the platelet membrane [12]. In contrast, it does not impair the initial step of adhesion and this suggests that the sequence on collagen which recognizes the platelet is located in another portion of peptide CB4.

The organization of a molecule of collagen, such as has been described, allows us to propose a scheme that shows how the active sequence is repeated all along the fibres (Fig. 8.2).

Incidence of Non-enzymatic Glycosylation of Collagen and its Effect on Aggregating Potency. Non-enzymatic glycosylation is an extracellular

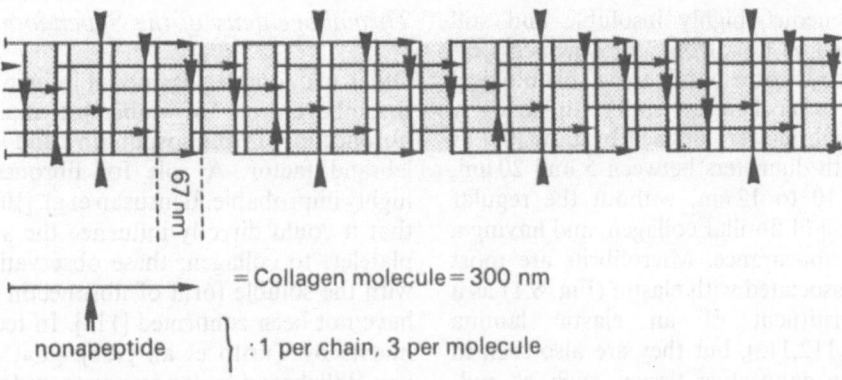

67 nm

⟶ = Collagen molecule = 300 nm

nonapeptide } : 1 per chain, 3 per molecule

Fig. 8.2. Schematic representation of a collagen fibre showing the recurrence of an "active" nine amino acid sequence reaching with platelets every 67 nm. In the scheme, each triple helix is represented by an *arrow*.

modification of collagen which is linked to ageing and strongly accelerated in diabetes [84,111]. Non-enzymatic glycosylation provokes an increased ability of collagen to aggregate platelets [85,86] which might contribute to the thrombotic tendency observed in diabetes. This is considered in more detail in Chap. 12.

Interaction Between Platelets and other Genetic Types of Collagen. Basement membranes interact weakly with platelets [58,59]. Some studies on the platelet–type IV collagen interaction seemed to indicate that this collagen does not react with platelets [3,117,124] unless artificially organized structures are reconstituted [4,5]. These structures are far from representing the network organization of type IV collagen in basement membrane, which has not been studied in that particular respect. For type V collagen, very few data have been reported; those that have been indicate a very poor reactivity [24,124]. This suggested to Parson et al. [104] that type V collagen provides a very weak thrombogenic surface for circulating blood once smooth-muscle cells have migrated into the intima during the repair process which follows vascular injury and de-endothelialization. Whatever its role may be, Barnes has also reconstituted "active" fibrillar forms from type V collagen and also from type VI collagen [5]. This indicates that all collagens bear particular elements in their primary structure which react with platelets, provided that an ordered tertiary and quaternary arrangement makes them accessible. The real question concerning these collagens is that of the physiological significance of this reactivity.

Interaction of Blood Platelets with Microfibrils

Microfibrils are difficult to study because they are heterogeneous, highly insoluble and still poorly defined, at least from a biochemical perspective. They correspond to a histological rather than a biochemical entity. In an early study, microfibrils were defined by Low [91] as filaments with diameters between 5 and 20 nm, most often 10 to 12 nm, without the regular cross-striation of fibrillar collagen, and having a beaded-like appearance. Microfibrils are most frequently associated with elastin (Fig. 8.1) as a second constituent of an elastic lamina [25,52,67,75,112,113], but they are also seen in several loose connective tissues such as pulmonary alveolae and placental villi in the vicinity of the trophoblastic basement membrane [20,55].

Microfibrillar structures are also seen near the basement membrane in subendothelium.

An important review by Cleary and Gibson [26] has recently collated the very diverse results from different laboratories working on the biochemical characterization of microfibrils, noting the source of the microfibrils (species, organs, age) and the conditions applied to their extraction. In their histochemical studies on rabbit aorta, Birembaut et al. [13] described the microfibrils as fibrillar (as they were visualized by tannic acid) and polyanionic (because they were stained by ruthenium red, a cationic stain) structures presenting glycanic determinants (revealed by peroxidase-labelled lectins): microfibrils were thus described as acidic glycoproteins. Another important property is their resistance to collagenase [13,26,45,79,121, mentioned earlier when we described the thrombogenicity of the collagenase-treated rabbit aortas. Fauvel et al. [45] have identified a major glycoproteic constituent in bovine aortic microfibrils, of molecular weight of 128 000 (GP128), which resists collagenase, is partly digested by pepsin, sensitive to chymotrypsin and involved in the interaction of microfibrils with platelets. GP 128 has been identified in a culture medium of endothelial cells and its endothelial origin was demonstrated by the use of a labelled precursor [46]. It has now been isolated from these culture media by high pressure liquid chromatography coupled with molecular sieve gel filtration and preparative iso-electrofocusing and identified as an acidic glycoprotein with a pH of 4.8. In its purified form, it inhibits the aggregation of platelets by microfibrils and does not affect the platelet–collagen interaction. Some very recent and preliminary data indicate that GP 128 has a thrombospondin-like antigenicity.

Are Other Macromolecules Involved in the Thrombogenicity of the Subendothelium?

Other macromolecules possibly involved in the thrombogenicity of subendothelium include fibronectin, thrombospondin and von Willebrand factor. A role for fibronectin seems highly improbable. Bensusan et al. [10] suggested that it could directly influence the adhesion of platelets to collagen; these observations, made with the soluble form of fibronectin in plasma, have not been confirmed [117]. In recent experiments by Turito et al. [126], goat anti-rabbit von Willebrand factor serum or an IgG fraction prepared from this serum, was incubated with vessel segments before exposure to blood in the annular perfusion chamber of Baumgartner; both preparations were shown to diminish the adhesion of platelets, suggesting that this factor could mediate platelet attachment to subendothelium in a manner similar to that of plasma von Willebrand factor.

The main origin of thrombospondin is in the alpha granules of platelets from which it is released during their activation by thrombin. Thrombospondin is a filamentous material of molecular weight 540 000, with a pH = 4.7, resulting from the association of three 150 000 subunits [48,65,80,93]. Endothelial cells, smooth-muscle cells and fibroblasts synthesize a material which has been immunologically identified as thrombospondin [64,94,97,107]. It has been shown by Mosher et al. [97] that 10^6 confluent endothelial cells secrete 49 μg thrombospondin within 24 hours. The exact role of tissue thrombospondin is not yet known. Lahav et al. [78] used cross-linking reagents to show that platelet-derived cells and endothelial cells secreted thrombospondin which could bind to fibronectin and they suggested the existence of a ternary complex between thrombospondin, fibronectin and collagen. This should be important in the organization of the extracellular matrix, though these authors discussed it mainly in terms of platelet adhesion.

Recent work has shown that platelets also adhere to laminin- and nidogen-coated surfaces [33,62]: these two glycoproteins are present in basement membrane where they act as anchoring proteins for the attachment of the endothelial cell to the matrix.

In Vivo Studies on the Thrombogenicity of Subendothelium

Interesting studies have been done in vivo to investigate the variability of the thrombogenicity of the subendothelium. Two factors are important: the time and the level in the vascular tree.

Variability as a Function of Time

Various authors studied the reactivity of the subendothelium exposed to blood after an endothelial injury caused by a balloon catheter [7,35,53,121]. They pointed out the rapidity of in vivo platelet deposition: Born and Richardson [17] calculated that the rate of incorporation of circulating platelets into a thrombus is approximately 10^4/s. Within a few minutes, a de-endothelialized surface is almost completely covered by a monolayer of activated platelets which then increases in thickness in some areas to form microthrombi. The formation of these microthrombi is maximal 10 min after the lesion is initiated. Between 30 min and 3 h after the lesion is initiated, the microthrombi disappear and the monolayer itself diminishes until only rare platelets are still seen in contact with the wall. The re-endothelialization then takes place progressively with only very few denuded areas observable: an interesting point is that this subendothelium has become thromboresistant.

What is the cause of this time-dependent variability? The apparent diminution of the thrombogenicity of the exposed subendothelial areas may be due to proteolytic enzymes (collagenase, elastase, cathepsins and so on) coming from leucocytes, from platelets or present in the wall itself. It is interesting to note that in addition to their elastase, platelets contain a heparitinase-like enzyme. Also, substances coming from the plasma and possibly from the cells can coat the wall or adsorb on it, conferring thromboresistance on the surface.

Platelets deposited on the wall are continuously renewed and there is a permanent and rapid exchange between circulating and deposited platelets. This exchange is observable after injection of indium-111 labelled platelets into an animal, and diminishes as the thrombogenicity of the exposed subendothelium diminishes.

Variability along the Vascular Tree

This is due to two main factors. Firstly, endothelial cells adapt their function following the level of the vascular tree. In particular, endothelium varies in its ability to synthesize subendothelial macromolecules and this may explain the variability observed from one vessel to another. Secondly, haemorheological factors also vary with the flow conditions, which depend on the diameter of the vessel and the distensibility of the wall. Flow differences between the various types of arteries and the veins must be mentioned in this respect.

Mechanism of Platelet Activation

The adhesion of platelets to subendothelium or to any isolated thrombogenic macromolecule triggers a series of phenomena which leads the

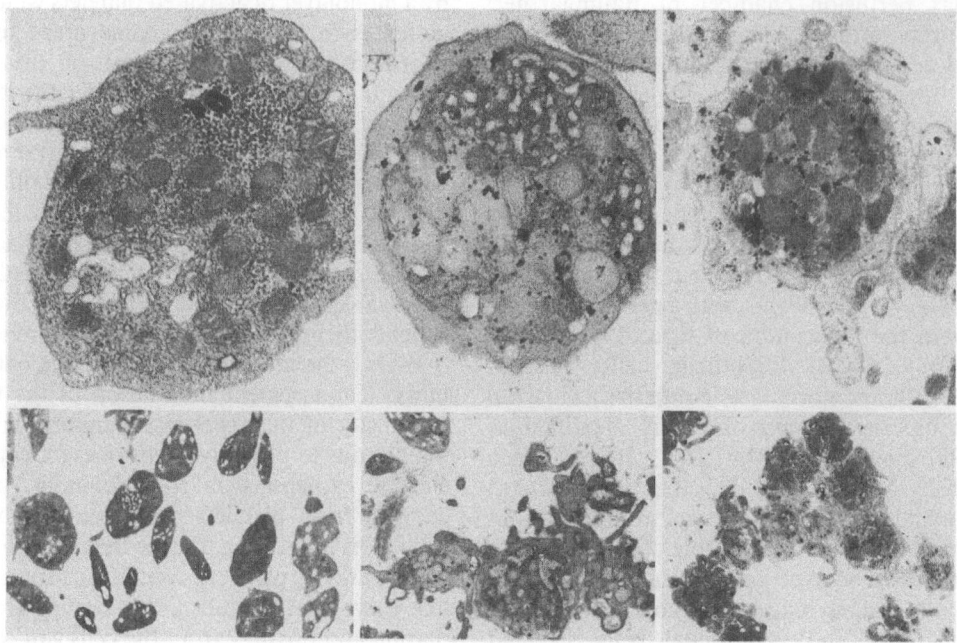

Fig. 8.3. Activation of platelets by microfibrils. The micrographs show the rapid emission of pseudopodia and the centralization of the intracellular organelles; small platelet aggregates are visible after 30 s.

platelets to build an aggregate and participate in the coagulation of the plasma. The main step of the platelet activation is the release of pro-aggregating substances and the expression of a pro-coagulant surface on their membrane together with receptors for substances which fill the role of aggregating cofactors: one of these is fibrinogen. The series of micrographs in Fig. 8.3 illustrates the various steps of the activation resulting from the stirring of a platelet-rich plasma with microfibrils in an aggregometer cuvette. Very rapidly platelets lose their discoid shape and put out pseudopodia. This immediately precedes the beginning of concentration of the organelles towards the centre of the platelets; degranulation is then observed as the platelets start to aggregate. This kind of observation can be made whatever the aggregating agent is; in particular, with collagen the photomicrograph would be identical.

Activation is a very complex phenomenon and is schematically represented in Fig. 8.4. The trigger is the binding of an agonist to its receptor on the membrane which is manifested by the adhesion of the platelets to the collagen fibre. This provokes a signal which induces the transmission of a message into the platelets, via particular second messengers such as diacylglycerol (DAG) and inositol 1-4-5 triphosphate (IP3). These second messengers originate from phos-

phatidyl inositol 4-5 biphosphate (PIP2) and result from the effect of the enzyme phospholipase C which hydrolyses PIP2 into 1-2-diacylglycerol which is then phosphorylated into phosphatidic acid, a direct precursor of phospho-inositides [61,127].

Despite its rapid conversion, DAG plays the role of second messenger by activating protein-kinase C which phosphorylates a protein of molecular weight 47 000. This protein is probably involved in the process of exocytosis during which the granules are released from the platelets and seems to facilitate a fusion of the membranes of the granules with the internal membrane. The internal membrane invaginates as an open canalicular system through which the organelles are extruded [69]. The other second messenger, IP3, releases calcium from the dense tubular system [49,50]: this, together with an influx of external calcium, increases the level of calcium in the cytoplasm and permits activation of a calmodulin-dependent myosin light-chain kinase which phosphorylates the 20 000 light chain of myosin involved in platelet shape change, emission of pseudopodia and granule centralization [31,32].

Calcium is also involved in the activation of a phospholipase, A_2 [110], which degrades phosphatidyl-choline in the membrane to produce lysophosphatidyl-choline, an intermediate in the

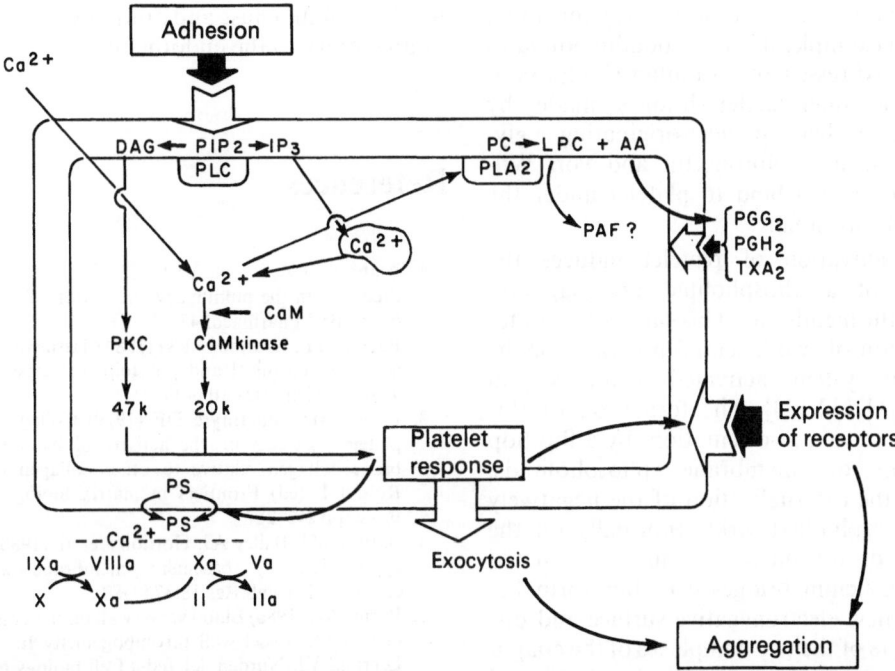

Fig. 8.4. Mechanism of platelet activation. In this scheme, the trigger is the adhesion of the platelet to collagen. Phospholipase C converts PIP2 (phosphatidyl inositol 4-5 biphosphate) into DAG (diacylglycerol) and produces IP3 (inositol 1-4-5 triphosphate). DAG activates protein kinase-C (PKC) which phosphorylates a protein of molecular weight 47 000 possibly involved in granule exocytosis. IP3 releases calcium from the dense tubular system and the binding of calcium to calmodulin (*CAM*) allows the activation of a CAM-dependent protein kinase (*CAM-kinase*) which phosphorylates a protein of molecular weight 20 000 which is the light chain of myosin involved in platelet shape change and granule centralization. Calcium also activates phospholipase A_2 which breaks down phosphatidyl-choline (*PC*) to produce lysophosphatidyl-choline (*LPC*), an intermediate in the PAF-acether pathway, and arachidonic acid, the precursor of pro-aggregating prostaglandins G_2 (PGG$_2$) and H_2 (PGH$_2$) and thromboxane A_2(TXA$_2$). (For details, see CHAP et al. (1986) Platelet phospholipids and their metabolites in the regulation of platelet response. In: Jolles P, Legrand YJ, Nurden AT (eds) Biology and pathology of the platelet-vessel wall interactions. Academic Press, New York, pp 267–287). As membrane modifications occur receptors for aggregtion cofactors released from the platelets (e.g. fibrinogen) become exposed, and a pro-coagulant surface is expressed secondarily to a surface of negatively charged phosphatidylserine (*PS*) which permits fixation of the factors of the complexes of the intrinsic coagulation, i.e. tenase complex (factors IXa and VIIIa, to activate factor X) and prothrombinase complex (factors Xa and Va to generate thrombin) via calcium bridges.

PAF-acether pathway, and arachidonic acid, the precursor of the pro-aggregating prostaglandins G_2 and H_2 and of thromboxane A_2. These derivatives find specific receptors on the platelet membrane.

All these complex phenomena are under the control of various regulatory mechanisms such as the levels of calcium and of adenylate cyclase/cyclic adenosine monophosphate (cAMP) phosphodiesterase. An increase in the cAMP level results in an inhibition of the activation of platelets. This can be due either to an activation of adenyl cyclase (this is the mode of action of prostacyclin and its stable analogues), or to an inhibition of phosphodiesterase (this is the mode of action of xanthines such as theophylline or of a pyrimido-pyrimidine such as dipyridamole).

What happens to platelet membrane during the activation process? At least two important events take place:

1. The activation leads to the expression of receptors for agents which play the role of aggregating cofactors. On the surface of unactivated platelets, these receptors are unavailable. The receptor for fibrinogen is borne by its glycoprotein complex IIb IIIa, which forms on the surface of the platelets when they are activated by ADP or by collagen [81,102]. It is known that fibrinogen bound to platelet permits the formation of intermembrane bridges which link two platelets together: this explains how fibrinogen is committed in platelet aggregation. The binding of other substances to the membrane also necessitates the expression of a recep-

tor, but this has been established for thrombin only. As an example, thrombospondin bound to a receptor expressed by thrombin should consolidate the interplatelet bridges made by fibrinogen. Other platelet-originating glycoproteins such as fibronectin and von Willebrand factor also bind to platelet under the influence of thrombin.

2. The activation of platelet induces the expression of a phospholipid pro-coagulant surface on the membrane. This surface facilitates the activation of two essential proteases of the coagulation system: activated factor X and thrombin [11,132,133]. The formation of this surface is due to a re-orientation, by a flip-flop mechanism, of membrane phospholipids, leading to the externalization of the negatively charged phosphatidyl serine (normally on the inner face of the membrane in unstimulated platelets). Calcium bridges can thus form between this new electronegative surface and different factors of the two complexes of the coagulation cascade: the tenase complex (activated factor VIII and factor X) and the pro-thrombinase complex (activated factor V and prothrombin). Bevers et al. [11] have shown that the re-orientation of phosphatidyl-serine occurs under the combined effect of collagen and thrombin.

Conclusion

Haemostasis is a drama which is played on an active stage. The scenery is the vessel wall itself which directly participates in the action and directly influences the behaviour of the actors. These actors are the cellular and molecular elements which will build or help to build a haemostatic plug, and participate in its consolidation and dispersion when it has finished its repair role at the site of vessel damage. We have described the regulatory effect of the endothelium haemostasis due to the properties of its surface and its synthetic activities. We have also emphasized the active role of platelets in the building of the aggregate and the activation of coagulation. The vessel wall is in fact fragile and many factors may provoke the detachment of the endothelial cell from its basement membrane and subendothelial support. This has very dramatic consequences, since the protective role that we described thus disappears. We will describe

in Chap. 14 the cause and dramatic consequences of any injury to the endothelium.

References

1. Alabaster DA, Bakhle YS (1973) The removal of noradrenaline in the pulmonary circulation of rat isolated lungs. Br J Pharmacol 47:325–333
2. Bachman F, Kruithof EKO (1984) Tissue plasminogen activator: chemical and physiological aspects. Semin Thromb Hemostas 10:6–17
3. Barnes MJ, MacIntyre DE (1979) Collagen induced platelet aggregation: the activity of basement membrane collagen relative to other collagen types. In: Robert L (ed) Frontiers in matrix biology. Karger, Basle, pp 246–287
4. Barnes, MJ, Bailey AJ, Gordon J et al. (1980) Platelet aggregation by basement membrane associated collagen. Thromb Res 18:375–380
5. Barnes MJ (1986) Blood vessel wall matrix components involved in vessel wall thrombogenicity In: Jolles G, Legrand YJ, Nurden AT (eds) Cell biology and pathology of the platelet vessel wall interaction. Academic Press, New York, pp 21–37
6. Baumgartner HR, Haudenschild C (1972) Adhesion of platelets to subendothelium. Ann NY Acad Sci 201:22–36
7. Baumgartner HR (1973) The sub-endothelial surface and thrombosin. In: Deutsch E, Brinkhous KM, Lechner K, Himmon S (eds) Pathogenesis and clinical trials. Thrombosis, Stuttgart, pp 91–105
8. Baumgartner HR, Tschopp TB, Meyer D (1980) Shear rate dependent inhibition of platelet adhesion and aggregation on collagenous surfaces by antibodies to human factor VIII/Von Willebrand factor. Br J Haematol 44:127–159
9. Becker CG, Harpel PC (1976) Alpha-2-macroglobulin on human vascular endothelium. J Exp Med 144:1–9
10. Bensusan HB, Koh TL, Henry KG et al. (1978) Evidence that fibronectin is the collagen receptor in platelet membranes. Proc Natl Acad Sci USA 75:5864–5868
11. Bevers, EM, Comfurius P, Zwaal RFA (1983) Changes in membrane phospholipid distribution during platelet activation. Biochim Biophys Acta 736:57–66
12. Bevers EM, Karniguian A, Legrand YJ et al. (1985) Collagen derived octapeptide inhibits platelet procoagulant activity induced by the combined action of collagen and thrombin. Thromb Res 37:365–370
13. Birembaut P, Legrand YJ, Bariety J et al. (1982) Histochemical and ultrastructural characterization of subendothelial glycoprotein microfibrils interacting with platelets. J Histochem Cytochem 30:75–80
14. Bjorck C, Larsson R, Olsson P et al. (1981) Uptake and inactivation of thrombin by the fresh, glutardialdehyde or heparin treated human umbilical cord vein endothelium. Thromb Res 21:603–610
15. Bolhuis PA, Sakariassen KS, Sander JH et al. (1981) Binding of factor VIII von Willebrand factor to human arterial sub-endothelium precedes increased platelet adhesion and enhances platelet spreading. J Lab Clin Med 47:568–577
16. Born GVR, Cross MJ (1963) The aggregation of blood platelets. J Physiol (Lond) 168:178–195

17. Born GVR, Richardson PD (1980) Activation time of blood platelets. J Membr Biol 57:87–90
18. Bounameaux Y (1959) L'accolement des plaquettes aux fibres sous-endothéliales. Bull Soc Biol 153:865–869
19. Brass LF, Bensusan HB (1976) On the role of the collagen carbohydrate residues in the platelet-collagen interaction. Biochim Biophys Acta 444:43–52
20. Bray BA, Hsu KC, Wigger JH et al. (1975) Association of fibrinogen and microfibrils with trophoblast basement membrane. Connect Tissue Res 3:55–71
21. Buonassini V, Venter JC (1976) Hormone and neurotransmitter receptors in an established vascular endothelial line. Proc Natl Acad Sci USA 73:1612
22. Cazenave JP, Blondowska D, Richardson M et al. (1979) Quantitative radioisotopic measurement and scanning electron microscopic study of platelet adherence to a collagen-coated surface and to sub-endothelium with a rotating probe device. J Lab Clin Med 93:60–70
23. Chesney C, Harper E, Colman RW (1972) Critical role of carbohydrate side chains of collagen in platelet aggregation. J Clin Invest 51:2693–2701
24. Chiang TM, Mainardi CL, Seyer JM et al. (1980) Collagen platelet interaction. Type V (AB) collagen induces platelet aggregation. J Lab Clin Med 95:99–107
25. Cleary EG, Fanning JC, Prosser I (1981) Possible role of microfibrils in elastogenesis. Connect Tissue Res 8:161–166
26. Cleary EG, Gibson MA (1983) Elastin associated microfibrils and microfibrillar proteins. Int Rev Connect Tissue Res 10:97–209
27. Collen D, Rijken DC, Van Damme J et al. (1982) Purification of human tissue type plasminogen activator in centigram quantities from human melanoma cell culture fluid and its conditioning for use in vivo. Thromb Haemost 48:294–296
28. Collen D (1980) On the regulation and control of fibrinolysis. Thromb Haemost 43:77–89
29. Comp PC, Jacoks RM, Rubenstein C et al. (1981) A lysine adsorbable plasminogen activator is elevated in condition associated with increased fibrinolytic activity. J Lab Clin Med 97:637–645
30. Cazervionke RL, Smith JB, Hoak JC et al. (1979) Use of a radio-immunoassay to study thrombin-induced release of PGI_2 from cultured endothelium. Thromb Res 14:781–786
31. Daniel JL, Holmsen H, Adelstern RS (1977) Thrombin stimulated myosin phosphorylation in intact platelets and its possible involvement in secretion. Thromb Haemost 38:984–989
32. Daniel JL, Molish IR, Holmsen H (1981) Myosin phosphorylation in intact platelets. J Biol Chem 256:7510–7514
33. De Groot PG, Ottenhoff-Rovers M, Dziadek M et al. (1985) Nidogen is present in the extra-cellular matrix of human endothelial cells and is involved in platelet adhesion to this matrix. Thromb Haemost 54:257 (abstract)
34. Dollery CT, Friedman LA, Hensby CN et al. (1979) Circulating prostacyclin may be reduced in diabetes. Lancet II:1365
35. Dosne AM, Drouet L, Dassin E (1976) Usefulness of [51]chromium platelet labelling for the measurement of platelet deposition on sub-endothelium. Microvasc Res 11:111–114
36. Drouet L, Efremidis A, Spaet TH (1979) Etude in vivo du dépôt au sous-endothélium aortique des plaquettes marquées à l'Indium 111. Nouv Rev Fr Hématol 21:101–109
37. Drouet L, Carrier JL, Rosa JP et al. (1983) Platelet deposition. In: Woolf N (ed) Biology and pathology of the vessel wall. Praeger, New York, pp 81–100
38. Drouet L, Baumann FCH, Baudin B, Dupuy E (1988) Seric angiotensin converting enzyme: a marker of venous endothelial cell functions? comparison to tissue plasminogen activator and von Willebrand factor in thromboembolic pathology. (Submitted for publication).
39. Emets JJ (1979) The vascular wall and fibrinolysis. Haemostasis 8:333–339
40. Engelberg H (1985) Heparin, heparin fractions, and the atherosclerotic process. Semin Thromb Hemost 11:48–55
41. Esmon CT, Owen WG (1981) Identification of an endothelial cell co-factor for thrombin-catalyzed activation of protein C. Proc Natl Acad Sci USA 78:2249–2252
42. Esmon CT, Esmon NL (1984) Protein C activation. Semin Thromb Hemost 10:122–130
43. Fauvel F, Legrand YJ, Caen JP (1978) Platelet adhesion to type I collagen and $\alpha_1(I)_3$ trimers: involvement of the C-terminal $\alpha_1(I)$ CB6 peptide. Thromb Res 12:273–285
44. Fauvel F, Grant ME, Legrand YJ et al. (1977) Interaction of blood platelets with an extract from adult bovine aorta: requirement for Von Willebrand factor. Proc Natl Acad Sci USA 80:551–554
45. Fauvel F, Legrand YJ, Bentz H et al. (1978) Platelet collagen interaction: adhesion of human blood platelets to purified CB_4 peptide from type III collagen. Thromb Res 12:841–850
46. Fauvel F, Campos-Oriola R, Leger D et al. (1984) Aortic endothelial cells in culture secrete glycoproteins reacting with blood platelets. Biochem Biophys Res Commun 123:114–120
47. Fishman JA, Ryan GB, Karnovsky MJ (1975) Endothelial regeneration in the rat carotid artery and the significance of endothelial denudation in the pathogenesis of myointimal thickening. Lab Invest 32:339–349
48. Gartner TK, Dockter ME (1983) Secreted platelet thrombospondin binds monovalently to platelets and erythrocytes in the absence of free Ca^{2+}. Thromb Res 33:19–30
49. Gerrard JM, Butler AT, White JG (1978) Calcium release from a platelet calcium sequestering membrane fraction by arachidonic acid and its prevention by aspirin. Prostaglandins 15:703–715
50. Gerrard JM, White JG, Petterson DA (1978) The platelet dense tubular system: its relationship to prostaglandin synthesis and calcium flux. Thromb Haemost 40:224–235
51. Glimelius B, Busch C, Hook M (1978) Binding of heparin in the surface of cultured human endothelial cells. Thromb Res 12:773–782
52. Greenlee TK, Ross R, Hartman JL (1966) The fine structure of elastin fibers. J Cell Biol 30:60–71
53. Groves HM, Kinlough-Rathbone RL, Richardson M et al. (1979) Platelet interaction with damaged rabbit aorta. Lab Invest 40:194–200
54. Gryglewski RJ (1979) Prostacyclin as a circulatory hormone. Biochem Pharmacol 28:3161–3166
55. Gutman N, Bardos P, Arbeille B et al. (1977) Isolement de glycoprotéines de structure du placenta humain. CR Acad Sci (Paris) 285:583–587
56. Higgs EA, Moncada S, Vane JR et al. (1978) Effect of prostacyclin (PGI_2) on platelet adhesion to rabbit arterial sub-endothelium. Prostaglandins 16:17–22

57. Hoyer LW, Rizza CR, Tuddenham EGD et al. (1983) Von Willebrand factor multimer patterns in Von Willebrand's disease. Br J Haematol 55:493–507

58. Huang TW, Lagunoff D, Benditt EP (1974) Non aggregative adherence of platelets to basal lamina in vitro. Lab Invest 31:156–162

59. Huang TW, Benditt EP (1978) Mechanism of platelet adhesion to basal lamina. Am J Pathol 92:99–108

60. Hugues J, Lapiere CM (1964) Nouvelles recherches sur l'accolement des plaquettes aux fibres de collagène. Thromb Haemost 11:327–345

61. Huzoor A, Navran SS, Chang J et al. (1982) Investigation on the effects of phospholipase C on human platelets: evidence that aggregation induced by phospholipase C is independent of prostaglandin generation, released ADP, and is modulated by cyclic AMP. Thromb Res 27:405–417

62. Ill CR, Engvall E, Ruoslahti E (1984) Adhesion of platelets to laminin in the absence of activation. J Cell Biol 99:2140–2145

63. Jaffe R, Deykin D (1974) Evidence for a structural requirement for the aggregation of platelets by collagen. J Clin Invest 53:875–883

64. Jaffe EA, Ruggiero JT, Leung LLK et al. (1983) Cultured human fibroblasts synthesize and secrete thrombospondin and incorporate it into extracellular matrix. Proc Natl Acad Sci USA 80:998–1002

65. Jaffe EA, Leung LLK, Nachman TL et al. (1982) Thrombospondin is the endogenous lectin of human platelets. Nature 295:246–247

66. Jelenska MM, Wegrzynowicz Z, Dancewicz AM et al. (1973) Interaction of platelets with collagen modified by irradiation or partial proteolysis. Thromb Res 2:41–48

67. Jones CJP, Sear CHJ, Grant ME (1980) Ultrastructural study of fibroblasts derived from bovine ligamentum nuchae and their capacity for elastogenesis in culture. J Pathol 131:35–53

68. Junod AF (1975) Metabolism, production and release of hormones and mediators in the lung. Ann Rev Respir Dis 112:93

69. Kaplan KL, Broekman MJ, Chernoff A et al. (1979) Platelet granule proteins: studies on release and subcellular localisation. Blood 53:604–618

70. Karniguian A, Legrand YJ, Caen JP (1982) Prostaglandins: specific inhibition of platelet adhesion to collagen and relationship with cAMP level. Prostaglandins 23:437–457

71. Karniguian A, Legrand YJ, Lefrancier P et al. (1983) Effect of a collagen derived octapeptide on different steps of the platelet/collagen interaction. Thromb Res 32:593–604

72. Kirchof B, Grunwald J (1982) Thrombin inhibition and thrombin generation by cultured aortic endothelial and smooth muscle cells from mini pig. Thromb Haemost 48:101–103

73. Kisiel W, Ericsson LH, Davie EW (1976) Proteolytic activation of protein C from bovine plasma. Biochemistry 15:4893–4900

74. Kjaerheim A, Hovig T (1962) The ultrastructure of haemostatic blood platelet plugs in rabbit mesenterium. Thromb Haemost 7:1–7

75. Krauhs JM (1983) Microfibrils in the aorta. Connect Tissue Res 11:153–167

76. Kronick P, Jimenez SA (1981) The size of collagen fibrils that stimulate platelet aggregation in human plasma. Biochem J 186:5–12

77. Kruithof EKO, Bachmann F (1982) Studies on the binding of plasminogen activator to fibrinogen and fibrin. In: Henschen A, Graeff H, Lottspeich F (eds) Fibrinogen – Recent biochemical and medical aspects. De Gruyter, Berlin, pp 377–387

78. Lahav J, Schwartz MA, Hynes RO (1982) Analysis of platelet adhesion with a radioactive chemical crosslinking reagent: interaction of thrombospondin with fibronectin and collagen. Cell 31:253–262

79. Lamberg SI, Poppke, DC, Williams BR (1980) Isolation of elastic tissue microfibrils derived from cultured cells of calf ligamentum nuchae. Connect Tissue Res 8:1–8

80. Lawler JW, Slayter HS, Colingan JE (1978) Isolation and characterization of a high molecular weight glycoprotein from human blood platelets. J Biol Chem 253:8609–8616

81. Legrand C, Dubernard V, Nurden AT (1985) Characteristics of collagen and ADP fibrinogen binding to human platelets. Biochim Biophys Acta 812:802–810

82. Legrand YJ, Fauvel F, Kartalis G et al. (1979) Specific and quantitative method for estimation of platelet adhesion to collagen. J Lab Clin Med 94:438–446

83. Legrand YJ, Karniguian A, Lefrancier F et al. (1980) Evidence that a collagen-derived nonapeptide is a specific inhibitor of platelet-collagen interaction. Biochem Biophys Res Commun 96:1579–1585

84. Le Pape A, Muh JP, Bailey AJ (1981) Characterization of N-glycosylated type I collagen in streptozotocin-induced diabetes. Biochem J 197:405–412

85. Le Pape A, Guitton JD, Gutman N et al. (1983) Non-enzymatic glycosylation of collagen in diabetes: incidence on increased normal platelet aggregation. Haemostasis 13:36–41

86. Le Pape A, Gutman A, Guitton JD et al. (1983) Non-enzymatic glycosylation increases platelet aggregating potency of collagen from placenta of diabetic human beings. Biochem Biophys Res Commun 111:602–610

87. Lewis PJ, Shepherd GL, Ritter J (1981) Prostacyclin and pre-eclampsia. Lancet I:559

88. Leytin VL, Misselwitz FV, Lyubimova EV et al. (1981) Automated technique for measuring platelet adhesion to a surface in the presence of plasma. Thromb Res 23:201–205

89. Lieberman GE, Lewis GP, Peters TJ (1977) A membrane-bound enzyme in rabbit aorta capable of inhibiting adenosine-diphosphate-induced platelet aggregation. Lancet II:330–332

90. Lijnen HR, Collen D (1982) Interaction of plasminogen activators and inhibitors with plasminogen and fibrin. Semin Thromb Hemost 8:2–10

91. Low FN (1982) Microfibrils, a small extra-cellular component of connective tissue. Anat Rec 142:131–137

92. Mannucci PM, Rota L (1980) Plasminogen activator response after dDAVP. A clinico-pharmacological study. Thromb Res 20:69–76

93. Margossian SS, Lawler JW, Slayters HS (1981) Physical characterization of platelet thrombospondin. J Biol Chem 256:7495–7500

94. McPherson J, Sage H, Bornstein P (1981) Isolation and characterization of a glycoprotein secreted by aortic endothelial cells in culture: apparent identity with platelet thrombospondin. J Biol Chem 256:11330–11336

95. Moncada S, Gryglewski RJ, Bunting S et al. (1976) An enzyme isolated from arteries transforms prostaglandin endoperoxyde to an unstable substance that inhibits platelet aggregation. Nature 263:663–665

96. Moncada S, Palmer R and Higgs EA (1986) Generation

of prostacyclin and endothelial cells. In: Jolles G, Legrand YJ, Nurden AT (eds) Cell biology and pathology of the platelet vessel wall interaction. Academic Press, London, pp 289–304

97. Mosher DF, Doyle MJ, Jaffe EA (1982) Synthesis and secretion of thrombospondin by cultured human endothelial cells. J Cell Biol 93:343–448

98. Muggli R, Baumgartner HR (1973) Collagen induced platelet aggregation: requirement for tropocollagen multimers. Thromb Res 3:715–731

99. Muggli R (1978) Collagen induced platelet aggregation: native collagen quaternary structure is not essential requirement. Thromb Res 13:829–843

100. Muggli R, Baumgartner HR, Tschopp TB et al. (1980) Automated microdensitometry and protein assay as a measure for platelet adhesion and aggregation on collagen-coated slides under controlled flow conditions. J Lab Clin Med 95:195–207

101. Mulvihill JN (1984) Interaction des plaquettes sanguines et des protéines plasmatiques avec des surfaces artificielles. Thèse Sc Phys, Strasbourg

102. Niewiarowski S, Kornecki E, Budzynski AZ et al. (1983) Fibrinogen interaction with platelet receptors. Ann NY Acad Sci 408:536–555

103. Noordhoek Hegt V (1976) Distribution and variation of fibrinolytic activity in the walls of human arteries and veins. Haemostasis 5:355–372

104. Parson TJ, Haycraft DL, Hoak JC, Sage H (1983) Diminished platelet adherence to type V collagen. Arteriosclerosis 3:589–598

105. Pennica D, Holmes WE, Kohr WJ et al. (1983) Cloning and expression of human tissue-type plasminogen activator cDNA. Nature 301:214–221

106. Pintigny D, Legrand C, Karniguian A et al. (1985) Evidence that a collagen derived octapeptide inhibits fibrinogen binding to platelet stimulated by collagen and not by ADP. Biochem Biophys Res Commun 128:547–553

107. Raugi GJ, Memby SM, Abbott-Brown D et al. (1982) Thrombospondin: synthesis and secretion by cells in culture. J Cell Biol 95:351–354

108. Rijken DC, Collen D (1981) Purification and characterization of the plasminogen activator secreted by human melanoma cells in culture. J Biol Chem 256:7035–7041

109. Rikjen DC, Hoylaerts M, Collen D (1982) Fibrinolytic properties of one-chain and two-chain human extrinsic (tissue-type) plasminogen activator. J Biol Chem 257:2920–2925

110. Rittenhouse-Simmons S, Deykin D (1978) The activation by Ca^{2+} of platelet phospholipase A_2. Effects of dibutyryl cyclic adenosine monophosphate and 8-(N,N diethylamino)-octyl-3,4,5 trimethoxybenzoate. Biochim Biophys Acta 543:409–515

111. Robert B, Szigetti M, Derouette JC et al. (1971) Studies on the nature of the "microfibrillar component" of elastic fibers. Eur J Biochem 21:507–516

112. Robins SP, Bailey AJ (1972) Age related changes in collagen: the identification of reductible lysine–carbohydrate condensation products. Biochem Biophys Res Commun 48:76–84

113. Ross R, Bornstein P (1969) The elastic fibers. I. The separation and partial characterization of its macromolecular components. J Cell Biol 40:366–381

114. Ruggeri ZM, Nilsson IM, Lombardi R et al. (1982) Aberrant multimeric structure of von Willebrand factor in a new variant of von Willebrand disease (type IIC).

J Clin Invest 70:1124–1127

115. Sakariassen KS, Bolhuis PA, Sixma JJ (1979) Human blood platelet adhesion to artery sub-endothelium is mediated by factor VIII-von Willebrand factor bound to the sub-endothelium. Nature 279:636–638

116. Sakariassen KS, Aarts Paam, De Groot PG et al. (1983) A perfusion chamber developed to investigate platelet interaction in flowing blood with human vessel wall cells, their extra-cellular matrix, and purified components. J Lab Clin Med 102:522–535

117. Santoro SA, Cunningham LW (1981) The interaction of platelets with collagen. In: Gordon JL (ed) Platelets in biology and pathology. Elsevier, Amsterdam, pp 249–264

118. Santoro SA (1983) Preferential binding of high molecular weight forms of von Willebrand factor to fibrillar collagen. Biochim Biophys Acta 156:123–126

119. Scheffel U, MacIntyre PA, Evatt B et al. (1977) Evaluation of ^{111}Indium as a new high photon yield gamma emitting physiological platelet label. John Hopkins Med J 140:285–297

120. Stel HV, Sakariassen KS, Scholte BJ et al. (1984) Characterization of 25 monoclonal antibodies to factor VIII-von Willebrand factor: relationship between ristocetin-induced platelet aggregation and platelet adherence to sub-endothelium. Blood 63:1408–1415

121. Stemerman MB, Baumgartner HR, Spaet TH (1971) The sub-endothelial microfibrils and platelet adhesion. Lab Invest 24:179–186

122. Stenflo J (1984) Structure and function of Protein C. Semin Thromb Hemost 10:109–121

123. Strum JM, Junod A (1972) Radioautographic demonstration of 5-hydroxytryptamine-^{3}H uptake by pulmonary endothelial cells. J Cell Biol 54:456

124. Trelstad RL, Carvalho AC (1979) Type IV and type "A-B" collagens do not elicit platelet aggregation or serotonin release reaction. J Lab Clin Med 93:499–505

125. Tschopp TB, Baumgartner HR, Silverbauer K et al. (1979) Platelet adhesion and platelet thrombus formation on sub-endothelium of human arteries and veins exposed to flowing blood in vitro. A comparison with rabbit aorta. Haemostasis 8:19–29

126. Turitto VT, Weiss HJ, Baumgartner HR (1980) The effect of shear rate on platelet interaction with sub-endothelium exposed to citrated human blood. Microvasc Res 19:352–365

127. Vickers JD, Kinlough-Rathbone RL, Mustard JF (1982) Changes in phosphatidylinositol-4-5-biophosphate 10 seconds after stimulation of washed rabbit platelets with ADP. Blood 60:1247–1259

128. Wallen P, Bergsdorf N et al. (1982) Purification and identification of two structural variants of porcine tissue plasminogen activator by affinity adsorption on fibrin. Biochim Biophys Acta 719:318–328

129. Weksler BB, Marcus AJ, Jaffe EA (1977) Synthesis of prostaglandin I2 (prostacyclin) by cultured human and bovine endothelial cells. Proc Natl Acad Sci USA 74:3922–3926

130. Weksler BB, Ley CW, Jaffe Ea (1978) Stimulation of endothelial cell prostacyclin production by thrombin, trypsine and the ionophore A23187. J Clin Invest 62:923–930

131. Zwaal RFA (1978) Membrane and lipid involvement in blood coagulation. Biochim Biophys Acta 515:163–205

132. Zwaal RFA, Hemker HC (1982) Blood cell membranes and haemostasis. Haemostasis 11:12–39

Chapter 9

The Vascular Wall and Hormonal Control of Vasomotor Function

F. Alhenc-Gelas and P. Corvol

The simplest concept of hormonal control of vascular tone would be that circulating vasoconstrictor and vasodilator factors act directly upon the smooth-muscle cells of the media after they have crossed or circumvented the endothelial cells and the fibroconnective matrix of the intima, without undergoing modification. It is, however, known that more complex biochemical processes take place in the vascular wall or on its surface and that the endothelium plays a major role in the hormonal modulation of vasomotor function, having the enzymatic equipment necessary for activating angiotensin I to become the vasoconstricting angiotensin II and for inactivating the vasodilator bradykinin. Pharmacologically, the endothelium has characterizable receptors for histamine and serotonin. It takes up vasoactive adenylated nucleotides and noradrenaline, and metabolizes certain vasoactive prostaglandins. Recently, the physiological importance of the interaction of vasoactive hormones with vascular endothelium has been clearly demonstrated by the fact that certain compounds like bradykinin and histamine activate endothelial phospholipase and stimulate the release of prostacyclin and prostaglandins E_2 and $F_{2\alpha}$. In addition, very elegant pharmacological experiments have shown that acetylcholine and bradykinin cannot dilate the de-endothelialized aorta. It is now known that many vasoactive compounds, peptides, autocoids, cyclic nucleotides and arachidonic acid exert their vasodilatory or sometimes vasoconstrictive effects mainly because their interaction with the endothelium results in the formation in the endothelial cell of compounds, whose nature is only partly known, which modulate the contractility of vascular myocytes. Similarly, the potentiation of noradrenaline vasoconstriction by anoxia seems to depend on the endothelium.

Most vasoactive compounds that interact with the endothelium are only released acutely and/or locally by an inflammatory reaction, allergy, platelet aggregation or anoxia. A noteworthy exception is angiotensin I which is continuously produced from renin in the systemic and pulmonary circulations [27]. By converting angiotensin I to angiotensin II, the vascular endothelium plays a permanent essential role in controlling local arterial pressure and blood flow. This role is particularly well illustrated by the haemodynamic consequences of the inhibition of the vascular converting enzyme, i.e. vasodilatation and lowering of arterial pressure. These consequences are observed when the renin–angiotensin system is stimulated in normal man, certain hypertensive patients and during animal experimentation [43] (Fig. 9.1). Renin itself is in fact synthesized in the vascular wall. The epithelioid cells of the afferent arteriole of the renal glomerulus which synthesize and secrete renin are vascular smooth-muscle cells with endocrine differentiation. Renin secretion thus appears to be an endocrine function of a specialized part of the arterial tree. Under certain circumstances, such as renal ischaemia, other arterial segments, e.g. the interlobular arteries, may synthesize renin.

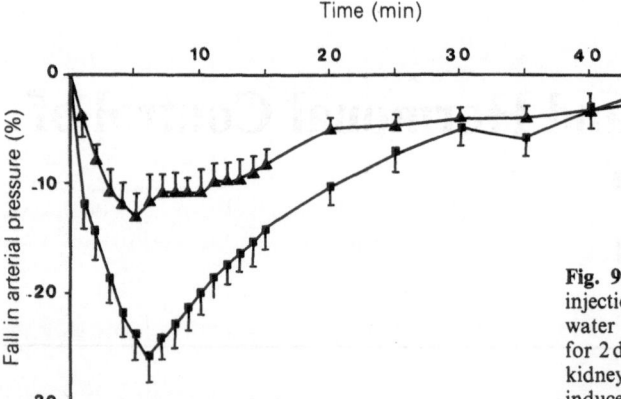

Fig. 9.1. Decrease in arterial pressure after intra-arterial injection of 1 mg/kg captopril in rats dehydrated by total water restriction for 48 h (▲), or by frusemide (40 mg/d for 2 d) (■). Dehydration stimulates renin secretion by the kidneys and the vascular converting enzyme blockage then induces hypotension (reproduced from [43]).

This chapter will therefore be divided into three parts. The first deals with the vascular endothelial metabolism of vasoactive peptides and endothelial peptidases, particularly angiotensin II production by the converting enzyme. The second part will attempt to summarize our present fragmentary knowledge of the interaction of peptide and non-peptide vasoactive compounds with endothelial cells and of the part played by the endothelium in modulating the pharmacological and physiological effects of vasoactive agents. The third part will deal with the endocrine function of the vascular muscle and describe the cellular and molecular mechanisms of renin biosynthesis.

Vasoactive Peptide Metabolism of the Vascular Endothelium

Converting Enzyme or Kininase II

The converting enzyme was discovered in horse plasma by Skeggs et al [101]. Renin hydrolyses angiotensinogen and releases angiotensin I, a decapeptide devoid of any known vascular activity at physiological concentrations. The activation of angiotensin I requires its hydrolysis by the converting enzyme, which splits off the histidyl–leucine dipeptide from its *C*-terminal extremity [34] (Fig. 9.2). Angiotensin II is an extremely potent vasoconstrictor in the systemic circulation. It acts directly on the contractile tone of the vascular smooth-muscle cells, and also indirectly, by facilitating noradrenergic transmission and inhibition of neuronal catecholamine re-uptake. It also stimulates aldosterone secretion by the corticoadrenals and exerts various pharmacological effects on the central nervous system. All these actions, like the peripheral activities of angiotensin II, combine to maintain arterial pressure and the sodium–water balance [94].

The converting enzyme is the only enzyme known to be able to activate angiotensin I to become angiotensin II in the circulation. Similarly, it activates de-aspartate-angiotensin I to become de-aspartate-angiotensin II (or angiotensin III), which is also a vasopressor stimulating aldosterone secretion (Fig. 9.2). It was

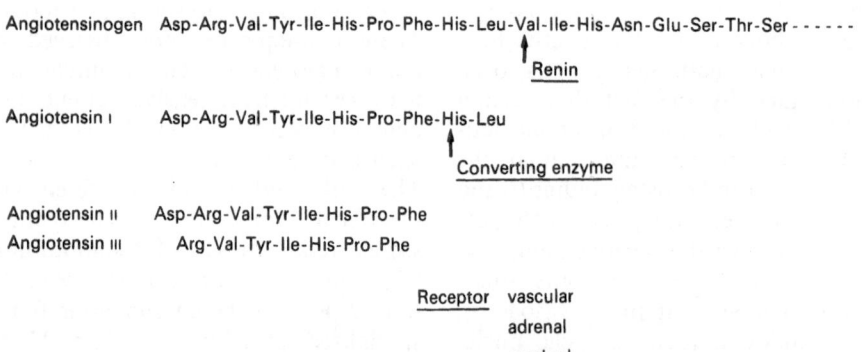

Fig. 9.2. The renin–angiotensin system.

thought after Skeggs' discovery that plasma alone was responsible for angiotensin I activation, but Ng and Vane noticed that the conversion of angiotensin I to angiotensin II proceeded too slowly in plasma alone to account for the in vivo formation of angiotensin II. They showed in the dog that in vivo conversion occurred mainly in the pulmonary circulation [76]. Biron and Huggins [11] independently obtained similar results. When angiotensin I was injected into the femoral or renal circulation, its conversion seemed negligible compared with that taking place in the pulmonary circulation [10]. The pulmonary vascular bed was thus responsible for the in vivo conversion of angiotensin I to angiotensin II. At the same time it had been observed that bradykinin, a potent vasodilating nonapeptide was rapidly inactivated during its passage through the pulmonary circulation [35]. Yang et al [11] showed that the converting enzyme was able, in pig plasma, lung and kidney, to inactivate bradykinin by sequential cleavage of phenylalanyl–arginine and seryl–proline dipeptides from its C-terminal extremity and to activate angiotensin I. The converting enzyme is thus kininase II or "dipeptidyl-carboxy-peptidase" (another carboxypeptidase – carboxypeptidase N or kininase I – which cleaves off the bradykinin terminal arginine residue, was previously discovered by Erdös in human plasma).

The precise location of the converting enzyme in the vascular wall was finally documented following two observations: firstly, kininase II was found in the microsomal fraction of lung and kidney homogenates, which indicated that it is at least partly a membrane-bound enzyme [8]. Subsequently, this enzyme was isolated from pig and rabbit and used to prepare antibodies. Ryan et al. [96] and Caldwell et al. [16] showed immunohistochemically that the converting enzyme was an ectoenzyme of the endothelial cell membrane. It is also found on the membrane of certain intracellular vesicles. This immunohistochemical approach showed that the converting enzyme was not only found in the endothelium of the pulmonary capillaries, but that it was also found in the aorta and large vessels. Finally, the importance of the pulmonary circulation for angiotensin I and bradykinin metabolism can only be explained by the extent of the capillary bed and by the fact that all the circulating blood has to traverse it. Splanchnic circulation also plays an important part in the in vivo metabolism of vasoactive peptides, as confirmed by the converting enzyme

diffusion along the extensive mesenteric capillary network [13].

Structure and Enzymatic Properties of the Vascular Converting Enzyme

In order to study the structure of the converting enzyme and to prepare specific antibodies, it was first necessary to isolate it from the membrane and purify it. It was purified from lung, plasma and kidney [3]. The converting enzyme isolated from lung comes from the capillary endothelial cells. The one isolated from kidney comes mainly from the brush border of the epithelial cells of the proximal convoluted tubule. The lung enzyme was isolated and purified from rabbit after solubilization with non-ionic detergents [25]. In the human lung and kidney, trypsin was used to solubilize the enzyme, which probably occurred by cleavage of a hydrophobic piece anchoring it to the membrane [77,105]. It was believed previously that only the trypsin-solubilized human enzyme was stable enough to be further purified. However the converting enzyme was recently isolated in its complete, hydrophobic form from human kidney using the zwitterionic non-denaturing detergent and a hydrophobic chromatography step [2]. It can also be isolated by solubilization with Triton X 100 [33]. The use of specific inhibitors immobilized on agarose carriers has made it possible to simplify the methods of purifying the solubilized enzyme by including an affinity chromatography step [2,14,88]. The plasma enzyme can be purified directly without extraction, since it circulates in soluble form; however, plasma is a relatively poor source of converting enzyme when compared with the lung or kidney [34].

The converting enzyme isolated from human lung is a glycoprotein apparently composed of a single peptidic chain of molecular mass 155 000, as determined on polyacrylamide gel [77]. Iwata et al. [52] observed that rabbit converting enzyme has a similar structure and that alkaline hydrolysis with concentrated ammonia splits the lung enzyme into two fragments of molecular weights 82 000 and 62 000 respectively. The 82 000 fragment has the same NH_2-terminal peptide sequence as the native enzyme; it thus constitutes the N-terminal part of the molecule. It also possesses enzymatic activity. These results throw light on the mechanisms that govern the functioning of the converting enzyme, since they indicate that the enzyme's active site is located on the N-terminal side of the molecule. The role of the C-terminal part is unknown but it seems logical to think that it helps to anchor the molecule on the membrane. In that case, the active site would protrude into the vascular lumen.

At the time of writing, the complete primary structure of the converting enzyme molecule is not known and its molecular cloning has not yet been reported. Thus only fragmentary information is available on the nature of the peptide sequences constituting the molecule, particularly those involved in the active site. An *N*-terminal peptide sequence of about 20 amino acids has been determined in rabbit, calf and man [2,52,97]. The nature of certain amino acids participating in the enzymatic activity was determined indirectly by using specific reagents that block the activity by interacting with a specific reactive group, carried by a given amino acid. This is how it is known that at least one arginine and one tyrosine residue are involved in the active site [115]. One glutamic acid residue is also necessary for the enzymatic activity. Recently Harris and Wilson [48] isolated a sequence of amino acids belonging to the active site of bovine enzyme after affinity labelling of that glutamic acid residue. This sequence exhibited a high degree of homology with a known sequence of the active site of carboxypeptidase A, which underscores that the converting enzyme belongs to a "family" of carboxypeptidases with analogous structures, at least at the active site. The mechanism whereby the molecule is anchored on the endothelial cell is not known. The trypsin-cleavable moiety is, however, certainly small, because on polyacrylamide gel the molecular mass of the trypsin-solubilized enzyme seems very close to that of the detergent-solubilized hydrophobic enzyme [2].

The carbohydrate content of the converting enzyme constitutes about 25% of its molecular mass. The lung enzyme (the capillary endothelial form) contains a form richer in sialic acid than the kidney enzyme, which is extracted from the tubular epithelial cell [77]. We shall see below the possible significance of this observation.

Polyclonal antibodies have been obtained against the converting enzyme purified from rabbit and rat lung, pig kidney and human lung and kidney [5,26,86]. These antibodies permitted immunological characterization of the enzyme and comparison of converting enzymes from vascular endothelium, plasma and kidney tubular epithelium. The enzymes extracted from these three sources appear to be immunologically identical [5]. A direct radioimmunological assay for the human enzyme has been devised and used to characterize and quantitate the enzyme in biological tissues and fluids [5]. Lastly, these antibodies allowed the study of the converting enzyme distribution in various vascular territories, especially in man (see below). Monoclonal antibodies directed against the bovine and rat enzymes have also been produced [6].

Enzymatic Activity

The converting enzyme is a dipeptidyl carboxypeptidase which cleaves off the *C*-terminal dipeptide (and sometimes the tripeptide) from several oligopeptides. It is relatively non-specific in vitro hydrolysing numerous natural or synthetic peptides. Bradykinin and its higher analogue lysylbradykinin are inactivated by successive cleavage of the dipeptides phenylalanyl–arginine and seryl–proline. Angiotensin I is activated to angiotensin II by cleavage of the dipeptide histidyl–leucine. However, angiotensin II is no longer sensitive to the converting enzyme, since the latter does not cleave upstream of a proline [34] (see the structure of angiotensin II in Fig. 9.1). Angiotensin II is thus formed according to a double enzymatic reaction: renin hydrolyses a leu–val linkage of angiotensinogen, releases angiotensin I and the converting enzyme hydrolyses a phe–his linkage of angiotensin I and releases angiotensin II (Fig. 9.2). Angiotensin II itself is protected from inactivation by the converting enzyme but is sensitive to other angiotensinases. 1-de-aspartate-angiotensin I is converted to the biologically active heptapeptide 1-de-aspartate-angiotensin II or angiotensin III. In vitro, the converting enzyme has an even higher affinity for 1-de-aspartate-angiotensin I than for angiotensin I [111].

The converting enzyme is in fact able to hydrolyse in vitro a large number of peptides, such as 5-leucine- and 5-methionine-enkephalin, substance P, neurotensin and the insulin B chain [3]. It also hydrolyses the hypothalamic decapeptide LH–RH and in this exceptional case splits off a tripeptide on both the *N*-terminal and the *C*-terminal sides [102]. However, except for angiotensin I and bradykinin, it is not certain that all these peptides are hydrolysed in vivo, since the affinity shown by the enzyme for them is weak compared with that for angiotensin I and bradykinin. Most of these peptides are produced in certain regions of the brain (where the brain converting enzyme may perhaps have a local role in their metabolism) but not in the circulation. It is therefore conceivable that the only substrates of the vascular converting enzyme are angiotensin I (and perhaps de-aspartate-angiotensin I) which is constantly produced in the circulation, and bradykinin which is released under certain circumstances.

Identification of the molecular mechanisms by which the converting enzyme interacts with angiotensin I and bradykinin is important for the preparation of specific inhibitors. However, these mechanisms are only partially known. Converting enzyme is a zinc-metalloprotein. Zinc binds to an anionic site of the enzyme active centre and is indispensable for the enzymatic activity; it constitutes one of the fixation sites of substrates and of peptide or non-peptide inhibitors. Chelating agents of divalent metals such as ethylenediaminetetra-acetic acid (EDTA) inhibit the enzyme by complexing the zinc atom and preventing the binding of the substrate molecules [15]. A study of the effect of chlorine on the enzymatic activity increases understanding of the mechanism of hydrolysis. This effect of chlorine can also have important consequences for the in vivo functioning of the enzyme. Chlorine increases the affinity for, and the hydrolysis rate of, certain substrates. Thus conversion of angiotensin I to angiotensin II depends on the presence of chlorine and hydrolysis of angiotensin I is considerably reduced in a chlorine-free medium. However, hydrolysis of bradykinin is only moderately affected by chlorine whose absence only reduces the hydrolysis rate by about half [34].

In endothelial and epithelial cells, where the converting enzyme is bound to certain intracellular membrane structures [13], its function might be very different from that on the cell surface since the chlorine concentration in the intracellular medium is only about 20 mM. Chlorine probably acts as an allosteric effector of the enzyme. It probably binds to a lysine residue close to the active site and thus modifies the configuration of the enzyme molecule [99]. Chlorine also plays an important part in the binding of certain inhibitors to the enzyme. In the absence of chlorine these inhibitors do bind, but the stability of the enzyme–inhibitor complex is greatly reduced [22]. Chlorine thus constitutes a true modulator of the activity and inhibition of the converting enzyme in vitro and maybe in the cell. The chlorine concentration is high in the vascular wall and in plasma, and varies little under different physiological conditions. Therefore, chlorine probably does not play an important part in regulating the enzyme's activity in circulating blood.

The Converting Enzyme in the Vascular Endothelium

The presence of the converting enzyme in a vascular territory can be schematically visualized in three different ways: (a) by studying the conversion of angiotensin I to angiotensin II in an isolated vascular territory in vivo; (b) by measuring the enzymatic activity in homogenates or subcellular fractions of vascular segments or isolated capillaries; and (c) by using specific antibodies and immunohistochemical techniques in optical or electron microscopy. In future it will also be possible to show the existence of messenger ribonucleic acids encoding the enzyme by using specific complementary nucleotidic probes.

The converting enzyme is an ectoenzyme of the endothelial cell. Histochemically, it is found at two different locations in this cell: on the luminal but not the basolateral membrane and on the membrane of endocytoplasmic vesicles (Fig. 9.3). Intracytoplasmic structures corresponding to the endoplasmic reticulum or to the Golgi apparatus are sometimes labelled by the antibodies, but these appearances, which illustrate the intracellular biosynthesis or maturation of the enzyme, are mainly visible only in kidney or intestine epithelial cells [13].

The converting enzyme is diffusely present in the endothelium of large vessels and capillaries

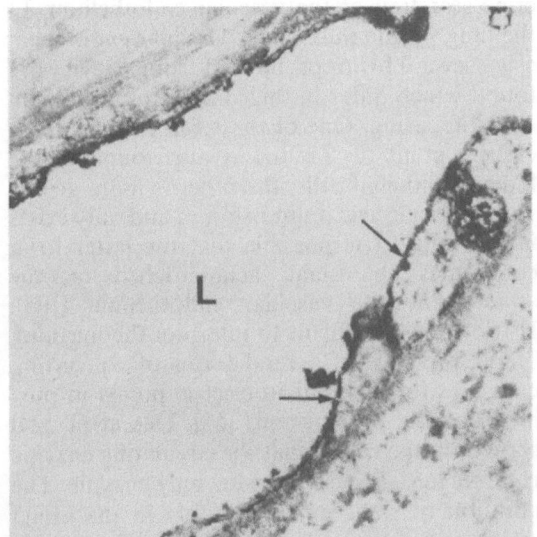

Fig. 9.3. Section of the luminal face of the plasma membrane, of vesicles (*arrows*) and of an endocytotic vacuole in human vascular endothelial cells. Ultrastructural immunohistochemistry using indirect immunoperoxidase with a human anti-serum converting anti-enzyme (reproduced from [13]). ×25 000

[13,16,96]. Since the arteriolo–capillary endothelial surface is far larger than that of the larger arteries, most of the conversion of angiotensin I to angiotensin II takes place in the arteriolo-capillary territory, especially in the pulmonary and mesenteric circulations. However, this conversion can also occur in other circulations such as the renal one [30]. Thus, angiotensin II has an important role in regulating certain local circulations in addition to its general action on the peripheral resistance and arterial pressure. This is the case for the circulations of the kidney (especially intraglomerular), the placenta and the brain.

The converting enzyme is found in the capillaries of the adult kidney floculus in the rat and rabbit [108]. In the human kidney it is found in those capillaries which invade the glomerulus early in foetal life during the early stages of glomerular differentiation. It persists throughout foetal life and might thus play an important part in the haemodynamics of the glomerular circulation and filtration of the foetus [73]. In neonatal and adult man, it seems that the converting enzyme is little expressed in the glomerular circulation and it is rather the circulating plasma enzyme that converts to angiotensin II at the glomerular level, the large quantities of angiotensin I produced by renin immediately after its secretion in the afferent arteriole. Under all circumstances in which renin secretion is increased by hypovolaemia or lowering of the perfusion pressure in the afferent arteriole, angiotensin II plays a major role in controlling glomerular filtration. Inhibition of the converting enzyme may therefore reduce glomerular filtration and cause renal insufficiency [27,45]. Angiotensin II constricts the glomerular efferent arteriole and maintains a high intravascular pressure in the glomerular circulation. According to this scheme the production of angiotensin I by renin in the afferent arteriole and its conversion to angiotensin II by the converting enzyme in the glomerulus are two essential mechanisms controlling the glomerular circulation.

The placental circulation is very rich in converting enzyme, which is located in the vessels of the villi [45]. In the foetus, the placenta seems to be the major site of angiotensin I conversion to angiotensin II, since the lung circulation is poorly developed before birth.

The circulation of the brain is another local circulation whose perfusion rate and pressure might be regulated by angiotensin II. In particular the cerebral microvessels seem to contain a large amount of converting enzyme [85].

Angiotensin II also participates in several physiological functions, such as the control of thirst, by acting directly on certain neurons particularly in the hypothalamus. These neurons also contain converting enzyme themselves and probably angiotensinogen. It is not known whether the vascular angiotensin II formed in the brain microvessels is implicated in neuronal function and in the central effects of the renin–angiotensin system.

Converting Enzyme Secretion by the Vascular Endothelium

Origin of the Plasma Converting Enzyme. As we have seen, the endothelial cell converting enzyme is both an ectoenzyme bound to the luminal plasma membrane, and a circulating protein, and its concentration in plasma, as determined by direct radioimmunological assay, is of the order of 10^{-9} M, which is relatively high for a circulating enzyme [5].

The circulating enzyme, also called "soluble", has enzymatic and immunological properties identical to those of the membrane enzyme but, as might be expected, it differs slightly in its structure from the enzyme of the endothelial cell membrane, because it probably does not carry the hydrophobic piece necessary for membrane anchorage and its carbohydrate content is different. Several arguments support the idea that the plasma converting enzyme is an endocrine secretion of the vascular endothelium. In the lung (in the pulmonary vascular endothelial cells) several forms of the converting enzyme are found which differ in their migration profile on electrofocusing. One of these forms has an isoelectric point of 5.2 and is also found in the kidney epithelial cells; the other is more acidic, with an isoelectric point below 5, and only exists in the lung. It is possible that the latter form constitutes the sialic acid-enriched enzyme secreted by the vascular endothelium [105]. However, it is difficult to interpret the meaning of the presence of several forms of converting enzyme with different isoelectric points in purified enzyme preparations [63]. Das et al. [24] showed in the rabbit that the circulating enzyme is richer in sialic acid than the lung enzyme. The addition of sialic acid, probably in the Golgi apparatus of the endothelial cell, to the soluble form of the converting enzyme, after its synthesis, stabilizes the enzyme in the circulation after its secretion and modifies its binding to liver lectins. More direct results have been obtained with cultures of bovine aorta endothelial cells:

these cells release converting enzyme into the culture medium and this release is blocked by cycloheximide, an inhibitor of protein synthesis. In this case too, the physicochemical characteristics of the enzyme secreted by the cells suggest that it is indeed the circulating plasma enzyme [49].

All these arguments suggest, albeit indirectly, that the plasma converting enzyme is secreted by the vascular endothelial cell; so does the observation that the enzyme content of plasma falls when there is extensive destruction of the lung vasculature [83]. Other cells in the organism, such as macrophages, are able to express the converting enzyme, at least under certain conditions, and perhaps to secrete it into the circulation. For instance, in sarcoidosis it is highly probable that the abnormal increase of the plasma converting enzyme is due to its secretion by activated macrophages [5].

Regulation of the Vascular Converting Enzyme

The factors responsible in vivo for the converting enzyme synthesis by the endothelium are not known. Successive measurements of the enzyme content of the plasma in a given individual have shown that it remains remarkably stable for periods of several months or years. The enzyme levels display no circadian variations and are not influenced by position changes, sex or ageing. There are, however, large interindividual differences in the levels of circulating converting enzyme. Taken together, all these observations might suggest the existence of a predominant genetic determination of the synthesis and release of the vascular converting enzyme. There is at least one published observation of a familial increase in the circulating converting enzyme [80]. The future demonstration of genetic determination of the regulation of the vascular enzyme would be of great interest in understanding the genetics of the endothelial proteins.

The only known endocrine influence on the synthesis and release of the vascular converting enzyme, which might be responsible for alterations observable when measuring the circulating enzyme, is that of thyroid hormones. Hyperthyroidism is accompanied by an increase in the circulating enzyme content [118]. Thyroid hormones stimulate in vitro biosynthesis of the converting enzyme in cultured endothelial cells [60]. This shows that these cells possess receptors for thyroid hormones and that they respond to these hormones by increasing the synthesis of

the converting enzyme and possibly of other endothelial proteins. However, in euthyroid subjects there is no correlation between the circulating enzyme and plasma T3, so that the latter does not seem to be an essential determinant of enzyme release in plasma in the normal subject [31].

Glucocorticoids in vitro also stimulate the synthesis of the endothelial converting enzyme, but in vivo, the level of circulating enzyme does not change during adrenal dysfunctioning [68]. Insulin seems to reduce in vitro enzyme biosynthesis by the endothelium [60]. Finally, it has been observed in rat and man that during prolonged chronic treatment with converting enzyme inhibitors, such as captopril or enalapril (as used in the treatment of arterial hypertension), the level of circulating converting enzyme increased [40]. This increase was found by both the enzymatic method and specific radioimmunological assay [98]. All this suggests converting enzyme synthesis is induced in the endothelium but its exact mechanism is still totally unknown.

Does a Complete Renin–Angiotensin System Exist in the Vascular Endothelium?

Angiotensin I, produced in the circulation by the hydrolysis of angiotensinogen by renin, is converted to angiotensin II by the converting enzyme present in plasma and on the surface of the vascular endothelium. This conversion step is generally considered as non-limiting for the production of angiotensin II. Angiotensin II probably reaches its specific receptors on the contractile vascular smooth-muscle cells by diffusing between the endothelial cells of the arterioles. Besides this very simple scheme which assumes that angiotensin II is formed in the plasma (this has been well established) there is also the possibility that angiotensin II is locally released in the vascular wall, especially in the endothelium [32,106]. This hypothesis rests on the demonstration, by enzyme activity measurements and by labelled amino acid incorporation into cultures of endothelial cells from several animal species, of the activity of a pseudo-renin, which might be authentic renin synthesized by the cells. Moreover, angiotensinogen was found in these cells and reacted with specific antibodies [32]. In addition, converting enzyme has been shown to be present inside the endothelial cell or at least on the membrane of certain intracellular vesicles [13]. It is thus possible that angiotensins I and II are produced inside the endothelial cell from renin and angiotensinogen synthesized in situ or originating from plasma. The smooth-muscle cells of the media might also be able to synthesize renin in culture [93], which is all the more likely as the epithelioid cells of the glomerulus afferent arteriole, which certainly synthesize renin, constitute a form of endocrine differentiation of vascular smooth-muscle cells (see below).

The amounts of renin, angiotensinogen and angiotensins I and II measured in the endothelial cells are in fact very small compared with the circulating levels. Moreover, it has not been possible to demonstrate convincingly the presence in the endothelium of specific messenger RNAs encoding renin and angiotensinogen [32]. It is thus impossible to conclude with certainty the existence of a complete intra-endothelial renin–angiotensin system, even if its existence could account for certain disturbing observations on the persistence of the anti-hypertensive effect of the renin–angiotensin system inhibitors despite their disappearance from plasma [112]. Similarly, it has been observed that perfusion of isolated rat hind leg with the tetradecapeptide, which constitutes the first 14 amino acids of angiotensinogen and is hydrolysable by renin, results in the formation of angiotensins I and II and in the vasoconstriction of the preparation [81]. This suggests that renin or a tetradecapeptide-cleaving enzyme able to release angiotensin I from circulating substrate exists in the vascular wall, even if in this type of experiment it is impossible to distinguish between local synthesis and intracellular uptake.

Other Peptidases of the Vascular Endothelium

Whereas the kidney and intestinal epithelial cells contain several vasoactive peptide-hydrolysing enzymes, it seems that the converting enzyme is the main, if not the only, membrane peptidase of the vascular endothelium. This assumption rests on the identification of the peptide fragments released by incubation of bradykinin with lung homogenates [19] and on the fact that hydrolysis of angiotensin I and bradykinin is inhibited in vivo by low concentrations of converting enzyme inhibitors [82]. However, the aminopeptidase activity found in cultured endothelial cells from human umbilical cord inac-

tivates angiotensin II [55]. This enzyme is intracellular and not membrane-bound, but its physiological role is unknown. It is not certain that it is to be found in all vascular territories, and this especially applies to the pulmonary circulation; neither is it certain that it would hydrolyse in vivo angiotensin II, whose inactivation seems to depend essentially on plasma angiotensinases [89].

Modulation by the Endothelium of the Physiological and Pharmacological Activity of Vasoactive Compounds

Modulation by the vascular endothelium of the activity of vasoactive compounds has only recently been fully recognized and studied. It has been observed in recent years that the endothelium metabolizes vasoactive peptides and certain autacoids, that it possesses receptors for several of them, that it synthesizes and releases prostacyclin and certain vasoactive prostaglandins and that this release can be stimulated by bradykinin, histamine, thrombin, ATP and arachidonic acid. Similarly, after an initial observation with acetylcholine, it is now known that many vasodilating compounds, and sometimes vasoconstricting ones, induce the formation of one or several compounds in the endothelial cell. The nature of these compounds is not completely known, but they are known to be responsible for the modifications in the contractility of the smooth-muscle cells of the media.

Apart from some of the vasoactive peptides described in the preceding chapter, the endothelium takes up and metabolizes adenylated nucleotides, noradrenaline and 5-hydroxytryptamine [100]. It is noteworthy that the endothelial cells inactivate these compounds and at the same time undergo physiologically important biochemical modifications under their influence, leading to the production of cyclic nucleotides and the release of prostacyclin and other factors which influence vascular tone. The best example of this complex interaction between vasoactive compounds and the endothelium is probably the inactivation of bradykinin by the surface of the endothelial cell. However, bradykinin strongly stimulates endothelial phospholipase A_2 and releases one or several vasodilating factors which act directly upon the smooth-muscle cells. For peptides and autocoids, a physiologically important equilibrium certainly exists between inactivation and physiological action; this equilibrium depends on the relative abundance of the receptors and inactivation enzymes, on their respective affinities and on the biochemical properties of the metabolites released.

Stimulation of Prostaglandin Formation by Vasoactive Agents in the Vascular Endothelium

Isolated arterial and venous segments or cultured endothelial cells synthesize certain prostaglandins and prostacyclin [4,72,110]. The extent of this synthesis and its physiological role in the absence of appropriate agonists are still a matter of controversy. However, it is well established that endothelial cells release large quantities of prostaglandins $F_{2\alpha}$ and E_2 and prostacyclin when they are stimulated by various vasoactive agents such as bradykinin, histamine and also thrombin [4,7,50,74]. Arachidonic acid, a substrate of cyclo-oxygenase, also stimulates the release of prostaglandins by the endothelium in vitro.

After being stimulated, human umbilical cord endothelial cells release prostacyclin (quantitated by measuring its stable metabolite 6-keto-

$Pg F_{1\alpha}$). They also release prostaglandins E_2 and $F_{2\alpha}$ and probably small quantities of thromboxane A_2 [4,74,110]. Cells of arterial origin release prostacyclin predominantly; those of venous origin prostacyclin and prostaglandin $F_{2\alpha}$. There may be some differences, depending on the vascular territory, between the amount and nature of the released cyclo-oxygenase products; these differences probably reflect differences in the enzyme equipment of the various vascular segments from both arteries and veins [4,7,50,74,103,110].

The mechanisms of prostaglandin release under the effect of bradykinin and histamine have been studied in cultured endothelial cells. Figure 9.4 shows the biosynthesis scheme for prostacyclin and prostaglandins in the vascular endothelium. Since arachidonic acid, a substrate of cyclo-oxygenase, is able to stimulate prostaglandin synthesis in the endothelium directly, it is highly probable that bradykinin and histamine act upstream to release arachidonic acid from cellular lipids. Arachidonic acid release is indeed known to be the rate-limiting step in prostaglandin synthesis. It is also known that prostaglandins are not stored but are secreted by the cell as soon as they are synthesized [72]. Bradykinin and histamine in fact stimulate an endothelial phospholipase and this action is abolished by mecaprin, a phospholipase A_2 inhibitor [4] (Fig. 9.5). Under the effect of these agents, phospholipase A_2 hydrolyses the cellular

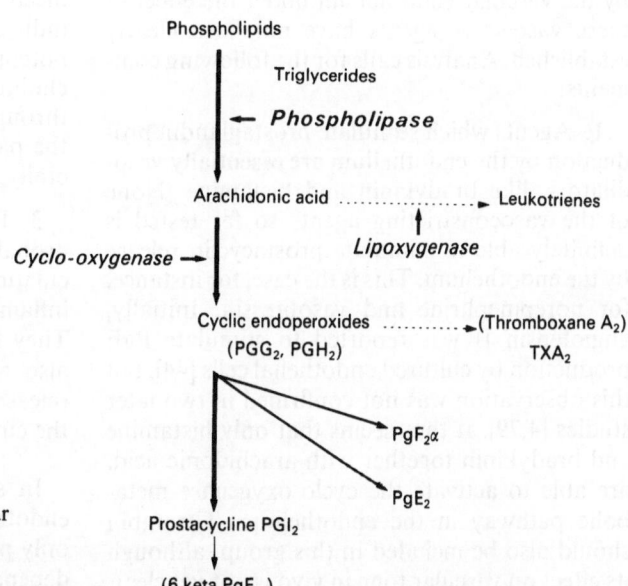

Fig. 9.4. Metabolism of arachidonic acid and pathways of prostaglandin synthesis in the vascular endothelium. The quantitative importance of the lipoxygenase pathway and of thromboxan A_2 synthesis is a matter of discussion.

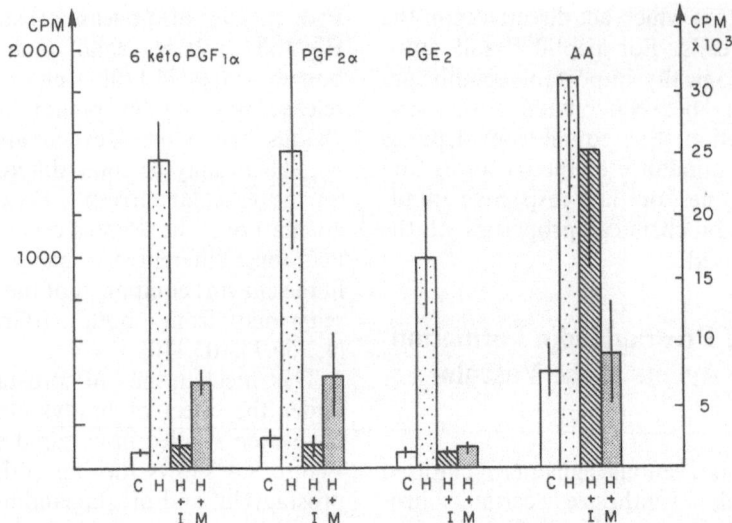

Fig. 9.5. Effect of 10^{-6} M indomethacin or 10^{-4} M mepacrin on release by 5×10^{-6} M histamine of ^{14}C-labelled arachidonic acid (*left scale*) and prostaglandins (*right scale*) from cultured human endothelial cells. Histamine (*H*) stimulates the release of arachidonic acid and prostaglandins. In the presence of indomethacin (*H* + *I*), an inhibitor of cyclo-oxygenase, histamine still releases arachidonic acid (which is no longer metabolized to prostaglandins). In the presence of mepacrin (*H* + *M*), an inhibitor of phospholipase A$_2$, histamine no longer has any effect upon arachidonic acid release (and therefore on prostaglandin release) (reproduced from reference 4).

lipids and releases arachidonic acid, which becomes available for the synthesis of prostacyclin and prostaglandins [72]. The effect of histamine is mediated by type H1 receptors [4,7,110]. Other results obtained, in particular with thrombin, suggest that the activation of endothelial phospholipases A$_2$ and C is itself dependent on intracellular calcium [1].

The physiological consequences of phospholipase activation and prostaglandin release by the vascular endothelium under the effect of these vasoactive agents have not been clearly established. Analysis calls for the following comments:

1. Agents which stimulate prostaglandin production by the endothelium are essentially vasodilators, like bradykinin and histamine. None of the vasoconstricting agents so far tested is definitely able to stimulate prostacyclin release by the endothelium. This is the case, for instance, for norepinephrine and vasopressin. Initially, angiotensin II was reported to stimulate PgE production by cultured endothelial cells [44], but this observation was not confirmed in two later studies [4,79]. It thus seems that only histamine and bradykinin together with arachidonic acid, are able to activate the cyclo-oxygenase metabolic pathway in the endothelium. Thrombin should also be included in this group, although its effect on vascular tone in vivo is not yet clear.

2. The physiological effects of bradykinin and histamine are to bring about arteriolar vasodilation and increase capillary permeability by acting at the venular level [53]. These effects are partly due to the prostaglandins they release, which thus probably participate in their action on the vascular tone. However, the bradykinin vasodilation is not always abolished by cyclo-oxygenase inhibitors. This indicates that, as described in the next paragraph, other possible mechanisms may account for the vasodilation induced by these compounds. Moreover, certain potent vasodilating compounds, like acetylcholine, always exert their physiological action through the endothelium but apparently without the participation of prostacyclin or, more generally, of cyclo-oxygenase products (see below).

3. Finally, bradykinin and histamine are probably not permanently present in the circulation, but are released acutely during the inflammatory reaction, fibrinolysis or allergy. They therefore cannot be responsible, and this also applies to thrombin, for any permanent release of prostacyclin and prostaglandins into the circulation.

In short, the formation and release of the endothelial cyclo-oxygenase products seem to be only partly responsible for vasomotor reactions depending on the vascular endothelium.

Vasomotor Reactions Depending on the Vascular Endothelium

In 1980 Furchgott and Zawadzki [39] observed that the vasodilating effect of acetylcholine on isolated vascular segments required an intact endothelium. While investigating why acetylcholine did not always induce a vasodilatory response, they noticed that when the endothelium in the preparation was injured, the vasodilatory effect was suppressed. When the endothelial cells remained intact there was always vasodilation. This muscarinic-type effect of acetylcholine was observed in vitro, when the vascular segment was "preconstricted" by, for example, noradrenaline. The hypothesis formulated by these authors was that, in response to acetylcholine, the endothelium produces a substance that relaxes the vascular smooth-muscle tone. This endothelium-dependent vasodilatory effect of acetylcholine was abolished by phospholipase inhibitors such as mecaprin, but not by cyclo-oxygenase inhibitors [39]. This suggested that the factor responsible was a product of the alternative arachidonic acid metabolic pathway, i.e. the lipoxygenase pathway. Anoxia also inhibits the endothelium-dependent vasodilatory response.

The same observation has been made with substance P [119] and the unknown factor(s) transmitted from the endothelium to the media was quickly baptized "endothelium-derived relaxing factor (EDRF)". Bradykinin, which in vivo is a potent vasodilator, is also able to relax the tone of arterial segments constricted by noradrenaline through an endothelium-dependent mechanism. In this case, however, things seem more complex since, in contrast with the effects of acetylcholine and substance P, the response varies according to species. Thus, in the dog, the endothelium is necessary for the relaxing effect of bradykinin and cyclo-oxygenase inhibitors have no effect. However, arterial segments from cat or rabbit respond to bradykinin even when the endothelium is destroyed, and this response can be inhibited by cyclo-oxygenase inhibitors. In the latter two species, bradykinin thus seems to exert a direct effect, which is independent of endothelial cells, and is probably mediated by prostaglandins [18].

Other compounds whose vasodilatory effect is endothelium-dependent, at least in some species, include ADP, ATP, arachidonic acid, thrombin, histamine, serotonin and perhaps a vasodilatory drug, hydralazine. However, the responses to adenosine, isoproterenol and other potent vaso-dilators such as nitroprussiate, nitrated derivatives or minoxidil, seem to be endothelium-independent [104,114].

The exact chemical nature of EDRF is unknown, although early experiments suggested that it might be an endothelial lipoxygenase product. It might even be that one or several vasodilatory factors, depending on the species and vascular territory, are transmitted from the endothelium to the smooth-muscle cells. In any case, arachidonic acid apparently exerts its vasomotor effects by transmitting a cyclo-oxygenase product from the endothelium to the smooth-muscle cells [28,69], which acetylcholine, for instance, does not do [39]. The difficulty of identifying EDRF is probably due to the small amounts of releasable EDRF in endothelial cells and perhaps to its short biological life span. However, the mechanism of action of EDRF on smooth-muscle cells is thought to involve the production of cyclic GMP and probably the phosphorylation of certain intracellular proteins [42,92]. Co-culture of endothelial cells with smooth-muscle cells, in which this accumulation of intracellular cGMP due to various agonists can be measured, will perhaps make progress possible in isolating and identifying the factor(s) responsible for the vasodilatory response mediated by endothelial cells [42].

The endothelium also seems to mediate some vasoconstrictory responses. This is the case for the potentiation through hypoxia, of the vasoconstrictory response to noradrenaline in isolated dog arterial and venous fragments, abolished by de-endothelialization [114]. The arachidonic acid-induced vasoconstriction in dog veins [69] is a further example.

Physiological and Physiopathological Consequences of the Role of the Endothelium in Controlling Vasomotor Activity

Most observations about the interaction of endothelial cells with vasoactive agents have been made either on isolated vascular segments or on cultured endothelial cells. The biochemical environment of endothelial cells, the fibroconnective matrix of the subendothelium and most important of all the haemodynamic conditions in vivo certainly play a role in the functioning of the endothelial cell.

This explains how certain physiologically interesting proteins such as the converting enzyme are expressed by the endothelial cell membrane only if this membrane has no contact with the subendothelial matrix; hence the luminal polarity of the converting enzyme present in the endothelium [75]. However, the conditions of wall tension and flow that affect the endothelium profoundly alter its ability to release prostacyclin. It has been shown that when cultured endothelial cells are submitted to either constant or pulsatile flows of medium, they respond by releasing prostacyclin, but that the release was greater with pulsatile flow [36]. This clearly shows that the haemodynamic conditions affect the biochemical properties of endothelial cells.

There are structural and physiological differences between arteries and veins. Arteries and arterioles have more converting enzyme than veins and venules [54] and their ability to release prostacyclin on stimulation appears to be quantitatively higher [4,103,110]. It may be that these differences, reflecting differences in enzymatic equipment, are at least partly due to the haemodynamic conditions of high and low pressure affecting the cells concerned.

Among the physiological circumstances which bring into action the endothelial control of vasomotor activity, the release of bradykinin or of histamine are acute local phenomena. This is probably also true of the presence of acetylcholine in the circulation. Hypoxia deserves special mention. During hypoxia, the conversion in vivo of angiotensin I to angiotensin II is slightly reduced, but this reduction seems to be due rather to the haemodynamic modifications of the pulmonary circulation than to the direct effect of hypoxia on the endothelial converting enzyme suggested earlier [84]. During hypoxia, the endothelium probably remains able to produce angiotensin II from angiotensin I. In fact, the most evident vascular response to hypoxia is a vasoconstriction, partly mediated by the release of catecholamines into the circulation. This hypoxic vasoconstriction seems at least partly dependent on the endothelium, as mentioned in the preceding paragraph, and is considerably reduced when the endothelial cells are suppressed [29]. Hypoxia is thus an exceptional circumstance during which the endothelium takes part in vasoconstriction.

It is obviously logical to look for possible perturbations of the mechanisms of control of muscle tone by the endothelium during arterial spasm, thrombosis or atheromatosis. Platelet aggregation in vitro on segments of previously constricted dog coronary artery induces vasodilation. This vasodilation depends on the in-

tegrity of the endothelium; injury to the endothelium causes vasoconstriction [20]. The dilating effect is probably due to the release by the platelets of adenylated nucleotides. Platelet serotonin does not take part in this effect [51]. These results are not yet transposable to man and to all vascular territories, but they suggest that endothelial injury favours not only the formation of a platelet rhrombus, but also vasoconstriction and vascular occlusion when the platelets aggregate.

Certain results suggest that in rabbits rendered atheromatous by a cholesterolaemia-increasing regimen, endothelial vasodilation and/or protection against vasoconstriction are reduced [47]. Similarly it has been observed that in aortic segments from hypertensive rats, the vasodilating ability of the endothelium was reduced [65, 66,116]. It is, however, difficult to extrapolate to the situation in vivo from results generally obtained with vascular segments in vitro and also to compare the pharmacological responses of healthy and pathological arteries, since the latter obviously also undergo deterioration of the media and smooth-muscle cells.

Endocrine Function of the Smooth-Muscle Cell: The Renin-Secreting Cell

Morphology of Renin-Secreting Cells

Ruyter [98] was the first to describe the presence of "epithelioid" cells in the juxtaglomerular apparatus. These cells are located in the media of the afferent arteriole of the glomerulus. Ruyter showed the existence of myofilaments and formulated the hypothesis that these cells display an intermediate morphology between that of endocrine cells and that of vascular smooth-muscle cells, hence the name "myoepithelioid cells".

The presence of renin inside the myoepithelioid cells of the juxtaglomerular apparatus was suspected from results obtained using biochemical techniques (renin extraction) and non-specific staining techniques (Bowie staining). The formal demonstration that these cells contain renin was made by immunohistochemistry [70,78,90] (Fig. 9,6). Renin synthesis was demonstrated by hybridization in situ [23]. Mouse or human renin cDNA specific-

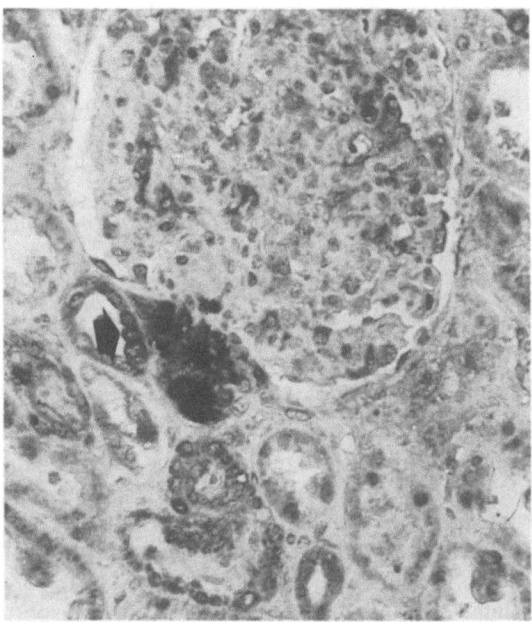

Fig. 9.6. Renin in the juxtaglomerular apparatus. Immunohistochemistry with human anti-renin polyclonal antibodies. The cells of the afferent arteriole are strongly positive. The *arrow* marks the macula densa. Peroxidase antiperoxidase. × 300

ally hybridizes with messenger RNAs from the juxtaglomerular cells of the afferent arteriole of mouse, rat and human kidney. Under certain circumstances (such as the physiopathological intervention of the renin–angiotensis system following sodium depletion, renal artery stenosis or renal infarct), other smooth-muscle cells are able to synthesize and store renin. Immunohistochemically, recruitment of juxtaglomerular cells can be observed inside the afferent arteriole and sometimes even in the interlobular arterioles in cases of intense stimulation of renin secretion [17,64,109]. In contrast with the rat efferent arteriole, the human efferent arteriole does not synthesize renin.

Renin is stored in myoepithelioid cells in granules of similar appearance to those in endocrine cells. Rhomboid protogranules are present which sometimes contain paracrystalline structures. These protogranules probably evolve into mature spherical electron-dense granules [9]. The presence of renin inside these granules has been confirmed by immunohistochemistry and electron microscopy [17,62,64]. Renin is probably stored in them in its mature form, although there is no definite evidence to this effect, since the antibodies used so far cannot discriminate between renin and prorenin.

Renin Structure and Renin Precursor Maturation

Thanks to cloning techniques, it has been possible to determine the cDNA sequence of mouse submaxillary renin [87] and of human kidney renin [56]. The amino acid sequence of the biosynthetic renin precursor could therefore be deduced. Renin is biosynthesized as preprorenin. A model for the maturation of preprorenin to renin in mouse submaxillary gland was proposed which compared the precursor sequence deduced from the renin cDNA with the primary renin sequence determined by conventional protein sequencing [71,87]. A first cleavage releases the prepeptide or signal peptide. This event takes place in the rough endoplasmic reticulum. The later maturation of prorenin to renin probably takes place in the Golgi apparatus. This maturation occurs after cleavage of a 43 amino acid N-terminal peptide. The renin so formed is then stored in the protogranules, and later in the mature granules. Prorenin is inactive wherereas renin is an aspartyl protease able to hydrolyse angiotensinogen to angiotensin I.

Another pathway for renin biosynthesis might co-exist with this scheme, which resembles that observed in numerous endocrine cells. This possibility applies in particular to the human kidney myoepithelial cell. In this alternative pathway, prorenin would not be activated to renin but would be exported into plasma as inactive prorenin [41]; 60%–80% of renin in human plasma is in the form of prorenin. The identities of this inactive plasma renin and of prorenin, the biosynthetic precursor of renin, were recently demonstrated using antibodies against peptides derived from the profragment of human renin; these antibodies recognize inactive renin, whether its origin is renal, amniotic or plasmatic [12,58]. Those two pathways might be regulated in different ways, as was found to be the case in other endocrine cells, including the ATt20 mouse cell synthesizing the adrenocorticotropic hormone (ACTH) precursor [41,46].

Assuming that the regulated pathway is the major one, prorenin would mature progressively and be stored inside secretory granules. The intragranular renin could be rapidly released under the influence of various second messengers, such as those produced by the activation of adenylate cyclase (beta-adrenergic receptor), or the movements of intracellular calcium (action of angiotensin II).

The constitutive pathway would lead to the production of prorenin and its release into plasma. Prorenin levels would not vary notably upon stimulation or acute blocking of the juxtaglomerular cell. Indeed, acute stimulation of the renin-secreting cell by diuretics or the inhibition of the converting enzyme lead to a rapid increase in the level of active renin without any variation in the level of circulating prorenin.

The primary sequence of renin shows the wide homology between this enzyme and the family of acid proteases (e.g. cathepsin D and pepsin) [87]. The organization of the renin and pepsin genes is also similar. These investigations support the hypothesis of Tang et al. [107] on the evolution of acid proteases; this hypothesis is based on crystallographic data obtained with some of these acid proteases. However, in contrast to other acid proteases which cleave a wide variety of substrates, renin hydrolyses only angiotensin I from angiotensinogen. Human renin only has slight affinity for rat and dog angiotensinogen which it hydrolyses very slowly.

Because of the analogy between renin and other acid proteases, it is possible to devise tridimensional structural models for human submaxillary renin. Such models may be helpful in the elaboration of specific renin inhibitors. The therapeutic interest of these inhibitors would lie in their inhibition of the first step in the renin–angiotensin system, i.e. the action of renin upon angiotensinogen [21].

Role of Renin Secretion

The renin–angiotensin system plays a major role in the control of arterial pressure and sodium–water metabolism. The renin-secreting cell occupies a critical position in the nephron, at the level of the glomerulus afferent arteriole, just before the opening of the glomerulus capillary bundle. It is in close contact with the macula densa and is richly innervated by beta-adrenergic fibres. These particular connections of the renin-secreting cell explain why it responds to different stimuli, including:

1. *The Perfusion in the Afferent Arteriole of the Juxtaglomerular Apparatus.* When the pressure in the renal artery decreases following artery stricture, hypovolaemia or orthostatism, renin secretion increases independently of any reduction in glomerular filtration or renal blood flow and in the absence of any renal hypoxia. Using Davis's non-filtrating kidney, it was demonstrated that the variations in afferent arteriole pressure modulate renin secretion independently of the macula densa and of the sympathetic innervation of the juxtaglomerular apparatus [38].

The fine mechanism of renin release outside the cell under the influence of variations in the stretching of the arteriolar wall (detected by stretch receptors) or in pressure on the wall (detected by baroreceptors) is unknown.

2. *Modifications of the Composition of Urine in the Macula Densa Before it Enters the Distal Tubule.* The macula densa is the initial part of the distal tubule marking the junction between the proximal part of the nephron, where the haemodynamic influences are strongest and its terminal part, where the fine hormonal control of sodium, potassium and water conservation is exerted. For the afferent arteriole, the macula densa cells constitute a chemoreceptor that senses the variations in the sodium, chloride and calcium content of urine when it enters the distal tubule. Renin release seems to be inversely correlated with the sodium load and transport through the macula densa [113], since the ion responsible for controlling renin secretion probably is not sodium but chloride. Several experiments have indeed shown that if chloride is replaced by other ions, the perfusing sodium salts are no longer able to lower renin secretion [59]. It is therefore clear that the variations in the composition of the urine electrolytes in the macula densa cells affect renin secretion, but it is not known whether the determining factor is the amount or the concentration of sodium (or chloride) in urine or the amount of sodium (or chloride) reabsorbed by the macula densa cells.

3. *Stimulation of the Adrenergic Nervous System and Renin Secretion.* The existence of adrenergic (and cholinergic) innervation of the juxtaglomerular apparatus was shown with silver staining, and then confirmed by electron microscopy. This innervation essentially affects the myoepithelial cells. The presence of a nerve ending on the macula densa has not been shown. Electrical stimulation of the perirenal sympathetic nerves increases renin secretion, whether the kidney is filtrating or not. Perfusions of adrenaline, noradrenaline or isoproterenol either in vivo or in the isolated perfused kidney, also stimulate renin secretion. Administration of the various known beta-blockers can block renin secretion induced by diverse stimuli. The existence in the juxtaglomerular apparatus of a beta-adrenergic receptor (beta 1 or beta 2) is thus certain [70,90]. Conversely, it is less evident that alpha-adrenergic receptor plays an important role in regulating renin secretion. Beta adrenergic agonists may well activate the adenylate

cyclase of the juxtaglomerular cell [61]. The events following the increase in intracellular cyclic AMP are less well known.

4. *Angiotensin II*. Angiotensins II and III and the hexapeptide reduce renin secretion independently of positive sodium balance, and angiotensin II is the most potent of these hormones. The importance of this short feedback mechanism for the renin secretion has been demonstrated by administering substances which block the renin–angiotensin system. Antagonists of angiotensin II, or inhibitors of the converting enzyme increase renin secretion by suppressing the negative feedback effect of angiotensin II on the juxtaglomerular cell.

5. *Calcium*. Calcium probably acts as a chemical mediator for stimulus–secretion coupling at the intrarenal baroreceptor, as well as in the mechanism of renin secretion induced by angiotensin II [91]. In addition, calcium by itself modifies renin secretion [37]: for instance, in the filtrating or non-filtrating kidney, intrarenal calcium gluconate perfusion reduces renin secretion; again, when isolated rat kidney is perfused with media containing various calcium concentrations, there is an inverse relationship between calcium content and renin release in the perfusate. Finally, perfusion with calcium inhibitors such as nifedipine increases renin secretion [67].

It is therefore likely that renin release and intracellular calcium levels are inversely related, that is the juxtaglomerular cell would behave in the opposite way to the endocrine cell, in which an increase in intracellular calcium usually induces hormone release.

In conclusion, the myoepithelioid cells of the juxtaglomerular apparatus are the only vascular smooth-muscle cells to act as endocrine cells. If the perfusion pressure in the arteriole where they are located drops, they can immediately release active renin from their granule storage. Circulating renin acts upon angiotensinogen to release inactive angiotensin I, which is then converted to angiotensin II by the plasmatic and endothelial converting enzyme. The final effect of angiotensin II is to re-establish normal arterial pressure by a direct vasoconstricting action on the vascular smooth muscle and indirectly, by increasing sodium reabsorption in the distal tubule by aldosterone secretion. In case of chronic stimulation, the smooth-muscle cells of the afferent arteriole are able to differentiate into endocrine cells; the "endocrine" recruitment can take place up to the level of interlobular arteries.

There is so far no proof that the other vascular smooth-muscle cells of the organism have the same capacity to differentiate into endocrine cells and to respond to pressure variations by releasing renin.

Note Added in Proof

Since this chapter was completed, important data have been reported in the field of endothelium-derived vasoactive agents. One endothelium-derived relaxing factor was identified as the compound nitric oxide, which relaxes vascular tissues and inhibits platelet adhesion and aggregation. Nitric oxide is produced in porcine endothelial cells challenged with bradykinin. The amino acid L-arginine is probably the source of nitric oxide in these cells (Palmer RMJ, Ferrige AG and Moncada S (1987) Nature 327: 524–526; Radomski MW, Palmer RMJ and Moncada S (1987) Br J Pharmacol 92: 639–646; Radomski MW, Palmer RMJ and Moncada S (1987) Biochem Biophys Res Commun 148:1482–1489; Palmer RMJ, Ashton DS and Moncada S (1988) Nature 33: 664–666).

A novel vasoconstrictor peptide, designated as endothelin, was discovered in porcine endothelial cells. Endothelin is a 21-amino acid cyclic peptide deriving from a 203-residues preproendothelin molecule. The mechanisms involved in preproendothelin maturation are still unknown at present. As segmental homologies were observed between endothelin and several peptidic neurotoxins acting as modulators of the voltage-dependent ion channel, and as the presence of calcium was recognized as necessary for the vasoconstrictor action of endothelin, this latter peptide may act directly on calcium channels in contractile cells (Yanagisawa M, Kurihara H, Kimura S et al. (1988) Nature 332:411–415).

References

1. Adams Brotherton AF, Hoak JC (1982) Role of Ca^{++} and cyclic AMP in the regulation of the production of prostaglandin by the vascular endothelium. Proc Natl Acad Sci USA 79:495–499
2. Alhenc-Gelas F, Cumin F, Yasui T et al. (1987) Purification and sequencing of the human kidney angiotensin-converting enzyme. (Submitted for publication)
3. Alhenc-Gelas F, Erdös EG (1986) The angiotensin I converting enzyme. In Robertson JIS (ed), Biochemistry of the renin system, Gower, London
4. Alhenc-Gelas F, Tsai SJ, Callahan KS, Campbell WB, Johnson RP (1982) Stimulation of prostaglandin formation by vasoactive mediators in cultured human

endothelial cells. Prostaglandins 24:723–742

5. Alhenc-Gelas F, Weare JA, Johnson RL Jr, Erdös EG (1983) Measurement of human converting enzyme level by direct radioimmunoassay. J Lab Clin Med 101:83–96

6. Auerbach R, Alby L, Grieves J et al. (1982) Monoclonal antibody against angiotensin converting enzyme: its use as a marker for murine, bovine and human endothelial cells. Proc Natl Acad Sci USA 79:7891–7895

7. Baenziger NL, Force LE, Becherer PR (1980) Histamine stimulates prostacyclin synthesis in culture human umbilical vein endothelial cells. Biochim Biophys Res Commun 92:1435–1440

8. Bakhle YS (1968) Conversion of angiotensin I to angiotensin II by cell-free extracts of dog lung. Nature 220:919–921

9. Barajas L (1966) The development and ultrastructure of the juxtaglomerular cell granule. J Ultrastruct Mol Struct Res 15:400–413

10. Biron P, Campeau L (1971) Pulmonary and extrapulmonary fate of angiotensin I. Rev Can Biol 30:27–34

11. Biron P, Huggins CG (1968) Pulmonary activation of synthetic angiotensin I. Life Sci 7:965–970

12. Bouhnik J, Fehrentz JA, Galen FX et al. (1985) Immunologic identification of both plasma and human renal inactive renin as prorenin. J Clin Endocrinol Metab 60:399–401

13. Bruneval P., Hinglais N, Alhenc-Gelas F et al. (1986) Angiotensin I converting enzyme in human intestine and kidney. Ultrastructural immunohistochemical localization. Histochemistry 85:73–80

14. Bull HG, Thornberry NA, Cordes EH (1986) Purification of angiotensin-converting enzyme from rabbit lung and human plasma by affinity chromatography. J Biol Chem 260:2963–2972

15. Bünning P, Riordan JF (1985) The functional role of zinc in angiotensin converting enzyme: implications for the enzyme mechanism. J Org Biochem 24:183–198

16. Caldwell PRB, Seegal BC, Hsu KC, Das M, Soffer RL (1976) Angiotensin-converting enzyme. Vascular endothelium localization. Science 191:1050–1051

17. Camilleri JP, Hinglais N, Nochy D, Phat VN, Bariety J (1982) Immunohistochemistry of renin in human diseased kidney. Clin Exp Hypertens [A] 5:1179–1190

18. Cherry PD, Furchgott R, Zawadzki JV, Jothianandan D (1982) Role of endothelial cells in relaxation of isolated arteries of bradykinin. Proc Natl Acad Sci USA 79:2106–2110

19. Cicilini MA, Caldo H, Berti JD, Camargo ACM (1977) Rabbit tissue peptidases that hydrolyse the peptide hormone bradykinin. Biochem J 163:433–439

20. Cohen RA, Shepherd JT, Vanhoutte PM (1983) Inhibitory role of the endothelium in the response of isolated coronary arteries to platelets. Science 221:273–274

21. Corvol P, Menard J (1986) Du gène de la rénine aux inhibiteurs. Ann Endocrinol (Paris) 47:156–166

22. Cumin F, Alhenc-Gelas F, Corvol P (1987) Binding of the specific tritiated inhibitor RU44403 to homogeneous human angiotensin converting enzyme. Direct evidence for a chloride effect in the stability of the enzyme inhibitor complex. J Hypertension (abstract) (in press)

23. Darby JA, Aldred P, Crawford RJ et al. (1985) Renin gene expression in vessels of the ovine renal cortex. J Hypertension 3:9–11

24. Das M, Hartley JL, Soffer RL (1977) Serum-angiotensin converting enzyme. Isolation and relationship to the pulmonary enzyme. J Biol Chem 252:1316–1319

25. Das M, Soffer RL (1975) Pulmonary angiotensin converting enzyme. Structural and catalytic properties. J Biol Chem 250:6762–6768

26. Das M, Soffer RL (1976) Pulmonary angiotensin-converting enzyme antienzyme antibody. Biochemistry 15:5088–5093

27. Davis JO, Freeman RH (1966) Mechanism regulating renin release. Physiol Rev 56:1–56

28. De Mey JG, Claeys M, Vanhoutte PM (1982) Endothelium dependent inhibitory effects of acetylcholine, adenosine triphosphate, thrombin and arachidonic acid in the canine femoral artery. J Pharmacol Exp Ther 222:166–173

29. De Mey JG, Vanhoutte PM (1983) Anoxia and endothelium dependent reactivity of the canine femoral artery. J Physiol 335:65–74

30. Di Salvo J, Peterson A, Montefusco C, Mentam (1971) Intrarenal conversion of angiotensin I to angiotensin II in the dog. Circ Res 29:398–406

31. Dollin C, Savoie F, Richard J, Alhenc-Gelas F Unpublished data

32. Dzau VJ (1986) Significance of the vascular renin–angiotensin pathway. Hypertension 8:553–559

33. Ehlers MRW, Maeder DL, Kirsch RE (1986) Rapid affinity chromatographic purification of human lung and kidney angiotensin-converting enzyme with the novel N-carboxyalkyl dipeptide inhibitor N[1(S)-carboxy-5-aminopentyl]glycylglycine. Biochim Biophys Acta 883:361–372

34. Erdös EG (1977) The angiotensin converting enzyme. Fed Proc 36:1760–1768

35. Ferreira SH, Vane JR (1967) The disappearance of bradykinin and eledoisin in the circulation and vascular bed of the cat. Br J Pharmacol Chemother 30:417–424

36. Frangos JA, Eskin SG, MacIntyre LV (1985) Flow effects on prostaglandin production by cultured human endothelial cell. Science 227:1477–1479

37. Fray JCS, Park CS (1979) Influence of potassium, sodium, perfusion pressure and isoprenaline on renin release induced by anticalcium deprivation. J Physiol 292:363–372

38. Freeman RH, Davis JO (1983) Factors controling renin secretion and release. In: Genest J (ed) Hypertension. McGraw-Hill, New York, pp 225–249

39. Furchgott RF, Zawadzki JV (1980) The obligatory role of endothelial cells in the relaxation of arterial smooth muscle by acetylcholine. Nature 288:373–376

40. Fyrquist F, Grönhagen-Riska, Hortling L, Forslund T, Tikkanen I, Klockars M (1983) The induction of angiotensin converting enzyme by its inhibitors. Clin Exp Hypertens [A] 5:1319–1330

41. Galen FX, Corvol MT, Devaux C et al. (1984) Renin biosynthesis by human tumoral juxtaglomerular cells. Evidence for a renin precursor. J Clin Invest 73:1144–1155

42. Ganz P, Davies PF, Leopold JA, Gimbrine MA Jr, Alexander RW (1986) Short and long term interactions of endothelium and vascular smooth muscle in coculture; effects on cyclic GMP production. Proc Natl Acad Sci USA 83:3552–3556

43. Gardes J, Gonzalez MF, Corvol P, Menard J (1986) Influence of converting enzyme inhibition on the hormonal and renal adaptation to hyper- and hyponatremic dehydration. J Hypertens 4:189–196

44. Gimbrone MA, Alexander RW (1975) Angiotensin II stimulation of prostaglandin production in cultured

human vascular endothelium. Science 189:219

45. Glance DG, Elder MG, Blowan DL (1984) The effects of the components of the renin–angiotensin system on the isolated perfused human placenta cotyledon. Am J Obstet Gynecol 149:450–454

46. Gumbiner B, Kelly RB (1982) Two distinct intracellular pathways transport secretory and membrane glycoproteins to the surface of pituitary tumor cells. Cell 28:51–59

47. Habib JB, Wells SL, Williams CL, Henry PD (1984) Atherosclerosis impairs endothelium dependent arterial relaxation. Circulation 70 [Suppl II]:123 (abstract)

48. Harris RB, Wilson IB (1985) Sequencing of an active site peptide of angiotensin I converting enzyme containing an essential glutamic acid residue. J Biol Chem 260:2208–2211

49. Hayes LW, Goguen CA, Ching SF, Slakey LL (1978) Angiotensin converting enzyme: accumulation in medium from cultured endothelial cells. Biochem Biophys Res Commun 82:1147–1153

50. Hong SL (1980) Effect of bradykinin and thrombin on prostaglandin synthesis in endothelial cells from calf and pig aorta and human umbilical cord vein. Thromb Res 18:787–795

51. Houston DS, Shepherd JT, Vanhoutte PM (1985) Adenine nucleotide, serotonin and endothelium dependent relaxations to platelets. Am J Physiol 248:4389–4395

52. Iwata K, Blacher R, Soffer RL, Lai CY (1983) Rabbit pulmonary angiotensin-converting enzyme: the NH_2 terminal fragment with enzymatic activity and its formation from the native enzyme by NH_4OH treatment. Arch Biochem Biophys 227:188–201

53. Johnson AR (1979) Effects of kinins on organ systems. In: Erdos EG (ed) Bradykinin, kallidin and kallikrein. Handbook of Experimental Pharmacology, vol. XXV [Suppl] pp 357–388

54. Johnson AR (1980) Human pulmonary endothelial cells in culture. Activities of cells from arteries and cells from veins. J Clin Invest 65:841–850

55. Johnson AR, Erdös EG (1977) Metabolism of vasoactive peptides by human endothelial cells in culture: angiotensin I converting enzyme (kininase II) and angiotensinase. J Clin Invest 59:684–695

56. Kageyama R, Ohkubo H, Nakanishi S, Murakami K (1983) Cloning and sequence analysis of cDNA for human renin precursor. Proc Natl Acad Sci USA 80:7405–7409

57. Keeton TK, Campbell WB (1980) The pharmacological alteration of renin release. Pharmacol. Rev 32:202

58. Kim SJ, Hirose S, Miyazaki H et al. (1985) Identification of plasma inactive renin as prorenin with a site-directed antibody. Biochem Biophys Res Commun 126:641–645

59. Kirchner KA, Kotchen TA, Galla JH, Luke RG (1978) Importance of chloride for inhibition of renin by sodium chloride. Am J Physiol 235:F444–F450

60. Krulewitz AH, Baur WE, Fanburg B (1984) Hormonal influence on endothelial cell angiotensin-converting enzyme activity. Am J Physiol 247:C163–C168

61. Kurz A, Pfeilschifter J, Bauer C (1984) Is renin secretion governed by the calcium permeability of juxtaglomerular cell membrane? Biochem Biophys Res Commun 124:359–366

62. Lacasse J, Ballak M, Mercure C et al. (1985) Immunocytochemical localisation of renin in juxtaglomerular cells. J Histochem Cytochem 33:323–332

63. Lanzillo JJ, Stevens J, Tumas J, Fanburg BL (1983) Spontaneous change of human plasma angiotensin I-converting enzyme isoelectric point. Arch Biochem Biophys 227:434–439

64. Lindop GBM (1987) Morphological aspects of renin synthesis, processing, storage and secretion. Kidney Int (in press)

65. Lockette W, Otsuka Y, Carretero OA (1986) The loss of endothelium dependent vascular relaxation in hypertension. Hypertension 8 [Suppl II]:61–66

66. Luscher TF, Vanhoutte PM (1986) Endothelium dependent responses to platelets and serotonin in spontaneously hypertensive rats. Hypertension 8 [Suppl II]:55–60

67. Marre M, Misumi J, Raemsch KD, Corvol P, Menard J (1982) Diuretic and natriuretic effects of nifedipine on isolated perfused rat kidney. J Pharmacol Exp Ther 233:263–270

68. Mendelsohn FAO, Lloyd CJ, Katchel C, Funder JW (1982) Induction by glucocorticoids of angiotensin converting enzyme production from bovine endothelial cells in culture and rat lung in vivo. J Clin Invest 70:684–692

69. Miller VM, Vanhoutte PM (1985) Endothelium dependent contractions to arachidonic acid are mediated by products of glycooxygenase. Am J Physiol 248:H433–H437

70. Minuth L, Hackenthal E, Poulsen K, Rix E, Taugner R (1981) Renin immunohistochemistry of the differentiating juxtaglomerular apparatus. Anat Embryol 162:173–181

71. Misono KS, Chang JJ, Inagami T (1982) Amino acid sequence of mouse submaxillary gland renin. Proc Natl Acad Sci USA 79:4858–4862

72. Moncada S, Vane JR (1979) Pharmacology and endogenous roles of prostaglandin endoperoxides, thromboxane A2 and prostaglandin. Pharmacol Rev 30:293

73. Mounier F, Hinglais N, Sich M et al. Ontogenesis of angiotensin I converting enzyme in human kidney. Kidney Int (in press)

74. Mullane KM, Moncada S (1980) Prostaglandin release and the modulation of some vasoactive hormones. Prostaglandins 20:125–149

75. Nakache M, Gaub ME, Schreiber AB, MacConnell HM (1986) Topological and modulated distribution of surface markers on endothelial cells. Proc Natl Acad Sci USA 83:2874–2878

76. Ng KKF, Vane JR (1967) Conversion of angiotensin I to angiotensin II. Nature 216:762–766

77. Nishimura K, Yashida N, Hiwada K, Ueda E, Kokubu T (1977) Purification of angiotensin I-converting enzyme from human lung. Biochim Biophys Acta 483:398–408

78. Nochy D, Barres D, Camilleri JP, Bariety J, Corvol P, Menard J (1983) Abnormalities of renin-containing cells in human glomerular and vascular renal diseases. Kidney Int 23:375–379

79. Ody C, Seillan C, Russo-Marie F, Duval D (1983) Angiotensin II does not elicit any specific prostaglandins secretion in piglet cultured endothelial cells. Thromb Res 31:219–231

80. Okabe T, Fusijawa M, Yatsumoto H, Takaku F, Lanzillo JJ, Fanburg BL (1985) Familial elevation of serum angiotensin converting enzyme. Q J Med 216:55–61

81. Oliver JA, Sciacla RR (1984) Local generation of angiotensin II as a mechanism of regulation of peripheral vascular tone in the rat. J Clin Invest 74:1247–1251

82. Ondetti MA, Cushman DW (1982) Enzymes of the renin angiotensin system and their inhibitors. Ann Rev Biochem 51:283–308

83. Oparil S, Low J, Koerner TJ (1976) Altered angiotensin I conversion in pulmonary disease. Clin Sci Mol Med 51:538–543

84. Oparil S, Winternitz S, Gould V, Baerwaldt M, Szidon P (1982) Effect of hypoxia on the conversion of angiotensin I to II in the isolated perfused rat lung. Biochem Pharmacol 31:1375–1379

85. Orlowski M, Wilk E (1978) Concentration of angiotensin converting enzyme and angiotensin degrading enzymes in brain microvessels. Fed Proc 37:602 (abstract)

86. Oshima G, Gecse A, Erdös EG (1974) Angiotensin I converting enzyme of the kidney cortex. Biochim Biophys Acta 350:26–37

87. Panthier JJ, Foote S, Chambraud D, Strosberg D, Corvol P, Rougeon F (1982) Complete amino acid sequence and maturation of the mouse submaxillary gland renin precursor. Nature 298:90–92

88. Pantoliano MW, Holmquist B, Riordan JF (1984) Affinity chromatographic purification of ACE. Biochemistry 23:1037–1042

89. Peach MJ (1977) Renin angiotensin system: biochemistry and mechanisms of action. Physiol Rev 57:313–370

90. Phat VN, Camilleri, JP, Bariety J et al. (1981) Immunohistochemical characterization of renin-containing cells in the human juxtaglomerular apparatus during embryonal and fetal development. Lab Invest 45:387–390

91. Pinet F, Mizrahi J, Laboulandine I, Menard J, Corvol P (1987) Regulation of renin secretion in cultured human transfected juxtaglomerular cells. J Clin Invest (in press)

92. Rapoport RM, Murad F (1983) Agonist-induced endothelium dependent relaxation in rat thoracic aorta may be mediated through cGMP. Circ Res 52:352–357

93. Re RN, Fallon JT, Dzau VJ, Quay S, Haber E (1982) Renin synthesis by canine aortic smooth muscle cells in culture. Life Sci 30:99–106

94. Reid IA (1984) Actions of angiotensin II on the brain: mechanism and physiologic role. Am J Physiol 246:F533–F543

95. Ruyter JHC (1925) Über einen merwürdigen Abschnitt der Vasa afferentia in der Mauseniere. Z Zellforsch Mikrosk Anat 2:242–248

96. Ryan US, Ryan JW, Whitaker C, Chiu A (1976) Localization of angiotensin-converting enzyme (Kininase II). Immunohistochemistry and immunofluorescence. Tissue Cell 8: 125–145

97. Saint-Clair DK, Presper KA, Smith PL, Stump DC, Heath EC (1986) Bovine angiotensin-converting enzyme: aminoterminal sequence analysis and preliminary characterization of a hybridiseation-selected primary translation product. Biochim Biophys Res Commun 141:968–972

98. Sassano P, Chatellier G, Billaud E, Alhenc-Gelas F, Corvol P, Menard J (1987) Treatment of mild to moderate essential hypertension with or without the converting enzyme inhibitor Enalapril. Results of a 6 months double blind trial. Am J Med (in press)

99. Shapiro R, Riordan JF (1983) Critical lysine residue at the chloride binding site of angiotensin converting enzyme. Biochemistry 22:5315–5321

100. Shepherd JT, Vanhoutte PM (1979) The human cardiovascular system. Facts and concepts, Raven Press, New-York

101. Skeggs LT, Kahn JR, Shumway NP (1956) The preparation and function of the hypertensin-converting enzyme. J Exp Med 103:295–299

102. Skidgel RA, Erdös EG (1985) Novel activity of human angiotensin I converting enzyme: release of the NH_2 and COOH terminal tripeptides from the luteinizing hormone-releasing hormone. Proc Natl Acad Sci USA 82:1025–1029

103. Skidgel RA, Printz MA (1978) PG 12 production by rat blood vessels: diminished prostaglandin formation in veins compared to arteries. Prostaglandins 16:1–16

104. Spolas EG, Folio G, Quilley J, Chander P, Margiff JC (1983) Endothelial mechanism in the vascular action of hydralazine. Hypertension 5 [suppl I]:107–111

105. Stewart TA, Weare JA, Erdös EG (1981) Purification and characterization of human converting enzyme (kininase II). Peptides 2:145–152

106. Swales JD (1979) Arterial wall or plasma renin in hypertension. Clin Sci 56:293–298

107. Tang J, James MNG, Hsu IN, Jenkins JA, Blundell TL (1978) Structural evidence for gene duplication in the evolution of acid proteases. Nature 271:618–621

108. Taugner R, Ganten D (1982) The localization of converting enzyme in kidney vessels of the rat. Histochemistry 75:191–201

109. Taugner R, Hackenthal E, Nobiling R, Harlacher M, Reb G (1981) The distribution of renin in the different segments of the renin atrial tree. Immunocytochemical investigation in the mouse kidney. Histochemistry 73:75–88

110. Terragno DA, Crowshaw K, Terragno NA, Macgiff JC (1978) Prostaglandins synthesis by bovine mesenteric arteries and veins. Circ Res 36 [suppl 1]:76

111. Tsai BS, Khosla MC, Peach MJ, Bumpus FM (1975) Synthesis and evaluation of Des-Asp₁-angiotensin I: a precursor for Des-Asp angiotensin II (A II). J Med Chem 18:1180–1183

112. Ungcr T, Ganten D, Lang RE, Schölkens BA (1984) Is tissue converting enzyme inhibition a determinant of the antihypertensive efficacy of converting enzyme inhibitors? Studies with the two different compounds Hoe 498 and MK 421 in spontaneously hypertensive rats. J Cardiovasc Pharmacol 6:872–880

113. Vander AJ (1967) Control of renin release. Physiol Rev 47:359–382

114. Vanhoutte PM, Rimeze TJ (1982) Role of endothelium in the control of vascular smooth muscle function. J Physiol (Paris) 78:681–686

115. Weare JA (1982) Activation inactivation of human angiotensin I converting enzyme (EC 3-4-15-1) following chemical modifications of amino groups near the active site. Biochem Biophys Res Commun 104:1319–1326

116. Winquist RJ, Bunting PB, Bskin EP, Wallace AA (1984) Decreased endothelium-dependent relaxation in New Zealand genetic hypertensive rats. J Hypertension 2:541–545

117. Yang HYT, Erdös EG, Levin Y. (1970) A dipeptidyl carboxypeptidase that converts angiotensin I and inactivates bradykinin. Biochim Biophys Acta 214:374–376

118. Yotsumoto H, Imai I, Kuzuya N, Uchimura H, Matsuzaki F (1982) Increased levels of serum-angiotensin converting enzyme activity in hyperthyroidism. Ann Int Med 96:326–328

119. Zawadzki JV, Furchgott RF (1981) Fed Proc 40:689 (abstract)

Section II

Pathophysiology of Experimental and Human Atherosclerosis

Chapter 10

The Lesions of Atherosclerosis

C. L. Berry

Atherosclerosis is defined as "a variable combination of changes of the intima of arteries consisting of the focal accumulation of lipids, complex carbohydrates, blood, and blood products, fibrous tissue and calcium deposits" [6]. This clearly defines established disease; this chapter considers how the changes in vessels develop.

There are three main classes of arterial lesions which have been considered to represent various stages of atherosclerosis:

1. The fatty streak and gelatinous elevation
2. The fibroatheromatous plaque
3. The complicated lesion

Although more complex classifications have been used there seems little justification for some of them and some have bizarre features such as that of Haust and More [2] which includes "microthrombi" as a form of early lesion. Recently, however, Stary [4,5] has implemented a seven-stage classification scheme (see Table 10.1) for reporting the findings of a study of 900 cases of individuals dying from atherosclerosis under the age of 30. It is likely that the type 1 lesions in this scheme are not the precursors of atheromatous lesions [5]. His extended scheme is useful in considerations of pathogenesis and makes the important point that the advent of haemorrhage in the lesion greatly increases fibrogenesis. Features of this work are incorporated into the account given below.

Early Changes, Fatty Streaks and Gelatinous Elevations

Lipid deposits may be found in the coronary vessels in infants, notably at the sites of eccentric intimal thickening opposite the flow dividers of coronary bifurcations. These deposits, in macrophages, disappear before fatty streaks appear at around 10 years of age. The streaks are composed of macrophage foam cells and show the accumulation of lipid in intimal smooth-muscle cells. There is also scattered extracellular lipid. The streaks cause a slight elevation of the intimal surface and are usually oval or linear in shape, with their long axis lying in the direction of blood flow in larger vessels.

It is clear that fatty streaks, which occur at high frequency in the proximal aorta and carotid arteries in the 2nd and 3rd decade, are not always the precursors of established lesions. In some races 35%–45% of the intimal surface may be involved by these streaks in the 3rd decade, as studies on American servicemen killed in action have shown, but longitudinal population surveys suggest that most of the streaks are reversible. This is also supported by the fact that they are probably commoner in young women than men, in marked contrast to what would be expected from the incidence of atherosclerotic vascular disease.

The gelatinous elevation is a greyish-white translucent area of intima, slightly raised and

Table 10.1. Definition of types of lesions in human coronary intima (Stary [4,5])

Lesion type	Descriptive terms used in this chapter	Descriptive terms used in the literature	Characteristic composition
I	Isolated MFC	None	MFC[a] in the intima as isolated cells; lipid does not accumulate extracellularly
II	Fatty streak	Fatty streak, dot	Layers of MFC; lipid-laden intimal SMC[b]; thinly scattered extracellular lipid particles
III	Pre-atheromatous lesion	Not previously described	All type II changes plus multiple small pools of extracellular lipid
IV	Atheromatous lesion	Atheroma, fibrolipid plaque, atherosclerotic plaque, fibrous plaque, complicated lesion	All type II changes plus massive confluent pool (core) of extracellular lipid replacing much of musculoelastic intima
V	Fibroatheroma		All type IV changes plus massive collagen layers (cap) above core; microhaemorrhages or thrombus may be present
VI	Ulcerated fibroatheroma	Complicated lesion	All type V changes plus ulceration of surface
VII	Fibrous plaque	Fibrous plaque	Intima massively thickened by collagen layers; intra- and extracellular lipid absent or negligible

[a] Macrophage foam cells
[b] Smooth-muscle cells.

usually elongated in the long axis of the artery. There is intimal oedema and fibrin insudation on sectioning, and dispersed lipid may be present in small amounts. The role of these lesions in atherogenesis is uncertain; finding this type of change at the margin of established lesions has been assumed by some to indicate that it represents a progressive process.

The Fibroatheromatous Plaque and Early Established Lesion

Fibrous plaques are clearly raised lesions, usually greyish-white in appearance. This colour is due to the presence of a thin cap of fibrous tissue and glycosaminoglycan (GAG) over a lipid mass, the whole covered by endothelium. Elastic tissue is often seen in the cap of larger lesions. Deep to this cap, lipid-laden smooth-muscle cells are found, but extracellular lipid (mainly cholesterol and its esters with some neutral fat) is abundant and may form a large pool containing cell debris. The presence of this mass in the deeper layers of the intima is associated with atrophic changes in the underlying media due to the pressure exerted by the blood in the artery.

The Complicated Lesion

Many fibrous lesions progress to complicated plaques. Ulceration of the endothelium as the lesion grows larger exposes the blood to intensely thromboplastic material and thrombosis may occur. In larger arteries this may not result in occlusion and the thrombus is then incorporated into the plaque and increases its rate of growth. Haemorrhage into the plaque, often from granulation tissue vessels in its base, may cause ulceration, displace atheroma thus causing embolism or result in occlusion due to displacement of the plaque into the lumen. Fibrogenesis is increased where haemorrhage has occurred. Calcification is common in the lesion and the surrounding, often atrophic, media.

The Pathogenesis of the Lesions

This section will look at the sequence of events leading to the production of these lesions. If the intima is damaged – experimentally this has been accomplished with balloons, painting with acid, mechanical abrasion, air drying and so on –

platelet adhesion follows and there is infiltration of lipid from the blood. These processes, together with the injury, result in the proliferation of smooth-muscle cells and their migration from the media into the intima, with subsequent production of GAGs and scleroproteins. This is a reversible lesion, although all of the connective tissue may not regress. Proliferation of smooth-muscle cells is brought about in part by platelet-derived growth factor (PDGF), a basic glycoprotein of molecular weight 32 000 which also stimulates collagen synthesis via an effect on protein synthesis generally. Low-density lipoprotein (LDL) also has a direct stimulatory effect on smooth-muscle cell division, an effect reversed by high-density lipoprotein (HDL).

The area of intimal thickening produced in this way may progress to become a fibrous plaque if lipid is not cleared or if intimal damage persists or is renewed. High lipid levels in the blood or hypertension will increase the amount of lipid "arriving" in the area; changes in the handling of lipid may prevent its mobilization. Thus a process of repair becomes pathological. The two fundamental processes in the generation of the lesions of atherosclerosis are thus seen to be smooth-muscle cell proliferation and the deposition of intra- and extra-cellular lipid. (See Ross and Glomset [3], Brown and Goldstein [1]). Details of their respective roles are covered in the following chapter and are supplemented by data on individuals with metabolic abnormalities.

The Role of Lipids

Lipids have previously been classified by their ultracentrifugal characteristics but current knowledge of their metabolism and interactions between lipids and lipoproteins suggests that this is an unsatisfactory method. For this discussion we will use a simplified classification (see Tables 10.2 and 10.3) which shows the major classes of lipid and their associated proteins. Disturbances of function may be related to the lipid or protein moiety. There are three functions of these proteins, the "apoproteins": they help to solubilize cholesterol and triglyceride by interacting with phospholipid; they regulate the reactions with enzymes such as lecithin cholesterol acyl transferase (LACT) and lipoprotein lipase; and they bind to surface receptors on cells, thus determining rates of uptake and metabolism of other lipoproteins.

The relationship between LDL and atherosclerosis is seen most clearly in type II hyperlipidaemia (Table 10.3) where there is a defect in the mechanism of receptor-mediated endocytosis, the method by which cells normally obtain essential lipids. Heterozygotes synthesize half the normal number of LDL receptors on the cell surface, homozygotes none. There is a feedback system controlling the expression of the gene encoding for the receptor so that when a cell has all the cholesterol it needs, fewer receptors are made and the cell extracts less lipid from surrounding fluids. Clearly, the reduction in levels of receptors, especially in the liver (the site

Table 10.2. Classification of lipids

Associated apoprotein	Lipid	Function of protein moiety
A1, A2	HDL	Activation of transferase
B	VLDL, LDL	Triglyceride transport. Binds to LDL receptor, with uptake into cells
C1, C2, C3	VLDL, HDL	Activate and inhibit lipoprotein lipase
D	HDL	
E	VLDL, HDL, IDL	Control of clearance of VLDL, chylomicron remnants

Table 10.3. Typical lipoprotein patterns in hyperlipidaemia

Type	Chylomicrons	VLDL	LDL	HDL
I	High	Normal or low	Normal or low	Normal
IIa	Normal	Normal	High	Normal
IIb	Normal	Raised	High	Normal
III	Normal		VLDL/LDL raised and abnormal in composition	
IV	Normal	Raised	Normal	Normal
V	Normal	Raised	Normal	Normal

of metabolism of cholesterol to bile acids), results in high circulating levels of lipid and thus in atherosclerosis. Plasma LDL is about doubled in heterozygotes and raised by up to sixfold in homozygotes. The increase in risk of a myocardial infarction in heterozygotes before age 60 is around 25 times that in the normal population and homozygotes usually have an infarction before the age of 20.

Certain authors (see Brown and Goldstein [1]) consider that in the population at large there may be individuals with a relative deficiency of LDL receptors, who therefore have a consequent rise in plasma LDL levels.

How do levels of LDL relate to atherosclerosis? The initiating event in this disease is damage to endothelium, which may occur for a number of reasons as described in Chaps. 11 and 14. The change in endothelial permeability which follows allows LDL particles and platelets to penetrate the deeper layers of the intima, where the disease begins. Following the release of PDGF, medial smooth-muscle cells divide and migrate into the affected area, as do other cells from the blood which provide the macrophage population. Both cell types take up LDL, meta-

bolize it and become foam cells. If large amounts of lipid are present, cholesterol accumulates in large quantities and various metabolic products and products derived from damaged and dead cells help to produce atheroma, the pultacious mess found in the damaged intima.

References

1. Brown MS, Goldstein JL (1984) How LDL receptors influence cholesterol and atherosclerosis. Sci 215:58–66
2. Haust MD, More RH '(1972) Development of modern theories on the pathogenesis of atherosclerosis. In: Wissler WR, Geer JC (eds) the pathogenesis of atherosclerosis. Williams and Wilkins, Baltimore
3. Ross R, Glomsett JA (1976) The pathogenesis of atherosclerosis. N Engl J Med 295:369–377
4. Stary HC (1985) Macrophage foam cells in the coronary artery intima of human infants. Ann NY Acad Sci 454:5–8
5. Stary HC (1987) Evolution and progression of atherosclerosis in the coronary arteries of infants and children. In: Bates SR, Gangloff E (eds) Atherosclerosis and ageing. Springer-Verlag, Berlin Heidelberg New York
6. WHO Study Group (1958) Classification of atherosclerotic lesions. WHO, Geneva (WHO technical report series: lipids in atherosclerosis)

Chapter 11

Atherosclerosis: Cellular Aspects

O. Kocher, I. Hüttner and G. Gabbiani

Introduction

Despite intensive clinical, experimental and epidemiological studies, the mechanisms leading to the formation of the lesions of atherosclerosis remain mysterious. In the last few years, however, significant progress has been made in understanding the roles played by arterial wall components, blood-borne cells and plasma constituents in the development of the atheromatous plaque [for reviews see 4,43,49,63,90,92,93,130]. This process appears more and more to depend on interactions among various causative factors, thus making the search for a prime cause particularly difficult and possibly pointless. In this chapter, we shall limit our discussion to the role of cellular elements, particularly those normally present in the vessel wall, in the development of the atheromatous plaque.

Role of Endothelium

The endothelial monolayer contributes to the maintenance of homeostasis in the vessel wall and in the vessel lumen via three main mechanisms: (a) production of metabolically active substances, (b) formation of an efficient barrier to plasma constituents and (c) prevention of platelet adhesion and blood coagulation (see Chap. 1). Normal endothelial cells synthesize numerous active substances such as prostacyclin

[72], a potent inhibitor of platelet aggregation, von Willebrand factor [56] and plasminogen activator [68]; they also contain receptors to vasoactive amines such as catecholamines, serotonin, histamine and peptides such as bradykinin and angiotensin, which in turn contribute to the regulation of the vascular tone [53].

Alteration of the endothelial barrier is obvious in small artery disease, particularly in hypertension and diabetes, and it has also been widely implicated in large artery disease [59,121,129]. Loss of continuity of the endothelial cell layer is the major cause of thrombosis because it exposes the highly thrombogenic subendothelial tissues to platelets. Furthermore, there is evidence that endothelial denudation leads to proliferation and migration of smooth-muscle cells into the intima as a result of the release of platelet-derived growth factor (PDGF) at sites where endothelial continuity is disrupted [90,92]. Non-denuding endothelial injuries leading to dysfunction of a structurally continuous but remodelled endothelial cell layer may, however, also contribute to smooth-muscle cell proliferation by altering transport, altering platelet interaction and by altering control of growth factor production [90,104]. The replication rate of endothelial cells is very low in normal adult animals [19,103] and not randomly distributed [102]. It correlates with the low rate of spontaneous cell death occurring in aortic endothelium under normal conditions [102]. The overall replication of aortic endothelium is increased in response to hypertension [19,103] which in turn results in extensive remodelling of the endothelial cell layer [19, see also below].

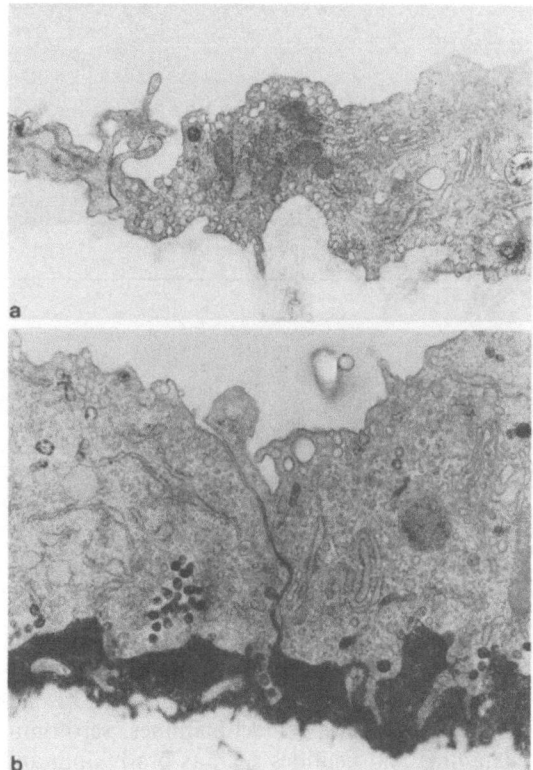

Fig. 11.1. Unstained ultra-thin sections of thoracic aorta endothelium in a normotensive rat **a** and hypertensive rat 7 d after aortic ligature **b**. The rats have been injected intravenously with horseradish peroxidase (5 ml/100 g body weight) 5 min before perfusion. In the hypertensive animal, the subendothelium is filled with the tracer, which is present also in the interendothelial cleft and in plasmalemmal vesicles. In the normotensive animal, horseradish peroxidase labels only occasional plasmalemmal vesicles. × 30 000

Modifications of the Endothelium in Pathological Conditions

Hypertension

Several laboratories have studied the effects of different types of hypertension on the endothelium [19,30,103]. Among the variables studied are changes in morphology, permeability, replication rate and cell density. Most of the work has been done in rat or rabbit aortas.

In the normal adult rat aorta, endothelial cells form a continuous monolayer and contain the usual cytoplasmic organelles, such as endoplasmic reticulum, mitochondria and Golgi apparatus [30]. They are interconnected by numerous tight and gap junctions. The endothelium is practically impermeable to a short perfusion (5 min) of a relatively small amount of

a tracer such as horseradish peroxidase (HRP) (Fig. 11.1a), and its replication rate is very low (Table 11.1) [30]. During hypertension, induced either by ligation of the aorta (an experimental model equivalent to the one-kidney Goldblatt model [88]) or by uninephrectomy and a sodium chloride-rich diet with or without subcutaneous administration of desoxycorticosterone acetate (DOCA), the morphology of the endothelial cells is modified; they bulge into the lumen of the blood vessels (Fig. 11.2) and are particularly rich in microfilament-containing cytoplasmic projections [19]. The endothelial cell volume and cell density are increased [19,54]. The lateral surface occupied by tight junctions increases in all types of hypertension, but there is no change in the lateral surface occupied by gap junctions [54]. However, gap junctions appear irregularly shaped in hypertensive animals compared to controls [54]. While there is an increase in tight-junction density in the aortic endothelium during hypertension, the percentage of discontinuous tight junctions (probably representing steps in assembly and/or disassembly) is also increased [54]. This may explain the increased permeability of the endothelial cell layer in certain hypertensive situations [30,54].

Table 11.1. Thymidine index and density of rat aortic endothelial cells during various hypertensive situations (de Chastonay et al. [19])

Treatment	BP	Thymidine index	Cell density
None	113/68	0.007 ± 0.0007	75 ± 1
Aortic ligature 7 d	166/118	0.120 ± 0.008	83 ± 3
Aortic ligature 40 d	182/124	0.002 ± 0.0006	91 ± 2
None	113/69	0.0039 ± 0.0005	79 ± 2
Sodium-rich diet 7 d*	115/77	0.0171 ± 0.0038	90 ± 1
Sodium-rich diet 40 d*	161–113	0.0046 ± 0.0005	85 ± 2
None	113/65	0.0039 ± 0.0003	75 ± 2
0.9% NaCl + DOCA 10 d*	167/120	0.0660 ± 0.004	103 ± 3
0.9% NaCl + DOCA 40 d*	197/128	0.0031 ± 0.0004	130 ± 2

* Unilaterally nephrectomized animals.

In the early stages of hypertension (7 d after the beginning of the experiment), permeability to HRP is increased in animals treated by aortic ligation or with uninephrectomy, a sodium chloride-rich diet and DOCA [30] (Fig. 11.1b). In chronic stages (40 d after the beginning of the experiment), increased permeability to HRP per-

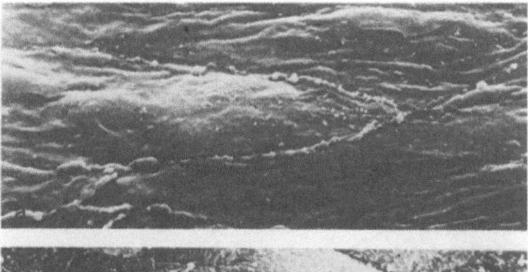
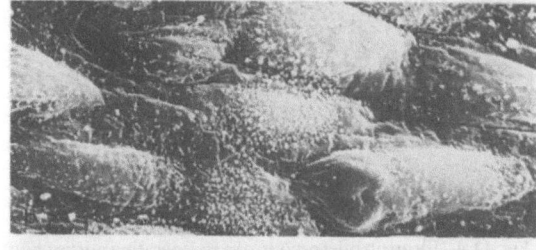
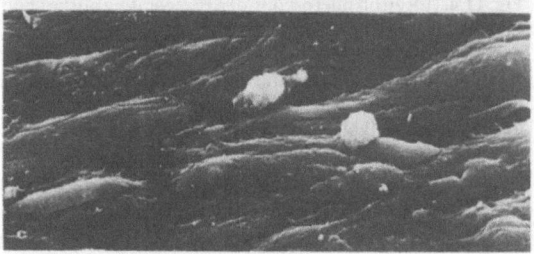

Fig. 11.2. Scanning electron micrographs comparing the luminal surfaces of endothelial cell layers in thoracic aorta of **a** a normotensive rat, **b** a hypertensive rat 7 d after aortic ligature and **c** a hypertensive rat 40 d after aortic ligature. In the normotensive rat, endothelial cells are flat and have well-defined boundaries. In the hypertensive rat 7 d after aortic ligature, the picture shows a cluster of endothelial cells bulging into the lumen; they are particularly rich in microvillous projections. In the hypertensive rat 40 d after aortic ligature, the surface of the endothelial cells is again relatively flat and cells lack surface projections [19]. × 2600

The endothelial replication rate (measured as a ^{3}H-thymidine index on "en face" preparations of the endothelial monolayer) is very low in normotensive adult animals and increases during the early phase of hypertension (Table 11.1)[19,103]; it decreases to control values in the chronic stage of hypertension (Table 11.1) [19]. The decrease in the ^{3}H-thymidine index taking place between the acute and chronic phases of experimental hypertension may be due either to a progressive adaptation of endothelial cells to the pathological state or to exhaustion of their replicative activity [19].

Some of the structural and functional changes induced by hypertension in the aortic endothelium have also been described in animal models testing the effect of risk factors of atherosclerosis other than hypertension. For instance, increased permeability of structurally continuous rabbit aortic endothelium exposed to moderate hypercholesterolaemia [114] and to nicotine [8] has been reported. Increased replication of endothelial cells in continuous rat aortic endothelium has been described in experimental endotoxaemia [85]. These data suggest that our observations on the aortic endothelium in various models of hypertension may be relevant to other pathological conditions and illustrate how risk factors of atherosclerosis may affect the structure and function of endothelium.

Endothelial Regeneration

Endothelial denudation using various mechanical devices, most commonly an inflated balloon catheter [3], is the most popular method of producing an experimental atheromatous plaque, and most useful in the study of the regeneration of the endothelium [33,61,105]. other models have been developed using a nylon filament to produce small areas of endothelial denudation [84]. These small lesions are re-endothelialized within 8 to 48 h, depending on the shape and location and, remarkably, do not produce intimal thickening [86]. The use of these models was based on the hypothesis that an early step in the development of atheromatosis is loss of endothelium, over a small area, followed by platelet adhesion [92]. Adhering platelets release mitogens such as PDGF, which stimulate migration and growth of smooth-muscle cells in the intima [94]. These experiments with a small defined endothelial lesion suggest that focal loss of endothelial cells is not necessarily followed by smooth-muscle activation [83].

sists in animals treated with uninephrectomy, a sodium chloride-rich diet and DOCA, but is no longer present in those treated by aortic ligation [30].

Aortic endothelial cells of hypertensive animals contain, in addition to the usual cytoplasmic organelles, well-developed microfilament bundles similar to the stress fibres present in cultured cells [29,30,33,128]. These stress fibres can be stained with anti-actin and anti-myosin antibodies [33,128,131]. Their role is probably to improve cellular adhesion to the substrate in conditions of haemodynamic stress. Similar stress fibres are present in newborn rat aorta [62] and also in regenerating endothelium after experimental endothelial injury [see below, 33]. They are very rare in the adult rat aortic endothelium except in areas of turbulent blood flow, e.g. at the branching of intercostal arteries [131].

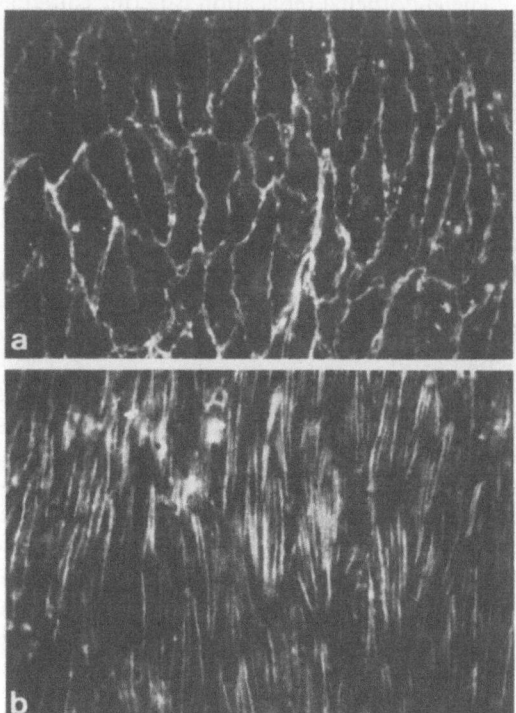

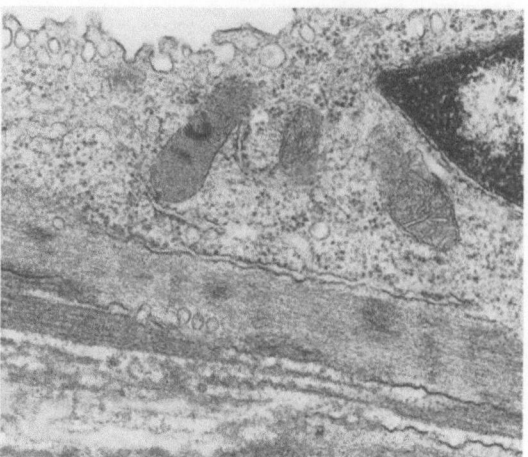

Fig. 11.4. Electron microscopy of a regenerating aortic endothelial cell showing a stress fibre located in the abluminal portion of the cytoplasm. Note extracellular matrix fibres which appear connected with microfilaments of the stress fibre across the plasma membrane, thus forming a fibronexus. × 47 000

Fig. 11.3. Immunofluorescent staining with anti-actin antibodies of endothelial cell "en face" preparations. **a** Normal endothelial cell layer with actin localized at the periphery of the cells. **b** Regenerating endothelial cell layer with actin organized into stress fibres oriented parellel to the blood flow [55]. × 800

of actin, actin-binding proteins and fibronectin with these areas of the membrane [10,117]. In any event, the lack of filipin binding shows that the fibronexus-forming plasmalemma is peculiar with respect to the remaining plasma membrane.

It is not known why regenerating endothelial cells develop large numbers of stress fibres

After denudation of large areas of endothelium, endothelial regeneration usually starts from collateral blood vessels, the endothelial cells dividing and migrating into the injured area, which has been covered by platelets [105]. These migrating endothelial cells differ from stationary cells in that they contain very well-developed stress fibres located always in the abluminal region of the cytoplasm [33] (Fig. 11.3). These are connected to extracellular fibronectin accumulations [55], thus forming a fibronexus [50] (Fig. 11.4). In regenerating endothelium, these structures persist long after cell replication has ceased. Treatment of the regenerating endothelium with filipin, an ultrastructural probe for membrane sterols, shows that the zones of interaction of the plasmalemmal membrane with stress fibres are resistant to the formation of filipin–sterol complexes [55] (Fig. 11.5). Rarity or lack of filipin–sterol complexes in the plasma membrane may indicate membrane domains with low cholesterol content or membrane domains which are too rigid to be easily deformed, probably due to the association

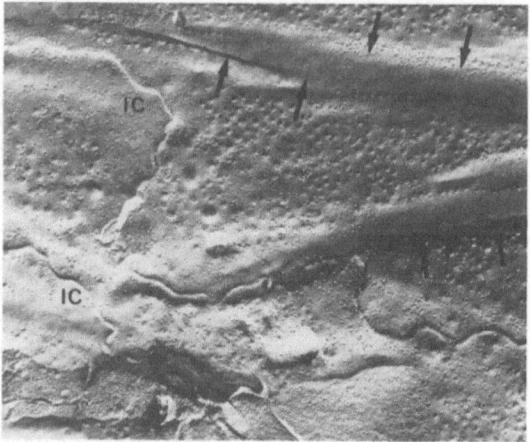

Fig. 11.5. Freeze-fracture electron micrograph of regenerating rat aortic endothelial cells treated with filipin. The plane of fracture exposes the P face of the abluminal plasma membrane of adjacent endothelial cells separated by typical interendothelial clefts (*IC*). Two band-like membrane regions appear devoid of vesicle stomata, and correspond to positions of insertion of intracytoplasmic stress fibres into the membrane (*arrows*) [55]. × 47 000

(similar to those seen in endothelial cells in culture conditions and also in hypertension in vivo) and what the functional significance is of the enhanced cell-to-matrix attachment (as reflected by the large number of stress fibres) in regenerating aortic endothelial cells. The large number of stress fibres in regenerating endothelium certainly indicates that in addition to haemodynamic stress (as in hypertension), altered composition of the connective tissue matrix may be an important factor resulting in the development of actin filament bundles that improve adhesion of endothelial cells to the matrix.

The complexity of tight junctions and the average width of gap junctions increase in regenerating endothelium 15 d after injury, when compared with normal endothelium [55]. However, in the very early phase of endothelial regeneration gap junctions are rare at the leading edge of the endothelial layers [112]. These observations are consistent with the view that endothelial cells move following injury while maintaining a continuous monolayer sheet; this probably requires a constant breakdown and rebuilding of their junctions.

The morphology of regenerating aortic endothelium is reminiscent of that of aortic endothelium in hypertension, supporting the view that increased rate of replication may be a mechanism responsible for the structurally and functionally modified (remodelled) endothelial cell layer in hypertension [19,53].

In the rat aorta, the endothelium regenerates completely over the injured area from the intercostal arteries [61,98]. This is not the case in rabbit aorta or in rat carotid artery where, at a certain point, the endothelium stops dividing for unknown reasons [17,33]. This model is very convenient for studying the biology of blood vessels in the absence of endothelium.

Endothelial Cell Death

In normal conditions, a small proportion of endothelial cells dies daily; these are replaced without loss of the monolayer continuity [46]. Dead endothelial cells have the ability to bind IgGs. Their frequency can therefore be counted in "en face" preparations of aortic endothelium after incubation with labelled IgG: 0.19% of the cells are positive under normal conditions [46]. The correlation of IgG binding with cell death has been further verified by injecting rats intravenously with chlorotetracycline, a fluorescent calcium-binding probe very efficient in showing intracellular calcium deposits which are known to represent an early event in cell death [22]. Ninety-five per cent of the IgG-labelled cells emitted chlorotetracycline fluorescence [46]. IgGs bind to the cytoskeleton, more particularly to vimentin, the endothelial cell intermediate filament protein, through the Fc fragment [47]. In addition, the Clq portion of the complement has been shown to bind to vimentin [67]. One can imagine that the IgG–complement cascade could stimulate proteolytic and membranolytic activities, as well as chemotactic factors, encouraging circulating monocytes to remove dead endothelial cells.

Basal replication in the normal endothelium being 0.15% [103], the residence time of dead endothelial cells has been estimated to be 24 h [46]. In vitro, their residence time goes up to 4 d; they are then fragmented and quick movement of neighbouring viable cells takes place to cover the empty area [46]. Rapid covering by viable cells of the empty area in vivo is essential to prevent platelet adhesion. Increased endothelial replication has been described in endotoxaemia [36,85] as well as during hypertension [19,103]. This could be due at least in part, to increased endothelial cell death, suggesting a common pathway whereby various risk factors of atherosclerosis may induce endothelial dysfunction: increased endothelial cell death, increased endothelial replication and a structurally and functionally modified endothelial cell layer.

Endothelial Dysfunction as a Predisposing Factor for the Development of Atherosclerosis

Since the discovery by several groups of PDGF [1,63,95,127], a factor which stimulates replication of smooth-muscle cells in vitro, and the production of experimental evidence that endothelial injury causes platelet adhesion to the denuded wall as well as smooth-muscle cell replication and migration from the media into the intima [115], the establishment of discontinuity of the endothelium has been considered to be a major factor in the development of atherosclerosis. Inhibition by heparin of intimal proliferation after endothelial injury supports this theory [16,45]. Endothelial cells have been shown to produce a heparin-like factor which inhibits smooth-muscle cell proliferation in vitro [12,13] and which may be important for the control of smooth-muscle cell proliferation in vivo. Remarkably, this factor has no effect on endothelial or fibroblastic proliferation in vitro [13].

The role of the endothelium as a thromboresistant surface and as a barrier regulating the transfer of serum constituents into the vessel wall is well accepted [53]. It is also known that endothelial permeability increases during pathological conditions such as hypertension [30] and hypercholesterolaemia [114], though the existence of a real discontinuity in the endothelium allowing platelet aggregation has never been demonstrated clearly. However, non-denuding endothelial injuries may also contribute to smooth-muscle cell proliferation in a number of ways: by altering transport, by altering platelet interaction and by altering control of growth-factor production [104]. Altered transport has been indicated by enhanced endothelial permeability in the absence of endothelial cell loss in rabbit aorta subjected to moderate hypercholesterolaemia [114]. As described above, we have found correlation between a structurally continuous but remodelled endothelial cell layer and increased permeability in various hypertensive models [19,30,54]. Altered transport across continuous endothelium presumably would not include the PDGF released from aggregated platelets. However, other mitogens may be available and may have access to the arterial wall at sites of increased permeability. It is also conceivable that endothelial cells, while maintaining continuity, may become altered in ways that allow altered platelet interactions without adherence. These could include contact with platelets followed by release, uptake and transport of platelet factors.

Another possible role for endothelial cells in stimulation of smooth-muscle growth is the production of endothelial cell-derived growth factors. Endothelial cells produce several growth factors in culture including a mitogen resembling PDGF [20,35]. It has been observed that both tumour promoters and endotoxin increase mitogen release by endothelium [25]. Thus, an important consideration in understanding the circumstances that lead to growth-factor formation by endothelial cells is their state in culture and in vivo. Using DNA hybridization, the levels of messenger RNA (mRNA) for the β chain of PDGF have recently been evaluated in endothelial cells in culture and in vivo. It was found that endothelial cells in vivo possess very low levels of mRNA for PDGF, whereas cultured endothelium forms high levels of PDGF on a continuous basis suggesting that endothelial cells

in culture may be in an abnormal or "injured" state.

Rat aortic endothelial cells in various hypertensive situations as well as during regeneration in vivo are similar to cultured endothelial cells in a number of morphological and functional aspects. It can be hypothesized that stimulated endothelium in vivo may release growth factors that can induce smooth-muscle cell migration and proliferation in conjunction with growth-factor release by macrophages, and possibly autogenous growth factor release by the stimulated smooth-muscle cells. This pathway may be important in hypertension, diabetes, cigarette smoking or other circumstances associated with an increased incidence of atherosclerosis [90].

Finally, the other way in which dysfunction of a continuous endothelial cell layer may contribute to atherosclerosis is via the accumulation of lipid. Accumulation of proteoglycans has been demonstrated at sites of endothelial regeneration. These sites are known to have a high affinity for low-density lipoprotein and thus may cause accumulation by increasing the affinity of the vessel wall for lipoproteins rather than increasing the entry rate [71,87].

Role of Smooth-Muscle Cells

The Arterial Smooth-Muscle Cell in Experimental Atherosclerosis

It is generally accepted that the migration of smooth-muscle cells from the arterial media into the intima is responsible for the development of atherosclerosis [4,49]. We have just seen that one of the major factors playing a role in this phenomenon is possibly endothelial cell dysfunction. Although endothelial dysfunction implies the presence of a structurally and functionally modified but continuous endothelial cell layer, it is evident that the best experimental model to study fast-developing intimal thickening is complete removal of endothelium from a selected vascular segment.

Different techniques have been described to produce endothelial loss. The most commonly used is endothelial removal by an inflated balloon catheter introduced in a given artery [3]. Other investigators dry the endothelial surface using an air stream [24]. Platelets adhere to the injured area and after 4–7 d, smooth-muscle cells begin to migrate from the media into the intima where they actively divide [11,17,24,48]. The appearance of intimal smooth-muscle cells is different from that of normal medial smooth-muscle cells in that they contain a more developed endoplasmic reticulum, but fewer microfilamentous bundles [61] (Fig. 11.6); i.e. they acquire a morphology similar to that of myofibroblasts [28,109]. After endothelial regeneration is completed, smooth-muscle cells stop dividing and recover an appearance similar to that of normal medial smooth-muscle cells [61]. Interestingly, this redifferentiation may also take place in the absence of endothelium [17, Hüttner et al., unpublished observations]. This modulation between a "contractile" and a "synthetic" phenotype has been also studied extensively in smooth-muscle cell culture [15]. In the rat carotid artery, the replication index of intimal smooth-muscle cell returns to normal 8 weeks after endothelial injury, even in the areas where the endothelium has not regenerated; however, the layer of smooth-muscle cells in contact with the blood stream still continues to replicate [17]. At this stage, platelets no longer adhere to the vessel wall [17].

Little is presently known about biochemical and/or histochemical differences between quiescent medial and atheromatous smooth-muscle cells. During the past years, the work of many laboratories has shown that cytoskeletal components can be used as markers of cell differentiation and dedifferentiation [for review see 98]. Three classes of cytoskeletal structures are present in most cells of vertebrates: microfilaments (containing mainly actin), microtubules (containing mainly tubulin), and intermediate filaments (whose composition depends on the embryological origin of the cell) [65,113]. Although actin is a very well-conserved protein, six isoforms have been identified in mammals [122]: α-cardiac muscle, α-skeletal muscle, α- and γ-smooth-muscle, and β- and γ-cytoplasmic. These two last isoforms are present in virtually all eukaryotic cells. Quantitative analysis of the different actin isoforms has been used as a tool to study actin divergence during differentiation and evolution [110,122]. The expression of intermediate filament proteins is also related to differentiation and provides information about cellular origins [76]. Antibodies against these proteins are extensively used for diagnostic purposes in tumour pathology [32,73, for review see 77,98]. We have attempted to define the cytoskeletal features of arterial smooth-muscle cells in normal conditions and during the evolution of experimental and human atheromatous plaques.

Smooth-muscle cells of the arterial wall are heterogeneous as far as their content of intermediate filament protein is concerned [31]. In normal adult rat aorta, double immunofluorescence staining with anti-vimentin and

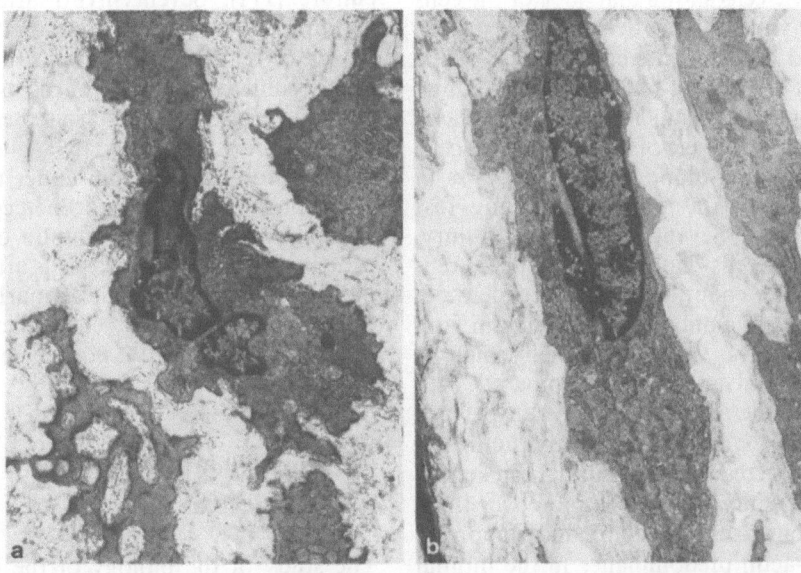

Fig. 11.6. Morphological features of a representative rat aortic medial smooth-muscle cell **a** and of an intimal smooth-muscle cell 15 d after endothelial injury **b**. A normal smooth-muscle cell **a** shows microfilaments distributed in bundles and the usual cytoplasmic organelles. An intimal smooth-muscle cell 15 d after endothelial injury **b** shows a well-developed endoplasmic reticulum and Golgi apparatus; microfilament bundles are decreased compared to normal medial smooth-muscle cells. × 5500

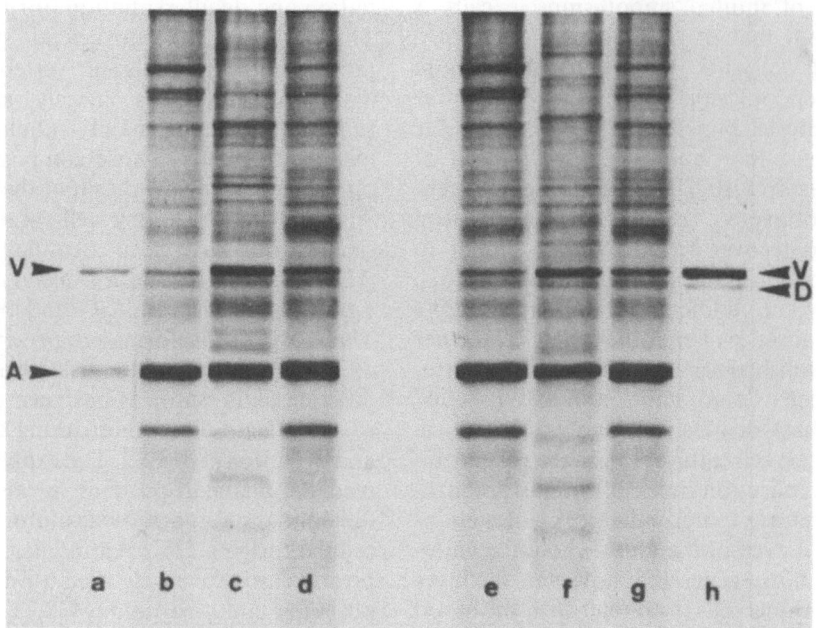

Fig. 11.7. SDS-PAGE of control rat aortic media (**b** and **e**), intimal thickening 15 d after endothelial injury (**c** and **f**) and intimal thickening 75 d after injury (**d** and **g**). Lanes **b**, **c** and **d** are loaded with the same amounts of total proteins, whereas **e**, **f** and **g** are loaded with the same amounts of DNA, reflecting the amounts of cellular proteins. Cytoskeletal preparations of cultured embryo lung fibroblasts **a** and cultured baby hamster kidney cells **h** serve as markers. *A*, actin; *V*, vimentin; *D*, desmin. Note that in addition to the changes of actin, vimentin and desmin, the tropomyosin band is weaker in intimal thickening after endothelial injury (**c** and **f**), compared with both normal media (**b** and **e**) and thickening 75 d after endothelial injury (**d** and **g**). [61]

anti-desmin antibodies on enzymatically isolated smooth-muscle cells shows that 48% of the cells contain both vimentin and desmin, 51% vimentin alone and only 1% desmin alone [61]. The percentage of the smooth-muscle cells containing the two intermediate filament proteins decreases to 21% as the percentage of smooth-muscle cells containing vimentin alone increases to 79% in replicating cells of the intimal thickening 15 d after the injury. Seventy-five d after injury, smooth-muscle cells have stopped their replication and the endothelium has regenerated; 50% of the cells contain both vimentin and desmin, and 50% vimentin alone, which is similar to normal aorta [61]. Compared with normal aorta, the amount of vimentin per cell increases, while the amounts of desmin, actin and tropomyosin decrease, in the intimal thickening 15 d after the injury (Fig. 11.7); moreover there is a switch from α-actin predominance in normal aorta to a β-actin predominance in the intimal thickening 15 d after the injury. Dividing intimal smooth-muscle cells share a number of similarities with fetal or newborn rat aortic smooth-muscle cells [62], those from human athero-

matous plaques [34,60] and those growing in culture [111]. Seventy-five d after injury, the endothelium is regenerated and intimal smooth-muscle cells have stopped replicating and undergone redifferentiation towards normal medial smooth-muscle cells, as far as their cytoskeletal composition is concerned [57]. Interestingly, the cytoskeleton of the media under the thickening does not show any change when compared with normal media. The modulation of cytoskeletal proteins in smooth-muscle cells after endothelial injury provides biochemical markers to differentiate smooth-muscle cells replicating and invading the intima from normal medial smooth-muscle cells.

Arterial Smooth-Muscle Cells in Human Atherosclerosis

The study of the biology of the human atheromatous plaque is complex because of its mixed cell population composed of at least smooth-muscle cells, macrophages and lymphocytes [97]. The relative significance of these three cell types varies according to the views of different authors and possibly depending on the pathogenesis of the atheroma: when the major stimulus is hyperlipidaemia, macrophages may already be present in the arterial intima in the early stages of the disease [38,39], whereas in hypertensive patients, smooth-muscle cells appear to be the earliest cell type involved [91]. Macrophages may be well represented in the fibrous cap and in the central core of complicated lesions [130].

Another problem arises from the morphological alterations of the involved cells (Fig. 11.8). For example, both macrophages and smooth-muscle cells have been described as precursors of "foam cells" [37] which contain accumulations of lipids; this makes the determination of the origin of single foam cells on morphological grounds alone very difficult.

Recent work [58] has shown that most cells composing complicated atheromatous plaques in human carotid arteries express the HLA-DR antigen, normally present on cells of the immune system, but only 60% of them are stained with a monoclonal antibody specific for macrophages and lymphocytes. Many of the remaining cells

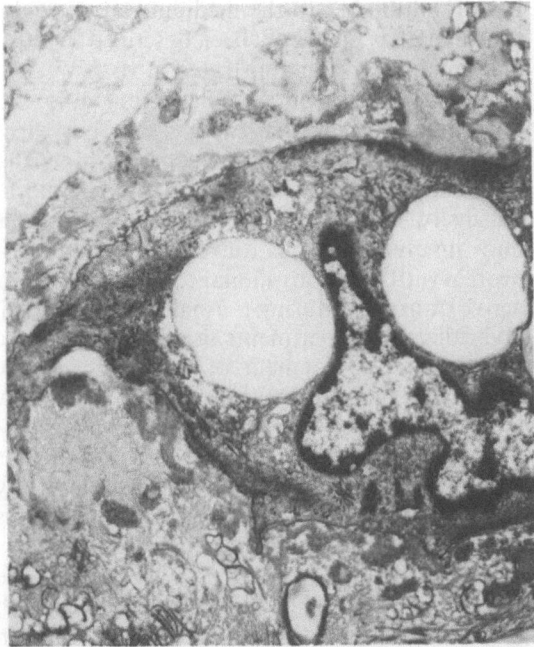

Fig. 11.8. Modified smooth-muscle cell in a human aortic atheromatous plaque. A large part of the cytoplasm is occupied by lipid accumulations. However, microfilament bundles are still numerous. The extracellular space is occupied by collagen, elastin and amorphous material. × 7500

express desmin, suggesting a muscular origin. The HLA-DR antigen is only rarely expressed by normal medial smooth-muscle cells [58]. This suggests that the HLA-DR expression is part of the phenotypic change of the atheromatous smooth-muscle cell.

The distribution of vimentin, desmin, actin and tropomyosin may vary in the normal media and fibrous or complicated atheromatous plaques of different arteries such as aorta, coronary artery and femoral artery [60]. Double immunofluorescence staining with anti-vimentin and anti-desmin antibodies shows that all smooth-muscle cells contain vimentin but only a varying percentage of them contain in addition desmin: 3% in the aorta, 11% in the coronary artery and 37% in the femoral artery. Thus, desmin-containing cells are more numerous in "muscular arteries" than in "elastic arteries", in accordance with previous work on the rat arterial tree [78]. In fibrous atheromatous plaques, desmin-positive cells are extremely rare or even absent (less than 3% of the total cell population even in the femoral artery), but reappear in complicated atheromatous plaques (12%–20% of the cell population), suggesting that, as in experimental intimal thickening after endothelial injury [61], atheromatous smooth-muscle cells are first dedifferentiated but tend to redifferentiate with time towards a more muscular phenotype [60].

Densitometric analysis of SDS-polyacrylamide electrophoretic gels shows that the content of vimentin per cell increases while actin and tropomyosin decrease in fibrous plaques when compared with normal media; moreover, fibrous plaques show a β-cytoplasmic actin predominance [60] (Fig. 11.9). These changes provide biochemical markers to differentiate atheromatous smooth-muscle cells (representing more than 80% of the total cell population in the samples of fibrous atheromatous plaques we have studied) from normal medial smooth-muscle cells. The β-actin predominance persists in complicated plaques but we do not know whether this depends on the persistence of this isoform in smooth-muscle cells, or whether this is due to a contamination by non-muscle cells which contain only β- and γ-cytoplasmic actins [60]. Human atheromatous smooth-muscle cells share many similarities with dividing rat intimal smooth-muscle cells 15 d after endothelial injury [61], newborn rat aortic smooth-muscle cells [62] and cultured rat aortic smooth-muscle cells [111]; thus they can be considered to be dedifferentiated cells.

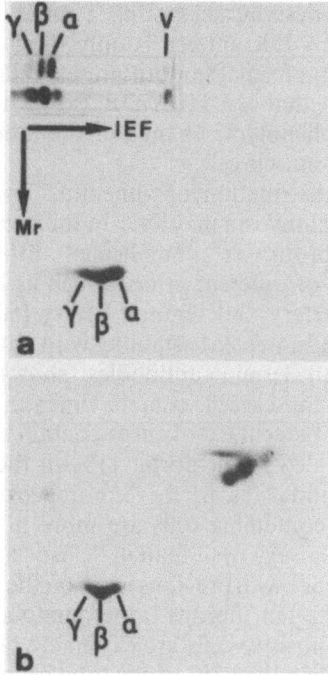

Fig. 11.9. Isoelectric focusing and bidimensional gel electrophoresis of total extracts from human aortic media **a** and fibrous atheromatous plaque **b**. In the media, the α-isoform of actin is predominant. In the fibrous plaque, the β-actin isoform becomes predominant. Additional unknown proteins, which are not present in the media, are present in the plaque. [60]

Growth Factors for Arterial Smooth-Muscle Cells

Under normal conditions, smooth-muscle cells in adult animals do not replicate. One of the main questions in the study of atherosclerosis concerns the identification of factors promoting smooth-muscle cell movement and replication. The discovery of PDGF which stimulates smooth-muscle cell replication and movement from the media to the intima [92] has furnished an answer to this question. PDGF is a heat-stable cationic protein with a molecular weight of 30 000, composed of two chains linked by disulphide bridges. It is localized in the α-granules of platelets together with platelet factor IV and β-thromboglobulin. It binds to a specific receptor on smooth-muscle cells and initiates DNA synthesis [9,51,94]. PDGF is chemotactic for smooth-muscle cells [44]. It stimulates LDL binding and degradation in cultured aortic monkey smooth-muscle cells by increasing the number of LDL receptors [14]. it increases the

motility and modifies the microfilament system organization in cultured glial cells [70]. PDGF disrupts the microfilament bundles of Balb/c 3T3 cells and modifies the distribution of vinculin normally localized in the adhesion plaques [7,52]. After 5 min of incubation of the cells with PDGF, vinculin disappears from the adhesion plaques and accumulates in the perinuclear region; this alteration in vinculin distribution is followed by a disruption of actin-containing stress fibres. After 60 min, vinculin reappears in the adhesion plaques, while stress fibres are not reconstituted. These changes suggest that the stimulation of cell growth induced by PDGF is associated with reorganization of cytoskeletal proteins [52].

Recent reports have shown that PDGF-like molecules are synthesized by various cultured cell types including endothelial cells [20] and smooth-muscle cells isolated from 15-day-old rat aorta [107] or rat carotid intimal thickening 15 d after endothelial injury [124].

A PDGF-like factor is also secreted by adult rat smooth-muscle cells growing in primary cultures which have been exposed to foetal calf serum for a short time [74]. Serum-free medium containing this factor initiates DNA synthesis in growth-arrested cultured smooth-muscle cells. This effect is neutralized by the addition of PDGF antibodies to the medium [74]. To date, such activities have only been described in vitro, and remain to be demonstrated in vivo.

Recently, important homologies of amino acid sequence between the simian sarcoma virus oncogene v-*sis* predicted translation product and PDGF have been reported [21,126]. The mechanisms by which this oncogene transforms cells may involve the constitutive expression of a protein with functions similar to those of a factor active transiently during normal cell growth [21]. Hybridization experiments using a DNA probe for the coding fragment of the *sis* gene have shown that this sequence is expressed at very low levels in cultured human umbilical vein endothelial cells and bovine aortic endothelial cells [2]. In addition, Collins et al. [18] have shown by cDNA cloning that the β-chain of the PDGF is expressed in cultured endothelial cells.

Besides PDGF, other growth factors are present in serum, including epidermal growth factor, somatomedins and insulin [94]. In addition, activated macrophages produce a growth factor, very similar structurally to PDGF [69,108], which stimulates fibroblasts and smooth-muscle cell replication in vitro [66]. Activated macrophages express the gene coding for the β-chain of PDGF [69, 108]. This factor could be of primary importance for the development of the atheromatous plaque in diseases such as hyperlipidaemia, where a number of macrophages accumulate in the intima [38]. Interestingly, replication of cultured smooth-muscle cells is stimulated by exposure to LDL from hyperlipidaemic animals [23]. Among the growth factors which may play a role in atheromatosis, PDGF is the better characterized. Work still needs to be done to understand the exact nature of the other factors and their eventual participation in the development of atherosclerosis.

Monoclonality of Atheromatous Smooth-Muscle Cells

In 1973, Benditt and Benditt [5] reported that in contrast to normal aortic medial smooth-muscle cells, smooth-muscle cells of atheromatous plaques were uniform as far as their content of glucose-6-phosphate dehydrogenase (G-6-PD) isoenzymes was concerned. On this basis, they suggested that atheromatous plaques are the result of monoclonal smooth-muscle cell proliferation. This observation has been confirmed by Pearson et al. [80–82] and Thomas et al. [120]. Later, however, Thomas and Kim [119] observed that among the human aortic plaque samples of their study, only 25% were monotypic for one isoenzyme. According to them, the monotypism observed in the lesions is the result of a phenotypic selection, which in fact already exists to some extent in normal tissues of heterozygotes. The same kind of selection has been observed in cultured human skin fibroblasts taken from heterozygote individuals [132], where two-thirds of the cultures became monotypic after several passages; the other third remained ditypic, but quickly became senescent and died. These considerations led Thomas and Kim [119] to favour a phenotypic selection rather than a monoclonal proliferation as responsible for the G-6-PD monotypism found in the atheromatous plaque.

The monoclonal hypothesis is compatible with a viral stimulus for atheromatous plaque formation and recently, evidence that viral mRNA can be expressed in atheromatous plaque smooth-muscle cells has been reported [6]. The study was performed on human aortic samples obtained from patients undergoing bypass surgery and on arteries from chickens infected with a variant of Marek disease virus, which belongs to the herpes virus group. The animals infected develop vascular lesions very similar to human atherosclerotic plaques. In situ hybridization experiments using DNA probes for herpes simplex virus on human samples showed positive cells mostly located in the intima in 13 of the 160 cases studied. Hybridization with probes for Epstein–Barr virus, cytomegalovirus and Marek disease virus remained negative. Positive hybridization was obtained with the probe for Marek disease virus in the intimal thickening of infected chicken arteries. The authors suggest that a viral infection of arterial smooth-muscle cells could in some cases be the trigger of intimal proliferation and produce genomic alterations responsible for clonal expansion of atherosclerotic lesions [6]. The identification of the fact that the simian sarcoma virus oncogene, v-sis predicted translation product, contains important sequence homologies with PDGF, relates atherosclerosis to tumour-like proliferative disorders [21,126].

The Role of the Macrophage

The role of macrophages in atherosclerosis is still controversial, and it is not clear whether they are responsible for the initiation, the progression or even the regression of the atheroma [89]. It has been well demonstrated that blood monocytes have the ability to adhere to the vessel wall and to migrate into the intima in various conditions of increased endothelial permeability, primarily during hyperlipidaemia [38,59,101]. Some evidence also exists for monocyte migration during hypertension [116; Kocher, unpublished observations] (Fig. 11.10).

In the intima, macrophages which possess a high-affinity receptor for acetylated LDL [42] actively phagocytose these molecules, and this results in intracellular accumulation of significant amounts of cholesterol esters. Lipid droplets appear in the cytoplasm of the macrophages, changing their morphology considerably. These modified macrophages are the major source of "foam cells", the lipid-laden cells in the atheroma [101]. Morphological distinction between monocyte- and smooth-muscle cell-derived foam cells is difficult, if not impossible. The monocyte-derived subpopulation can be recognized by several characteristics which include: (a) surface binding and phagocytosis of antibody-coated or complement-coated erythrocytes to detect specific surface receptors for the Fc portion of the IgGs [100,118] or the third

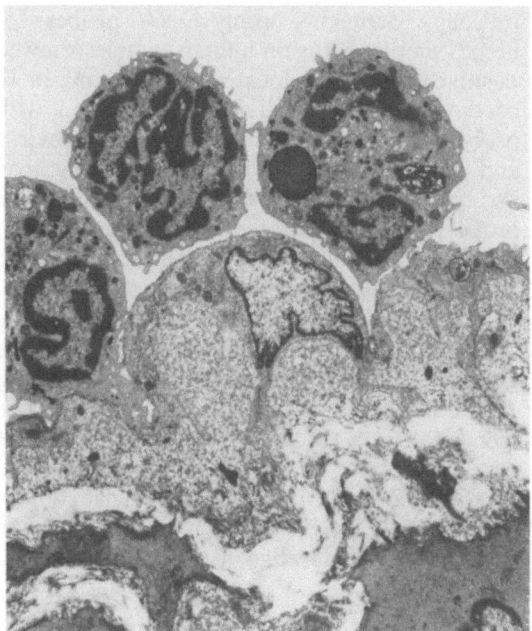

Fig. 11.10. Monocytes adhering to the endothelium in an area of increased permeability of a rat thoracic aorta during the early stage of experimentally induced hypertension. × 7000

component of the complement [101]; (b) presence of acid lipase activity [99] and intracellular lysozyme-like antigen [101]; and (c) rapid adhesion to glass [101]. In addition, blood monocytes exhibit both non-specific esterase and peroxidase activities; peroxidase activity is lost when monocytes enter the tissues and become actively phagocytic [38]. Foam cells derived from smooth-muscle cells do not show any of these characteristics.

More recently, monoclonal antibodies specific for macrophages have been used to study the role and behaviour of these cells in the vessel walls of hypercholesterolaemic rabbits [125]. During the very early phases of hyperlipidaemia, circulating monocytes adhere to the endothelium before any lesion is visible and migrate into the intima. Macrophage-derived foam cells are practically the only component of the early intimal thickening, as shown by means of macrophage antibody staining. However, in advanced lesions, foam cells which do not react with the antibody, presumably derived from smooth-muscle cells, are predominant in the deep portions of the plaques, whereas the superficial layer still contains macrophage-derived foam cells [125]. Seventy-four weeks after termination of the

cholesterol-rich diet, the number of cells with specific macrophage staining is markedly diminished in the plaques.

Watanabe and collaborators [125] consider the macrophage as a scavenger cell which prevents lipid accumulation in the intima and progression of the atherosclerotic lesions. Once lipid-laden, foam cells migrate back into the bloodstream by crossing the arterial endothelium. in the late fatty lesions, the ratio of penetrating monocytes to foam cells leaving the lesion is close to one, and the lesion ceases to progress [39]. During hyperlipidaemia, the macrophage appears to clear lipids and to prevent their accumulation in the artery wall. According to Gerrity [39], the migration of smooth-muscle cells in advanced fibrosclerotic plaques is the result of a relative failure of macrophages to clear sufficient lipids. Thus, in chronic hyperlipidaemia, fatty streak lesions could be precursors of atheromatous plaques. However, the presence of macrophages in the atheromatous plaque is perhaps not only related to a scavenger activity. In vitro experiments have shown that these cells produce a growth factor for smooth-muscle cells and fibroblasts [96]. Macrophage-conditioned medium added to wound fibroblasts cultured in medium containing plasma-derived serum induces DNA synthesis [66]. Recent experiments have shown that this macrophage-derived growth factor is structurally very similar to PDGF [108], By secreting growth factors, the macrophage could be involved in the progression of the atheromatous plaque.

Conclusions

We have presented a selection of recent findings relevant to the understanding of the pathogenetic mechanisms leading to atheromatous plaque formation. In conclusion, we would like to stress some points which we feel would be useful for the planning of future experimentation.

First, it appears that discontinuity of the endothelial cell layer is not a necessary prerequisite for the formation of the atheromatous plaque. This negative conclusion derives from careful observations (not always easy to achieve) of early human atheromatous lesions [49,130], of experimental lesions either during hyperlipidaemia or hypertension [19,41,59] and from elegant studies on the influence of the size of experimental endothelial injury in the production of smooth-muscle cell proliferation [83]. If this conclusion is true, the role of platelet adhesion to arterial wall as a pathogenetic factor for atherosclerosis is less important than has been thought since the first description of PDGF [91,92]. If the pathogenetic role of platelet adhesion becomes less relevant, the role of PDGF itself remains important, particularly since the discovery that PDGF can be present in several cell types that are important for the development of atheromatous lesions, such as macrophages, endothelial cells and smooth-muscle cells themselves. Endothelial cells appear to produce PDGF in culture but only in very small amounts in vivo (2). It would be of interest to study the production of PDGF by regenerating or dividing endothelial cells in vivo, since these acquire several features similar to those observed in endothelial cells during culture [33].

Macrophages appear to be good candidates for the role of stimulators of smooth-muscle cell growth in view of the observation that they may be present very early in arterial intima during experimental hyperlipidaemia [39,41,59]. Further studies of the distribution of these cells during situations other than hyperlipidaemia which predispose to atheromatous lesions will clarify this point.

The central question of atherosclerosis research is what stimulates movement and replication of smooth-muscle cells. Factors stimulating these activities may be present in plasma, either originating from circulating cells (PDGF) or as plasma components (lipoproteins). In all these cases, variations in endothelial permeability may play a permissive role in the accumulation of mitogenic substances in the intima. Hence the study of the barrier function of endothelial cells under experimental or pathological conditions should give important information concerning the mechanisms predisposing to atheroma formation.

The work of Benditt and Benditt [5], of Thomas et al. [120] and of other laboratories [34,61,106] concerning the clonal selection of atheromatous smooth-muscle cells raises the question of whether all smooth-muscle cells are similar or whether a small population in the arterial wall is more growth susceptible. This subpopulation would thus be selectively responsible for the onset of the atheromatous response. It is generally accepted that the morphological features of arterial smooth-muscle cells evolve with the age of the individual, but remain homogeneous within the smooth-muscle cell population of a given vessel in individuals of the same

age [40,62,75]. However, previous work has shown that a certain degree of heterogeneity exists within a population of arterial smooth-muscle cells of any given animal. This heterogeneity concerns the content of angiotensin receptors [79], the antigenic expression of myosin at different levels of the arterial tree [64] and the presence of vimentin or desmin as the main intermediate filament component of smooth-muscle cells, either at different levels of the arterial tree [78] or within the same arterial segments [26,27].

The structural variability of arterial smooth-muscle cells under normal conditions may be due to the different mechanical properties and hormone responsiveness of different arterial vessels [123]. Changes in the cytoskeletal composition of smooth-muscle cells are also observed during development [62], in response to experimental endothelial injury [61], in formation of atheromatous lesions in man [60] and in tissue culture [111]. To date, no population of smooth muscle cells selectively involved in atheroma formation has been identified but further studies in this area may provide useful information on the mechanisms of atheroma development [106].

In conclusion, although the pathogenesis of atheroma remains mysterious, many recent and stimulating observations have indicated new areas for research which appear promising for the solution of one of the most important problems in human pathology.

Acknowledgements. This chapter has been supported in part by the Swiss National Science Foundation, Grant no. 3.107-0.85. We thank Mrs. M.M. Rossire for typing the manuscript.

References

1. Antoniades HN, Stathakos D, Scher CD (1975) Isolation of a cationic polypeptide from human serum that stimulates proliferation of 3T3 cells. Proc Natl Acad Sci USA 72:2635–2639
2. Barrett TB, Gajdusek CM, Schwartz SM et al. (1984) Expression of the *sis* gene by endothelial cells in culture and in vivo. Proc Natl Acad Sci USA 81:6772–6774
3. Baumgartner HR, Studer A (1966) Folgen des Gefässkathetarismus am normo- und hypercholesterinaemischen Kaninchen. Pathol Microbiol 29:393–405
4. Benditt EP (1977) The origin of atherosclerosis. Sci Am 236:74–85

5. Benditt EP, Benditt JM (1973) Evidence for a monoclonal origin of human atherosclerotic plaques. Proc Natl Acad Sci USA 70:1753–1756

6. Benditt EP, Barrett T, McDougall JK (1983) Viruses in the etiology of atherosclerosis. Proc Natl Acad Sci USA 80:6386–6389

7. Bockus BJ, Stiles CD (1984) Regulation of cytoskeletal architecture by platelet-derived growth factor, insulin and epidermal growth factor. Exp Cell Res 153:186–197

8. Booyse FM, Osikowicz G, Quarfoot AJ (1981) Effects of chronic oral consumption of nicotine on the rabbit aortic endothelium. Am J Pathol 102:229–238

9. Bowen-Pope D, Ross R (1982) Platelet-derived growth factor. II. Specific binding to cultured cells. J Biol Chem 257:5161–5171

10. Brown D, Montesano R, Orci L (1982) Patterns of filipin-sterol complex distribution in intact erythrocytes and intramembrane particle-aggregated ghost membranes. J Histochem Cytochem 30:702–706

11. Burns ER, Spaet TH, Stemerman MB (1978) Response of the arterial wall to endothelial removal: an autoradiographic study. Proc Soc Exp Biol Med 159:473–477

12. Castellot Jr JJ, Addonizio ML, Rosenberg R et al. (1981) Cultured endothelial cells produce a heparin-like inhibitor of smooth muscle cell growth. J Cell Biol 90:372–379

13. Castellot Jr JJ, Favreau LV, Karnovsky MJ et al. (1982) Inhibition of vascular smooth muscle cell growth by endothelial cell-derived heparin. Possible role of a platelet endoglycosidase. J Biol Chem 257:11256–11260

14. Chait A, Ross R, Albers JJ et al. (1980) Platelet-derived growth factor stimulates activity of low density lipoprotein receptors. Proc Natl Acad Sci USA 77:4084–4088

15. Chamley-Campbell J, Campbell GR, Ross R (1979) The smooth muscle cell in culture. Physiol Rev 59:1–61

16. Clowes AW, Karnovsky MJ (1977) Suppression by heparin of smooth muscle cell proliferation in injured arteries. Nature 265:625–626

17. Clowes AW, Reidy MA, Clowes MM (1983) Kinetics of cellular proliferation after arterial injury. I. Smooth muscle growth in the absence of endothelium. Lab Invest 49:327–333

18. Collins T, Ginsburg D, Boss JM et al. (1985) Cultured human endothelial cells express platelet-derived growth factor B chain: cDNA cloning and structural analysis. Nature 316:748–750

19. De Chastonay C, Gabbiani G, Elemer G et al. (1983) Remodelling of the rat aortic endothelial layer during experimental hypertension. Changes in replication rate, cell density, and surface morphology. Lab Invest 48:45–52

20. DiCorleto PE, Bowen-Pope DF (1983) Cultured endothelial cells produce a platelet-derived growth factor-like protein. Proc Natl Acad Sci USA 80:1919–1923

21. Doolittle RF, Hunkapiller MW, Hood LE et al. (1983) Simian sarcoma virus onc gene, v-sis, is derived from the gene (or genes) encoding a platelet-derived growth factor. Science 221:275–277

22. Farber JL, Chien KR, Mittnacht S (1981) The pathogenesis of irreversible cell injury in ischemia. Am J Pathol 102:271–281

23. Fischer-Dzoga K, Wissler RW (1983) The latest evidence, in vitro and in vivo, on controlling smooth muscle cell proliferation and arterial wall cholesterol ester accumulation by controlling LDL concentration. In: Hauss WH, Wissler RW (eds) Clinical implications of recent research results in arteriosclerosis. Westdeutscher Verlag, Opladen, p 369

24. Fishman JA, Ryan GB, Karnovsky MJ (1975) Endothelial regeneration in the rat carotid artery and the significance of endothelial denudation in the pathogenesis of myointimal thickening. Lab Invest 32:339–351

25. Fox PL, DiCorleto PE (1984) Regulation of a production of a platelet-derived growth factor-like protein by cultured bovine aortic endothelial cells. Cell Physiol 121:298–308

26. Frank ED, Warren L (1981) Aortic smooth-muscle cells contain vimentin instead of desmin. Proc Natl Acad Sci USA 78:3020–3024

27. Gabbiani G, Kocher O (1983) Cytocontractile and cytoskeletal elements in pathologic processes. Pathogenetic role and diagnostic value. Arch Pathol Lab Med 107:622–625

28. Gabbiani G, Ryan GB, Majno G (1971) Presence of modified fibroblasts in granulation tissue and their possible role in wound contraction. Experientia 27:549–550

29. Gabbiani G, Badonnel MC, Rona G (1975) Cytoplasmic contractile apparatus in aortic endothelial cells of hypertensive rats. Lab Invest 32:227–234

30. Gabbiani G, Elemer G, Guelpa CH et al. (1979) Morphologic and functional changes of the aortic intima during experimental hypertension. Am J Pathol 96:399–422

31. Gabbiani G, Schmid E, Winter S et al. (1981) Vascular smooth muscle cells differ from other smooth muscle cells: predominance of vimentin filaments and a specific alpha-type actin. Proc Natl Acad Sci USA 78:298–302

32. Gabbiani G, Kapanci Y, Barazzone P et al. (1981) Immunochemical identification of intermediate-sized filaments in human neoplastic cells. A diagnostic aid for the surgical pathologist. Am J Pathol 104:206–216

33. Gabbiani G, Gabbiani F, Lombardi D et al. (1983) Organization of actin cytoskeleton in normal and regenerating arterial endothelial cells. Proc Natl Acad Sci USA 80:2361–2364

34. Gabbiani G, Kocher O, Bloom WS et al. (1984) Actin expression in smooth muscle cells of rat aortic intimal thickening, human atheromatous plaque, and cultured rat aortic media. J Clin Invest 73:148–152

35. Gajdusek CM, DiCorleto P, Ross R et al. (1980) An endothelial cell derived growth factor. J Cell Biol 85:467–472

36. Gaynor E (1971) Increased mitotic activity in rabbit endothelium after endotoxin: An autoradiographic study. Lab Invest 24:318–320

37. Geer JC, Haust DM (1972) Smooth muscle cells in atherosclerosis. Karger, Basel (Monographs on atherosclerosis, vol 2)

38. Gerrity RG (1981) The role of the monocyte in atherogenesis. I. Transition of blood-borne monocytes into foam cells in fatty lesions. Am J Pathol 103:181–190

39. Gerrity RG (1981) The role of the monocyte in atherogenesis. II. Migration of foam cells from atherosclerotic lesions. Am J Pathol 103:191–200

40. Gerrity RG, Cliff WJ (1975) The aortic tunica media of the developing rat. I. Quantitative stereologic and biochemical analysis. Lab Invest 32:585–600

41. Gerrity RG, Naito HK, Richardson M et al. (1979) Dietary induced atherogenesis in swine: morphology of

the intima in prelesion stages. Am J Pathol 95:775–786

42. Goldstein JL, Ho YK, Basu SK et al. (1979) Binding site on macrophages that mediates uptake and degradation of acetylated low density lipoprotein, producing massive cholesterol deposition. Proc Natl Acad Sci USA 76:333–337

43. Gotlieb AI (1982) Smooth muscle and endothelial cell function in the pathogenesis of atherosclerosis. Can Med Assoc J 126:903–908

44. Grotendorst GR, Seppa HEJ, Kleinman HK et al. (1981) Attachment of smooth muscle cells to collagen and their migration toward platelet-derived growth factor. Proc Natl Acad Sci USA 78:3669–3672

45. Guyton JR, Rosenberg RD, Clowes AW et al. (1980) Inhibition of rat arterial smooth-muscle cell proliferation by heparin. In vivo studies with anticoagulant heparin. Circ Res 46:625–634

46. Hansson GK, Schwartz SM (1983) Evidence for cell death in the vascular endothelium in vivo and in vitro. Am J Pathol 112:278–286

47. Hansson GK, Starkenbaum GA, Benditt EP et al. (1984) Fc-mediated binding of IgG to vimentin-type intermediate filaments in vascular endothelial cells. Proc Natl Acad Sci USA 81:3103–3107

48. Hassler O (1970) The origin of the cells constituting arterial intima thickening. An experimental autoradiographic study with the use of H^3-thymidine. Lab Invest 22:286–293

49. Haust MD (1978) Light and electron microscopy of human atherosclerotic lesions. Adv Exp Med Biol 104:33–59

50. Heggenes MH, Ash JF, Singer SJ (1978) Transmembrane linkage of fibronectin in intracellular actin-containing filaments in cultured human fibroblasts. Ann NY Acad Sci 312:414–417

51. Heldin CH, Westermark B, Wasteson A (1979) Platelet-derived growth factor: purification and partial characterization. Proc Natl Acad Sci USA 76:3722–3726

52. Herman B, Pledger WJ (1985) Platelet-derived growth factor-induced alterations in vinculin and actin distribution in BALB/c-3T3 cells. J Cell Biol 100:1031–1040

53. Hüttner I, Gabbiani G (1983) Vascular endothelium in hypertension. In: Genest J, Kuchel O, Hamet P, Cantin M (eds) Hypertension, 2nd edn. McGraw-Hill, New York, pp 473–488

54. Hüttner I, Mo Costabella P, De Chastonay C et al. (1982) Volume, surface, and junctions of rat aortic endothelium during experimental hypertension. A morphometric and freeze fracture study. Lab Invest 46:489–504

55. Hüttner I, Walker C, Gabbiani G (1985) Aortic endothelial cell during regeneration. Remodeling of cell junctions, stress fibers, and stress fiber-membrane attachment domains. Lab Invest 53:287–302

56. Jaffe EA, Hoyer LW, Nachman RL (1974) Synthesis of von Willebrand factor by cultured human endothelial cells. Proc Natl Acad Sci USA 71:1906–1909

57. Jellinek H, Nagy Z, Hüttner I et al. (1969) Investigations of the permeability changes of the vascular wall in experimental malignant hypertension by means of a colloidal iron preparation. Br J Exp Pathol 50:13–16

58. Jonasson L, Holm J, Skalli O et al. (1985) Expression of class II transplantation antigen on vascular smooth muscle cells in human atherosclerosis. J Clin Invest 76:125–131

59. Joris I, Zand T, Nunnari JJ et al. (1983) Studies on the pathogenesis of atherosclerosis. I. Adhesion and migration of mononuclear cells in the aorta of hypercholesterolemic rats. Am J Pathol 113:341–358

60. Kocher O, Gabbiani G (1986) Cytoskeletal features of human normal and atheromatous arterial smooth muscle cells. Hum Pathol 17:875–880

61. Kocher O, Skalli O, Bloom WS et al. (1984) Cytoskeleton of rat aortic smooth muscle cells. Normal conditions and experimental intimal thickening. Lab. Invest 50:645–652

62. Kocher O, Skalli O, Cerutti D et al. (1985) Cytoskeletal features of rat aortic cells during development. An electron microscopic, immunohistochemical, and biochemical study. Circ Res 56:829–838

63. Kohler N, Lipton A (1974) Platelets as a source of fibroblast growth-promoting activity. Exp Cell Res 87:297–301

64. Larson DM, Fujiwara K, Alexander RW et al. (1984) Heterogeneity of myosin antigenic expression in vascular smooth-muscle in vivo. Lab Invest 50:401–407

65. Lazarides E (1980) Intermediate filaments as mechanical integrators of cellular space. Nature 283:249–256

66. Leibovitch SJ, Ross R (1976) A macrophage-dependent factor that stimulates the proliferation of fibroblasts in vitro. Am J Pathol 84:501–514

67. Linder E, Helin H, Chang CM et al. (1983) Complement-mediated binding of monocytes to intermediate filaments in vitro. Am J Pathol 112:267–277

68. Loskutoff DJ, Edington TS (1977) Synthesis of a fibrinolytic activator and inhibitor by endothelial cells. Proc Natl Acad Sci USA 74:3903–3907

69. Martinet Y, Bitterman PB, Mornex JF et al. (1986) Activated human monocytes express the c-sis proto-oncogene and release a mediator showing PDGF-like activity. Nature 319:158–160

70. Mellström K, Höglund AS, Nister M et al. (1983) The effect of platelet-derived growth factor on morphology and motility of human glial cells. J Musle Res Cell Motil 4:589–609

71. Minick CR, Stemerman MB, Insull W Jr (1977) Effect of regenerated endothelium on lipid accumulation in the arterial wall. Proc Natl Acad Sci USA 74:1724–1728

72. Moncada S, Vane JR (1979) Arachidonic acid metabolites and the interactions between platelets and blood-vessel walls. N Engl J Med 300:1142–1147

73. Nagle RB, McDaniel KM, Clark VA et al. (1983) The use of antikeratin antibodies in the diagnosis of human neoplasms. Am J Clin Pathol 79:458–466

74. Nilsson J, Sjölund M, Palmberg L et al. (1985) Arterial smooth-muscle cells in primary culture produce a platelet-derived growth factor-like protein. Proc Natl Acad Sci USA 82:4418–4422

75. Olivetti G, Anversa P, Melissari M et al. (1980) Morphometric study of early postnatal development of the thoracic aorta in the rat. Circ Res 47: 417–424

76. Osborn M, Weber K (1982) Intermediate filaments: cell-type-specific markers in differentiation and pathology. Cell 31:303–306

77. Osborn M, Weber K (1983) Biology of disease. Tumor diagnosis by intermediate filament typing: a novel tool for surgical pathology. Lab Invest 48:372–394

78. Osborn M, Caselitz J, Weber K (1981) Heterogeneity of intermediate filament expression in vascular smooth-muscle: a gradient in desmin positive cells from the rat aortic arch to the level of the arteria iliaca communis.

Differentiation 20:196–202

79. Paglin S, Stukenbrok H, Jamieson JD (1984) Characterization of rabbit aortic smooth muscle cells grown in culture. J Cell Biol 99:28a

80. Pearson TA, Wang A, Solez K et al. (1975) Clonal characteristics of fibrous plaques and fatty streaks from human aortas. Am J Pathol 81:379–387

81. Pearson TA, Dillman JM, Solez K et al. (1978) Clonal characteristics in layers of human atherosclerotic plaques. Am J Pathol 93:93–102

82. Pearson TA, Dillman JM, Solez K et al. (1978) Clonal markers in the study of the origin and growth of human atherosclerotic lesions. Circ Res 43:10–18

83. Reidy MA (1985) A reassessment of endothelial injury and arterial lesion formation. Lab Invest 53:513–520

84. Reidy MA, Schwartz SM (1981) Endothelial regeneration. III. Time course of intimal changes after small defined injury to rat aortic endothelium. Lab Invest 44:301–308

85. Reidy MA, Schwartz SM (1983) Endothelial injury and regeneration. IV. Endotoxin – a non-denuding injury to aortic endothelium. Lab Invest 48:25–34

86. Reidy MA, Clowes AW, Schwartz SM (1983) Endothelial regeneration. V. Inhibition of endothelial regrowth in arteries of rat and rabbit. Lab Invest 49:569–575

87. Richardson M, Ihnatowycz I, Moore S (1980) Glycosaminoglycan distribution in rabbit aortic wall following balloon catheter deendothelialization: an ultrastructural study. Lab Invest 43:509–516

88. Rojo-Ortega JM, Genest J (1968) A method for production of experimental hypertension in rats. Can J Physiol Pharmacol 46:883–885

89. Ross R (1981) Atherosclerosis: a problem of the biology of arterial wall cells and their interactions with blood components. Arteriosclerosis 1:293–311

90. Ross R (1986) The pathogenesis of atherosclerosis – an update. N Engl J Med 314:488–500

91. Ross R, Glomset JA (1973) Atherosclerosis and the arterial smooth muscle cell. Science 180:1332–1339

92. Ross R, Glomset JA (1976) The pathogenesis of atherosclerosis (first of two parts). N Engl J Med 295:369–377

93. Ross R, Harker L (1976) Hyperlipidemia and atherosclerosis. Science 193:1094–1100

94. Ross R, Vogel A (1978) The platelet-derived growth factor. Cell 14:203–210

95. Ross R, Glomset J, Kariya B et al. (1974) A platelet-dependent serum factor that stimulates the proliferation of arterial smooth muscle cells in vitro. Proc Natl Acad Sci USA 71:1207–1210

96. Ross R, Kariya B, Vogel A et al. (1979) Cell proliferation: platelet- and macrophage-derived growth factor. Adv Inflam Res 1:183–187

97. Ross R, Wight TN, Strandness E et al. (1984) Human atherosclerosis. I. Cell constitution and characteristics of advanced lesions of the superficial femoral artery. Am J Pathol 114:79–93

98. Rungger-Brändle E, Gabbiani G (1983) The role of cytoskeletal and cytocontractile elements in pathologic processes. Am J Pathol 110:359–392

99. Schaffner T, Bekermeier M, Fischer-Dzoga K (1976) Histochemical localization of acid lipase in developing primate atheroma. Circulation 54 [Suppl 2]:85

100. Schaffner T, Vesselinovitch DV, Wissler RW (1979) Macrophages in experimental and human atheromatous lesions: immunomorphologic identification. Fed Proc 38:1076

101. Schaffner T, Taylor K, Bartucci EJ et al. (1980) Arterial foam cells with distinctive immunomorphologic and histochemical features of macrophages. Am J Pathol 100:57–80

102. Schwartz SM, Benditt EP (1976) Clustering of replicating cells in aortic endothelium. Proc Natl Acad Sci USA 73:651–653

103. Schwartz SM, Benditt EP (1977) Aortic endothelial cell replication. I. Effects of age and hypertension in the rat. Circ Res 41:248–255

104. Schwartz SM, Gajdusek CM, Selden SC III (1981) Vascular wall growth control: the role of the endothelium. Arteriosclerosis 1:107–126

105. Schwartz SM, Haudenschild CC, Eddy EM (1978) Endothelial regeneration. I. Quantitative analysis of intimal stages of endothelial regeneration in rat aortic intima. Lab Invest 38:569–580

106. Schwartz SM, Reidy MR, Clowes A (1985) Kinetics of atherosclerosis: a stem cell model. Ann NY Acad Sci 454:292–304

107. Seifert RA, Schwartz SM, Bowen-Pope DF (1984) Developmentally regulated production of platelet-derived growth factor-like molecules. Nature 311:669–671

108. Shimokado K, Raines EW, Madtes DK et al. (1985) A significant part of macrophage-derived growth factor consists of at least two forms of PDGF. Cell 43:277–286

109. Skalli O, Gabbiani G (1988) The biology of the myofibroblast in relation to wound contraction and fibrocontractive diseases. In: Clark RAF, Henson P (eds) The molecular and cellular biology of wound repair. Plenum, New York, pp 373–402

110. Skalli O, Gabbiani G, Vandekerckhove J et al. (1984) The fibroblastic nature of myofibroblasts suggested by their actin isoform pattern. Fed Proc 43:802

111. Skalli O, Bloom WS, Ropraz P et al. (1986) Cytoskeletal remodelling of rat aortic smooth muscle cells "in vitro": Relationships to culture conditions and analogies to "'in vivo" situations. J Submicrosc Cytol 18:481–493

112. Spagnoli LG, Pietra GG, Villaschi S et al. (1982) Morphometric analysis of gap junctions in regenerating arterial endothelium. Lab Invest 46:139–148

113. Steinert PM, Jones JCR, Goldman RD (1984) Intermediate filaments. J Cell Biol 99:22s–27s

114. Stemerman MB (1981) Effects of moderate hypercholesterolemia on rabbit endothelium. Arteriosclerosis 1:25–32

115. Stemerman MB, Ross R (1972) Experimental arteriosclerosis: Fibrous plaque formation in primates, an electron microscope study. J Exp Med 136:769–789

116. Still WJS (1968) The pathogenesis of the intimal thickenings produced by hypertension in large arteries in the rat. Lab Invest 19:84–91

117. Tamm SL, Tamm S (1983) Distribution of sterol-specific complexes in a continually shearing region of a plasma membrane and at procaryotic–eucaryotic cell junctions. J Cell Biol 97:1098–1106

118. Taylor K, Schaffner T, Wissler RW et al. (1979) Immunomorphologic identification and characterization of cells derived from experimental atherosclerotic lesions. Scan Electron Microsc 3:815–822

119. Thomas WA, Kim DN (1983) Atherosclerosis as a hyperplastic and/or neoplastic process. Lab Invest 48:245–255

120. Thomas WA, Reiner JM, Janakidevi K et al. (1979) Population dynamics of arterial cells during athero-

genesis. X. Study of monotypism in atherosclerotic lesions of black women heterozygous for glucose-6-phosphate dehydrogenase (G-6-PD). Exp Mol Pathol 31:367–386

121. Thorgeirsson G, Robertson AL (1978) The vascular endothelium. Pathologic significance: a review. Am J Pathol 93:803–848

122. Vandekerckhove J, Weber K (1978) At least six different actins are expressed in higher mammals: an analysis based on the amino acid sequence of the amino-terminal tryptic peptide. J Mol Biol 126:783–802

123. Vanhoutte PM (1978) Heterogeneity in vascular smooth-muscle. In: Kaley G, Altura BM (eds) Microcirculation, vol II. University Park Press, Baltimore, pp 181–309

124. Walker LN, Bowen-Pope DF, Reidy MA (1984) Secretion of platelet-derived growth factor (PDGF)-like activity by arterial smooth-muscle cells is induced as a response to injury. J Cell Biol 99:416a

125. Watanabe T, Hirata M, Yoshikawa Y et al. (1985) Role of macrophages in atherosclerosis. Sequential observations of cholesterol-induced rabbit aortic lesion by the immunoperoxidase technique using monoclonal antimacrophage antibody. Lab Invest 53:80–90

126. Waterfield MD, Scrace GT, Whittle N et al. (1983) Platelet-derived growth factor is structurally related to the putative transforming protein p28^{sis} of simian sarcoma virus. Nature 304:35–39

127. Westermark B, Wasteson A: A platelet factor stimulating human normal glial cells. Exp Cell Res 98:170–174

128. White GE, Gimbrone Jr MA, Fujiwara K (1983) Factors influencing the expression of stress fibers in vascular endothelial cells in situ. J Cell Biol 97:416–424

129. Wiener J, Giacomelli F (1980) Morphogenesis of hypertensive vascular disease. In: Kaley G, Altura BM (eds) Microcirculation, vol 3. University Park Press, Baltimore, pp 187–222

130. Wissler RW (1984) Principles of the pathogenesis of atherosclerosis. In: Braunwald E (ed) Heart disease: a textbook of cardiovascular medicine, 2nd Edn. Saunders, Philadelphia, pp 1183–1204

131. Wong AJ, Pollard TD, Herman IM (1983) Actin filament stress fibers in vascular endothelial cells in vivo. Science 219:867–869

132. Zavala C, Herner G, Fialkow PJ (1978) Evidence for selection in cultured diploid fibroblast strains. Exp Cell Res 117:137–144

Chapter 12

Extracellular Matrix Changes in Atherosclerosis

T. N. Wight

A number of major events characterize the genesis of the atherosclerotic lesion. These events include the proliferation of intimal smooth-muscle cells, accumulation of lipid and extracellular matrix, calcification and thrombosis. Although the initiating factors are still not completely understood, it is clear that the deposited extracellular matrix affects all of these events. There are two fundamental questions concerning the role of the extracellular matrix components in the genesis of the atherosclerotic plaque. The first is whether specific sets of extracellular matrix (ECM) molecules regulate certain processes associated with the genesis of the lesion. The second question is why extracellular matrix accumulates within the lesion as the disease progresses. These two questions, which deal with reciprocal events, are the focus of much of the intense research that is being done to define further the nature of this major disease of man.

Collagens

Collagens constitute a major class of protein within blood vessels and accumulate during the formation of the atherosclerotic plaque [2,85]. They represent approximately one-third of the dry weight and 60% of the total protein content of the advanced atherosclerotic plaque and are a prominent morphological component (Fig. 12.1). The major collagen type present is type I (approximately 60% of the total) with lesser amounts of types III, IV and V [see reviews

2,53]. It appears that type I is the major collagen type in normal blood vessels [27,58], indicating that significant shifts in the proportions of collagens I and III do not occur during atherogenesis as was originally thought [54]. Type V collagen has also been shown to increase in atherosclerosis [58,66]. This collagen type has been described as a pericellular collagen surrounding arterial smooth-muscle cells [24] and it is intimately associated with type I and III collagen fibrils [52]. Increases in type V collagen have also been observed during wound healing and in inflammation suggesting that this extracellular matrix change may be indicative of a response to injury [19,63]. It is interesting to note in this regard that arterial basement membranes, present below endothelial cells and surrounding smooth-muscle cells, are markedly thickened in atherosclerotic arteries (Fig. 12.2), although no increases in basement-membrane-specific molecules such as type IV collagen have been observed. The functional significance of hypertrophic basement membranes in atherosclerosis is still not understood.

The causes of collagen accumulation within the atherosclerotic plaque are unknown but several "atherogenic factors" have been shown to influence arterial wall collagen metabolism. Increased arterial collagen deposition has been observed in hypertension [64], during immunological injury [76] and in experimental injury, either mechanical (as with a balloon catheter) or chemical (as in chronic lipid feeding) [2,18,22,44,55]. In addition, cell-culture studies have demonstrated increased collagen synthesis when arterial smooth-muscle cells are stimulated

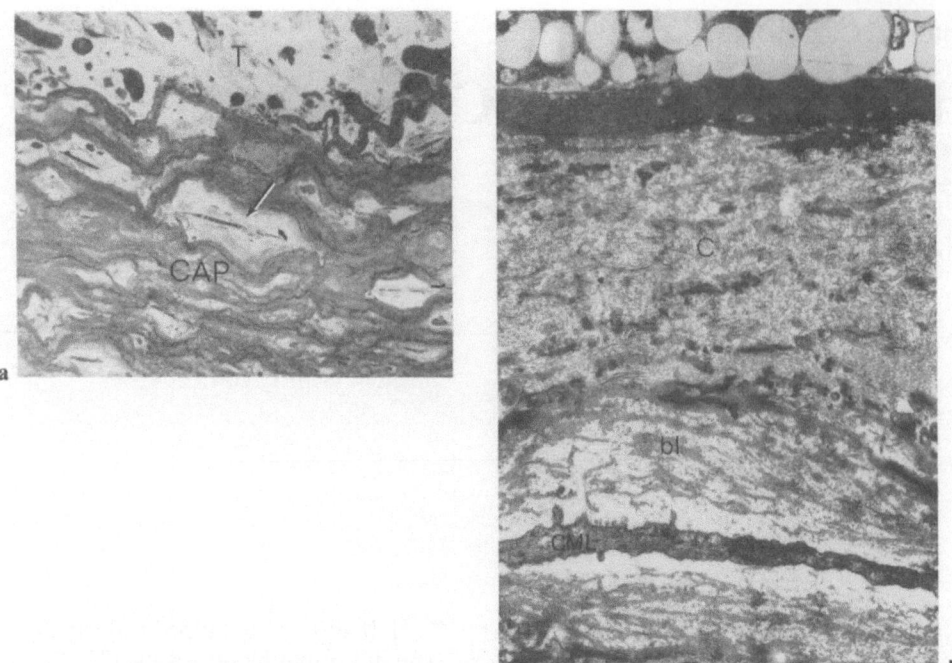

Fig. 12.1. a A light micrograph showing the interface between a partially organized thrombus (*T*) and the adjacent fibrous cap (*CAP*) of a human atherosclerotic lesion. The cap contains a number of elongated cells surrounded by connective tissue (*arrow*). × 170. **b** An electron micrograph of a similar area to that shown in **a**, demonstrating that the interface between the fibrous cap and thrombus consists of dark-staining amorphous material and dense connective tissue consisting primarily of collagen fibrils (*C*). A portion of a smooth-muscle cell (*CML*) surrounded by extensive basal lamina-like material (*bl*) is present in the fibrous cap. × 10 000

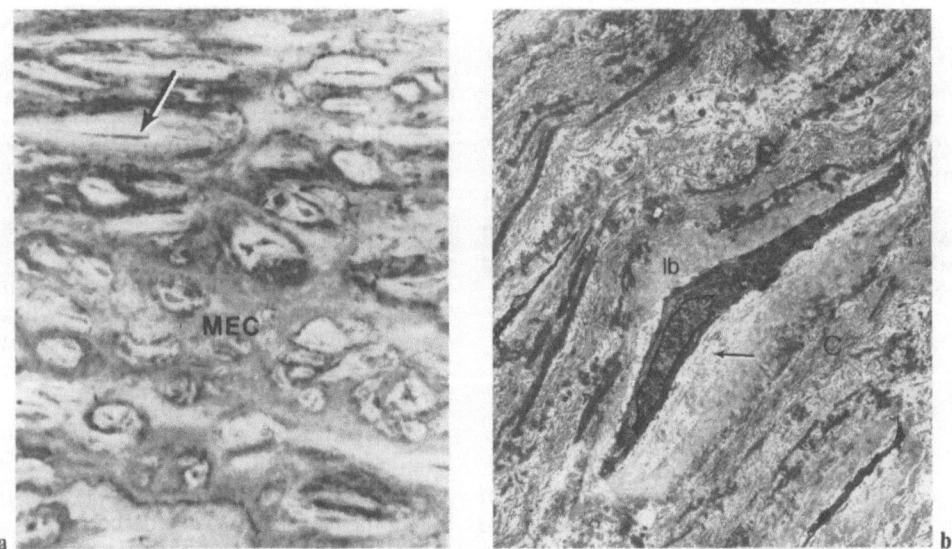

Fig. 12.2. a A light micrograph of a portion of a fibrous cap from a human atherosclerotic lesion of the superficial femoral artery. The "cap" consists of a number of elongated cells (*arrows*), lying in a cavity or lacuna space. The cells are separated by dense extracellular matrix (*MEC*). × 125. **b** An electron micrograph of a region of the cellular cap of a human superficial femoral artery lesion illustrating that the elongated cells are smooth-muscle cells surrounded by an extensively thickened basal lamina (*lb*). The cavity in which the cells sit also contains a number of proteoglycan granules (*arrow*). Dense connective tissue is present between cells and consists of collagen fibrils (*C*) and dark-staining, partially fragmented elastic fibres (*E*). × 53 000

to divide, indicating that the proliferative phase of atherosclerosis may be accompanied by increased collagen synthesis [8,28,49]. Cultured smooth-muscle cells from atherosclerotic rabbit aortas were also found to have increased collagen biosynthesis when compared to smooth-muscle cells from normal aortas [70]. Additional studies are needed to determine the molecular mechanism(s) responsible for altered arterial collagen metabolism during events associated with the genesis of the atherosclerotic lesion.

The consequence of arterial collagen accumulation within atherosclerotic arteries are severalfold. First, an increase in this fibrous protein can potentially lead to an increase in arterial wall "stiffness" due to changes in mechanical properties (see Chap. 3). An increase in vascular resistance could lead to hypertensive conditions and may predispose the arterial wall to rupture and/or ulceration. In addition, the activation of platelets by the fibrillar collagens (I and III) which occurs during the normal process of wound healing [1] may have adverse consequences within the arterial wall. Platelets and their products have been implicated in the early fibromuscular phases of lesion formation as well as in later events leading to thrombosis [for reviews, 77,78]. For example, the release of platelet mitogens such as platelet-derived growth factor (PDGF) in response to collagen could cause medial smooth-muscle cells to migrate from the media to the intima since PDGF has been shown to be both chemotactic and mitogenic for arterial smooth-muscle cells [25,77,78]. Exposure of circulating platelets to arterial wall collagen could occur as a result of interruption of the endothelial cell layer caused either by endothelial injury and desquamation [82] or by the splitting of the arterial wall as a consequence of excessive collagen accumulation. Fibrillar collagen is abundant within the fibromuscular cap adjacent to developing thrombi (Fig. 12.1b).

In addition to influencing platelet deposition and thrombus formation, collagen may also influence the amount of lipid deposited within atherosclerotic arteries. For example, accelerated atherosclerotic disease in coronary, cerebral and peripheral arteries occurs frequently in patients with diabetes mellitus, even though most of these patients have plasma cholesterol and low-density lipoprotein (LDL) levels that are within the normal range [5]. It is known that as glucose builds up within diabetic tissue, it can bind non-enzymatically to collagenous proteins. These glycoprotein end products are capable of reacting with additional amino groups on other proteins such as LDL to form intermolecular cross-links [7]. Such cross-linking may result in the trapping of LDL within the collagen-rich matrix via hyperglycaemia-induced glycosylation end products.

Proteoglycans

The major families of arterial proteoglycans, chondroitin sulphate proteoglycan (CSPG), dermatan sulphate proteoglycan (DSPG) and heparan sulphate proteoglycan (HSPG), accumulate in both spontaneous and experimental atherosclerosis (see reviews 3,9,91). Generally, these studies demonstrate that proteoglycans accumulate only in the early lesions of blood vessels and specifically within the intima (Figs 12.3 and 12.4) but decrease in

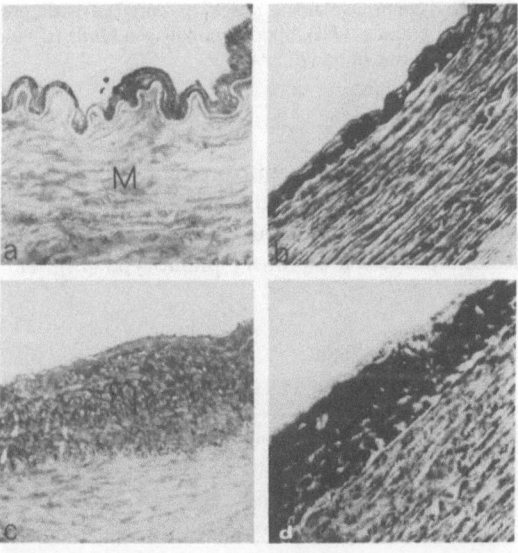

Fig. 12.3. a This light micrograph illustrates that the narrow intima of a normal blood vessel stains more intensely with alcian blue, indicative of proteoglycan, than the underlying medial layer (*M*). × 250. **b** This section was immunostained with a monoclonal antibody against aortic CSPG using an indirect peroxidase method. Note that the normal narrow intima is more intensely stained than the underlying medial layer. × 250. **c** This section was taken from a fibromuscular hyperplastic intima (early atherosclerosis) and demonstrates intense alcian blue staining in the thickened intima (*I*). × 250. **d** This section was taken from an artery comparable to that shown in **c** and immunostained with a monoclonal antibody against aortic CSPG as described for **b**. These sections also demonstrate an enrichment of proteoglycan in the thickened intima. Reproduced with permission [92a]. × 250

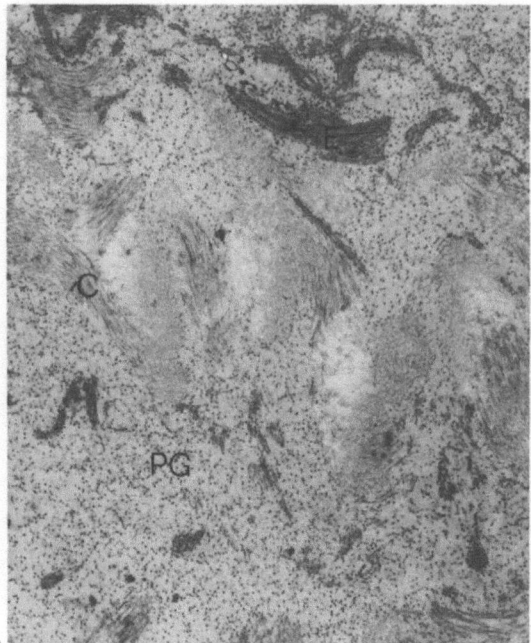

Fig. 12.4. An electron micrograph taken of the intima of an early developing human atherosclerotic lesion demonstrating that the "connective tissue space" is enriched with proteoglycan granules (*PG*), bundles of collagen fibrils (*C*) and immature elastic fibres (*E*). × 8600

[14]). Proteoglycan accumulation is thought to predispose the arterial wall to further complications such as lipid accumulation, calcification and thrombosis due to the ability of these molecules to bind to a variety of other molecules involved in these processes. The literature is confusing as to whether there are changes in specific types of proteoglycans during atherogenesis. Several studies indicate specific increases in DSPG while other studies indicate changes specifically in CSPG. The reasons for these apparent discrepancies may lie with a number of factors such as use of different methods for extraction and identification, species differences, site of the vasculature analysed, nature of the inciting stress and severity of the lesion. However, it is clear, as with collagen, that several "atherogenic factors" influence proteoglycan deposition.

For example, normal aorta from species susceptible to atherosclerosis (chicken, pigeon and rabbit) contains greater amounts of aortic glycosaminoglycans than does aorta from atherosclerosis-resistant species (rat, mouse and hamster) [17,50]. Experimental injury created either by physical trauma such as balloon catheterization or chronic lipid feeding leads to increased content of proteoglycans within experimental lesions [81,92]. Two other conditions associated with the rapid onset of atherosclerosis – diabetes and hypertension – are characterized by excessive accumulation of proteoglycans within affected arteries [6,16,29,72,73,83,84]. It should be stressed that

concentration in the fibrous plaque and complicated lesion. It is also interesting to note that the arterial smooth-muscle cells in the early lesions possess a secretory phenotype (Fig. 12.5

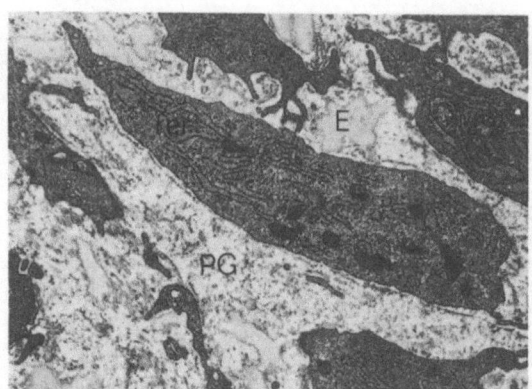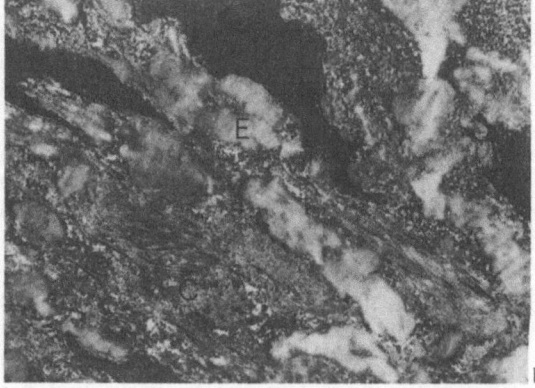

Fig. 12.5. a An electron micrograph of a portion of a thickened intima taken from a rat carotid artery 2 weeks following experimental injury. Note that the smooth-muscle cells have extensive rough endoplasmic reticulum (*rer*), characteristic of a "secretory phenotype". The extracellular space is occupied principally by elastin (*E*) and amorphous granular material (*PG*) thought to represent proteoglycan. × 12000. **b** An electron micrograph of a portion of a thickened intima from a rat carotid artery 3 months after experimental injury. Note that the extracellular space now contains a large number of collagen fibrils (*C*) and elastic fibres (*E*) but little proteoglycan. Reproduced with permission [14]. × 12000

not only are quantitative differences obvious in these states, but the physical–chemical nature of the proteoglycans may also be modified. For example HSPGs isolated from lesions of diabetic rats have a reduced sulphate content [39] indicating alteration in charge characteristics. These modifications may influence the way in which proteoglycans interact with other extracellular matrix molecules and with cells. For example, CSPG isolated from human fatty fibrous plaques exhibits less tendency than normal to aggregate with hyaluronic acid and the aggregates present in fatty fibrous plaques are smaller than those present in normal aorta [90]. Altered aggregate formation may have some bearing on arterial calcification, a complication of advanced atherosclerosis, since a number of studies indicate that the degree of proteoglycan aggregation influences calcification in cartilage [71].

Although it is clear that proteoglycans accumulate within lesions of atherosclerosis, the reasons for this accumulation are not clear. Cell-culture studies have revealed that proteoglycan synthesis is increased when arterial smooth-muscle cells are stimulated to proliferate [93] and endothelial cells are stimulated to migrate [38]. Both of these events are associated with the onset of atherosclerosis. Interestingly, arterial smooth-muscle cells, when in contact with collagen gels, switch from making a chondroitin sulphate-rich extracellular matrix to depositing a dermatan sulphate-rich extracellular matrix, indicating that the nature of the matrix surrounding the cells influences their biosynthetic capacity [46]. Other possible "atherogenic factors" shown to influence proteoglycan synthesis by arterial cells include cyclic stretching [47], oxygen content [69], prostaglandins [70a], interleukins [57], lipids, lipoproteins [68,95,96] and cell age [45].

As with collagens, there are several different mechanisms by which changes in arterial proteoglycan metabolism may influence the atherogenic process. Several studies indicate that proteoglycans within the vessel wall not only serve as structural links binding components of the extracellular matrix to each other and to cells, but also influence several arterial properties fundamental to the atherosclerotic process. Their presence within the major layers of the arterial wall is essential for maintaining visco-elastic properties by virtue of their ability to bind water and maintain compressibility [15]. In addition, the HSPGs of the endothelium are involved in maintaining blood fluidity at the blood–tissue interface. This is due to their ability to bind coagulant and anti-coagulant factors [51]

as well as to influence the metabolism of lipid-containing chylomicron particles by virtue of their ability to bind lipoprotein lipase at the endothelial cell surface [40]. It has been postulated that the actions of this enzyme may generate cholesterol-rich degradation products which could gain access to the vessel wall and promote lipid deposition [98].

In addition to influencing the lipid content of the arterial wall via lipoprotein lipase, a number of studies indicate that proteoglycans can interact with lipoproteins in the presence of calcium ions to form insoluble complexes [9]. This interaction is believed to be one of the main reasons why lipoproteins accumulate in the arterial wall during the formation of the atherosclerotic plaque. Experimental studies reveal that regions of blood vessels enriched in proteoglycans, such as the re-endothelialized region following balloon injury, exhibit a high propensity to accumulate lipoproteins [20,92]. These complexes form via electrostatic forces involving the positive charge of the protein moiety of the lipoprotein and the negative charges present on the glycosaminoglycan chains of the proteoglycans. Also Ca^{2+} may act as a bridge between negative charge groups of the two sets of molecules. In addition, non-coulombic forces may be involved since ionic detergents are frequently required to dissociate the complexes. Such interactions may explain why lipids accumulate outside of cells, but proteoglycans may also be involved in intra-cellular lipid accumulation. A number of studies have shown that LDL complexed to proteoglycan is recognized by the modified LDL receptor of the macrophage and the catabolism of this complex is greatly diminished [21]. This may be a partial explanation of the mechanism by which macrophages turn into lipid-filled foam cells. These studies point to the importance of proteoglycans in arterial lipid accumulation – a hallmark of the atherosclerotic process.

Finally, arterial wall proteoglycans and/or their breakdown products have been implicated in growth control within the vessel wall. For example, it has been demonstrated that fragments of heparan sulphate glycosaminoglycan derived from HSPGs synthesized by endothelial and smooth-muscle cells are potent inhibitors of arterial smooth-muscle cell growth [10,11,12,23,51]. These studies illustrate that not only are the structural properties of the extracellular matrix important in the atherogenic process but also the components of the matrix may regulate specific cell processes associated with the progression of vascular disease.

Elastin

Elastin appears to be quantitatively the second major extracellular macromolecule of the fibrous plaque (see review [41]). Biochemical analysis of alkali-isolated elastin fractions from atherosclerotic lesions revealed a much higher content of polar amino acids than that found for elastin from normal arteries, and this abnormality appears to increase with increasing severity of the disease [37,42]. This compositional difference may be due to the binding of other proteins to the elastic fibre. Examples of such proteins include the microfibrillar proteins, proteoglycans and their associated glycoproteins such as link proteins, collagens and elastase or lysyl oxidase [41]. Another possibility is that these secondary proteins may be derived from low-density (LDL) and very low-density lipoproteins (VLDL).

Several workers have performed in vivo and in vitro studies which suggest a specific role for elastin in the accumulation of lipid in the atherosclerotic plaques [42,43,65,88]. Normal aortic elastin appears to consist of a protein–lipid complex with the lipids presumably bound by hydrophobic stacking to hydrophobic sites in the elastin molecule [31]. While the lipid moiety of normal elastin is small, it increases markedly in elastin isolated from human atherosclerotic plaques [42], as well as in atherosclerotic arteries from monkeys and rabbits [43]. Morphological studies [26] have demonstrated that the "perifibrous lipid" first described by Smith et al. [86] and present in ageing human arteries is deposited adjacent to and within elastic fibres (Fig. 12.6).

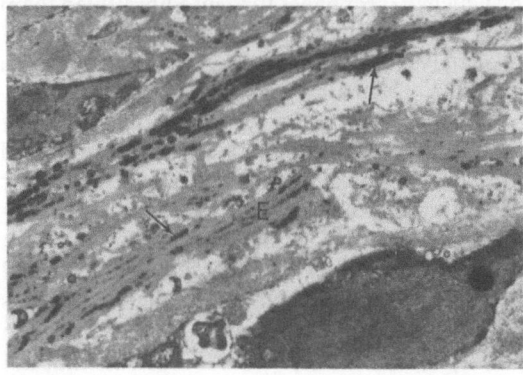

Fig. 12.6. This electron micrograph is of a section of a human atherosclerotic lesion which has been stained for lipid. Numerous lipid droplets (*arrows*) are present within and adjacent to the elastic fibres (*E*). (Courtesy of Dr. John Guyton, Baylor College of Medicine) × 12 600

The binding of lipids to elastin could interfere with new elastic fibre formation because the peptide chains would not associate properly and cross-linking would become randomized. In addition, since elastic recoil is somewhat dependent upon exposure of the lipophilic groups of the native elastin molecule, lipid binding to these groups would remove part of the restoring force and prevent the return of the fibre to a fully functional state [89]. Another facet of lipid binding to elastin is the observation that naturally occurring hydrophobic ligands such as free fatty acids bind to elastin [34] and in turn increase the binding of pancreatic elastase causing elastolysis [35,36].

Fragmentation or splitting of elastic fibres is a frequently observed event in the ageing and atherosclerotic process [75]. Destruction of elastic fibres markedly compromises the rheological properties of the vessel wall since elastin confers a high degree of elasticity and allows the blood vessel to recoil during systole-diastole. Thus, breakdown of the elastic fibre could lead to increased arterial wall stiffness which would result in haemodynamic stress within the vessel wall itself. Such stress may be a source of "injury", promoting the disease process. Several different elastases have been identified and these include those from pancreas, leucocytes, platelets, macrophages and smooth-muscle cells; all cells involved in atherosclerotic disease [4,75]. Elastase is a calcium-dependent enzyme and it is interesting that several reports suggest that calcium interacts with elastin and that elastin is a calcium "sink" [4,80,89].

Arterial calcification is a prominent event in the evolution of the atherosclerotic plaque [97], and calcium deposition appears to occur in the elastic fibre [41,89] as well as within lipid deposits (Fig. 12.7). The mechanism(s) by which calcium associates with elastin is not clear, i.e. there may be direct binding of Ca^{2+} through ionic interaction with the protein backbone of elastin or association may be via some accessory protein(s) [80,89]. Conformational changes in the elastin molecule promote calcium binding [79] and it is interesting in this regard that organic solvents induce conformational changes in the elastin which significantly increase calcium binding [80]. These observations have led to the suggestion that elastin–lipid interactions promote deposition of calcium, predisposing the tissue to advanced atherosclerosis. The importance of calcium in the atherogenic process is further emphasized by the finding that calcium agonists such as lanthanum have been shown to sig-

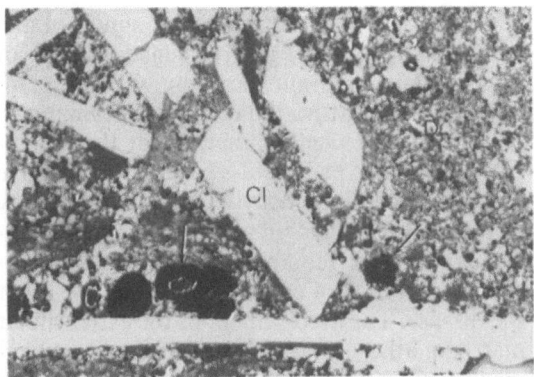

Fig. 12.7. An electron micrograph of a portion of the necrotic core of a human atherosclerotic plaque. Lipid debris (*D*), cholesterol clefts (*Cl*) and calcium spheres (*arrows*) are characteristic of this phase of the disease.

nificantly retard the development of experimental atherosclerosis [41].

Glycoproteins

Besides the fibrous proteins and proteoglycans, glycoproteins appear also to increase within developing atherosclerotic lesions [3,74]. Two structural glycoproteins that have been shown

to increase are fibronectin and thrombospondin (Fig. 12.8). Fibronectin is a polymorphic glycoprotein of plasma and is present in an insoluble form in a variety of tissues including blood vessels [30]. Recent studies have demonstrated that a variety of cells including vascular endothelial and smooth-muscle cells synthesize fibronectin in cell culture [33,48]. These observations have led to the suggestion of a dual origin of fibronectin, an insoluble form in tissues and a soluble form in plasma. Since this glycoprotein is capable of multiple molecular interactions, fibronectin is thought to be an important structural element at sites where plasma and tissue come into contact, especially at sites of tissue damage. An example of this would be the case where blood platelets interact with the fibronectin present within basement membranes and connective tissue that are exposed when the endothelial layer lining the blood vessel is injured.

Fibronectin is also released from platelets during their aggregation and soluble fibronectin potentiates the action of plasminogen activator. Fibronectin also can covalently bind to fibrin via its specific binding site and thus promote events involved in the thrombotic process. In addition, fibronectin is concentrated in areas of intense cellular activity and studies indicate that this extracellular matrix macromolecule appears essential for cellular attachment, migration and

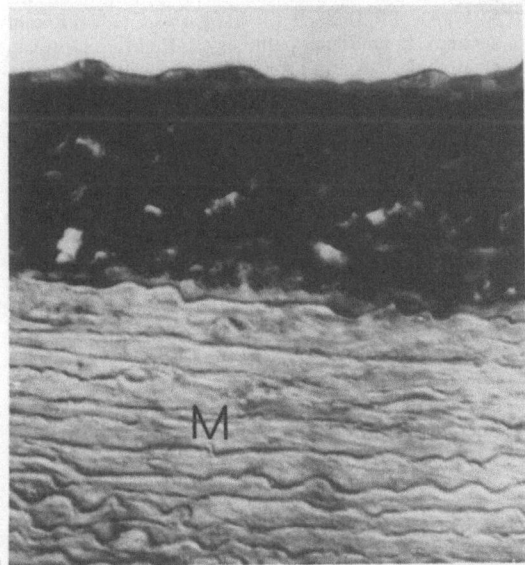
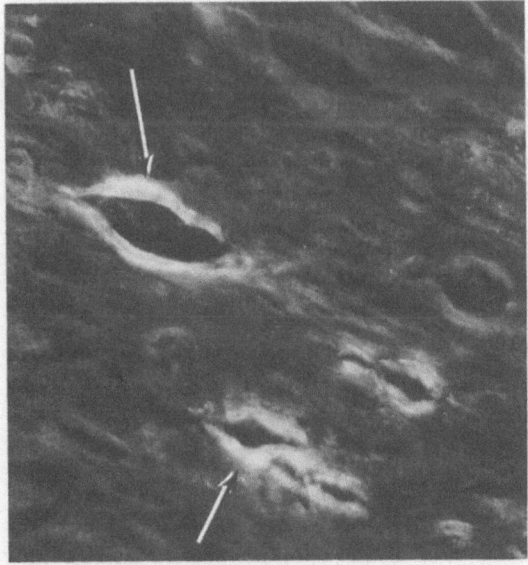

Fig. 12.8. a A portion of a thickened intima from a fat-fed rabbit, immunostained with anti-sera against fibronectin. Intense staining is confined to the thickened intima medial layer (*M*). ×325. **b** This section was taken from the fibrous cap of a human atherosclerotic aorta immunostained with anti-sera against thrombospondin. Staining was pronounced surrounding clef-like areas (*arrows*). Reproduced with permission [94]. ×260

proliferation (see review, [49]). Immuno-cytochemical studies reveal that fibronectin is enriched in developing lesions (Fig. 12.8a) as well as within advanced lesions of atherosclerosis [67,87]. Thus, fibronectin is important for two reasons; one is that it promotes events associated with the clotting mechanism (i.e. haemostasis and thrombosis) and second, it provides a substrate for cell attachment, proliferation and migration. At this point it is not clear whether fibronectin within lesions arise principally from leakage of plasma fibronectin into the plaque or whether resident cells of the plaque synthesize and deposit fibronectin. Arterial wall fibronectin accumulation may very well be a combination of both of these possibilities [13,87].

Another glycoprotein that is present in blood vessels and appears to be of dual origin is thrombospondin. Like fibronectin, thrombospondin is also present within platelets and thought possibly to be a platelet lectin. It is released upon the attachment and aggregation of platelets [32]. In addition, a number of studies have shown that both endothelial and smooth-muscle cells synthesize thrombospondin in culture [59,61] and that synthesis is elevated when arterial smooth-muscle cells are stimulated to proliferate. Immunostaining of normal and diseased blood vessels indicates a basement membrane or basal lamina localization [94] (Fig. 12.8b). Binding studies [62] also suggest interaction between type V collagen and thrombospondin and this interaction may partially explain the accumulation of these two matrix molecules within developing atherosclerotic lesions. The functional significance of this accumulation in the vessel wall is not understood and more studies are needed in this area. It is noteworthy that thrombospondin potentiates the action of certain mitogens such as epidermal growth factor (EGF), indicating a potential role for this molecule in growth regulation in the arterial wall [49].

Summary

Current theories of atherosclerosis must account for the events that lead to the formation of the atherosclerotic plaque. Important factors include endothelial damage due to injury, adhesion of formed elements of the blood as well as deposition of coagulation and anti-coagulation factors, migration of medial smooth-muscle cells to the intima and their subsequent proliferation, ccumulation of lipid and compounds of the xtracellular matrix, and calcification. All of hese events are influenced by various components of the extracellular matrix. It remains to be determined whether the role of these components in these processes is permissive, adaptive or in some cases, instructive. In addition, many of these events contribute to the changing extracellular matrix composition during atherogenesis and in turn contribute to the severity of the disease. The involvement of the extracellular matrix in atherogenesis must therefore be examined in two ways; firstly to determine which events lead to increased extracellular matrix deposition and the fibrotic response and secondly to determine how the extracellular matrix influences the events that form the basis for the development of the atherosclerotic plaque. The extracellular matrix can no longer simply be considered an inert structural element but rather a mixture of molecules that reside outside of cells and that are capable of modifying cellular behaviour and metabolism.

References

1. Barnes MJ (1982) The collagen–platelet interaction. In: Weiss JB, Jayson MN (eds) Collagen in health and disease. Churchill Livingstone, Edinburgh, pp 179–197
2. Barnes MJ (1985) Collagens in atherosclerosis. Collagen Rel Res 5:65–97
3. Berenson GS, Radhakrishnamurthy B, Srinivasan SR, Vijayagopal P, Dalferes ER Jr, Sharma C (1984) Recent advances in molecular pathology: carbohydrate-protein macromolecules and arterial wall integrity – a role in atherogenesis. Exp Mol Pathol 41:267–287
4. Bieth J (1978) Elastases: structure, function, and pathological role. In: Robert L (ed) Frontiers of matrix biology, vol 6. Karger, Basel
5. Briones ER, Mao SJT, Palumbo PJ, O'Fallon WM, Chenoweth W, Kottke BA (1984) Analysis of plasma lipids and apolipoproteins in insulin-dependent and non-insulin dependent diabetics. Metabolism 33:42–49
6. Brown DM, Klein DJ, Michael AF, Oegema TR (1982) ^{35}S-glycosaminoglycan and ^{35}S-glycopeptide metabolism by diabetic glomeruli and aorta. Diabetes 31:418–425
7. Brownlee M, Vlassara H, Cerami A (1985) Non-enzymatic glycosylation products on collagen covalently trap low density lipoprotein. Diabetes 34:938–941
8. Burke JM, Ross R (1977) Collagen synthesis by monkey arterial smooth muscle cells during proliferation and quiescence in culture. Exp Cell Res 107:387–395
9. Camejo G (1982) The interaction of lipids and lipoproteins with the intercellular matrix of arterial tissue: its possible role in atherogenesis. Adv Lipid Res 19:1–53
10. Castellot JJ Jr Addonizio ML, Rosenberg R, Karnovsky MJ (1981) Cultured endothelial cells produce a heparin-like inhibitor of smooth muscle cell growth. J Cell Biol 90:372–379

11. Castellot JJ Jr, Beeler DL, Rosenberg RD, Karnovsky MJ (1984) Structural determinants of the capacity of heparin to inhibit the proliferation of vascular smooth muscle cells. J Cell Physiol 120:315–320

12. Castellot JJ Jr, Rosenberg RD, Karnovsky MJ (1984) Endothelium, heparin and the regulation of vascular smooth muscle cell growth. In: Jaffee EA (ed) Biology of endothelial cells. Martinus Nijhoff, Boston, pp 118–128

13. Clark RAF, Quinn JH, Winn HJ, Lanigan JM, Dellepella P, Colvin RB (1982) Fibronectin is produced by blood vessels in response to injury. J Exp Med 156:646–651

14. Clowes AW, Reidy MA, Clowes MM (1983) Mechanisms of stenosis after arterial injury. Lab Invest 49:208–215

15. Comper WD, Laurent TC (1978) Physiological function of connective tissue polysaccharides. Physiol Rev 58:255–315

16. Crane WA (1962) Sulfate utilisation and mucopolysaccharide synthesis by the mesenteric arteries of rats with experimental hypertension. J Pathol 84:113–122

17. Curwen KD, Smith SC (1977) Aortic glycosaminoglycans in atherosclerosis-susceptible and resistant pigeons. Exp Mol Pathol 27:121–133

18. Ehrhart LA, Holderbaum D (1980) Aortic collagen, elastin, and non fibrous protein synthesis in rabbits fed cholesterol and peanut oil. Atherosclerosis 37:423–432

19. Ehrlich HP, Brown H, White BS (1982) Evidence for type V and I trimer collagens in Dupuytren's contracture palmer fascia. Biochem Med 28:273–284

20. Falcone DJ, Hajjar DP, Minick CR (1980) Enhancement of cholesterol and cholesteryl ester accumulation in re-endothelialized aorta. Am J Pathol 99:81–104

21. Falcone DJ, Mateol N, Shio H, Minick CR, Fowler SD (1984) Lipoprotein-heparin-fibronectin-denatured collagen complexes enhance cholesteryl ester accumulation in macrophages. J Cell Biol 99:1266–1274

22. Fischer GM, Swain ML, Cherian K (1980) Increased vascular collagen and elastin synthesis in experimental atherosclerosis in the rabbit. Atherosclerosis 35:11–20

23. Fritze LM, Reilly C, Rosenberg RD (1985) An anti-proliferative heparan sulfate species produced by post-confluent smooth muscle cells. J Cell Biol 100:1041–1049

24. Gay S, Martinez-Hernandez A, Rhodes RK, Miller EJ (1981) The collagenous exocytoskeleton of smooth muscle cells. Collagen Rel Res 1:377–384

25. Grotendorst GR, Seppa, HE, Kleinman HK, Martin GR (1981) Attachment of smooth muscle cells to collagen and their migration toward platelet derived growth factor. Proc Natl Acad Sci USA 78:3669–3672

26. Guyton JR, Bacon TM, Schifani TA (1985) Quantitative ultrastructural analysis of perifibrous lipid and its association with elastin in non-atherosclerotic human aorta. Arteriosclerosis 5:644–652

27. Hanson AN, Bentley JP (1983) Quantitation of type I to type III collagen ratios in small samples of human tendon, blood vessels and atherosclerotic plaque. Ann Biochem 130:32–40

28. Holderbaum D, Ehrhart LA (1984) Modulation of types I and III procollagen synthesis at various stages of arterial smooth muscle cell growth in vitro. Exp Cell Res 153:16–24

29. Hollander W, Kramsch D, Farmelant M, Madoff IM (1968) Arterial wall metabolism in experimental hypertension of coarctation of aorta of short duration. J Clin Invest 47:1221–1229

30. Hynes RD (1981) Fibronectin and its relation to cellular structure and behavior. In: Hay E (ed) Cell biology of extracellular matrix. Plenum Press, New York pp 295–334

31. Jacotot B, Beaumont JL, Monnier G, Szigeti M, Robert B, Robert L (1973) Role of elastic tissue in cholesterol deposition in the arterial wall. Nutr Metabol 15:46–58

32. Jaffe E, Leung LL, Nachman RL, Levin RI, Mosher DF (1982) Thrombospondin is the endogenous lectin of human platelets. Nature 295:246–248

33. Jaffe EA, Mosher DF (1978) Synthesis of fibronectin by cultured human endothelial cells. J Exp Med 147:1779–1791

34. Jordan RE, Hewitt N, Lewis W, Kagan, H, Franzblau G (1974) Regulation of elastase-catalyzed hydrolysis of insoluble elastin by synthetic and naturally occurring hydrophobic ligands. Biochemistry 13:3497–3503

35. Kagan HM, Lerch RM (1976) Amidated carboxyl groups in elastin. Biochim Biophys Acta 434:223–232

36. Kagan HM, Milbury PE, Kramsch DM (1979) A possible role for elastin ligands in the proteolytic degradation of arterial elastic lamellae in the rabbit. Circ Res 44:95–103

37. Keeley FW, Partridge SM (1974) Amino acid composition and calcification of human aortic elastin. Atherosclerosis 19:287–296

38. Kinsella MG, Wight TN (1986) Modulation of sulfated proteoglycan synthesis by bovine aortic endothelial cells during migration. J Cell Biol 102:679–687

39. Kjellen L, Bielefeld D, Höök M (1982) Reduced sulfation of liver heparan sulfate in experimentally diabetic rats. Diabetes 32:337–342

40. Klinger MM, Margolis RU, Margolis RK (1985) Isolation and characterization of the heparan sulfate proteoglycans of the brain. Use of affinity chromatography on lipoprotein lipase–agarose. J Biol Chem 260:4082–4090

41. Kramsch DM (1981) Biochemical changes of the arterial wall in atherosclerosis with special reference to connective tissue: promising experimental avenues for their prevention. In: McDonald TF, Bleakly Chandler A (eds) Connective tissues in arterial and pulmonary disease. Springer-Verlag, Berlin Heidelberg New York, pp 95–151

42. Kramsch DM, Franzblau C, Hollander W (1971) The protein and lipid composition of arterial elastin and its relationship to lipid accumulation in the atherosclerotic plaque. J Clin Invest 50:1666–1677

43. Kramsch DM, Hollander (1973) The interaction of serum and arterial lipoproteins with elastin of the arterial intima and its role in the lipid accumulation in atherosclerotic plaques. J Clin Invest 52:236–247

44. Langner RO, Modrak JB (1981) Alteration of collagen synthesis in different tissues of the atherosclerotic rabbit. Artery 9:253–261

45. Lark MW, Perigo S, McFarlane S, Wight TN (1987) Influence of cell age on proteoglycan metabolism by arterial smooth muscle cells. (Submitted for publication)

46. Lark MW, Wight TN (1986) Modulation of proteoglycan metabolism by aortic smooth muscle cells grown on collagen gels. Arteriosclerosis 6:638–650

47. Leung DY, Glagov S, Mathews MB (1976) Cyclic stretching stimulates synthesis of matrix components by arterial smooth muscle cells in vitro. Science 191:475–477

48. Macarak EJ, Kriby E, Kirk T, Kefalides NA (1978) Synthesis of cold insoluble globulin by cultured calf endothelial cells. Proc Natl Acad Sci 75:2621–2625

49. Majack RA, Bornstein (1986) Biosynthesis and modulation of extracellular matrix components by cultured vascular smooth muscle cells. CRC Press, Boca Raton, Florida

50. Mancini M, Rossi GB, Oriente P (1965) Possible relationship between aortic acid mucopolysaccharides and species susceptibility to experimental atherosclerosis. Nature 207:1206

51. Marcum JD, Reilly CF, Rosenberg RD (1987) Heparan sulfate species and blood vessel wall function. In: Wight TN, Mecham EJ (eds) Biology of the extracellular matrix, proteoglycans. Academic Press, New York

52. Martinez-Hernandez A, Gay S, Miller EJ (1982) Ultrastructural localization of type V collagen in rat kidney. J Cell Biol 92:343–349

53. Mayne (1984) Vascular connective tissue: normal biology and derangement in human disease. In: Uitto J, Perejda (eds) Diseases of connective tissue: the molecular pathology of the extracellular matrix. Marcel Decker, New York.

54. McCullagh KA Balian G (1975) Collagen characteristization and cell transformation in human atherosclerosis. Nature 258:73–75

55. McCullagh KG, Ehrhart LA (1974) Increased arterial collagen synthesis in experimental canine atherosclerosis. Atherosclerosis 19:13–28

56. McCullagh KG, Ehrhart LA (1977) Enhanced synthesis and accumulation of collagen in cholesterol-aggravated pigeon atherosclerosis. Atherosclerosis 26:341–352

57. Montesano R, Orci L, Vassalli P (1985) Human endothelial cell cultures: phenotypic modulation by leukocyte interleukins. J Cell Physiol 122:424–343

58. Morton LF, Barnes MJ (1982) Collagen polymorphism in the normal and diseased blood vessel wall. Investigations of collagen types I, III, and V. Atherosclerosis 42:41–51

59. Mosher DF, Doyle MJ Jaffe EA (1982) Synthesis and secretion of thrombospondin by cultured human endothelial cells. J Cell Biol 93:343–348

60. Mourao PA, Pillai S, Di Ferrante (1981) The binding of chondroitin 6 sulfate to plasma low density lipoprotein. Biochim Biophys Acta 674:178–187

61. Mumby SM, Abbott-Braun D, Raugi GJ, Bornstein P (1984) Regulation of thrombospondin secretion by cells in culture. J Cell Physiol 120:280–288

62. Mumby SM, Raugi GJ, Bornstein P (1984) Interactions of thrombospondin with extracellular matrix proteins: selective binding to type V collagen. J Cell Biol 98:646–652

63. Narayanan AS, Engel LD, Page RC (1983) The effect of chronic inflammation on the composition of collagen types in human connective tissues. Collagen Rel Res 3:232–334

64. Nissen R, Cardinale GJ, Udenfriend S (1978) Increased turnover of arterial collagen in hypertensive rats. Proc Natl Acad Sci USA 75:451–453

65. Noma A, Hirayama T, Yachi A (1983) Studies on the binding of plasma low density lipoproteins to arterial elastin. Connect Tissue Res 11:123–133

66. Ooshima A (1981) Collagen alpha B chain: increased proportion in human atherosclerosis. Science 213:666–668

67. Orekhov AN, Andreeva ER, Shekhonin BV, Tertov VV, Smirnov VN (1984) Content and localization of fibronectin in normal intima, atherosclerotic plaque and underlying media of human aorta. Atherosclerosis 53:213–219

68. Pietila K (1982) Long-term effect of hyperlipidemic serum on the synthesis of glycosaminoglycans and on the rate of growth of rabbit aortic smooth muscle cells in culture. Atherosclerosis 42:67–75

69. Pietila K, Jaakkola O (1984) Effect of hypoxia on the synthesis of glycosaminoglycans and collagen by rabbit aortic smooth muscle cells in culture. Atherosclerosis 50:183–190

70. Pietila K, Nikkari T (1980) Enhanced synthesis of collagen and total protein by smooth muscle cells from atherosclerotic rabbit aortas in culture. Atherosclerosis 37:11–9

70a. Pietila K, Moilanen T, Nikkari T (1980) Prostaglandins enhance the synthesis of glycosaminoglycans and inhibit the growth of aortic smooth muscle cells in culture. Artery 7:509–518

71. Poole AR, Rosenberg LC (1987) Proteoglycans, chondrocalcin and the calcification of cartilage matrix in endochondral ossification. In: Wight TN, Mecham RP (eds) Biology of the extracellular matrix proteoglycans. Academic Press, New York

72. Reynertson RH, Roden L (1986) Proteoglycans and hypertension. II. [^{35}S] sulfate incorporation into aorta proteoglycans of spontaneously hypertensive rats. Collagen Rel Res 6:103–120

73. Reynertson RH, Parmley RT, Roden L, Oparil SL (1986) Proteoglycans and hypertension. I. A biochemical and ultrastructural study of aorta glycosaminoglycans in spontaneously hypertensive rats. Collagen Rel Res 6:77–103

74. Robert L (1978) Structural glycoproteins of connective tissue. In: Biochemistry of normal and pathological connective tissue. Collaque du CNRS No. 287, vol 2, pp 189–194

75. Robert L, Jacob MP, Frances C, Godeau G, Hornbeck W (1984) Interaction between elastin and elastases and its role in the aging of the arterial wall, skin and other connective tissues. A review. Mech Ageing Dev 28:155–166

76. Rokosova B, Bentley JP (1979) Biosynthesis of aortic collagen and glycosaminoglycan following immunological injury. Atherosclerosis 32:359–365

77. Ross R, Glomset JA (1976) The pathogenesis of atherosclerosis. N Engl J Med 295:369–377, 420–425

78. Ross R (1986) The pathogenesis of atherosclerosis – an update. N Engl J Med 314:488–500

79. Rucker RB, Ford D, Riemann WG, Tom K (1974) Additional evidence for the binding of calcium ions to elastin at neutral sites. Calcif Tissue Res 14:317–325

80. Sandberg LB, Soskel NT, Leslie JG (1981) Elastin structure, biosynthesis, and its relation to disease states. N Engl J Med 304:566–579

81. Saxena ID Nagchaudhuri J (1970) Effect of high fat diets on acid mucopolysaccharides of the aorta in rabbits. Ind J Exp Biol 8:15–18

82. Schwartz SM (1980) Role of endothelial integrity in atherosclerosis. Artery 8:305–314

83. Sirek OV, Sirek A, Cukerman E (1980) Arterial glycosaminoglycans in diabetic dogs. Blood Vessels 17:271–275

84. Sirek OV, Sirek A, Cukerman E (1981) Intermittent hyperinsulinaemia and arterial glycosaminoglycans in dogs. Diabetologia 21:154–159

85. Smith EB (1965) The influence of age and atherosclerosis on the chemistry of aortic intima. Part 2, collagen and mucopolysaccharides. Atherosclerosis 5:224–228

86. Smith EB Evans PH, Downham MD (1967) Lipid in the aortic intima. The correlation of morphological and chemical characteristics. Atherosclerosis 7:171–186

87. Stenman S, von Smitten K, Vaheri A (1980) Fibronectin and atherosclerosis. Acta Med Scand [Suppl] 642:165–170
88. Tokita K, Kanno K, Ikeda K (1977) Elastin sub-fraction as a binding site for lipids. Atherosclerosis 28:111–119
89. Urry DW (1978) Molecular perspectives of vascular wall structure and disease: the elastic component. Perspect Biol Med 21:265–295
90. Wagner WD, Hardingham T, Edwards I (1983) Decreased amount and size of artery chondroitin sulfate proteoglycan – hyaluronic acid aggregate in atherosclerosis. Arteriosclerosis 3:471a
91. Wight TN (1980) Vessel proteoglycans and thrombogenesis. In: Spaet Th (ed) Progress in hemostasis and thrombosis. Grune and Stratton, New York, pp 1–39
92. Wight TN, Curwen KD, Litrenta MM, Alonso DR, Minick CR (1983) Effect of endothelium on glycosaminoglycan accumulation in injured rabbit aorta. Am J Pathol 113:156–164
92a. Wight TN, Kinsella M, Lark M, Potter-Perigo S (1987) CIBA Foundation symposium 124. Functions of the proteoglycans. J Wiley, Chichester
93. Wight TN, Kinsella MG, Potter-Perigo S (1985) Proteoglycans synthesized and secreted by cultured vascular cells. In: Reddi (AH) Extracellular matrix: structure and function. Alan R Liss, New York
94. Wight TN, Raugi GJ, Mumby SM, Bornstein P (1985) Light microscopic immunolocation of thrombospondin in human tissues. J Histochem Cytochem 33:295–302
95. Wosu L, McCormick S, Kalant N (1984) Interaction of high and low density lipoprotein on glycosaminoglycan secretion by human vascular smooth muscle cells and fibroblasts. Can J Biochem Cell Biol 62:984–990
96. Wosu L, Parisella R, Kalant N (1983) Effect of low density lipoproteins on glycosaminoglycan secretion by cultured human smooth muscle cells and fibroblasts. Influence of serum concentration and cell proliferation rate. Atherosclerosis 48:205–220
97. Yu SY (1974) Calcification processes in atherosclerosis. In: Wagner WD, Clarkson TB (eds) Arterial mesenchyme and atherosclerosis. Plenum Press, New York, pp 403–425
98. Zilversmit DB (1973) A proposal linking atherogenesis to the interaction of endothelial lipoprotein lipase with triglyceride-rich lipoprotein. Circ Res 33:633–638

Chapter 13

Lipid Accumulation in the Vessel Wall

S. Moore

Introduction

Since the early 1970s experiments have demonstrated that lesions of the arterial wall, morphologically and chemically resembling those seen in atherosclerosis in man, can be induced by various forms of injury to the inner lining of the vessel [69]. The lesions occur without dietary manipulation of the blood lipid content. These observations have led to new theories concerning the pathogenesis and mechanisms of lipid deposition or accumulation in lesions. Inevitably they have also led to a re-examination of the parts played by hyperlipidaemia and by alterations of lipoprotein metabolism in the genesis and progression of lesions. In addition, study of the development, progression and regression of lesions induced by differing forms of vessel-wall injury may give further insights into the mechanisms involved [72].

It is also well established that lesions resembling those found in the human disease can be induced in a variety of animals by dietary lipid supplement. In these experiments very high blood lipid levels are induced [29,30]. It would be expected therefore that the combination of hyperlipidaemia and injury would be particularly effective in inducing lesions experimentally and there is in fact very good evidence for the synergy of hyperlipidaemia and injury, particularly immunological injury [64].

The important question of which factors are critical to the genesis of lesions is shrouded by our ignorance concerning the occurrence and nature of "early" lesions in man. Longitudinal studies provide increasing evidence that many of the lesions found in vessels in early life are transient and may be repaired, a factor of great importance in considering lipid accumulation in arteries (see Chap. 10).

The cellular composition of lesions induced by dietary manipulation differs from that of injury-induced lesions. In one form of injury-induced lesion, that resulting from removal of the endothelial layer, the intimal thickening is composed exclusively of smooth-muscle cells which later acquire lipid droplets in their cytoplasm [78]. This contrasts with diet-induced lesions where the initial and main cellular element appears to be monocytes [39], which at first attach to the endothelial surface, then migrate into the intima and become lipid-laden foam cells. When these lesions eventually ulcerate, platelets are deposited and smooth-muscle cells appear in the deeper parts of the lesions [30]. The relative proportion of macrophages to smooth-muscle cells can be altered by manipulation of the diet. Certain diets, for example those containing peanut oil, cause lesions which are mainly composed of smooth-muscle cells, whereas diets rich in oleic acid cause lesions in which macrophages predominate [105]. The macrophage-rich lesions induced by diets containing oleic acid show little evidence of smooth-muscle cell proliferation, collagen synthesis or proteoglycan content. The peanut oil lesions show abundant smooth-muscle cells with an increase in interstitial connective tissue containing collagen and pro-teoglycans, but they also have a component of lipid-laden macrophages.

Similar lesions are induced by repeated injury caused mechanically by an indwelling aortic catheter [67] or by repeated injections of lym-phocytotoxic serum into a temporarily isolated segment of a carotid artery in rabbits [35]. These lesions, unlike those resulting from removal of the endothelium with a balloon catheter, are composed of two cell types – smooth-muscle cells and lipid-filled macrophages, the latter tending to occur in the superficial part of raised lesions (towards the lumen) which is covered by platelet–fibrin thrombus. As the lesions regress in size and lipid content, following removal of the injury stimulus, the macrophages also disappear leaving lesions consisting of smooth-muscle cells.

From the foregoing brief summary of the cellular composition of various lesions induced either by diet or by injury in normal-fed animals, it is apparent that manipulation of either hyper-lipidaemic diets or the mechanisms of injury can alter the cellular content of lesions. It is also apparent that there are some similarities between injury-induced and diet-induced lesions. It may therefore be useful to examine more closely the sequence of events in the formation of lesions induced by injury and by dietary hyper-lipidaemia.

Lesions Induced by Injury

As indicated above there are two types of lesion which can be induced by injury. Although there are many similarities between them, there are also striking differences in terms of their pro-gression and regression [70].

In the course of experiments designed to in-vestigate the relationship between embolism of platelet aggregates into the renal microcircula-tion and the development of hypertension and nephrosclerosis [66], two observations were made which indicated that vascular injury by itself, without dietary lipid supplement, could lead to the accumulation of lipid in vessel walls. Polyethylene catheters containing segments of magnesium–aluminium wire were passed into the aorta of rabbits by way of a femoral artery. The upper part of the catheter, lying in the tho-racic aorta, contained short segments of the wire and holes were cut in the polyethylene, using a dental drill, to expose small areas of the wire. On

these, platelets aggregated and were discharged distally as small emboli. The emboli lodged in the intertubular arterioles of the renal cortex where they could be found easily in relation to the leakage into the vessel walls of colloidal carbon, injected intravenously before killing the animals. Endothelial disruption occurred in relation to those platelet aggregates and mono-cytes accumulated.

After several days a marked neo-intimal thick-ening occurred in which both monocytes and smooth-muscle cells were prominent [60]. An unexpected finding was the presence of abundant lipid in the walls of these vessels, revealed by staining frozen sections of the tissue with oil-red O [74]. This became apparent 48 h after the placement of a source of emboli in the aorta and indicated that lipid deposition could occur as a response to injury in the arterial system, in the absence of dietary lipid supplement.

Injury from Indwelling Aortic Catheter

When catheters had been left in place in the aorta for weeks to months raised yellow lesions were found on the intimal surface, apparently due to repeated contact between the catheter and the vessel wall [67]. When these lesions were sectioned, the histological appearance was ident-ical to that of complicated human athero-sclerosis. The surface of the lesions showed platelet–fibrin thrombus. The shoulders of the plaques were generally covered by endothelium, and the surface area of the plaques often showed a "fibrous cap", composed mainly of smooth-muscle cells but with some admixture of macro-phages where the thrombus deposition was abundant. Calcification and even ossification occurred in the lesions.

When the catheter was removed the lesions rapidly regressed [77]. This involved a marked reduction in size and loss of lipid. Eventually, in weeks to months following removal of the catheter, the lesions took the form of intimal thickenings composed of smooth-muscle cells without stainable lipid. The capacity of these lesions to regress was remarkable. For example, heavy deposits of calcium disappeared from the lesions.

Immunological Injury

Similar lesions were induced by immunological injury. The method used by O'Connell and

Mowbray was adopted [80]. They had shown that immunological damage to rabbit carotid arteries in the presence of a hyperlipidaemic diet induced intimal thickenings containing lipid. The immunological insult was the placement of human serum into a segment of carotid artery for several minutes. An attempt was made to determine whether repeated injury of this sort would cause lipid-containing atherosclerotic lesions in the absence of dietary lipid supplement. It was found that human serum which was cytotoxic for rabbit lymphocytes was effective [35]. Following four injections into a temporarily isolated segment of rabbit carotid artery at weekly intervals, raised lipid-containing lesions were induced.

Again, following cessation of the injury stimulus, the lesions rapidly regressed so that by 4 weeks following the last of four injections, intimal thickenings composed of smooth-muscle cells, without stainable lipid, were the only lesions observed [36].

In both of these experiments employing repeated injury to induce atherosclerotic lesions in normal-fed rabbits, fatty streaks and oedematous (gelatinous) lesions occurred fleetingly. In a study of regression of lesions which had been induced by the placement of an indwelling aortic catheter, fatty streaks were observed in a narrow temporal window from 7 to 9 d following removal of the catheters, after they had been in place for 2 weeks [77]. These fatty streaks took the form of lipid-filled cells in a thickened intima. Interestingly, the cells, when they could be identified by electron microscopical examination, were of smooth-muscle cell type. Less commonly oedematous intimal thickenings were observed without stainable lipid but with the cells widely separated, presumably by fluid. Similar lesions were occasionally seen after two or three injections of lymphocytotoxic human serum into the rabbit carotid artery.

Chemical analysis of the lesions induced by the placement of an indwelling aortic catheter showed considerable increase in free cholesterol and cholesteryl ester [24].

Inhibition of Lesion Development

These experiments demonstrated that repeated or continuous injury to the endothelial surface of arteries of normal-fed rabbits induced lesions identical to those of human atherosclerosis. Two important conclusions could be drawn from this. One is that injury alone, in the absence of hyper-

lipidaemia or hypercholesterolaemia, could induce atherosclerosis. The other was that either injury or the thrombosis associated with injury were causally related to lesion development. To determine whether injury or thrombosis was the factor essential for lesion development, we asked whether, if platelets were removed from the system, lesions would develop normally [76]. This question seemed even more relevant in view of the recently described role of a material released from platelets in stimulating the proliferation of smooth-muscle cells [85].

In an attempt to answer the question, rabbits were made severely thrombocytopenic before the insertion of indwelling aortic catheters, by the intravenous injection of anti-platelet serum, raised in a sheep to rabbit platelets. Controls were given normal sheep serum. When the platelet count was markedly reduced (4000 or less) there was marked inhibition or prevention of lesion development [85]. When thrombocytopenia was induced after the catheter injury had occurred, even for as short a time as 1 h, there was no inhibition of lesion development [34].

Platelet-Derived Chemotactic Factor

This experiment clearly demonstrated that platelets or material released from platelets were/was essential for lesions to develop. With the description of platelet-derived growth factor (PDGF) it seemed likely that this was the material responsible. However, as well as there being proliferation of smooth-muscle cells, it was apparent that the cells migrated from the media to the intima. In experiments designed to seek a chemotactic factor for smooth-muscle cells in platelets it was shown that platelet releasate stimulated the migration of smooth-muscle cells across a filter of 8.0 μm pore size. The filter had to be coated with gelatin for migration to occur [54,55]. The subsequent demonstration by Grotendorst and colleagues of the necessity for collagen in the system confirmed the finding and showed that the material responsible was PDGF [43]. However, for migration the factor acted alone without, as in the case of proliferation, the requirement for other serum growth factors [101,53].

It is now established that endothelial cells [26] and activated monocytes produce growth factors [40] which may be identical to PDGF. It has also been shown that smooth-muscle cells themselves can, under certain conditions, produce the factor. This was first shown in studies of aortic explants from rat pups [87]. The findings may explain why the intimal thickening that develops following balloon removal of the endothelium is progressive.

Balloon-Catheter Removal of the Endothelium

The type of lesion, described above, resulting from injury to the aorta by an indwelling catheter or by repeated injections of serum, differs in a number of respects from the lesions induced by the removal of aortic endothelium by a balloon catheter.

When the endothelium is stripped off, platelets rapidly cover the denuded surface [7]. Initially there may be small platelet aggregates occurring at intervals along the surface, which is totally covered by a single cell layer composed of platelets [6]. Later, only the monolayer remains. The uptake of platelets is maximal within 30 min after removal of the endothelium. The platelets remain for several days, then gradually are lost from the surface so that at 4 d many platelets have come off and at 7 d the surface is virtually platelet free [44]. Smooth-muscle cells begin to migrate from the media about 48 h after injury and are proliferating to form a neo-intima at 4 d. The intimal thickening is progressive in the sense that for as long as we have observed it (24 months) the number of layers of smooth-muscle cells increases [78]. Thickening is greater in the abdominal than in the thoracic aorta and is greater in the areas which have become covered by endothelium regenerating from the branch vessels [78].

The process is progressive and unlike the indwelling catheter or immunological injury does not regress. Like the lesion induced by repeated injury it is preventable or can be markedly inhibited by inducing profound thrombo-cytopenia with anti-platelet serum [37].

Lipid Accumulation in Lesions

Lipid appears in the thickened intima more slowly than in the lesions induced by an indwelling catheter, but persists and seems to increase in amount [78]. The morphological impression of abundant lipid deposition gained from viewing frozen sections of the tissue stained with oil-red

O is substantiated by chemical analysis which shows at 6, 12 and 24 months after a single balloon removal of the endothelium a marked increase in lipid including free cholesterol but especially cholesteryl ester, which is increased 40–50 times above the concentration in normal intima-medial tissue [4].

These experiments demonstrate that lipid-containing lesions similar to those of human atherosclerosis can be induced simply by removing the endothelial lining in normally fed animals. The lipid accumulates exclusively or preferentially in areas that have become re-endo-thelialized. The lesions are progressive in the sense of continuing to increase in thickness and to accumulate lipid.

This is in contrast to the lesions induced by repeated mechanical or immunological injury which regress in size and content of lipid when the injury stimulus is removed [71].

Lipid-Containing Lesions Following Vascular Injury in Man

Although it is clear that lesions identical to those of human atherosclerosis can be induced in normal-fed rabbits and swine with "normal" blood lipid levels by various mechanisms of endothelial or intimal injury, it is relevant to ask if this has any meaning or implication for the development of atherosclerosis in man. There are in fact a number of examples of athero-sclerosis and thrombosis associated with injury to vessels in human pathology.

Coronary venous bypass grafts develop in-timal thickening and features comparable with atherosclerosis [13]. These changes are associ-ated with damage to the endothelium and throm-bus formation early after graft placement and later, and may be susceptible to modification by drugs [38,18].

Homocysteinaemia in subjects with inherited abnormalities of methionine metabolism is associated with the precocious development of atherosclerosis and thromboembolic compli-cations [12]. Continuous intravenous infusion of homocysteine thiolactone in baboons caused patchy endothelial loss in arteries followed by focal intimal thickening [48]. Similar experiments in pigs gave negative results [82]. Exposure of rabbits to homocysteine thiolactone caused intimal thickening in some [62] but not in all such studies (27). More recently there has been speculation that moderate homocysteinaemia due to deficiency of the enzyme cystathionine synthetase may be a risk factor for arterio-sclerotic cerebrovascular disease, coronary heart disease and peripheral vascular disease [79].

Radiation injury to arteries causes intimal thickening with lipid deposits, including cholesterol crystals [8].

In late rejection of transplanted organs lipid-containing intimal thickenings occur in arteries of the grafted organ [81]. In heart transplants this intimal thickening tends to be concentric [98] and in this way mimics the lesion induced by the synergy of immunological injury and cholesterol feeding [64]. The development of this lesion is a significant cause of graft failure.

The insertion of catheters through the umbilical artery into the aorta in neonates to monitor blood gases induces arterial lesions similar to those produced experimentally in rabbits by the placement of a polyethylene catheter in the aorta [97].

The placement of a shunt between an artery and a vein to provide access for dialysis causes thrombosis in the vein just distal to the shunt connection [57]. Eventually changes occur which are similar to those which have been produced experimentally by the creation of arteriovenous fistulas [91]. They are presumably due to disruption of the endothelium of the vein with the subequent laying down of thrombus material. In man, the development of these lesions frequently leads to failure of the shunt. The tissue removed at the time of shunt revision shows a spectrum of atherosclerotic changes [92]. These always include intimal thickening, often with lipid deposition and sometimes, at the base of the plaque, cholesterol crystals are observed [68].

Re-stenosis after percutaneous transluminal coronary angioplasty occurs in approximately 30% of cases [45]. Experimental [32] and clinical studies [96] support the use of aspirin in inhibiting the development of this complication, suggesting that endothelial injury and platelet deposition are important in its causation. However, it is not evident why re-stenosis does not occur in every case since the procedure must destroy the endothelium in the segment which is dilated. This may be partly due to the relative paucity of smooth-muscle cells in the advanced atherosclerotic plaque and possibly their inability to replicate further. A more likely reason is destruction of the viable medial coat of the vessel by overdistension. In experimental induction of atherosclerosis in rabbits by balloon-catheter removal of the aortic endothelium, we observed that overdistension of the balloon caused extensive medial necrosis accom-

panied by a very attenuated development of intimal thickening and lipid accumulation [78].

The vascular changes in the placenta in diabetes mellitus, commonly referred to as "atherosis", are essentially foam-cell collections in the intima. However, whether they can be attributed to an "injury" mechanism or relate to the disordered lipid metabolism in diabetes or to some other factor or factors, is unknown.

These examples indicate that injury to vessels is associated with the development of atherosclerotic lesions in man. The injury, whether it involves arteries or veins, is associated with endothelial destruction and thrombus formation. The improved results of coronary artery bypass grafts in a second series compared to a first in the same institution are probably attributable to more careful handling and preparation of the vein grafts to ensure better survival of the endothelium [42]. This factor is probably also responsible for the better results using internal mammary artery grafts instead of venous bypass grafts [42].

These clinical examples of injury-induced or related atherosclerosis provide support for the experimental evidence that vessel injury, especially endothelial injury, provides a stimulus for plaque formation.

In these experiments, as in most of the clinical examples cited above, it is clear that extensive endothelial damage and disruption occur, and set in motion a progressive series of events culminating in the development of lipid-rich atherosclerotic plaques.

Relationship of Injury to "Early" Human Atherosclerotic Lesions

Whether such injury occurs "spontaneously" in man is unknown but seems very unlikely. In recent experimental studies of endothelial injury, both in vivo and in vitro, it has been clearly demonstrated that as single cells die they are extruded from the surface without loss of continuity [83,106]. This is accomplished by the neighbouring endothelial cells spreading under the dead cell. It is not known if such "non-denuding" endothelial injury results in intimal thickening though the evidence available is that it does not. Could repeated injury of single cells or small groups of cells serve as an appropriate stimulus? We do not know the answer to this crucial question. There are several lesions which have been characterized as early lesions of atherosclerosis which may be the hallmark or evidence of such injury. These are spontaneously occurring mural thrombi, oedematous (gelatinous) lesions and fatty streaks. It is of interest that all of these lesions have been seen in atherosclerosis caused by an indwelling aortic catheter or by repeated immunological injury to a segment of carotid artery in rabbits [73]. Could such lesions represent responses to localized or focal injury to the vessel wall? This speculation leads to another consideration. We do not know what causes the diffuse intimal thickening which occurs in the arteries of man and is present from an early age. Could these evidences of focal injury (early lesions) be the precursors of diffuse intimal thickening [73]?

Mechanisms of Lipid Accumulation in Injury-Induced Lesions

Following removal of the aortic endothelium with a balloon catheter, stainable lipid occurs preferentially or exclusively in areas where endothelial regrowth has occurred [78]. The areas which remain uncovered by endothelium show little or no stainable lipid. These findings are supported by chemical analysis of intima–media tissue which shows increases in lipid content including free cholesterol and cholesteryl ester in the endothelium-covered neo-intima. The increase in cholesteryl ester is particularly marked. These elevations in lipid content persist for 24 months following the injury [4]. Since there is no increase in plasma lipid levels the accumulation of lipid must depend on alterations within the vessel wall.

Such alterations might include increased entry of lipid, decreased hydrolysis, increased synthesis or increased retention. One of the striking features of the neo-intima, covered by regenerated endothelium (white areas), is the high concentration of proteoglycans as demonstrated by ruthenium red staining of the glycosaminoglycan elements of the proteoglycan molecule in material examined in the electron microscope [84]. By contrast there is very much less proteoglycan in the areas of neo-intima uncovered by regenerated endothelium (blue areas).

Proteoglycans and Lipid Accumulation

These findings suggest one mechanism by which lipid might accumulate preferentially in the endothelium-covered neo-intima, namely by low-density lipoprotein binding to the glycosaminoglycans. This mechanism was outlined by Iverius in 1972 [56]. The complexing of lipid to glycosaminoglycans has been supported by histochemical observations [107,65], chemical analysis of atherosclerotic tissue [90,95] and by in vitro studies of the interaction between arterial glycosaminoglycans and lipoproteins [10,9]. Camejo and colleagues have isolated a lipoprotein-complexing moiety of proteoglycan from human aorta [15]. In the neo-intima there are both qualitative and quantitative changes in the glycosaminoglycan components of the proteoglycan formed in response to balloon-catheter removal of the endothelium [2,3]. In particular there is an increase in chondroitin sulphates and the appearance of chondroitin-4 sulphate, not normally present in tissue of the intima–media.

The proteoglycan whose synthesis is increased in the neo-intima is retained in the part covered by endothelium but rapidly disperses into the medium from the tissue denuded of endothelium when incubated in tissue culture [2]. Presumably a similar equilibration occurs in vivo. If lipoprotein has complexed to the proteoglycan the diffusion of lipoprotein–glycosaminoglycans complexes into the bloodstream might explain the low concentrations of lipid (cholesteryl ester and free cholesterol) in these areas compared to the areas covered by regenerated endothelium. The kinetics of lipoprotein entry and removal from the neo-intima parallel these events [1]. While it is clear that lipoprotein entry into the endothelium-covered neo-intima is increased over the normal [25], the accumulation of lipid in these areas as compared to the areas remaining denuded of endothelium indicates some mechanism for retention of lipoprotein. Binding to glycosaminoglycans is a likely mechanism.

The part played by sterol ester hydrolase activity in the two areas in relation to lipid accumulation is debated. Hajjar et al. showed that this was reduced in the white areas compared to the blue areas [46], whereas Day and colleagues showed a higher level of cholesteryl ester hydrolase activity in the white areas [25].

Of the mechanisms proposed, that of binding of low-density lipoprotein (LDL) to proteoglycan together with an increased rate of entry of LDL, seem to have the most experimental support.

The concept is not a new one. Virchow in 1856 speculated that injury to the luminal surface of an artery leads to an exudation of "mucous" substances, which cause fatty deposits to occur [99]. It is of interest in this regard that Duff, in studying dietary atherosclerosis in rabbits, identified a "change in the ground substance" which preceded the appearance of stainable lipid in the intima [28]. It may be that proteoglycans are involved in lipid deposition in the vessel wall, whether the initial stimulus is injury [69] or hyperlipidaemia [84a]. Increased content of proteoglycans and alteration of proteoglycan synthesis are associated with lipid deposition in response to injury [72], dietarily induced atherosclerosis [51,104] and the spontaneously occurring atherosclerosis in White Carneau pigeons [20,102].

In the injury models of atherosclerosis described above we have examined the concentration of glycosaminoglycans as identified by ruthenium red staining of tissue prepared for examination in the electron microscope. Following balloon-catheter de-endothelialization there is an increased concentration of large (20–50 nm) ruthenium red particles [84]. These represent dermatan and/or chondroitin sulphate. The increased concentration of these substances is observed in the areas of the neo-intima which have been covered by regenerating endothelium. The areas not covered have a decreased concentration of large ruthenium red stained particles. The increased concentration corresponds to increased lipid deposition, as observed in oil-red O stained sections, in the endothelium-covered areas, as compared to very little or no stainable lipid in the areas remaining uncovered by endothelium. Similarly, lesions produced by an indwelling aortic catheter in rabbits [75] and rats [52] show an increased concentration of large ruthenium red particles in relation to lipid accumulation. They decrease before the lipid begins to disappear during regression.

Chemical analysis of the tissue shows a marked increase in lipid content, especially cholesteryl ester, in the endothelium-covered areas with very little increase above control values in the areas uncovered by endothelium [24,4]. These changes persist at 6 months, 1 year and 2 years following removal of the endothelium. During this time the thickness of the neo-intima continues to increase, especially that of the endothelium-covered portion. The difference in thickness, comparing the two areas, may be due to endothelial-cell derived growth factor (EDGF). It is probable that the main component of an endothelial stimulus to growth of mesenchymal cells is similar to platelet-derived growth factor as shown by competitive inhibition of PDGF by EDGF for the receptors

on smooth-muscle cells, fibroblasts and 3T3 cells [21].

Modification of Low-Density Lipoprotein

Endothelial cells have an important influence on the metabolism of LDL by smooth-muscle cells [23]. This can be shown in systems employing co-culture of the cells or medium conditioned by the growth of endothelial cells [21,23]. In these conditions the amount of LDL bound, internalized and degraded by smooth-muscle cells is increased. High-affinity LDL receptors increase greatly in number, and increase further when more endothelial cells are added. The metabolism of LDL by smooth-muscle cells plateaus when the numbers of smooth-muscle cells and endothelial cells are approximately equal. This increased uptake and metabolism may be related to the metabolic requirements of the cells to proliferate since it has been demonstrated that exposure of smooth-muscle cells or fibroblasts to mitogens such as PDGF or FGF increases LDL metabolism co-incident with the stimulation of cells into the growth cycle [17,22]. These in vitro findings are supported by in vivo studies which have shown increased degradation of LDL in areas of endothelial regeneration, following balloon-catheter removal of the endothelium [93]. These studies have used radio-iodinated tyramine cellobiose which remains in the tissue following degradation of LDL and serves as a measure of the amount of LDL which has been taken up by the vessel wall. They are consistent also with the studies by Vlodavsky and Fielding showing increased uptake of LDL by non-confluent (proliferating) endothelial cells compared to quiescent (stationary) endothelial cells [100].

Qualitative Change in Proteoglycans and Lipoprotein Binding

It is apparent that selective extracts of proteoglycan from the arterial wall have more affinity for LDL than other proteoglycan elements [16]. Camejo has described a lipoprotein-binding proteoglycan fraction extracted from normal arterial wall [15]. We have observed that such extracts from neo-intima, collected following balloon de-endothelialization, remove much more cholesterol from solution than extracts of normal aortic wall [5].

Alterations of Lipoprotein and Vessel Wall Uptake

It is likely that particular lipoproteins bind more actively to glycosamingoglycans. The modification of Lp(a) by sulphated polysaccharides which promotes its uptake by mouse peritoneal macrophages has been cited as a reason for its "atherogenicity" [58].

The recognition that people with normal LDL–cholesterol levels but increased levels of LDL apolipoprotein B (hyperapobetalipoproteinaemia) often experience premature coronary artery disease [89] is consistent with trapping of apo B by glycosaminoglycans. The mechanism may also go some way towards explaining the benefit of high levels of HDL [63] relative to LDL since HDL can disassociate the complexes formed between glycosaminoglycans and LDL [9].

The interactions of lipoprotein with the vessel wall include modifications of lipoprotein structure by cells of the vessel wall and by macrophages. While smooth-muscle cells and macrophages in atherosclerotic plaques can accumulate lipid to the extent that the phenotype of a particular lipid-laden cell is no longer identifiable, these cells in tissue culture do not take up cholesterol, even in the presence of high concentrations of LDL [88,103]. Moreover the presence of high concentrations of LDL down regulates the high-affinity LDL receptors so that LDL is not readily internalized. The high concentration of LDL also inhibits intracellular synthesis of cholesterol. When LDL is modified by exposure to acetic anhydride [41], malondialdehyde [33] or diketene [61] the scavenger receptor of the macrophage recognizes it and uptake occurs.

Endothelial cells [50] and smooth-muscle cells [49] can modify LDL to make it recognizable to the scavenger receptors of monocyte/macrophages. The mechanism may involve an intrinsic LDL oxidative process [21]. The modification that occurs in vivo to facilitate uptake is unknown. However, lipoproteins containing apo B isolated from human aortas stimulate cholesterol esterification in macrophages [19]. It is known that macrophages take up LDL–proteoglycan complexes [86]. This uptake enhances cholesteryl ester accumulation in the macro-phages. Similarly the formation of complexes of LDL with heparin and fibronectin or heparin–fibronectin and collagen increases the uptake and stimulates the synthesis of cholesteryl oleate [31].

Summary

From the foregoing it is clear that alterations of the proteoglycan molecule during the formation of neo-intimal tissue might favour the binding of LDL to the glycosaminoglycans, arranged in such a combination as to represent the lipoprotein complexing proteoglycan described by Camejo [15]. Moreover, it is apparent that various modifications of lipoprotein may favour its binding to proteoglycans normally present in the vessel wall. The part played by endothelium in possibly modifying LDL to make it acceptable to macrophages and possibly to smooth-muscle cells for phagocytosis may be important in lipid accumulation in the vessel wall. It is clear that regenerated endothelium allows more LDL to pass into the intima than does normal undamaged endothelium. The mechanism by which this occurs is still not understood but may in part be receptor mediated as in the studies of Vlodavsky and Fielding [100] showing that proliferating (regenerating) endothelium binds and takes up more LDL than normal undamaged endothelium.

The kinds of experimental situations we have been considering involve either repeated or extensive damage to the endothelium. There is no evidence in man that extensive endothelial denudation occurs in response to any stimulus or injury apart from the examples cited above. If a single endothelial cell is damaged so that it dies and detaches, neighbouring cells spread rapidly beneath the dying cell so that actual exposure of the subendothelial tissue to the bloodstream does not occur. It has recently been reported that one form of endothelial injury, that caused by air drying in the rabbit carotid artery, presents a surface which remains inert to the deposition of platelets for several hours after the endothelial loss [14]. This may mean that small areas of injured endothelium may be recovered by migrating and proliferating endothelium before a stimulus to the migration and proliferation of smooth-muscle cells is delivered to the vessel wall.

These considerations make it likely that non-denuding endothelial injury is more likely to occur in response to stimuli, as, for example, that due to endotoxin which causes endothelial cell turnover without disruption.

Stemerman has shown that in rabbits there are focal areas of increased permeability to low-density lipoprotein in the aortic endothelium [94]. Whether these areas develop because of focal endothelial injury or dysfunction is unknown. Additionally, "spontaneously" occurring endothelial damage has been observed in normal aortas of various species. This has been shown by uptake of vital dyes by dead endothelial cells [11] and by the accumulation of IgG in damaged or dying endothelial cells [47].

Possibly accentuation of these "normal" cycles of cell death and replacement may set the stage for other injurious agents to initiate lesion development. This "non-denuding" endothelial injury may, possibly by release of endothelial cell growth factor, stimulate the migration and proliferation of smooth-muscle cells and the formation of an intimal thickening or plaque. Once this has happened then the stage is set for accumulation of lipid by the mechanisms discussed above. These might include increased uptake by regenerating endothelium, modification of lipoprotein by endothelium, trapping of LDL by proteoglycan synthesized by the proliferating intimal cells and possibly by other mechanisms.

Such a formulation marries the concepts of the disease as a response to injury and to hyperlipidaemia. The relative importance of the part played by these agencies in the usual setting of human atherosclerosis remains to be determined and is clearly of great importance in devising means to modify or control the disease process.

References

1. Alavi M, Moore S (1984) Kinetics of low density lipoprotein interactions with rabbit aortic wall following balloon catheter deendothelialization. Arteriosclerosis 4:395–402
2. Alavi M, Moore S (1985) Glycosaminoglycan composition and biosynthesis in the endothelium-covered neointima of de-endothelialized rabbit aorta. Exp Mol Pathol 42:389–400
3. Alavi M, Moore S (1987) Proteoglycan composition of rabbit arterial wall under conditions of experimentally induced atherosclerosis. Atherosclerosis 63:65–74
4. Alavi M, Dunnett CW, Moore S (1983) Lipid composition of rabbit aortic wall following removal of endothelium by balloon catheter. Arteriosclerosis 3:413–419
5. Alavi M, Weigensberg BI, Haqqee R et al. (1986)

Affinity of proteoglycans of normal and de-endothelialized rabbit aortic tissue for cholesterol. Arteriosclerosis 6:556a (abstract)
6. Baumgartner HR (1972) Platelet interaction with vascular structures. Thromb Haemost [Suppl] 51:161–176
7. Baumgartner HR, Studer A (1963) Gezielte überdehung der Aorta abdominalis am normo und hypercholesterinaemischen Kaninchen. Pathol Microbiol 26:129–148
8. Benson EP (1973) Radiation injury to large arteries. III. Further examples with prolonged asymptomatic intervals. Radiology 106:195–197
9. Bihari-Varga M (1978) Influence of serum high density lipoproteins on the low density lipoprotein–aortic glycosaminoglycan interactions. Artery 4:504–511
10. Bihari-Varga M, Begh M (1967) Quantitative studies of the complex formed between acid mucopolysaccharides and serum lipoproteins. Biochim Biophys Acta 144:202–210
11. Bjorkerud S, Bondjers G (1972) Endothelial integrity and viability in the aorta of normal rabbit and rat as evaluated by dye exclusion tests and interference contrast microscopy. Atherosclerosis 15:285–300
12. Boers GHJ, Smals AGH, Trijbels FJM et al. (1985) Heterozygosity for homocystinuria in premature peripheral and cerebral occlusive arterial disease. N Engl J Med 313:709–715
13. Bourassa MG, Enjalbert M, Campeau L et al. (1984) Progression of atherosclerosis in coronary arteries and bypass grafts: ten years later. Am J Cardiol 53:102C–107C
14. Buchanan MR, Richardson M, Hass TA et al. (1986) Vascular subendothelium is not thrombogenic: influence of 13-HODE. Circulation 74:II–98 (abstract 389)
15. Camejo G, Ponce E, Lopez F et al. (1983) Partial structure of the active moiety of a lipoprotein complexing proteoglycan from human aorta. Atherosclerosis 49:241–254
16. Camejo G, Lopez A, Lopez F et al. (1985) Interaction of low density lipoproteins with arterial proteoglycans: the role of charge and sialic acid content. Atherosclerosis 55:93–105
17. Chait A, Ross R, Albers JJ et al. (1980) Platelet derived growth factor stimulates activity of LDL receptors. Proc Natl Acad Sci USA 77:4084–4088
18. Cheseboro JH, Fuster V, Elveback LR et al. (1984) Effect of dipyridamole and aspirin on late vein-grafts patency after coronary bypass operations. N Engl J Med 314:209–214
19. Clevidence BA, Morton RE, West G et al. (1984) Cholesterol esterification in macrophages: stimulation by lipoprotein-containing Apo B isolated from human aortas. Arteriosclerosis 4:196–207
20. Curwen KD, Smith SC (1977) Aortic glycosaminoglycans in atherosclerosis-susceptible and -resistant pigeons. Exp Mol Pathol 27:121–133
21. Davies PF (1986) Biology of disease: vascular cell interactions, with special reference to the pathogenesis of atherosclerosis. Lab Invest 55:5–24
22. Davies PF, Kerr L (1982) Modification of low density lipoprotein metabolism by growth factors in cultured vascular cells and human skin fibroblasts: dependance on duration of exposure. Biochim Biophys Acta 712:26–32
23. Davies PF, Truskey GA, Warren HB et al. (1985) Metabolic cooperation between vascular endothelial cells and smooth muscle cells in co-culture: changes in low

density lipoprotein metabolism. J Cell Biol 101:871–879

24. Day AJ, Bell FP, Moore S et al. (1974) Lipid composition and metabolism of thrombo-atherosclerotic lesions produced by continued endothelial damage in normal rabbits. Circ Res 34:467–476

25. Day AJ, Alavi M, Moore S (1985) Influx of (^{3}H,^{14}C) cholesterol-labelled lipoproteins into re-endothelialized and de-endothelialized areas of ballooned aortas in normal-fed and cholesterol-fed rabbits. Atherosclerosis 55:339–351

26. Dicorleto PE, Bowen-Pope DF (1983) Cultured endothelial cells produce a platelet derived growth factor-like protein. Proc Natl Acad Sci USA 80:1919–1923

27. Donohue S, Sturman JA, Saull G (1974) Arteriosclerosis due to homocyst(e)inemia. Failure to reproduce the model in weanling rabbits. AM J Pathol 77:167–174

28. Duff GL (1935) Experimental cholesterol atherosclerosis and its relationship to human atherosclerosis. Arch Pathol Lab Med 20:80–124, 259–304

29. Duff GL, McMillan GC (1951) Pathology of atherosclerosis. Am J Med 11:92–108

30. Faggioto A, Ross R, Harker L (1984) Studies of hypercholesterolemia in the nonhuman primate. I. Changes that lead to fatty streak formation. Arteriosclerosis 4:323–340

31. Falcone DJ, Mated N, Shio H et al. (1984) Lipoprotein–heparin–fibronectin–denatured collagen complexes enhance cholesteryl ester accumulation in macrophages. J Cell Biol 99:1266–1274

32. Faxon DP, Sanborn TA, Haudenschild CC et al. (1984) Effect of antiplatelet therapy on restenosis after experimental angioplasty. Am J Cardiol 53:72C–76C

33. Fogelman AM, Schechter I, Seager J et al. (1980) Malondialdehyde alteration of LDL leads to cholesteryl ester accumulation in human monocyte-macrophages. Proc Natl Acad Sci USA 77:2214–2218

34. Friedman RJ, Burns ER (1978) Role of platelets in the proliferative response of the injured artery. In Spaet TH (ed) Progress in hemostasis and thrombosis, vol 4. Grune and Stratton, Orlando, Florida, pp 249–278

35. Friedman RJ, Moore S, Singal DP (1975) Repeated endothelial injury and induction of atherosclerosis in normolipemic rabbits. Lab Invest 30:404–415

36. Friedman RJ, Moore S, Singal DP et al. (1976) Regression of injury-induced atheromatous lesions in rabbits. Arch Pathol Lab Med 100:189–195

37. Friedman RJ, Stemerman MB, Wenz B et al. (1977) The effect of thrombocytopenia on arteriosclerotic plaque formation in rabbits. I. Smooth muscle cell proliferation and re-endothelialization. J Clin Invest 60:1191–1201

38. Fuster V, Cheseboro JH (1985) Aortocoronary artery vein graft disease: an experimental and clinical approach for the understanding of the role of platelets and platelet inhibitors. Circulation [Suppl V] 72:65–70

39. Gerrity R (1981) The role of the monocyte in atherogenesis. I. Transition of blood-bone monocytes into foam cells in fatty lesions. Am J Pathol 103:181–190

40. Glenn K, Ross R (1981) Human monocyte derived growth factor(s) for mesenchymal cells: activation of secretion by endotoxin Concanavalin A. Cell 25:603–615

41. Goldstein JL, Ho YK, Basu SK et al. (1979) Binding site on macrophages that mediates uptake and degradation of acetylated LDL, producing massive cholesterol deposition. Proc Natl Acad Sci USA 76:333–337

42. Grondin CM, Campeau L, Lespérance J et al. (1984) Comparison of late changes in internal mammary artery and saphenous vein grafts in two consecutive series of patients 10 years after operation. Circulation 70 I:208–212

43. Grotendorst GR, Chang T, Seppa HEJ et al. (1982) Platelet-derived growth factor is a chemoattractant for vascular smooth muscle cells. J Cell Physiol 113:261–266

44. Groves HM, Kinlough-Rathbone RL, Richardson M et al. (1979) Platelet interaction with damaged rabbit aorta. Lab Invest 40:194–200

45. Gruentzig A (1982) Results from coronary angioplasty and implications for the future. Am Heart J 103:779–783

46. Hajjar DP, Falcone DJ, Fowler S et al. (1981) Endothelium modifies the altered metabolism of the injured aortic wall. Am J Pathol 102:28–39

47. Hansson GK, Holm J, Kral JG (1984) Accumulation of IgG and complement factor C3 in human arterial endothelium and atherosclerotic lesions. Acta Pathol Microbiol Immunol Scand [A] 92:429–435

48. Harker LA, Ross R, Slichter SJ et al. (1976) Homocystine-induced arteriosclerosis: the role of endothelial cell injury and platelet response in its genesis. J Clin Invest 58:731–741

49. Heinecke JW, Rosen H, Chait A (1984) Iron and copper promote modification of low density lipoproteins by human arterial smooth muscle cells in culture. J Clin Invest 74:1890–1894

50. Henriksen T, Mahoney EM, Steinberg D (1981) Enhanced macrophage degradation of LDL previously incubated with cultured endothelial cells: recognition by receptors for acetylated LDL. Proc Natl Acad Sci USA 78:6499–6503

51. Hollander W (1976) Unified concept on the role of acid mucopolysaccharide and connective tissue protein in the accumulation of lipid and lipoproteins and calcium in atherosclerotic plaque. Exp Mol Pathol 25:106–120

52. Huang W, Richardson M, Alavi M et al. (1984) Proteoglycan distribution in rat aortic wall following indwelling catheter injury. Atherosclerosis 51:59–74

53. Ihnatowicz IO, Cazenave JP, Mustard JF et al. (1979) The effect of a platelet-derived growth factor on the proliferation of rabbit arterial smooth muscle cells in tissue culture. Thromb Res 14:477–487

54. Ihnatowicz IO, Winocur PD, Moore S (1979) A platelet-derived factor chemotactic for rabbit arterial smooth muscle cells. Thromb Haemost 42:202

55. Ihnatowicz IO, Winocur PD, Moore S (1981) A platelet-derived factor chemotactic for rabbit arterial smooth muscle cells in culture. Artery 9:316–327

56. Iverius PH (1972) The interaction between human plasma lipoprotein and connective tissue glycosaminoglycans. J Biol Chem 247:2607–2613

57. Kaegi A, Pineo GF, Shimazu A et al. (1974) Arteriovenous shunt thrombosis: prevention by sulfinpyrazone. N Engl J Med 290:304–306

58. Krempler F, Kostner GM, Roscher A et al. (1984) The interaction of Apo B-containing lipoproteins with mouse peritoneal macrophages: a comparison of Lp(a) with LDL. J Lipid Res 25:283–287

59. Loop FD, Lytle BW, Cosgrove DM et al. (1968) Influence of the internal mammary artery graft on 10-year survival and other cardiac events. N Engl J Med 314:1–6

60. Lough J, Moore S (1972) Platelet embolic injury: the

healing process. Exp Mol Pathol 17:144
61. Mahley RW, Innerarity TL, Weisgraber KH et al. (1979) Altered metabolism (in vivo and in vitro) of plasma lipoproteins after selective chemical modification of lysine residues of the apoproteins. J Clin Invest 64:743–750
62. McCully KS, Wilson RB (1970) Production of arteriosclerosis by homocystinemia. Am J Pathol 61:1–12
63. Miller GI, Miller NE (1975) Plasma high-density-lipoprotein concentration and development of ischemic heart disease. Lancet I:16–19
64. Minick CR, Murphy GE, Campbell WG, Jr (1966) Experimental induction of atheroarteriosclerosis by the synergy of allergic injury to arteries and lipid-rich diet. J Exp Med 124:635–652
65. Minick CR, Alonso DR, Littrenta MM et al. (1977) Regenerated endothelium and intimal proteoglycan accumulation. Abstracts of the 31st annual meeting of the Council on Arteriosclerosis, Miami Beach, Florida, 28–30 Nov 1977
66. Moore S (1967) Hypertension and nephrosclerosis: A re-appraisal and a new theory of renal ischemia. Am Heart J 74:730–732
67. Moore S (1973) Thromboatherosclerosis in normolipemic rabbits: A result of continued endothelial damage. Lab Invest 29:478–487
68. Moore S (1975) Clinical correlations. Thromb Hemost 33:417–425
69. Moore S (1981) Injury mechanisms in atherogenesis. In: Moore S (ed) Vascular injury and atherosclerosis. Marcel Dekker Inc, New York, pp 131–148
70. Moore S (1983) Atheroma. In: Harrison MJG, Dyken ML (eds) Neurology, vol 3, cerebro-vascular disease, Butterworth, pp 1–24 (Butterworths International Medical Reviews)
71. Moore S (1983) Lipid accumulation in the rabbit aortic wall in response to injury. In: Schettler G, Gotto AM, Middelhoff G, Habenict AJR, Jurutka KR (eds) Atherosclerosis. Springer-Verlag, Berlin Heidelberg New York, pp 135–139
72. Moore S (1985) Pathogenesis of atherosclerosis. Metabolism [Suppl 1] 34: 13–16
74. Moore S, Lough J (1978) Lipid accumulation in renal arterioles due to platelet aggregate embolism. Am J Pathol 58:283–293
75. Moore S, Richardson M (1985) Proteoglycan distribution in catheter-induced aortic lesions in normolipidemic rabbits. Atherosclerosis 55:313–330
76. Moore S, Friedman RJ, Singal DP et al. (1976) Inhibition of injury-induced thromboatherosclerotic lesions by antiplatelet serum in rabbits. Thromb Haemost 35:70–81
77. Moore S, Friedman RJ, Gent M (1977) Resolution of lipid-containing atherosclerotic lesions induced by injury. Blood Vessels 14:193–203
78. Moore S, Belbeck LW, Richardson M et al. (1982) Lipid accumulation in the neointima formed in normally fed rabbits in response to one or six removals of the aortic endothelium. Lab Invest 47:37–42
79. Mudd SH (1985) Vascular disease and homocystine metabolism. N Engl J Med 313:751–753
80. O'Connell TX, Mowbray JF (1973) Effects of humoral transplantation antibody on the arterial intima of rabbits. Surgery 74:145–152
81 Porter KA, Rendall JM, Solinski C et al. (1966) Light and electron microscopical study from 33 human renal allografts and isografts 1 3/4–2 1/2 years after transplantation. Ann NY Acad Sci 129:615–636
82. Reddy GSR, Wilcken DEL (1982) Experimental homocystinemia in pigs: comparison with studies in 16 homocystinemic patients. Metabolism 31:778–783
83. Reidy MA, Schwartz SM (1983) Endothelial injury and regeneration. IV. Endotoxin: a nondenuding injury to aortic endothelium. Lab Invest 48:25–34
84. Richardson M, Ihnatowycz IO, Moore S (1980) Glycosaminoglycan accumulation in rabbit aortic wall following balloon catheter de-endothelialization: an ultrastructural study. Lab Invest 43:509–516
84a. Richardson M, Gerrity R, Alavi MZ, Moore S (1982) Proteoglycan distribution in areas of differing permeability to Evans blue dye in the aorta of young pigs. An ultrastructural study. Arteriosclerosis 2:369–379
85. Ross R, Glomset JA, Kariya B et al. (1974) A platelet dependent serum factor that stimulates the proliferation of arterial smooth muscle cells in vitro. Proc Natl Acad Sci USA 71:1207–1210
86. Salisbury BGJ, Falcone DJ, Minick CR (1985) Insoluble low density lipoprotein–proteoglycan complexes enhance cholesteryl ester accumulation in macrophages. Am J Pathol 120:6–11
87. Seifert RA, Schwartz SM, Bowen-Pope DF (1984) Developmentally regulated production of platelet-derived growth factor-like molecules. Nature 311:669–671
88. Shechter I, Fogelman AM, Haberland ME et al. (1981) The metabolism of native and malondialdehyde–altered LDL by human monocyte-macrophages. J Lipid Res 22:63–71
89. Sniderman A, Shapiro S, Marpole D et al. (1980) Association of coronary atheroclerosis with hyperapobetalipoproteinemia (increased protein but normal cholesterol levels) in human plasma low density B-lipoproteins. Proc Natl Acad Sci USA 77:604–608
90. Srinavsasan SR, Dolan P, Radhakrishnamurthy B et al. (1975) Lipoprotein-acid mucopolysaccharide complexes of human atherosclerotic lesion. Biochim Biophys Acta 338:58–70
91. Stehbens WE (1974) Haemodynamic production of lipid deposition, intimal tears, mural dissection and thrombosis in the blood vessel wall. Proc R. Soc London (Biol) 185:357–373
92. Stehbens WE, Karmody AM (1975) Venous atherosclerosis associated with arteriovenous fistulas for haemodialysis. Arch Surg 110:176–180
93. Steinberg D, Pittman RC, Carew RC (1985) Mechanisms involved in the uptake and degradation of low density lipoprotein by the artery wall in vivo. Ann NY Acad Sci 454:195–206
94. Stemerman MB, Morrel EM, Burke KR et al. (1986) Local variation in arterial wall permeability to low density lipoprotein in normal rabbit aorta. Arteriosclerosis 6:64–69
95. Stevens RL, Colombo M, Gonzales JJ et al. (1976) The glycosaminoglycans of human artery and their changes in atherosclerosis. J Clin Invest 58:470–478
96. Thornton MA, Gruentzig AR, Hollman J et al. (1984) Coumadin and aspirin in prevention of recurrence after transluminal coronary angioplasty: a randomized study. Circulation 69: 721–727
97. Tyson JE, Desa DJ, Moore S (1976) Thrombo-atheromatous complications of umbilical arterial catheterization in the newborn period: a clinico-pathological study. Arch Dis Child 51:744–754
98. Uys CJ, Rose HG (1982) Cardiac transplantation: aspects of pathology. Pathol Annu 17:147–178
99. Virchow R (1856) Phlogose und thrombose im gefas-

system. In: Gesammelte Abhandlungen zur wissenschaftlichen medizin. Medinger Sohn, Frankfurt, p458

100. Vlodavsky I, Fielding PE, Fielding CJ et al. (1978) Role of contact inhibition in the regulation of receptor-mediated uptake of low density lipoprotein in cultured vascular endothelial cells. Proc Natl Acad Sci USA 75:356–360

101. Vogel A, Raines E, Kariya B et al. (1978) Co-ordinate control of 3T3 cell proliferation by platelet derived growth factor and plasma components. Proc Natl Acad Sci USA 25:2810–2814

102. Wagner WE, Nohlgren SR (1981) Aortic glycosaminoglycans in genetically selected WC-2 pigeons with increased arteriosclerosis susceptibility. Arteriosclerosis 1:192–201

103. Weinstein DB, Carew TE, Steinberg D (1976) Uptake and degradation of LDL by swine arterial smooth muscle cells with inhibition of cholesterol biosynthesis. Biochim Biophys Acta 424:404–421

104. Wight TN, Ross R (1975) Proteoglycans in primate arteries. I. Ultrastructural localization and distribution in the intima. J Cell Biol 67:660–671

105. Wissler RW, Frazier LW, Hughes RH et al. (1962) Atherogenesis in the Cebus monkey. I. A comparison of three food fats under controlled dietary conditions. Arch Pathol 74:312–322

106. Wong MK, Gotlieb AI (1984) In-vitro re-endothelialization of a single cell wound: role of microfilament bundles in filipodia-mediated wound closure. Lab Invest 51:75–81

107. Zugibe FT (1963) Histochemical studies of human coronary atherogenesis – comparison with aortic and cerebral atherogenesis. Circ Res 13:401–409

Chapter 14

Role of Platelets in Atherosclerosis and its Complications

Y. J. Legrand and L. O. Drouet

Introduction

The data presented in Chap. 8 clearly establish that the endothelium protects the vessel wall from being injured by various factors in blood cells and plasma. One of the main factors in atherosclerotic disease is proliferation of the smooth-muscle cells in the intima, which follows their migration from the media through the internal elastic lamina. Although it is not likely that the *only* cause of atherosclerosis is an alteration in the properties of the endothelial cells or desquamation of the endothelium, it is obvious that any injury of the endothelium plays a part in triggering the atherosclerotic and thrombotic processes. The results from experiments using in vivo or in vitro models of de-endothelialization, summarized in Chap. 8, support this statement. However, they are perhaps an insufficient demonstration; only in a very few animal models has it been shown that endothelial cell lesions directly precede the first manifestation of atherosclerosis.

The only animal species in which this sequence of events has been demonstrated is the white Carneau pigeon, which is presented as an animal with high cardiovascular risk. Structural investigations on its vascular tree revealed a high frequency of endothelial lesions in areas where the early signs of arterial alteration develop later. These lesions have been directly related to changes in the endothelial equipment of enzymes which are directly involved in lipid and cholesterol metabolism [37]. They facilitate the activation of circulating thrombocytes, as shown

by a rise in the level of a factor in the plasma resembling platelet factor 4. Even though the white Carneau pigeon is the only animal species in which a direct cause–effect relationship can be demonstrated between a primary endothelial injury and the occurrence of vascular lesions, it is important to define clearly the main risk factors of these endothelial injuries. Their description will precede that of the role of blood platelets in the atherosclerotic process.

Factors in Endothelial Injury

Very many factors provoke endothelial lesions; the effects of some of these are additive. Some are simple in effect such as the mechanical factors which are applied in many experimental models: for example, the model described by Moore [25] has a catheter floating freely in a rabbit artery, continuously scraping the endothelium. Other factors causing lesions will now be summarized. The incidence of risk factors such as diabetes on hypertension [9,18,22] will not be discussed here.

Metabolic Factors

A good example of a metabolic factor involved in lesion formation is provided by hyperlipidaemia. Several studies have shown the occurrence of endothelial desquamation preceding the formation of atherosclerotic lesions in animals on a hyperlipidaemic and hypercholesterolaemic diet [8,27,29,33]. It seems that not only cholesterol but also hydroxylation derivatives resulting from its metabolism, are toxic. In particular, dihydroperoxides derived from cholesterol products are particularly toxic for endothelial cells. In general, every lipid-derived peroxide formed secondarily to the production of free radicals inhibits prostacyclin synthetase and as a consequence, the formation of prostacyclin; they therefore destroy the homeostatic balance between prostacyclin and thromboxane.

Many authors have emphasized endothelial cell alterations which rapidly follow an overload in dietary cholesterol in animals. One of these alterations is an increase in the number of lysosomes present in an endothelial cell [13]; from this observation it may be proposed that, following dietary overload of cholesterol, lytic enzymes could be released in higher quantities than normal and thus degrade the protein and glycoproteins which anchor endothelial cells to the vascular matrix. This is speculative. It has also been shown that cholesterol overload provokes an increase in the number of the cytoplasmic microfilaments involved in endothelial cell contraction [35], and of the Weibel–Palade bodies which merge with the endothelial membrane [38]. It should also be noted that during the induction of a hyperlipidaemia, leucocytes attach to the endothelium, incorporate lipids and then migrate into the subendothelium, as endothelial lesions start to form (see Chap. 11).

Toxic Factors

Many epidemiological and experimental studies have demonstrated clearly the toxicity of tobacco for the endothelium; its mode of action is probably not unique. Rats and rabbits inhaling cigarette smoke show swelling of the cells in endothelium in which denuded areas with microthrombi are observed, mainly in small arteries submitted to a relatively low shear rates. One toxic agent in smoke is carbon monoxide; it has not been established clearly whether its effect on the endothelium is direct or secondary, being linked to the increase of various other substances in the plasma of smokers, for example the free fatty acids, very low-density lipoprotein (VLDL), LDL and various hormones such as cortisol, epinephrine and cyclic guanine monophosphate (GMP). Nicotine is also cytotoxic, adding to their effect.

Another toxic agent in man is homocystine which induces endothelial damage and vascular lesions observable from early childhood. Homocystine has therefore been used to induce experimental endothelial lesions in baboons [15].

Immunological Factors

Circulating immune complexes can provoke endothelial lesions either directly (producing auto-antibodies against the endothelium) or indirectly, secondarily to their own toxicity. These lesions caused by immunological factors are the most frequent in areas of disturbed flow. This is explained by the fact that flow conditions can modify the expression of surface antigens of the endothelium, so that the binding of circulating antibodies or immune complexes could be facilitated: this is the case in disseminated lupus erythematosis, a disease in which the binding of IgG to cultured endothelial cells has been demonstrated [4]. Several factors explain why the resulting alterations of the endothelium destroy its thromboresistance: exhaustion of prostacyclin-determined resistance, occurring rapidly after inhibition of the synthesis of tPA, and direct cytolytic effects. Endothelial lesions also occur following organ transplantation; vascular thrombosis is one of the main complications and is initiated by endothelial detachment followed by adhesion of platelets to the denuded subendothelial areas [31]. Arterial lesions with early de-endothelialization may be the consequence of allergic phenomena, in response to different antigens such as microorganisms, vaccines, antibiotics or certain drugs; cigarette smoke would also produce endothelial desquamation in hypersensitized smokers, as a response to an allergen induced by an IgE antibody [2].

Infectious and Viral Factors

Antibodies against the microfilaments of the endothelial cytoskeleton have been found in association with several infections and viral diseases [20], such as infectious mononucleosis, polyarthritis and diseases induced by cytomegaloviruses. The direct effects of the viral infections on the endothelial cells have been studied mainly in animals; it seems that endothelial cells are less affected by viruses than are smooth-muscle cells. For example, a proliferation of these cells has been noted in normocholesterolaemic chickens submitted to infection by herpes type viruses [24]. Nevertheless, repeated endothelial lesions have been noted in cases of prolonged viral infection.

Role of Mononuclear Cells

Gerrity and colleagues [12], feeding animals on hyperlipidaemic diets, have shown that activated macrophages derived from circulating monocytes adhere to the endothelium, spread along its surface and form small aggregates. This is of great importance in the genesis of arterial lesions, since macrophages contain various proteolytic enzymes (elastase etc.) which can be released into the endothelium and into the subjacent layers of the wall. Some of these cells can pass through the endothelium into the intima and thus provide a clearance system for lipids coming from plasma at this level [12]. The foam cells observed in the intima during experimentally induced atherosclerosis may derive from macrophages. Another factor which is relevant to the origin of the arterial lesion is monocyte-derived growth factor (MGDF), described by Ross et al. [34] as one of the many factors implicated in the migration and proliferation of smooth-muscle cells.

The origins of the endothelial lesions are thus multiple and varied. They are aggravated locally by the haemodynamic conditions: all the factors discussed are more important in turbulence zones (see Chap. 7) in the initiation of thrombosis. For this reason, endothelial lesions are mainly observed in arteries. If an injury is unique, it will evolve and be followed by endothelial repair; if the injury is repeated, an atherosclerotic lesion may develop. In both situations platelets play an important role, releasing different factors within the wall.

Role of Platelets in Atherosclerosis

Injured endothelium loses its protective effect in three ways:

1. It stops protecting the vessel wall: the selective and specific sieve that it provides for plasma is altered; plasma factors such as cholesterol carried by the LDL enter the wall freely.
2. It is no longer protected from blood cells: polymorphonuclear monocytes and platelets

adhere to the exposed subendothelial surfaces.
3. It no longer offers protection against plasma clotting factors: the extrinsic and intrinsic coagulation pathways are activated, either by the exposure of an endothelium-linked tissue factor (extrinsic pathway) or secondarily to the expression of a pro-coagulant surface on the platelet membrane (see Chap. 8). This results in the generation of thrombin in plasma [42,43] which participates in both coagulation and platelet activation.

We will briefly summarize the different steps of the activation process which results from the adhesion of platelets to the subendothelial thrombogenic macromolecules.

The first step is activation of membrane phospholipases which leads to the production of phosphatidic acid and thromboxane A_2, both of which are involved in the mobilization of calcium from the dense tubular system [10,11]. Calcium is indispensable to the phosphorylation of proteins which are implicated in the centralization of the granules and the secretion of their contents. Among the released substances [19], platelet-derived growth factor (PDGF) is particularly important in the development of the atherosclerotic process.

A Few Data on PDGF and its Role

One of the principal manifestations of atherosclerosis is the proliferation of smooth-muscle cells in the intima as a repair phenomenon when the endothelium has been injured. A large variety of growth factors have now been identified in various types of cells.

PDGF is certainly the best characterized of all the growth factors so far identified. In 1974 Ross and colleagues [34] showed that serum derived from whole blood stimulated in vitro mesenchymatous cell growth, in contrast with serum derived from platelet-poor plasma which did not stimulate this effect. This observation was the first step in experiments which led to the identification, isolation and characterization of PDGF as a cationic glycoprotein (Pi = 10.2). It exists in two different molecular forms with different carbohydrate contents, but with the same biological activities: PDGF 1 has a molecular weight of 51 000 and contains 7% carbohydrate; PDGF 2 has a molecular weight of 28 000 and contains 4% carbohydrate.

PDGF is produced by the megakaryocyte and localized in the alpha granules of platelets [3,19]: each platelet contains 1000 molecules of PDGF. It is released during the aggregation which follows the adhesion of platelets to subendothelium and collagen. PDGF is a potent mitogen: it mainly acts on mesenchymatous cells such as fibroblasts, glial cells and smooth-muscle cells. In contrast, it does not affect endothelial and epithelial cells, hepatocytes and splenic lymphocytes. Its action is not felt by cells which are normally exposed to blood flow: its effect is manifest in situ, when platelets have been activated by contact with subendothelium.

The first step of the stimulation of a target cell by PDGF is its binding to a receptor on the membrane, which initiates a series of events leading to the increased synthesis of DNA and cell proliferation [33]. The binding is specific, with a Kd between 10^{-10} and 10^{-9} M. 3T3 cells offer 400 000 receptors for PDGF on their surface. These receptors are phosphorylated on their tyrosyl residues via a protein tyrosine kinase bound to the cell membrane [1,6,27,40]. This is a common mechanism for all growth factors; they all stimulate, on their target cell, protein tyrosine kinase activity which phosphorylates a protein of molecular weight 180 000, which is the receptor of PGDF. This receptor itself bears protein tyrosine kinase activity and is therefore able to be autophosphorylated by PDGF. The mode of action of PDGF has been reviewed recently by Deuel and Huang [6]. We will only comment that PDGF is a potent metabolic stimulant and also that the biosynthetic activities stimulated by PDGF are the same as those which are stimulated in virus-transformed cells. In this regard, it must be underlined that, in retrovirus-stimulated cells, a protein kinase is also responsible for the cell transformation.

This stimulative mechanism explains the variety of effects of PDGF:

1. In repair and wound-healing processes, because of its mitogenic properties: stimulation of cells secondarily stimulates the synthesis of connective tissue macromolecules. PDGF is also a potent chemotactic agent for a variety of cells involved in repair and inflammatory processes. It also increases the production of collagenase which degrades those fibres of collagen disrupted before the tissue repair processes begin.

2. In the genesis of the atherosclerotic lesions: this effect is due to its chemotactic effect on the smooth-muscle cells which migrate from the media into the intima where they proliferate. It has also been shown that PDGF stimulates the synthesis of cholesterol esters and phospholipids and increases the number of LDL receptors on the same cells[1]. This facilitates an enhanced uptake of cholesterol by smooth-muscle cells.

3. In oncogenesis: the amino acid sequence of PDGF presents important homologies with that of certain transforming proteins identified in Simian sarcoma virus or SV 40 transformed cells, as well as in human osteosarcoma cells. This shows that PDGF is an important regulator of transformed cell growth.

Progression of the Atherosclerotic Lesion

The release of PDGF at the site of an endothelial lesion is thus one of the main factors in atherosclerosis. In unique and isolated endothelial injuries, the arterial lesion is reversible. The damaged surface is coated by a platelet layer which rapidly becomes non-thrombogenic; these platelets will release PDGF which will stimulate smooth-muscle cells to proliferate and to secrete the elements of a neo-intima. This is less thrombogenic and will be rapidly covered by a new endothelium.

Arterial lesions will be serious if endothelial injuries are repeated or become chronic. In this case, the lesion will progress and be slowly and inexorably complicated; the smooth-muscle cells will proliferate, extracellular matrix macromolecules will be synthesized and the elastic lamina will divide into fragments. Elastase-type enzymes secreted by polymorphonuclear leucocytes and blood platelets [17] and also present in the wall itself [16] are probably implicated in parietal elastinolysis. The resulting diminution of arterial elasticity is aggravated by increased synthesis and secretion of collagen from proliferating smooth-muscle cells. This results in a fibromusculoelastic lesion, which is the first step of the process leading to the constitution of an atherosclerotic plaque.

We have mentioned that PDGF helps to establish the lipid overload by stimulating cholesterol ester synthesis and endocytosis. Cholesterol is carried by the oleic-acid-rich LDL, internalized and released in its free form inside the cells where it is re-esterified by a cholesterol ester synthetase: this enzyme preferentially utilizes linoleic acid which is thereafter released in the extracellular parietal matrix as cholesterol linoleate. This forms lipid deposits which will bind to elastin and glycosaminoglycans by hydrophobic interactions (see Chaps 12 and and 13). The athero-sclerotic plaques will then calcify, forming hydroxyapatite crystals [36]. The evolution of the process will then lead to those thrombotic accidents which mark the beginning of the clini-cal manifestations of the disease. The athero-sclerotic plaques rich in lipids and calcium will reach a critical size for the vascular lumen and for the wall thickness; they will then become unstable and start fissuring. There are two conse-quences of this fissure formation: – it exposes blood to large altered areas which will activate haemostasis and coagulation factors and it exposes the media to high pressure which may give rise to dissection.

Direct involvement of platelets in the for-mation of a lesion has been demonstrated by the fact that in animals made thrombocytopenic by the use of anti-platelet antibodies, endothelial injuries did not cause severe vascular damage. However, in hyperlipidaemia, atherosclerotic lesions may form in which platelets are not impli-cated. Within a week of the onset of hyper-lipidaemia, monocytes attach to the endothelium and migrate towards the subendothelium where a fat overload gives rise to foam cells [11]. A monocyte-derived growth factor (MDGF) has been implicated, as well as an endothelial-cell-derived growth factor (ECDGF). These prob-ably participate in cell proliferation.

Involvement of Haemostatic Factors in the Complication of the Arterial Lesions

There are three complications of arterial lesions in which haemostatic factors are involved: thrombosis, venous occlusion and spasm.

Thrombosis

Thrombosis results from the simultaneous acti-vation of platelets and plasma coagulation factors; its mechanism is similar to that of hae-mostasis. The resulting thrombus is composed of a platelet aggregate consolidated in a fibrin network within the vessel lumen.

Platelet Activation

Important steps in platelet activation are the secretion of granules, more particularly of aden-osine diphosphate (ADP) from dense granules,

and activation of phospholipases, and these steps lead to the production of thromboxane A_2 (TxA_2). These two steps are particularly import-ant, as both ADP and TxA_2 recruit circulating platelets to aggregate to those which are adhered to damaged vessel. This process necessitates rapid mobilization of plasma and platelet fibrinogen and binding of the latter to its receptor expressed on the glycoprotein complex GP IIb/IIIa. Bound fibrinogen then forms inter-platelet bridges in the presence of calcium [26], and the process is possibly consolidated by the binding of other co-factors to their receptors (such as thrombospondin).

Activation of Plasma Coagulation Factors

This is due to generation of thrombin at the level of a parietal injury. Thrombin generation results from:

1. The activation of the coagulation extrinsic pathway under the effect of tissue factors from the endothelium;

2. The activation of the intrinsic pathway initiated by the activation of factor XII [5] by the injured endothelial cells [41] or collagen [15]. We have already mentioned the part platelets play in this pathway, providing a pro-coagulant surface which facilitates the activation of factor X and pro-thrombin (see Chap. 6, ref. 14).

The resulting thrombin has two main effects:

1. It stimulates the aggregation of platelets as it increases the production of TxA_2 and expresses receptors for different co-factors of aggregation (fibronectin, thrombospondin) on the membrane [30];

2. It converts fibrinogen into fibrin and acti-vates factor XIII which stabilizes the fibrin clot. Plasma coagulation is the formation of an insol-uble clot with which fibronectin and throm-bospondin associate. Platelets themselves participate in the organization of the clot: they incorporate into its meshwork by establishing intermolecular bridges between extracellular fibrillar proteins present in the clot and intra-cellular contractile proteins via transmembrane glycoproteins and calcium ion. Pro-coagulant and pro-aggregating mechanisms are thus closely related.

Fate of the Thrombus

A thrombus is rarely obstructive in arteries. Its fragmentation, accelerated by the activation of the fibrinolytic system, is mainly due to blood flow. Resulting fragments are carried on by the flow and will embolize in the distal circulation. The thrombus may also remain where it has formed, be invaded by parietal smooth-muscle cells and thus be incorporated into the wall. In this instance, it will participate in the aggravation of the atherosclerotic plaque and contribute to the intimal thickening.

Thromboembolic Occlusion

Obstruction of small vessels by emboli resulting from fragmentation of a mural thrombus may have dramatic consequences. At first this obstruction will damage or even destroy the vascular bed downstream. There may be cerebrovascular accidents, acute ischaemia of the lower limbs or renal or more rarely myocardial infarction. However, it is clear that there is usually a primary thrombotic origin to myocardial infarction. This is shown by anatomopathological observations revealing occlusive thrombi in patients dying after acute infarction, and by radioisotopical observations with ^{111}In-labelled platelets [29] and ^{125}I fibrinogen. Angiographical examinations have revealed that in 90% of the patients who have an acute infarction, formation of an occlusive thrombus precedes the acute episode [7]. It is important to note that the distal embolization of fragmented thrombi is not the only cause of obstructive accidents in atherosclerosis and that fissuring of plaques releases fragments which also embolize.

Arterial Spasm

This is largely due to the vasoconstrictor effect of thromboxane A_2 produced by platelets during their activation. Angiography has shown frequent association of coronary spasm with myocardial ischaemia which is reversed when a relaxation of the vessel wall is observed [21,23,39]. It is evident that as arterial spasm narrows the vascular lumen it facilitates its obstruction by transient platelet deposits or emboli which circulate in the bloodstream. Arteries which present atherosclerotic lesions are more sensitive to spasm than normal arteries. If the spasm is transitory, its effect is limited; even if it provokes occlusion, the relaxation which follows brings the circulation back to normal and helps in the dissolution of any mural thrombus

which may have been formed at the site. If the spasm is prolonged and if an occlusive thrombus has been formed, it remains following vessel relaxation and the artery remains obstructed, unless the thrombus is dislodged by the blood flow and embolizes distally.

Sudden death and unstable angina are two of the dramatic consequences of the arterial spasm. It is therefore important to emphasize the role played by platelets in its genesis: when platelets are activated by turbulence provoked by an atherosclerotic plaque, they release TxA_2 locally in an amount which is sufficient to induce a disquieting spasm. Moreover, normal endothelium receives messages from the circulation which it transmits to the underlying layers of the wall. Endothelial injury modifies this transmission and the result is that smooth-muscle cells become more sensitive to vasoconstrictor agents such as TxA_2.

Conclusion

The relevance of the numerous factors involved in the formation and complication of arterial lesions must be clearly defined. In the middle of the nineteenth century, Virchow listed the three factors he believed were important in coagulation: vessel-wall injury, changes in the coagulability of blood and changes in flow. Normally, the interaction of endothelium, blood constituents and haemodynamics leads to the repair of any break in the wall. Haemostasis and thrombosis are firmly bound together and this point must be clearly emphasized. In both phenomena, there is interdependence between vascular, platelet and plasma factors; their interaction leads to the formation of a thrombus and to its dissolution within the framework of a series of events which includes the adhesion and activation of platelets. This process leads not only to their aggregation, but also helps in the generation of thrombin and the coagulation of fibrinogen, producing the fibrin clot. There is thus a combined effect of vessel, platelets and plasma. One should not omit to mention changes occurring in the wall, which have significance in repair. These can be exaggerated and then become risk factors for vessel injury. The role of the factors involved in primary haemostasis and in coronary atherosclerosis must therefore be clearly defined. This is important not only in terms of mechanisms, but also for diagnostic and prevention purposes.

References

1. Antoniades HN, Williams LT (1976) Human platelet-derived growth factor: structure and function. Thromb Res 9:575–583
2. Becker CG, Levi R, Zavecz JH (1979) Induction of IgE antibodies to antigen isolated from tobacco leaves and from cigarette smoke condensate. Am J Pathol 96:249–255
3. Chernoff A, Levine RF, Goodman DS (1980) A growth promoting activity derived from guinea pig mega-caryocytes. J Clin Invest 65:926–930
4. Cines DB, Lyss AP, Reeper M et al. (1984) Presence of complement-fixing anti-endothelial cell antibodies in systemic lupus erythematosus. J Clin Invest 73:611–625
5. Cochrane CG, Griffin JH (1979) Molecular assembly in the contact phase of the Hageman factor system. Am J Med 67:657–664
6. Deuel TF, Huang JS (1984) Platelet-derived growth factor: structure, function and roles in normal and transformed cells. J Clin Invest 74:669–676
7. Dewood MA, Spores J, Noiske R et al. (1980) Prevalence of coronary occlusion during the early hours of transmural myocardial infarction. N Engl J Med 303:897–902
8. Freidman RJ, Moore S, Singal DP (1975) Repeated endothelial injury and induction of atherosclerosis in normolipemic rabbits by human serum Lab Invest 32:404–415
9. Gabbiani G, Elemer G, Guelpa CH et al. (1979) Morphological and functional changes of the aortic intima during experimental hypertension. Am J Pathol 96:399–414
10. Gerrard JM, Butler AT, White JG (1978) Calcium release from a platelet calcium sequestering membrane fraction by arachidonic acid and its prevention by aspirin. Prostaglandins 15:703–715
11. Gerrard JM, White JG, Petterson DA (1978) The platelet dense tubular system: its relationship to prostaglandin synthesis and calcium flux. Thromb Haemost 40:224–235
12. Gerrity RG, Naito HK, Richardson M et al. (1979) Dietary induced atherogenesis in swine: morphology of the intima in pre-lesion stages. Am J Pathol 95:775–784
13. Goldfischer S, Schiller B, Wolinsky H (1975) Lipid accumulation in smooth muscle cell lysosomes in primate atherosclerosis. Am J Pathol 78:497–504
14. Goode TB, Davies PF, Reidy MA et al. (1977) Aortic endothelial cell morphology observed in situ by scanning electron microscopy during atherogenesis in the rabbit. Atherosclerosis 27:235–251
15. Harker LA, Ross R, Slichter SJ et al. (1976) Homocystine-induced arteriosclerosis: the role of endothelial cell injury and platelet response in its genesis. J Clin Invest 58:731–741
16. Hornebeck W, Brechemier RD, Bourdillon MC et al. (1981) Isolation and partial characterization of an elastase-like protease from rat aorta smooth muscle cells: possible role in the regulation of elastin biosynthesis. Connect Tissue Res 8:245–249
17. Hornebeck W, Pignaud G, Legrand YJ et al. (1984) Differentiation of the elastase-type protease of platelets from other elastases. Clin Physiol Biochem 2:166–175
18. Huttner I, Badonnel MC, Elemer G et al. (1979) Aortic intima of the rat in various phases of hypertension. Exp Mol Pathol 31:191–200
19. Kaplan KL, Broekman MJ, Chernoff A et al. (1979) Platelet granule proteins: studies on release and subcellular localisation. Blood 53:604–618
20. Linder EM, Hormia VP, Letho T et al. (1981) Identification of cytoskeletal intermediate filaments of vascular endothelial cells as targets for autoantibodies in patient sera. Clin Immunol Immunopathol 21:217–227
21. Luchi RJ, Chahine RA, Raizner AE (1979) Coronary artery spasm. Ann Intern Med 91:441–449
22. Majack RA, Bhalla RC (1980) Endothelial alterations and colloidal carbon permeability in the peripheral vasculature of the spontaneously hypertensive rat. Exp Mol Pathol 32:201–215
23. Maseri A, D'Abbate A, Baroldi G et al. (1978) Coronary vasospasm as a possible cause of myocardial infarction. A conclusion derived from the study of "preinfarction" angina. N Engl J Med 299:1271–1277
24. Minick CR, Fabricant CG, Fabricant J et al. (1979) Atheroarteriosclerosis induced by infection with a herpes virus. Am J Pathol 96:673–700
25. Moore S (1979) Endothelial injury and atherosclerosis. Exp Mol Pathol 31:182–190
26. Niewiarowski J, Kornocki E, Budzynski AZ et al. (1983) Fibrinogen interaction with platelet receptors. Ann NY Acad Sci 408:536–555
27. Nishimura J, Huang JS, Deuel TF (1982) Platelet-derived growth factor stimulates tyrosine-specific protein kinase activity in Swiss mouse 3T3 cell membranes. Proc Natl Acad Sci USA 79:4303–4307
28. Pittilo RM, Mackie IJ, Rowles PM et al. (1982) Effects of cigarette smoking on the ultrastructure of rat thoracic aorta and its ability to produce prostacyclin. Thromb Haemost 48:173–176
29. Pledger WJ, Stiles CD, Antoniades HN, Scher CD (1977) Induction of DNA synthesis in BALB/c 3T3 cells by serum components: re-evaluation of the commitment process. Proc Natl Acad Sci USA 74:4481–4485
30. Plow EF, Srouji AH, Meyer D et al. (1984) Evidence that three adhesive proteins interact with a common recognition site on activated platelets. J Biol Chem 259:5388–5391
31. Ritchie JL, Stratton JR, Thiele B et al. (1981) Indium 111 platelet imaging for detection of platelet deposition in abdominal aneurysms and prosthetic arterial grafts. Am J Cardiol 47:882–889
37. Ross R (1981) Atherosclerosis: a problem of the biology of arterial wall cells and their interactions with blood components. Arteriosclerosis 1:293–311
33. Ross R, Harker L (1976) Hyperlipidemia and atherosclerosis: chronic hyperlipidemia initiates and maintains lesions by endothelial cell desquamation and lipid accumulation. Science 193:1094–1100
34. Ross, R, Glomset J, Kariba B et al. (1974) A platelet-dependent serum factor that stimulates the proliferation of arterial smooth muscle cells in vitro. Proc Natl Acad Sci USA 71:1207–1210
35. Shimamoto T, Hidaka H, Moriya K et al. (1976) Hyper-reactive arterial endothelial cells: a clue for the treatment of atherosclerosis. Ann NY Acad Sci 275:266–285
36. Stemerman MB, Ross R (1972) Experimental atherosclerosis. I. Fibrous plaque formation in primates, an electron microscopic study. J Exp Med 136:769–777
37. Subbiah RMT, Unni KK, Kottite BA (1976) Arterial

and metabolic changes during the critical period of spontaneous sterol accumulation in pigeon aorta. Exp Mol Pathol (1979) 24:287–301

38. Trillo AA, Prichard RW (1979) Early endothelial changes in experimental primate atherosclerosis. Lab Invest 41:294–302

39. Vincent GM, Anderson JL, Marshal HW (1983) Coronary spasm producing coronary thrombosis and myocardial infarction. N Engl J Med 309:220–223

40. Westermark B, Wasteson A, Heldin CH (1982) Stimulation of tyrosine-specific phosphorylation by platelet-derived growth factor. Nature 295:419–420

42. Wiggings RC, Loskutoff DJ, Cochrane GG et al. (1980) Activation of rabbit Hageman factor by homogenates of cultured rabbit endothelial cells. J Clin Invest 65:197–206

42. Zwall RFA (1978) Membrane and lipid involvement in blood coagulation. Biochim Biophys Acta 515:163–205

43. Zwaal RFA, Hemker HC (1982) Blood cell membranes and haemostasis. Haemostasis 11:12–39

Chapter 15

Mechanical Factors in the Pathogenesis, Localization and Evolution of Atherosclerotic Plaques

S. Glagov, C. K. Zarins, D. P. Giddens and D. N Ku

Mechanical stresses associated with blood flow and pressure have been associated with the pathogenesis of atherosclerosis in the human arterial tree. Observations that plaques tend to occur preferentially in relation to branch ostia, bifurcations and bends of the major arteries have led to hypotheses that alterations in flow which occur at transitions in geometric configuration potentiate the development of intimal lesions. The proposed fluid dynamic localizing features include increases in flow velocity [38,199] and wall shear stress [67], low flow velocity and reduced wall shear stress [25,66,219], flow separation [64,134,172] and departures from uniform unidirectional laminar flow such as turbulence [58,94,206] and variations in flow direction and velocity [46,68,123,185]. Regions of branching and curvature are associated also with alterations in the distribution of mural tensile stresses. These forces influence artery wall metabolism and correspond to local modifications of artery wall thickness [215], composition [60,83,91], microarchitecture [35,214] and compliance [40] and may therefore be expected to play a role in determining local susceptibility to atherosclerosis [82].

Several specific arteries or artery segments are especially prone to the development of plaques [158,159,208]. These include the proximal internal carotid artery, the coronary arteries, the distal aorta and the arteries supplying the lower extremities [55,180]. Other major vessels, such as the pulmonary, renal, mesenteric and internal mammary arteries as well as the thoracic aorta, are largely spared despite the presence of both

branches and bends and the occurrence of advanced disease at nearby susceptible locations [84,85]. Despite the high incidence of lesions in the arteries which are at high risk, individual differences in the relative extent and clinical consequences of plaque formation in each of these regions are common [48,85]. Thus, individuals with severe coronary artery disease may have little involvement at the carotid bifurcation, while others with symptomatic carotid disease have little or no clinical evidence of significant coronary artery disease. Selective involvement of specific vessels and discrepancies in relative degrees of involvement of the particularly susceptible vessels in the same individual have also been attributed to mechanical effects associated with local differences in flow and geometric configuration [15,166,174,199]. In the present chapter, we review evidence for the role of mechanical factors in plaque distribution, initiation and evolution and consider the manner in which these factors may account for differences in involvement of the major arteries.

The Mechanical Forces

Blood Flow

Several brief considerations of the mechanical stresses associated with blood flow in arteries will serve to identify the physical forces in question. Mean velocity of flow at any location in a straight vessel without branches and under a given head of pressure is determined by its cross-sectional area. If the volume flow is constant, velocity will be greatest where the vessel is narrowest and least where the vessel is widest. Because of frictional resistance at the inner surface, flow velocity is not uniform across the lumen of an artery. The velocity is greater in midstream than at the blood–endothelium interface. The gradient of the velocities from centre-stream to wall will be symmetrical across the vessel lumen at any given level so long as there is radial and axial symmetry of the vessel, that is, so long as any narrowing results from a symmetrical taper or constriction and the vessel does not bifurcate or bend. When the narrowing is eccentric or where the vessel bifurcates, the velocity profile, i.e. the distribution of velocities across the lumen, is altered and may no longer be the same in all directions. If the parent and branch vessels are again straight and unbranched beyond the division or bend, the symmetry of the flow profile is restored. In general, the flow divider, which is the distal edge of a branch ostium or the junctional ridge at a bifurcation, intercepts the flow profile at some distance from the wall towards the centre of the lumen. It is therefore a region of higher flow velocity than that which prevails at the wall. The relative diameters of the parent vessel and its divisions and the angle of the bifurcation also determine the degree of displacement of the flow profile.

Flow is considered to be laminar where the fluid displacement proceeds along predictable, stable lines of flow. Departures from unidirectional, laminar flow may occur under a number of circumstances. Complex paths, such as vortices and eddys, may form, particularly about branch points and distal to stenoses. Transient retrograde flow may also occur. Such flow profiles are not necessarily turbulent, for turbulence implies random movement of elements in the flow field. The conditions which determine whether flow will be laminar or turbulent are a function of flow velocity, vessel diameter and blood viscosity. The Reynolds number, which takes these factors into account, is an expression which characterizes the tendency of flow to remain stable. It is given by the expression:

$$Re = \frac{Vd}{v}$$

where V is the mean value of the time-averaged velocity across the vessel, d is the vessel diameter and v is the kinematic viscosity. For a given

system, a Reynolds' number exceeding a certain value predicts the occurrence of turbulent flow. Extreme or abrupt changes in geometry which occur in relation to severe stenoses, projecting edges or other obstacles in the flow stream may also cause focal turbulent disturbances regardless of general stable conditions.

When the flow profile is displaced, the gradient of the velocities near the wall is altered. On the side towards which the higher midstream velocities are displaced, for example in the region just behond a flow divider at a branch site, the velocity gradient is increased. Where the flow velocity profile has been skewed away from the artery wall the velocity gradient is decreased and a zone of flow separation is said to occur. The gradient of the velocities near the artery wall determines the wall shear stress. This force, transmitted to the endothelium by the flowing blood, tends to displace the endothelium and the inner layers of the artery wall over the immediately subjacent layers in the direction of flow. It is given by the expression:

$$\tau_w = \mu \ \partial u / \partial y_w$$

where μ is the viscosity of blood (poise) and $\partial u / \partial y$ is the velocity gradient at the wall. For a parabolic flow profile the expression is:

$$\tau_w = \frac{4\mu Q}{r^3}$$

where Q is the volume flow (cm^3/s) and r is the radius (cm). It is noteworthy that radius appears in the denominator to the third power. Thus, small changes diameter will result in marked changes in wall shear stress. The reader is referred to recent texts for detailed treatment of the relevant fluid dynamic principles [28,72,198].

Pressure and Wall Tension

The distending force due to intraluminal pressure acts as a compressive and convective force, perpendicular to the endothelial lining. It may also be considered as the force tending to tear the vessel open and acting in a circumferential direction trangential to the wall and perpendicular to the axial trans-section of the vessel. The tangential or circumferential tension produced in relation to intraluminal pressure is well approximated by the Law of Laplace which relates tension, pressure and radius. If the thickness of the wall is negligible compared with the radius, the expression is:

$$T = Pr$$

where T is the tension (dynes/cm), P is the pressure (dynes/cm^2) and r is the radius (cm). Although the force due to distending pressure has a radial compressive component, an axial elongation component and a circumferential or tangential component [50], it is the latter which is most closely related to the thickness, composition and structure of the arterial wall. The relationship between tangential tension and circumference as pressure increases may therefore be considered to be a measure of the elasticity of the artery wall [22]. The close structural association between collagen and elastin in the artery wall results in a modulus of elasticity lower than that of collagen alone and higher than elastin [60,88]. As collagen fibres are drawn taught with increasing distension, distensibility decreases as the effective resultant modulus of elasticity increases [214]. At distending pressures below usual diastolic levels the modulus is closer to that of elastin [157]. Between diastolic and systolic pressures the modulus approaches that of collagen. Since collagen is relatively inextensible, there is little change in distensibility with pressure increases above systolic levels.

The tangential tension is actually supported by the full thickness of the artery wall. Tangential tensile stress on the wall or the force per unit axial trans-sectional area is expressed by:

$$S = Pr/d$$

where d is the wall thickness (cm) and S is the stress (dynes/cm^2). At bends and bifurcations, the effective radius is modified and cannot be considered to be equal to the radius of the artery itself [22]. At the inner or concave side of a bend, for example, the curvature of the artery cylinder is opposite to the curvature imposed by the bend, tending to "flatten" the artery wall and thereby to increase the effective radius. Thus, the tensile stress is elevated in this region compared to the straight portion of the vessel. Conversely, at the outer or convex aspect the two curvatures are in the same direction. The effective radius is therefore decreased in this region and tensile stress is lower than if the vessel were straight. Similar relationships exist in relation to changes in effective radius at more complex configurations about bifurcations. The effects of distending pressure may be dampened or modified by surrounding adventitial and peri-adventitial tissues which may furnish support and tend to limit vessel

excursions, depending upon the density and rigidity of the tissues. Tensile forces also tend to modify vessel length as pressure varies and may therefore alter the configuration of bends and bifurcations. Positional changes of organs and movement of extremities also alter vessel configurations, while tethering at branches tends to limit these effects. The resulting stresses are in general relatively small compared to those imposed by intraluminal distending pressure.

Pulsatile Flow

Arteries are normally subjected to pulsatile flow and therefore to excursions in flow velocity and distending pressure in the course of the cardiac cycle and to modulation of these features in relation to changing levels of peripheral resistance. The mechanical stresses related to these variations may be expected to elicit both short-term and long-term reactions. Although marked local haemodynamic differences are demonstrable about bifurcations in the presence of steady flow [135,219], recent evidence suggests that differences in both flow dynamics and cyclical wall motion associated with pulsatile flow may play a critical role in atherogenesis [66,68,123]. The effect of pulsation may also have a greater impact in some regions than in others [7,93], while the number of pulsations over time may have a marked bearing on both the degree and localization of intimal lesions in all susceptible regions [82,138].

In summary, the artery wall is normally exposed to two major mechanical forces, to wall shear stresses acting principally at the blood–endothelium interface and related directly to the flow velocity profile, and to tensile stresses, acting across the vessel wall and related directly to pressure and radius. It should be noted that these forces co-exist throughout the arterial tree. At locations of changing configuration, where the distribution of plaques may be correlated with the distribution of mechanical forces, regions of altered wall shear stress are also often regions of altered tensile stress.

The Physical State of the Plaque

Mechanical forces may determine plaque localization, affect plaque composition and influence the consequences of lesion formation at each of the recognized stages of lesion development. There is general agreement at present that the fully developed atherosclerotic lesion consists of an accumulation, mainly in the intima, of smooth-muscle cells [74,98,210] and macrophages [76,77,152,169,181], intracellular and extracellular lipids in several physical and chemical states and interstitial macromolecules including collagen and elastin fibres as well as glycosaminoglycans and fibrin. Calcification is frequent and is at times a prominent feature. In the characteristic established lesion, the components are organized to form a plaque in which extracellular lipids and tissue debris form a central core or necrotic centre. This region is usually separated from the artery lumen by a well-organized denser subendothelial layer composed of cells and connective tissue fibres (Fig. 15.1). This "fibrous cap" is more or less rigid but not hard, while the necrotic interior tends to be soft, grumous or semi-fluid. Calcification tends to make the plaque brittle.

The intima of those arteries which are rarely the site of plaque formation usually consists of endothelial cells which rest almost directly on the internal elastic lamina and are attached by peripheral dense bodies to the elastin at points of relatively tenacious adhesion [200]. In other locations, several layers of cells, largely smooth muscle, and connective tissue fibres occupy the intimal region between the endothelium and the internal elastic lamina. This appearance may be focal, particularly at branches and bends where configurational changes occur [186], but may also be diffuse. Regions of fibrocellular intimal thickening tend in general to be the same regions in which lipid-containing plaques occur [16, 61,144,185,192,201], but a transition from lipid-free, well-organized, intimal fibrocellular thickenings to typical lipid-containing atherosclerotic plaques remains to be demonstrated. Focal intimal thickening, characterized mainly by relatively uniform cellular hyperplasia, may also occur and persist in response to focally altered haemodynamic conditions. Often present at anastomotic sites, these deposits contain abundant ground substance, rich in proteoglycans, but lipid deposition, necrosis, fibrosis and calcification are not characteristic features. Mechanical factors such as disordered or turbulent flow, vibration or compliance mismatch of the anastomosed components have been proposed as possible stimuli for this presumably reactive intimal proliferative reaction [104,142]. Although deposits of this type may modify flow, they remain pliable and are not subject to disruption.

Whether intimal fibrocellular thickening is a necessary precursor state for human atherosclerosis or merely identifies a region of mesenchymal activation and predilection for plaque formation should specific atherogenic conditions such as hyperlipidaemia be present, remains to be clarified. Nevertheless, the terms "intimal thickening", "atherosclerosis" and "arteriosclerosis" are often used interchangeably in both human and experimental reports dealing with atherogenesis, including those studies which deal with mechanical predisposing factors. Reservations concerning the relationship between intimal fibrocellular thickening and atherosclerosis should be kept in mind when evaluating such studies.

Precursor states for atherosclerosis consisting of intimal swelling, corresponding to increased accumulations of plasma constituents but with little cellular reaction, have been proposed by some observers [97,177], while others have assigned the central initiating role to smooth-muscle proliferation [10,163,210]. The unequivocal identification of lipid-laden cells and extra-

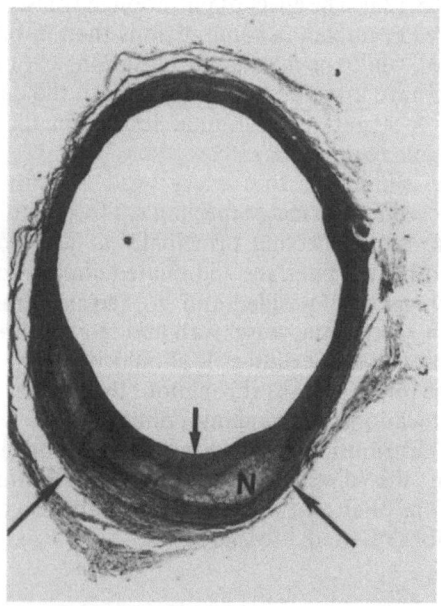

Fig. 15.1. Typical atherosclerotic plaque in a human superficial femoral artery which was fixed while distended at 100 mmHg. The plaque has been modelled such that the lumen remains circular. The plaque bulges outward (*two arrows*) and the necrotic centre (*N*) is sequestered from the lumen by a fibrous cap (*single arrow*) which is similar in thickness to the adjacent uninvolved artery wall and is organized in circumferentially oriented layers.

cellular lipids in grossly apparent flat intimal streaks and dots in young individuals has resulted in a widely held consensus that these are the initial recognizable lesions. Definitive evidence that any of these changes are necessary or irreversible precursors of the characteristic advanced plaque is lacking [141]. On the contrary, there is evidence that many fatty streaks may be evanescent and that fatty streaks and fibrous plaques do not always occur in precisely the same locations [173]. The drastic obstructive complications of atherosclerosis are not associated with these presumed precursor states but with stenoses, disruptions and thromboses which occur in relation to the fibrous, "raised", fully developed, complex plaques which contain the elements and organizational features described above. The physical properties of advanced, complex lesions which may be important determinants of the tendency to disrupt and expose thrombogenic plaque material [108] to the bloodstream, remain to be investigated [150].

If the relationship between fibrocellular intimal thickening and the development of atherosclerosis is not clear, neither is the transition from either intimal swelling by plasma constituents or from the fatty streak to the fully developed fibrous, complex plaque [141]. Nor is there as yet strong evidence that the various manifestations of intimal change are related to different specific features of the flow profile or to differential effects of tensile stress. Nevertheless, the organization of the characteristic atherosclerotic plaque suggests a close relationship to both fluid dynamic and tensile forces associated with the circulation. The luminal surface of most atherosclerotic lesions is concave, in conformity with the curvature of the artery as a whole, tending to maintain a regular circular or slightly ovoid luminal cross-section. Uncomplicated plaques do not bulge into the lumen on transverse sections [89]. In suitably prepared specimens, lesions tend instead to bulge outward, deforming the underlying media and adventitia (Figs. 15.1 and 15.2). The fibrous cap is usually a well-organized multi-layered fibrocellular structure containing various proportions of collagen, elastin and smooth muscle. It is often comparable in thickness to the adjacent uninvolved sector of the artery wall and is covered by an intact endothelial lining with an immediately subjacent newly formed internal elastic lamina [89]. Thus, the configuration of a typical plaque indicates that modelling, reflecting the influence of mechanical factors, occurs in the course of atherogenesis in a manner which results in

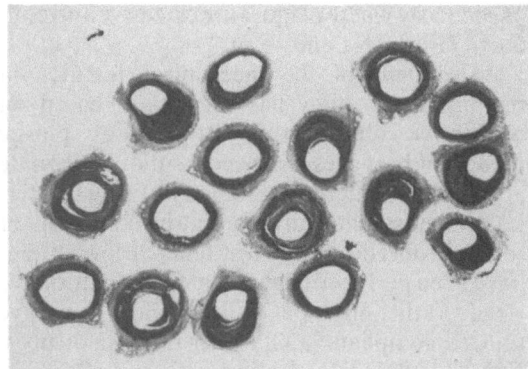

Fig. 15.2. Sequential transverse sections through adjacent atherosclerotic plaques of a human superficial femoral artery fixed while distended at physiological intraluminal pressure. There is considerable variation in lesion configuration and composition but most lesions are eccentric. The severity of stenosis varies with the level of the section but the lumen is round or oval thoughout.

sequestration of the necrotic centre of the plaque away from the lumen, while the fibrous cap forms an effective new artery wall over the lesion and preserves a regular, nearly circular lumen.

Mechanical Stresses and the Distribution of Intimal Plaques

Methodological Problems

Conclusions regarding the haemodynamic determinants of plaque localization have been drawn mainly from correlations of the distribution of grossly visible human and experimental intimal plaques with presumed or demonstrated qualitative features of the corresponding flow fields. Lesion distribution has usually been determined by "en face" examination of vessels which have been removed, opened axially and flattened intimal side up. The precise location of lesions with respect to bifurcations, bends and flow dividers has been difficult to define with such preparations due to deformations introduced by collapse and retraction of vessels upon removal, and to flattening at ostia of branches [36,89,214,220,222]. Furthermore, few investigators have quantitated actual plaque volume or composition. Reliance exclusively on estimates of the extent of surface area covered by plaques may result in misleading conclusions when significant differences in plaque thickness and composition are not taken into account, particularly where plaques are extensive and confluent (Figs. 15.1 and 15.2).

Assessment of the mechanical features in regions under study has often been based on qualitative observations in glass or plastic models of branches and bends without adequate attention to the precise geometry of actual arteries, to realistic flow divisions at birfurcations or to the properties of the test fluids. Flow studies have been performed mainly under conditions of steady rather than pulsatile flow and few have utilized compliant [132] rather than rigid tubes. Experimental data dealing with the effects of flow velocity and direction on the orientation of endothelial cells [62] have been used to support preconceptions with regard to the precise location of lesions and the role of fluid dynamics and have given rise to assertions that elevated flow velocity and high levels of shear stress are major determinants of lesion localization. Recent studies of flow field characteristics have overcome some of the methodological difficulties by utilizing casts of actual vessels or precise glass or plastic models [66,135] constructed to scale and based on measurements from angiograms

[12,13]. Laser-doppler anemometry [2,13] and particle tracking [115,117,195] in fluids of suitable viscosity, using flow divisions in the physiological range and under conditions of pulsatile flow [121,123], have now provided accurate representations of low profiles [66]. Attempts are being made to assess the effects of compliant walls on near-wall flow profiles [132]. Such data have been correlated in several studies with determinations of plaque localization, area, thickness and composition on histological sections at standard levels in corresponding artery specimens fixed while distended at physiological mean pressure in order to restore in situ configurations [219,222].

Fluid Dynamic Characteristics

Quantitative correlative studies utilizing haemodynamic measurements in accurate models of the human carotid bifurcation under conditions of steady flow and sections of controlled pressure fixed postmortem specimens [219] have revealed that both intimal thickening and characteristic

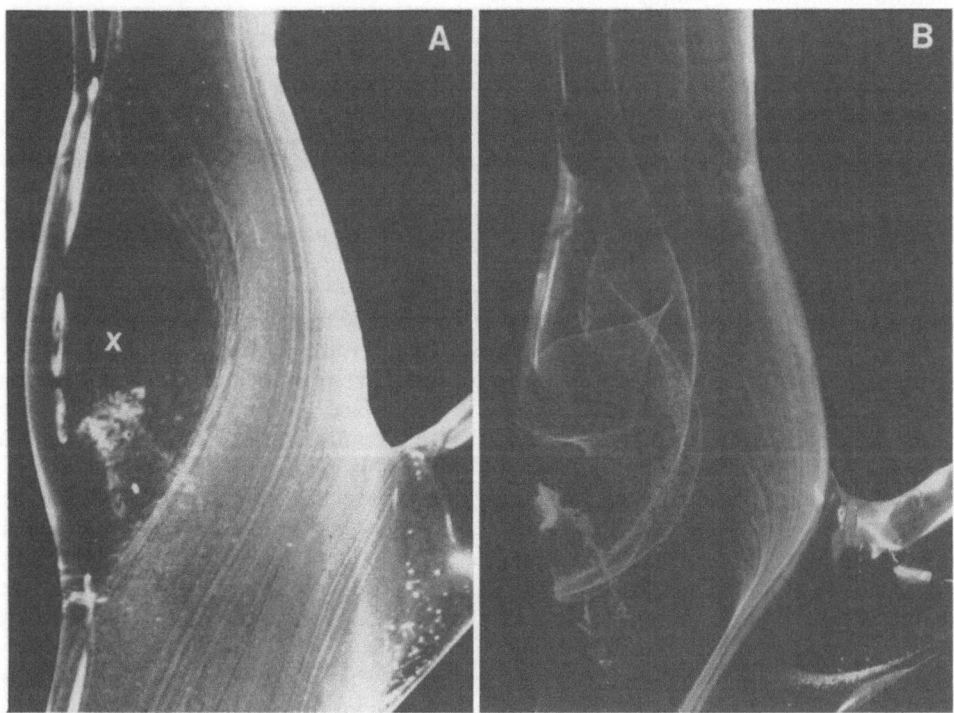

Fig. 15.3. Flow field visualization in an accurate scale model of the human carotid bifurcation. Flow is visualized by the passage of hydrogen bubbles in the plane of the bifurcation angle (**A**) and in a plane perpendicular to the bifurcation angle (**B**). A zone of separation (*X*) is clearly visible at the lateral (outer) wall in **A**, while a complex helical flow pattern is evident in the same region in **B**. The flow remains linear and axial in the region of the flow divider in both planes.

plaque formation are greatest at the outer wall of the sinus of the internal carotid artery, opposite the ostium of the external carotid artery (Figs. 15.3 and 15.4). Maximum involvement occurs where wall shear stress is lowest, in the region of flow separation, and in relation to complex flow patterns and vortices which form near the lateral wall at the bifurcation and extend into the sinus (Figs. 15.5 and 15.6). At the flow divider, where wall shear stress is highest, the vessel tends to be spared (Fig. 15.7). These findings reinforce earlier observations that human plaques about the ostia of the major visceral branches as they arise from the aorta occur primarily at the inlet or proximal rim of the orifice and not at the distal edge on the flow divider [26]. The proximal inlet is presumably a region of flow separation and low wall shear, while the flow divider is a region of relatively high wall shear [25].

Low flow induced by obstruction to venous outflow resulted in plaque formation in the otherwise spared renal artery of the cholesterol-fed rabbit [151]. The absence of plaque formation within channels created by the placement of constricting aortic bands in experimental animals

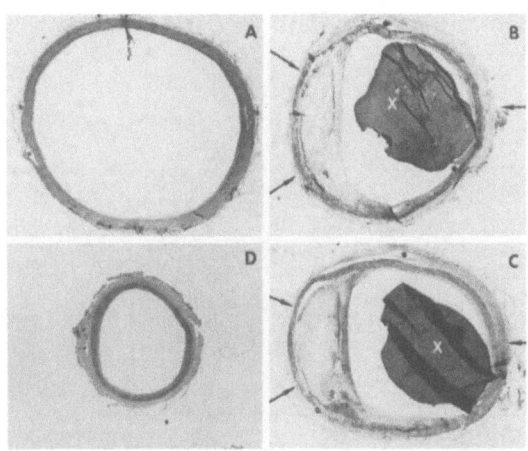

Fig. 15.5. Transverse sections of a controlled pressure fixed human carotid bifurcation at the levels (*A*,*B*,*C* and *D*) indicated in Fig. 15.4. No lesions are evident in the common carotid artery (**A**) or the distal internal carotid (**D**) where the flow profiles tend to be parabolic and symmetrical. At the entrance to the sinus (**B**) and in mid-sinus (**C**), the plaque is at the lateral or outer wall (*two arrows*) where the gradient of near-wall velocities is lowest, while the region at the flow divider or inner wall (*single arrow*) is spared, i.e. where the near-wall velocity gradient is greatest. The dark material (*X*) in the lumen at levels B and C is gelatin used in preparation of the specimens.

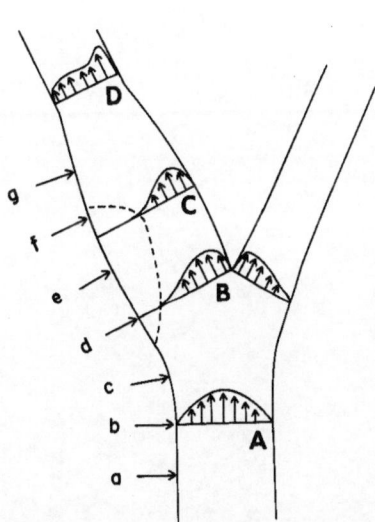

Fig. 15.4. Velocity profiles in the human carotid bifurcation. The profile is parabolic and symmetrical in the common carotid artery (*level A*). The high centre-line velocities are skewed towards the flow divider as the flow enters the carotid sinus (*level B*) and remains displaced towards the inner wall at the mid-sinus level (*level C*). As flow leaves the sinus (*D*) the original symmetrical, parabolic profile tends to be restored. The lower case letters (*a* through *g*) indicate positions along the lateral wall for which the shear stresses are plotted in Fig. 15.6. The interrupted line marks the region of separation and complex flow.

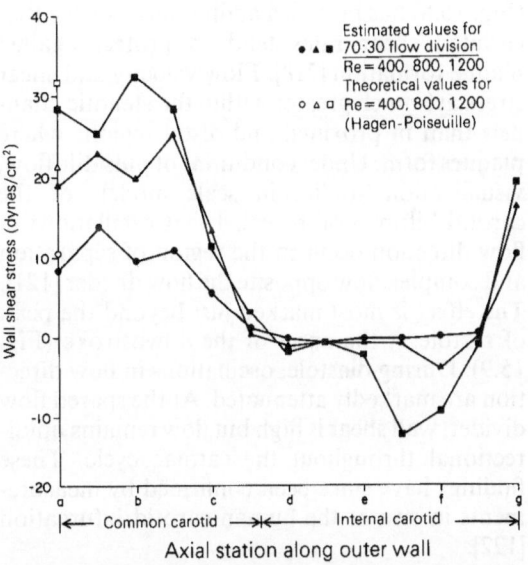

Fig. 15.6. Shear stress distribution along the outer wall where the plaque usually forms. Positions correspond to the markers at the outer wall in Fig. 15.4. Plots are shown for Reynolds numbers which include the range of values which occur in the course of the cardiac cycle. At the level of the spared common carotid artery (*a*,*b*) shear stress is relatively high. In the sinus (*c*,*d*,*e*,*f*) where plaque is thickest, shear stress falls. In the distal sinus and in the distal internal carotid artery (*g*) where plaque formation is minimal, shear stress is again elevated.

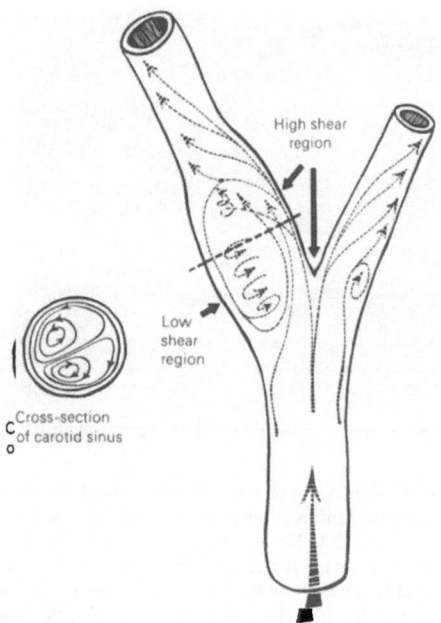

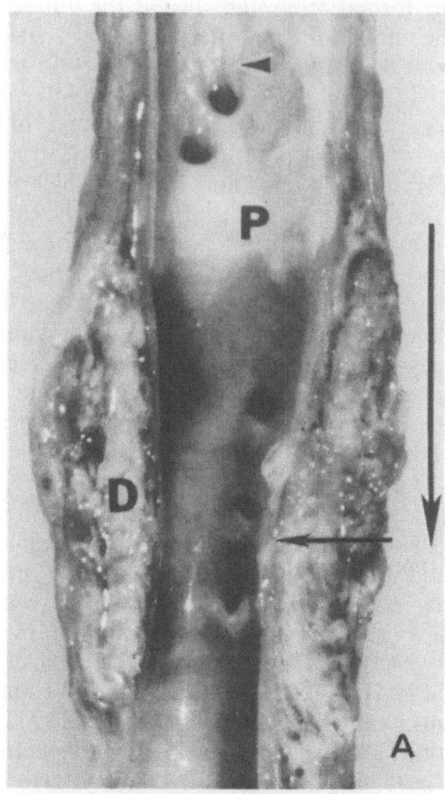

Fig. 15.7. Flow directions at the human carotid bifurcation. Low flow velocity and decreased shear stress occur at the outer walls. High flow velocity and elevated shear stress prevail at the inner wall in the region of the flow divider.

(Fig. 15.8) has provided additional evidence that elevated shear rates tend to protect against plaque formation [218]. Flow velocity and shear stress are much greater within the stenotic channels than in proximal and distal regions where plaques form. Under conditions of pulsatile flow, visualization studies in scale models of the carotid bifurcation revealed that oscillations in flow direction occur in the region of separation and complex flow opposite the flow divider [121]. The effect is most marked just beyond the peak of systole at the onset of the downstroke (Fig. 15.9). During diastole, oscillations in flow direction are markedly attenuated. At the spared flow divider, wall shear is high but flow remains unidirectional throughout the cardiac cycle. These findings have since been confirmed by measurements in situ at the human carotid bifurcation [122].

An analogous geometric situation prevails at the human distal aortic bifurcation [66]. Corelations of intimal thickening with flow characteristics in casts of the distension-fixed postmortem human abdominal aortic bifurcation have revealed that lateral locations, opposite the central flow divider at the origin of the common iliac arteries, are preferential sites of plaque formation and are regions of low wall

Fig. 15.8. A Experimental stenosis of the aorta of a cholesterol-fed monkey. Plaque forms proximal to the stenosis (*P*), but the stenotic channel (*horizontal arrow*), where flow velocity is elevated, is spared as is the immediately distal region of turbulent flow. At a branch ostium in the proximal region, a plaque has formed at the inlet or low shear side (*arrowhead*). The *long arrow* indicates the direction of flow. The dacron sleeve used to constrict the artery is marked *D*. **B** Scanning electron micrograph of the distal edge of the proximal plaque reveals a sharp transition between the lesion and the spared region.

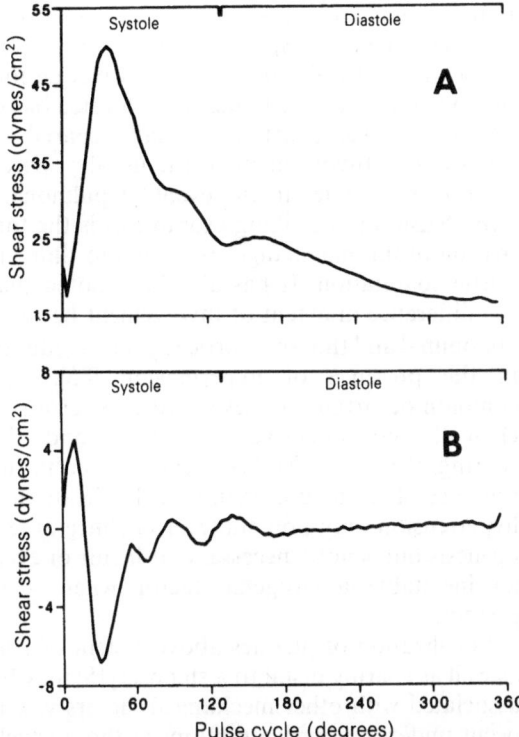

Fig. 15.9. Variation in shear stress in the course of the cardiac cycle under conditions of pulsatile flow in the carotid sinus. At the inner wall (**A**), where lesions do not occur, shear stress is elevated and does not reverse in direction. At the lateral or outer wall (**B**), where plaques are common, shear stress is low and oscillates in direction. The oscillation in shear stress direction at the outer wall occurs mainly during systole.

shear stress. Some intimal thickening is however also noted along the flow divider. Although this change is less prominent than at the lateral walls, it seems to precede the changes at the lateral wall. Intimal thickening associated with the flow divider region of high shear stress may be an early and relatively slow and limited reaction, possibly associated with the increasing shear stresses associated with altered blood flow velocity [61] and with changing mural tensile stress as the vessel enlarges during growth. Intimal changes with the characteristic features of atherosclerosis, which are found at the lateral walls in the regions of low shear, probably begin when manifest plaques are developing elsewhere and may continue to enlarge. Further studies based on numbers of specimens sufficient to take these factors into account should reveal the time-course of changes at the flow divider and of the progressive atherosclerotic changes at the lateral walls.

Probability maps of plaque localization in the proximal human anterior decending coronary artery place the development of plaques mainly opposite the flow divider of the ostium of the left circumflex coronary artery [193]. With respect to branch geometry this location corresponds to that noted for the proximal human internal carotid artery and also with conditions at the aortic bifurcation [63]. It is therefore likely to be a region of low and oscillating wall shear stress. Comparison of blood flow velocity distribution in the human right coronary artery, as determined by angiography, with the localization of plaques in postmortem specimens revealed that lesions formed preferentially on the inner wall where flow velocity was low [164].

As noted above, the coronary arteries have been shown, in postmortem studies, to be selectively and extensively involved by atherosclerosis, while vessels with less complex non-oscillating systolic flow patterns, such as the renal and mesenteric arteries in the same individuals tend to be spared in regions beyond their ostia [84,85]. If oscillations in flow direction play a role in plaque localization and such oscillations occur mainly during systole [121], the coronary arteries should be expected to be particularly vulnerable, since the cyclic variations in flow in these vessels include a bimodal excursion during systole [12]. Exposure of the coronary arteries to two changes in direction of flow during systole may account for preferential involvement of these vessels in comparison to other major arteries where flow does not reverse in direction. A potentiating role for elevated heart rate was suggested initially on the basis of a comparative study of atherosclerotic localization in the human aorta, and its coronary and renal artery branches and considerations of the nature of coronary and renal blood flow [85]. Recent experiments in which heart rate has been correlated with diet-induced atherosclerosis in non-human primates have indicated that decreased heart rate protects against both coronary artery [7] and carotid sinus [8] atherosclerosis.

Experimental studies designed to assess the effects of behavioural stress on atherogenesis in the same species also revealed a positive correlation between pulse rate and coronary artery and carotid involvement [114,138]. Individual differences in human heart rate would be expected to exert a marked long-term effect on exposure of these vessels to systolic flow reversals, for the difference in pulse rate between 60 and 80 beats per min corresponds to a difference of more than 10 million beats per year. The effect is further accentuated by the fact that systolic time is relatively independent of heart rate, while diastolic time is shortened as heart rate increases. Differences in total heart beats during 24 h of usual activity among clinically healthy men appear to be a reflection of parallel differences in rate at various levels of activity and are consistent over extended time intervals [86]. Physical fitness is characterized by a relatively low basal heart rate, a rapid return to the basal rate after exertion and relatively lower rates at standard levels of activity [107]. The salutory effects of exercise on the cardiovascular system may in part be related to these effects.

Turbulent and disordered flow have received attention as possible determinants of plaque localization [46,58,94,185]. Although focal cyclic departures from unidirectional laminar flow are demonstrable in various regions of the arterial tree, turbulent or random flow patterns do not occur under normal conditions. If the distinction between complex flow patterns and turbulence is made, turbulence has not been shown to correlate with the localization of atherosclerotic lesions in human specimens or in arteries of experimental animals. Turbulent and disordered flow conditions have been induced by establishing abrupt transitions in vessel dimensions by means of arteriovenous anastomoses [47, 184,188], sharp bends [90], aneurysms [187] or high-grade stenoses [124,170,190]. Intimal thickenings with some of the features of atherosclerosis have been produced at positions of presumed or induced disturbed or turbulent flow. On the other hand, in regions immediately distal to tight stenoses, where turbulence has been shown to occur [131], arteries are usually spared of plaques [18,39,119]. When direct measurements of turbulence velocity have been compared with plaque localization distal to experimental aortic coarctations, turbulent flow did not correlate positively with plaque formation, despite higher than normal blood pressure levels distal to the stenoses [124].

Blood Pressure and Tensile Stress

Elevated blood pressure is closely associated with symptomatic atherosclerotic disease [32,113,160] and there is evidence, both in human vessels and in experimental models, for a relationship between blood pressure and the localization of plaques [20,85,100,160]. In general, postmortem studies have revealed that hypertension is associated with an increase in both extent and severity of atherosclerosis. Although the effect appears to be most marked in locations already predisposed to plaque formation, lesions can be induced by elevated blood pressure in some arteries usually spared of disease [84]. Involvement of the usually spared pulmonary arteries in the event of pulmonary hypertension is a striking example as is the formation of intimal changes proximal to thoracic aortic coarctation. It has also been noted that the difference in extent of involvement between abdominal and thoracic aortic regions is reduced in the presence of hypertension [85]. The common occurrence of severe atherosclerosis in clinically normotensive individuals and the sparing of vessels distal to stenoses even in the presence of elevated pressure [124] indicate that hypertension may potentiate or enhance atherogenesis but is not a necessary co-factor or even an inevitable atherogenic factor when it is present.

Localization of plaques above a stenosis [99] as well as sparing distal to a stenosis [18] may be associated with other mechanical factors which occur under the same conditions as those which result in differences in blood pressure. It has been shown for example that aortic wall motion, i.e. the change in diameter between diastole and systole, is markedly decreased distal to an experimental coarctation [136]. Wall motion above the stenosis may not be significantly increased. Diet-induced atherosclerosis may not differ markedly in extent from controls proximal to coarctation but is practically absent distal to the stenosis. Elevations in pressure and increases in diameter also result in increases in mural tensile stress. In the course of the cardiac cycle, vessels such as the coronary arteries would be subjected to more complex and wider oscillations in tensile stress than visceral arteries because of the special configuration of the cyclic pressure variation in the coronary circulation [85]. Local geometric configurations also result in differences in the distribution of tensile stresses associated with pulsation. Regions such as the inner or concave aspects of bends and the outer walls of bifurcations would also be exposed to different variations in mural tensile stress than regions of greater curvature on the convex side of bends or adjacent to flow dividers. The presumed distribution of tensile stresses and of oscillations in tensile stress at such sites corresponds to both medial and intimal thickness and to susceptibility to plaque formation. Plaques form preferentially at the inner curvature of tortuous vessels such as the splenic artery. Experiments designed to isolate cyclic changes in diameter from cyclic changes in pressure and flow dynamics may help to isolate the relevant variables and determine their relative contributions to plaque location.

Proposed Mechanisms for Plaque Initiation

Investigations dealing with mechanisms by which mechanical stresses may contibute to plaque formation have attempted to identify features which may account for the increased entry into and/or accumulation of circulating atherogenic substances in the artery intima or which may result in the induction of one or more of the changes presumed to be indicative of early plaque formation. Although findings in correlative studies tend to establish close relationships between plaque localization and geometric configuration, the mechanisms which underlie the putative role of mechanical factors in atherogenesis are not clear and the relative contributions, direct or indirect, of the forces associated with flow and tensile stress to plaque initiation, progression or complication remain to be clarified.

Endothelial Disruption

According to the recently proposed version of the endothelial injury hypothesis [163], thrombocytes accumulate at regions of endothelial disruption where focal denudation of the lumen surface has occurred. As a result, a platelet-derived growth factor is released inducing medial smooth-muscle cell proliferation. This proliferative reaction may be sustained by repeated endothelial injury and by the penetration of circulating atherogenic materials into the intima across the endothelial defect. In this view, endothelial disruption and smooth-muscle proliferation are presumed to be the critical early events in atherogenesis. This mechanism of plaque initiation gained support from experiments in which intimal thickening by smooth-muscle cell proliferation occurred after extensive mechanical removal of endothelium by means of a balloon catheter or after focal injuries made with specially designed instruments [14]. The hypothesis that focal disruption of the endothelial surface is a major initiating event in atherogenesis suggested to some investigators that high shear rate could be one of the conditions favouring such disruption, for experiments utilizing mechanical devices designed to increase shear stress indicated that elevated shear stress could injure endothelium, with detachment of cells occurring at shear rates exceeding 400 dynes/cm^2 [67].

The demonstration that endothelial cell orientation is sensitive to shear stress [49] and that cells are usually aligned in the direction of flow [34,62,126,147,155], has also been interpreted as evidence that elevated shear stresses may cause endothelial deformation and injury in vivo. Appearances on scanning electron micrographs of flow dividers suggesting endothelial deformation were also offered as evidence for such an effect [154]. Recent experimental evidence however indicates that endothelial removal by balloon scraping also injures the inner media [153]. Thus, the proliferative healing response is not necessarily a demonstration that endothelial desquamation is the critical precursor of smooth-muscle proliferation. Furthermore, the endothelium has not been found to be disrupted before experimental diet-induced atherosclerotic lesions form in susceptible locations [53,196]. Evidence has actually been forthcoming that an endothelial lining or the restoration of recently removed endothelium is required for experimental lesions to form and that plaques do not form where the endothelium is absent [31,56,143,179]. In addition, preparations of aortic branch ostia fixed while distended at physiological pressure with minimal subsequent manipulation fail to reveal endothelial cell degeneration or deformation at flow dividers or in other regions normally exposed to relatively elevated wall shear stresses [220]. Where susceptible locations have been studied under conditions of adequate pressure fixation, endothelial cells tend to be rounded or at least not markedly aligned or clearly oriented [130].

In vitro studies confirm effects of shear stress on endothelial orientation and function [49,81] but fail to reveal evidence of gross injury or denudation even at markedly elevated levels [125]. More recent formulations of the endothelial injury hypothesis do not emphasize an essential role for endothelial disruption or detachment, at least during lesion initiation, but consider endothelial injury in the broader sense of departures from normal function which could result in altered permeability or loss of factors which control thrombogenicity or cell shape [54,162].

Wall Shear Stress

The mechanisms by which low wall shear stress and oscillations in flow direction might favour atherogenesis remain to be identified. Flow visualization studies in the scale model of the human carotid bifurcation reveal that several pulse cycles are required to clear circulating markers from the susceptible region of low and oscillating shear, while transport is not delayed in the relatively spared region of high shear and unidirectional flow at the flow divider [121]. Thus, the duration of exposure of the endothelium to atherogenic lipoproteins and/or to substances favouring the entrapment of atherogenic particles in the intima would presumably be increased in the low shear regions predisposed to atherosclerosis [80]. Permeability is increased in susceptible locations as indicated by increased passage of albumin-bound dyes into the intima [3,9,71,78,79,103,139]. In addition to the increased residence time of atherogenic particles, flow reversal or complex flow patterns could conceivably loosen or elevate overlapping endothelial cell junctions and facilitate ingress of such materials, thereby favouring the initiation or acceleration of atherogenesis. Regions of unidirectional flow and high shear rate, where cells are aligned in the direction of flow, tend to be spared, while regions in which cells are not aligned [130], presumably where wall shear is oscillatory and low, are at higher risk. It is conceivable that moderately elevated levels of shear stress may induce formation of tighter intercellular attachments [148] and reduce transendothelial permeability in regions resistant to atherosclerosis. Low and oscillating shear could also induce alterations in endothelial cell metabolism [45,80]. Delayed transport of erythrocytes could render such locations relatively hypoxic. Delayed access of nutritive compounds as well as a reduced rate of transfer away from the lumen surface of catabolic products could modify the elaboration and secretion of regulatory substances associated with vasoactivity, coagulation or permeability.

Circulating cells are considered to play a role in lesion induction [110,169] and the accumulation of monocytes, thrombocytes and granulocytes on endothelial lumen surfaces may also be affected by wall shear rate. Low wall shear rate favours margination and attachment of these elements [116]. Monocytes may enter the intima [111], incorporate lipids and return to the bloodstream [76,77]. Platelets and granulocytes elaborate agents which interact with the endo-

thelium to alter permeability and thrombogenicity, and induce smooth-muscle cells to proliferate and contract [45,54]. The degree to which the adhesive or chemotactic forces [78] which affect circulating cells may be enhanced by metabolic alterations associated with focally increased endothelial exposure to circulating substances or to injury has not been established [101,109]. The extent to which these attractive forces may be counteracted by the mechanical forces associated with flow velocity [52] and shear rate [6,116] remains to be subjected to quantitative study in regions susceptible and resistant to atherosclerosis. Since endothelial cell turnover may be increased in regions prone to develop plaques, it has been suggested that in regions of reactive endothelial proliferation after a denuding injury, intercellular junctions may be "leaky" [205]. Openings may favour the selective entry into the artery wall of circulating atherogenic plasma components such as low-density lipoproteins. Clear links have not been established among these interactions, mechanical factors and atherogenesis. Although the mechanisms by which low shear stress and/or departures from unidirectional flow favour plaque initiation remain to be elaborated, plaques do not appear initially in regions of high wall shear rate, and endothelial disruption or desquamation do not appear to be immediate events in plaque initiation.

Transport Through the Artery Wall

The state of the arterial wall, in particular of the media, has also been associated with mechanisms favouring the accumulation of atherogenic substances in the intima and inner media [27,70,178,204]. Although substances may diffuse through the wall in both directions, the distending pressure acts to promote entry of circulating materials into the wall by convection [23,27,197]. Transit through the wall towards the adventitia may be determined in large part by the density of the wall as well as the gradient of chemical activity across the wall, water flux [69] and local differences in ultrastructural specialization of endothelial cells and medial components [204]. Localization of plaques in the major arteries rather than in small arteries or arterioles could be associated with the relatively slower convection through the relatively thick and fibrous walls of the larger vessels. The accumulation of connective tissue fibres [57] and the changes in mechanical properties with age

[59,128,149] in large arteries predisposed to atherosclerosis would tend to render these vessels more dense and therefore less compliant and increasingly resistant to convection across the wall. Differences in configuration and organization of the media and in the continuity of such structures as the internal elastic lamina may also influence both convection and diffusion of materials into and across the wall.

Although intimal thickening which appears to precede plaque formation in many locations may be a non-atherosclerotic response to local stresses, it may also play a role in atherogenesis by creating a relatively thicker and denser wall [96]. Active tension associated with smooth-muscle contraction in the media would also be expected to increase wall density [171] particularly in vessels subject to spasm. It has been proposed that the effects of vasoactive stimuli such as nicotine [209] may act by inducing spasm, thereby increasing resistance to transmural convection [24]. Arteries with increased perivascular tissue support tend to be spared. This occurs for example when arteries such as the internal carotid artery pass through boney canals, or when segments of the coronary artery loop into the myocardium and then re-emerge [75]. The intraosseous carotid and occasional intramyocardial coronary segments are spared while plaques may be prominent in both proximal and distal regions. These observations suggest that reduced transmural pressure gradients may tend to affect the role of convective forces.

The smooth-muscle cells of the media are concerned mainly with the development of active tension and with the biosynthesis and degradation of matrix components in response to tensile stress [129]. Arteries subjected to hypertension elaborate connective tissue and develop a corresponding increase in wall thickness and cross-sectional area in relation to imposed tensile stress [11,41,161,189,212]. This adaptive reaction may also render the wall increasingly dense and tend to retard transmural transport. The alterations in mural metabolism induced by hypertension may be modified by exposure to atherogenic lipids [217] and conversely, hypertension may modify the metabolic processes involved in the incorporation and processing of lipids [99,213,217]. According to this conception the balance among plasma lipid concentration, near-wall haemodynamic conditions, convectional forces, wall metabolism and wall density determines the rate at which lipids can be cleared from the wall. Differences in the depth of penetration of vasa vasorum from the adventitia into the media and the state of the perivascular lymphatic channels [106] could also influence the rate of passage of atherogenic substances through the artery wall. The highly susceptible abdominal aortic segment is devoid of medial vasa vasorum, while the more resistant thoracic segment is permeated by vasa vasorum to within 0.5 mm of the intima [216]. Deformation and compression of the media by hypertension may also act to compress intramural vasa vasorum and interfere with medial nutrition and transmural clearance [165]. Lipid trapped in the intima is apparently dealt with initially by macrophages. In the event that this response cannot keep pace with the rate of accumulation due to alterations in the wall or to modifications of the lipids, the artery wall reacts to sequester and isolate the material. This reaction requires the participation of medial smooth-muscle proliferation and the elaboration of connective tissue fibres to form a fibrous cap. The composition of the fibrous cap would be modified in accordance with the tensile stresses in the artery wall.

Mechanical Factors in Lesion Progression
Secondary Disruption

Once initiated and enlarged by the continued accumulation of cells, lipids and matrix macromolecules in the intima, plaques initially develop irregular or bosselated endothelial surfaces (Fig. 15.10), corresponding to the contours of the immediately subjacent foam cells [196]. Such surfaces may engender focal instabilities of flow and modifications of wall compliance may favour disruption of the endothelial cells which have been attenuated over the underlying intimal elements. Penetration through the endothelium by foam cells and exposure of fibres have been demonstrated in carefully prepared specimens of multilayered lesions [54,196], but demonstration that flow disturbances due to lesion surface irregularities result in endothelial disruption has not been shown. Lowered shear stress, departures from unidirectional flow and local flow instabilities may however favour the adhesion and deposition of thrombocytes and monocytes at sites of secondary endothelial surface penetrations where lesion components are exposed to the bloodstream. The accretion of blood platelets and fibrin at these locations may then contribute to plaque enlargement and complexity by

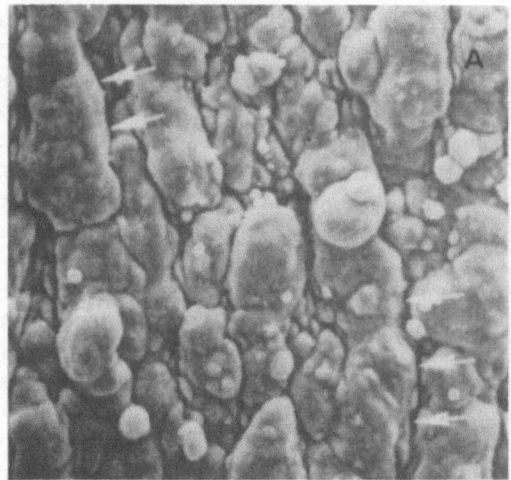

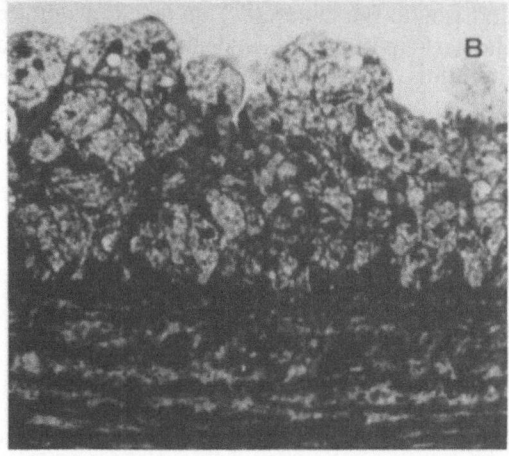

Fig. 15.10. A Scanning electron micrograph of the surface of an early experimental foam-cell lesion similar to an early human fatty streak. **B** Transmission electron micrograph of a cross-section of the same plaque. The surface is irregular consisting of mounds and crevices (*white arrows*) as the intact endothelial cells conform to the contours of the underlying lipid-filled foam cells, but the endothelial surface is intact.

release of growth factors and the formation of thrombi [145,162]. The deposition of thrombus on plaques [29] has been well documented and many observations indicate that accretion of thrombotic material may be organized and incorporated into plaques contributing directly to plaque growth and to the complexity of plaque morphology [30,51]. Advanced plaques with well-organized fibrous caps would be expected to present smooth and regular lumen surfaces to the bloodstream but may contain elements such as collagen which are highly thrombogenic when exposed to the circulation.

Although turbulence does not occur under normal circumstances, changes in vessel con-figuration associated with the development of atherosclerotic plaques may create conditions favouring the development of instabilities and turbulence. The likelihood of such changes is further enhanced when vessels enlarge and become tortuous with age, when multiple plaques occur in close axial proximity in the same vessel, when there are marked variations in peripheral resistance or when there are abrupt changes in pulse rate or pressure. Although regions distal to very tight experimental stenoses tend to be spared of plaques compared to controls [18], artery segments distal to less severe strictures tend to develop plaques of greater complexity than those of controls [17]. The features in these regions, consisting of calcification, haemorrhage, thrombosis and necrosis, suggest that flow instabilities distal to stenosis which are not tight enough to prevent plaque formation may nevertheless be associated with plaque disruption and with modifications in plaque morphogenesis. Haemodynamic changes resulting from alterations in configuration due to plaque formation may then favour atherogenesis in regions which are normally spared.

Artery Adaptation

Recent experimental studies have revealed that arteries undergo what appears to be compensatory adaptation as plaques deposit [19]. On the basis of determinations made on experimental artery cross-sections, the area delineated by the internal elastic lamina has been shown to enlarge as plaque area increases such that the original lumen cross-sectional area tends to be maintained [4]. As the process continues, growth of the plaque may eventually exceed the adaptive possibilities of the artery. Adaptive enlargement has also been demonstrated in human coronary arteries [87]. These data suggest that when the lesion area exceeds about 40% of the area encompassed by the internal elastic lamina, the absolute lumen area is progressively diminished. The mechanism by which the adaptive process is achieved is not clear. Possible explanations include effects of the plaque directly upon the subjacent wall, as well as effects of altered flow on the opposite intact portion of the wall.

Early uncomplicated plaques tend to be eccentric with lumen surfaces conforming to the general curvature of the vessel while the artery lumen maintains a circular or slightly oval cross-section [89]. Plaques are therefore crescentic on cross-section. As the lesion is sequestered from the lumen, the underlying wall is usually thinned and both plaque and wall tend to bulge outward [42,89]. This outward extension could be the result of compressive forces exerted by the developing plaque or of atrophy of the underlying wall due to interference with diffusion from the lumen into the media. Alternatively, the intact artery wall opposite the developing plaque may respond by circumferential extension, thereby enlarging the artery. Such a reaction could be induced by the increased flow velocity resulting from narrowing of the lumen by the enlarging plaque. Increased flow in an artery, which occurs for example proximal to an arteriovenous fistula, results in enlargement of artery diameter, while decreased flow distal to a stenosis results in artery narrowing.

In a quantitative experimental study of the enlargement of the iliac artery proximal to an anastomosis to the iliac vein, artery diameter was found to increase until the shear stress was restored to the same level as that in the contralateral unanastomosed side [221]. Immediately after establishment of the anastomosis, wall shear stress was estimated to be approximately 100 dynes/cm^2 on the operated side. After 6 months, shear stress on both sides proved to be approximately 15 dynes/cm^2, the same level of shear stress which has been reported for normal arteries of several mammalian species in several locations. Similar adjustments in artery diameter [112,207], including reduction when flow is reduced [95], may also result in the re-establishment of normal levels of shear stress in other experimental models [112]. Adjustment of the uninvolved wall in the case of the developing plaque may be mediated by an endothelium-derived factor released in response to altered flow velocity [176] and similar in its effect to the factor which mediates the relaxation response of the wall to acetylcholine [73]. Narrowing in response to reduced flow is also dependent on the presence of endothelium [127]. It is especially noteworthy that endothelium-dependent relaxation appears to be lost in the presence of experimental hypertension [133].

Marked local dilatation of atherosclerotic vessels, leading to the formation of aneurysms, results in the accretion of thrombus composed of layers at various stages of organization. The lumen however, remains circular on cross-section, despite tortuosity of the dilated segment, and its diameter approaches that of normal aorta. Thus it appears that both the modelling of the plaque and of the mural thrombus as well as the maintenance of an adequate lumen area

in relation to enlarging atherosclerotic plaques are adaptive reactions designed to restore and maintain optimal conditions for adequate, laminar flow for as long as possible. The reactions appear to be closely related to flow field characteristics. The development and organization of the fibrous cap would appear also to be regulated by tensile stresses associated with distending pressure.

Since diseased vessels often proceed to obstruction or excessive dilatation, the adaptive processes are limited. Individual differences in the degree and nature of tissue responses to mechanical stresses related to flow, pressure and geometric configuration and in relation to age, sex and nutritional and metabolic status are likely to determine the morphogenesis and consequences of plaque formation and the repercussions on the artery wall. In the presence of continued and overwhelming exposure to atherogenic stimuli, and/or if there is exhaustion or modification of adaptive tissue reactions, eventual symptomatic circulatory compromise occurs. Growth of the plaque from its site of origin into the lumen and extension around the circumference may lead to encirclement of the lumen by the lesion. In keeping with the possible mechanisms which underlie the adaptive process, involvement of the entire wall circumference may prevent further artery enlargement. The lesion would then proceed to encroach on the lumen. Flow disturbances associated with a stenosis, particularly with the eccentric configuration typical of advanced plaques, could also contribute to modifications in the rate of plaque development, the direction of plaque extension and the composition of the plaque. On the other hand, stenoses would result in increased flow velocity at the narrowing and could conceivably be associated with focal slowing of the atherogenic process. It has already been noted that arteries may become narrowed and atrophic [18] distal to a tight stenosis but still be relatively spared of atherosclerotic disease. The precise relationship between the rate of plaque growth and changes in haemodynamic status remains to be clarified.

Mechanical Factors in Plaque Complication

Since atherosclerotic vessels appear to enlarge to some degree in many instances and flow velocity increases in regions narrowed by plaques, atherosclerosis tends to remain clinically quiescent until late in the development of the disease. Clinical symptoms occur only when the disease reaches a stage when blood flow in the region of a plaque is inadequate to meet the needs of the supplied tissues or when an artery is rapidly or suddenly occluded or ruptured. Thus, the extent of the process, defined as the proportion of an artery lumen surface involved by atherosclerotic plaques, is not identical with the severity of the disease [89]. Severity is a measure of the degree of actual or potential interference with flow regardless of the extent of involvement. Morbidity or mortality may supervene as the result of one or a few critically located and rapidly stenosing plaques. Plaque complication, regardless of the severity of stenosis, is frequently the critical factor determining the consequences to the patient and may also be associated with local mechanical conditions. These complications include the closely related phenomena of plaque ulceration or disruption, plaque haemorrhage and thrombus formation [211]. Plaques differ in consistency in relation to their composition. Some are relatively soft and friable, while others are densely sclerotic and calcific. Some have well-formed fibrous caps, mimicking the architecture and width of a normal artery wall, while others are covered by a narrow zone of loose connective tissue or by endothelium alone.

Haemorrhagic disruption of fractured or fissured plaques by entry of blood from the lumen, and the deposition of thrombi on disruptions have both been shown to underlie sudden occlusion [44,55,102,156]. Susceptibility to rupture, fracture, fissuring or plaque haemorrhage [105] resulting from mechanical stresses due to sudden changes in pressure, flow or pulse rate, or from torsion and bending in relation to organ movements, is likely to be closely related to plaque consistency [37,65,89]. Thrombus formation is likely to depend both on exposure of plaque constituents after disruption of the plaque and on modifications of the flow profile in the region of plaque formation. Research concerned with the determination of the physical properties of atherosclerotic vessels and of individual plaques is in progress [150]. The goals of these investigations are to establish relationships between plaque composition and configuration and the tendency for plaques to crack, form surface fissures or disrupt and to identify the haemodynamic or other mechanical circumstances under which plaques may be fractured. Plaques also contain blood vessels, usually originating by ingrowth of capillaries from adventitial vasa vasorum. Haemorrhagic disruption of these vessels in relation to changes in plaque consistency and/or physical stresses has been proposed as another mechanism for plaque disruption and rapid enlargement.

Direct Interventions on Plaques

In the light of the development of several direct physical interventions designed to remove plaques, reduce plaque size or create new channels through plaques, information concerning the mechanical properties of plaques and of atherosclerotic arteries is of increasing practical significance.

In endarterectomy the plaque is removed by surgical dissection, usually in a plane beneath the lesion which removes the entire intima. Often a portion of the underlying media is removed as well, but neither haemorrhage nor dissection of blood into the remaining artery wall occurs [191]. Presumably integrity of the wall is maintained mainly by the adventitia which may be quite fibrotic and mechanically strong beneath advanced plaques [93]. Enlargement of the lumen is associated with increased tangential tension which may stimulate thickening of the wall until a suitable wall thickness is achieved.

Percutaneous transluminal angioplasty is a widely used procedure. It consists of passing a balloon-tipped catheter into the lumen, to the level of a stenosing plaque and inflating the balloon rapidly, in order to force enlargement of the narrowed channel. The desired effect is brought about for the most part by fracture and partial separation of the plaque from the underlying wall [137,168,182]. The plaque may at times be fragmented as a result of this intervention but usually it retains its general shape. It is not compressed or herniated into the wall. The separation from the wall creates an undermining cleft around the plaque (Fig. 15.11), thereby enlarging the effective lumen cross-sectional area [137]. The plaque usually remains attached at its central, presumably oldest and most fibrous portion. In many instances the artery wall is also disrupted, but the adventitia at the site of the break is apparently adequate to maintain the integrity of the vascular channel and a pseudoaneurysm rarely develops. The effective radius of the lumen is however enlarged by this procedure, probably resulting in a cor-

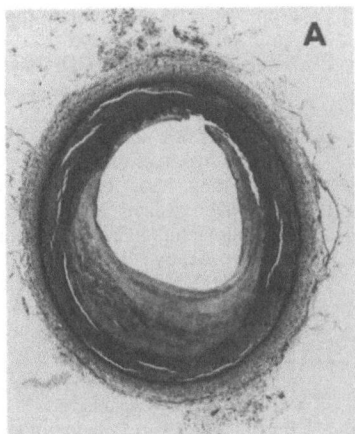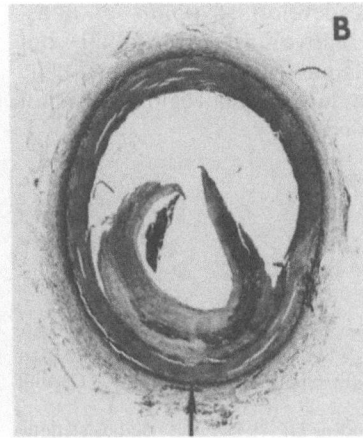

Fig. 15.11. Experimental transluminal angioplasty done under conditions of distension at physiological pressure performed in a human superficial femoral artery obtained at autopsy. **A** Intact plaque. **B** Angioplasty region. The plaque is separated from the underlying wall at its lateral edges and remains firmly attached in its central thickest portion (*arrow*). Disruption of the plaque and the media, increased diameter and eventual modelling–retraction of the plaque edges contribute to the increase in lumen area.

responding increase in tangential tensile stress which may act to keep the vessel open to its new diameter. The healing process includes remodelling, blunting and retraction of the projecting sharp plaque edges, presumably leaving a larger lumen than when the plaque was intact. Thrombus may also collect beneath the raised plaque edge and should eventually organize to restore a lumen of adequate size and shape. Many of the vessels so treated re-obstruct in several months. It is not yet clear whether gradually accumulating thrombus, intimal proliferation [5] or recurrent atherosclerosis is at the basis of this complication.

In laser beam searing, a multichannel catheter is passed to the lesion and a laser beam is directed at the lesion under direct visualization. Thermal lysis disrupts portions of the plaque [1] and the resulting particles may be aspirated by a third channel [33]. Laser beams may also be used to heat a probe at the end of a catheter [43,167]. The probe is then passed into the artery to burn out substantial regions of the plaque. There are as yet few data concerning the short-term or long-term effects of these laser beam procedures.

The atherotome is a tool consisting of a rotatable blade within a partially open capsule. Layers of plaque may be shaved off from the lumen side with the device [140]. The long-term effects on the plaques have not been defined.

A drill catheter consists of a rotating or vibrating drill bit at the end of a catheter [118]. It is designed to create an opening in a markedly stenosed vessel by boring out a new enlarged channel.

Revascularization with grafts, is increasingly used. This operative procedure employs autologous veins or arteries as well as synthetic materials to re-establish blood flow by bypassing stenosing obstructions or as substitutes for vessels which are aneurysmal and subject to rupture. The most common vessel in use for bypass grafting is the saphenous vein. If flow is initially inadequate, the bypassing vessel may be occluded by thrombus. With poor flow the bypass vessel may be occluded gradually by progressive intimal fibrocellular thickening [202]. Although nearly all bypass grafts show some evidence of intimal thickening [21,194], the process is apparently self-limiting if flow remains at an adequate level. This finding is in keeping with the adaptive response to tensile stress [223] or to shear stress [221] which seems to prevail in arteries, i.e. low flow rates induce lumen narrowing until a normal level of flow and wall shear stress is established. New plaques have been shown to occur in bypass grafts [146,203] but the relative roles of flow dynamics, wall tension or clinical risk factors in this development are not yet clear. Synthetic graft materials include woven fibre conduits. If these are of sufficiently large calibre, patency may be maintained for long periods. Organizational features including capillary ingress and smooth muscle and connective tissue modelling [36,92] resemble those which are noted in vein grafts. Intimal hyperplasia, at times obstructive, may develop at points of anastomosis to the native artery [104,142], but atherosclerotic plaques rarely occur in synthetic grafts. Similar conduits of small calibre have not been used with success.

References

1. Abela GS, Normann S, Cohen D et al. (1982) Effects of carbon dioxide Nd-YAG and argon laser radiation on coronary atheromatous plaque. Am J Cardiol 50:1109–1205

2. Ahmed SA, Giddens DP (1984) Pulsatile poststenotic flow studies with laser doppler anemometry. J Biomech 17:695–705

3. Anthony A, Bacon T, Doebler et al. (1981) Ribonucleic acid changes in medial cells in Evans blue positive regions of the dog aorta. Exp Mol Pathol 35:347–358

4. Armstrong ML, Heistad DD, Marcus ML et al. (1985) Structural and hemodynamic responses of peripheral arteries of Macaque Monkeys to atherogenic diet. Arteriosclerosis 5:336–346

5. Austin GE, Ratliff NB, Hollman J et al. (1985) Intimal proliferation of smooth muscle cells as an explanation for recurrent coronary artery stenosis after percutaneous transluminal coronary angioplasty. J Am Coll Cardiol 6:369–375

6. Badimon L, Badimon JJ, Galvez et al. (1986) Influence of arterial damage and wall shear rate on platelet deposition. Arteriosclerosis 6:312–320

7. Beere PA, Glagov S, Zarins CK (1984) Retarding effect of lowered heart rate on coronary atherosclerosis. Science 226:180–182

8. Beere PA, Stankunivicius R, Ku DN et al. (1986) Low heart rate retards carotid atherosclerosis. Arteriosclerosis 6:524 (abstract)

9. Bell FP. Gallus A, Schwartz CJ (1974) Aortic endothelial permeability to albumin: focal and regional patterns of uptake and transmural distribution of ^{131}I-albumin in the young pig. Exp Mol Pathol 20:57–68

10. Benditt EP (1974) Evidence for a monoclonal origin of human atherosclerotic plaques and some implications. Circulation 50:650–652

11. Berry CL, Greenwald SE (1976) Effect of hypertension on the static mechanical properties and chemical composition of the rat aorta. Cardiovasc Res 10:437–451

12. Bharadvaj BK, Mabon RF, Giddens DP (1982) Steady flow in a model of the human carotid bifurcation. I. Flow visualization. J Biomech 15:348–362

13. Bharadvaj BK, Mabon RF, Giddens DP (1982) Steady flow in a model of the human carotid bifurcation. II. Laser doppler anemometer measurements. J Biomech 15:363–378

14. Bjorkerud S, Bondjers G (1973) Arterial repair and atherosclerosis after mechanical injury. V. Tissue response after induction of a large superficial transverse injury. Atherosclerosis 18:235–255

15. Blumenthal HT (1967) The pathology of human arteriosclerosis. In: Blumenthal HT (ed) Cowdry's arteriosclerosis, a survey of the problems, 2nd edn. CC Thomas, Springfield, Illinois, pp 227–414

16. Bocan TMA, Guyton JR (1985) Human aortic fibrolipid lesions. Progenitor lesions for fibrous plaques, exhibiting early formation of the cholesterol-rich core. Am J Pathol 120:193–206

17. Bomberger RA, Zarins CK, Glagov S (1981) Subcritical arterial stenosis enhances distal atherosclerosis. J Surg Res 30:205–212

18. Bomberger RA, Zarins CK, Taylor KE et al. (1980) Effect of hypotension on atherogenesis and aortic wall composition. J Surg Res 28:401–409

19. Bond MD, Adams MR, Bullock BC (1981) Complicating factors in evaluating coronary artery atherosclerosis. Artery 9:21

20. Breterton KN, Day AJ, Skinner SL (1977) Hypertension-accelerated atherogenesis in cholesterol-fed rabbits. Atherosclerosis 27:79–87

21. Bulkley BH, Hutchins GM (1977) Accelerated "atherosclerosis". A morphologic study of 97 saphenous vein coronary artery bypass grafts. Circulation 55:163–169

22. Burton AC (1954) Relation of structure to function of the tissues of the wall of blood vessels. Physiol Rev 34:619–642

23. Caro CG (1973) Transport of material between blood and wall in arteries. In: Ciba Foundation Symposium on atherogenesis: initiating factors. Associated Scientific Publishers, Amsterdam

24. Caro CG, Fish PJ, Jay M et al. (1986) Influence of vasoactive agents on arterial hemodynamics: possible relevance to atherogenesis. Biorheology 23:197 (abstract)

25. Caro CG, Fitzgerald JM, Schroter RC (1971) Atheroma and arterial wall shear. Observation, correlation and proposal of a shear dependent mass transfer mechanism for atherogenesis. Proc R Soc Lond (Biol) 117:109–159

26. Caro CG, Fitzgerald JM, Schroter RC (1969) Arterial wall shear and distribution of early atheroma in man. Nature 223:1159–1161

27. Caro CG, Lever MJ, Laver-Rudich Z et al. (1980) Net albumin transport across the wall of the rabbit common carotid artery perfused in situ. Atherosclerosis 37:497–511

28. Caro CG, Pedley TJ, Schroter RC et al. (1978) The mechanics of the circulation. Oxford University Press, Oxford.

29. Chandler AB (1970) Thrombosis and the development of atherosclerotic lesions. In: Jones RJ (ed) Atherosclerosis: proceedings of the second international symposium. Springer-Verlag, Berlin Heidelberg New York pp 88–93

30. Chandler AB, Pope JT (1975) Arterial thrombosis in atherogenesis: a survey of the frequency of incorporation of thrombi into atherosclerotic plaques. In: Hautvast JGAJ, Hermus RJJ, van der Haar F (eds) Blood and arterial wall in atherogenesis and arterial thrombosis. EJ Brill, Leiden, pp 110–118

31. Chidi CC, Klein L, DePalma R (1979) Effect of regenerated endothelium on collagen content in the injured artery. Surg Gynecol Obstet 148:839–843

32. Chobanian AV (1983) The influence of hypertension and other hemodynamic factors in atherogenesis. Cardiovasc Dis 26:177–196

33. Choy DSJ, Stertzer SH, Rotterdam HZ et al. (1982) Laser coronary angioplasty: experience with 9 cadaver hearts. Am J Cardiol 50: 1209–1211

34. Clark JM, Glagov S (1976) Luminal surface of distended arteries by scanning electron microscopy: eliminating configurational artifacts. Br J Exp Pathol 57:129–135

35. Clark JM, Glagov S (1985) Transmural organization

of the arterial wall: the lamellar unit revisited. Arteriosclerosis 5:19–34

36. Clowes AW, Gown AM, Hanson SR et al. (1985) Mechanisms of arterial graft failure. I. Role of cellular proliferation in early healing of PTFE prostheses. Am J Pathol 118:43–54

37. Constantinides P (1984) Atherosclerosis a general study and synthesis. Surv Synth Pathol Res 3:477–498

38. Cornhill JF, Roach MR (1976) A quantitative study of the localization of atherosclerotic lesions in the rabbit aorta. Atherosclerosis 23:489–501

39. Coutard M, Osborne-Pellegrin MJ (1983) Decreased dietary lipid deposition in spontaneous lesions distal to a stenosis in the rat caudal artery. Artery 2:182–198

40. Cox RH (1977) Carotid artery mechanics and composition in renal and DOCA hypertension in the rat. Cardiovasc Med 2:761–766

41. Cox RH (1982) Changes in arterial wall properties during development and maintenance of renal hypertension. Am J Physiol 242:H477–H484

42. Crawford T, Levene CI (1955) Medial thinning in atheroma. J Path Bact 66:19

43. Cumberland DC, Taylor DI, Welsh CL et al. (1986) Percutaneous laser thermal angioplasty: Initial clinical results with a laser probe in total peripheral artery occlusions. Lancet II:1457–1459

44. Davies MJ, Thomas AC (1985) Plaque fissuring – the cause of acute myocardial infarction, sudden ischaemic death and crescendo angina. Br Heart J 53:363

45. Davies PF (1986) Biology of disease. Vascular cell interactions with special reference to the pathogenesis of atherosclerosis. Lab Invest 55:5–24

46. Davies PF, Remuzzi A, Gordon EJ et al. (1986) Turbulent fluid shear stress induces vascular endothelial cell turnover in vitro. Proc Natl Acad Sci USA 83:2114–2117

47. Davis PF, Stehbens WE (1985) The biochemical composition of haemodynamically stressed vascular tissue. I. The lipid, calcium and DNA concentrations in experimental arteriovenous fistulae. Atherosclerosis 56:27–37

48. DeBakey ME, Lawrie GM, Glaeser DH (1985) Patterns of atherosclerosis and their surgical significance. Ann Surg 201:115–131

49. Dewey CF, Bussolari SR, Gimbrone MA et al. (1981) The dynamic response of endothelial cells to fluid shear stress. J Biochem Eng 103:177–185

50. Doyle JM, Dobrin PB (1973) Stress gradients in the walls of large arteries. J Biomech 16:631–639

51. Duguid JB (1949) Pathogenesis of atherosclerosis. Lancet II:925–927

52. Ernst E, Weihmayr M, Schmid M et al. (1986) Cardiovascular risk factors and hemorrheology, physical fitness, stress and obesity. Atherosclerosis 59:263–269

53. Faggiotto A, Ross R, Harker L (1984) Studies of hypercholesterolemia in the nonhuman primate. I. Changes that lead to fatty streak formation. Arteriosclerosis 4:323–340

54. Faggiotto A, Ross R (1984) Studies of hypercholesterolemia in the nonhuman primate. II. Fatty streak conversion to fibrous plaque. Arteriosclerosis 4:341–356

55. Falk E (1983) Plaque rupture with severe pre-existing stenosis precipitating coronary thrombosis: characteristics of coronary atherosclerotic plaques underlying fatal occlusive thrombi. Br Heart J 50:127

56. Falcone DJ, Hajjar DP, Minick CR (1984) Lipoprotein and albumin accumulation in reendothelialized and deendothelialized aorta. Am J Pathol 114:112–120

57. Feldman SA, Glagov S (1971) Transmedial collagen and elastin gradients in human aortas: reversal with age. Atherosclerosis 13:385–394

58. Ferguson GG, Roach MR (1976) Flow conditions at bifurcations as determined in glass models with reference to the focal distribution of vascular lesions. In: Bergel DH (ed) Cardiovascular fluid dynamics. Academic Press, New York and London pp 141–157

59. Fischer GM (1976) Effects of spontaneous hypertension and age of arterial connective tissue in the rat. Exp Gerontol 11:209–215

60. Fischer GM, Llaurado JG (1966) Collagen and elastin content in canine arteries selected from functionally different vascular beds. Circ Res 19:394–399

61. Flaharty JT, Ferrans VJ, Pierce JE et al. (1972) Localizing factors in experimental atherosclerosis. In: Siegel BL, Likoff W et al. (eds) Atherosclerosis and coronary heart disease. Grune and Stratton, New York, pp 40–83

62. Flaherty JT, Pierce JE, Ferrans VJ et al. (1972) Endothelial nuclear patterns in the canine arterial tree with particular reference to hemodynamic events. Circ Res 30:23–33

63. Fox B, James K, Morgan et al. (1982) Distribution of fatty and fibrous plaques in young human coronary arteries. Atherosclerosis 41:337-347

64. Fox JA, Hugh AE (1976) Static zones in the internal carotid artery: correlation with boundary layer separation and stasis model flows. Br J Radiol 43:370–376

65. Friedman M (1971) The coronary thrombus – its origin and fate. Hum Pathol 40:153

66. Friedman MH, Hutchins GM, Bargeron CB et al. (1981) Correlation between intimal thickness and fluid shear in human arteries. Atherosclerosis 39:425–436

67. Fry DL (1973) Responses of arterial wall to certain physical factors. In: Ciba Foundation Symposium on atherogenesis: initiating factors. Associated Scientific Publishers, Amsterdam, pp 93–125

68. Fry DL (1976) Hemodynamic forces in atherogenesis. In: Steinberg P (ed) Cerebrovascular diseases. Raven Press, New York, pp 77–95

69. Fry DL (1985) Steady-state macromolecular transport across a multilayered arterial wall. Math Model 6:353–368

70. Fry DL (1985) Mathematical models of arterial transmural transport. Am J Physiol 248:H240–H263

71. Fry DL, Mahley RW, Weisgraber KH et al. (1977) Simultaneous accumulation of Evans blue dye and albumin in the canine aortic wall. Am J Physiol 233:H66–H79

72. Fung YC (1984) Biodynamics: circulation. Springer-Verlag, Berlin Heidelberg New York

73. Furchgott RF (1983) Role of endothelium in responses of vascular smooth muscle. Circulation 53:557–573

74. Geer JC, Haust MD (1972) Smooth muscle cells in atherogenesis. Karger, Basel, pp 1–140

75. Geiringer E (1951) The mural coronary. Am Heart J 41:359

76. Gerrity RG (1981) The role of the monocyte in atherogenesis. I. Transition of blood-borne monocytes into foam cells in fatty lesions. Am J Pathol 103:181–190

77. Gerrity RG (1981) The role of the monocyte in atherogenesis. II. Migration of foam cells from atherosclerotic lesions. Am J Pathol 103:191–200

78. Gerrity RG, Goss JA, Soby L (1985) Control of monocyte recruitment by chemotactic factor(s) in lesion-prone areas of swine aorta. Arteriosclerosis 5(1):55–66

79. Gerrity RG, Richardson M, Somer JB et al. (1977) Endothelial cell morphology in areas of in vivo Evans blue uptake in the aorta of young pigs. Am J Pathol 89:313–334

80. Gessner FB (1973) Hemodynamic theories of atherogenesis. Circ Res 33:259–266

81. Gimbrone MA, Dewey CF, Davies PF et al. (1981) Hemodynamic shear stress and endothelial cell function: in vitro studies. Thromb Hemost 46:153

82. Glagov S (1972) Hemodynamic risk factors: mechanical stress, mural architecture, medial nutrition and the vulnerability of arteries to atherosclerosis. In: Wissler RW, Geer JC (eds) Pathogenesis of atherosclerosis. Williams and Wilkins, Baltimore

83. Glagov S (1984) Microarchitecture of arteries and veins. In: Abramson D, Dobrin P (eds) Blood vessels and lymphatics. Academic Press, Orlando, Florida pp 3–16

84. Glagov S, Ozoa AK (1960) Significance of the relatively low incidence of atherosclerosis in the pulmonary, renal and mesenteric arteries. Ann NY Acad Sci 149:940

85. Glagov S, Rowley DA, Kohut R (1961) Atherosclerosis of human aorta and its coronary and renal arteries. Arch Pathol 72:558–571

86. Glagov S, Rowley DA, Cramer DB et al. (1970) Heart rates during 24 hours of usual activity for 100 normal men. J App Physiol 29:799–805

87. Glagov S, Weisenberg E, Kolletis G et al. (1986) Compensatory enlargement of human atherosclerotic coronary arteries. N Engl J Med 316:1371–1375

89. Glagov S, Zarins CK (1983) Quantitating atherosclerosis: problems of definition. In: Bond MG, Insull W, Glagov S, Chandler AB and Cornhill F (eds) Clinical diagnosis of atherosclerosis: quantitative methods of evaluation. Springer-Verlag, Berlin Heidelberg New York, pp 11–35

90. Greenhill NS, Stehbens WE (1987) Haemodynamically-induced intimal tears in experimental U-shaped arterial loops as seen by scanning electron microscopy. Br J Exp Pathol 66:557–584

91. Greenwald SE, Berry CL (1978) Static mechanical properties and chemical composition of the aorta of spontaneously hypertensive rats: a comparison with the effects of induced hypertension. Cardiovasc Res 12: 364–372

92. Greisler HP, Kim DU, Price JB et al. (1985) Arterial regenerative activity after prosthetic implantation. Arch Surg 120:315–323

93. Gryska PF (1961) The physical properties of arteries after endarterectomy. Surgery 113:227–229

94. Gutstein WH, Farrell GA, Armellini C (1973) Blood flow disturbance and endothelial cell injury in pre-atherosclerotic swine. Lab Invest 29:134–149

95. Guyton JR, Hortley CJ (1985) Flow restriction of one carotid artery in juvenile rats inhibits growth of arterial diameter. Am J Physiol 248:H540–H546

96. Hartman JD (1977) Structural changes within the media of coronary arteries related to intimal thickening. Am J Pathol 89:13–34

97. Haust MD (1971) The morphogenesis and fate of potential and early atherosclerotic lesions in man. Hum Pathol 2:1–29

98. Haust MD, More RH (1963) Significance of the smooth muscle cell in atherogenesis. In: Jones RJ (ed) Evolution of atherosclerotic plaque. University of Chicago Press, Chicago, pp 51–65

99. Hollander W, Madoff I, Paddock J et al. (1976) Aggravation of atherosclerosis by hypertension in a sub-

human primate model with coarctation of the aorta. Circ Res [Suppl 2] 38:63

100. Hollander W, Prusty S, Kirkpatrick B et al. (1977) Role of hypertension in ischemic heart disease and cerebral vascular disease in the cynomolgus monkey with coarctation of the aorta. Circ Res [Suppl 1] 40:70–83

101. Hoover RL, Folger R, Haering WA et al. (1980) Adhesion of leukocytes to endothelium: roles of divalent cations, surface charge, chemotactic agents and substrate. J Cell Sci 45:73–86

102. Horie T, Sekiguchi M, Hirosawa K (1978) Coronary thrombosis in pathogenesis of acute myocardial infarction: histopathological study of coronary arteries in 108 necropsied cases using serial sections. Br Heart J 40:153

103. Imai H, Scott RF, Thomas WA (1982) Evans Blue dye: its accumulation in extracellular space in relation to endothelial cell basement membrane in normal and atherosclerotic areas of abdominal aorta. Arch Pathol Lab Med 106:186–191

104. Imparato AM, Bracco A, Kim GE et al. (1972) Intimal and neointimal fibrous proliferation causing failure of arterial reconstructions. Surgery 72:1007–1017

105. Imparato AM, Riles TS, Mintzer R et al. (1983) The importance of hemorrhage in the relationship between gross morphologic characteristics and cerebral symptoms in 376 carotid artery plaques. Ann Surg 197:195–203

106. Jellinek H, Fuzesi S, Solti F et al. (1986) Ultrastructural study of canine aortic damage caused by disturbance of transmural transport. Exp Mol Pathol 44:67–75

107. Jennings G, Nelson L, Nestel P et al. (1986) The effects of changes in physical activity on major cardiovascular risk factors, hemodynamics, sympathetic function, and glucose utilization in man: a controlled study of four levels of activity. Circulation 73:30–40

108. Jeynes BJ, Warren BA (1981) Thrombogenicity of components of atheromatous material. Arch Pathol Lab Med 105:353–357

109. Jorgensen L, Packham MA, Rowsell HC et al. (1972) Deposition of formed elements of blood on the intima and signs of intimal injury in the aorta of rabbit, pig, and man. Lab Invest 27:341–350

110. Joris I, Majno G (1979) Inflammatory components of atherosclerosis. In: Weissmann G, Samuelson B, Paoletti R (eds) Advances in inflammation research, vol 1. Raven Press, New York, pp 71–85

111. Joris I, Zand T, Nunnari JJ et al. (1983) Studies on the pathogenesis of atherosclerosis. I. Adhesion and emigration of mononuclear cells in the aorta of hypercholesterolemic rats. Am J Pathol 113:341–358

112. Kamiya A, Togawa T (1980) Adaptive regulation of wall shear stress to flow change in the canine carotid artery. Am J Physiol 239:H14–H21

113. Kannel WB, Schwartz MJ, McNamara PM (1969) Blood pressure and risk of coronary heart disease: the Framington Study. Dis Chest 56:43–52

114. Kaplan JR, Clarkson TB, Manuck SB (1984) Pathogenesis of carotid bifurcation atherosclerosis in cynomolgus monkeys. Stroke 15:994–100

115. Karino T (1987) Microscopic structure of disturbed flows in the arterial and venous systems, and its implication in the localization of vascular diseases. Int Angiol 5:297–313

116. Karino T, Goldsmith HL (1979) Aggregation of human platelets to collagen on the walls distal to a tubular expansion. Microvasc Res 17:217–237

117. Karino T, Kwong HHM, Goldsmith HL (1979) Particle

flow behaviour in models of branching vessels. I. Vortices in 90° T-junctions. Biorheology 16:231–248

118. Kensey K, Nash J, Abrahams C, Lake K, Zarins CK (1986) Recanalization of obstructed arteries using a flexible rotating tip catheter. Circulation [Suppl] 74(2):(abstract)

119. Khalifa AMA, Giddens DP (1981) Characterization and evolution of poststenotic flow disturbances. J Biomech 14:279–296

120. Klocke FJ, Mates RE, Canty JM et al. (1985) Coronary pressure-flow relationships. Controversial issues and probable implications. Circ Res 56:310–323

121. Ku DN, Giddens DP (1983) Pulsatile flow in a model carotid bifurcation. Arteriosclerosis, 3:31–39

122. Ku DN, Giddens DP, Phillips DJ et al. (1985) Hemodynamics of the normal human carotid bifurcation: in vitro and in vivo studies. Ultrasound Med Biol 1:13–26

123. Ku DN, Zarins CK, Giddens DP et al. (1985) Pulsatile flow and atherosclerosis in the human carotid bifurcation: positive correlation between plaque localization and low and oscillating shear stress. Arteriosclerosis 5:292–302

124. Ku DN, Zarins CK, Giddens DP, Glagov S (1986) Reduced atherogenesis distal to stenosis despite turbulence and hypertension. Circulation [suppl] 74(2):abstract

125. Langille BL (1984) Integrity of arterial endothelium following acute exposure to high shear stress. Biorheology 21:333–346

126. Langille BL, Adamson SL (1981) Relationship between blood flow direction and endothelial cell orientation at branch sites in rabbits and mice. Circ Res 48:481–488

127. Langille BL, O'Donnell F (1986) Reductions in arterial diameter produced by chronic diseases in blood flow are endothelial-dependent. Science 231:405–407

128. Learoyd BM, Taylor MG (1966) Alterations with age in the viscoelastic properties of human arterial walls. Circ Res 18:278–292

129. Leung, DYM, Glagov S, Mathews MB (1976) Cyclic stretching stimulates synthesis of matrix components by arterial smooth muscle cells in vitro. Science 191:475–477

130. Lewis JC, Taylor RG, Norman BS et al. (1982) Endothelial surface characteristics in pigeon coronary artery atherosclerosis. I. Cellular alterations during the initial stages of dietary cholesterol challenge. Lab Invest 46:133–138

131. Lieber BB (1985) Ordered and random structures in pulsatile flow through constricted tubes. Ph.D. thesis, Georgia Institute of Technology, Atlanta

132. Liepsch D, Moravec S, Zimmer R (1983) Pulsating flow in distensible models of vascular branches. Fluid dynamics as a localizing factor for atherosclerosis. In: Schettler G et al. (eds) Atherosclerosis VI. Springer-Verlag, Berlin Heidelberg New York, pp 46–70

133. Lockette W, Otsuka Y, Carretero O (1986) The loss of endothelium-dependent vascular relaxation in hypertension. Hypertension 8 [Suppl II]:II61–II66

134. LoGerfo FW, Nowak MD, Quist WC et al. (1981) Flow studies in a model carotid bufurcation. Arteriosclerosis 1:235–241

135. LoGerfo FW, Nowak MD, Quist WC (1985) Structural details of boundary layer separation in a model human carotid burfurcation under steady and pulsatile flow conditions. J Vasc Surg 2:263–269

136. Lyon RT, Hass A, Davis HR et al. (1987) Protection from atherosclerotic lesion formation by reduction of artery wall motion. J Vasc Surg 5:59–67

137. Lyon RT, Zarins CK, Lu CT et al. (1981) Arterial wall disruption by balloon dilatation: quantitative comparison of normal, stenotic and occluded vessel. Surg Forum 32:326–328

138. Manuck SB, Kaplan JR, Clarkson TB (1983) Behaviorally induced heart rate reactivity and atherosclerosis in cynomolgus monkeys. Psychosom Med 45:95–108

139. Markle RA, Hollis TM (1981) Influence of locally altered in vivo shear stress on aortic histamine-forming capacity and aortic albumin uptake. Blood vessels 18:45–57

140. Martin F, McAuley BJ, Zimmerman JJ et al. (1986) Transluminal atherectomy: hemodynamic effects in occlusive vascular disease by doppler analysis. Circulation [Suppl] 74(2):abstract

141. McGill HC (1984) Persistent problems in the pathogenesis of atherosclerosis. Arteriosclerosis 4:443–451

142. Megerman J, Abbot WM (1983) Compliance in vascular grafts. In: Creighton BW (ed) Vascular grafting: clinical applications and techniques, John Wright and Sons Ltd, Bristol, pp 344–364

143. Minick CR, Stemerman MB, Insull W (1979) Role of endothelium and hypercholesterolemia in intimal thickening and lipid accumulation. Am J Pathol 95:131–158

144. Montenegro MR, Eggen DA (1968) Topography of atherosclerosis in the coronary arteries. Lab Invest 18:586–593

145. More RH, Movat HZ, Haust MD (1957) Role of mural fibrin thrombi of the aorta in the genesis of arteriosclerotic plaques. Arch Pathol 63:612–620

146. Neitzel GF, Barboriak JJ, Pintar K, Qureshi I (1986) Atherosclerosis in aortocoronary bypass grafts: morphologic study and risk factor analysis 6 to 12 years after surgery. Arteriosclerosis 6:594–600

147. Nerem RM, Levesque MJ, Cornhill JF (1981) Vascular endothelial morphology as an indicator of the pattern of blood flow. J Biomech Eng 103:171–176

148. Nerem RM, Levesque MJ, Sato M (1986) Mechanical properties of endothelial cells. Biorheology 23:230

149. Newman DL, Lallemand RC (1978) The effect of age on the distensibility of the abdominal aorta in man. Surg Gynecol Obstet 147:211–214

150. Owens C, Richardson PD (1985) Mechanical properties of carotid artery walls. Proceedings 11th annual NE bioengineering conference, 1985, pp 12–15

151. Ozoa AK, Glagov S (1963) Induction of atheroma in renal arteries of rabbits. Arch Pathol 76:667

152. Poole JCF, Florey HW (1958) Changes in the endothelium of the aorta and the behaviour of macrophages in experimental atheroma of rabbits. J Path Bact 75:245–252

153. Reidy MA (1985) Biology of disease: a reassessment of endothelial injury and arterial lesion formation. Lab Invest 53:513–520

154. Reidy MA, Bowyer DE (1977) Scanning electron microscopy of arteries. The morphology of aortic endothelium in haemodynamically stressed areas associated with branches. Atherosclerosis 26:181–194

155. Reidy MA, Langille BL (1980) The effect of local blood flow patterns on endothelial cell morphology. Exp Mol Pathol 32:276–289

156. Ridolfi RL, Hutchins GM (1977) The relation between coronary artery lesions and myocardial infarcts: ulceration of atherosclerotic plaques precipitating coronary thrombosis. Am Heart J 93:468

157. Roach MR, Burton AC (1957) The reason for the shape

of the distensibility curves of arteries. Can J Biochem Physiol 35:681–690

158. Roberts JC, Moses C, Wilkins RH (1959) Autopsy studies in atherosclerosis. I. Distribution and severity of atherosclerosis in patients dying without morphologic evidence of atherosclerotic catastrophe. Circulation 20:511

159. Roberts JC, Wilkins RH, Moses C (1959) Autopsy studies in atherosclerosis. II. Distribution and severity of atherosclerosis in patients dying with morphologic evidence of atherosclerotic catastrophe. Circulation 20:520

160. Robertson WB, Strong JP (1969) Atherosclerosis in persons with hypertension and diabetes mellitus. Lab Invest 18:538

161. Rodbard S (1970) Negative feedback mechanisms in the architecture and function of the connective and cardiovascular tissues. Perspect Biol Med 13:507–527

162. Ross R (1986) The pathogenesis of atherosclerosis – an update. N Engl J Med 314:488–500

163. Ross R, Glomset J (1976) The pathogenesis of atherosclerosis. N Engl J Med 295:369

164. Sabbah HN, Khaja F, Brymer JF et al. (1984) Blood velocity in the right coronary artery: relation to the distribution of atherosclerotic lesions. Am J Cardiol 53:1008–1012

165. Sacks AH (1975) The vasa vasorum as a link between hypertension and arteriosclerosis. Angiology 26:385–390

166. Saltissi S, Webb-Peploe MM, Cortart DJ (1979) Effect of variation in coronary artery anatomy on distribution of stenotic lesions. Br Heart J 42:186–191

167. Sanborn TA, Faxon DP. Haudenschild CC, Ryan TJ (1985) Experimental angioplasty: circumferential distribution of laser thermal energy with a laser probe. J Am Coll Cardiol 5:934–938

168. Sanborn TA, Faxon DP, Haudenschild CC, Gottsman SB, Ryan TJ (1983) The mechanism of transluminal angioplasty: evidence for formation of aneurysms in experimental atherosclerosis. Circulation 68:1137–40

169. Schaffner T, Taylor K, Bartucci EJ et al. (1980) Arterial foam cells exhibit distinctive immunomorphologic and histochemical features of macrophages. Am J Pathol 100:57–80

170. Schneiderman G, Ellis CG, Goldstick TK (1979) Mass transport to walls of stenosed arteries: variation with Reynolds number and blood flow separation. J Biomech 12:869–877

171. Schneiderman G, Pritchard WF, Ramirez CA, Colton CK, Smith KA, Stemerman MB (1983) Rabbit aortic medial thickness under relaxed and specified simulated in vivo conditions. Am J Physiol 245:H623–H627

172. Scharfstein H, Gutstein WH, Lewis L (1963) Changes of boundary layer flow in model systems, implications for initiation of endothelial injury. Circ Res 13:580–584

173. Schwartz CJ, Mitchell JRA (1962) Observations on localization of arterial plaques. Circ Res 11:63–73

174. Sharp WV, Donovan DL, Teague PC et al. (1982) Arterial occlusive disease: a function of vessel bifurcation angle. Surgery 91:680–684

175. Small DM (1970) The physical state of lipids of biological importance: cholesteryl esters, cholesterol, triglyceride. In: Blank M (ed) Surface chemistry of biological systems. Plenum Press, New York, pp 55–83 (Advances in experimental medicine and biology, vol 7)

176. Smiesko V, Kozik J, Dolezel S (1985) Role of endo-

177. Smith EB, Ashall C (1984) Compartmentalization of water in human atherosclerotic lesions. Changes in distribution and exclusion volumes for plasma macromolecules. Arteriosclerosis 4:21–27

178. Smith EB, Staples EM (1980) Distribution of plasma proteins across the human aortic wall. Atherosclerosis 37:579–590

179. Smith EB, Staples EM, Dietz HS, Smith RH (1979) Role of endothelium in sequestration of lipoprotein and fibrinogen in aortic lesions, thrombi and graft pseudointimas, Lancet II:812–816

180. Solberg LA, McGarry PA, Moossy J et al. (1968) Severity of atherosclerosis in cerebral arteries, coronary arteries, and aortas. Ann NY Acad Sci 149:956–973

181. Stary HC (1983) Macrophages in coronary artery and aortic intima and in atherosclerotic lesions of children and young adults up to age 29. In: Schettler FG, Gotto AM, Middelhoff G et al. (eds) Atherosclerosis VI. Proceedings of the sixth international symposium. Springer-Verlag, Berlin Heidelberg New York, pp 462–466

182. Steele PM, Chesebro JH, Stanson AW et al. (1985) Balloon angioplasty. Natural history of the pathophysiologic response to injury in pig model. Circ Res 57:105–112

183. Stehbens WE (1965) Intimal proliferation and spontaneous lipid deposition in the cerebral arteries of sheep and steers. J Athero Res 5:556–568

184. Stehbens WE (1973) Experimental arteriovenous fistulae in normal and cholesterol-fed rabbits. Pathology 5:311–324

185. Stehbens WE (1975) The role of hemodynamics in the pathogenesis of atherosclerosis. Prog Cardiovasc Dis 18:89–103

186. Stehbens WE (1981) Intimal cushions (pads): structure, location and functional significance. In: Schwartz CJ et al. (eds) Structure and function of the circulation, vol 2. Plenum, New York, pp 603–634

187. Stehbens WE (1981) Predilection of experimental arterial aneurysms for dietary-induced lipid deposition. Pathology 13:735–747

188. Stehbens WE, Karmody AM (1975) Venous atherosclerosis associated with arteriovenous fistulas of hemodialysis. Arch Surg 110:176–180

189. Still WJS (1968) The pathogenesis of the intimal thickenings produced by hypertension in larger arteries in the rat. Lab Invest 19:84–91

190. Subbiah MTR, Kottke IA, Kottke, BA, Bale LK, Grabau C (1980) Regional differences in cholesterol content of aorta in response to experimental coarctation in spontaneously atherosclerosis susceptible pigeons. Basic Res Cardiol 75:583–589

191. Sumner DS, Hokanson E, Strandness DE (1969) Arterial walls before and after endarterectomy. Arch Surg 99:606–611

192. Suzuki K (1967) Experimental studies on morphogenesis of arteriosclerosis, with special reference to relation between hemodynamic change and developments of cellulo-fibrous intimal thickening and atherosclerosis. Gunma J Med Sci 16:185–243

193. Svindland A (1983) The localization of sudanophilic and fibrous plaques in the main left coronary bifurcation. Atherosclerosis 48:139–145

194. Szilagyi DE, Elliott JP, Hageman JH et al. (1973) Biologic fate of autogenous vein implants as arterial substitutes. Surgery 178:232

195. Talukder N, Giddens DP (1983) Quantitative flow visu-

alization studies in a carotid artery bifurcation model. Proceedings of ASME. Biomech Symposium. AMD 56:165–168

196. Taylor K, Glagov S, Lamberti J et al. (1978) Surface configuration of early atheromatous lesions in controlled-pressure, perfusion-fixed monkey aortas. In: Johari O (ed) Scanning electron microscopy. Scanning Electron Microscopy Inc., Chicago, pp 459–464

197. Tedgui A, Lever MJ (1985) The interaction of convection and diffusin in the transport of ^{131}I-albumin within the media of the rabbit thoracic aorta. Circ Res 57:856–863

198. Tennekes H, Lumley JL (1972) A first course in turbulence. Massachusetts Institute of Technology Press, Cambridge, Mass.

199. Texon M (1960) The hemodynamic concept of atherosclerosis. Bull NY Acad Med 36:263–274

200. Ts'ao CH, Glagov S (1970) Basal endothelial attachment: tenacity of cytoplasmic dense zones in the rabbit aorta. Lab Invest 23:510–516

201. Velican C, Velican D (1976) Intimal thickening in developing coronary arteries and its relevance to atherosclerotic involvement. Atherosclerosis 23:345–355

202. Vlodaver A, Edwards JE (1971) Pathologic changes in aortic–coronary arterial saphenous vein grafts. Circulation 44:719–728

203. Walton KW, Slaney G, Ashton F (1985) Atherosclerosis in vascular grafts for peripheral vascular disease. Atherosclerosis 54:49–64

204. Weinbaum S, Caro CG (1976) A macromolecule transport model for the arterial wall and endothelium based on the ultrastructural specialization observed in electron microscopic studies. J Fluid Mech 74:611–640

205. Weinbaum S, Pfeffer R, Chien S (1986) Lipoprotein transport across the endothelium and its possible relationship to atherosclerosis. Biorheology 23:194

206. Wesolowski SA, Fries CC, Sabini AM, Sawyer PN (1965) The significance of turbulence in hemic systems and in the distribution of the atherosclerotic lesion. Surgery 57:155

207. Whitmore RL (1968) Rheology of the circulation. Pergamon Press, Oxford, pp 90–108

208. Wilkins RH, Roberts JC, Moses C (1959) Autopsy studies in atherosclerosis. III. Distribution and severity of atherosclerosis in the presence of obesity, hypertension, nephrosclerosis and rheumatic heart disease. Circulation 20:527

209. Winniford MD, Wheelan KR, Kremers MS et al. (1986) Smoking-induced coronary vasoconstriction in patients with atherosclerotic coronary artery disease: evidence for adrenergically mediated alterations in coronary artery tone. Circulation 73:662–667

210. Wissler RW (1967) The arterial medial cell, smooth muscle, or multifunctional mesenchyme? Circulation 36:1–4

211. Wissler RW (1985) The evolution of the atherosclerotic plaque and its complications. In: Connor WE, Bristow JD (eds) Coronary heart disease. Lippincott, Philadelphia, pp 193–214

212. Wolinsky H (1972) Long-term effects of hypertension on the rat aortic wall and their relations to concurrent aging changes. Circ Res 30:301

213. Wolinsky H, Fowler S (1978) Participation of lysosomes in atherosclerosis. N Engl J Med, 299:1173–1178

214. Wolinsky H, Glagov S (1964) Structural basis for the static mechanical properties of the aortic media. Circ Res 14:400

215. Wolinsky H, Glagov S (1967) A lamellar unit of aortic medial structure and function in mammals. Circ Res 20:99

216. Wolinsky H, Glagov S (1969) Comparison of abdominal and thoracic aortic medial structure in mammals: deviation from the usual pattern in man. Circ Res 25:677

217. Wolinksy H, Goldfischer S, Daly MM et al. (1975) Arterial lysosomes and connective tissue in primate atherosclerosis and hypertension. Circ Res 36:553

218. Zarins CK, Bomberger RA, Glagov S (1981) Localization of stenosis: increased flow velocity inhibits atherogenesis. Circulation 64 [Suppl II]:221–227

219. Zarins CK, Giddens DP, Bharadvaj BK et al. (1983) Carotid bifurcation atherosclerosis: quantitation of plaque localization with flow velocity profiles and wall shear stress. Circ Res 53:502–514

220. Zarins CK, Taylor KE, Bomberger RA et al. (1980) Endothelial integrity at aortic ostial flow dividers. In: Johari O (ed) Scanning electron microscopy, III. Scanning Electron Microscopy Inc, Chicago, pp 249–254

221. Zarins CK, Zatina MA, Giddens DP et al. (1987) Shear stress regulation of artery lumen diameter in experimental atherogenesis. J Vasc Surg 5:413–420

222. Zarins CK, Zatina MA, Glagov S (1983) Correlation of post-mortem angiography with pathologic anatomy: quantitation of atherosclerotic lesions. In: Bond MG et al. (eds) Clinical diagnosis of atherosclerosis: quantitative methods of evaluation. Springer-Verlag, Berlin Heidelberg New York, Chap. 13

223. Zwolak RM, Adams MC, Clowes AW (1987) Kinetics of vein graft hyperplasia: association with tangential stress. J Vasc Surg 5:126–136

Chapter 16

Progression, Topographical Aspects and Regression of Atherosclerosis

H. Bouissou, M. T. Pieraggi and M. Julian

Introduction

The term atherosclerosis, first used by Marchand in 1904, is applied to severe arterial sclerosis associated with atheromatous lesions in varying degrees of evolution and coalescence. Macroscopically there are three classical stages: stage I, where there are fatty streaks only; stage II, where pustules and atherosclerotic plaques are separable; and stage III, with coalescent plaques, usually calcified, cartilaginous and ulcerated (see below). We will describe vessel changes in terms of these classical stages, but longitudinal studies suggest that pathogenetic consideration may require their reclassification if newer concepts are upheld (see Chap. 10). From stage II onwards, the artery outside the area covered by plaque undergoes distinct arteriosclerosis, with fibrosis and loss of elasticity of the arterial wall (see Chap. 4). Atherosclerosis and arteriosclerosis thus diminish the resistance of the wall and favour ectasia. These changes account for the macroscopic appearance of the artery involved: the variations in diameter; the rigid, non-expandable, hard wall with its sometimes almost stony consistency; and the pale yellow colour of the normal or thickened intima which contrasts with the cartilaginous opalescent yellow plaques and the whitish fibrosis of the fibrous plaques. The characteristic grinding sound scissors make when they crush such an artery is due to calcification in the form of scales of varying thickness; these scales exude a yellowish lipid material under pressure.

The distribution and clinical effects of atherosclerotic lesions permit classification of the disease [13,14]. Type I disease is where the predominant site at the lesions is in the coronary arterial bed; type II disease is when the lesions are present predominantly in the major branches of the aortic arch; in type III disease the lesions occur in the visceral branches of the abdominal aorta (coeliac, superior mesenteric and renal arteries); and type IV disease is where the lesions are predominantly in the terminal abdominal aorta and its major branches. Types IV and I represent about two-fifths and one-third of cases respectively, type III only 3%. Arterial stenoses are usually segmental and proximal; less commonly, the occlusive disease occurs in distal arteries. As a general rule, the distal arteries of the arm and the leg and the intraparenchymal arteries are not affected. Patients with disease types I and III are younger and are significantly more likely to be female than patients with other types. Patients with type II disease often go on to develop type IV and vice versa. More rapid rates of progression can be observed in patients with disease types II and IV.

Natural History and Progression of Atherosclerosis

Atherosclerosis is a progressive arterial disease. The concept of lesion progression with increasing age has been supported by extensive autopsy

242   Diseases of the Arterial Wall

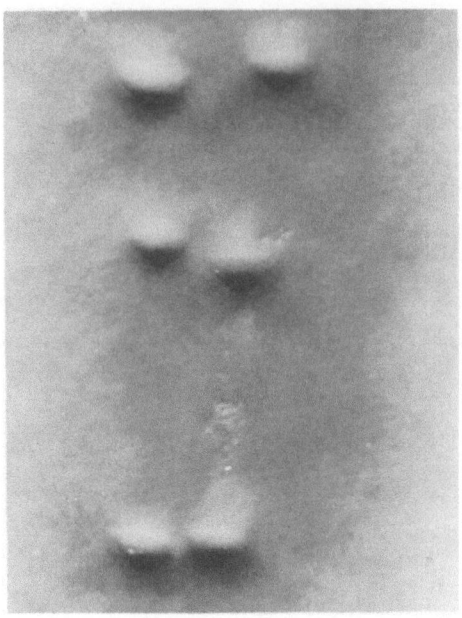

Fig. 16.1. The aorta showing a dot (gross appearance) between the entrances to the intercostal arteries.

studies. Ideally, it is important to differentiate arterial lesion progression from clinical progression which involves end-organ impairment [37,38].

Progression

Three lesions may be considered as precursors of the atherosclerotic plaque: the dot, the streak and the gelatinous elevation [2,5,7,9,29,30,43, 49, 56, 59].

The fatty dot (Figs 16.1 and 16.2) is yellowish, well delimited and flat or slightly raised. It has a

diameter of up to 0.5 mm. Schwartz et al. [58] observed it in 43% of children aged less than 1 year; it is present in the aorta in all cases after this age. Made up of a small number of more or less scattered foam cells laden with esterified cholesterol, this lesion may be resorbed, remain stable or lease very discrete intimal fibrosis. The respective roles of monocyte-macrophages and smooth-muscle cells [23,55,58] in the progression of the fatty dot is discussed elsewhere (see Chap. 11).

The streak (Figs. 16.3 and 16.4) can be seen macroscopically as a yellowish slightly elevated, trail. It is usually located with its long axis aligned along the vessel in the direction of blood flow and it measures from a few millimetres to 1–5 cm in length. Streaks may be numerous and can form a meshwork, notably in the thoracic aorta at the origin of intercostal arteries [47]. They are composed of one or several trails of foam cells massed together in the intima with extracellular matrix (see Chap. 12). The total intima lipid content in the streak is markedly increased compared with that in normal intima [63]. The content of glycosaminoglycans (GAGs) is high, but as the streak progresses the levels of dermatan sulphate and heparan sulphate tend to decrease [64]. The streak may also be resorbed (see Chap. 10).

The mechanisms of lesion progression are a matter of debate [46]. It is possible that high plasma levels of LDL, which may result from a relative deficiency of LDL receptors [8,25,35] or abnormalities of lipoprotein particles and changes in endothelial permeability, allow LDL particles to penetrate into cells. Thus a large number of cells (macrophages and smooth-muscle cells) become incapable of metabolizing cholesterol and become overloaded. It has been

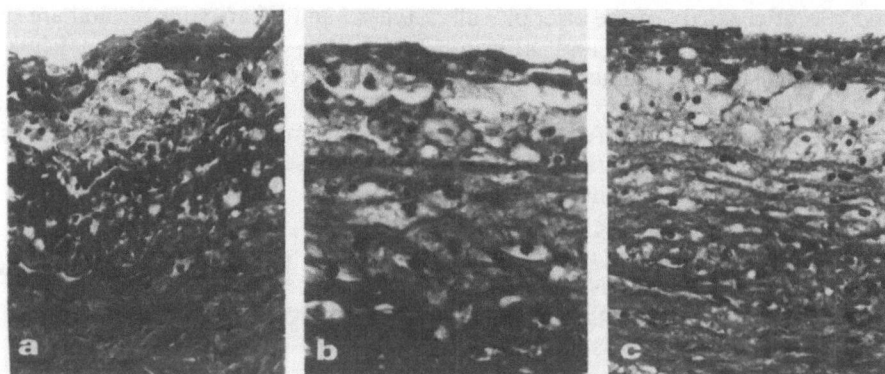

Fig. 16.2 a–c. The aorta showing the microscopic appearance of a dot and a streak. Intimal foam cells clumped at **a** and **b**, evolving towards the streak **c**. Masson's stain, ×25; ×40; ×25

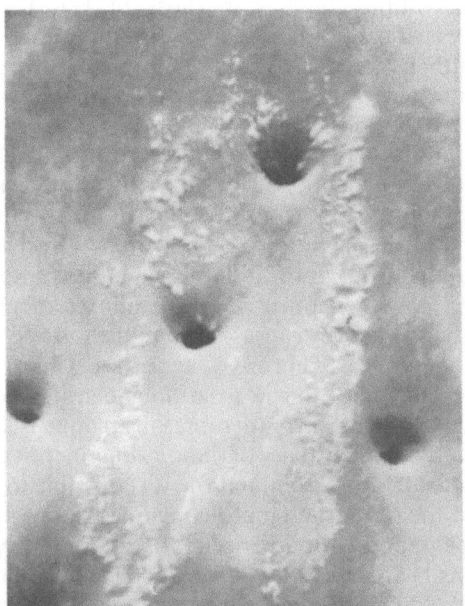

Fig. 16.3. The aorta showing the gross appearance of a streak. Mesh-like aspect: thin lipidic lines parallel to the axis of the aorta.

shown that both types of cells can take up chemically modified cholesterol via a pathway which is not mediated by LDL receptors. The result is foam cells, the numbers of which depend on extracellular lipid content [26,28].

It has been suggested that cell necrosis could be the critical event in progression of the lesion and initition of lipid-rich core formation [21,29,53]. Following cell necrosis, the residual cholesterol and its esters become extracellular

and insoluble, tend to crystallize and may lead to the arrival of new cells or to fibrosis [1]. A new influx of GAGs and lipids may lead to the formation of foam cells. The existing lesion thus increases in size. Another hypothesis is that cell necrosis may not be responsible for the extracellular lipids. Ultrastructural evidence supports the idea that direct interactions between collagen, proteoglycans and elastin with permeating LDLs could initiate the extracellular deposition of lipids [26,32] (see Chap. 12). Concerning the cell types involved in the process, most electron microscopy studies have indicated the presence of both macrophages and smooth-muscle cells with abundant myofilaments, lipid vacuoles and rough endoplasmic reticulum (see Chap. 11). In addition, LDLs promote smooth-muscle cell proliferation [20] and biosynthesis of fibrous proteins and GAGs [24]. Thus the streak, which is now irreversible, is covered by varying degrees of intimal sclerosis. If external conditions are not modified, the streak becomes relatively stable. Following the arrival of further constituents of the plasma, including lipids, the lesions may increase in size in successive bursts. The role played by streaks in the atherosclerotic process remains unclear. It is possible that various clinico-pathological entities are important, for example the age, the ethnic group and the location of the patient [45,51,53].

The gelatinous elevation (Figs 16.5, 16.6 and 16.7) is a greyish area, usually translucent but sometimes opaque, coating the intima to varying degrees. Isolated, limited, round or oval, it usually has a diameter of 0.8 mm to 1 cm but may be larger. It is often associated with other

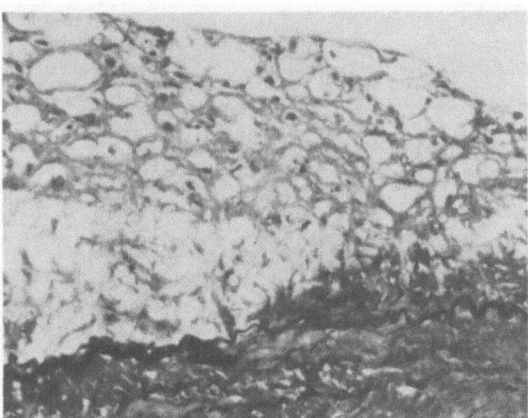

Fig. 16.4. Iliac artery. A developed streak, irreversible due to modification of the elastic layer, which is in contact with deposits of extracellular lipids. Foam cells on the surface. Verhoeff's stain, × 25

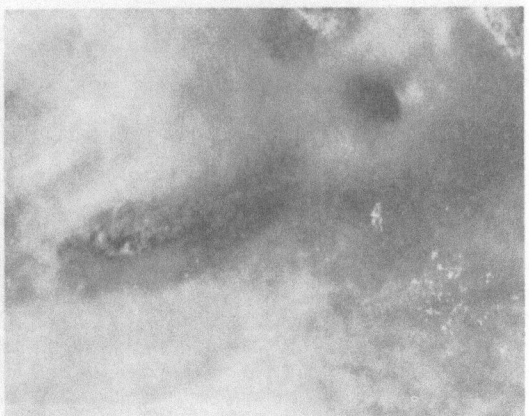

Fig. 16.5. Aorta showing the gross appearance of a gelatinous plaque at the entrance to an intercostal artery. Translucid elevation of the intima.

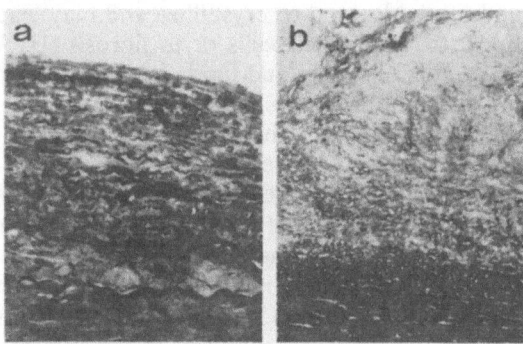

Fig. 16.6 a,b. Gelatinous plaque in the aorta. **a** Reversible, taking up the whole surface of the intima, but the internal elastic lamina is intact. Verhoeff's stain, × 25. **b** Irreversible, penetrating the intima–media area and encroaching on the media, with lysis of elastic layers. Verhoeff's stain, × 10

potential or established atherosclerotic lesions. Formed of an abundance of plasmatic substances [29,62], the lesion is seen to consist of focal oedema separating the elements of the subendothelial connective tissue. A variety of cells can be found within it, for example smooth-muscle cells and histiocytes. The smallest lesions affect only the innermost layers of the intima, but may sometimes take over the entire intima. In advanced lesions, fragmentation and distortion of the elastic and collagen fibres often occurs. Protein precipitates become visible, together with fibrin deposits in the form of fine streaks, bands or stars. Ultramicroscopy confirms the microscopic findings: the oedematous intercellular spaces contain fibrin, small granulations and filamentous material and there are few lipid droplets or osmiophilic bodies. Our

observations agree with those of Haust [29]. Biochemically, according to several studies by Smith between 1963 and 1983 [60,61], the gelatinous elevation contains twice as much water and fewer proteoglycans than a normal aorta wall, and very few lipids. The gelatinous elevation is not mentioned by Ross [55], nor by Small and Shipley [59]. It may be likened to the "focal myxoid degeneration" described by Orcel and Chomette [49]. If the lesion is slight, it is resorbable. A succession of marked influxes of plasmatic substances causes intimal fibrosis and the formation of a fibrous plaque (see Chap. 4). The possibility that a gelatinous elevation can develop into a fibroatheromatous plaque is suggested by co-existence of these lesions in the same aorta. On gross examination, it is possible to see that the fibroatheromatous plaque is greyish in appearance, with a peripheral scattering of yellowish dots. Under microscopic and ultramicroscopic examination, it resembles the streak, but with additional oedema [Haust's mixed gelatinous fatty streak [29]]. The lipids present are trapped by peripheral fibrosis, which penetrates more deeply in the increasingly damaged intima–media area. The extension of the lesion is accompanied by accentuation and lysis of the pre-existing internal elastic fibres.

The fibrolipidic lesions of atherosclerosis, that is, atheromatous "pustules" or "plaques", are irreversible lesions and are likely to evolve progressively [66,67]. The two differ only in their degree of evolution and extension [5,6,49].

The atheromatous pustule (Figs 16.8 and 16.9) is lenticular, barely elevated, smooth and shiny, whitish-yellow with a pearly lustre at its centre or over its whole surface. Its diameter is 0.5 cm

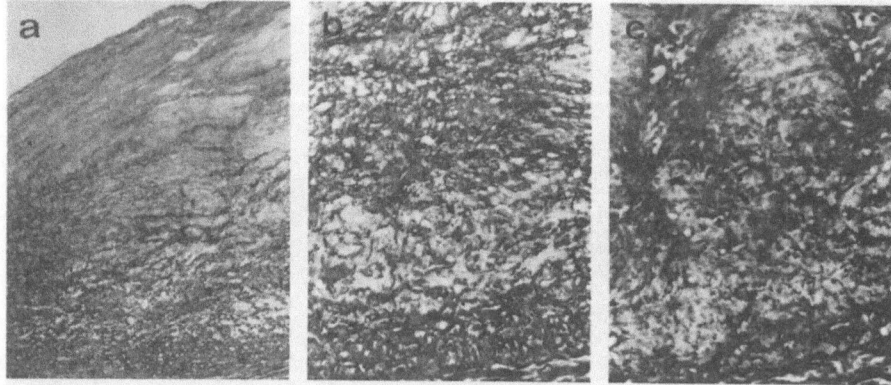

Fig. 16.7 a–c. Aorta showing a gelatinous plaque evolving towards an atheromatous pustule with intra- and extracellular lipid infiltration. **a** General appearance: thickened intima, influx of GAG, destruction of the elastic layers of the intima–media area, where lipids can be seen. Verhoeff's stain, × 10. **b** and **c** Intima–media zone, showing lipids and destruction of elastic layers. Verhoeff's stain, × 25

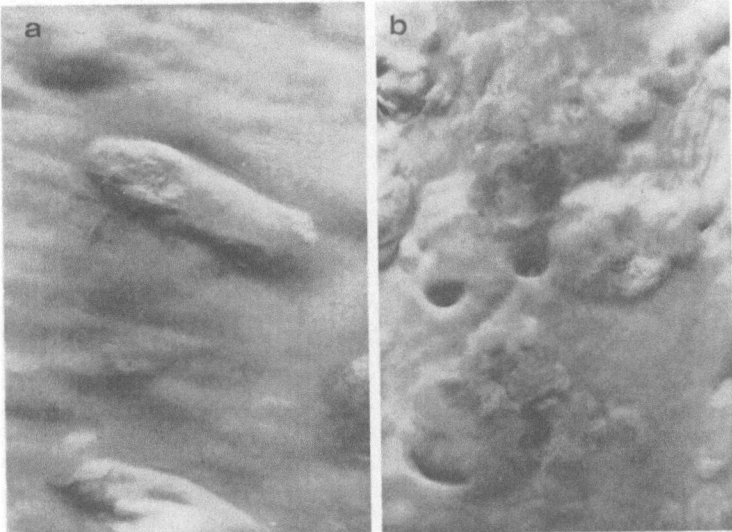

Fig. 16.8 a, b. Pustule and atheromatous plaques of varying dimensions with and without ulceration. The intima around the plaques is fibrous, cartilaginous and thickened. The entrance regions of the collateral arteries are infiltrated, and the lumen is narrowed.

and its thickness 2 mm. It is found in the intima and overlapping into the intima–media zone, and is surrounded by varying degrees of fibrosis. Its centre contains extracellular and intracellular lipids, cellular debris, fibrin and GAGs. Tiny calcified deposits may be observed.

The atheromatous plaque (Figs 16.8 and 16.10) develops from the pustule. Nummular, it has a diameter of 1–3 cm and a thickness of 2–4 mm. It is of distinctly different consistency from the arterial wall surrounding it. A cross-section (Fig. 16.11) shows a butter yellow layer

between a thickened intima and a thin media, from which it appears to be separated by a narrow whitish sclerotic band. Situated in the deep part of the intima, which forms a fibrous cap, it occupies the entire intima–media zone and

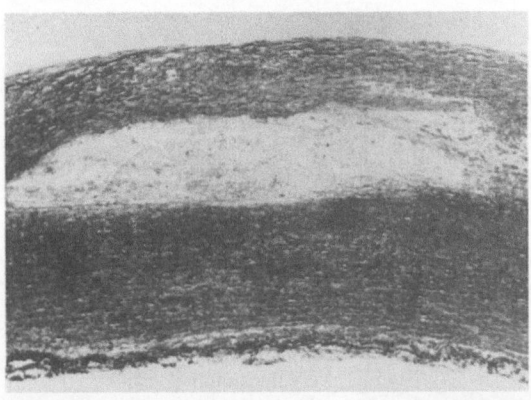

Fig. 16.9. Microscopic appearance of an atheromatous pustule: clump of extracellular lipids with some cholesterol crystals, enclosed between a thickened and sclerotic intima and a fibrous media, in the fragile intimalization area. Verhoeff's stain, × 10

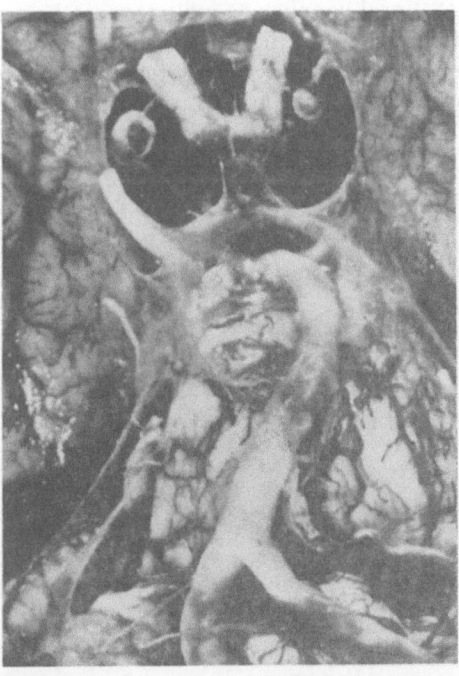

Fig. 16.10. Basilar artery. Atheromatous plaques clearly showing through.

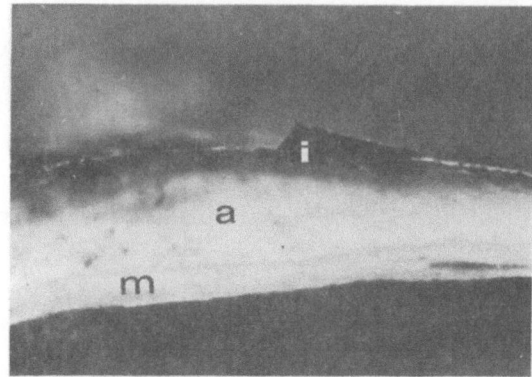

Fig. 16.11. Section of an atheromatous plaque. The atheromatous material (*a*) is visible between the intima (*i*) where a thrombus can be seen, and the thin media (*m*).

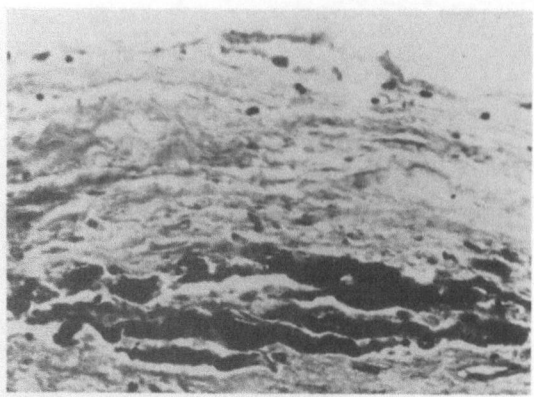

Fig. 16.13. Aorta. Mural thrombus incorporated in a fibrous intima. Masson's stain, × 25

impinges upon the media to varying degrees (Fig. 16.12). In the centre is lipid which has a necrotic, eosinophilic and grainy appearance, with many lanceolated crystals whose number increases with the age of the lesion. Clumps of fibrin and of cellular debris are also seen. The centre of the plaque is surrounded by a fibrous cap, the thickness of which varies from one plaque to another. The structures of the centre and the periphery are similar to those of the pustule. Newly formed vessels are numerous at the base of the lesion and result either from vasa vasorum or from organized mural thrombi (see Fig. 16.13). Focal lymphocyte infiltrates can be seen at the periphery of the lesion [34,47,64]. Ultrastructural evidence confirms the light microscopic findings and shows variable numbers of

lipid-laden macrophages and smooth-muscle cells, as well as peripheral fibrosis and a lipid-rich core with cell debris and calcium deposits [3,22].

The established lesion becomes detectable using angiography, as it elevates the intima. This becomes rigid and sometimes calcified. The lumen may slowly be narrowed by the development of one or more plaques. Relative ischaemia then occurs in the area involved. Plaques may become stable or develop further following the formation of new endothelial lesions which cause accretion of thrombi and an influx of GAGs and lipids which further generates new lesions and brings in smooth-muscle cells. The volumes and surface areas of the lesions are thus increased. The plaque becomes more elevated within the

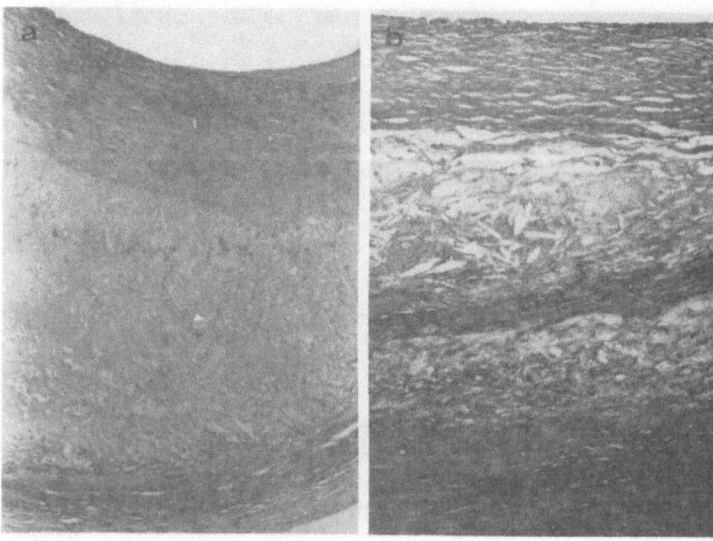

Fig. 16.12. a An atheromatous plaque almost occluding the left coronary artery (common trunk). Very thick intimal "cap"; atheromatous material; badly damaged internal elastic lamina; thin fibrous media. Verhoeff's stain, × 10. **b** Aortic atheromatous plaque, showing the same modifications as in **a**. Verhoeff's stain, × 25

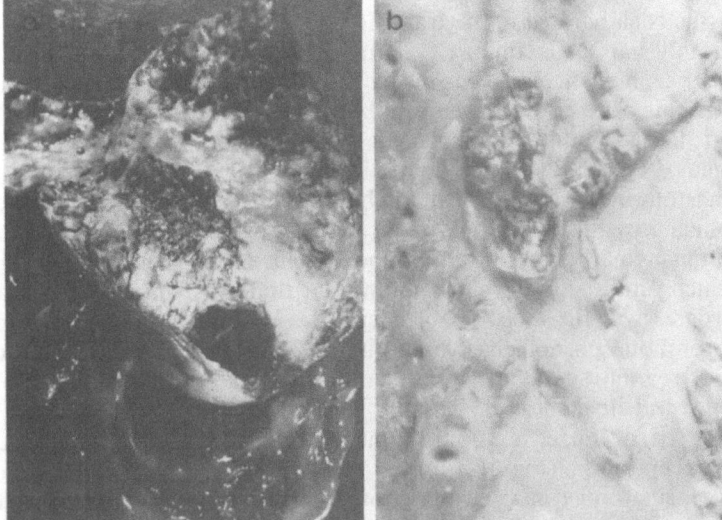

Fig. 16.14. Aorta. **a** Ascending aorta. Central ulceration of a vast atheromatous patch showing the classic "paved aorta" appearance. **b** Abdominal aorta. Atheromatous plaques of which one is ulcerated.

lumen and the fibrous shell of the intima is accentuated by re-endothelialized, fibrotic thrombi (this is particularly noticeable in musculo-elastic or elastic arteries). This irregular development results in the proliferation and coalescence of the atheromatous plaques, and in the progressive destruction of the arterial media which may now consist only of a few muscular cells or a very reduced distorted musculo-elastic framework. This damage and the presence of variable quantities of trapped lipids account for the appearance of these advanced lesions on gross examination, in particular the extensive hypertrophic or atrophic sclerosis and the classical "paved" appearance of the intimal surface of the vessel deformed by the plaques (Fig. 16.14). Between the plaques, the arterial walls appear more or less sclerotic or distended [8]. It is at this stage that lesions which have long remained silent become a problem. Atherosclerosis may now be revealed by a complication.

Complications

Calcification is usually considered as one of the complications of atherosclerosis. This is not entirely justified because plaques are almost always calcified and the calcifications are not directly responsible . for the clinical manifestations of atherosclerosis [9]. The calcification may be situated at the periphery or in the atheromatous pulp.

Ulceration (Fig. 16.14) plays a key role in the evolution of the plaque and the development of thrombus [12,18,27,64]. Ulceration is a rupture of the endothelium and fibrous cap, and can result from the following mechanisms; (a) tensile stresses between the rigid plaque and the more supple arterial wall surrounding it; (b) ischaemic processes leading to the formation of an ulcerated necrotic area; or (c) rupture of newly formed vessels responsible for intraplaque haematoma (see below). Healing can occur by re-endothelialization of mural thrombi. These ulcerations may be the source of distal atheromatous emboli, which are formed of cholesterol crystals, fibrin and platelets. Most of them are multiple and clinically silent, but they can be responsible for distal, limited, sometimes multiple, ischaemic lesions. They can be recognized in kidney biopsy specimens. These areas of ulceration are a constant target for platelet aggregation and thrombosis (Fig. 16.14a) [33].

Thrombosis is one of the major complications of atherosclerosis. Thrombi of various ages, formed successively by repeated mural deposits and accompanied by intermittent thrombus fragmentation, are probably the basis of the disease. Another factor is intraplaque haemorrhage (discussed below). Arterial occlusion can occur and is often the first serious clinical event of the disease. Usually, numbers of thrombi increase in successive waves. Their composition and expression vary according to the arteries involved. In the aorta and the large trunks, the thrombi are hyaline, granular, thin and disorganized, while in the muscular arteries they appear massive and stratified, poor in cells and very asymmetrical, being roughly crescentic in shape. With each new thrombus, new elastic tissue is elaborated in the transitional zone and the internal elastic lamina is split (in the coronary arteries in particular). The media is fibrous and atrophic. These successive thrombi may become calcified or infiltrated by small quantities of lipids. In general, a clot is formed so rapidly that anastomotic circulation cannot be established, which explains the more frequent and severe parenchymal lesions in fibrin–blood occlusions, which almost always cause infarction, than in progressive atherosclerotic stenosis. These local thrombotic phenomena play an important role in the evolution of the plaque, its occlusive tendency and its clinical expression. The thrombus may be dislodged and/or fragmented causing distal embolism (Fig. 16.14b).

Haematomas may occur within the plaque (intraplaque haemorrhage). The plaque suddenly increases in size and becomes stenotic, protruding into the lumen [12,19,27,31,39]. These intraplaque haemorrhages may result from rupture of newly formed vessels but they are more often considered to be a consequence of ulceration which allows the blood to penetrate into the wall and dissociates the plaque from the underlying media. This event can be critical, notably in the carotids [31,37]. Resorption of haematomas and organization of mural thrombi explains the presence of fibrin deposits and haemosiderin within the lesion. Formation of new vessels increases the risk of haemorrhage and ulceration.

Stenosis may be caused by a large plaque, or more often by an ulcerated plaque accompanied by thrombosis (this is frequent in the coronary arteries). Areas of stenosis located close together (Fig. 16.15) will give rise to ischaemia (cerebral and myocardial); total stenosis results in infarction (intestinal, renal, cerebral or myocardial).

Aneurysms are evidence of the dislocation of the medial elastic framework and of the fragility of the atherosclerotic arterial wall. They are most commonly found in the lower abdominal artery between the renal arteries and the bifurcation, though they may occur in any artery. They are usually fusiform, with no individual border, and are lined by thrombus which is a possible source of distal embolism. An atheromatous plaque may contribute to the formation of a dissecting aneurysm since in the atherosclerotic media marked sclerotic and necrotic lesions co-exist [6]. Aneurysms are covered in detail in Chap. 31.

Specific Findings

Knowledge of topographical aspects of the human circulation is essential to an understanding of angiographic and ultrasound images.

The aorta is a large-calibre elastic vessel where two major complications are frequently seen: parietal thrombi, causing embolism and sometimes occlusion, and aneurysms, with their risk of rupture. There is a third and less common complication; dissecting aortic aneurysms. In general, aortic atheromatous deposits are fairly well tolerated and the discovery of severely damaged aortas in subjects whose death was due to a variety of other causes is common. The

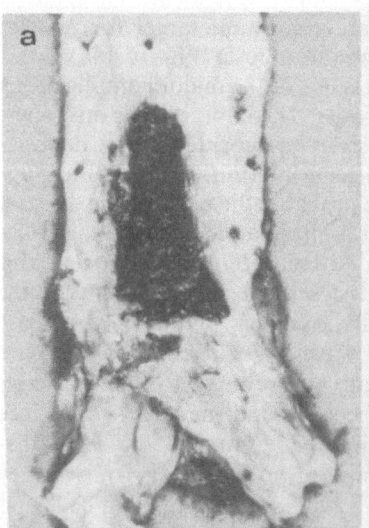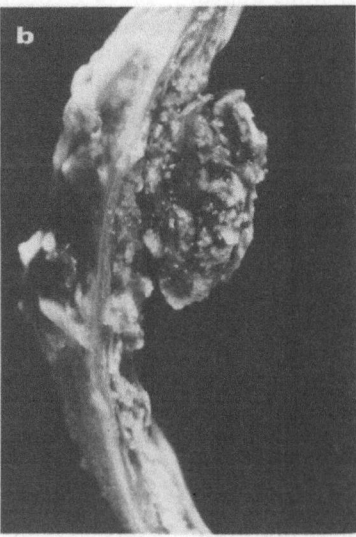

Fig. 16.15. Aorta. **a** Atheromatous material entering the arterial lumen through an ulcerated plaque: a cause of lipid embolism. **b** Blood clot above the iliac junction, on an ulcerated plaque; a cause of stenosis.

plaques are situated on the ascending intra-pericardial arch, at the ostia of the coronary arteries, at the site of closure of the ductus arteriosus, at the orifices of the large trunks and at the ostia of the intercostal arteries. Atherosclerosis is generally more advanced in the abdominal aorta and around the orifices of the coeliac axis and the mesenteric and renal arteries. It is in this area that the lesions are the most severe, long-established and damaging.

The following findings are made in aortas where there are extensive, coalescent, ulcerated plaques:

1. Aneurysms are almost always fusiform (see Chap. 31). When ectasia commences, it always progresses in accordance with Laplace's Law; the tangential tension in the wall increases as a product of the radius. The aneurysmal pocket becomes covered with successive thrombi and embolization may follow. Bacteria may sometimes colonize these thrombi.

In the aortic arch, the aneurysm may involve one of the three segments. Aneurysms in either segments I or III are likely to remain limited to that segment; aneurysms in segment II overlap with part or all of the adjacent segments. Dissection may occur. Abdominal aneurysms are the most common form but about a quarter of these go unrecognized and are only found at autopsy.

2. Thrombi are generally mural but may result in occlusion at a bifurcation. They commonly originate either in the common iliac artery or the distal portion of the aorta, and may extend retrogradely to the kidneys (Fig. 16.14).

The "syndrome of the aortic arch" describes stenotic or occlusive lesions of the supra-aortic branches in their thoracic portion, clinically expressed by varying degrees of ischaemia in the brachiocephalic areas. The most obvious sign is decrease or disappearance of the arterial pulses in the upper limbs and the neck. The lesions may involve one or several trunks. Isolated lesions may often remain clinically silent as obstruction is compensated for by other pathways; diversion during exercise of the upper limbs of blood destined for the brain is characteristic, as seen in subclavian steal syndrome, but blood may also be diverted through the brachiocephalic trunk.

Lesions of the carotid artery are situated, in order of decreasing frequency, in the bulb and the first centimetre of the internal carotid extending to the entrance of the external carotid, in the cavernous portion of the internal carotid and lastly in the common carotid (atheroma of the common carotid is rare). Single or multiple, these lesions are often bilateral (66% of cases) and may be symmetrical or asymmetrical. They affect the cerebral circulation by decreasing flow and by embolism. The plaques are often of limited width, but are raised and often ulcerated [16,18,48]. There is significant correlation between the incidence of stroke and the presence and extent of ulceration [52].

The vertebrobasilar axis may be affected at three points: the orifice, the endocranial segment of the vertebral artery and the basilar trunk. As in the carotid axes, these lesions are stenotic but rarely calcified. The wide anatomical and functional variations in the vertebral arteries and their connections account for the great variability in clinical expression. In the basilar trunk, lesions decrease in extent with distance from the fusion of the vertebral arteries, and are mostly ventral. The arteries of the base of the brain are evidently rigid, studded with yellowish, sometimes calcified and often coalescent, stenotic plaques.

Atherosclerosis frequently involves the coronary arteries, and in man it is mainly in the proximal parts of the three main branches that are affected [10]. In an aged artery the lesions are often visible on the epicardial surface. They consist of sinuous or straight threads which are indurated, rigid and white or yellowish. The common trunk of the left coronary artery and the intraventricular artery is most frequently involved, followed by the right trunk and the circumflex artery. The result is occlusion of the lumen, by progressive development of one or several increasingly stenotic atheromatous plaques, or by the sudden additional effect of thrombosis. The consequences of occlusion of a coronary artery are variable [44]. The chances of infarction are increased if the occlusion involves a large-calibre arterial trunk and lesion development is rapid. The existence of prior occlusions also makes the consequences more serious. Lesions vary greatly from one individual to another and from trunk to trunk. They resemble those of the aorta, but associated thrombosis is more common, aneurysms are exceptional and fibrosis is always present, causing ectasia or major concentric or eccentric narrowing of the lumen. As a general rule these stenoses only affect the coronary reserve when the artery has been narrowed by 50%–75%. There is thus a long period of clinical silence. It is widely accepted that organization of successive mural thrombi and intermittent fragmentation with distal embolism may explain progression [19]. The role of spasm [44] and/or platelet aggregation [33] remains a matter of debate.

The coeliac trunk and the mesenteric arteries are often affected by atherosclerosis. Atheroma is ostial or proximal, affecting the first 2 or 3 cm of the artery. The distal network is normal and stenosis tends to be concentric. The result is chronic or severe intestinal ischaemia (mesenteric infarction). There may be aneurysms of the trunk of the superior mesenteric artery.

The renal and splenic arteries are often markedly affected along the whole of their length, but the lesions are more marked near the aorta. The splenic artery may be very sinuous and calcified. Renal atherosclerosis may remain latent or cause arterial hypertension which is often associated with progressive renal failure if both arteries are affected. Stenosis of the main renal artery or of its branches is the main cause of renovascular hypertension and is particularly common in men over the age of 50. It should be distinguished from fibromuscular hyperplasia (particularly in women) (Chap. 29). Dissection of the wall may occur.

The arteries of the lower limbs as far as the trifurcation are also liable to stenosis. Atheromatous deposits are extensive in areas exposed to haemodynamic trauma, for example the iliac bifurcation, the lower femoral and higher popliteal arteries and the tibial arch crossing the bony ligament. Elsewhere, the lower limb arteries show significant lesions of the intima and the internal elastic lamina, but these are simple arteriosclerotic lesions and are practically without atheromatous deposits.

Although atherosclerosis of the pulmonary arteries is mostly secondary to pulmonary hypertension from any cause, some patches can be observed near bifurcations in individuals over 40–50 years of age (see Chap. 35).

Polyarterial involvement is very frequent, as shown by clinical and anatomical data. The Framingham study showed that 15%–20% of patients presenting with a symptom indicating one site of atherosclerosis were in fact affected in two major sites. The risk increases with age. The associations of lesions are not fortuitous, but have not been clearly explained: coronary artery disease is often associated with symptomatology related to the abdominal aorta or the carotid artery, for example. Coronary atherosclerosis associated with arterial disease of the lower limbs is in our opinion more frequent than is generally thought; the term "arterial disease of the lower limbs" is vague, and arterios-

clerotic lesions are too often confused with atheromatous lesions. In fact, the frequency of polyarterial lesions in atherosclerosis is difficult to determine, and is probably underestimated.

Problems of Regression

In recent decades, many studies have been carried out on the regression of atheromatous lesions in humans and animals. Various methods have been used separately or in association, including angiography, morphological, biochemical and haemodynamic tests and clinical observation.

Recent publications [2,11,36,40,41,42,57,65] show that although our knowledge of the development of atherosclerosis has progressed, it is still impossible with our present methods of investigation to assess precisely the regression of the atheromatous lesion, particularly in man. Ideally, regression is the return of the arterial wall to a normal state. This does not occur however, and the term regression is used instead to designate any modification (biochemical or morphological) which brings about repair or beneficial remodelling of a lesion and which generally results in an increased diameter of the arterial lumen. To base a diagnosis of lesional regression on an angiographical image may be inaccurate. If an angiography shows an increase in the diameter of the lumen, this may indeed correspond to an improvement in the atheromatous condition, but it may also result from disease-linked modifications and merely indicate remodelling, repair, lysis or recanalization of the thrombus, or even progression of the lesions resulting in ectasia through thinning of the media. Though repeated angiography is, in theory, an excellent way of appreciating the regression or progression of atherosclerosis, in practice and in order to be of any value the comparison of images taken at different times must take into account numerous factors which are difficult to control, such as the state of the patient (arterial tension, vascular tonicity, spasm), the thickness of the soft tissues, angle of incidence of the rays and so on. For the same reasons, during programmed studies, angiographic data cannot provide definite information on the beneficial effect of regimens or treatments, in spite of sophisticated computerized methods.

Certain data provided by clinical observation and retrospective epidemiological studies indicate that regression is possible in man: during the Second World War, food restrictions were followed by a reduction in the number of ischaemic cardiac events; in the United States and in Scandinavian countries, after low-lipid diets became fashionable and drug administration improved, a distinct decrease in mortality due to coronary heart disease was noted. In fact, as multiple factors together with associated thrombosis contribute to this ischaemia, it is difficult to affirm that the improvements observed are indeed due to regression of the atherosclerotic lesion, and not to stabilization in the absence of haemodynamic or thrombotic complications. However, certain studies indicate undeniable regression in atherosclerosis. Such studies are rare, however, and should be commented on. Their results are often based on intervention in moderate atherosclerosis, on "hypertreated" atherosclerosis (ileal bypass, plasma exchange, multifactorial reductions of risk) or on patients with familial hyperlipidaemia. The latter are far from common, and present in fact lipid insudation into the arterial wall and not true atheromatous lesions, at least in the early stages of their disease. Plasma lipid changes probably play a role in angiographic regression. Reduced plasma cholesterol and HDL levels, as well as increased HDL levels, are probably involved in regression of atherosclerosis (see Chap. 19).

In animals, in particular in non-human primates (rhesus and cynomolgus monkeys) in whom the atheromatous lesions most resemble those of man, experimental methods are more effective in assessing possible regression of atherosclerosis. The literature shows that purely lipid lesions (dots and fatty streaks) disappear and leave no scar. It also shows that developed lesions (plaques) may be modified morphologically in favour of regression or stabilization, that is, there may be a decrease or no change in the surface area involved, a lessening or stabilization of thickening of the intima, less evident or similar medial lesions and a lesser quantity of intra- and extracellular lipids, correlating with a decrease in foam cells and cell proliferation. Biochemically, there are lower levels of total, free and esterified cholesterols, of phospholipids and of triglycerides and an increase in collagen levels, while GAG levels remain the same. Calcium levels increase.

However, two remarks, applicable to all other experimentation, qualify these results. Firstly, the basic experimental schema is usually as follows: during a certain period, atheroma is induced by various hyperlipidaemic diets, and the monkeys are then divided into homogeneous groups according to their serum lipid values. One group continues to be fed the hyperlipidaemic diet while others continue without it or are treated with various drugs. However, it is well known that considerable variations exist in the extent and severity of atheroma from one monkey to another, even if they have high and comparable cholesterol blood levels. Our own study of cynomolgus monkeys receiving a hyperlipidaemic diet for 2 years showed that atherosclerosis of the coronary arteries ranged from severe fibrolipidic lesions to minimal or nonexistent atheroma [58]. Moreover, it is not possible to evaluate the gravity, the morphology and the biochemical constitution of lesions at the time when an atherogenic diet is discontinued, yet these are compared with terminal lesions whose area, depth, morphology and biochemistry are perfectly accessible to analysis. How valid is a comparison between two different groups of animals when it is known that, like man, they do not respond identically to a given lipid overload? Is it possible to compare, in the same animals, preliminary "fictitious" lesions and final "actual" lesions?

Finally, it is extremely difficult to establish valid tests for the regression of atherosclerosis, since the causes inducing and influencing the evolution of this disease are so varied and so variable, and the individual reactions so diverse. It is also very difficult to affirm that the reduction of cardiovascular mortality is indeed related to regression of the lesions, and not to stabilization. Complications might be more often avoided if detection and follow-up of high-risk subjects in families with hypercholesterolaemia, diabetes, arterial hypertension or cardiovascular disease are regularly carried out by means of blood and skin tests [5,50].

From the experimental viewpoint studies are easier to perform in animals than in man, but it is necessary to find a means of assessing precisely the initial degree of atherosclerosis. Lesion regression studies demand further perfection of non-invasive methods, of which nuclear magnetic resonance (NMR) is certainly one. This will lead to a better understanding of the regression of atheroma and also of the processes of its deposition. Animal experimentation already enables us to assess the overall degree of atheroma by comparing treated and untreated groups matched as homogeneously as possible. However, coronary and aortic angiography should always be included as selection tests before and at the end of experimentation; soon NMR may be systematically used.

Will morphological examination, and in particular measurement of cholesterol and apoprotein B in skin biopsies, make it possible to trace the regression of atheroma? Current studies give promising results when performed in man on two groups of subjects, those with normal and with pathological coronary x-ray results [15], and in cynomolgus monkeys fed a hyperlipidaemic diet for 2 years [4]. They demonstrate that the best marker of coronary atheroma is the cutaneous level of apoprotein B. Experimental studies on the regression of atheroma will no doubt further our knowledge of this condition.

References

1. Adams CW (1971) Lipids, lipoprotein and atherosclerotic lesions. Proc R Soc Med 64:902–906
2. Barndt R, Blankenhorn DH, Crawford DW (1977) Regression and progression of early femoral atherosclerosis in treated hyperlipoproteinemia patients. Ann Int Med 86:139–146
3. Bocan T, Schifani A, Guyston JR (1986) Ultrastructure of the human aorta: fibrolipid lesion formation of the atherosclerotic lipid-rich care. Am J Pathol 123:413–424
4. Bouissou H, De Graeve J, Aouidet A, Kokolo J, Julian M, Thiers J (1988) Coronary atheroma in the cynomolgus monkey: protective value of serum and cutaneous lipoprotein B measurement. Virchows Arch [A] (in press)
5. Bouissou H, Julian M, Pieraggi MT, Aouidet A, Marie W (1984) Microangiopathie et atteinte conjonctive dermiques du sujet jeune avec ou sans antécedent direct cardiovasculaire ou diabétique. Nouv Presse Med 46:2815–2816
6. Bouissou H, Pieraggi MT, Alzieu J et al. (1978) Les dissections aortiques aigues communicantes (essai pathogénique) paroi artérielle. Arterial Wall 4:135–141
7. Bouissou H, Pieraggi MT, Julian M (1981) L'atherome. Ann Pathol 1:5–20
8. Brown MS, Goldstein JL (1984) How LDL receptors influence cholesterol and atherosclerosis. Sci Am 215:58–66
9. Capron L (1983) Athérosclérose: description et mécanismes. Rev Neurol 139:167–176
10. Crawford T (1977) Pathology of ischaemic heart disease. Butterworth, London
11. Daoud AJ, Fritz KE, Jarmolych J, Wiener B (1983) Measurement of regression in atherosclerosis. In: Schettler G, Gotto AM (eds) Atherosclerosis VI Springer-Verlag, Berlin Heidelberg New York, pp 155–159
12. Davies MO, Thomas AC (1985) Plaque fissuring, the cause of acute myocardial infarction, sudden ischaemic death and crescendo angina. Br Heart J 53:363–373
13. Debakey ME (1978) Pattern of atherosclerosis and rates of progression. In: Paoletti R, Gotto AM Jr (eds) Atherosclerosis reviews; vol 3. Raven Press, New York, pp 1–56
14. Debakey ME, Lawrie GM, Glaeser DH (1985) Patterns of atherosclerosis and their surgical significance. Am Surg 201:115–131
15. De Graeve J, Bouissou H, Thiers JC, Fouet J, Valdiguie P (1984) Is cutaneous apoprotein B a better discriminator than serum lipoproteins for atherosclerosis? Atherosclerosis 52:301–308
16. Dixon S, Tais SO, Raviola C et al. (1982) Natural history of non-stenotic, asymptomatic ulcerative lesion of the carotid artery Arch Surg 117:1493–1498
17. Eikelboom BC, Riles TS, Minter R et al. (1983) Inaccuracy of angiography in the diagnosis of carotid ulceration. Stroke 14:882–885
18. Falk E (1983) Plaque rupture with severe pre-existing stenosis precipitating coronary thrombosis: characteristics of coronary atherosclerotic plaques underlying fatal occlusive thrombi. Br Heart J 50:127–134
19. Falk E (1985) Unstable angina with fatal outcome: dynamic coronary thrombosis leading to infarction and/or sudden death. Circulation 71:699–708
20. Fless GM, Kirchausen T, Fischer-Dzoga et al. (1982) Serum low-density lipoprotein with mitogenic effects on cultured aortic smooth muscle cells. Thrombosis 41:171–183
21. Geer JC, Haust MD (1972) Smooth muscle cells in atherosclerosis. Karger, Basel (Monographs in atherosclerosis 2)
22. Gendre P (1970) Ultrastructure de la paroi artérielle humaine saine et athéroscléreuse. Pathol Eur 5:283–306
23. Gerrity MG (1981) The role of the monocytes in atherogenesis. II. Migration of foam cells from atherosclerotic lesions. Am J Pathol 103:191–200
24. Ghidoni JJ, O'Neal RM (1967) Recent advances in molecular pathology: a review. Ultrastructure of human atheroma. Exp Molec Pathol 7:378–400
25. Goldstein JL, Kita T, Brown MS (1983) Defective lipoprotein receptors and atherosclerosis. N Engl J Med 309:288–293
26. Goldstein JL, Brown MS (1985) Les récepteurs LDL, le cholesterol et l'athérosclérose. Pour la Science 87:62–71
27. Gore I (1963) Ulceration of and embolization by atheromato. In: Jones R (ed) Evolution of the atherosclerotic plaque. University of Chicago Press, Chicago, pp 316–329
28. Haberland ME, Fogelman AM, Edwards PA (1982) Specificity of receptor mediated recognition of malondialdehyde-modified low density lipoproteins. Proc Natl Acad Sci USA 79:1712–1716
29. Haust MD (1971) The morphogenesis and fate of potential and early atherosclerotic lesions in man. Hum Pathol 2:1–29
30. Haust MD (1983) Derivation and progression of atherosclerotic plaques. In: Schetten G, Gotto AM (eds) Atherosclerosis VI. Springer-Verlag, Berlin Heidelberg New York, pp 350–357
31. Imparato AM, Riles TS, Mintzer R et al. (1983) The importance of hemorrhage in the interrelationship between gross morphologic characteristics and cerebral symptoms in 376 carotid artery plaques. Ann Surg 197:195–203
32. Jacotot B (1982) L'athérosclérose: aspects fondamentaux, aspects cliniques. Editions Phil, Paris (collection Lepetit)
33. Jorgensen L, Rowsell HC, Hovig T et al. (1967) Adenosine diphosphate-induced platelet aggregation and myocardial infarction in serine. Lab Invest 17:616–622
34. Kohchi K, Takebayashi S, Hirok T et al. (1985) Significance of adventitial inflammation of the coronary

artery in patients with unstable angina: results at autopsy. Circulation 71:709–718

35. Kovanen PT (1985) Le contrôle du cholestérol. La Recherche 16:1472–1480

36. Kramer JR, Kitazume H, Proudfit WL (1983) Progression and regression of coronary atherosclerosis: relation to risk factors. Am Heart J 105:134–144

37. Kramer JR, Kitazume H, Proudfit WL et al. (1983) Segmental analysis of the rate of progression in patients with progressive coronary atherosclerosis. Am Heart J 106:1427

38. Kramer JR, Matsuda Y, Mulligan JC et al. (1981) Progression of coronary atherosclerosis. Circulation 63:519–524

39. Lusby RJ, Ferrell LD, Ehrenfeld WK, Stoney RJ, Wylie EJ (1982) Carotid plaque hemorrhage: its role in production of cerebral ischemia. Arch Surg 117:1479–1488

40. Malinow MR (1981) Regression of atherosclerosis in humans: fact or myth? Circulation 64:1–3

41. Malinow MR (1983) Experimental models of atherosclerosis regression. Atherosclerosis 48:105–118

42. Malinow MR (1983) Potential for atherosclerosis regression in perspective. In: Schettler G, Gotto AM (eds) Atherosclerosis VI. Springer-Verlag, Berlin Heidelberg New York, pp 207–211

43. Malinow MR (1984) Atherosclerosis: progression, repression and resolution. Am Heart J 108:1523–1537

44. Mautner RK, Cooper MD, Philips JH (1983) Catheter induced coronary artery spasm: an angiographic manifestation of vasospastic angina. Am Heart J 106:659

45. McGill HC Jr (1968) The geographic pathology of atherosclerosis. Williams and Wilkins, Baltimore, pp 1–93

46. McGill HC Jr (1984) Persistent problems in the pathogenesis of atherosclerosis. Arteriosclerosis 4:443–451

47. Mitchell JRA, Schwartz CJ (1965) Study of cardiovascular disease at necropsy in arterial disease. Blackwell Scientific Publications, Oxford, pp 377–396

48. Moore WS, Boree C, Malone JM et al. (1978) Natural history of non-stenotic asymptomatic ulcerative lesions of the carotid artery. Arch Surg 113:1352–1359

49. Orcel L, Chomette G (1978) Anatomie pathologique vasculaire: athérosclerose. Flammarion Médecine Sciences, Paris, pp 160–161

50. Pieraggi MT, Bouissou H, Julian M, Aouidet A, Marie W (1984) Relations entre les lésions du derme et le "terrain artérioscléreux". Nouv Presse Med 13, 41:2518–2519

51. Restrepo C, Tracy RE (1975) Variation in human aortic fatty streaks among geographic locations. Atherosclerosis 21:179–193

52. Ricota JJ, Schenk EA, Ekholm SE, Deweese JA (1986) Angiographic and pathologic correlates in carotid artery disease. Surgery 99:284–292

53. Ross R, Glomset JA (1973) Atherosclerosis and the arterial smooth muscle cell. Science 180:1332–1339

54. Ross R (1978) L'athérosclérose. La Recherche 9:131–137

55. Ross R (1981) Atherosclerosis: a problem of the biology of arterial wall cells and their interactions with blood components. Atherosclerosis 1:293–311

56. Ross R et al. (1986) Human atherosclerosis. I. Cell constitution and characteristics of advanced lesions of the superficial femoral artery. Am J Pathol 114:79–93

57. Saint-Clair RW (1984) Regression de l'athérosclérose en expérimentation animale: conceptions actuelles sur les mécanismes intracellulaires et biochimiques. Acquisit nouv pathol cardiovasc 26:153–181

58. Schwartz CJ, Ardlie NG, Carter RF, Paterson JC (1967) Gross aortic sudanophilia and hemosiderin deposition. A study of infants, children and young adults. Arch Pathol 83:325–336

59. Small DM, Shipley GG (1974) Physical-chemical basis of lipid deposition in atherosclerosis. Science 185:222–229

60. Smith B (1983) Identification of the gelatinous lesions. In: Schetter G, Gotto AM (eds) Atherosclerosis VI. Springer-Verlag, Berlin Heidelberg New York, pp 170–173

61. Smith EB, Slater RS (1973) Relationship between plasma lipids and arterial tissue lipids. Nutr Metab 15:17–26

62. Tracy RE, Devane Y, Kissling G (1985) Characteristics of the plaque under a coronary thrombus. Virchows Arch [A] 405:411–427

63. Velican C, Velican D (1978) Heterogeneity in composition and aggregation patterns of coronary intima acid mucopolysaccharides (glycoaminoglycans). Atherosclerosis 29:141–159

64. Vlodaver Z, Edwards JE (1972) Anatomie pathologique de l'athérosclérose coronaire. Acquisition nouvelles en pathologie cardiovasculaire 14:283–302

65. Wissler RW (1983) Chairman's introduction to workshop on regression. In: Schettler G, Gotto AM (eds) Atherosclerosis VI. Springer-Verlag, Berlin Heidelberg New York pp 183–186

66. Woolf N (1982) Pathology of atherosclerosis. Butterworth, London

67. World Health Organisation (1958) Classification of atherosclerotic lesions. WHO, Geneva (Technical report series no 143)

Section III

Management of Patients with Atherosclerosis

Chapter 17

A Pharmacological Approach to Degenerative Alterations of Arterial Walls

A. Simon, J. Levenson and M. Safar

The pharmacological approach to the arterial system in man has received little attention until recently. The lack of investigation of arteries is due in part to the pulsatility of the physiological phenomena which take place on the arterial side of the circulation. Additionally, the study of arteries has been impaired for a long time by the absence of appropriate quantitative methods. Today, two important factors allow us to re-evaluate the place of the arterial system in clinical and pharmacological investigations: (a) rapid developments in the field of cardiovascular pharmacology and (b) progress in non-invasive techniques. Hypertension, which produces degenerative changes in the whole circulatory system, represents a good model for investigation of the arteries and their response to pharmacotherapy.

The Primary Effects of Drugs on Arteries

Role of Exaggerated Constriction of Arteries in Vascular Disease

An exaggerated constriction of arteries is associated with many vascular diseases. The purpose of pharmacological intervention is to reverse this vasoconstriction which has adverse long-term effects on the circulatory function.

Hypertension

Small Arteries. The elevation of peripheral resistance is the hallmark of sustained essential hypertension; it is due to the reduction of the global inner cross-section of the small arteries and of the arterioles [10,11] and depends on structural (increased wall to lumen ratio) and/or to functional changes of the arterial walls. Such changes induce an increased reactivity to vasoactive agents and some impairment in the capacity of relaxation of arterioles [11].

Large Arteries. Hypertension affects the large arteries by increasing their calibre and decreasing their compliance (or distensibility) [54, 58]. Such alterations are not only a secondary phenomenon to the increased distending pressure of the arterial walls; they may also result from functional and/or structural changes of the arteries [5,6,50]. These latter consist of smooth-muscle hypertrophy, an accumulation of scleroproteins with inversion of the elastin to collagen ratio, and waterlogging [63].

Heart Failure

Although the neurogenic vasomotor tone has been reported to be normal in patients at rest with moderate heart failure, the adreno-sympathetic response to exercise is increased [8]. Moreover, the sensitivity of arterial and atrial mechanoreceptors is reduced in heart failure, which causes a lack of reduction of the sympathetic hyperactivity in response to exercise, especially in the non-metabolic circulation [8]. Lastly, the renin–angiotensin system is abnormal and the sodium content of the arterial walls is increased in heart failure, so contributing to excessive arterial rigidity and hyper-responsiveness of the arteries to vasoactive agents [8].

Atherosclerosis

Atherosclerotic disease may disrupt the physiological balance between constriction and dilation of large arteries; indeed, experimental studies have shown that the relaxation of isolated large arteries by acetylcholine is reduced in the presence of atherosclerotic lesions of their walls [14]. Such relaxation is mediated by production and release of relaxing substances from endothelial cells in response to muscarinic stimulation by acetylcholine [14], and endothelial denudation, which initiates the atherosclerotic process, might alter this relaxation phenomenon; moreover, some studies suggest that the existence of stenosis in large conducting arteries limits the vasodilation reserve of the arterioles distal to the stenosis (i.e. reduces the ratio between flow at maximal vasodilation and flow at rest [45].

Vascular Spasm

Vascular spasm or vasospasm is defined as the abnormal or exaggerated constriction of a vessel. It may be generalized or localized and it is not obligatorily related to the vessel size. Physiopathological interest in vasospasm mainly concerns the peripheral musculo-elastic arteries such as the digital vessels in Raynaud's disease; however, other arteries such as conducting, cerebral and coronary arteries and lower limb arteries can be also affected by vasospasm [40]. Indeed, spasm of large cerebral arteries has been reported in cerebral infarction and may play a role in the pathogenesis of certain transient ischaemic attacks [40]. In addition to the classical description of Printzmetal, recent clinical reports suggest that coronary vasospasm is frequent and may be superimposed on partial atherosclerotic occlusion [40]. Lastly, radiological evidence of spasm has been reported in patients with atherosclerosis obliterans of the lower limbs [40].

Response of Arteries to Acute Pharmacological Interventions

Essential hypertension is a good model for studying the arterial effects of pharmacological interventions. These can be classified according to their main effect on the smooth muscle of arteries (Table 17.1).

Neurogenic Action

Alpha-Adrenergic Blockade. Intravenous administration of nicergoline, a potent post-synaptic alpha-adrenergic blocking drug, greatly reduces the total peripheral resistance and increases systemic arterial compliance of hypertensive patients [21]. Intravenous administration of urapidil, another alpha-adrenergic blocker, reduces the forearm vascular resistance, does not change the diameter and compliance of the brachial artery and decreases the brachial-to-radial pulse wave velocity [22].

Thus, in hypertension, acute alpha–adrenergic blockade produces dilation of the small arteries due to smooth-muscle relaxation, an effect previously demonstrated with other classical alpha-adrenergic blockers such as prazosin [41]. In contrast, the effects of alpha-adrenergic blockade are not marked in the large arteries, since brachial artery calibre and compliance are not affected by urapidil [22]; however, the reduction in arterial pulse wave velocity observed with urapidil [22] and the increase in systemic arterial compliance reported with nicergoline [21] can be considered to reflect some improvement in the distensibility of large arteries. This phenomenon is probably secondary to the mechanical relaxation of the arterial walls induced by decrease in systemic pressure [18,21,22] but it is also likely that alpha-adrenergic blockade produces isometric relaxation of the smooth muscle of the brachial artery [4].

Beta-Adrenergic Blockade. Intravenous injection of propranolol, a non-cardioselective beta-adrenergic blocker, increases total peripheral resistance and decreases systemic arterial compliance of patients over 50 years of age with arteritis or systolic hypertension. Such an effect cannot result from mechanical pressure changes since systemic mean pressure is not modified acutely by propranolol [24]. It is likely that the blockade of peripheral beta adrenoceptors by propranolol unmasks unopposed alpha vasoconstriction which is responsible for an increased smooth-muscle tone in the arterial walls [60]. In contrast, propranolol injection in young systolic hypertensive patients does not modify peripheral resistance and systemic arterial compliance [24,51]. The decreased number of beta adrenoceptors found with advancing age, concomitant with the unchanged number of alpha adrenoceptors, may explain how the balance between the two populations of receptors is more

Table 17.1. Systemic and forearm arterial effects of anti-hypertensive drugs classified according to their mechanism of action

Main mechanism of action	Type of drug	Type of pathology	Arterial effects	
			Vascular resistance (S, systemic; F, forearm)	Arterial compliance
Neurogenic	Alpha Blockers			
	IV nicergoline (21)	Sustained hypertension	Decrease (S)	Increase (S)
	IV urapidil (22)	Sustained hypertension	Decrease (S,F)	No change (F)
	Beta blockers			
	IV propranolol (23)	Arteritis	Increase (S)	Increase (S)
	IV propranolol (25)	Borderline hypertension	Increase (S)	—
Hormonal	Converting enzyme inhibitors			
	Oral captopril (52)	Sustained hypertension	Decrease (S,F)	Increase (S,F)
	Oral enalapril (19)	Normotension	Decrease (S)	Increase (S)
Cellular	Calcium antagonists			
	Oral nifedipine (26)	Sustained hypertension	Decrease (S)	Increase (S)
	IV diltiazem (48)	Sustained hypertension	Decrease (S,F)	Increase (S)
	Oral nitrendipine (28)	Sustained hypertension	Decrease (F)	Increase (F)
	Oral nicardipine (27)	Sustained hypertension	Decrease (F)	Increase (F)
Other	Nitrate derivatives (low doses)			
	IV sodium nitroprusside (51)	Systolic hypertension	No change (S)	Increase (S)
	IV nitroglycerin (29)	Arteritis	No change (S,F)	Increase (S,F)
	Direct acting vasodilators			
	IV dihydralazine (58)	Sustained hypertension	No change (F)	Decrease (F)

IV, intravenous

easily disrupted by propranolol in older than in younger patients [7,24].

Acute intravenous administration of non-cardioselective propranolol and selective primidolol in two groups of young borderline hypertensive patients produces contrasting effects on forearm arterial circulation [25]. Propranolol reduces the hyperkinetic pattern (increased blood velocity and flow), while primidolol has no comparable effect [25]. Thus, only non-selective beta-adrenergic blockade (propranolol) can normalize the brachial artery circulation of labile hypertensive patients which suggests that beta adrenoceptors are implicated in the exaggerated vasodilation of small arteries in these young patients.

These observations provide evidence that acute beta-adrenergic blockade exerts different arterial effects which are a function of the physiopathological pattern of hypertension; older subjects whose beta adrenoceptor population is reduced are extremely sensitive to the non-selective beta blockade which unmasks an unopposed alpha vasoconstriction and aggravates their arterial alterations [23,24,51]. In contrast, younger hypertensive patients with increased beta-adrenergic activity respond to beta-adrenergic blockade by reducing arteriolo-dilation of the forearm. This tends to normalize the hyper-

kinetic circulation in their peripheral arteries [25].

Hormonal Action

The hormonal actions of drugs are illustrated by the angiotensin converting enzyme inhibitors whose main effect is to prevent the vasoconstrictive action of angiotensin II [2]. Arterial effects have been studied both in the systemic arterial circulation and in the forearm arterial circulation of patients with sustained essential hypertension.

Systemic Arterial Circulation. Single oral doses of captopril increase systemic arterial compliance and decrease total peripheral resistance in patients with mild to moderate hypertension [52]. Enalapril has comparable effects in mildly sodium-depleted normotensive subjects [19]. When comparing the arterial effects of captopril given either in a single dose or for 5 d, the increase in systemic arterial compliance is similar for the two methods of administration, but the reduction in peripheral resistance is more pronounced after 5 d of treatment than after a single dose [52]. Such an observation on the kinetics of arterial effects of captopril suggests that converting enzyme inhibition acts more rapidly on the large arteries than on the arterioles [52,53].

Forearm Arterial Circulation. Acute oral captopril increases the diameter and compliance of the brachial artery and decreases the forearm vascular resistance of hypertensive patients [53]. Thus, acute converting enzyme inhibition produces a global arterial vasodilation which affects both the large and the small arteries of the forearm [53]. Brachial artery vasodilation can be considered to be an active phenomenon because it counteracts the mechanical reduction in calibre which must result from the blood pressure drop induced by captopril [53]. This active vasodilation of arteries implies that the mechanism responsible is the inhibition of formation of angiotensin II in plasma and/or inside the arterial walls, as suggested by recent experimental findings in the rat [3].

Cellular Action

Calcium antagonists, which interact with the excitation–contraction coupling of the smooth-muscle cell by blocking calcium entry, are good examples of drugs with cellular action. The effects of calcium antagonists have been studied in the systemic and forearm arteries of hypertensive patients.

Systemic Arteries. A single oral dose of nifedipine and intravenous infusion of diltiazem both bring about a rapid decrease in systemic blood pressure by reducing total peripheral resistance in moderate permanent hypertension [26]. Moreover, nifedipine increases systemic arterial compliance [26].

Forearm Arteries. Acute oral administration of nifedipine [26], nitrendipine [28] and nicardipine [27] and the intravenous infusion of diltiazem [48] increase brachial artery diameter and blood velocity and flow, and decrease the forearm vascular resistance of hypertensive patients. Moreover, acute oral administration of nitrendipine increases brachial artery compliance, reduces brachial-to-radial pulse wave velocity and decreases the characteristic input impedance of the brachial artery [28,34]. Thus, in essential hypertension calcium antagonists dilate both the large and the small arteries of the forearm. Brachial artery dilation is an active phenomenon which occurs despite the systemic blood pressure fall which accompanies administration of calcium antagonists and contributes to increased

arterial compliance. Calcium antagonists also improve the distensibility of the arterial wall as assessed by the reduction of pulse wave velocity and characteristic impedance observed with nitrendipine [28,38]. The common mechanism of these arterial effects of calcium antagonists is related to the calcium entry blockade in the smooth-muscle cells which produces smooth-muscle relaxation.

Other Actions

Nitrate Derivatives. Intravenous infusion of low doses of sodium nitroprusside or nitroglycerin does not modify total peripheral resistance in older patients with arteritis or systolic hypertension, but strongly increases systemic arterial compliance [29,51]. This increase in systemic arterial compliance follows a selective decrease in systolic pressure without change in diastolic level [29,51]. Low doses of intravenous nitroglycerin also dilate the brachial artery of middle-aged moderate hypertensive patients, but do not change brachial artery blood velocity and flow, and forearm vascular resistance [49]. In addition, the input impedance spectrum of the brachial artery is modified by nitroglycerin (Fig. 17.1): input impedance of a large artery represents a measure of the opposition to pulsatile inflow in this artery which is due to the physical properties of the downstream arterial circulation (arterial distensibility and small artery resistance) and to the properties of the blood. Input impedance is calculated as the instantaneous ratio of pressure and flow waves in the brachial artery; practical calculation of this pressure–flow ratio requires the transformation of pressure and flow pulses into a sum of sinusoidal waves or harmonics according to the Fourier theorem, by means of an analogue-to-digital converter and a computer. The harmonics of impedance (impedance spectrum) are represented as a plot of moduli (amplitude of pressure to flow ratio) and phases (timing in relation to other harmonics) against the frequency of their harmonics [49].

Thus, under perfusion of low doses of nitroglycerin, the moduli curve becomes flatter and shifts its minimum towards the left while the phase curve becomes positive for lower harmonic frequencies than in the control state [49]. The plateau of the moduli curve at the higher frequencies represents the characteristic impedance, i.e. the impedance to flow inside the brachial artery in the absence of wave reflections against the small arteries of the hand or the forearm [42]. Moreover it has been shown that characteristic

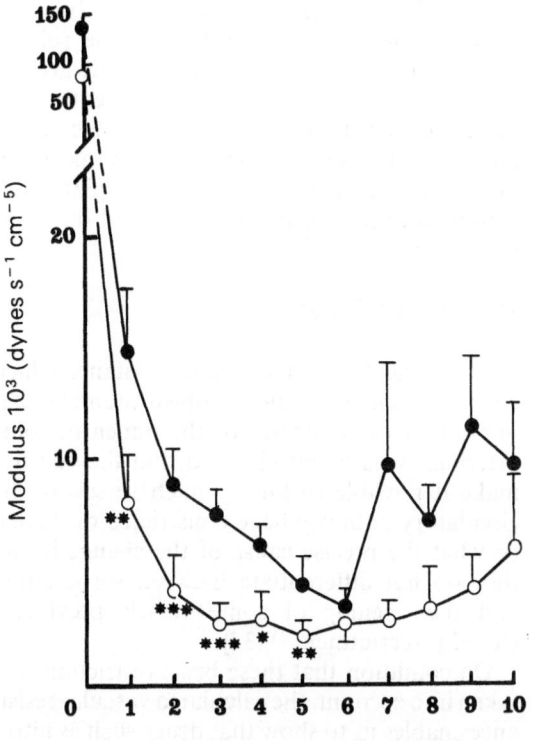

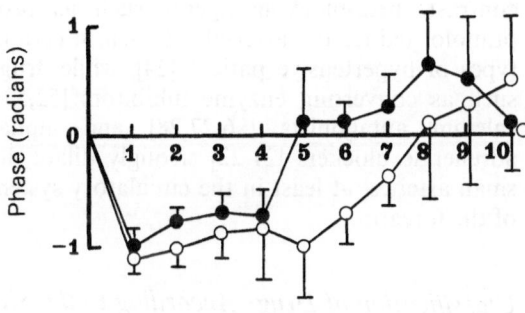

● Control (*n* = 6)
○ Nitroglycerin

Fig. 17.1. Input impedance spectrum of the brachial artery before (●) and after (○) intravenous administration of low doses of nitroglycerin (six patients). * $P < 0.05$, ** $P < 0.01$, *** $P < 0.001$. (Simon et al. [49])

impedance is inversely related to the cross-sectional area of the artery and directly related to pulse wave velocity along this artery according to the formula $Zc = \rho \, C/a$ where Zc is the characteristic impedance, ρ the blood density, C the pulse wave velocity and a the cross-section of the artery. Thus, the fact that nitroglycerin decreases the characteristic impedance indicates that this drug has a predominant effect on the large arter-

ies of the forearm and of the hand, probably by a dilating effect on their cross-section [49].

Dihydralazine. Intravenous administration of dihydralazine in moderate hypertensive patients decreases the forearm vascular resistance; such a decrease can be considered as a local reflection of the systemic arteriolodilation induced by this drug [58]. In contrast, dihydralazine reduces the brachial artery diameter and does not change the forearm arterial compliance from its baseline value [58]. It is likely that the baroreflex sympathetic discharge produced by dihydralazine explains the vasoconstriction of the brachial artery and the lack of increase in arterial compliance [58].

Differences in the Response of Large and Small Arteries to Pharmacological Interventions

Response of Large Arteries

The smooth-muscle content of large artery walls increases progressively from the aorta to the peripheral conducting arteries where smooth muscle is organized in rings around the vessel lumen [39]. It follows that the functional characteristics of peripheral arteries may be modified by the activation of their smooth muscle. Thus, smooth-muscle relaxation increases arterial compliance providing that the intra-arterial pressure remains unchanged; in contrast, smooth-muscle relaxation tends to reduce arterial compliance when arterial calibre is kept constant [4]. These observations indicate that interpretation of the response of large arteries to acute pharmacological action must take into account the modification of several parameters such as arterial calibre, arterial distensibility and arterial compliance [31].

Arterial Calibre. Pharmacological modification of calibre directly reflects the vasoactivity of arteries [55]. Thus, the increased diameter of the brachial artery observed following administration of angiotensin converting enzyme inhibitors [52,53], calcium antagonists [28] and nitrate derivatives [49], demonstrates that these drugs have the capacity to dilate large arteries. This vasodilation results from a decrease in smooth-muscle tone induced by the pharmacological effects of the drug, such as the reduction of angiotensin II (in plasma and/or in the arterial wall) or the calcium entry blockade in the smooth-muscle cell [9].

Arterial Distensibility. Distensibility (%dV/dP) [17] represents the capacity for distension of arterial walls independent of intra-arterial volume. Its evaluation is based on the pulse wave velocity measurement which is related to Young's elastic modulus (E) according to the classical Moens–Korteweg [20] formula:

$$C = Eh/2R\rho$$

where C represents pulse wave velocity, E the circumferential elastic modulus, h and R the wall thickness and radius of the artery and ρ the blood density. If the ratio h/R remains unchanged, the modifications of pulse wave velocity and elastic modulus are strongly correlated and arterial distensibility can be calculated from pulse wave velocity. However, the h/R ratio may change following acute pharmacological intervention; converting enzyme inhibitors [52,53], calcium antagonists [26,27,28] and nitrate derivatives [49] greatly increase arterial radius and consequently decrease the h/R ratio because of the incompressibility of the arterial walls. Thus, the only condition where pulse wave velocity measurements allow the evaluation of the elastic modulus of the arterial walls is when arterial radius does not change; this is the case following administration of urapidil [22]. The reduction in pulse wave velocity observed after its administration reflects an improvement of the elastic properties of the arterial walls; however, it must be questioned whether this increased distensibility is due to the reduction in distending pressure or to isometric relaxation of the arterial smooth muscle induced by the alpha-adrenergic blockade with urapidil [22].

Arterial Compliance. Arterial compliance (dV/dp) represents the global capacity of an elastic artery to increase its volume by a rise in intra-arterial pressure. Arterial compliance is equal to the product of the intra-arterial volume (V) and the distensibility of the arterial wall (% dV/dP) so that these two factors are implicated in the response of arterial compliance to pharmacological interventions. The modifications of arterial volume (or arterial calibre) are generally parallel to the changes in arterial compliance, since drugs which dilate the brachial artery [converting enzyme inhibitors [52,53], calcium antagonists [26,27,28] and nitrate derivatives [47] increase arterial compliance, whereas those which do not change [urapidil [22]] or even

decrease [dihydralazine [58]] arterial calibre have no effect on arterial compliance. The role of distensibility of the arterial walls in the arterial response to pharmacological interventions is more difficult to evaluate because the determination of arterial distensibility by means of pulse wave velocity depends on complex interactions between elastic modulus, wall thickness and arterial radius [48].

Response of Small Arteries

The modifications of vascular resistance which occur in response to the administration of drugs imply that the diameter of the lumen of some arteriolar vessels has changed, but they do not make it possible to know which vessels in the circulatory pathway have constricted or dilated or what the precise cause of the change is and they do not differentiate between vasodilation and the opening of some vessels previously closed ("recruitment") [37].

On condition that these basic restrictions are taken into account, the calculated vascular resistance enables us to show that drugs such as nitroglycerin [49] and nitrate derivatives, given at low doses [51], have no effect on the arterioles; in contrast, beta-blocking agents such as propranolol reduce the arteriolar lumen in certain types of hypertensive patient [24], while drugs such as converting enzyme inhibitors [52,53], calcium antagonists [26,27,28] and alpha-adrenergic blockers [21,22] strongly dilate the small arteries, at least in the circulatory system of the forearm.

Classification of Drugs According to the Site of the Arterial Response

Evidence of a pharmacological effect on large arteries in humans is an original finding. It is commonly thought that only the small arteries are vasoactive while large arteries behave in a purely mechanical and passive manner. In fact, drugs may have a preferential or combined action on the large and/or small arteries, so defining three pharmacological groups of drugs active in hypertension (Table 17.1).

The first group includes alpha-adrenergic blockers [21,22] and dihydralazine [58] which exert an exclusive arteriolodilation effect without having a significant effect on large arteries.

The second group includes converting enzyme inhibitors [52,53] and calcium antagonists [26,27,28] which produce a combined vasodilation of small and large arteries of the systemic and forearm circulations, and markedly increase large artery compliance.

A third group is represented by nitrate derivatives administered at low doses [49,50,51] which predominantly dilate the brachial artery without modifying the forearm arterioles, and increase both systemic and forearm arterial compliance [49].

The mechanisms responsible for the differential action of drugs on small and large arteries are complex. Firstly, the pathological processes of arterial walls may modify the pharmacological reactivity of the different sides of the arterial circulation. Secondly, the pharmacological action on smooth muscle is non-uniform and may differ according to the regional circulation and the type of vessel. Finally, the response of large arteries to drugs may be influenced by physiological regulation phenomena such as baroreflex interactions.

Arterial Consequences of the Physiological Counter-regulations Induced by Drugs

Three main factors may influence direct pharmacological effects on arteries: change in blood pressure, baroreflex activation and structural alterations of the arterial walls.

Blood Pressure Changes

Variations in blood pressure affect the calibre and compliance of large arteries mechanically; however, the volume–pressure relationship in arteries is curvilinear so that the capacity to increase arterial volume with increments in pressure tends to decrease with elevation in pressure. This results from the non-uniform composition of human arteries [4]. Thus, the mechanical effect of blood pressure changes on diameter and compliance of large arteries may contribute to minimize pharmacological action: this is the case for the captopril-induced dilation of the brachial artery. Thus, the greater the blood pressure fall following administration of captopril, the less the arterial dilation [52,55]; indeed the blood pressure fall tends to counterbalance mechanically the active vasodilatory action of captopril. The modifications of calibre of large arteries observed following drug administration thus represent the net effect on the arterial wall

of the blood pressure change and of the smooth-muscle pharmacological response.

Baroreflex Activation

Any pharmacological intervention in the arterial tree may change baroreflex activation by modifying the distending pressure [46], the calibre [46], the smooth-muscle tone and the distensibility of those segments of the arterial wall containing mechanoreceptors, such as the carotid sinus or aortic arch. Baroreflex activation follows neurogenic modifications which are superimposed peripherally on vascular pharmacological activity. Such baroreceptor actions will be considered only as they modify short-term drug effects; however, the long-term interaction between baroreceptor re-adjustments and pharmacological action is an important and difficult problem which requires further investigation and the development of new methodological approaches in clinical investigation. This is true for the action of dihydralazine in hypertension where systemic blood pressure reduction induces a strong baroreflex sympathetic discharge [58] which increases the smooth-muscle tone of the large arteries (which are not responsive to the vasodilatory effect of dihydralazine) and may explain the decrease in brachial artery calibre observed following administration of this drug [58].

Baroreflex activation also explains the differences observed between younger and older hypertensive patients in the arteriolar response to intravenous infusion of low doses of nitroglycerin [30]. Younger patients exhibit an increase in diastolic pressure and total peripheral resistance due to a sympathetic baroreflex discharge on the systemic arterioles; such a sympathetic effect is related to the rapid intravenous entry of nitroglycerin into the circulation [35] and to the high sympathetic tone of younger hypertensive subjects. In contrast, older hypertensive patients do not show arteriolo-constriction following administration of nitroglycerin because their sympathetic tone and their baroreflex sensitivity are lower than in younger subjects [30].

Structural Alterations of Vascular Systems

Structural changes of arteries in hypertension may modify the response to pharmacological stimulations. Indeed, the smooth-muscle hypertrophy of the small arteries in the hypertensive state responds to reduction in smooth-muscle tone by a greater decrease in vascular resistance than in the normotensive state [11]. Similarly the alteration of beta-adrenoceptors which occurs with advancing age might explain the vasoconstricting effect of propranolol on arteries of older patients with systolic hypertension [24] and in atherosclerosis obliterans of the lower limbs [23].

Long-term Effects of Drugs on Arteries

Chronic action of drugs on the arterial system induces, in addition to the smooth-muscle tone modifications of the arterial walls, other physiological changes, such as the reversibility of structural alterations of arteries induced by the trophic direct action of drugs and/or the indirect effect of pressure changes on the arterial walls.

Beta-Adrenoceptor Blocking Drugs

Three months of treatment with propranolol decreases systemic blood pressure without modifying the arterial properties of the forearm of patients with mild to moderate essential hypertension [56] (Fig. 17.2). In contrast, 3 months' treatment with pindolol decreases systemic blood pressure and concomitantly increases calibre, blood flow and compliance of the brachial artery [35]. The reasons for the differences of effects of these two beta blockers on the forearm arteries are unclear; the strong partial agonist activity of pindolol on the beta-2 adrenoceptors of the arterial walls might explain its dilatory effect on arteries [35].

Vasodilating Drugs

Nicardipine, a new calcium antagonist, induces a large increase in brachial artery diameter and a decrease in forearm vascular resistance in moderately hypertensive patients [27]. Enalapril, a well-known angiotensin converting enzyme inhibitor gives similar results and increases arterial compliance of the forearm after 6 months of treatment [57] (Fig. 17.2). These arterial responses to drugs are similar to those obtained in acute administration and are probably due to a direct action on the arterial walls related to specific pharmacological properties such as the calcium entry blockade or the inhibition of angiotensin II formation.

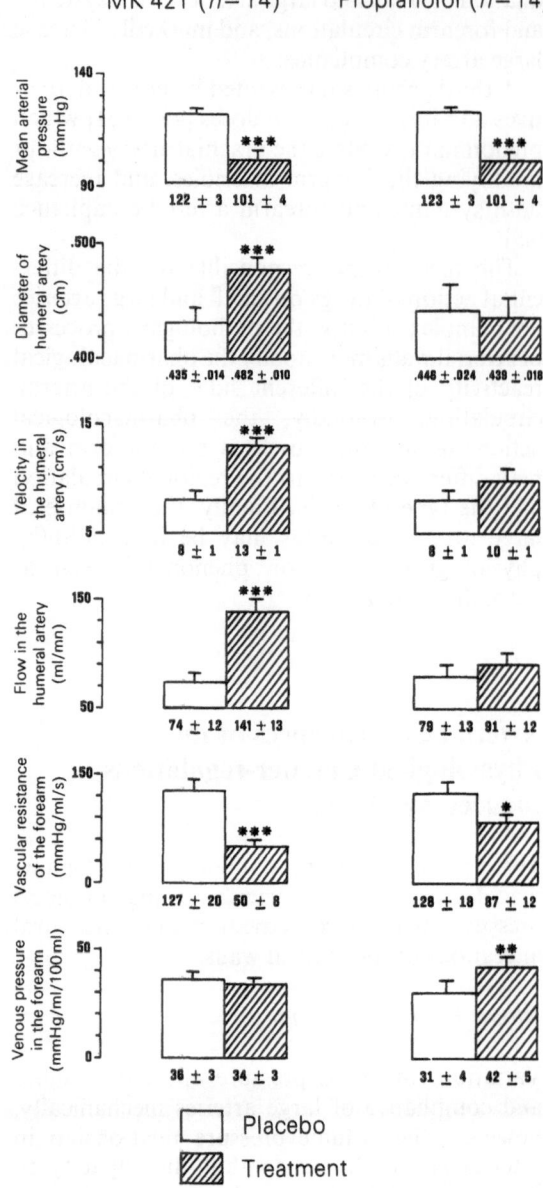

Fig. 17.2. Changes in forearm circulation after 3 months of treatment with MK 421 (enalapril) and propranolol. Values are mean ±SEM. *$P<0.05$, **$P<0.01$, ***$P<0.001$. (Simon et al. [56])

Implications for Prognosis

Chronic pharmacological modifications of arterial systems may have important prognostic implications. In hypertension, a decrease in blood pressure does not prevent degenerative athero-arteriosclerotic lesions of large arteries [13]. The reversibility of those arterial changes due to hypertension, produced by certain antihypertensive drugs, may represent a new way for a better prevention of the degenerative process of the arterial system [12].

Secondary Effects of Drugs on the Function of Arteries

The primary action of drugs on arteries induces important consequences for the three main functions of arterial systems, i.e. the conducting, the buffering and the baroreflex functions.

Conducting Function

The conducting function of arteries consists of delivering blood to peripheral organs [42,59]. Since blood flow in conducting arteries is equal to the product of arterial cross-sectional area and blood velocity [20], any change in these two parameters may affect the conducting function [1,55].

Role of Large and Small Arteries

The increase in blood velocity observed inside the lumen of large arteries after drug administration is always related to a distal arteriolodilation and constitutes a major mechanism for increasing arterial blood flow; increased blood velocity has been observed in the brachial artery after acute and chronic administration of several drugs such as converting enzyme inhibitors [52,53,56] and calcium antagonists [26,27,28]. The increase in arterial blood velocity may also contribute to an increase in the pressure drop between two different sites in conducting arteries; under baseline conditions the mean pressure drop between axillary and radial arteries is 2 to 4 mmHg [61], but limb vasodilation such as that induced by post-ischaemic reactive hyperaemia can increase the brachio-radial pressure drop to 25 mmHg [61]. This observation demonstrates that pressure gradient lost in a large artery is directly related to the blood velocity; this physiological concept is embodied in the Navier–Stokes equation, whose simplified form is [16]:

$$dp/dz = \rho/g.dV/dt + aV$$

where dp is the pressure gradient which exists between two sites of one artery at distance dz apart, ρ is blood density, g the gravitational

acceleration, dV/dt the acceleration of the blood, V the blood velocity and a is a friction coefficient of blood against the arterial walls.

The increase in calibre of large arteries may also participate in increasing arterial blood flow, as demonstrated for the brachial artery with several drugs including converting enzyme inhibitors [52,53,56] and calcium antagonists [26,27,28]. However, vasodilation of large arteries enables an increased flow only if blood velocity does not decrease; this happens with low doses of nitroglycerin. These low doses do not increase arterial flow despite marked vasodilation of the brachial artery as arterial blood velocity has been reduced concomitantly [49].

Physiological Implications of Drug-Mediated High Flow

The high arterial flow produced by pharmacological intervention is not always beneficial. In heart failure, drugs which increase blood flow in certain ischaemic organs are considered to be beneficial, especially in exercise conditions [44]. In contrast, in hypertension most regional flows are normal except for renal blood flow [36], thus the need to increase peripheral flows pharmacologically is questionable. Lastly, in the case of large artery stenosis marked arteriolodilation may induce a steal phenomenon, i.e. an excessive perfusion of non-ischaemic tissues resulting in low perfusion in ischaemic regions, as demonstrated by the effect of dypiridamol on coronary circulation [62]. Nevertheless, the dilation of a large artery before and after a stenosis may represent a means of increasing flow, on condition that pressure distal to the stenosis does not fall [62].

Buffering Function

The buffering function of large arteries involves damping the pulsatility of pressure and flow generated by the intermittent cardiac pump [42,43]. Indeed, during ventricular ejection, the distension of the elastic walls of the aorta and peripheral large arteries allows the storage of a fraction of the blood ejected, which is then transmitted to the peripheral arterial bed during diastole because of the elastic contraction of arteries. The storage of blood during systole in the large arteries, and its restitution during diastole acts as a hydraulic filter which buffers the pulsatile phenomena inside the arterial tree.

Physiological Approach to the Buffering Function of Arteries

Arterial compliance represents the main factor determining buffering function. Any drug which decreases blood pressure must theoretically increase arterial compliance because of the non-linear elastic behaviour of arteries [18]. However, despite markedly reducing systemic pressure, drugs such as alpha-adrenergic blockers (urapidil) [22] or arteriolar vadodilators (dihydralazine) [58], do not increase arterial compliance. In contrast, arterial compliance increases strongly in response to the administration of nitrate derivatives [49,51], calcium antagonists [26.27.28], converting enzyme inhibitors [52,53,56] and certain beta blockers with intrinsic sympathetic activity [53]. Moreover, drugs that increase arterial compliance, also decrease systolic pressure and pulse pressure; these can be considered to be the two simplest indices of arterial pressure pulsatility. Thus, acute oral administration of nifedipine decreases systolic pressure more significantly than diastolic pressure [26]; acute oral administration of nitrendipine strongly decreases pulse pressure [28]; chronic oral administration of enalapril decreases systolic pressure to a greater extent than diastolic pressure and also decreases pulse pressure [53]; and nitrate derivatives specifically decrease systolic pressure without changing diastolic pressure [51]. In contrast, nicergolin [21], dihydralazine [58] and urapidil [22], which do not change arterial compliance, decrease systolic and diastolic pressures proportionately without a clear reduction of pulse pressure.

Arterial Pulsatility and the Degenerative Process in Arteries

The wide-ranging pulse pressure which results from the loss of distensibility of large arteries in hypertension might be a major factor in the degenerative process in arterial walls; indeed, experimental studies in vitro show that cyclic stress of arteries stimulates the proliferation of smooth-muscle cells. Moreover, pulsatile strains on bioelastomers, similar to those contained in the arterial walls, accelerate their degenerative changes and ultimately cause fatigue and fracture [15,33]. It follows from these observations that any pharmacological increase in arterial compliance, by reducing the pulsatility of arterial pressure, might contribute to slowing of the degenerative process in arteries and help to prevent athero-arteriosclerotic complications.

Baroreflex Function

Laplace's Law

Intra-arterial baroreceptors are stimulated by arterial wall tension. The circumferential tension (T) of a cylindrical artery, which represents the major component of the total wall tension, is equal, according to Lame's law (derived from the Laplace's law) to the product of pressure (P) multiplied by radius (R), divided by the wall thickness (h) of the artery [46]:

$$T = PR/h$$

During acute pharmacological intervention, the arterial wall thickness may be considered to be unchanged, and the modification of arterial tension depends only on pressure and radius of the artery and on their variations:

$$T = R\Delta P + P\Delta R$$

In contrast, in chronic drug treatment, structural changes of the arterial walls may modify wall thickness which then becomes an important variable to be taken into account in the interpretation of Lame's law.

Role of Vasodilation of Large Arteries

Lame's law indicates that both calibre and pressure of large arteries play a part in the regulation of the stretch of intra-arterial baroreceptors. Such a concept is supported by a comparison of the haemodynamic effects of two vasodilating agents, diltiazem and dihydralazine, administered intravenously to hypertensive patients [48]. Dihydralazine produces a strong tachycardia while diltiazem does not change heart rate, despite a blood pressure fall comparable to that produced by dihydralazine [48]. In addition, no correlation was observed between change in pressure and change in heart rate in the overall population of patients receiving diltiazem and dihydralazine [48]. In contrast, a strong correlation was found between change in tangential tension and change in heart rate in diltiazem- and dihydralazine-treated patients [48]. Thus, for a similar blood pressure fall, tangential tension is reduced by drugs that decrease large artery calibre (dihydralzine), but remains unchanged by drugs which vasodilate the large arteries (calcium antagonists, converting enzyme inhibitors, nitrate derivatives) [32,53]. Such a phenomenon might partly explain the different heart rates induced by the different vasodilators (Fig. 17.3).

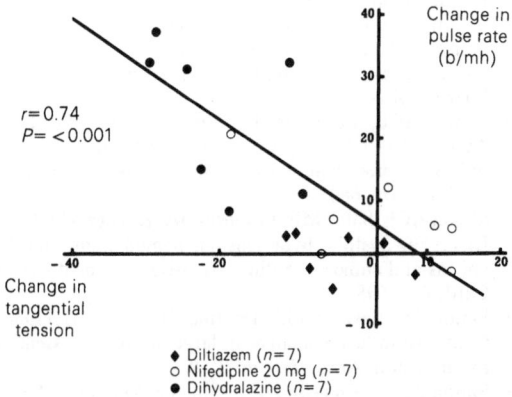

Fig. 17.3. Anti-hypertensive vasodilators: relationship between change in heart rate and change in tangential tension observed in three groups of patients. (o) dihydralazine, $n = 7$; (o) nifedipine, $n = 7$; ($\diamond$) diltiazem, $n = 7$. (Levenson et al. [31])

References

1. Benetos AM, Simon ACh, Bouthier JD, Levenson JA, Lagneau PL, Safar ME (1986) Pulse doppler: an evaluation of diameter, blood velocity and blood flow of the common carotid artery in patients with isolated unilateral stenosis of the internal carotid artery. Stroke 16:969–972

2. Collis MG, Keddie JR (1981) Captopril attenuates adrenergic vasoconstriction in rat mesenteric arteries by angiotensin dependent and independent mechanism. Clin Sci 61:281–286

3. Cohen ML, Kurz KD (1982) Angiotensin converting enzyme inhibition in tissues from spontaneously hypertensive rats after treatment with captopril or MK 421. J Pharmacol Exper Ther 220:63–69

4. Cox RH (1978) Passive mechanics and connective tissue compositions of canine arteries. Am J Physiol 234:H533–H541

5. Dobrin PB, Rovich AA (1969) Influence of vascular smooth muscle on contractile mechanics and elasticity of arteries. Am J Physiol 217:1644–1652

6. Dobrin PB (1978) Mechanical properties of arteries. Physiol Rev 58:397–460

7. Drayer JL, Weber MA (1982) Antihypertensive agents which inhibit sympathetic activity: potentially adverse effects of combination treatment. Am Heart J 104:660–664

8. Flain SF, Zelis R (1981) The effects of nitroglycerin on regional blood flow during exercise in heart failure. In: Vanhoulec PM, Leusen I (eds) Vasodilatation. Raven Press, New York, pp 469–475

9. Fleckenstein A (1977) Specific pharmacology of calcium in myocardium: cardiac pacemakers, and vascular smooth muscle. Ann Rev Pharmacol Toxicol 17:149

10. Folkow B, Neil E (1971) Circulation. Oxford University Press, New York, pp 36–56

11. Folkow B (1982) Physiological aspects of primary hypertension. Physiol Rev 62:347–504

12. Folkow B (1983) Structural factors: the vascular wall consequences of treatment. Hypertension 5 [Suppl III]:58–62

13. Freis ED (1982) The Veterans' trial and sequelae. Br J Clin Pharmacol 13:67–72

14. Furchgott RF, Davidson D, Lin CI (1979) Conditions which determine whether muscarinic agonists contract or relax rabbit aortic strips. Blood Vessels 16:213–214

15. Gent AN (1972) Fracture of elastomers. In: Liebowitz H (ed) Fracture. Academic Press, New York

16. Greenfield JC (1966) Pressure gradient technique. Rushmer RF (ed), Yearbook Medical Publishers, Chicago, pp 83–93

17. Gribbin B, Pickering IG, Sleight P (1979) Arterial distensibility in normal and hypertensive men. Clin Sci 56:413–417

18. Hallock P, Benson IC (1937) Studies on the elastic properties of human isolated aorta. J Clin Invest 16:595–602

19. Ibsen H, Egan B, Osterziel K, Vander H, Julius S (1983) Reflex hemodynamic adjustments and baroreflex sensitivity during converting enzyme inhibition with MK 421 in normal humans. Hypertension 5:184–191

20. Korteweg DJ (1878) Über die Fortpflanzungsgeschwindigkeit des Schalles in elastischen Röhren. Ann Phys Chem Neue Foige 5:225

21. Levenson JA, Chelly J, Payen D, Simon ACh, Safar ME (1981) Vasodilating antihypertensive drugs: effect on arterial compliance. A preliminary report. Symposium sur les alpha-bloquants, pharmacologie expérimentale et clinique. Editions Masson pp 271–276

22. Levenson JA, Simon ACh, Bouthier JD, Benetos AM, Safar ME (1984) Post-synaptic alpha-blockade and brachial artery compliance in essential hypertension. J Hypertens 2:37–41

23. Levenson JA, Simon ACh, Fiessinger JN, Safar ME, London GM, Housset EM (1982) Systemic arterial compliance in patients with arteriosclerotis obliterans of the lower limbs. Observations on the effect of intravenous propranolol. Arteriosclerosis 2:266–271

24. Levenson JA, Simon ACh, Maarek BC et al. (1983) Comparative effects of beta-adrenergic blockade in systolic hypertension according to age. Eur Heart J 4 [Suppl G]:51–54

25. Levenson J, Simon A, Safar M, Bouthier J, London G (1985) Elevation of brachial arterial blood velocity and volumic flow mediated by peripheral beta adrenoreceptors in patients with borderline hypertension. Circulation 71(4):663

26. Levenson JA, Safar ME, Simon ACh, Bouthier JD, Griener JL (1983) Systemic and arterial hemodynamic effects of nifedipine (20 mg) in mild to moderate hypertension. Hypertension 5 [Suppl V]:V57–V60

27. Levenson JA, Simon ACh, Bouthier JD, Maarek BC, Safar ME (1985) The effect of acute and chronic nicardipine therapy on forearm arterial hemodynamics in essential hypertension. Br J Clin Pharmacol 20:107s–113s

28. Levenson JA, Simon ACh, Safar ME, Bouthier JD, Maarek BC (1984) Large arteries in hypertension: acute effect of a new calcium entry blocker nitrendipine. J Cardiovasc Pharmacol 6[Suppl 7]:1006–1010

29. Levenson JA, Simon ACh, Safar ME, Fiessinger JN, Housset EM (1982) Systolic hypertension in arteriosclerosis obliterans of the lower limbs. Clin Exp Hypertens [A] 4:1059–1072

30. Levenson JA, Simon ACh, Safar ME, Chelly JE, Gitelman RCh (1980) Effect of intravenous nitroglycerin on arterial compliance and baroreflex sensitivity in hypertension. In: Sleight PS (ed) Arterial baroreceptors and

hypertension. Oxford University Press, Oxford, pp 510–520

31. Levenson JA, Simon ACh, Safar ME (1985) Vasodilatation of small and large arteries in hypertension. J Cardiovasc Pharmacol 7 [Suppl 2]:115–120

32. Maarek B, Simon AC, Levenson J, Bouthier J, Safar ME (1986) Chronic effects of pindolol on the arterioles, large arteries and veins of the forearm in mild to moderate essential hypertension. Clin Pharmacol Ther 39:403–408

33. Leung DY, Glagov MS, Mathews MB (1976) Cyclic stretching stimulates synthesis of matrix components by arterial smooth muscle cells in vitro. Science 191:475–477

34. McDonald DA (1974) Blood flow in arteries, 2nd edn. Edward Arnold, London, pp 309–350

35. Mason DT (1978) Afterload reduction and cardiac performance. Physiologic basis of systemic vasodilators as a new approach in treatment of congestive heart failure. Am J Med 65:106

36. Messerli FH, De Carvalho JGR, Christie B, Frohlich ED (1978) Systemic and regional hemodynamics in low, normal and high cardiac output borderline hypertension. Circulation 58:441

37. Milnor WR (1981) Steady flow and properties of the vascular wall. In: Milnor WR (ed) Hemodynamics. Williams and Wilkins, Baltimore, pp 11–48, 56–91

38. Noble MIM (1979) The load system–aortic input impedance and pressure gradient. In: Noble MIM (ed) The cardiac cycle. Blackwell Scientific Publications, Oxford, pp 163–192

39. Noordergraaf A (1978) Circulatory system dynamics. Academic Press, New York, pp 137–139

40. Nveten JM, Vanhoutte PM (1981) Calcium entry blockers and vasospasm. In: Vanhouette PM, Leusen I (eds) Vasodilatation. Raven Press, New York, pp 459–468

41. Oates HF, Graham RM, Strokes GS (1977) Mechanism of the hypotensive action of prazosin. Arch Int Pharmacodyn Ther 227:41–48

42. O'Rourke MF (1982) Arterial function in health and disease. Churchill Livingstone, Edinburgh

43. O'Rourke MF (1976) Pulsatile arterial hemodynamics in hypertension. Aust NZ J Med 6:40–48

44. Packer M, Le Jemtel TH (1982) Physiologic and pharmacologic determinants of vasodilator response: a conceptual frame-work for rational drug therapy for chronic heart failure. Proc Cardiovasc Dis 24:275–292

45. Roth AC, Young DF, Cholving R (1976) Effect of collateral and peripheral resistance on blood flow through arterial stenosis. J Biomed 9:367

46. Rushmer RK (1970) Cardiovascular dynamics. Saunders, Philadelphia, pp 163–167

47. Randall US, Elser MD, Bulloch GF et al. (1976) Relationship of age and blood pressure to baroreflex sensitivity and arterial compliance in man. Clin Sci 51:357–360

48. Safar ME, Simon ACh, Levenson JA, Cazor JL (1983) Hemodynamic effects of diltiazem in hypertension. Circ Res 52 [Suppl 1]:169–173

49. Simon ACh, Levenson JA, Levy BY, Bouthier JE, Peronneau PA, Safar ME (1982) Effect of nitroglycerin on peripheral large arteries in hypertension. Br J Clin Pharmacol 14:241–246

50. Simon ACh, Levenson JA, Bouthie JD, Safar ME, Avolio AP (1985) Evidence of early degenerative changes or large arteries in human essential hypertension. Hypertension 7:675–680

51. Simon ACh, Safar ME, Levenson JA, Kheder AD, Levy BI (1979) Systolic hypertension: hemodynamic mechanisms and choice of antihypertensive treatment. Am J Cardiol 44:505–511

52. Simon AC, Levenson JA, Bouthier JE, Safar ME (1984) Captopril induced changes in large arteries in essential hypertension. Am J Med 76:71–75

53. Simon ACh, Levenson JA, Bouthier J, Maarek B, Safar ME (1985): Effects of acute and chronic angiotensin-converting enzyme inhibition on large arteries in human hypertension. J Cardiovasc Pharmacol 7[Suppl I]:45–51

54. Simon ACh, Safar ME, Levenson JA, London GM, Levy BI, Chau NP (1979) An evaluation of large arteries compliance in man. Am J Physiol 237:H550–H554

55. Simon ACh, Safar ME, Levenson JA, Bouthier JD, Benetos AM (1983) Action of vasodilating drugs on small and large arteries of hypertensive patients. J Cardiovasc Pharmacol 5:626–631

56. Simon ACh, Levenson JA, Bouthier JD et al. (1984) Comparison of oral MK 421 and propranolol in mild to moderate essential hypertension and their effects on arterial and venous vessels of the forearm. Am J Cardiol 53:781–787

57. Simon ACh, Levenson JA, Safar AM, Bouthier JD, Safar ME (1984) ACE inhibition and brachial artery haemodynamics in hypertension. Br J Clin Pharmacol 18:243s–246s

58. Simon ACh, Laurent S, Levenson JA, Bouthier JE, Safar ME (1983) Estimation of forearm arterial compliance in normal and hypertensive men from simultaneous pressure and flow measurements in the brachial artery, using a pulsed Doppler device and a first-order arterial model during diastole. Cardiovasc Res 17:331

59. Smith JJ, Kampire JP (1980) Blood and the circulation: general features. In: Smith JJ, Kampire JP (eds) Circulatory physiology – the essentials. Williams and Wilkins, Baltimore, pp 1–18

60. Ulrych M, Frohlich ED, Dustan HP, Page IH (1968) Immediate hemodynamic effects of beta-adrenergic blockade with propranolol in normotensive and hypertensive man. Circulation 37:411–416

61. Wallace JM, Stead EA (1959) Fall in pressure in radial artery during reactive hyperemia. Circ Res 7:876–879

62. Winbury MM (1982) Proximal and distal coronary arteries. In: Santamore WP, Bove AA (eds) Coronary artery disease. Urban and Schwarzenberg, Baltimore, pp 63–78

63. Wolinsky H (1980) Response of the rat aortic media to hypertension. Morphological and chemical studies. Circ Res 26:507

The Arterial Wall: Relationships Between Haemodynamic and Structural Aspects in Man

M. Safar, A. Simon and J. Levenson

The relationships between the arterial wall and haemodynamic factors have been investigated many times in animal experiments. For evident methodological reasons, comparable studies are difficult to carry out in man. Many indices of cardiovascular function can be studied extensively but structural modifications of the vascular wall are more difficult to demonstrate. Autopsy observations are of little help in physiopathological terms because of the advanced nature of the lesions found. In vivo biopsies are difficult to obtain and interpret owing to the concomitant modifications in the geometry of the arterial walls. Recently, ultrasonographic techniques in man [20] have been developed and applied to the non-invasive measurement of the internal diameter of superficial peripheral arteries, and linear changes have been studied in vivo [11,21,27]. These techniques allow a more precise analysis of the relationship between structure and function of large arteries in humans. After a brief methodological recapitulation, the present review proposes studying the relationship between haemodynamic factors and the arterial wall in certain peripheral arteries studied under three conditions in man: in the normal state, in arterial hypertension and in atherosclerosis of the lower limbs.

Methods of Studying the Vascular System in Man

Study of the vascular system in man is based on the analysis of the pressure–flow relationship. Since flow rate and pressure depend on cardiac pulsation, their variations are highly non-linear [16,18], so that utilization of all the curves entails complex calculations. The problem is that of evaluating vascular impedance, which would allow the simultaneous appraisal of the resistance, elastic and inertial components of the high-pressure system [16,18]. This requires a Fourier analysis of simultaneously measured pressure waves and flow. Evaluation of the relationship between pressure and flow has been carried out for each harmonic and reveals a spectrum of modules and phases. However, this method of study, although it is very precise, requires invasive exploration and a sophisticated mathematical approach. For this reason, only linear models of the circulation are used in current practice. In such models, the resistance is calculated as the ratio between mean arterial pressure and mean flow, permitting a semiquantitative and indirect estimation of the calibre of the arterioles. The compliance, defined as the ratio between the variation of volume and the variation of intra-arterial blood pressure, provides a measure of the elasticity of the arterial system [8].

Linear Model of the Vascular System

Figure 18.1 shows the essential variables of circulatory function [8], these are "mean" haemodynamic values. Within the system, the total blood volume (TBV) is distributed in three principal reservoirs: the arteries, the veins and the heart. The heart, considered as a continuous pump, allows circulation of the blood. Since the distensibility of the veins is greater than that of the arteries, the pressure in the veins is lower than that in the arteries (Table 18.1). Two resistances (the one arterial and the other venous) counteract the flow in the circuit. The capillaries are situated at the arterio-venous junction. In the model, P designates the pressure, C the compliance, V the volume, R the resistance and Q the flow. Furthermore, a designates the arteries, c the capillaries, v the veins and RA the right auricle. The symbol o, when it is added to V again, indicates the volume at zero pressure, which is also termed the "unstressed volume"

Table 18.1. Characteristics of the different parts of the vascular system [8,16,18]

Vessel	Mean pressure (mmHg)	Amount of elastic tissue	Primary function
Aorta	90–110	+ + + +	Conductive
Large arteries	80–90	+ + + +	Resistive
Arterioles	40–60	+ + (mainly arterial smooth muscle)	Resistive
Capillaries	15–25	0	Exchange
Small veins	5–10	0 or +	Capacitive
Large veins	0–2	+ + +	Resistive

for a given segment of the circulation. This unstressed volume is difficult to measure under physiological conditions in humans. Its measurement constitutes an essential problem in interpreting the modifications in the function of the cardiovascular system [8]. Taking this limitation into account, it is nevertheless possible to analyse the pressure–volume relationship and the pressure–flow relationship in the human arterial system.

Figure 18.2 shows schematically the pressure–volume relationship in the system of the large

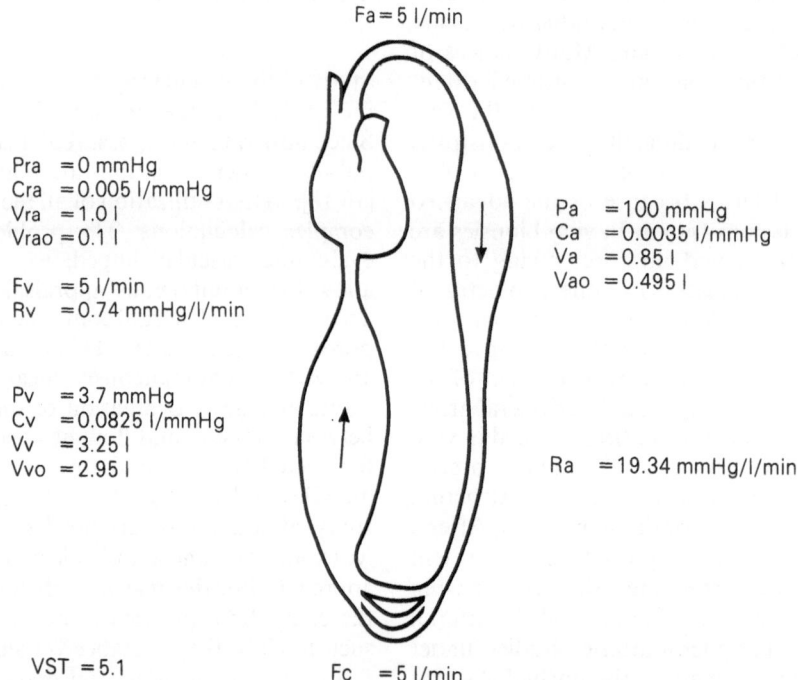

Fig. 18.1. Example of a linear model of the vascular circulation [8]. (For abbreviations see text.)

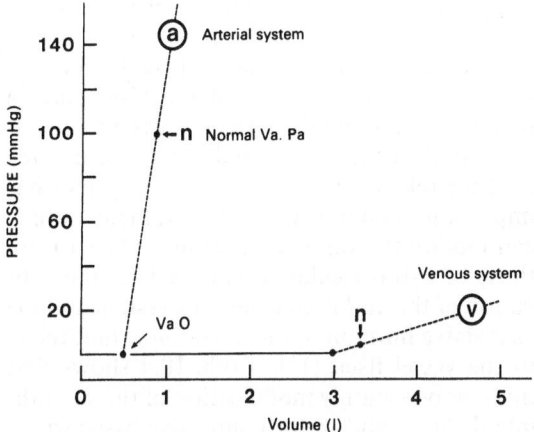

Fig. 18.2. Pressure–volume relationship in an artery of large calibre. The pressure–volume relationship in the venous system is shown for comparison [8]. (For abbreviations see text.) *n* represents the equilibrium point of the normal state for the arterial system (*a*) and the venous system (*v*).

Table 18.2. Systemic and brachial arterial compliance in normal subjects and patients with hypertension. The mean age is 38 years [25,27]

	Normal subjects	Sustained systolo-diastolic hypertension
Age (years)	37 ± 2	39 ± 1
Systolic arterial pressure (mmHg)	126 ± 3	190 ± 4***
Diastolic arterial pressure (mmHg)	71 ± 3	103 ± 2***
Carotid-femoral pulse wave velocity (m/s)	8.4 ± 0.25	10.98 ± 0.27***
Systemic arterial compliance (ml/mmHg/m^2)	1.26 ± 0.4	0.88 ± 0.02***
Brachial arterial compliance (arbitrary units)	1.22 ± 0.12	0.87 ± 0.8**

± 1, standard deviation of the mean; * $P < 0.05$; ** $P < 0.01$; *** $P < 0.001$ (compared to normal subjects)

Table 18.3. Systemic arterial compliance in normal subjects and in patients with isolated systolic hypertension and arteriopathy of the lower limbs [12,13,26]

	Normal subjects	Isolated systolic hypertension	Normal subjects	Arteriosclerosis of the lower limbs
Age (years)	51 ± 1	51 ± 2	51 ± 2	53 ± 2
Systolic arterial pressure (mmHg)	152 ± 2	183 ± 4***	126 ± 4	152 ± 6***
Diastolic arterial pressure (mmHg)	85 ± 2	85 ± 2	68 ± 3	65 ± 2
Systemic arterial compliance	2.13 ± 0.16 (ml/mmHg)	1.41 ± 0.08**	1.06 ± 0.07 (ml/mmHg/m^2)	0.87 ± 0.04**

± 1, standard deviation of the mean; * $P < 0.05$; ** $P < 0.01$; *** $P < 0.001$ (compared to normal subjects)

arterial trunks. The curve is derived from measurements carried out in dogs and extrapolated to humans [8]. Up to an arterial volume of about 500 ml, the arterial pressure is zero (*Vao*). Any increase in volume above 500 ml gives rise to a rapid and pronounced increase in arterial pressure: when the volume increases by 0.85 l, the arterial pressure reaches 100 mmHg. The inverse of the slope of the curve represents the arterial compliance (dV/dP: increase in volume per unit of pressure). Because of the elastic and collagenous content of the large arteries the curves are sigmoid, so that the compliance decreases when the pressure increases. The pressure–volume curve is modified by age: the older the subject, the greater the displacement of the curve to the right, and the greater the decrease in the slope of the curve (the compliance) [18]. Thus, two principal factors determine arterial compliance: the level of pressure and the morphological state of the arterial wall.

Arteries of large calibre (especially the aorta) mainly act as hydraulic filters [16,18]. Thanks to their visco-elastic properties, they allow the pulsatile flow of the heart to be converted into a continuous flow in the periphery, thus permitting capillary perfusion. Indeed, during systole, part of the ventricular ejection volume is stored in large arteries. It is then restored to the periphery during diastole. This property is called the Windkessel effect. This simple model of the arterial circulation can be utilized in the diastolic part of the arterial pressure curve. Compliance can then be calculated as the relationship between the time constant of the diastolic decay of the pressure curve and the peripheral resistance [14,26,27]. A validation of this measure was carried out in humans both for the systemic circulation and for the circulation of the forearm [25,27]. The normal values of compliance are shown as a function of age in Tables 18.2 and 18.3.

As to the relationship between pressure and flow in the arterial system, the simplest model is represented by liquid draining away in a cylindrical tube. There is a linear relationship between the pressure gradient and the flow rate of the liquid. The slope of the curve for pressure (abscissa) against flow rate (ordinate) depends on the cross-section of the tube and the consistency of the liquid. For a given pressure gradient, the flow rate is higher the less viscous the liquid and the greater the cross-section of the tube (Poisseuille's law) [8,16,18]. The inverse of the slope of the pressure–flow curve ($\triangle P / \triangle Q$) represents the flow resistance, which is also termed the resistance of the circulation. Figure 18.3 shows several different pressure–flow curves [7]. Curve A represents the physical system described above. This case is not frequently found in the arterial system. Curve B represents the case of a vascular bed responding passively to an increase in pressure with an arterial dilatation (resulting in a lowering of resistance). Curve C represents the case of a vascular bed responding actively to an increase in pressure by a reduction in the calibre of the vessels (resulting in an increase in resistance). This last variation of the response, which is termed autoregulation, is a homeostatic process by which a vascular bed maintains its constant flow despite changes in perfusion pressure. This phenomenon of autoregulation, which occurs in the majority of the central organs, such as the brain, the kidney and the heart, is an important factor to be considered in any analysis of the structure–function relationship.

The study of the distribution of pressures in the vascular system (Table 18.1) indicates that flow resistance mainly occurs at the level of the arterioles. Thus, the arteriolar calibre plays a major role in the modifications of vascular resistance. In the context of a study on the structure–function relationships of the vascular system, it is important to analyse multiple parameters which can modify the vascular resistance. Apart from those of extravascular origin, such as the activation of the autonomic nervous system and the vasoactive hormones, the most important relate to the vessel itself [17]. Table 18.4 shows that this involves mainly modification of the smooth-muscle fibres and/or the connective tissue which may be due to either structural abnormalities or to disorders in excitation–contraction coupling. The present review is devoted to these two principal aspects. Other factors, such as the role of the endothelium and the perivascular nerves, are dealt with in the pertinent sections of this book.

Table 18.4. Vascular factors affecting vascular resistance [17]

	Structural factors	Modifications of excitation–contraction coupling	Other factors
Reduction of the arteriolar lumen	Connective tissue	Increase of the sensitivity	Endothelium
	Mass of arterial smooth muscle	Alteration of intracellular calcium metabolism	Peripheral nerves
Decrease in the number of arterioles	Reduction in the number of vessels	Recruitment phenomenon	

Measuring the Diameter of Arteries of Large Calibre

Whereas the functions of the vascular system (resistance, compliance) have been studied for a long time in humans, the geometry of the arterial system is less well known, especially in the case of intact human arteries analysed in situ. Ultrasonic techniques have recently resulted in a complete transformation of the study of this problem. Using Doppler methods, it is possible to evaluate the diameter of peripheral and straight superficial arteries, such as the brachial artery and the common carotid artery [2,11,21]. In parallel, the

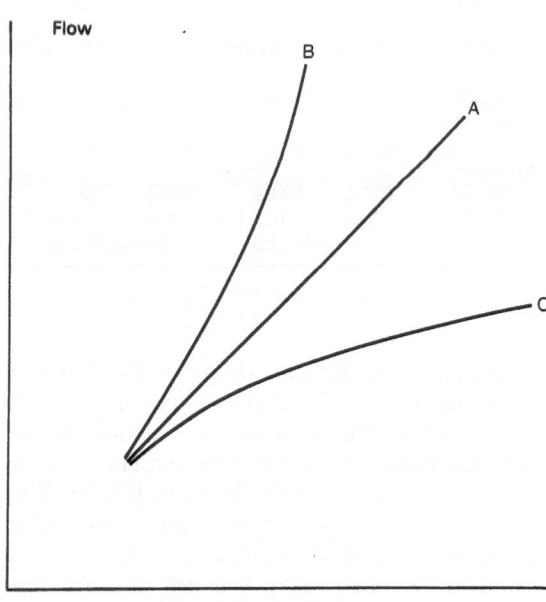

Fig. 18.3. Different forms of the pressure–flow relationship in the arterial system [7]. (For abbreviations see text.)

mean velocity of the blood can be measured and the blood flow, resistance and the compliance can be calculated on the basis of these determinations. These methods can be rationalized and validated in humans, in particular for the study of the circulation of the forearm [2,11,21,25].

With Doppler techniques, transcutaneous measurement of the arterial diameter encounters two obstacles: (a) a continued Doppler emission instrument does not allow the distance between target particles to be determined; this makes it impossible to appraise the diameter of circulating columns of blood, and (b) the Doppler effect is proportional to the incident angle of the ultrasonic beam; this angle is difficult to establish correctly on a transcutaneous basis. Bidimensional pulse Doppler with a double chamber has allowed these problems to be solved: (a) a regulatable reception system allows the distance between the target particles to be calculated in relation to the sensor, and (b) bidimensional registration by means of a double probe provides an exact value for the incidence angle [11].

The adjustable reception makes it possible to localize the red blood corpuscles. Each pulse emission of very short duration (0.5 to 2 μs) is separated by an interval of 32 to 64 μs. In this interval, the sensor becomes the receptor. An electronic gate allows regulation of the reception at any value between 0 and 32 or 64 μs. However, the regulation can only be allowed down to a minimum level of 0.5 μs. The reception is thus defined in relation to the emission by two parameters: its delay (t) and its duration (r). These reception times can be converted to distances by an elementary echographic procedure: as the reception time t corresponds to the time for the ultrasound to pass to the sensor and return from the sensor to the target, half of this time multiplied by the velocity (c) of the ultrasound is equal to the sensor–target distance: $Q = t/2 \times c$. Under these conditions, the interval between two emissions (32 or 64 μs) represents the maximum depth of exploration of the sensor, i.e. 2.5 to 5 cm for a speed of the ultrasound in the tissue of 1540 m/s; the adjustment interval (0.5 μs) measures 0.4 mm. The delay and the duration of reception respectively represent the depth and the width of the sample volume. In other words, it is possible to evaluate a maximum distance of 2.4 or 5 cm from the sensor and to alter the depth and the width of this exploration arbitrarily. It is thus possible to measure local velocities of blood and to apply these measures to the estimation of the diameter of a vessel and also of

blood flow. With this objective, one thus reduces the duration of reception to its minimum value (1 μs). This reception is then progressively advanced from the surface to the deep tissue in steps of 0.5 μs and thus it traverses the lumen of the vessel.

The first wave of perceived velocity (depth t_1) and the last perceived wave (depth t_2) correspond respectively to the proximal wall and the distal wall of the vessel: the apparent diameter is thus $(t_2 - t_1)$. However, the real diameter is equal to its projection on the cross-section of the vessel, i.e. $(t_2 - t_1)$ multiplied by sine θ, θ being the angle of incidence of the ultrasound beam. Once this diameter has been determined, the depth of reception is regulated in order to measure the global velocity of the entire blood column, i.e. the cross-sectional velocity. The product of velocity (V) multiplied by the cross-section of the vessel (inferred from the diameter) gives the vascular flow (Q), according to the formula:

$$Q = \pi D^2/4 \times V$$

However, all the calculations are only valid if the angle of incident ultrasound is exactly known. A system of two sensors forming an angle alpha between them allows this difficulty to be surmounted. The velocity of the blood is thus calculated as the vectoral sum of the two components V_1 and V_2. When V_1 and V_2 are equal in absolute value, the incident angle of each ultrasonic beam is geometrically equal to half of the angle alpha [20]. In practice, it is sufficient to adjust the position of the probe in relation to the artery in such a way that the velocities registered on each sensor are equal. For this, it is necessary that the angle thus determined remains stable during the entire manipulation. To this end, stereotactic fixation apparatus allows the probe to be immobilized on the skin in the correct position. Thus, the double-probe system and its fixation device on the skin allow any uncertainty associated with the measurement of the incident angle of the ultrasonic beam to be eliminated.

This ultrasound system has been utilized clinically in normal subjects [11] and in hypertensive patients [21] as well as in patients suffering from arteriosclerosis of the lower limbs [12,13]. Tables 18.5 and 18.6 indicate the diameter values of the brachial artery and the common carotid artery as a function of age in normal and pathological subjects.

Simultaneous measurement of pressure and flow in the brachial artery by pulsed Doppler allows the brachial arterial compliance to be

Table 18.5. Diameter of the brachial artery and the carotid artery in normal subjects and in patients with chronic hypertension. Mean age is 35 years [12,21]

	Normal subjects	Sustained systolo-diastolic hypertension
Age (years)	36 ± 1	35 ± 2
Systolic arterial pressure (mmHg)	120 ± 2	164 ± 4***
Diastolic arterial pressure (mmHg)	78 ± 2	105 ± 3***
Diameter of the brachial artery (cm)	0.422 ± 0.011	0.482 ± 0.013**
Diameter of the carotid artery (cm)	0.651 ± 0.015	0.639 ± 0.014*

± 1, standard deviation of the mean; * $P < 0.05$; ** $P < 0.01$; *** $P < 0.001$ (compared to normal subjects)

Table 18.6. Diameter of the brachial and the carotid arteries in normal subjects and in patients with sustained systolo-diastolic hypertension and arteriosclerosis obliterans of the lower limbs. The mean age is 50 years [2,12,13,21]

	Normal subjects	Chronic systolo-diastolic hypertension	Arteriosclerosis obliterans of the lower limbs
Age (years)	50 ± 1	50 ± 2	50 ± 2
Systolic arterial pressure (mmHg)	124 ± 3	178 ± 3***	151 ± 7***
Diastolic arterial pressure (mmHg)	80 ± 2	102 ± 3***	69 ± 3
Diameter of the brachial artery (cm)	0.436 ± 0.013	0.517 ± 0.013***	0.452 ± 0.021
Diameter of the carotid artery (cm)	0.653 ± 0.011	0.665 ± 0.018*	—

± 1, standard deviation of the mean; ** $P < 0.05$; *** $P < 0.001$ (compared to normal subjects)

determined [27]; this represents the volume distensibility of the brachial artery and its branches. Its evaluation is based on diastolic analysis of the pressure curves and the brachial blood flow by means of a mathematical model of first order in the forearm circulation [27]. The Fourier breakdown of the pressure waves and the brachial arterial flow into elementary sinus waves allows the ratio between the pressure and the flow rate to be calculated at each harmonic [16,18]. One thus obtains an impedance spectrum consisting of modules and phases of the pres-

sure–flow relationship at each harmonic. The impedance spectrum expresses the phenomena which impede the pulsatile progression of the blood distal to the brachial artery; it thus reflects the physical properties of the arterial system of the brachial artery [23].

Apart from enabling the geometry of the arterial system to be studied, the Doppler methods allow the significance of the stress mechanics of the wall to be evaluated for the first time in a non-invasive fashion in humans [16,18,23]. The tangential tension can be measured as the product of the mean arterial pressure (P) and the arterial radius (R). Measurement of the relation between the product PR and the thickness of the vessel, is still not possible. Its measurement would be essential for establishing the structure–function relationship of the vessels in response to different chemical or pathophysiological stresses [23].

Haemodynamic Consequences of Diseases of the Large Arterial Trunks

Two diseases which originally affect the structure of the large arterial trunks in humans are known to cause systemic haemodynamic repercussions [24]: systolic hypertension in elderly subjects which results from vascular ageing, and arteriopathy of the lower limbs which results from atherosclerosis. In the latter case, the anomalies of the structure–function relationship are particularly evident since the analysis is no longer made in terms of continuous circulation but in terms of the pulsatile model of the arterial system [18].

Reduction of Arterial Compliance

Arterial compliance is reduced in old subjects with systolic hypertension [26] and patients with arteriosclerosis of the lower limbs [12,13] (Table 18.3). This reduction is independent of age and the level of mean arterial pressure. As the arterial compliance is a product of volume and arterial distensibility [16,18], the contribution of arterial volume must be considered first. This is possibly important in the case of systolic hypertension of elderly subjects since the aortic diameter increases with age [18]. It is more difficult to

Table 18.7. Systolic and diastolic arterial pressure in 105 subjects with amputation of the lower limbs following trauma compared to 184 control subjects [10]

Age range (years)	40–49	50–59	60–69	70–79	80–89
Systolic arterial pressure (mmHg):					
Normal subjects	132 ± 4	138 ± 2	141 ± 2	150 ± 2	154 ± 3
Subjects with amputation	$148 \pm 5*$	$148 \pm 4*$	$151 \pm 3**$	160 ± 3	$167 \pm 3*$
Diastolic arterial pressure (mmHg):					
Normal subjects	77 ± 3	79 ± 1	80 ± 1	80 ± 1	81 ± 2
Subjects with amputation	83 ± 2	82 ± 2	83 ± 1	82 ± 2	82 ± 3

$* P < 0.05$; $** P < 0.01$; $*** P < 0.001$

evaluate in arteriopathy of the lower limbs owing to the prominence of the arterial stenoses observed [12,13]. However, the mechanism is certainly of little importance as reduction in arterial compliance can be demonstrated even in vascular regions which are apparently undamaged [13]. The reduction in arterial compliance is thus probably associated with structural modification of the vascular wall such as is observed classically in atherosclerosis and vascular ageing and which is attested to by the frequency of calcification of the large arterial trunks observed in these cases [24].

Increase in Systolic Pressure

Both in arteriosclerosis of the lower limbs and arterial hypertension of elderly subjects, the systolic pressure is abnormally elevated, whereas the diastolic pressure is often normal or reduced. In the absence of modifications of ventricular ejection, the increase in systolic pressure is attributable to two principal mechanisms: the reduction of arterial compliance, which increases the speed of propagation of the pulse wave, and the modification in the timing of reflected waves. The latter may have a major role in the pathogenesis of arteriopathy of the lower limbs [12,24]. The systolic pressure results from the summation of two waves, an incident wave and a reflected wave from the periphery after impact of the incident wave. The reflected wave is more pronounced the closer the points of reflection. Under normal conditions, these points are generally situated in the periphery of the circulatory system where the arterioles originate [18]. In arteriopathy of the lower limbs, the importance of lesions of the aortic bifurcation makes them a zone of predilection for the points of reflection, thus giving rise to an abnormal increase in reflected waves and to a pre-

dominance of systolic hypertension [16,18]. A similar phenomenon is observed from the age of 50 in patients with traumatic amputation of the limbs (Table 18.7) [10].

Reversibility and Aggravation of Haemodynamic Disorders

Despite the significance of the structural modifications of the vessel wall usually observed, isolated systolic hypertension of elderly subjects and systolic hypertension in patients with atherosclerosis of the lower limbs are pharmacologically reversible after acute administration of nitrate derivatives [9,20]. These drugs are able to increase arterial compliance and to lower systolic pressure, and this without significant modification of diastolic pressure. Thus functional factors are able to modify the damping function of the large arterial trunks of elderly atherosclerotic subjects. Among the principal factors which can exacerbate the reduction in arterial compliance observed in the aged subject, one must include the perfusion of isotonic saline (Table 18.8) [13]; intravenous injection of propranolol [12,13,24] and administration of neurohumoral agents such as angiotensin and serotonin [16,18].

Restructuring of the Vascular System in Permanent Essential Systolo-Diastolic Arterial Hypertension

Various physiological and pathological considerations suggest that the level of arterial pres-

Table 18.8. Arteriosclerosis obliterans of the lower limbs: aggravation of the reduction in arterial compliance after administration of 2 l of isotonic saline in 2 h [13]

	Before perfusion of saline	After perfusion of saline
Systolic arterial pressure (mmHg)	151 ± 7	$174 \pm 5^+$
Diastolic arterial pressure (mmHg)	69 ± 3	78 ± 4
Diameter of the brachial artery (cm)	0.452 ± 0.021	0.456 ± 0.019
Blood flow of the brachial artery (ml/min)	100 ± 13	83 ± 9
Vascular resistance of the forearm (mmHg/ml/s)	74 ± 13	$93 \pm 13^*$
Compliance of the brachial artery (10^{-4} ml/mmHg)	102 ± 13	$73 \pm 12^{**}$

± 1, standard deviation of the mean; * $P < 0.05$; $^+$ $P < 0.02$; ** $P < 0.01$

sure is intimately associated with two important functions of the cardiovascular system, i.e. maintenance of blood flow and an effective capillary filtration pressure, both necessary for oxygen to be conveyed to the tissues. In essential hypertension in the rat, it is probable that these two parameters are kept normal [6,17]. In human hypertension, the same is probably true and requires adaptive changes of the cardiovascular system. These changes affect the small arteries and the heart but also the large arteries. The damping function of the artery is altered and a reduction of the arterial compliance is likewise observed [27]. The diminution of compliance is similar in moderate hypertension and severe hypertension, thus reflecting intrinsic modifications of the walls of the large arterial trunks of the hypertensive subject [27] (Table 18.2).

The arterial modifications in the cardiovascular system of the hypertensive subject may be functional (associated with neurohumoral phenomena) or structural or they may depend on both factors simultaneously. In this context, it is necessary to recall that the product of pressure × radius is increased in any hydraulic system in which there is an increase of pressure. In view of this, the only factor allowing maintenance of parietal pressure is, according to Laplace's principle, an increase in the thickness of the wall [6] and such changes are to be discussed in the case of human hypertension. In the present review, both arterial and arteriolar changes will be analysed.

Arguments in Favour of Structural Modifications in the Cardiovascular System of the Hypertensive Subject

In the hypertensive subject, the increase of systolic stress favours the development of cardiac hypertrophy, a phenomenon which is known and recognized [6,16,18]. Echocardiographic data have documented the structural modifications of the heart in the hypertensive.

The existence of a hypertrophy of the vessels, and especially of the arterial media, continues to be discussed. Although established in animals [1], indirect proof has been provided in humans by computer studies allowing the introduction in systemic models of the circulation of haemodynamic measurements carried out in the hypertensive patient [4]. In order to establish compatibility between clinical measurements and the characteristics of the model, modification of certain coefficients is necessary. In particular, it is necessary to change the coefficient characterizing the structure of the vessels, i.e. the variables reflecting their number and the thickness of their walls. However, the confirmation of these data on the basis of the investigations carried out in humans is more difficult, both at the level of the arterioles and at that of the arteries of large calibre.

At the arteriolar level, the presence of structural anomalies is suggested by the fact that the vascular resistance of the hypertensive subject is increased even under conditions of maximum vasodilatation [6,9,28]. Indeed, as Folkow has shown [6], a variation of 5% in the internal diameter of the resistance vessels is capable of giving rise to an increase of 25% in peripheral resistance. The increase of the pressor response to diverse physiological and pharmacological stimuli is also indirect evidence in favour of structural modifications of arterioles. The histopathological arguments are more difficult to prove. Autopsy specimens show vascular alterations, but the stage in which they are demonstrated makes it doubtful whether they have a role in the pathophysiological mechanism of hypertension. However, recent studies concerning the pre-capillary sector of the microvessels in human hypertension have shown a hypertrophy of the media relative to the internal diameter of the arterioles, and this is in proportion to the level of arterial pressure [5,29].

In arteries of large calibre, the evidence for the presence of structural anomalies is based solely on indirect arguments. At the level of the forearm circulation, the reduction of arterial compliance is independent of the level of mean arterial pressure and is not modified by blockade of the autonomic nervous system [24]. Similar observations have been made in spontaneous hypertension in the rat where the existence of structural alterations of the wall has been demonstrated [14,24]. At the level of the carotid circulation, the diameter of the common carotid artery is normal, and this despite the chronic increase in arterial pressure [2]. Recent studies [19] have shown that there is an increase in the mass of smooth-muscle fibres with a raised frequency of tetraploid cells and consequently a rise in the content of deoxyribonucleic acid in the human carotid artery. The thickening of the wall of the large vessels of the hypertensives may be secondary, either to the increase in mean arterial pressure or to the increase in pulsatile pressure or to the two factors in combination [18]. The exact contribution of these two mechanisms is difficult to establish precisely in human hypertension. However, it is admitted that the bioelastomers of the arterial wall are easier to alter experimentally during pulsatile stress than during continuous stress [18].

The existence of structural modifications in the cardiovascular system in the hypertensive human does not allow us to exclude the role of functional factors acting on the wall of the vessels. These may be operative at several levels. On the one hand, increased neurohumoral activity (coming in particular from the sympathetic nervous system or from the renin–angiotensin system) may play a role in the initial phase of the hypertensive disease and may, in the long term, have a trophic influence [6]. On the other hand, any structural modification in the wall of the vessels is accompanied by functional modification directly associated with medial hypertrophy of the wall. The hyper-reactivity of the vascular system to different stimuli, in particular to angiotensin and noradrenaline, is one of the classical features of hypertension [6].

Reversibility of Structural and Haemodynamic Alterations of the Cardiovascular System in Hypertensive Subjects

To the extent that it is admitted that the cardiovascular system adapts structurally to haemodynamic stress, the problem of the reversibility of lesions during treatment carried out in humans may be discussed [30]. The most classic

cases are those produced by regression of myocardial hypertrophy observed after healing of valvular lesions. Another example is that of the reversibility of the structural modifications of the cardiovascular system demonstrated after pharmacological treatment of human arterial hypertension. At the myocardial level, echocardiographic studies have shown that cardiac hypertrophy may regress in parallel to the lowering of arterial pressure. However, the reversibility of the lesions depends in part on the medication used. Certain anti-hypertensive agents induce little regression in myocardial hypertrophy for the same degree of lowering of arterial pressure at which other drugs give rise to regression [30]. The activation of the sympathetic nervous system may play a role in the persistence of hypertrophy. Comparable phenomena may exist at the level of the large arterial trunks [22,23].

The problem of the reversibility of structural lesions at the arteriolar level has been discussed a great deal. The fall of vascular resistance under conditions of maximum vasodilatation may be indirect evidence for reversibility [6]. The question is of interest since it involves the problem of cardiovascular coupling in the treated hypertensive [18]. In agreement with various experimental data [3], it appears that certain anti-hypertensive drugs give rise to a simultaneous regression of the hypertrophy of the heart and the vessels whereas other drugs result in divergent changes [3,30]. Such a problem necessitates meticulous investigations in the course of treatments of long duration and therefore requires further study.

References

1. Berry CL, Henrichs J (1982) Morphometric investigations of hypertrophy in the arteries of DOCA hypertensive rats. J Pathol 136:85–94
2. Bouthier J, Benetos A, Simon A, Levenson J, Safar M (1985) Pulsed Doppler evaluation of diameter, blood velocity and blood flow of common carotid artery in sustained essential hypertension. J Cardiovasc Pharmacol 7:S99–S104
3. Carlier P, Rorive G (1985) Pathogenesis and reversibility of the aortic changes in experimental hypertension. J Cardiovasc Pharmacol 7:S46–S51
4. Chau NPH, Safar ME, London GM, Weiss YA (1979) Essential hypertension: an approach to clinical data by the use of models. Hypertension 2:87–97
5. Furuyama M (1962) Histometrical investigations of arteries in reference to arterial hypertension. Tokohu J Exp Med 75:388–414

6. Folkow B (1982) Physiological aspect of primary hypertension. Physiol Rev 62:347–503

7. Green JF (1977) Mechanical concepts in cardiovascular and pulmonary physiology. Lea and Febiger, Philadelphia, pp 25–54

8. Guyton AC (1980) Circulatory physiology III. Arterial pressure and hypertension. Saunders, Philadelphia pp 293–306

9. Korner PL, Shaw J, Uther B, West MJ, McRitchie RJ, Richards JG (1973) Autonomic and non-autonomic circulatory components in essential hypertension in man. Circulation 68:107–117

10. Labouret G, Achimastos A, Benetos A, Safar M, Housset E (1983) L'hypertension artérielle systolique des amputés traumatiques. Presse Med 21:393–400

11. Levenson JA, Peronneau PP, Simon ACH, Safar ME (1981) Pulsed Doppler: determination of diameter, blood flow velocity and volume flow of brachial artery in man. Cardiovasc Res 15:164–170

12. Levenson JA, Simon ACH, Safar ME, Fiessinger JN, Housset EM (1982) Systolic hypertension in arteriosclerosis obliterans of the lower limbs. Clin Exp Hypertens [A] 4:1059–1072

13. Levenson JA, Simon ACH, Maarek BE, Gitelman GJ, Fiessinger JN, Safar ME (1985) Regional compliance of brachial artery and saline infusion in patients with arteriosclerosis obliterans. Arteriosclerosis 5:80–87

14. Levy BI, Benessiano J, Poitevin P, Lubin L, Safar ME (1985) Systemic arterial compliance in normotensive and hypertensive rats. J Cardiovasc Pharmacol [Suppl 2] 7:22–27

15. London GM, Levenson JA, Safar ME, Simon ACH, Guerin AP, Payen D (1983) Hemodynamic effects of head-down tilt in normal subjects and sustained hypertensive patients. Am J Physiol 245:H195–H202

16. Milnor WR (1982) Hemodynamics. Williams and Wilkins, London, pp 56–91, 189–231

17. Mulvany MJ (1984) Pathophysiology of vascular smooth muscle in hypertension. J Hypertens [Suppl 3] 2:413–420

18. O'Rourke MF (1982) Arterial function in health and disease. Churchill Livingstone, Edinburgh, pp 53–66, 153–169, 196–252

19. Owens GK, Rabinovitch PS, Schwartz SM (1982) Smooth muscle cell hypertrophy versus hyperplasia in hypertension. Proc Natl Acad Sci USA 78:7759–7765

20. Peronneau PP, Leger F (1977) Applications of pulsed Doppler systems. Echocardiology. Martinus Nijhoff, The Hague, pp 223–232

21. Safar ME, Peronneau PP, Levenson JA, Totomoukouo JA, Simon ACH (1981) Pulsed Doppler: diameter, blood flow velocity and volumic flow of the brachial artery in sustained essential hypertension. Circulation 63:393–400

22. Safar ME, Simon ACH, Levenson JA, Cazor JL (1983) Hemodynamic effect of Diltiazem in hypertension. Circ Res [Suppl I] 52:169–173

23. Safar ME, Bouthier JA, Levenson JA, Simon AC (1983) Peripheral large arteries and the response to antihypertensive treatment. Hypertension [Suppl III] 5:63–68

24. Safar ME, Simon ACH, Levenson JA (1984) Structural changes of large arteries in sustained essential hypertension. Hypertension [Suppl III] 6:117–121

25. Simon ACH, Safar ME, Levenson JA, London GM, Levy BI, Chau NPh (1979) An evaluation of large arteries compliance in man. Am J Physiol 237:H550–H554

26. Simon ACH, Safar ME, Levenson JA, Kheder AD, Levy BI (1979) Systolic hypertension: hemodynamic mechanisms and choice of antihypertensive treatment. Am J Cardiol 44:505–511

27. Simon ACH, Laurent S, Levenson JA, Bouthier J, Safar ME (1983) Estimation of forearm arterial compliance in normal and hypertensive men from simultaneous pressure and Doppler flow measurements. Cardiovasc Res 17:331–338

28. Sivertsson R (1970) The hemodynamic importance of structural vascular changes in essential hypertension. Acta Physiol Scand [Suppl] 343:1–56

29. Suwa N, Takahashi T (1971) Morphological and morphometrical analysis of circulation in hypertension and ischemic kidney. Urban and Schwarzenberg, Berlin

30. Tarazi RC, Fouad FM (1984) Reversal of cardiac hypertrophy in humans. Hypertension [Suppl III] 6:140–146

Chapter 19

The Epidemiology of Atherosclerosis

J. L. Richard

Clinical complications of atherosclerosis are quite frequent in later life and are a major health problem. However, the site and the extent of lesions and the age when first clinical symptoms appear are very variable. Such a variability is observed among individuals, and also among populations; it is a noteworthy characteristic of these diseases and it suggests that their development is conditioned, at least partly, by determinants with an unequal distribution among individuals and among populations. The search for factors predisposing to atherosclerosis is therefore a necessary prerequisite for prevention. A logical and deliberate epidemiological research programme carried out in the United States since 1950, has contributed much to a better understanding of these diseases, particularly coronary heart disease and, to a lesser degree, stroke.

Atherosclerosis in Human Populations

Death rates from coronary heart disease vary a great deal; for example, the standardized mortality rates among middle-aged men in 1969–1971 in 26 industrialized countries varied from one to more than six (Fig. 19.1): a particular geographical distribution of the disease becomes clear [44]. The highest rates are observed in Finland and in the Anglo-Saxon countries and the lowest ones in the Mediterranean countries and in Japan. In Europe, mortality clearly increases from South to North. Furthermore, it has been demonstrated that such mortality is low or very low in developing countries.

Mortality due to stroke varies also, but its geographical distribution differs from that of coronary heart disease: the highest death rates are reported in Japan and in developing countries; death rates are lower and less variable in industrialized countries other than Japan. In general, there is a certain negative relationship between these two mortalities.

Studies of coronary heart disease and stroke morbidity can only take account of limited population samples. Generally speaking, the geographical variability of measured rates is in agreement with mortality variability: the incidence of coronary heart disease is high in Northern European countries and is low in Japan and in some Mediterranean and developing countries. Table 19.1 gives the mean annual incidence of the three main diseases of atherosclerosis according to sex and age as well as the incidence of stroke measured in some recent studies. It is noteworthy that, among men, the American annual incidence of coronary heart disease increases from about 1% at 50 years of age to above 2% around 60 years of age. It is very infrequent among women before the menopause but the incidence increases quickly afterwards, the difference between sexes lessening with age. Within the age limits of the study, the incidences of cerebral infarction and of intermittent claudication are lower. In the Paris Study the incidence of coronary heart disease among men is about half of the Framingham incidence.

These data, already rather old (no recent data being available) and observed in specific populations, cannot be extrapolated without reservation. In the French study, the lower incidence is in agreement with other data (mortality, registers etc.), strongly suggesting a mod-

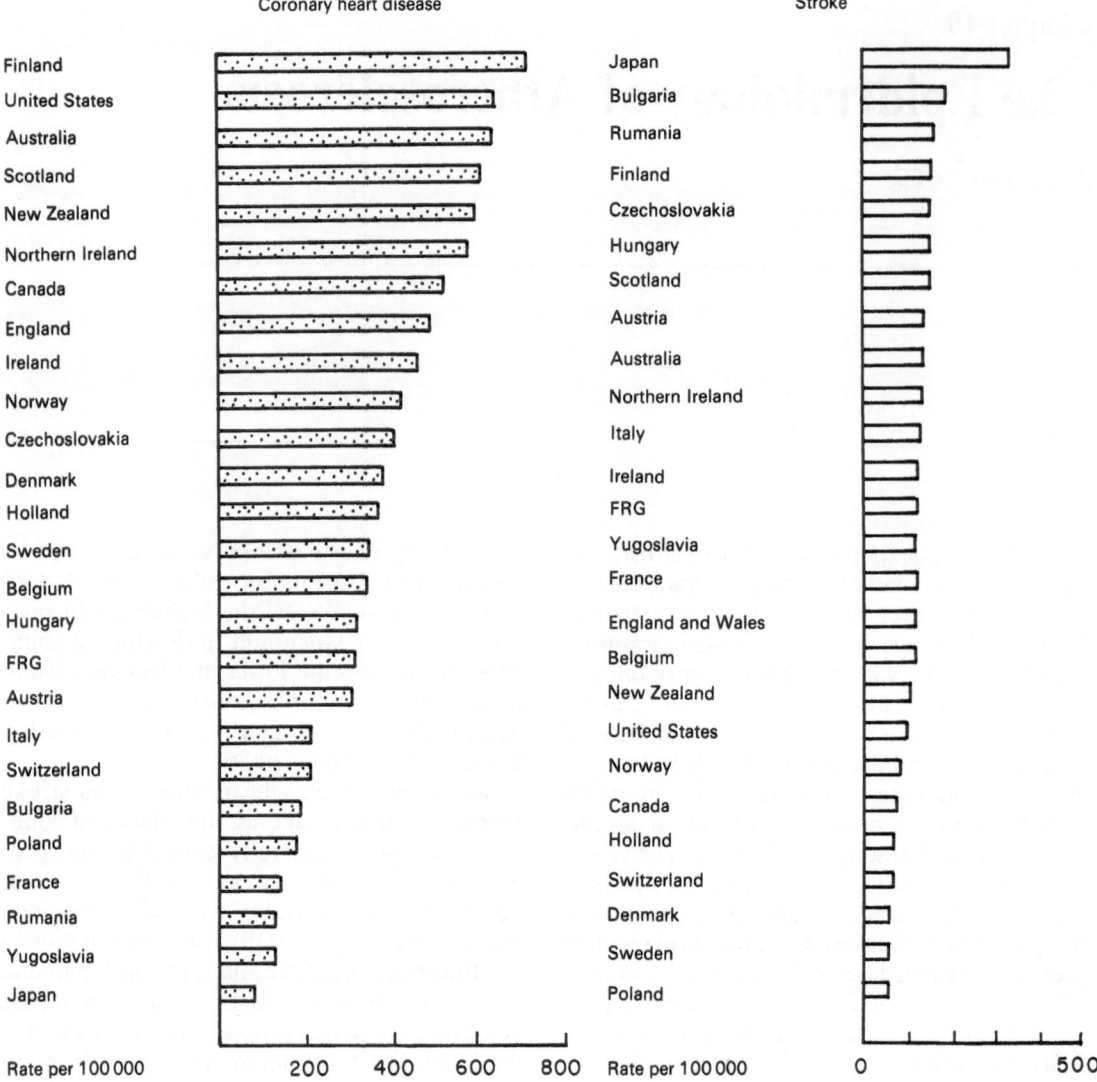

Fig. 19.1. Mortality in 26 industrialized countries, 1969–1971. Males, 40–69 years old, standardized rates.

erate frequency of coronary heart disease [18]. In the case of stroke, the incidence increases very much with age and is, as a whole, slightly lower in females. There is no very clear geographical variation in incidence of stroke, but in the Ni-Hon-San Study, the incidence is much higher in Japan than in Hawaii (standardized rates 7.4 and 2.7 per 1000 respectively).

Some surveys which have compared male groups in different geographical locations have confirmed the variability in the incidence of coronary heart disease (Table 19.2) and also the strong correlation between incidence and mortality rates as shown clearly in the myocardial infarction registers [14,19,24,26,38,47].

Because of their frequency, these diseases are a major public health problem in industrialized countries. Table 19.3 gives the proportions of deaths from coronary heart disease and stroke occurring in men between the ages of 40 and 69 in the industrialized countries shown in Figs 19.1 and 19.2. The proportion of male deaths from coronary heart disease in 1969–1971 is notably high in some countries, for example Finland, the United States, Scotland and Australia. In the Framingham Study, after 14 years of follow-up, 12% of the 40–44-year-old men, 20% of those between 50 and 54 years and 25% of those over 55 years had some clinical features of coronary heart disease. In the Paris Prospective Study,

Table 19.1. Mean annual incidence of atherosclerosis per 1000 in some population studies in males (M) and females (F)

Age	Coronary heart disease Framingham M	Coronary heart disease Paris M	Coronary heart disease Framingham F	Intermittent claudication Framingham M	Intermittent claudication Framingham F	Cerebral infarction Framingham M	Cerebral infarction Framingham F	Stroke Ni-Hon-San Study Japan M	Stroke Ni-Hon-San Study Hawaii M	Göteborg (Sweden) M	Göteborg (Sweden) F	Oxford (England) M	Oxford (England) F	Melbourne (Australia) M	Melbourne (Australia) F	South Alabama (United States) M	South Alabama (United States) F	Washington (United States) M	Washington (United States) F
25–34														0.12	0			0.04	0.08
35–44	3.8[a]		0.6	0.6	0.3	0.1	0							0.23	0.24	0.49[d]	0.35	0.40	0.23
45–54	9.9	3.5 / 5.0 / 7.3	2.6	1.9	0.7	1.0	0.9	1.7	1.8	0.97	0.65			1.46	0.98			1.35	1.14
55–64	21.0	10.6	9.5	5.3	1.8	3.0	2.3	4.1 / 7.2[c]	1.3 / 2.3	3.16	1.74	3.18	2.92	4.70	2.24	2.74	2.33	4.3	2.29
65–74	19.4		14.0	6.1	5.4	2.1	4.6					5.95	6.89	13.23	7.90	5.41	3.65	8.4	7.57
75–84												17.82	14.21	22.51	21.39	13.85	7.19	20.69	17.57
85+														13.67	32.62				

[a] 30–44 years old.
[b] Estimated incidence respectively at 45, 50, 55 and 60 years.
[c] 65–69 years old.
[d] 20–54 years old.

Table 19.2. Comparative incidence of coronary heart disease in population studies (standardized rates per 1000)

1. Seven Country Study and Paris Prospective Study. Males, 40–59 years old. Five-year incidence

East Finland	43.0
US Railroad	32.6
Zutphen, Holland	31.1
Slavonia[a]	17.0
Paris Study	16.8
Rural Italy	16.6
West Finland	14.9
Rome	12.3
Dalmatia[a]	10.3
Greek Islands	8.2
Serbia[a]	5.3

2. Pooling Project[b] and Paris Prospective Study. Males, 45–49 years old. 8.6-year incidence

Pooling Project	93.9
Paris Study	29.7

[a] Yugoslavia.
[b] Pooled analysis of five American prospective studies.

Table 19.3. Relative likelihood (in %) of coronary heart disease or stroke being the cause of male premature mortality (40–69 years)

Coronary heart disease		Stroke	
New Zealand	41.0	Japan	26.4
Australia	40.3	Bulgaria	15.8
Canada	38.8	Rumania	12.0
United States	38.6	Hungary	10.2
Northern Ireland	37.8	Czechoslovakia	9.6
Finland	37.3	Italy	9.4
Norway	36.7	Austria	8.9
Scotland	35.5	Scotland	8.7
Sweden	33.7	Australia	8.7
England	32.7	Ireland	8.6
Denmark	32.0	Northern Ireland	8.4
Ireland	31.5	Finland	8.3
Holland	30.3	Yugoslavia	8.3
Czechoslovakia	24.9	France	8.2
Belgium	21.9	Federal Republic	
Hungary	21.2	of Germany	8.1
Federal Republic		England	8.0
of Germany	20.4	New Zealand	7.6
Austria	19.6	Belgium	7.6
Switzerland	17.3	Norway	7.5
Italy	15.7	Sweden	6.6
Bulgaria	15.3	United States	6.3
Poland	11.7	Holland	6.0
France	9.7	Switzerland	5.9
Rumania	9.5	Canada	5.9
Yugoslavia	8.8	Denmark	5.7
Japan	6.6	Poland	4.0

after 6.6 years of follow-up, 4.6% of a male population with a mean age of 47 years at entry into the study had some clinical complications of atherosclerosis, among which 3.4% had coronary heart diseases, 0.8% peripheral vascular diseases and 0.4% stroke [7].

Temporal Variations

The frequency of atherosclerosis can also vary a great deal across time. It is probable that the frequency of coronary heart disease increased considerably in the United States between the two World Wars, and in Europe after the Second World War. Recent variations of mortality are better known (Fig. 19.2). Stroke mortality is decreasing in many countries, but the recent increase observed in Eastern European countries is in disagreement with this general trend.

The interpretation of these contemporary trends is difficult. They may be due to the improvement of health statistics, changes in populations' life styles, consequences of preventive measures, effects of modern treatments or inadequacy of health services in some countries. Some new data on coronary heart disease suggest that the decrease in mortality observed in the United States could be due to a decrease of the incidence itself following the preventive measures adopted by the population and simultaneously to a decrease of the lethality of acute clinical events [31].

The Search for Associations

Secular and Geographical Relationships

The high variability between human populations in the frequency of atherosclerosis has led to a search for associated factors. The incidence of coronary heart disease is linked to variables related to the way of life or to the socio-economic development of populations. Thus, death rates from coronary heart disease are highly correlated with the gross national product, with energy consumption and also with the consumption of cigarettes, meat, animal proteins, saturated lipids and refined sugar. However, a negative correlation is observed with the consumption of complex carbohydrates, fibre and alcohol (Fig. 19.3).

Such relationships have led to many attempted analyses. Thus, from statistics in 20 countries, an index taking into account the consumption of saturated lipids, polyunsaturated lipids and cholesterol was found to be significantly associated with coronary death rates after adjustment for consumption of fibre and refined sugar [22]. However, such relationships cannot reasonably

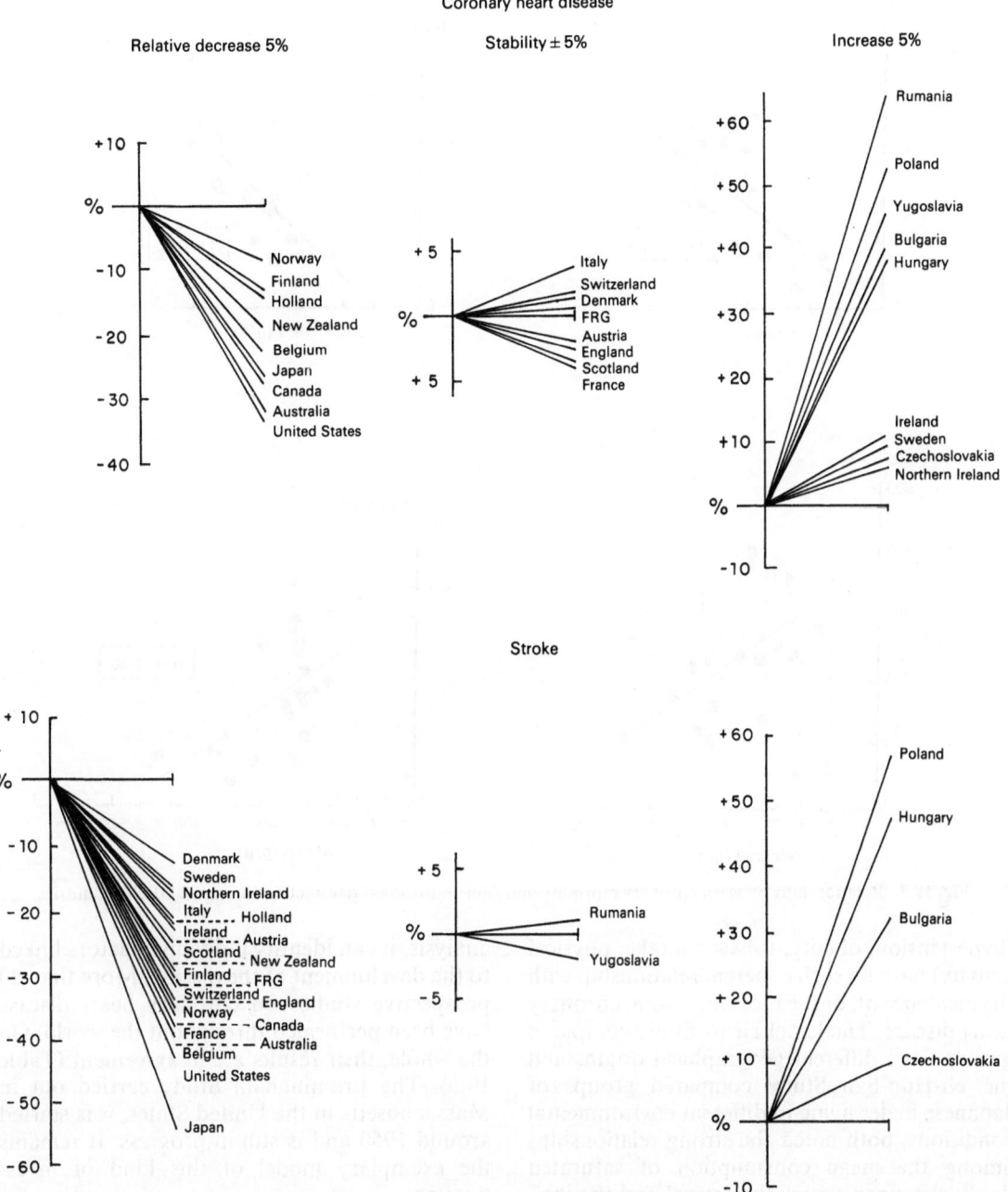

Fig. 19.2. Recent trends of mortality in 26 industrialized countries between 1969–1971 and 1980–1982. Males, 40–69 years old, standardized rates.

be interpreted as causal, mainly because all of the variables considered are highly intercorrelated.

Group Comparisons

Comparisons of populations with different coronary heart disease mortality or morbidity rates have confirmed previous associations. Three studies are exemplary on this account. In the Seven Countries Study, the coronary heart disease 5-year incidence and 10-year mortality are seen to be strongly associated with the populations' mean serum cholesterol and also with the percentage of calories provided by the saturated fatty acids of the diet, these two variables being strongly intercorrelated (Table 19.4). It is worthwhile noting that the levels of other risk factors

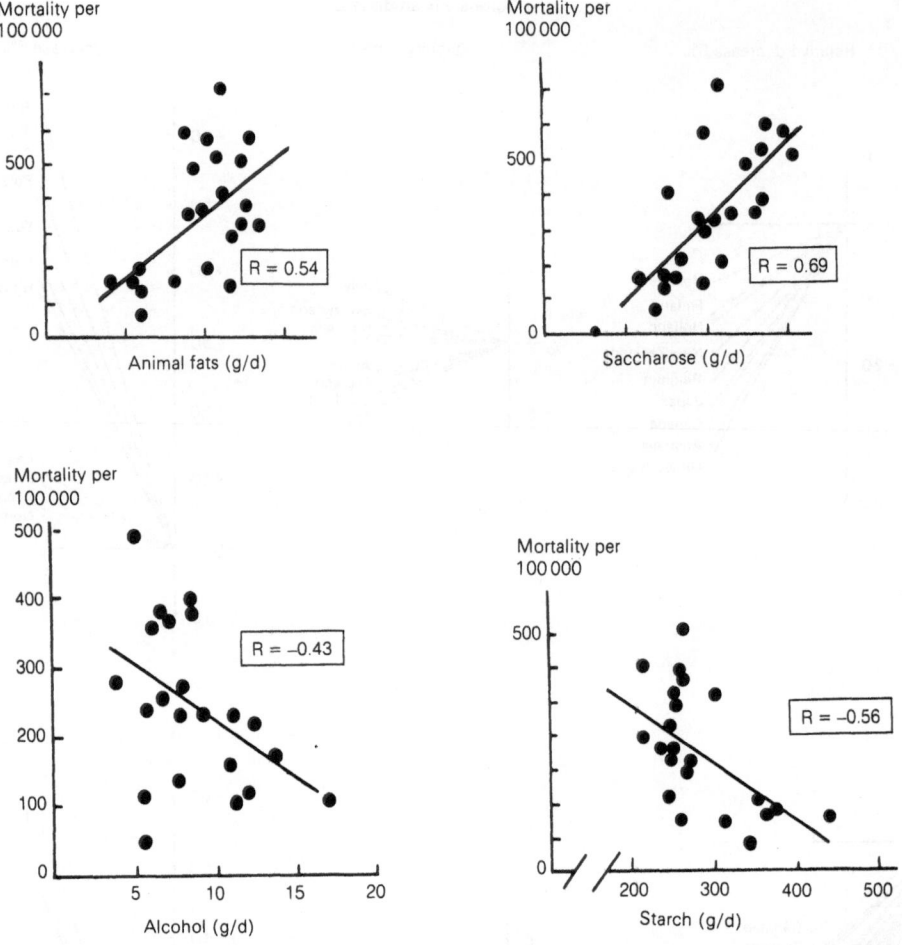

Fig. 19.3. Relationships between coronary mortality and some nutritional parameters among 22 OECD countries.

(hypertension, obesity, tobacco intake, physical activity) have no such coherent relationship with the incidence of, or the mortality from, coronary heart disease. The Israel Heart Study compared groups from different geographical origins and the Ni-Hon-San Study compared groups of Japanese males living in different environmental conditions; both noted the strong relationships among the mean consumption of saturated lipids, the mean serum cholesterol and the incidence of coronary heart disease (Table 19.4).

The Search for Causes

This main stage of epidemiological research needs planned studies. The prospective study is the methodology most suited to such research. It requires a long follow-up of a large number of subjects but, using techniques of multivariate analysis, it can identify individual factors linked to the development of the disease. More than 20 prospective studies on coronary heart disease have been performed throughout the world. On the whole, their results are in agreement (Table 19.5). The Framingham Study carried out in Massachusetts in the United States, was started around 1950 and is still in progress. It remains the exemplary model of this kind of investigation.

The Classical Risk Factors

Some biological or environmental factors have a significantly greater effect than others on the development of coronary heart disease: the incidence of the disease is increased in the presence of qualitative factors (for example diabetes) and with the level of quantitative factors; this last relationship is a continuous one, with no obvious threshold which allows us to discriminate

Table 19.4. Comparison of populations: relationship between saturated lipid content of diets, mean serum cholesterol and coronary incidence or mortality

	Saturated lipids (percentage of calories)	Cholesterol (mg/dl)	Incidence or mortality/1000
Seven Countries Study			
Finland	21	261	45.5[a]
United States	18	240	42.4
Holland	19	235	31.7
Italy	10	204	20.3
Yugoslavia	10.5	187	14.3
Greece	8	204	6.6
Japan	3	148	6.0
Israel Heart Study[b]			
Western Europe	11.3	219	10.0[c]
Eastern Europe	10.8	218	9.8
Southern Europe	9.9	210	9.0
Israel	9.3	212	9.0
Middle East	8.5	203	7.4
North Africa	8.1	195	8.0
Ni-Hon-San Study			
California	26.2	226	4.3[c]
Hawaii	23.3	219	2.9
Japan	6.7	176	1.4

[a] Mortality in 10 years.
[b] According to the geographical origin.
[c] Annual incidence.

between the subjects who are healthy and those who have developed the disease: the higher the level of the factor, the higher the incidence of the disease.

An authentic scientific consensus now recognizes, without dispute, eight independent risk factors for coronary heart disease: sex, age, serum cholesterol, blood pressure, tobacco consumption, diabetes, lack of exercise and type A personality. The relationships of the disease with age and sex are clearly illustrated in Table 19.1. Figures 19.4 and 19.5, based on data from the Paris Prospective Study and the French–Belgian Project, illustrate these relationships, except that with physical activity which has been observed in the Framingham and the Whitehall studies [2,28]. Table 19.5 summarizes the main independent risk factors in the prospective studies conducted among middle-aged men and using a multivariate analysis technique [3].

The concept of independent risk factors is of great theoretical importance for the following reasons:

1. They are supposed to be "aetiological" in the disease and also to be the targets of prevention. Other factors, for example triglycerides,

uric acid and obesity, whose univariate association with the disease disappears when other factors are taken into account, probably do not have any direct and important aetiological relationship with coronary heart disease.

2. They allow a quantitative estimation of the multivariate risk from an index which considers the relative part each factor plays in that risk. The incidence calculated from that index for a population of subjects is in very good agreement with the incidence observed. Table 19.6 gives an estimation of the risk of developing coronary heart disease in 5 years for a 50-year-old man, as a probability per 100 according to the level or the presence of classical risk factors. It is worth noting that the relative risk increases from 1 to 86 between the lowest level of exposure (cholesterol 160 mg/dl and blood pressure 120 mmHg in a non-smoking and non-diabetic man) and the highest level (cholesterol 360 mg/dl and blood pressure 220 mmHg in a smoking and diabetic man). This table, theoretically referring only to the men of the Paris Prospective Study between 45 and 54 years of age, can reasonably be applied to middle-aged Frenchmen.

3. The concept of risk factor does not involve an unquestionable causal relationship with the

Table 19.5. Multivariate analysis of the main independent risk factors among middle-aged men

Risk factors	Coronary heart disease			Stroke			Peripheral vascular disease		
	Total number of studies	Significant	Not significant	Total number of studies	Significant	Not significant	Total number of studies	Significant	Not significant
Age	15	13	2	8	7	1	—	—	—
Blood pressure	16	14	2	11	11	0	3	2	1
Cholesterol	16	13	3	10	2	7	3	2	1
Tobacco	15	12	3	9	4	5	3	3	0
Diabetes	6	4	2	3	0	3	2	1	1
Triglycerides	6	1	5	3	1	2	2	0	2
Obesity	12	1	11	9	0	9	1	1[a]	0
Physical activity	9	5[a]	4	—	—	—	—	—	—
Haematocrit	10	2	8	3	1	2	1	1	0
Alcohol	4	2[a]	2	5	2	3	—	—	—
Pattern A	4	4	0	—	—	—	—	—	—

[a] Negative relationship

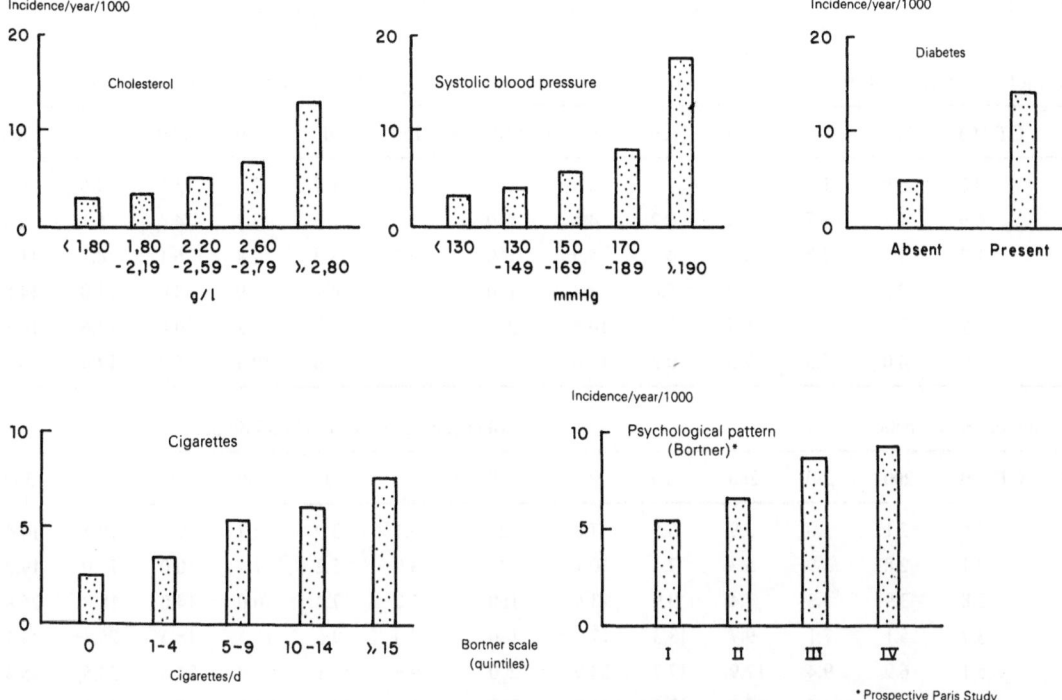

Fig. 19.4. Incidence of coronary heart disease and classical risk factors.

disease. The term "risk factor" has been proposed precisely in order to remove any ambiguity and to express only the observed relationship and the allowed probabilistic prediction.

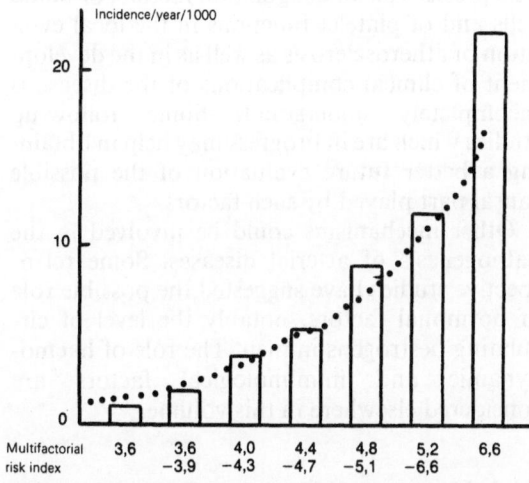

Fig. 19.5. Multifactorial risks (age, cholesterol, blood pressure, cigarettes and diabetes) in coronary heart disease. Each column gives the observed incidence. The *dotted line* gives the expected theoretical incidence. (Based on data from the Paris Prospective Study.)

Other Risk Factors

Instead of listing the 246 risk factors recently noted, it is more interesting to stress a few essential points [12].

The risk estimated from the level of total serum cholesterol is in fact the result of the risk linked to the fraction of cholesterol (about 2/3) carried by low-density lipoproteins (LDL) and of the negative risk linked to the fraction of cholesterol (about 1/4) carried by high-density lipoproteins (HDL). This protective factor has been identified in both sexes even in advanced age [9]. The ratios taking into account these lipid variables (for instance, total cholesterol/HDL cholesterol) give a better estimation than a measure of total cholesterol alone. The levels of the main apoproteins of the various lipoproteins (Apo B for LDL and Apo[1] for HDL) are also linked to the disease; these links are in good agreement with the links observed with the levels of the lipoproteins themselves. However, the epidemiological data concerning these proteins are still too limited to allow a reasonable use of their levels in the prediction of coronary heart disease.

A family history, especially a paternal history, of coronary heart disease or hypertension is

Table 19.6. Probability of a 50-year-old man developing coronary heart disease in 5 years according to the level or the presence of risk factors: Paris Prospective Study

Non-smoker and non-diabetic							Smoker (20 cigarettes/day) and non-diabetic						
SBP	CT 160	200	240	280	320	360	SBP	CT 160	200	240	280	320	360
120	0.7	0.9	1.3	1.7	2.4	3.3	120	1.3	1.7	2.5	3.2	4.6	6.3
140	0.9	1.2	1.7	2.3	3.2	4.4	140	1.7	2.3	3.2	4.4	6.1	8.4
160	1.2	1.7	2.3	3.2	4.3	5.9	160	2.3	3.2	4.4	6.1	8.2	11.2
180	1.6	2.2	3.1	4.2	5.8	7.8	180	3.0	4.2	5.9	8.0	11.0	14.8
200	2.2	3.0	4.1	5.6	7.7	10.4	200	4.2	5.7	7.8	10.6	14.6	19.8
220	2.9	4.0	5.5	7.5	10.2	13.8	220	5.5	7.6	10.4	14.2	19.4	26.2

Non-smoker and diabetic							Smoker (20 cigarettes/day) and diabetic						
SBP	CT 160	200	240	280	320	360	SBP	CT 160	200	240	280	320	360
120	1.6	2.1	3.0	3.9	5.5	7.6	120	3.1	3.9	5.7	7.4	10.5	14.4
140	2.1	2.8	3.9	5.3	7.4	10.1	140	3.9	5.2	7.4	10.0	14.0	19.2
160	2.8	3.9	5.3	7.4	9.9	13.6	160	5.2	7.4	10.0	14.0	18.8	25.8
180	3.7	5.1	7.1	9.7	13.3	17.9	180	7.0	9.6	13.5	18.3	25.3	34.1
200	5.1	6.9	9.4	12.9	17.7	23.9	200	9.6	13.1	17.9	24.5	33.6	45.4
220	6.7	9.2	12.6	17.2	23.5	31.7	220	12.7	17.5	24.0	32.8	44.6	60.3

The subjects of the Paris Prospective Study have a mean serum cholesterol of 222 mg/dl and a mean systolic blood pressure of 141 mmHg; among the 68% male smokers the mean daily cigarette consumption is 15.7, and 1.5% have a definite diabetes.

SBP, systolic blood pressure in mmHg; CT serum cholesterol in mg/dl.

linked to the development of the disease independently of classical risk factors, suggesting the possible pathogenic role of genetic factors other than those which control the level of cholesterol or the values of blood pressure [4]. Likewise, the level of circulating plasma insulin and the insulin/fasting glycaemia ratio were independent predictive factors in three prospective studies conducted in Paris, Helsinki and Büsselton in New Zealand [6]. Also, the number of leucocytes can predict the development of a myocardial infarction independently of other risk factors; for example, tobacco consumption increases leucocytosis [49]. The level of the electrophoretic fraction albumin-alpha1-globulin of serum proteins is linked negatively to the development of coronary heart disease in the Paris Prospective Study; this very strong relationship is independent of classical risk factors [40]. A moderate alcohol consumption (under half a litre of wine per day) has a negative association with the disease [39].

Three prospective studies (Framingham, Göteborg in Sweden and Northwick Park in England) have recently shown a relationship which is independent of classical risk factors between the level of plasma fibrinogen and the development of a coronary disease [16,25,45]. The precise role of coagulation factors, of blood cells and of platelet functions in the local evolution of atherosclerosis as well as in the development of clinical complications of the disease is incompletely understood. Some follow-up studies which are in progress may help in obtaining a better future evaluation of the possible causal part played by such factors.

Other mechanisms could be involved in the pathogenesis of arterial diseases. Some retrospective studies have suggested the possible role of hormonal factors, notably the level of circulating oestrogens in men. The role of haemodynamic and immunological factors are considered elsewhere in this volume.

Risk Factors of the Other Locations of Atherosclerosis

Table 19.5 shows very clearly that independent risk factors are not perfectly identical at the various locations of atherosclerosis. Only blood pressure and age are linked very strongly to stroke in all studies. The association with cholesterolaemia is observed only in some Western populations; in the Japanese studies, the association is low or nil with cerebral infarctions and negative with cerebral haemorrhages. The relationship between stroke and tobacco is also inconstant. Various publications, with some contradictions, have suggested a relationship between stroke and alcohol consumption, and between stroke and use of oral contraceptives in women [17,46].

In the case of peripheral arterial disease, tobacco is the most significant risk factor in the three mentioned studies. Blood pressure and serum cholesterol are significantly linked to the disease in two of the three studies and diabetes in one study. A recent publication of the Framingham Study confirmed the risk variations according to age and sex as well as the main risk factors linked to peripheral arterial disease [15].

Conclusions and Hypotheses

The possible aetiological role of each of the main risk factors seems more or less probable after taking into account the available clinical, experimental, biological and epidemiological data. A multifactorial pathogenesis of atherosclerosis is now well recognized and seems reasonable. In the case of coronary heart disease, a sequence of causal relationships has been proposed between the role of saturated fats in the diet, the level of serum cholesterol and the development of a coronary disease: this is the diet heart hypothesis or lipid hypothesis [35,37].

The level of serum cholesterol is partly dependent on the diet, and particularly on the lipid consumption and on the respective proportions of calories given by saturated and unsaturated fatty acids. A basal and physiological cholesterolaemia which is genetically determined would be close to 150–160 mg/dl. A supplementary fraction, labile and variable between people, might be under the influence of environmental factors (notably the diet) and might depend on complex interactions between these factors and individual metabolic peculiarities, most of them of genetic origin. Such an interpretation is in agreement with the great variability of serum cholesterol seen among individuals as well as among populations.

Among all classical risk factors, serum cholesterol is the factor most clearly linked to the incidence of coronary heart disease in human populations. A clear relationship is also seen with the content of saturated lipids in the diet. However, other risk factors (blood pressure, tobacco consumption, physical activity) do not have the same discriminative power among populations. Such facts appear very clearly in the analysis of the Seven Countries Study. Although the measure of individual consumption of saturated lipids is very imprecise, some prospective studies have shown a relationship between this consumption and the future development of coronary heart disease [42].

A different significance for aetiological factors is likely in other localizations of atherosclerosis. The diet hypothesis does not seem to be applicable to stroke for which hypertension is the prevailing risk factor; it could play a certain part in peripheral vascular diseases in which tobacco is a demonstrated causal factor.

Preventive Trials

Descriptive and analytic epidemiology can only show associations and a strict causal interpretation needs experimentation in man. This would involve preventive trials with a random distribution of people in two groups (intervention and control). Theoretically, in the controls all the factors other than the treatment(s) whose effects must be evaluated should be matched; this new step in epidemiological research allows the recognition of the factors playing a causal role in the development of the disease and brings about a better understanding of preventive measures.

The trials must comply with exacting restraints, in particular with strict planning and a sufficient sample size. These restraints have a scientific base and cannot be eluded.

Since the mid-1970s, many primary preventive trials have increased our knowledge. They were often considerable undertakings, notable for their extent, their rigour and their scientific results.

Treatment of Hypertension

The benefit of anti-hypertensive treatment in preventing the development of stroke and of noncoronary cardiac complications was quickly demonstrated by the first clinical trials conducted among severe hypertensive subjects, often

Table 19.7. Preventive trials: severe hypertension

Trials	Sample size	Initial blood pressure (mmHg)	Relative effect of treatment (%)[a]				
			Stroke	Heart failure	Hard coronary disease	All events	Total mortality
Hamilton	61	$\geqq 110$	− 57	− 75	—	− 70**	—
Wolff	87	$\geqq 175.3/108.3$	—	− 100	—	− 69***	—
Veterans adm.	523	90–129	− 75	− 100	− 31	− 72***	− 65
	143	115–129	− 75	− 100	—	− 93***	− 100
	210	105–114 }	− 70	− 100	− 15	{ − 77***	− 59
	170	90–104 }				{ − 33	− 65

*P < 0.05; **P < 0.01; ***P < 0.001
[a] In the intervention group compared with the control group.

with complications (Table 19.7). The magnitude of the benefit was proportional to the blood pressure level, to the decrease of that level and to the permanency of that decrease, this benefit disappearing after the withdrawal of the treatment and the subsequent increase in the blood pressure level. However, the benefit of the treatment was not proved in the moderate hypertension blood pressure ranges; moreover, the incidence of coronary heart disease did not seem to be much reduced [8].

Later trials involved large sample sizes in order to clarify the benefits of treatment (Table 19.8). They have confirmed the spectacular reduction

of stroke and have shown an improvement of non-coronary cardiac pathology following antihypertensive treatment [10,13,23,27,43]. These reductions are observed until the moderate hypertension blood pressure ranges, in both sexes and until a rather advanced age [1]. These effects seem largely independent of the type of treatment. The results are discordant and not very conclusive in the case of coronary heart disease. A clear answer on the efficacy of antihypertensive treatments for coronary disease probably requires a trial with a very large sample size and it is to be feared that such an answer might never be given.

Table 19.8. Preventive trials: moderate hypertension

Trials	Sample size	Initial blood pressure (mmHg)	Mean relative reduction (%)	Relative effect of treatment (%)[a]			
				Stroke	Coronary heart disease		Total mortality
					Hard	Mortality	
USPHS	389	90–114	− 10.0	− 100	+ 17	0	—
Oslo	785	150–179 (<110)	− 10.0	− 64	+ 29	+ 300	+ 11
Australia	3427	95–109	− 5.6	− 45*	0	− 55	− 29*
HDFP	11940	$\geqq 90$	− 6.0	− 35**	− 16**	− 21	− 17**
	1063	$\geqq 115$	− 5.6	− 45	− 32	− 14	− 7
	3052	105–114	− 7.0	− 35	− 19	+ 13	− 13
	7825	90–104	− 5.3	− 32	− 12	− 15	− 20**
MRC	17354	90–109	—	− 45**	—	+ 9	− 2
EWPHBP[b]	840	90–119	—	− 32	—	—	− 9

* P < 0.05; ** P < 0.01; *** P < 0.001
[a] In the intervention group compared with the control group.
[b] Significant reduction of total cardiovascular mortality (− 27%, P = 0.037) and cardiac mortality (− 38%, P = 0.036).

Table 19.9. Preventive trials: hypocholesterolaemic treatment

Trials	Sample size	Initial serum cholesterol (mg/dl)	Mean relative reduction (%)	Relative effect of treatment (%)[a]			Total mortality
				Coronary heart disease			
				Non-fatal infarction	Total morbidity	Morbidity	
Clofibrate	5331	248	− 9	− 25*	− 20*	+ 8	+ 22*
Cholestryamine	3806	292	− 8.5	− 19	− 19*	− 24	− 7

* $P < 0.05$; **$P < 0.01$; *** $P < 0.001$
[a]In the intervention group compared with the control group.

Nutritional Prevention

The effects of trying to reduce cholesterolaemia by dietary change have been evaluated in many trials whose methodological flaws are now quite clear, in particular their insufficient sample sizes. An American study, the National Diet Heart Study, has increased the precision of the methodology of trials of that type; the necessary sample size (more than 50 000 subjects) has led to an abandonment of this kind of research and here one may express the fear that a scientific evaluation of the efficacy of nutritional prevention may not be provided [30]. However, these trials and studies have introduced the feasibility of reducing, through a diet, the serum cholesterol of groups of individuals and of populations. Moreover, the Oslo trial, listed with the multifactorial trials, included a dietary intervention among hypercholesterolaemic subjects along with an anti-smoking campaign. The coronary morbidity and mortality are significantly reduced in the intervention group whose serum cholesterol is on average 13% lower than that of controls. The benefit is proportional to the decrease of serum cholesterol and seems to depend for the most part on that decrease, whereas the effect of a cigarette consumption decrease seems to be much more limited [11].

Hypocholesterolaemic Treatments

Two other trials have suggested a significant reduction of coronary morbidity following a hypolipidaemic treatment with clofibrate or with cholestyramine (Table 19.9) [21,34]. In both cases, the observed reduction seems to be dependent on the extent of the cholesterol level decrease; the trial using cholestyramine showed that the decrease of the LDL cholesterol was the main variable linked to the reduction of inci-

dence whereas the variations of the level of HDL cholesterol or of the triglycerides only had a modest effect. However, the trial using clofibrate was accompanied by a significant increase ($P < 0.05$) of all-cause mortality which disappeared after withdrawal from the treatment. This suggests a possible side-effect of the long-term use of the drug which had not been observed with cholestyramine. One can note that these two trials and the Oslo trial are strong arguments in favour of an aetiological role for serum cholesterol and for the diet hypothesis.

Multifactorial Prevention

Multifactorial preventive trials are summarized in Table 19.10. The Oslo trial has been quoted previously. The MRFIT trial, a considerable and rigorous enterprise, does not show any noticeable decrease of the coronary mortality in spite of a 22% multifactorial risk reduction in the intervention group [29]. This negative result could be due to (a) an absence of effect of the prevention, (b) sample fluctuations, (c) the absence of noticeable differences between the cholesterolaemia in both groups during the whole trial or (d) the excess mortality observed in some subgroups of hypertensives treated with large quantities of diuretics. It is notable that, in this trial, hypercholesterolaemic non-hypertensive subjects, similar to those of the Oslo trial, have a 49% reduction in their coronary mortality. The absence of a relative reduction of serum cholesterol and the excess mortality of some hypertensives in the treated group are probably responsible for the negative results of that trial.

In the World Health Organization (WHO) cooperative trial where the multifactorial risk decreased by 11%, neither the mortality nor the morbidity reductions were significant at the 5%

Table 19.10. Preventive trials: multifactorial prevention

Trials	Sample size	Initial values			Relative reduction of multifactorial risk (%)	Relative effect of treatment (%)[a]			
		Cholesterol (mg/dl)	Blood pressure (mmHg)	Cigarettes		Coronary heart disease			Total mortality
						Non-fatal infarction	Total morbidity	Mortality	
Oslo	1232	329	<150	13	—	−39	−45*	−55*	−31
MRFIT	12 686	240	91	21.6	−22	—	—	−7	+2
WHO	49 781	216	138	11.3	−11.1	—	−3.9	−7.4	−2.7
Projects English	18 210	216	140	8.6	−3.9	—	+5	+8	+14
Belgian	19 409	229	141	12.5	−15.8	−26*	−24*	−21	−17*
Italian	6027	220	133	12.0	−28.2	—	−14	−30	−6
Polish	6135	188	126	13.5	−4.2	—	−20	−20	−22
North Karelia	the country	274	149	10	−17.7	—	—	−4.8	—

* $P < 0.05$; ** $P < 0.01$; *** $P < 0.001$

[a] In the intervention group compared with the control group.

level, but the power of the trial was reduced because of a final sample size smaller than expected [48]. This international cooperative trial, in which the random allocation into two groups concerned collectivities (factories) and not individuals, covered several national projects, among which two (the British and the Belgian projects) had rather big sample sizes [20,41]. The British project, with no modification either in the multifactorial risk or in the complications, contrasts with the Belgian project in which the multifactorial risk is reduced by 15.8% and each of the considered complications is reduced by about 20% ($P<0.05$). The Italian project, with a smaller sample size, has similar results.

Besides these planned studies, cardiovascular community control programmes have been set up in different countries and sometimes evaluated; the most elaborate one is the North Karelia Project in Finland which led to a higher decrease of coronary mortality than was observed in a neighbouring district and in the rest of country where no community control programme existed [32,33].

The information obtained so far from the primary preventive trials can be summed up in a few statements:

1. Anti-hypertensive treatment is effective in both sexes until a rather advanced age and in a broad range of blood pressure values including moderate hypertension. This treatment is effective in the prevention of myocardial damage and of stroke; its effectiveness in the prevention of coronary disease is not proved. Adverse side-effects of some treatments (diuretics in rather large doses) are suggested among hypertensives with ECG abnormalities.

2. The dietary and/or pharmacological decrease of the serum cholesterol level, notably of its fraction carried by the LDL, is followed by a reduction of coronary heart disease incidence which is proportional to the amplitude of the decrease. Such a relationship is observed in a very large range of cholesterol levels, even in the range of moderate hypercholesterolaemias. Side-effects of some long-term drug treatments are observed.

3. The decrease of cigarette consumption has a complementary effect independent of that due to the decrease of serum cholesterol.

4. The multifactorial prevention is effective towards cardiovascular mortality and coronary heart disease incidence; this last effectiveness might only appear after a certain reduction of the cholesterol level.

5. The feasibility of changing whole populations' life styles in a suitable direction for prevention of atherosclerosis has been demonstrated. It is very likely that the recent decrease of coronary mortality observed in some countries is secondary to a moderate reduction of the mean level of risk factors.

Prevention: A Public Health Strategy

The coherence of descriptive or analytic epidemiological data with those obtained in preventive trials or experiments is very impressive. This coherence as a whole is a very strong argument in favour of the diet hypothesis being the main aetiological factor in coronary atherosclerosis and in favour of hypertension being the main aetiological factor for the adverse effects of cerebral atherosclerosis. It provides a basis for cardiovascular disease prevention which is now becoming a public health strategy. The overall goal is to develop a primary prevention strategy which could control one of the calamities of modern societies [36].

Prevention remains the main goal of coming years. A preventive strategy includes screening and the treatment of high-risk subjects but also, and maybe above all, a community-wide approach which will help the population to respect the rules of prevention. In particular, dietary changes aimed at reducing the mean serum cholesterol below 2 g/l do not seem to be possible before the launching of modified foods with good nutritive qualities, but also with characteristics meeting the requirements of nutritional prevention. This control of lipidaemia, the correct treatment of hypertension, the control of weight, a moderate alcohol consumption, a decrease in tobacco consumption and an increase of physical activity might have a major effect on premature cardiovascular mortality and morbidity.

Scepticism in the face of such evidence may not now be justified. Cardiovascular disease prevention might be one of the major changes and one of the main medical achievements in the next century.

References

1. Amery A, Birkenhager W, Brixko P (1985) Mortality and morbidity results from the European working party

on high blood pressure in the elderly trial. Lancet I:1349–1370

2. The Belgian-French Pooling Project (1984) Assessment of type A behaviour by the Bortner scale and ischaemic heart disease. Eur Heart J 5:440–446

3. Bloch C, Richard JL (1985) Les facteurs de risque dans l'Etude Prospective Parisienne. I. Comparaison avec les études étrangères. Rev Epidem Santé Publ 33:108–120

4. Cambien F, Richard JL, Ducimetiere P (1980) Etude des antécédents familiaux de cardiopathies ischémiques et d'hypertension artérielle en liaison avec la prévalence des facteurs de risque et l'incidence des cardiopathies ischémique. L'Etude Prospective Parisienne. Rev Epidem Santé Publ 28:21–37

5. Castelli WP (1984) Epidemiology of coronary heart disease. The Framingham study. Am J Med 76:4–12

6. Ducimetiere P, Eschwege E, Papoz L, Richard JL, Claude JR, Rosselin G (1980) Relationship of plasma insulin levels to the incidence of myocardial infarction and coronary heart disease mortality in a middle-aged population. Diabetologia 19:205–210

7. Ducimetiere P, Richard JL, Claude JR, Warnet JM (1982) Les cardiopathies ischémiques – incidence et facteurs de risque. L'Etude Prospective Parisienne. INSERM, Paris

8. Freis ED (1967) Effects of treatment on morbidity in hypertension. JAMA 202:1028–1034. 1970, 213:1143–1150

9. Gordon T, Castelli WP, Hjortland MC, Kannel WB, Dawber TR (1977) High density lipoprotein as a protective factor against coronary heart disease. The Framingham study. Am J Med 62:707–714

10. Helgeland A (1980) Treatment of mild hypertension: a five-year controlled drug trial. The Oslo Study. Am J Med 69:725–732

11. Hjermann I, Byre KV, Holme I, Leren P (1981) Effect of diet and smoking intervention on the incidence of coronary heart disease. Lancet II:1303–1310

12. Hopkins PN, Williams RR (1981) A survey of suggested coronary risk factors. Atherosclerosis 40:1–52

13. Hypertension Detection and Follow-up Program Cooperative Group (1979) Five-year findings of the hypertension detection and follow-up program. I. Reduction in mortality of persons with high blood pressure, including mild hypertension. II. Mortality by race, sex and age. JAMA 242:2562–2577

14. Kagan A, Harris BR, Winkelstein W et al. (1974) Epidemiologic studies of coronary heart disease and stroke in Japanese men living in Japan, Hawaii and California: demographic, physical, dietary and biochemical characteristics. J Chronic Dis 27:345–364

15. Kannel WB, McGee DL (1985) Update on some epidemiologic features of intermittent claudication: The Framingham study. J Am Geriatr Soc 33:13–18

16. Kannel WB, Castelli WP, Neeks SL (1985) Fibrinogen and cardiovascular disease J Am Coll Cardiol 5:517 (abstract)

17. Longstreth WT Jr, Koepsell TD, Yerby MS, Van Belle G (1985) Risk factors for subarachnoid hemorrhage. Stroke 16:377–385

18. Keys A (1970) Coronary heart disease in seven countries. Circulation 41:Suppl 1

19. Keys A (1980) Seven countries – a multivariate analysis of death and coronary heart disease. Harvard University Press, Cambridge, Mass

20. Kornitzer M, Debacker G, Dramaix M et al. (1983) Belgian heart disease prevention project: incidence and mortality results. Lancet I:1066–1070

21. Lipid Research Clinics Program (1984) The Lipid Research Clinics coronary primary prevention trial results. I. Reduction in incidence of coronary heart disease. II. The relationship of reduction in incidence of coronary heart disease to cholesterol. JAMA 251:351–374

22. Liu K, Stamler J, Trevisan M, Moss D (1982) Dietary lipids, sugar, fiber, and mortality from coronary heart disease. Bivariate analysis of international data. Arteriosclerosis 2:221–227

23. Management Committee (1980) The Australian therapeutic trial in mild hypertension. Lancet I:1261–1267

24. McGill HC Jr (1968) The geographic pathology of atherosclerosis. Williams and Wilkins, Baltimore

25. Meade TW, North WRS, Chakrabarti R et al. (1980) Haemostatic function and cardiovascular death: early results of a prospective study. Lancet I:1050–1054

26. Medalie JH, Kahn HA, Neufeld HN, Riss E, Goldbourt U (1973) Five-year myocardial infarction incidence. II. Association of single variables to age and birth-place. J Chronic Dis 26:329–349

27. Medical Research Council Working Party (1985) MRC trial of treatment of mild hypertension: principal results. Br Med J 291:97–104

28. Morris JN, Everitt MG, Pollard R, Chave SPW, Semmence AM (1980) Vigorous exercise in leisure-time: protection against coronary heart disease. Lancet II:1207–1210

29. Multiple Risk Factor Intervention Trial Research Group (1982) Multiple risk factor intervention trial (MRFIT): risk factor changes and mortality results. JAMA 248:1465–1477

30. National Diet Heart Study Research Group (1968) Final report. Circulation 37:Suppl I

31. Pell S, Fayerweather WE (1985) Trends in the incidence of myocardial infarction and in associated mortality and morbidity in a large employed population, 1957–1983. N Engl J Med 312:1005–1011

32. Puska P, Nissinen A, Salonen JT, Tuomilehto J (1983) Ten years of the North Karelia project; results with community-based prevention of coronary heart disease. Scand J Soc 11:65–68

33. Puska P, Salonen JT, Nissinen A et al.(1983) Change in risk factors for coronary heart disease during 10 years of a community intervention program (North Karelia project). Br Med J 287:1840–1844

34. Report from the Committee of Principal Investigators (1978) A cooperative trial in the primary prevention of ischemic heart disease using clofibrate. Br Heart J 40:1069–1118

35. Richard JL (1980) Lipides alimentaires, cholestérolémie et cardiopathies ischémiques. Rev Epidem Santé Publ 28:461–484

36. Richard JL (1983) Les essais de prévention primaire multifactorielle des cardiopathies ischémiques. Rev Epidem Santé Publ 33:121–133

37. Richard JL (1983) Nutrition et cardiopathies ischémiques. Cah Nutr Diet 18:17–28

38. Richard JL (1983) Peculiarities of coronary heart diseases in the French population. In: Schettler G, Gotto AM (eds) Atherosclerosis VI. Springer-Verlag, Berlin Heidelberg New York, pp 821–825

39. Richard JL (1984) Athérosclérose et alcool. Med Cardiovasc 49:13–21

40. Richard JL, Ducimetiere P, Warnet JM (1983) Albumin α I-globulins as a negative risk factor for hard coronary disease. Eur Heart J 4:7

41. Rose G, Tunstall-Pedoe HD, Heller RF (1983) UK heart

disease prevention project – incidence and mortality results. Lancet I:1062–1065

42. Shekelle RB, Shryock AM, Paul O, Lepper M, Stamler J, Raynor WJ (1981) Diet, serum cholesterol, and death from coronary heart disease. The Western Electric study. N Engl J Med 304:65–70

43. Smith WF (1977) Treatment of mild hypertension – results of a ten-year intervention trial by the US Public Health Service Hospitals Cooperative Study Group. Circ Res [Suppl 1] 40:98–105

44. Uemura K, Piza Z (1985) Recent trends in cardiovascular disease mortality in 27 industrialized countries. World health statistics quarterly. World Health Organization, Geneva, 38(2):142–162

45. Wilhemsen L, Svardsudd K, Korsan-Bengtsen K et al. (1984) Fibrinogen as a risk factor for stroke and myocardial infarction. N Engl J Med 311:501–505

46. Wolf PA (1985) Risk factors for stroke. Stroke 16:359–360

47. World Health Organization (1976) Myocardial infarction community registers. Public Health in Europe, No.5, Copenhagen

48. World Health Organization European Collaborative Group (1983) Multifactorial trial in the prevention of coronary heart disease. 3. Incidence and mortality results Eur Heart J 4:141–147

49. Zalokar JB, Richard JL, Claude JR (1981) Leukocyte count, smoking, and myocardial infarction. N Engl J Med 304:465–468

Chapter 20

Risk Factors for Atherosclerosis: A Clinician's Pragmatic Appraisal

L. Capron

Ignorance and Concern

Cardiovascular diseases are the major cause of adult mortality in industrial countries. Yet our understanding of atherosclerosis remains poor: we still do not know exactly how atherosclerosis develops and progresses, and what causes it.

For researchers these uncertainties are stimulating. They aim at understanding mechanisms, and are mostly concerned with clarifying the pathogenesis of the condition. In the present volume many pages are devoted to the results of such investigations, and to the hypotheses that have stemmed from them.

For clinicians these gaps in our knowledge are worrying rather than stimulating. They have to deal daily with the appalling consequences of a disease which they are unable to explain and control. Practising physicians are rather more concerned with the progression and the prevention of atherosclerosis. Nevertheless, it is an uneasy feeling not to understand that disease which one has to diagnose and treat so often. That is one reason why reassuring explanatory theories of atherosclerosis have been built up and adopted by a majority of disquieted clinicians. However, anxiety should not be alleviated at the expense of lucidity: uncertainty should lead neither to naiveté nor to dogmatism; however attractive they may look, the proffered answers should be regarded as hypotheses which must be rigorously tested before they are accepted or rejected.

Concerning the progression and complications of atherosclerosis, almost everybody agrees that thrombosis is the usual link between an atherosclerotic plaque and the occurrence of clinical manifestations. But several questions have not yet been solved: Why and how does thrombosis develop? Which factors modulate the ensuing ischaemia? The modest efficiency of the available anti-thrombotic treatments (anti-platelet agents, anti-coagulants, thrombolytic drugs) is fair proof of the weaknesses of our present knowledge.

The aetiology of atherosclerosis is thought by many to depend on multiple risk factors. The present chapter deals with these and tries to examine them in the spirit that I have just defined. Mine is not the viewpoint of an epidemiologist or of a pure experimenter, but rather the viewpoint of a consumer, i.e. a physician who has to care for atherosclerotic patients, and who, because of that, has become interested in the advances of our knowledge about the epidemiology of atherosclerosis, and arterial wall biology.

Genesis of Risk Factors

Soon after the Second World War, in the United States, it appeared that the causes of coronary heart disease (CHD) would be identified through long-term population studies, rather than through experimental and transverse clinical studies: the first large-scale epidemiological projects concerned with atherosclerosis were then organised. In the words of Ancel Keys [59], their aim was to "find individual characteristics in apparently healthy middle-aged men related to future tendency to develop CHD, an aim based on the conviction that the physico-chemical characteristics of the individual should have predictive value. It was believed that the discovery of predictive variables – what we now call *risk factors* – should point out the way to preventive efforts". In the early 1960s, the first European epidemiological studies of coronary heart disease (Scandinavia, Yugoslavia) and the Seven Countries Study [59] (United States, Yugoslavia, Japan, Finland, Italy, Netherlands, Greece) were undertaken. The first French study began in the late 1960s [32]. This short historical review is a reminder that: (a) risk factors were initially a working hypothesis and (b) their study has mainly been concerned with coronary heart disease, the most prevalent manifestation of atherosclerosis.

Definition and Nature of Risk Factors

By generalizing and restating the words of Ancel Keys, one can define a risk factor for a given disease as a constitutional state (e.g. sex, age, heredity), a pathological state (e.g. high blood pressure, dyslipidaemia) or a life habit (e.g. cigarette smoking, diet, poor physical activity) which is associated with an increased incidence of that disease.

For atherosclerosis, the number of possible risk factors is quite large: in 1981 Hopkins and Williams [49] analysed more than 800 publications and counted 246 potential risk factors. Such an overabundance of information about the determinants of atherosclerosis came from various sources: experimental work on the mechanisms of atherosclerosis; morphological (pathological, angiographic) studies of arteries; clinical cross-sectional studies of the case–control type; confrontations of statistics from various countries and times; and, of course, prospective epidemiological studies which have followed up large numbers of people for long periods of time. However, by taking into account only this last source of information (which undoubtedly is the most reliable), and only those risk factors that appear to act independently from other risk factors, one can greatly reduce the number of likely risk factors for atherosclerosis. Table 20.1 lists the main candidates for a role as "major" risk factors for atherosclerosis.

Table 20.1. Major risk factors for atherosclerosis

Constitutional states
Age
Male sex
Atherosclerotic heredity
Personality ("type A")
Pathological states
High blood pressure
Plasma lipids and lipoprotein abnormalities
Diabetes mellitus
Obesity
Life habits
Cigarette smoking
Diet (rich in cholesterol and in saturated fats)
Sedentary life

Meaning of Risk Factors

At this point a pragmatic description of the risk factors for atherosclerosis we must ask ourselves, what is their exact significance? This crucial step of reasoning is often omitted by those who think that as long as one has identified the risk factors, one has discovered the causes of atherosclerosis. A good part of our debates about atherosclerosis and of our failure to efficiently prevent it may have come from a misunderstanding of the exact meaning of risk factors. Yet, to avoid that error, one has simply to remember that risk factors are identified by establishing statistical correlations between the presence of the considered risk factor and the occurrence of a clinical manifestation of atherosclerosis.

Correlation and Causation

Correlation is not causation. This simple principle of statistics cannot be overlooked: a statistical correlation between a factor A and a disease B does not necessarily mean that a cause effect relationship exists between A and B. Epidemiological studies, however excellent they may be, yield data that are only correlations. It is the interpretation of these correlations (often supported by information from other fields) that may eventually conclude that a claim of causation is justified; that a risk factor has been promoted to the rank of causative factor [9]. The passage from correlation to causation is so thorny and so difficult that strict rules have been set up to constrain it. Even so, impassioned arguments still take place [10,67].

Table 20.2 lists the criteria believed to identify a risk factor for atherosclerosis firmly [100], and

Table 20.2. Criteria believed to identify a risk factor for atherosclerosis [100]

Criteria to establish that a phenomenon is an actual risk factor for atherosclerosis [100]

1. The association between the factor and the risk of an atherosclerotic event (e.g. coronary heart disease) must be statistically strong
2. The association must be graded in nature, i.e. such that any increase in the factor must be associated with a proportional increase in risk
3. The association must have the necessary temporal relationship, i.e. the risk factor must precede the disease
4. The association must be generally consistent in most of the studies which have assessed it among various populations
5. The association must be independent from any other risk factor. For instance, the association between salt consumption and arterial diseases is not independent of hypertension; salt is an indirect risk factor, whereas hypertension is a direct (or independent) risk factor
6. The association must have an accurate predictive value: application of data on risk factor from one population yields good prediction of risk in another population and the individuals in it
7. The association must be coherent: the epidemiological data are consistent with findings from other research methods, and reasonable pathogenic mechanisms are known

Criteria to establish that a risk factor is a cause of atherosclerosis

1. Controlling the risk factor must lower the incidence and prevalence of the disease
2. The risk factor is a necessary and sufficient condition for the development of the disease

---

to establish it as a cause or the cause of the disease. By sticking to these rules, the list of established risk factors for atherosclerosis becomes even shorter than that in Table 20.1. There only remain: age, male sex, high blood pressure, modifications of plasma cholesterol (increased concentrations of total cholesterol and of cholesterol carried by low-density lipoproteins and reduced concentration of cholesterol carried by high-density lipoproteins), diabetes mellitus, cigarette smoking and a diet rich in cholesterol and saturated fat. When dealing with causes, hypercholesterolaemia is the only factor which may lay claim to be a (but not the) *cause* of atherosclerosis.

Arterial Diseases and Atherosclerosis

All major risk factors in Table 20.1, as identified by epidemiology, are not, stricto sensu, risk factors for atherosclerosis. They actually are risk factors for that arterial disease whose diagnosis has led to their identification. It is an extra-

polation to talk about "risk factors for atherosclerosis" when in fact we deal with factors about which we can only say that they increase the risk of coronary heart disease, of ischaemic strokes or of peripheral (lower limb) artery disease; the majority of epidemiological inquiries have studied coronary heart disease (myocardial infarction, angina pectoris, sudden cardiac death).

In order to get rid of this ambiguity many pathological studies have analysed the association between the past existence of risk factors and the presence of atherosclerotic lesions at autopsy. They have been either retrospective (and thus liable to bias) or prospective, depending on whether the information about the risk factors had been collected after or before death. From a critical review of all such studies Solberg and Strong [99] have concluded that a firm relationship has been established between atherosclerotic lesions and the following risk factors: increased plasma cholesterol, decreased HDL cholesterol and increased systolic and diastolic blood pressures. Solberg and Strong have estimated that all these combined risk factors explain about 25% of the individual variations in atherosclerotic lesions. This rate may seem pretty low and lead to the belief that: (a) a large part (up to 75%) of the individual variation in atherosclerotic lesions remains unexplained and may be modulated by influences other than the risk factors known to date; and (b) risk factors may enhance atherosclerotic arterial disease other than by acting directly upon the development of atherosclerosis.

Triglycerides and Atherosclerosis

Plasma triglycerides yield an example which illustrates the remarks I have so far made on the complexity of the links between risk factors and atherosclerosis. In 1959 Albrink and Man [3] reported for the first time on an association between plasma triglycerides and coronary heart disease. Since then about 30 articles have confirmed such a correlation. Hulley et al. [50] have reviewed those which could be analysed and found that in all but two the association was indirect. In particular, high levels of triglycerides are associated with a lowered concentration of HDL cholesterol. Of the two studies against which this criticism could not be levelled, the Stockholm Prospective Study [20] was the most important. Its authors, although they have adhered to their initial conclusions [21], have agreed that they would never be able to incorporate HDL cholesterol into their analysis: "the fact is that the fuse to the deep freeze where we had all 6464 plasma samples broke after two years and we did not find out until 1966" [18]. In spite of that, Carlson et al. have hypothesized that triglycerides might interfere with coronary heart disease in a selective manner: they may be an independent risk factor for myocardial infarction, but not for angina pectoris, [1,19] which would lead us to think that they might enhance less the progression of atherosclerosis itself than thrombosis. Such an influence has also been ascribed to triglycerides in venous thrombosis [40]. Nevertheless, with the present state of our knowledge, we must consider that triglycerides are neither a cause, nor an independent risk factor for atherosclerosis: lowering their plasma concentration through diet or drugs is of no demonstrated benefit in the control or prevention of atherosclerosis [79]. To conclude otherwise, one has to show that triglycerides act independently from HDL cholesterol and other risk factors (smoking, obesity, high blood pressure). A carefully performed study that establishes an association between coronary artery lesions and triglycerides cannot be held as convincing if its statistical analysis has not taken into account HDL cholesterol [11].

Possible Mechanisms of Action for Risk Factors

Our knowledge on atherosclerosis has come from two main sources. On one side, pathological and experimental studies have inspired various theories of atherogenesis [12]. On the other, epidemiological studies have discovered several risk factors for arterial diseases. I shall examine the links between those two sources, that is, the various ways in which a risk factor can act upon the chain of events in atherosclerosis in order to promote the occurrence of clinically recognizable manifestations (angina pectoris, myocardial infarction, ischaemic stroke, intermittent claudication etc.).

Development of Atherosclerotic Plaques

We do not completely understand the sequence of events that transforms part of a healthy arterial wall into a mature atherosclerotic plaque which may then become complicated. The list of potential mechanisms is therefore quite long. To shorten it, I have confined myself to the currently prevailing theories of pathogenesis. They cite the endothelium (endothelial injury), blood platelets, smooth-muscle cells and the monocyte–macrophages as important [12]. Risk factors may affect endothelial injury; platelet adhesion, activation and secretion; penetration of large molecules (e.g. LDL) into the arterial intima; migration and proliferation of smooth-muscle cells; diapedesis of monocytes into the intima; disturbances of lipoprotein metabolism in smooth-muscle cells or monocyte–macrophages; overproduction of fibrous proteins by smooth-muscle cells; and degeneration of foam cells. All these possibilities are an inexhaustible source of inspiration for experimental work. One must, however, remain cautious and not overinterpret those studies that suggest a relationship between a risk factor and a cellular event supposedly playing a role in atherogenesis.

Regarding this point, insulin is a useful example to examine. The first direct suspicions for an association between insulin and atherosclerosis emerged from works on experimental "atherosclerosis" in non-diabetic chickens [101,103] and diabetic dogs [27]. Later on three epidemiological studies [30,89,112] reported on a correlation, apparently independent of other risk factors, between plasma insulin concentration (as measured in the fasted state and after an oral glucose load) and the occurrence of coronary heart disease. They brought support to the conclusions of many other clinical studies which had used less reliable methods [105]. It is on this basis that the idea that insulin is probably a risk factor for atherosclerosis depends, together with the idea that hyperinsulinaemia may be the missing link between diabetes and atherosclerosis [15,105,106]. Following this hypothesis many studies have documented those effects of insulin upon the arterial wall that may enhance atherogenesis. The major findings have been that:

1. Insulin increases lipid synthesis by the arterial wall in diabetic [16] and non-diabetic [17,102,104] animals;

2. Insulin stimulates the migration [77] and, through an interaction with well-characterized membrane receptors [60,86], the proliferation [85,87,107] of cultured arterial smooth-muscle cells;

3. Diabetes reduces the activity of lysosomal enzymes in aortic smooth-muscle cells, and that insulin treatment reverses this inhibition [114].
4. In vitro production of prostacyclin, a powerful vasodilating and platelet anti-aggregating eicosanoid, by the arterial wall is reduced by insulin [63,64];
5. In diabetic dogs, insulin treatment promotes the accumulation of glycosaminoglycans in the arterial wall [98].

These experimental observations provide a reasonable basis for the epidemiological association between insulin and atherosclerosis provided that two conditions are fulfilled. Firstly, one has to admit that lipogenesis, migration, proliferation and enhanced lysosomal activity of smooth-muscle cells, reduced synthesis of prostacyclin by the arterial wall and accumulation of glycosaminoglycans in the arteries are all critical mechanisms in atherosclerosis. Secondly, one has to accept several extrapolations that are still pretty daring: from arterial lipogenesis to atherogenesis; from the reactions of smooth-muscle cells isolated in a plastic culture dish to those of the same cells placed in their natural environment as part of a living arterial wall submitted to complex haemodynamic forces; and from the characteristics of an experimental animal artery to those of a human artery.

So far, neither of these two conditions has been fully satisfied. Furthermore, there is no definite proof that hyperinsulinaemia is an entirely independent risk factor: some studies have shown its epidemiological association with high blood pressure [74,96] and with lipid abnormalities regarded as atherogenic [83]. Like many other risk factors, hyperinsulinaemia appears to be suspected, but not definitely guilty, of playing a direct, causal role in atherosclerosis. It would thus be premature to add the control of hyperinsulinaemia as another objective to the treatment of diabetes. However, it is necessary further to explore the links between diabetes and atherosclerosis with due regard to the fact that they may proceed through intermediates other than hyperinsulinaemia [38].

Complications of Atherosclerotic Plaques

The major complication of atherosclerosis is thrombosis: atherosclerotic plaques grow mainly through the incorporation of thrombi. Athero-

sclerotic thrombosis provokes arterial occlusions that are responsible for a great majority of the clinical manifestations of atherosclerosis [76]. Oddly enough, few investigations have been devoted to finding out why a thrombus comes to develop on an atherosclerotic plaque. Rupture of a plaque leading to its ulceration [28,34,44] and haemorrhages inside the plaque [4,51,71,97,111] probably play a role in thrombosis. But, due to the lack of relevant experimental models, we still understand only poorly the mechanisms and consequences of these crucial events. Most of our precious knowledge has come from a small number of enlightening pathological observations of human diseased arteries.

A "risk factor" for atherosclerosis may, without necessarily interfering with atherogenesis itself, act to initiate the formation and growth of thrombi, but also the ulceration and haemorrhage in atherosclerotic plaques. In addition, as is the case with venous thrombi [80], the growth and evolution of arterial thrombi is likely to depend upon the fibrinolytic and thrombolytic capacities of the plasma and vascular wall. Some risk factors might interfere with the subtle balance between activation and inhibition of plasminogen [36]. Clues for such an action exist for the harmful effect of diabetes [7,66] and of lipoprotein abnormalities [35,71] and for the beneficial effects of physical training [113] and of moderate alcohol consumption [65]. Paradoxically, cigarette smoke would rather appear to stimulate fibrinolysis in vitro [37]. However, a similar interaction may not take place in vivo, in the arteries of smokers.

Consequences of Arterial Obstructions due to Atherothrombosis

Not all arterial occlusions, even when they involve a main trunk, always cause ischaemia, i.e. clinical symptoms. Asymptomatic occlusions of the superficial femoral [25], left coronary [29], or internal carotid [46] arteries can occur, which underlines the crucial role of collateral circulations [45]. Some risk factors may act by hampering the development of collateral channels or by hindering their efficiency. No such hypothetical mechanism has so far been demonstrated for any of the established risk factors. We indeed know very little about the build-up and regulation of vicarious blood supplies [62]. Progress in this field will certainly be useful for the better understanding of arterial diseases as well as improving our therapeutic capabilities.

Other Unsolved Issues

Do Risk Factors Exert the Same Influence in all Populations?

The incidence and prevalence of coronary heart disease vary widely among populations. In the Seven Countries Study [59] (16 cohorts, adding up to 12 000 men initially free of coronary heart disease, followed-up for 10 years) the average coronary heart disease mortality rate was 246 per 10 000 with a wide range of variation, from 0 in Crete to 681 per 10 000 in East Finland. Such a variability in incidence was partially accounted for by the variability in three risk factors: arterial pressure, plasma cholesterol and dietary habits. The other risk factors assessed in this study (smoking, obesity, physical activity) did not contribute significantly. An analysis of coronary heart disease epidemiology in France [31,91] provided even stronger evidence that variations in risk factors cannot explain the variations in coronary heart disease incidence fully. In men initially aged 35–45 and followed up for 8.6 years in the Paris Prospective Study [32] coronary heart disease incidence was 29.7 per 1000. In the United States, according to the Pooling Project [109], the corresponding incidence was 93.9 per 1000. However, the average level of the major risk factors (cholesterol, blood pressure, smoking) was not definitely lower in France than in North America, and could not account for the impressive difference in coronary heart disease incidence.

Risk factor discrepancies explain only a small part of the discrepancy in coronary heart disease incidence between the Paris Prospective Study and either the American cohort of the Seven Countries Study or the Pooling Project. With an equal score of risk in 50–year-old men, the American coronary heart disease incidences were respectively 1.6 and 2.2 times higher than the Parisian incidence. Influences other than the risk factors assessed in these comparisons probably act upon the development of coronary heart disease. We can but speculate about their nature: genetics [55,78,95,108] viral [6,9,33] or as yet undefined.

Are Risk Factors Similar for all Arterial Localizations of Atherosclerosis?

It is difficult to answer this question because so few data are available. In fact, only one large-scale prospective epidemiological investigation, the Framingham Study [42], has simultaneously and reliably compared the risk factors for the three main localizations of atherosclerosis: coronary arteries through the occurrence of coronary heart disease; abdominal aorta and lower limb arteries through intermittent claudication [57]; and cerebral arteries through atherothrombotic brain infarction [52]. This study allows us to draw two conclusions. Firstly, the total plasma cholesterol influences all three localizations in a similar way [42], but the effect of lipoproteins (LDL and HDL) varies between localizations [43]. Secondly, the four major risk factors (cholesterol, high blood pressure, smoking, diabetes) have about the same impact upon coronary heart disease, whereas smoking and diabetes are the most important risk factors for intermittent claudication and hypertension clearly heads the bill for the risk of cerebrovascular disease [42]. In addition, age does not exert a uniform influence: clinical manifestations of cerebral atherosclerosis tend to occur 10 to 20 years later than those of coronary atherosclerosis [13,70]. Several hypotheses should be considered to account for these discrepancies [14].

Firstly, the inequality of risk factor influence might be apparent rather than real, resulting from a statistical artefact. Two major biases may operate: the bias of small numbers and the bias of misdiagnosis.

In spite of the size of the Framingham Study, the number of non-coronary arterial events has been comparatively small: by the 16th year of that study [42], which follows 5209 adults, 533 coronary events had been recorded compared with only 122 cases of intermittent claudication and 81 atherothrombotic brain infarctions. The smaller the number of events the greater the risk of error in estimating the influence of risk factors. This bias of small numbers would explain the fact that for the less frequent events (ischaemia of the lower limbs and brain) an artificially exaggerated influence is attributed to one or two risk factors, whereas for the commonest coronary heart disease all factors exert an almost uniform influence, the mathematical basis for this being more secure.

Secondly, the bias of diagnostic errors may account for the marked impact of hypertension upon cerebral risk. Some cerebral infarctions, for example cerebral embolism of cardiac origin and many subarachnoid and cerebral haemorrhages, are not directly or solely caused by atherosclerosis; other arterial and cardiac diseases are involved which, like atherosclerosis, are enhanced by hypertension. Finally, high blood pressure promotes all types of cerebrovascular accidents [13]. Epidemiological studies have not routinely used cerebral CAT (computerized axial tomography) scanning and thus have not always been able to assess the exact nature of the neurological events recorded. They may have grossly overestimated the influence of hypertension upon what they have wrongly considered as exclusive manifestations of atherosclerosis.

Statistical biases can thus partially account for the inequalities in the influence of risk factors. However, to rely solely upon this explanation would lead, for instance, to the refutation of such a well-established fact as the discrepant influence of age. It is, therefore, reasonable to think that at least part of the inconsistency is real. Four explanations can then be considered.

Firstly, the clinical manifestations of atherosclerosis are not caused by the same mechanisms in all arterial territories. They depend not only upon the site and extent of the atherosclerotic lesions themselves, but also upon the resistance of the organ involved to ischaemia and upon the local capacity for the development of an efficient collateral circulation. These physiological differences are still poorly understood. The susceptibility to a given risk factor, depending on its mode of action, could therefore vary from one arterial site to the other.

Secondly, non-atherosclerotic arterial diseases may, when associated with atherosclerosis, participate in the clinical sequelae from which epidemiology extracts its data. For instance, apart from atherosclerosis, arteriosclerosis, medial calcification and arteriolosclerosis contribute to lower limb arteriopathy [2,4]. Some factors influencing these particular arterial involvements (age, hypertension, diabetes) may be identical to some risk factors for atherosclerosis. In the various arterial territories, they may appear from epidemiological studies to have a variable strength, depending upon the respective role of non-atherosclerotic arterial diseases in the occurrence of the clinical manifestations.

Thirdly, the nature (the morphology and possibly the pathogenesis) of atherosclerotic lesions might vary from one localization to another. Atherosclerosis is a pleomorphic disease: the respective contributions of the atheroma and sclerosis are uneven. Femoral artery atherosclerosis is more sclerotic and less atheromatous than atherosclerosis of coronary arteries and aorta [82,92] Atherosclerosis may conceivably result from mechanisms and influences that differ slightly from one type of lesion or from one localization to another.

Fourthly, we wrongly estimate the relevance of risk factors to atherosclerosis. At best, variations in the known risk factors can account statistically for only 50% of the variability in cardiovascular risk. In a more conservative estimate, their influence in atherosclerosis might be much less [26]. If such were the case, it would not be surprising that the action of these weak risk factors may appear to vary between arterial territories, due to differing biases. In other words, by not concentrating our attention upon the leading actor (because we do not know who he is), we would organize those supporting roles which we scrutinize too closely into a futile hierarchy, which ultimately has less importance in the plot we are attempting to unravel.

Is Amending Risk Factors Controlling Atherosclerosis?

This is currently the most important and most controversial issue about atherosclerosis. The answer we give depends (a) from a theoretical viewpoint, on our opinion of the causal role of risk factors in atherosclerosis and (b), from a practical viewpoint, on our opinion of the possibility and strategies for an efficient prevention of atherosclerosis through an offensive action against risk factors. I shall only consider primary prevention, which aims at avoiding the occurrence of a disease in individuals who are apparently still free from that disease, and not secondary prevention, which aims at avoiding the progression of a chronic disease in patients who already are ill with that disease.

In this regard, hypercholesterolaemia has been the most thoroughly studied amenable risk factor. Several large-scale intervention trials have been carried out to determine whether lowering plasma cholesterol (with diet, drugs or both) significantly reduces the incidence of coronary heart disease. The Lipid Research Clinic's Coronary Prevention Trial [68,69] was the first study to demonstrate convincingly that treating hypercholesterolaemia (plasma cholesterol over 2.65 g/l) with diet and cholestyramine lessens coronary mortality and myocardial infarction incidence in 35–60-year-old men. According to this trial, the expected rate of decrease in coronary risk is twice the observed rate of decrease in plasma cholesterol: a 25% cholesterol reduction predicts a halving of the incidence of coronary heart disease. In December 1984, less than a year after these results had been published, a consensus conference of the National Institute of Health (NIH) [23] concluded that hypercholesterolaemia is an established cause of coronary heart disease, and that cholesterol lowering treatment provides an efficient means of preventing coronary heart disease. Practical recommendations for such prevention followed.

However, controversies have emerged [61,88] about the statistical validity of the Lipid Research Clinic's results for this trial and, even more clearly, about the extrapolations made from them: there is enough doubt to believe that a causal role for cholesterol in atherosclerosis has not been established unequivocally. In addition, the effect becomes quite modest when the results of the Lipid Research Clinic's trial are transposed into daily clinical practice. A medical practitioner would have to treat 200 patients for 7 years with cholestyramine resin (six 4-g packets per day) to avoid a coronary event in three of them [90]. The estimated cost (in drug expenses alone) for preventing one myocardial infarction or one coronary death would amount to US $775 600 [47]. Using these arguments and others, some experts [2,81] have criticized the optimistic, although stringent, conclusions and recommendations of the NIH consensus on prevention of coronary heart disease through cholesterol lowering.

For the other risk factors (high blood pressure, diabetes, smoking etc.) proofs of efficient prevention are lacking or unconvincing. Two large-scale controlled trials have tested the efficiency of multifactorial primary prevention of coronary heart disease. The larger (nearly 13 000 high-risk men, 35–57 years old, followed up for a mean of 7 years) has analysed the effects of a combined intervention against hypertension (drugs), cholesterol (diet) and smoking [75]. The results of this Multiple Risk Factor Intervention Trial have been negative: neither total mortality nor coronary heart disease incidence decreased in the treated group. Interpretations of this disappointing outcome have varied between: (a) concluding that the inefficiency of such "preventive" measures is demonstrated; (b) pointing out some methodological weaknesses of the trial; and (c) suggesting the presence of an unexpected untoward factor which would have acted against the treated group (for instance, antihypertensive treatment may have worsened coronary heart disease in some participants). The

other trial, run in Norway, has had positive results [48]. It involved 1232 hypercholesterolaemic men aged 40–49, followed up for 5 years with active intervention against cholesterol (diet) and tobacco smoking. A modest but significant effect was found on the incidence of coronary events. But here also, statistical criticism has cast doubt upon the validity of the result [88].

Taken together, these prevention trials are disappointing. Risk factor intervention is either inefficient or of such marginal efficiency that its reality and practical usefulness are uncertain. The weakness of these demonstrations may be in part due to the age of the trial subjects (more than 35) and to the relatively short duration of the studies (7 years or less). It could be that in order to be efficient, the prevention of atherosclerosis should be initiated at an earlier age, and be sustained for a longer time. If one sticks to what has been firmly demonstrated, it seems that, apart from diabetes and hypertension, the control of risk factors does not increase life expectancy. In addition, the benefits of antidiabetic and anti-hypertensive treatments do not seem to include a reduction in the manifestations of atherosclerosis.

Apart from formal therapeutic trials, epidemiological statistics have shown that in several industrialized countries the incidences of coronary heart disease and of cerebral arterial diseases have been declining since the mid 1960s [39,41,84,110]. An explanation for this encouraging trend may be the simultaneous reduction of the risk factor level in the population (improved efficiency and generalization of antihypertensive and anti-diabetic treatments, modifications in dietary habits, decreased tobacco consumption, increased physical activity). But this interpretation lacks a convincing demonstration, and the statistics from some countries, such as Sweden, do not support it. Other explanations are possible: causal risk factors for cardiovascular disease may have changed without being noticed because they still escape scrutiny.

All these conflicting studies do not provide a scientifically sound basis for conviction. The debate goes on with, on the one side, those who believe in the efficiency of prevention and promote it to the medical community and the lay public, and, on the other, those who remain sceptical enough not to encourage mass preventive intervention.

Hypercholesterolaemia: A Nearly Perfect Example of a Risk Factor

Hypercholesterolaemia is another example of a risk factor. Aside from age, a factor whose power is not doubted by anyone, but against which no action will ever be possible, hypercholesterolaemia is the most solidly anchored risk factor in the minds of biologists, physicians and even the lay public: it has been the most thoroughly studied determinant of atherosclerosis. There are several reasons for this. The measurement of plasma cholesterol has been easy for a long time; historically, in the early 20th century the first animal model of "atherosclerosis" was established by feeding rabbits a cholesterol-enriched diet [5]. Also, with familial hypercholesterolaemia (type IIa hyperlipoproteinaemia), nature has provided us with a spectacular demonstration of the role played by cholesterol in human arterial disease.

Increased total plasma cholesterol is associated with an increased risk of coronary heart disease but, as is now well known, the influence of cholesterol depends upon its plasma lipoprotein carrier: coronary heart disease risk is positively correlated with LDL cholesterol concentration, but negatively with HDL cholesterol concentration [22]. The characterization of cholesterol as a risk factor has evolved along with the progress in lipoprotein biochemistry: we are now able to estimate risk according to the nature of plasma apolipoproteins [8,73], apo A-I for HDLs and apo B for LDLs; and we shall soon be able to identify the abnormalities in apolipoprotein genes routinely [58,94]. In spite of such impressive advances in molecular biology, measurement of total plasma cholesterol is not yet considered to be a completely obsolete investigation [53]. Total plasma cholesterol remains an established risk factor [54] and we shall now see how it fulfils the requirements listed in Table 20.2.

Agreement

As far as coronary heart disease is concerned, cholesterol meets the criteria listed in the first part of Table 20.2, rather well in terms of the following points:

1. The association between cholesterol and coronary heart disease is strong. In the Framingham Study the predictive value of cholesterol ranked second among the four major pathological risk factors, immediately after hypertension, and ahead of smoking and diabetes [42].

2. The association is graded in nature: in the Framingham Study, the risk starts increasing from a threshold concentration (around 2.2 g/l before, and 2.6 g/l after 50 years of age).

3. All prospective epidemiological trials have shown that the association is temporally consistent: cholesterolaemia measured at the outset of the study in individuals free from coronary heart disease correlated with the incidence of coronary heart disease events that occurred later on.

4. No published epidemiological investigation run anywhere in the world has ever failed to find an association between cholesterol and coronary heart disease. In this regard the Seven Countries Study has been especially demonstrative [59]

5. Cholesterol is an independent risk factor although, as already stated, the evaluation of risk can be sharpened by separating out the various lipoprotein fractions of plasma cholesterol.

6. The correlation can be transferred from one population to another without losing its predictive power: the results in the Seven Countries Study [59] remained consistent among populations in which mean cholesterolaemia varied between 1.6 g/l (Japan, Yugoslavia) and 2.5 g/l (Finland). However risk factors, including hypercholesterolaemia, do not seem to exert an equal influence in the USA and in France [31,91]; but this analysis did not take into account the possible confounding role of HDL cholesterol in such an intriguing discrepancy.

7. The association between coronary heart disease and cholesterol has been confirmed by patho-anatomical surveys [99], and although less eloquently so, by cross-sectional angiographic studies [73,93]. In addition to the seminal observations of Anitschkow and Chalatow in 1913 [5], a huge number of publications have confirmed the link between cholesterol and experimental atheroma-like lesions in animals.

For the most critical part of Table 20.2, the second part, only the first criterion can be considered as fulfilled. We have already seen that, according to the results of the Lipid Research Clinic's trial, the treatment of hypercholesterolaemia can reduce coronary heart disease incidence [68,69]. Even if cholesterol is *a* cause of coronary heart disease, one cannot state that it is *the* cause (criterion no. 2 in the second part of Table 20.2). Increased plasma cholesterol is a state that is neither necessary nor sufficient for the development of coronary heart disease, which may develop in subjects whose plasma cholesterol has always been normal (according to the presently accepted criteria), and may not develop in hypercholesterolaemic individuals.

Disagreement

In spite of that, it is not totally unreasonable to think that cholesterol does not play a perfect role as a risk factor for atherosclerosis.

As we have just seen, cholesterol is a well-established risk factor for coronary heart disease, which does not automatically mean that it is also a risk factor for atherosclerosis in general. The association of hypercholesterolaemia with atherosclerosis of cerebral arteries (athero-thrombotic brain infarction) and of lower limb arteries (intermittent claudication) has been much less well investigated than that with coronary heart disease. Here, only the Framingham Study allows valid comparisons. A first report [42] in 1972 stated that "serum cholesterol seems to have about the same degree of association (on the average) to all three major atherosclerotic outcomes". Another epidemiological investigation which was cross-sectional instead of longitudinal, but which relied upon much more sophisticated diagnostic criteria, found an association between hyperlipidaemia (hyper-cholesterolaemia, hypertriglyceridaemia or pre-scription of a lipid-lowering treatment) and an increased prevalence of large-vessel peripheral arterial disease of the lower limbs; however, this increase (29%) was not statistically significant [25]. A second report from Framingham [56] in 1974 concluded that the risk of atherothrombotic brain infarction was correlated with chole-sterolaemia, but only before age 60, that is before the age for the maximal incidence of this com-plication. Finally, doubt arises mainly from a third report from Framingham produced in 1981 [43]. An analysis comparing the influences of LDL and HDL cholesterols led to the following conclusions:

1. Coronary heart disease incidence and mor-tality are positively correlated with LDL con-centration, but negatively with HDL concentration, which confirms many other observations from other sources.

2. For intermittent claudication, the risk profile is much less straighforward: the enhanc-ing effect of LDLs and the protective effect of HDLs, although still present, become weak and non-significant.

3. For atherothrombotic brain infarctions the protective role of HDLs is also non-significant; most of all, the enhancing action of LDLs com-pletely disappears, and reverses, i.e. the relation-ship with LDLs is negative, and significantly so in women. Thus the LDLs theoretically exert either no effect or a protective effect.

All these data certainly need to be repeated and confirmed. With the present state of our know-ledge we cannot consider that cholesterol abnor-malities (hypercholesterolaemia, high LDLs, low HDLs) are an established and equally important risk factor for all the clinical sequelae, i.e. all the localizations, of atherosclerosis. I have already given some explanations which may account for this discrepancy: methodological bias in the epid-emiological surveys cannot be held as solely responsible.

The belief for a causal role of cholesterol in coronary heart disease mainly relies upon the results of one long-term, large-scale therapeutic trial. The statistical validity of these results and of the inferences drawn from them has been debated [61,88]. Furthermore, no study has established similar benefits of cholesterol low-ering for the non-coronary localizations of atherosclerosis.

The vast amount of available information con-cerning the association between cholesterol and vascular diseases, of which I have analysed but a small sample, allows us to conclude that hyper-cholesterolaemia:

1. Is certainly a risk factor for coronary heart disease;

2. Is probably a risk factor for atherosclerosis in general; and

3. Is likely to be a cause of atherosclerosis, but a doubtful one whose influence may be minute if one relies strictly upon the available evidence, without extrapolating from it.

Conclusion

The reference criteria of pragmatism are: to func-tion realistically and to succeed in practice. From the author's clinical point of view, critical evalu-ation of risk factors causes them to lose part of their generally acknowledged sturdiness. None of them can be promoted to the rank of established cause for atherosclerosis in general. The one we know best, hypercholesterolaemia, is a cause, and probably not the main cause, of coronary heart disease. One may comment in two ways upon this disappointing state of affairs.

The initial way of thinking may be that, after 40 years of hard work, we can only begin to assess the exact influence of the risk factors we know; additional evidence will accumulate which will confirm that hypertension, hypercholesterolaemia, diabetes and smoking are indeed the four major causes of atherosclerosis; by going on ploughing this same furrow, undisputable proof will certainly be forthcoming. The most optimistic, i.e. the majority, even believe that there is no longer any need to wait for such evidence: the time has already come to engage in an action against the risk factors so as to defeat atherosclerosis, and to study the effects of the risk factors upon the arterial wall to unravel the enigma of atherosclerosis.

A second more considered way of thinking is that risk factors may play a role, but probably are modest actors in the scene which produces atherosclerosis. The major causes of the disease remain hidden: few improvements in the atherosclerotic state of many patients are to be expected from the control of risk factors; we must go on searching along imaginative paths, clinically as well as experimentally; no project should be regarded as eccentric at first sight, even if it does not follow the hypothesis that risk factors are the causes of atherosclerosis.

The dilemma is set: no one can solve it today, and due attention should be paid to it.

References

1. Åberg H, Lithell H, Selinus I et al. (1985) Serum triglycerides are a risk factor for myocardial infarction but not for angina pectoris. Results from a 10-year follow-up of Uppsala primary preventive study. Atherosclerosis 54:89–97
2. Ahrens EH Jr (1985) The diet–heart question in 1985: has it really been settled? Lancet 1:1085–1087
3. Albrink MJ, Man EB (1959) Serum triglycerides in coronary artery disease. Arch Intern Med 103:4–8
4. Ammar AD, Wilson RL, Travers H et al. (1984) Intraplaque hemorrhage: its significance in cerebrovascular disease. Am J Surg 148:840–843
5. Anitschkow N, Chalatow S (1983) On experimental cholesterin steatosis and its significance in the origin of some pathological processes. Arteriosclerosis 3:178–182
6. Benditt EP, Barrett T, McDougall JK (1983) Viruses in the etiology of atherosclerosis. Proc Natl Acad Sci USA 80:6386–6389
7. Brownlee M, Vlassara H, Cerami A. (1983) Nonenzymatic glycosylation reduces the susceptibility of fibrin to degradation by plasmin. Diabetes 32:680–684
8. Brunzell JD, Sniderman AD, Albers JJ et al. (1984) Apoproteins B and A-1 and coronary artery disease in humans. Arteriosclerosis. 4:79–83
9. Burch PRJ (1980) Ischaemic heart disease: epidemiology risk factors and cause. Cardiovasc Res 14:307–338
10. Burch PRJ (1983) The Surgeon General's "epidemiologic criteria for causalty". A critique. J Chronic Dis 36:821–836
11. Cabin HS, Roberts WC (1982) Relation of serum total cholesterol and triglyceride levels to the amount and extent of coronary arterial narrowing by atherosclerotic plaque in coronary heart disease. Quantitative analysis of 2,037 five mm segments of 160 major epicardial coronary arteries in 40 necropsy patients. Am J Med 73:227–234
12. Capron L (1983) Athérosclérose. Description et mécanismes. Rev Neurol (Paris) 139:167–176, 239–250
13. Capron L (1984) Epidémiologie et facteurs de risque des accidents artériels cérébraux. Rev Neurol (Paris) 140:161–170
14. Capron L, Cristol R (1985) Facteurs de risque selon les territoires artériels. In: Prévention des accidents vasculaires cérébraux. Académie Nationale de Médecine, Paris, pp 53–57
15. Capron L, Fiessinger JN, Housett E (1982) Paroi artérielle, insuline, diabète et athérosclérose. In: Rathery M (ed) Journées annuelles de diabétologie de l'Hôtel-Dieu. Flammarion Médécine-Sciences, Paris, pp 247–255
16. Capron L, Philippe M, Fiessinger JN et al. (1984) Diabetes and insulin: actions and interactions upon the glucose metabolism of rat aorta. Diabete Metab 10:78–84
17. Capron L, Phillippe M, Guilmot JM et al. (1981) Effects of insulin exposure upon the rat aortic media: influence of hydrostatic forces. Arteriosclerosis 1:345–352
18. Carlson LA (1972) Citation classic. Commentary. Lancet I:865–868
19. Carlson LA, Åberg H (1985) Serum triglycerides: an independent risk factor for myocardial infarction but not for angina pectoris. N Engl J Med 312:1127
20. Carlson LA, Böttiger LE (1972) Ischaemic heart disease in relation to fasting values of plasma triglycerides and cholesterol: Stockholm prospective study. Lancet I:865–868
21. Carlson LA, Böttiger LE (1980) Triglycerides and coronary heart disease. N Engl J Med 303:1061–1062
22. Castelli WP, Abbott RD, McNamara PM (1983) Summary estimates of cholesterol used to predict coronary heart disease. Circulation 67:730–734
23. Consensus Conference (1985) Lowering blood cholesterol to prevent heart disease. JAMA 253:2080–2086
24. Cormier JM, Fiessinger JN, Capron L (1984) Athérosclérose oblitérante des membres inférieurs. Encyclopédie Médico-chirurgicale, Cœur-Vaisseaux, 1316 A[10] Paris, p 22
25. Criqui MH, Fronek A, Barrett-Connor E et al. (1985) The prevalence of peripheral arterial disease in a defined population. Circulation 71:510–515
26. Crouse JR (1984) Progress in coronary heart disease risk-factor research: what remains to be done? Clin Res 30:1125–1127
27. Cruz AB, Amutuzio DS, Grande F et al. (1961) Effect

of intra-arterial insulin on tissue cholesterol and fatty acids in alloxan-diabetic dogs. Circ Res 9:39–43

28. Davies MJ, Thomas AC (1985) Review: plaque fissuring – the cause of acute myocardial infarction, sudden ischaemic death, and crescendo angina. Br Heart J 53:363–373

29. DePace NL, Kimbiris D, Iskandrian AS et al. (1983) Total occlusion of left main coronary artery without angina pectoris. Arch Intern Med 143:1064–1065

30. Ducimetière P, Eschwege E, Papoz L et al. (1980) Relationship of plasma insulin levels to the incidence of myocardial infarction and coronary heart disease mortality in a middle-aged population. Diabetologia 19:205–210

31. Ducimetière P, Richard JL, Cambien F et al. (1980) Coronary heart disease in middle-aged Frenchmen. Comparisons between Paris Prospective Study, Seven Countries Study, and Pooling Project. Lancet I:1346–1350

32. Ducimetière P, Richard P, Claude JR et al. (Groupe d'Etude sur l'Epidémiologie de l'Athérosclérose) (1981) Les cardiopathies ischémiques. Incidence et facteurs de risque. L'Etude Prospective Parisienne. INSERM, Paris, p 149

33. Fabricant CG (1981) Herpes virus-induced atherosclerosis. Diabetes [Suppl 2] 30:29–31

34. Falk E (1983) Plaque rupture with severe pre-existing stenosis precipitating coronary thombosis. Characteristics of coronary atherosclerosis plaques underlying fatal occlusive thrombi. Br Heart J 50:127–134

35. Fletcher A, Alkjaersig N, Schonfeld G et al. (1981) Fibrinogen catabolism in patients with type II and type IV hyperlipidemia. Effects of dietary and clofibrate treatment on laboratory findings. Arteriosclerosis 1:202–209

36. Franzén J, Nilsson B, Johansson BW et al. (1983) Fibrinolytic activity in men with acute myocardial infarction before 60 years of age. Acta Med Scand 214:339–344

37. Galanakis DK, Laurent P, Janoff A (1982) Cigarette smoke contains anticoagulants against fibrin aggregation and factor XIIIa in plasma. Science 217:642–645

38. Ganda OP (1980) Pathogenesis of macrovascular disease in the human diabetic. Diabetes 29:931–942

39. Garraway WM, Whisnant JP, Furlan AJ et al. (1979) The declining incidence of stroke. N Engl J Med 300:449–452

40. Gennes JL de, Ozanne P, Turpin G (1982) Les accidents thrombo-emboliques veineux au cours des hyperlipidémies idiopathiques. Nouv Presse Med 11:1777–1780

41. Gillum RF, Folsom AR, Blackburn H (1984) Decline in coronary heart disease mortality. Old questions and new facts. Am J Med 76:1055–1065

42. Gordon T, Kannel WB (1972) Predisposition to atherosclerosis in the head, heart, and legs. The Framingham study. JAMA 22:661–666

43. Gordon T, Kannel WB, Castelli WP et al. (1981) Lipoproteins, cardiovascular disease, and death. The Framingham study. Arch Intern Med 141:1128–1131

44. Gore I (1983) Ulceration of and embolization by atheromata. In: Jones RJ (ed) Evolution of the atherosclerotic plaque, The University of Chicago Press, Chicago, pp 315–329

45. Gregg DE, Patterson RE (1980) Functional importance of the coronary collaterals. N Engl J Med 303:1404–1406

46. Hennerici M, Aulich A, Sandmann W et al. (1981) Incidence of asymptomatic extracranial arterial disease. Stroke 12:750–755

47. Himmelstein DU, Woolhandler S (1984) Free care, cholestyramine, and health policy. N Engl J Med 311:1511–1514

48. Hjermann I, Holme I, Velve Byre K et al. (1981) Effects of diet and smoking intervention on the incidence of coronary heart disease. Lancet II:1303–1310

49. Hopkins PN, Williams RR (1981) A survey of 246 suggested coronary risk factors. Atherosclerosis 40:1–52

50. Hulley SB, Rosenman RH, Bawol RD et al. (1980) Epidemiology as a guide to clinical decisions. The association between triglyceride and coronary heart disease. N Engl J Med 302:1383–1389

51. Imparato AM, Riles TS, Mintzer R et al. (1983) The importance of hemorrhage in the relationship between gross morphologic characteristics and cerebral symptoms in 376 carotid artery plaques. Ann Surg 197:195–203

52. Kannel WB (1971) Current status of the epidemiology of brain infarction associated with occlusive arterial disease. Stroke 2:295–318

53. Kannel WB, Castelli WP (1979) Is the serum total cholesterol an anachronism? Lancet II:950–951

54. Kannel WB, Castelli WP, Gordon T et al. (1971) Serum cholesterol, lipoprotein, and the risk of coronary heart disease. The Framingham study. Ann Intern Med 74:1–12

55. Kannel WB, Feinleib M, McNamara PM et al. (1979) An investigation of coronary heart disease in families. The Framingham study. Am J Epidemiol 110:281–290

56. Kannel WB, Gordon T, Dawber TR (1974) Role of lipids in the development of brain infarction: the Framingham study. Stroke 5:679–685

57. Kannel WB, Skinner JJ Jr, Schwartz MJ et al. (1970) Intermittent claudication. Incidence in the Framingham study. Circulation 41:875–883

58. Karathanasis SK, Zannis VI, Breslow JL (1984) A DNA insertion in the apolipoprotein A-I gene of patients with premature atherosclerosis. Nature 305:823–825

59. Keys A (1984) Seven Countries. A multivariate analysis of death and coronary heart disease, Harvard University Press, Cambridge

60. King GL, Goodman AD, Buzney S et al. (1985) Receptors and growth-promoting effects of insulin and insulinlike growth factors on cells from bovine retinal capillaries and aorta. J Clin Invest 75:1028–1031

61. Kronmal RA (1985) Commentary on the published results of the Lipid Research Clinics coronary primary prevention trial. JAMA 253:2091–2093

62. Kumars S, West D, Shahabuddin S et al. (1983) Angiogenesis factor from human myocardial infarcts. Lancet II:364–367

63. Lasché EM (1984) The effect of high insulin concentrations on prostacyclin production as measured by 6-keto PGF$_{1\alpha}$ determination. Prostagland Leukot Med 14:181–184

64. Lasché EM, Larson RE (1982) Interaction of insulin and prostacyclin production in the rat. Diabetes 31:454–458

65. Laug WE (1983) Ethyl alcohol enhances plasminogen activator secretion by endothelial cells. JAMA 250:772–776

66. Leet H, Paton RC, Passa P et al. (1981) Fibrinogen binding and ADP-induced aggregation in platelets

from diabetic subjects. Thromb Res 24:143–150

67. Lilienfeld AM (1983) The Surgeon General's "epidemiologic criteria for causality": a criticism of Burch's critique. J Chronic Dis 36:837–845

68. Lipid Research Clinics Program (1984) The Lipid Research Clinics coronary primary prevention trial results. I. Reduction in incidence of coronary heart disease. JAMA 251:351–364

69. Lipid Research Clinics Program (1984) The Lipid Research Clinics coronary primary prevention trial results. II. The relationship of reduction in incidence of coronary heart disease to cholesterol lowering. JAMA 251:365–374

70. Louis S, McDowell F (1970) Age: its significance in nonembolic cerebral infarction. Stroke 1:449–453

71. Lowe GDO, Mc Ardle BM, Stromberg P et al. (1982) Increased blood viscosity and fibrinolytic inhibitor in type II hyperlipoproteinemia. Lancet I:472–475

72. Lusby RJ, Ferrell LD, Ehrenfeld WK et al. (1982) Carotid plaque hemorrhage. Its role in production of cerebral ischemia. Arch Surg 117:1479–1488

73. Maciejko JJ, Holmes DR, Kottke BA et al. (1983) Apolipoprotein A-1 as a marker of angiographically assessed coronary-artery disease. N Engl J Med 309:383–389

74. Modan M, Halkin H, Almog S et al. (1985) Hyperinsulinemia. A link between hypertension, obesity and glucose intolerance. J Clin Invest 75:809–815

75. Multiple Risk Factor Intervention Trial Research Group (1982) Multiple risk factor intervention trial. Risk factor changes and mortality results. JAMA 248:1465–1477

76. Mustard JF, Roswell HC, Murphy EA et al. (1963) Intimal thrombosis in atherosclerosis. In: Jones RJ (ed) Evolution of the atherosclerotic plaque. The University of Chicago Press, Chicago, pp 183-203

77. Nakao J, Ito H, Kanayasu T et al. (1985) Stimulatory effect of insulin on aortic smooth muscle cell migration induced by 12-L-hydroxy-5,8,10,14-eicosatetraenoic acid and its modulation by elevated extracellular glucose levels. Diabetes 34:185–191

78. Neufeld HN, Golbourt U (1983) Coronary heart disease: genetic aspects. Circulation 67:943–954

79. NIH consensus development conference (1984) Summary: treatment of hypertriglyceridemia. Arteriosclerosis 4:296–301

80. Nilsson IM, Ljungnér H, Tengborn L (1985) Two different mechanisms in patients with venous thrombosis and defective fibrinolysis: low concentration of plasminogen activator or increased concentration of plasminogen activator inhibitor. Br Med J 290:1453–1456

81. Oliver MF (1985) Consensus or nonsensus conferences on coronary heart disease. Lancet I:1087–1089

82. Orcel L, El Salem Ch, Natali J et al. (1979) Aspects anatomo-pathologiques des artériopathies chroniques oblitérantes des membres. Sem Hop Paris 55:743–747

83. Orchard TJ, Becker DJ, Bates M et al. (1983) Plasma insulin and lipoprotein concentrations: an atherogenic association? Am J Epidemiol 118:326–337

84. Pell S, Fayerweather WE (1985) Trends in the incidence of myocardial infarction and in associated mortality and morbidity in a large employed population 1957–1983. N Engl J Med 312:1005–1011

85. Pfeifle B, Ditschuneit H. (1981) Effects of insulin on growth of cultured human arterial smooth muscle cells. Diabetologia 20:155–158

86. Pfeifle B, Ditschuneit H (1983) Receptors for insulin and insulin-like growth factor in cultured arterial smooth muscle cells depend on their growth state. J Endocrinol 96:251–257

87. Pfeifle B, Ditschuneit HH, Ditschuneit H (1980) Insulin as a cellular growth regulator of rat arterial smooth muscle cells in vitro. Horm Metab Res 12:381–385

88. Pocock SJ (1985) Current issues in the design and interpretation of clinical trials. Br Med J 290:39–42

89. Pyörälä K (1979) Relationship of glucose tolerance and plasma insulin to the incidence of coronary heart disease: results from two population studies in Finland. Diabetes Care 2:131–141

90. Rahimtoola SH (1985) Cholesterol and coronary heart disease: a perspective. JAMA 252:2094–2095

91. Richard JL, Cambien F, Ducimetière P (1981) Particularités épidémiologiques de la maladie coronarienne en France. Nouv Presse Med 10:1111–1114

92. Ross R, Wight TN, Strandness E et al. (1984) Human atherosclerosis. I. Cell constitution and characteristics of advanced lesions of the superficial femoral artery. Am J Pathol 114:79–93

93. Salel AF, Fong A, Zelis R et al. (1977) Accuracy of numerical coronary profile. Correlation of risk factors with arteriographically documented severity of atherosclerosis. N Engl J Med 296:1447–1450

94. Scott J, Knott TJ, Priestley LM et al. (1985) High density lipoprotein composition is altered by a common DNA polymorphism adjacent to apoprotein A II gene in man. Lancet I:771–773

95. Shea S, Ottman R, Gabrieli C et al. (1984) Family history as an independent risk factor for coronary artery disease. J Am Coll Cardiol 4:793–801

96. Singer P, Godicke W, Voigt S et al. (1985) Postprandial hyperinsulinemia in patients with mild hypertension. Hypertension 7:182–186

97. Singh RN (1984) Progression of coronary atherosclerosis. Clues to pathogenesis from serial coronary arteriography. Br Heart J 52:451–461

98. Sirek OV, Sirek A, Cukerman E (1981) Intermittent hyperinsulinemia and arterial glycosaminoglycans in dogs. Diabetologia 21:154–159

99. Solberg LA, Strong JP (1983) Risk factors and atherosclerotic lesions. A review of autopsy studies. Arteriosclerosis. 3:187–198

100. Stamler J (1979) Research related to risk factors. Circulation. 60:1575–1587

101. Stamler J, Pick R, Katz LN (1960) Effect of insulin in the induction and regression of atherosclerosis in the chick. Circ Res 8:572–576

102. Stout RW (1968) Insulin-stimulated lipogenesis in arterial tissue in relation to diabetes and atheroma. Lancet II:702–703

103. Stout RW (1970) Development of vascular lesions in insulin-treated animals fed a normal diet. Br Med J iii:685–687

104. Stout RW (1975) The effect of insulin on the incorporation of D-glucose-U-^{14}C into the lipids of the rat aorta in vivo. Horm Metab Res 7:31–34

105. Stout RW (1979) Diabetes and atherosclerosis – the role of insulin. Diabetologia 16:141–150

106. Stout RW (1981) The role of insulin in atherosclerosis in diabetics and nondiabetics. A review. Diabetes [Suppl 2] 30:54–57

107. Stout RW, Bierman EL, Ross R (1975) Effect of insulin on the proliferation of cultured primate arterial smooth muscle cells. Circ Res 36:319–327

108. Swonden CB, McNamara PM, Garrison RJ et al. (1982) Predicting coronary heart disease in siblings. A

multivariate assessment. The Framingham heart study. Am J Epidemiol 115:217–222

109. The Pooling Project Research Group (1978) Relationship of blood pressure, serum cholesterol, smoking habit, relative weight and ECG abnormalities to incidence of major coronary events: final report of the Pooling Project. J Chronic Dis 31:201–206

110. Thom TJ, Kannel WB (1981) Downward trend in cardiovascular mortality. Ann Rev Med 32:427–434

111. Virmani R, Roberts WC (1983) Extravasated erythrocytes, iron, and fibrin in atherosclerotic plaques of coronary arteries in fatal coronary heart disease and their relation to luminal thrombus: frequency and significance in 57 necropsy patients and in 2958 five mm segments of 224 major epicardial coronary arteries. Am Heart J 105:788–797

112. Welborn TA, Wearne K (1979) Coronary heart disease incidence and cardiovascular mortality in Busselton with reference to glucose and insulin concentrations. Diabete Care 2:154–160

113. Williams RS, Loge EE, Lewis TL et al. (1980) Physical conditioning augments the fibrinolytic response to venous occlusion in healthy adults. N Engl J Med 302:987–991

114. Wolinsky H, Goldsfischer S, Capron L et al. (1978) Hydrolase activities in the rat aorta. I. Effects of diabetes mellitus and insulin treatment. Circ Res 42:821–831

Chapter 21

Diagnostic Principles and Therapeutic Management of Patients with Atherosclerosis

Lower Limb Arteriopathies

J. N. Fiessinger

Atherosclerosis of the lower limbs may be diagnosed when symptomatic arteriopathy is present, or in association with other diseases or an epidemiological survey. The investigation of symptomatic arteriopathies will be examined here.

Diagnosis

Clinical Diagnosis

Clinical diagnosis is based on the functional symptoms of arterial insufficiency. Intermittent claudication is an exertional muscular pain which disappears rapidly after the effort is stopped. It is almost pathognomonic of arterial insufficiency [2]. However, if arterial insufficiency is diagnosed only on the basis of intermittent claudication the diagnosis is wrong in 50% of cases [12]. Rest pains are felt first in the big toe and the fore foot. These pains are relieved when their legs are allowed to dangle, and are highly suggestive of arterial insufficiency. However, these posture pains may also be caused by neuropathy, infected venous trophic ulceration and hypertensive ischaemic ulcers.

Symptoms must be confirmed by physical examination. Here again, however, errors are frequent; few surveys have assessed the diagnostic value of clinical data [12,13,36,42]. The dorsalis pedis pulse is congenitally absent in 4%–12% of patients and this sign has no diagnostic

value [13]. A decreased or absent posterior tibial pulse is considered as more specific but even so, only half of patients [13] with an abnormal posterior tibial pulse actually have an arteriopathy. Moreover, 30% of patients with arteriopathy have a clinically normal posterior tibial pulse [13].

Auscultation of arteries from the navel to the popliteal space at rest, after exertion and after compression [8] is useful in diagnosis. However, bruits may be present in normal patients. Almost half of patients with symptomatic diabetic arteriopathy have no bruits on physical examination [36]. Other investigations are thus necessary to diagnose arteriopathy conclusively.

Investigations

Further investigations aim to confirm a clinically suspected arteriopathy, to determine the importance of ischaemia, to document the location of arterial lesions, to contribute to aetiological diagnosis (i.e. atherosclerosis) and to detect other arterial sites involved by atherosclerosis. With these multiple aims and techniques of investigation, a strategy for diagnosis has to be planned.

Measurement of Systolic Ankle Pressure

This is the first step in diagnosis [5,45]. A distal pressure below 95% of the humeral pressure indicates arterial disease [4]. The value of the distal pressure is related to the degree of ischaemia. It is an objective criterion for follow-up of patients [14]. A low distal pressure is specific but not sensitive: intermittent claudication may be associated with normal pressure if arterial lesions are proximal or limited below the popliteal artery. Measuring pressure after exertion or hyperaemia [3,6] increases the sensitivity of this method. The time it takes to recover basal pressure after treadmill exercise allows a functional assessment of arterial insufficiency [46]. Medial calcifications in some patients (diabetic, old patients) do not allow the measurement of systolic pressure because arteries are incompressible. In such patients, if measurement is

possible it is probably overvalued. The use of several cuffs placed along the lower limb allows assessment of the location of lesions [22] but given the poor specificity of this method, it is of limited value [34]. Plethysmography allows us to measure pressure at the big toe and to calculate a pressure gradient between ankle and toe. It is useful in diagnosing distal arteriopathies and is accurate in predicting healing of gangrene or amputation [7,33].

Measurements of flow are less sensitive than measurements of pressure in assessing ischaemia.

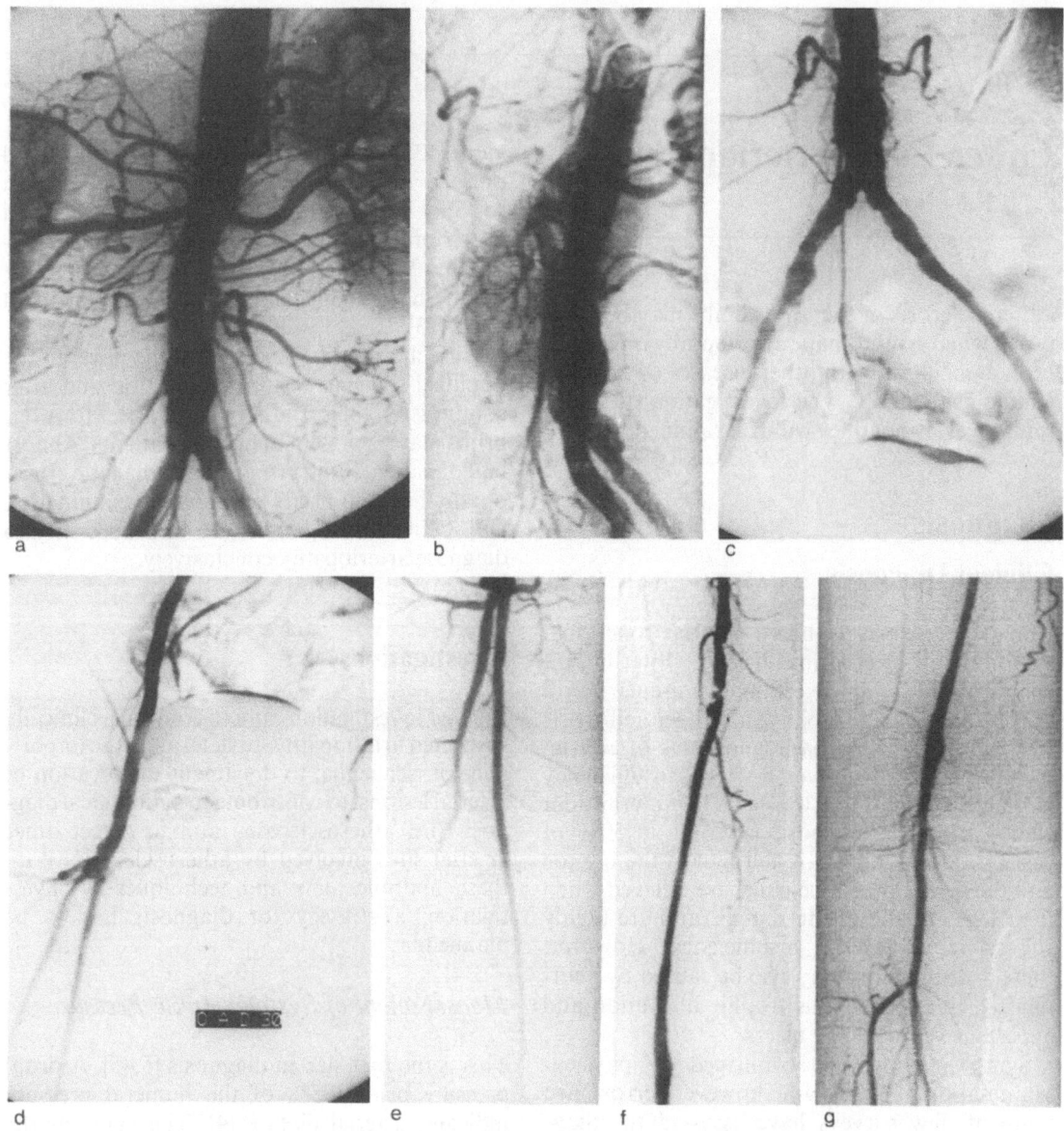

Flow measurements are made using plethysmography or isotopic clearance and are not required in clinical practice [45].

Irrigraphy is a rheoplethysmographic method used in assessing quantitative and functional characteristics of arteriopathies [23]. It determines the sites of lesions and the value of the collateral circulation. The transcutaneous measurement of O_2 partial pressure can be used to assess the microcirculatory and tissue consequences of ischaemia. This new method seems to be useful in determining the prognosis of severe ischaemia [40,50].

Assessment of the Topography of Arterial Lesions

This is the second step in the investigation of suspected arteriopathy. For many years it depended on arteriography, physical examination and roentgenography [2]. Antero-posterior and lateral roentgenograms of the abdomen are still used in the investigation of ischaemia of the lower extremities. Calcification of the aortic wall is visualized using roentgenography and allows assessment of the subrenal aorta. It is almost always present in aneurysms, suggesting atherosclerosis. Use of a directional doppler detector allows assessment of arterial lesions but is not of as much value for lower limbs as for aortic arch branches. Ultrasound scanning used in association with directional Doppler studies constitutes a major advance in the investigation of arteriopathies.

The results of ultrasound scanning adds significance to the results of doppler studies done by the same physician at the same time. These two techniques allow accurate study of lower limb arteries, and digestive and renal artery ostia. They also contribute to the determination of the aetiology of arteriopathy [24,45,53]. Pulsed Doppler techniques allow accurate measurement of flow [37].

Angiography is the key to assessing the topography of lesions. It also contributes to the aetiological diagnosis. Angiography is necessary in the surgical management which follows the antero-posterior and lateral visualization of the subrenal aorta, renal and digestive arteries and lower extremity arteries as far as the sole of the foot. Aortography has a morbidity of 0.1% and a mortality of 0.05% [46]. This is an argument in favour of restrictive use of aortography; it is only indicated if surgical treatment is envisaged [35]. The recent development of computerized angiography decreases the risks involved. After intravenous injection of contrast media, this method can be used to assess arteriopathy [41]. However, the quality of imaging is often poor, particularly in obese and old patients. For these reasons, computerized aortography following intra-arterial injection of contrast media should be preferred. When compared with conventional aortography, it allows reduction of the amount of contrast medium, saves films and is less painful [31]. Risks still have to be evaluated, but it may replace conventional aorto-arteriography (Fig. 21.1a).

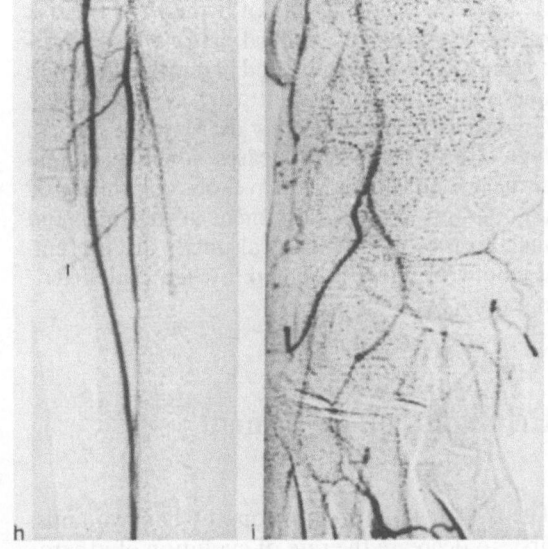

Fig. 21.1 a–i. Computerized angiography of the abdominal aorta and the lower limb arteries through a 5 F Seldinger catheter in the left femoral artery in a patient with ischaemia of the right lower limb. Ten 30-ml injections of Hexabrix diluted to 50% were given. **a, b** Antero-posterior and lateral views of the aorta. Mild stenosis of the right renal artery. Ulcerated plaque in the subrenal aorta. Aneurysm of the lower part of the aorta. **c** Iliac arteries are atheromatous, but without significant stenosis. **d, e** The right femoral bifurcation is normal. **f** Severe narrowing and ulceration of the right superficial femoral artery. **g** The right popliteal artery is normal. Mild narrowing of the origin of the anterior tibial artery. **h, i** The anterior tibial artery supplies the plantar arcade. The other leg arteries are patent. (Courtesy Dr JM Pernes, Vascular Radiology Service and Prof JC Gaux, Hospital, Broussais, Paris)

Coronary Involvement

This is the main prognostic factor in athero-sclerotic patients [11,15,42], emphasizing the importance of the assessment of coronary risk. Systematic treadmill ECG [15,38] reveals coronary insufficiency in 30% of asymptomatic patients. Alternatively, myocardial thallium scan after dipyridamole injection allows assessment of coronary risk without any exercise [4]. Systematic coronary angiography is normal in only 8% of patients with lower limb arteriopathy and shows lesions amenable to surgical treatment in 30% of cases [27].

The Strategy of Investigation

The measurement of systolic ankle pressure is systematic. Investigations are completed by taking on ECG during treadmill exercise and by measuring the pressure at the big toe or the transcutaneous or partial pressure of oxygen. Ultrasound scanning–Doppler studies are however the main method used when investigating lower limb arterial involvement and extension of atherosclerosis to other sites.

Coronary risk assessment is of great importance. Intermittent claudication disturbs the treadmill ECG and will be apparent following myocardial thallium scan. If these investigations are positive or if there is a history of coronary artery disease, coronary angiography is indicated strongly.

Arteriography is rarely indicated to determine the aetiology of arterial disease, but is required for surgical treatment, the indication for which depends on the possibility of bypass and the risk of development of arterial lesions [14]. Aortography is usually indicated in patients in whom anatomical surgical treatment is considered. Computerized angiography decrease the risk. In high-risk patients or in cases where proximal surgery is considered, intravenous computerized angiography allows assessment of the aorta and iliac arteries and, if femoral pulses are present, it could be complemented by femoral arteriography.

Principles of Treatment

The treatment of an arteriopathy has two aims: first, to decrease the rate of evolution of athero-sclerosis; second, to diminish the arterial insufficiency.

Treatment of Atherosclerosis

The evolution of atherosclerotic lesions depends on their site [10]: aneurysms often develop in proximal lesions, whereas stenoses are frequently observed in peripheral lesions [9]. The factors responsible for plaque ulceration and healing are unknown. Usually, preventative treatment consists of the control of risk factors of atherosclerosis. Many retrospective or uncontrolled studies suggest that the prognosis improves if patients stop smoking [47,51]. This is supported by a prospective study of 200 patients with 5-year follow-up [47]. The level of plasma carboxyhaemoglobin is significantly higher in patients with failure of surgical bypass than in patients with good surgical results.

Lipid abnormalities are not strongly correlated with peripheral arteriopathies [2,10]. In patients with both hyperlipidaemia and arterial involvement, a study showed that those treated for their lipid abnormality have a decreased progression rate of angiographically assessed lesions [17]. In diabetes, the risk of arteriopathy and amputation is increased but it has not been shown that control of the diabetes influences the prognosis of arterial involvement; hypertension makes the treatment of arteriopathy more difficult. Indeed, some systolic hypertension may compensate for the decreased arterial compliance [32]. The normalization of blood pressure decreases distal pressures and may aggravate ischaemia, thus beta blockers probably have a deleterious effect [29].

Thrombosis is a major complication of atherosclerosis and heparin is widely used in the early phase of an acute arterial occlusion. It prevents thrombophlebitis and may prevent the extension of arterial thrombosis. Although good results were reported with anti-vitamin K agents [48] they have been supplanted by platelet anti-aggregant agents. These have two theoretical advantages: first, platelets have a major role in arterial thrombosis and second, iatrogenic accidents are less frequent with anti-aggregant agents. In coronary and carotid lesions, studies suggest that aspirin is effective [52]. In lower limb arterial lesions, a prospective random study showed that arteriographically assessed lesions evolve less rapidly in patients treated with aspirin and dipyridamole, whereas aspirin alone is not efficient [28].

Surgery may eliminate an arterial lesion known to present a further risk, as with aortic and popliteal aneurysms [10].

Treatment of Ischaemia

There are three aims in the treatment of ischaemia: to restore the blood supply, to improve the flow and to act on the vasomotricity and the function of collaterals.

Restoration of the Blood Supply

This is the main aim in the treatment of ischaemia. Surgery is of primary importance, but angioplasty is now also indicated [10]. Thromboendarterectomy and bypasses are both surgical methods used in the treatment of ischaemia. Thromboendarterectomy removes the lesion by dissection through the outer part of the media. It allows the conservation of collateral branches, but aneurysms of the aorta and iliac arteries and restenoses of external iliac and femoral arteries are potential late complications.

Bypass grafts are made of venous or prosthetic material. They follow an anatomical or extra-anatomical (femoro-femoral, axillo-femoral) course. There is a risk of infection and false aneurysms may develop at the anastomosis between the prosthesis and the arterial wall. In the aorta up to the external iliac artery, bypass surgery or thromboendarterectomy share the same indications, and which one is chosen will depend on the extent of lesions, their level and the habits of the surgeon. Below the external iliac artery, bypass surgery is usually indicated. Below the popliteal artery, a venous graft bypass is the best technique, allowing femoro-leg bypass. Transluminal angioplasty is mainly indicated in isolated iliac stenoses, with 80% patency after 2 years [1,26,55]. In segmental lesions of the superficial femoral artery, long-term patency is not so good, so that the complications of the technique and the relatively favourable prognosis of these lesions should be taken into account.

Intravenous streptokinase treatment, although it is efficient in 30% of early arterial occlusions, can lead to major complications which necessitates the use of in situ thrombolysis [19]. However intra-arterial thrombolysis often results in haemorrhagic complications [25,43]. In regimens where systemic fibrinolysis is absent or low, thrombosis may occur in catheters [18,20].

Thus thrombolysis is indicated in severe ischaemia when surgery is not indicated.

Treatment Affecting Blood Flow

Treatment acts on three components of blood viscosity – fibrinogen, the haematocrit and red cell deformability. Anti-fibrinogen agents are not efficient in the treatment of ischaemia [49]. Although haemodilution and abundant intravenous perfusions [10,54] increase blood flow in ischaemic areas, their clinical efficiency has not yet been demonstrated. Recently it has been suggested that drugs acting on blood flow could improve the claudication distance [21,30,39].

Actions on Vasomotricity and Collateral Branches

The results in this field are limited or controversial [16,51,44], whether due to surgical intervention (sympathectomy) or produced through vasoactive drugs.

References

1. Abbott WM (1980) Percutaneous transluminal angioplasty. Surgeon's view. AJR, 135:917–920
2. Allen-Barker-Hines (1980) Peripheral vascular diseases, 5th edn. Juergens JL, Spittell JA, Fairbairn JF (eds). Saunders, Philadelphia
3. Baker JD (1978) Post-stress doppler ankle pressures. A comparison of treadmill exercise with two other methods of induced hyperemia. Arch Surg 113:1171–1173
4. Boucher CA, Brewster DC, Darling RC et al. (1985) Determination of cardiac risk by dipyridamole–thallium imaging before peripheral vascular surgery. N Engl J Med 312:389–384
5. Carter SA (1969) Clinical measurement of systolic pressures in limbs with arterial occlusive disease. JAMA 207:1879–1874
6. Carter SA (1972) Response of ankle systolic pressure to leg exercise in mild or questionable arterial disease. N Engl J Med 287:578–582
7. Carter SA (1973) The relationship of distal systolic pressures to healing of skin lesions in limbs with arterial occlusive disease with special reference to diabetes mellitus. Scand J Clin Lab Invest. (Suppl 128) 31:239–243
8. Carter SA (1981) Arterial auscultation in peripheral vascular disease. JAMA 246:1682–1686
9. Chilvers AS, Thomas ML, Browse NL (1974) The progression of arteriosclerosis, a radiological study. Circulation 50:402–408
10. Cormier JM, Fiessinger JN, Capron L (1984) Athérosclérose oblitérante des membres inférieurs. Encycl Med Chir (Paris), Cœur-Vaisseaux, 11316 A10, 12–1984
11. Criqui MH, Coughlin SS, Fronek A (1985) Non-invasively diagnosed peripheral arterial disease as a pre-

dictor of mortality; results from a prospective study. Circulation 72:768–773

12. Criqui MH, Fronek A, Barrett-Connor E et al. (1985) The prevalence of peripheral arterial disease in a defined population. Circulation 71:510–515

13. Criqui MH, Fronek A, Klauber MR et al. (1985) The sensitivity, specificity and predictive value or traditional clinical evaluation of peripheral arterial disease: results from noninvasive testing in a defined population. Circulation 71:516–522

14. Cronenwett JL, Warner KG, Zelenock GB et al. (1984) Intermittent claudication current results of nonoperative management. Arch Surg 119:430–436

15. Cutler BS, Wheeler HB, Paraskos JA, Cardullo PA (1981) Applicability and interpretation of electrocardiographic stress testing in patients with peripheral vascular disease. Am J Surg 141:501–506

16. De Cree J, Leempoels J, Geukens H, Verhaegen H (1984) Placebo-controlled double-blind trail of ketanserin treatment of intermittent claudication. Lancet II:775–779

17. Duffield RG, Lewis B, Miller NE et al. (1983) Treatment of hyperlipidaemia retards progession of symptomatic femoral atherosclerosis. A randomised controlled trial. Lancet II:639–652

18. Eskridge JM, Becker GL, Rabe FE et al. (1983) Catheter-related thrombosis and fibrinolytic therapy. Radiology 149:429–432

19. Fiessinger JN, Aiach M (1983) Utilisation des thrombolytiques dans les affections artérielles. Agressologie 24:573–577

20. Fiessinger JN, Vitoux JF, Pernes JM et al. (1986) Complications of local thrombolytic therapy in treatment of peripheral arterial occlusions. AJR 146:157–159

21. Forconi S. Pieragalli D, Acciavatti A et al. (1984) Positive effect of oral buflomedil on exercise-induced haemorhological damage and on claudication distance in peripheral obliterative arterial disease patients. J Int Med Res 12:188–192

22. Francfort JW, Bigelow PS, Davis JT, Berkowitz HD (1984) Noninvasive techniques in the assessment of lower-extremity arterial occlusive disease. The advantages of proximal and distal thigh cuffs. Arch Surg 119:1145–1148

23. Franco A, Morzol B, Piquard JF et al. (1978) Valeur de l'irrigraphie dans l'évaluation de la fonction fémoropoplitée et distale. J Mal Vasc 3:188–193

24. Glover JL, Bendick PJ, Jackson VP et al. (1984) Duplex ultrasonography digital subtraction angiography and conventional angiography in assessing carotid atherosclerosis. Arch Surg 119:664–669

25. Graor RA (1985) Local thrombolysis in the treatment of thrombosed arteries, bypass grafts and arteriovenous fistulas. Vasc Surg 2:406–414

26. Guidet B. Angel C. Seurot M et al. (1983) Angioplastie endoluminale percutanée des artères iliaques. Résultats immédiats et à distance. Press Med 12:507–511

27. Hertzer NR, Beven EG, Young JR et al. (1984) Coronary artery disease in peripheral vascular patients. A classification of 1000 coronary angiograms and results of surgical management. Ann Surg 199:223–233

28. Hess H, Miestaschk A, Deischsel G (1985) Drug-induced inhibition of platelet function delays progression of peripheral occlusive arterial disease. A prospective double-blind arteriographically controlled trial. Lancet I:415–419

29. Ingram DM, House AK, Thrompson GH et al. (1982) Beta-adrenergic blockade and peripheral vascular disease. Med J Aust 1:509–511

30. Jones NAG, De Haas H, Zahavi J. Kakkar VV (1982) A double blind trial of suloctudil v. placebo in intermittent claudication. Br J Surg 69:38–40

31. Kaufman SL, Chang R, Kadir S et al. (1984) Intraarterial digital subtraction angiography in diagnostic arteriography. Radiology 151:323–327

32. Levenson JA, Simon AC, Fiessinger JN et al. (1982) Systemic arterial compliance in patients with arteriosclerosis obliterans of the lower limbs. Observations on the effect of intravenous propranolol. Arteriosclerosis 2:266–271

33. Lezack JD, Carter SA (1973) The relationship of distal systolic pressures to the clinical and angiographic findings in limbs with arterial occlusive disease. Scand J Clin Lab Med (Suppl 31) 31:97–101

34. Lynch TG, Hobson RW, Wright CB et al. (1984) Interpretation of doppler segmental pressures in peripheral vascular occlusive disease. Arch Surg 119:465–467

35. Macpherson DS, James DC, Bell PRF (1980) Is aortography abused in lower-limb ischaemia, Lancet II:80–82

36. Marinelli MR, Beach KW, Glass MJ et al. (1979) Noninvasive testing vs clinical evaluation of arterial disease. A prospective study. JAMA 241:2031–2034

37. Peronneau P, Xhaard M, Diebold B et al. (1976) Débitmétrie transcutanée par vélocimétrie ultrasonore Doppler à émission pulsée. Premières applications cliniques. Nouv Presse Med. 5:2547

38. Petiau JC, Lancelin B. Guermonprez JL et al. (1978) Artériopathie des membres inférieurs: intérét de l'épreuve d'effort. Nouv Presse Med 7:2074

39. Porter JM, Cutler BS, Lee BY et al. (1982) Pentoxifylline efficacy in the treatment of intermittent claudication: multicenter controlled double-blind trial with objective assessment of chronic occlusive arterial disease patients. Am Heart H 104:66–72

40. Ratliff DA, Chant ADB, Webster JHH (1984) Predicting of amputation wound healing: the role of transcutaneous p02 assessment. Br J Surg 71:219–222

41. Raynaud A, Seurot M, Angel C et al. (1983) Indications actuelles de l'angiographie numérique par voie veincise dans l'étude des axes ilio-fémoraux. Press Méd 12:1641–1644

42. Reunanen A, Takkunen H, Aromaa A. (1982) Prevalence of intermittent claudication and its effect on mortality. Acta Med Scand 21:249–256

43. Sicard GA. Schier JJ, Totty WG et al. (1985) Thrombolytic therapy for acute arterial occlusion. Vasc Surg 2:65–78

44. Sorlie D, Myhre K (1978) Effects of physical training in intermittent claudication. Scand J Clin Lab Invest 38:217–222

45. Strandness DE (1983) Noninvasive evaluation of arteriosclerosis. Comparison of methods. Arteriosclerosis 3:103–116

46. Szilagyi DE, Smith RF, Elliot JP, Hageman JH (1977) Translumbar aortography. A study of its safety and usefulness. Arch Surg 112:399–408

47. Thomas M (1981) Smoking and vascular surgery. Br J Surg 68:601–604

48. Tillgren C (1965) Obliterative arterial disease of the lower limbs. IV. Evaluation of long-term anticoagulant therapy. Acta Med Scand 168:203–219

49. Tonnesen KH, Sager PH, Gormsen J (1978) Treatment of severe foot ischaemia by defibrination with ancrod: a randomized blind study. Scand J Clin Lab Invest 38:431–435

50. Vayssairat M, Mathieu JF, Priollet P et al. (1984) Mesure transcutanée de la pression partielle d'oxygène. Une nouvelle méthode d'exploration fonctionnelle en pathologie vasculaire. Presse Med 13:1683–1686

51. Verstraete M (1982) Current therapy for intermittent claudication. Drugs 24:240–247

52. Warlow C (1985) Transient ischaemic attacks, current treatment concepts. Drugs 29:474–482

53. Wetzner SM, Kiser LC, Bezreh JS (1984) Duplex ultrasound imaging: vascular applications. Radiology 150:507–514

54. Yates CIP, Berent A, Andrews V, Dormandy JA (1979) Increase in leg blood-flow by normovolaemic haemodilution in intermittent claudication. Lancet II:166–168

55. Zeitler E, Rochter EL. Roth FJ, Schoop W. (1983) Results of percutaneous transluminal angioplasty. Radiology 146:57–60

Arteriosclerotic Lesions of Carotid Arteries

C. F. Degos

Arterial thrombosis is a frequent cause of cerebral ischaemia. In fact, it is estimated that 75% of ischaemic episodes occurring in the carotid territory are due to underlying atheromatous vascular narrowing. Carotid artery thrombosis gives rise clinically to contralateral sensory and/or motor deficits that appear rapidly if not abruptly. Associated risk factors include old age [79] and known atherosclerotic vascular disease.

Faced with evidence of cerebral ischaemia, it is important for clinicians to understand the underlying pathophysiological mechanisms in order to institute appropriate and preferably definitive therapy.

Haemodynamic Factors

Pathophysiology of Cerebrovascular Thrombosis

Atheromatous obstruction develops gradually. It occurs preferentially at bifurcations in the arterial tree, particularly where the internal and external carotid arteries arise from the common carotid. When a 75% proximal stenosis of the internal carotid artery is detected, a similar lesion is present in the opposite carotid in three out of ten cases. Stenoses of the petrous portion of the internal carotid artery or of the proximal middle cerebral artery are rare, and in any case, not amenable to surgery. Extracranial stenoses of the internal carotid artery are generally well-tolerated until the degree of luminal obstruction exceeds 80%. Beyond this degree of narrowing, distal blood flow is significantly compromised. However, adequate cerebral profusion may still be obtained by reversal of flow in the ophthalmic artery which is supplied by the external carotid artery, or through anastomoses within the circle of Willis (anterior and posterior communicating arteries).

Thus, it is unusual to see clinical evidence of obstruction at the origin of the internal carotid artery. For it to be evident, circulation in the external carotid artery and in the circle of Willis would have to be impaired. Such impairment could result from the extension of thrombosis into these collateral vessels. Alternatively, an atheromatous plaque in the proximal part of the internal carotid artery could fragment under the pressure of collateral flow, giving rise to emboli that occlude useful anastomoses. This event is fortunately rare.

Neurological abnormalities generally appear when arteries distal to the internal carotid become occluded.

Mechanism of Vascular Occlusion

The growth rate of atheromatous plaques is difficult to predict and varies from individual to individual. As a general rule, stenosis lesions increase in size slowly over a period of several years. The risk of cerebral ischaemia is ever present, and is related at least in part to the topography of the occluding lesion. Ulcerated atheromatous plaques are probably more frequent than current imaging techniques would lead us to believe, and provide fertile ground for platelet deposition. Such platelet thrombi are friable and may give rise to emboli that lodge in distal arteries causing transient cerebral ischaemia or frank cerebral infarction. Actual fragments of atheromatous plaques may also embolize in this fashion.

The Diagnosis of Cerebral Ischaemia Due to Thrombosis

Clinical Features

Whereas transient ischaemic attacks (TIA) are almost always caused by cerebrovascular thrombosis, it may be more difficult to pinpoint the cause of hemispheric infarction. Cerebrovascular accidents (CVA) due to thrombosis generally evolve over a period of several hours, may occur during sleep and may be characterized by intermittent or stuttering progression (stroke in evolution). Cerebral haemorrhage, in contrast, is abrupt in onset, often accompanies physical exertion and is associated with headache, nausea and impaired consciousness. Haemorrhagic infarction generally occurs in individuals with antecedent hypertension, but this is not always a useful clue given the high frequency of arterial hypertension in individuals with atherosclerosis. In practice, all these clinical features provide only presumptive evidence for the nature of the underlying lesion. Physical examination may be equally uninformative as both haemorrhage and thrombosis may give rise to the identical constellation of neurological findings in the involved territory (see Table 21.1).

Table 21.1. Symptoms and signs of various arteriosclerotic lesions of carotid arteries

Deep hemispheric infarction
Contralateral hemiplegia proportional to infarct size
No sensory abnormalities
No hemianopsia
Verbal aphasia if dominant hemisphere is involved

Superficial hemispheric infarction
Contralateral hemiplegia predominantly of the face and arm
Sensory disturbances notably astereognosis
Lateral homonymous hemianopsia
Dominant hemisphere: Broca's aphasia and ideational and motor apraxia
Non-dominant hemisphere: asomatognosia and anosognosia

Massive hemispheric infarction
Severe hemiplegia proportional to infarct size
Hemianaesthesia
Hemianopsia
Marked disturbances in symbolic thinking

Anterior cerebral artery occlusion
Hemiplegia principally affecting the lower limb
Frontal signs (grasping, indifference, euphoria, stereotyped behaviour)

Anterior choroidal artery occlusion
Massive contralateral hemiplegia
Sensory disturbances
Hemianopsia

Carotid Bruits

Neurological examination must include auscultation of the neck and orbits [69]. The presence of a bruit at the angle of the jaw indicates stenosis at the carotid bifurcation. However, very tight stenoses may be silent, and the absence of a bruit is not necessarily reassuring.

The Role of Computerized Axial Tomography

Experience has shown that the clinical picture may be misleading in cases of cerebral ischaemia [65], and computerized axial tomography is often necessary to distinguish between haemorrhagic and thrombotic infarctions. Early haemorrhagic lesions are hyperdense and span a region not necessarily corresponding to a discrete vascular territory. In contrast, thrombotic infarctions are isodense initially, and affect a distinct cortical or subcortical region. Ultimately, these lesions become hypodense, but visualization may be enhanced with the infusion of contrast agents. Occasionally, a lesion may be seen at the junction of two territories supplied by different arteries ("watershed infarction"), both affected by thrombosis.

Differential Diagnosis

Even if carotid atherosclerosis is present, other possible causes of cerebral ischaemia must not be overlooked. Thus, a careful cardiac examination is indicated including, if necessary, electrocardiography, echocardiography and Holter monitoring to rule out the possibility of embolization from a cardiac source. Benign cerebral tumours (meningioma) or malignant tumours (glioma, metastases) may be associated with neurological symptoms suggesting cerebrovascular disease, but these lesions are generally identified by tomodensitometry. So-called "complicated migraine headaches" or episodes of hypoglycaemia may also mimic cerebrovascular accidents. The use of oral contraceptives particularly by women over 35 with a history of migraine headaches or with other risk factors for vascular disease is associated with an increased incidence of cerebral thrombosis [63,79].

Transient Ischaemic Attacks

Initial Approach

Although the neurological signs associated with transient ischaemic attacks (TIA) are happily regressive, it is in this clinical context that urgent investigations are warranted. Once a neurological deficit is established, it is usually too late for successful surgical intervention.

An individual who has suffered one or more TIA is at risk for the development of frank cerebral infarction. The magnitude of the risk is difficult to estimate, and many factors are probably involved. In approximately 50% of cases of cerebral infarction due to internal carotid artery stenosis, patients report one or more antecedent TIAs. However, the exact frequency of cerebral infarction following transient ischaemia in the territory of the internal carotid artery is unknown. The incidence ranges between 13% and 40% in different studies.

In practice, all TIAs are considered ominous, and every attempt should be made to localize the lesion responsible and to estimate its size. It is also important to identify any other supra-aortic stenoses and to evaluate the patency of distal vessels as well as the degree of collateral circulation.

Clinical Aspects and the Limits of Classification: Prolonged Reversible Ischaemic Neurological Deficits (PRIND) and Lacunar Infarction

TIAs give rise to a variety of symptoms and signs depending on the vascular territory involved (see Table 21.1): contralateral sensory and/or motor deficits may occur, as may speech deficits or mono-ocular blindness due to ophthalmic artery occlusion.

By definition, TIAs regress within 24 h, but this designation is not entirely satisfactory. Certain deficits associated with small cerebral infarctions may regress completely in a matter of days (PRIND) and these episodes should be handled like TIAs. Furthermore, early tomodensitometry has confirmed that true TIAs can often be associated with frank anatomical lesions. It is clear that serial scanning is necessary in cases of ischaemia in order to evaluate parenchymal integrity.

The term "mini-infarct" is often applied to lacunae or small cerebral infarctions. However, lacunar infarction refers to a specific entity associated with hypertension and atherosclerosis, characterized by fibrinoid necrosis of small, penetrating cerebral arteries. The resulting infarctions are small and there is a paucity of clinical signs that often regress over a period of several days: pure hemisensory deficits, hemiplegia, dysarthria and awkwardness of the hand, all associated with round, hypodense cerebral lesions. Therapy should be supportive and directed towards control of arterial hypertension and elimination of other risk factors. Co-existing carotid artery atherosclerosis should not be overlooked.

Ancillary Investigations

In experienced hands, Doppler flow studies are indispensable in the investigation of carotid artery stenosis [66]. This technique permits the localization of areas of narrowing in the internal carotid artery in the neck, and also provides valuable information about the external carotid artery, the direction of flow in the ophthalmic artery and the integrity of the circle of Willis. A Doppler examination that identifies a "tight stenosis" with echogenic material present in the arterial lumen signals a situation of extreme urgency. Even moderate stenoses without impairment of distal blood flow may be significant since non-obstructive but fissured or ulcerated atheromatous plaques invite thrombus formation.

Ultrasound studies [73,88] are non-invasive, devoid of risk and are thus useful in the initial assessment of patients with suspected carotid artery stenosis. If the clinical situation points to the desirability of surgery, this pre-operative arteriography is the investigation of choice. This technique provides information about the structure of stenoses, atheromatous plaques and about the general state of the intracranial vasculature. Arteriography is performed under anaesthesia. All four axes are injected. The route of infection is variable: either both brachial arteries and the left carotid are used, or selective angiography is performed after femoral artery catheterization.

Digital angiography is also becoming popular [82,83,89].

Treatment

Prior to surgery, parenteral anti-coagulant therapy with heparin may be instituted if the risk of embolization is considered great. Definitive therapy is surgical endarterectomy. This may occasionally require the use of a venous patch (generally taken from the saphenous vein) to enlarge the arterial lumen. In some cases, an intraoperative shunt is created if collateral flow is judged potentially inadequate. Following surgery, oral anti-coagulants are administered for a period of several months, and serial ultrasound examinations are performed to assess vascular patency.

Two clinical situations merit special attention. When bilateral carotid artery obstruction is present, the risk from surgery is great even when an intraoperative shunt is created. It may be advisable to avoid surgery in these circumstances. Bypass surgery has been attempted (temporo-sylvian anastomosis), but is waning in popularity since restoring blood flow does not reduce the risk of embolization.

Asymptomatic carotid artery stenoses also pose a therapeutic problem. Little is known about the natural history of such lesions, and most neurologists agree that surgery is not indicated in this situation [81]. Asymptomatic plaques may represent lesions with a smooth or endothelialized surface (ulcerated plaques are thought to heal after approximately 2 months). Asymptomatic stenoses may also reflect the existence of adequate collateral circulation. Medical therapy for prevention of recurrence is considered only in cases where surgery is contraindicated. Anti-platelet agents are used in preference to oral anti-coagulants because the risk of haemorrhage is great in elderly individuals with atherosclerosis and often hypertension.

Other treatment measures are often important [62] including blood pressure control [74,75], weight reduction [72], control of diabetes and hyperlipidaemia [77] and correction of polycythaemia [71,78,85]. The consumption of tea, coffee and alcohol does not seem to be important, and curiously smoking [64] does not constitute a significant risk.

Cerebrovascular Infarction

Diagnosis

In the early hours after a cerebral ischaemic event, it is difficult to predict whether neurological deficits will regress or persist. The distinction is nonetheless important since a less aggressive attitude is warranted in cases of cerebral infarction as opposed to cases of transient ischaemia without anatomical sequelae. Clinical examination and early tomodensitometry are generally unhelpful in making this distinction,

but nuclear magnetic resonance may become the most sensitive technique for the detection of early infarction [81]. In practice, patients are generally seen several hours after the onset of symptoms, and the passage of time often helps to differentiate between the two possibilities.

The clinical syndromes characteristic of cerebral infarction are easily recognized in the case of major artery occlusion (see Table 21.1). More discrete signs are the rule in occlusion of small distal vessels.

Apart from axial tomography, the only other tests warranted on an emergency basis are cardiac assessment and determination of serum glucose and electrolyte concentrations. No abnormalities are visualized by tomodensitometry in the first hours after cerebral thrombotic infarction, but within 24–48 h, a hypodense lesion is usually evident. In exceptional cases, lesions may be seen as early as 4 h after infarction. The degree of perilesional oedema may be estimated by scanning when there is displacement of neighbouring or of midline structures. This oedema generally disappears within a month. The injection of contrast medium "enhances" visualization of areas of infarction and attests to a disruption of the blood–brain barrier. Finally, tomodensitometry even without infusion of contrast medium can identify areas of haemorrhagic infarction.

Clinical Course

Over a period of 3 weeks following frank cerebral infarction, a number of histological changes occur in the ischaemic territory. The area of compromised circulation tends to diminish as blood flow is diverted from adjacent vascular territories and from arterioles within the pia mater. Local acidosis promotes vasodilatation in the infarcted area, but oxygen consumption remains low (superfluous perfusion).

Initial Medical Therapy

Initial therapy should include correction of electrolyte imbalances and control of extreme hypertension (BP > 200/100) using diuretics. Nursing care is also important, particularly the prevention of bed sores. In cases where cerebral oedema is marked, osmotic agents such as intravenous mannitol or oral glycerol may be used. It is generally agreed that corticosteroids should not be used systematically as was once the practice because these agents can aggravate hypertension and diabetes, and promote infection.

The use of anti-coagulants is controversial. There is a small risk of haemorrhage (6%) within a previously thrombotic infarction when anti-coagulation therapy is instituted. The more important question, however, is whether any benefit can be derived from anti-coagulation therapy. Some authors [68] feel that thrombus extension and formation of emboli can be reduced. The only clearcut indication for anti-coagulation therapy is the so-called "stroke in evolution" [84]. In this condition deterioration is twice as frequent in untreated patients.

There does not appear to be any justification for the use of vasodilators. Generally, vessels within infarcted areas are already dilated, and further vasodilatation may result in circulatory steal from other cerebral territories.

The Role of Surgery in Carotid Artery Stenosis

The indications for surgery are very limited. Apart from the rare instance of significant carotid obstruction in the neck diagnosed within the first few hours after the onset of ischaemic symptoms, surgery is contraindicated. Immediate reperfusion can in fact aggravate cerebral infarction, and surgery is generally deferred for at least 1 month or until tomodensitometry indicates substantial infarct resolution.

Long-Term Medical Therapy

Long-term therapy includes control of arterial hypertension and of other risk factors for the progression of atherosclerosis [72]. Anti-platelet agents may also be of benefit [58,60,61,67,70,87]. There is as yet no justification for the use of vasodilators or other medications that enhance glucose or oxygen consumption.

Physiotherapy and speech therapy are important, if sometimes arduous. Rehabilitation of this nature is not always adequately subsidized by insurers.

References

56. Anon. (1984) Accidents artériel cérébraux. Rev Prat 34:1091–1154

57. Anon. (1977) Accidents cérébraux ischémiques transitoires. Rev. Prat 27:1–76

58. Bousser MG, Eschwege E, Hagueneau M et al (1981, 1982, 1983) Essai controlé AICLA. Prévention secondaire des accidents ischémiques cérébraux liés à l'athérosclérose par l'aspirine et le dipyridamole. Rev Neurol 137:333–341, 138:1–5, 139: 335–348

59. Cambier J, Masson M, Dehen H (1985) Neurologie, 5th edn. Masson, Paris

60. Canadian Cooperative Study Group (1978) A randomised trial of aspirine and sulfinpyrazone in threatened stroke. N Engl J Med 299:53–59

61. Candelise L, Lanli G, Perrone P et al. (1982) A randomised trial of aspirine and sulfinpyrazone in patients with TIA. Stroke 13:175–179

62. Capron L (1984) Epidémiologie et facteurs de risque des accidents artériels cérébraux. Rev Neurol 140:162–170

63. Collaborative group for the study of stroke in young women (1973) Oral contraception and increased risk of cerebral ischemia or thrombosis. N Engl J Med 288:871–878

64. Dawber TR, Wolf PA, Thomas HE et al (1975) Epidemiology of cerebral accidents due to atherosclerosis. In: Castaigne P, Lhermitte F, Gauthier JC (eds) Maladies vasculaires cérébrales. Conferences de la Salpétrière. Baillière, Paris

65. Derouesne C (1983) Pratique neurologique. Flammarion Paris

66. Dutreix JL, Genre O, Monegier du Sorbier C et al. (1985) Corrélations ultrasoniques, artériographiques et anatomo-pathologiques dans 59 cas d'athérosclérose carotidienne. Rev Neurol 141:128–136

67. Fields WS, Lemak AN, Frankowski RF, Hardy RJ (1977, 1978) Controlled trial aspirine in cerebral ischemia. Stroke 8:301–316, 9:309–319

68. Garde A, Samuelson K, Fahlgren H et al. (1983) Treatment after transient ischemic attacks: a comparison between anticoagulant drug and inhibition of platelet aggregation. Stroke 14:677–681

69. Gautier JC, Rosa A, Lhermitte F (1975) Auscultation carotidienne. Corrélation chez 200 patients avec 322 angiographies. Rev Neurol 131:175–184

70. Guiraud-Chaumeil B, Rascol A, David J et al. (1982) Prévention des récidives des accidents vasculaires cérébraux ischémiques par les anti-agrégants plaquettaires. Rev Neurol 138:367–385

71. Harrison MJG, Pollock S, Kendall BE, Marshall J (1981) Effect of haematocrit on carotid stenosis and cerebral infarction. Lancet II:114–115

72. Hubert HB, Feinleib M, McNamara PM, Castelli WP (1983) Obesity as an independant risk factor of cardiovascular disease: a 26 year follow-up of participants in Framingham study. Circulation 67:968–977

73. Humber PR, Leipold GR, Wickbom IG, Bernstein EF (1980) Ultrasonic imaging of the carotid arterial system. AM J Surg 140:199–202

74. Hypertension detection and follow-up program Cooperative Group (1982) Five year findings of the hypertension detection and follow-up program. III. Reduction of stroke incidence among persons with high blood pressure. JAMA 247:633–638

75. Kannel WB, Dawber TR, Sorlie P, Wolf PA (1976) Components of blood pressure and risks of atherothrombotic brain infarction: the Framingham study. Stroke 7:327–331

76. Kannel WB, Gordon T (1978) Evaluation of cardiovascular risk in the elderly: the Framingham study. Bull NY Acad Med 54:573–591

77. Kannel WB, Gordon T, Dawber TR (1974) Role of lipids in the development of brain infarction: the Framingham study. Stroke 5:679–685

78. Kannel WB, Gordon T, Wolf PA, McNamara P (1972) Hemoglobin and the risk of cerebral infarction: the Framingham study. Stroke 3:679–685

79. Kannel WB, Wolf PA (1983) Epidemiology of cerebrovascular disease. *In:* Ross RP (ed) Vascular diseases of the central nervous system, Churchill Livingstone, Edinburgh, pp 1–24

80. Kistler JP, Buonanno FS, Dewitt LD et al. (1983) Application de la résonance magnétique nucléaire à l'étude des maladies cérébro-vasculaires. Presse Med, 48:3086–3090

81. Laplane D, Josse MO (1983) La chirurgie des sténoses carotidiennes asymptomatiques est-elle légitime? Ann Med Interne 134:663–675

82. Little JR, Furlan AJ, Modic MT, Weintein MA (1982) Digital substraction angiography in cerebrovascular disease. Stroke 13:557–566

83. Manelfe C, Ducos de Lahitte M, Marc-Vergnes JP et al. (1982) Investigation of extra-cranial cerebral arteries by intravenous angiography: report of 1000 cases. Am J Neuroradiol 3:287–293

84. Millikan CH, McDowell FH (1981) Treatment of progressing stroke. Stroke 12:397–409

85. Moulias R, Congy F, Mercier M et al (1982) l'élévation de l'hématocrite est-elle in facteur de risque vasculaire cérébral? Etude prospective chez 150 sujets ayant eu un accident vasculaire cérébral et chez 150 témoins appariés. Nouv Presse Med II:567–570

86. Ramollissements cérébraux. Rev Prat 26:595–730

87. Sorensen PS, Pedersen H, Marquarsden J et al. (1982, 1983) Acetyl salicylic acid in the prevention of stroke in patients with reversible ischemic attacks. Acta Neurol Scand 65:176–177, Stroke 14:15–22

88. Wolverson MK, Heiberg E, Sundaram M et al. (1982) Carotid atherosclerosis: high-resolution real-time sonography correlated with angiography. Am J Neuroradiol 3:601–607

89. Zwiebel WL (1982) Comparison of ultrasound and digital substraction arteriography. *In:* Mistretta CA, Crummy AB, Strother CM, Sacketts JF (eds). Digital substraction arteriography: an application of computerized fluoroscopy. Year Book Medical Publishers, Chicago, pp 45–52

Atheromatous Stenosis of Renal Arteries

J.-B. Michel, G. Chatelier and J. Ménard

Introduction

While the frequency of arterial hypertension in a determined population is easy to establish, the frequency of atheromatous stenosis of the renal arteries in a population of hypertensive patients is more difficult to define. The reported fre-

quency of renovascular hypertension varies from 1% to 10% in populations of hypertensive patients. Variation is due essentially to the diagnostic screening methods used but also to the age of the patient population screened [92,93, 94,105,109]. In a retrospective study on 1122 intravenous pyelographies of renal arteries performed for the aetiological diagnosis of hypertension, we have shown that the frequency of renal artery stenosis was around 8.6%, and moreover that, on this sample of hypertensive patients, the frequency of renal artery stenosis increased with age [119] (Fig. 21.2). The importance of renal artery stenosis in hypertensive patients needs to be established by a prospective epidemiological study. This is now possible because of the high sensitivity and the low morbidity associated with digital intravenous angiography used in screening.

The association of atheromatous stenosis of a renal artery with a high blood pressure level does not mean that a direct relationship exists between the two. During autopsies [95,129] or analysis of abdominal aortographies obtained from arteritic patients [102,104,136], it is possible to show that a significant degree of stenosis of a renal artery can be associated with a normal blood pressure. Moreover, studies have clearly shown that hypertension commonly coexists with atheroma, speeding up the evolution of atheromatous disease regardless of the territory (coronaries, carotids, renal artery). The relationship between renal artery stenosis and hypertension is thus complex; an atheromatous stenosis could be the cause and/or the consequence of hypertension.

These concepts explain why it is necessary though difficult to establish a direct relationship between stenosis of the renal artery and high

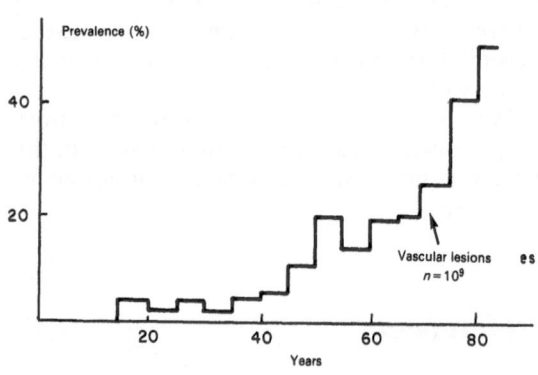

Fig. 21.2. Progressive increase in the frequency of atheromatous stenosis of renal arteries in relation to age in hypertensive patients [117].

blood pressure. The methods used to investigate the cause of renovascular hypertension have been based on and continue to be based on the experimental approach previously described by Golblatt in 1934 [106].

The evolution of atheromatous stenosis of the renal artery may lead to complete occlusion, which seems to be much more frequent when the stenosis is tight [128,142]. This in turn may lead to renal insufficiency of vascular origin. Although stenosis of the renal artery can be asymptomatic or lead to renal insufficiency, the most frequent symptom of renovascular disease is hypertension. We develop this aspect in the following section.

Strategy of Diagnosis

Diagnosis of Renovascular Hypertension

The investigation of the aetiology of reno-vascular hypertension is necessary if a specific vascular disease is to be treated effectively and without morbidity.

Renal artery stenosis has no specific clinical symptoms [131]. Statistically, atheromatous stenosis of renal arteries is more frequent in hypertensive patients over the age of 45, when hypertension is severe, when there is no family history of hypertension (only one-third of reno-vascular hypertensive patients have a family history and when patients are white and smokers. Three clinical findings are particularly relevant for the diagnosis of renovascular hypertension: recently apparent or recently aggravated, hypertension resistance to anti-hypertensive treatment or hypertension which is particularly sensitive to inhibitors of the renin–angiotensin system [139]. The specificity of an abdominal bruit is very low under these circumstances because of the extent of atheroma in the aorta and its visceral branches. Hypokalaemia related to secondary hyperaldosteronism may be taken as indirect evidence of renovascular hypertension.

The visceral consequences of this hypertension are often strongly related to the blood pressure level and the visceral action of vasoconstrictive hormones. Retinopathy, cardiac hypertrophy (detected by the Sokolov index on ECG, the cardiothoracic ratio and echocardiography) renal disease (detected by proteinuria and a decrease in creatinine clearance) may all follow the rise in pressure.

Diagnosis of the Stenosis

A positive diagnosis of renal artery stenosis may be made using intravenous digital angiography showing both the renal arteries and aorta [126]. This radiographical method is used more often than intravenous pyelography in the screening of renal artery disease. If specificity of functional intravenous pyelography (IVP) is good (0.98) in the diagnosis of renal artery stenosis, sensitivity of IVP is low (0.48) [119]. Digital intravenous angiography may be used in association with functional IVP with two objectives:

1. Diagnosis of the consequence of unilateral renal artery stenosis on exocrine function of the kidney can be appreciated by studying delayed secretion and wash out of the stenosed kidney. The macroscopic morphology can be appreciated by comparing the length of the two kidneys and by observing the ratio between the excretory cavities and the cortex. Echography also permits measurement of the length of the kidneys, determination of the ratio between cortex and medulla, possible dilatation of excretory cavities and presence of cysts in the diseased kidney.

2. IVP also permits the diagnosis of associated uropathy, uni- or bilateral nephropathy, renal lithiasis, pyelonephritis, congenital hypoplasia of the kidney, tumour and partial renal infarct.

Renal Consequence of the Stenosis

The consequence of the stenosis on the endocrine function of the kidney is determined by analysis of the renin–angiotensin system. Analysis should include measurement of levels of enzymatic activity and of the molecular concentration [117] of renin in the peripheral blood, measurement of the same levels in the renal veins and the analysis of the clinical and hormonal effects of the pharmacological blockade of the renin–angiotensin system in these patients. The enzymatic activity of renin is measured by the radioimmunoassay of angiotensin I generated in the plasma of the patient. The direct measurement of molecular concentrations of renin is now possible by direct solid phase immunoassay [117].

Theoretically, in renovascular hypertension there is always a discrepancy between renin secretion by the kidney and homeostasis of salt. There are a lot of practical problems in the exact determination of the relationship of plasma renin activity (PRA) to salt homeostasis in renovascular hypertension. Difficulties arise in choosing the physiological conditions under which to take the sample. Posture, stress, exercise, diuretic treatments, salt intake, plasma volume etc, may all be varied. Each group of specialists has its own method.

The results obtained by the groups in Montreal and New York are representative of the literature. Messerli et al. [118] have shown an improvement of hypertension and an elevated level of plasma renin activity in 19 of 32 patients treated by surgical revascularization of the kidney. By contrast, the 13 patients in which surgery failed to improve hypertension, only one had a high level of plasma renin activity. For Vaughan et al. [138], 13 of 18 patients improved by surgery had increased plasma renin activity, whereas PRA was increased in only one of the six patients not improved by surgery. The measurement of PRA in peripheral blood samples is thus an indicator for therapeutic intervention [127], but cannot be proposed as a screening test because of its low sensitivity and specificity in detecting renovascular hypertension [107].

The fall in blood pressure obtained by inhibition of the renin–angiotensin system and the associated increase in the circulating renin level are also possible indicators of the participation of the renin–angiotensin system in hypertension. Before the appearance of converting enzyme inhibitors, intravenous injection of saralazine, a competitive antagonist of angiotensin II, had been used. Possibly this drug has some agonist properties and mimics angiotensin II in positive salt homeostasis [97,99].

Converting enzyme inhibitors have no agonist properties [98]. In 15 patients with renovascular hypertension, with a normal salt intake and without any drug intake during the week before investigation, acute converting enzyme inhibition induced a 15% decrease of blood pressure and a 457% increase of the PRA [135]. Nevertheless, the sensitivity and the specificity of this pharmacological test to distinguish between renovascular and essential hypertension do not permit its use in the screening of hypertension. Other studies have shown that the fall in blood pressure obtained by chronic converting enzyme inhibition treatment has a better predictive value than the acute test in the prognosis of renal revascularization [91].

The simultaneous measurement of plasma renin activity in both renal veins and in vena cava is probably the most accurate way of demonstrating the stimulation of renin secretion in the stenosed kidney and its freination on the opposite side. Sealey et al. [130] have measured plasma renin activity in blood samples from both renal veins, and from vena cava or aorta in 43 essentially hypertensive patients. They demonstrated that the plasma renin activity in the renal vein was always 24% higher than plasma renin activity in peripheral blood samples. Vaughan et al. [138] studied the value of this test for predicting the level of blood pressure which resulted from renal revascularization in relation to the separated function of both kidneys and to data obtained in essential hypertension. On the stenosed side a ratio (PRA in renal vein: PRA in vena cava) higher than 0.48 is related to a stimulation of renin secretion, whereas in the opposite kidney, a ratio lower than 0.24 indicates an arrest of secretion or a retention of renin by a healthy kidney. The prognosis of hypertension after renal revascularization depends on both the degree of stenosis of the diseased kidney (inappropriate secretion of renin) and the absence of disease in the contralateral kidney (arrest of renin secretion). Several studies have been proposed to potentiate the sensitivity of this test by pharmacological stimulation. Presently, the acute administration of a single dose of a converting enzyme inhibitor before the catheterization of renal veins is used.

More recently, studies of glomerular filtration in the kidney distal to the stenosis, using measurement of the clearance of DPTA before and during inhibition of converting enzyme, have permitted the demonstration of the role of the renin angiotensin system in the preservation of glomerular filtration in the stenosed kidney [140] and seem to have predictive value for the outcome of revascularization.

It is clear that there is no ideal method for screening renovascular hypertension. Digital intravenous angiography is probably the best screening test for renovascular disease, but there are no pathophysiological tests which predict with certainty the relationship between hypertension and stenosis of the renal artery, and which thus predict the outcome of renal revascularization. Nevertheless, in our experience, failure of perfect renal revascularization to improve hypertension in renovascular disease is rare.

Diffusion of the Disease

The systemic consequences of renovascular hypertension related to atheromatous stenosis of renal arteries are of two types: organic consequences of hypertension and the effect of atherosclerosis on other organs. These two aspects are always inter-related but hypertension is not simply an additional risk factor of atheroma and conversely, diffuse atheroma alone is not a cause of hypertension.

Three organs are particularly sensitive to renovascular hypertension: the heart, the brain and the kidney. In renovascular hypertension, when the peripheral resistance and the impedance of the aorta increase in relation to the high level of circulating angiotensin, the myocardium, and particularly the left ventricle cardiac muscle, is progressively adapted to this new afterload by concentric hypertrophy and by decreased contractility. The contraction of the left myocardium generates more force but less velocity in an attempt to maintain systolic wall stress and the energy efficiency of contraction within normal limits. Later on, these adaptive mechanisms can be overtaken and cardiac insufficiency may result. These mechanisms are not specific to renovascular disease but are specific to hypertension. In contrast, a deficit in potassium associated with secondary hyperaldosteronism, which may be aggravated by diuretic prescription, is more specific for the pathophysiology of renovascular hypertension. Hypokalaemia may carry an arrhythmogenic risk which is potentiated by cardiac hypertrophy, fibrosis and inotropic drugs [121,123]. As we have seen, the cardiac consequences of hypertension are demonstrated by the cardiothoracic ratio on thoracic radiography, by ECG and by echocardiography.

As the myocardium increases its mass in relation to increase in afterload, the walls of the arteries also increase their mass by hyperplasia of muscle cells. In arterioles, this structural increase in wall thickness, called hypertensive microangiopathy, may be responsible for distal occlusion leading to focal ischaemia. This type of lesion has been described in the contralateral unstenosed kidney and in the brain. In the latter, lacuna infarcts can be symptomatic or asymptomatic.

In hypertension of renovascular origin the consequences of high blood pressure (microangiopathy, lacunae, nephronic ischaemia and glomerulosclerosis) are only seen in the contralateral unstenosed kidney, [108,141] unless hypertension existed before the onset of renovascular disease or the stenosed kidney has been revascularized without normalization of blood pressure, in which case glomerulosclerosis can be seen in the stenosed kidney. The stenosed kidney is however, usually protected against hypertension and does not develop this type of lesion, but can show lesions with more diffused ischaemia related to the stenosis. Increase in thickness of the arterial wall is reversed by normalization of blood pressure, whereas glomerulosclerosis is probably irreversible. The latter can be prevented by normalization of perfusion pressure.

Renovascular hypertension can be induced during atherogenesis by three distinct mechanisms:

1. Physical mechanisms: the increase in perfusion pressure in large arteries leads to an increase in the transudation of macromolecules, particularly of lipoproteins, from the arterial lumen to the arterial wall.

2. Structural mechanisms, secondary to the increase in the medial thickness due to smooth-muscle cell hyperplasia in the arterial wall.

3. Functional mechanisms, secondary to the contractile effect of angiotensin II on endothelial and smooth-muscle cells of large arteries. The relationships between hypertension and atherosclerosis are thus complex and multifactorial.

At the time of evaluation of the organic consequences of hypertension, evaluation of atherosclerosis in other vascular territories should be carried out, for example in coronary arteries, carotids, the abdominal aorta, and so on. These associated localizations should be studied when symptoms such as angina pectoris, transitory ischaemic attacks or intermittent claudication are present. A complete clinical examination of all the vascular territories should include pulse palpation, a search for bruits and measurement of blood pressure differences between both arms and between arms and legs. This clinical investigation may be completed by complementary non-invasive vascular study such as doppler examination of the carotids and echography of the abdominal aorta. Also, the pathological state of the abdominal aorta should be evaluated by angiography. On 1122 arterial visualizations performed because of the presence of hypertension, there were 123 arterial lesions in 109 patients including 103 examples of renovascular disease, 13 significant aortic stenoses, five aneurysms of the abdominal aorta and two aneurysms of the splenic artery [119]. The presence of both renovascular hypertension and an aneurysm of the abdominal aorta should indicate the urgency of obtaining complete normalization of blood pressure, because high blood pressure level increases the risk of rupture. The association of renovascular hypertension with chronic arteriopathy of the lower extremities should indicate treatment without beta blockers which aggravate the functional symptoms of lower extremity arteriopathy.

In conclusion, the visceral consequences of hypertension and the extent of atherosclerosis are often important in renovascular hypertension. For example, when cardiac insufficiency exists in patients with angina pectoris and renovascular hypertension, it is difficult to determine the role respective of hypertension, the activation of the renin–angiotensin system and the diffusion of the disease.

Therapeutic Approach

The treatment of hypertension which is secondary to atheromatous stenosis of the renal arteries may be medical or surgical. Surgical treatment may involve nephrectomy or revascularization of the kidney. The more logical treatment is to produce regression of the activation of the renin–angiotensin system and this can be achieved by removing the ischaemic kidney, by removing the stenosis or by using drugs which inhibit the renin–angiotensin system.

This secondary hypertension is often resistant to conventional anti-hypertensive medical therapy [103,115]. Moreover, Hunt and Strong [113] have shown that the morbidity and the mortality were greater in a group of patients treated medically than in a group of patients treated by revascularization of the kidney. This study [113] is unique because of the number of patients included and the duration of the survey. Although prospective, it was not randomized and the most severely affected patients were not surgically treated. Dean [100] raised some questions about the long-term effects of anti-hypertensive therapy on the function of the stenosed kidney. In renovascular hypertension, converting enzyme inhibitors are particularly efficient in decreasing blood pressure, and may have no effect on plasma creatinine when the contralateral kidney is untouched [110,111]. However, they could impair renal function when a solitary kidney is stenosed or when the stenosis is bilateral [90,112]. In fact, when the stenosis is unilateral and when the contralateral kidney is

untouched, the consequences of the inhibition of converting enzyme in the stenosed kidney are similar, but cannot be detected by the level of peripheral plasma creatinine. To analyse the consequences of the inhibition of the renin–angiotensin system it is also necessary to perform studies on the separate function of both kidneys [110,140]. Moreover these types of treatment seem to increase the frequency of complete occlusion of the stenosed renal artery [111,143]. In experimental rats with renovascular hypertension we have demonstrated that chronic treatment with converting enzyme inhibitors decreased the weight and increased the frequency of ischaemic lesions in the kidney distal to the stenosis [120]. In man, the long-term effects of converting enzyme inhibition on the morphology of the stenosed kidney are not known.

The first type of surgery used to cure renovascular hypertension was nephrectomy [96]. The indications for nephrectomy in renovascular hypertension are now limited to a small and non-functional kidney (length <8 cm) in high-risk patients with associated renal parenchymatous disease and iatrogenic occlusion or dissection of the renal artery. Partial nephrectomy may be indicated in segmental infarctions or when arterial disease is very distal and localized.

Renal revascularization is the most suitable method for treatment of renovascular hypertension. Revascularization of the kidney decreases blood pressure by suppressing the abnormal activation of the renin–angiotensin system and permits the conservation of the exocrine function of the kidney. The surgical results are improved by good selection of patients and by progress in the materials used: the obsolescence of dacron grafts and the use of saphenous vein grafts or direct arterio-arterial anastomosis when possible [116,124] have improved results which are predominantly related to the technological quality of the graft. The most frequent causes of failure are thrombosis, restenosis or dilatation of venous grafts [133]. In renovascular hypertension of atheromatous origin, the majority of people do not recover spontaneously (i.e. complete normalization of blood pressure is not obtained without drugs) but rather a significant improvement (i.e. decrease in blood pressure level and a decreased intake of anti-hypertensive drugs) is associated with a regression of visceral consequences of hypertension.

Since 1978, a new method of revascularization by endoluminal dilatation of the stenosis has become available for the treatment of renovascular hypertension. In this case technological progress (slenderness of the catheter, quality of balloons) has also lead to improvement of results.

The majority of atheromatous stenoses of renal arteries can be treated by endoluminal dilatation; by contrast ostial diseases associated with atheromatous plaques of the aortic wall are not amenable to endoluminal dilatation [132,137]. This method is less invasive than surgery but, as with surgery, carries a risk of dissection and of complete occlusion of the renal artery. The long-term effects of the renal revascularization method on atheromatous stenoses and on hypertension are still unknown.

Two specific points should be emphasized. The possibility of asymptomatic renal artery stenosis arises when a stenosis is shown on an angiography performed for other problems as aneurysm of the aorta or arteriopathy of the lower limb, in a normotensive patient. If surgery is to be carried out on the abdominal aorta, preventive revascularization of the kidney can be performed simultaneously [141], although no prospective study has justified this surgical intervention. In renal insufficiency of vascular origin, renal revascularization can be of very significant value when the renal tissue distal to the stenosis or occlusion is still living and functional [101,114,125]. The prognostic criteria for revascularization performed for renal insufficiency are the length of the kidney, the conservation of the ratio between medulla and cortex and the absence of acquired cystic disease in the kidney. These can be checked by echography. The existence of an occlusion (more than a simple stenosis) of the renal artery and the results of a kidney biopsy can also help to predict the outcome of renal revascularization on kidney function [122].

References

90. Aldigier JC, Plouin PF, Guyenne TT et al. (1982) Comparison of the hormonal and renal effects of captopril in severe essential and renovascular hypertension. Am J Cardiol 49:1447–1452
91. Atkinson AB, Brown JJ, Cumming AMM et al. (1982) Captopril in the management of hypertension with renal artery stenosis: its long term effect as a predictor of surgical outcome. Am J Cardiol 49:1460–1466
92. Ayers CR, Slaughter AR, Smallwood MD et al. (1973) Standards for quality care of hypertensive patients in office and hospital practice. Am J Cardiol 32:533–545
93. Bech K, Hilden T (1975) The frequency of secondary hypertension. Acta Med Scand 197:65–69
94. Berglund G, Ove A, Wilhelsen L (1976) Prevalence of primary and secondary hypertension: studies in a random population sample. Br Med J ii:554–556
95. Berthaux P, Beck H, Polet D et al. (1962) Les sténoses des artères rénales chez le vicillard. Sem Hôp Paris 61:3451–3454
96. Butler AM (1937) Chronic pyelonephritis and arterial hypertension. J Clin Invest 16:889–897
97. Carey RM, Vaughan ED, Ackerley JA et al. (1978) The immediate pressor effect of saralasin in man. J Clin Endocrinol Metab 46:36–43
98. Case DB, Wallace JM, Keim HA et al. (1976) Estimating renin participation in hypertension: superiority of converting enzyme inhibitor over saralasin. Am J Med 61:790–796
99. Case DB, Wallace JM, Keim HJ et al. (1976) Usefulness and limitations of saralasin, a weak competitive agonist for angiotensin II. for evaluating the renin and sodium factors in hypertensive patients. Am J Med 60:825–836
100. Dean RH, Kieffer RW, Smith BM et al. (1981) Renovascular hypertension. Anatomic and renal function changes during drug therapy. Arch Surg 116:1408–1415
101. Dean RH, Shack RB, Rhamy RK et al. (1977) The effect of renal revascularization on kidney function. J Surg Res 22:443–448
102. Dustan HP, Humphries AW, De Wolfe VG, Page IH (1964) Normal arterial pressure in patients with renal arterial stenosis. JAMA 187:1028–1029
103. Dustan HP, Page IH, Poutasse EF, Wilson (1963) An evaluation of treatment of hypertension associated with occlusive renal arterial disease. circulation 27:1018–1027
104. Eyler WE, Clark MD, Garman JE et al. (1962) Angiography of the renal areas including a comparative study of renal arterial stenosis in patients with and without hypertension. Radiology 78:879–891
105. Gifford RW (1969) Evaluation of the hypertensive patient with emphasis on detecting curable causes. Milbank Q 47:170–186
106. Golblatt M, Lynch J, Hanzal RF, Summerville WW (1984) Studies on experimental hypertension. I. The production of persistent elevation of systolic blood pressure by means of renal ischaemia. J Exp Med 59:347–379
107. Grim CE, Luft FC, Weinberger NH, Grim CM (1979) Sensitivity and specificity of screening test for renovascular hypertension. Ann Intern Med 91:617–622
108. Heptinstall RH, Hill GS (1967) Steroid induced hypertension in the rat. A study of the effects of renal artery constriction on hypertension caused by doxycorticosterone. Lab Invest 16:751–767
109. Hillestad L (1968) Systemic arterial hypertension. Aspect of etiology and pathogenesis in a retrospective study of an hospital material. Acta Med Scand 184:225–229
110. Hodsman GP, Brown JJ, Cumming AMM et al. (1984) Enalapril in treatment of hypertension with renal artery stenosis. Changes in blood pressure renin, angiotensin I and II, renal function and body composition. Am J Med 20:52–60
111. Hollenberg NF (1983) Medical therapy of renovascular hypertension: efficacy and safety of captopril in 259 patients. Cardiovasc Rev Rep 3:673–676
112. Hricik DE (1983) Captopril induced functional renal insufficiency in patients with bilateral renal artery stenosis in a solitary kidney. N Engl J Med 308:273–27
113. Hunt JC, Strong CS (1973) Renovascular hypertension:

mechanisms, natural history and treatment. Am J Cardiol 32:562–574

114. Jamieson CG, Clarkson AR, Wookroffe AJ, Faris I (1984) Reconstructive renal vascular surgery for chronic renal failure. Br J Surg 71(32):338–340

115. Kjellbo H, Hund N, Bergentz SE, Hood B (1970) Renal artery stenosis and hypertension. Scand J Urol Nephrol 4:43–47

116. Lagneau P, Michel JB (1981) Arterial reconstructive surgery for renovascular hypertension. Arch Surg 116:999–1002

117. Menard J, Guyenne TT, Corvol P et al. (1985) Direct immunometric assay of active renin in human plasma. J Hypertension 3:275–278

118. Messerli FF, Genest J, Nowaczynski W et al. (1975) Hypertension with renal artery stenosis: humoral, hemodynamic and histopathologic factors. Am J Cardiol 36:702–707

119. Michel JB, Chastang C, Girard J (1982) Apport de la visualisation vasculaire associée à l'urographie intraveineuse dans le diagnostic de l'hypertension, artérielle. ADELF, Hôpital Necker, Paris

120. Michel JB, Dussaule JC, Choudat et al. (1986) Renal and hormonal effects of antihypertensive treatment in one clip two kidney hypertension in rats. Kidney Internat 29:1011–1020

121. Michel JB, Menard J, Dussaule JC et al. (1984) Can the inhibition of the renin angiotensin system have a cardioprotective effect. J Cardiovasc Pharmacol 7:S75–S79

122. Michel JB, Nussaume O, Valverde JP et al. (1985) Revascularisation renale tardive. J Mal Vasc 10:221–226

123. Michel JB, Salzmann JL, Ossondo Nlom N et al. (1986) Morphometric analysis of collagen network and plasma perfused capillary bed in the myocardium of rats during evolution of cardiac hypertrophy. Basic Res Cardiol 81:142–155

124. Noble MJ, Novick AC, Straffon RA, Stewart BA (1979) Aortorenal reimplantation in treatment of renovascular hypertension. Urology 14:566–569

125. Novick AC, Textor SC, Bodie B, Khauli RB (1984) Revascularization to preserve renal function in patients with atherosclerotic renovascular disease. Urol Clin North Am 11:477–490

126. Osborne RW, Goldstone J, Hillman BJ (1981) Digital videosubtraction angiography: screening technique for renovascular hypertension. Surgery 90:932–939

127. Pickering TG, Sos TA, Vaughan ED et al. (1984) Predictive value and changes of renin secretion in patients undergoing successful renal angioplasty. Am J Med 76:398–404

128. Schreber MJ, Pohi MA, Novick AC (1984) The natural history of atherosclerotic and fibrous renal artery disease. Urol Clin North Am 11:383–392

129. Schwartz CF, White IA (1964) Stenosis of the renal artery: an unselected necropsy study. Br Med J ii:1415–1421

130. Sealey JE, Buhler ER, Vaughan ED, Laragh JH (1973) The physiology of renin secretion in essential hypertension: estimation of renin secretion rate and renal plasma flow from peripheral and renal vein renin levels. Am J Med 55:391–401

131. Simon N, Franklin SS, Bleifer KH, Maxwell MH (1972) Clinical characteristics of renovascular hypertension. JAMA 220:1209–1218

132. Sos TA, Pickering TG, Saddekni S et al. (1984) The current role of renal angioplasty in the treatment of renovascular hypertension. Urol Clin North Am 11:503–513

133. Stanley JC, Ernst CB, Fry WJ (1973) Fate of 100 aortorenal vein grafts: characteristics of late graft expansion, aneurysmal dilatation and stenosis. Surgery 74:931–944

134. Thibonnier M, Joseph A, Sassano P et al. (1984) Improved diagnosis of unilateral renal artery lesions after captopril administration. JAMA 251:56–60

135. Thibonnier M, Sassano P, Plouin PE et al. (1982) diagnostic value of a single dose of captopril in renin and aldosterone dependent surgically curable hypertension. Cardiovasc Rev Rep 3:1659–1668

136. Van Velzer DA, Burge CM, Morris GC (1961) Atherosclerotic narrowing of renal arteries associated with hypertension. AJR 86:807–818

137. Vaughan ED (1985) Renovascular hypertension. Kidney Int 27:811–827

138. Vaughan ED, Buhler ER, Laragh JH et al. (1973) Renovascular hypertension: renin measurements to indicate hypersecretion and controlateral suppression, estimate renal plasma flow, and score for surgical curability. AM J Med 55:402–414

139. Vaughan ED, Case DB, Pickering TG et al. (1984) Clinical evaluation for renovascular hypertension and therapeutic decisions. Urol Clin North Am 11:393–408

140. Wenting GJ, Tan Tjong HL, Derkx FMM et al. (1984) Split renal function after captopril in unilateral renal artery stenosis. Br Med J 288:886–890

141. Wilson C, Byrom FB (1939) Renal changes in malignant hypertension experimental evidence. Lancet I:136–139

142. Wollenveber J, Sheps SG, David DG (1968) Clinical course of atherosclerotic renovascular disease. Am J Cardiol 21:60–71

143. Ying CY, Tifft CP, Gavras H, Chobanian AV (1984) Renal revascularization in the azotemic hypertensive patient resistant to therapy. N Engl J Med 31:1070–1075

Coronary Arteries

J. Bonnet

Reliable diagnosis of coronary atherosclerosis is hindered by the intrathoracic location of the small-diameter coronary arteries, themselves attached to a mobile organ. It is practically impossible to demonstrate lesions in these arteries using my direct non-invasive method, and the diagnosis normally rests on evaluation of the functional effects of the lesions. Often, these are only apparent late in the course of the disease, or are detected by incidental radiological investigations. The latter rely on invasive techniques, and thus pose both practical and ethical problems.

The progressive but variable course of coronary atherosclerosis means that the more serious conditions are the easiest to diagnose.

Diagnosis

Clinical Findings

The clinical diagnosis of coronary athero-sclerosis can only be made late in the course of the condition, whose end result is angina pectoris or myocardial infarction [144,147]. The essential feature is gripping chest pain of variable intensity, radiating towards the lower jaw, shoulders and arms. It tends not to be affected by breathing in one lying down. Proper evaluation of the circumstances triggering the symptoms as well as their duration and any associated signs is essential, since the gravity and instability of the angina can indicate the presence of myocardial infarction. This is strongly suggested by onset of angina at rest which lasts at least 20 min even after administration of glycerol trinitrate, associated with vagal symptoms including sweating, nausea, malaise and occasionally dyspnoea. Angina pectoris, on the other hand, is indicated by shorter duration of chest pain which is relieved by glyceryl trinitrate within 20 min, which is reliably triggered by exercise or equivalent factors such as digestion, excitement, anger, nightmares, cold, wind and awakening and which ceases in the absence of exertion. Onset of chest pain at rest or during the night may be indicative of unstable angina.

However, the history is sometimes not conclusive. The pain may be atypical, of variable duration and with an onset seemingly unrelated to the cardinal triggering factors. The pain may be restricted to the epigastrium with limited radiation, disappearing after warming, or appearing only at night or only after exercise. Critical attention should be paid to pain radiating to the back, which in the patient with high blood pressure may indicate aortic dissection.

Negative signs on clinical examination are also significant. Rales indicating left ventricular insufficiency may be found on chest examination, which can also rule out pulmonary pathology. Blood pressure is often elevated at the onset of anginal pain, or in the initial phase of myocardial infarction, and then falls in the ensuing stages.

Auscultation is usually normal, although pericardial rub or left-sided gallop rhythm may indicate early complications of myocardial infarction such as pericarditis or left ventricular insufficiency. The presence of a cardiac murmur is particularly significant, especially in the presence of exercise-induced angina, since it indicates aortic reduction or obstructive heart disease. In myocardial infarction, it may also be suggestive of mechanical complications such as mitral dysfunction or rupture of the septum. Particular attention should be paid to signs of possible aortic dissection such as the murmur of aortic insufficiency, interscapular murmur, and inequality of pulse pressure between the two arms.

The presence of hepatic pain or hepatomegaly, and hepato-jugular reflux is indicative of right ventricular pathology. Oedema of the lower limbs should be absent; if present and unilateral, it may indicate venous thrombosis, and point towards a diagnosis of pulmonary embolus.

Although a diagnosis of angina pectoris or myocardial infarction cannot be made with certainty on clinical examination, the clinical signs are highly suggestive. Acute pulmonary pathology, pericarditis, aortic dissection and pulmonary embolus can usually be ruled out because of the absence of positive signs on clinical examination.

Complementary Investigations

Both the gravity and initial complications of angina pectoris and myocardial infarction can be assessed by complementary investigations. They may provide further evidence in favour of coronary pathology of an atherosclerotic nature [146,150,151,155,159]. Myocardial infarction can usually be confirmed on the ECG. It is characterized initially by an elevation of the ST segment and T wave (Fig. 21.3) and progresses rapidly to a subepicardial ischaemia in the infarcted area with the appearance of a Q wave (Fig. 21.4). A mirror image subendocardial lesion indicates the myocardial nature of the lesion, and can rule out the possibility of pericarditis.

However, the ECG diagnosis of acute myocardial necrosis is not altogether straightforward, since the Q wave is not always observed. Three types of electrical signal may be observed: isolated elevation of the ST segment, an abnormal T wave or a depressed ST segment. The elevation of levels of myocardial enzymes can help confirm the presence of myocardial necrosis. In the absence of the Q wave, and in view of the poor prognosis of these types of infarct, especially the subendocardial ones, such factors should not be overlooked. During an attack, the ECG can confirm angina pectoris due to ischaemia or a subendocardial or subep-

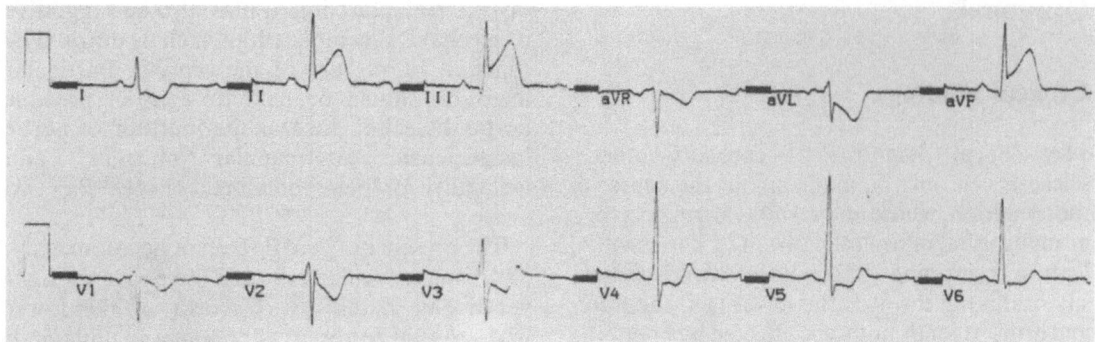

Fig. 21.3. Mr N. Recent inferior infarct, raised ST segment in DII, DIII and aVF; mirror image ST depression in V2, V3, V4, V5, V6.

icardial lesion (Fig. 21.5). At other times, however, the ECG tends to be normal.

The sequelae of previous myocardial necrosis, either recognized or otherwise, are indicated by the presence of a Q wave. However, in the absence of a well-defined acute myocardial necrosis, the presence of the Q wave should be interpreted with caution. The Q wave found in leads D3, VF, V2, V5 and V6 may also be due to heart muscle hypertrophy or the Wolf–Parkinson–White syndrome, which may thus be confused with previous myocardial necrosis. In between attacks, the presence of ischaemia, localized subepicardial lesions or isolated alteration in the T wave are indicative of permanent ischaemia. This tends to confirm the anginal nature of the pain (Fig. 21.6).

The ECG may also be frankly abnormal in association with disorders of repolarization, but the characteristic signs of myocardial ischaemia will be absent, Left ventricular hypertrophy, heart muscle hypertrophies (particularly in the apex) and metabolic or drug-induced disorders can also lead to marked alterations in the ST segment and especially in the T wave.

Thus the ECG, which is a simple and specific investigation during an anginal attack or the acute phase of a myocardial infarction, may be negative or non-specific at other times. It cannot really be used to confirm, or even in some cases detect, the presence of coronary artery disease.

Acute myocardial infarction is confirmed by an assay of circulating myocardial enzymes released from areas of necrosis. Levels of creatine phosphokinase (CPK) and their myocardial isoenzymes, the creatine kinase/MB isoenzyme (CK/MB), are elevated within the first 6 h after the infarction. They return to normal over the next 3 to 5 d. A rise in the level of lactic dehydrogenase (LDH) is observed later in the course of the infarction and it takes longer to return to normal than the levels of other enzymes.

Acute myocardial infarction is thus readily diagnosed from the clinical, ECG and biochemical findings. However, early diagnosis is required if prompt and aggressive treatments such as the administration of thrombolytic agents are to be instigated. Unfortunately the presence of a Q wave is not a reliable indicator, and there is too great a time lag before the rise

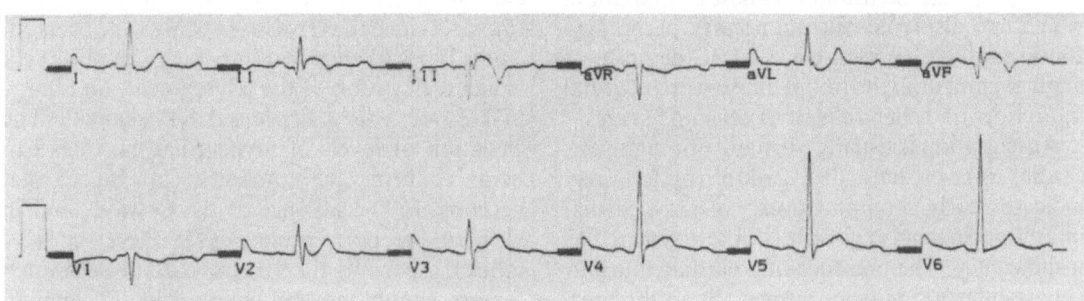

Fig. 21.4. Mr N. Inferior infarct: appearance of Q wave in DII, DIII and aVF with subepicardiac ischaemic persistence in DII, DIII and aVF.

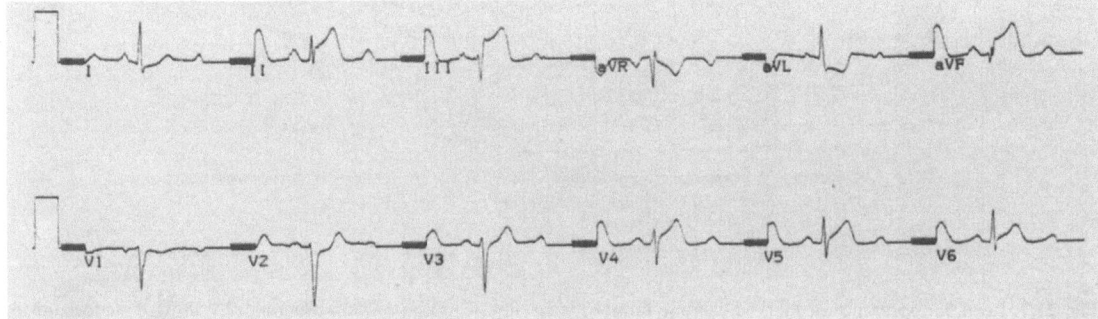

Fig. 21.5. Spontaneous angina with elevated ST segment recorded during an anginal attack at rest in hospital. Marked elevation of ST segment in DII, DIII and aVF, less marked in V4, V5 and V6, associated with slight depression in DI, aVL and V2: inferior and lateral Prinzmetal angina.

in circulating myocardial enzymes is detectable. The diagnosis thus depends on clinical criteria and the disorders of repolarization observed on the ECG.

Echocardiography can be of help in the early diagnosis of myocardial infarction, although its main value is in assessing ventricular involvement and other complications. Two-dimensional echocardiograms of akinetic or segmentally dyskinetic areas can be a valuable adjunct for early diagnosis of acute myocardial infarction before the appearance of the Q wave or the elevation of circulating myocardial enzymes. Echocardiography is particularly valuable for estimation of the haemodynamic effects of acute myocardial infarction by evaluation of left ventricular function and determination of the ejection fraction. Follow-up echograms can also detect onset of recent mechanical complications such as rupture of the ventricles or septum, mitral incompetence, intraventricular thrombosis or aneurysm. In the absence of myocardial infarction, echocardiography provides valuable

evidence for a diagnosis of coronary artery disease. It may detect stenosis of the left coronary branch, and in patients with atypical pain and no obvious ECG abnormalities, a picture of segmental dyskinesia is highly suggestive of coronary artery disease. Confirmation depends on the observation of ischaemia during exercise.

The exercise test carried out on an ergometric bicycle or treadmill in graded steps of 3 min duration can detect effort-induced ischaemia by a downsloping or negative depression of the ST segment of more than 1 mm. Alterations in the Q wave and amplitude of the R wave are also indicators of exercise-induced ischaemia. A depression of more than 4 mm with a fall in blood pressure at the end of the test, together with a disorder in ventricular rhythm, are indicative of serious pathology, either a stenosis of the coronary stem or three vessel disease. In these cases, early investigation by coronary angiography is required (Fig. 21.7) Nevertheless, in some cases the exercise test is inconclusive. It may be negative in patients with obvious symp-

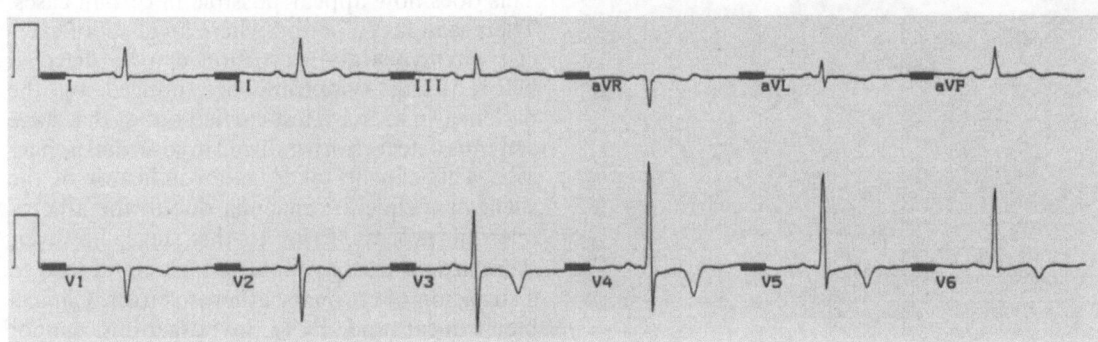

Fig. 21.6. Exercise-induced angina with persistence of ECG abnormalities in the absence of the anginal attacks, with negative T wave in DI. aVL. V2, V3, V4, V5 and V6 and subepicardial ischaemia with slight depression of the ST segment in V3, V4 and V5: pre-infarctional state in the anterior region.

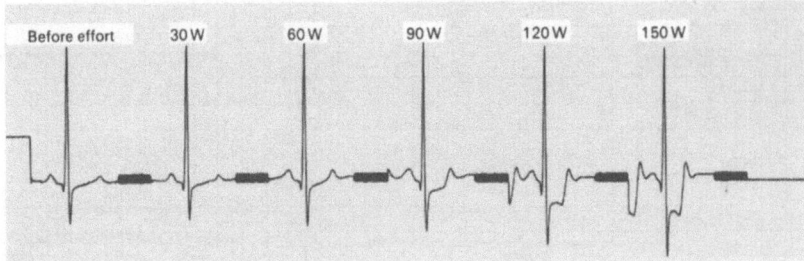

Fig. 21.7. Lead 5: appearance of marked exercise-induced ischaemia. Coronary artery angiography showed stenosis of the main left coronary vessel.

toms, or it may be positive in patients with atypical pain, or in whom atherosclerosis is unlikely, for example young physically fit patients and premenopausal women in the absence of risk factors.

Thallium-201 scintigraphy can make up for the lack of sensitivity of the exercise test. This isotope, injected at the end of the exercise test, does not localize initially in the poorly perfused areas, although over the following 3 h the tracer becomes redistributed in the ischaemic areas. Such zones can thus be discriminated from areas of necrosis (Fig. 21.8). The presence of exercise-induced ischaemia in a patient with anginal pain clinches the diagnosis of angina pectoris, and therefore of coronary atherosclerosis. The presence of risk factors and evidence of atherosclerotic damage in other vessels provides further confirmation of the diagnosis.

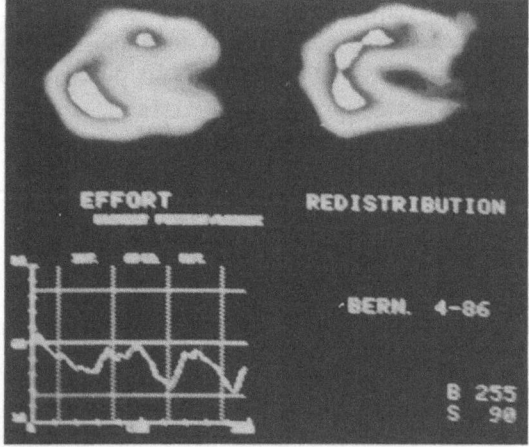

Fig. 21.8. Thallium-201 scintigram: reduced apical binding of thallium during exercise with redistribution of isotopic tracer in the ischaemic zone 3 h later: suggestive of stenosis of the anterior interventricular branch, confirmed by angiography.

Arguments in Favour of a Diagnosis of Coronary Atherosclerosis

Accurate diagnosis of coronary atherosclerosis is not a simple matter. It can only be established definitively from radiological evidence of atherosclerotic lesions in the coronary tree. Unfortunately this evidence can only be provided by invasive methods with their ethical and practical implications. In practice, clinical examination and complementary investigations should provide enough evidence for a diagnosis of coronary atherosclerosis with almost 100% reliability before recourse to such methods. Three main lines of evidence support a clinical diagnosis of coronary atherosclerosis.

The first depends on determination of the presence of coronary artery disease as demonstrated by myocardial infarction or angina pectoris. As mentioned above, the diagnosis of these conditions is not always straightforward. Coronary atherosclerosis is insidious in onset, and only when it can be detected with around 50% certainty does it have functionally observable effects. This indicates the need for a way of detecting the condition in its infraclinical stage. This does now appear possible in certain cases. There is in fact a period where an effect of exercise on myocardial perfusion can be detected before frank symptoms are noticed by the patient. An exercise test carried out at this stage can thus detect effort-induced myocardial ischaemia. This can be taken as an indicator of the silent or incipient ischaemia due to the atherosclerotic process. Prior to this stage, however, there are no tests available which could lead to a suspicion of coronary atherosclerosis. Clinical biochemical and ECG investigations cannot identify individuals with incipient coronary atherosclerosis with any degree of reliabiity. Only the development of specific markers for atherosclerosis will settle the matter.

The second line of evidence comes from the presence of predisposing factors, i.e. the classical risk factors for heart disease: diabetes, hypercholesterolaemia, high blood pressure and smoking. The presence of a single one of these factors can provide a basis for suspicion of coronary disease, particularly when associated with environmental factors such as stress or a positive family history for the condition.

The third line of evidence derives from the presence of atherosclerotic lesions in other parts of the arterial system, either above the aorta or in the lower limbs. Slight neurological signs, intermittent claudication, abnormalities in pulses and blood pressure or during auscultation of the course of arteries, as well as ECG abnormalities are all indications for further investigations. These include doppler or computerized tomography (CT) scans of the lower limbs or supra-aortic trunks. Detection of a lesion with haemodynamic effects can help in reaching a diagnosis in patients with slight chest symptoms or those with atypical ECG results.

Proof of the existence of coronary atherosclerosis can only be provided by coronary angiography [148,161]. This is an invasive technique which is not without risk. Using Judkin's technique there is a mortality rate of 1 per 1000, and so the indications for this investigation should be considered carefully. Nevertheless, in view of the grave implications for the patient, the diagnosis of myocardial infarction or angina pectoris must be backed up by demonstration of the presence of coronary lesions.

In patients with myocardial infarction, the use of early and aggressive treatments such as intracoronary fibrinolysis or transluminal dilatation may necessitate coronary angiography within the first hours or days after the attack. However, the indications and value of such treatments have not been established unequivocally, and coronary angiography during the initial phase of myocardial infarction is not employed universally. It can, however, detect total thrombosis of the coronary artery, although after removal of the obstrucion by thrombolytic treatment and dilatation, the underlying atherosclerosis will be revealed. However, in 5%–10% of cases, early coronary angiography can detect stenosis of the coronary artery in the absence of total occlusion or obvious thrombotic processes. The pathophysiology of the infarct is unclear. It is thought to be due to arterial spasm in parallel with a thrombotic process, immediately followed by endogenous fibrinolysis.

In patients with angina pectoris, coronary angiography is only indicated if there is a tendency towards an unstable state. This is seen in patients with major ECG abnormalities at rest, signs of marked impairment in the exercise test or a lack of response to treatment.

In practice, coronary angiography is mainly carried out to reach a decision with respect to treatment rather than to establish a diagnosis of angina pectoris or coronary atherosclerosis per se. For example, selective angiography of the coronary arteries can determine the presence, extent and severity of the lesions (Figs. 21.9 and 21.10). Current techniques of digital angiography can in fact provide quantitative estimates of the stenosis, and hence give an idea of the extent of the atherosclerotic process. Such methods also highlight the seeming lack of a relationship between the clinical manifestations and the extent of the lesions, ranging from the existence of three-vessel atherosclerosis with marked haemodynamic effects in an almost symptom-free patient to the total absence of atherosclerosis in patients with typical anginal symptoms and ECG abnormalities such as those observed in the X syndrome.

Thus, the occurrence of myocardial infarction or the presence of angina pectoris together with a positive exercise test in a patient with vascular risk factors is a sure sign of the existence of coronary atherosclerosis. However, it must be stressed that direct radiological observation of the lesions is the only sure proof of the existence

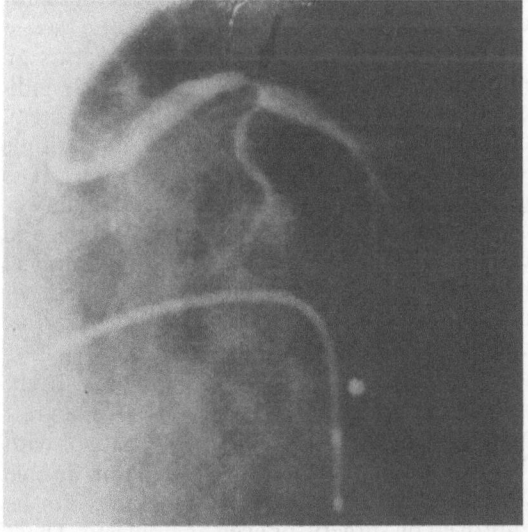

Fig. 21.9. Isolated, long and narrowed stenosis on the left coronary trunk.

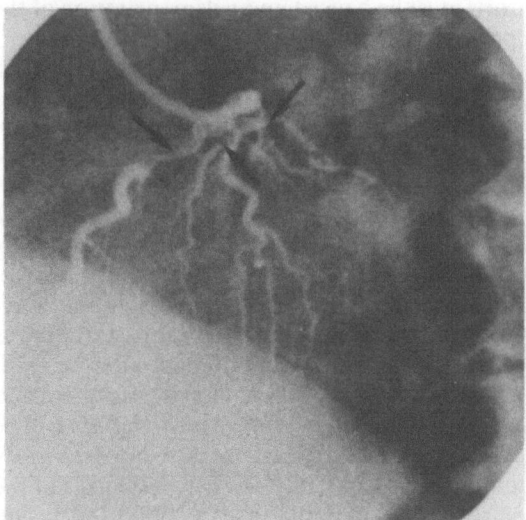 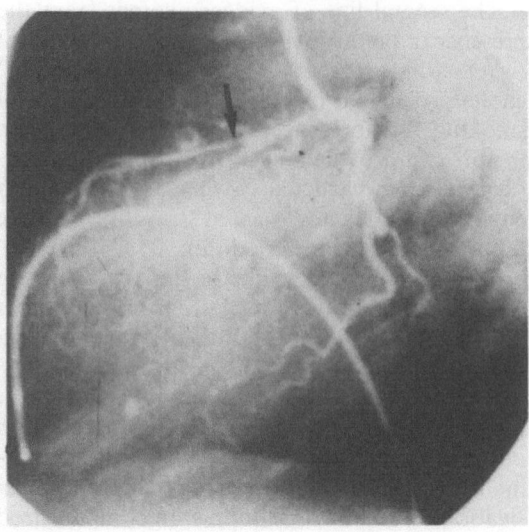

Fig. 21.10. Diffuse atherosclerotic lesions with long narrowed stenoses on the first course of the anterior interventricular branch and with a stenosis on the circumflex branch.

Fig. 21.11. Long distal stenosis in the anterior interventricular branch.

of coronary atherosclerosis. In spite of these considerations, this investigation does not form part of the battery of routine investigations required for diagnostic purposes, and it tends to be reserved for decisions concerning treatment.

Treatment

The treatment of angina pectoris can in theory be considered to prevent myocardial infarction. The means employed must be matched to the severity of the condition [145,152,160]. Drugs are the first line of treatment. Four main types of agent are used, either alone or in combination: beta blockers, calcium blockers, slow release nitrites or nitrates and their equivalents, molsidamine, vastarel and amiodarone. The choice of drug depends on the state of the patient, especially with respect to heart rate, overall cardiac function, arrhythmia, blood pressure and presence of other arterial disease. Persistence or deterioration of the anginal symptoms after combined drug treatment is an indication for more direct treatments such as transluminal angioplasty or coronary surgery. In such cases, coronary angiography is invaluable for determination of the type and extent of the lesions.

The indications for transluminal angioplasty depend on the facilities available and the experi-

ence of the treatment team. It is preferable to surgery in patients with isolated or double stenoses of the coronary arteries, and in cases where surgery is contraindicated. It is often carried out after diagnostic coronary angiography. However, in 20%–25% of cases, stenoses have been found to reform within the 6-month period following the dilatation procedure. These patients either require further dilatation or surgical intervention. The risk of complications such as thromboses or the very occasional dissection requires the proximity of facilities for cardiac surgery (Figs. 21.11–21.13).

Coronary bypass surgery also represents an important therapeutic option. It is clearly indicated in cases of stenosis of the main stem, although in patients with isolated or multiple stenoses, angioplasty may be preferable. It is generally accepted, however, that proximal two- or three-vessel impairment or eccentric and ulcerating plaque-type lesions do require surgical treatment. The frequency of early thromboses of the graft (10%) and the possibility of graft disease or atherosclerosis in saphenous bypasses has led to use of the internal mammary artery instead of the saphenous vein as a source of graft material. On the other hand, in the patient with few symptoms and with only single or double vessel involvement, overall survival may not be improved by surgery, or even angioplasty. These treatments do, however, reduce the consumption of drugs and improve the overall quality of life.

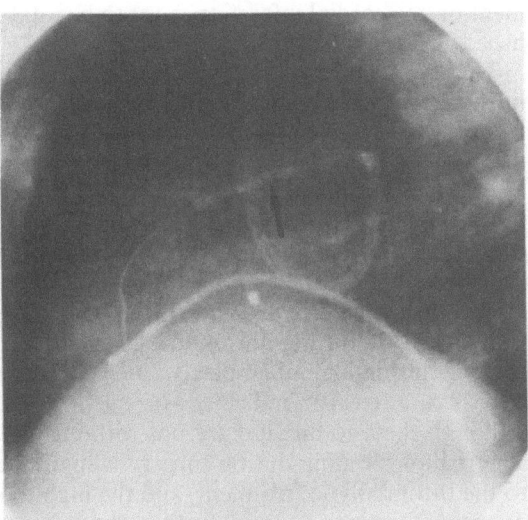

Fig. 21.12. Positioning of the angioplasty balloon catheter in the stenosed region. Onset of dilatation with clear imprint of the stenosis on the balloon.

When treating the patients with angina, the problems posed by unstable angina must be discriminated from cardiac muscle disorders due to severe ischaemia. Unstable angina requires treatment in hospital because of the risk of myocardial infarction. Stabilization is usually achieved by complete bed rest combined with anti-coagulant and anti-anginal medication such as intravenous glyceryltrinitrate, beta blockers,

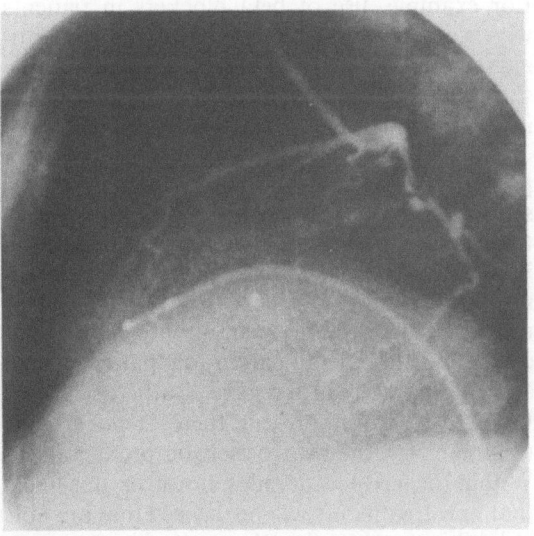

Fig. 21.13. Angiogram after dilatation: complete disappearance of stenosis from the anterior interventricular branch (smooth bore and absence of stenosis).

calcium blockers and amiodarone. Coronary angiographic investigations can also be carried out at this time, and the patient can then be conveniently treated by transluminal angioplasty or bypass surgery. If the angina does not stabilize, more aggressive treatments such as the administration of thrombolytic agents or aortic counterpulsation are indicated.

Severe cardiac muscle ischaemia with marked impairment of cardiac function and ejection fraction, usually in the presence of major atherosclerotic lesions of the coronary vessels, seems to be beyond current therapeutic control. For young patients, this incapacity is particularly dramatic. Now that graft rejection can be partially suppressed by administration of cyclosporins, heart transplantation may represent a viable option for such patients.

It can be seen that the treatment strategy to be adopted for the patient with angina is not a cut and dried issue. It should depend on the severity of the condition and be based on functional considerations rather than with respect to the underlying causal process, namely coronary atherosclerosis. Specific treatment of the atherosclerotic lesions is by angioplasty and aorto-coronary bypass surgery. A specific treatment aimed at reducing the atherosclerotic lesions in the coronary vessels has yet to be found.

The treatment of myocardial infarction seems to have undergone almost annual upheavals since the mid-1970s. Increased understanding and control of the initial complications of the infarction have led to a reduction in mortality. There has also been a complete revision of classical methods of treatment of the infarct itself, the medium- and long-term evolution of which depends on its initial size. Two approaches have been employed. The first is the use of pharmacological treatments which act at the periphery of the necrotic zone in order to reduce the size of the infarction. These peripheral areas are thought to be at risk from spread of the necrotic process. The second is represented by attempts at early vascularization by thrombolytic treatment.

At present, although there is some agreement on the overall strategy to adopt, many different treatments are currently in use for the treatment of myocardial infarction [149,153,154,157,158, 162]. The classical medical treatment is still the basis of therapy for the uncomplicated infarction when seen in the later stages. Bed rest aimed at limiting myocardial oxygen consumption is the first recommendation. There is a trend to earlier mobilization, although a period of 6 to 10 days' rest and normalization of circulating myocardial enzymes are generally thought to be required before allowing the patient to get up. Although initial treatment of pain and reduction in stress are mandatory, preventive anti-arrhythmic treatment such as administration of lignocaine during the acute phase is still a matter for debate. The frequency of thrombo-embolic complications during the acute phase requires the aministration of anti-coagulants. Low doses of heparin or low molecular weight heparinoids are given to low-risk patients with a small area of necrosis, while high doses are required for patients at a high risk of thrombosis.

Pharmacological reduction of the infarcted area is now attempted systematically. Various agents are used, although their real influence on the size of the infarct or on long-term survival has not yet been clearly established in prospective studies. Nitrites and nitrates are now used routinely for their anti-anginal action and their haemodynamic effects, especially in patients with left ventricular insufficiency and increased pulmonary pressure. Use of hyaluronidase or GIK solution (glucose, insulin, potassium) is not widespread, as their effectiveness has not been clearly demonstrated. On the other hand, beta blockers (atenolol, metoprolol) can reduce the size of the infarct and improve the long-term survival in certain cases, namely those less than 65 years old, with an infarction of recent origin, with no evidence of major left ventricular insufficiency, with a systolic pressure above 100 mmHg and a heart rate above 60, without obstructive lung disease and without auriculo-ventricular block. Some calcium blockers appear to reduce the size of the infarct, although their effect on long-term survival is less well established.

At present, the principle treatment for myocardial infarction is to attempt revascularization of the damaged tissue. Early perfusion of the necrosed area with thrombolytic agents can reduce the size of the infarct and improve overall contractility of the left ventricle. It also reduces early mortality and may also reduce longer term mortality. Streptokinase is the most frequently used agent and is administered either in a single bolus dose or by slow infusion. A number of new fibrinolytic agents have been developed. For example tPA has been found to be more effective in the short term, although at the expense of an increased tendency to reobstruct. APSAC and ScuPA are variants of tPA, and agents chemically coupled to monoclonal anti-fibrin antibodies have also been devised. These new thrombolytics have a specific action on the thrombus itself, although their place in the treat-

ment of myocardial infarction is yet to be established.

These agents can be administered either intravenously or directly into the coronary arteries. Both routes seem equally effective, and the intravenous route is usually adopted for reasons of convenience and expediency.

Whatever the thrombolytic agent used, early post-infarction perfusion may dissolve the thrombus, but the stenosis due to the underlying coronary atherosclerosis nevertheless remains. The threat of further thromboses is therefore real. Transluminal angioplasty and coronary bypass are clearly indicated in such cases, although these techniques are not without risk. The risk of bleeding during surgery is high due to the thrombolytic treatment, and the high frequency of restenoses and further thromboses after angioplasty is evidence of the limited effectiveness of this procedure.

At present it is still not clearly established whether all patients treated with thrombolytics and revascularized require investigation and specific treatment of residual stenosis. In cases where this may be appropriate, the interval between the infarction and the administration of fibrinolytics is still a matter of debate. After the acute phase of the infarction, the long-term treatment of the patient poses a number of problems. The patient is at risk not only from further infarction, but also from arrhythmia, cardiac insufficiency and sudden death. Only a few treatments have been demonstrated clearly to be effective in the prevention of these complications. For example, use of beta blockers in patients with cardiac dysfunction, or of platelet anti-aggregants and anti-vitamin K agents has been shown to reduce the incidence of subsequent infarction and improve long-term survival.

Thus, myocardial infarction, which in 50% of patients is the first sign of coronary atherosclerosis, reflects an irreversible impairment of the heart musculature with grave short-term implications. In many cases, the long-term prognosis is also not particularly favourable. This unsatisfactory situation can only really be remedied by the identification of patients with asymptomatic, progressive atherosclerosis. Unfortunately, at present there is no specific treatment for the atherosclerotic process itself. Within the next few decades, however, it is likely that new treatments will be devised that are able to inhibit or reduce the atherosclerotic processes. Their use will nevertheless depend on finding a method for selecting patients who would benefit from such treatment in the first place.

References

144. Alpert JS, Braunwald E (1980) Pathological and clinical manifestations of acute myocardial infarction. In: Braunwald E (ed) Heart disease: a textbook of cardiovascular medicine. Saunders, Philadelphia, p 1309
145. Boulay F (1987) Traitement de l'insuffisance coronaire chromique. In: Gilgenkrantz JM, Royer RJ, Zannad F (eds) Thérapeutique en pathologie cardiovasculaire. Flammarion, Paris, p 565
146. Bruce RA, Horstein TIR (1969) Exercise testing in evaluation of patients with ischemic heart disease. Prog Cardiovasc Dis 11:371
147. Cohn PF, Braunwald E (1980) Chronic coronary artery disease. In: Braunwald E (ed) Heart disease: a textbook of cardiovascular medicine. Saunders, Philadelphia, p 1387
148. De Wood MA, Spores J, Notske R. (1980) Prevalence of total coronary occlusion during the early hours of transmural myocardial infarction. N Engl J Med 303:897
149. GISSI (1986) Effectiveness of intravenous thrombolytic treatment in acute myocardial infarction. Lancet I:397
150. Goldman MJ (1973) Maladie coronarienne; l'infarctus du myocarde. In: Goldman MJ (ed) Electrocardiographie clinique. Vigot, Paris, p 160
151. Goldman S, Tselos S, Cohn K (1976) Marked depth during treadmill exercise testing: indication of severe coronary artery disease. Chest 69:729
152. Gruntzig A, Senning A, Siegenthalerwe L (1979) Non operative dilatation of coronary artery stenosis: percutaneous transluminal coronary angioplasty. N Engl J Med 301:61
153. ISIS–I Collaborative Group (1986) Randomised trial of intravenous atenolol among 16027 cases of suspected acute myocardial infarction: ISIS I. Lancet I:57
154. Norwegian Multicenter Study Group (1981) Timolol induced reduction in mortality and reinfarction in patients surviving acute myocardial infarction. N Engl J Med 304:801
155. Ogawa H, Hiramori K, Haze K et al. (1985) Classification of non-Q-wave myocardial infarction according to electrocardiographic changes. Br Heart J 54:473
156. The Persantine-Aspirin Reinfarctus Study Research Group (1980) Persantine and aspirin in coronary heart disease. Circulation 62:449
157. Sixty Plus Reinfarction Study Research Group (1980) A double-blind trial to assess long-term oral anticoagulant therapy in elderly patients after myocardial infarction. Lancet II:989
158. Sobel BE, Braunwald E (1980) The management of acute myocardial infarction. In: Braunwald E (ed) Heart disease, a textbook of cardiovascular medicine. Saunders, Philadelphia, p 1353
159. Strauss HW, Harrison K, Langan JK (1975) Thallium-201 for myocardial imaging. Relation of Thallium-201 to regional myocardial perfusion. Circulation 51:641
160. Surgery in the treatment of coronary artery disease. Circulation, 1985, Part II, V:72
161. Wilson RF, Holda MD, White CW (1986) Quantitative angiographic morphology of coronary stenoses leading to myocardial infarction or unstable angina. Circulation 73:286
162. Zannad F (1987) Limitation pharmacologique de la taille de l'infarctus du myocarde (thrombolyse exclue). In: Gilgenkrantz JM, Royer RJ, Zannad F (eds) Thérapeutique en pathologie cardiovasculaire, Flammarion, Paris, p 594

Section IV

The Arterial Wall in Diabetes

Chapter 22

Clinical and Physiopathological Features in Diabetic Vascular Disease

A. Mossaz and J. P. Assal

Links Between Diabetes and Arterial Disease

It is usually agreed that endothelial injury is the primary lesion in the development of atherosclerosis [11,13]. This lesion, at first reversible, may evolve by platelet aggregation, smooth-muscle cell proliferation (under mitogenic influence), lipid deposition and modification of the vascular matrix composition. The pathogenesis of atherosclerosis does not differ essentially in diabetic and non-diabetic patients. In the diabetic, however, macrovascular disease appears prematurely and with greater severity than in the non-diabetic. Diabetic woman may suffer from vascular problems before the menopause [20], whereas atherosclerosis in non-diabetic women is rare before this period.

Calcification of the vessels is more frequent in diabetics. There are two types: (a) calcium deposition in atheromatous lesions and (b) medial calcification [16]. The atheromatous lesions are more peripheral and often preferentially affect the leg arteries.

Diabetic microangiopathy is to be distinguished from macrovascular disease. Its characteristics are, from the histological viewpoint, thickening of the basement membrane of the capillaries and extravascular deposition of glycoprotein material. While the microangiopathy plays a part in renal and ocular lesions, in diabetic foot lesions its role is reversed [38].

Pathogenic Factors

Hyperglycaemia

The link between vascular disease and hyperglycaemia is difficult to establish in epidemiological studies, the problem being to determine the duration of hyperglycaemia, particularly in type II diabetes. There is, however, an increase in the prevalence of macroangiopathy in patients suffering from asymptomatic hyperglycaemia [19,27] and a relationship between the duration of diabetes and the appearance of vascular disease [42]. Nevertheless some patients with poor metabolic control may not develop premature atherosclerosis. Diabetic heterogeneity as well as genetic and environmental factors may also account for the development of vascular disease [4,49]. Hyperglycaemia seems not to be the only essential factor in the appearance of atherosclerosis in diabetics: disturbance of lipid metabolism, obesity and hypertension are other risk factors frequently encountered.

Hyperlipidaemia

Hyperlipidaemia is often linked with diabetes and is considered to be a determining factor in the development of atherosclerosis [3,44]. Diabetics, especially those with poor metabolic control, frequently have high levels of VLDL. On the other hand, a decrease in the HDL-cholesterol level and an increase in the LDL level may also be seen in diabetic patients, particularly when they are obese and sedentary.

Rheological and Haemostatic Factors

Platelet function studies are of great interest when studying the links between diabetes and arterial diseases. Data concerning the platelet adhesion and aggregation do not always coincide. It nevertheless appears that the premature endothelial injury seen in diabetics is an important stimulus of platelet aggregation. Platelet survival is reduced [11,13]; prostacyclin synthesis is diminished in the two major forms of diabetes; and fibrinolytic activity is also decreased [13,23]. Thrombocytes seem to synthesize more thromboxane A_2 [13]. An increase in plasma concentration of von Willebrand factor, also synthesized by the endothelial cell, is found in most of the studies.

Hyperinsulinism

Insulin probably contributes to the development of macrovascular complications in diabetes [11]. Some of the effects of insulin may favour the appearance of atheromatous lesions, for example stimulation of smooth-muscle cells and the incorporation and synthesis of lipids in the artery wall. Due to treatment of peripheral resistance to endogenous insulin, some diabetics are in a state of chronic hyperinsulinism [48].

Epidemiological Factors and Prevention

General Remarks

Since the introduction of insulin therapy and antibiotic therapy, vascular complications are now the main cause of morbidity and mortality in the diabetic population [15,26,55]. Diabetic life expectancy is still less than that of the general population [45]. Cardiovascular diseases are responsible for the higher mortality in diabetic groups, particularly in non-insulin-treated patients [18]. In insulin-treated patients the difference in mortality rates is due mainly to acidocedotic and nephrotic complications

[12,15]. Increase in mortality is more important in the younger groups, with the greater difference in the 25–39 year age group [18,23,28]; this disparity then decreases but mortality still remains higher than in the general population. Diabetic women have the same mortality rate as diabetic men [25]. Certain factors have been suggested (greater increase of cholesterol and beta-lipoproteins, hypertension, obesity) to try to explain why the female protection against vascular illness disappears in diabetics; there is as yet no satisfactory answer to this question [21].

Coronary Heart Disease: Mortality and Glucose Intolerance

Coronary heart disease is the major cause of death in diabetics. Myocardial infarction is two to three times more frequent [25,34] and leads to a higher mortality rate in the weeks following the infarction (35% in the diabetic group versus 18% in the control group) [47]. Coronary risk seems to be particularly high in the insulin-treated female group [21]. Duration of the disease for longer than 10 years represents an additional risk [45].

Mortality due to coronary heart disease and glucose intolerance appear to be related. Age, systolic blood pressure values and electrocardiographic abnormalities are prior indicators of mortality [19].

Peripheral Arterial Disease: Morbidity

The prevalence of this condition is related to the fasting blood glucose levels [4]. However, cigarette smoking is among the other most important risk factors that determine the development of peripheral vascular insufficiency [4]. Obesity, in contrast, is of minor importance. Generally, the prevalence of atherosclerosis obliterans in the diabetic population is about 30% [4] and is associated with an increased risk of amputation (50%–70% of non-traumatic amputations are performed on diabetic patients). More diabetics suffer from lower limb ischaemia than do control patients [24,42].

Cerebrovascular Disease: Mortality

Diabetics have twice as many fatal cerebrovascular accidents than do the rest of the population. Mortality is related to age and duration of the disease [28,55].

Specific Organic Involvement

Coronary Heart Disease

Clinical Factors

Apart from the classical presentation of coronary heart disease as effort angina, unstable angina, myocardial infarction and sudden death, ischaemic accidents may arise in atypical patterns in diabetics. Silent infarcts supervene two to six times more frequently in diabetics than in non-diabetics. Physicians must be particularly aware of the indicators, which include thoracic or gastric atypical pain, unexplained fatigue, indisposition, palpitations, nausea or vomiting. Acute cardiac failure or metabolic disturbance may also be signs of an ischaemic accident. Autonomic neuropathy could be responsible for the decrease in pain during myocardial infarction [34].

Three points must be stressed concerning myocardial infarction in a diabetic population: (a) the mortality rate is high after acute attacks; (b) recurrences are more frequent [47]; and (c) long-term prognosis is worse, often due to major cardiac dysrhythmia, such as ventricular fibrillation [34.

Glucose intolerance may often be seen in the normal subject just after a myocardial infarct, but it appears that a greater number of these people later develop true diabetes (up to 40% according to some studies) [21].

Mortality rate is higher in diabetics who need to be treated either with oral agents or with insulin, compared with those treated by diet only [34]. The major cause of death is ventricular pump failure associated with cardiogenic shock and valvular rupture.

Diagnosis

Diagnosis of coronary heart disease is based on the following: a typical history of thoracic pain, electrocardiographic signs of ischaemia or earlier infarct, a positive stress test or isotopic scintigraphy and angiographic evidence of coronary arterial thrombosis. Where surgery or percutaneous transluminal angioplasty may be performed, coronary angiography is justified to determine the localization and extent of lesions.

Treatment

Treatment of coronary heart disease does not differ in diabetic and non-diabetic patients.

Nitrous agents, calcium antagonists and beta-blockers may be used, in the latter case with due caution as they reduce the cardiac output increase which occurs in response to hypoglycaemia. A surgical approach, angioplasty or fibrinolysis may be chosen depending on the anatomical lesions. Elimination of risk factors remains a very important aspect of prophylaxis.

Non-atheromatous Cardiopathy

In diabetics there seems to be a myocardial microangiopathy resembling those encountered in the eye and the kidney [34,49]. These modifications are not accompanied by a significant obstruction of the vessels and their importance remains hypothetical.

Diabetic Foot and Peripheral Arterial Disease

Diabetic foot problems often lead to amputation. The genesis of trophic lesions may be explained by simultaneous or independent action of macroangiopathy, neuropathy or infection [36].

Peripheral Arterial Disease

Clinical Aspects. Peripheral arterial disease is a health-threatening complication in diabetes. Atherosclerosis is more frequent and of greater severity in the diabetic population than in the non-diabetic, leading to a higher rate of gangrene and amputation. Bell's series of autopsies demonstrated that atherosclerotic gangrene was 38 times more frequent in diabetic men than in control subjects [5]. Atheromatous lesions are more diffuse and localized to smaller arteries, favouring particularly the leg and preventing development of effective collateral circulation [50].

Intermittent claudication is a frequently encountered complaint in diabetics. It is characterized by pain on walking for a distance which quickly disappears after stopping. Pain is most often located in the calf but may also affect the foot, the thigh or the buttock. Rapid fatigue in the limbs, cold feet and difficulties in the healing of wounds may also be attributable to chronic ischaemia. On examination the skin is dry, smooth and thin, the hair disappears and subcutaneous tissue becomes atrophied (Fig. 22.1). The foot is pale or decubitus and may show

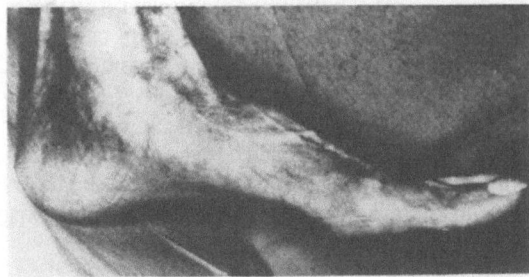

Fig. 22.1. Trophic lesions in a diabetic foot in a case of chronic ischaemia.

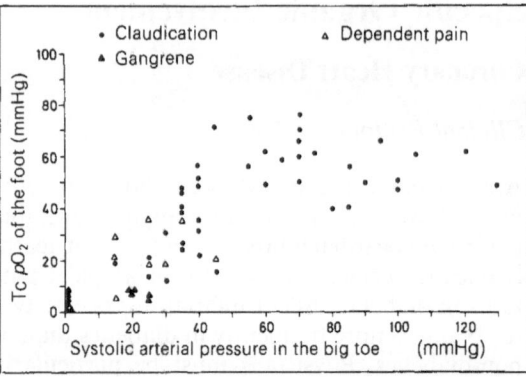

Fig. 22.3. Pedal transcutaneous oxygen pressure (tcpO₂) and toe systolic blood pressure (SBP) in 59 patients suffering from lower limb arterial occlusive disease.

redness due to vasodilatation. Bruits and the absence of popliteal or pedal pulse must be systematically searched for in the vascular area to try to localize the stenosis.

Multiple calcifications may be seen on standard x-rays (Fig. 22.2). In more severe cases of ischaemia, rest pain occurs in the toes. The final stage is represented by trophic lesions. Dry gangrene may then follow, due to either embolism or digital occlusion. This may be exacerbated by pre-existing stenosis or local infection [36].

Diagnosis. Clinical examination helps to determine the diagnosis of occlusive arterial disease.

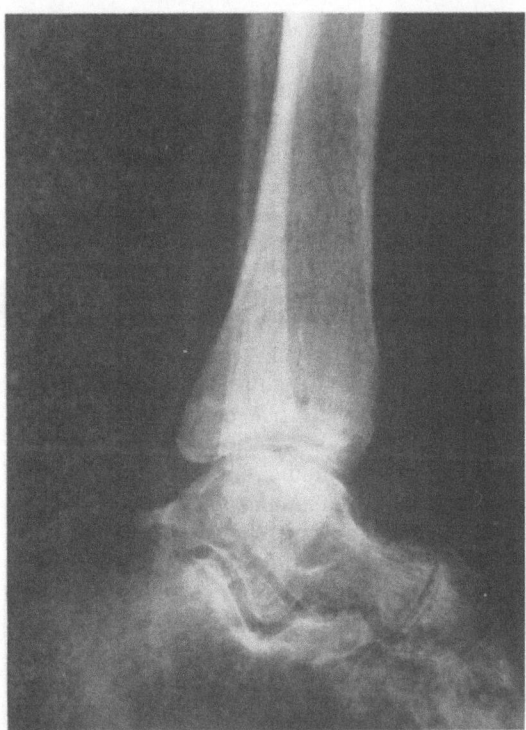

Fig. 22.2. Ankle x-ray showing extensive calcification of the vessels in a 76-year-old type I diabetic woman.

Confirmation is obtained using such techniques as Doppler ultrasonography [32] and measurement of toe systolic blood pressure [33]. It must be emphasized that vascular calcification, which is frequent in diabetics, provokes vessel rigidity that sometimes leads to an overestimation of the systolic pressure [57]. Transcutaneous oxygen pressure (tcpO₂), which is related to local blood flow, is a good index of the nutritional circulation and correlates well with the clinical state. TcpO₂ has been proven useful in determining the gravity of the ischaemia [8]; patients presenting with ischaemic tissue necrosis have pedal tcpO₂ values less than 10 mmHg (Fig. 22.3) [30].

The angiographic features of atherosclerosis obliterans in diabetics are as follows: tortuous arteries, ulcer lesions, calcifications and multiple level stenosis or obstructions (Fig. 22.4). Abdominal aorta occlusion, if it occurs, is usually situated below the renal arteries [14].

Treatment. The therapeutic measures against intermittent claudication include daily walking exercises to encourage collateral vessel development (often of much benefit). Vasodilator agents sometimes lead to subjective improvement; however, according to data produced in some studies, anti-platelet drugs seem, for the moment, to be of theoretical interest only [29]. Percutaneous transluminal angioplasty may eliminate short non-calcified stenosis.

Vascular surgery is the ideal treatment in the case of rest pain and ischaemic trophic lesions. This includes aorto-iliac endarterectomy, aorto-iliac or aorto-femoral graft or, more distally, femoral-leg graft, which all produce good clinical and haemodynamic results. Vasodilator drugs are a second choice in this situation, but sometimes improvement following their use is

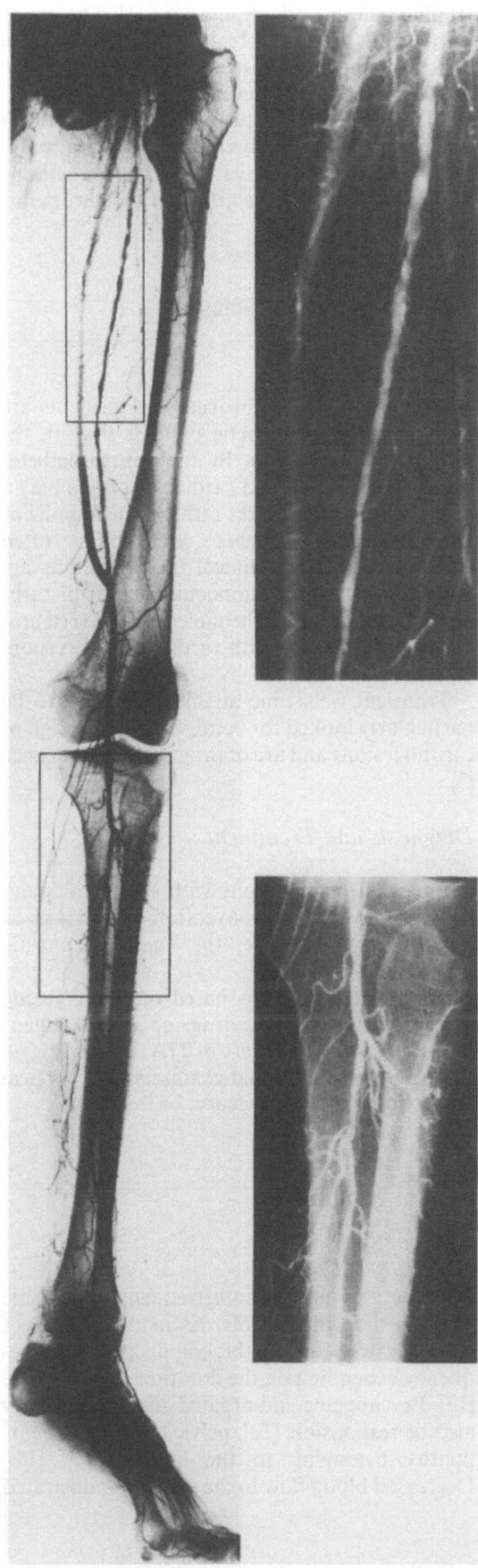

Fig. 22.4. Left lower limb arteriography. Multiple femoral, fibular and anterior tibial artery stenoses with femoro-popliteal bypass in a 62-year-old type II diabetic woman already amputated on the right side.

observed, with an increase in pedal tcpO$_2$ values [8]. Lumbar sympathectomy in patients suffering from rest pain or trophic lesions is rarely performed unless reconstructive surgery is impracticable. Results are poor in the case of autonomic neuropathy; before thinking of this type of intervention, it must also be pointed out that sympathectomy is contraindicated when pedal systolic blood pressure is below 50 mmHg [54]. Following bilateral sympathectomy unwanted effects, such as impotence, are not rare.

For the prevention of atherosclerosis, particularly in diabetics, obtaining the best possible metabolic control, avoiding smoking and treating hypertension are essential prophylactic measures [23].

Role of Peripheral Neuropathy in the Genesis of Diabetic Foot Lesions

Atherosclerosis obliterans, even though more frequent in diabetics, cannot alone account for all the foot lesions. Many are due to peripheral neuropathy [41]. Because of a loss of sensitivity to pain, lesions may appear without the patient being aware. The skin is typically dry and cracked because of decreased perspiration due to autonomic nervous system injury, with a tendency to fissures leading to secondary infection. Plantar pressure ulcers and Charcot's neuroarthropathy are the result of static disorders [41,46] (Fig. 22.5). Even if the neuropathic foot often appears warm in the presence of arteriovenous shunt, the possibility of a co-existing distal arterial insufficiency must be born in mind [9].

Role of Infection

The diabetic foot is often infected. This is exacerbated by microtraumas that are all the more dangerous, being associated with loss of sensitivity and ischaemia. Metabolic disorders, such as a decrease in effective granulocytic function due to hyperglycaemia, are predisposing factors. Mixed bacterial flora (staphylococcus, streptococcus, *E. coli,* pseudomonas etc.) are most often encountered together with anaerobic microorganisms. Osteomyelitis may be superimposed on the plantar pressure ulcer.

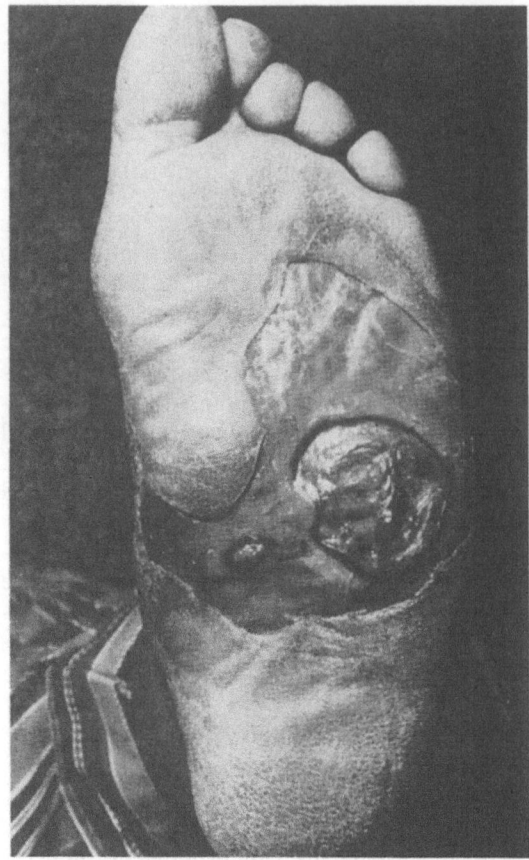

Fig. 22.5. Neuro-arthropathic plantar ulcer.

Diabetic Microangiopathy

Diabetic microangiopathy is often said to be responsible for trophic lesions, but studies on pedal $tcpO_2$ have recently shown that, when not associated with macrovascular disease, the microangiopathy is not able to provoke ischaemia [31,57]. In the absence of haemodynamically significant atheromatous lesions, peripheral neuropathy is undoubtedly the most important factor leading to trophic lesions [38]. As in cardiac failure, lower limb oedema is a predisposing factor [37].

Prophylactic Measures and Treatment

The prevention of arteriopathy and/or neuropathy is of primordial importance in the diabetic foot. Educating patients in daily foot care (choice of footwear, avoidance of sharp instruments, use of hydrating and anti-fungal creams etc.) leads to a decrease in the incidence of

trophic lesions and amputations [1]. The use of intravenous anti-microbial agents in the event of infection or vascular reconstructive surgery in major ischaemia is nevertheless often mandatory. When these measures fail, and the limb or the patient's life is threatened, amputation is indicated. Strict observation of the remaining limb is imperative since arterial and neuropathic diseases are bilateral.

Cerebrovascular Disease

Clinical Aspects

As stated earlier, cerebrovascular accidents are more frequent in diabetic patients than in the rest of the population. In addition to diabetes itself, hypertension and cardiac disorders play a role. Thrombotic strokes rather than embolic or haemorrhagic accidents are most often encountered. Usual clinical manifestations are aphasia, dysarthria, monocular visual disturbance, hemiparesis or hemiplegia and vertebrobasilar insufficiency with vertigo, double vision, ataxia or drop attack.

Transient ischaemic attacks (TIA) are to be particularly looked for because they are a sign of carotid lesions and are of prognostic significance [2].

Diagnosis and Treatment

Doppler ultrasonography with spectrum analysis [6], angioscan and conventional or digitalized angiography [51] are the main diagnostic methods.

Medical treatment is based on prophylactic measures. Anti-platelet drugs or anti-coagulant agents are used in the case of TIA. Depending on the results of additional examinations, surgical correction may be necessary.

Sexual Impotence

Aetiology

Impotence is much more frequent in diabetic patients than in controls. As many as 50% of diabetics may develop this complication, the frequency depending on the duration of the disease [17]. Psychogenic and organic disturbances may also be responsible [22], pelvic autonomic neuropathy belonging to the latter group [17]. Decreased blood flow to the penis, demonstrated by refined diagnostic procedures, is more frequent than previously thought. In the presence of associated risk factors (cigarette smoking, dyslipidaemia, hypertension), impotence of arterial origin must be suspected strongly [52]. Hypoandrogenism and undesired effects of certain drugs may also be responsible.

The Leriche syndrome due to obstruction of the terminal part of the aorta is often associated with erectile disorders [35]. More distal lesions also have been shown to provoke such disturbances [39].

Diagnosis and Treatment

Doppler velocymetry [40] and nocturnal phalloplethysmography [10] are non-invasive diagnostic methods and may be used in association with selective angiography [39] when a surgical procedure is envisaged. Reconstructive surgery appears to offer a promising therapeutic alternative [53].

Conclusion

Macroangiopathic complications are still one of the principal causes of morbidity and the major cause of mortality in diabetic patients. Atheromatous disease is often more diffuse than in the general population. Correction of metabolic disorders and the suppression of associated risk factors (particularly hypertension and cigarette smoking) are imperative in the prevention of atherosclerosis.

Medical and surgical means may lower the incidence of serious ischaemic accidents, but these are still more frequent in diabetic subjects than in the general population. Further studies are required to ascertain the role of biological modifications in the genesis of atherosclerosis if a better understanding and better treatment of this disease are to be achieved.

References

1. Assal JP, Ekoe JM, Lacroix A (1984) L'enseignement au malade sur sa maladie et son traitement. Un succès thérapeutique, un échec du corps médical. Journées Annuelles de Diabétologie de l'Hôtel Dieu, Flammarion, Paris, pp 193–207
2. Barnett HJM (1979) The pathophysiology of transient cerebral ischemic attacks: therapy with platelet anti-aggregants. Med Clin North Am 63:649–679
3. Beach KW, Brunzell JD, Conquest LL et al. (1979) The correlation of arteriosclerosis obliterans with lipoproteins in insulin-dependent and non-insulin-dependent diabetes. Diabetes 28:836–840
4. Beach KW, Brunzell JD, Strandness E (1982) Prevalence of severe arteriosclerosis obliterans in patients with diabetes mellitus: relation to smoking and form of therapy. Arteriosclerosis 2:275–280
5. Bell ET (1950) Incidence of gangrene of the extremities in non-diabetic and in diabetic persons. Arch Pathol 49:469–473
6. Blackshear WM, Philipps DJ, Thiele BL et al. (1979) Detection of carotid occlusive disease by ultrasonic imaging and pulsed doppler spectrum analysis. Surgery 86:698–706
7. Bongard O, Krahenbuhl B (1984) Pedal blood flow and transcutaneous pO_2 in normal subjects and in patients suffering from severe arterial occlusive disease. Clin Phys 4:393–401
8. Borzykowski J, Krahenbuhl B (1981) measurement of pedal transcutaneous oxygen tension to follow up lower limbs arterial occlusive disease. Vasa 10:137–140
9. Boulton AJ, Scarpello JH, Ward JD (1982) Venous oxygenation in the diabetic neuropathic foot: evidence of arteriovenous shunting? Diabetologia 22:6–8
10. Casey WC (1979) Phallography: technique and results of nocturnal tumescence monitoring. J Urol 122:752–753
11. Chait A, Bierman EL, Brunzell JD (1985) Diabetic macroangiopathy. In: Alberti KG, Krall LP (eds) The diabetes annual, Elsevier, Amsterdam pp 323–349
12. Christlieb AR, Warram JH, Krolewski AS et al. (1981) Hypertension: the major risk factor in juvenile-onset insulin-dependent diabetics. Diabetes [Suppl 2] 30:90–96
13. Colwell JA, Winocour PD, Lopes-Virella M et al. (1983) New concepts about the pathogenesis of atherosclerosis in diabetes mellitus. Am J Med 75:67–80
14. Costello P, Hoar C, Wheelock FC (1984) Arteriography. In: Kozak GP (ed) Management of diabetic foot problems. Saunders, Philadelphia, pp 122–132
15. Dorman JS, Laporte RE, Kuller LH et al. (1984) The Pittsburgh insulin-dependent diabetes mellitus (IDDM) morbidity and mortality study. Diabetes 33:271–276
16. Edmonds ME, Morrison N, Laws JW et al. (1982) Medial arterial calcification and diabetic neuropathy. Br Med J 284:928–930
17. Ellenberg M (1980) Sexual function in diabetic patients. Ann Intern Med 92:331–333
18. Entmacher PS, Root HF, Marks HH (1964) Longevity of diabetic patients in recent years. Diabetes 13:373–377
19. Fuller JH, Shipley MJ, Rose G et al. (1980) Coronary-heart-disease risk and impaired glucose tolerance. The Whitehall study. Lancet II:1373–1376
20. Ganda OP (1984) Pathogenesis of accelerated atherosclerosis in diabetes. In: Kozak GP (ed) Management of diabetic foot problems. Saunders, Philadelphia, pp 17–26
21. Garcia MJ, McNamara PM, Gordon T et al. (1974) Morbidity and mortality in diabetics in the Framingham population. Diabetes 23:105–111
22. Gfeller R, Assal JP, Ekoe JM (1981) Les problèmes psychologiques des diabétiques. Rev Ther 38:1069–1074
23. Jarrett RJ, Keen H, Chakrabarti R (1982) Diabetes, hyperglycemia and arterial disease. In: Keen H, Jarrett RJ (eds) Complications of diabetes. Edward Arnold, London, pp 179–203

24. Kahn HA, Herman BJ, Medalie JH et al. (1971) Factors related to diabetes incidence: a multivariate analysis of two years observation on 10,000 men. J Chronic Dis 23:617–629

25. Kannel WB, McGee DL (1979) Diabetes and cardiovascular disease. The Framingham study. JAMA 241:2035–2038

26. Kannel WB, McGee DL (1979) Diabetes and cardiovascular risk factors: the Framingham study. Circulation 59:8–13

27. Keen H, Jarrett RJ, Fuller JH et al. (1981) Hyperglycemia and arterial disease. Diabetes [Suppl 2], 30:49–53

28. Kessler II (1971) Mortality experience of diabetic patients. Am J Med 51:715–724

29. Krahenbuhl B (1983) Traitment de l'insuffisance artérielle périphérique. In: Fabre J (ed) Thérapeutique médicale. Flammarion, Paris, pp 708–713

30. Krahenbuhl B, Dubas JM (1981) Transcutaneous oxygen pressure on the foot of normal subjects and patients suffering from arterial occlusive disease. In: Jageneau AH (ed) Noninvasive methods in cardiovascular haemodynamics. Elsevier Biomedical, Amsterdam, pp 469–474

31. Krahenbuhl B, Mossaz A (1985) On vascular non-disease of the foot in diabetes. N Engl J Med 312:1190 (letter)

32. Lassen NA, Tonnensen KH, Holstein P (1976) Distal blood pressure. Scand J Clin Lab Invest 36:709–712

33. Lassen NA, Krahenbuhl B, Hirai M (1977) Occlusion cuff for routine measurement of digital blood pressure and blood flow. Am J Physiol 232:338–340

34. Leland OS (1982) Diabetes and the heart. In: Kozak GP (ed) Clinical diabetes mellitus. Saunders, Philadelphia, pp 302–316

35. Leriche R, Morel A (1948) The syndrome of thrombotic obliteration of the aortic bifurcation. Ann Surg 127:193–206

36. Levin ME (1980) The diabetic foot. Angiology 31:375–385

37. Lithner F, Tornblom N (1980) Gangrene localized to the lower limbs in diabetics. Acta Med Scand 208:315–320

38. Logerfo FW, Coffman JD (1984) Vascular and microvascular disease of the foot in diabetes. Implications for foot care. N Engl J Med 311:1615–1619

39. Michal V, Kovac J, Belan A (1984) Arterial lesions in impotence: phalloarteriography. In Angiol 3:247–254

40. Nath RL, Menzoian JO, Kaplan KH et al. (1981) The multidisciplinary approach to vasculogenic impotence. Surgery 89:124–131

41. Oakley W, Catterall RC, Martin MM (1956) Aetiology and management of lesions of the feet in diabetes. Br Med J iv:953–957

42. Ostrander LD, Francis T, Hayner NS et al. (1965) The relationship of cardiovascular disease to hyperglycemia. Ann Intern Med 62:1188–1198

43. Ostrander LD, Lamphiear DE, Carman WJ et al. (1981) Blood glucose and risk of coronary heart disease. Arteriosclerosis 1:33–37

44. Santen RJ, Willis PW, Fajans SS (1972) Atherosclerosis in diabetes mellitus. Correlations with serum lipid levels, adiposity and serum insulin level. Arch Intern Med 130:833–843

45. Shenfield GM, Elton RA, Bhalla IP et al. (1979) Diabetic mortality in Edinburgh. Diabete Metab 5:149–158

46. Sinha S, Munichoodappa CS, Kozak GP (1972) Neuro-arthropathy (Charcot joints) in diabetes mellitus. Medicine 51:191–210

47. Soler NG, Pentecost BL, Benett MA et al. (1974) Coronary care for myocardial infarction in diabetics. Lancet I:475–477

48. Stout RW (1979) Diabetes and atherosclerosis – the role of insulin. Diabetologia 16:141–150

49. Stout RW (1981) Blood glucose and atherosclerosis. Arteriosclerosis 1:227–234

50. Strandness DE, Priest Re, Gibbons GE (1964) Combined clinical and pathologic study of diabetic and non-diabetic peripheral arterial disease. Diabetes 13:366–372

51. Turnipseed WD, Sackett JF, Strother CM et al. (1982) A comparison of standard cerebral arteriography with non-invasive doppler imaging and intravenous angiography. Arch Surg 117:419–421

52. Virag R, Bouilly P, Frydman D (1985) Is impotence an arterial disorder? A study of arterial risk factors in 440 impotent men. Lancet I:181–184

53. Virag R, Zwang G, Dermange H et al. (1980) Exploration et traitement chirurgical de l'impuissance vasculaire. J Mal Vasc 5:205–209

54. Walker PM, Johnston KW (1980) Predicting the success of a sympathectomy: a prospective study using discriminant function and multiple regression analysis. Surgery 87:216–220

55. West KW, Ahuja MMS, Bennett PH et al. (1983) The role of circulating glucose and triglyceride concentrations and their interactions with other "risk factors" as determinants of arterial disease in nine diabetic population samples from the WHO multinational study. Diabetes Care 6:361–369

56. Wheelock FL, Gibbons GW: (1984) Arterial reconstruction. Femoral-popliteal-tibial. In: Kozak GP (ed) Management of diabetic foot problems. Saunders, Philadelphia, pp 173–187

57. Wyss CR, Matsen FA, Simmons CW (1984) Transcutaneous oxygen tension measurements on limbs of diabetic and non-diabetic patients with peripheral vascular disease. Surgery 95:339–345

Chapter 23

Diabetic Retinopathy

A. Garner

Although the ophthalmoscopic appearance of the diseased retina in diabetes mellitus was first described by von Jaeger [62] in 1855, detailed information about the histopathological changes was not forthcoming, despite a careful account by Ballantyne and Loewenstein [13] based on conventional sections, until the vessel injection studies of Ashton [4,5] and the retinal digest preparations introduced by Kuwabara and Cogan [72] allowed a more complete appreciation. Subsequent investigations, some using fluorescein angiography, have enhanced yet further our understanding of the pathophysiology of the diabetic retina.

Predisposing Factors

Metabolic Disturbance

The importance of the metabolic and hormonal dysfunctions associated with diabetes in causing retinal pathology is highlighted by the development of retinopathy, albeit infrequently, in patients who develop diabetes as a secondary consequence of pancreatitis [40,92] or haemochromatosis [53] or following pancreatectomy [100]. It has been pointed out that the retinal complications are relatively infrequent in secondary diabetes when compared with their incidence in spontaneous diabetes but this may be a reflection of the generally shorter duration of the metabolic upset, since Mohan et al. [87] found that 13 of 40 patients with tropical pancreatic diabetes, a condition compatible with prolonged survival, had demonstrable retinopathy. Reti-

nopathy can also develop in animals with artificially induced diabetes [20,63]. Moreover, the therapeutic benefit of maintaining normoglycaemia, while open to qualification, is widely attested to [23,47,79,109]. Engerman and Kern [43] have produced an experimental retinopathy, comparable to that seen in diabetes, in dogs fed high levels of galactose, which suggests that the effect is attributable to hexoses in general. It would be naive, nevertheless, to imagine that the relationship is direct or unifactorial, since others have found that patients with mild diabetes controlled by dietary means were still at risk of developing retinopathy if growth hormone levels were increased and there was hypoinsulinaemia [15].

Genetic Considerations

The observation that the vascular basement membrane thickening of diabetic microangiopathy may occasionally be seen in patients with disease of very short duration [91] or even precede the onset of diabetes [95] might suggest that the microvascular changes are part of the primary genetic defect of essential diabetes. Conflicting with this conclusion, however, is evidence of the production of similar basement membrane thickening in animals with experimentally induced diabetes [19,89] and the previously noticed development of retinopathy in patients with secondary diabetes.

Even so, genetic factors cannot be discounted totally as there is evidence of an enhanced risk that patients with primary insulin-dependent diabetes and the histocompatibility antigen HLA-DR4 will have retinal disease [38]. A susceptibility to retinopathy has also been noted in insulin-dependent diabetics in the presence of histocompatibility antigens HLA-B8 and HLA-B15 [14]. Identical twins who develop non-insulin-dependent diabetes are usually equally likely to develop retinopathy whereas the same is not true of monozygotic twins with the insulin-dependent form of the disease [76].

Arterial Disease

It is, perhaps, relevant when considering retinal vasculopathy to examine the effects of alterations in the flow of blood reaching the eye, especially since atherosclerosis is conspicuous in the dia-

betic state and affects the ophthalmic and central retinal arteries. A postmortem study of the ophthalmic artery in diabetics with and without retinopathy revealed that the opening from the internal carotid artery was generally narrower in the former group [50] (Fig. 23.1). There were, however, numerous individual exceptions where the artery was markedly stenosed on one side and relatively uninvolved on the other in the presence of bilateral and equal retinopathy: conversely, not all diabetics with narrowed ophthalmic artery openings had evidence of retinopathy. Taken as a whole, these findings led to the conclusion that ophthalmic artery atheroma may aggravate an established retinopathy but is unlikely to be an initiating factor [50].

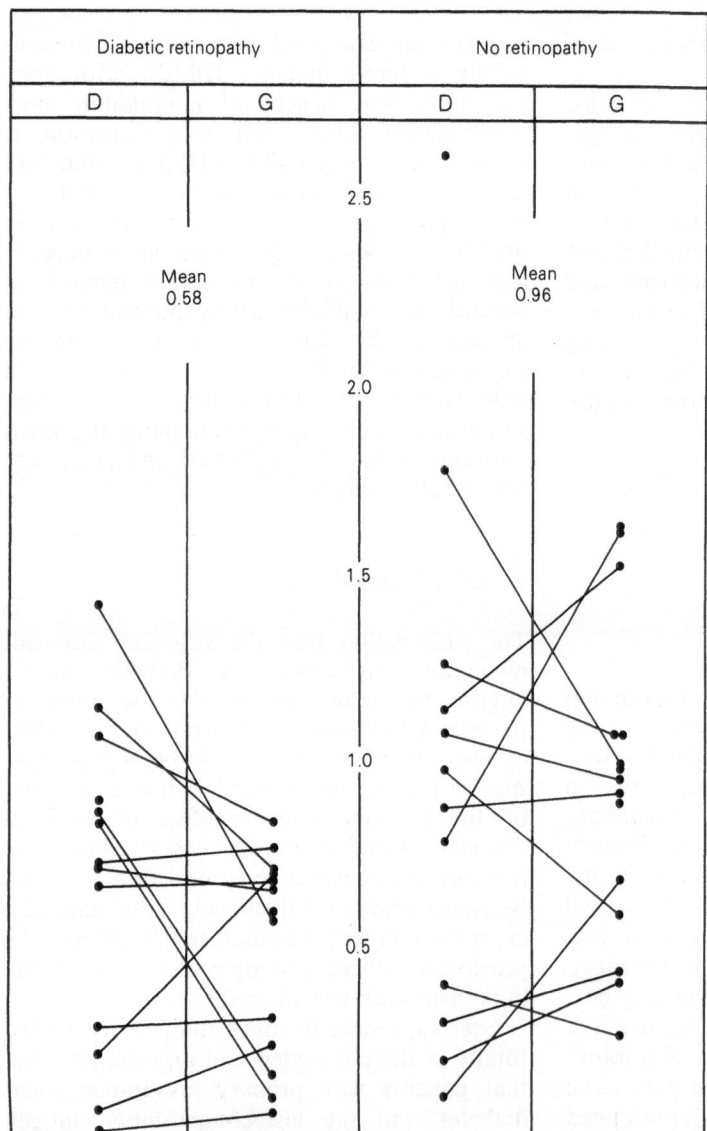

Fig. 23.1. Calibre of ophthalmic artery opening in diabetic patients at autopsy with and without evidence of retinopathy. (Reproduced with permission from Garner and Klintworth [51].)

There is no convincing evidence that co-existing systemic hypertension adds to the risk of developing diabetic retinopathy [68].

Duration

The incidence of retinopathy increases with the age of the patient. Thus, the onset of diabetes in adult life is associated with an earlier and more rapid progression of retinopathy than is the case in childhood [68,95]. A higher risk of developing retinal disease is even more closely linked with the duration of the diabetic state, one reported series of 173 patients with juvenile onset diabetes having an incidence of 1% up to 4 years after diagnosis, 25% between 5 and 9 years and 67% from 10 to 16 years [48]. It is probable that all diabetics, given time, would develop retinopathy.

Pre-retinopathy

Even before there is clinically recognizable retinopathy, functional abnormalities may be present in the form of hyperpermeability of the blood–retinal barriers and increased blood flow.

Breakdown of the Blood–Retinal Barriers

In the healthy state the extravascular compartment of the retina is isolated from the circulating blood so that molecules of the size of lipids and proteins are prevented from passing from one compartment to the other. There are two barriers. The first, the inner blood–retinal barrier, is provided by the endothelium lining the retinal blood vessels, while the second, the outer blood–retinal barrier, is provided by the monolayer of pigment epithelium between the retina and the underlying choroid. At an ultrastructural level the barriers depend on the presence of tight encircling junctions (zonulae occludens) between adjacent cells.

One of the earliest changes in the diabetic state is an increase in permeability of these barriers such that water and even proteins and lipids can leak into the retina [30]. The leakage can be detected in vivo by observing the fate of fluorescein dye injected into the circulation. Evidence of the dye reaching the vitreous gel was originally construed as indicating breakdown of the vascular endothelium [31] but later evidence, supplemented by data from spontaneously diabetic rats, suggests that the leakage may be occurring simultaneously or even preferentially from the choroid across the retinal pigment epithelium [66,103]. The precise cause of the increase in permeability in human diabetes is not known, although some separation of the tight junctions has been described in dogs with alloxan-induced diabetes [106]. There is no conclusive evidence concerning the immediate stimulus to increased permeability but the beneficial effect of normoglycaemia in reducing intravitreal dye accumulation suggests that metabolic factors are involved [24].

Increased Blood Flow

A second early abnormality in increased blood flow through the retina which at first may reflect reduced transit time rather than vascular dilatation [71], although venous dilatation is one of the earliest ophthalmoscopic findings in diabetic retinopathy [74]. The increased flow may be an autoregulatory response to either tissue hypoxia caused in part, perhaps, by glycosylation of the circulating haemoglobin with consequent reduction in oxygen delivery to the tissues [78] or to the accumulation of metabolites [68]. The exacerbating effect of anaemia on the retinopathy in diabetes might also be explicable in terms of tissue hypoxia [94]. There is also evidence from laboratory experiments that hyperglycaemia per se can provoke a rise in retinal blood flow [55]. Some clinical reports of adverse effects of continuous subcutaneous insulin infusion may be explicable in terms of inhibited autoregulation: diabetics with proliferative retinopathy often respond badly to carefully and evenly regulated normoglycaemia by developing multiple cotton-wool spots [75]. This could be because the acute ischaemia required to create the retinal 'infarct' or cotton-wool spot is precipitated by abolishing the compensatory vasodilatation induced by hyperglycaemia.

Background Retinopathy

Some time after the appearance of a defective blood–retinal barrier and increased blood flow, clinically demonstrable changes can be expected in the retina. Such changes constitute background retinopathy and include venous dilatation, oedema, exudates, haemorrhages and microaneurysms. Fluorescein angiography reveals foci of unperfused capillaries and histological study provides evidence of pericyte degeneration and basement membrane thickening in addition to the clinically recognizable phenomena. None of the clinical findings are specific for diabetic retinopathy but when seen in combination in the posterior fundus they constitute a diagnostically identifiable picture.

Venous Dilatation

Both generalized and focal venous dilatation with some increase in tortuosity are well-recognized but little understood features of diabetic retinopathy. The emergence of venous dilatation early in the course of the retinopathy suggests that it is primarily a functional adaptation, especially as it can be reversed by good metabolic control [74], but whether it occurs in response to increased blood flow or tissue hypoxia or both is not known. Ultimately the dilatation tends to become permanent as sclerotic changes supervene.

Oedema

Oedema, should it involve the macular region of the retina, can have serious consequences for vision but the term has tended to be used somewhat indiscriminately to describe leakage of any kind that causes generalized induration of the retina. Used in the strict sense of a transudate with little protein or lipid, retinal oedema is most often associated with the increased transmural pressure that occurs in arterial hypertension and venous obstruction: provided the basic condition is treated the oedema resolves completely. However, the oedema seen in the diabetic retina is more often characterized by residual "hard" exudates, indicating that the fluid was not a pure transudate as might have been supposed [68]. The reason for the confusion is that true oedema fluid and plasma exudates are indistinguishable when they first form since both diffuse freely through the retina to create an overall thickening. Nevertheless, diffusion of pure transudate can occasionally occur, giving rise to cystoid spaces through the pooling of extracellular colourless fluid in the outer and inner plexiform layers of the retina [46].

Exudates

Retinal exudates represent the leakage of plasma from the circulating blood and in the context of diabetes the leakage is confined to the capillary vessels. This implies some loss of structural integrity in the endothelial lining of the affected vessels since molecules the size of proteins and lipids would not be expected to cross the capillary wall due to increased hydrostatic or osmotic pressure. There is no evidence of either increased pinocytotic activity or defective intercellular junctions in rats with spontaneous diabetes [17]: instead leakage of horseradish peroxidase tracer appears to be a consequence of endothelial cell degeneration. Some authors, however, have described evidence of transport across the tight junctions in streptozotocin- and alloxan-induced diabetes in rats [60] and dogs [106] respectively. Irrespective of the mechanism there is experimental evidence that control of the metabolic defect can reverse the permeability [63]. There is evidence in humans of accelerated endothelial cell death in response to hyperglycaemia [81] and a gradual loss of both endothelial cells and pericytes can be observed by histological examination of the retinal capillaries. Loss of integrity of the endothelial lining of microaneurysms is also common, these being one of the more potent sources of exudate.

Initially the exudate, as noted previously, is ophthalmoscopically indistinguishable from oedema fluid. As it diffuses through the retina there is a prevailing drift towards the choroid engendered by the intraocular pressure [45]. However, the junctional complexes between the terminal regions of the Müller cells and the adjacent photoreceptor cells, the so-called external limiting membrane of the retina, prevent deeper penetration by the lipid and protein component of the exudate while allowing water and electrolytes to pass unhindered. This means that the exudate becomes progressively inspissated until, eventually, semi-solid residues are deposited in the outer plexiform layer of the retina (Fig. 23.2). The deposits are circumscribed, if irregular in shape, and by this time are recognizable clinically as "hard" exudates. The lipid content can impart a shiny appearance, particularly if the diabetic state is complicated by hyperlipidaemia, when the exudates may be termed "waxy". Lateral diffusion from a single focus of leakage in association with progressive inspissation can result in an annular or circinate pattern of exudation. The distribution of exudates forming around the macula is determined

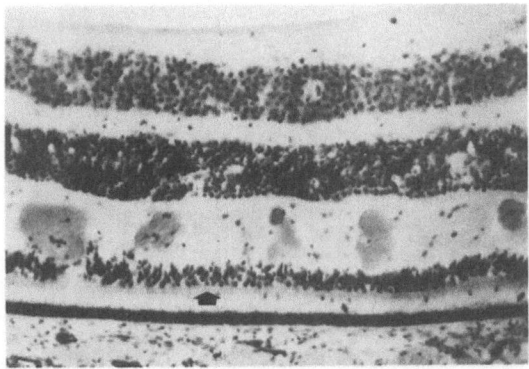

Fig. 23.2. Section through the posterior retina of a diabetic patient, showing circumscribed "hard" exudates in the outer plexiform layer. Although the exudates derive from vessels in the inner retina they migrate towards the choroid only to be held back by the external limiting membrane (*arrow*) (Reproduced with permission from Garner and Klintworth [91].) H & E, × 180

by the radial outflow of the axons in Henle's layer which creates a stellate appearance.

Subsequent resolution of the exudates may take some months, especially when the lipid component is prominent. Clofibrate therapy can accelerate the resorption, possibly through reducing the initial cholesterol content, but would not be expected to prevent the development of exudates [39].

Haemorrhages

Intraretinal haemorrhages in diabetes usually originate in the capillary bed, microaneurysms being a particularly potent source. When microaneurysms are the source, the extravasated blood pools around the aneurysm ("dot-and-blot" haemorrhages) before resolving some weeks or months later [68]. Sometimes the haemorrhages can, if sufficiently severe and close to the surface of the retina, spill into the potential space between the inner limiting membrane of the retina and the condensation of collagen fibres which constitute the posterior face of the vitreous. Rarely, extension into the vitreous may occur. Pre-retinal haemorrhages may be complicated by fibrous tissue proliferation which occurs as part of the process of organization, but those limited to the retrohyaloid space usually clear within 2 or 3 months. Resolution in the latter situation may be facilitated by a greater tendency for the haemorrhage to remain fluid [28]. Bleeding from the larger intraretinal vessels

is unusual but not unknown and, since the vessels are located within the nerve fibre layer of the retina, presents clinically in a linear or flame-shaped pattern. This type of haemorrhage may be a complication of concomitant systemic hypertension but has also been described in normotensive diabetic patients [68].

Microaneurysms

Microaneurysms are focal distensions in the walls of capillary blood vessels. They are an early and almost invariable, but not specific, feature of diabetic retinopathy. Commonly they develop in clusters at the margins of focal areas of unperfused capillaries, predominating on the venous side of the capillary bed. They persist for several months and then disappear, after which new crops of aneurysms form elsewhere in the retina [67]. The apparent resolution of the microaneurysms is attributable to gradual encrustation of the sac with thrombus and the eventual occlusion of the affected capillary: such aneurysms will not be seen by ophthalmoscopy or by fluorescein angiography.

From a morphological standpoint two types of microaneurysm can be recognized. The first type is a saccular distension of part of the wall of a capillary, whereas the second appears to be a sequel to the formation of distended loops or kinks in the affected capillary [5] (Fig. 23.3).

Saccular microaneurysms are seen in vascular retinopathies caused by a variety of diseases

Fig. 23.3. Diabetic retinopathy. The patient vessels are outlined by Indian ink injection. Clusters of microaneurysms are seen in relation to areas of hypoperfusion and while some are saccular there is also evidence of preceding loop formation, the affected capillary segments being both kinked and dilated. × 70

including systemic hypertension, sickle-cell disease, retinal vein occlusion and macroglobulinaemia, a possible common denominator being increased intraluminal tension due to raised perfusion pressure, hyperviscosity or obstructed outflow. Simple distension due to increased intraluminal pressure does not account, however, for the focal nature of the lesions and additional explanations are necessary. One of the more popular theories involves the pericytes in the wall of the affected capillary, especially as degeneration of these cells is a distinctive feature of diabetic retinopathy (vide infra). Selective degeneration of these cells, which are the counterpart of the smooth-muscle component of larger vessels, has been considered to provoke a focal loss of capillary tone and aneurysmal distension [26]. Although there is some statistical correlation between pericyte loss and microaneurysm formation [113], the histological correlation is far from precise. Also, a contractile function for the pericyte in the retina has not been demonstrated. It is, nevertheless, possible that loss of part of the cellular content of the capillary wall would, irrespective of any functional loss, predispose the affected segment to dilatation in the presence of some other distending factor. Some authors have attempted to associate microaneurysm formation with basement membrane changes [58] and traction by the argyrophilic fibres that accumulate within the diabetic retina [112] but these theories have not attracted much support.

Aneurysms that are preceded by loop formation are particularly common in diabetes. The kinking that preceded the loop is associated with an apparent elongation and dilatation in a segment of the affected capillary and it is possible that the process is one of focal endothelial cell hyperplasia. This could mean that limited neovascularization is occurring, especially as the loops are usually directed towards areas of capillary closure. The formation of loops may also be abetted by the proliferation of reticulin fibres at the base of each loop [8].

Capillary Non-perfusion

Foci of capillary closures are another early feature of diabetic retinopathy [70] (Fig. 23.4) and microscopical examination shows them to be completely acellular [7,26]. Cotton-wool spots, which are associated with disrupted axoplasmic transport and degeneration in the nerve fibre

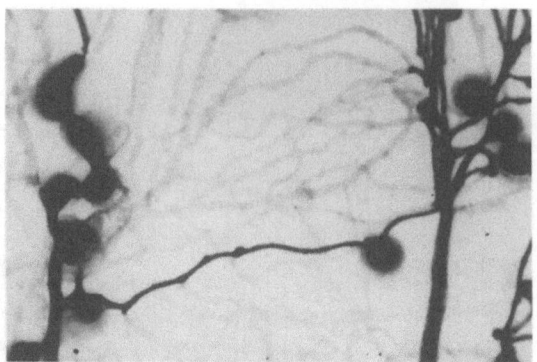

Fig. 23.4. In this diabetic retina the functioning vessels are perfused with Indian ink and subsequent trypsin digestion of the neuronal components has allowed visualization of other defunct, largely acellular, capillaries. Numerous microaneurysms and irregular venular dilatations are also seen. × 170

layer as a consequence of acute ischaemia [9,83], may be concomitant finding [69]. The number and size of the avascular areas increases with the duration of the retinopathy and it is possible that occlusion of the feeding arteriole is the principal cause [4]. Whether this is because of changes in the vessel wall or in the circulating blood is uncertain. Hyalinization of the walls of the retinal arterioles is often seen, although never to the extent of complete stenosis. The role of endothelial cell hyperpermeability has already been stressed and it is possible that the abnormal vessel lining may also predispose to thrombosis, especially as a reduction in prostacyclin metabolites has been noted [36]. Changes in the circulating blood conducive to microthrombosis are also well documented, for example platelet abnormalities [36,90] and hyperviscosity due to increases in acute phase plasma protein levels [84]. Trauma resulting in haemoconcentration and increased blood viscosity was considered to be the cause of early emergence of florid retinopathy in two diabetic patients reported by Alexander et al. [3]. Evidence of microthrombi in the retinal capillaries of rats with streptozotocin-induced diabetes has been presented by Ishibashi et al. [60] but convincing evidence of a similar process in humans has yet to be presented.

A further factor which may contribute to capillary closure is tissue oedema [6]. Siphoning of blood away from some capillaries because of dilatation of others has also been postulated [25] but the evidence provided by fluorescein angiographs lends little support to this suggestion [70].

Pericyte Degeneration

Although not peculiar to diabetes, in no other retinopathy is pericyte degeneration such a conspicuous finding. The recognition that there is a selective deficiency of pericytes relative to endothelial cells in the diabetic retina came from the study of retinal digest preparations, a technique involving immersion of the whole retina in a proteolytic enzyme solution until the neurons fall away to leave an intact vascular bed [73]. Mounting and staining the vessels with haematoxylin and eosin showed a relative absence of viable pericytes, the pericyte nuclei failing to stain with the basic dye and taking up eosin instead [7] (Fig. 23.5). Called, for want of a better name, eosinophilic degeneration, such pericyte change is associated with both empty, ghost nuclear outlines in preparations stained by the periodic acid-Schiff sequence and with electron microscopical evidence of both nuclear and cytoplasmic necrosis.

The same eosinophilic degeneration is not demonstrable in brain capillaries even though it may be conspicuous in the retina [1] suggesting that the process is limited to the retinal circulation. The cause of the degeneration is obscure although pericytes are thought to be more vulnerable to hypoxia than are endothelial cells [11,88]. How this could account for limitation of the degeneration in the retina is not clear unless the excessive metabolic demands of the retina expose any latent hypoxia in advance of that in other tissues. Some recent studies suggest that the retinal selectivity may be more apparent than real. Labelling with tritiated thymidine of the

capillaries in rats with streptozotocin-induced diabetes suggests that the deficit in the pericyte population is due to a low rate of regeneration in response to an initial accelerated degeneration [93]: in other tissues the regeneration rate matches equally the degenerative process [101]. But still the problem remains as to why pericyte regeneration should be so much less efficient in the retina.

Basement Membrane Thickening

Retinal capillaries, like those in many other tissues in diabetics, acquire thickened basement membranes [18,102]. Haematogenous elements including lipid, fibrin and haemosiderin have been described in the thickened membrane [8] which may reflect the characteristic increase in endothelial cell permeability, especially as it has been shown that the basement membrane changes in rats with streptozotocin-induced diabetes are preceded by a permeability increase in the vascular endothelium [59]. However, leakage is probably a minor factor in the thickening and increased synthesis is likely to account for most of the thickening. In the context of basement membrane thickening in general it is possible that there is both increased synthesis of the peptide precursor units and glycosylation in response to hyperglycaemia [2,97]. An observation of some fascination was that the capillary basement membranes in the swim-bladders of eels kept in cold water were considerably thickened compared with those of eels kept in water at 15°–20°C: in parallel with these differences was

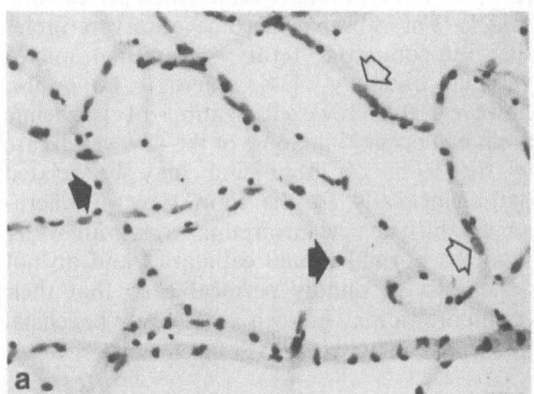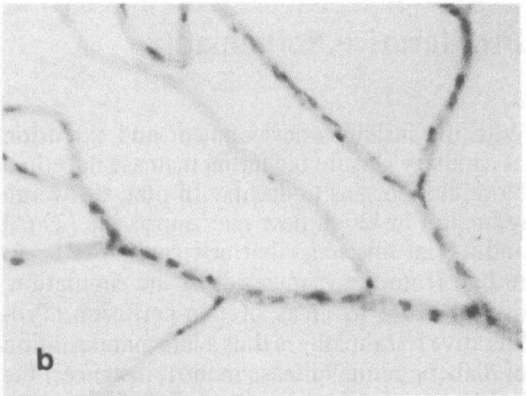

Fig. 23.5a,b. Retinal digest preparations stained with haematoxylin and eosin. **a** Normal retina. The capillaries show both the elongated nuclei of endothelial cells (*open arrows*) and the rounded, darker staining nuclei of pericytes (*solid arrows*). × 70. **b** Diabetic retina. The endothelial cell nuclei are essentially unchanged and are even increased in number but the pericyte nuclei are reduced in number and most of those that remain appear pale (having taken up eosin rather than haematoxylin). × 180

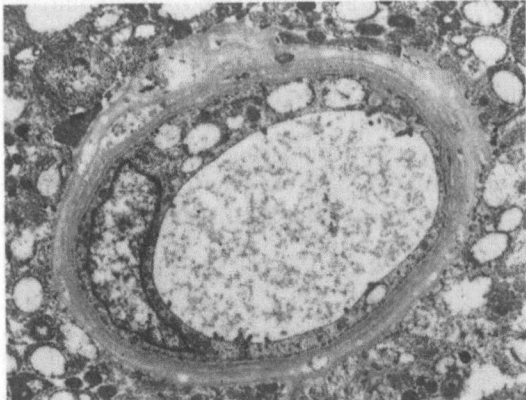

Fig. 23.6. Electron micrograph of a capillary in diabetic retinopathy showing a thickened, laminated basement membrane with some entrapped cellular debris, possibly the result of pericyte degeneration. Uranyl acetate/lead citrate. × 18 600

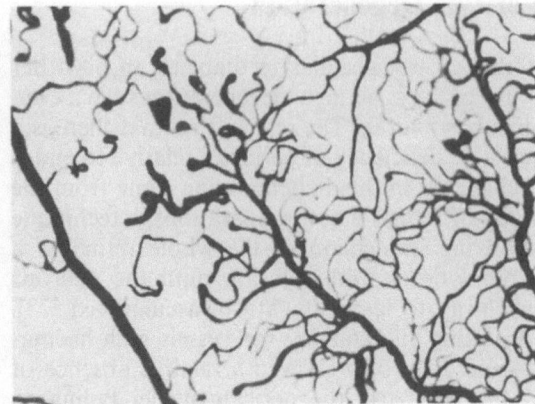

Fig. 23.7. Diabetic retina injected with Indian ink to show several dilated and tortuous "new" capillaries in relation to an area of underperfusion. × 70

the finding of marked hyperglycaemia in the eels maintained at low temperature [16].

Sometimes the basement membrane is seen to be laminated (Fig. 23.6) and this has given rise to the concept of accelerated endothelial cell degeneration, with each successive wave of renewal being associated with the formation of a new membrane [104,105].

Argyrophilic reticulin fibres are normally present in the retina, particularly around the blood vessels, but their number is greatly increased in diabetic retinopathy [10]. These fibres blend with the vascular basement membranes which might infer similarities in terms of both composition and pathogenesis.

Proliferative Retinopathy

With the insidious development and extension of capillary closure the initial increase in retinal blood flow begins to decline. In time an overall reduction in blood flow may supervene [29,68] and it is at this stage that new capillaries begin to bud from the venous side of the circulation, usually close to areas of non-perfusion. Proliferative retinopathy is thus a late manifestation of diabetic retinal disease in most instances, the incidence in one series increasing from 1.5% after 16 years from the onset of insulin-dependent diabetes to 18% after 26 years [77]. It also appears that diabetics with juvenile onset insulin-dependent disease are more likely to develop pre-

retinal neovascularization than are those with the late onset non-insulin-dependent form [21].

Intraretinal Vascular Proliferation

Although the term proliferative retinopathy usually infers vascular proliferation on the surface of the retina, new vessels can also develop within the retina. The new vessels are seen as dilated, tortuous capillaries extending into areas of failed perfusion (Fig. 23.7). On purely morphological grounds it can be difficult to distinguish between vessels that are newly formed, i.e. that represent the re-endothelialization of defunct capillary basement membrane tubes [105], and those that are remodelled pre-existing vessels. For this reason some observers prefer the non-committal term "intraretinal microvascular anomaly" [34]. There is no doubt, however, that revascularization of ischaemic areas can occur [54]. Some of the new vessels are in the form of loops and may be related pathogenetically to the loop type of microaneurysm. The new intraretinal vessels are wider than the normal retinal capillaries and do not appear to be unduly permeable so that their proliferation may be seen as a largely beneficial attempt to revitalize areas of ischaemia.

Pre-retinal Fibrovascular Proliferation

The proliferation of new vessels in front of the retina is most conspicuous in the posterior fundus and over the optic disc. In some instances the choroid can be the source of the neo-vascularisation on the optic disc [68] and exceptionally the retinal arteries have been shown to be the source, but usually the new capillaries arise from the venous side of the retinal circulation.

At first, having broken through the inner limiting membrane, the new vessels tend to lie flat on the surface of the retina (Fig. 23.8) where they proliferate to form delicate arcades and fronds. Forward extension of the growing vessels is facilitated by retraction of the vitreous since it is usual for them to be attached to the condensation of collagen fibres responsible for the posterior face of the hyaloid or vitreous gel. Collapse and retraction of the vitreous gel is a poorly understood but frequent phenomenon in patients with proliferative diabetic retinopathy [33]: possibly the structure of the gel is destroyed by the leakage of charged electrolyte ions from the hyper-permeable new vessels [28]. The traction exerted by the retreating vitreous body can tear the delicate new vessels and make them bleed, with serious consequences for vision.

Some degree of fibrous tissue proliferation is also usual, tending to be more conspicuous in the later stages of angiogenesis, and the consequences can be even more disastrous than those attending bleeding from the new vessels, particularly as there is no capacity for resolution. The harmful effects of the fibrotic component are attributable in part to the formation of an optically opaque membrane and the risk of retinal tearing and tractional detachment from the contractile action of the actin-rich "myo-fibroblasts" [55,107]. Before significant fibrosis

has developed, the vascular proliferations may either regress spontaneously or be encouraged to do so by laser treatment. The not entirely appropriate label of retinitis proliferans (there is no evidence of inflammation) has been given to the irreversible late fibrotic stage of proliferative retinopathy. The source of the fibroblasts is unknown. The pigment epithelium of the retina is a potential source [85] through a capacity for metaplasia but direct evidence that such cells are involved in the proliferative retinopathy associated with diabetes has not been presented. Nor does there seem to be any reason to implicate the glial component of the retina [57]. Metaplasia of the vascular endothelium is another possibility, conceivably in an indirect manner via pericytes which have been shown to incorporate [¹⁴C] proline and form collagen in vitro [27]. There is also evidence from tissue culture experiments that hyperglycaemia can be a stimulus to collagen synthesis [78].

When they first develop, the new capillary buds appear to consist entirely of endothelial cells [72] with pericytes developing later [35]. The endothelial cells may be fenestrated [98,107]; the fenestrations measuring 40–80 nm with a covering unit membrane diaphragm [107] together with evidence of incomplete tight junctions in the immature capillaries and even open inter-endothelial clefts [86] may account for the characteristic leakiness of the new vessels.

The nature of the mechanism underlying neo-vascularization has attracted considerable attention in recent years. Based on the observation that the new vessels usually originate from the edges of areas of capillary closure in the retina (Fig. 23.9), the idea grew that the relationship

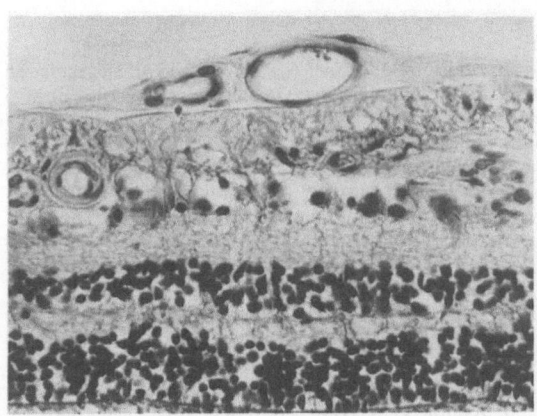

Fig. 23.8. Proliferative retinopathy. Newly formed capillaries are seen lying on the inner surface of the retina. H & E, ×450

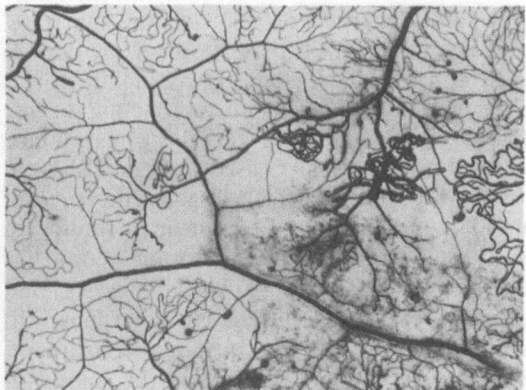

Fig. 23.9. Proliferative diabetic retinopathy. A leash of new vessels, previously identified as being pre–retinal, is seen overlying an area of underperfused retina. Indian ink injection, ×50

might be causal [111]. A similar connection between angiogenesis and reduced retinal perfusion in retrolental fibroplasia had previously been noted by Ashton et al. [12], and comparable reasoning can be advanced with respect to most vascular retinopathies giving rise to new vessel formation. Nevertheless, it appears that the retina needs to remain viable for the relationship to hold since occlusion of the central retinal artery, which gives rise to retinal infarction, is only rarely complicated by neovascularization.

On the basis of such circumstantial evidence it was postulated that retinal tissue subjected to reduced perfusion to render it hypoxic but not anoxic might release or activate a factor with angiogenic properties. The molecular size of the factor would be such as to allow it to diffuse across the inner limited membrane of the retina into the vitreous where it would accumulate until present in sufficient quantity to stimulate the underlying vessels to proliferate towards the source of the stimulus. It is only in recent years, however, that the existence of the putative factor has been demonstrated [49]. A fraction with angiogenic properties has been isolated from the vitreous of patients with both diabetic [52] and nondiabetic proliferative retinopathy [99], but it is not present in biologically significant concentrations in normal vitreous [44]. The active component appears to be an anionic moiety of extremely small size, i.e. with a molecular weight of 300–600, bound to a carrier protein [42]. Other workers have identified a substance of larger molecular size [32] which may represent contamination of the carrier protein or, more likely, mean that the moiety of molecular weight 300–600 is the active fraction of a larger peptide chain. The angiogenic factor stimulates capillary but not large vessel endothelium and then only if the cells are grown on a collagenous substrate, and has no effect on pericytes in tissue culture [43]. It is probable that the retinal factor is similar, if not identical, to tumour angiogenic factor since antibodies raised against factor derived from the Walker ascites sarcoma showed affinity for the angiogenic retinal fraction present in feline retina [65]. Experimental studies in the kitten indicate that the normal retina has angiogenic capacity [64] which suggests that pathological proliferation is a response to a quantitative increase in a normal constituent. Evidence to support this was forthcoming following the discovery that the angiogenic fraction not only stimulates endothelial cell mitosis and migration but also a pro-collagenase which, in turn, mobilizes latent collagenase in the tissues

[110]. Using the latter property as a means of quantifying angiogenic activity, that is, measuring the amount of labelled collagen degraded by a given amount of extract, it has been possible to show that ischaemic retina from kittens subjected to vaso-obliterative doses of oxygen was at least three times more active than extracts of normal retina [99].

There is also evidence of a vaso-inhibitory substance in the normal vitreous [22,82]. This could mean that the baseline angiogenic activity in the retina required to maintain the existing vasculature is normally prevented from provoking intravitreal vascularization by virtue of a preponderance of inhibitory factor in the vitreous. For pre-retinal angiogenesis to occur it is necessary for this inhibitory behaviour to be overcome [61].

Conclusion

Diabetic retinopathy, like the parent disease, is a complex and multifactorial condition. Metabolic factors, especially hyperglycaemia, have been shown to be of major importance in its pathogenesis and there is evidence that functional derangements precede the morphological tissue changes. Intraretinal disturbance develops first but may give way to pre-retinal proliferative phenomena as retinal ischaemia intensifies, the latter being associated with the release of a fraction with angiogenic properties.

References

1. Addison DJ, Garner A, Ashton N (1970) Degeneration of intramural pericytes in diabetic retinopathy. Br Med J i:246–266
2. Albert KGMM, Hockaday TDR (1975) The biochemistry of the complications of diabetes mellitus. In: Keen H, Jarrett J (eds) Complications of diabetes. Edward Arnold, London, pp 221–264
3. Alexander WD, Kearns M, Kohner EM et al. (1979) Trauma and severe proliferative retinopathy in diabetes mellitus. Br Med J ii:831
4. Ashton N (1953) Arteriolar involvement in diabetic retinopathy. Br J Ophthalmol 37:282–292
5. Ashton N (1958) Diabetic micro-angiopathy. Adv Ophthalmol 8 (Bibl Ophthalmol Fasc 52) :1–84
6. Ashton N (1959) Diabetic retinopathy, a new approach. Lancet II:628–630
7. Ashton N (1963) Studies of the retinal capillaries in relation to diabetic and other retinopathies. Br J Ophthalmol 47:521–538

8. Ashton N (1974) Vascular basement membrane changes in diabetic retinopathy. Br J Ophthalmol 58:344–366

9. Ashton N, Dollery CT, Henkind P et al. (1966) Focal retinal ischaemia; ophthalmoscopic, circulatory and ultrastructural changes. Br J Ophthalmol 50:281–384

10. Ashton N, Tripathi RC (1975) Perivascular and inter-vascular reticular fibres of the retina. Am J Ophthalmol 80:337–359

11. Ashton N, Tripathi RC (1977) The problem of selective pericyte injury. 9th European conference on microcirculation, Antwerp, 1976 Bibl Anat, 16:19–23

12. Ashton N, Ward B, Serpell G (1954) Effect of oxygen on developing retinal vessels with particular reference to the problem of retrolental fibroplasia. Br J Ophthalmol 38:397–432

13. Ballantyne AJ, Loewenstein A (1943) Diseases of the retina. I. Pathology of diabetic retinopathy. Trans Ophthalmol Soc UK 63:95–115

14. Barbosa J, Ramsay RC, Knobloch WH et al. (1980) Histocompatibility antigen frequencies in diabetic retinopathy. Am J Ophthalmol 90:148–153

15. Barnes AJ, Kohner EM, Johnston DG et al. (1985) Severe retinopathy and mild carbohydrate intolerance; possible role of insulin deficiency and elevated circulating growth hormone. Lancet I:1465–1468

16. Bendayan M, Rasio SA (1981) Hyperglycemia and microangiopathy in the eel. Diabetes 30:317–325

17. Blair NP, Tso MOM, Dodge JT (1984) Pathologic studies of the blood-retinal barrier in the spontaneously diabetic BB rat. Invest Ophthalmol Vis Sci 25:302–311

18. Bloodworth JMB (1963) Diabetic microangiopathy. Diabetes 12:99–114

19. Bloodworth JMB, Engerman RL, Powers KL (1969) Experimental diabetic microangiopathy. I. Basement membrane statistics in the dog. Diabetes 18:455–458

20. Bloodworth JMB, Molitor DL (1965) Ultrastructural aspects of human and canine diabetic retinopathy. Invest Ophthalmol 4:1037–1048

21. Bodansky HJ, Cudworth AG, Whitelocke RAF et al. (1982) Diabetic retinopathy and its relation to the type of diabetes; review of a retinal clinic polulation. Br J Ophthalmol 66:496–499

22. Brem S, Preis I, Langer R et al. (1977) Inhibition of neovascularization by an extract from vitreous. Am J Ophthalmol 84:323–328

23. Cahill GF Jr, Etzwiler DD, Freinkel N: (1976) "Control" and diabetes. N Engl J Med 294:1004–1005

24. Christacoupoulos PD, Soeldner JS, Gleason RF (1972) Variation in retinal haemodynamics in different stages of diabetes. Excerpta Medica, Amsterdam, pp 180–181 (International Congress Series 280)

25. Cogan DG, Kuwabara T (1963) Capillary shunts in the pathogenesis of diabetic retinopathy. Diabetes 12:293–300

26. Cogan DG, Toussaint D, Kuwabara T (1961) Retinal vascular patterns. IV. Diabetic retinopathy. Arch Ophthalmol 66:366–378

27. Cohen MP, Frank RN, Khalifa AA (1980) Collagen production by cultured retinal capillary pericytes. Invest Ophthalmol Vis Sci 19:90–94

28. Constable IJ (1975) Pathology of vitreous membranes and the effect of haemorrhage and new vessels on the vitreous. Trans Ophthalmol Soc UK 95:382–386

29. Cunha Vaz JG (1978) Pathophysiology of diabetic retinopathy. Br J Ophthalmol 62:351–355

30. Cunha Vaz JG (1983) Studies on the pathophysiology of diabetic retinopathy: the blood-retinal barrier in diabetes. Diabetes [Suppl 2] 32:20–27

31. Cunha Vaz JG, Abreu JR, Campos AJ et al. (1975) Early breakdown of the blood-retinal barrier in diabetes. Br J Ophthalmol 89:649–656

32. D'Amore PA (1982) Purification of a retina-derived endothelial cell mitigen/angiogenic factor. J Cell Biol 95:9055 (abstract)

33. Davis MD (1965) Vitreous contraction in proliferative diabetic retinopathy. Arch Ophthalmol 74:741–751

34. Davis MD, Norton EWD, Myers FL (1969) The airline classification of diabetic retinopathy. In: Goldberg MF, Fine SL (eds) Treatment of diabetic retinopathy. United States Public Health Service, Washington DC, Publ. No. 1890, pp 7–22

35. Deruaz JP (1969) Etude histo-pathologique de la vascularisation rétinienne par la methode de la digestion enzymatique. Doc Ophthalmol 25:282–359

36. Dobie JG, Kwaan HC, Colwell JA et al. (1973) The role of platelets in pathogenesis of diabetic retinopathy. Trans Am Acad Ophthalmol Otolaryngol 77:43–46

37. Dollery CT, Friedman LA, Hensby CN et al. (1979) Circulating prostacyclin may be reduced in diabetes. Lancet II:8156–8157

38. Dornan TL, Ting A, McPherson CK et al. (1982) Genetic susceptibility to the development of retinopathy in insulin-dependent diabetics. Diabetes 31:226–231

39. Duncan LJP, Cullen SF, Ireland JT et al. (1968) A three year trial of atromid therapy in exudative diabetic retinopathy. Diabetes 17:458–467

40. Duncan LJP, Macfarlane A, Robson JS (1958) Diabetic retinopathy and nephrology in pancreatic diabetes. Lancet I:822–826

41. Elston SF, Kissun RD, Schor A et al. (1987) A low molecular mass angiogenic fraction in vitreous from patients with extraretinal neovascularisation. (in press)

42. Elston SF, Schor AM, Weiss JB (1985) Bovine retinal angiogenesis factor is a small molecule (molecular mass <600). Invest Ophthalmol Vis Sci 26:74–79

43. Engerman RL, Kern RS (1984) Experimental galactosemia produces diabetic-like retinopathy. Diabetes 33:97–100

44. Felton SM, Brown GC, Felberg NT et al. (1979) Vitreous inhibition of tumour neovascularization. Arch Ophthalmol 97:1710–1713

45. Foulds WS (1976) Clinical significance of trans-scleral fluid transfer. Trans Ophthalmol Soc UK 96:290–308

46. Frangieh GT, Green WR, Engel HM (1981) A histopathologic study of macular cysts and holes. Retina 1:311–336

47. Frank RN (1984) On the pathogenesis of diabetic retinopathy Ophthalmology 91:626–634

48. Frank RN, Hoffman WH, Podgor MJ et al. (1982) Retinopathy in juvenile-onset type 1 diabetes of short duration. Diabetes 31:874–882

49. Garner A (1986) Ocular angiogenesis. Int Rev Exp Pathol 28:249–306

50. Garner A, Ashton N (1972) Ophthalmic artery stenosis and diabetic retinopathy. Trans Ophthalmol Soc UK 92:101–110

51. Garner A, Klintworth GK (1982) Pathobiology of ocular disease. Marcel Dekker, New York

52. Glaser BM, D'Amore PA, Michels RG et al. (1980) The demonstration of angiogenic activity from ocular tissues. Ophthalmology 87:440–446

53. Griffiths JD, Dymock IW, Davies EWG et al. (1971) Occurrence and prevalence of diabetic retinopathy in haemochromatosis. Diabetes 20–766–770

54. Hamilton AM, Kohner EM, Rosen D et al. (1974) Experimental venous occlusion. Proc R Soc Med 67:1045–1948

55. Hill DW, Atherton HA (1979) Experimental studies of the retinal circulation relating to diabetic retinopathy. Trans Ophthalmol Soc UK 99:4–7

56. Hiscott PS, Grierson I, Hitchins CA et al. (1983) Epiretinal membranes in vitro. Trans Ophthalmol Soc UK 103:89–102

57. Hiscott PS, Grierson I, Trombetta CJ (1984) Retinal and epiretinal glia; an immunohistochemical study. Br J Ophthalmol 68:698–707

58. Hogan MJ, Feeney L (1963) Ultrastructure of retinal blood vessels. II. The small vessels. J Ultrastruct Res 9:29–46

59. Ishibashi T, Tanaka K, Taniguchi Y (1979) An experimental study on the role of vascular permeability in the development and progression of diabetic retinopathy. Acta Soc Ophthalmol Jap 83:766–782

60. Ishibashi T, Tanaka K, Taniguchi Y (1980) Disruption of blood retinal-barrier in experimental diabetic rats; an electron microscopic study. Exp Eye Res 30:401–410

61. Jacobson B, Basu PK, Hasany SM (1984) Vascular endothelial cell growth inhibitor of normal and pathologic human vitreous. Arch Ophthalmol 102:1543–1545

62. Jaeger E (1955) Beitrage zur Pathologie des Auges. Kaiserlich-Koniglichen Hof-und Staatsdruckerci, Wien, pp 33–36

63. Kernell A, Arnqvist H (1983) Effect of insulin treatment on the blood-retinal barrier in rate with septozotocin-induced diabetes. Arch Ophthalmol 101:968–970

64. Kissun RD, Garner A (1977) Vasoformative properties of normal and hypoxic retinal tissue. Br J Ophthalmol 61:394–398

65. Kissun RD, Hill CR, Garner A et al. (1982) A low molecular weight angiogenic factor in cat retina. Br J Ophthalmol 66:165–169

66. Kohner EM (1980) Fluorescein in diabetic retinopathy. In: Friedman EA, L'Esperance FA Jr (eds) Diabetic renal-retinal syndrome. Grune and Stratton, New York, pp 103–112

67. Kohner EM, Dollery CT (1970) The rate of formation and disappearance of microaneuryons in diabetic retinopathy. Eur J Clin Invest 1:167–171

68. Kohner EM, Dollery CT (1975) Diabetic retinopathy. In: Kean H, Jarrett J (eds) Complications of diabetes. Edward Arnold, London, pp 7–98

69. Kohner EM, Dollery CT, Bulpitt CJ (1969) Cotton-wool spots in diabetic retinopathy. Diabetes 18:691–704

70. Kohner EM, Dollery CT, Paterson JW et al. (1967) Arterial fluorescein studies in diabetic retinopathy. Diabetes 16:1–10

71. Kohner EM, Hamilton AM, Saunders SJ et al. (1975) The retinal blood flow in diabetes. Diabetologia 11:27–33

72. Kuwabara T, Cogan DG (1960) Studies of retinal vascular patterns. I. Normal architecture. Arch Ophthalmol 64:904–911

73. Kuwabara T, Cogan DG (1963) Retinal vascular patterns. VI. Mural cells of the retinal capillaries. Arch Ophthalmol 69:492–502

74. Larsen HW (1960) Diabetic retinopathy. Acta Ophthalmol [Suppl] 60:1–89

75. Lawson PM, Champion PC, Canny C et al. (1982) Continuous subcutaneous insulin infusion (CS11) does not prevent progression of proliferative and pre-proliferative retinopathy. Br J Ophthalmol 66:762–766

76. Leslie RDG, Pyke DA (1982) Diabetic retinopathy in identical twins. Diabetes 31:19–21

77. Lestradet H, Papoz L, Hellouin De Menibus C (1981) Long term study of mortality and vascular complications in juvenile-onset (type 1) diabetes. Diabetes 30:175–179

78. Li W, Shen S, Khatami M et al. (1984) Stimulation of retinal capillary pericyte protein and collagen synthesis in culture by high-glucose concentration. Diabetes 33:785–789

79. Liang JC, Goldberg MF (1980) Treatment of diabetic retinopathy. Diabetes 29:841–851

80. Little HL (1981) Pathogenesis. In: L'Esperance FA Jr, James WA Jr (eds) Diabetic retinopathy, clinical evaluation and management. Mosby, St. Louis, pp 58–88

81. Lorenzi M, Cagliero E, Toledo S (1985) Glucose toxicity for human endothelial cells in culture; delayed replication, disturbed cell cycle, and accelerated death. Diabetes 34:621–627

82. Lutty G, Thompson DC, Gallup JG et al. (1980) Inhibition of neovascularization by bovine vitreous in the CAM assay. Invest Ophthalmol Vis Sci [ARVO Suppl]:138

83. McLeod D (1976) Retinal ischaemia, disc swelling and axoplasmic transport. Trans Ophthalmol Soc UK 96:313–316

84. McMillan DE (1976) Plasma protein changes, blood viscosity, and diabetic microangiopathy. Diabetes [Suppl 2] 25:858–864

85. Machemer R, Horn D, Aaberg TM (1978) Pigment epithelial proliferation in human retinal detachment with massive periretinal proliferation. Am J Ophthalmol 85:181–191

86. Miller H, Miller B, Zonis S et al. (1984) Diabetic neovascularization; permeability and ultrastructure. Invest Ophthalmol Vis Sci 25:1338–1342

87. Mohan R, Rajendran B, Mohan V et al. (1985) Retinopathy in tropical pancreatic diabetes. Arch Ophthalmol 103:1487–1489

88. Oliveira F de (1966) Pericytes in diabetic retinopathy. Br J Ophthalmol 50:134–143

89. Osterby R, Seyer-Hansen K, Gundersen HJG et al. (1978) Growth hormone enhances basement membrane thickening in experimental diabetes. Diabetologia 15:487–489

90. Porta M, Hilgard P, Kohner EM (1980) Platelet shape change abnormalities in diabetic retinopathy. Diabetologia 18:217–221

91. Sabour MS, MacDonald MK, Robson JS (1962) An electron microscopic study of the human kidney in young diabetic patients with normal renal function. Diabetes 11:291–295

92. Sevel D, Bristow JH, Bank S et al. (1971) Diabetic retinopathy in chronic pancreatitis. Arch Ophthalmol (1985) 86:245–250

93. Sharma NK, Gardiner TA, Archer DB (1985) A morphologic and autoradiographic study of cell death and regeneration in the retinal microvasculature of normal and diabetic rats. Am J Ophthalmol 100:51–60

94. Shorb SR (1985) Anemia and diabetic retinopathy. Am J Ophthalmol 100:434–436

95. Siperstein MD, Unger RH, Madison LL (1968) Studies of muscle capillary basement membranes in normal subjects, diabetic and prediabetic patients. J Clin Invest 47:1973–1999

96. Soler NG, Fitzgerald MG, Malins JM et al. (1969)

Retinopathy at diagnosis of diabetes, with special reference to patients under 40 years of age. Br Med J iii:567

97. Spiro RG (1976) Investigations into the biochemical basis of diabetic basement-membrane alterations. Diabetes [Suppl 25]:909–913

98. Taniguchi Y (1976) Ultrastructure of newly formed blood vessels in diabetic retinopathy. Jpn J Ophthalmol 20:19–31

99. Taylor CM, Weiss JB, Kissun RD et al. (1986) Effect of oxygen tension on the quantities of procollagenase-activating angiogenic factor present in the developing kitten retina. Br J Ophthalmol 70: 162–165

100. Tiengo A, Segato T, Briani G et al. (1983) The presence of retinopathy in patients with secondary diabetes following pancreatectomy or chronic pancreatitis. Diabetes Care 6:544–570

101. Tilton RG, Hoffmann PL, Kilo C et al. (1981) Pericyte degeneration and basement membrane thickening in skeletal muscle capillaries of human diabetics. Diabetes 30:326–334

102. Toussaint D, Dustin P (1963) Electron microscope studies of the retinal capillaries in normal and diabetic subjects. Bull Mem Acad R Med Belg 3:95–121

103. Tso MOM, Cunha-Vaz JG, Shih C-Y et al. (1980) Clinico-pathologic study of blood-retinal barrier in experimental diabetes mellitus. Arch Ophthalmol 98:2032:2040

104. Vracko R (1974) Basal lamina layering in diabetes mellitus; evidence of accelerated rate of cell death and cell regeneration. Diabetes 23:94–104

105. Vracko R, Benditt EP (1970) Capillary basal lamina thickening, its relationship to endothelial cell death and replacement. J Cell Biol 47:281–285

106. Wallow IHL, Engerman RL (1977) Permeability and potency of retinal blood vessels in experimental diabetes. Invest Ophthalmol Vis Sci 16:447–461

107. Wallow IHL, Geldner PS (1980) Endothelial fenestrae in proliferative diabetic retinopathy. Invest Ophthalmol Vis Sci 19:1176–1183

108. Wallow IHL, Greasner ML, Stevens TS (1981) Actin filaments in diabetic fibrovascular preretinal membrane. Arch Ophthalmol 90:2175–2181

109. Waltman SR, Santiago J, Krupin T et al. (1979) Vitreous fluorophotometry and blood-sugar control in diabetics. Lancet II:8151

110. Weiss JB, Hill CR, Davis RJ et al. (1983) Activation of a procollagenase by low molecular weight angiogenesis factor. Biosci Rep 3:171–177

111. Wise GN (1956) Retinal neovascularization. Trans Am Ophthalmol Soc 54:726–826

112. Wolter JR (1962) The nature of capillary microaneurysms in diabetic retinopathy. Diabetes 11:126–131

113. Yanoff M (1966) Diabetic retinopathy. N Engl J Med 274:1344–1349

Chapter 24

Diabetic Nephropathy

J. Bariéty

The hallmark of diabetic nephropathy is glomerulosclerosis, the generalized microangiopathy seen in patients with diabetes mellitus. Diabetic nephropathy is probably responsible for 25% of cases of end-stage renal failure [149], and an increasing number of diabetics require either dialysis or transplantation [59]. Type I insulin-dependent diabetics generally have a more severe and rapidly progressive nephropathy than do type II initially insulin-independent diabetics [2,54,102]. Almost half of all type I diabetics develop nephropathy leading to uraemia after 10 to 30 years of overt diabetes mellitus [2,44,52,66,102,143,194]. Type II diabetics may also suffer renal damage and develop renal insufficiency, but mortality in this population is generally due to cardiovascular rather than renal disease [122]. The lesions of diabetic nephropathy occur early, preceding any clinical signs of frank nephropathy [45].

Anatomopathological Features

Gross Anatomy

Half of all diabetics with renal insufficiency have symmetrical kidneys of normal or increased size, while the remaining half have evidence of moderate renal atrophy. The tendency towards nephromegaly probably reflects underlying hypertrophy of some nephrons which counterbalances the retractile fibrosis generally seen in chronic renal failure. In kidneys of increased size, the cortical zone is widened and the arteries at the corticomedullary junction are patent. Type I diabetics have kidneys of increased size and

volume long before there are any clinical or structural manifestations of renal disease [69]. This nephromegaly may persist when glomerulopathy becomes overt.

Light Microscopy

In the so-called pre-clinical phase of diabetic nephropathy, no specific renal abnormalities are evident apart from an increase in glomerular size in type I diabetes [135]. Once diabetic nephropathy is advanced, however, glomerular arteriolar and arterial lesions are evident.

Glomerular Lesions

Two forms of glomerular lesions are characteristically seen: diffuse glomerulosclerosis and nodular glomerulosclerosis. Both result from an increase in mesangial matrix and from thickening of the glomerular basement membrane. Other lesions referred to as "hyaline", "insudative" or "exudative" are less frequent and include the "fibrin cap" and "capsular drop" glomerular lesions.

Nodular glomerulosclerosis. Nodular glomerulosclerosis, originally described by Kimmelstiel and Wilson [90], is characterized by the presence of centrilobular, round, homogeneous nodules generally found at a distance from the vascular pole of the glomerulus (Fig. 24.1). The number and size of these nodules vary from one glomerulus to another. A single glomerulus may contain a very large solitary nodule, or a number of nodules of variable size. The nodules are acel-

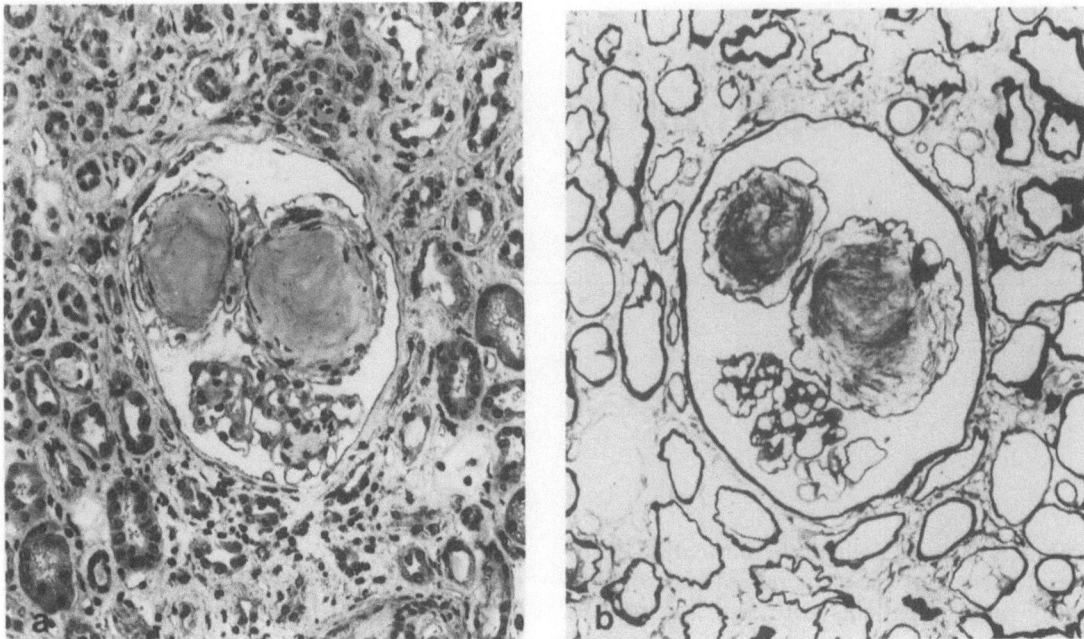

Fig. 24.1a, b. Diabetic nodular glomerulosclerosis. **a** Masson's trichrome: note the paucicellular rounded centrilobular nodules. The tubules are atrophic and the tubular basement membranes are enlarged. × 225. **b** Silver staining (Marinozzi), same glomerulus. The centrilobular nodules are argyrophilic and show a lamellated pattern. × 225

lular but may be surrounded by a cellular rim. Red blood cells may sometimes be trapped within these nodules. Adjacent capillaries are often dilated, occasionally aneurysmal and have thickened walls without the double contour seen in lobular or membranoproliferative glomerulonephritis. Glomerular nodules are PAS positive, and staining of small nodules is especially intense. Green and occasionally partial red staining is seen with Masson's trichrome. The nodules stain faintly red with sirius red but are not birefringent under polarized light. However, Bowman's capsules and interstitial collagen are highly birefringent using this method [23]. Staining of nodules with silver methenamine is patchy and sometimes lamellar. The glomeruli or glomerular lobules that are nodule-free exhibit diffusely increased PAS and silver staining of mesangial areas associated with a variable degree of mesangial hypercellularity. Thickening of Bowman's capsules may result in the formation of fibrous crescents.

Diffuse Glomerulosclerosis. Diffuse glomerulosclerosis, seen more frequently than the nodular form, is characterized by an increase in mesangial matrix and by glomerular basement membrane thickening (Fig. 24.2) [13,173]. The latter

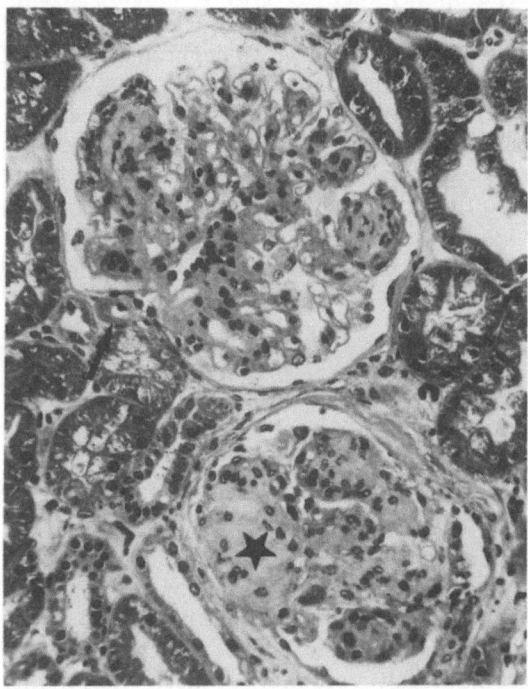

Fig. 24.2. Diabetic diffuse glomerulosclerosis. The mesangial matrix is diffusely enlarged with a centrilobular nodule (*stern*). The number of mesangial cells is increased. Arteriolar hyalin deposits are seen (*arrow*). Masson's trichrome, × 200

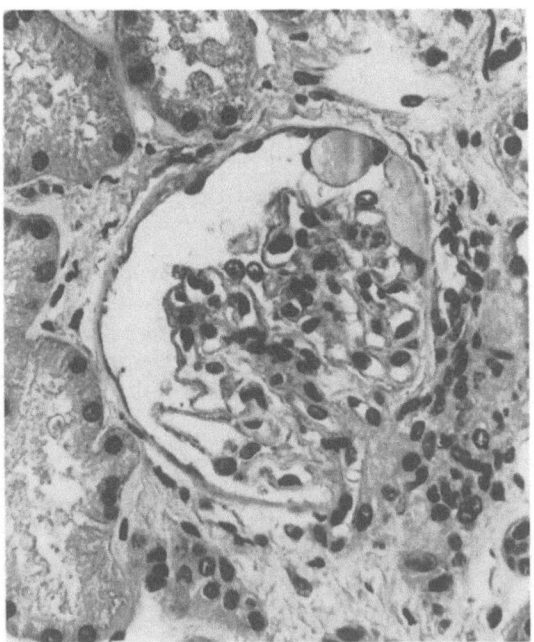

Fig. 24.3. Diabetic diffuse glomerulosclerosis. Hyalin capsular drop lesion between epithelial cells and the basement membrane of the Bowman's capsule. × 375

is sometimes visible only by electron microscopy. Mesangial hypercellularity may be present and is particularly frequent early in the course of diabetic nephropathy, suggesting a possible involvement of these cells in the production of excess mesangial matrix.

"Fibrin Cap" Lesion.   Fibrin cap lesions [9] are eosinophilic, homogeneous, hyaline deposits found in dilated capillary lumina, generally at the periphery of a glomerular lobule. These lesions share the staining characteristics of fibrin, and may sometimes contain lipids and foam cells. Podocytes adjacent to fibrin cap lesions may become altered leading to adhesion between the glomerular basement membrane and the Bowman's capsule, and to formation of focal and segmental glomerular hyalinosis. This appearance is non-specific but frequently encountered in diabetes.

"Capsular Drop" Lesion. Capsular drop lesions [9] are rounded, eosinophilic deposits situated between epithelial cells and the basement membrane of a Bowman's capsule (Fig. 24.3). Like fibrin cap lesions, capsular drops stain red with trichrome dyes and are not argyrophilic.

Tubular and Interstitial Lesions

Tubular basement membranes are thickened which may be in contrast with the underlying tubular atrophy. Interstitial fibrosis may be evident as may be cellular infiltration by lymphocytes, plasmocytes, neutrophils and eosinophils in the absence of urinary tract infection. Interstitial fibrosis is particularly marked in cases associated with arterial and arteriolar lesions [5].

Vascular Lesions

Vessels of all sizes are sclerotic and this is probably one of the major factors contributing to the development of renal insufficiency. The intima and media of arterioles are frequently disrupted by large hyaline deposits. Glomerular afferent and efferent arterioles may both be involved. The degree of arteriohyalinosis does not correlate with the severity of glomerular lesions [13]. Arteriolar aneurysms can be visualized in histological sections and after microangiography [127].

Electron Microscopy

An early sign of diabetic nephropathy is an increase in the amount of mesangial matrix seen by electron microscopy [42]. The matrix can aggregate into nodules with time. The glomerular basement membrane is segmentally thickened initially and eventually can attain a width up to ten times greater than normal. Marked glomerular basement membrane thickening is particularly associated with nodular glomerulosclerosis (Fig. 24.4) [91]. The material responsible for mesangial expansion is finely granular. When nodules are present, they contain bundles of 10–12 nm fibrillar hollow microtubules that are irregularly oriented. Within a given bundle, the microtubules are generally parallel [23,170] and resemble the structures described in other forms of glomerulopathy [60,81,100]. Studies of the size, arrangement and biochemical composition of these microtubules suggest that they are glycoproteinaceous [88].

Glomerular basement membrane thickening may reflect a variety of ultrastructural changes, including a homogeneous widening of the lamina densa, subendothelial accumulation of microfibrillar material similar to that seen within glomerular nodules and/or heterogeneous aggregation of membrane-like material in the subepithelial space. The mesangial matrix and the glomerular basement membrane contain cellular

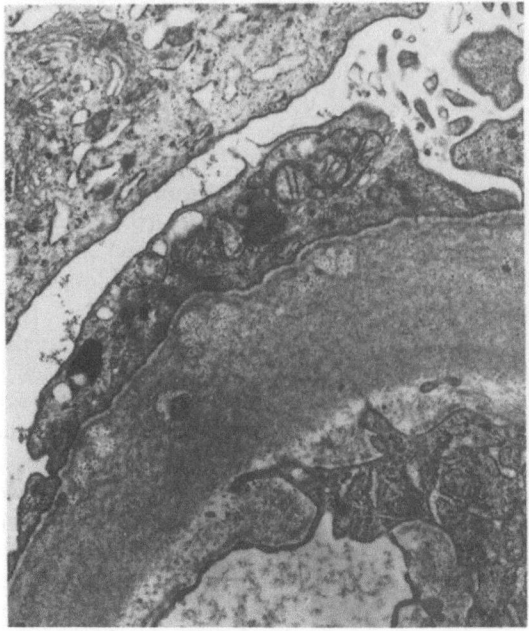

Fig. 24.4. Diabetic glomerulosclerosis. The glomerular basement membrane is considerably enlarged with subepithelial irregular layers surrounding punctuated clear areas. The subendothelial space contains numerous microfibrils. Uranyl lead, × 8900

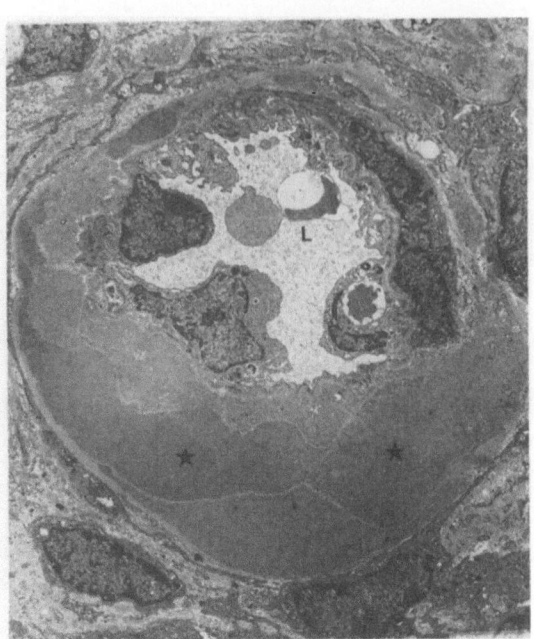

Fig. 24.6. Juxtaglomerular arteriole. Voluminous hyalin deposits are seen in the arteriolar wall (stern). *L*, lumen. Uranyl lead, × 2100

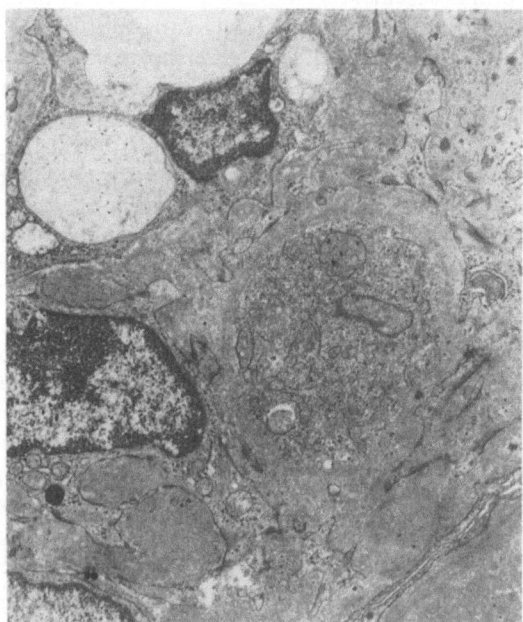

Fig. 24.5. Diabetic glomerulosclerosis. Numerous cell remnants are embedded in the enlarged mesangial matrix. Uranyl lead, × 7200

remnants called "striated membranous structures" [8] and "round extracellular particles" (Fig. 24.5) [7]. These remnants are not unique to diabetic nephropathy but are particularly marked in glomerulosclerosis. Podocytes may degenerate and detach themselves from the underlying glomerular basement membrane [38]. Capsular drop and fibrin cap lesions are composed of finely granular, electron dense material. This material is found between the epithelium and the basement membrane of the Bowman's capsules in the case of the capsular drop lesions, whereas it is located in the capillary subendothelial and the mesangial areas in fibrin cap lesions. Similar material can be found in the subendothelium and occasionally in the media of arterioles (such as the juxtaglomerular arterioles) (Fig. 24.6).

Immunohistochemistry

Circulating proteins such as IgG, IgM, albumin, ceruloplasmin and fibrin may be found within

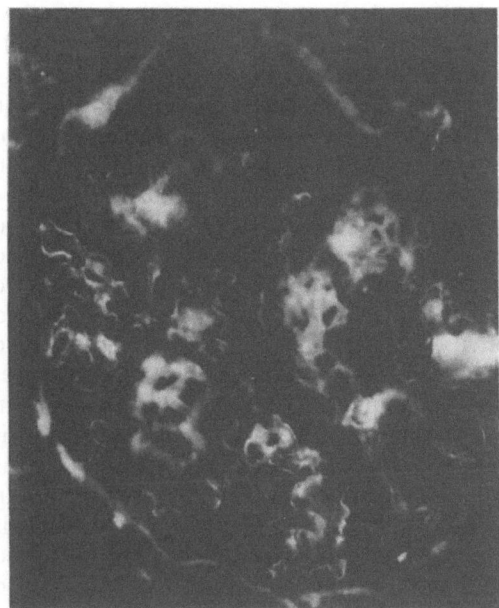

Fig. 24.7. Immunofluorescence. Staining of the glomerular basement membrane and of the enlarged mesangial matrices by the fluoresceinated anti-IgG

the glomerular basement membrane, the tubular basement membrane and the basement membrane of Bowman's capsules, and to a lesser extent in glomerular nodules (Fig. 24.7) [112,196]. Anti-glomerular basement membrane antibodies have not been found either in serum or in renal eluates [61]. Any of the proteins detected in renal tissue appear to be trapped non-specifically by the kidney. This may be favoured by the non-enzymatic glycolysation of proteins which often occurs in diabetes [163]. In this process, glucose molecules attached to matrix proteins may provide carbonyl groups as linkage sites for circulating plasma proteins [21]. The terminal complement activation complex C5b-9 has been detected in diabetic kidneys [55,74]. Immunoperoxidase staining of sections examined by electron microscopy has further localized C5b-9 to the cellular remnants within connective tissue matrices [74]. Immunomorphological data suggest that glomerulosclerosis is a consequence of the accumulation of matrix protein. The mesangium and glomerular basement membrane in glomerulosclerosis stain intensely for fibronectin, laminin, type IV collagen and for a proteoglycan containing heparan sulphate [23,55,56,161]. Actomyosin [160,162] has also been found in mesangial areas. The presence of this substance and of cellular remnants points to a possible impairment in the clearance of cellular

debris. Fibrin cap and capsular drop lesions contain high molecular weight proteins such as immunoglobulins, fibrinogen, complement components and lipoproteins. The mechanism for the formation of these lesions is unknown. C3 and often IgM and C1q are found in hyaline arteriolar deposits. This is thought to be the consequence of slow, spontaneous complement activation by the alternate pathway [62]. C3b generated in this fashion diffuses into the arteriolar subendothelial space and becomes linked to hyaluronic acid. C3b is rapidly inactivated by factors I and H to form iC3b. This is apparently linked by esterification to the hydroxide groups on the disaccharide subunits of hyaluronic acid. Factor H is also a member of this complex. The IgM present in hyalin arteriolar lesions appears to be an autoantibody (immunoconglutinin) directed against epitopes of iC3b. The resulting antigen–antibody interaction may then elicit classical complement activation. It is interesting to note that this phenomenon is not accompanied by inflammation in the form of angiitis.

Specificity of Diabetic Glomerulosclerosis

The constellation of findings including diffuse or nodular glomerulosclerosis, arteriolar hyalinosis, capsular drop lesions and linear staining for IgG of non-basement membrane specificity is highly suggestive of diabetic nephropathy. Additional immunofluorescent studies and examination of tissue by electron microscopy generally suffice to distinguish diabetic nephropathy from conditions having a similar light microscopic appearance: glomerulosclerosis seen in hypertension or old age [72], or in certain hepatic conditions [128]; lobular forms of membranoproliferative glomerulonephritis type I or II [166]; and light chain glomerulopathy [23,60].

Superimposed Renal Lesions

Although the existence of glomerulonephritis in diabetics is thought to be purely coincidental, the frequency of the association may in fact be of significance [49]. The following types of glomerulonephritis have been described in diabetics: membranous glomerulonephritis [1,14,29,36,125,148,164,189,190,193], endocapillary proliferative glomerulonephritis

[29,53,130,167], "mesangial" [29,73] or membranoproliferative [29,190] glomerulonephritis, nephrotic syndrome in association with minimal glomerular lesions [22,29,155,178] and rapidly progressive glomerulonephritis [25]. It is conceivable that the functional and structural lesions of diabetic nephropathy, and the frequency of infections may predispose diabetics to various forms of glomerulonephritis [1,14,102].

Papillary Necrosis

Papillary necrosis may occur in association with a variety of clinical conditions including diabetes mellitus in 50% of cases [72,97]. It is generally a late complication of diabetes [72] reflecting microvascular disease of the vessels nourishing the renal papillae. Urinary tract infection may be a precipitating factor [154].

Clinical Manifestations

The detection of proteinuria by the usual methods (e.g. Albustix) indicates the presence of the clinical phase of diabetic nephropathy. This clinical phase is preceded by a pre-clinical phase known to exist in type I insulin-dependent diabetes but less well documented in type II diabetes.

Pre-clinical Phase

Subtle renal anomalies may be present very early in the course of type I diabetes, although the exact proportion of affected individuals is unknown.

The pre-clinical phase of diabetic nephropathy is characterized by a state of hypertrophy and hyperfunction. Renal size increases in this phase of the disease in humans, and this phenomenon is also observed within days of the chemical induction of diabetes in experimental animals [117,157,165,179]. In diabetic rats, renal protein, RNA and DNA content increases, and ribosomes have been shown to incorporate more amino acids than in normal rats [144]. Glomerular size is also increased [134]. Strict glucose control using insulin can reduce nephromegaly in both humans and animals [117,150,157]. Glomerular filtration rate (GFR) [30,31, 50,114,117,174] is also elevated in pre-clinical

diabetic nephropathy as is renal plasma flow (RPF) [30,114]. These anomalies generally subside after several days to weeks of insulin therapy [114,117,157,179]. Similar elevations in GFR and RPF are seen in streptozotocin- or alloxan-treated rats, pancreatectomized rats and spontaneously diabetic mice [48,63,77,94,111].

A number of factors have been invoked and probably act in concert to cause the observed elevations in GFR and RPF. In addition to extracellular volume expansion, decreased activity of vasoconstrictor mechanisms and increased production of vasodilatory substances may contribute to the renal hyperperfusion seen in early diabetes. Plasma volume expansion is known to elicit renal vasodilatation [43] and has been demonstrated in human [16,77] and in experimental [33,82] diabetes mellitus. This volume expansion may result from enhanced sodium reabsorption [24,113] which is coupled with the enhanced proximal tubular reabsorption of glucose seen in early diabetes mellitus [96,177,195]. Decreased plasma renin activity in response to volume expansion [33] and a reduced number of glomerular angiotensin II receptors [6] have been noted in alloxan-treated diabetic rats. In man, plasma renin activity is abnormally low in diabetics [32,34], and an elevated circulating concentration of inactive hormone, probably prorenin [101], has been described in some cases. The latter may result from the inhibition by high glucose concentrations of cathepsin B [39], a substance involved in the conversion of inactive renin to active renin.

Prostaglandin may also play a role in glomerular hyperfiltration as treatment with indomethacin in experimental situations can decrease GFR and RPF [93]. In fact, enhanced renal prostaglandin synthesis has been detected in glomeruli isolated from diabetic rats [159]. A hepatic hormone, glomerulopressin, which may also augment GFR has been identified in diabetic dogs [48]. Vascular reactivity to catecholamines is slightly decreased in insulin-dependent diabetes [35], alloxan-induced diabetes [33] and after pancreatectomy [65]. Glucose and certain glucoregulatory hormones may have direct vasodilatory effect on the renal circulation. Nevertheless, this effect is probably minimal as glucose infusion into normal subjects does not markedly increase GFR [17,115,119,157]. Furthermore, normal individuals treated with glucagon or growth hormone show only a slight elevation in GFR when compared to subjects with type I diabetes and comparable circulating hormonal concentrations [139,158]. Insulin deficiency alone does not appear to cause increases in GFR and RPF as insulin infusion does not correct these anomalies if hyperglycaemia is deliberately maintained [156]. Haemodynamic factors such as increased glomerular plasma flow [77,94,111], increased transglomerular hydrostatic pressure [77,94] and perhaps alterations in the surface area for filtration [75] account for hyperfiltration in diabetes.

Of diabetics, 30%–45% [186] excrete abnormal quantities of albumin detectable only by radioimmunoassay [116,138]. Such microalbuminuria is generally associated with poor glucose control or with physical activity [95,180,181,187]. Effective insulin therapy can abolish microalbuminuria [28,180,182,185]. However, once albuminuria exceeds 30 $\mu g/min$, the effect of insulin is variable [58,186]. Albuminuria of this magnitude is thought to presage the development of overt, irreversible nephropathy [123,141,183]. If microalbuminuria persists in spite of adequate glucose control, a state of incipient diabetic nephropathy is said to be present [120,122].

Microalbuminuria in pre-clinical diabetic nephropathy reflects a defect in glomerular filtration and not in tubular reabsorption. Indeed, the urinary excretion of ß$_2$-microglobulin (molecular weight 20 000), normally freely filtered across the glomerular capillary wall and largely taken up by the tubule, is not modified [180]. Glomerular filtration of proteins is influenced by molecular size and charge, and by the haemodynamic forces operating across the glomerular capillary wall. A simple size-selective defect does not appear to account for microalbuminuria as the clearance of neutral dextran, which is a molecule similar in size to albumin, is not altered in early diabetes [126]. It is unlikely that the early microalbuminuria can be attributed to changes in charge selectivity as elevated rates of both albumin and IgG excretion have been observed [182]. Also, the rapidity with which insulin therapy and cessation of physical activity abrogate proteinuria is incompatible with changes in charge selectivity.

Thus, microalbuminuria appears to be linked to haemodynamic alterations favouring the passage of albumin across the filtration barrier, for example elevated hydrostatic pressure and flow as soon with physical exertion [118] and with hyperglycaemia [180].

The reasons for the abnormalities seen in pre-clinical diabetic nephropathy have yet to be fully understood. However, these anomalies herald the development of overt renal disease and predict early mortality [122]. Microalbuminuria may occur in type II diabetes but its significance is unclear. Hyperfiltration has not been reported in this patient population, and abnormal albumin excretion may be due to factors different from those invoked in type I diabetes.

Clinical Phase

This phase is characterized by persistent proteinuria, hypertension and a drop in GFR. These abnormalities supervene on average 15–20 years after the onset of type I insulin-dependent diabetes [102], and generally do not appear earlier than 10 years after the onset of clinical hyperglycaemia [92]. Renal disease in the clinical phase is slowly progressive [85,184]. However, it must be remembered that this syndrome affects only 40%–45% of type I diabetics; the remaining individuals may have retinopathy but are generally spared any renal involvement [57,120].

Once diabetes has been established for more than 30 years, the incidence of new cases of nephropathy is virtually zero [92]. Thus, some individuals may not develop renal disease even after decades of hyperglycaemia, ostensibly similar in degree to that suffered by patients who do develop nephropathy [110]. The reasons for this paradox are unknown [45,121]. It is unclear as yet whether strict metabolic control can prevent the advent of diabetic nephropathy, and it is likely that other genetic and environmental factors as well as arterial hypertension may play an important role.

Proteinuria

The first sign of overt diabetic nephropathy is persistent proteinuria detectable by conventional tests. Renal function declines on average over a 5-year period after the onset of proteinuria, but the rate of decline varies from individual to individual and may be rapid or slow. Prognosis is roughly correlated with the degree of proteinuria [191], although some cases of significant glomerular disease without proteinuria have been reported [64,71,191].

In this phase, a variety of mechanisms may be responsible for abnormal protein excretion. Glomerular size selectivity appears to be altered as the fractional clearance of neutral dextrans over 46 Å in diameter is abnormally elevated [126]. Charge selectivity has not been directly studied. Nevertheless, decreased glomerular capillary sialic acid concentration [41,89, 188,197] and heparan sulphate concentration [37,136] have been reported. These molecules normally provide fixed negative charges that repel anionic molecules from the filtration barrier. Haemodynamic factors such as increased transcapillary hydrostatic pressure may also play a role, and may be especially marked in kidneys that have adapted to progressive nephron loss [15,78]. The appearance of proteinuria has in fact been correlated with antecedent hyperfiltration [122].

Arterial Hypertension

Hypertension accompanies proteinuria in 50%–75% of cases [51,83,137,140,191], and is ubiquitous in those patients who develop renal insufficiency [142]. However, hypertension has on occasion been reported to precede proteinuria and decreased GFR [123]. Physical exertion may elevate blood pressure considerably [86,120] and this phenomenon may be observed at the stage of incipient diabetic nephropathy. Hypertension may accelerate the rate of decline of renal function in diabetics with overt proteinuria [70].

The pathogenesis of hypertension in diabetic nephropathy is poorly understood [153]. Plasma volume expansion due to the enhanced tubular sodium reabsorption seen in association with hyperglycaemia may play a role, as may high extracellular glucose concentrations [24, 96,177,195]. It has also been suggested that exogenous insulin or high levels of endogenous insulin as seen in type II diabetes may directly increase tubular reabsorption of sodium [47].

Glomerular Filtration

Decline in GFR is inexorable, and in poorly controlled diabetics averages 1 ml/min/month. The rate of decline is variable but is linear in any given individual [122]. GFR declines most rapidly in those patients with poor metabolic control and arterial hypertension [26,129]. RPF has been shown to diminish in diabetic nephropathy, and this may account in part for the reduction in GFR [199]. The glomerular surface area available for filtration is also diminished [67,108,199] thereby depressing GFR by an alteration in the ultrafiltration coefficient. Glomerular capillary hydrostatic pressure increases with progressive nephron loss as seen in diabetic rats after unilateral nephrectomy [80]. Such adaptive changes may lead to further renal damage.

Nephropathy and Diabetic Retinopathy

Individuals with diabetic nephropathy almost invariably have associated retinopathy [27,76] although the converse is not true. This suggests that the retina and the kidney are differentially susceptible to the development of microangiopathy. Furthermore, proteinuria in a diabetic without retinopathy is distinctly unusual and should prompt a search for other causes of renal disease (see Chap. 23).

Microscopic Haematuria

Microscopic haematuria as a sign of diabetic nephropathy is a subject of controversy: some authors consider it rare [87] while others claim that it is present in 30% of cases [132].

End-Stage Renal Failure

Renal insufficiency is often poorly tolerated by diabetics. The gastrointestinal complications of uraemia may be accentuated by diabetic neuropathy [168]. The advent of hyporeninaemic hypoaldosteronism may increase the risk for the development of hyperkalaemia and acidosis [46]. Sudden glucose elevations may expand plasma volume and precipitate congestive heart failure and pulmonary oedema [4].

The clinical manifestations of renal disease may be identical in type I and type II diabetes mellitus but progression of the disease is certainly slower in the latter condition [122]. Cardiovascular complications overshadow renal involvement in type II diabetes and mortality is primarily related to cardiovascular disease [124].

Pathogenesis

It is impossible at present to outline clearly and coherently the pathogenesis of diabetic nephropathy. The following is a critical review of the putative factors involved.

Genetic Factors

It has been suggested that genetic susceptibility to the development of microangiopathy underlies the development of end-organ disease in diabetes [169]. Indeed, basement membrane thickening of capillary walls in muscle is a frequent finding in diabetes of whatever duration and may be present in pre-diabetics without hyperglycaemia [169]. Other authors have disputed these findings [68,198]. However, genetic factors may be at play in determining the propensity to renal disease of certain diabetics.

Role of Insulin Deficiency and Hyperglycaemia

A wealth of data indicates that diabetic nephropathy arises in a perturbed metabolic environment. The renal lesions of diabetes occur not only in primary cases of diabetes, but also in chemically induced experimental diabetes [108,176] and in cases of secondary diabetes in man.

The characteristic lesions of glomerular sclerosis are observed only after several years of clinical diabetes [133]. Normal kidneys transplanted into diabetic patients [105,109] or into hyperglycaemic rats also develop the typical lesions of diabetic nephropathy. In experimental animals, strict glucose control can prevent the appearance and evolution of renal lesions [18,151], and glomerular lesions have been reported to regress after transplantation of islets of Langerhans [104,175]. It is much more difficult to establish whether strict metabolic control can prevent the late complications of diabetes mellitus in man [19,145].

The precise mechanisms whereby the diabetic state leads to nephropathy are unknown. There is no evidence for an immunological mechanism involving insulin as a "planted" glomerular antigen or as a substance eliciting pathogenetic anti-insulin antibodies. The role of non-enzymatic glycosylation of connective tissue is discussed in Chap. 22. The infusion of glycosylated plasma proteins in mice [103] has led to glomerular basement membrane thickening, but this was not observed with glycosylated albumin in rats [84]. An increase in glomerular basement membrane hydroxylysine concentration with attachment of the disaccharide glucose–galactose due to heightened glucosyl transferase activity has been reported [10,171,172] but not confirmed [11,41,89,152,188,197]. Decreased glomerular basement membrane sialic acid content [41,89,188,197] and heparan sulphate content [136] have also been reported. Contradictory evidence exists concerning glomerular basement membrane turnover. Decreased turnover was observed in rats injected with tritiated amino acids or radiolabelled proline [20,146]. However, the bioavailability of uridine diphosphate sugars is increased in diabetic rats, and these molecules act as carbohydrate donors, and are involved in the production of glycoproteins and glucosaminoglycans [40].

Haemodynamic Factors

Altered haemodynamic forces such as increased hydrostatic pressure and hyperfiltration appear to play an important role in the development of diabetic nephropathy. Experimental procedures such as renal ablation that induce hyperfiltration [78,79] can provoke structural changes in the remaining renal tissue similar to those seen in diabetic nephropathy [131]. It has also been demonstrated that unilateral renal artery stenosis in man [12] and in diabetic animals [106] can prevent the development of renal lesions in the hypoperfused kidney.

Elevated transcapillary pressures may have an effect on the number and size of adjacent mesangial cells as seen in hypertension [3,192]. Collagen production [99] and deposition of circulating proteins in the mesangium and glomerular basement membrane may also be enhanced [147]. Such deposition may be facilitated by non-enzymatic glycosylation of circulating proteins that may then become attached to glomerular collagen and glycoproteins. A defect in macromolecular clearance has been demonstrated in the mesangium of diabetic rats [107].

The pathogenesis of diabetic nephropathy cannot be entirely explained by haemodynamic perturbations. Experimentally induced hyperfiltration is associated with renal lesions not completely identical to those seen in diabetic nephropathy. Hyperfiltration after unilateral nephrectomy is associated with segmental and focal glomerular hyalinoses but not with diffuse glomerulosclerosis [200]. Furthermore, hyperfiltration is not seen in type II diabetes yet renal lesions in this disease may resemble those seen in type I diabetes [121]. In some individuals hyperfiltration may be present for more than 20 years in association with only minimal or moderate renal damage [110].

Rational therapeutic decisions will have to await a better understanding of the pathogenesis of diabetic nephropathy. The abnormal metabolic milieu of the diabetic is certainly a major factor and it is clear that glucose control is beneficial in experimental models of diabetes. It is to be hoped that the role of strict metabolic control in humans will soon be elucidated.

References

1. Ainsworth SK, Hirsch HZ, Brackett NC, Brissie RM, Williams AV Jr, Hennigar GR (1982) Diabetic glomerulopathy. Histopathologic, immunofluorescent and ultrastructural studies of 16 cases. Hum Pathol 13:470–478

2. Andersen AR, Christiansen JS, Andersen JK, Kreiner S, Deckert T (1983) Diabetic nephropathy in type I (insulin-dependent) diabetes: an epidemiologic study. Diabetologia 25:496–501

3. Ausiello DA, Kreisberg JI, Roy C, Karnovsky MJ (1980) Contraction of cultured rat glomerular cells of apparent mesangial origin after stimulation with angiotensin II and arginine vasopressin. J Clin Invest 65:754–760

4. Axelrod L (1975) Response of congestive heart failure to correction of hyperglycemia in the presence of diabetic nephropathy. N Engl J Med 298:1243–1245

5. Bader R, Bader H, Grund KE, Mackensen-Haen S, Christ H, Bohle A (1980) Structure and function of the kidney in diabetic glomerulosclerosis: correlations between morphological and functional parameters. Pathol Res Pract 167:204–216

6. Ballerman BJ, Skorecki KL, Brenner BM (1984) Reduced glomerular angiotensin II receptor density in early untreated diabetes mellitus in the rat. Am J Physiol 274:F110–F116

7. Bariety J, Callard P, Appay MD, Grossetete J, Mandet C (1972) Ultrastructural study of some frequent and poorly known intraglomerular structures. In: Hamburger J, Crosnier J, Maxwell MH (eds) Advances in nephrology, vol 3. Year Book Medical Publishers, Chicago, pp 153–172

8. Bariety J, Callard P (1975) Striated membranous structures in renal glomerular tufts. An electron microscopy study of 340 human renal biopsies. Lab Invest 32: 636–641

9. Barrie HJ, Askanasy CL, Smith GW (1952) More glomerular changes in diabetics. Can Med Assoc J 66:428–431

10. Beisswenger PJ, Spiro MJ (1972) Studies on the human glomerular basement membrane. Composition, nature of the carbohydrate units and chemical changes in diabetes mellitus. Diabetes 22: 180–193

11. Beisswenger PJ (1976) Glomular basement membrane. Biosynthesis and chemical composition in the streptozotocin diabetic rat. J Clin Invest 58:844–852

12. Berkman J, Rifkin H (1975) Unilateral nodular diabetic glomerulosclerosis (Kimmelstiel-Wilson). Metabolism 22:715–722

13. Bell ET (1953) Renal vascular disease in diabetes mellitus. Diabetes 2:376–389

14. Bertani T, Olesnicky L, Abu-Regiaba S, Glasberg S, Pirani CL (1983) Concomitant presence of three different glomerular diseases in the same patient. Report of a case and review of the literature. Nephron 34:260–266

15. Brenner BM, Meyer, TW, Hostetter TH (1982) Dietary protein intake and the progressive nature of kidney disease: the role of hemodynamically mediated glomerular sclerosis in aging, renal ablation, and intrinsic renal disease. N Engl J Med 307:652–659

16. Brotchner-Mortensen J (1973) Glomerular filtration rate and extracellular fluid volumes during normoglycemia and moderate hyperglycemia in diabetes. Scand J Clin Lab Invest 32:311–316

17. Brotchner-Mortensen J (1973) The glomerular filtration rate during moderate hyperglycemia in normal man. Acta Med Scand 194:31–37

18. Brown DM, Andres GA, Hostetter TH, Mauer SM, Price R, Venkatachalam AM (1982) Kidney complications. Diabetes [Suppl 1] 3:71–81

19. Brownlee M, Cahill GF Jr (1979) Diabetic control and vascular complications. Atherosclerosis Rev. 4:29–34

20. Brownlee M, Spiro RG (1979) Glomular basement membrane metabolism in the diabetic rat. Diabetes 28:121–125

21. Brownlee M, Pongor S, Cerami A (1983) Covalent attachment of soluble proteins by non-enzymatically glycosylated collagen. J Exp Med 158:1739–1744

22. Brulles A, Caralps A, Vilardell M (1977) Nephrotic syndrome with minimal glomerular lesions (lipoid nephrosis) in an adult diabetic patient. Arch Pathol Lab Med 101:270–271

23. Bruneval P, Foidart JM, Nochy D, Camilleri JP, Bariety J (1985) Comparative evaluation of glomerular matrix proteins in nodular glomerulosclerosis associated with light chain deposition disease and diabetes mellitus. Hum Pathol 16:477–484

24. Carney SL, Wong NLM, Dirks JH (1979) Acute effects of streptozotocin diabetes on rat renal function. J Lab Clin Med 98:950–960

25. Carstens SA, Hebert LA, Garancis JC (1982) Rapidly progressive glomerulonephritis superimposed on diabetic glomerulosclerosis. JAMA 247:1453–1457

26. Cataland S, O'Dorisio TH (1983) Diabetic nephropathy. Clinical course in patients treated with the subcutaneous insulin pump. JAMA 249:2059–2061

27. Chahaz PS, Kohner EM (1983) The relationship between diabetic retinopathy and nephropathy. Diabet Nephropath 2:4–5

28. Chiasson JL, Ducros F, Poliquin-Hamet M, Lopez D, Lecavalier L, Hamel P (1984) Continuous subcutaneous insulin infusion (Mill-Hill infuser) versus multiple injections (Medi-jector) in the treatment of insulin-dependent diabetes mellitus and the effect of metabolic control on microangiopathy. Diabetes Care 7:331–337

29. Chihara J, Takebayashi S, Taguchi T, Yokoyama K, Harada T, Naito S (1986) Glomerulonephritis in diabetic patients and its effect on the prognosis. Nephron 43:45–49

30. Christiansen JS, Gammelgaard J, Frandsen M, Parving HH (1981) Increased kidney size, glomerular filtration rate and renal plasma flow in short-time insulin-dependent diabetics. Diabetologia 20:451–456

31. Christiansen JS, Gammelgaard J, Tronier B, Svendsen PA, Parving HH (1982) Kidney function and size in diabetics before and during initial insulin treatment. Kidney Int 21:683–688

32. Christlieb AR, Munichoodapa C, Braaten JT (1974) Decreased response of plasma renin activity to orthostatic hypotension. Diabetes 23:835–840

33. Christlieb AR (1974) Renin, angiotensin, and norepinephrine in alloxan diabetes. Diabetes 23:962–970

34. Christlieb AR, Kaldany A, D'Elia VA (1976) Plasma renin activity and hypertension in diabetes mellitus. Diabetes 25:969-974

35. Christlieb AR, Janka HV, Kraus B et al. (1976) Vascular reactivity to angiotensin II and to norepinephrine in diabetic subjects. Diabetes 25:268–274

36. Churg J, Ehrenreich T (1973) Membranous nephropathy. In: Kincaid-Smith P et al. (eds) Glomerulonephritis, Morphology, natural history and treat-

ment. Wiley, New York, pp 443–448

37. Cohen MP, Surma ML (1981) Sulfate incorporation into glomerular basement membrane is decreased in experimental diabetes. J Lab Clin Med 98:715–722

38. Cohen AH, Manpaso F, Zamboni I (1977) Glomerular podocyte degeneration in human renal disease. An ultrastructural study. Lab Invest 37:40–42

39. Coradello H, Pollak A, Pagano, Leban J, Lubec G (1981) Non-enzymatic glyclosylation of cathepsin B: possible influence on conversion of proinsulin to insulin. J Med Sci 9:766–767

40. Cortes P, Dumler F, Sury Sastry KS, Verghese CP, Levin NW (1982) Effects of early diabetes on uridine diphosphosugar synthesis in the rat renal cortex. Kidney Int 21:676–682

41. Cruz A, Moreau Lalande H (1978) Biochemical studies on glomerular basement membrane in human diabetic microangiopathy. Pathol Biol 26:411–417

42. Dachs S, Churg J, Mautner W, Grishman E (1964) Diabetic nephropathy. Am J Pathol 44:155–168

43. Daugharty TM, Veki IF, Nicholas DP, Brenner BM (1973) Renal response to chronic intravenous salt loading in the rat J Clin Invest 52:21–31

44. Deckert T, Poulsen JE, Larsen M (1978) Prognosis of diabetics with diabetes onset before the age of thirty one. Diabetologia 14:363–370

45. Deckert T, Parving HH, Andersen AR et al. (1982) Diabetic nephropathy. A clinical and morphometric study. In: Eschwege E (ed) Advances in diabetes epidemiology. INSERM Symposium number 22, Jena, East Germany, Elsevier Biomedical Press, Amsterdam, pp 235–243

46. Defronzo RA (1980) Hyperkalemia and hyporeninemic hypoaldosteronism. Kidney Int 17:118–134

47. Defronzo RA (1981) The effect of insulin on renal sodium metabolism. Diabetologia 21:165–171

48. Del Castillo E, Fuenzalida R, Uranga J (1977) Increased glomerular filtration rate and glomerulopressin activity in diabetic dogs. Horm Metab Res 9:46–53

49. Ditscherlein G (1985) Renal histopathology in hypertensive diabetic patients. Hypertension 7:29–31.

50. Ditzel J, Schwartz M (1967) Abnormally increased glomerular filtration rate in short-term insulin-treated diabetic subjects. Diabetes 16:264–267

51. Drury PL (1983) Diabetes and arterial hypertension. Diabetologia 24:1–9

52. Entmacher PS, Root HF, Marks HH (1964) Longevity of diabetic patients in recent years. Diabetes 13:373–377

53. Epstein SE, Becker EL (1961) Acute glomerulonephritis in an adult with long-standing diabetes mellitus. Ann Intern Med 54:97–103

54. Fabre J, Balant LP, Dayer PG, Fox HM, Vernet AT (1982) The kidney in maturity onset diabetes mellitus: a clinical study of 510 patients. Kidney Int 21:730–738

55. Falk RJ, Dalmasson AP, Kim Y et al. (1983) Neoantigen of the polymerized ninth component of complement. J Clin Invest 72:560–573

56. Falk RJ, Scheinman JI, Mauer SM, Michael AF (1983) Polyantigenic expansion of basement membrane constituents in diabetic nephropathy. Diabetes (Suppl 2) 32:34–39

57. Feldman JN, Hirsch SR, Beyer MM, James WA, L'Esperance FA, Friedman EA (1982) Prevalence of diabetic nephropathy at time of treatment for diabetic retinopathy. In: Friedman EA, L'Esperance FA (eds) Diabetic renal–retinal syndrome. Grune and Stratton,

New York, pp 9–20

58. Feldt-Rasmussen B, Mathiesen ER, Hegedus L, Deckert T (1986) Kidney function during 12 months of strict metabolic control in insulin-dependent diabetic patients with incipient nephropathy. N Engl J Med 314:665–670

59. Friedman EA (1982) Diabetic nephropathy. Strategies in prevention and management. Kidney Int 21:780–791

60. Gallo GR, Feiner HD, Katz LA et al. (1980) Nodular glomerulopathy associated with an amyloidotic Kappa light chain deposits and excess immunoglobulin light chain synthesis. Am J Pathol 99:621–644

61. Gallo GR (1970) Elution studies in kidneys with linear deposition of immunoglobulin in glomeruli. Am J Pathol 61:377–394

62. Gamble CN (1986) The pathogenesis of hyaline arteriosclerosis. Am J Pathol 22:410–420

63. Gartner K (1978) Glomerular hyperfiltration during the onset of diabetes mellitus in two strains of diabetic mice. Diabetologia 15:59–63

64. Gellman DD, Pirani CI, Soothill JF, Muehrcke RC, Kark RM (1959) Diabetic nephropathy: a clinical and pathologic study based on renal biopsies. Medicine 38:321–367

65. Goldberg E, Rosenblum I (1966) Reduction in cardiovascular and metabolic responses to phenylephrine in acutely pancreatectomized dogs. Am Heart J 72:483–488

66. Goldstein HH (1974) Discussion. The problem of end-stage diabetic nephropathy. Kidney Int [Suppl 1] 6:21–26

67. Gundersen HJG, Osterby R (1977) Glomerular size and structure in diabetes mellitus. II. Late abnormalities. Diabetologia 13:43–48

68. Gundersen HJ, Osterby R, Lundbaek K (1978) The basement membrane controversy. Diabetologia 15:361–363

69. Gundersen HJG, Morgensen CE (1981) The relationship between kidney size and function in short-term diabetic patients Diabetologia 21:498–499

70. Hasslacher C, Stech W, Wain P, Ritz E (1983) Normal history of nephropathy in type I diabetes – roles of metabolic control and hypertension. In: Proceedings of the 16th annual meeting of the American Society of Nephrology, 1983, p 29A

71. Hatch FE, Watt MF, Kramer NC, Parrisch AE, Howe JS (1961) Diabetic glomerulosclerosis: a long term follow-up study based on renal biopsies: Am J Med 31:216–230

72. Heptinstall RH (1983) Diabetes mellitus and gout. In: Heptinstall RH: Pathology of the kidney, 3rd ed. Little, Brown and Co, Boston, pp 1397–1453

73. Herlihy NG, Nordquist JA, Kandal AK, Llach F (1981) Diabetic nephropathy associated with fibrin formation. Hum Pathol 12:658–660

74. Hinglais N, Kazatchkine MD, Bhakdi S et al. (1986) Immunohistochemical study of the C5b-9 complex of complement in human kidneys. Kidney Int 30:399-410

75. Hirose K, Tsuchida H, Osterby R, Gundersen HJG (1980) A strong correlation between glomerular filtration rate and filtration surface in diabetic kidney hyperfunction lab Invest 43:434–437

76. Honey GE, Pryse-Davies J, Roberts DM (1962) A survey of nephropathy in young diabetics. Q J Med 31:473–483

77. Hostetter TH, Troy JL, Brenner BM (1981) Glomerular hemodynamics in experimental diabetes. Kidney Int 19:410–415

78. Hostetter TH, Olson JL, Rennke HG, Venkatachalam MA, Brenner BM (1981) Hyperfiltration in remnant nephrons: a potentially adverse response to renal ablation. Am J Physiol 241:F85–F93

79. Hostetter TM, Rennke HG, Brenner BM (1982) The case for intrarenal hypertension in the initiation and the progression of diabetic and other glomerulopathies. Am J Med 73:375–380

80. Hostetter TH, Meyer TW, Rennke HG, Brenner BM (1983) Influence of strict control of diabetes on intrarenal hemodynamics. Kidney Int 23:215 (abstract)

81. Hsu HC, Churg J (1979) Glomerular microfibrils in renal disease: a comparative electron microscopic study. Kidney Int 16:497–504

82. Illstrup KM, Keane WF, Michels LD (1981) Intravascular and extracellular volumes in the diabetic rat. Life Sci 29:717–724

83. Jarrett RJ, Keen H, Chakrabari R. (1982) Diabetes, hyperglycemia and arterial disease. In: Keen H, Jarrett J (eds) Complications of diabetes, 2nd edn. Edward Arnold, London, pp 179–203

84. Jeraj KP, Michael AF, Mauer SM, Brown DM (1983) Glycosylated and normal human or rat albumin do not bind to renal basement membranes of diabetic and control rats. Diabetes 32:380–382

85. Jones RH, Hayakawa H, McKay JD, Parsons V, Warkins PJ (1979) Progression of diabetic nephropathy. Lancet I:1105–1106

86. Karlefors T (1966) Circulatory studies during exercise with particular reference to diabetics. Acta Med Scand [Suppl 449] 180:1–87

87. Kasinath BS, Mujais SK, Spargo BH, Katz AI (1983) Non diabetic renal disease in patients with diabetes mellitus. Am J Med 75:613–617

88. Kefalides NA (1972) Biochemical studies of the glomerular basement membrane in the normal kidney. Adv Nephrol 2:3–24

89. Kefalides NA (1974) Biochemical properties of human glomerular basement membrane in normal and diabetic kidneys. J Clin Invest 53:403–407

90. Kimmelstiel P, Wilson C (1936) Intercapillary lesions in glomeruli of kidney. Am J Pathol 12:83–105

91. Kimmelstiel P, Osawa G, Beres J (1966) Glomerular basement membrane in diabetics. Am J Clin Pathol 45:21–31

92. Knowles HC Jr (1974) Magnitude of the failure problem in diabetic patients. Kidney Int [Suppl 1] 6:2–7

93. Koch-Jensen P, Steven K, Christiansen JS, Parving HH (1983) The effects of indomethacin on glomerular hemodynamics in experimental diabetes. Clin Res 31:431 (abstract)

94. Koch-Jensen P, Christiansen JS, Steven K, Parving HH (1981) Renal function in diabetic rats. Acta Endocrinol (Suppl 242) 97:25

95. Koivisto VA, Hurrenen NP, Vierikko P (1981) Continuous subcutaneous insulin infection corrects exercise-induced albuminuria in juvenile diabetes. Br Med J 282:778–779

96. Kokko JP (1973) Proximal tubule potential difference. Dependance on glucose, HCO3, and amino acids. J Clin Invest 52:1362–1367

97. Lauler DP, Schreiner GE, David H (1960) Renal medullary necrosis. Am J Med 29:132–156

98. Lee CS, Mauer SM, Brown DM, Sutherland DER, Michael AF, Najarian JS (1974) Renal transplantation in diabetes mellitus in rats. J Exp Med 139:783–800

99. Leung DY, Glagov S, Mathews MB (1976) Cyclic stretching stimulates synthesis of matrix components by arterial smooth muscle cells in vitro. Science 191:475–477

100. Linder J, Croker BP, Vollmer RT, Shelburne J (1983) Systemic Kappa light chain deposition: an ultrastructural and immunohistochemical study. Am J Surg Pathol 7:85-93

101. Luetscher JA, Kraemer FB, Wilson DM, Schwartz HC, Bryer-Ash M (1985) Increased plasma inactive renin in diabetes mellitus. A marker of microvascular complications. N Engl J Med 312:1412–1417

102. McCrary RF, Pitts TO, Puschett JB (1981) Diabetic nephropathy: natural course, survivorship and therapy. Am J Nephrol 1:206–218

103. McVerry BA, Hopp A, Fischer C, Huehns ER (1980) Production of pseudo-diabetic renal glomerular changes in mice after repeated injections of glycosylated proteins. Lancet I:738–740

104. Mauer SM, Steffes MW, Sutherland DER, Najarian JS, Michael AF, Brown DM (1974) Studies of the rate of regression of the glomerular lesions in diabetic rats treated with pancreatic islet transplantation. Diabetes 24:280–285

105. Mauer SM, Barbosa J, Verner RL, et al. (1976) Development of diabetic vascular lesions in normal kidneys transplanted into patients with diabetes mellitus. N Engl J Med 295:916–920

106. Mauer SM, Steffes MW, Azar S, Sandberg SK, Brown DM (1978) The effects of Goldblatt hypertension on development of the glomerular lesions of diabetes mellitus in the rat. Diabetes 27:738–744

107. Mauer SM, Steffes MW, Chern M, Brown DH (1979) Mesangial uptake and processing of macromolecules in rats with diabetes mellitus. Lab Invest 41:401–406

108. Mauer SM, Steffes MW, Brown DM (1981) The kidney in diabetes. Am J Med 70:603–612

109. Mauer SM, Steffes MW, Connett J, Najarian JS, Sutherland ER, Barbosa J (1983) The development of lesions in the glomerular basement membrane and mesangium after transplantation of normal kidneys to diabetic patients. Diabetes 32:948–952

110. Mauer SM, Steffes MW, Ellis EN, Brown DM (1984) Can the insulin-dependent diabetic patient be managed without kidney biopsy? In: Robinson RR (ed) Nephrology, vol II. Springer-Verlag, Berlin Heidelberg New York, pp 1103–1108

111. Michels LD, Davidman M, Keane WF (1981) Determinants of glomerular filtration and plasma flow in experimental diabetic rats. J Lab Clin Med 98:869–885

112. Miller K, Michael AF (1976) Immunopathology of renal extracellular membranes in diabetes mellitus: specificity of tubular basement membrane immunofluorescence. Diabetes 25:701–708

113. Mogensen CE (1971) Maximum tubular reabsorption capacity for glucose and renal hemodynamics during rapid hypertonic glucose in normal and diabetic subjects. Scand J Clin Lab Invest 28:101–109

114. Mogensen CE (1971) Glomerular filtration rate and renal plasma flow in short-term and long term juvenile diabetes mellitus. Scand J Clin Lab Invest 28: 91–100

115. Mogensen CE (1971) Glomerular filtration rate and renal plasma flow in normal and diabetic man during elevation of blood sugar levels. Scand J Clin Lab Invest 28:177–182

116. Mogensen CE (1971) Urinary albumin excretion in early and long-term juvenile diabetes. Scand J Clin Lab Invest 28:183–193

117. Mogensen CE, Andersen MJF (1973) Increased kidney size and glomerular filtration rate in early juvenile diabetes. Diabetes 22:706–712

118. Mogensen CE, Vittinghus E, Solling K (1979) Abnormal albumin excretion after two provocative renal tests in diabetes: physical exercise and lysine injection. Kidney Int 16:385–393

119. Mogensen CE, Steffes MW, Deckert T, Sandahl-Christiansen J (1981) Functional and morphological renal manifestations in diabetes mellitus. Diabetologia 21:89–93

120. Mogensen CE, Christiansen CK, Vittinghus E (1983) Changes in renal function and blood pressure control in diabetes mellitus. With special reference to exercise-induced changes in albumin excretion and blood pressure. In: Friedman EA, L'Experance FA (eds) Diabetic renal – retinal syndrome II: prevention and management. Grune and Stratton, New York, pp 41–58

121. Mogensen CE (1983) A complete screening of urinary albumin concentration in an unselected diabetic outpatient clinic population (1082 patients). Diabetic Nephrop 2:11–18

122. Mogensen CE (1984) Clinical and renal functional studies of diabetic nephropathy in humans. In: Robinson RR (ed) Nephrology, vol II. Springer-Verlag, Berlin Heidelberg New York, pp 1053–1073

123. Mogensen CE, Christiansen CK (1984) Predicting diabetic nephropathy in insulin dependent patients. N Engl J Med 311:89–93

124. Mogensen CE (1984) Microalbuminuria predicts clinical proteinuria and early mortality in maturity-onset diabetes. N Engl J Med 310:356–360

125. Murphy WM, Deodhar SD, McCormack LJ, Osborne DG (1973) Immunopathologic studies in glomerular disease with membranous lesions. Am J Clin Pathol 60:364–376

126. Myers BD, Winetz JA, Chiu F, Michaels AS (1982) Mechanisms of proteinuria in diabetic nephropathy. Kidney Int 21:633–641

127. Nakamoto Y, Takazakura E, Hayakawa H et al. (1980) Intrarenal microaneurysms in diabetic nephropathy. Lab Invest 42:433–439

128. Nochy D, Druet P, Bariety J (1984) IgA nephropathy in chronic liver disease. Contrib Nephrol 40:268–275

129. Nyberg G, Blohme G, Björck S, Norden G (1985) Progression of diabetic nephropathy in relation to plasma glucose control and blood pressure control. Diabetic Nephrop 4:53–54

130. Olivero J, Suki WN (1977) Acute glomerulonephritis complicating diabetic nephropathy. Arch Intern Med 137:732–734

131. Olson JL, Hostetter TH, Rennke HG, Brenner BM, Venkatachalam MA (1982) Altered glomerular permselectivity and progressive sclerosis following extreme ablation of renal mass. Kidney Int 22:112–126

132. O'Neill WM, Wallin JD, Walker PD (1983) Hematuria and red cell casts in typical diabetic nephropathy. Am J Med 74:389–395

133. Osterby R, Gundersen HJG (1975) Glomerular size and structure in diabetes mellitus. Diabetologia 11:225–229

134. Osterby R, Gundersen HJG (1980) Fast accumulation of basement membrane material and the rate of morphological changes in acute experimental diabetic glomerular hypertrophy. Diabetologia 18:493–500

135. Osterby R (1983) Basement membrane morphology in diabetes mellitus. In: Ellenberg M, Rifkin H (eds) Diabetes mellitus. Theory and practice 3rd edn. Medical Examination Publishing Co, New York, pp 323–341

136. Parthasarathy N, Spiro RG (1982) Effect of diabetes on the glycosaminoglycan components of the human glomerular basement membrane. Diabetes 31:738–741

137. Parving HH, Jensen HAE, Mogensen CE, Evrin PE (1974) Increased urinary albumin-excretion rate in benign essential hypertension. Lancet I:1190–1192

138. Parving HH, Noer I, Decker T (1976) The effect of metabolic regulation on microvascular permeability to small and large molecules in short-term diabetes. Diabetologia 12:161–166

139. Parving HH, Christiansen JS, Noer I, Tronier B, Mogensen CE (1980) The effect of glucagon infusion on kidney function in short term insulin-dependent juvenile diabetes. Diabetologia 19:350–384

140. Parving HH, Smidt UM, Frisberg B, Bonnevie-Nielsen V, Andersen AR (1981) A prospective study of glomerular filtration rate and arterial blood pressure in insulin-dependent diabetics with diabetic nephropathy. Diabetologia 20:457–461

141. Parving HH, Oxenboll B, Svendsen PA, Sandahl-Christiansen J, Andersen AR (1982) Early detection of patients at risk of developing diabetic nephropathy. Acta Endocrinol 100:550–555

142. Parving HH, Andersen AR, Smidt UM, Christiansen JS, Oxenboll B, Svendsen PAA (1983) Diabetic nephropathy and arterial hypertension. The effect of antihypertensive treatment. Diabetes [Suppl 2] 32:83–97

143. Paz-Guavara AT, Hsu TH, White P (1975) Juvenile diabetes mellitus after forty years. Diabetes 24:559–565

144. Peterson DT, Greene WC, Reaven GM (1971) Effect of experimental diabetes mellitus on kidney ribosomal protein synthesis. Diabetes 20:649–654

145. Pirart J (1978) Diabetes mellitus and its degenerative complications: a prospective study of 4,400 patients observed between 1947 and 1973. Diabetes Care 1:168–252

146. Price RG, Spiro RG (1977) Studies on the metabolism of the renal glomerular basement membrane. Turnover measurements in the rat with the use of radiolabeled amino acids. J Biol Chem 252:8597–8602

147. Raij L, Keane WF, Osswald H, Michael A (1979) Mesangial function in ureteral obstruction in the rat. J Clin Invest 64:1204–1212

148. Rao KV, Crosson JT (1980) Idiopathic membranous glomerulonephritis in diabetic patients. Arch Intern Med 140:624–627

149. Rao TKS, Friedman EA (1982) Diabetic nephropathy in Brooklyn. In: Friedman E, L'Esperance FA(eds) Diabetic renal–retinal syndrome, vol 2. Grune and Stratton, New York, pp 3–8

150. Rasch R (1979) Prevention of diabetic glomerulopathy in streptozotocin diabetic rats by insulin treatment. Diabetologia 16:125–128

151. Rasch R (1979) Prevention of diabetic glomerulopathy in streptozotocin diabetic rats by insulin treatment. Glomerular basement thickness. Diabetologia 16:319–324

152. Risteli J, Koivisto VA, Akerblom HK, Kivirikko K (1976) Intracellular enzymes of collagen biosynthesis in rat kidney in streptozotocin diabetes. Diabetes 25:1066–1070

153. Ritz E, Hasslacher C (1984) Genesis and treatment of hypertension in diabetes mellitus. Diabetic Nephrop 3:2–11

154. Robbins SL, Mallory GK, Kinney TD (1946) Necrotizing renal papillitus: a form of acute pyelonephritis. N Engl J Med 231:885–893

155. Robinson GC, McConnel D (1961) Simultaneous onset diabetes mellitus and the nephrotic syndrome. Can Med Assoc J 85:80–81

156. Sandahl-Christiansen J, Frandsen M, Parving HH (1981) The effect of intravenous insulin infusion on kidney function in insulin dependent mellitus. Diabetologia 20:199–204

157. Sandahl-Christiansen J, Gammelgaard J, Frandsen M, Parving HH (1981) Increased kidney size, glomerular filtration rate and renal plasma flow in short-term insulin-dependent diabetics. Diabetologia 20:451–456

158. Sandahl-Christiansen J, Gammelgaard J, Orkov H, Andersen AR, Telmer S, Parving HH (1981) Kidney function and size in normal subjects before and during growth hormone administration for one week. Eur J Clin Invest 11:487–490

159. Schambelan M, Blake S, Sraer J et al. (1985) Increased prostaglandin by glomeruli isolated from rats with streptozotocin-induced diabetes mellitus. J Clin Invest 75:404–412

160. Scheinman JI, Fisch AJ, Michael AF (1974) The immunohistopathology of glomerular antigens. The glomerular basement membrane, collagen, actomyosin antigens in normal and diseased kidneys. J Clin Invest 54:1144–1154

161. Scheinman JI. Fisch AJ, Michael AF (1978) The immunohistopathology of glomerular antigens. II. The glomerular basement, actomyosin and fibroblast surface antigens in normal, diseased and transplanted human kidneys. Am J Pathol 90:71–88

162. Scheinman JI, Steffes MW, Brown DM, Mauer SM (1978) The immunohistopathology of glomerular antigens. III. Increased mesangial actomyosin in experimental diabetes in the rat. Diabetes 27:632–637

163. Schober E, Pollak A, Coradello H, Lubec G (1982) Glycosylation of glomerular basement membrane in type I (insulin-dependent) diabetic children. Diabetologia 25:485–487

164. Schreiner GE (1971) The nephrotic syndrome. In: Strauss MB, Welt LG (eds) Disease of the kidney. Little, Brown and Co., Boston, chap. 16

165. Seyer Hansen K (1976) Renal hypertrophy in streptozotocin-diabetic rats. Clin Sci Mol Med 51:551–555

166. Seymour AE, Spargo BH, Penkas R (1971) Contributions of renal biopsy studies to the understanding of disease. Am J Pathol 65:550–598

167. Sharma HM, Yum MN, Kleit S (1974) Acute glomerulonephritis with diabetes mellitus. Report of a case. Arch Pathol 97:152–154

168. Shideman JR, Buselmeier TJ, Kjellstrand CM (1976) Hemodialysis complications in insulin dependent diabetic patients accepted for transplantation. Arch Intern Med 136:1126–1130

169. Siperstein MD, Unger RH, Madison II (1968) Studies of muscle capillary basement membranes in normal subjects, diabetic and prediabetic patients. J Clin Invest 47:1973–1999

170. Sohar E, Ravid M, Ben-Shaul Y, Reshef T, Gafni J (1970) Diabetic fibrillosis. A report of three cases. Am J Med 64:49–51

171. Spiro RG, Spiro MJ (1971) Studies on the biosynthesis of the hydroxylysine-linked disaccharide unit of basement membranes and collagens. J Biol Chem 246:4899–4909

172. Spiro RG, Spiro MJ (1971) Effect of diabetes on the biosynthesis of the renal glomerular basement membrane studies on the glycosyl-transferase. Diabetes 20:164–167

173. Spuhler O, Zollinger Hu (1944) Die diabetische glomeruloskle-rose. Helv Med Acta 11:27–30

174. Stalder G, Schmid R (1959) Severe functional disorders of glomerular capillaries and renal hemodynamics in treated diabetes mellitus during childhood. Ann Paediatr 193:129–138

175. Steffes MW, Brown DM, Basgen JM, Mauer SM (1980) Amelioration of mesangial volume and surface alterations following islet transplantation in diabetic rats. Diabetes 29:509–515

176. Steffes MW, Buchwald H, Wigness BD et al. (1982) Diabetic nephropathy in the uninephrectomized dog: microscopic lesions after one year. Kidney Int 21:721–724

177. Ullrich KJ, Rumrich G, Kloss S (1974) Specificity and sodium dependance of the active sugar transport in the proximal convolution of the rat kidney. Pflugers Arch 351:35–49

178. Urizar RE, Schwartz A, Top F, Vernier RL (1969) The nephrotic syndrome in children with diabetes mellitus of recent onset. Report of five cases. N Engl J Med 281:173–181

179. Puig JG, Anton FM, Grande C, et al. (1981) Relation of kidney size to kidney function in early juvenile diabetes. Diabetologia 21:363–367

180. Viberti GC, Pickup JC, Jarrett RJ, Keen H (1979) Effect of control of blood glucose on urinary excretion of albumin and B2—microglobulin in insulin-dependent diabetes. N Engl J Med 300:638–641

181. Viberti GC, Pickup JC, Bilous RW, Keen H, Mackintosh D (1981) Correction of exercise-induced microalbuminuria in insulin-dependent diabetes after 3 weeks of subcutaneous insulin infusion. Diabetes 30:818–823

182. Viberti GC, Mackintosh D, Bilous RW, Pickup JC, Keen I (1982) Proteinuria in diabetes mellitus: role of spontaneous and experimental variation of glycaemia. Kidney Int 21: 714–720

183. Viberti GC, Hill RD, Jarrett RJ, Argyropoulos A, Mahmud U, Keen H (1982) Microalbuminuria as a predictor of clinical nephropathy in insulin-dependent diabetes mellitus. Lancet I:1430–1432

184. Viberti GC, Bilous RW, Mackintosh D, Keen H (1983) Monitoring glomerular function in diabetic nephropathy. Am J Med 74:256–264

185. Viberti GC, Keen H (1984) The patterns of proteinuria in diabetes mellitus. Relevance to pathogenesis and prevention of diabetic nephropathy. Diabetes 33:686–692

186. Viberti GC, Wiseman MJ (1984) Early markers of diabetic nephropathy: a road of prevention. In: Robinson RR (ed) Nephrology, vol II. Springer-Verlag, Berlin Heidelberg New York, pp 1094–1102

187. Vittinghus E, Mogensen CE (1982) Graded exercise and protein excretion in diabetic man and the effect of insulin treatment. Kidney Int 21:725–729

188. Wahl P, Deppermann D, Hasslacher C (1982) Biochemistry of glomerular basement membrane of the normal and diabetic human. Kidney Int 21:744–749

189. Warms PC, Rosonbaum BF, Michelis MF, Haas JE (1973) Idiopathic membranous glomerulonephritis occuring with diabetes mellitus. Arch Intern Med 132:735–738

190. Wass JAH, Waltkins PJ, Dische FE, Parson V (1978) Renal failure, glomerular disease and diabetes mellitus. Nephron 21:289–296

191. Watkins PJ, Blainey JD, Brewer DB et al. (1972) The

natural history of diabetic renal disease. Q J Med 164:437–456

192. Webb C, Bohr DF (1981) Recent advances in the pathogenesis of hypertension. Consideration of structural, functional, and metabolic vascular abnormalities resetting in elevated arterial resistance. Am Heart J 102:251–264

193. Wehner H, Bohle A (1974) The structure of the glomerular basement membrane in diabetes mellitus with and without the nephrotic syndrome. Virchows Arch [A] 364:303–309

194. Weiner RB (1982) Observations of diabetic, hypertensive physician following renal transplantation. Diabetic Nephrop 1:20–21

195. Weinman EJ, Suki WN, Eknoyan G (1976) D-glucose enhancement of water reabsorption in proximal tubule

of the rat kidney. Am J Physiol 231:777–780

196. Westberg MG, Michael AF (1972) Immunohistopathology of diabetic glomerulosclerosis. Diabetes 21:163–174

197. Westberg MG, Michael AF (1973) Human glomerular basement membrane: chemical composition in diabetes mellitus. Acta Med Scand 134:39–47

198. Williamson J, Kilo C (1977) Current studies of capillary basement membrane disease in diabetes mellitus. Diabetes 26:65–73

199. Winetz JA, Golbetz HV, Spencer RJ, Lee JA, Myers BD (1982) Glomerular function in advanced human diabetic nephropathy. Kidney Int 21:750–756

200. Zucchelli P, Cagnoli L, Casanova S, Donini U, Pasquali S (1983) Focal glomerulosclerosis in patients with unilateral nephrectomy. Kidney Int 24:649–655

Chapter 25

Vascular Connective Matrices and Haemostasis in Diabetic Angiopathy

J. P. Muh and Y. J. Legrand

Glycoprotein Biosynthesis Modification

Diabetic angiopathy is characterized by two types of alterations. The first concerns glyco-protein metabolism: an increase in synthesis or a decrease in catabolism leads to the accumulation of "glycoprotein material" [98], the chief consequence of which is the thickening of the vascular basal lamina [13]. The second type of alteration consists of non-enzymatic glyco-sylation of proteins [84].

Increase in the Biosynthesis of Glycoproteins

This involves an increase in the number of gly-canic units (hydroxylysyl glycosides) bound to collagen chains [5,11,114] and an increase of lysine and proline hydroxylation which affects the basement membrane collagen synthesis rate [20,29,106]. There is also evidence of an increase in lysine hydroxylase activity in the diabetic rat [28]. It has been demonstrated that hyper-glycaemia plays a direct role in these modifications in the diabetic rat [67] and that insulin is important for correcting them [30].

It has been shown that in the case of strep-tozotocin-induced diabetes, lysine and proline hydroxylation was increased after 12 weeks [99], while the glucosyl transferase activity was normalized with insulin therapy [113]. In KK mice, kidney glucosyl transferase determinations are normal for some [26] and increased for others between the 25th and 55th days of life [127], and there is an associated plasma rise in UDP galactosyl transferase [73]. In the C57 BL db/bd mouse, kidney galactosyl transferase activity is lowered from the 13th month of life [72]. These enzymatic variations must be related to other changes which occur in diabetes as kidney protein synthesis increases [98]; slowing of their catabolism and transferase modifications are induced by hyperglycaemia [112].

Since the mid-1970s, basing their arguments on the enzymatic perturbations just described, many authors have proposed a genetic and molecular foundation for the complications of diabetes mellitus. the discovery of non-genetic, non-enzymatic glycosylation processes [55] challenged these proposals. The HbA_1C rates study on diabetic twins demonstrates the significance of these processes [119].

Decrease of Glycoprotein Degradation

Lysosomal hydrolases play a major part in glycoconjugate degradation [83]. Betaglucuroni-dase and glucosaminidase levels are increased [12,14] while glucosidase activity is reduced [48]. Moreover, Sternberg described an inhibi-tion of hydroxylysyl-glucoside α-glycosi-dase activity in basal lamina collagen [115]. That experimental result favours slowed catab-olism. Nicoll was able to demonstrate a poly-nuclear collagenolytic activity reduction in diabetics [88].

It appears that the elevation of the activity of plasma hydrolases of lysosomal origin is connected with a decrease of α and β (kidney) glycohydrolases [49]. The consequence of these enzymatic changes is the accumulation of disaccharide units associated with the hydroxylysine of glomerular basement membrane collagen [11].

Metabolic Deviation: The Polyalcohol Pathway

To these enzymatic disturbances must be added the consequences of a deviation in glucose metabolism on the polyalcohol pathway [51]. The particular locations of aldose reductase in eyes, kidney and peripheral nerves probably explains its important metabolic effect on the vascular system [50,80]. Thus, this intracellular accumulation of polyalcohol is associated with those other enzymatic disturbances within "target organs" which are affected by diabetic disease.

Non-enzymatic Glycosylation: Molecular and Cellular Consequences

Non-enzymatic Glycosylation (NEG) of Proteins

The non-enzymatic fixation of glucose to proteins is a post-translational modification first described by Allen et al. [2], who described haemoglobin A_1C in 1958. The precise identification of this modification was made by Holmquist [63] and Bookchin [16]. In 1969, Rahbar [96] demonstrated that this form of a haemoglobin was greatly increased in diabetic red blood cells. Since that time, HbA_1C determination has been used in therapeutic follow-up since its level is proportional to levels of blood glucose. Glucose fixation to proteins depends on the fixation of that molecule to amines.

The glucose molecule is first condensed, because of its semi-aldehyde function, with the α amino group of the N-terminal amino acid or with "ε amino groups" of lysines and hydroxylysines located on the peptide chains. There is a rapid reversible formation of Schiff's base [62]. Then, synthesized aldimin is slowly transformed into a stable cetoamine by Amadori's rearrangement into the β pyranose type configuration [45,87] (see Fig. 25.1).

The main functional consequence of glucose fixation on the N-terminal valine of the haemoglobin β chain is the formation of a more stable $Hb-O_2$ complex, and the subsequent decrease of P_{50} in peripheral tissues [8] (Fig. 25.1).

The most widely used method for the determination of non-enzymatically bound glucose is based on its conversion in 5-hydroxymethylfurfural which then reacts with thiobarbituric acid [46]. Another assay using ionic exchange chromatography after reduction with tritiated borohydride can determine the quantity of glucose fixed on crystalline proteins [27] or on collagen [74]. These methods have shown that non-enzymatic glycosylation is a general

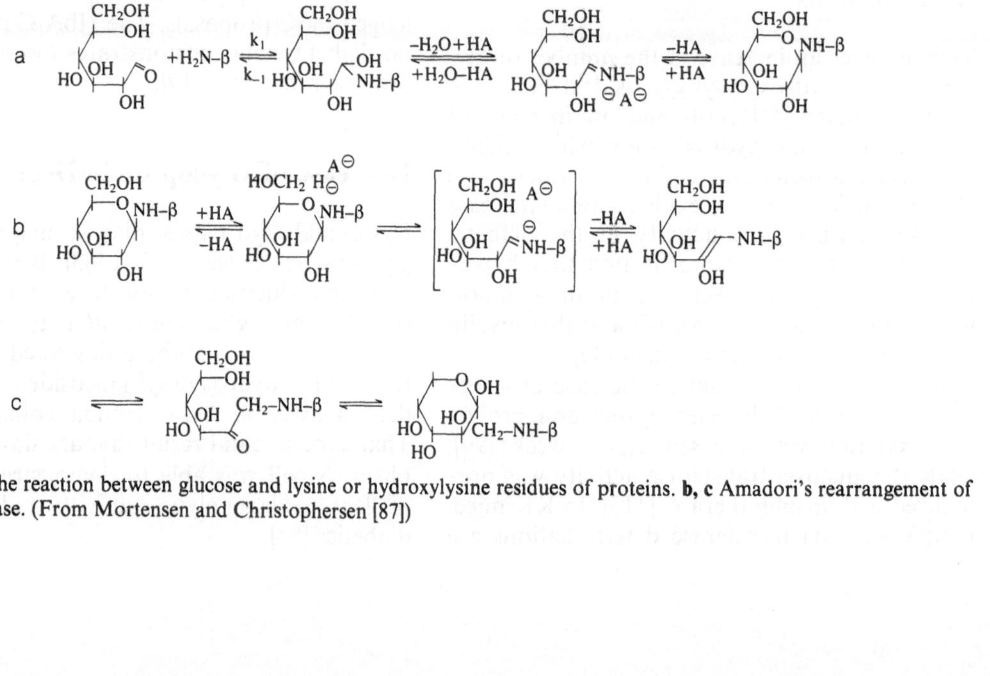

Fig. 25.1. a The reaction between glucose and lysine or hydroxylysine residues of proteins. **b, c** Amadori's rearrangement of the Schiff's base. (From Mortensen and Christophersen [87])

phenomenon affecting a number of proteins [21], including the following:

1. Circulating proteins and secreted proteins: serum albumin, anti-thrombin III, fibrinogen, LDL, HDL, pancreatic ribonuclease and fibronectin (cIg).
2. Cellular proteins: haemoglobin, red blood cell stroma proteins, ferritin, cathepsin B, β-Nac-D glucosaminidase, membrane proteins of endothelial cells, tubulin and myelin.
3. Connective tissue proteins: aortic wall collagens, coronary arteries, glomerular basal lamina, type I, II and III collagens, crystalline, proteins, crystalline capsules and fibronectin.

Incidence in the Vascular Connective Matrix

Collagens

The initial demonstration of the formation of the products of lysine–carbohydrate condensation at the collagen level was reported by Robins and Bailey [100] in 1972. These results were confirmed by Tanzer [120].

Arterial collagens, especially type III and capillary type IV, are proteins suitable for non-enzymatic glycosylation: they are rich in lysine and hydroxylysine, their half-life is long (about $2\frac{1}{2}$ years [69]) and their extracellular position exposes them directly to variations in blood glucose. Non-enzymatic glycosylation of diabetic vessels was demonstrated in aortic [103] and basal laminae [31,75,122]. These results were obtained following work on interstitial collagen metabolism in diabetic patients and in animals with experimentally induced diabetes [54,60,82]. It has been shown that the connective tissue of diabetics undergoes an accelerated ageing. The measurement of levels of cetoamine-bound sugars shows they are increased in the rat glomerular basement membrane [75] and in type IV human collagen [93]; it was found that 25.9 nmol of hexose were bound with 1 μmol of hydroxyproline [123]. Type IV collagen in diabetics contains one or two glucose molecules non-enzymatically bound to a triple helix molecule the molecular weight of which is 380 000 [121]. This glomerular basement membrane collagen is hyperglycosylated in diabetes while the number of cross-links decreases by about 40% [75]. These modifications can explain the genesis of glomerular basement membrane denseness in part, since non-enzymatic glycosylation occurs of lysine and hydroxylysine involved in fibril stabilization [57]. However, glucose acts directly as a cross-linking agent for proteins, as demonstrated by Eble [43], Pongor [95] and Kent [66] in their work on glycosylated collagen. Glycosylated collagen is composed of high molecular weight elements which are not very soluble and are stable at high temperatures in in vitro conditions. These molecular characteristics explain the increased resistance to collagenases [126] and the basement membrane thickness [68]. Glucose can still act as a cross-linking agent between non-enzymatically glycosylated collagen and other circulating proteins such as albumin and immunoglobulin G [20]. This phenomenon can explain the linear deposition of anti-albumin and anti-IgG fluorescent antibodies on the diabetic basal laminae [85].

In short, collagen non-enzymatic glycosylation leads to the formation of insoluble collagen molecules of high molecular weight which are resistant to collagenase and are able to fix some circulating proteins.

These modifications associated with the cross-linking characteristics of glucose can explain the alteration of the properties and the quality of arterial type I and III collagens, and the type IV collagen and glycoprotein material accumulation within basement membranes.

Basement Membrane Glycoproteins

Type IV collagen, the major component in basement membrane, is associated with a number of proteins and proteoglycans [56]. Fibronectin is one of these components and plays a major role in the adhesion of endothelial cells to collagen [104,128].

Tarsio et al. have demonstrated the non-enzymatic glycosylation of fibronectin and have studied its functional incidence [118]. Non-enzymatically glycosylated fibronectin in the diabetic is two to three times more abundant than in the non-diabetic. That increase is proportional to the blood glucose level. In in vitro conditions, the authors have demonstrated a decrease in its affinity for heparin and collagen. For the other proteins involved in the interactions between cells and the extracellular matrix, the information is not available, except for laminin whose rate of interaction is increased [102].

Proteoglycans

In an experimental animal model C57 BL.Ksj db/db mice carrying EhS tumours Rohrbach et al. have demonstrated a decrease of heparin sulphate BM1 [102]. Parthasarathy has confimed this result in human glomerular basement membrane [91]. Since glomerular proteoglycans are implicated as "anionic filters" this decrease could explain in part the proteinuria observed during the development of diabetic glomerulopathy.

Functional Incidence

We have described the main modifications induced by non-enzymatic glycosylation in the extracellular matrix of vessels. These modifications, for the most part, induce functional alterations which represent a molecular basis for diabetic angiopathy.

The main alterations are as follows:

1. Anomalies of the fibrillogenesis and of the polymerization of the interstitial collagens (types I and III) which lead to an increase in their resistance to collagenases and to modifications of their mechanical properties.
2. At the site of the basal laminae, accumulation of type IV collagen and laminin occurs, and there is modification of the adhesion capacity of fibronectin and a decrease of heparin sulphate. We have also noted the presence of glycosylated collagen–serum albumin–immunoglobulin G complexes.
3. Related to the interactions between blood cells and the vessel wall is an increase and acceleration of the platelet aggregation process. All these alterations lead to an angiopathy which favours thrombotic phenomena.

Collagen Fibrillogenesis

Molecular alterations in collagen may be in part responsible for arterial wall lesions (rigidity), the consequent modification of vascular reactivity [70] and an increase of protein transfer [92]. Chang's study on diabetic collagen solubility argues for the existence of an intermolecular cross-links modification [24].

In 1981, Andreassen [6] demonstrated that the thermal and mechanical stability of diabetic collagen fibres was increased. During a kinetic study on the formation of type I collagen fibrils in vitro, Guitton demonstrated competition between glycosylamines and cross-link formation [57]. These results were confirmed by Lien in 1984 [77]. Instability due to fibril hyperhydration, associated with glucose cross-

linking properties [43], induces a formation of complex derivative molecule [95]. This molecular reorganization is accompanied by an increased resistance to proteolysis [79, 107].

Platelet Aggregation

When an endothelial lesion forms, blood comes into contact with the subendothelium and the platelet–collagen interactions begin [9, 34]. If we place healthy platelets in contact with non-enzymatically glycosylated type I collagen, we notice an increase in platelet aggregation: there is an increase in velocity and intensity of effect with shortening of the latent period [76].

Endothelial Cells, Platelets and Haemostasis in Diabetic Angiopathy

Endothelial Cells

Chapter 2 of this book describes endothelial/blood relationships. Most of the knowledge acquired about the functions of endothelial cells is the result of in vitro cellular culture studies. The extrapolation to phenomena occurring in vivo is consequently hazardous but is nevertheless possible in some cases. It is practicable for those metabolic activities of the endothelial cell that are related to a haemostasis or thrombosis. These phenomena may be involved in vascular alterations associated with diabetes mellitus, for instance, the synthesis of plasma or tissue factors promoting or regulating molecular or cellular interactions in primary haemostasis, coagulation or fibrinolysis. Variations of these factors may indicate alterations of the endothelium itself rather than disturbances of metabolite synthesis.

Endothelial Alterations

These alterations affect some endothelial properties, for example they may increase permeability for dyes [7] or for lipoproteins [132]. One of the most striking manifestations is the increase of circulating von Willebrand's factor [38] measured either by immunological methods [32, 34, 90, 97] or by its co-factor activity for ristocetine [15, 81]. There is controversy surrounding the correlation or non-correlation between von Willebrand's factor levels and the appearance of vascular lesions or retinopathy [38]. In any case, the endothelium is involved, as it appears that a rise in von Willebrand's factor must be the cause or a consequence of vascular lesions.

Functional Anomalies

These are related to the fact that diabetic endothelial cells are exposed to glucose toxicity, as recently reported by Lorenzi et al. [78]. Glucose toxicity is manifested as changes in cycles, such as a slowing down in multiplication of endothelial cells or their accelerated death rate. Metabolic abnormalities mainly affect prostacyclin $PG1_2$ and the tissue plasminogen activator (TPA); most of the comparative data were collected by Colwell et al. [38] in 1983.

Diabetes is associated with a striking reduction of prostacyclin production and of its 6 ceto $PGF_{1\alpha}$ metabolite by endothelial, arterial or venous cells, in both diabetic humans [61,117,129] and in animals with streptozotocin- or alloxan-induced diabetes [53,61,100]. It seems that this reduction strikes mainly insulin-dependent patients, and the problem is corrected by insulin [31,101] or by transplantation of islets of Langerhans in animals [53].

Diabetes is associated with a reduction of the fibrinolytic actvity of plasma as described and this reduced actvity can be considered to be the result of the effect of tissue plasminogen activator and endothelial fibrinolysis inhibitors. Tissue plasminogen activator (TPA) production is measured by the increase of plasma fibrinolytic activity induced by anoxaemia provoked with a pneumatic cuff, a system that Almer et al. have used to demonstrate a decrease of TPA activity in diabetes [3]. In fact, a decrease in TPA activity may be more important in the venous system than the arterial: this may explain why other authors have found an increase of TPA production in the aorta of diabetic rats [109]. Nevertheless, most of the studies show a decrease either of TPA or plasma fibrinolytic activity which can be linked to endothelial alterations.

Platelet Function and Coagulation Factors in Diabetic Microangiopathy

Lesions of diabetic angiopathy developed in the whole vascular system are described, as are macro- and microangiopathies. Microroangiopathy appears more specific in its lesions and semiology; it is associated with alterations of microvessels (arterioles, capillaries), and with an increase of adhesiveness and rigidity of red blood cells. Platelet function and the coagulation system are also disturbed.

We will focus mainly on platelet function and coagulation factor changes because they are implicated in the genesis of microthrombi.

Platelet Function in the Diabetic State

The platelets of diabetics are more sensitive to aggregant agents than those of controls. Colwell et al. [37] reported that platelet hyperactivity is apparent before the appearance of vascular lesions in insulin-dependant diabetic patients without retinopathy. Capillary lesions have multiple causes such as endothelial change, hypertension, obesity, lipid disturbances and hyperglycaemia.

Activated platelets which have attached to injured endothelium spread out and produce platelet-derived growth factor (PDGF). PDGF secretion initiates proliferation of fibroblasts and smooth-muscle cells. This phenomenon is an early step in the atherogenic process (see Chap. 14).

In Vitro Platelet Function. Aggregation variables have been studied, such as reactivity induced by ADP, collagen, epinephrin or thrombin and as arachidonic acid catabolism and granular content secretion. Some studies have demonstrated platelet hypersensitivity to aggregating agents [37,59,105,108]. However, in the absence of vascular lesions, that hypersensitivity is not observed [35,37]. Some studies do not show any significant differences between diabetic and normal platelets [39,94]. In most of the studies concluding that there is hyperaggregability of platelets, an increase of arachidonic acid metabolism has been noticed: this was also noticed in patients without vascular lesions. Thus, we can describe an increased secretion of pro-aggregating prostaglandins such as PGE_2 [58], thromboxane A_2 (TxA2) [130], $PGF_{2\alpha}$ [25] and malondialdehyde [116]. This increase is more significant in cases with vascular lesions. We know that endoperoxides and thromboxane A_2 can occupy bond sites available for fibrinogen on the platelet surface. In a recent study, Di Mino et al. [41] compared fibrinogen–platelet surface interactions induced by collagen or ADP. The results obtained show that platelets from diabetics without retinopathy bind more fibrinogen than normal platelets, because far more receptors are exposed on the platelet surface. This has been linked to the increased production of TxA_2 and endoperoxide stimulated by the effect of ADP and collagen, and more precisely to intraplatelet ADP being increased under the effect of endoperoxide and of TxA_2 raising its production.

The fact that the anomaly observed in diabetes has no bearing on the glycoprotein complex GpIIb IIa (receptor site carrier of fibrinogen), is demonstrated by the use of a monoclonal antibody specific to that complex which does not bind to diabetic platelets or normal platelets. Taking into account that fibrinogen binding to platelets is one of the most important co-factors in platelet aggregation, the augmentation of its link to membrane is in perfect concordance with the observations showing a hypersensitivity to aggregating agents and a hyperproduction of endoperoxide, TxA_2 and malondialdehyde. The increase is also in accordance with other observations concerning the secretion of granules, including dense granules: Di Mino et al. [41] have demonstrated that collagen, ADP and a stable analogue of TxA_2 (U 46619) provoked an increased secretion of adenhylic nucleotides from platelets of diabetics without retinopathy when compared with normal platelets. This is also true for alpha granule: some studies have related the site of circulating plasma beta thromboglobulin to platelet activation found in in vivo; beta thromboglobulin levels have been found to be high in patients without vascular lesions [18,22,37] which means that platelet activation can occur in vivo before the vascular lesions appear. Nevertheless, it is noticeable that the highest levels we observed were in patients with lesions.

Two important points must be made. Firstly we must note Kwan's description [71] of increased normal platelet aggregation due to a factor in the plasma of a diabetic patient who died of disseminated intravascular coagulation. This factor has been called PAEF (platelet aggregation enhancing factor); another has been compared to soluble immune complexes.

Secondly, Le Pape et al. [76] have commented on the reactivity of non-enzymatic glycosylated collagen towards normal platelets.

In Vivo Platelet Function. We know that plasma beta thromboglobulin levels are related to the activation of platelets in the bloodstream.

Other studies have shown that platelet survival is decreased [1,40] in diabetic patients with or without patent vascular alterations. Platelet life shortening is not constant: Jones et al. only demonstrated in diabetic patients with severe retinopathy [65].

Coagulation Factors in Diabetes

The study of the coagulation system in diabetes has produced divergent results. Some of them show hypercoagulability with platelet function enhancement. Diabetic hypercoagulability is linked to a significant increase of fibrinogen [17,64], mainly in those patients with vascular lesions. Other factors are also enhanced, including factor V and factor VIII_c [17,47] for instance. The foregoing suggests that high levels of fibrinogen are related to diabetic angiopathy.

The hypercoagulability of diabetics is associated with a significant decrease of fibrinolytic activity and of fibrin sensitivity to plasmin. These changes are found in non-insulin-dependent patients with or without vascular lesions. These results, reported by Brownlee et al. [19], are the consequence of non-enzymatically glycosylated fibrinogen and may induce the deposition of glycosylated fibrin at the site of lesions. They may thus play a major part in the development of diabetic complications.

The controversy about diabetic hypercoagulability can be linked to doubtful matching between control and patient groups as well as to purely technical problems.

There are some contradictory data concerning anti-thrombin III, which is the main circulating inhibitor of the coagulation enzymes: a state of hypercoagulability should be associated with a decrease of anti-thrombin III. However, analysis of the literature shows the level of anti-thrombin III can either be decreased [10,110], normal [53,131] or increased [44,47]. These findings are independent of vascular lesions although Borsey et al. [17] observed the anti-thrombin III increase to be pronounced in patients with retinopathy. Ceriello et al. [23] observed a decrease of anti-thrombin III activity associated with hypercoagulability; hypothetically they connected this with an alteration of the characteristics of this inhibitor, induced by non-enzymatic glycosylation. Its concentration in plasma did not differ significantly from that in controls.

Differences are observed when the biological activity of anti-thrombin III or its plasma concentration are studied.

Conclusion

Is diabetes associated with a state of hyper-coagulability? Many authors think so, even if there is some variation in the results and interpretation. It seems that thrombogenic factors are increased, while the activity of factors inhibiting thrombosis are decreased.

Increased thrombogenic activity may be due to the increased sensitivity of platelets to aggregation inducers, the increased reactivity of collagen under the influence of non-enzymatic glycosylation, the increase in the level of von Willebrand's factor and the presence of plasma factors inducing aggregation in plasma.

The decrease in activity of thrombosis-regulating factors occurs when the production of platelet prostacyclin, the release of tissue plasminogen activator and the fibrinolytic activity of plasma are lowered.

Associated with all these alterations are rheological modifications characterized by slowing of blood flow [42] and reduction of erythrocyte deformability and these are corrected by insulin. An increase of platelet aggregation [124] and a hyperadhesivity of red blood cells to endothelium are found, correlated with the severity of vascular complications [40,86].

Are these alterations implicated in the triggering of vascular lesions or are they the consequence of them? It is not easy to answer that question and only a chronological study of the alterations occurring in vascular walls and a study of primary haemostasis may answer it. The important point is to know if the hyperglycaemia responsible for the non-enzymatic glycosylation of vessel-wall proteins is the main factor causing vascular lesions, or whether the liberation of platelet factors is the initial cause of these vascular lesions. This debate is not concluded.

Alterations of circulating coagulation factors, platelet hyperactivity, endothelial activity and subendothelial non-enzymatic glycosylation together make capillaries and arteries prime sites for the production of diabetic angiopathy.

References

1. Abrahamsen AF (1968) Platelet survival studies in man with special reference to thrombosis and atherosclerosis. Scand J Haematol [Suppl] 3:7–53
2. Allen DN, Shroeder WA, Balog J (1958) Observation of the chromatographic heterogeneity of normal adult and foetal hemoglobin: a study of the effect of cristallization and chromatography on the heterogeneity and isoleucine content. J Am Chem Soc 80:1698–1734
3. Almer LD, Pandolfi M, Nilsson ID (1975) Diabetic retinopathy and the fibrinolytic system. Diabetes 24:529–534
4. Almer LD, Pandolfi M (1976) Fibrinolysis and diabetic retinopathy. Diabetes [Suppl 2] 25:807–810
5. Anderson JW (1975) Metabolic abnormalities contributing to diabetic complications. I. Glucose metabolism in insulin insensitive pathways. Am J Clin Nutr 28:273–280
6. Andreassen TT, Seyer-Hansenk, Bailey AJ (1981) Thermal stability, mechanical properties and reducible crosslinks of rat tail tendon in experimental diabetes. Biochim Biophys Acta 677:313–317
7. Arquilla ER, Weringer EJ, Nakajo M (1976) Wound healing: a model for the study of diabetic angiopathy. Diabetes [Suppl] 25:811–819
8. Artuson G, Garby L, Robert M et al. (1974) Oxygen affinity of whole blood in vivo and under standard conditions in subjects with diabetes mellitus. Scand J Clin Lab Invest 34:19–22
9. Bailey AJ (1981) The non enzymatic glycosylation of proteins. Horm Metab Res [Suppl] 11:90–94
10. Banerjee RN, Sahni Al, Kumere U et al. (1974) Antithrombin III deficiency in maturity onset diabetes mellitus and atherosclerosis. Thromb Diath Haemost 31:339–345
11. Beisswenger PJ, Spiro RG (1970) Human glomerular basement membrane: chemical alterations in diabetes mellitus. Science 168:596–598
12. Belfiore F, Lo Vecchio L, Napoli E (1972) Serum beta-glucuronidase activity in diabetic patients as related to vascular complications and degree of glucose metabolic disorder. Am J Med Sci 264:457–466
13. Belfiore F (1980) Diabetic microangiopathy. In: Belfiore F: Enzyme regulation and metabolic diseases. Karger, Basel, pp 532–539
14. Belfiore F, Napoli E, Lo Vecchio L (1982) Serum-N-acetyl-beta-glucosaminidase activity in diabetic patients. Diabetes 21:1168–1172
15. Bensoussan D, Le Levy-Toledano S, Passa P et al. (1976) Platelet hyperaggregation and increased plasma level of Von Willebrand factor in diabetic retinopathy. Diabetologica 11:307–312
16. Bookchin RM, Gallop PM (1968) Structure of hemoglobin A. Nature of the N-terminal beta chain blocking group. Biochim Biophys Res Commun 32:86–93
17. Borsey DQ, Prowse CV, Gray RS et al. (1984) Platelet and coagulation factors in proliferative diabetic retinopathy. J Clin Pathol 37:659–664
18. Borsey DQ, Dawes J, Fraser et al. (1980) Plasma beta-thromboglobulin in diabetes mellitus. Diabetologia 18:353–357
19. Brownlee M, Vlassara H, Cerami A (1983) Non enzymatic glycosylation reduces the susceptibility of fibrin to degradation by plasmin. Diabetes 32:630–637
20. Brownlee M, Pongor S, Cerami A (1983) Covalent attachment of soluble proteins by non enzymatically glycosylated collagen: role in the in situ formation of immune complexes. J Exp Med 158:1739–1744
21. Brownlee M, Vlassara H, Cerami A (1984) Non enzymatic glycosylation and the pathogenesis of diabetic complications. Ann Intern Med 101:527–537
22. Burrows AW, Chavin SI, Hockaday TDR (1978) Plasma thromboglobulin concentrations in diabetes

mellitus. Lancet I:235–237

23. Ceriello A, Dello-Russo P, Zuccoti C et al. (1983) The decreased antithrombin III activity in diabetes may be due to non enzymatic glycosylation. A preliminary report. Thromb Haemost 50:633–634

24. Chang K, Uitto J, Rowald EA et al. (1980) Increased collagen non-linkage in experimental diabetes. Reversal by beta-aminopropionitrile and D-Penicillamine. Diabetes 29:778, 781

25. Chase HP, Williams RL, Dupont J (1979) Increased prostaglandin synthesis in childhood diabetes mellitus. J Pediatr 94:185–189

26. Chen M, Velasco C, Camerini-Davalos Ra (1975) Renal glucose utilization in genetically diabetic microangiopathy. Experimentia 31:1130–1132

27. Chiou SH, Chylak LTT, Bunn HF (1981) Non-enzymatic glycosylation of bovine lens cristallins. J Biol Chem 256:5176–5180

28. Cohen MP, Vogt CA (1972) Evidence for enhanced basement membrane synthesis and lysine hydroxylation in renal glomerulus in experimental diabetes. Biochem Biophys Res Commun 49:1542–1546

29. Cohen MP, Vogt CA (1975) The effect of diabetes on renal lysine utilization. Horm Metab Res 7:338–342

30. Cohen MP, Khalifa A (1977) Effect of diabetes and insulin on rat renal glomerular protocollagen hydroxylase activities. Biochim Biophys Acta 496:88–94

31. Cohen MP, Wu VY (1981) Identification of specific amino acids in diabetic glomerular basement membrane collagen subjected to non enzymatic glycosylation in vivo. Biochem Biophys Res Commun 10:1549–1554

32. Coller BS, Frank R, Milton RC et al. (1978) Plasma cofactors of platelet function: correlation with diabetic retinopathy and hemoglobin A$_{1c}$. Ann Intern Med 88:311–316

33. Colwell JA, Nai RMG, Halushka PV et al. (1979) Platelet adhesion and aggregation in diabetes mellitus. Metabollism 28:394–400

34. Colwell JA, Halushka PV (1980) Platelet function in diabetes mellitus. Br J Haematol 44:521–526

35. Colwell JA, Lopez-Virella M, Haluscha PV (1981) Pathogenesis of atherosclerosis. Diabetes Care 4:121–133

36. Colwell JA, Von Zile O, Kilpatrick JM et al. (1981) Plasma factors and platelet aggregation in diabetes mellitus. Horm Metab Res [Suppl] 11:1–6

37. Colwell JA, Winocour PD, Haluska PV (1983) Do platelets have anything to do with diabetic microvascular disease? Diabetes [Suppl] 32:14–15

38. Colwell JA, Winocour PD, Lopez-Virella M et al. (1983) New concepts about the pathogenesis of atherosclerosis in diabetes mellitus. In: New perspectives in non insulin dependant diabetes mellitus and the role of glitizide in its treatment. Am J Med 75:67–80

39. Corbella E, Miragliota G, Masper R et al. (1979) Platelet aggregation and antithrombin III levels in diabetic children. Haemostasis 8:30–37

40. Dassin E, Najean Y, Poirier O et al. (1978) In vivo platelet kinetics in thirty one diabetic patients. Correlation with the degree of vascular impairment. Thromb Haemost 40:83–88

41. Di Mino G, Silver MJ, Cerbone AM et al. (1985) Increased binding of fibrinogen to platelets in diabetes: the role of prostaglandins and thromboxane. Blood 65:156–162

42. Dintenfass L (1979) Haemorheology of diabetes mellitus. Adv Microcirc 8:14–36

43. Eble AJ, Thorpe SR, Baynes JW (1983) Non enzymatic glucosylation and glucose dependant cross-linking of protein. J Biol Chem 258:9406–9412

44. Elder GE, Mayne EE, Daly JG et al. (1980) Antithrombin III activity and other coagulation changes in proliferative diabetic retinopathy. Haemostasis 9:288–296

45. Fischer RW, Winterhalter KA (1981) The carbohydrate moiety in hemoglobin A$_{1c}$ is present in the ring form. FEBS Lett 145:145–147

46. Fluckiger R, Winterhalter KH (1976) In vitro synthesis of hemoglobin A$_{1c}$. FEBS Lett 71:356–360

47. Fuller JH, Keew H, Jarrett RJ et al. (1979) Haemostatic variables associated with diabetes and its complications. Br J Med 2:964–966

48. Fushimi H, Tarui S (1976) Beta-glycosidases and diabetic microangiopathy. I. Decreases of beta-glucosidase activities in diabetic rat kidney. J Biochem (Tokyo) 79:265–270

49. Fushimi H, Tarui S (1976) Beta-glycosidases and diabetic microangiopathy. II. An insulin dependent isozyme of beta-N-acetylglucosaminidase. J Biochem (Tokyo) 79:271–275

50. Gabbay KH (1968) The sorbitol pathway: enzyme localisation and content in normal and diabetic nerve and cord. Diabetes 17:239–243

51. Gabbay KH (1975) Hyperglycemia, polyol metabolism, and the complications of diabetes mellitus. Ann Rev Med 26:521–536

52. Gandolfo GM, De Angelis A, Torresi MV (1980) Determination of antithrombin III activity by different methods in diabetic patients. Haemostasis 9:15–19

53. Gerrard JM, Stuart MJ, Rao GHR (1980) Alteration in the balance of prostaglandin and thromboxane synthesis in diabetic rats. J Lab Clin Med 95:950–958

54. Golub LM, Genwald RA, Zebrowski EJ et al. (1979) The effect of experimental diabetes on the molecular characteristics of soluble rat tail tendon collagen. Biochem Biophys Acta 585:481–487

55. Gonen B, Rubenstein AH (1978) Haemoglobin A$_{1c}$ and diabetes mellitus. Diabetologia 15:1–8

56. Grant ME, Heathcote JG, Orkins RW (1981) Current concepts of basement-membranes: structure and function. Bioscience Rep 1:819–842

57. Guitton JD, Le Pape A, Sizaret PY et al. (1981) Effects of in vitro N-glucosylation on type I collagen fibrillogenesis. Bioscience Rep 1:945–954

58. Halushka PV, Lurie D, Colwell JA (1977) Increased synthesis of prostaglandin E like material by platelets from patients with diabetes mellitus. N Engl J Med 297:1306–1310

59. Halushka PV, Roger RC, Loadholt LB et al. (1981) An increased platelet thromboxane synthesis in diabetes mellitus. J Lab Clin Med 97:87–96

60. Hamlin CR, Kohn RR, Luschin JH (1975) Apparent accelerated ageing of human collagen in diabetes mellitus. Diabetes 29:902–904

61. Harrison MF, Johnson M (1981) Vascular prostacyclin release and metabolic derangement in diabetes. Horm Metab Res [Suppl] 11:43–49

62. Higgins PJ, Bunn HF (1981) Kinetic analysis of the non enzymatic glycosylation of hemoglobin. J Biol Chem 256:5204–5208

63. Holmquist WR, Schroder WA (1966) A new N-terminal blocking group involving a Schiff base in hemoglobin A$_{1c}$. Biochemistry, 5:2489–2503

64. Hughes A, McVerry BA, Wilkinson L et al. (1983). Diabetes, a hypercoagulable state? Haemostatic vari-

ables in newly diagnosed type 2 diabetic patients. Acta Haematol 69: 254–259

65. Jones RL, Paradise C, Peterson CM (1981) Platelet survival in patients with diabetes mellitus. Diabetes 30:486–489

66. Kent MJC, Light ND, Bailey AJ (1985) Evidence for glucose-mediated covalent cross-linking of collagen after glycosylation in vitro. Biochem J 225:745–752

67. Khalifa A, Cohen MP (1975) Glomerular protocollagen lysyl hydroxylase activity in streptozotocin diabetes. Biochim Biophys Acta 386:332–339

68. Kilo C, Vogler N, Williamson JR (1972) Muscle capillary basement membrane changes related to aging and to diabetes mellitus. Diabetes 21:881–905

69. Kvirikko KI (1973) Urinary excretion of hydroxyproline in health and disease. Int Rev Connect Tissue Res 5:93–161

70. Kohner EM (1975) Dynamic changes in the microcirculation of diabetics as related to diabetic microangiopathy. Acta Med Scand [Suppl] 578:41–47

71. Kwaan HC, Colwell JA, Suwandela N (1972) Disseminated intravascular coagulation in diabetes mellitus with reference to the role of platelet aggregation. Diabetes 21:108–113

72. Lacord-Bonneau M, Muh JP, Gutman N et al. (1983) Glucosyl and galactosyl transferase activities of diabetic (db/db) and obese (ob/ob) mice kidneys. Int J Biochem 15:759–762

73. Lee LPK, Prasad A, Bolton KJ et al. (1977) Serum UDP galactose: glycoprotein galactosyltransferase in diabetics with microangiopathy. Clin Biochem 10:111–117

74. Le Pape A, Muh JP, Bailey AJ (1981) Characterization of N-glucosylated type I collagen in streptozotocin induced diabetes. Biochem J 197:405–412

75. Le Pape A, Guitton JD, Muh JP (1981) Modification of glomerular basement membrane cross-links in experimental diabetic rats. Biochem Biophys Res Commun 100:1214–1221

76. Le Pape A, Guitton JD, Gutman N et al. (1983) Non enzymatic glycosylation of collagen in diabetes: incidence on increased normal platelet aggregation. Haemostasis 13:36–41

77. Lien YH, Stern R, Fu JCC et al. (1984) Inhibition of collagen fibril formation in vitro and subsequent crosslinking by glucose. Science 225:1489–1491

78. Lorenzi M, Cagliero E, Toredo S (1985) Glucose toxicity for human endothelial cells in culture. Delayed replication, disturbed cell cycle and accelerated death. Diabetes 34:621–627

79. Lubec G, Pollak A (1980) Reduced susceptibility of non enzymatically glucosylated glomerular basement membrane to proteases in thickening of diabetic glomerular basement membranes due to reduced proteolytic degradation. Renal Physiol 3:4–8

80. Ludvigson MA, Sorenson RL (1980) Immunohistochemical localization of aldose reductase. II. Rat eye and kidney. Diabetes 29:450–459

81. Lufkin EG, Fass DN, O'Fallon WN et al. (1979) Increased von Willebrand factor in diabetes mellitus. Metabolism 28:62–66

82. Madia AM, Bozovski SJ, Kagan HM (1979) Changes in lung lysyloxydase activity in streptozotocin diabetes and in starvation. Biochem Biophys Acta 585:481–487

83. Mahadevan S, Dillard CJ, Tappel AL (1979) Degradation of polysaccharides, mucopolysaccharides, and glycoproteins by lysosomal glycosidases. Arch Biochem Biophys 129:525–533

84. Means GE, Chang MK (1982) Non enzymatic glycosylation of proteins. Structure and function changes. Diabetes [Suppl] 31:1–4

85. Miller K, Michael AF (1976) Immunopathology of renal extra cellular membranes in diabetes: specifity of tubular basement membrane immunofluorescence. Diabetes 25:701–708

86. Monnier VM, Vishwanath V, Franck KE et al. (1986) Relation between complications of type I diabetes mellitus and collagen-linked fluorescence. N Engl J Med 314:403–408

87. Mortensen HB, Christophersen C (1982) Glycosylation of human hemoglobin A, kinetics and mechanisms studied by isolelectric focusing. Biochim Biophys Acta 707:154–163

88. Nicoll GA, Gollapudi GM, Ramamurthy NS et al. (1981) Suppressed collagenolytic activity in polymorphonuclear leucocytes from diabetic humans. Experientia 37:315–317

89. Orkin RW, Gehron P, McGoodwin EB et al. (1977) A murine tumor producing a matrix of basement membrane. J Exp Med 145:204–220

90. Pandolfi M, Almer L, Holmberg L (1974) Increased von Willebrand-antihemophilic factor A in diabetic retinopathy. Acta Ophthalmol 52:823–828

91. Parthasarathy N, Spiro RG (1982) Effect of diabetes on the glycosaminoglycan component of the human glomerular basement membrane. Diabetes 31:738–741

92. Parving HH (1976) Increased microvascular permeability to plasma proteins in short and longterm juvenile diabetics. Diabetes [Suppl] 25:884–889

93. Perejda AJ, Uitto J (1982) Non enzymatic glycosylation of collagen and other proteins: relationship to development of diabetic complications. Coll Relat Res 2:81–88

94. Peterson HD, Gormsen J (1978) Platelet aggregation in diabetes mellitus. Acta Med Scand 203:125–130

95. Pongor S, Ulrich PC, Bencsath PA et al. (1984) Aging of proteins: isolation and identification of a fluorescent chromophore from the reaction of polypeptides with glucose. Proc Natl Acad Sci USA 81:2684–2688

96. Rahbar S, Paulsen E, Ranney HM (1986) Studies of hemoglobin in patients with diabetes mellitus. Diabetes 18:332–335

97. Rak K, Beck P Udvardy M et al. (1983) Plasma level of beta-thromboglobulin and factor VIII related antigen in diabetic children and adults. Thromb Res 29:155–162

98. Reddi AS (1978) Diabetic microangiopathy. I. Current status of the glomerular basement membrane. Metabolism 27:107–124

99. Risteli J, Koivisto VA, Akerblom HK et al. (1976) Intracellular enzymes of collagen biosynthesis in rat kidney in streptozotocin diabetes. Diabetes 25:1066–1070

100. Robins SP, Bailey AJ (1972) Age related changes in collagen: the identification of reductible lysine-carbohydrate condensation products. Biochim Biophys Res Commun 48:76–84

101. Rogers SP, Larkins RG (1981) Production of 6-oxo prostaglandin F 1 alpha by rat aorta; influence of diabetes, insulin treatment and cationic deprivation. Diabetes 30:935–939

102. Rohrbach DH, Hassell JR, Kleinman HK et al. (1982) Alterations in the basement membrane (heparan sulfate) proteoglycan in diabetic. Diabetes 31:185–188

103. Rosenberg H, Modrak JP, Hassing JM et al. (1979) Glycosylated collagen. Biochem Biophys Res Commun 91:498–501

104. Ruoslahti E, Engvall E, Haymann EG (1981) Fibronectin: current concepts of its structure and function. Coll Relat Res 1:95–128

105. Sagel J, Colwell JA, Crook L et al. (1975) Increased platelet aggregation in early diabetes mellitus. Ann Int Med 82:733–738

106. Sato T, Munakata H, Yoshinaga K et al. (1975) Comparison of the chemical composition of glomerular and tubular basement membranes obtained from human kidneys of diabetics and non diabetics. Clin Chim Acta 61:145–150

107. Schnider SL, Kohn RR (1981) Effects of age and diabetes mellitus on the solubility and non enzymatic glucosylation of humans skin collagen. J Clin Invest 67:1630–1635

108. Silberbauer K, Schernthaner G, Sinzinger H et al. (1981) Platelet aggregation and reversible platelet aggregates in type I diabetes staged by retinal fluorescein angiography. Atherosclerosis 40:81–90

109. Smokovitis A, Auerswald W, Muller K et al. (1982) Plasminogen activator activity in the aorta of strains of rats with genetically determined different pattern of lipids in the blood: induction of increased activity by streptozotocin. Haemostasis 12:256–261

110. Sowers JR, Tuck ML, Sowers DK (1980) Plasma antithrombin III and thrombin generation time: correlation with haemoglobin A1 and fasting serum glucose in young diabetic women. Diabetes Care 3:655–658

111. Spiro RG (1969) Glycoproteins: their biochemistry, biology and role in human disease. N Engl J Med 281:991–1001 and 1043–1056

112. Spiro RG (1970) Chemistry and metabolism of the basement membrane. In: Ellenbert M, Rifkin H (eds) Diabetes mellitus: theory and practice. McGraw Hill, New York, pp 210–214

113. Spiro RG, Spiro MJ (1971) Effect of diabetes on the biosynthesis of the renal glomerular basement membrane. Studies on the glucosyltransferase. Diabetes 20:641–648

114. Spiro RG, Spiro MJ (1971) Studies on the biosynthesis of the hydroxylysine-linked disaccharide unit of basement membrane and collagens. III. Tissue and subcellular distribution of the glycosyltransferases and the effect of various conditions on the enzyme levels. J Biol Chem 246:4919–4925

115. Sternberg M, Andre J (1982) Glucose inhibition of the alpha glucosidase specific for basement membrane and collagen disaccharide units. FEBS Lett 139:53–56

116. Stuart MJ, Elrad H, Graeber JC et al. (1979) Increased synthesis of prostaglandin endoperoxide and platelet hyperfunction in infants of mothers with diabetes mellitus. J Lab Clin Med 94:12–17

117. Stuart MJ, Sunderji SG, Allen JB (1981) Decreased prostacyclin production in the infant of diabetic mother. J Lab Clin Med 98:412–416

118. Tarsio JF, Wigness B, Rhode TD et al. (1985) Non enzymatic glycation of fibronectin and alterations in the molecular association of cell matrix and basement membrane components in diabetes mellitus. Diabetes 34:477–484

119. Tattersall RB, Pyke DA, Ranney JH et al. (1975) Hemoglobin components in diabetes mellitus: studies in identical twins. N Engl J Med 293:1171–1173

120. Tanzer ML, Fairweather R, Gallop PM (1972) Collagen cross links: isolation of reduced N hexosyl-hydroxylysine from borohydride reduced calf skin insoluble collagen. Arch Biochem Biophys 151:137–141

121. Trueb B, Fluckiger R, Winterhalter KH (1984) Non enzymatic glycosylation of basement membrane collagen in diabetes mellitus. Coll Relat Res 4:239–251

122. Uitto J, Grant GA, Perejda AJ et al. (1980) Glycosylation of human glomerular basement membrane collagen (GBMC): increased non enzymatic glucosylation in diabetes. Fed Proc 39:1722

123. Uitto J, Perejda AJ, Grant GA et al. Glycosylation of human glomerular basement membrane collagen: increased content of hexose in ketoamine linkage and unaltered hydroxylysine-O-linked glycosides in patients with diabetes. Connect Tissue Res 10:287–296

124. Vague P, Juhan I (1983) Red cell deformability, platelet aggregation and insulin action. Diabetes [Suppl] 32:88–91

125. Van Zile J, Kilpatrick M, Laimins M et al. (1981) Platelet aggregation and release of ATP after incubation with soluble immune complexes purified from the serum of diabetic patients. Diabetes 30:575–579

126. Vater CA, Harris ED, Siegel RC (1979) Native cross-links in collagen fibrils induce resistance to human synovial collagenase. Biochem J 181:639–645

127. Velasco C, Opperman W, Marine N et al. (1974) Effect of genetic diabetes on kidney glucosyltransferase. Horm Metab Res 6:427

128. Wautier JL, Wautier MP, Pintigny D et al. (1983) Factors involved in cell adhesion to vascular endothelium. Blood Cells 9:221–234

129. Yamada KM (1983) Cell surface interactions with extracellular materials. Ann Rev Biochem 52:761–799

130. Ylikorkala O, Kaila J, Viinikka L (1981) Prostacyclin and thromboxane in diabetes. Br Med J 283:1148–1150

131. Ziboh VA, Maruta H, Lord J et al. (1979) Increased biosynthesis of thromboxane A2 by diabetic platelets. Eur J Clin Invest 9:223–226

132. Zucker ML, Comperts ED, Russel D (1979) Antithrombin functional activity after saturated and unsaturated fatty meals and fasting in normal subjects and some diseases states. Thromb Res 15:37–48

133. Zweifach BW (1980) Integrity of vascular endothelium. Acta Microcirc 9:206–225

Section V

Inflammatory Diseases of the Arterial Wall

Chapter 26

Immunological Processes in the Vascular Wall

M. D. Kazatchkine, P. S. Seifert and U. E. Nydegger

Many mechanisms may be involved in the mediation of immunologically induced vascular injury. These mechanisms can be classified as humoral or cellular (Table 26.1). Humoral mechanisms involve deposition or in situ formation of immune complexes and the intravascular activation of complement. Cellular mechanisms lead to granulomatous reactions. This chapter will review the pathophysiology of vascular injury mediated by immune complexes, complement and specific and non-specific cellular effector mechanisms. Some of the functional and immunological characteristics of endothelial cells will also be considered, since the clinical features of inflammatory vascular injury depend to a large extent on the absence (as in post-capillary venules and glomerular capillaries) or presence (as in arteries) of an intact non-fenestrated continuous layer of endothelial cells.

Effector Mechanisms in Immunologically Mediated Vascular Damage

Immune Complexes and Antibodies to Vascular Structures

Immune complexes associated with vasculitis involve antigens of exogenous origin (e.g. bacterial, viral, parasitic, fungal and tumour antigens and those induced by drugs or food allergens) or an endogenous auto-antigen, indicating the essential role of the Fc portion of complexed antibody in the mediation of tissue injury. Complexed antigens modulate effector

Table 26.1. Immunological mechanisms of vascular damage

Humoral immune mechanisms
  Deposition of circulating immune complexes
  In situ formation of immune complexes
    Structural vascular antigens
    Non-native "planted" antigens bound to vascular structures
  Intravascular activation of complement
Granulomatous vascular lesions
  Direct lymphocyte-mediated cytotoxicity (?)

functions of antibodies in the immune complex by expressing variable epitopic density. These epitopes are recognized by antibodies of various affinities and by other microenvironmental characteristics such as their electric charge and size. Immune complexes may be: (a) soluble, circulating immune complexes; (b) soluble extravascular immune complexes in inflammatory exudates; (c) deposited or in situ tissue-bound immune complexes.

The transient occurrence of circulating immune complexes must be regarded as a physiological phenomenon. Antigens are constantly removed and destroyed following formation of complexes with specific antibodies and clearance of the complexes by cells of the reticuloendothelial system. Complexed antigens may also be transported to germinal centres of lymph nodes where they become trapped on the surface of follicular dendritic cells, a process that allows the antigen to escape from its catabolic fate and to be presented to lymphocytes for immune recognition and generation of specific memory cells.

When circulating or tissue-bound immune complexes are formed in response to exogenous or endogenous antigens, a dynamic relationship exists between antigen and antibody in free and complexed form: $Ag + Ab \rightleftharpoons [AgAb]$. The status of this equilibrium is dependent on many factors including the relative concentrations of antigen and antibody available for immune complex formation, antigen valence, the class and avidity of specific antibodies for the antigen, the interaction of complexes with anti-complexed antibodies and complement and the rate at which immune complexes are cleared.

Prolonged persistence of antigens following infection with weakly cytopathogenic viruses (e.g. hepatitis B virus) or entrapment of bacterial antigens in chronic inflammatory sites results in repeated stimulation of the immune response and chronic formation of immune complexes. Antigen excess in immune complexes results in saturation of antibody sites, so that cross-linking is prevented and only "simple" immune complexes are formed regardless of the valence of antigen. In contrast, maximal cross-linking occurs in immune complexes formed at equivalence where the valence of the antigen becomes a major determinant of the size of the complexes. At equivalence and in antibody excess, multivalent antigens form large soluble immune complexes or tend to precipitate, whereas oligovalent antigens will form small soluble immune complexes. Once formed, antigen–antibody complexes become the target of numerous humoral and cellular recognition systems which profoundly influence the composition and the fate of the nascent immune complexes (Fig. 26.1):

1. The interaction of immune complexes with complement leads to ionic fixation of Clq to the $C\gamma_2$ domain of IgG or the $C\mu_4$ domain of IgM, to covalent binding of C4b to the Fd region of the heavy chain and to covalent binding of C3b to amino acid residues of Fd (opsonization) (see below);

2. Fibronectin may bind to immune complexes via the globular heads or the collagen-like domains of Clq.

3. Rheumatoid factors and anti-idiotypic antibodies may bind to complexed antibody molecules; rheumatoid factors react with the Fc portion of complexed IgG, whereas anti-idiotypic antibodies recognize antigenic determinants in the variable region of antibodies. As seen in animal models of polyclonal B cell activation and even in normal human sera, anti-idiotypic antibodies may form idiotype–anti-idiotype antigen–antibody complexes with free antibody molecules. Each type of protein which binds to immune complexes significantly alters the size of the complexes and thus changes their

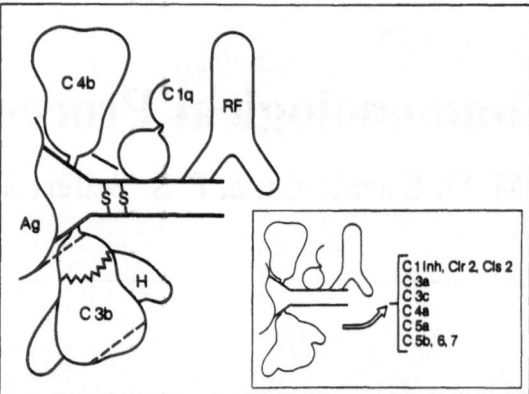

Fig. 26.1. Schematic representation of an antigen–antibody complex. The complexed antibody carries a number of complement proteins. Clq is bound to the $C\gamma_2$ domain of the heavy chain. C4b and C3b are covalently bound to the Fab; IgG-rheumatoid factor is also shown. *Inset* at lower right corner of figure: interaction with complement leads to formation of a number of complement activation products.

complement-activating properties, mechanism and rate of clearance from the circulation and ultimate capacity to produce an inflammatory reaction in tissues [75].

Once immune complexes have interacted with complement and have been opsonized, their clearance from the circulation depends on their interaction with C3b receptors (CR1) on erythrocytes and on the functional state of the reticuloendothelial system (Table 26.2). Erythrocyte CR1 is the major source of receptors in whole blood. CR1 endows the cells with the capacity for repeated uptake and release of C3b-bearing immune complexes which allows the erythrocytes to transport the complexes, deliver them to the sites of removal and, possibly, prevent their interaction with other structures such as vascular endothelium [33]. In vivo studies in monkeys demonstrated that injected preformed immune complexes bind to erythrocytes and that erythrocyte-bound complexes are delivered through the portal circulation to the liver, where the complexes are stripped from the cells and transferred to Kupffer cells [19]. An impaired clearance of immune complexes in the liver has been found in hypocomplementaemic animals in whom complexes appear widely distributed throughout the body.

Transfer of complexes to macrophages lining organs of the reticuloendothelial system is possible because of the synergistic function of CR1 and Fc receptors on these cells and because CR1 serves as a co-factor for cleavage of immune-complex-bound C3b into C3bi and C3dg. The

Table 26.2. Systems involved in removal of antigen and immune complexes and potential consequences of their failure

System	Function	Consequences of failure
Lymphocytes	Production of specific antibodies	Lack of efficient removal of antigen
Classical complement pathway	Binding of C3b and C4b to immune complexes; inhibition of immune complex precipitation	Defective clearance of immune complexes. Unprocessed complexes may form or be deposited in tissues
Alternative complement pathway	Binding of C3b to immune complexes; solubilization of pre-formed complexes	Defective clearance of immune complexes. Large unprocessed immune complexes may form or be deposited in tissues
Total mass of erythrocytes bearing C3b receptors (CR1)	Transport of immune complexes in the circulation	Defective clearance of immune complexes
Reticuloendothelial system	Removal of immune complexes	Accumulation of immune complexes in the circulation favours deposition of complexes, vasculitis and tissues lesions

latter C3 fragments can bind to CR3 and possibly to another distinct type of C3 receptor on phagocytes (CR4) (Fig. 26.2). The number of CR1 on erythrocytes is a phenotypic characteristic of each individual, and is decreased by approximately 40% in patients with systemic lupus erythematosus (SLE) because of both genetic and acquired factors, including complement activation and anti-receptor auto-antibodies [115]. Expression of CR1 is also decreased on neutrophils and B lymphocytes of patients with SLE and is lost on glomerular podocytes of patients with type IV SLE nephritis [58]. Expression of CR1 on erythrocytes is also defective in patients with acquired immunodeficiency syndrome (AIDS) but is not altered in patients

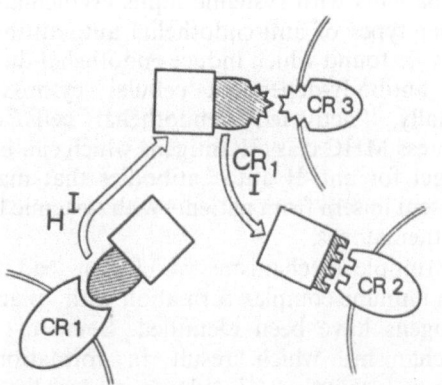

Fig. 26.2. Complement receptors for C3 fragments interact with immune complexes carrying C3b (*lower left*), C3bi (*upper drawing*) and C3dg (*lower right*). Co-factors for C3b modification on immune complexes are drawn near the arrows: I and H transform C3b into C3bi; CR1 and I transform C3bi into C3dg.

with certain immune complex diseases and hypocomplementaemia such as rheumatoid arthritis. It is likely that a reduced number of CR1 on erythrocytes limits the capacity of the cells to transport and clear immune complexes from the circulation. Decreased clearance of immune complexes may also result from defective function of cells belonging to the reticuloendothelial system (RES). Function of the RES can be investigated in vivo by studying the clearance of autologous erythrocytes coated with anti-rhesus IgG (Fc receptor function of the spleen) or with anti-rhesus IgM and complement (C3 receptor function of the liver) [37]. Using these assays, clearance of immune complexes was found to be decreased in patients with SLE and Sjögren's syndrome (Fc receptor function), primary biliary cirrhosis (C3 receptor function) and other diseases, and to be improved by plasma exchange therapy in several conditions.

The route by which macromolecular and complexed IgA is cleared in humans is unknown. In rodents, a mechanism independent of reticuloendothelial system macrophages has been described which involves the binding of IgA to a secretory component expressed on the surface of hepatocytes and epithelial cells of the bile canaliculi and secretion of complexed IgA as secretory IgA in the bile. IgA-containing immune complexes which do not activate the classical pathway may bind relatively few C3b molecules and thus poorly interact with erythrocyte CR1 or CR3 on liver cells. Ineffective clearance of macromolecular IgA in the presence of an intact complement system may be responsible for the accumulation of IgA aggregates in the mesangium and for systemic IgA deposits occurring in a number of human systemic vasculitides (Table 26.3).

Table 26.3. Diseases involving macromolecular IgA

Schönlein–Henoch purpura
Other systemic vasculitides with vascular IgA deposits
Primary IgA nephropathy
Liver disease
Dermatitis herpetiformis
Some cases of systemic lupus erythematosus with glomerular IgA deposits
Some cases of rheumatoid arthritis with vascular IgA deposits
Behcet's disease

Besides the systemic factors that have just been discussed, a number of local factors will determine whether circulating immune complexes will eventually deposit in vessels and tissues. Factors which determine trapping of circulating immune complexes in glomerular capillaries have been extensively studied in animal models [21]. Some of these factors are listed in Table 26.4. Such determinants are probably of critical importance for immune complex deposition since, although defective clearance increases availability of immune complexes for deposition, patients with high levels of circulating complexes do not necessarily have systemic deposits in vessels and tissues. The latter observation, as well as information obtained from experimental models, has suggested alternative ways in which insoluble immune complexes may form in tissues. These include in situ immune complex formation following: (a) interaction of an inhaled or subcutaneously injected antigen with circulating antibody in a pre-sensitized host (Arthus reaction); (b) interaction of circulating antibody with intrinsic vascular structures [e.g. anti-glomerular basement membrane (GBM) disease]; and (c) interaction of circulating antibody with nonnative "planted" antigens in vascular structures.

Table 26.4. Factors which regulate deposition of circulating immune complexes in glomeruli (mesangial and/or subendothelial)

Modification of circulating immune complexes: complement, anti-antibodies, erythrocyte CR1, function of phagocytic cells of the reticuloendothelial system
Haemodynamic factors: renal blood flow, haemodynamic changes in glomeruli (e.g. changes in hydrostatic pressure and filtration fraction)
Mesangial function
Characteristics of immune complexes: size, antigen/antibody ratio, charge

The latter mechanism is considered operative predominantly in the formation of subepithelial immune complexes in glomeruli.

The Arthus reaction, resulting in formation of insoluble immune complexes within the vessel wall at the site of antigen penetration, provides a model for hypersensitivity pneumonitis in "farmer's lung" disease and for certain cutaneous necrotizing vasculitides [16]. Complement activation by large immune aggregates at the site of their formation results in local deposition of C3b, release of chemotactic C5a, influx of activated granulocytes and leucocyte-mediated tissue injury.

Anti-GBM disease represents an example of an immune complex disease in which an antibody reacts to a native (fixed) antigen on the capillary wall [114]. A direct role for antibodies targetted against vascular structures has also been suggested in the pathogenesis of vascular rejection in patients with renal transplants and of vasculitis in patients with systemic lupus erythematosus. Kidney transplant recipients may have circulating anti-endothelial cell antibodies that react with umbilical cord vein endothelial cells and endothelium of peritubular capillaries, and which are not directed against class I or ABO blood group antigens [76]. IgG anti-endothelial antibodies have been found in sera from patients with active systemic lupus erythematosus. The sera may also contain IgG complexes that bind to endothelial cells. In vitro experiments demonstrated that binding of these antibodies to endothelial antigens triggers complement activation, induces disruption of the monolayer in endothelial cell cultures and secretion of prostacyclin and causes platelet adherence. In patients with systemic lupus erythematosus, other types of anti-endothelial auto-antibodies may be found which induce endothelial damage by antibody-dependent cellular cytotoxicity. Finally, "activated" endothelial cells often express MHC class II antigens which can be the target for anti-II auto-antibodies that may be present in sera from patients with systemic lupus erythematosus.

Multiple mechanisms which can lead to in situ immune complex formation with "planted" antigens have been identified. Some of these mechanisms which result in formation of immune complexes in glomeruli are listed in Table 26.5 [21]. Immune-complex-mediated damage following in situ complex formation occurs in a way similar to that which occurs in some drug-induced (e.g. penicillin) immune cytopenias. As mentioned above, in situ immune

Table 26.5. In situ formation of immune complexes with non-native "planted" antigens in glomeruli

Subepithelial immune complex formation
  Charge-dependent mechanisms (e.g. cationic antigens or antibodies)
  Direct binding of antigen or of IgG to the capillary wall by undefined charge-independent mechanisms
Subendothelial immune complex formation
  Charge-dependent mechanisms (e.g. large cationic antigens)
  Affinity of antigen for the glomerular basement membrane (e.g. DNA)
Mesangial uptake of macromolecular antigens
Secondary rearrangement of pre-formed complexes (independent of their localization)
  By interaction with rheumatoid factors, anti-idiotypic antibodies and complement proteins

complex formation is likely to be of particular relevance in the pathogenesis of membranous nephritis and probably the predominant pathway for immune complex formation in human vasculitides with immune deposits. Analysis of immune complex and complement deposits in membranous nephritis revealed that local complement activation by immune complexes results in little C3 deposition and in no activation of the terminal complement sequence during the early phase of this disease and that complement activation is insufficient to prevent immune complex formation. A possible explanation for the latter phenomenon may be that classical pathway activation is more efficient at preventing immediate aggregation of immunoglobulins, which occurs in the fluid phase by Fc–Fc interactions, than lattice formation, which follows "true" antigen–antibody reactions [94]. In situ immune complex formation may be

relevant to the pathogenesis of some forms of vasculitis and glomerulonephritis in patients with systemic lupus erythematosus. For example, DNA may bind directly to the glomerular basement membrane by charge-independent mechanisms and initiate glomerular immune complex formation with trapped circulating anti-DNA antibodies [51]. Animal experiments have suggested that this mechanism may be operative in vivo.

Whatever the primary mechanism of immune complex formation in a tissue may be, tissue-bound complexes can be secondarily rearranged and processed in a similar fashion to circulating immune complexes when they become accessible to plasma proteins [75]. In this respect, deposited or in situ formed immune complexes may themselves become the target for in situ secondary immune complex formation by interaction of free antigen valences with circulating antibodies, free antibody valences with antigen or of immune complexes with anti-antibodies, e.g. rheumatoid factors and/or anti-idiotypes. As mentioned in the next section, activation of the complement system may at least partly resolubilize tissue-bound complexes at a time when it also initiates a local inflammatory reaction.

Complement

The complement system comprises 19 plasma proteins which represent 5% of the plasma protein content. Upon activation, complement components interact within distinct and finely regulated functional units (Fig. 26.3) [60]. The

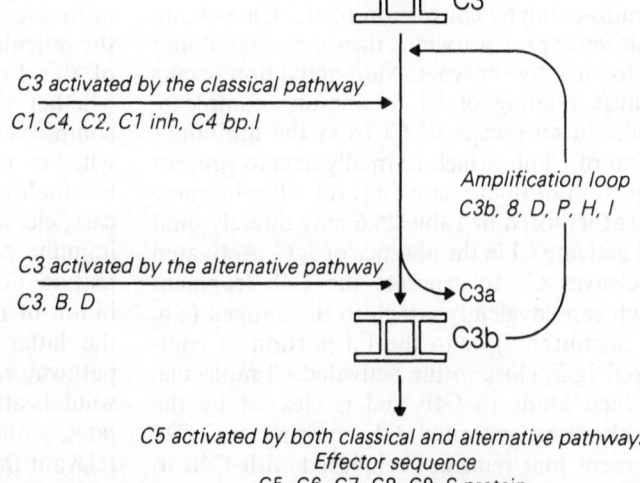

Fig. 26.3. Complement activation. Cleavage of C3 may result from the assembly of the classical pathway C3 convertase, the "priming" alternative pathway convertase or the amplification C3 convertase. Generation of C3b triggers formation of C5 convertases and assembly of the C5b-9 effector sequence.

C3

C3 activated by the classical pathway
C1, C4, C2, C1 inh, C4 bp.I

Amplification loop
C3b, B, D, P, H, I

C3 activated by the alternative pathway
C3, B, D

C3a

C3b

C5 activated by both classical and alternative pathways
Effector sequence
C5, C6, C7, C8, C9, S protein

classical and the alternative pathways of activation both form specific enzymatic complexes termed C3 convertases that cleave C3 and generate the major cleavage fragment C3b. A single amplification pathway exists that augments C3 cleavage once initial C3b has been generated and there is a common effector sequence comprised of C3b and components C5 to C9 that generates the opsonizing, vasoactive, leucocyte-attracting, immune regulatory and cytolytic activities of complement. The classical pathway of activation comprises the C1 complex formed by one molecule of C1q, two molecules of C1r and two molecules of C1s, the component proteins C2 and C4 and the regulatory proteins C1 inhibitor (C1inh), C4-binding protein (C4BP) and I. The alternative and amplification pathways involve the components C3, B and D, and three regulatory proteins P, H and I. When not engaged in the assembly of a membrane-bound C5b-9 complex, components C5–C9 bind the serum S protein to form a fluid-phase cytolytically inactive SC5b-9 complex. Phagocytic cells and lymphocytes express specific receptors for some complement proteins and their fragments, which upon interaction with their ligands elicit the various cellular responses that result in the inflammatory process [33].

The interaction of immune complexes with complement occurs in the fluid phase, on cell surfaces and in the interstitial spaces. It may proceed via either the classical or alternative pathways. Classical pathway activation is initiated by binding of the C1q subunit of C1 to at least two closely positioned IgG molecules or to complexed IgM. The role of antigen in the antigen–antibody complex is to aggregate IgG molecules in order to facilitate multivalent binding of C1q. The binding of C1q is followed by auto-catalytic conversion of the C1r subunit to an active protease which then converts subunit C1s to an active enzyme. Auto-activation occurs because binding of C1 to immune complexes results in an escape of C1 from the inhibitory action of C1inh, which normally acts to prevent activation of fluid phase C1 [118]. Non-immune activators listed in Table 26.6 may directly bind and activate C1 in the absence of IgG. Activated C1 cleaves C4 to generate the C4b fragment which can covalently attach to the antigen (e.g. cell membranes) or to the Fd portion of complexed IgG, close to the activated C1 molecule. C2 then binds to C4b and is cleaved by the neighbouring activated C1, yielding the C2a fragment that remains complexed with C4b to form the bimolecular C4b,2a classical pathway

Table 26.6. C1 activators

Fc portion of complexed immunoglobulins
    IgG1, IgG2, IgG3
    IgM
Non-immune activators
    RNA viruses
    Some strains of Salmonella, *E. coli* and *Neisseria*
    Heparin–protamine complexes
    Heart mitochondrial membranes
    Monosodium urate crystals

C3 convertase. The next step in complement activation is the cleavage of C3 by the immune-complex-bound C3 convertase, which results in release of the anaphylatoxin C3a and generation of C3b. Like C4b, nascent C3b transiently expresses a reactive site that allows formation of a covalent bond with hydroxyl groups or amino groups on immune complexes and/or on bystander surfaces [65].

Binding of multiple C3b molecules to the activating immune complex changes the specificity of the C3 convertase to a C5 convertase and triggers activation of the terminal sequence C5–C9. Function of the classical pathway C3 and C5 convertase is controlled by the regulatory proteins C4BP and I. Complement activation by immunoglobulin in immune complexes may also occur through the alternative pathway as evidenced by the capacity of aggregated immunoglobulins, including IgA, to activate this pathway, and by the enhancing role of specific antibodies on alternative pathway activation by bacteria, parasites, fungi and virus-infected cells.

A critical function of complement with regard to the pathogenesis of vasculitis is its ability to prevent immune complex precipitation and to prepare optimally solubilized complexes for their intravascular transport to the sites of removal in the reticuloendothelial system [94]. In the study of this function in vitro, it is of importance whether the immune complexes interact with complement at the time of their formation or whether complement is added to pre-formed insoluble (precipitated) complexes. In the former case, classical pathway components bind to the immune complexes simultaneous to their formation and prevent their precipitation (inhibition of immune precipitation) (Fig. 26.4). In the latter case, activation of the alternative pathway results in disruption of the lattice and solubilization of pre-formed immune precipitates, a phenomenon which may be particularly relevant for the local rearrangement of immune complexes deposited in tissues. Complement-

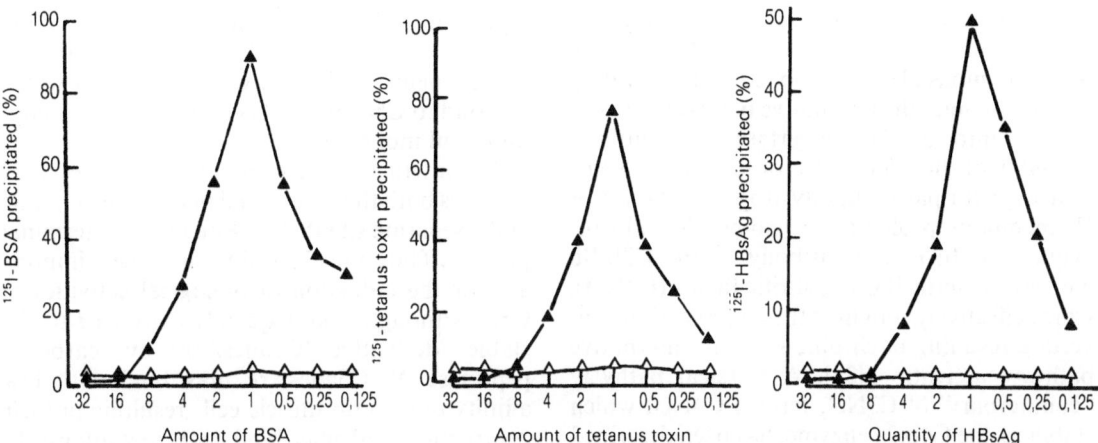

Fig. 26.4. Formation of insoluble immune complexes at equivalence is prevented in these experiments when antigen and antibody are reacted in the presence of fresh human serum rather than in heat-inactivated, complement-depleted serum. The abscissa shows increasing amounts of antigen [bovine serum albumin (BSA), tetanus toxoid and hepatitis B virus surface antigen] which were added to constant amounts of specific antibodies raised in rabbits. The ordinate indicates the percentage of radiolabelled antigen precipitated.

dependent inhibition of immune complex precipitation is more effective in slight antibody excess, most probably because binding of C4b and a large number of molecules of C3b to IgG prevents Fc–Fc interactions. The Fc–Fc interactions are responsible for the immediate aggregation of complexes that precede formation of a stable lattice. Failure of classical-pathway-dependent solubilization and opsonization of immune complexes probably accounts for the frequent association of C4, C2 and C1 deficiencies with immune complex diseases and vasculitis (Table 26.7) [92,93].

Activation of the alternative pathway represents a natural system of defence against infection as it may occur in the absence of specific antibodies [59]. It is initiated by the binding to surfaces of C3b molecules that have been generated in plasma. C3b molecules that are bound

Table 26.7. Deficiencies of classical pathway complement components associated with immune complex diseases (adapted from Schifferli and Peters [92])

Deficient component	Disease	% of patients with immune disease among homozygous deficient patients
C1q	Glomerulonephritis, SLE-like syndrome	94
C1r	Glomerulonephritis	75
C1s	Glomerulonephritis, SLE	75
C4	SLE	87
C2	SLE	57

Table 26.8. Activators of the alternative pathway

Many strains of gram-negative and gram-positive bacteria
LPS from gram-negative bacteria
Fungi and yeast cell walls
Some viruses; virus-infected cells
Parasites
Some tumour cells
Cholesterol crystals
Aggregated human IgG, IgA and IgE

to "activating surfaces" (Table 26.8) interact with factors B and D to form the amplification C3 convertase which cleaves C3 and deposits C3b molecules on the target surface. Bound C3b is rapidly inactivated by membrane inhibitors, i.e. "decay accelerating factor" (DAF) and the C3b receptor (CR1), and by the plasma control proteins H and I when it is bound to membranes of autologous cells or to "non-activating" surfaces of the alternative pathway. DAF, an intrinsic membrane glycoprotein of molecular weight 70 000 [74] is present on endothelial cells, erythrocytes and leucocytes and probably represents the major protective factor on these cells for resisting complement activation and lytic damage [1].

Two membrane-associated families of molecules, sialic acid [31] and sulphated mucopolysaccharides (e.g. heparan sulphate) [57] modulate the ability of cell surfaces to activate the alternative pathway by enhancing the interaction between surface-bound C3b and the regulatory protein H. On "activating" surfaces, the alternative pathway C3 convertase C3b, Bb may be stabilized by properdin (P) which delays the spontaneous decay of the enzyme. Binding of multiple molecules of C3b to the activating surface changes the alternative pathway C3 convertase into a C5 convertase and initiates assembly of the C5b-9 effector sequences. Some patients with partial lipodystrophy or type I or II membranoproliferative glomerulonephritis, acquire an IgG auto-antibody to the C3b,Bb convertase, termed C3 nephritic factor (C3Nef), which effectively stabilizes the amplification convertase resulting in chronic systemic alternative pathway activation [24]. A classical pathway "counterpart" of C3Nef, termed C4Nef, which stabilizes the C4b,2a enzyme, has been described in certain patients with post-streptococcal glomerulonephritis and in patients with SLE.

Cleavage of C5 by the classical and alternative pathway C5 convertases releases C5a and generates C5b. Nascent C5b can form a stable bimolecular C5b,6 complex with C6. The C5b,6 complex can reversibly bind to cell membranes. With the addition of C7–C9, it can lyse unsensitized bystander cells, a process known as reactive lysis. Several protective mechanisms contribute to the inhibition of autologous cell lysis by complement. These include: the membrane-associated factors which control function of C3 and C5 convertases on autologous cell surfaces (i.e. CR1 and DAF); plasma high-density lipoproteins which inhibit insertion of C9 in cell membranes bearing C5b-8 complexes [85]; and a recently identified species-specific protein which prevents binding of C9 to homologous C5b-8 [45]. Furthermore, nucleated cells are relatively resistant to membrane attack by complement because of cell membrane renewal and a shorter life span of the complement lytic channels in membranes of nucleated cells [82]. Thus, the finding of C5b-9 neo-antigens in a pathological tissue does not necessarily mean that locally formed m(C5b-9) complexes are directly causing tissue lesions and cell death [7,22,49]. It is important to note that C5b-9 complexes also exert several biological effects at sublytic concentrations, such as stimulation of arachidonic acid metabolism and release of leukotrienes from leucocytes [97].

The anaphylatoxins C3a, C4a and C5a generated after C3/C5 convertase formation [50] do not remain at the site of activation but are released into plasma and extracellular fluids. C5a, the most potent anaphylatoxin, induces a variety of biological effects which are important for the initiation of an inflammatory reaction, by reacting with specific receptors on leucocytes.

Approximately $1–3 \times 10^5$ high-affinity receptors $(0.3–1 \times 10^9 \text{ M}^{-1})$ have been demonstrated on human neutrophils [12]. Other cell types which respond to C5a are mast cells, basophils, eosinophils and monocytes.

The interaction between C5a and its receptor involves both the C-terminal part of the molecule and a separate site that is close to the N-terminal portion. Thus, the C-terminal arginine is important for the induction of biological activities of C5a. Although C5a is quickly converted to the stable derivative C5adesArg by carboxypeptidase N, C5adesArg may bind with low affinity to smooth-muscle cells resulting in their contraction and may bind with low affinity to leucocytes to induce chemotaxis, enhanced adhesiveness and lysosomal degranulation. C5a is chemotactic at nanomolar concentrations, causing directed migration of neutrophils, eosinophils, basophils and monocytes against the gradient of concentration which occurs when the peptide diffuses away from the site of complement activation [34] (Table 26.9). Neutrophil stimulation by C5a leads to the respiratory burst and increased membrane expression of C3b receptors (CR1) which indirectly facilitates phagocytosis [32]. C5a enhances neutrophil turnover of arachidonic acid and stimulates cellular production of the inflammatory mediators 5-HETE and leukotriene B4. Both of these lipids are known to be neutrophil chemotactic factors themselves [100]. C5a and C5adesArg also enhance the adhesiveness of neutrophils to foreign surfaces and endothelial cells and reversibly aggregate neutrophils in vitro and in vivo [23,108]. The latter effect is secondary to enhanced expression of the adhesion-promoting molecule Mol [2]. It mediates intravascular aggregation of leucocytes and leucopenia in patients undergoing haemodialysis with complement-activating membranes or car-

Table 26.9. Effects of binding of C5a and C5adesArg to C5a receptors on neutrophils

Chemotaxis
Enhanced expression of C3b receptors (CR1) and enhanced phagocytic capacity
Enhanced expression of Mol and enhanced adhesiveness; reversible aggregation of neutrophils; enhanced adhesion of neutrophils to endothelial cells
Increased oxygen consumption
Oxygen burst
Stimulation of arachidonate metabolism (cellular production of 5-HETE and LTB4)
Specific granule secretion

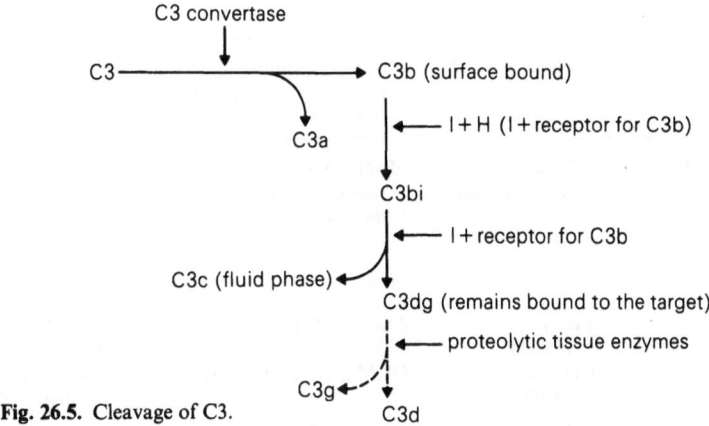

Fig. 26.5. Cleavage of C3.

diopulmonary bypass [14,111]. This mechanism is probably essential in the pathogenesis of pulmonary endothelial cell damage in the adult respiratory distress syndrome (ARDS) [84]. Experimental evidence for a pathogenic link between complement-mediated pulmonary sequestration of neutrophils and lung endothelial cell injury has been provided by the observation that intravenous injection of C5a or complement-activated autologous plasma induces sequestration of neutrophils in the lung and that C5a causes neutrophils to bring about release of Cr from labelled umbilical vein endothelial cells in culture [90]. Recent data indicate that oxygen-derived free radicals (the iron-catalysed H_2O_2 conversion product OH⁻) released from complement-sequestered neutrophils play a major role in endothelial cell damage [107]. In addition, complement and neutrophils have been shown to exaggerate endothelial cell injury initiated by endotoxin. Lipopolysaccharide (LPS) appears to enhance neutrophil-mediated damage induced by C5a in the lungs of experimental animals in vivo and human endothelial cells in vitro. In as much as LPS can activate complement, large amounts of LPS might produce injury by a combination of direct effects on endothelial cells and indirect effects on neutrophils activated by C5a.

Complement fragments generated during complement activation and certain complement proteins bound to the targets of complement activation may interact with receptors on effector cells to trigger phagocytosis and/or other specific cellular responses which contribute to the inflammatory process [33]. These fragments include the C3b, C3bi, C3dg and C3d cleavage fragments of C3 (Fig. 26.5) and Clq and H. The binding specificities and cell distribution of the

receptors CR1, CR2 and CR3, which bind the C3b, C3bi and C3dg/C3d fragments of C3 covalently attached to surfaces, are summarized in Table 26.10.

CR1 is a polymorphic glycoprotein with a molecular weight of approximately 200 000. CR1 on erythrocytes transports opsonized immune complexes to the sites of their removal in the reticuloendothelial system. As mentioned earlier, the role of erythrocytic CR1 in the clearance of immune complexes may provide an explanation for the association of certain immune complex diseases with deficiencies of CR1. The main function of CR1 on neutrophils and monocytes is to enhance phagocytosis of IgG-coated particles and mediate internalization of small ligands bearing C3b. Binding of fibronectin or other connective tissue components to C3b-coated particles confers the ability to ingest the opsonized targets in the absence of antibody on both monocytes and C5a-stimulated neutrophils [81]. Interaction of bound C3b with CR1 has also been shown to induce enzyme release from neutrophils and to trigger the oxidative metabolism in neutrophils and monocytes. A role for CR1 on B lymphocytes in B cell differentiation has been suggested by the observations that F(ab¹)2 anti-CR1 enhanced polyclonal antibody secretion by peripheral blood mononuclear cells in the presence of pokeweed mitogen.

The predominant or exclusive ligand of CR3 is C3bi. The receptor is present on neutrophils, monocytes, large granular lymphocytes and follicular dendritic cells. The expression of CR3 is up-regulated by chemo-attractants and decreased by interferons. CR3 consists of two polypeptide chains, an alpha chain with a molecular weight of 165 000 that expresses the Mac1/0KM1 and Mol determinants and a beta

Table 26.10. Human C3 receptors

Receptor	Ligand*	Molecular weight	Monoclonal antibodies specific for epitopes on receptors	Cell distribution
CR1	C3b, C4b,C3bi	200 000 (polymorphic)	Monoclonal antibodies raised against the purified molecule	Erythrocytes, neutrophils, eosinophils, monocytes, B lymphocytes, some T (CD4 +) lymphocytes, follicular dendritic cells, glomerular podocytes
CR3	C3bi	α chain: 165 000	Anti-Macl, anti-Mol	Neutrophils, eosinophils, monocytes, large granular lymphocytes, follicular dendritic cells
	C3dg?	ß chain: 90 000	OKM1, OKM10, Leu15	
CR2	C3dg C3d C3bi,C3b	145 000	Anti-B2, HB-5, OKB7	B lymphocytes, follicular dendritic cells
CR4(?)	C3dg	?	?	Neutrophils

* The underlined ligand is that which binds to the receptor with highest affinity.

chain of molecular weight 90 000 that is shared with two other leucocyte membrane proteins, LFA1 and p 150,95 [91]. Binding of C3bi-coated particles to CR3 on phagocytes is considerably enhanced in the presence of divalent cations. CR3 is probably the most important C3 receptor for phagocytosis of opsonized bacteria, since C3b is rapidly converted in serum into C3bi. CR3 binds with high affinity to surface-bound C3bi; it also binds to unopsonized yeast and to some bacterial surfaces; and it can trigger phagocytosis of C3bi-coated particles in the presence or absence of IgG. The roles of CR3 in phagocytosis and of the Mol alpha chain in promoting adherence of neutrophils to surfaces explain the complex defects in neutrophil functions reported in patients with recurrent bacterial infections whose cells are deficient in the Macl/Mol antigen and/or in the common beta chain of the Macl-LFA1-p150,95 family of molecules [35].

CR2, the C3d and Epstein–Barr virus receptor, is present only on B lymphocytes and follicular dendritic cells. It plays an important role in antigen presentation and B cell proliferation. Thus C3 receptors are essential for phagocytosis and elimination of antigens, the clearance of immune complexes and immune regulation. As it has briefly been discussed above, both their acquired or genetic deficiency (e.g. CR1 deficiency) and their enhanced expression (e.g. the Mol alpha chain of CR3) participate in the pathogenesis tissue damage.

Experimental and in vivo evidence discussed above indicates that complement may mediate immunologically induced vascular inflammatory injury by either of the two following pathways:

1. Complement may directly induce vascular damage without involvement of inflammatory cells by generating m(C5b-9) complexes and inserting them into lipid bilayers of target cell membranes. Evidence for this pathogenic mechanism in human vasculitides is still indirect, based on the well-established cytolytic properties of m(C5b-9) and on the finding of C5b-9 neo-antigens in tissue lesions [7,22]. C5b-9 deposits are present in association with immune deposits along glomerular capillary walls and in the mesangium in a variety of glomerular diseases, including all forms of nephritis in systemic lupus erythematosus, membranous nephropathy, anti-GBM nephritis, acute post-streptococcal glomerulonephritis, IgA glomerulonephritides and membranoproliferative nephritis [49]. C5b-9 deposits are found in association with IgG and C1 at the dermal–epidermal junction of involved skin in systemic lupus erythematosus [8] and in vessels with acute experimental endothelial injury induced by in situ formation of IgA-containing immune complexes in the lung [52]. It should be noted however, that C5b-9 neo-antigens in a tissue may represent either "true" m(C5b-9) complexes or passively absorbed non-cytolytic SC5b-9 complexes. Furthermore, m(C5b-9) may be pathogenic independently of its cytolytic capacity, by reversibly altering cell metabolism and inducing release of inflammatory mediators such as eicosanoids, interleukin-1 and reactive oxygen species [6].

2. Complement may induce vascular injury in an indirect fashion by generating chemotactic peptides which can activate leucocytes inducing their adherence to target tissues and the release of inflammatory mediators and toxic oxygen products. This mechanism may be initiated by local or systemic complement activation. It involves neutrophils when systemic complement activation or IgG-dependent activation are the initiating events. It predominantly involves monocytes/macrophages when the phlogistic complex contains IgA rather than IgG [30]. One of the differences between IgG- and IgA-containing immune complexes is that macromolecular IgA only activates the alternative pathway and generates relatively less C5a than IgG-containing complexes. Generated C5a would then primarily stimulate local resident cells (i.e. macrophages) rather than diffuse away from the site of its generation so as to attract and activate circulating neutrophils as occurs when complement is activated by IgG-containing immune complexes. The relatively poor complement-activating capacity of IgA may account for the occurrence (e.g. in mesangial nephropathy) of IgA immune deposits and complement deposits without evidence of infiltrating leucocytes. IgG immune complexes or immunoglobulin-independent systemic complement activation generate large amounts of C5a which attracts neutrophils and stimulates their effector functions.

In addition to C5a, C3 fragments and complexed immunoglobulins contribute to cell-mediated complement-dependent damage to tissues. For example, stimulation of CR3 and of Fc receptors induces degranulation of prostaglandins, thromboxane and leukotrienes. Finally, as mentioned above, sublytic amounts of C5b-9 may also activate neutrophils and monocytes to release mediators of inflammation.

Cellular Effector Mechanisms

Effector cells of the immune system are involved in at least three different pathways of immunologically mediated damage of the vascular wall: (a) neutrophils and monocytes are recruited in the amplification and effector phase of immune-complex-mediated damage; (b) neutrophils are primarily involved in vascular damage induced by systemic complement activation; (c) cytotoxic effector subsets of lymphocytes and monocytes/macrophages that have been activated by factors derived from specific-

ally activated lymphocytes determine cell-mediated immunity reactions against allogeneic or other uncharacterized antigens. In all instances, a complex dynamic interplay between cells and soluble effector or regulatory molecules focuses effector mechanisms on the target(s) of immune recognition so as to optomize their removal or destruction and limit adverse reactions to the host.

Monocytes

Mononuclear phagocytes are found circulating in the blood (monocytes) and fixed in tissues (tissue macrophages) [43]. Circulating monocytes originating from the bone marrow spend approximately 24 h in the bloodstream before emigrating to the extravascular pool. Giant cells and epithelial cells derive from tissue macrophages. During inflammation, there is an increase in monocyte production, an increase in the influx and accumulation of monocytes at the site of the inflammatory reaction. Monocytes and macrophages may be triggered to become effector cells following the interaction of soluble or particulate ligands with specific receptors or "recognition units" on the cell membrane. These include receptors for the Fc part of immunoglobulins, receptors for cleavage fragments of complement proteins, receptors for lymphokines, receptors for several plasma proteins and "recognition units" for particles of defined biochemical composition (Table 26.11).

Table 26.11. Ligands for surface receptors expressed by monocytes/macrophages

Fc portion of immunoglobulins:
  IgG
  IgE
Complement fragments:
  C3b (CR1)
  C3bi (CR3)
  C5a, C5adesArg
  C3a, C3adesArg
Gamma interferon, other lymphokines (MIF)
Colony-stimulating factor (CSF)
Insulin
Plasma proteins:
  transferrin, fibronectin, fibrinogen, lipoproteins, alpha-2 macroglobulin–protease complexes
Mannose (yeast cell wall)
Activating surfaces of the alternative complement pathway

Fc gamma receptors on human monocytes bind the Cγ3 domain of complexed or aggregated IgG1 or IgG3. The Cγ2 domain and the hinge region of the immunoglobulin are required for stable conformational binding. Interaction of Fc receptors with IgG molecules bound by their Fab arms to a foreign particle and the subsequent cross-linking of the receptors on the cell surface is the most potent signal for phagocytosis by monocytes. The IgG-coated particle that has adhered to the monocyte is surrounded by cytoplasmic extensions of the cell, leading to formation of a vesicle (phagosome) that separates from the plasma membrane and fuses with primary and secondary lysosomes. Ingested organisms are killed and degraded in lysosomes. Fc receptors that have been cleared from the macrophage surface during formation of the phagosome may be recycled back to the plasma membrane of the cell.

Although binding of IgG to Fc receptors is sufficient to trigger phagocytosis by monocytes, the process is usually facilitated by the fact that biological targets are co-opsonized with other molecules such as C3 fragments or fibronectin which enhance adhesion of the targets and/or trigger their ingestion. Fc gamma receptors on monocytes also mediate antibody-dependent cellular cytotoxicity (ADCC) by which IgG-bearing opsonized cells are killed by the Fc receptor-bearing effector cell; Fcε receptors on monocytes mediate ADCC against IgE-coated parasitic targets.

Human monocytes express four complement receptors: CR1, CR3, C5a and C3a receptors. The interaction of CR1 and CR3 with C3b and C3bi on opsonized targets will cause particles to adhere to monocytes. This interaction is not sufficient to trigger phagocytosis on its own unless CR1 and CR3 have been "activated" by exposure of the monocytes to a chemotactic factor, to specific T cell-derived lymphokines, to phorbol esters or to fibronectin or laminin [42,80,116]. The enhancing effect of connective tissue proteins on phagocytosis may be particularly relevant to the pathogenesis of vascular lesions where extracellular matrices become exposed. Surface-bound or soluble fibronectin induces phagocytosis of C3b-coated particles in the absence of fibronectin on the particle to be phagocytosed. CR1 on monocytes also mediate absorbtive endocytosis by these cells. Absorbtive endocytosis differs from phagocytosis in that it involves receptor-mediated ingestion of soluble ligands and the formation of small-size clathrin-coated vesicles. Expression of CR1 and of CR3 on monocytes is enhanced by IL-1 and by chemo-attractants including C5a and C5adesArg. As mentioned earlier, CR3 is a member of a family of three structurally related surface glycoproteins, CR3, LFA1 and p150,95 which function in adhesion events. All three of these receptors recognize a fungal pathogen, *Histoplasma capsulatum* CR3 mediates binding to monocytes of C3bi-coated particles and of unopsonized zymosan, and may function as a receptor for beta glucans [86,87]. By enhancing expression of CR3 on monocytes, C5a and C5adesArg increase adhesiveness of the cells to surfaces of pathogens, to artificial surfaces and also to vascular endothelium [108].

The adherence of monocytes to the endothelium has been shown to be involved in experimentally induced atherosclerosis and in some animal models of immune-complex-mediated vascular injury involving macromolecular IgA. Further consequences of the binding of C5a/C5adesArg to C5a receptors on monocytes include chemotaxis and stimulation of the oxygen metabolism and degranulation. C5a-dependent migration of monocytes is enhanced by factors produced by fibroblasts and fragments of collagen (Table 26.12) and is inhibited by lymphokines (MIF) and the Bb fragment generated from factor B during activation of the alternative complement pathway. Receptors for C3a have not yet been biochemically characterized on monocytes, although their existence is indicated by the observation that C3a and C3adesArg trigger IL-1 secretion by human cultured monocytes [64].

Table 26.12. Chemotactic factors for human monocytes

Complement fragments: C5a; C5adesArg; C5b,6, 7
Lymphokines
LTB4, 5-HETE
Fragments of fibronectin
FMLP

Interaction of monocyte receptors with various ligands, irrespective of engaging phagocytosis, triggers the respiratory burst which involves increased oxygen consumption by the cells and production of reactive oxygen species such as superoxide anion, hydrogen peroxide, hydroxyl radicals and singlet oxygen [3]. Extracellular and intraphagosomal superoxide anion derivatives are microbicidal. Their antimicrobial effects are augmented by halide ions in the presence of myeloperoxidase in monocytes and of catalase in mature macrophages. Reactive oxygen species contribute to tissue damage by oxidizing proteins, lipids and nucleic acids.

Secretory products of macrophages include enzymes, enzyme inhibitors, coagulation factors, fibronectin, plasma proteins, IL-1, tumour necrosis factor, growth-promoting factors and arachidonic acid derivatives (Table 26.13). Complement proteins are secreted by macrophages at all stages of activation. Secretion of enzymes, neo-synthesis of arachidonic metabolites and production of IL-1 require engagement of specific membrane receptors or exposure of the cells to membrane-active pharmacological agents. Among the enzymes that are secreted by macrophages are acid hydrolases which are stored in primary and secondary lysosomes. They may be released into the extracellular milieu following stimulation by immune complexes. Neutral proteases secreted by monocytes/macrophages include plasminogen activator, elastase and collagenase, all of which are capable of degrading collagens, elastin and proteoglycans in connective tissue. Lysozyme present in lysosomal vesicles selectively degrades some bacterial cell walls and may contribute, in association with other enzymes, to the tissue damage produced by stimulated macrophages [43].

Macrophages produce large amounts of prostaglandins and leukotrienes following stimu-

Table 26.13. Secretory products of macrophages

Enzymes
  Acid hydrolases
    Proteases and peptidases
    Glycosidases
    Phosphatases
    Lipases
  Neutral proteases
    Collagenases (specific for collagens types I, II, III; specific for collagen type V)
    Plasminogen activator
  Angiotensin-converting enzyme
Enzyme inhibitors
  Alpha-2 microglobulin and other protease inhibitors
Plasma proteins
  Fibronectin
  Complement components C1, C4, C2, C3, C5, B, D, P, H and I
  Pro-coagulant proteins: tissue factor, factors V, VII, IX and X
Reactive oxygen species
Arachidonic acid metabolism derivatives
  Leukotrienes
  Prostaglandins, thromboxane B_2
Interleukins
  IL-1
  Interferon
CSF
  And other factors promoting cell proliferation (including growth factors for fibroblasts, smooth-muscle cells and endothelial cells)

Table 26.14. Stimuli for IL-1 release from monocytes

Gram-negative bacteria, LPS, the polysaccharidic moiety of LPS
Gram-positive bacteria, cell walls, muramyl dipeptide
Yeast cell walls
Viruses
C5a, C5adesArg, C3adesArg, C3b
Surface-bound IgG
Activated T cells
Phorbol myristate acetate
Silica
Urate crystals

lation of membrane phospholipase A_2 and release of arachidonic acid. Prostaglandins (PGE_2) affect macrophage function and exert suppressive effects on the immune response. Among the leukotrienes, LTB4 and 5HETE are particularly relevant to the inflammatory process due to their potent chemotactic activity and aggregating effects on leucocytes. Recently identified lipoxines favour lysosomal enzyme release and production of oxygen radicals by monocytes. Stimulated monocytes also release platelet-activating factor (PAF-acether), a potent mediator that attracts leucocytes, triggers the respiratory burst and induces a variety of systemic effects including hypotension and contraction of smooth-muscle cells [66].

Monocytes/macrophages are the primary source of IL-1 whose synthesis is induced by a wide variety of stimuli (Table 26.14) [26,28]. Little is known of the precise mechanisms by which monocytes react to IL-1 inducers; however, ligand–receptor interactions are generally involved, e.g. when cells are exposed to LPS or to C5a/C5adesArg. Following contact of the inducing agent with the monocyte membrane, activation of protein kinase C and phosphorylation of intracellular proteins take place which lead to gene activation and generation of two different types of RNA specific for IL-1. mRNA specific for IL-1 appears by 2 h after stimulation. Two large IL-1 precursor proteins (molecular weight approximately 33 000) are then cleaved into peptides of molecular weight 17 500 prior to membrane expression and to extracellular release [68]. IL-1 has a wide range of biological activities (Table 26.15) [26]. It is involved in the regulation of the immune response due to its ability to induce (a) secretion of the T cell growth factor interleukin-2 (IL-2) and (b) the expression of IL-2 receptors by T cells. IL-1 is involved in the initiation of the immune response since its membrane form participates with class II antigens in antigen pres-

Table 26.15. In vivo effects of IL-1

Fever
Neutrophilia
Synthesis of acute phase reactants
Proliferation of fibroblasts
Increased protein degradation
Immune regulation

entation by macrophages. Two different T cell-dependent pathways have been described for stimulating membrane IL-1 expression in macrophages. One involves cell–cell contact and the other involves a lymphokine secreted by activated T cells. IL-1 also behaves as a key mediator of the inflammatory reaction, acting as a chemoattractant. It induces the synthesis of various proteins including collagen and proteinases, supporting the concept that it is involved in tissue injury and induction of sclerosis (e.g. in synovitis and in hepatic or pulmonary fibrosis) [95]. IL-1 can act as a circulating hormone, generating fever and responsible for the production of acute-phase reactants by hepatocytes. Locally secreted IL-1 may recruit and mediate monocyte adhesion to the endothelium by increasing the expression of specific leucocyte adherence molecules on the surface of endothelial cells; IL-1 is also secreted by injured or stimulated smooth-muscle cells so that a self-amplifying and perpetuating inflammatory process is triggered at the site of vascular injury.

Neutrophils

Neutrophils represent 50%–70% of the total number of leucocytes in peripheral blood. The cells do not circulate continually since part of the population is located along the margins of blood vessels. The half-life of neutrophils in the blood is about 7 h. As mentioned earlier, neutrophils are the primary effector cells in IgG immune-complex-induced injury and in tissue injury initiated by systemic complement activation. Specific receptors for chemotactic molecules are present on the neutrophil cell membrane (Table 26.16).

The most potent chemo-attractant for neutrophils is C5a [100]. The relative importance of

Table 26.16. Receptors for chemotactic factors on neutrophils

C5a: 80 000 receptors/cell (Kd = 2×10^9 M^{-1})
FMLP: 70 000 receptors/cell
LTB4: 25–40 000 receptors/cell
Chemotactic lymphokines:?

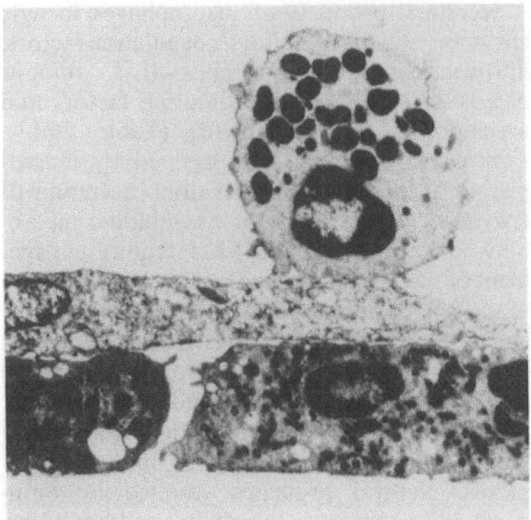

Fig. 26.6. Electron photomicrograph of a neutrophil adherent to a cultured endothelial cell.

C5a as a mediator of neutrophil influx has been shown in several experimental models. Serum from C5-deficient mice generates no chemotactic activity when treated with immune complexes and C5-deficient mice exhibit a delay or abrogation of neutrophil accumulation in the pulmonary vasculature following intratracheal administration of pre-formed immune complexes [63]. Derivatives of the lipoxygenase pathway are also chemotactic for neutrophils. The most potent lipid mediator is LTB4 and the mono-HETE are less active. Most biologically active leukotrienes are rapidly inactivated so that their effects are temporally limited. Lymphokines that are chemotactic for several types of leucocytes or specifically chemotactic for neutrophils have recently been described. Activated neutrophils adhere to endothelial cells and usually migrate through the junctional complexes between the cells (Fig. 26.6). Erythrocytes and various soluble inflammatory mediators enhance neutrophil migration and adhesion of neutrophils to the endothelium. The two most prominent mechanisms by which neutrophils mediate tissue injury are release of granule-contained enzymes and the production of oxygen radicals. The cytoplasm of neutrophils contains a large number of granules of two types, primary (azurophil) granules, which are the lysosomes, and secondary (specific) granules which predominantly contain alkaline phosphatase (Table 26.17) [61].

Secretion of enzymes is induced by a variety of stimuli. During phagocytosis, intracellular

Table 26.17. Content of neutrophil granules

Primary granules (lysosomes)
  Acid hydrolases: beta glucuronidase, arylsulphatase, cathepsin
  Microbicidal enzymes: myeloperoxidase, lysozyme
  Neutral proteases: elastase, collagenase, proteases

Secondary granules
  Microbicidal enzymes: lysozyme
  Neutral proteases: collagenase
  Lactoferrin
  Vitamin B_{12}-binding protein

Table 26.18. Oxygen radicals derived from neutrophils

Peroxidase-dependent
Peroxidase-independent
Superoxide anion
Hydrogen peroxide
Hydroxyl radical
Singlet oxygen

granules fuse with the plasma membrane surrounding the phagosome and discharge their contents of enzymes into the phagocytic vacuole. The degranulation which follows phagocytosis involves the primary and secondary granules, with release of acid hydrolases, myeloperoxidase and lysozyme. Other stimuli or degranulation include interaction with IgG-coated or C3b- and C3bi-coated surfaces which have not been phagocytosed, and interaction with the same molecules that induce chemotaxis. Many of the lysosomal enzymes are optimally active only at acidic pH. Whether the pH in inflammatory lesions is sufficiently decreased to allow a significant contribution by the acid hydrolases remains unknown. Neutral proteases, which are active at neutral and alkaline pH, may be more important in tissue destruction [48]. These enzymes may also facilitate emigration of leucocytes through the basement membrane. In contrast, acid proteases would primarily function within lysosomes to digest phagocytosed microorganisms. The secretory function of neutrophils is also important in controlling vascular permeability at inflammatory sites and regulating the cooperation between neutrophils and other cells of inflammation. The effects of leucocyte degranulation are inhibited by enzyme inhibitors which are released simultaneously or are present in the extracellular fluid.

Neutrophils respond to phagocytosable stimuli, to antigen–antibody complexes and to a number of soluble agents (including C5a) with a burst of oxygen consumption. The oxygen consumed is converted to a number of highly reactive oxygen derivatives: the superoxide anion (O^{2-}); hydrogen peroxide (H_2O_2); hydroxyl radicals (OH^-); and singlet oxygen (Table 26.18) [62]. Reactive oxygen derivatives are either dependent on or independent of peroxidase. The neutrophil peroxidase (myeloperoxidase) is present in high concentrations in primary granules. Myeloperoxidase can use the H_2O_2 generated by the oxidative burst to oxidize halides

and form powerful toxic molecules [62]. Reactive species of the peroxidase system may attack cell surfaces by halogenation and oxidation. The role of peroxidase-dependent products in the extracellular toxicity of neutrophils is still debated. Peroxidase-independent products of the respiratory burst may be released extracellularly: they include superoxide, hydrogen peroxide and hydroxyl radicals. Of these products only OH^- is toxic enough to kill bacteria. Superoxide anions may reduce metalloproteins which are critical for oxygen-dependent metabolism. Multiple factors are involved in protection against injury by reactive oxygen species. Enzymatic detoxification of superoxide and hydrogen peroxide by superoxide dismutase and catalase is highly efficient.

Eicosanoids and platelet-activating factor are synthesized de novo within neutrophils and released extracellularly when the cells are activated. Other cells also produce these mediators including monocytes and basophils. Eicosanoids are metabolites of arachidonic acid which is derived from the degradation of membrane phospholipids by phospholipase A_2 following cell activation. The arachidonic acid may be converted into prostaglandins and thromboxanes via the cyclo-oxygenase pathway or into hydroxyperoxy (HETE) hydroxyeicosatetrahoic acids (HPETE) and leukotrienes B4, C4, D4 and C4(SRS-A) by lipoxygenase enzymes [25]. Some effects of representative lipid mediators are listed in Table 26.19.

Table 26.19. Role of representative lipid mediators in inflammation

Leukotriene B4
  Chemotaxis of neutrophils; increase in vascular permeability in the presence of PGE_2
Leukotriene D4
  Smooth-muscle cell contraction; increase in vascular permeability
Prostaglandin E_2
  Vasodilatation; increase in vascular permeability when acting in synergy with a chemotactic agent
PAF
  Platelet release reaction; neutrophil aggregation; stimulation of neutrophil release and neutrophil respiratory burst; increase in vascular permeability

Effector Lymphocytes

Effector lymphocytes include cytotoxic T cells and T cells involved in delayed-type hypersensitivity [83]. Cytotoxic T cells are generated in response to allo-antigens and in host defence against viruses. Activation of T cells by allo-antigens involves recognition of class II molecules by CD4$^+$ cells and their subsequent activation in the presence of IL-1. Activated CD4$^+$ cells then co-operate in the differentiation of a subset of cytotoxic CD8$^+$ effector cells through generation of IL-2 and other as yet poorly identified factors. Cytotoxic T cells recognize class I molecules on allografts. Effector T cells of delayed-type hypersensitivity are activated upon specific recognition of the antigen. They secrete a variety of lymphokines, some of which activate macrophages (Table 26.20). Most of the infiltrating cells in granulomas are macrophages activated by T cell products. The most important of these T cell-derived mediators include: (a) macrophage inhibitory factor (MIF) which prevents the macrophages from leaving the site of cellular immune reactions; (b) macrophage activating factor; (c) chemotactic lymphokines; and (d) gamma interferon. In addition to its antiviral effects, gamma interferon has a number of biological effects that may be relevant to tissue injury during granuloma formation and in graft rejection [109]. These include alterations in cell phenotype with enhanced expression of class II molecules and Fc receptors [102], inhibition of growth of T and B cells, enhanced phagocytosis, the spreading and cytotoxic properties of macrophages, enhanced expression of adhesion-promoting proteins on leucocytes, enhanced cytotoxicity of lymphocytes and positive feedback production of interferon.

Table 26.20. Properties of lymphokine-activated macrophages

Attachment to surfaces
Increased number of lysosomes
Release of reactive oxygen radicals
Release of neutral proteases
Enhanced expression of surface Fc receptors
Enhanced expression of class II molecules

Endothelial Cells

The endothelium participates in various haemostatic, metabolic and endocrine processes that have been discussed in Chap. 1. This section will focus on immunological aspects of endothelial reactions such as expression of surface antigens by resting and stimulated endothelial cells and interactions of leucocytes with the endothelium.

ABH blood group antigens are expressed on the endothelium throughout the vascular system. Endothelial cells also express class I antigens of the major histocompatibility complex (MHC). Antigenic expression of the three loci (A, B and C) coding for class I molecules is increased following treatment of endothelial cells with gamma interferon. MHC class I antigens are targets for cytolytic T lymphocytes in allograft rejection. MHC class II antigens are encoded by three distinct loci in the HLA-D region, the DP, DQ and DR loci. These antigens stimulate allogeneic T helper cells in graft rejection and are essential for the presentation of antigen in the induction of a normal immune response. Lymphokines can promote expression of MHC class II antigens on antigen-presenting cells and induce synthesis and expression of MHC class II antigens by cells which do not express the molecules in their unstimulated state such as endothelial cells and fibroblasts [78].

Expression of class II antigens (but not of factor VIII-related antigen on skin, glomerular and peritubular endothelial cells) in animals is suppressed when lymphokine release is suppressed in vivo by treatment with cyclosporin A. Gut and skin endothelium which are normally Ia negative express MHC class II molecules during local infections and graft versus host reactions [44]. Using sections of fresh-frozen tissue, class II antigens have been found on the endothelium of most capillary beds in man except for those of the brain and placenta. Expression in larger vessels varies depending on anatomical location and within a single tissue. It may be that small vessel endothelium exhibits a greater sensitivity to lymphokines and that basal circulating levels are sufficient to maintain expression of class II antigens, whereas arterial endothelium requires a local immunological reaction that would generate larger amounts of lymphokines.

In vitro, less than 1% of the cells in human umbilical vein endothelial cell cultures are II positive, and these cells cannot be depleted with monoclonal antibodies directed against monocytes. When cultured cells are stimulated with gamma interferon, HLA-DR antigens are expressed on all cells, whereas HLA-DP antigens are expressed to a lesser degree and HLA-DQ antigens are only expressed on a minority of cells [79]. The finding that endothelial cells are capable of expressing Ia antigens substantiated the hypothesis that this cell type may function as an antigen-presenting cell. Cultured endothelial cells may stimulate allogeneic T cell proliferation, conversely, supernatants from endothelial and T cell co-cultures and conditioned medium from allo-antigen-stimulated T cells stimulate MHC class II antigen expression by endothelial cells. Endothelial cells can also substitute for monocytes in the proliferative response of purified T cells to antigen and mitogen in reactions that are restricted by class II antigens [99]. Furthermore, as will be discussed later, endothelial cells can secrete IL-1 which provides in addition to antigen and MHC class II determinants, the second signal by an antigen-presenting cell for the induction of T cell activation.

The in vivo significance of these interactions is uncertain. It is conceivable that endothelial cells could present antigens from blood-borne pathogens to T cells, support a local T cell response in the vessel, promote T cell adherence, facilitate subsequent emigration of the cells into the vascular wall and form perivascular infiltrates. Increased expression of MHC class II antigens could also induce autoimmune responses against endothelial cell surface antigens that are not normally presented to T cells, i.e. in the context of MHC class II determinants, in a way similar to the induction of autoimmunity against thyroid epithelial cells and cells of the islets of Langerhans. Furthermore, MHC class II antigens may represent the target for anti-Ia autoantibodies in some patients with systemic lupus erythematosus [15].

As discussed above, expression of MHC class I and class II antigens provides a stimulus for the activation and cytolytic functions of T cells in allogeneic transplantation. It has been noticed however, that recipients of renal transplants from related, HLA-identical living donors occasionally rejected their transplant hyperacutely, suggesting that endothelial cells were the main target. This observation led to the discovery of a non-HLA-related system of endothelial cell antigens encoded by genes located on chromosome 6. At least some of these antigens are shared by monocytes. The presence of preformed antibodies to donor endothelial cell/monocyte antigens generally leads to graft rejection in HLA-identical combinations, whereas pre-formed antibodies to monocytic antigens alone appear to be benign. In a recent randomized prospective study, 7% of the recipients of HLA-identical kidneys were found to have antibodies against monocytic determinants of the donor [11].

Normal human endothelial cells in culture do not express Fc and C3 receptors. Indirect evidence suggests that they may express receptors for Clq. Cultured endothelial cells infected with *Herpes simplex* virus (HSV) type I exhibit Fc- and C3b-binding activities. Both the Fc and C3b receptors are viral glycoproteins. Simultaneous expression of both receptors occurs at 4 h postinfection, yet they are distinct receptors in that one ligand does not inhibit binding of the other. Passive adsorption of viral glycoproteins is not responsible since UV-inactivated virus, which is capable of adsorption but not of replication, had no receptor-inducing effects. Biochemical analysis of cell surface proteins revealed that the Fc receptor is HSV glycoprotein E and the C3b receptor, HSV glycoprotein C [98]. Thus, infection with HSV I may lead in vivo to the expression of endothelial Fc- and C3b-binding molecules which could facilitate binding of circulating immune complexes to infected cells. Viral antigens have been detected in the vessel walls in vivo in several forms of vasculitis and HSV infection has been implicated in the development of erythema multiforme in some patients. Fixed immune complexes are stimulatory for leucocytes and could activate complement, two situations that are potentially harmful to the endothelium and which could represent mechanisms for viral-associated immune vascular injury.

A number of other antigens associated with resting and/or stimulated endothelial cells are probably involved in various forms of immunologically mediated vasculitis. Of particular interest may be: (a) endothelial cell antigenic determinants, cross-reactive with alkaline phosphatase of neutrophils, which are recognized by circulating auto-antibodies in patients with Wegener's granulomatosis and some patients with other forms of microscopic polyarteritis; (b) neoantigenic determinants expressed on cultured endothelial cells stimulated with IL-1, which are formed in vivo in vessels of some patients with inflammatory systemic diseases.

Interactions between leucocytes and endothelial cells are essential in the initiation of the inflammatory reaction since leucocytes must adhere to and emigrate through the endothelium in order to perform their effector functions at extravascular sites [46]. Neutrophils continuously interact with endothelial cells physiologically. During their approximately 7-h life time in the circulation, neutrophils are in dynamic equilibrium between a freely circulating and a marginating pool. The marginating neutrophils adhere to endothelial cells (mainly in small venules) and emigrate from the blood vessels, moving between the endothelial cells to reach the extravascular tissue. Associated with acute inflammatory states there is a characteristic rapid increase in neutrophil adhesion and emigration, apparently due to chemotactic agents which are generated extravascularly and diffuse across the vessel wall. Maximal neutrophil adherence occurs shortly after endothelial contact and is cation-dependent. Since both cell types exhibit a negative surface charge, cations may aid in reducing the electrostatic repulsion. A reduction in surface charge occurs after neutrophil stimulation with chemotactic factors. Several molecules have been implicated in adhesion, including lactoferrin, members of the LFA1, MAC1/Mol, GP 150,95 family and ICAM-1 [96].

The LFA1, Mac1/Mol, GP 150,95 family, also termed the CD 18 complex, is made up of glycoproteins defined by monoclonal antibodies and consists of a common beta subunit and a unique alpha subunit [91]. The alpha subunit appears to confer the adhesivity. Mac1, which also functions as a CR3 receptor, is present on monocytes and granulocytes. The GP 150,95 molecule may also bind iC3b (the ligand for CR3) and is found on granulocytes, while LFA1 permits homotypic adhesion of lymphocytes. The expression of these molecules is modulated by LPS, IL-1, TNF and GM-CSF. In addition to their expression on cell surface, Mac1 and GP 150,95 are stored intracellularly and are rapidly mobilized to the surface under the influence of certain physiological triggers. The intercellular adhesion molecule (ICAM-1), also defined by monoclonal antibodies, appears to be widely distributed, occurring on lymphocytes, endothelial cells and fibroblasts [89]. ICAM-1 is hypothesized to be a ligand for LFA1 and monoclonal antibodies to either can inhibit adhesivity. Interleukin-1 and several other cytokines are stimulatory for ICAM-1 expression.

Several soluble molecules are stimulators of leucocyte–endothelial cell adhesion. Interleukin-1, which can be synthesized by monocytes and endothelial cells in vitro, promotes monocyte and neutrophil adhesion to endothelial cells [4]. The stimulatory effect is maximal at 4–6 h and requires protein synthesis. In contrast, molecules such C5a and FMLP induce maximal adhesion within 5 min. Thrombin exhibits a similar time course for induction of adhesion. However, it appears that it does so via stimulation of platelet-activating factor (PAF) synthesis. PAF stimu-

lation of endothelial cell is distinct from that brought about by C5a and FMLP since prior exposure to PAF does not desensitize the cells to C5a or FMLP and vice versa. Also, PAF remains associated with the endothelial cell surface, suggesting that it brings about an increase in endothelial cell stickiness, whereas C5a and FMLP appear to act on the neutrophil and monocyte. In this regard IL-1 was reported to exert its effect(s) only on the endothelial cell. Leukotriene B4, an arachidonic acid metabolite of granulocytes, can stimulate adhesion of granulocytes to endothelial cells as well. Thus, it is apparent that many independent routes for leucocyte–endothelial cell adhesion are available.

Peripheral blood lymphocytes continually emigrate from the bloodstream to enter lymphoid tissues. Tissue lymphocytes in turn emerge from these sites to re-enter the circulation (via the lymphatics and ultimately the thoracic duct), thus establishing a circuit which probably assists in immune surveillance. Lymphocytic emigration is regulated by recognition of specialized endothelium in post-capillary venules within secondary lymphoid organs. The localization and morphology of this endothelium characterizes high endothelial venules (HEV). Compared to the thin flat endothelium found elsewhere in the microvasculature, endothelial cells of the HEV are large and columnar. Morphological and functional characteristics of HEV appear to be mediated by local T cell traffic since HEV flatten when T cells are congenitally absent or experimentally depleted, and regenerate when T cells are repleted. Furthermore, HEV are associated with the localized accumulation of lymphocytes at sites of chronic inflammation, suggesting that this phenotype is inductible. Specific membrane receptors are present on lymphocyte subpopulations which allow them to recognize determinants on HEV at different locations. Thus, for example, circulating B cells preferentially bind to gut-associated Peyer's patch HEV while T lymphocytes exhibit a stronger affinity for peripheral node HEV. The ability of lymphocytes to recognize HEV is present in mature migratory phase cells but not their sessile precursors in the bone marrow and thymus.

The interaction between endothelial cells and lymphocytes occurs by lymphocyte membrane attachments within the "pits" of the HEV microvilli. It is an energy- and calcium-dependent process which requires the participation of microfilaments in the lymphocytes. As with neutrophils, emigration is believed to occur predominantly via intercellular routes although intracellular passage has also been observed. At present, LPS, IL-1 and gamma-interferon are known to increase binding of T and B cells to the endothelium.

Following adhesion, leucocytes may be released back into the circulation or emigrate into the surrounding tissues. Since adhesion and emigration are events triggered by tissue injury or infection, molecules which stimulate adhesion may be involved in other functional responses such as chemotaxis, opsonization, lysosomal degranulation and superoxide anion production. Passage of leucocytes through the endothelium can proceed via cell junctions or transcellularly and does not necessarily in itself induce injury or increase permeability. However, it has been observed that neutrophil degranulation can induce endothelial cell detachment in vitro and that superoxide anion derivatives can produce endothelial cell functional changes as well as cell death. The p 150,95 adhesion molecule has been shown to be important in endothelial cell detachment in vitro after it has encountered activated neutrophils. Often, the soluble mediators of cell adhesion and emigration also participate directly or indirectly in increasing vascular permeability and stimulating lysosomal degranulation and superoxide anion production. Thus, leucocyte adhesion to endothelium may result in endothelial cell damage due to mediators exhibiting multiple stimulating effects.

Endothelial cells under normal in vitro culture conditions are reported to secrete molecules with IL-1 like activity and IL-1 inhibitory activity simultaneously. Three agents identified to date which can stimulate endothelial cells to produce IL-1 are lipopolysaccharide (LPS), tumour necrosis factor (TNF) and thrombin [69,72,103]. The synthesis of IL-1 is blocked by cycloheximide and with LPS stimulation was shown to be enhanced by pre-treating endothelial cells with gamma interferon for 24–48 h. Maximal intracellular IL-1 accumulation in stimulated endothelial cells in vitro occurs at 24 h but begins as early as 1 h after stimulation.

Interleukin-1 is now known to modulate the haemostatic state of endothelial cells. For example, it stimulates prostacyclin synthesis in endothelial cells which have been cultured for more than 6 h [88]. In contrast, other prostacyclin inducers, such as thrombin and arachidonic acid, act within minutes. IL-1 induces a decrease in tissue plasminogen activator and an increase in the inhibitor of plasminogen activator in vitro. These effects reach a plateau at about

24 h and persist for 72 h. IL-1 also induces a procoagulant tissue factor activity in endothelial cells in vitro and in vivo within 3–5 h which returns to normal pre-stimulation values by 24 h. Thus, IL-1 mediates an early pro-coagulant state of the endothelium as well as a switch-over to a later anti-coagulant state [5, 73].

Endotoxin can also stimulate endothelial cell pro-coagulant activity and bring about a decrease in thrombomodulin, a cell surface glycoprotein which when bound to thrombin activates circulating protein C, which in turn acts as an anti-coagulant by degrading factor Va and factor VIIIa [71]. The kinetics of this response are similar to those mediated by IL-1. Hence, clinical states of hypercoagulability related to immune or inflammation events may operate via mechanisms involving the endothelial cell and soluble mediators.

Tumour necrosis factor (TNF) is a protein of molecular weight 17 000, synthesized by activated mononuclear phagocytes. It is directly cytotoxic for certain tumour cells in vitro and causes necrosis and regression of many solid tumours in vivo. When applied to endothelial cells in vitro, there is an increase in neutrophil adhesivity which is maximal at 4 h, requires protein and RNA synthesis and can result after only 5 min stimulation with TNF. Treatment of neutrophils with TNF likewise increases their adhesivity to endothelial cells (contrary to the situation with IL-1). However, the maximal response occurs at 5 min and does not require protein or RNA synthesis [39]. TNF increases the expression of HLA-A and B MHC class I antigens on cultured umbilical vein endothelial cells [18]. After TNF treatment, the expression of these antigens on endothelial cells, increases ninefold. This increase is time- and dose-dependent, is maximal at 4 d following administration and requires 7 d to return to baseline levels after TNF withdrawal. The increased expression of these antigens appears to result from an increased mRNA expression although this latter increase is proportionately greater than the increase of class I antigens on the cell surface (hence there may be storage of antigens). MHC class II antigens, which are normally absent on cultured human umbilical cord endothelial cells, were not induced by TNF treatment.

The effects of TNF on class I antigen expression are similar to those produced by gamma interferon. Cycloheximide, when added at the same time as interferon, leads to superinduction of class I mRNA transcripts and to the appearance of class II mRNA antigens.

Cycloheximide by itself has no such effect nor does it modulate an increase in mRNA induced by type II interferon. On the other hand, cycloheximide blocks the increase in class I mRNA induced by TNF. Hence, TNF appears to act on MHC gene expression through a recently described synthesized protein intermediate. Modifications in endothelial cell morphology and function have been noted in post-capillary venules in delayed-type hypersensitivity reactions and in post-capillary HEV of lymph nodes after antigen stimulation. Crude lymphokine preparations were shown to induce endothelial cell shape changes in vitro and TNF and gamma interferon incubations alone have been able to induce this response [104]. Under TNF and/or gamma interferon stimulation the cells assume a fibroblast-like growth pattern in which individual cells become elongated, a process which is discernible at 24 h and reaches a maximum at 72–96 h. It is thus clear that endothelial cells respond in a wide variety of ways (e.g. by HLA antigen expression, leucocyte adhesivity, IL-1 and prostacyclin production, morphological changes and expression of factors involved in haemostasis) to a variety of soluble mediators in a defined temporal sequence. Recently, a monoclonal antibody was developed which recognizes a cell surface protein on endothelial cells stimulated by IL-1, LPS and TNF but not on unstimulated cells or cells stimulated by alpha, beta or gamma interferons. The expression of this protein is maximal 4–6 h after stimulation in vitro and declines to basal levels by 24 h. The monoclonal antibody also binds to human skin microvascular endothelium in delayed-type hypersensitivity reactions, further demonstrating the dynamic changes which endothelial cells are involved in during immune and inflammatory vascular reactions.

Interactions Between Immunological Effectors and the Coagulation and Fibrinolytic Systems

Leucocytes, immune complexes and complement directly or indirectly contribute to thrombogenesis as was recognized in animal studies of the endotoxin-induced Shwartzman reaction. As was discussed in the previous section, cytokines from stimulated cells can induce both pro-coagulant and fibrinolytic activities on endothelial cells and thereby modulate the haemostatic state. Independent of the nature of the immunological process that leads to vascular damage, exposure of subendothelium and medial layers of the vessel wall results in platelet adhesion and activation of blood coagulation and complement.

Although platelets are predominantly involved in haemostasis they also express Fc-gamma receptors. Cross-linking of the receptor with complexed or aggregated IgG leads to shape change, aggregation and release of mediators from cytoplasmic granules, and may result in an active energy-dependent uptake of complexed IgG by the cells. Interactions between platelets and the complement system include: (a) binding of the collagen part of Clq to the platelet membrane; (b) specific interaction of platelets with C3a/C3adesArg; (c) modulation of thrombin-dependent activation by factor D since both proteases may compete for the same binding site on the platelet membrane; (d) enhancement of the thrombin-induced release reaction and platelet aggregation by uptake or assembly of C5b-9 complexes on the platelet surface; and (e) modulation of convertase formation by DAF. Finally, human (primate) platelets do not express CR1 whereas those from rabbit and guinea pig do. Platelets stimulated with immune complexes, thrombin or leucocyte-derived PAF release granule, contents, eicosanoids, serotonin and lysosomal enzymes. Platelet alpha granules contain Clinh which inhibits activation of C1 and kallikrein. Serotonin, a vasoactive amine, is also stored in alpha granules and metabolized by endothelial cells. Platelets also contain a well-defined growth factor, PDGF, which is mitogenic and chemotactic for vascular smooth-muscle cells. It may play an important role in vessel wall repair through its effects on smooth-muscle cells. Lipoxygenase-dependent arachidonic acid metabolites released by platelets may be processed by leucocytes into leukotrienes. Several of the enzymes that may be released by platelets, such as collagenase, elastase and heparinase, act on vascular structures. Activated platelets express membrane-associated procoagulant activity and fibronectin which triggers adhesion of monocytes.

Complex interactions occur between complement and proteins of the coagulation and fibrinolytic systems, as recently reviewed by Sundsmo and Fair [105]. Many of these interactions have been studied in purified protein systems and in general only occur at relatively high concentrations, thus raising the question of physiological relevance. However, since biological surfaces often provide catalytic sites for these reactions and since a local concentration of plasma proteins may occur during inflammatory response, the possibility exists for local complement–coagulation interactions in vivo.

Hereditary angioneurotic oedema is a genetic disease in which afflicted individuals are C1-inhibitor (C1inh) deficient. Kallikrein is unregulated in this disease since its functioning is governed by the presence of C1inh. In vitro plasmin generation may directly act on C1 to activate the complement system or inactivate C1inh, thereby releasing C1 from inhibition. Whether this is the actual mode of the tissue-factor-induced complement activation in vivo is unknown. Plasmin can also cleave C3 and C5 but in vitro this process is inefficient compared to the action of C3 convertase. Clot lysis and retraction are inhibited in blood de-complemented by the alternative pathway, indicating a connection between complement and fibrinolysis. This may involve kallikrein activation of factor B to produce Bb which may in turn activate plasminogen. Kallikrein appears to cleave rabbit C5 and human C3 but its relative efficiency has not been reported. In most cases (e.g. thrombin-mediated C3/C5 cleavage) the cleavage efficiency is very low compared to that normally obtained with physiological enzymes.

The mechanisms by which leucocytes contribute to thrombogenesis are poorly understood. Tissue trauma and ischaemia may lead to release of agents which promote increased vascular permeability (e.g. kinins, prostaglandins and histamine) and to exposure of factor XII to collagen. Activated factor XII promotes generation of chemotactic factors which can potentiate leucocyte aggregation, adhesion to endothelium and migration into the vessel wall and underlying tissues. One of these chemotactic factors, LTB4, enhances neutrophil adhesion and emigration both directly and through its ability to increase vascular permeability. Leucocytes may generate significant pro-coagulant activity following prolonged incubation in vitro. The activity is restricted to monocytes and macrophages which have been activated, for example by reaction with specifically sensitized lymphocytes [29]. In vitro stimulation of peripheral blood lymphocytes with LPS or immune complexes results in a T cell-dependent lymphokine-mediated stimulation of tissue factor production by monocytes/macrophages. Tissue factor, a lipoprotein expressed on the cell surface, can initiate the extrinsic pathway of coagulation. Monocytes/macrophages also synthesize and assemble FVII/VIIa on their surface. FVII/VIIa binds to tissue factor in a calcium-

dependent process to form a complex capable of converting factor X to its active form.

Proteases from leucocytes may be the source of proteolytic activity which initiates intrinsic coagulation and cleavage of C3. Leucocyte synthesis of eicosanoids and release of lysosomal enzymes from activated polymorphs can both be stimulated or inhibited by different eicosanoids. Leukotrienes formed in the lipoxygenase pathway, prostaglandins and thromboxanes are powerful mediators of the inflammatory reaction and also interact with platelets. Sensitized basophils and macrophages can elaborate platelet-activating factor (PAF), which is a potent inflammatory agent [66].

Several mechanisms are probably responsible for limitation of thrombus growth by leucocytes. Activated neutrophils generate a variety of oxidative species which impair platelet aggregation. In addition, synthesis of prostaglandin E_1 and inactivation of ADP by neutrophils inhibits platelet aggregation. Neutrophil proteases may inactivate coagulation factors and thus limit fibrin formation. During the later stages of thrombogenesis, neutrophils (and monocytes), through plasmin-dependent and plasmin-independent fibrinolysis, play a role in the resolution of thrombi and in addition phagocytose fibrin strands, platelet aggregates and cell debris. Neutrophil granules are rich in highly charged cationic proteins, as well as glycosaminoglycans which resemble chondroitin sulphate. These promote the formation of complexes of soluble fibrin monomers with fibrinogen or fibrin degradation products. Lastly, C3a and C5a can stimulate IL-1 release from monocytes which can induce endothelial cell pro- and anti-coagulatory activities.

Experimental Models of Immunologically Mediated Vascular Lesions

This section will review briefly experimental models of immune-complex-induced and complement-induced vascular injury. As mentioned earlier, mechanisms by which antigen–antibody complexes may form and/or deposit in the vascular wall include: (a) local tissue interaction between antigen and pre-formed antibody (Arthus reaction); (b) deposition of circulating immune complexes; and (c) interaction of circulating antibody with an antigen in situ, whether the antigen is a constitutive vascular wall antigen or whether it is an exogenous "planted" antigens.

The Arthus reaction was initially observed after intradermal injection of an antigen into appropriately sensitized animals, the result being a localized acute necrotizing vasculitis [16]. For induction of this type of vasculitis, relatively large amounts of antigen–antibody precipitates must form in the vessel wall and perivascular areas. One of the reactants, either antigen or antibody, circulates in the peripheral blood and is distributed in extravascular spaces. The other reactant is injected locally. Within a few hours, the injected antigen (active Arthus reaction) or the injected antibody (passive Arthus reaction) diffuses from the site of injection forming a decreasing concentration gradient. Within the area of diffusion, immune precipitates will form at the site where the antigen/antibody ratio is optimal. Locally formed immune complexes will react with complement and induce release of anaphylatoxins which enhance vascular permeability and attract (via C5a, C5adesArg) polymorphonuclear leucocytes to the site of lesions. Upon contact with the immune complexes, polymorphonuclear leucocytes liberate lysosomal enzymes which contributes to necrosis at the site of their accumulation.

Serum sickness reactions provide a model of vascular damage induced by deposition of circulating immune complexes [17,40]. Acute serum sickness is produced in rabbits by administration of a single large dose of heterologous protein (e.g. bovine serum albumin). The fate of the injected antigen may be followed by using a radiolabelled protein. Following an initial period in which the antigen equilibrates in extravascular spaces, a slow linear logarithmic decrease in antigen plasma concentration occurs due to the physiological catabolism of the protein. Then, an accelerated clearance of the antigen is observed coincident to the appearance of antibody. During this period, soluble antigen–antibody complexes form (because of antigen excess) which activate complement and localize within glomerular capillaries and other capillary beds producing vasculitis. The antigen and antibody can be detected within circulating complexes and in glomerular capillaries in association with C3 fragments. The vasculitis is characterized by infiltration of neutrophils whereas glomerular deposits are primarily associated with a mononuclear cell infiltrate. After day 12, only free antibody is detected in the circulation coincident to a transient increase in immunoglobulin and complement deposition in glomeruli.

Chronic serum sickness is induced by repeated injections of a heterologous protein. The amount of injected antigen is calculated to ensure antigen excess. The site of immune complex deposition within the glomeruli depends on size, charge and other characteristics of the complexes. Thus, small antigen–antibody complexes tend to form subepithelial deposits in glomeruli whereas larger complexes tend to localize in mesangial areas. Chronic serum sickness is not associated with the occurrence of extraglomerular vascular deposits of immune complexes. Several animal models have been characterized in which the occurrence of glomerulonephritis and vasculitis is associated with circulating complexes in antigen excess. For example, infection of neonatal mice with lymphochoriomeningitic virus induces a state of partial tolerance to the antigen, persistent viraemia, formation of viral–anti-viral immune complexes and a chronic form of glomerulonephritis in which specific anti-viral antibodies and viral antigen deposits are found. This model is in many ways similar to human postviral systemic vasculitis seen in persistent infection with HBV. The virus replicates continuously in hepatocytes where it exerts mild to no cytopathic effects because replication is strictly intracellular. Patients have systemic vasculitis with circulating immune complexes containing HBs antigen, IgM, IgG cryoprecipitating complexes and deposits of HBs antigen anti-HBs antibodies in vessel walls [27].

Some strains of mice (female F1 hybrids from NZB and NZW strains, male BXSB mice and MRL/1 mice) spontaneously develop an autoimmune disease that shares many immunopathological features with human SLE, including hypergammaglobulinaemia, antinuclear antibodies, anti-T cell antibodies, haemolytic anaemia, rheumatoid factors and occurrence of glomerulonephritis and vasculitis (which is more severe in MRL/1 animals) with extensive deposits of IgG and fragments of C3 [106]. Circulating immune complexes containing DNA and anti-DNA antibodies, the viral antigen gp 70 and anti-gp 70 antibodies can be detected in the circulation before immune complexes are found in glomeruli. The pathogenic role of circulating complexes in these animals is suggested by the finding of DNA, gp 70, anti-DNA and anti-gp 70 antibodies in eluates from renal biopsies, and by the fact that administration of denatured DNA to young BW hybrids increases the severity of nephritis. Immune-complex-mediated glomerular and vascular lesions similar to those found in autoimmune mice may be induced in animals given polyclonal B cell activators such as trypanosomes or LPS wherein auto-antibodies of various specificities develop [51]. The same mechanism operates in mice during expreimentally induced graft versus host reaction obtained by injecting spleen cells from a parent to an F1 hybrid. In the latter model, polyclonal activation of B cells is dependent on semi-allogenicity of T cells from the donor [41].

Animal models of in situ antigen–antibody complex formation have generally examined mechanisms of immune complex formation in glomerular capillaries rather than models of immune-complex-mediated systemic vasculitis. The lack of systemic vascular deposits in animals with chronic serum sickness suggests, however, that in situ immune complex formation is the major pathway by which immune deposits may form in systemic small vessel vasculitides. As discussed earlier, in situ immune complex formation may occur between circulating antibodies and insoluble antigens fixed in vessels (i.e. glomerular basement membrane) or between circulating antibodies and a variety of soluble endogenous and exogenous antigens that first become localized in their free form ("planted") in the capillary.

The classical model of anti-GBM disease is that of heterologous anti-GBM nephritis oberved in rats injected with rabbit anti-rat glomerular basement membrane antibodies [114]. In the heterologous phase of the disease, rabbit antibodies rapidly bind to the glomerular basement membrane in the kidney and to the basement membrane in the lungs where they can be detected as linear deposits in association with C3 by indirect immunofluorescence. If sufficient amounts of antibody have been injected, proteinuria appears. The histological picture of glomerular lesions is that of a diffuse endocapillary and extracapillary proliferative glomerulonephritis. Neutrophils, and at a later stage, monocytes, accumulate in the capillaries. Complement is required for leucocyte infiltration and proteinuria to appear. The "autologous" phase of the disease occurs after 8–10 d, following the binding of rat anti-rabbit immunoglobulin antibodies to the fixed anti-GBM antibodies. Experimental models in which mice were injected with specific heterologous anti-laminin or anti-type IV collagen antibodies revealed morphological changes in the glomerular basement membrane and the occurrence of proteinuria. Autoimmune models of anti-

GBM disease are induced in sheep by immunizing the animals with heterologous glomerular basement membrane in adjuvant, which leads to production of anti-GBM antibodies and to severe anti-GBM nephritis [101]. Repeated injections of HgC12 or of D-penicillamine induce anti-GBM auto-antibodies in susceptible strains of rats, emphasizing the role of genetic factors in the susceptibility to autoimmune disease. Another model involving in situ immune complex formation is Heymann nephritis which occurs when rats are immunized with tubular brush border antigens (FxIA). Immunohistochemical analysis of diseased kidney demonstrates the presence of subepithelial IgG deposits. The disease was first believed to be mediated by circulating immune complexes until a passive model was developed in which rats were injected with heterologous antibodies to FxIA which bind to a cross-reacting antigen expressed on the basal part of glomerular epithelial cells [10]. Proteinuria in this model is complement-dependent but cell-independent, in that no inflammatory cells participate in the lesions.

Several models have been developed in which subepithelial and subendothelial in situ formation of immune complexes in glomeruli can be triggered by initial localization of antigen or antibody or by trapping of pre-formed immune complexes: (a) subepithelial immune deposit formation may be induced by initial localization in the glomerular capillary wall of large cationic proteins, including BSA, IgG, ferritin and the native cationic protein lyzozyme [9,38,113]; (b) in situ immune complex formation can also occur with anionic antigens. Thus, free DNA may bind to the glomerular basement membrane and initiate in situ formation of immune complexes [51]. How anionic molecules deposit in the anionic glomerular capillary wall remains unclear. One possibility (as suggested in a model of SLE nephritis) is that a loss of anionic sites occurs in the capillaries, which precedes localization of complexes. A variety of proteins involved in immune reactions, such as neutrophil cationic proteins or cationic products of complement activation may have such charge-dependent effects. Alternatively, binding of cationic antibodies to anionic antigens could serve to localize the antigen in a vessel wall by a charge-dependent mechanism. This mechanism could account for subepithelial immune deposition observed after alternate infusion of BSA and anti-BSA antibodies in animals. It also suggests that pre-formed antibodies could initiate formation of immune complexes in an antibody excess situation. Finally, as shown with cationized bovine gammaglobulin–anti-bovine gammaglobulin complexes, initial trapping of immune complexes may be the primary event which triggers immune complex formation [36]. Subsequent interaction of complexes with antigen, antibodies, complement and/or anti-antibodies such as anti-idiotypes and rheumatoid factors contributes to local rearrangement of the complexes and alterations in their phlogistic potential [75].

Although the primary causative antigen is unknown in most human vasculitides, these models are probably relevant for the pathogenesis of most human vasculitides and glomerulonephritides with immune deposits.

Immunologically Mediated Vascular Lesions in Man

Within the spectrum of human vascular diseases, the following disorder categories may be considered to be dependent on one or several of the immunological mechanisms described above: small vessel vasculitides, polyarteritis nodosa and related syndromes, pulmonary endothelial cell injury in adult respiratory distress syndrome (ARDS) and vascular rejection of transplants. Table 26.21 lists a number of clinical circumstances in which evidence suggestive of immunologically mediated vascular damage is available. Such evidence may be direct or indirect and includes:

1. Identification of a causative antigen and demonstration of a systemic and/or local specific immune response against the antigen.
2. Clinical (chronological) association between the occurrence of the disease and exposure to infectious antigens or drugs.
3. Clinical and histopathological similarity of the disease to experimental models of immunologically mediated vasculitis (e.g. serum sickness, immune complex vasculitis in animals with persistent viral infections, vasculitis in autoimmune mice with SLE-like syndromes).
4. Immunohistochemical evidence of immunoglobulin deposits and/or deposition of complement proteins; presence of vascular and/or perivascular mononuclear cell infiltrates.

Table 26.21. Immunologically mediated vascular lesions in humans

Systemic small vessel vasculitides
  Immune-complex-mediated diseases
    Immune complex vasculitides (small vessel vasculitides) involving foreign antigens:
      Heterologous proteins: serum sickness
      Viruses: hepatitis B virus; cytomegalovirus; mononucleosis; rubella; mycoplasma
      Bacteria: bacterial endocarditis; shunt nephritis; leprosy
      Parasites: malaria
      Drugs: penicillamine; propylthiouracil; antibiotics
    Putative immune complex vasculitides (small vessel vasculitides):
      Systemic lupus erythematosus
      Rheumatoid arthritis
      Sjögren's syndrome
      Cryoglobulinaemia
      Hypocomplementemic vasculitis
    Diseases involving IgA-containing immune complexes and/or macromolecular IgA (small vessel vasculitides):
      Schönlein–Henoch purpura
      Systemic lupus erythematosus (some)
      Dermatitis herpetiformis
  Small vessel vasculitides with granulomatous changes with/without evidence for immune-complex-mediated damage:
    Wegener's granulomatosis
    Lymphoid granulomatosis
    Behcet's disease
Systemic necrotizing vasculitides of the polyarteritis group
  Classic polyarteritis nodosa
  Churg and Strauss granulomatosis
Giant cell arteritides
  Temporal arteritis and Takayashu's disease (immunological mechanisms are only presumptive)
Pulmonary endothelial cell injury mediated by intravascular complement activation
  Extracorporeal circulation
  Adult respiratory distress syndrome (ARDS)
Vascular injury in transplant rejection

5. Systemic immunological abnormalities including hypergammaglobulinaemia, cryoglobulinaemia, rheumatoid factors, circulating immune complexes and acquired hypocomplementaemia.
6. Clinical improvement of the disease associated with plasma exchange, steroid and/or immunosuppressive therapy.

The clinical spectrum of vasculitides is discussed in Chaps. 27 and 28. Within this group, systemic small vessel vasculitides and medium-size vessel vasculitides of the polyarteritis group are heterogeneous disorders that are generally believed to be caused by immune complexes. Deposits of complement and immunoglobulins are often found at the site of vascular lesions. Circulating immune complexes may be found which relate to the activity of the disease. Small vessel vasculitis may occur in an idiopathic form, in association with autoimmune diseases such as systemic lupus erythematosus, rheumatoid arthritis and cryoglobulinaemia, and in association with well-characterized bacterial, parasitic and viral diseases. In subacute bacterial endocarditis, immune complexes have been found in patients with a disease of long duration and with extravalvular vascular spread [55]. In meningococcal disease, vasculitis, iritis and neuropathy are thought to be due to the formation and deposition of immune complexes in tissues. Complexes of antigen and specific antibodies are also found in patients with disseminated gonococcal infection. The erythematous skin rash observed in this condition is evidence of a transient form of vasculitis. The vasculitis that may complicate acute rheumatic fever could also be due to type A beta-haemolytic streptococcus-containing immune complexes. Another well-documented form of immune complex vasculitis accompanying bacterial disease is *Staphylococcus albus* shunt nephritis [47].

In leprosy, the infective agent is relatively non-toxic (patients with early lepromatous leprosy may have many millions of bacilli per gram of skin tissue without overt clinical symptoms) so that the contribution of host factors appears to be of major importance in the evolution of this disease. Erythema nodosum leprosum (ENL), a reactive state which occurs usually towards the end of the 1st year of treatment, is characterized by tender subcutaneous nodules, which are histologically characterized by polymorphonuclear leucocyte infiltration and deposits of IgG and complement. There is evidence of an immune complex involvement in the pathogenesis of ENL based on clinical and immunopathological tissue studies [110]. This type of lesion, which is sometimes accompanied by albuminuria, resembles an Arthus reaction. Immune complexes have most frequently been found in the circulation of patients with this particular phase of the disease. Incubation of certain immune-complex-containing sera with *Mycobacterium leprae* decreased the amount of ^{125}I–C1q binding material to about half the control levels suggesting that *M. leprae* is an important antigen in these complexes. Antigen–antibody systems other than those involving mycobacterial antigens may also be present in leprosy sera. Mixed cryoglobulins with an IgM component and exhibiting anti-IgG activity have been found in some sera. The finding of increased levels of C3d, a catabolic fragment of C3, in 70% of erythema nodosum leprosum plasma samples provides

additional evidence that complement is activated in vivo, possibly by circulating immune complexes.

In parasitic disease, the most comprehensive data implicating immune complexes in the pathogenicity of vascular lesions relates to lesions in the glomerular tuft and the choroid plexus. A glomerulonephritis with immunohistopathological features similar to those occurring during acute serum sickness is observed during the course of acute *Plasmodium falciparum* infection. In man, deposits of immunoglobulins, complement and *P. malariae* antigens are found in glomerular capillary walls. A study of antibodies eluted from renal biopsies demonstrated the presence of antibodies specific for *P. malariae* in the majority of cases. In human and murine malaria, circulating immune complexes have been detected by two different techniques. Evidence that such complexes could be deposited or formed in situ in tissues comes from the finding by immunofluorescence of IgG in choroid plexus.

Persistent viral infections occurring in a variety of hosts are associated with virus-induced immune complex deposits and disease. Since viruses are self-replicating agents, they provide an important supply of antigenic material which in most instances induces an immune response. After production of anti-viral antibodies by the host and liberation of infectious viral particles by the infected cells, immune complexes can be formed. Immune complexes are also formed on virus-infected cells when infections are caused by budding viruses such as measles, mumps and *Herpes* in man. As mentioned earlier, an example of chronic viral disease associated with immune complex vasculitis is persistent infection with hepatitis B virus (HBV) [67]. The association of persistent HB antigenaemia and generalized necrotizing vasculitis is now well recognized. Particular attention was drawn to this association by the discovery of HBsAg and of HBs–anti-HBs circulating immune complexes in the sera of patients with polyarteritis and the demonstration of HbsAg in vascular lesions of a patient who developed polyarteritis 3 weeks after transfusion of HbsAg-positive blood. Deposition or in situ formation of hepatitis B virus-containing immune complexes in the vessel walls of susceptible persons could lead to activation of complement and initiation of the inflammatory vascular response characteristic of polyarteritis nodosa. It is reasonable to suspect that circulating immune complexes of HBV–anti-HBV are pathogenic in this syndrome even though it may be difficult, in a given patient, to detect directly such complexes due to their transient appearance. There is no apparent relationship, however, between onset and activity of vasculitis and onset and activity of liver disease. Thus, all combinations may be observed, including vasculitis as the initial clinical illness and expression of severe vasculitis together with subclinical liver disease.

Laboratory Assessment of Immunologically Mediated Vascular Lesions

In Situ Analysis

In situ analysis of immune components may provide direct evidence for the involvement of immune mechanisms in the pathogenesis of vascular lesions. Immunohistochemical techniques (i.e. immunofluorescence and immunoperoxidase staining) allow the detection of antibodies, antigens and complement proteins, and the determination of phenotypic characteristics of infiltrating cells in the lesions. Thus, IgG deposits are usually found in vessel wall biopsies from patients with active SLE, and IgA deposits are found in biopsies from patients with Schönlein–Henoch purpura. Anti-light chain antibodies and antibodies against isotypic determinants may be evidence of monoclonal immunoglobulin deposits in vasculitis associated with monoclonal B cell proliferation. The availability of anti-sera against C3 fragments and complement activation products (e.g. neo-antigens expressed by the C5b-9 terminal complex of complement) allows direct assessment of in situ complement activation in tissue lesions (Table 26.22) (Fig. 26.7) [49].

Using a combination of antibodies, in vivo cleavage of C3 could be demonstrated at the site of immune complex formation in the early stage of epimembranous nephritis with the finding of superimposable patterns of staining with anti-C3, anti-C3d and anti-H antibodies. No C5b-9 neo-antigens were detected in C3 deposits indicating that in situ cleavage of C3 was effectively regulated by control proteins. However, at later stages of the disease, C3 cleavage products and C5b-9 neo-antigens were found in the same location in glomerular capillary walls indicating that secondary modifications of immune complexes had enhanced their complement-activating potential thus circumventing regulatory

Table 26.22. Assessment of complement activation in tissues using antibodies to complement antigens

	No evidence for complement activation	C3 cleavage but no recruitment of terminal sequence	Activation of the whole complement sequence
Anti-C3 (anti-C3c antigen)	−	+	+
Anti-C3d	−	+	+
Anti-C3g	−	+	−/+
Anti-H	−	+	−/+
Anti-(C5b-9) neo-antigens	−	−	+

mechanisms and allowing cleavage of C5 and activation of the terminal effector sequence. Complement deposits may also be found in vessels independently of the presence of immunoglobulins. In large vessels, C5b-9 neo-antigens have been localized in fibrous plaques but not in atherosclerotic-free intimas, suggesting a pathogenic involvement of complement in the progression of the atherosclerotic lesion [112].

The use of monoclonal antibodies has allowed precise identification of the surface phenotype of infiltrating cells in granulomas and perivasular infiltrates in some forms of vasculitis and vascular transplant rejection [77].

Cell surface antigens on the major subsets of infiltrating lymphocytes are listed in Table 26.23.

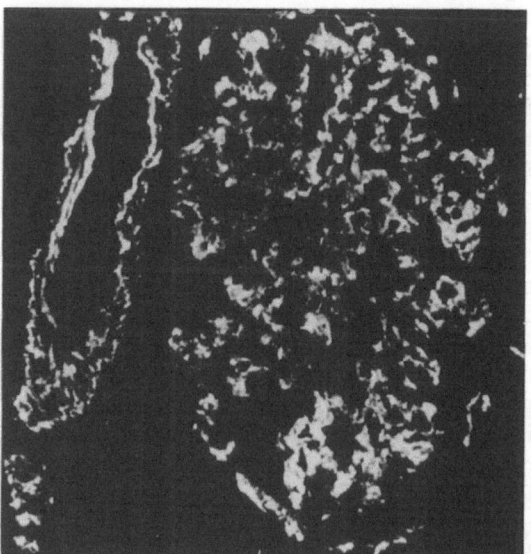

Fig. 26.7. Nephritis during systemic lupus erythematosus. Staining of adjacent sections of the same glomerulus and artery with anti-(C5b-9)-neo-antigens. × 450

Table 26.23. Surface antigens of major mononuclear cell lines

Cluster of differentiation	CD3 (OKT3-Leu4)	CD4 (OKT3-Leu4)	CD8 (OKT8-Leu2)	CD19 (BR Leu12)	CD16 (Leu11)	CD11b (OKM1,Mol)
Helper T cell	+	+	−	−	−	−
Suppressor T cell	+	−	+	−	−	−
Cytotoxic T cell	+	−	+	−	−	+
B cells	−	−	−	+	−	−
NK cells	−	−	−	−	+	−/+
Monocytes	−	−	−	−	−	+

Assessment of Humoral Abnormalities Associated with Immunologically Mediated Vasculitis

These assessments include the detection of soluble immune complexes, the detection of cryoglobulins and the assessment of complement activation and of complement receptors on circulating cells (Table 26.24). Currently used immune complex assays are designed to detect an immunoglobulin or complement component in antigen–antibody complexes, but not the antigen itself [117]. Identification of specific antigens in complexes is therefore often impossible, except in a few cases when the antigen could be identified as the HBs antigen for example. Knowledge of the composition of circulating complexes would provide support for their pathogenic importance in those cases in which immune complexes of identical composition are

Table 26.24. Laboratory assessment of the interaction of immune complexes with complement

Principle	Assays
Detection of immune complexes with complement reactants	C1q-binding assay Anti-C3 assays Conglutinin-binding assay Raji cell assay
Simultaneous demonstration of complement activation: decreased levels of complement components	CH50 C4 (immunochemical) C4, C2 (haemolytic)
Products generated by complement activation	C3dg (immunochemical) C3a, SC5b-9 (radioimmunoassay)
Capacity of patient's serum to process immune complexes	Inhibition of immune complex precipitation
Capacity of patient's cells to process immune complexes	Clearance of antibody-coated autologous cells; enumeration of C3b receptors

found in tissue deposits. Determination of the size and composition of complexes may be important since current evidence suggests different mechanisms of clearance from the circulation for soluble complexes of different composition or size. In most current assays however, the size of reacting material is not well known. The availability of assays capable of discriminating between large (>21 S) and small (<21 S) IgA-containing immune complexes in serum have allowed the detection of large IgA macromolecular complexes in patients with Schönlein–Henoch purpura [54].

Among the numerous assays that have been described for the detection of soluble immune complexes, those using complement as a recognition system are of value for detecting complexes that could activate complement in vivo, for example the C1q-binding assay, the anti-C3 assays, the conglutinin-binding assay and the Raji cell assay. Although the conglutinin-binding assay and the Raji cell assay are held to be technically reliable, they detect C3bi deposited on immune complexes, which is only one of the breakdown products of C3 in the proteolytic C3 degrading sequence. Hence, with a single type of immune complex assay only a selected population of complexes is revealed. The C1q-binding assay has proven its reliability in the follow-up of patients with SLE. The presence of circulating complexes cannot, however, be considered as a diagnostic criterion. The incidence of a positive C1q-binding assay in patients with renal lupus is higher than in patients without renal SLE, but the sensitivity and specificity of this finding is too low to be of predictive value in individual patients. Different glomerular lesions in lupus nephritis have been associated with different types of immune complexes, especially of different sizes, but such findings in serum cannot replace histological examination of renal tissue.

The simultaneous use of multiple assays for detecting immune complexes may be helpful in differentiating between various forms of vasculitis, such as cutaneous versus systemic and acute versus chronic or between different forms of systemic small vessel vasculitis [53]. In patients with Schönlein–Henoch purpura, IgA-containing immune complexes are present during the acute phase and absent during remission. Furthermore, the C1q-binding assay is usually negative, whereas it is often positive in patients with other types of systemic small vessel vasculitis. Thus, the use of two immune complex assays may help to differentiate between various forms of small vessel vasculitis. In diseases such as Wegener's granulomatosis, periarteritis nodosa and giant cell arteritis the presence of circulating immune complexes does not have any diagnostic significance.

Many studies have appeared concerning the relationship between the presence of soluble complexes and disease activity or the outcome of treatment (immunosuppressive and/or plasma exchange). At best the results of immune complex assays can be used in combination with other clinical parameters to monitor disease activity. It is important to note that several studies have shown that acute renal involvement in systemic small vessel vasculitis may initially be associated with the presence of circulating immune complexes. However, in patients in whom the renal manifestations of small vessel vasculitis led to progressive renal failure, circulating immune complexes were no longer detectable [56]. This suggests that soluble complexes are not involved in the later phase of glomerulonephritis. In patients with acute cutaneous vasculitis, circulating immune complexes are usually absent, in contrast to patients in whom the cutaneous manifestations are part of a systemic vasculitis disorder.

Circulating immune complexes have been described in many bacterial, viral and parasitic infections. In patients with infective endocarditis, 63% will exhibit a positive C1q-binding assay [55]. This incidence is significantly higher than among patients with septicaemia without endocardial infection in whom the C1q-binding assay is usually negative. The use of the C1q-binding assay may therefore be relevant for the diagnosis of infective endocarditis. In infections with hepatitis B virus, circulating immune complexes have been demonstrated during the prodromal, acute and chronic phases. Several studies have shown the presence of hepatitis B antigen and antibody in circulating complexes as well as in tissue deposits suggesting an important pathogenic role for immune complexes. Their presence is not however of diagnostic or prognostic importance.

In parallel with the detection of soluble immune complexes, their complement-activating capacity may be readily assessed by measuring functional activity of individual complement components in serum [75]. However, normal levels of total complement or of single components do not rule out consumption of complement components if such consumption is masked by hypersynthesis, which often occurs in chronic inflammatory diseases. Decreased complement levels, for example of CH50 and C4, do not allow a distinction to be made between complement consumption or primary deficiency which can both be associated with immune complex disease. Therefore, assessment of breakdown products generated during complement activation is used to obtain direct evidence for in vivo complement activation. Such tests include: (a) the immunochemical quantitation of $C1rC1s$ $(C1$ $inh)_2$ released from activated C1 on immune complexes following classical pathway activation: (b) the sensitive radioimmunochemical assessment of generated C3a anaphylatoxin [13]; (c) immunochemical quantitation of the C3dg/C3d fragment in polyethyleneglycol-precipitated serum. Levels of C3d were shown to correlate significantly with the presence of C1q-binding immune complexes in extracellular fluids of patients with systemic lupus erythematosus (Fig. 26.8); and (d) radioimmunochemical determination of SC5b-9 in patient's plasma [70]. The latter assay may provide direct evidence for the capacity of circulating complexes to activate the terminal sequence, which most biological effects of activated complement depend on. It is of interest that preliminary results indicate a relatively poor capacity of circulating complexes in patients with systemic lupus erythematosus to cleave C5 and generate SC5b-9.

The capacity of patients' sera to keep immune complexes in solution can be assessed by evaluating the capacity of the test serum to prevent precipitation of heterologous bovine serum albumin (BSA)–anti-BSA complexes or by evaluating the capacity of such serum to solubilize immune complexes pre-formed in slight antibody excess [93]. In patients with systemic lupus erythematosus the results of the latter test were shown to correlate significantly with the functional capacity of the alternative pathway and with levels of factor B.

Survival studies of isoagglutinin-bearing and chromium-labelled erythrocytes can be used as

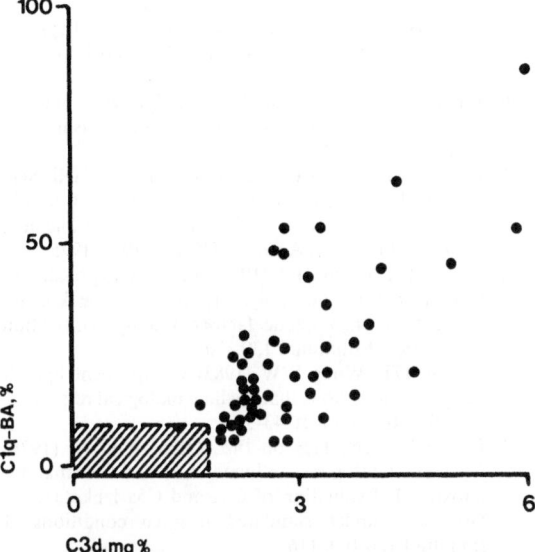

Fig. 26.8. Correlation between C1q-binding assay (ordinate) and serum C3d levels (abscissa) in patients with systemic lupus erythematosus. Shadowed area indicates range found in 30 normal blood donors.

a tool to investigate "C3b-receptor-mediated clearance" of immune complexes [37]. The clearance was found to be retarded in patients with primary biliary cirrhosis and in some patients with systemic lupus erythematosus, whereas it was normal in patients with other liver diseases and in patients having undergone splenectomy. Defective clearance of autologous erythrocytes sensitized with IgG ("Fc receptor function") was found in patients with systemic lupus erythematosus, in whom the clearance was improved following treatment with plasma exchange, and in patients with Sjögren's syndrome and DRW3 individuals.

The number of C3b receptors on erythrocytes can be measured using radiolabelled antibodies to CR1 [115]. In normal blood donors, assuming the existence of two codominant alleles coding for the number of receptors per erythrocyte, approximately 54% of subjects are homozygous for the high receptor allele, 12% homozygous for the low number of receptors and 34% are heterozygous. In systemic lupus erythematosus, CR1 numbers are decreased so that approximately 60% of patients are homozygous for the low receptor number phenotype, and only 5% are homozygous for high CR1 numbers. A role for inheritance in the decreased numbers of CR1 on erythrocytes of SLE patients has been suggested since healthy relatives of the patients also had low numbers of CR1 on erythrocytes.

Acquired factors also contribute to decreased expression of CR1 in SLE, as indicated by the following observations: (a) numbers of CR1 on erythrocytes in some patients with systemic lupus erythematosus were lower than would have been predicted by inheritance; (b) no relationship between CR1 numbers on erythrocytes was found between patients and offspring in some studies of the families of patients with SLE; (c) higher numbers of CR1 sites on erythrocytes have been observed in patients with remission than in patients with active disease. No correlation was found however between CR1 numbers on erythrocytes and the occurrence or immunohistopathological type and severity of lupus nephritis; and (d) three patients with SLE and low numbers of CR1 on erythrocytes have been found to have antibodies to the receptor. Independent of the mechanism by which cellular expression of CR1 may be decreased in SLE, the role of CR1 in the processing of circulating immune complexes strongly suggests that a deficiency in CR1 could predispose to impaired clearance of immune complexes and immune-complex-mediated vasculitis in SLE patients.

References

1. Asch AS, Kinoshita T, Jaffe EA, Nussenzweig V (1986) Decay-accelerating factor is present on cultured human umbilical vein endothelial cells. J Exp Med 163:221–226
2. Arnaout MA, Hakim RM, Todd RF III, Dana N, Colten HR (1985) Increased expression of an adhesion-promoting surface glycoprotein in the granulocytopenia of hemodialysis. N Engl J Med 312:457–462
3. Babior BM (1980) Oxygen-dependent microbial killing by phagocytes. N Engl J Med 298:659–721
4. Bevilacqua MP, Pober JS, Wheeler ME et al. (1985) Interleukin-1 acts on cultured human vascular endothelium to increase the adhesion of polymorphonuclear leukocytes, monocytes and related leukocyte cell lines. J Clin Invest 76:2003–2011
5. Bevilacqua MP, Schleef RR, Gimbrone MA et al. (1986) Regulation of the fibrinolytic system of cultured human vascular endothelium by interleukin-1. J Clin Invest 78:587–591
6. Bhakdi S, Tranum-Jensen J (1983) Membrane damage by complement. Biochim Biophys Acta 737:343–372
7. Biesecker G, Katz S, Koffler D (1981) Renal localization of the membrane attack complex in systemic lupus erythematosus nephritis. J Exp Med 154:1779–1794
8. Biesecker G, Lavin L, Ziskind M et al. (1982) Cutaneous localization of the membrane attack complex in discoid and systemic lupus erythematosus. N Engl J Med 306:264–270
9. Border WA, Ward HJ, Kamil ES et al. (1982) Induction

of membranous nephropathy in rabbits by administration of an exogenic cationic antigen: demonstration of a pathogenic role for electrical charge. J Clin Invest 69:451–461

10. Camussi G, Brentsens JR, Kerjashchki D et al. (1985) Antibody-induced redistribution of Heyman's antigen on the surface of visceral glomerular epithelial cells in Lewis rats. Kidney Int. 27:207 (abstract)

11. Cerilli J, Brasile L, Galousis T et al. (1985) The vascular endothelial cell antigen system. Transplantation 39(3):286–289

12. Chenoweth DE, Hugli TE (1978) Demonstration of a specific C5a receptor on intact human polymorphonuclear leukocytes. Proc Natl Acad Sci USA 75:3943–3947

13. Chenoweth DE, Hugli TE (1982) Assays for chemotactic factors and anaphylatoxins. In: Nakamura RM, Dito WR, Tucker ES III (eds) Immunologic analysis: recent progress in diagnostic laboratory immunology. Masson, New York, pp 227–237

14. Chenoweth DE, Cheung AK, Ward DM et al. (1983) Anaphylatoxin formation during hemodialysis: effects of different dialyzer membranes. Kidney Int 24:770–774

15. Cines DB, Lyss AP, Reeber M et al. (1983) Presence of complement-fixing anti-endothelial cell antibodies in systemic lupus erythematosus. J Clin Invest 73:611–625

16. Cochrane CG, Weigle WO (1958) The cutaneous reaction to soluble antigen–antibody complexes. J Exp Med 108:591–607

17. Cochrane CG, Koffler D (1973) Immune complex disease in experimental animals and man. Adv Immunol 16:185–215

18. Collins T, Lapierre LA, Fiers W et al. (1986) Recombinant human tumor necrosis factors increases mRNA levels and surface expression of HLA-A,B antigens in vascular endothelial cells and dermal fibroblasts in vitro. Proc Natl Acad Sci USA 83:446–450

19. Cornacoff JB, Hebert LA, Smead WL et al. (1983) Primate erythrocyte immune complex clearing mechanism. J Clin Invest 71:236–247

20. Cotran RS, Gimbrone MA Jr, Bevilacqua MP et al. (1986) Induction and detection of a human endothelial activation antigen in vivo. J Exp Med 164:661–666

21. Couser WG (1985) Mechanisms of glomerular injury in immune-complex disease. Kidney Int 28:569–583

22. Couser WG, Baker PJ, Adler S (1985) Complement and the direct mediation of immune glomerular injury: a new perspective. Kidney Int 28:879–890

23. Craddock PR, Hammerschmitt DE, White JG et al. (1977) Complement (C5a)-induced granulocyte aggregation in vitro. J Clin Invest 60:260–264

24. Daha MR, Van Es LA (1979) Further evidence for the antibody nature of C3 nephritic factor (C3Nef). J Immunol 123:755–758

25. Davies Ph, Bailey PJ, Goldenberg MM (1984) The role of arachidonic acid oxygenation products in pain and inflammation. Ann Rev Immunol 2:335–358

26. Dinarello CA (1984) Interleukin-1 and the pathogenesis of the acute phase response. N Engl J Med 311:1413–1418

27. Duffy J, Lidsky MD, Sharp JJ et al. (1976) Polyarthritis, polyarteritis and hepatitis B. Medicine 55:19–35

28. Durum SK, Schmidt JA, Oppenheim JJ (1985) Interleukin-1: an immunological perspective. Ann Rev Immunol 3:263–287

29. Edwards RL, Ewan VA, Rickles FR (1986) Macrophage procoagulants, fibrin deposition and the inflammatory response. In: Phillips SM, Escobar MR (eds) Hypersensitivity, the reticuloendothelial system, a comprehensive treatise. Plenum, New York, pp 233–266

30. Emancipator SN, Lamm ME (1986) Pathways of tissue injury initiated by humoral immune mechanisms. Lab Invest 54:475–478

31. Fearon DT (1978) Regulation by membrane sialic acid of beta-1-H-dependent decay association of amplification C3 convertase of the alternative complement pathway. Proc Natl Acad Sci USA 75:1971–1975

32. Fearon DT, Collins LA (1983) Increased expression of C3b receptors on polymorphonuclear leukocytes induced by chemotactic factors and by purification procedures. J Immunol 130:370–375

33. Fearon DT, Wong WW (1983) Complement ligand-receptor interactions that mediate biological responses. Ann Rev Immunol 1:243–272

34. Fernandez HN, Henson PM, Otani A et al. (1978) Chemotactic response to human C3a and C5a anaphylatoxins. I. Evaluation of C3a and C5a leukotaxis in vitro and under simulated in vivo conditions. J Immunol 120:109–116

35. Fischer A, Seger R, Durandy A et al. (1985) Deficiency of the adhesive protein complex lymphocyte function antigen 1, complement receptor type 3, glycoprotein p150,95 in a girl with recurrent bacterial infections. Effects on phagocytic cells and lymphocyte functions. J Clin Invest 76:2385–2392

36. Ford PM, Kosatka I (1981) A mechanism of enhancement of immune complex deposition following in situ immune complex formation in the mouse glomerulus. Immunology 43:433–438

37. Frank MM, Lawley T, Hamburger MI et al. (1983) Immunoglobulin G Fc receptor-mediated clearance in autoimmune disease. Ann Intern Med 98:206–218

38. Gallo GR, Caulin-Glaser T, Lamm ME (1981) Charge of circulating immune complexes as a factor in glomerular basement membrane localization in mice. J Clin Invest 67:1305–1313

39. Gamble JR, Harlan JM, Klebanoff SJ et al. (1985) Stimulation of the adherence of neutrophils to umbilical vein endothelium by human recombinant tumor necrosis factor. Proc Natl Acad Sci USA 82:8667–8671

40. Germuth FG Jr, Rodriguez E (1973) Immunopathology of the renal glomerulus. Immune complex deposit and antibasement membrane disease. Little, Brown, Boston

41. Gleichmann E, Issa P, Vanelven E et al. (1978) The chronic graft-versus-host reaction: a lupus erythematosus like syndrome caused by abnormal T-B cell interactions. Clin Rheum Dis 4:587–602

42. Griffin JA, Griffin FM Jr (1979) Augmentation of macrophage complement receptor function in vitro. J Exp Med 150:653–675

43. Griffin FM (1982) Mononuclear cell phagocyte mechanisms and host defense. In: Gallin JI, Fauci AS (eds) Advances in host-defense mechanisms. Raven Press, New York pp 31–55

44. Groenwegen G, Buurman WA, Vander Unden CJ (1985) Lymphokine dependence of in vivo expression of MHC class II antigens by endothelium. Nature 316:361–363

45. Hansch GM, Hammer CH, Vanguri P et al. (1981) Homologous species restriction in lysis of erythrocytes by terminal complement proteins. Proc Natl Acad Sci USA 78:5118–5121

46. Harlan JM (1985) Leukocyte-endothelial interactions. Blood 65:513–525

47. Harkiss GD, Brown DL, Evans DB (1975) Longitudinal study of circulating immune complexes in a patient with staphylococcus albus-induced shunt nephritis. Clin Exp Immunol 24:396–400

48. Havemann K, Janoff A (1978) Neutral proteases of human polymorphonuclear leukocytes. Urban and Schwarzenberg, Baltimore and Munich

49. Hinglais N, Kazatchkine MD, Bhakdi S et al. (1986) Immunohistochemical study of the C5b-9 complex of complement in normal and diseased human kidneys: diversity in localization and potential role in tissue damage. Kidney Int 30:399–410

50. Hugli TE (1986) Biochemistry and biology of anaphylatoxins. Complement 3:111–127

51. Izui S, Lambert PH, Fournie GJ et al. (1977) Features of systemic lupus erythematosus in mice injected with bacterial lipopolysaccharides: identification of circulating DNA and renal localization of DNA-anti-DNA complexes. J Exp Med 145:1115–1123

52. Johnson KJ, Wilson BS, Till GO et al. (1984). Acute lung injury in rat caused by immunoglobulin A immune complexes. J Clin Invest 74:358–369

53. Kauffmann RH, Herrmann WA, Meyer CJLM et al. (1980) Circulating and tissue-bound immune complexes in allergic vasculitis: relationship between immunoglobulin class and clinical features. Clin Exp Immunol 41:459–470

54. Kauffmann RH, Herrmann WA, Meyer CJLM et al. (1980) Circulating IgA-immune complexes in Henoch-Schönlein purpura. Am J Med 69:859–866

55. Kauffmann RH, Thompson J, Valentijn RM et al. (1981) The clinical implications and the pathogenetic significances of circulating immune complexes in infective endocarditis. Am J Med 71:17–25

56. Kauffmann RH, Valentijn RM (1982) The spectrum of circulating immune complexes in various forms of systemic small vessel vasculitis. Neth J Med 25:193–201

57. Kazatchkine MD, Fearon DT, Silbert JE et al. (1979) Surface-associated heparin inhibits zymosan-induced activation of the human alternative pathway by augmenting the regulatory action of the control proteins on particle-bound C3b. J Exp Med 150:1202–1215

58. Kazatchkine MD, Fearon DT, Appay C et al. (1982) Immunohistochemical study of the human glomerular C3b receptor in normal kidney and in seventy five cases of renal diseases: loss of C3b receptor antigen in focal hyalinosis and in proliferative nephritis of SLE. J Clin Invest 69:900–912

59. Kazatchkine MD, Nydegger UE (1982) The human alternative complement pathway: biology and immunopathology of activation and regulation. Prog Allergy 30:193–234

60. Kazatchkine MD, Nydegger UE (1986) Complement mediated injury. In: Phillips SM, Escobar MR (eds) The reticuloendothelial system, a comprehensive treatise, Plenum, New York, pp 173–196

61. Klebanoff SJ, Clark RA (1978) The neutrophil. North-Holland, Amsterdam

62. Klebanoff SJ (1982) Oxygen dependent cytotoxic mechanisms of phagocytes. In: Gallin JI, Fauci AS (eds) Advances in host-defense mechanisms. Raven Press, New York, pp 111–162

63. Larsen GL, Mitchell BC, Henson PM (1981) The pulmonary response of C5 sufficient and deficient mice to immune complexes. Annu Rev Respir Dis 123:434–439

64. Laude M, Haeffner-Cavaillon N, Fischer E et al. (1985) Human C3 and C3a stimulate IL-1 secretion by cultured human monocytes. Complement 2:48 (abstract)

65. Law SK, Levine RP (1977) Interaction between the third complement protein and cell surface macromolecules. Proc Natl Acad Sci USA 74:2701–2704

66. Lynch JM, Worthen GS, Henson PM (1984) Platelet-activating factor. In: Buckle DR, Smith H (eds) Development of anti-asthma drugs. Butterworth, London, pp 73–88

67. Madalinski K, Bragiel I (1979) HbsAg immune complexes in the course of infection with hepatitis B virus. Clin Exp Immunol 36:371–378

68. March CJ, Mosley B, Larsen A et al. (1985) Cloning sequence and expression of two distinct human interleukin-1 complementary DNAs, Nature 315:641–647

69. Miossec P, Ziff M (1986) Immune interferon enhances the production of interleukin 1 by human endothelial cells stimulated with polysaccharide. J Immunol 137:2848–2852

70. Mollnes TE, Lea T, Mellbye OJ et al. (1986) Complement activation in rheumatoid arthritis evaluated by C3dg and the terminal complement complex. Arthritis Rheum 29:715–721

71. Moore KL, Andreoli SP, Esmon NL et al. (1987) Endotoxin enhances tissue factor and suppresses thrombomodulin expression of human vascular endothelium in vitro. J Clin Invest 59:124–130

72. Nawroth PP, Bank I, Handley D et al. (1986) Tumor necrosis factor/cachectin interacts with endothelial cell receptors to induce release of interleukin-1. J Exp Med 163:1363–1375

73. Nawroth PP, Handley DA, Esmon CT et al. (1986) Interleukin-1 induces endothelial cell procoagulant while suppressing cell-surface anticoagulant activity. Proc Natl Acad Sci USA 83:3460–3464

74. Nicholson-Weller A, Burge J, Fearon DT et al. (1982) Isolation of a human erythrocyte membrane glycoprotein with decay-accelerating activity for C3 convertases of the complement system. J Immunol 129:184–189

75. Nydegger UE, Kazatchkine MD (1986) Modulation by complement of immune complex processing in health and disease in man. Prog Allergy 39:361–392

76. Paul LC, Carpenter CB (1980) Antibodies against renal endothelial alloantigens. Transplant Proc 12:43–48

77. Platt JL, Lebien TW, Michael AF (1982) Interstitial mononuclear cell populations in renal graft rejection. Identification by monoclonal antibodies in tissue sections. J Exp Med 155:17–30

78. Pober JS, Collins T, Gimbrone MA et al. (1986) Inducible expression of class II major histocompatibility complex antigens and the immunogenicity of vascular endothelium. Transplantation 41(2):141–146

79. Pober JS, Gimbrone MA Jr, Cotran RS et al. (1983) Ia expression by vascular endothelium is inducible by activated T cells and by human gamma interferon. J Exp Med 147:1339–1353

80. Pommier CG, Inada S, Fries LG et al. (1983) Plasma fibronectin enhances phagocytosis of opsonized particles by human peripheral blood monocytes. J Exp Med 157:1844–1854

81. Pommier CG, O'Shea, Chuseo T et al. (1984) Studies on the fibronectin receptors of human peripheral blood leucocytes. J Exp Med 159:137–151

82. Ramm LE, Whitlow MB, Koski CL et al. (1983) Elimination of complement channels from the plasma membranes of U937, a nucleated mammalian cell line:

temperature dependence of the elimination rate. J Immunol 131:1411–1415

83. Reinherz EL, Schlossmann SF (1980) The differentiation and function of human T lymphocytes: a review. Cell 19:821–825

84. Rinaldo JE, Rodgers R (1982) Adult respiratory distress syndrome: changing concepts of lung injury and repair. N Engl J Med 396:900–909

85. Rosenfeld SI, Packman CH, Leddy JP (1983) Inhibition of the lytic action of cell bound terminal complement components by human high density lipoproteins and apoproteins. J Clin Invest 795–808

86. Ross GD, Cain JA, Lachmann PJ (1985) Membrane complement receptor type three (CR3) has lectin-like properties analogous to bovine conglutinin and functions as a receptor for zymosan and rabbit erythrocytes as well as a receptor for C′C3b. J Immunol 134:3307–3315

87. Ross GD, Cain JA, Myones BL (1987) Specificity of membrane complement receptor type three (CR3) for beta-glucans. Complement 4:61–74

88. Rossi V, Brevario F, Ghezzi P et al. (1985) Prostaglandin synthesis induced in vascular cells by interleukin 1. Science 229:174–176

89. Rothlein R, Dustin ML, Marlin SD et al. (1986) A human intracellular adhesion molecule (ICAM-1) distinct from LFA1. J Immunol 137:1270–1274

90. Sachs T, Moldon CF, Craddock PR et al. (1978) Oxygen radical mediates endothelial cell damage by complement stimulated granulocytes. An in vitro model of immune vascular damage. J Clin Invest 61:1161–1167

91. Sanchez-Madrid F, Nagy JA, Robbins P et al. (1983) A human leukocyte differentiation antigen family with distinct alpha subunits and a common beta subunit. J Exp Med 158:1785–1803

92. Schifferli JA, Peters DK (1983) Complement, the immune complex lattice and the pathophysiology of complement deficiency syndromes. Lancet II:957–959

93. Schifferli JA, Steiger G, Hauptmann G et al. (1985) Formation of soluble immune complexes by complement in sera of patients with various hypocomplementemic states: difference between inhibition of immune precipitation and solubilization. J Clin Invest 76:2127–2133

94. Schifferli JA, Peters KD (1986) The role of complement and its receptor in the elimination of immune complexes. N Engl J Med 315:488–495

95. Schmidt JA, Mizel SB, Cohen D et al. (1982) Interleukin-1, a potential regulator of fibroblast proliferation. J Immunol 128:2177–2182

96. Schwartz BR, Ochs HD, Beatty PG et al. (1985) A monoclonal antibody defined membrane antigen complex is required for neutrophil–neutrophil aggregation. Blood 65:1553–1556

97. Seeger W, Suttorp N, Hellwig A et al. (1986) Noncytolytic terminal complement complexes may serve as calcium gates to elicit leukotriene B4 generation in human polymorphonuclear leukocytes. J Immunol 137:1286–1293

98. Smiley ML, Hoxie JA, Friedman HM (1985) Herpes simplex virus type I injection of endothelial, epithelial and fibroblast cells induces a receptor for C3b. J Immunol 134:2673–2678

99. Shore A, Leary P, Teitel JM (1986) Comparison of accessory cell functions of endothelium cells and monocytes: IL2 production by T cells and PFC generation. Cell Immunol 100:210–223

100. Snyderman R, Pike MC (1984) Chemoattractant receptors on phagocytic cells. Annu Rev Immunol 2:257–281

101. Steblay RW (1962) Glomerulonephritis induced in sheep by injection of heterologous glomerular basement membrane and Freund's complete adjuvant. J Exp Med 116:253–261

102. Steeg PS, Moore RN, Johnson HM (1982) Regulation of murine Ia antigen expression by a lymphokine with immune interferon activity. J Exp Med 156:1780–1786

103. Stern DM, Bank I, Nawroth PP et al. (1985) Self-regulation of procoagulant events on the endothelial cell surface. J Exp Med 162:1223–1235

104. Stolden AM, Guinan EC, Fiers W et al. (1986) Recombinant tumor necrosis factor and immune interferon act singly and in combination to reorganize human vascular endothelial cell monolayers. Am J Pathol 123:16–24

105. Sundsmo JS, Fair DS (1983) Relationship among the complement, kinin, coagulation and fibrinolytic systems. Springer Semin Immunopathol 6:231–258

106. Theofilopoulos AN, Dixon FJ (1981) Etiopathogenesis of murine systemic lupus erythematosus. Immunol Rev 55:179–216

107. Till G, Ward PA (1986) Systemic complement activation and acute lung injury. Fed Proc 45:13–18

108. Tonnesen MG, Smedly LA, Henson PM (1984) Neutrophil-endothelial cell interactions. Modulations of neutrophil adhesiveness induced by complement fragments C5a and C5adesArg and formyl-methionyl-leucyl-phenylalanine in vitro. J Clin Invest 74:1581–1592

109. Trinchieri G, Perussia B (1985) Immune interferon: a pleiotropic lymphokine with multiple effects. Immunol Today 6:131–136

110. Valentijn RM, Faber WR, Lai A et al. (1982) Immune complexes in leprosy patients from an endemic and a nonendemic area and a longitudinal study of the relationship between complement breakdown products and the clinical activity of erythema nodosum leprosum. Clin Immunol Immunopathol 22:194–202

111. Van Oeveren W, Kazatchkine MD, Descamps-Latscha B et al. (1985) Deleterious effects of cardiopulmonary bypass. A prospective study of bubble versus membrane oxygenation. J Thorac Cardiovasc Surg 888–899

112. Vlaicu R, Niculescu F, Rus HG et al. (1985) Immunohistochemical localization of the terminal C5b-9 complement complex in human aortic fibrous plaque. Atherosclerosis 57:163–170

113. Vogt A, Rohrbach R, Shimuzu F et al. (1982) Interaction of cationized antigen with rat glomerular basement membrane. Kidney Int 22:27–35

114. Wilson CB, Dixon FJ (1981) The renal response to immunological injury. In: Brenner BM, Rector FC Jr (eds) The kidney. Saunders, Philadelphia, pp 1237–1350

115. Wilson JG, Wong WW, Schur PH et al. (1982) Mode of inheritance of decreased C3b receptors on erythrocytes of patients with systemic lupus erythematosus. N Engl J Med 307:981–986

116. Wright SD, Silverstein SC (1982) Tumor promoting phorbol esters stimulate C3b and C3bi receptor-mediated phagocytosis in cultured human monocytes. J Exp Med 156:1149–1161

117. Zubler RH, Lambert PH (1978) Detection of immune complexes in human diseases. Prog Allergy 24:1–48

118. Ziccardi RJ (1983) The first component of human complement (C1): activation and control. Springer Semin Immunopathol 6:213–230

Chapter 27

The Vasculitis Syndromes in Small- and Medium-Sized Vessels

J. Bariéty and C. Jacquot

Part I. Lesions and Mechanisms

The hallmark of angiitis is necrosis and infiltration of the vascular wall by inflammatory cells. Arteries, arterioles, capillaries and venules may be affected. The lesions may be widely disseminated or may involve a region such as the skin or a single organ such as the appendix or gall bladder. Clinical manifestations are thus diverse and depend upon the area of involvement. Although the histological diagnosis of angiitis is made when inflammation and necrosis are found in a vascular wall, this does not provide adequate information about the type of angiitis, its cause, pathogenesis, spread or prognosis. It is particularly difficult to identify small-vessel angiitis because the characteristic vascular lesions are not always evident. Perivascular aggregation of inflammatory cells is not a proof of angiitis. Regardless of the cause of inflammation, infiltrating cells migrate from the lumen to the interstitium of small vessels.

Giant cell arteritis is discussed elsewhere (Takayasu's disease p. 453 and temporal arteritis p. 463) as will be Kawasaki's disease (p. 474), Behcet's syndrome (p. 473) and Liebow's disease (p. 592).

Lesions

The lesions may be found in medium- and small-sized arteries or in venules, capillaries and arteri-

oles. Lesions in a combination of these sites is a frequent finding.

Medium- and Small-Sized Arteries

The lesions evolve through three stages: the acute, the subacute and the scarring stages. Lesions in each of these phases may co-exist in the same patient.

Acute Phase

In this stage the arterial wall is necrotic and filled with inflammatory cells. The normal arterial wall architecture is obscured by material which has the staining characteristics of fibrin. Such fibrinoid necrosis involves the whole or part of the arterial circumference and may affect only the internal layers or the entire thickness of the arterial wall. The internal elastic lamina is destroyed. Numerous cells (neutrophils, mononuclear cells, eosinophils) infiltrate the necrotic zone and sometimes pass beyond this area into the perivascular space (Fig. 27.1). This picture inspired the term periarteritis. In any one section, perivascular cellular infiltration may be the only visible lesion, but serial sections allow the identification of the characteristic areas of necrosis and cellular infiltration of the vascular wall. Although disseminated along the arterial tree, these lesions tend to be located in particular segments, often at points of bifurcation. Skipped areas are common and biopsies may be normal and therefore misleading. When fibrinoid necrosis involves only a part of the arterial cir-

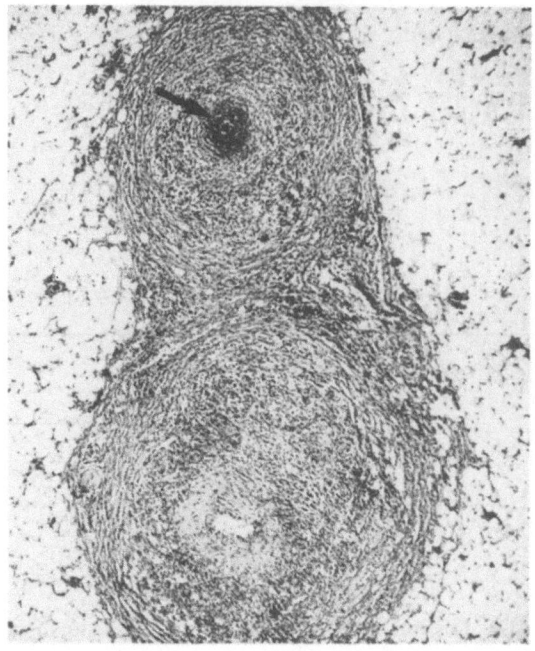

Fig. 27.1. Polyarteritis nodosa. Acute lesion of angiitis in an arterial bifurcation. The three layers of the arterial wall are infiltrated by inflammatory cells. The lumen is severely narrowed. There is a recent thrombus in one arterial lumen (*arrow*). The media is necrotic. Inflammatory infiltrate spreads in the perivascular tissue. H & E, × 90

cumference, distension of this region may lead to the formation of a small aneurysm (Fig. 27.2). These aneurysms give a typical angiographic picture and are responsible for the clinical nodular pattern. Luminal thrombosis may occur in the involved vessel and in contiguous aneurysms, leading in some cases to distal ischaemia. Red blood cells may extravasate into the vascular wall and cause haemorrhage.

Subacute Phase

Fewer inflammatory cells are present at this stage; often only mononuclear cells are present. Necrotic areas are invaded by fibroblasts or smooth-muscle cells. These cells are responsible for considerable concentric thickening of the intima. Any thromboses are beginning to organize.

Scarring

Inflammatory cells have disappeared by this stage. Haemosiderin is present in areas of previous red blood cell extravasation. Zones of medial destruction are replaced by fibrous tissue and intimal fibrous thickening may cause a considerable reduction in luminal diameter. The rup-

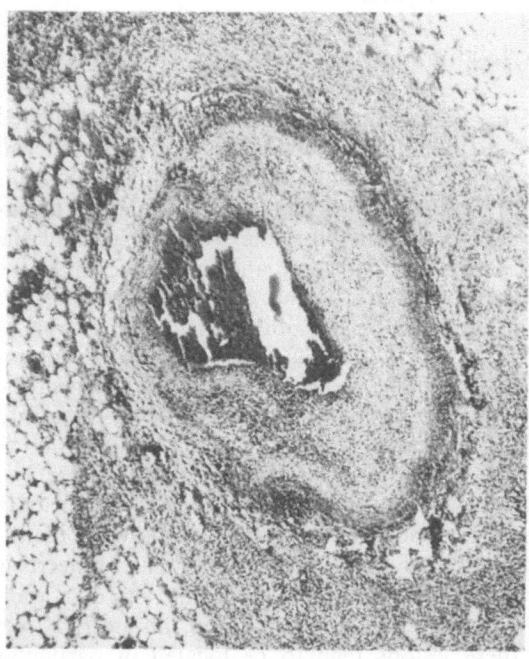

Fig. 27.2. Polyarteritis nodosa. Focal lesion with thinning of the arterial wall which could lead to an aneurysm. H & E, × 90

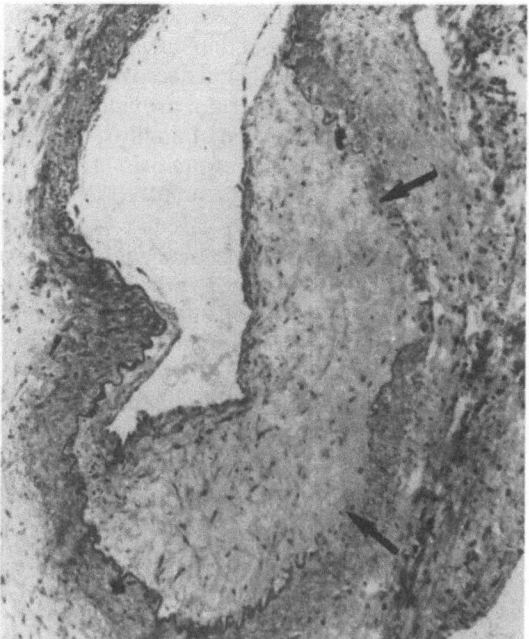

Fig. 27.3. Polyarteritis nodosa. Scar lesion with focal intimal thickening and disruption of the media. The internal elastic lamina is blurred (*arrows*). H & E, × 50

tured internal elastic lamina is not reconstructed, but is replaced by collagen. However, strands of elastic tissue are often seen within the thickened intima (Fig. 27.3).

The Microcirculation

Cutaneous angiitis is the prototype of small-vessel angiitis. This condition is easy to diagnose and biopsy specimens are readily obtained. Arterioles, capillaries and venules may be affected. Sections of 1 μm have shown that venules are more frequently involved than arterioles. Generally, all the lesions are of the same age. The entire vascular wall is the seat of fibrinoid necrosis, and is invaded by variable numbers of neutrophils and mononuclear cells (Fig. 27.4). Many of these cells contain lysed nuclei, hence the term leucocytoclastic angiitis (Fig. 27.5). Lesions have been described that are either neutrophil-predominant or alternatively are lymphocyte-predominant [274]. In patients with hypocomplementaemia [196], neutrophils apparently prevail, whereas lymphocytes prevail in normocomplementaemic subjects [275] (Fig. 27.6). The nature of the cellular infiltrate may change with time. In some cases of drug hyper-

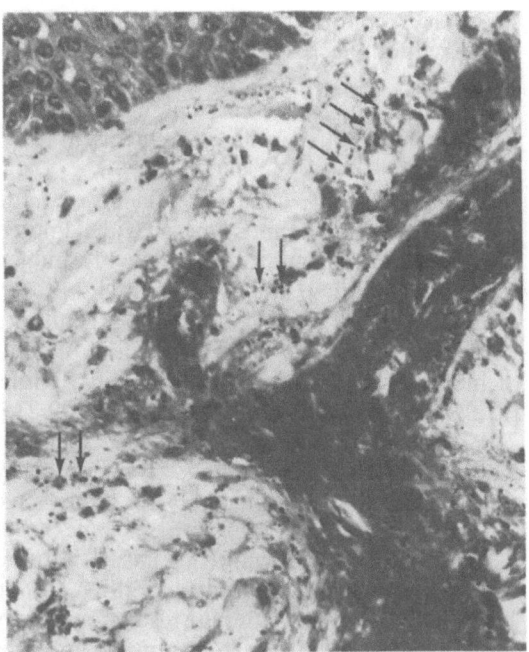

Fig. 27.5. Leucocytoclastic angiitis. The small dermal vessels show diffuse necrosis with thrombosis and polymorphonuclear infiltrates. There are numerous nuclear remnants (*arrows*). Masson's trichrome, × 380

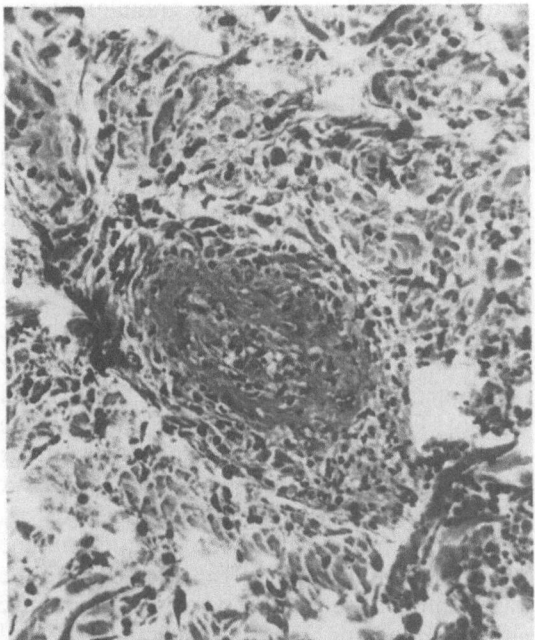

Fig. 27.4. Cutaneous angiitis. Fibrinoid necrosis in the wall of a venule in the lower dermis. It is associated with a polymorphonuclear infiltrate. H & E, × 150

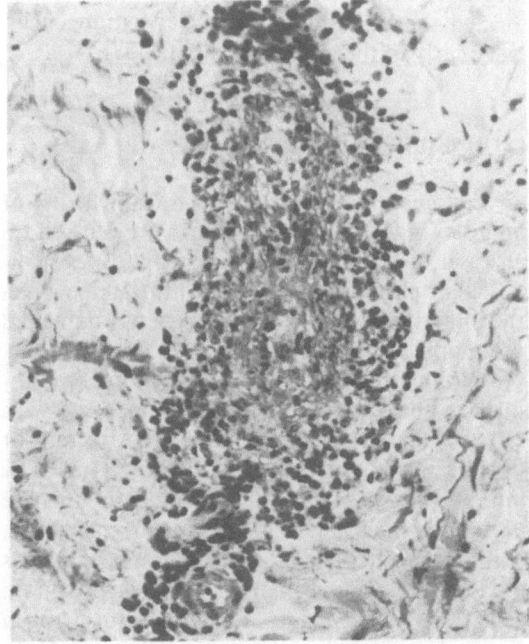

Fig. 27.6. Cutaneous angiitis. The perivenular infiltrate is mainly composed of mononuclear cells. H & E, × 160

sensitivity, mononuclear and eosinophilic infilt-ration without necrosis has been described [218].

Electron Microscopy. This has not been par-ticularly informative in the study of angiitis.

Immunofluorescence and Immunoperoxidase Staining. These methods have provided limited data which are difficult to interpret. Fibrinogen is usually found in the vascular walls [27,99, 128,155,166,204,230,286]. Numerous authors have reported that immunoglobulin and/or com-plement deposits are present in angiitis lesions [75,87,98,121,131,134,138,162,194,204,209,215, 230,256], but this is by no means a constant feature. IgM has been found more frequently than IgG or IgA. Vascular IgG deposits have been reported in patients with systemic lupus erythematosus [22,39] while IgA has been found in cases of rheumatoid purpura [87,281,287]. Immunofluorescent staining results are very often negative, at least for immunoglobulins [27,42,98,198,305]. The presence of immuno-globulins does not necessarily signify immune complex deposition, but may simply represent immunoglobulin trapping in damaged vessel walls. IgM and C3 are often found in the arterioles of apparently normal kidneys. The absence of immunoglobulin deposits is sig-nificant only early in the evolution of angiitic lesions since immune complexes can be rapidly degraded in vessel walls. Skin biopsies have con-firmed that immunoglobulin deposition occurs very early in the course of angiitis [37].

Granulomas

Granulomas may accompany certain forms of angiitis (Fig. 27.7). These lesions are centred on a vessel or may sometimes appear to be extra-vascular, although this is difficult to judge when the underlying vessel walls are severely damaged or completely destroyed.

A granuloma is a compact collection of mono-nuclear phagocytes sometimes interspersed with other inflammatory cells. Necrosis and/or fibrosis may or may not be present [1,3]. The mononuclear phagocytes [277,292] include immature macrophages, mature macrophages, epithelioid cells and giant cells. The macrophages are derived from monocytes, and the epithelioid cells from macrophages. Giant cells result from the fusion of either of these cell types [1,3]. Granuloma formation involves two compulsory phases and one optional phase: (a) the influx and

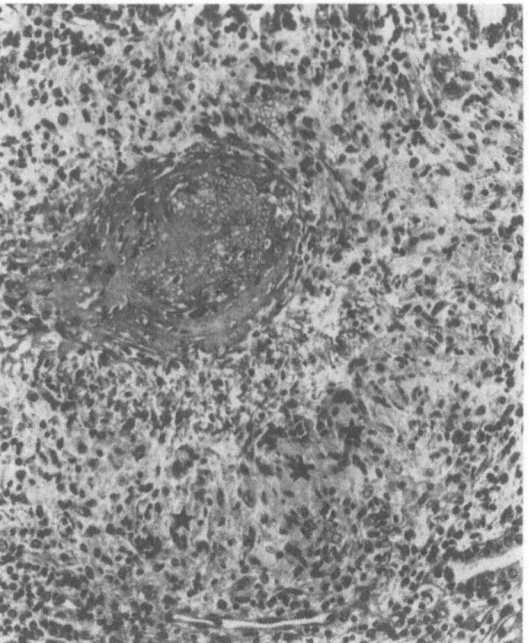

Fig. 27.7. Granulomatous angiitis. The granuloma, made of epithelioid cells (*stars*), spreads outside the vascular wall. The media and the intima show fibrinoid necrosis. The lumen is occluded by fibrin and red blood cells. H & E, × 190

accumulation of mononuclear phagocytes in the vascular lesion, (b) the aggregation and organ-ization of young mononuclear phagocytes and mature macrophages (mature granuloma) and (c) the eventual development of epithelioid cells (epithelioid granuloma). Granulomas usually persist until the original inciting factors are degraded or destroyed [1,3].

The mechanisms of granuloma formation and the significance of this lesion are still incom-pletely understood. Immune factors may be at play. Cell-mediated immunity involves sen-sitization of immunocompetent cells by antigen and the release by these cells of various lym-phokines, including macrophage migration inhi-bition factor. These soluble mediators favour macrophage accumulation and activation [1,3,36]. Immune complexes can also activate macrophages through an interaction between the immunoglobulin Fc portion and the Fc receptor of mononuclear phagocytes. Nevertheless, granulomas may form for entirely non-immune reasons which is not surprising considering the vast number of substances which exist that have either chemotactic [273,297,304] or maturational [3,278] effects on macrophages.

Two mononuclear phagocytic activities are pronounced within granulomas: (a) the liberation of a variety of substances [3,83] including oxygen radicals, lysozyme, neutral proteases, lysosomal enzymes, complement factors, pyrogens, prostaglandins, interferons and other factors which can modulate the activity of other cells; and (b) endocytosis of particles smaller than 10 nm in diameter (phagocytosis). The plasma membranes of mononuclear phagocytes bear receptors for the Fc portion of immunoglobulins, for C3b and for certain glycoproteins. These cells can also bind certain tumour cells and parasites [3,293,294] as well as denatured particles or molecules [269]. Such binding facilitates phagocytosis. Once a foreign particle is internalized in an endocytotic vesicle, fusion with lysosomes occurs and the particle is rapidly digested by various hydrolases [269]. Epithelioid cells apparently have greater secretory capacity than macrophages, but are less able to phagocytose [227]. This point remains controversial.

Certain features may distinguish immune from non-immune granulomas [288]. Those of immunological origin apparently contain epithelioid cells and lymphocytes, and are markedly fibrotic. The fibrosis is attributed to the release of a specific fibroblast-activating factor by epithelioid cells. Granulomas of non-immunological origin contain only phagocytic macrophages without epithelioid cells and are not fibrotic. Since very few studies have been performed using specific markers to classify the macrophages within granulomas, it is wise to await further information before drawing any conclusions about the respective roles of mature macrophages and epithelioid cells.

The appearance of a granuloma may be modified by the presence of necrosis or by non-mononuclear phagocytic cell infiltration. Necrosis is probably due to the release of acid hydrolases, neutral proteases and oxygen radicals by activated macrophages [2,82]. Neutrophils are usually found in newly formed or necrotic granulomas, and eosinophils, basophils, and T and B lymphocytes may also be seen. The factors which attract these cells have not been completely identified.

Glomerular Lesions

Glomerular lesions are quite often seen in the context of angiitic syndromes. These lesions are easily detected because of their clinical manifestations and can be studied at an early stage by renal biopsy.

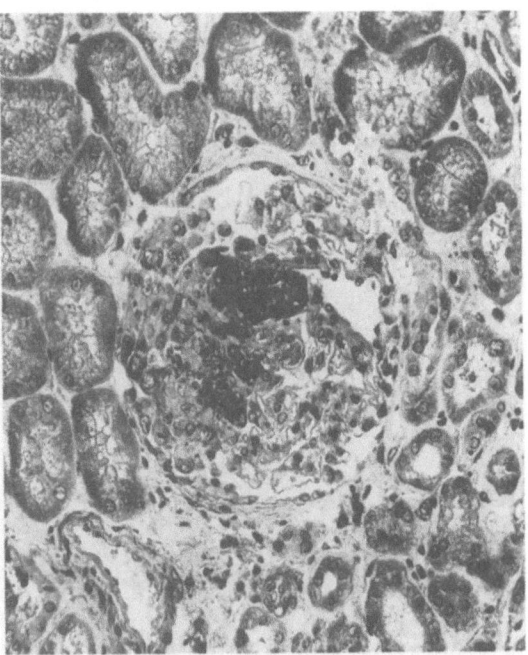

Fig. 27.8. Microscopic polyarteritis nodosa. Focal glomerular necrotic lesion with an epithelial crescent in Bowman's space. Masson's trichome, × 250

Angiitis is associated with two basic glomerular lesions: variably extensive necrosis of the glomerular tuft, and extracapillary cellular proliferation within Bowman's space (Fig. 27.8). Glomerular capillary necrosis is often focal and is especially well seen in silver-stained sections showing basement membrane disruption by a fibrinoid material. The severity of the lesions in a single glomerulus and the number of glomeruli affected vary from case to case. Glomerulonephritis may be focal and segmental with discrete areas of necrosis in involved glomeruli eliciting local epithelial crescent formation. Alternatively, a diffuse extracapillary glomerulonephritis may occur in which all glomeruli are progressively effaced by a ring of extracapillary cells occupying Bowman's space. Necrosis is evident only in relatively young glomerular lesions. Its mechanism is poorly understood. The extracapillary cells include macrophages and proliferating epithelial cells of Bowman's capsules. This makes it difficult to decide whether the cells invading Bowman's space originate from the glomerular capillary tuft or from the interstitium. Occasionally, granulomatous lesions containing epithelioid cells may surround the vascular tuft in the periglomerular interstitium [42,202] (Fig. 27.9).

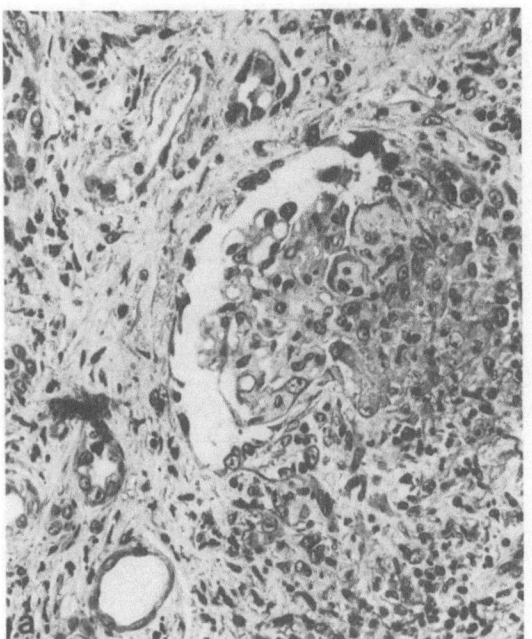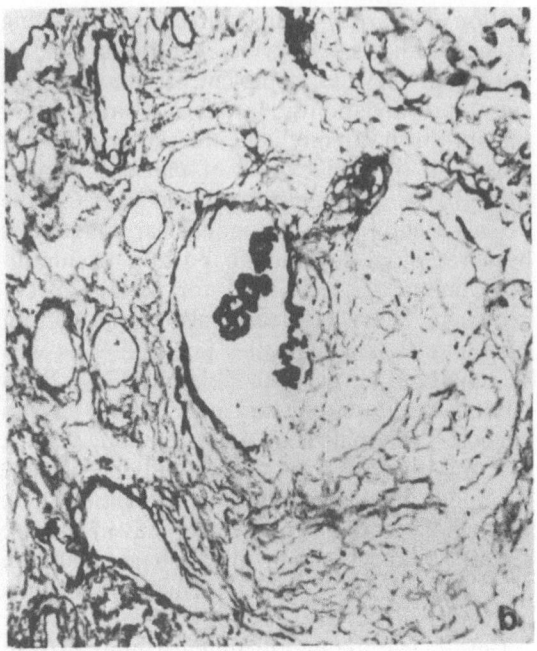

Fig. 27.9 a,b. Wegener's disease. **a** Subtotal necrosis of the glomerular tuft associated with cell proliferation in the Bowman's space and the periglomerular interstitial area. Masson's trichrome, × 275. **b** Using silver staining the focal disruption of the glomerular tuft and of the Bowman's capsule is evident. Silver methenamine stain, × 275

Other glomerular lesions may accompany these basic lesions depending on the exact cause or on the type of angiitis.

Immunofluorescent staining invariably reveals fibrin in early lesions especially in areas of extra-capillary proliferation with Bowman's space. Complement and immunoglobulins tend not to be found. However, glomerular IgA and fibrin deposits are always present in cases of rheumatoid purpura, and IgG deposits are characteristic of evolving lupus nephritis [132].

Mechanisms

Immune mechanisms are invariably cited to explain the pathogenesis of human angiitis (see Chap. 26). However, non-immune mechanisms may play a role often in association with classical immune mechanisms.

Immune Mechanisms

Angiitis in humans closely resembles many experimental animal models of hypersensitivity angiitis. However, factors inducing hypersensitivity are rarely if ever identified in clinical practice.

Accumulation of Immune Complexes (Type III Hypersensitivity)

Experimental Pathology. Existing experimental models point to the importance of immune complexes in the pathogenesis of angiitis. These complexes may be formed in situ as in the Arthus model, or may originate in the circulation as in acute serum sickness [73,92,194].

The Arthus reaction is a localized form of angiitis which occurs when animals are re-challenged by a subcutaneous injection of an antigen with which they have been previously immunized. The walls of postcapillary venules are invaded by neutrophils and haemorrhagic necrosis ensues. Immune complex formation in this model occurs in situ when the re-injected antigen encounters circulating specific antibodies. The further development of the vascular lesion is complement- and neutrophil-dependent [63]. Such a mechanism may be responsible for localized forms of angiitis in areas of the body to which antigens have ready access such as the skin and the respiratory and digestive tracts.

In serum sickness, immune complexes form in the circulation and are then deposited in vascular walls. This implies that at least some immune complexes are not destroyed by mononuclear phagocytes in the circulation, and that these complexes have properties favouring their deposition in vascular walls.

Acute serum sickness is classically induced by the injection of a large dose of a foreign protein, such as bovine serum albumin, into rabbits [92]. These animals develop acute glomerulonephritis and arteritis within 10 to 14 days. These abnormalities appear during the immune antigen elimination phase when immune complexes are present in conditions of antigen excess. Deposition of these immune complexes in vascular walls depends upon several incompletely understood factors [63,256,306]: complex size, the nature of the antibody response, the ability of the complexes to fix complement, the release of vasoactive substances during the immune response, the clearance of the complexes by phagocytes and various haemodynamic factors. Local factors may also promote immune complex deposition [253]: turbulent flow at vascular bifurcations, elevated blood pressure (hypertension, predominance of lesions in susceptible areas), local spasm, cold, pressure and punctures.

An inflammatory reaction is triggered by the presence of immune complexes in the vascular wall: the binding and activation of complement then lead to the release of factors chemotactic for neutrophils. This in turn ensures phagocytosis or proteolytic destruction of the immune complex deposits. However, vascular damage also occurs as a result of the action of lysosomal enzymes and oxygen radicals [119]. Monocyte macrophages may also be activated via their Fc receptors and may contribute to the vascular damage. Prostaglandins play a regulatory role in this process. The typical arterial lesion is segmental. There is necrosis, rupture of the internal elastic lamina and infiltration by neutrophils and later by mononuclear cells. Immunofluorescent staining reveals small deposits of BSA, immunoglobulins and complement in the arterial and glomerular lesions. It is interesting to note that the lesion of acute serum sickness can regress despite the local accumulation of immunoglobulins and complement, suggesting that these molecules can lose their inflammatory potential [306].

In chronic serum sickness, angiitis is uncommon even in the face of persistently high circulating immune complex levels. In the rabbit model where chronic serum sickness is induced by repeated injection of foreign protein over several weeks, glomerulonephritis [93] and immune deposition in small vessels and elsewhere [38] are frequent, but angiitis is very rare. Severe angiitis without glomerulonephritis [128] can however be induced by antigen injection after an interval of several weeks.

The following hypothesis has been advanced to explain the differences between acute and chronic serum sickness [56]. In acute serum sickness, large quantities of antigen are injected and allowed to diffuse out of the circulation, particularly into vascular walls. Antibodies are subsequently released into the circulation and are prompted to form in situ immune complexes with the previously deposited antigen. Injections of small doses of antigen on a daily basis as in the chronic serum sickness model leads to immune complex formation in the circulation and these complexes are presumably more readily eliminated.

Injection of pre-formed immune complexes occasionally provokes arteritis [193], but immune complex deposition, especially in venules, without angiitis is much more common [62]. Thus, the simple presence of immune complexes in vessel walls does not necessarily lead to angiitis.

Human Pathology. The scattered reports of immunoglobulin and/or complement deposition in angiitis lesions [75,87,98,134,138,194,204, 215,256] do not provide formal proof of the presence of immune complexes. Unfortunately, it is generally impossible to elute sufficient material from vascular lesions to demonstrate the presence of specific antigen–antibody complexes. In many cases, there is no clue to the nature of the antigens involved. Hepatitis B surface antigen is the only specific antigen that has been convincingly identified in angiitic lesions thus far [20,91,99,134,143,215]. Some streptococcal, staphylococcal and mycobacterial antigens have also been reported in cutaneous vascular lesions [228,229]. Patients with cutaneous vasculitis have been studied hourly after intradermal histamine injection: the lesions induced probably involve immune complex formation [138] as suggested by the early and invariable presence of immunoglobulins and complement even before cellular infiltration occurs. The cutaneous vessels of patients with mixed cryoglobulinaemia may contain immunoglobulins of the same classes as in the circulating cryoglobulins [129]. In rheumatoid arthritis or mixed cryoglobulinaemia, the

IgG which is sometimes detected in vascular walls may act as an antigen combined with rheumatoid factor or with anti-idiotypic antibodies [125]. The absence of immunoglobulins in vasculitic lesions is significant only early in their evolution since immune complexes in this location may be rapidly degraded.

Other Immunological Mechanisms

Type I Hypersensitivity. Type I hypersensitivity refers to the release of mediators from IgE-bearing mastocytes when these cells are re-challenged with an antigen to which they have been previously sensitized. This phenomenon has not been implicated directly in the pathogenesis of human angiitis. Nevertheless, when this process occurs, vascular permeability is increased and this could favour in situ formation of immune complexes in vascular walls, the deposition of circulating immune complexes, or the deposition of other phlogistic non-immune material. The angiitis and allergic granulomatosis characteristic of the Churg–Strauss syndrome [58] are accompanied by asthma, hypereosinophilia and eosinophilic tissue infiltration, e.g. in vascular lesions, granuloma and renal interstitium. Type I hypersensitivity may play a role in this syndrome.

Type II Hypersensitivity. Cytotoxic antibodies that fix complement are responsible for the necrotic angiitis with neutrophilic infiltration seen in hyperacute and certain forms of acute allograft rejection in both experimental animals and humans. This phenomenon occurs in human renal transplant rejection. It is possible that a similar process occurs in Wegener's granulomatosis [290]. In this condition, auto-antibodies to a cytoplasmic antigen of monocytes, neutrophils and endothelial cells have been described [188,189,291]. The role of these antibodies in pathogensis remains to be established. Similarly, the anti-glomerular basement membrane antibodies seen in Goodpasture's syndrome and in related experimental models are directed against a vascular constituent. These antibodies are thought to play a pathogenetic role.

Type IV Hypersensitivity. The existence of granulomatous angiitis, certain lymphocyte- and monocyte-predominant forms of cutaneous angiitis [274] and Wegener's granulomatosis [127] suggests that cell-mediated immune mechanisms may be at play. T lymphocytes sensitized to antigens of the vascular wall would thus be responsible for a type IV hypersensitivity reaction leading to angiitis [194]. Experimental evidence points to the involvement of smooth-muscle cell antigens in this process [148,217]. Cell-mediated immune mechanisms also play a role in the development of vascular and glomerular lesions during renal allograft rejection in both animals and humans.

In addition to the structural antigens of the vascular wall, certain deposited immune complexes, especially when formed in conditions of antibody excess, may contain insoluble antigens that can elicit a cell-mediated granulomatous response [130].

Very few immunomorphological studies have been performed that look at the exact phenotype and function of the mononuclear cells within angiitic lesions. Many varieties of cytotoxic cells may be involved theoretically: these sensitized T cells can interact directly with antigens located in the vascular wall. T cells might be prompted to attack the vascular wall if this structure has been modified in any way, for example, by drugs, toxins or viruses. Antibody-dependent cell-mediated cytoxicity may also be involved in a classical interaction between antibody and the Fc receptor for immunoglobulin present on a variety of cells (B lymphocytes, "null" cells, large granular lymphocytes, certain activated T cells, natural killer cells and macrophages). Finally it is possible that certain natural killer cells may act even in the absence of antibody.

The characterization of the cells involved in the pathogenesis of angiitis is difficult not only for methodological but also for physio-pathological reasons. The cells which are specifically responsible for the induction of vascular lesions cannot be distinguished easily from those recruited during the inflammatory process. Numerous soluble mediators are released in this context (macrophage migration inhibition factor, interleukin-2, gamma interferon) [235], and all these substances attract many different kinds of cells in a non-specific fashion.

Thus, no definitive information exists about the cause and pathogenesis of most forms of human angiitis. Although immunological mechanisms are classically invoked, certain non-immunological mechanisms may be involved and should be considered.

Non-immunological Mechanisms

Certain a priori non-immunological conditions are known to provoke or to facilitate the development of angiitis.

Pulmonary hypertension [283] and re-perfusion of the mesenteric bed after surgical correction of aortic coarctation [97] are both associated with the development of angiitis in these specific territories. Malignant arterial hypertension in man usually leads to vascular necrosis without cellular infiltration. This condition should therefore be excluded when cases of putative angiitis present. In the rat, arterial hypertension, regardless of its causes, provokes necrosis with cellular infiltration in vessel walls [155].

Angiitis can also be induced by bacterial, fungal, rickettsial [144,296] or viral [266] infection of vascular walls. It is unclear whether these agents directly cause angiitis or whether they do so by the liberation of endotoxin or by the mechanism of immune complex formation [26]. Angiitis is frequently seen in appendicectomy specimens. These lesions may be due to direct effects of the intestinal flora or to in situ immune complex formation [171,222].

Cholesterol emboli are frequently associated with typical angiitic lesions. Although rare, destruction of arterial wall is seen in cases of tissue embolization from atrial myxomas [180].

Drug addicts are known also to develop angiitis, and vasoactive drugs have been incriminated [59]. In such cases, other drugs or microbial or viral inoculation, or particularly infection with hepatitis B virus, may be responsible.

Medications are frequently cited as the cause of angiitis [147]. This concept has been popular since Rich [244] reported fatal cases of angiitis in patients treated with sulphonamides and by serotherapy. It is probable that many of these patients died of the infections for which they were being treated, and that their angiitis was due to classical serum sickness. Thus the role of the various medications thought to cause angiitis is still unclear. In many situations, suspect medication may have been administered after and not before the development of angiitis [248]. There is no experimental evidence that medications can induce angiitis or necrotic glomerulonephritis by immune complex formation, although they can theoretically act as haptens. It would seem that medications are primarily responsible for small-vessel, especially cutaneous, angiitis [218,274,275]. A hypersensitivity mechanism is possibly involved, given the skin eruptions, eosinophilia and unexplained fever seen in these forms of angiitis.

Certain lymphoproliferative syndromes are associated with "angiitis". In Liebow's syndrome [185], lymphocytoid and plasmacytoid cells are known to involve and destroy vessels in multiple organ sites.

Classification

There is no satisfactory classification of angiitis. No single criterion such as aetiology, pathogenesis, the type of vessel involved, the nature of the cellular infiltrate or the associated clinical or laboratory data can provide the basis for an adequate classification. The distribution of lesions and the type of vessel affected may differ from case to case even in the same form of angiitis, and similar lesions may be seen in different forms of angiitis. The cause and the pathogenesis of most forms of angiitis are incompletely understood and difficult to study. Clinical and laboratory features are often ubiquitous and non-discriminatory.

Semantic difficulties also add to the confusion. Different terms are often used to designate the same entity, and different entities are sometimes lumped together under the same name. Some commonly used terms bear no relationship to reality. For instance, periarteritis nodosa is more appropriately called panarteritis or polyarteritis nodosa [113]. Connective tissue disorders and the various collagenoses are largely conditions which do not specifically involve either connective tissue or collagen. The term hypersensitivity angiitis was coined by Zeek to describe the disseminated, necrotic vascular lesions seen in the small arteries, arterioles, venules and capillaries (including those of the splenic and pulmonary beds) and occasionally in the glomeruli of some patients [309]. Zeek [307,308] compared this form of angiitis with polyarteritis nodosa which affects larger vessels, particularly medium-sized arteries, and which spares pulmonary arteries and splenic arterioles. The term hypersensitivity angiitis was later adopted to denote cutaneous forms of angiitis without apparent visceral involvement (75). The use of the term hypersensitivity angiitis when there is small-vessel involvement is misleading since this suggests that a hypersensitivity mechanism is not involved in polyarteritis nodosa affecting medium-sized vessels but is always present in small-vessel disease. Neither of these statements is necessarily true. Davson et al. [85] hoped to avoid all confusion by classifying angiitis on purely morphological grounds. Two forms are defined by these authors: classical angiitis corresponding to the condition of polyarteritis nodosa originally described by Kussmaul and Maier [175], and microscopic angiitis affecting small vessels, including capillaries and glomeruli. Numerous "overlap" syndromes exist, however. It is also to be noted that the microscopic angiitis of Davson corresponds to hypersensitivity angiitis as described by Zeek.

Despite all of the pitfalls of any one classification system, some system if desirable for didactic reasons and to some extent in order to define prognosis and response to treatment. The classification proposed in the second part of this chapter is advanced for this purpose although it is certainly somewhat artificial and incomplete.

Part II. Clinical and Nosological Aspects

Vasculitis occurs in a variety of clinical contexts [194]. Often its cause is unknown or poorly understood; this is true for polyarteritis nodosa (PAN), the Churg–Strauss syndrome, Wegener's granulomatosis and for Schönlein–Henoch purpura. Predisposing factors are sometimes suspected, for example drugs, viral or bacterial infection or malignancy, particularly of haematological origin. The same form of vasculitis, PAN for example, may be primary or caused by an aetiological agent such as hepatitis B virus. Conversely, the same agent may be associated with different forms of vasculitis: hepatitis B virus may cause macroscopic or microscopic PAN. Clinical features of several types of vasculitis may co-exist in the same patient [178]. Vasculitis may also arise in association with illnesses uncommonly characterized by angiitis such as systemic lupus erythematosus (SLE), mixed cryoglobulinaemia, rheumatoid arthritis (RA), Sjögren's syndrome [33], dermatomyositis [156], bacterial endocarditis, sepsis due to certain organisms, rickettsial infections and after surgical repair of aortic coarctation [5,97].

The Vasculitis Syndrome

Vasculitis may affect virtually any organ. In any given organ, however, angiitis is characterized by necrosis and inflammation giving rise to a clinical picture common to many forms of vasculitis. Thus, organ-specific clinical syndromes can be described. These are generally associated with constitutional signs of inflammation.

Age, Sex and Constitutional Signs

With the exception of Kawasaki's syndrome which rarely affects adults [43], vasculitis can occur in subjects of all ages. The mean age of onset is approximately 45 years [74,106]. Slightly more men than women develop Wegener's granulomatosis (1.3 to 2:1) [106]. The sex ratio is less marked in other forms of vasculitis. Constitutional signs such as rapid weight loss and fever herald the development of vasculitis in 50% of cases, and eventually affect all untreated patients [64,74,106,135]. The fever pattern is variable and chills may occur. Often the cause of fever remains elusive until other signs of angiitis appear. Fever may disappear with the development of severe renal failure, thus obscuring the clinical picture.

Laboratory findings are non-specific: variable elevations in erythrocyte sedimentation rate (ESR) (> 60 mm in 1 h), neutrophil-predominant hyperleucocytosis, moderate thrombocytosis, hyper-alpha 2-globulinaemia and hyper-gammaglobulinaemia. Hypereosinophilia is characteristic of the Churg–Strauss syndrome, but may also be seen in PAN [74,106,135]. Total haemolytic complement and the various complement factors are decreased only in certain aetiological circumstances. Certain forms of vasculitis when associated with severe renal disease may be accompanied by signs of disseminated intravascular coagulation [72].

Dermatological Signs

Four lesions which are easily recognized and biopsied are frequent in vasculitis: purpura, nodules, livedo and urticarial plaques.

Purpura is often petechial and sometimes infiltrated. A bullus may precede the development of a purpuric lesion, or such a lesion may develop around a necrotic zone millimetres to several centimetres in diameter. Necrotic lesions often leave residual scars. Purpura is due to angiitis of small dermal vessels. Limb gangrene is rare and related to thrombosis of large calibre arteries.

Subcutaneous nodules may be found anywhere but are most frequently seen on arms and legs. The lesions are usually well-localized, but in some cases of systemic lupus erythematosus, may become confluent, giving rise to "subacute nodular panniculitis" [212]. Purpura and nodules may be spontaneously painful or painful in response to pressure. These lesions are often transient and should be biopsied as soon as they are recognized. Biopsy of a nodule is generally more informative than biopsy of a purpural lesion. Livedo reticularis or racemosa is a non-specific inflammatory lesion frequently encountered in cases of vasculitis.

Peripheral Nervous System Signs

Amyotrophic sensorimotor polyneuritis [216] of abrupt or erratic onset is relatively characteristic of the vasculitis syndrome. Cranial nerves are rarely involved. Subclinical cases may be detected by systematic nerve conduction studies [300]. When polyneuritis is suspected, a nerve–muscle biopsy is indicated for diagnostic confirmation.

Central Nervous System Signs

The signs of cerebral angiitis are extremely diverse and include generalized or focal seizures, focal deficits, intracerebral haematomas, progressive encephalopathy with or without specific neurological deficits, and meningeal haemorrhage. These usually late manifestations must be distinguished from complications of treatment. In some cases, central nervous system involvement may herald the development of angiitis, and may be responsible for the only clinical signs of vasculitis. Invasive testing is then required. Cerebral angiography reveals a variety of more or less widespread lesions including segmental narrowing sometimes alternating with sausage-like dilatations, and thromboses. Cerebral biopsy, despite its dangers, may be necessary in fulminant cases of suspected vasculitis before therapy can be instituted [79,216,250,303].

Renal Involvement

Renal manifestations including proteinuria, microscopic or macroscopic haematuria, renal insufficiency and arterial hypertension are frequent.

Lesions of medium-sized arteries, such as those seen in macroscopic PAN, may cause severe or malignant hypertension, arterial aneur-

ysms, papilloedema, unilateral or bilateral loin pain, macroscopic haematuria and oligo-anuric or high-output acute renal failure. Alternatively, these lesions may be associated with few or no symptoms. Renal arteriography or digital angiography may be used for anatomical diagnostic confirmation. Small saccular aneurysms are generally found in grape-like clusters within the renal parenchyma. These lesions are persistently opacified. This distinguishes them from normal vascular curves. Arteriovenous aneurysms may be seen. Other radiological abnormalities are frequent and include segmental narrowing alternating at times with dilatation, arterial obliteration, infarcts and intra- or peri-parenchymal haematomas. Such anomalies can also affect the liver, the spleen, the mesentery and the brain [52,94,108,135,155,282].

Microscopic renal lesions may occur either alone or in association with macroscopic lesions. The associated signs are those of extracapillary glomerulonephritis; that is proteinuria, microscopic or macroscopic haematuria, oligo-anuric or high-output rapidly progressive renal failure [166,176,267,302]. Renal biopsy is classically contraindicated by the presence of renal aneurysms [24,135], but the risk of biopsy is probably overestimated [10]. Surgical or transcutaneous biopsy should be performed when angiography is contraindicated because of renal failure or is unrevealing, and when renal tissue is needed to assess the diagnosis and the reversibility of lesions. Transcutaneous biopsies should only be performed in centres equipped for carrying out arterial embolization should this become necessary.

Urological Manifestations

Ureteric stenosis has been described in PAN [136,157,208], Wegener's granulomatosis [47,58,246] and Schönlein–Henoch purpura [150]. It may be latent or discovered by intravenous urography, or it may be responsible for urinary or retroperitoneal symptoms. Bilateral stenosis leading to anuria has been described in a few cases. Testicular involvement is classical [16,122].

Cardiac Manifestations

Left ventricular failure or global cardiac failure may occur in PAN [64,122,135] and more frequently in the Churg–Strauss syndrome. Cardiomegaly and repolarization abnormalities are frequently observed [16,161,172]. Angina pectoris is rare. Asymptomatic pericardial effusions are common. In the Churg–Strauss syndrome, right ventricular failure may be due to acute or chronic respiratory insufficiency. Arterial hypertension may be responsible for left ventricular failure, particularly in PAN or in cases of vasculitis with advanced renal failure and volume overload.

Gastrointestinal Manifestations

Vasculitis can affect any part of the gastrointestinal tract. Haemorrhage and perforation usually occur in the small intestine but may rarely involve the stomach and the duodenum. In the latter case, fibroscopy is a useful diagnostic procedure. Perforations may also occur in the appendix or gall bladder. A rare complication is intestinal infarction following occlusion of a mesenteric artery. Very often the clinical picture is confusing, consisting of constant or paroxysmal abdominal pain, transit problems or haemorrhage of variable severity [44,64,71,74,135,223,255,264,310].

Although these manifestations are evocative of vasculitis, diagnosis is very difficult and rests upon evidence provided by arteriography or by exploratory laparotomy and biopsy.

In known cases of vasculitis, it is sometimes difficult to distinguish symptoms due to new areas of involvement from those due to complications of therapy. For example, acute pancreatitis is rare and the contribution of vascular lesions versus other factors such as corticosteroid therapy is difficult to appreciate. Liver disease is almost invariably associated with the presence of hepatitis B antigen [52,135]. Nevertheless, hepatic angiitis may exist and be responsible for pain, elevation of alkaline phosphatase levels [70] and intra- or peri-hepatic haemorrhage [6]. Splenic infarcts can occur, are often asymptomatic and are readily visualized by CAT scan [15,71].

Pulmonary Manifestations

Pulmonary involvement is more or less prominent depending upon the type of vasculitis.

Asthma occurs in 100% of cases of the Churg–Strauss syndrome and in 0% [74] to 30% of cases of PAN [106,135,264), but never in Wegener's granulomatosis [14]. Severe, steroid-dependent

asthma culminating in respiratory failure is the general rule [135]. In PAN, asthma may precede other signs of vasculitis by several years. The advent of such signs may coincide with worsening of the asthmatic condition. Conversely, respiratory infections or attempts at desensitization may trigger additional signs of vasculitis [135]. In the Churg–Strauss syndrome, patients often have a history of cutaneous or respiratory allergic manifestations in the absence of a family history of atopy. The evolution of the asthma is unpredictable. It may regress in favour of other locations of angiitis or persist when all other lesions disappear. Asthma is often accompanied by fever, hyperleucocytosis and hypereosiniphilia. The frequency of the latter is difficult to estimate since there is no clearcut border between the Churg–Strauss syndrome and PAN [74,106,108,135]. Small, eosinophil-rich, inflammatory pleural effusions are frequent. Chest pain, cough, mucopurulent expectoration and haemoptysis may occur. Massive haemoptysis is encountered only in Wegener's granulomatosis [47,133,152,286].

Radiological findings are diverse. Diffuse and fleeting bilateral pulmonary infiltrates may be seen in association with asthma-evoking Loeffler's syndrome. Diffuse interstitial opacities are rare. Irregular nodular infiltrates may be seen in all forms of vasculitis, and this often leads to diagnostic confusion with malignancy. Regional adenopathy is exceptional. Cavitation occurs in 50% of cases of Wegener's granulomatosis but is rare in PAN [75,76,106,107,135]. Tomography may sometimes be necessary in order to visualize cavities. The evolution of pulmonary infiltrates is unpredictable; they may regress spontaneously, leaving only a faint residual radiological image in the same or in a different area. This is especially true in Wegener's granulomatosis. Large infiltrates may persist for several months. This invariably poses a diagnostic dilemma since immunosuppressive therapy may be complicated by infections giving similar radiological signs.

Pulmonary functional abnormalities co-exist with radiological lesions and have been particularly well studied in Wegener's granulomatosis [76,107]. Residual abnormalities in pulmonary function are usually observed even after clinical cure.

Ocular Lesions

These include choroiditis, uveitis, aneurysmal dilatations, retinal detachment, multiple stenoses with or without obliteration of retinal arteries, and hypertensive retinopathy [135].

Articular and Muscular Lesions

Peripheral, fleeting arthralgia occurs in 30%–60% of cases of vasculitis. Often an early clinical sign, arthralgia may evolve into frank inflammatory arthritis. Myalgia is as common as arthralgia, but actual myositis is rare [64, 74,75,76,114,135,264].

Diagnostic Considerations

The diagnosis of vasculitis is suspected when constitutional symptoms are associated with signs of multi-system disease. Often several weeks, months or years may elapse before the diagnosis becomes evident. Some cases are recognized only at autopsy after fulminant illness or after a fatal complication. Delays in diagnosis may have dire consequences but they are not uncommon because of the diversity of clinical manifestations of vasculitis. As soon as the diagnosis of vasculitis is suspected, in presence of skin lesions for example, the extent of the disease should be evaluated for two reasons: (a) in order to assess the involvement of crucial organs such as the kidney, the lungs, the heart, the brain and the peripheral nerves; and (b) in order to select appropriate sites for confirmatory biopsies.

Appropriate investigations include anteroposterior and lateral chest radiographs, sinus x-rays, ENT and opthalmological examinations, evaluation of renal function (serum creatinine, 24-h protein excretion, examination of the urinary sediment) and occasionally fibreoptic bronchoscopy. Even slight microscopic haematuria and/or proteinuria should not be neglected and in themselves justify renal biopsy. Nerve conduction studies and electromyography may be useful in detecting latent disease and in selecting biopsy sites. Gastrointestinal endoscopy, CAT scanning and conventional or digital angiography may be indicated.

These investigations permit the selection of appropriate biopsy sites. Tissue samples should be sufficiently large and numerous in order to detect the often scattered and segmental lesions of vasculitis. Easily accessible organs are generally biopsied first, but pulmonary or cerebral biopsies may sometimes be necessary.

While documenting the extent of the disease, historical evidence should be collected to help pinpoint the cause or the type of vasculitis and should include otorhinolaryngological symptoms, allergic manifestations, hepatitis, the use of medications, serotherapy, vaccinations, drug abuse, cold intolerance etc. These data may prompt other investigations such as blood cultures, echocardiography, hepatitis B serology, rickettsial serodiagnosis, a search for cryoglobulins, anti-nuclear antibodies or rheumatoid factors, serum and urinary protein electrophoresis or immunoelectrophoresis and measurement of serum IgA levels, total haemolytic complement and complement fractions.

"Primary" Vasculitis

Angiitis Affecting Medium-Sized Arteries with or without Small Vessel Involvement

Classical Polyarteritis Nodosa (Kussmaul Maier Disease, Macroscopic Polyarteritis Nodosa)

PAN can occur at any age and affects males and females with equal frequency. In two-thirds of cases the onset of disease is marked by fever and weight loss. These may be the only symptoms for a long period of time, and the advent of other signs may be necessary before the appropriate diagnosis can be made. In the series reported by Godeau et al. [135], myalgia with or without myositis, arthralgia and polyneuritis were observed in 60% of cases, cutaneous lesions in 36% and pulmonary abnormalities in 21%. Asthma is said to be present in 0% [74] to 30% of cases [106,135,264], and in the series of Bletry et al. [32], differences exist between necrotizing angiitis with asthma and PAN without asthma. Hypereosinophilia is seen in 90% of cases with asthma but in only 6% of cases without asthma, cutaneous signs of vasculitis in 20% versus 60% of cases respectively, arthralgia in 4% versus 33% of cases and pulmonary infiltrates in 15% versus 4% of cases. The reported frequency of renal involvement is variable and undoubtedly depends upon patient recruitment. Renal insufficiency occurred in 21% of the cases reported by Godeau et al. [135]. Both macroscopic and/or microscopic vascular lesions may be responsible. Possible complications include renal infarction and sub-capsular haematoma. Arterial hypertension, seen in 40% of cases of PAN, generally supervenes during the 1st year of illness. Hypertension may be moderate and easy to control, or highly malignant when associated with renal artery stenosis.

With continuing improvements in anti-hypertensive therapy, the influence of hypertension on the prognosis of PAN should diminish. Hypertension persists after the cure of PAN [122,135]. The other signs of PAN and their frequency are listed in Table 27.1. Ureteric involvement and orchitis may mark the onset of illness [135]. Ultimately, the diagnosis of PAN rests upon angiographical and histological evidence. In 60% of cases of PAN [40], intraparenchymal aneurysms of medium-sized arteries are found. These are uncommon in Wegener's granulomatosis [21],

SLE [108], atrial myxoma [180], familial Mediterranean fever [95], Schönlein–Henoch purpura [178] and drug-induced angiitis [289]. Arterial aneurysms may go undetected if thrombosed. Biopsy specimens should be obtained whenever possible from sites of suspected involvement. Skin biopsies, especially of dermo-hypodermal nodules, and nerve–muscle biopsies are generally highly diagnostic [135].

PAN can be rapidly fatal when digestive and cerebral complications occur. Oligoanuria is often associated with PAN and iatrogenic complications may also occur, especially in cases of paroxysmal illness poorly controlled by therapy [76,135]. In most cases, however, PAN occurs in episodes that respond well to therapy. After a 1- to 2-year treatment period, drug dosages can be decreased or therapy stopped. Peripheral neuropathies usually regress with few sequelae [64,135]

Table 27.1. Clinical manifestations in polyarteritis nodosa (1169 cases) (From Even et al. [106])

Cases	Mowrey (1954) GR607	Nuzum (1954) GR175	Rose (1957) AC98	Frohnert (1967) AC130	Leib (1979) AC64	Guillevin (1980) AC95
Age		41		6–75(55)		
Sex ratio		4	1.6	2		
Allergic background	12			30	30	
Fever	68	81		76	36	74
Weight loss	58	46		71		63
Myalgias	50	35		30	18	49
Articular involvement			27	58	52	38
Polyarthritis				33	22	
Joint pain				25	30	
Neuropathies	66	54	36	55	72	68
Cardiac and/or coronary involvement			48	10	30	
Pericarditis				2	3	
Arterial thrombosis of limbs				2	3	
Phlebitis				3		
Arterial hypertension	58	54	56	56	55	37
Nephropathies						27
Proteinuria		71		75	45	
Haematuria				4	21	
Renal insufficiency	27	26		30	33	
Mesenteric involvement			70			
Abdominal pains	48	62			42	23
GI tract infarction					9	
Acute cholecystitis					17	
GI tract ulcers		2		14		
Hepatomegaly					13	
Jaundice	12					
Cutaneous lesions				58	28	
Indurated purpura	21	35		28	11	
Cutaneous nodules	25	16	27	18	8	53
Orchioepididymitis				2	2	
Asthma			12		21	
Respiratory involvement	36	47	33	38	47	39

GR, general review
AC, author cases

after several months of physiotherapy but central nervous system involvement may result in permanent neurological deficits. Small areas of myocardial infarction may also persist. Asthma, especially if present before the onset of other signs of vasculitis, may continue to be bothersome. Gastrointestinal perforations can occur even when PAN appears to be in remission [64,135,310]. Nephropathy can regress, especially if therapy was instituted early in the course of renal insufficiency. Chronic renal failure can occur and remain stable for several years; renal protection depends upon perfect control of arterial blood pressure. Haemodialysed or transplanted patients have the same prognosis as any other patients with chronic renal failure provided that PAN is maintained in remission. The poor prognosis formerly associated with anuria [166,176] can be improved with intensive therapy in a properly equipped centre.

The mainstay of therapy in PAN is high-dose corticosteroids which should be administered for at least 2 months, and then slowly withdrawn as clinical and laboratory signs of inflammation regress. The concurrent use of cyclophosphamide permits lower dose steroid therapy and this regimen apparently has a better 5-year prognosis [109,179,311]. However, some clinicians have reported that iatrogenic complications are more frequently seen in association with use of cyclophosphamide, and that the use of this drug does not appreciably improve prognosis [141]. The effects of plasmapheresis and bolus intravenous methylprednisolone therapy [164] are currently being assessed. It would appear that such therapy and/or bolus injections of cyclophosphamide may be beneficial, especially in cases where oral steroid therapy is contraindicated or insufficiently effective [30,245] (the treatment of nephropathy will be discussed in the section on "microscopic" PAN). In some cases, PAN may have a prolonged course with multiple relapses and visceral involvement and this poses difficult therapeutic problems. When treatment of PAN has been successfully stopped for more than 1 year, relapses are rare [135]. The 5-year survival rate when PAN persists has improved from less than 13% to over 50% with currently available therapy [64,135,179].

Churg–Strauss Syndrome ("Allergic Granulomatous Angiitis")

This syndrome, first described in 1951 [58] is rare and only 200 cases have been reported in the world literature, of which 15 occurred in France.

This syndrome is characterized by clinical signs and histological lesions similar to those seen in PAN, but with the following peculiarities: (a) severe steroid-dependent asthma with peripheral hypereosinophilia in 60%–100% of cases; (b) angiitis affecting arteries and veins of both medium and small calibre; (c) vascular and extravascular granulomas; and (d) eosinophilic infiltration of involved tissue [32,57,106]. Serological evidence of hepatitis B infection is infrequent in this syndrome [32]. Very often, affected individuals have a personal atopic history. Asthma precedes other signs of disease by an average of 8 years (0 to 30 years) and persists after clinical cure. Renal involvement is half as frequent as in PAN [57,106]. Prognosis is poorer than in PAN, and death often ensues due to left, right or biventricular heart failure, although Chumbley [57] does not mention this.

It is often difficult to make a diagnosis of the Churg–Strauss syndrome with absolute certainty [106]. The reasons for this are many: asthma and hypereosinophilia alone are insufficient criteria and characteristic lesions of granulomatous angiitis rich in eosinophils must be found; PAN may be accompanied by hypereosinophilia [74,135]; vascular aneurysms may be a feature of the Churg–Strauss syndrome [108,179] and noneosinophilic extravascular granulomas may be seen in PAN [58].

In fact, a recent series of 27 patients said to have the Churg–Strauss syndrome on the basis of cutaneous extravascular necrotizing granulomas included patients with illnesses as diverse as Wegener's granulomatosis (three cases), PAN (one case), SLE (two cases), rheumatoid arthritis (four cases), haemopoietic malignancy (three cases) including multiple myeloma (one case), subacute endocarditis (one case), chronic active hepatitis (two cases), ulcerative colitis (one case), three cases of unclassified vasculitis, and only seven cases of the Churg–Strauss syndrome per se as defined above [117]. This is why many authors include this syndrome in the diagnostic category of classical PAN, specifying that it is characterized by granulomatous polyarteritis with pulmonary involvement [248].

Others, like Fauci, use the term "overlap syndrome" [108]. It is possible that the only difference between the two entities is the portal of entry of the putative triggering antigen, airborne in the Churg–Strauss syndrome, and digestive in the case of PAN. The treatment of the Churg–Strauss syndrome is similar to that of PAN.

Necrotizing Angiitis Involving Vessels of Small Calibre

In 1952 Zeek proposed a classification of necrotizing angiitis based on a complete review of the literature since the original article of Kussmaul and Maier in 1866 and on personal experience with 50 patients as well as experimental data. She distinguished five types of angiitis: hypersensitivity angiitis, allergic granulomatous angiitis, rheumatic arteritis, periarteritis nodosa and temporal arteritis [308]. This classification excluded some forms of angiitis.

Hypersensitivity angiitis referred to a small-vessel vasculitis affecting venules and capillaries more often than arterioles. This condition was characterized by necrotizing glomerulonephritis, and pulmonary, splenic, hepatic and cardiac involvement, and was uniformly fatal in days to weeks if left untreated. The term hypersensitivity angiitis was chosen for the following reasons: In 1937, Clark and Kaplan described cases of vasculitis associated with serum sickness [60]; in 1942, Rich and Gregory confirmed this association and described another sulphonamide-related case [244], and Zeek herself found necrotizing angiitis at autopsy in numerous patients treated with sulphonamides [309] although a causal relationship was not demonstrated. Hypersensitivity angiitis corresponds to what is commonly called microscopic PAN. Zeek compared this form of angiitis to the macroscopic PAN described by Kussmaul and Maier. In the former, the lesions are said to be all of the same age, whereas lesions of different ages can co-exist in macroscopic PAN. In practice, while it is clear that macroscopic PAN specifically affects medium-sized arteries, the other putative histological differences seem theoretical and difficult to demonstrate in restricted biopsy specimens.

Small vessel necrotizing angiitis occurs in a very heterogeneous group of clinical syndromes. For reasons of clarity, it is useful to separate cutaneous or predominantly cutaneous syndromes from cutaneo-visceral syndromes. The former may be referred to as hypersensitivity angiitis secondary to a known agent, although this deviates from the definition of this disorder proposed by Zeek. The graver cutaneo-visceral syndromes would fall into the category of microscopic PAN. Thus, Godeau et al. use the terms hypersensitivity angiitis or Zeek's angiitis to refer to an essentially benign cutaneous syndrome which is acute and heralded by purpura, and which may occasionally affect muscles, joints, the heart, the lungs or the kidney [135]. The causative agent, be it microbial, heterologous serum or medications, is usually identifiable. The condition usually resolves in a few weeks, although the prognosis may be less favourable in cases with visceral involvement. Hypersensitivity angiitis will be discussed in more detail in the section on secondary angiitis.

Other apparently primary forms of small vessel angiitis occur in Schönlein–Henoch purpura and in hypocomplementaemic, urticarial angiitis defined by strict laboratory criteria.

Cutaneous Forms

Pure Cutaneous Forms. The term Gougerot's triad (purpura, nodules, livedo) is commonly used in the French literature to refer to purely cutaneous vasculitis [135]. The lesions previously described above may be found anywhere but are often found in the most dependent portions of the lower limbs. Erythematous rosettes are relatively characteristic. The illness usually evolves in irregular spurts lasting 1 to 4 weeks, and may last for several years without resulting in cutaneous necrosis or in an alteration of general wellbeing. All gradations of illness ranging from Gougerot's triad to a severe form of microscopic PAN may be seen. It is therefore important to be vigilant in order to rule out visceral angiitis or hypocomplementaemic angiitis. Some purely cutaneous forms of angiitis may be characterized by chronic, indurated urticaria or paroxysmal urticaria occurring at 1- to 30-day intervals. Patients complain of a burning sensation rather than of pruritis. All of these elements are fleeting and usually last less than 3 d. Hyperpigmentation may persist but there is no residual scarring.

Hypocomplementaemic Urticarial Angiitis or MacDuffie's Syndrome. This syndrome was first described by Agnello in 1971 [4], and is characterized by chronic urticaria and hypocomplementaemia. Women are more often affected and the condition tends to be episodic. The chronic urticaria, generally non-pruritic but sometimes painful, is accompanied by venular leucocytoclastic angiitis with neutrophilic infiltration.

The hypocomplementaemia occurs as a result of classical pathway activation [205,213,257, 261,271,275]. Complement abnormalities include a profound decrease of C1q together with low C3 and C4 levels whereas levels of C1r and C1s and the components of the alternative pathway are normal. The late complement components may be decreased [111]. In some patients, circulating anti-C1q precipitins are found [205,213]. Between episodes or during treatment with corticosteroids, C1q levels may increase. C1 esterase inhibitor levels are normal [111,205,213]. There is no cryoglobulinaemia and no anti-nuclear antibodies are found [257,261]. In addition to urticaria, the following may be present: purpura, arthralgia and arthritis, abdominal pain in the absence of a specific lesion, Quincke's oedema, adenopathy, uveitis and episcleritis. In the series of 16 patients reported by Schwartz and MacDuffie, eight of the ten smokers developed chronic obstructive lung disease which was severe in three cases [261]. Pseudotumour cerebri and myositis have been described [111,192]. A quarter of all patients have an associated glomerulopathy [111,213]. Proliferative lesions are seen with granular deposits of IgG and complement [111,192, 241,260,271], resembling the picture seen in SLE [213]. Focal necrosis occurs in some cases.

Low-dose corticosteroid therapy is recommended by MacDuffie et al., who reserve high-dose therapy for cases of where there is renal involvement. MacDuffie's syndrome and SLE may share certain features. Circulating anti-Sm antibodies have been reported in MacDuffie's syndrome [260] and urticaria may occur in SLE [257]. Some cases of cutaneous hypocomplementaemic vasculitis with a partial C3 deficit have been described [200].

Cutaneo-Visceral Forms

Microscopic Polyarteritis Nodosa. Microscopic PAN is a severe systemic illness affecting vessels of small calibre. The clinical signs, treatment and prognosis are often those of macroscopic PAN but there is no aneurysm and arterial hypertension. Necrotizing glomerulonephritis is frequently seen in microscopic PAN. Diagnosis may be difficult, particularly if the clinical picture includes only rapidly progressive necrotizing glomerulonephritis without immunoglobulin deposition, associated with vague systemic signs such as arthralgia, myalgia and fleeting erythema [64,69,153,201,219]. Renal prognosis without treatment is poor, but general outcome is usually favourable [201,219,249]. It is debatable whether such isolated glomerulopathies should be classified as localized forms of microscopic PAN.

Two therapeutic regimens have been shown to ensure a cure rate of 50%–75%: bolus methylprednisolone (30 mg/kg) on three occasions followed by oral prednisone (1 mg/kg/d) and in some cases cyclophosphamide [34,35,249], or oral prednisone (1 mg/kg/d), cyclophosphamide (2–3 mg/kg/d) and azathioprine (1 mg/kg/d) [239,258]. Plasmapheresis may be useful in cases associated with dialysis-dependent renal insufficiency [239]. The earlier treatment is instituted, the better are the renal and the general prognosis [160,302]. Corticosteroid therapy at a moderate dose should be continued for at least 12 months to reduce the risk of relapse [280].

Overlap between the syndromes of macroscopic and microscopic PAN is frequent [74,80,108,135].

It would seem that no clear boundaries exist between the various conditions in which apparently idiopathic small vessel vasculitis may occur (i.e. Gougerot's triad, microscopic PAN and hypersensitivity angiitis). When confronted with small vessel vasculitis, it is probably more useful for treatment and prognosis to itemize the organs involved and the possible aetiological factors than to try to classify the angiitis in any one diagnostic category.

Overlap Syndromes. Many authors since Zeek [309] have pointed out that certain patients develop a mixture of the symptoms seen in various forms of vasculitis such as microscopic or macroscopic PAN, the Churg–Strauss syndrome and hypersensitivity angiitis [73,108,135]. Leavitt and Fauci have recently described ten such apparently idiopathic cases of so-called "polyangiitis overlap syndrome" [178].

In the above-mentioned series, a combination of macroscopic PAN and the Churg–Strauss syndrom was seen in three cases, temporal arteritis and macroscopic PAN in one case, macroscopic PAN and Takayasu's syndrome in one case and giant cell arteritis and the Churg–Strauss syndrome or Wegener's granulomatosis in one other case: two other patients had a nonspecific syndrome with myositis. Overlap syndromes are seen in a younger population of patients than in macroscopic PAN (an average age of onset of 25 versus 41 years). There is no association with hepatitis B antigen. The overlap syndrome is generally severe enough to warrant treatment with prednisone and often with cyclophosphamide. Leavitt and Fauci insist that severe forms of angiitis that cannot be placed into the classical categories of vasculitis be designated "polyangiitis overlap syndrome" to avoid therapeutic delay.

Rheumatoid Purpura or Schönlein–Henoch Purpura. Schönlein–Henoch purpura is a cutaneous and systemic form of vasculitis which affects joints, the digestive tract and the kidney. It is a relatively common condition; over 1500 cases have been reported including approximately 100 cases in adults [18]. It is characterized by the presence of IgA deposits and occasionally C3 in the vascular walls of normal and involved skin, and in the glomerular mesangium [87,194]. Other immunoglobulins may also be present but usually in smaller quantities [194]. Some cases of angiitis clinically identical to Schönlein–Henoch purpura have been described in which IgA deposits were absent. Mesangial IgA deposits, in contrast, may be seen in other conditions not associated with angiitis such as Berger's disease [154] and SLE [158]. When circulating immune complexes in which IgA predominates are detected, especially early in the course of disease, this is an important argument for the pathogenicity of these complexes [168]. Other mechanisms have been proposed including the formation of in situ immune complexes composed of circulating IgA and a selectively localized or intrinsic mesangial antigen [191]. Complement activation by the alternative pathway and occasionally by the classical pathway appears to play an important role [183]. The number of circulating lymphocytes bearing surface IgA is increased in Schönlein–Henoch purpura. The spontaneous production of IgG, IgA and IgM is also increased both in Schönlein–Henoch purpura and in SLE [18,61, 104].

Upper respiratory infection precedes the onset of Schönlein–Henoch purpura by a few weeks in one-third to more than one-half of cases. This suggests that a mucosal IgA reaction to a bacterial or viral antigen may be involved in the pathogenesis of Schönlein–Henoch purpura. Alimentary antigens may trigger a similar reaction in the digestive tract which is especially rich in IgA [18,23,104,183]. Different medications (tetracycline, sulphonamides, aspirin, thiazides, quinidine derivatives, indomethacin) have been cited as possible aetiological agents but appropriate immunohistochemical studies were not performed in every case [18,23,25,206]. A few cases linking skin or glomerular IgA deposits to treatment with trimethoprim-sulphamethoxazole [65], quinidine derivatives [268] or propylthiouracil [139] have been described. We have ourselves seen two cases of Schönlein–Henoch purpura in patients with rheumatoid arthritis treated with non-steroidal anti-inflammatory drugs and with anti-malarials. A variety of other aetiological or predisposing factors have been cited [18,183] including influenza vaccinations,

primary tuberculous infection, tonsillectomy, insect bites, infections due to mycoplasma, yersinia or shigella [251], smallpox vaccination, cold exposure and monoclonal IgA gammopathy [96]. Schönlein–Henoch purpura is more frequent in children than in adults. Young adults are primarily affected. Males are more frequently affected than females regardless of age. The most comprehensive pediatric series deal with about 100 cases (in 18). The generally older series describing adult cases include an average of less than 30 patients each. Many of these studies omit a search for IgA in histological sections [18,23,25,173,183,270].

Schönlein–Henoch purpura may occur at any time, but cases tend to cluster in autumn and winter. The classical cutaneous signs have been described on p. 432. Purpura which is sometimes necrotic occurs preferentially in dependent parts. The legs are most frequently involved, but other skin areas may be affected. Erythematous or urticarial lesions (particularly of the scalp) may precede the characteristic purpuric lesions [18,135]. No cases of livedo have been described to our knowledge. Joint and gastrointestinal manifestations generally accompany the cutaneous signs. Colicky abdominal pain, diarrhoea and/or vomiting or haematemesis are common features. These may assume the proportions of a surgical emergency and may necessitate a period of parenteral feeding. Rarely, the following severe complications may occur: ileo-ileal intussusception (especially in boys), intestinal perforation, duodenal stenosis and acute convulsions, hypertensive encephalopathy, cerebral haemorrhage, pulmonary involvement, hepatomegaly, orchitis, parotiditis and ureteric stenosis [18,150].

The reported frequency of renal involvement varies from 22% to 92% depending upon patient recruitment and the stringency of diagnostic criteria; in a general paediatric ward, the reported frequency was 45%. Nephropathy is an early manifestation occurring in the 1st month of illness in 80% of cases, and within 3 months in 94% of cases. Nephropathy rarely develops any later, but when it does, it generally accompanies a new episode of purpura. In a few cases, nephropathy preceded the other signs of Schönlein–Henoch purpura by 5 to 11 months [18,86, 173,183,207].

The hallmark of renal involvement is haematuria which is macroscopic in 86% of cases. Haematuria occurs alone (5% of cases) or in association with proteinuria (50% of cases), or with proteinuria and the nephrotic syndrome

(30% of cases), and with renal insufficiency (15% of cases). Arterial hypertension is most frequent in severe cases of nephropathy and affects 30% of patients with the nephrotic syndrome and renal insufficiency [18,183,207,270]. The prognosis of Schönlein–Henoch purpura depends upon the severity of renal involvement. Generally, the nephropathy assumes a stuttering course lasting a few months to less than 2 years [183]. Renal insufficiency does not occur if the only renal signs are haematuria and proteinuria <1g/24h. However, end-stage renal failure ensues in 40%–60% of patients with renal insufficiency and/or severe nephrotic syndrome after a delay of 2 months to 8 years. Complete cure is possible in 25% of these patients. Renal prognosis is not linked to the severity of extrarenal manifestations. When moderate functional renal sequelae are present, these may disappear after several years, remain stable or gradually lead to end-stage renal failure. Renal biopsy is of considerable prognostic value. End-stage renal failure does not develop in patients with minimal, discrete lesions.

The incidence of end-stage renal failure increases with the percentage of glomerular crescentic involvement: when less than 50% of glomeruli contain crescents, the incidence is 4%; this incidence increases to 25% and to 66% when 50%–75% or greater than 75% of glomeruli are involved, respectively. These estimates have been confirmed in a large number of paediatric series [18,68,183,220,221]. The outcome in adults is similar although the general illness in adults may be more severe and even life-threatening [18,25,86,207,270]. Nevertheless, the renal prognosis in the majority of cases of Schönlein–Henoch purpura is excellent at least in the first few years after the onset of illness [173]. Even severe cases characterized by renal insufficiency and crescentic glomerulonephritis affecting more than half of all glomeruli may regress with few or no functional sequelae [220,221]. It is difficult to assess the value of commonly used therapeutic regimens since no prospective, randomized trials have been carried out [18]. Anti-coagulants and anti-platelet agents have been responsible for haemorrhagic complications [18,207]. Two retrospective studies conducted in paediatric nephrology centres suggest that treatment with bolus methylprednisolone, oral prednisone and in a few cases with cyclophosphamide can diminish the frequency of renal sequelae including chronic renal failure [220,221,249]. Plasmapheresis in association with drug therapy has seemed to be beneficial in a few cases [19,151,169,245]. Antibiotics are indicated only in cases of documented infection. Corticosteroids and non-steroidal anti-inflammatory agents have a beneficial effect on the cutaneous, digestive and articular manifestations of Schönlein–Henoch purpura.

Dialysis-dependent patients with Schönlein–Henoch purpura should be considered for renal transplantation. IgA deposition may occur in the renal allograft but is generally asymptomatic. However, it is reasonable to wait for several months to 1 year after the last attack of Schönlein–Henoch purpura before attempting transplantation [18,19].

Wegener's Granulomatosis

Wegener's granulomatosis is a clinical and histological entity characterized by granulomatous lesions of the upper and/or lower respiratory tract, angiitis and necrotizing glomerulonephritis [76].

The condition is very narrowly defined and as such is distinct from other forms of angiitis. A special therapeutic approach is also indicated. Diagnosis is based upon specific clinical and histological criteria, and often several months or years may elapse before the diagnosis becomes evident [107].

The first cases of this form of vasculitis were reported in 1931, and Wegener gave his name to this condition in 1936 [301]. A total of several hundred cases have been described including 85 cases in the largest published series of Fauci in 1983 [107]. Men are more often affected than women with a sex ratio of 3:2. The average age at onset is 41 years with a range from 14 to 71 years. Very few cases have been reported in blacks.

Wegener's granulomatosis is frequently preceded by ENT or pulmonary abnormalities (40% and 23% respectively.) This observation has led to a fruitless search for an inhaled bacterial or viral triggering antigen [149]. Even though such an agent has not been identified, this hypothesis should not be completely discarded in view of the recent report by Remee et al. who successfully treated six patients with trimethoprim-sulphamethoxazole alone [89]. Nalidixic acid is also thought to have a beneficial therapeutic effect. Finally, Wegener's granulomatosis is associated with the HLA antigens DR2 and B8 [103,167]. Almost any organ may be affected. The lungs, sinuses and kidneys are almost invariably involved (85% to 94% of

Diseases of the Arterial Wall

cases). Joints, the eyes, ears and nasopharynx are affected in 58%–67% of cases, the skin and the mucous membranes in 45% of cases, the nervous system in 22% of cases and the heart in 12% of cases. Gastrointestinal involvement is exceptional [226] as is central nervous system involvement [224,231]. Weight loss and fever occur in 75% of cases.

The illness is almost always heralded by respiratory symptoms that occur episodically over a period of several weeks to months. These symptoms may regress spontaneously in the initial phase of illness. Superinfection is very frequent. Common manifestations include chronic rhinitis with nasal obstruction, maxillary sinusitis (95% of cases), recurrent serous otitis, mastoiditis, laryngitis with or without associated dyspnoea [48], cough, atypical chest pain, haemoptysis which is rarely severe [53,286] and pleural effusions [107]. Regardless of the extent of involvement, asthma never occurs and respiratory insufficiency is rare. In 40% of cases, polyarthralgia and occasionally frank arthritis occur, but joint destruction is unusual. Ocular complications generally occur later in the disease and include proptosis, conjunctivitis, episcleritis, scleritis, corneal ulceration, angiitis affecting the optic nerve and rarely uveitis or obstruction of the nasolacrymal duct [47,76,107].

Cutaneous signs are non-specific and rarely occur at the onset of Wegener's granulomatosis [49,76,107]. Renal involvement is the inaugural sign in only 11% of cases [107].

Laboratory investigations are not very helpful in making a diagnosis. Generally, non-specific signs of inflammation are found in association with increased serum IgG and IgA levels but not IgM levels [107,284]. Complement levels are normal. Twenty per cent to 50% of patients have circulating rheumatoid factors and type III cryoglobulinaemia [107,247]. Two groups of authors have recently detected anti-cytoplasmic antibodies during acute attacks of Wegener's granulomatosis [140,291]. These antibodies are non-specific as they can be found in cases of microscopic PAN [188,189]. Patients with Wegener's granulomatosis are hepatitis B antigen-negative [107,247].

Radiographs of the upper respiratory tract may reveal a variety of non-specific lesions. Pulmonary abnormalities are more characteristic and include dense, ill-delimited infiltrates that are scattered throughout the lungs and that have a tendency to cavitate. These infiltrates may be single or multiple, pseudomiliary or confluent, and are of variable size (1–5 cm along their

longest axis). The lesions may regress or disappear without treatment between acute episodes. This makes diagnosis difficult, and it is not surprising that an average of 8.3 ± 1.5 months elapse between the first signs of illness and definitive diagnosis. The latter rests upon the identification of granulomatous angiitis in biopsy specimens [106,107,120]. Upper respiratory tract biopsies are often performed for reasons of easy accessibility and innocuity. The diagnostic yield is poor, however, since non-specific cellular infiltration and tissue necrosis without evidence of angiitis are usually seen. Laryngeal, tracheo-bronchial, orbital or cutaneous biopsies [48,107] are much more revealing. Transbronchial pulmonary biopsies are rarely useful. Often open lung biopsy is required [120]. The value of systemic renal biopsy is unknown, and cases of Wegener's granulomatosis without renal involvement have been described [46,54,55]. However, in Fauci's series, even though most patients had clinically minor renal disease, biopsies were useful for the detection of at least focal and segmental necrotizing lesions [107]. In the absence of granulomas, these lesions in the appropriate clinical context are sufficiently evocative to confirm the diagnosis. Renal biopsy should therefore be considered before open lung biopsy if other less invasive techniques are unsuccessful. This is especially true if proteinuria or haematuria are present in association with an unexplained ENT or respiratory condition.

Renal manifestations occur in 85% of cases of Wegener's granulomatosis [107]. They appear 2 months to over 3 years after the onset of illness and may sometimes be overlooked. Their discovery often helps in making the appropriate diagnosis [290]. Renal involvement may progress in a few days or a few weeks to renal insufficiency necessitating dialysis. The advent of renal failure represents a turning point which may have severe functional and even life-threatening consequences. Signs of renal involvement include unilateral or bilateral lumbar pain, murky or frankly bloody urine, oedema and oligoanuric renal insufficiency. Proteinuria and at least microscopic haematuria are always present. The nephrotic syndrome may occur. Blood pressure is generally normal in the absence of volume overload [17,149,153,237,267,302]. Renal pseudotumours [285], ureteral stenosis [21,246] and aneurysms [21,247] have been described in rare cases. Finally, a recent autopsy series has underlined the frequency of papillary necrosis [298]. Renal biopsy reveals lesions of variable severity ranging from focal, segmental necrosis with or without crescent formation to global necrosis and circumferential crescent formation affecting all glomeruli. These glomerular lesions may be associated with evidence of small vessel angiitis, interstitial or perivascular granulomas or peri- and intraglomerular granulomas destroying Bowman's capsules. These associated lesions are rarely present in biopsy specimens according to Fauci et al. [107].

When present, granulomas are polymorphic and contain lymphocytes, plasmocytes, macrophages and giant cells [155]. OKT4-positive cells predominate among the lymphocytes [284]. Fibrinogen is detected by immunofluorescence in the extracapillary space and in necrotic zones. Immunoglobulin and complement deposits are absent or discrete [107,247]. Rarely, lesions of thrombotic microangiopathy are superimposed [165]. These lesions, especially when unaccompanied by granulomas, are non-specific and can be found in other forms of vasculitis. They assume greater significance when associated with respiratory lesions. There is no correlation between initial renal function, glomerular necrosis or the extent of crescent formation and renal prognosis in treated patients [28,118, 149,237,302]. However, prognosis worsens with increasing glomerular scleroris and ischaemia especially if treatment was delayed [302].

Cyclophosphamide has been shown to be unquestionably superior to prednisone in the treatment of Wegener's granulomatosis. Fauci et al. recommend an initial treatment regimen consisting of cyclophosphamide (2 mg/kg/d) and prednisone (1 mg/kg/d) for 1 to 2 months [107]. Prednisone is then withdrawn progressively, as the clinical state allows, over a period of several months to 1 year. Cyclophosphamide should be continued for at least 1 year after complete clinical remission and the resolution of all laboratory signs. In some cases, relapses occur as treatment is tapered or stopped, and this generally prompts the re-institution of therapy for several years if not indefinitely. Ralapses often follow bacterial or viral infection [238]. If minor and involving discrete sites, some relapses may regress spontaneously without therapeutic modification. Treatment with cyclophosphamide should be monitored or stopped if the white blood count drops below 3000/mm³. In some cases, azathioprine has been successfully substituted for prednisone. Maintenance therapy with prednisone alone is more frequently associated with relapse than is therapy with cyclophosphamide alone or in combination with prednisone [149].

The main complication of cyclophosphamide therapy is haemorrhagic cystitis. Preventive measures include abundant fluid intake and the administration of cyclophosphamide in the morning.

In the 85 patients treated in this fashion by Fauci and followed for an average of 51±4.3 months, 75 patients survived and only six of the reported deaths could be directly attributed to Wegener's granulomatosis or to its treatment [107]. One case of malignant lymphoma occurred in a patient treated continuously for 9 years with cyclophosphamide [7]. These relatively favourable results may be due at least in part to patient selection since very few patients in this study had severe renal involvement.

Wegener's granulomatosis with severe renal insufficiency is a distinct clinical and histological entity with a different prognosis. This condition needs emergency therapy. There may be associated ocular involvement (77% of cases), muscular involvement (83% of cases), peripheral nervous system and/or cutaneous involvement (66% of cases) and central nervous system involvement (44% of cases). Granulomas are seen in one-third of all renal biopsies. Twenty-seven per cent to 38% of patients die in the first 2 months of treatment from pulmonary or gastrointestinal haemorrhage or sepsis. An additional 10% die in the following months. Only half of the patients originally requiring dialysis can eventually be weaned from this treatment. If serum creatinine remains above 400 μmol/l after the first 4 months of treatment, renal function usually deteriorates over a period of several months or years to end-stage renal failure, even if the condition appears to be otherwise in remission. In a few cases, however, haemodialysis was successfully stopped after more than 6 months of oligoanuria. These results are culled from several nephrological series [149,174,237,284,290,302]. In the series of 44 patients reported by Haworth [149], all the patients were initially treated with prednisone, azathioprine and cyclophosphamide, and a few also underwent plasmapheresis. The value of the latter cannot be assessed using currently available data. Six patients received no maintenance therapy; three are alive after 3 to 6 years of follow-up, and two died because of disease relapse even though treatment was re-instituted. Maintenance therapy in this series excluded cyclophosphamide. The relapse rate was 80% in patients treated with corticosteroids alone, and 25% in those treated with corticosteroids and azathioprine.

The efficacy of bolus methylprednisolone and bolus cyclophosphamide therapy is currently under investigation. The role of trimethoprim-sulphamethoxazole therapy, either alone or in association with immunosuppressive treatment, also warrants clarification [89]. Patients with end-stage renal failure may be successfully transplanted, and in the case of relapse, azathioprine may be substituted for cyclophosphamide [107,149,174,279].

Some cases of midline facial granuloma associated with necrotizing glomerulonephritis arising soon or a long time after remission of the initial episode have been recorded in the literature, and have all the features of Wegener's granulomatosis [285].

Angiitis Secondary to Known Aetiological Agents

Certain agents are known to cause angiitis which is clinically and histologically indistinguishable from primary angiitis. These agents include viruses, bacteria, parasites, medications etc. Vasculitis may also arise in association with neoplasms.

Viruses

Hepatitis B Virus

Hepatitis B infection may be associated with several types of vasculitis. In the prodromal phase of hepatitis, Gougerot's triad or hypersensitivity angiitis may arise, only to resolve when hepatitis becomes overt.

Markers of hepatitis B infection have been found in 0%–45% of cases of macroscopic PAN, and the role of HBs antigen in triggering this condition in some cases is now beyond doubt [24,32,56,64,99,135,267]. This is especially true in cases of angiitis not associated with asthma [32]. HBs antigen may be found in cryoprecipitates of patients with angiitis and mixed cryoglobulinaemia [12,182,199]. Hepatitis B virus is responsible for most of the alterations in hepatic function seen in the course of PAN. Chronic hepatitis and cirrhosis may ensue. The presence of hepatitis B surface antigen does not affect short-term prognosis, but in the long run, cure is rarely obtained [135]. Immunosuppressive therapy with corticosteroids or other agents may

retard seroconversion, and may even favour viral replication as determined by the measurement of DNA polymerase levels. It is for this reason that therapeutic regimens combining anti-viral agents and plasmapheresis are currently under investigation [142].

Other Viruses

An aetiological role has been suggested but not demonstrated for Herpes virus [236], EBV, CMV, influenza and arbovirus [75,84]. One case of relapse of necrotizing angiitis has been observed 10 d after anti-influenzal vaccination with killed myxovirus [45].

Bacterial and Parasitic Infections

Numerous bacterial infections may be accompanied by angiitis. The best known of these is streptococcal subacute bacterial endocarditis. A few rare cases of post-streptococcal glomerulonephritis and angiitis have also been described [50]. Fulminant acute rheumatic fever may also be associated with vasculitis [309]. In these conditions, immune complexes containing bacterial antigens play a probable but uncertain role. Glomerular involvement usually consists of IgG and complement deposition [154]. Other bacterial infections [232] and parasitoses may be associated with exclusively cutaneous vasculitis: pseudomonas septicaemia, rickettsioses [296, 303], gonococcal and meningococcal infections [240] and toxoplasmosis [101]. Many infectious agents are known to be responsible for erythema nodosum: streptococci, certain yersinia, Koch's bacillus, Hansen's bacillus and chalmydia [8,88]. Tuberculous meningo-encephalitis includes lesions of cerebral angiitis [79]. Finally, isolated angiitis in surgical appendectomy or cholycystectomy specimens is thought to be due to bacterial superinfection [44,110,186].

Angiitis Secondary to Therapeutic Agents

Hypersensitivity angiitis seems to be the most appropriate term for drug- or serum-induced vasculitis. In practice, this type of angiitis may be either confined to the skin or widely disseminated.

The cutaneous forms are benign. They are restricted to maculopapular or vesicular eruptions predominantly affecting the extremities. Nodules are not seen. Systemic signs are minimal or absent as is hypereosinophilia. Vasculitis usually develops 8 to 10 d after initial introduction of the causative agent, or more rapidly in the case of re-introduction. Delayed reactions after several weeks or months are possible [135].

Administration of heterologous serum elicits the classical picture of acute serum sickness including fever, urticaria, arthralgia and adenopathy. The same clinical syndrome may be elicited by non-protein medications. The syndrome resolves a few days after cessation of therapy [108].

The *disseminated forms* with visceral involvement are rare and the clinical picture, treatment and prognosis are those of microscopic PAN [135,218]. Mortality (84% in a series of 19 patients [218]) is high in cases where the diagnosis has been missed. Otherwise, the syndrome resolves rapidly upon withdrawal of the causative agent or under steroid therapy instituted in cases of life-threatening visceral involvement [135]. A number of medications may be responsible for crescentic necrotizing glomerulonephritis including chlorpropamide, D-penicillamine and phenylbutazone [100,181].

A few isolated cases of macroscopic PAN have been described in association with penicillin therapy [in 170,289].

Drug addicts may develop angiitis but the causative agent appears to be hepatitis B virus rather than any illicit drug [52,59,81]. Six cases of macroscopic PAN and of the Churg–Strauss syndrome have been described after desensitization for asthma and allergic rhinitis [234].

Angiitis in Cryoglobulinaemia

The principal constituents of cryoglobulins are immunoglobulins, usually of the IgG or IgM class (116). Type I cryoglobulins are composed of a single monoclonal immunoglobulin, usually IgG, with no known antibody activity. They are almost always seen in association with haematopoietic malignancy such as multiple myeloma or B cell lymphoma, and in Waldenström's macroglobulinaemia. Type II cryoglobulins consist of a monoclonal immunoglobulin, generally IgM but occasionally IgG or IgA, directed against a polyclonal IgG. Type II cryoglobulins are either primary or secondary to Waldenström's macroglobulinaemia, a chronic B cell lymphoproliferative disorder, Sjögren's syndrome or rarely "autoimmune" illness. Type III cryoglobulins are composed of two or more classes of polyclonal immunoglobulins. They are

never seen in cases of multiple myeloma or of Waldenström's macroglobulinaemia, but they are frequently present in small quantities in various haematological disorders, and in "autoimmune" disorders such as SLE, PAN [41], Schönlein–Henoch purpura [124] and Wegener's granulomatosis [107,247]. They are also seen in association with certain neoplasms and with systemic bacterial, fungal [123], parasitic or viral infections including hepatitis A [163] and B [182,199]. Some primary, idiopathic cases have been reported.

Type II and III cryoglobulins are circulating immune complexes that can fix complement. The IgM of mixed IgM–IgG cryoglobulins has rheumatoid factor activity [211].

Cryoglobulins may be responsible for factitious hypercytosis [41,116] and for a deceptive lowering of the erythrocyte sedimentation rate.

The clinical manifestations of cryoglobulinaemia have been described by Meltzer and Franklin [211]. Cases appear to be relatively uncommon, and the largest available series deals with 86 patients [41]. Symptoms are due to either hyperviscosity or to direct vascular wall damage. Hyperviscosity may be responsible for superficial, especially cutaneous, manifestations brought on by the cold.

Cryoglobulins can also penetrate vascular walls, activate complement and induce angiitis. In cases of secondary cryoglobulinaemia, the clinical picture is also characterized by signs of the underlying disease. The diagnosis of cryoglobulinaemia-related angiitis rests upon biopsy, generally of the skin, upon the discovery of cryoglobulins and complement in involved vessels [41,123,210] and an evidence of a cryoprecipitate in the serum, the composition of which has yet to be determined.

The signs of cryoglobulinaemia are relatively homogeneous. In two-thirds of cases, vasomotor phenomena are observed (Raynaud's syndrome and/or cutaneous lesions), and cold hypersensitivity occurs in less than 50% of cases. Renal insufficiency, mucosal bleeding, visual disturbances, abdominal pain or peripheral neuropathy are rare. The most frequent and the earliest manifestations involve the skin. All the lesions described on p.438 may be seen. Purpura is much more frequent in type II (60%) and in type III (70%) cryoglobulinaemia than in type I (15%). Purpura occurs in episodes lasting 1 to 2 weeks which alternate with periods of remission lasting several weeks or months. Distal necrosis is seen only in types I (40%) and II (20%) cryoglobulinaemia. Leg ulcers occur in less than 5%

of cases. Livedo and cold-induced urticaria are sometimes seen in types I and II cryoglobulinaemia. Twelve per cent of patients develop Gougerot's triad. Joint involvement is more frequent in type III cryoglobulinaemia (58% of cases) whereas neurological involvement is more frequent in types I (15%) and II (25%). Peripheral neuropathy is generally distal and asymmetric and may be abrupt in onset. The onset of motor signs follows that of sensory signs by several months to several years. Regression may occur slowly over a period of years. Medullary and central nervous system angiitis are rare.

Renal manifestations occur in 21% of cases. They include microscopic or macroscopic haematuria, proteinuria and renal insufficiency. An acute nephritic syndrome may be seen. Renal insufficiency with normal or decreased urine output may progress over a few weeks to a few months to a state requiring dialysis. In other cases, partial or complete remission occurs spontaneously or following treatment. In still other cases, the condition may worsen progressively over a period of several years [41, 299]. Renal involvement is of great prognostic importance: in 21 patients with nephropathy, five died because of renal failure and four others, still alive after 2 to 12 years of follow-up, developed chronic renal failure [41]. Renal lesions are more frequent in types I (25% of cases) and II cryoglobulinaemia (35% of cases) than in type III (12% of cases). There is no apparent correlation in the series of Brouet et al. [41] between the type of cryoglobulinaemia and the lesions seen by light microscopy. However, in a review of the literature, Heptinstall distinguishes the histological manifestations of type I from those of type II cryoglobulinaemia [159]. In type I, the usual lesions are those of membrano-proliferative glomerulonephritis with monocytic and neutrophilic infiltration, and C3 and cryoglobulin deposition. This picture is often seen in type II cryoglobulinaemia but glomerular deposits usually contain IgG, IgM and C3. These subendothelial and mesangial deposits are sometimes accompanied by occlusive capillary thrombi. The presence of thrombi is considered to be an indication of severity. Crescentic, necrotizing glomerulonephritis and arteriolar angiitis may occur.

Cordonnier et al. have reported four cases of type II (IgG–IgM) cryoglobulinaemia in which specific tubular and rosette-shaped crystalline deposits were seen by electron microscopy within the cryoprecipitate and in the glomerular sub-

endothelial deposits [66]. Such deposits can also be found in intraglomerular monocytes.

Gastrointestinal and retinal involvement have also been described. Ophthalmological examination may reveal granular deposits in retinal and conjunctival vessels and retinal vein thromboses.

The treatment of cryoglobulinaemia is that of the underlying cause. When life-threatening lesions or severe functional impairment are present, plasmapheresis with or without cryofiltration is indicated. This must be performed cautiously, especially if the serum cryoglobulin level is high and if the cryoglobulin is known to precipitate at a relatively elevated temperature [66,105,115,187,214,245,299]. Plasmapheresis may have to be performed in a room heated to 37°C. Exchanges are carried out three times a week for 2 weeks, and their frequency is then decreased until there is clinical improvement and a stable reduction in the cryocrit is obtained. In essential cryoglobulinaemia, it seems logical to institute immunosuppressive therapy for a few months in order to temper B cell hyperactivity [126]. Monthly injections or daily oral doses of cyclophosphamide or chlorambucil are commonly chosen. Moderate daily doses of prednisone alone are ineffective, but bolus methylprednisolone therapy is reputedly beneficial when used for the same indications as plasmapheresis, and when followed by conventional immunosuppressive maintenance therapy [90]. In non-threatening clinical situations, plasmapheresis and bolus methylprednisolone may be avoided in lieu of conventional drug therapy. Very satisfactory remissions that may last for years after therapy is discontinued are obtained in over 50% of cases [41].

Hypercholesterolaemic Angiitis

Hypercholesterolaemic angiitis (HCA) is induced by cholesterol emboli that originate from atheromatous plaques in the aorta or rarely in one of its branches. Diagnosis rests upon a positive biopsy of an affected organ [9,31, 135,195]. The histological lesions are characteristic and involve arterioles less than 200 μm in diameter. Needle-shaped, biconvex ghosts of cholesterol crystals, dissolved by the techniques of fixation, are seen. The underlying intima is thickened and macrophages, giant cells, neutrophils and occasionally eosinophils invade the often necrotic vascular wall [31,135,195]. This picture justifies the term HCA. The inflammatory reaction may give way to late, non-specific fibrosis. C3 and occasionally C4 levels are decreased, suggesting that the embolized atheromatous material may activate complement, and that this may play a pathogenetic role. Hypereosinophilia and moderate thrombopenia have been described [11,67,145]. Cholesterol levels are often normal or only slightly elevated at the time of diagnosis.

The incidence of HCA has been derived from autopsy series and is 4% in the general population, 15%–30% in patients with severe aortic atherosclerosis or abdominal aortic aneurysms and 77% of patients who have undergone surgery for an abdominal aortic aneurysm [31,135,272]. The sex ratio is 1 : 1. The high incidence at autopsy contrasts with the rarity of clinical manifestations which are generally only seen in males over the age of 50 [31,135]. Cholesterol emboli often arise a few days or weeks after aortic surgery or arteriography, or after anti-coagulant therapy, especially with anti-vitamin K derivatives [51,135,272]. Very often the underlying atherosclerosis is severe and has already been responsible for episodes of ischaemia. The aortic lesions are evident on plain abdominal radiographs and on echography.

The clinical manifestations depend upon the site of origin of the embolus. Most emboli originate in the abdomen. Iliac atheroma gives rise to lower limb emboli, while a plaque in the ascending aorta can shower all territories with emboli, including the coronary arteries. When atherosclerosis is widespread, the disease may mimic PAN or polymyositis; thus 4% of patients thought to have PAN in the series of Godeau et al. in fact had HCA [135].

Cutaneous signs most often provoke the correct diagnosis. The most classical is the "purple toe". This refers to the presence of well-delimited, violet patches on the plantar surface of one or more toes. The entire toe may be involved. This cutaneous sign may be preceded or accompanied by pain which is relieved when the limb is placed in a dependent position. The lesion may be transient or may lead to necrosis in the skin. Livedo reticularis frequently affects the lower limbs, the skin of the pelvic girdle and the trunk, but rarely the upper limbs. The lesion may predominate below the waist and may vary with time. There may be associated nodules, infiltrated and eventually necrotic plaques and necrotic petechial purpura.

Gastrointestinal emboli are responsible for fluctuating abdominal symptoms that may culminate in haemorrhage, ischaemic perforation of a hollow organ, acute pancreatitis or in generally asymptomatic splenic infarction [9,51,135,272]. Renal emboli are responsible for a stuttering form of renal insufficiency evolving over a period of months. Mild or moderate proteinuria, microscopic or macroscopic haematuria and arterial hypertension are common. Renal insufficiency may progress rapidly over several days to several weeks. If the characteristic cutaneous signs are absent, a renal biopsy may help to confirm the diagnosis of HCA. Of course PAN, emboli of cardiac origin, endocarditis, atrial myxoma and contrast-induced renal failure must be ruled out [51,197,272]. Three cases of crescentic glomerulonephritis associated with HCA have recently been described [137,146,243]. Renal insufficiency may regress at least partially even after the institution of dialysis. This regression may take several weeks to more than 6 months [13,197,272].

HCA evolves in a paroxysmal manner with intervening periods of spontaneous remission. The condition may be life-threatening in ill, aged patients with severe, generalized disease. Treatment is essentially preventive and designed to minimize the risk of embolization during arteriography or aortic surgery. Once embolization has occurred, treatment is symptomatic and may include dialysis [197,243,272]. Anti-coagulants are not indicated. Anti-platelet agents are sometimes used. Colchicine or moderate-dose corticosteroid therapy is said to have a favourable effect on the inflammatory syndrome and on the general state of wellbeing. In exceptional circumstances when the age, general state of health and the clinical picture permit, surgery may be considered to ablate well-localized atheromatous plaques [31,135,197].

Necrotizing Angiitis in Connective Tissue Disease and in Malignant Disorders

Connective tissue diseases may be associated with clinical and histological manifestations of angiitis [5,56,108,155].

Rheumatoid Arthritis (RA)

Vasculitis lesions are common in RA but are generally restricted to rheumatoid nodules.

Arteries and veins of small calibre may be affected [276]. Certain fulminant cases of RA mimic PAN and may be life-threatening. Vasculitis may supervene in highly sero-positive cases of RA, or may complicate relatively quiescent RA, sometimes when drug therapy is diminished [5,254,265]. A recent report of six cases of RA with vasculitis resembling PAN is particularly remarkable for the rapid development of angiitis (within 2 years of diagnosis of RA), the prevalence of pulmonary angiitis (three patients) and the severity of illness. All these patients died after an average of 24.5 months [177].

Bolus cyclophosphamide and methyl-prednisolone therapy followed by oral corticosteroids may improve the prognosis in this previously catastrophic condition [262]. Plasmapheresis may also have a beneficial if ephemeral effect [263].

Angiitis and Acute Systemic Lupus Erythematosus

Cutaneous and neurological vasculitis may occur in SLE [102,108,212]. Necrotic purpura is the usual sign and this may be accompanied by extensive cutaneous necrosis and digital gangrene. The neurological manifestations of SLE include motor, visual or sensory deficits, chorea, convulsions, psychiatric problems, myelopathy, cranial nerve palsy or peripheral neuropathy, and all these manifestations are linked to angiitis [102,212]. Medium-sized arteries or intracerebral arterioles are generally involved. The arteriographic picture may include multiple irregularities, stenoses and thromboses [78,102,112,259]. Digestive angiitis may cause ischaemia and lead to the perforation of a hollow organ [310]. The diagnosis of abdominal vasculitis complicating SLE is difficult to establish and is often delayed. The condition has a 50% mortality rate especially if there is involvement of other viscera. Coronary artery involvement may be responsible for severe myocardial ischaemia and heart failure [172]. IgG and complement deposits have been found in the vasculitic lesions but also in normal, healthy tissue. Renal angiitis is usually confined to glomuli and other renal vessels are spared [155]. Large aortic branch vessels may be involved [212]. The advent of vasculitis in SLE is a bad prognostic sign [22,29,177], especially when the kidney is involved. Treatment should probably include bolus methylprednisolone, cyclophosphamide

therapy and perhaps plasmapheresis in addition to corticosteroid therapy.

Other Connective Tissue Diseases and Angiitis

Clinical signs and histological manifestations of PAN arise in rare cases of scleroderma, dermatomyositis [156], atrophic polychondritis and mixed connective tissue disease [135]. Temporal arteritis and PAN have been reported to co-exist [135,225]. This association has been encountered in three out of 95 patients in the series of Godeau et al. [135].

Cutaneo-visceral vasculitis has been described in a few patients with alpha 1 anti-trypsin deficiency [184].

Angiitis and Malignancy

Macroscopic PAN has been reported in association with certain visceral and haematological malignancies but these cases are rare and may be coincidental [135]. Five cases of hairy cell leukaemia associated with angiitis have been described. Vasculitis ensues 6 months to 1 year after the diagnosis of haemopathy. Two patients with Hodgkin's disease out of a series of 1000 patients also had PAN. A few instances of central nervous system and medullary granulomatous angiitis apparently preceding the diagnosis of Hodgkin's disease were also cited. Associated viral infection was thought to be a predisposing factor, but this was not proven [77].

Microscopic PAN may be associated rarely with Hodgkin's disease, non-Hodgkin's lymphomas [77,233], multiple myeloma [203,242, 295] and other myeloproliferative syndromes [190]. Evidence of angiitis precedes the discovery of neoplasias by several months or years. Immune complexes containing tumour antigens may be involved but this remains to be demonstrated. Mixed cryoglobulinaemia may in fact account for the angiitis.

Localized Angiitis

Localized forms of angiitis exist which affect medium- and small-sized vessels. Some are idiopathic and others are associated with illnesses such as SLE. Thus, macroscopic PAN may be limited to the skin [73,135], to one or two muscle groups [114], to the gall bladder, to the appendix [44,110,186] or to an obstructed ureter [136, 157,201].

Localized central nervous system angiitis is rare but extremely serious [79,216,250,303]. The angiitis is often associated with concomitant herpetic infection, pyogenic meningitis, tuberculous or fungal meningitis, syphilis, coeliac disease [252] or SLE [78]. Primary angiitis strictly localized to the central nervous system is thus exceptional [79]. The onset of illness is marked by violent headaches and confusion followed by multifocal neurological deficits. There is generally no fever, myalgia or arthralgia. Arteriography provides the diagnosis in most but not all cases. Aneurysms are rare. A meningeal and cerebral biopsy may be indispensable. Corticosteroid therapy alone is effective but does not prevent relapse. A combination of oral cyclophosphamide and prednisone therapy is recommended by Cupps et al. who were able to induce complete and prolonged remission in four cases described recently [79].

References

1. Adams DO (1976) The granulomatous inflammatory response. Am J Pathol 84:163–192
2. Adams DO, Kao KJ, Farb R, Pizzo SV (1980) Effector mechanisms of cytolytically activated macrophages. II. Secretion of a cytolytic factor by activated macrophage and its relationship to secreted neutral proteases. J Immunol 124:293–300
3. Adams DO (1983) The biology of the granuloma. In: Loachim HL (ed) Pathology of granulomas. Raven Press, New York, pp 1–20
4. Agnello V, Koffler D, Eisenberg JW et al. (1971) C1q precipitins in the sera of patients with systemic lupus erythematosus and other hypocomplementic states: characterization of high and low molecular weight types. J Exp Med 134:228s–241s
5. Alarcon-Segovia D (1980) Classification of the necrotizing vasculitides in man. In: Alarcon-Segovia D (ed) Clinics in rheumatic diseases, vol 6. Saunders, Philadelphia, pp 223–231
6. Alleman MJA, Janssens AR, Spoelestra P et al. (1986) Spontaneous intrahepatic hemorrhages in polyarteritis nodosa. Ann Intern Med 105:712–713
7. Ambrus JL, Fauci AS (1984) Diffuse histiocytic lymphoma in a patient treated with cyclophosphamide for Wegener's granulomatosis. Am J Med 76:745–747
8. Amor B, Laoussadi S, Sansonetti P (1985) Microbiologie et virologie. In' Kahn MF, Peltier AP (eds) Maladies systémiques. Flammarion, Paris, pp 109–133
9. Anderson WR (1965) Necrotizing angiitis associated with embolization cholesterol. Case report with emphasis on the use of the muscle biopsy as a diagnostic aid. Am J Clin Pathol 43:65–71
10. Anonymous (1985) Systemic vasculitis. Lancet I:1252–1254
11. Anonymous (1985) The complement in atheroembolic disease. Lancet II:136
12. Anonymous (1985) Hematuria, proteinuria and hypertension in a patient with multiple organ system disease. Am J Nephrol 5:217–227
13. Anonymous (1981) Progressive renal failure with hematuria in a 62-year-old man. Clinicopathologic conference. Am J Med 71:468–474
14. Anonymous (1984) Wegener's granulomatosis. Lancet I: 260–261
15. Anonymous (1977) Postoperative fever, leukocytosis and acute renal failure. Clinicopathologic conference. Am J Med 63:421–429
16. Anonymous (1975) Fever, obtundation and acute renal failure. Clinico-pathologic correlations and long-term course. Am J Med 59:553–562
17. Appel GB, Gee B, Kashgarian M et al. (1981) Wegener's granulomatosis. Clinical-pathological correlations and long-term course. Am J Kidney Dis 1:27–37
18. Austin III HA, Balow JE (1983) Henoch–Schönlein nephritis: prognostic features and the challenge of therapy. Am J Kidney Dis 5:512–520
19. Bachman U, Biava C, Amend W et al. (1986) The clinical course of IgA nephropathy and Henoch–Schönlein purpura following renal transplantation. Transplantation 42:511–515
20. Baker AL, Kaplan MM, Benz WM et al. (1972) Polyarteritis associated with Australia antigen positive hepatitis. Gastroenterology 62:105–110
21. Baker SB, Robinson DR (1978) Unusual renal manifestations of Wegener's granulomatosis. Report of two cases. Am J Med 64:883–889
22. Baldwin DS, Gluck MC, Lowenstein J et al. (1977) Lupus nephritis. Clinical course as related to morphologic form and their transitions. Am J Med 62:12–30
23. Ballard HS, Eisenger, RP, Gallo G (1970) Renal manifestations of the Henoch–Schönlein syndrome in adults. Am J Med 49:328–335
24. Balow JE (1985) Renal vasculitis. Nephrology forum. Kidney Int 27:954–964
25. Bar-on H, Rosenmann E (1972) Schönlein–Henoch syndrome in adults. A clinical and histological study of renal involvement. Isr J Med Sci 8:1702–1715
26. Bayer AS, Theofilopoulos AN, Eisenberg R et al. (1976) Circulating immune complexes in infective endocarditis. N Engl J Med 295:1500–1505
27. Berger J, Yaneva H, Hinglais N (1971) Immunohistochemistry of glomerulonephritis. In: Hamburger J, Crosnier J, Maxwell MH (eds) Advances in nephrology. Year Book, Chicago, p 11
28. Berry S, Greene III, J, Park HS et al. (1981) Return of renal function after renal insufficiency with cyclophosphamide therapy in Wegener's granulomatosis. Arch Intern Med 141:544
29. Bhuyan UN, Malavtya AN, Malhotra KK (1983) Prognostic significance of renal angiitis in systemic lupus erythematosus (SLE). Clin Nephrol 20:109–111
30. Bletry O, Bussel A, Badelon I et al. (1982) Intérêt des échanges plasmatiques au cours des angéites nécrosantes: 11 cas. Nouv Press Med 11:2827–2831
31. Bletry O, Frances C, Kieffer E et al. (1986) Les embolies de cholstérol. Rev Prat 36:1301–1308
32. Bletry O, Guillevin L, Dubernet T et al. (1984) Angéites

necrosantes avec et sans asthme. Différences cliniques, biologiques, étiologiques. Rôle du virus B de l'hépatite. Presse Med 13:2245–2248

33. Bloch KT, Buchanan WW, Wohl MJ et al. (1965) Sjögren's syndrome. A clinical, pathological and serological study of sixty two cases. Medicine (Baltimore) 44:187–231

34. Bolton WK (1984) Use of pulse methylprednisolone in primary and multisystem glomerular diseases. In: Robinson RR. Nephrology, vol 2. Springer-Verlag, Berlin Heidelberg New York, pp 1464–1473

35. Bolton WK, Sturgill BC (1986) Pulse methylprednisolone therapy of rapidly progressive glomerulonephritis. 10 years experience. Kidney Int 29:180

36. Boros DL (1978) Granulomatous inflammation. Prog Allergy 24:184–267

37. Braverman IM, Yen A (1975) Demonstration of immune complexes in spontaneous and histamine-induced lesions in normal skin of patients with leukocytoclastic angiitis. J Invest Dermatol 64:105–112

38. Brentjens JR, O'Connell DW, Albini B, Andres GA (1975) Experimental chronic serum sickness in rabbits that received daily multiple and high doses of antigen: a systemic disease. Ann NY Acad Sci 254:603–613

39. Brentjens J, Ossi E, Albini B (1977) Disseminated immune deposits in lupus erythematosus. Arthritis Rheum 20:962–968

40. Bron KM, Strott CA, Shapiro AP (1965) The diagnostic value of angiographic observations in polyarteritis nodosa. A case of multiple aneurysms in the visceral organs. Arch Intern Med 116:450–454

41. Brouet JC, Clauvel JP, Danon F et al. (1974) Biologic and clinical significance of cryoglobulins. A report of 86 cases. Am J Med 57:775–788

42. Burkholder P (1968) Immunology and immunohistopathology of renal diseases. In: Becker EL (ed) Structural basis of renal disease. Harper and Row, New York, pp 197–218

43. Burnstein F, Metson R, Colman MF, Canalis R (1984) Kawasaki disease in adult. Arch Otolaryngol 110:543–545

44. Camilleri M, Pusey CD, Chadwick VS et al. (1983) Gastrointestinal manifestations of systemic vasculitis. QJ Med 206:141–149

45. Cannata J, Guesta V, Peral V et al. (1981) Reactivation of vasculitis after influenza vaccination. Br Med J 283:526

46. Carrington CB, Liebow AA (1966) Limited forms of angiitis and granulomatosis of Wegener's type. Am J Med 41:497–527

47. Case records of the Massachusetts General Hospital (1979) Case record 47–1979. N Engl J Med 300:1378–1385

48. Case records of the Massachusetts General Hospital (1979) Case 24–1979. N Engl J Med 300:1378–1385

49. Case records of the Massachusetts General Hospital (1979) Case 34–1979. N Engl J Med 301:421–428

50. Case records of the Massachusetts General Hospital (1983) Case 22–1983. N Engl J Med 308:1343–1353

51. Case records of the Massachusetts General Hospital (1983) Case 4–1983. N Engl J Med 310:244–253

52. Case records of the Massachusetts General Hospital (1985) Case 36–1985. N Engl J Med 313:622–631

53. Case records of the Massachusetts General Hospital (1986) Case 12–1986. N Engl J Med 314:834–844

54. Cassan P, Bergue A, Brochard C et al. (1976) Granulomatose de Wegener pulmonaire. Une observation. Nouv Press Med 34:2234–2236

55. Cassan SM, Coles DT, Harrison EG (1970) The concepts of limited forms of Wegener's granulomatosis. Am J Med 49:366–379

56. Christian CL, Sergent JS (1976) Vasculitis syndromes: clinical and experimental models. Am J Med 61:385–392

57. Chumbley LC, Harrison EG, De Remee RA (1977) Allergic granulomatosis and angiitis (Churg–Strauss syndrome). Report and analysis of 30 cases. Mayo Clin Proc 52:477–484

58. Churg J, Strauss L (1951) Allergic granulomatosis, allergic angiitis and periarteritis nodosa. Am J Pathol 27:277–301

59. Citron BP, Halpern M, McCarron M et al. (1970) Necrotizing angiitis associated with drug abuse. N Engl J Med 283:1003–1011

60. Clark E, Kaplan B (1937) Endothelial arterial and other mesenchymal alterations associated with serum disease in man. Arch Pathol 24:458–475

61. Clarkson AR, Woodroffe AJ, Bannister KM et al. (1984) The syndrome of IgA nephropathy. Clin Nephrol 21:7–14

62. Cochrane CG (1971) Mechanisms involved in the deposition of immune complexes in tissues. J Exp Med 134:75s–89s

63. Cochrane GC (1979) Immune complex-mediated tissue injury. In: Cohen S et al. (eds) Mechanisms of immunopathology. John Wiley and Sons, New York

64. Cohen RD, Conn DL, Ilstrup DM (1980) Clinical features, prognosis and response to treatment of polyarteritis. Mayo Clin Proc 55:146–155

65. Coquin Y, Modai J (1978) Vascularite cutanée bénigne lors d'un traitement par triméthoprime-sulfaméthoxazole. Nouv Presse Med 7:3145

66. Cordonnier D, Vialtel P, Renversez JC et al. (1982) Lésions rénales chez 18 malades porteurs de cryoglobulines mixtes IgM–IgG de type II. In: Actualités néphrologiques de l'Hôpital Necker. Flammarion, Paris, pp 219–241

67. Cosio FG, Zager RA, Sharma HM (1985) Atheroembolic renal disease causes hypocomplementaemia. Lancet II:118–121

68. Counahan R, Winterborn MH, White RHR et al. (1977) Prognosis of Henoch–Schönlein nephritis in children. Br Med J 2:11–14

69. Couser WG (1982) Idiopathic rapidly progressive glomerulonephritis Am J Nephrol 2:57–69

70. Cowan RE, Thomas GE, Mallinson CN et al. (1977) Polyarteritis nodosa of the liver: a report of two cases. Postgrad Med J 53:89–93

71. Coward RA, Gibbons CP, Brown CB et al. (1985) Gastro-intestinal haemorrhage complicating Wegener's granulomatosis. Br Med J 291:865–866

72. Crummy CS, Perlin E, Moquin RB (1971) Microangiopathic hemolytic anemia in Wegener's granulomatosis. Am J Med 51:544–548

73. Cupps TR, Fauci AS (1981) Pathophysiology of vasculitis. In: Smith LH Jr (ed) The vasculitides: major problems in internal medicine, vol 21. Saunders, Philadelphia, pp 6–19

74. Cupps TR, Fauci AS (1981) Systemic necrotizing vasculitis of the polyarteritis nodosa group. In: Smith LH Jr (ed) Major problems in internal medicine, vol 21. Saunders, Philadelphia, pp 26–49

75. Cupps TR, Fauci AS (1981) Hypersensitivity vasculitis. In: Smith LH Jr (ed) The vasculitides: major problems in internal medicine. Saunders, Philadelphia, pp 50–71

76. Cupps TR, Fauci AS (1981) Wegener's granulomatosis.

In: Smith LH Jr (ed) Major problems in internal medicine, vol 21. Saunders, Philadelphia, pp 72–87

77. Cupps TR, Fauci AS (1981) Vasculitis and neoplasm. In: Smith LH Jr (ed) Major problems in internal medicine, vol 21. Saunders, Philadelphia, pp 116–122

78. Cupps TR, Fauci AS (1981) Central nervous system vasculitis. In: Smith LH Jr (ed) Major problems in internal medicine, vol 21. Saunders, Philadelphia, pp 123–132

79. Cupps TR, Moore PM, Fauci AS (1983) Isolated angiitis of the central nervous system. Prospective diagnostic and therapeutic experience. Am J Med 74:95–105

80. D'Agati V, Chandler P, Nash M et al. (1986) Idiopathic microscopic polyarteritis nodosa: ultrastructural observation on the renal vascular and glomerular lesions. Am J Kidney Dis 7:95–110

81. Dally B, Guillevin L, Maindenberg M et al. (1979) Périartérite noueuse chez un toxicomane atteint d'hépatite B. Ann Med Interne 130:649–652

82. Dannenberg AM Jr, Sugimoto M (1976) Liquefication of caseous foci in tuberculosis. Am Rev Respir Dis 113:257–259

83. Davies P, Bonney RJ (1979) Secretory products of mononuclear phagocytes. A brief review. J Reticuloendothel Soc 26:37–48

84. Davies DJ, Moran JE, Niall JF et al. (1982) Segmental necrotizing glomerulonephritis with antineutrophil antibodies: possible arbovirus aetiology. Br Med J 285:606

85. Davson J, Ball J, Platt R (1948) The kidney in periarteritis nodosa. QJ Med 17:175–202

86. De Groc F, Beaufils H, Bendia N et al. (1985) Nephropathies glomérulaires du purpura rhumatoide de l'adulte. Néphrologie 6:82–83

87. De La Faille-Kuyper EHB, Kater L, Kooiker CJ, Mees EDJ (1973) IgA-deposits in cutaneous blood vessel walls and mesangium in Henoch–Schönlein syndrome. Lancet I: 892–896

88. Denneberg T, Friedberg M, Samuelsson T et al. (1981) Glomerulonephritis in infections with Yersinia enterocolitica O serotype 3. Acta Med Scand 209:97–101

89. De Remee RA, McDonald TJ, Weiland LH (1985) Wegener's granulomatosis: observations on treatment with antimicrobial agents. Mayo Clin Proc 60:27–32

90. De Vecchi A, Montagnino G, Pozzi C et al. (1983) Intravenous methylprednisolone pulse therapy in essential mixed cryoglobulinaemia nephropathy. Clin Nephrol 19:221–227

91. Dienstag JL (1981) Hepatitis B as an immune complex disease. Semin Liver Dis 1:45–50

92. Dixon FJ, Vazquez JJ, Weigle WO, Cochrane CG (1958) Pathogenesis of serum sickness. Arch Pathol 65:18–28

93. Dixon FJ, Feldman JD, Vazquez JJ (1961) Experimental glomerulonephritis: the pathogenesis of a laboratory model resembling the spectrum of human glomerulonephritis. J Exp Med 113:899–920

94. D'Izarn JJ, Boulet CP, Convard JP et al. (1976) L'artériographie dans la périartérite noueuse. A propos de 15 cas; J Radiol Electrol 57:505–509

95. Dor JF, Clauvel JP, Degos L et al. (1979) Hematome périrénal spontané au cours de la maladie périodique. 3 observations. Nouv Presse Med 8:1927–1929

96. Dosa S, Cairns SA, Mallick NP et al. (1980) Relapsing Henoch–Schönlein syndrome with renal involvement in a patient with an IgA monoclonal gammopathy. Nephron 26:145–148

97. Downing DF, Grotzinger PJ, Weller RW (1958) Coarctation of the aorta: the syndrome of necrotizing arteritis of the small intestine following surgical therapy. Maj Dis Child 96:711–719

98. Droz D, Noel LH, Leibowitch M, Barbanel C (1979) Glomerulonephritis and necrotizing angiitis. In: Hamburger J et al. (eds) Advances in nephrology, vol 8. Year Book, Chicago, p 343

99. Drueke T, Barbanel C, Jungers P et al. (1980) Hepatitis B antigen-associated periarteritis nodosa in patients undergoing long-term hemodialysis. Am J Med 68:86–90

100. Druet P, Jacquot C, Baran D et al. (1987) Immunologically mediated nephritis induced by toxins and drugs. In: Bach, PH, Lock EA (eds) Nephrotoxicity in the experimental and the clinical situation. Martinus Nijhoff, Dordrecht, Boston, Lancaster, pp 727–770

101. Dry J, Leynadier F, Henault S et al. (1979) Toxoplasmose aiguë et vascularite nécrosante. Ann Med Interne 130:401–404

102. Dubois EL (1974) The clinical picture of systemic lupus erythematosus. In: Dubois EL (ed) Lupus erythematosus, 2nd edn. University of Southern California Press, Los Angeles, pp 232–437

103. Elkon KB, Sutherland DC, Rees AJ et al. (1983) HLA antigen frequencies in systemic vasculitis: increase in HLA-DR2 in Wegener's granulomatosis. Arthritis Rheum 26:102–105

104. Emancipator SN, Gallo GR, Lamm ME (1985) IgA nephropathy: perspectives on pathogenesis and classification. Clin Nephrol 24:161–179

105. Evans TW, Nicholls AJ, Ward AM (1976) Acute renal failure in essential mixed cryoglobulinaemia: precipitation and reversal by plasma exchange. Clin Nephrol 21:287–293

106. Even P, Sors H, Venet A et al. (1985) Angéite granulomateuse allergique de Churg et Strauss, granulomatose de Wegener et vascularites pulmonaires apparentées. In: Khan MF, Peltier AP (eds) Maladies systémiques, Flammarion, Paris, pp 446–483

107. Fauci AS, Haynes BF, Katz P et al. (1983) Wegener's granulomatosis: prospective clinical and therapeutic experience with 85 patients for 21 years. Ann Intern Med 98:76–85

108. Fauci AS, Haynes BF, Katz P (1978) The spectrum of vasculitis: clinical, pathologic and therapeutic considerations. Ann Intern Med 89:660–676

109. Fauci AS, Katz P, Haynes BF et al. (1979) Cyclophosphamide therapy of severe systemic necrotizing vasculitis. N Engl J Med 301:235–238

110. Fayemi AD, Ali M, Braun EJ (1972) Necrotizing vasculitis of the gallbladder and the appendix. Am J Gastroenterol 67:608–612

111. Feig PU, Soter NA, Yager HM et al. (1976) Vasculitis with urticaria, hypocomplementemia and multiple system involvement. JAMA 236:2065–2068

112. Feinglass EJ, Arnett F, Dorsch AL et al. (1976) Neuropsychiatric manifestations of systemic lupus erythematosus. Diagnosis, clinical spectrum, and relationship to other features of the disease. Medicine 55:323–339

113. Ferrari E. (1903) Uber polyarteritis acuta nodosa (sogenannte Periarteritis nodosa), und ihre Beziehungen zur polymyositis und polyneuritis nodosa. Beitr Pathol Anat 34:350–386

114. Ferreiro JE, Saldana MJ, Azevedo SJ (1986) Polyarteritis manifesting as calf myositis and fewer. Am J Med 80:312–315

115. Ferri C, Moriconi L, Gremignai G et al. (1986) Treatment of the renal involvement in mixed cryoglobulinaemia with prolonged plasma exchanges. Nephron 43:246–253

116. Feuilhade P (1984) Cryoglobulinémies. In: Dreyfus B (ed) Hematologie. Flammarion, Paris, pp 605–608

117. Finan MC, Vinkelmann RK (1983) The cutaneous extravascular necrotizing granuloma (Churg–Strauss granuloma) and systemic disease: a review of 27 cases. Medicine 62:142–158

118. Fischer E, Blumberg A (1978) Prolonged anuria in Wegener's granulomatosis. Recovery of renal function. JAMA 240:1174–1175

119. Fligiel SEG, Ward PA, Johnson KJ et al. (1984) Evidence for a role of hydroxyl radical in immune-complex induced vasculitis. Am J Pathol 115:375–382

120. Flye MW, Mundinger GH, Fauci AS (1979) Diagnostic and therapeutic aspects of the surgical approach to Wegener's granulomatosis. J Thorac Cardiovasc Surg 77:331–337

121. Freedman P, Peters JH, Kark RM (1960) Localization of gamma globulin in the diseased kidney. Arch Intern Med 105:524–535

122. Frohnert PP, Sheps SG (1967) Long-term follow up study of periarteritis nodosa. Am J Med 43:8–14

123. Gamble CN, Ruggles SW (1978) The immunopathogenesis of glomerulo-nephritis associated with mixed cryoglobulinemia. N Engl J Med 299:81–84

124. Garcia-Fuentes M, Chantler C, Williams DG (1977) Cryoglobulinemia in Henoch–Schönlein purpura. Br Med J 2:163–165

125. Geltner D, Franklin EC, Frangione B (1980) Antiidiotypic activity in the IgM fractions of mixed cryoglobulins. J Immunol 125:1530–1535

126. Geltner D, Kohn RW, Gorevic P et al. (1981) The effects of combination therapy (steroids, immunosuppressives and plasmapheresis on 5 mixed cryoglobulinemia patients with renal neurologic and vascular involvement). Arthritis Rheum 24:1121–1127

127. Gephardt GN, Ahmad M, Tubbs RR (1983) Pulmonary vasculitis (Wegener's granulomatosis): immunohistochemical study of T and B cell markers. Am J Med 74:700–704

128. Germuth FG Jr, Heptinstall RH (1957) The development of arterial lesions following prolonged sensitization to bovine gamma globulin. Bull John Hopkins Hosp 100:58–70

129. Giannetti A, Serri F, Bernasconi C (1976) Immunofluorescence studies of the skin in mixed cryoglobulinemia and Schönlein–Henoch purpura. Acta Derm Venereol (Stockh) 56:211–216

130. Ginsburg C, McCluskey RJ, Nepon JT et al. (1982) Antigen and receptor-driven regulatory mechanisms. X. The induction and suppression of hapten-specific granulomas. Am J Pathol 106:421–431

131. Gitlin D, Craig JM, Janeway CA (1957) Studies on the nature of fibrinoid in the collagen diseases. Am J Pathol 33:55–77

132. Glassock RJ, Cohen AH, Adler SG, Ward HJ (1986) Secondary glomerular disease. In: Brenner BM, Rector FC (eds) The kidney, Saunders, Philadelphia, pp 1014–1084

133. Glassock RJ, Feinstein EI, Kitt D et al. (1985) Hemoptysis and acute renal failure in a young man Am J Nephrol 5:64–70

134. Gocke DJ, Hsu K, Morgan C et al. (1970) Association between polyarteritis and Australia antigen. Lancet II:1149–1153

135. Godeau P, Guillevin L (1985) Periartérite noueuse systémique. In: Kahn MF, Peltier AP (eds) Maladies systémiques, Flammarion, Paris, pp 414–445

136. Godeau P, Guillevin L, Lenne Y et al. (1979) Périartérite noueuse avec sténoses urétérales. Deux observations. Nouv Press Med 8:2269–2271

137. Goldman M, Thoua Y, Dhaene M et al. (1985) Necrotizing glomerulonephritis associated with cholesterol microemboli. Br Med J 290:205–206

138. Gower RG, Sams WM Jr, Thorn EG et al. (1977) Leucocytoclastic vasculitis: sequential appearance of immunoreactants and cellular changes in serial biopsies. J Invest Dermatol 69:477–484

139. Griswold WR, Mendoza SA, Johnston W et al. (1978) Vasculitis associated with propylthiouracil. Evidence for immune complex pathogenesis and response to therapy. West J Med 128:543–546

140. Gross WL, Ludemann G, Kieffer G et al. (1986) Anticytoplasmic antibodies in Wegener's granulomatosis. Lancet, I:806

141. Guillevin L, Le Thi Huong DU, Bussel A et al. (1985) Treatment of polyarteritis nodosa with plasma exchange: a randomized study in 60 patients. Plasma Ther Transfus Technol 6:483–486

142. Guillevin L, Trepo C, Aouate JM et al. (1986) Polyarteritis nodosa related to hepatitis B virus treatment with plasma exchange and vidarabine. Preliminary report. In: Nosé Y, Kjellstrand L, Yvanovich P (eds) (1985) Progress in artificial organs. ISAO Press, Cleveland, pp 827–829

143. Gupta RC, Kohler PJ (1984) Identification of HBs Ag determinants in immune complexes from hepatitis B virus-associated vasculitis. J Immunol 132:1222–1228

144. Hamashima Y, Tasaka K, Hoshino T et al. (1982) Mite associated particles in Kawasaki disease. Lancet, II:266

145. Hammerschmidt DE, Greenberg CS, Yamada O et al. (1981) Cholesterol and atheroma lipids activate complement and stimulate granulocytes. A possible mechanism for amplification of ischemic injury in atherosclerotic states. J Lab Clin Med 98:68–77

146. Hannedouche T, Godin M, Courtois H et al. (1986) Necrotizing glomerulonephritis and renal cholesterol embolization. Nephron 42:271–272

147. Hannedouche T, Godin M, Fillastre JP. (1986) Vascularites d'hypersensibilité d'origine médicamenteuse. Ann Med Interne 137:57–64

148. Hart MN, Tassel SK, Sadewasser KL et al. (1985) Autoimmune vasculitis resulting from in vitro immunization of lymphocytes to smooth muscle. Am J Pathol 119:448–455

149. Haworth SJ (1983) Renal involvement in Wegener's granulomatosis. The Hammersmith experiment. In: d'Amico G, Colasanti G (eds) Nephrology 83. Wichtig Editore, Milano, pp 33–43

150. Hayat P, Sonsino E, Bompard Y et al. (1978) Sténose ischémique de l'uretère au cours d'un purpura rhumatoide. Nouv Presse Med 7:3913–3920

151. Hene RJ, Kater L (1983) Plasmapheresis in nephritis associated with Henoch–Schönlein purpura in primary IgA nephropathy. Plasma Ther Transfus Technol 4:165–173

152. Hensley MJ, Feldman NT, Lazarus JM et al. (1979) Diffuse pulmonary hemorrhage and rapidly progressive renal failure. An uncommon presentation of Wegener's granulomatosis. Am J Med 66:894–898

153. Heptinstall RH (1983) Crescentic glomerulonephritis. In: Heptinstall RH (ed) Pathology of the kidney, 3rd edn. Little, Brown, Boston, pp 443–477

154. Heptinstall RH (1983) Focal glomerulonephritis. In: Heptinall RH (ed) Pathology of the kidney, 3rd edn. Little, Brown, Boston, pp 557–600
155. Heptinstall RH (1983) Polyarteritis (periarteritis) nodosa and other forms of vasculitis and rheumatoid arthritis. In Heptinstall RH (ed) Pathology of the kidney 3rd edn. Little, Brown, Boston, pp 793–838
156. Herreman, G, Godeau P, Marteau R et al. (1977) Dermatomyosite, fibrose pulmonaire et nécroses cutanées. Une nouvelle entité. A propos de deux observations. Ann Med Interne 128:773–779
157. Herreman G, Puech H, Galezowki N et al. (1985) Périartérite noueuse urétérale localisée. Guérison chirurgicale. Presse Med 14:48
158. Hill GS (1983) Systemic lupus erythematosus and mixed connective tissue disease. In: Heptinstall RH (ed) Pathology of the kidney, 3rd edn. Little, Brown, Boston, pp 839–906
159. Hill GS (1983) Multiple myeloma, amyloidosis. Waldenström's macroglobulinemia, cryoglobulinemia and benign monoclonal gammopathies. In: Heptinstall RH (ed) Pathology of the kidney 3rd edn. Little, Brown, Boston, pp 993–1067
160. Hind CRK, Lockwood CM et al. (1983) Prognosis after immuno-suppression of patients with crescentic nephritis requiring dialysis. Lancet I:263–265
161. Holsinger DR, Osmundson PH, Edwards JE (1962) The heart in periarteritis nodosa. Circulation 25:610–618
162. Hu CH, O'Loughlin S, Winkelmann RK (1977) Cutaneous manifestations of Wegener's granulomatosis. Arch Dermatol 113:175–182
163. Inman RD, Hodge M, Johnston MEA et al. (1986) Arthritis, vasculitis and cryoglobulinemia associated with relapsing hepatitis. A virus infection. Ann Med Interne 105:700–703
164. Jarrousse B, Guillevin L (1986) Intérêt de la methylprednisolone dans le traitement des vascularites nécrosantes systémiques. Résultats chez 15 patients. Ann Med Interne 137:23–26
165. Juncos LI, Alexander RW, Marbury TC (1979) Intravascular clotting preceding crescent formation in a patient with Wegener's granulomatosis and rapidly progressive glomerulonephritis. Nephron 24:17–20
166. Kanfer A, Sraer JD, Feintuch MH et al. (1976) Insuffisance rénale aigue au cours de la périartérite noueuse. Nouv Presse Med 5:1883–1888
167. Katz P, Alling DN, Haynes BF et al. (1979) Association of Wegener's granulomatosis with HLA-B8. Clin Immunol Immunopathol 14:268–270
168. Kauffmann RH, Herrmann WA, Meyer CJLM et al. (1980) Circulating IgA-immune complexes in Henoch–Schönlein purpura. A longitudinal study of their relationship to disease activity and vascular deposition of IgA. Am J Med 69:859–866
169. Kauffmann RH, Houwert DA (1981) Plasmapheresis in rapidly progressive Henoch–Schönlein glomerulonephritis and the effect on circulating IgA immune complexes. Clin Nephrol 16:155–160
170. Kleinknecht D, Kanfer A, Morel-Maroger et al. (1978) Immunologically mediated drug-induced acute renal failure. Contrib Nephrol 10:42–52
171. Kopaniak MM, Movat HZ (1982) Kinetics of acute inflammation induced by E. coli in rabbits. Am J Pathol 110:13–29
172. Korbert SM, Schwartz MM, Lewis EJ (1984) Immune complex deposition and coronary vasculitis in systemic lupus erythematosus. Am J Med 77:141–146
173. Koskimies O, Mir S, Rapola J et al. (1981) Henoch–Schönlein nephritis: long term prognosis of unselected patients. Arch Dis Child 56:482–484
174. Kuross S, Davin T, Khellstrand CM (1981) Wegener's granulomatosis with severe renal failure: clinical course and results of dialysis and transplantation. Clin Nephrol 16:172–180
175. Kussmaul A, Maier E. (1866) Uber eine nicht beschriebene eigenthümliche Arterienerkrankung (periarteritis nodosa), die mit Morbus Brightii und rapid fortschreitender allgemeiner Muskellähmnug einhergeht. Dtsch Arch Klin Med 1:484
176. Ladefoged J, Nielsen B, Raaschou F et al. (1969) Acute anuria due to polyarteritis nodosa. Am J Med 45:827–831
177. Lakhanpal S, Conn DL, Lie JT (1984) Clinical and prognostic significance of vasculitis as an early manifestation of connective tissue disease syndromes. Ann Intern Med 101:743–748
178. Leavitt RY, Fauci AS (1986) Polyangiitis overlap syndrome. Classification and prospective clinical experience. Am J Med 81:79–85
179. Leib ES, Restivo C, Paulus HE (1979) Immunosuppressive and corticosteroid therapy of polyarteritis nodosa. Am J Med 67:941–947
180. Leonhart ETG, Kullenberg KPG. Bilateral atrial myxomas with multiple arterial aneurysms. A syndrome mimicking polyarteritis nodosa. Am J Med 62:792–794
181. Leung ACT, McLay A, Dobbie JW et al. (1985) Phenylbutazone induced systemic vasculitis with crescentic glomerulonephritis. Arch Intern Med 145:685–687
182. Levo Y, Gorevic PD, Kassab HJ et al. (1977) Association between hepatitis B virus and essential mixed microglobulin. N Engl J Med 296:1501–1504
183. Levy M, Broyer M, Arsan A et al. (1976) Glomerulonéphrites du purpura rhumatoide chez l'enfant. Histoire naturelle et étude immunopathologique. In: Acutalités néphrologiques de l'Hôpital Necker. Flammarion, Paris, pp 174–226
184. Lewis M, Kallenbach J, Zaltman M et al. (1985) Severe deficiency of alpha 1-antitrypsine associated with cutaneous vasculitis, rapidly progressive glomerulonephritis and colitis. Am J Med 79:489–494
185. Liebow AS, Carrinton CRB, Friedman PH (1972) Lymphomatoid granulomatosis. Hum Pathol 3:457–558
186. Livolsi VA, Persin KH, Porter M (1973) Polyarteritis nodosa of the gallbladder presenting as acute cholecystitis. Gastroenterology 65:115–123
187. Lockwood CM (1979) Lymphoma, cryoglobulinemia and renal disease (Nephrology Forum). Kidney Int 16:522–530
188. Lockwood CM, Bakes D, Jones S, et al. (1987) Association of alkaline phosphatase with an autoantigen recognized by circulating anti-neutrophil antibodies in systemic vasculitis. Lancet I:716–719
189. Lockwood CM, Jones S, Bakes D, Savage GOS (1986) Antibodies to neutrophile cytoplasmic antigens in systemic vasculitis. In: Proceedings of XXIIIrd congress of EDTA-European renal association 28 June–3 July at the XVth annual conference of EDTA-European care association. Budapest, Hungary, 1–4 July 1986, p 93
190. Longley S, Caldwell JR, Panusch RS (1986) Paraneoplastic vasculitis. Unique syndrome of cutaneous angiitis and arthritis associated with myeloproliferative disorders. Am J Med 80:1027–1030

191. Lowance DC, Mullins JD, McPhaul JJ (1973) Immunoglobulin A (IgA) associated glomerulonephritis. Kidney Int 3:167–176

192. Ludovico CL, Allen RM, Maurer K (1979) Hypocomplementemic urticarial vasculitis with glomerulonephritis and pseudotumor cerebri. Arthritis Rheum 22:1024–1028

193. McCluskey RT, Benecerraf B (1959) Localization of colloidal substances in vascular endothelium: a mechanism to tissue damage. II. Experimental serum sickness with acute glomerulonephritis induced passively in mice by antigen–antibody complexes in antigen excess. Am J Pathol 35:275–295

194. McCluskey RT, Fienberg R (1983) Vasculitis in primary vasculitides, granulomatoses, and connective tissue diseases. Hum Pathol 14:305–315

195. MacDonel Richards A, Eliot RS, Kanjuh VI et al. (1965) Cholesterol embolism. A multiple-system disease masquerading as polyarteritis nodosa. Am J Cardiol 15:696–707

196. MacDuffie FC, Sams WM, Madonado JE et al. (1973) Hypocomplementemia with cutaneous vasculitis and arthritis. Mayo Clin Proc 48:340–348

197. MacGowan JA, Greenberg A (1986) Cholesterol atheroembolic renal disease. Report of 3 cases with emphasis on diagnosis by skin biopsy and extended survival. Am J Nephrol 6:135–139

198. McIntosh RM, Tingloff B, Kaufman D (1971) Immunohistology in renal disease. QJ Med 40:385–390

199. McIntosh RM, Koss MN, Gocke DJ (1976) The nature and incidence of cryoproteins in hepatitis B antigen (HBs Ag) positive patients. QJ Med 45: 23–38

200. McLean RH, Weinstein A, Chapitis J et al. (1980) Familial partial deficiency of the third component of complement (C3) and the hypocomplementemic cutaneous vasculitis syndrome. Am J Med 68:549–558

201. McLeish KR, Yun MN, Luft FC (1978) Rapidly progressive glomerulonephritis in adults: clinical and histologic correlation. Clin Nephrol 10:43–50

202. McManus JFA, Hornsby AT (1951) Granulomatous glomerulonephritis associated with polyarthritis. Report of a case. Arch Pathol 32:84–90

203. McMillen JJ, Krueger SK, Dyerg A (1986) Leukocytoclastic vasculitis in association with immunoglobulin A myeloma. Ann Intern Med 104:709–710

204. Magil A, Calderon L, Gershon S (1978) Demonstration of immune complex by electron microscopy and immunofluorescence in a fatal case of hypersensitivity angiitis. Can Med Assoc J 119:751–752

205. Marder RJ, Rent E, Choi EYC et al. (1976) C1q deficiency associated with urticarial like lesions and cutaneous vasculitis. Am J Med, 61:560–565

206. Marsh FP, Almeyda JR, Levy IS (1971) Non-thrombocytopenic purpura and acute glomerulonephritis after indomethacin therapy. Ann Rheum Dis 30:501–505

207. Meadow SR (1979) The prognosis of Henoch–Schönlein nephritis. Clin Nephrol 9:87–90

208. Melin JP, Lemaire P, Aubert L et al. (1982) Polyarteritis nodosa with bilateral ureteric involvement. Nephron 32:87–89

209. Mellors RC, Ortega LG (1956) Analytical pathology. III. New observations on the pathogenesis of glomerulonephritis, lipid nephrosis, periarteritis nodosa and secondary amyloidosis in man. Am J Pathol 32:455–499

210. Meltzer M, Franklin EC (1966) Cryoglobulinemia: a study of twenty nine patients. I. IgG and IgM cryoglobulins and factors affecting cryoprecipitability. Am J Med 40:828–836

211. Meltzer M, Franklin EC, Elias K et al. (1966) Cryoglobulinemia: a clinical and laboratory study. II. Cryoglobulins with rheumatoid factor activity. Am J Med 40:837–856

212. Meyer O, Margulis J, Kahn MF (1985) Lupus erythémateux disséminé. In: Kahn MF, Peltier AP (eds) Maladies systémiques. Flammarion, Paris, pp 202–295

213. Meyrier A, Francais P, Lesavre P et al. (1984) Vascularite urticarienne hypocomplémentémique avec glomérulopathie et veinulite rénale. Néphrologie 5:1–7

214. Meyrier A, Simon P, Mignon F et al. (1984) Rapidly progressive ("crescentric") glomerulonephritis and monoclonal gammapathies. Nephron 38:156–162

215. Michalak T (1978) Immune complexes of hepatitis B surface antigen in the pathogenesis of periarteritis noda. A study of seven necropsy cases. Am J Pathol 90:619–632

216. Moore PM, Fauci AS (1981) Neurologic manifestations of systemic vasculitis. A retrospective and prospective study of the clinico-pathologic features and response to therapy in 25 patients. Am J Med 71:517–524

217. Meyer CF (1984) The role of vascular smooth muscle cells in experimental autoimmune vasculitis. I. The initiation of delayed type hypersensitivity angiitis. Am J Pathol 117:380–390

218. Mullick FG, McAllister HA Jr, Wagner BM (1979) Drug-related vasculitis: clinicopathologic correlations in 30 patients. Hum Pathol 10:313–325

219. Neild GH, Cameron JS, Ogg CS et al. (1983) Rapidly progressive glomerulonephritis with extensive glomerular crescent formation. QJ Med 207:395–416

220. Niaudet P (1987) Personal communication

221. Niaudet P, Levy M, Broyer M et al. (1984) Clinicopathologic correlations in severe forms of Henoch–Schönlein purpura nephritis based on repeated biopsies. Contrib Nephrol 40:250–254

222. Nydegger UE, Lambert PH. (1980) The role of immune complexes in the pathogenesis of necrotizing vasculitides. In: Alarcon-Segovia D (ed) The necrotizing vasculitides. Saunders, Philadelphia, pp 255–278 (Clinics in rheumatic diseases, vol 6)

223. Offenstadt G, Pinta P, Morel–Maroger L et al. (1978) Angéite nécrosante révélée par une cholecystite aigüe et une insuffisance rénale aigüe. Ann Med Interne 129:269–271

224. Oimoni M, Suehiro I, Mizuno N et al. (1980) Wegener's granulomatosis with intracerebral granuloma and mammary manifestation. Arch Intern Med 140:853–854

225. O'Neil WM, Hammar SP, Bloomer A (1976) Giant cell arteritis with visceral angiitis. Arch Intern Med 136:1457–1160

226. Ozenne G, Heliot P, Lemercier JP et al. (1983) Vascularites et granulomatoses pulmonaires. Commentaires à propos d'un cas avec localisation colique. Rev Med Int 4:27–33

227. Papadimitriou JM, Spector WG (1971) The origin, properties and fate of epithelioid cells. J Pathol 105:187–203

228. Parish WE, Rhodes EL (1967) Bacterial antigens and aggregated gamma globulin in the lesions of nodular vasculitis. Br J Dermatol 79:131–147

229. Parish WE (1971) Studies on vasculitis. I. Immunoglobulins. BIC. C-reactive protein and bacterial antigens in cutaneous vasculitis lesions. Clin Allergy 1:97–109

230. Paronetto F, Strauss L (1962) Immunocytochemical observations in periarteritis nodosa. Ann Intern Med 56:289–296

231. Payton CD, Boulton Jones JM (1985) Cortical blindness complicating Wegener's granulomatosis. Br Med J 290:676

232. Peltier AP (1985) Vascularites nécrosantes: Généralités. In: MF Kahn, AP Peltier. Maladies systémiques. Flammarion, Medecine-Sciences, Paris. pp 409–413

233. Petzel RA, Brown DC, Staley NA et al. (1979) Crescentic glomerulonephritis and renal failure associated with malignant lymphoma. Am J Clin Pathol 71:728–732

234. Phanuphak P, Kohler PF (1980) Onset of polyarteritis nodosa during allergic hyposensitization treatment. Am J Med 68:479–485

235. Phillips SM, Carpentier CB, Merrill JP (1972) Cellular immunity in the mouse. II. Correlation of in vivo and in vitro phenomena. Cell Immunol 5:249–263

236. Phinney PR, Fligiel S, Bryson YJ et al. (1982) Necrotizing vasculitis in a case of disseminated neonatal herpes simplex infection. Arch Pathol Lab Med 16:64–67

237. Pinching AJ, Lockwood CM, Pussel BA et al. (1983) Wegener's granulomatosis: observations on 18 patients with severe renal disease. QJ Med 208:435–460

238. Pinching AJ, Rees AJ, Pussel BA et al. (1980) Relapse in Wegener's granulomatosis: the role of infection. Br Med J 281:836–838

239. Pusey CD, Lockwood CM (1984) Plasma exchange for glomerular disease. In: Robinson RR (ed) Nephrology vol 2. Springer-Verlag Berlin Heidelberg New York, pp 1474–1485

240. Rainford DJ, Woodrow DF, Sioper JC et al. (1978) Post-meningococcal acute glomerular nephritis. Clin Nephrol 9:249–253

241. Ramirez G, Saba S, Espinoza L (1987) Hypocomplementemic vasculitis and renal involvement. Nephron 45:148–150

242. Raper RF, Ibels L (1985) Osteosclerotic myeloma complicated by diffuse arteritis, vascular calcification and extensive cutaneous necrosis. Nephron 39:389–392

243. Remy P, Jacquot C, Nochy D et al. (1987) Cholesterol atheroembolic renal disease with necrotizing glomerulonephritis. Am J Nephrol 7:164–165

244. Rich AR (1942) Role of hypersensitivity in periarteritis nodosa as indicated by 7 cases developing during serum sickness and sulfonamide therapy. Bull Johns Hopkins Hosp 71:123–140

245. Rifle G, Chalopin JM, Tanter Y et al. (1983) Echanges plasmatiques en néphrologie. Etude clinique. In: Actualités néphrologiques de l'Hôpital Necker. Flammarion, Paris, pp 243–313

246. Ronco P, Mignon F, Lanoe Y et al. (1982) Ureteral stenosis in Wegener's granulomatosis. Nephron 30:201–204

247. Ronco P, Verroust P, Mignon F et al. (1983) Immunopathological studies in polyarteritis nodosa and Wegener's granulomatosis: a report of 43 patients with 51 renal biopsies. QJ Med 206:212–223

248. Rose GA, Spencer H (1957) Polyarteritis nodosa. QJ Med 26:43–81

249. Rose GM, Cole BR, Robson AM (1981) The treatment of severe glomerulopathies in children using high dose intravenous methylprednisolone pulses. Am J Kidney Dis 1:148–156

250. Rothenberg R (1985) Isolated angiitis of the brain.

251. Roza M, Galbe M, Gonzalez Baschwitz C et al. (1983) Henoch-Schönlein purpura after shigellosis. Clin Nephrol 20:269

252. Rush PJ, Inman R, Bernstein M (1986) Isolated vasculitis of the central nervous system in a patient with celiac disease. Am J Med 81:1092–1094

253. Ryan TJ (1979) Vasculitis. Immunology and localization. A review. J R Soc Med 72:527–529

254. Ryckewaert A (1985) Polyarthrite rhumatoide. In: Kahn MF, Peltier AP (eds) Maladies systémiques. Flammarion, Paris, pp 135–168

255. Sairanen E, Wasauberna C (1972) Periarteritis nodosa. A ten year follow-up study of ten cases. Acta Med Scand 191:501–504

256. Sams WM Jr, Claman HN, Kohler PF et al. (1975) Human necrotizing vasculitis: immunoglobulins and complement in vessel walls of cutaneous lesions and normal skins. J Invest Dermatol 64:441–446

257. Saurat JH, Sohier J (1985) Urticaires systémiques. In: Kahn MF, Peltier AP (eds) Maladies systémiques. Flammarion, Paris, pp 564–570

258. Savage CO, Wineares CG, Evans DJ et al. (1985) Immunosuppressive treatment of 34 patients with microscopic polyarteritis. Proc Eur Dial Transplant Assoc Eur Ren Assoc. Davison AM, Guillou PJ (eds) Baillière Tindall, London, 22:720–724

259. Scharre D, Petri M, Engman E et al. (1986) Large intracranial arteritis with giant cells in systemic lupus erythematosus. Ann Intern Med 104:661–662

260. Schultz DR, Perez GO, Volanakis JE et al. (1981) Glomerular disease in two patients with urticaria-cutaneous vasculitis and hypocomplementemia. Am J Kidney Dis 1:157–165

261. Schwartz HR, MacDuffie FC, Black LF et al. (1982) Hypocomplementemic urticarial vasculitis. Association with chronic obstructive pulmonary disease. Mayo Clin Proc 57:231–238

262. Scott DGI, Bacon PA (1984) Intravenous cyclophosphamide plus methylprednisolone in treatment of systemic rheumatoid vasculitis. Am J Med 76:377–384

263. Scott DGI, Bacon PA, Bothamley J et al. (1981) Plasma exchange in rheumatoid arthritis. J Rheumatol 8:433–439

264. Scott DGI, Bacon PA, Elliott PJ et al. (1982) Systemic vasculitis in a district general hospital 1972–1980: clinical and laboratory features, classification and prognosis of 80 cases. QJ Med 203: 292–311

265. Scott DGI, Bacon PA, Tribe CR (1981) Systemic rheumatoid vasculitis: a clinical and laboratory study of 50 cases. Medicine 60:288–297

266. Sergent JS (1980) Vasculitides associated with viral infections. In: Alarcon-Segovia D (ed) The necrotizing vasculitides. Saunders, Philadelphia, pp 339–350 (Clinics in rheumatic diseases 6)

267. Serra A, Cameron JS, Turner DR et al. (1984) Vasculitis affecting the kidney: presentation histopathology and long-term outcome. QJ Med 210:181–207

268. Shalit M, Fluegelman MY, Harats N et al. (1985) Quinidine-induced vasculitis. Arch Intern Med 145:2051–2052

269. Silverstein SC, Steinmann RM, Cohn ZA (1977) Endocytosis. Annu Rev Biochem 46:669–722

270. Sinniah R, Feng PH, Chen BTM et al. (1978) Henoch–Schönlein syndrome a clinical and morphological study of renal biopsies. Clin Nephrol 9:219–228

271. Sissons JGP, Williams DN, Peters DK et al. (1974)

Case of a renal transplant recipient. Am J Med 79:629–632

Skin lesions, angio-oedema and hypocomplementaemia. Lancet II:1350–1352

272. Smith MC, Ghose MK, Henry AR (1981) The clinical spectrum of renal cholesterol embolization. Am J Med 71:174–180

273. Snyderman R, Mergenhagen SE (1976) Chemotaxis of macrophages. In: Nelson DS (ed) Immunobiology of the macrophage. Academic Press, New York

274. Soter NA, Nihm MC Jr, Gigli J et al. (1976) Two distinct cellular patterns in cutaneous necrotizing angiitis. J Invest Dermatol 63:344–350

275. Soter NA (1976) Clinical presentations and mechanisms of necrotizing angiitis of the skin. J Invest Dermatol 67:354–359

276. Soter NA, Austen KF, Gigli I (1974) The complement system in necrotizing angiitis of the skin. Analysis of complement component activities in serum of patients with concomitant collagen vascular diseases. J Invest Dermatol 63:219–226

277. Spector WG (1974) The macrophage: its origins and role in pathology. Pathobiol Annu 4:33–64

278. Steinman RM, Cohn ZA (1974) The metabolism and physiology of the mononuclear phagocytes. In: Zweifach BW, Grant L, McCluskey RT (eds) The inflammatory process, vol 1, 2nd edn. Academic Press, New York pp 449–510

279. Steinman TI, Jaffe BF, Monaco AP et al. (1980) Recurrence of Wegener's granulomatosis after kidney transplantation. Am J Med 68:458–460

280. Stevens ME, Bone JM (1984) Follow-up prednisolone dosage in rapidly progressive crescentic glomerulonephritis successfully treated with pulse methylprednisolone of plasma exchange. Proc Eur Dial Transplant Ass Eur Ren Assoc 21:594–599

281. Stevenson JA, Leong LA, Cohen AH, Border WA (1982) Henoch–Schönlein purpura. Simultaneous demonstration of IgA deposits in involved skin, intestine and kidneys. Arch Pathol Lab Med 106:192–195

282. Stockigt JR, Topliss DJ, Hewett MJ (1979) High-renin hypertension in necrotizing vasculitis. N Engl J Med 300:1218

283. Symmers W StC (1952) Necrotizing pulmonary arteriopathy associated with pulmonary hypertension. J Clin Pathol 5:36–41

284. Tenberge IJM, Wilmink JM, Meyer CJL et al. (1985) Clinical and immunological follow-up of patients with severe renal disease in Wegeners's granulomatosis. Am J Nephrol 5:21–29

285. Thomas B, Houssin A, Saint Andre JP et al. (1983) Forme tumorale d'une granulomatose de Wegener succédant à un granulome malin facial. Intérêt nosoligique. Rev Med Interne 4:155–158

286. Thomashow BM, Fleton CP, Navarro C (1980) Diffuse intrapulmonary hemorrhage, renal failure and a systemic vasculitis. A case report and review of the literature. Am J Med 68:299–304

287. Touchard G, Marie P, Beauchant M et al. (1983) Vascular IgA and C3 deposition in gastrointestinal tract of patients with Henoch–Schönlein purpura. Lancet I:771–772

288. Turk JL, Parker D (1983) Immunological aspects of granuloma formation. In: Gordon CG, Hubbard R, Parke DV (eds) Immunotoxicology. Academic Press, London, pp 251–259

289. Vahanian A, Mignon F, Chavaz A et al. (1977) Angéite aigüe après pénicillinothérapie. A propos d'une observation avec regression artériographique des lésions anéurysmales intrarénales. J Urol Nephrol 83:285–289

290. Van Der Woude FJ, Hoornstje SJ, Weening JJ et al. (1982) Renal involvement in Wegener's granulomatosis. Nephron 32:185–187

291. Van Der Woude FJ, Lobatto S, Permin H et al. (1985) Antibodies against neutrophils and monocytes: tools for diagnosis and marker of disease activity in Wegener's granulomatosis. Lancet I:425–429

292. Van Furth R, Cohn ZA, Hirsch JG et al. (1972) The mononuclear phagocyte system. A new classification of macrophages, monocytes and their precursor cells. Bull WHO 46:845–852

293. Van Fruth R (1981) Mononuclear phagocytes: functional aspects. Martinus-Nijhoff. The Hague.

294. Van Fruth R (1982) Current view on the mononuclear phagocyte system. Immunobiology 161:178–185

295. Viard JP, Gherardi R, Salama J et al. (1986) Multinérite par vascularite au cours d'un plasmocytome solitaire récent. Presse Med 15:167

296. Walker DH, Mattern WD (1980) Rickettsial vasculitis. Am Heart J 100:896–906

297. Ward PA, Becker EL (1977) Biology of leukotaxis. Rev Physiol Biochem Pharmacol. 77:125–148

298. Watanabe T, Nagafuchi Y, Yoshikawa et al. (1983) Renal papillary necrosis associated with Wegener's granulomatosis. Hum Pathol 14:551–557

299. Weber M, Kohler H, Fries J et al. (1985) Rapidly progressive glomerulonephritis in IgA/IgG cryoglobulinemia. Nephron 41:258–261

300. Wees SJ, Sunwoo IL, Oh SJ (1981) Sural nerve biopsy in systemic necrotizing vasculitis. Am J Med 71:525–532

301. Wegener F (1939) Uber eine eigenartige rhinogene Granulomatose mit besonderer Beteiligung des Arteriensystems und der Nieren. Beitr Pathol Anat 102:36–68

302. Weiss MA, Crissman JD (1985) Segmental necrotizing glomerulonephritis: diagnostic, prognostic and therapeutic significance. Am J Kidney Dis 6:199–211

303. Wenzel RP, Hayden FG, Groschel DH et al. (1986) Acute febrile cerebrovasculitis: a syndrome of unknown perhaps rickettsial cause. Ann Intern Med 104:606–615

304. Wilkinson PC (1977) Chemotaxis and inflammation. Churchill-Livingstone, Edinburgh

305. Williams RC Jr (1980) Immune complexes in clinical and experimental medicine. Harvard University Press, Cambridge, p 231

306. Wilson CB, Dixon FJ (1986) The renal response to immunological injury. In: Brenner BM, Rector FC (eds) The kidney. Saunders, Philadelphia, pp 800–889

307. Zeek PM, Smith CC, Weeter JC (1948) Studies on periarteritis nodosa. III. The differentiation between the vascular lesions of periarteritis nodosa and of hypersensitivity. Am J Pathol 24:889–917

308. Zeek PM (1952) Periarteritis nodosa: critical review. Am J Pathol 22:777–790

309. Zeek PM (1953) Periarteritis nodosa and other forms of necrotizing angiitis. N Engl J Med 248:764–772

310. Zizic TM, Classen JN, Stevens MB (1982) Acute abdominal complications of systemic lupus erythematosus and polyarteritis nodosa. Am J Med 73:525–531

311. Zweiman B (1979) A new therapeutic strategy in systemic vasculitis. N ENgl J Med 301:266–267

Chapter 28

The Vasculitis Syndromes in Aorta and Large Arteries

Pathological Aspects

J.-P. Camilleri and P. Bruneval

Takayasu's Disease

Takayasu's disease is a chronic inflammatory arteriopathy which principally involves the aorta and its main branches. The terms "pulseless disease" [223], "non-specific aortoarteritis" [227], "aortic arch syndrome" [122] and "occlusive thromboaortopathy" [123] are acceptable synonyms for Takayasu's disease.

It seems likely that this syndrome was first documented in a young woman in 1856 by Savory [222]; postmortem examination revealed stenosis and thrombosis of the proximal branches of the aortic arch. Takayasu in 1908 [241] described the retinal vessel changes. Forty years later Shimuzu and Sano [223] characterized a more definitive syndrome among the various diseases with aortic arch involvement and named it "pulseless" disease. Although the first cases reported involved predominantly the aortic arch and the brachiocephalic branches, cases involving the abdominal aorta, the renal arteries [60,227,251] or the splanchnic arteries have been described more recently. Pulmonary artery involvement has been described by Luppi-Herrera [153]. Finally, large series affecting patients from countries other than Japan have been reported, notably in Asia and Africa [9,75,92,154,214,227,243,251].

The disease occurs predominantly in women according to nearly all series which have been published [60,123,154]; the sex ratio is about four females to one male. At the time of the diagnosis the mean age is around 30 years, and the onset of the disease always occurs before the age of 40 years.

Because of the unknown aetiology and the various clinical presentations of this nosologic entity, it has been necessary to define criteria for diagnosis [123,154,178]. Among these criteria, pathological features seemed likely to be essential [73,75]. Recent development of vascular surgery has provided numerous peroperative biopsy specimens and has allowed pathological examination to be included in the diagnosis and therapeutic approaches to the patient.

Pathological Features

The pathological features of Takayasu's disease were described from 21 autopsy cases by Nasu [181], and later refined by Sun et al. [227].

Retrospective analysis of the records of a series of 114 revascularizations performed on 107 patients allowed us to confirm the value of pathological data in identifying the disease entity.

Macroscopic Findings

Takayasu's disease is characterized by a focal stenotic process, involving the aorta and the proximal segments of its main branches [44,64]. The macroscopic findings may allow a peroperative diagnosis of this entity. The affected vessel appears very thick walled, shortened in length and rigid with marked perivascular sclerosis and adhesion to surrounding tissues

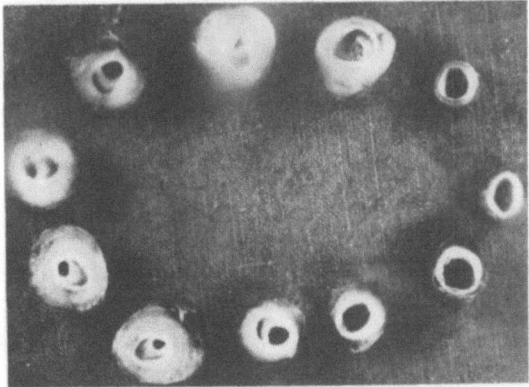

Fig. 28.1. Gross aspect of serial cross-sections of subclavian artery. The lesion consists of segmental thickening of the arterial wall with marked stenosis of the lumen.

(Fig. 28.1). The inner surface is thickened, greyish white and protrudes towards the lumen. At the level of collateral trunks, complete occlusion of the lumen may be found. Thrombosis is uncommon. Fibrosis may be found in hypertrophied regional lymph nodes. The affected parts of the vessel are clearly demarcated from the uninvolved healthy segments.

When compared with other pathological conditions of the aorta, one of the most characteristic hallmarks of this entity is the stenotic appearance without changes in the external diameter of the vessel. Aneurysmal dilatation is infrequent, but stenotic and post-stenotic dilatation of the aorta are highly characteristic of Takayasu's disease. Arterial rupture appears to be less frequent than in aneurysms of other types

(see Chap. 31). Arterial dissection is very uncommon.

Histopathological Criteria

Takayasu's disease is characterized histopathologically by a chronic inflammatory arteriopathy which progresses from the outer part of the vessel wall and leads to luminal occlusion (Figs. 28.2 and 28.3). Most changes are found in the adventitial and medial layers. Adventitial sclerosis consists of dense collagenous tissue, including thickened vasa vasorum and nerves, which penetrates into the outer part of the media. The media appears to be destroyed, with foci of elastic fragmentation and neovascularization.

The inflammatory cellular component may be productive, consisting predominantly of large areas of infiltration by lymphocytes and histiocytes with or without multinucleated giant cells (Fig. 28.4). Sometimes plasma cells may predominate (Fig. 28.5). Typical fibrinoid necrosis is rarely seen. Most often the cellular infiltration remains discrete, composed of small foci of mononuclear cells found at the junction between the media and the adventitia (Fig. 28.3). Sometimes no cell infiltrates can be found.

The intima shows marked thickening by loose connective tissue without newly formed vessels or haemosiderin deposits (Fig. 28.6). Usually there are no inflammatory cells. This change is responsible for the translucent appearance of the intimal plaques. Mural thrombosis may result from intimal abnormalities and the haemodynamic changes, but remains uncommon. It is clear that the alterations in the intima are

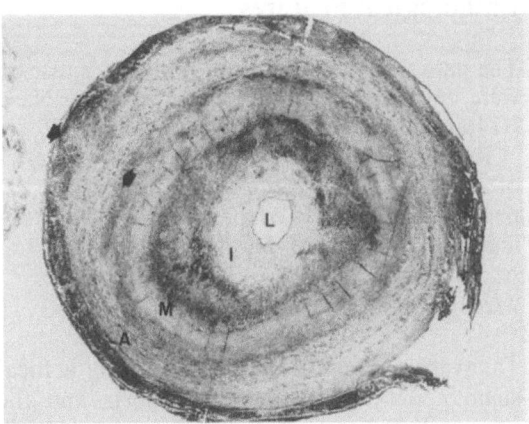

Fig. 28.2. Biopsy cross-section of subclavian artery showing marked luminal narrowing due to severe transmural fibrosis with adventitial thickening (*arrows*). *L*, lumen; *I*, intima; *M*, media; *A*, adventitia. Masson's trichrome stain, × 17

Fig. 28.3. Section of aorta showing dense collagenous thickening of the adventitia with small mononuclear cells at the medial–adventitial junction (*arrows*). *M*, media; *A*, adventitia. Haematoxylin–eosin–safran, × 375

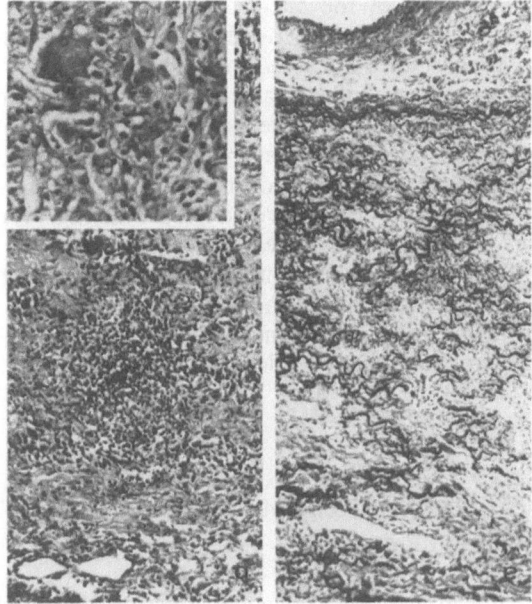

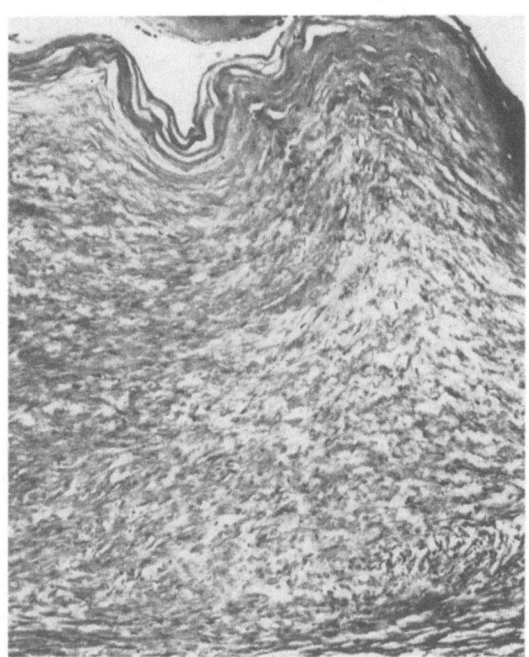

Fig. 28.4a, b. Detail of full-thickness biopsy section of aorta. **a** Trichrome stain showing transmural mononuclear infiltration. **b** Orcein stain showing marked disruption of elastic laminae. × 225

Fig. 28.6. Intimal thickening by paucicellular loose connective tissue in subclavian artery. Note the absence of newly formed vessels. Masson's trichrome stain, × 225

secondary to the changes in the outer coats. Superimposed atheromatous lesions and subsequent calcifications have been noted, mostly in Western countries.

Although the most striking histopathological changes have been described in the aorta and the main elastic branches (subclavian, primitive carotid and iliac arteries), the lesions can involve visceral muscular arteries such as the coeliac trunk, mesenteric or renal arteries. The most common sites of visceral artery involvement are the ostium and the proximal segment, which are trapped by the periaortic fibrotic tissue. Associated intimal thickening is present, but the media

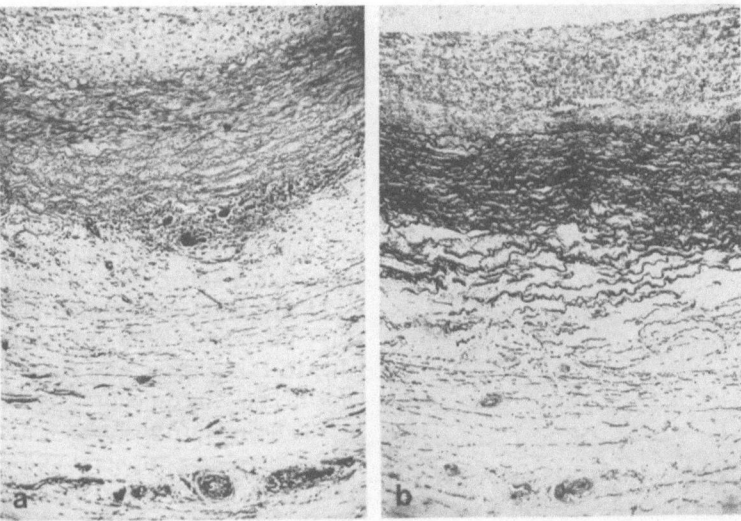

Fig. 28.5a,b. Detail of transmural cellular infiltration in active form. **a** Granulomatous pattern with giant cells. **b** Lymphoplasmacytic cellular infiltration. Haematoxylin–eosin–safran, × 45

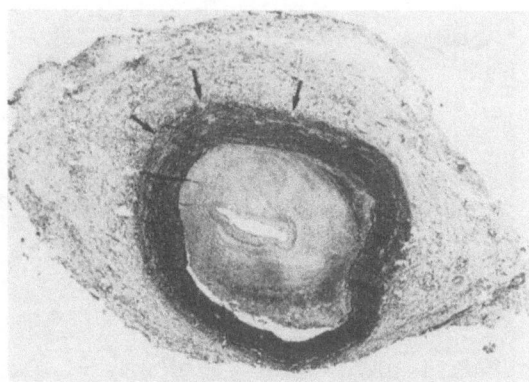

Fig. 28.7. Cross-section of subclavian artery showing inactive, scarring fibrous thickening of the intimal as well as the adventitial layers. The outer part of the media shows small foci of disruption (*arrows*). Orcein stain, × 19

may be nearly normal. It would be difficult to differentiate these lesions from adventitial dysplastic arteriopathy. Aortic wall biopsy may be useful to identify the inflammatory process.

With respect to the activity of the disease, the histopathological changes can be divided into three types: (a) granulomatous, (b) productive with lymphocytes and plasma cell infiltration and (c) scarring characterized by thickening and

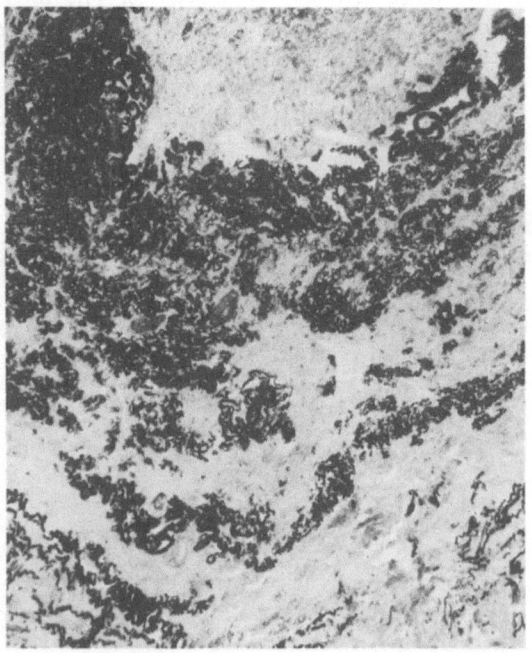

Fig. 28.8. Involvement of the subclavian artery at the healed stage. The lumen is completely occluded. The medial elastic laminae are markedly disrupted by dense acellular scarring fibrosis. Orcein stain, × 120

shortening of the affected vessel. About two-thirds of the cases operated on during the occlusive phase of the disease show fibrosis and scarring without inflammatory cellular infiltration (Fig. 28.7). In this healed stage, the diagnosis of Takayasu's disease may be strongly supported by the presence in the aortic wall of these medial–adventitial foci of destruction and scarring, even if cellular infiltrates are absent (Fig. 28.8).

Careful examination of para-aortic lymph nodes usually reveals non-specific sinusal histocytosis, and sometimes granulomatous epithelioid and giant cell lesions [142].

Sites of Involvement and Visceral Consequences

Clinical symptoms depend on the sites of arterial lesions.

The aorta is involved in about half of the patients with Takayasu's disease. In the reported Japanese cases, the arch is the most commonly affected region [123,178,181]. Associated aortic regurgitation due to dilatation of the ascending aorta and thickening of the valve cusps has been reported [249]. Other series, however, showed predominant involvement of the thoraco-abdominal segment [75,154,227,243,257]. Association with retroperitoneal fibrosis has been noted [219]. A constricting lesion occurs most frequently in the descending aorta. Atypical coarctation of the abdominal aorta, sometimes classified as a congenital change, may indicate stenosing Takayasu's arteritis. Secondary renal hypertension and ischaemic symptoms secondary to vascular insufficiency can be associated typically with such a stenotic lesion. Such coarctation has been reported in association with several well-defined syndromes, e.g. congenital hypercalcaemia, rubella, neurofibromatosis, fibromusclar dysplasia and radiation aortitis. Total occlusion is unusual in the aorta. Associated aneurysmal dilatation can be found [123]. Isolated aneurysms remain unusual and the differential diagnosis between aneurysms due to this disease and those due to giant cell aortitis may be difficult.

In the majority of cases, aortic involvement may extend to peripheral and visceral branches (usually in the proximal segment). Clinical symptoms may be the consequence of regional ischaemic processes or secondary to systemic manifestations such as hypertension.

The supra-aortic trunks are involved in almost 50% of all cases. The post-vertebral subclavian artery is most commonly affected. The common carotid arteries may also be affected, showing long stenoses which usually stop at the level of the bifurcation, rarely extending within the internal carotid artery. Humeral involvement has been occasionally reported, either isolated or in association with involvement of the subclavian artery.

Intra-abdominal visceral artery involvement results in ostial and/or proximal stenoses, usually asssociated with involvement of the abdominal aorta. Ischaemic manifestations can be seen in the intestinal tract when coeliac and mesenteric arteries are damaged, but are rarely observed because of the major collateral arterial supply in this area. Renal artery involvement can lead to severe renovascular hypertension, a major cause of mortality in Takayasu's disease [1,156]. Although the pathophysiology of this hypertension is predominantly renovascular [243], other factors may be at work, e.g. atypical aortic coarctation, aortic regurgitation or a decrease in the elastic properties of the diffusely affected aortic wall. Takayasu's disease restricted to the abdominal aorta and the proximal portion of its branches seems to be an important cause of renovascular hypertension in childhood [60,258].

Iliac vessels may be involved but the disease remains clinically uncommon beyond the bifurcation of the aorta. The most peripheral site noted in our series was a popliteal artery involvement. Claudication of the lower extremities is very rarely the initial symptom of the disease. In the young, fortuitous demonstration of aneurysms on arteriograms suggests either inflammatory or dysplastic arteriopathy.

Associated involvement of the coronary arteries has been reported [10,50,51,214,215]. Obstructive changes are usually found to involve the ostia of the coronary arteries, extending from the aorta [50,51,215]. Multifocal segmental coronary arteritis as the only manifestation of the disease is very uncommon [215]. Aneurysms have been noted [214]. Coronary insufficiency is rarely the presenting symptom [215], and patients can have symptoms typical of angina pectoris without obstructive lesions in the coronary arteries [92]. Associated endocardial lesions have been reported [46].

Pulmonary artery involvement in Takayasu's arteritis can be recognized not only by postmortem examination but also by arteriographic measurement of pulmonary arterial pressure and by pulmonary perfusion scanning [123]. Associ-ated pulmonary arterial changes were found in about 50% of all cases in the series reported by Lupi-Herrera et al. [153,154] and in that reported by Ishikawa [123]. Nasu previously described 12 cases with pulmonary artery involvement out of 21 autopsies reported [181]. In Europe, pulmonary involvement seems less frequent [42]. It appears that the pulmonary trunk and the main branches can be involved with arteritis (Fig. 28.9). The right artery is more often affected than the left. Various degrees of rigidity, irregularities of the vascular wall and narrowing can be observed on angiograms. In most series, patients are asymptomatic with respect to pulmonary involvement. Severe pulmonary hypertension with cyanosis and right heart failure, requiring a surgical approach to the lesion, remains rare [171]. Small intrapulmonary vessels are usually spared. There is only one reported case with small pulmonary arteriopathy very similar to the angiomatoid lesions seen in plexogenic arteriopathy [213].

Multifocal forms, with the intervening areas uninvolved, are frequent. Diffuse extensive forms an relatively uncommon. According to Ueno's classification, modified by Lupi-Herrera [154], four types of patients can be distinguished: in type I patients only the aortic arch and its main branches are affected; in type II patients the lesions are restricted to the thoracic descending and abdominal aorta; type III patients show

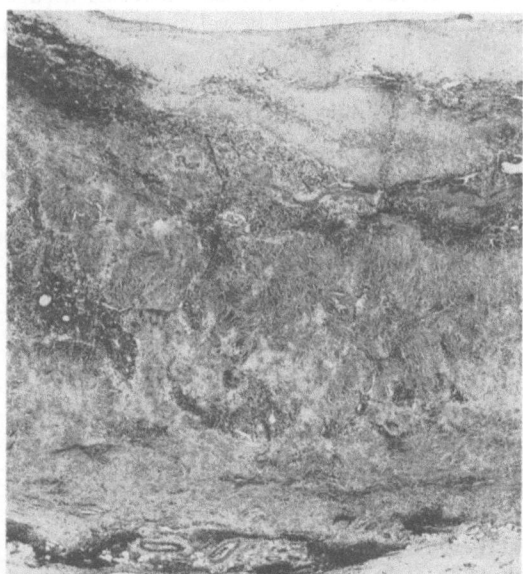

Fig. 28.9. Involvement of the pulmonary artery in a patient operated on for severe pulmonary hypertension. Masson's trichrome stain, × 19

features of the types I and II; and type IV patients show pulmonary artery involvement. The type III patients show the most common pattern of the disease (60%) [123,154,227]. At peroperative or postmortem examination, the anatomical pattern of the lesions always exceeds that which is suggested by the clinical manifestation. Treatment requires a complete examination of the patient in order to detect clinically inapparent lesions [75,141,142].

Involvement of venous trunks, particularly the vena cava, has been reported rarely, generally associated with thrombosis [48,224].

Associated Findings

Cardiomegaly with or without cardiac failure may be linked to systemic hypertension or aortic regurgitation. Rarely, associated primary cardiac involvement resembling dilated cardiomyopathy has been reported [49]. It is possible that both diseases share aetiological factors. There is one reported case with giant cell myocarditis [209].

Joint involvement has been described. At the end of the pre-occlusive phase of the disease, joint pain is frequent, particularly in the young. Associated findings of polyarthritis [42,70] or ankylosing spondylitis [83] have been reported occasionally. Association with Still's disease has been noted [264].

Several cases of Takayasu's arteritis associated with inflammatory bowel disease have been described [41,48,191]. The associated bowel disease may be Crohn's disease [78], ulcerative colitis or unclassified colitis. The vascular disease may be recognized at the same time as the bowel disease, or, more often, several years after the onset of the bowel disease.

Associated cases of cutaneous pyoderma gangrenosum are known [108]. Involvement of the kidney results usually from ischaemia or systemic hypertension. Associated findings of focal glomerulonephritis suggest similar immunological mechanisms in the two diseases [240,269]. Retroperitoneal fibrosis has been described [219].

Aetiological Factors

Takayasu's disease (non-specific aortoarteritis) accounts for most of the non-atherosclerotic lesions of the aorta and its main branches, except

for those found in association with syphilis, ankylosing spondylitis, rheumatoid athritis, rheumatic fever, systemic lupus erythematosus and allied syndromes and arteriopathies of a dysplastic and familial nature.

The incidence has been estimated at 2.6 cases per million inhabitants per year in the USA [93], so is ten times less common than temporal arteritis. If we take into account the asymptomatic forms of the disease, the actual incidence is probably higher than reported above.

The systemic expression of the disease, the constitutional signs of the pre-occlusive phase and the non-specific inflammatory syndrome have led to the hypothesis that it is an autoimmune disease. Circulating anti-aorta antibodies were detected by Ueda et al. [248] and Ito et al. [124], but controversial data showed that there was no obvious evidence for circulating autoimmunity [8,48]. However, it is likely that most cases under study represent the end stage of the process and that serological analysis of cases in the early stages might reveal elevated levels of anti-aorta antibodies. The circulating immune complex (CIC) level has been reported to be high in some cases when compared with that of normal controls, and immune deposits have been inconsistently observed in injured tissue [183,248]. These discrepancies and the absence of correlation between the immunological findings and the activity of the disease suggest that these complexes are not a primary aetiological factor [90].

In most series a previous history of tuberculosis is frequent and tuberculosis has been implicated in the pathogenesis of this condition [73,131,154,214,227]. Demonstration of tuberculous or non-specific lymphadenitis in association with segmental arterial lesions involving predominantly the external part of the arterial wall, suggests that lymphadenitis may cause aortitis. This hypothesis has been supported by Sen et al. [227] who were able to induce inflammatory lesions of the arterial wall by injection of tuberculous bacilli in the adventitia of the aortic wall of rabbits. They could induce adventitial inflammation by injection of antigens derived from the tuberculous bacilli (Freund adjuvant, BCG) in previously sensitized animals. By contrast, Morooka [173] did not succeed in inducing inflammatory lesions of the aortic wall by intravenous injection of serine protease. As far as the histopathological aspect is concerned, the adventitial granulomatous chronic inflammation suggests reaction to persistent antigenic irritants [250]. Antigenic fractions of mycobacteria can initiate a chronic granulomatous response. Fibrosis is always associated with the cicatricial stage of this type of tissue reaction. The associated finding of inflammation, bowel disease or immune glomerulonephritis may support such hypothesis. Finally, the existence of cases with familial history [179], and the analysis of HLA antigens have suggested the role of a congenital factor in the pathogenesis of the disease [180].

Temporal Arteritis (Temporal Arteritis/Polymyalgia Rheumatica Syndrome)

Temporal arteritis was first described by Hutchinson in 1890 [119]. Since the work of Horton et al. [113], temporal arteritis has been recognized as a clinicopathological entity characterized as granulomatous temporal arteritis of the elderly [118]. Gilmour in 1941 [84] and Cooke et al. in 1946 [55] showed the systemic distribution of the disease. In addition, polyarthritis of the limb girdles, called pseudopolyarthritis [77] or polymyalgia rheumatica [93,95,116,259], appeared to be a closely related condition affecting the same population. It is widely accepted that the two diseases are parts of the same syndrome. There is a weak predilection for women and the syndrome seems to be very rare in people under 50 years of age [26]. The incidence is probably higher in Northern European countries. The disease rarely occurs in black populations.

Pathological Aspects

Temporal arteritis is an occlusive segmental panarteritis which affects mainly the media and the internal elastic lamina of the vessel wall [150,196]. The terms "granulomatous arteritis" and "giant cell arteritis" are synonyms for a lesion characterized histopathologically by mononuclear and giant cells with disrupted elastic lamina. Associated findings of thrombus formation and occlusion of the lumen are usual. The disease affects the cranial arteries predominantly. The superficial temporal artery is the preferential site for biopsy and according to most authorities, the proper diagnostic approach to the disease is based on long segment biopsies (several millimetres) of artery and multiple sections [134].

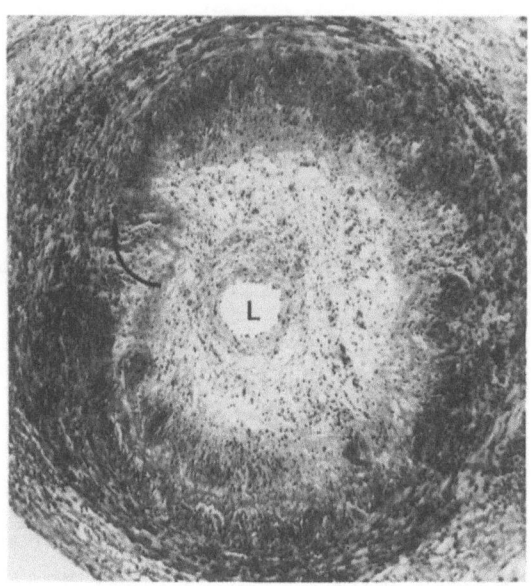

Fig. 28.10. Temporal arteritis. Marked stenosis of the lumen (*L*) due to intimal proliferation. Typical distribution of inflammatory infiltrates within intimal and medial layers, extending to the adventitia. Giant cells are visible close to the fragmented internal elastic lamina (*arrow*). Masson's trichrome stain, × 95

Macroscopically, the affected arteries are larger, sometimes nodular.

Typically the diagnosis is easy when three criteria are present (Figs. 28.10 and 28.11): (a) predominant involvement of the intimal–medial junction; (b) mononuclear granulomatous infiltrate with multinucleated giant cells; and (c) disruption of the internal elastic lamina. In middle-

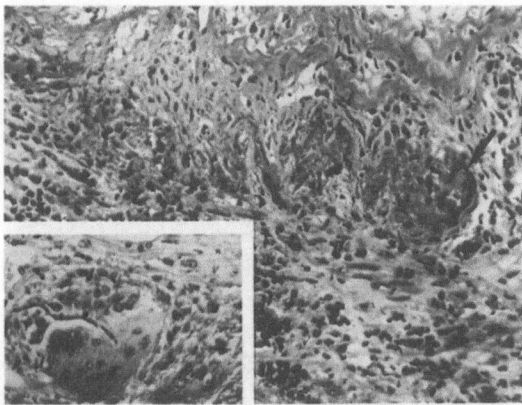

Fig. 28.11. Granulomatous infiltrate of the intimal–medial junction with multinucleated giant cells (*arrow*). Disruption of the internal elastic lamina. Masson's trichrome stain, × 375. *Inset*. Elastic fragment present within giant cells. × 750

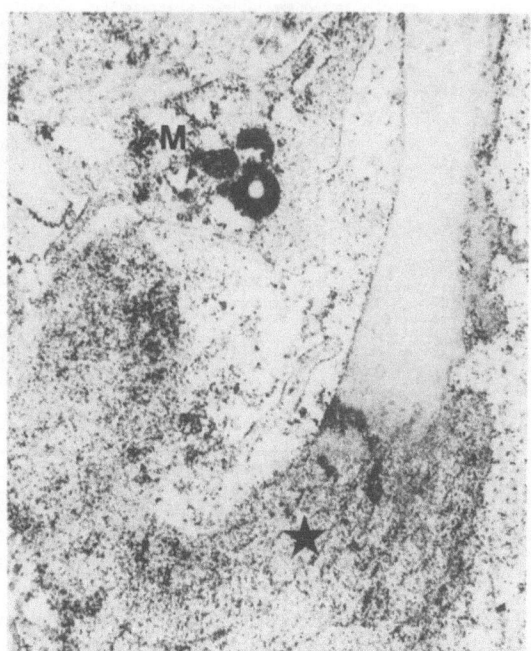

Fig. 28.12. Temporal arteritis. Electron micrograph showing focal granular appearance of elastin. *M*, macrophage. Uranyl lead, × 18 000

sized arteries, the internal elastic lamina can be missing or fragmented and have altered staining properties. Fibrinoid necrosis is rare. The intimal thickening may be concentric or eccentric, often leading to complete occlusion of the lumen. The adventitia is fibrotic and collections of lymphocytes may be seen around the vasa vasorum. Ultrastructural observations may identify a

variety of mononuclear cells, including lymphocytes, macrophages and ephithelioid-like cells [196,234]. Such cell changes have been noted [205] but seem to be neither specific nor primary. The infiltrating granulomatous cells are closely associated with elastic lamina fragments. Some segments of fragmented elastic lamina may exhibit a finely granular appearance and associated microfibrils may be markedly diminished (Fig. 28.12).

However, biopsy of the temporal artery fails to show histological evidence of giant cell arteritis in 60% of clinically genuine cases (b). Two types of histological changes may be found: (a) atypical or inactive healed forms characterized by intimal and medial fibrosis with defects of the elastic lamina and rare clusters of mononuclear cells hardly distinguishable from mobilized myocytes; (b) non-specific intimal fibrosis which appears similar to senile arteriosclerosis (Fig. 28.13). Fragmentation of internal elastic lamina is often observed and microcalcification may be closely opposed to fragments of elastin. Neither elastic lamina defects, nor rare giant cells, if viewed as solitary phenomena, can be regarded as pathognomonic of temporal arteritis.

There are risks in overdiagnosing temporal arteritis and subjecting aged patients to needless long-term corticosteroid therapy, as well as missing the diagnosis, with the inherent risk of tragic complications such as stroke or blindness. In such cases the biopsy specimens have to be carefully examined on multiple sections to avoid sampling error and on effort made to recognize small cluster of macrophages with or with-

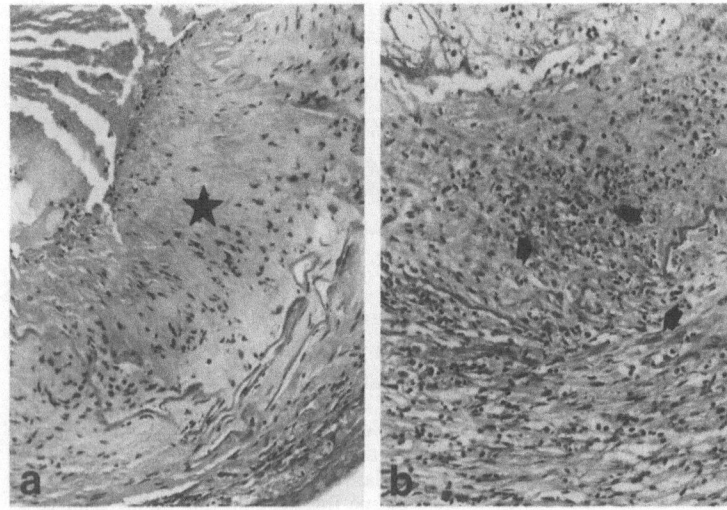

Fig. 28.13a, b. Biopsy specimen of temporal artery. **a** Focal non-specific intimal fibrosis (*star*). Haematoxylin–eosin–safran, × 165. **b** Same case. Focal infiltration of mononuclear cells without giant multinucleated cells. The internal elastic lamina is fragmented (*arrows*). × 210

out giant cells associated with fragmented and degenerative internal elastic lamina (Fig. 28.13b).

The focal nature of the histological changes has been highlighted by numerous authors [27,93,98,107]. Skip lesions are characterized by the presence of giant cell arteritis in one or more sections of a biopsy specimen while other sections of the same specimen disclose no or nonspecific lesions [134]. This characterization is, however, controversial. Cohen and Smith [53] suggested that the skip lesions noted by others were artefactual. By contrast, Klein et al. [134] from the Mayo Clinic found skip lesions in 17 patients out of a series of 60 patients; segments showing active disease were as short as 300 to 400 μm. This could explain the frequency of false negative results in most series [6]. These authors have also shown that unilateral temporal arteritis can occur; they suggest that fresh frozen sections might be reliable in detecting arteritis, and recommended preparation of multiple sections and the biopsy of the contralateral temporal artery of frozen section examination of the first side was negative. All the teams have stressed the need for long biopsy specimens, measuring at least 1–2 cm. After surgical excision the segment of temporal artery is cut transversely into fragments from 3 to 5 mm long; the end of each fragment is marked with india ink before paraffin embedding and multiple sections are performed.

Immunopathological Aspects

Direct immunofluorescence study of temporal artery biopsies has been proposed as a procedure of diagnostic value in recognizing temporal arteritis. Sauerbruck et al. [224] described immunoglobulin and complement deposits within the temporal artery wall during giant cell arteritis. Extracellular granular and/or linear deposits in active forms of the disease have been identified by Liang et al. [146], Plouvier et al. [202] and Park and Hazleman [195]. In addition, intracytoplasmic immunoglobulins in both plasma cells and macrophages have been noted [146,221]. These findings may suggest that humoral immune mechanisms play a pathogenetic role in the development of temporal arteritis [193]. Nevertheless Gallagher and Jones [81], using an immunoperoxidase method, failed to find extracellular complement. Such discrepancies may be due to different technical approaches, and the rapid degradation of

immune complexes within tissues might explain the fact that identifiable immunoglobulins or complement could only be detected in the cytoplasm of mononuclear cells. Furthermore there is no evidence that immunoglobulin and complement deposits result from primary deposition of antibodies specific to antigens of the arterial tissue or from plasma diffusion through porous endothelium. From a diagnostic point of view, it is necessary to stress the following points: (a) immunoglobulin and complement deposits are not specific and IgM and/or C3 deposits can sometimes be found in senile arteriosclerosis; (b) these deposits are usually absent in healed lesions and in normal segments of the temporal arteries with the skip lesions. Thus it seems likely that routine immunofluorescent investigation of biopsy specimens would not enhance the diagnostic value of temporal artery biopsy in this disease.

Immunolabelling of the cell populations involved in the granulomatosis process has become feasible since the recent availability of highly specific monoclonal antibodies directed against human cell surface antigens. Banks et al. in 1983 [12] showed the prevalence of the helper/inducer T lymphocyte subset in the mononuclear infiltrate of the arterial wall (Fig. 28.14). This finding, as well as the presence of histiocytes/macrophages, sometimes of epithelioid appearance in the lesions [196] and of circulating immunoblasts in the peripheral blood [66], are consistent with the role of a cell-mediated reaction in producing granulomatous lesions. In 1970 Papajiannis et al. [194] showed that subcutaneous implants of heterologous elastic tissue can apparently induce a giant cell granulomatous reaction. Nevertheless investigations of the response of peripheral blood lymphocytes to arterial or muscular antigens have produced controversial results [101,192,268]. The role of elastin, altered by ageing, in inducing the immune process has been evoked [101]. The systemic nature of the disease and the prominent site of damage in the "exposed" superficial temporal artery which shows signs of accelerated ageing, are not necessarily contradictory. It can be assumed that a systemic process might be provoked by primary localized tissue damage. The concept of elastolysis as an early event in temporal arteritis/polymyalgia rheumatica syndrome has been suggested recently by O'Brien and Argyle [184]; these authors proposed that temporal arteritis could be regarded as the florid expression of a pathological process that is common in ageing. Nevertheless there is still no

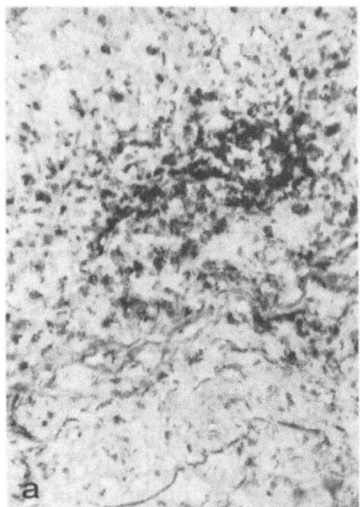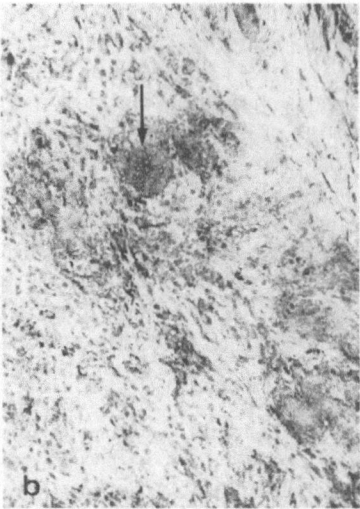

Fig. 28.14a,b. Temporal arteritis. Immunohistochemical identification of cell populations in a temporal artery biopsy specimen. Frozen section stain APAAP. **a** Large number of mononuclear cells reacting with anti-T11 monoclonal antibody (anti-pan T). **b** Macrophage-like cells reacting with EBM 11 monoclonal antibody (Dako Laboratories). Note the positivity of the giant multinucleated cells.

obvious evidence to support this hypothesis. In the context of autoimmune disease, associated findings of rheumatoid arthritis, thyroiditis, lymphoma and progressive systemic sclerosis have been reported [117,136,159].

Finally, attention has been focused recently on genetic factors [87,130,145]. Although a definite association with antigens of the major histocompatibility complex has not been demonstrated clearly [34,100,217], two separate teams have reported an increase in HLA-DR4 in patients with temporal arteritis [14,38]. These findings, which have been confirmed by Lowenstein et al. [152], and familial aggregation are consistent with a genetic predisposition.

Patterns of Arterial Involvement

The superficial temporal, occipital, facial and ophthalmic arteries are those most often involved in the disease. Visual symptoms and blindness represent the most dreaded manifestations of this medical emergency. Ischaemic manifestations within the area depending on the external carotid artery may lead rarely to gangrene of the scalp and of the tongue. Every branch of the carotid artery may be involved, as well as the vertebral and basilar arteries [246]. It is generally accepted that the anatomical pattern of the disease is probably more diffuse than expected from clinical symptoms [4,27,115,135,164,261]. The exceptional involvement of the intracranial arteries might be due to the paucity of elastin in the wall of these arterial segments. Arteriography can play a key role in

the diagnostic approach and can guide the biopsy.

Involvement of great vessels in the temporal arteritis/polymyalgia rheumatica syndrome is well documented [73,96,115,237,238,246]. Klein

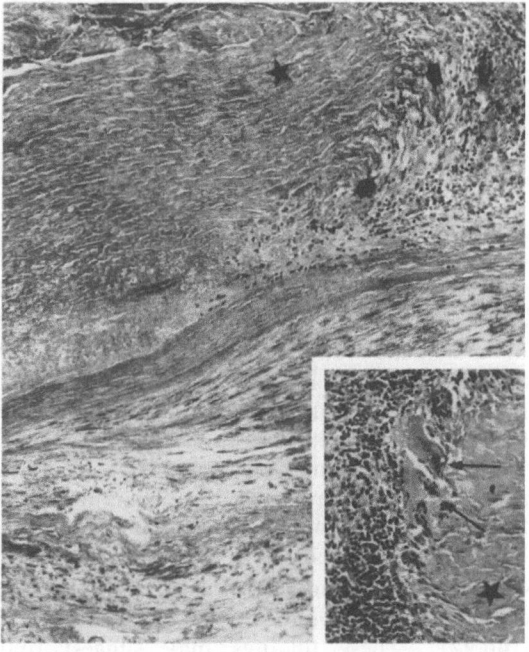

Fig. 28.15. Involvement of the subclavian artery in a patient with temporal arteritis. A large infarct like area can be seen within the media (*star*). Granulomatous cellular infiltration underlines the necrotic area (*thick arrows*). Masson's trichrome stain, × 165. *Inset.* Detail of the cellular infiltrate showing multinucleated giant cells (*thin arrows*) in close contact with the necrotic area (*star*). Masson's trichrome stain, × 280

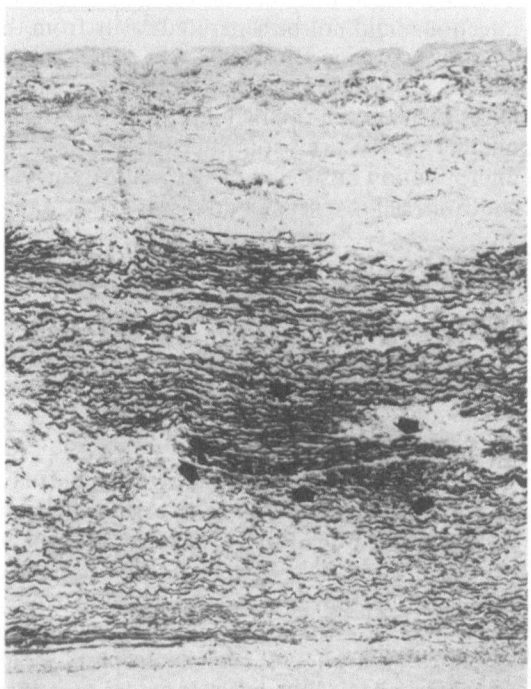

Fig. 28.16. Aortic involvement in a patient with temporal arteritis. The intimal layer is thickened. The medial elastic tissue is markedly disrupted by areas of elastolysis. In contrast, a necrotic area can be seen showing easily recognizable densely packed elastic laminae (*arrows*). Orcein stain, × 135

et al. in 1975 [135] noted 34 cases with great vessel involvement out of 248 patients. Involvement of the aorta may be asymptomatic and recognized only at angiography or postmortem examination. It may be responsible for aortic arch syndrome, aortic regurgitation, aneurysms [204] or dissection [158]. An associated finding of myocarditis has been reported [47]. These complications may reveal giant cell aortitis and represent a possible cause of sudden death by aortic rupture [190]. Histological aspects include marked fragmentation of elastic tissue, extensive giant cell granulomatous infiltration, marked intimal proliferation and thrombosis. The presence within the media of infarcted necrotic foci surrounded by macrophage and giant cell infiltrate supports the diagnosis of Horton's disease (Fig. 28.15). On staining these foci show recognizable remaining elastic tissue, in contrast to the neighbouring areas where there has been elastolysis (Fig. 28.16). Elastolysis in this disease may explain the propensity to medial dissection, aneurysms or rupture of the vascular wall. In addition to aortic involvement, the subclavian arteries are frequently involved [73,239] (Fig. 28.17). Other sites of involvement have been reported, e.g. peripheral arteries of the lower limbs [17], mesenteric arteries [96] and hepatic [186] or coronary [198] arteries. Although the renal arteries can be involved, renovascular hypertension remains very rare in this disease.

Such major involvement of the aorta and its branches can be recognized in the presence of symptoms of vascular insufficiency, vascular bruits or markedly diminished large artery pulses at a time of active temporal arteritis. Whatever the clinical presentation, aortography may show changes consistent with arteritis. In the event that bilateral biopsy of the superficial temporal

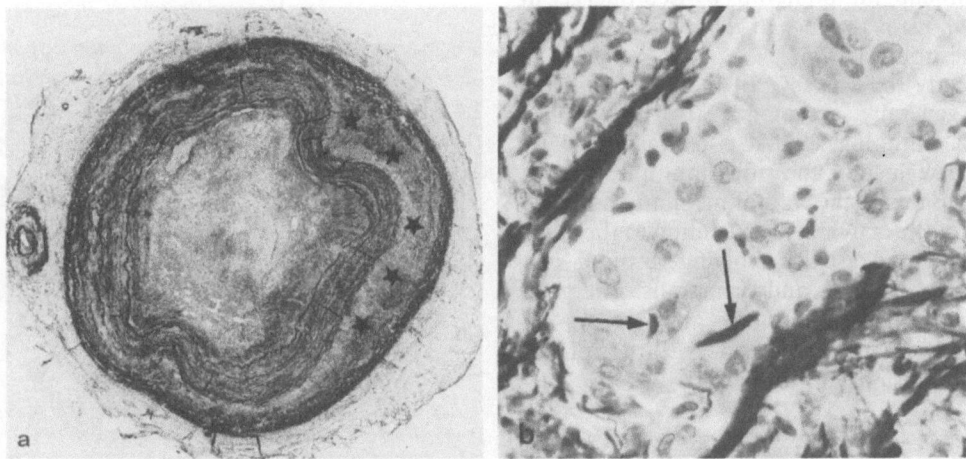

Fig. 28.17. a Severe arteritis of the subclavian artery in a patient with temporal arteritis. The lumen is completely occluded. The medial layer is dissociated by heavy cellular infiltration (*stars*). Orcein stain, × 20. **b** Detail of the same case as in **a**. Elastic laminae are disrupted by giant multinucleated cells showing elastic fragments within the cytoplasm (*arrows*). Orcein stain, × 510

artery might be negative, surgical exploration of the affected artery may be indicated to enable a diagnosis of Horton's disease. Theoretically, the histological pattern can provide evidence for its differentiation from the inflammatory diseases of the aorta and its major branches, particularly Takayasu's disease [73]. Moreover, associated constitutional and biological features are more often different in both diseases. However, the differential diagnosis can be very difficult. The onset of both diseases can present as isolated ischaemia of a limb; Takayasu's disease can occur later in life than is usually expected, and the histological pattern may be atypical in great vessels, notably in lesions with aneurysmal dilatation. One should bear in mind that elastolysis or giant cell infiltration, if viewed as solitary features, may not be regarded as pathognomonic of temporal arteritis and can be observed in Takayasu's disease. However, if various features are properly synthesized, it is generally possible to arrive at a specific diagnosis.

Associated findings of hepatic symptoms with non-specific and reversible histological lesions on liver biopsy [65,201], and hyperthyroidism [244] have been reported.

Thromboangiitis Obliterans (Von Winiwarter–Buerger's Disease)

Ever since the first descriptions of the disease by Von Winiwarter in 1879 and its characterization as thromboangiitis obliterans by Leo Buerger in 1908 [35], its acceptance as a specific entity has been questioned [86,256].

Currently the existence of this disease seems to be generally accepted [2,24,31,32,74,160, 163,175]. Pathologically, it is characterized by occlusive segmental and often multiple, inflammatory lesions in the wall of medium and small arteries of the extremities, extending to the veins and nerves and usually evolving through successive exacerbations and remissions. Buerger's disease remains infrequent and occurs in approximately 3% of patients with peripheral arterial occlusions in Europe [32].

Pathological Features

In previous descriptions based on the examination of amputated limbs, secondary findings due to extension of thrombosis and secondary infection could not be separated easily from the possibly specific features of the disease [76,125,132,138,231]. Nevertheless amputated limbs provide the opportunity to study a large number of sections of affected arteries; the interpretation of small biopsy specimens may be controversial in a segmental disease.

Arterial Changes

Macroscopically, the lesion is occlusive and segmental with intervening normal segments and relatively clear-cut demarcation of the involved segments. Sometimes periarterial fibrosis is marked.

Histopathologically, the lumen is occluded by thrombus at a variable stage of organization. In early cases, the features considered peculiar to Buerger's disease are a cellular thrombus and cellular infiltration of the vascular wall with preservation of its structure. (Fig. 28.18). Of course cellular infiltration by neutrophil polymorphonuclear cells and/or macrophages may represent non-specific reaction to thrombosis but, nevertheless the cellularity appears unusual. Neutrophils will often predominate and are often associated with lymphoplasmacytic infiltration. Rarely (in 2 of 113 cases collected by Kurozumi and Tanaka [138]) giant cells can be seen close to the denuded intima as well as within the thrombus (Fig. 28.19). Cellular immunotyping shows the macrophage nature of these giant cells. Clusters of prominent and partly detached endo-

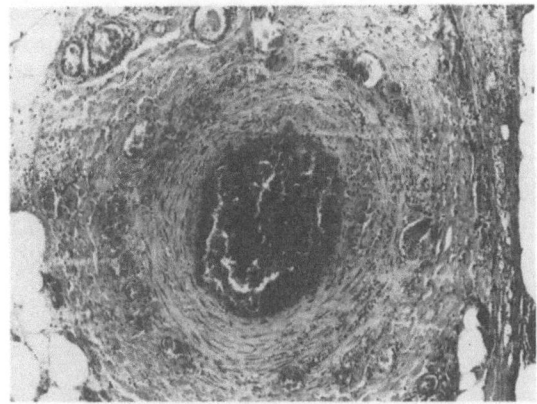

Fig. 28.18. Buerger's disease. Cross-section of a medium-sized artery showing complete occlusion of the lumen by fresh thrombus. The medial and adventitial layers show moderate cell infiltration with marked neovascularization. Haematoxylin–eosin–safran, × 120

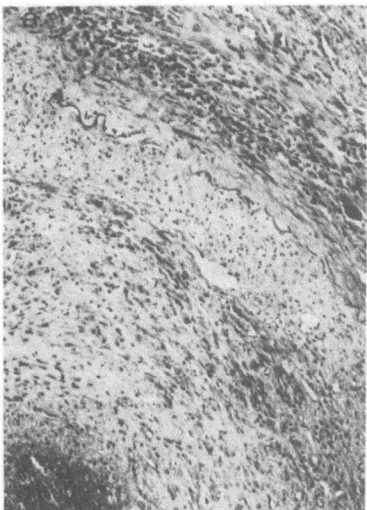

Fig. 28.19. Buerger's disease. Arterial occlusion. Granulomatous cellular infiltration with giant cells within the thrombus. Haematoxylin–eosin–safran, × 120

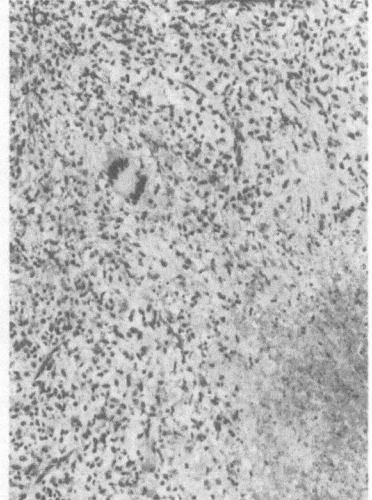

Fig. 28.20. Small-vessel involvement in a case of Buerger's disease. Small venule showing mononuclear and polymorphonuclear cell infiltration in a skin biopsy specimen. Haematoxylin–eosin–safran, × 410

thelial cells are rare associated findings. Necrosis is usually absent. The thickened intima can be involved by the cellular infiltration but the internal elastic lamina is well preserved. Most often medial changes remain discrete, showing sclerosis and neovascularization.

However, in most cases no distinctive feature can be seen. The pattern consists of organized thrombus with vascularization (Fig. 28.20). The media and internal elastic lamina are nearly normal without unusual cellular infiltration. Focal calcifications can be seen. Excessive adventitial and periadventitial fibrosis with involvement of veins and nerves is usually present. On the basis of arterial changes, it is impossible in these cases, to separate thromboangiitis obliterans from other peripheral occlusive arterial diseases or thromboembolism.

Venous Changes

Associated findings in veins are essential for characterizing Buerger's disease. Recurrent migratory superficial thrombophlebitis is noted in about 40% of cases and subcutaneous nodules resembling erythema nodosum can be the initial symptom before the symptoms of arterial disease are present.

The biopsy of an early lesion can show fresh thrombus including a large number of polymorphonuclear cells, macrophages and sometimes giant multinucleated cells (Fig. 28.21).

Such a pattern may suggest the diagnosis even in patients with mild or absent ischaemic symptoms. In more advanced lesions, the histopathological picture includes fibrous organizing lesions with intimal thickening and perivenous fibrosis. Such lesions can not be distinguished from common thrombophlebitis. Extensive fibrosis involving all components of the neurovascular bundle is well recorded.

In addition, cellular infiltration of small vessels, namely venules, can be seen in the vicinity of arterial or venous changes (Fig. 28.22).

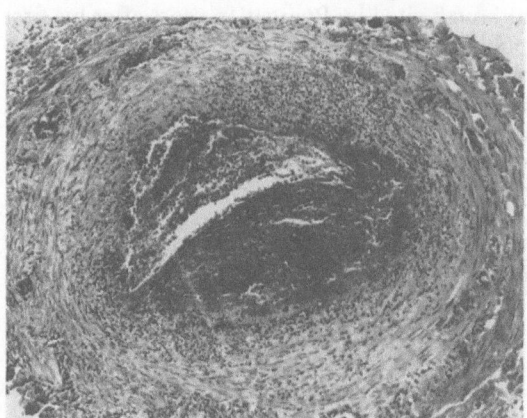

Fig. 28.21. Thrombophlebitis showing dense cellular infiltration and fresh thrombus. Haematoxylin–eosin–safran, × 135

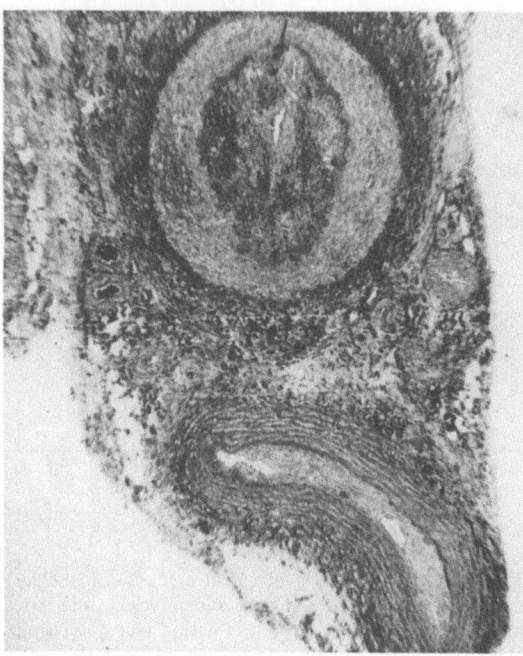

Fig. 28.22. Venous and arterial involvement in Buerger's disease. The artery is occluded by an organized thrombus. The elastic laminae are well preserved. The vein shows thickening. Cellular infiltration is scarce. Haematoxylin–eosin–safran, × 35

Pattern of Arterial Involvement

Medium-sized and small arteries are predominantly involved in Buerger's disease (tibial, radial, palmar and digital arteries). Both lower and upper extremities can be affected, but the consequences are more severe in the lower limbs. The disease can affect the major vessels secondarily and also the visceral arteries. Involvement of splanchnic and cerebral arteries, although cited by Buerger 16 years after his first publication [36], remains uncommon. Nevertheless involvement of the gastrointestinal tract can appear at any time in the course of the disease and present as an initial event prior to patent peripheral ischaemic disease [63,216]. Cerebral involvement with neurological expression was reported by Lindenberg and Spatz in 1939 [148] and by others since then [24,28,270], with an incidence of 0.5 to 10 per 100 [216].

Aetiological Factors

Thromboangiitis obliterans affects predominantly young men, usually under 40 years of age. In recent years there seems to have been a relative increase in the incidence of the disease among women. Incidence from 2.5% to 10% or 20% have been reported [32,137]. Tobacco smoking is a strong contributory factor but its relationship to the disease remains controversial. The prevalence of HLA-A9 and HLA-B5 antigens in affected persons [161] and occasional familial occurrence may suggest a genetic predisposition. The greater incidence of the disease in Israel [133], India [132] and Asia [110,231,232] compared with that in America or Europe, is in agreement with this hypothesis.

Recent immunological data seem to favour an autoimmune process [3,31,235]. The presence of specific anti-elastin antibodies [112] or a significantly higher stimulation index for cell-mediated sensitivity to human type I and type III collagens which are normal components of the arterial wall [3] may be found. The fact that patients with atherosclerosis have lower indexes does not confirm the hypothesis that these are secondary phenomena which might be a response to non-specific reparative processes in the arterial wall.

Infectious Aortitis and Arteritis

Since antibiotics have been developed and rheumatic fever largely prevented, the pathological and bacteriological features of infectious arteritides have changed dramatically. Syphilitic and tuberculous lesions are exceptional and will be considered separately. Rickettsial arteritides are no longer observed in Western Europe [16]. Viral vasculitis due to *Herpes zoster* and *varicella* involves small-sized vessels. In intrauterine rubella, foci of intimal hyperplasia are observed in large- and medium-sized arteries [39].

From physiopathological considerations infectious arteritides can be classified as follows [69,166,197,263]:

1. Mycotic aneurysms: first described by Osler, they result from arterial infection through septic embolism from endocarditis. Since the treatment of endocarditis has become more efficient their incidence has decreased. Bacteria responsible for subacute endocarditis are usually *Staphylococcus*, *Streptococcus* and more rarely gram-negative bacilli. An acute endocarditis, especially in intravenous drug abusers may give rise to the lesion. Fungal infections are excep-

tional, and are usually reported in immu-nocompromised patients or drug addicts [176]. In the course of endocarditis, mycotic aneurysms are frequently latent and are often diagnosed when they are complicated by arterial rupture [23] (Fig. 28.23).

2. Infected aneurysms: these are primary infectious aneurysms [57]. They represent the infection of any pre-existing aneurysm whatever its aetiology, occurring during bacteraemia [22]. They are more frequent in the abdominal aorta, like atherosclerosis itself which is the most fre-quent pre-existing lesion. They may be com-plicated by rupture [262]. *Salmonella* accounts for most of the cases [265,267], followed by *Staphylococcus* and *Escherichia coli* [162]. The primary site of infection is rarely found. Gram-negative bacteria have also been isolated from 15% of abdominal aortic aneurysms submitted to systematic bacterial culture. However, in these cases, infectious symptoms are lacking and post-operative septic risk is not increased [68].

3. Microbial aortitis and arteritis: infection of the arterial wall occurs during bacteraemia and is promoted by pre-existing arterial lesions, most frequently non-aneurysmal atherosclerosis in the aorta [13] and limbs [56]. The infection may develop in aortic coarctation [30]. Cerebral arter-itis has been reported rarely in meningitis [120]. Bacteriological features are the same as in infected aneurysms.

4. Contiguous arterial infections: although arterial walls are resistant to infection, they may be involved by surrounding septic foci, such as osteomyelitis and mediastinal and pulmonary abcesses [263].

5. Traumatic and post-operative infections are more frequent than they used to be. The infections may be secondary to penetrating arterial trauma, illicit drug injections, catheters [11,229] and cardiac surgery [5]. They are dis-cussed, in relation to vascular grafts, in Chap. 32.

In infections spreading haematogenously, thrombosis occurs early and inflammation is prominent in the innermost arterial layer with disruption of the internal elastic lamina. In infec-tions spreading contiguously or by colonization of vasa vasorum, inflammation is prominent in the outermost arterial tunica. These two mech-anisms result in destruction of the arterial wall more or less rapidly according to the virulence of the germ. The arterial wall is replaced by granulation tissue with suppurative necrosis. Arterial rupture may occur and produce life-threatening haemorrhage or false aneurysm.

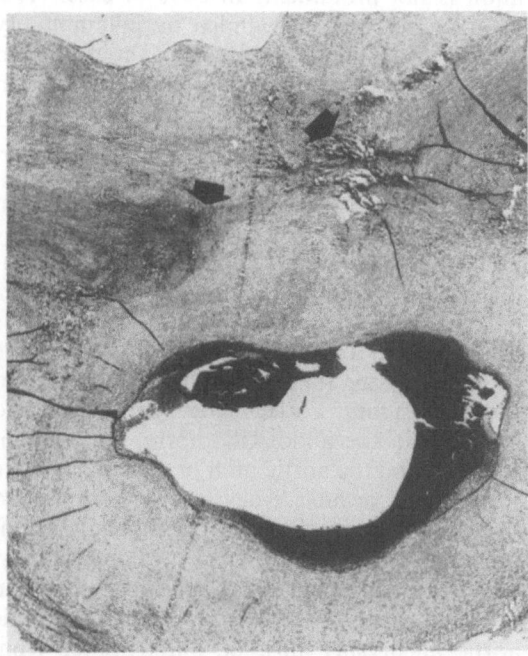

Fig. 28.23. False aneurysm due to mycotic aortitis in patient with coarctation of the aorta. The medial coat is ruptured (*arrows*) and covered by intimal thickening. The false aneur-ysm is bordered by organized connective tissue. Haema-toxylin–eosin–safran, × 23

Syphilitic Aortitis

Syphilitic involvement of the aorta occurs during the tertiary stage of the disease. In the past it was frequent, accounting for most of the aortitides. It then became less frequent so that it was observed in less than 1% of autopsies between 1950 and 1960 [102,103,105]. Syphilitic aortitis is now exceptional due to prevention and curative treat-ment with penicillin, and is rarely diagnosed, clinically. Serology is negative in 28% of syphi-litic aortitides [102].

Classification of the Lesions in Syphilitic Aortitis

Uncomplicated Syphilitic Aortitis. This com-prises syphilitic aortitis without aneurysm, aortic insufficiency or coronary ostial stenosis. Thirty-six per cent of cases in Heggtveit's series [102] had uncomplicated syphilitic aortitis. Aortic involvement is prominent in the ascending aorta, but can be observed in the subrenal abdominal aorta in 7% of cases [102]. No stenosing lesion of the branches of the aortic arch has been reported.

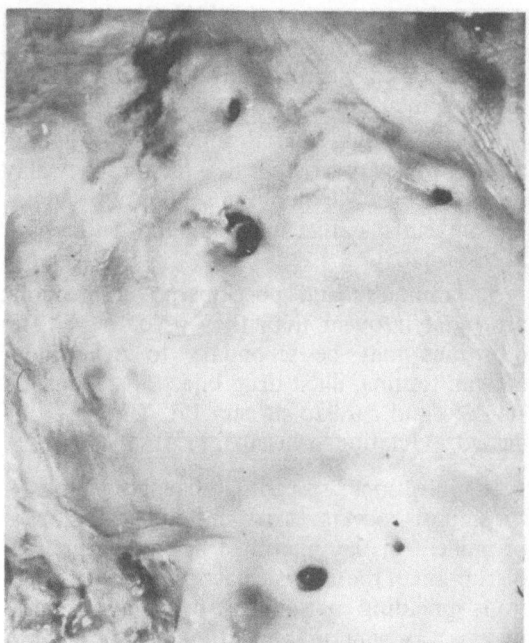

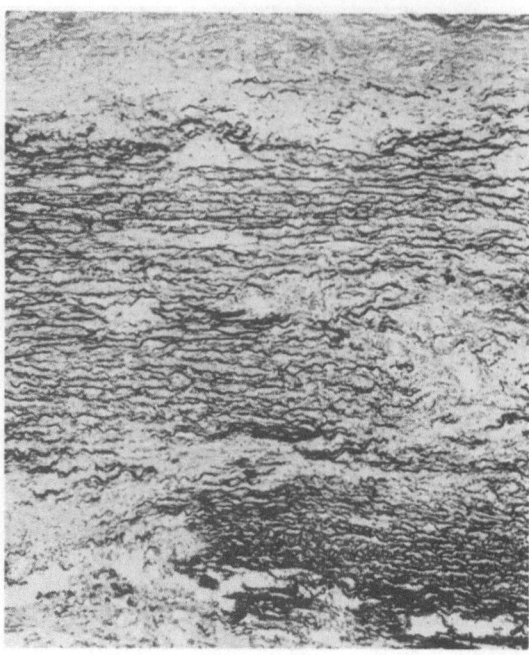

Fig. 28.24. Syphilitic aortitis. Gross aspect showing areas of whitish intimal thickening.

Fig. 28.26. Syphilitic aortitis. Areas of the media are replaced by foci of fibrosis and cellular infiltration. Orcein stain, × 70

Macroscopically, the aortic intima exhibits areas of whitish and shiny thickening. The intimal plaques are depressed by grooves with a "tree-bark" appearance (Fig. 28.24). Atherosclerosis is often an associated finding. Histology shows lymphocytic and plasmacytic inflammatory infiltrates around the vasa vasora which show obliterative endarteritis (Fig. 28.25). In the aortic media, smooth-muscle cells and elastic fibres are replaced by foci of inflammation and fibrosis

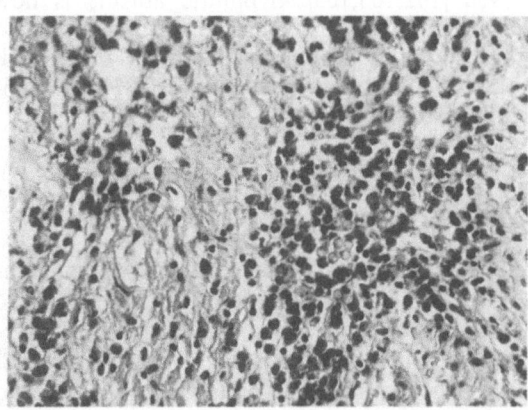

Fig. 28.25. Syphilitic aortitis. Plasma cell infiltration of the media and the adventitia. Haematoxylin–eosin–safran, × 200

(Fig. 28.26). In the intimal thickening, inflammation is not prominent. In 20% of cases [102] areas of necrosis surrounded by inflammatory cells (microgummas) are seen in the media. In rare instances spirochaetes are demonstrated in gumma with Warthin-Starry stain [252]. Giant cells are observed in 3% of cases [102].

Syphilitic Aortic Aneurysms. These result from destruction of the media by the inflammatory process. They are the most frequent complication of syphilitic aortitis, observed in 40% of cases [102]. Seven per cent of these aneurysms are multiple. They are located mainly in the ascending aorta, followed by the aortic arch and the descending aorta. Aneurysms are also observed in the abdominal aorta, the sinus of Valsalva and the aortic arch branches. They are fusiform or saccular. They may be complicated by thrombosis, erosion of vertebral bodies, compression of the superior vena cava and, mainly, rupture. Fatal aneurysmal rupture is found in 14% of cases [102]. Histologically, syphilitic aortic aneurysms must be differentiated from inflammatory infiltration occurring in some atherosclerotic aneurysms [18].

Syphilitic Aortic Insufficiency. This occurs in 25% of cases [102]. Aortic insufficiency results from two mechanisms: (a) dilatation of the aortic root by involvement of the sinus of Valsalva; (b) widening and thickening of the valve commissures with fibrosis of the cups and rolling of the free margin.

Syphilitic Coronary Ostial Stenosis. The stenosis of one or two ostia occurs in 28% of cases [102]. Ostial stenosis results from aortic intimal thickening, while a true syphilitic coronary arteritis has been observed only in one case [102]. Associated coronary atherosclerosis increases myocardial ischaemia induced by syphilitic ostial stenosis.

Tuberculous Aortitis and Arteritis

Tuberculous involvement of the aorta and large-sized arteries is now exceptional so that a review of the literature in the early 1960s reported only 100 cases [97,233]. The tuberculous infection spreads contiguously from lymphadenitis and from spondyloarthritis with Pott's cold abcess [72,253]. Haematogenous dissemination seems possible; tuberculous aortitis, pulmonary miliary

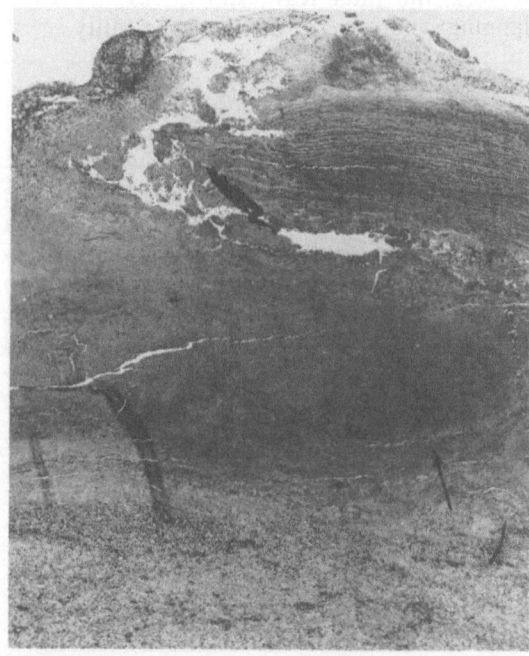

Fig. 28.27. Tuberculous aortitis developing in close contact with mediastinal adenopathy. Large area of caseous necrosis of the aortic wall, bordered by dense granulomatous and sclerosing reaction of the adventitia. Origin of the necrotic process appears to take place within the periaortic connective tissue (*arrow*). Haematoxylin-eosin-safran, × 32

tuberculosis [72] and colonization in the femoral artery [253] have been described. The abdominal and thoracic aorta are most frequently involved [253]. Involvement of the branches of the aortic arch has been observed.

Tuberculous infection produces granulomatous lesions, including epithelioid and giant cells in the arterial wall [182]. Acid-fast positive bacilli are sometimes demonstrated. Caseating necrosis may promote arterial rupture and false aneurysm formation (Fig. 28.27) [72,188,253]. Sometimes tuberculous aortitis does not exhibit aneurysmal lesions [233]. Stenosing lesions with periarterial fibrosis have been described.

A possible tuberculous aetiology for Takayasu's disease has been proposed [80]. Rasmussen's aneurysms develop from a pulmonary artery in an excavated pulmonary lesion. In tuberculous meningitis, cerebral arteritis is frequent [149]. In tuberculous aortitis surgical resection associated with anti-tuberculous chemotherapy prevents arterial ruptures [208].

Aortitis and Arteritis in Systemic Diseases and Miscellaneous Conditions

Behcet's Syndrome

Behcet's syndrome is characterized by relapsing buccal (98% of cases) and genital (88% of cases) ulcers, and iridocyclitis (66% of cases). Thrombophlebitis (37% of cases) and central nervous system and joint involvement (29% of cases) can be associated findings [19,40,230]. The aetiology is unknown. Immunogenetic factors are related to Behcet's syndrome: the $HLA-B_5$ phenotype is associated with ocular involvement and $HLA-B_{12}$ in cutaneo-mucous involvement [144]. Immune complexes have been implicated in the physiopathology of Behcet's syndrome [82]. The main morphological feature of Behcet's disease is a systemic vasculitis involving small-sized vessels, with inconstant fibrinoid necrosis [185,230]. However, large-sized vessel involvement is not infrequent and is mainly represented by thrombophlebitis. In contrast, large- and medium-sized arteries are rarely involved [40,230]. The involvement of large-sized vessels is via vasculitis of the vasa vasora [230]. Arterial involvement consists of thrombosis and aneurysm formation [40,139,230]. Histology shows a mononuclear inflammatory infiltrate in the three layers of involved arteries, with intimal thickening and disruption of the internal elastic lamina [207]. The arterial involvement is located (a) in the aorta [40,109,139,151], within the sinus of Valsalva [54]; (b) in the arteries of the limbs [43,67,139,207,230]; (c) in coronary arteries [33,139,230]; (d) in renal arteries [139,230]; (e) in the carotids [139]; and (f) in the pulmonary arteries [62,139].

Cogan's Syndrome

Cogan's syndrome is characterized by relapsing interstitial keratitis and vestibulo-auditory dysfunction in young adults. By 1980, 100 cases of this rare syndrome had been reported [99]. The aetiology is unknown. Ten per cent of cases are complicated by aortitis and life-threatening aortic insufficiency [45,99]. Histology shows polymorphic inflammatory infiltrates in the three layers of the aorta [99] and aortic cusps [45]. In typical Cogan's syndrome, there is no systemic vasculitis and vasculitis is restricted to ocular and auditory areas [99]. In contrast, in atypical Cogan's syndrome, systemic vasculitis is patent in several organs [45] and overlap with other vasculitides (periarteritis nodosa, Wegener's syndrome) is questioned [59].

Kawasaki Disease

Kawasaki disease or mucocutaneous lymph node syndrome was first described in Japan in the mid-1960s [129]. It was then observed in Western countries, the USA [21] and Western Europe [88]. Kawasaki disease occurs in young children, usually below 5 years old, with the highest frequency during the 2nd year of life. The incidence is 1.5 times higher for boys than for girls. Kawasaki disease is characterized by an acute febrile episode, a conjunctival infection, an oral mucous membrane involvement, an erythema of the extremities, a diffuse rash and non-suppurative lymphadenopathies [129]. The aetiology is still unknown [20].

Cardiovascular involvement is life-threatening, with a fatality rate of between 1% and 2% [20,21]. All the fatal cases present cardiovascular involvement [242]. This may be manifested by pericarditis, myocarditis [242,266], mitral insufficiency [121] or especially by coronary involvement [25]. Coronary lesions are observed in 20% of cases of Kawasaki disease [21,88,126,127,165,189]. They are characterized by aneurysms, thrombosis and stenosis [177]. Coronary lesions appear within the 1st month of the disease [79,111]. They may be complicated by myocardial infarction. Left coronary artery involvement has a more severe prognosis than right [111]. About 50% of coronary aneurysms disappear [126,220]. Late cardiac complications may occur [187,203].

Macroscopically, coronary aneurysms exhibit bead-like features [143,242]. The arterial media is replaced by fibrosis or granulation tissue in the early stages. The intima is thickened and sometimes inflammatory. The vascular lumen is occluded by thrombosis. Apart from coronary aneurysms, one rarely finds necrotizing arteritis characterized by fibrinoid necrosis of the media and polymorphic inflammatory infiltrates. Vasculitis of small intramyocardial vessels has been observed in very early stages of Kawasaki disease [79]. In addition to cardiac vascular involvement, most fatal cases present a more diffuse arteritis in iliac, gastrointestinal, renal and pulmonary territories [121,143,242]. Thus the topographic and microscopic features of the arteritis resemble periarteritis nodosa [143,242]. It remains unclear whether infantile periarteritis nodosa and fatal Kawasaki disease are overlapping syndromes or are the same entity.

Rheumatic Fever

Aortitis is not frequent in rheumatic fever, the incidence of which is nowadays low. The aorta can be involved in short segments or more extensively, with the abdominal aorta most likely to be affected [103,104,105,140]. Histologically, mononuclear inflammatory infiltrate is observed in the three arterial layers, with inconstant fibrinoid necrosis [105]. Aschoff's nodules are lacking. In relapsing or prolonged rheumatic fever, adventitial fibrosis and medial fibrous replacement may develop. Pulmonary arterial involvement is often associated with aortitis [105].

Rheumatoid Arthritis

Clark et al. [52] and Zvaifler and Weintraub [271] emphasized aortic involvement in rheumatoid arthritis (and in ankylosing spondylitis). In rheumatoid arthritis aortitis is most often associated with cardiac involvement. Aortic involvement is prominent in the aortic root and is combined with aortic insufficiency [104]. Unlike the situation in ankylosing spondylitis, aortic involvement may occur early in the course of rheumatoid arthritis but is haemodynamically moderate [140]. Epicardial coronary arteries can be involved, exhibiting thrombosis and aneurysm [104,174]. Histology shows mononuclear inflammatory infiltrates in the three aortic layers, yielding panaortitis [52,104,210,271]. Inflammation is prominent in the adventitia where obliterans endarteritis of vasa vasorum may be found, mimicking syphilitic aortitis [105]. Rarely rheumatoid nodules are observed in the aortic wall [104,210,271] and the latter may exhibit fibrinoid necrosis and polynuclear inflammatory infiltrates which give rise to "microabcesses" [104]. A differential diagnosis must be made between such severe rheumatoid arthritides with necrotizing vasculitis and polyarthritis nodosa.

Ankylosing Spondylitis

Aortitis is a well-known complication of ankylosing spondylitis [7,52,61,271] and is located in the aortic root. Aortic insufficiency occurs late in the course of ankylosing spondylitis, often after more than 10 years [140,271]. Among all the aetiologies of aortic insufficiency, ankylosing spondylitis is implicated in 5% of cases [271]. In contrast to the situation in rheumatoid arthritis, in ankylosing spondylitis aortic insufficiency is often severe [140,271]. Histological features are similar to those observed in rheumatoid arthritis [52,140,271], except that a subaortic fibrous ridge has been described in ankylosing spondylitis [37]. Aortic insufficiency is life-threatening in ankylosing spondylitis and should by systematically searched for by ultrasound [247].

Reiter's Syndrome

Reiter's syndrome is characterized by non-gonococcal urethritis, conjunctival injection and arthritis [260] and occurs in young males. Aortitis is present in 2%–5% of cases of Reiter's syndrome [58]. Involvement is located in the aortic root and may be complicated by aortic insufficiency [58,199,211] which sometimes worsens rapidly [29]. The histology of the aortic involvement is similar to that of ankylosing spondylitis [199]. Occasionally aortic involvement is distal, appearing as a stenosis near to the aortic bifurcation and associated with periaortic fibrosis [172].

Relapsing Polychondritis

Relapsing polychondritis is characterized by an inflammatory involvement of cartilages of nose, ears, trachea and joints [114,200,226]. Recurring attacks of the disease destroy cartilagenous structures and respiratory involvement is life-threatening. Aortic involvement occurs in 10% of cases [91] and is characterized by major lesions of the extracellular matrix [114]. These lesions may exhibit a pattern of cystic medial necrosis [226] and may be complicated by aneurysm and aortic dissection [91,114,226]. Inflammatory infiltration is observed around the vasa vasora [91].

Periaortitis or Inflammatory Aneurysms of the Aorta

Since their first description by Walker et al. [254], many clinical [15,85] and pathological reports [71,167,168,169,212] have dealt with inflammatory aneurysms of the aorta. These reports have contributed to the characterization of a new entity which is related to inflammatory involvement of the aorta.

Gross morphology allows diagnosis during surgery [85,254] or before operation by means of computerized tomography and ultrasonography [15,71]. The aneurysmal wall is thick, measuring 1–3 cm. Extensive fibrosis surrounds the aneurysm. The fibrosis may reach the duodenum, the mesentery, the inferior vena cava and the ureters which can be obstructed [85,169,212,254]. The abdominal part of the aorta is involved predominantly. A few cases present with pericoronary involvement contiguous to periaortitis [170].

Histology [71,168,169,212] shows atherosclerosis often complicated by mural thrombus. The media is atrophic or more often indistinguishable in fibrosis. The fibrous thickening of the adventitia is characteristic of these inflammatory aneurysms of the aorta. Nerves, periaortic lymphatics, vasa vasora and arterial branches are coated by fibrosis. (Fig. 28.28). In rare instances, venous thrombosis and arteritis have been reported in the periaortic fibrosis [71]. This fibrosis exhibits variable inflammatory infiltrates. Sometimes inflammation is discrete and localized around vasa vasora which do

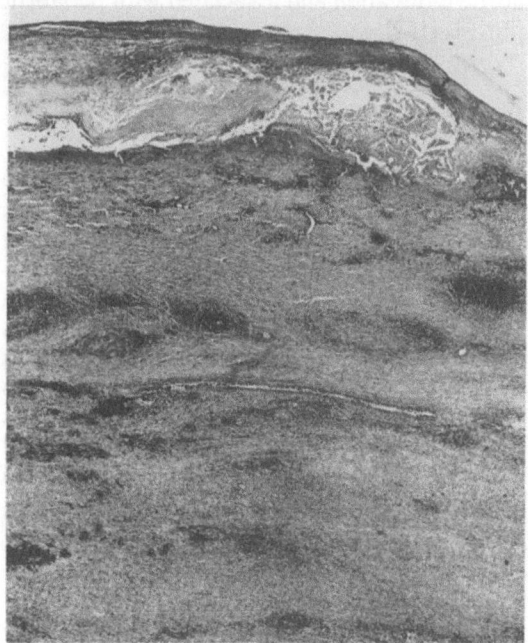

Fig. 28.28. Inflammatory aneurysms of the aorta with periaortitis.

not show endarteritis obliterans. Sometimes inflammatory infiltrates are dense throughout the aortic wall. The inflammatory cells are mononuclear cells, mainly polyclonal plasma cells [71]. Haemosiderin-loaded macrophages are rare.

Inflammatory aneurysms or aneurysms with periaortitis are considered to be a distinctive feature of atherosclerotic aneurysms [169,212], though they actually have the same topography and the same epidemiological background. The rupture of inflammatory aneurysms seems to be less frequent [254]. Their incidence ranges from 2% [71] to 10% [169,212,254] of operated aneurysms.

Inflammatory aneurysms have to be distinguished from inflammatory phenomena which are sometimes associated with atherosclerosis [18,225]. In the latter, adventitial fibrosis is not a prominent feature.

The pathogenesis of the periaortitis is unknown. Hypersensitivity to atheromatous material due to the thinned media has been hypothesized [169]. An associated inflammatory syndrome has sometimes been reported [254]. However, no infectious disease like syphilis, nor systemic disease, nor Takayasu's disease could be demonstrated [71,212,254]. Most cases of retroperitoneal or mediastinal fibrosis are associated with an aortic involvement very similar to

that of inflammatory aneurysms [167,168,169]. In this way both diseases widely overlap [228].

Treatment with corticosteroids has been proposed [15]. Surgical treatment of these aneurysms has a higher morbidity and mortality than surgical treatment of the usual atherosclerotic aneurysms [15].

Miscellaneous Conditions

A few reports deal with single or limited cases of aortitis and large-sized vessel arteritis in miscellaneous conditions.

Roth and Kissane [218] reported a case of scleroderma (progressive systemic sclerosis) evolving in a female patient and complicated by panaortitis with aortic valve involvement, as in ankylosing spondylitis.

The same pattern of aortic involvement has been observed in Crohn's disease and psoriasis [271].

In sarcoidosis, granulomatous aortitis may occur [155,245].

In angioimmunoblastic lymphadenopathy, a case with gangrene and popliteal and brachial arteritis has been reported [255].

Frusemide has been suspected of being responsible for aortitis disclosed at autopsy [236].

Non-specific Aortitis

Sometimes it is impossible to classify aortitis using either histological criteria (such as giant cell arteritis), or the clinical and biological features (such as rheumatoid or infectious diseases). Series of aortitides include a number of these cases [105,106,140,206].

References

1. Abe K, Miyazaki K (1982) Clinical aspect of aortitis syndrome with special influence to the relation between prognosis and hypertension. Jpn Circ J 40:190–193
2. Adar R (1974) Buerger's disease: the need for diagnostic criteria. Surgery 76:848
3. Adar R, Papa MZ, Halpern Z et al. (1983) Cellular sensitivity to collagen in thromboangitis obliterans. N Engl J Med 308:1114–1116
4. Ainsworth RW, Gresham GA, Balmforth GV (1961) Pathological changes in temporal arteries removed from unselected cadavers. J Clin Pathol 14:115–119
5. Allevato PA, Eisses JF, Mezger E et al. (1985) Nocardia asteroides aortitis with perforation of the aorta. Hum Pathol 16:743–745
6. Allsop CJ, Gallagher PJ (1981) Temporal artery biopsy

of giant-cell arteritis: a reappraisal. Am J Surg Pathol 5:317–323

7. Ansell BM, Bywaters EGL, Doniach I (1958) The aortic lesion of ankylosing spondylitis. Br Heart J 20:507–515
8. Asherson RA, Asherson GL, Schrire V (1968) Immunological studies in arteritis of the aorta and great vessels. Br Med J iii:589–591
9. Ask-Upmark E (1954) On the pulseless disease outside of Japan. Acta Med Scand 149:161–178
10. Aufderheide A, Henke B, Parker E (1981) Granulomatous coronary arteritis. Arch Pathol Lab Med, 105:647–649
11. Baker WH, Moran JM, Dorner DB (1979) Infected aortic aneurysm following arteriography. J Cardiovasc Surg 20:373–377
12. Banks PM, Cohen MD, Ginsburg WW et al. (1983) Immunohistologic and cytochemical studies of temporal arteritis. Arthritis Rheum 26:1201–1207
13. Bardin JA, Collins GM, Devin JB, Halasz NA (1981) Nonaneurysmal suppurative aortitis. Arch Surg 116:954–956
14. Barrier J, Bignon JP, Soulillou JP et al. (1981) Increased prevalence of HLA-DR4 in giant cell arteritis. N Engl J Med 305:104–105
15. Baskerville PA, Blakeney CG, Young AE, Browse NL (1983) The diagnosis and treatment of periaortic fibrosis (inflammatory aneurysms). Br J Surg 70:381–385
16. Bastin R, Bricaire F, Frottier J et al. (1983) Arteriopathies infectieuses. Ann Med Interne 134:475–478
17. Bastin R, Godeau P, Lavernhe J et al. (1969) Pseudopolyarthérite rhizomalique: de l'artérite temporale à l'artérite des membres inférieurs. Presse Med 77:1649–1651
18. Beckman EN (1986) Plasma cell infiltrates in atherosclerotic abdominal aortic aneurysms. Am J Clin Pathol 85:21–24
19. Behcet's disease Research Committee of Japan (1974) Behcet's disease: guide to diagnosis of Behcet's disease. Jpn J Ophthalmol 18:291–297
20. Bell DM (1985) Kawasaki syndrome: still a mystery after 20 years. JAMA 254:801–802
21. Bell DM, Morens DM, Holman RC et al. (1983) Kawasaki syndrome in the United States. Am J Dis Child 137:211–214
22. Bennett DE, Cherry JK (1967) Bacterial infection of aortic-aneurysms: a clinico-pathologic study. Am J Surg 113:321–326
23. Berkmen YM (1986) Medical aspects of infectious aortitis. In: Lande A, Berkmen YM, McAllister HA (eds) Aortitis; clinical, pathological and radiographic aspects, vol 1. Raven Press, New York, pp 161–172
24. Berlit P, Kessler C, Reuther P, Krause KH (1984) New aspects of thromboangiitis obliterans (Von Winiwarter-Buerger disease). Eur Neurol 28:394–399
25. Berry CL (1983) Kawasaki's disease. Pediatr Cardiol 4:233–234
26. Bethlenfalvay NC, Nusynowitz ML (1964) Temporal arteritis. A rarity in the young adult. Arch Intern Med 114:487–489
27. Bevan TA, Dunnill MS, Harrison MSG (1968) Clinical and biopsy findings in temporal arteritis. Ann Rheum Dis 27:271–277
28. Biller J, Asconate J, Challa VR et al. (1981) A case of cerebral thromboangiitis obliterans. Stroke 12:686–689
29. Block SR (1972) Reiter's syndrome and acute aortic insufficiency Arthritis Rheum 15:218–220
30. Bodner SJ, McGee ZA, Killen DA (1973) Ruptured

myotic aneurysm complicating coarctation of the aorta. Ann Thorac Surg 15:419–426
31. Bollinger A, Hollmann B, Schneider E et al. (1979) Thromboangiitis obliterans: Diagnose und Therapie im licht neuer immunologischer Befnüde. Schweiz Med Wochenschr 109:537–543
32. Bollinger A, Piquerez MJ, Largiarder J et al. (1983) Maladie de Buerger: concepts diagnostiques et thérapeutiques actuels. Ann Med Interne 134:436–440
33. Bowles CA, Nelson AM, Hammill SC, O'Duffy JD (1985) Cardiac involvement in Behcet's disease. Arthritis Rheum 28:345–348
34. Bridgeford PH, Lowenstein M, Bocanegra TS et al. (1980) Polymyalgia rheumatica and giant cell arteritis: histocompatibility typing and hepatitis-B infections studies. Arthritis Rheum 516–518
35. Buerger L (1908) Thromboangiitis obliterans: a study of the vascular lesions leading to presenile spontaneous gangrene. Am J Med Sci 136:567–580
36. Buerger L. (1924) The circulatory disturbances of the extremities including gangrene, vasomotor and trophic disorders. Saunders, Philadelphia
37. Bulkley BH, Roberts WC (1973) Ankylosing spondylitis and aortic regurgitation. Description of the characteristic cardiovascular lesion from study of eight necropsy patients. Circulation 48:1014–1027
38. Calamia KT, Moore SB, Elveback LR (1981) HLA-DR locus antigens in polymyalgia rheumatica and giant cells arteritis. J Rheumatol 8:993–995
39. Campbell PE (1965) Vascular abnormalities following maternal rubella. Br Heart J 27:134
40. Chajek T, Fainaru M (1975) Behcet's disease: report of 41 cases and a review of the literature. Medicine 54:179–196
41. Chapman R, Dawe C, Whorwell JP et al. (1978) Ulcerative colitis in association with Takayasu's disease. Dig Dis Sci 23:660–662
42. Chauvaud S, Gillon JF, Valois M et al. (1983) Localisation pulmonaire atypique de la maladie de Takayasu. Ann Cardiol Angeiol 32:55–57
43. Chavatzas D (1974) Popliteal artery thrombosis in Behcet's syndrome. A new manifestation of a very little known condition. Angiology 25:773–776
44. Cherigie E. Chermet J (1975) Maladie de Takayasu. Aspects angiographiques habituels et inhabituels. A propos de 15 observations arteriographiques. J Radiol Electrol 56:21–33
45. Cheson DB, Bluming AZ, Alroy J (1976) Cogan's syndrome: a systemic vasculitis. Am J Med 60:549–555
46. Chhetri MK, Pal NC, Neelkanthan C et al. (1970) Endocardial lesion in case of Takayasu's arteriopathy. Brit Heart J 32:859–861
47. Chomette G, Facquet J, Grosgogeat Y et al. (1975) Deux observations anatomocliniques singulières d'aortite à cellules géantes. Dissection aortique. association à une myocardite. Ann Med Int 126:573–581
48. Chopra P, Datta RK, Dasgupta A et al. (1983) Non specific aortoarteritis (Takayasu's disease). An immunologic and autopsy study. Jpn Heart J 24:549–556
49. Chopra P, Singhal V, Nayak NC (1978) Aortoarteritis and cardiopathy. A heretofore undescribed association. Jpn Heart J 19:358–365
50. Chun P, Jones R, Robinowitz M et al. (1980) Coronary aortal stenosis in Takayasu's arteritis. Chest 78:330–331
51. Cipriano PR, Silverman JF, Perlroth MG et al. (1977) Coronary arterial narrowing in Takayasu's aortis. Am J Cardiol 39:744–750

52. Clark WS, Kulka JP, Bauer W (1957) Rheumatoid aortitis with aortic regurgitation: an unusual manifestation of rheumatoid arthritis (including spondylitis). Am J Med 22:580–592

53. Cohen DN, Smith TR (1974) Skip areas in temporal arteritis: myth versus fact. Trans Am Acad Ophthalmol Otolaryngol 78:772–783

54. Comess KA, Zibelli LR, Gordon D, Fredrikson SR (1983) Acute, severe aortic regurgitation in Behcet's syndrome. Ann Intern Med 99:639–640

55. Cooke WT, Cloake PLP, Govan ADT et al (1946) Temporal arteritis: a generalized vascular disease. QJ Med 15:47–75

56. Cormier JM, Leguillou M (1969) Anévrysmes infectieux primaires des membres. Presse Med 77:1415–1417

57. Crane AR (1937) Primary multilocular mycotic aneurysm of the aorta. Arch Pathol 24:634–641

58. Csonka GW, Litchfield JW, Oates JK, Willcox RR (1961) Cardiac lesions in Reiter's disease. Br Med J i:243–247

59. Cupps TR, Fauci AS (1981) Cogan's syndrome. In: Smith LH (ed) The vasculitides, vol 1. Saunders, Philadelphia, pp 149–151

60. Danaraj TJ, Wong HO (1959) Primary arteritis of abdominal aorta in children causing bilateral stenosis of renal arteries and hypertension. Circulation 20:856–863

61. Davidson R, Baggensttos AH, Slocumb CH, Daugherty GW (1963) Cardiac and aortic lesions in rheumatoid spondylitis. Mayo Clin Proc 38:427–435

62. Davies JD (1973) Behcet's syndrome with haemoptysis and pulmonary lesions. J Pathol 109:351–356

63. Deitch EA, Sikkema WW (1981) Intestinal manifestations of Buerger's disease. Case report and literature review. Am Surg 47:326–328

64. Deutsch V (1974) Takayasu's arteritis: an angiographic study with remarks on ethnic distribution in Israel. AJR 122:13–28

65. Dickson ER, Maldonado JE, Sheps SG et al. (1973) Systemic giant cell arteritis with polymyalgia rheumatica, reversible abnormalities of liver function. JAMA 11:1496–1498

66. Eghtedari AA, Esselinckx X, Bacon PA (1976) Circulating immunoblasts in polymyalgia rheumatica. Ann Rheum Dis 35:158–162

67. Enoch BA, Catillo-Olivares JL, Kloo TCL et al. (1968) Major vascular complications in Behcet's syndrome. Postgrad Med J 44:453–459

68. Ernst CB, Campbell HC Jr. Daugherty ME, Sachatello CR (1977) Incidence and significance of intra-operative bacterial cultures during abdominal aortic aneurysmectomy. Ann Surg 185:626–630

69. Ewart JM, Burke ML, Bunt TJ (1983) Spontaneous abdominal aortic infections. Essentials of diagnosis and management. Am Surg 49:37–50

70. Falicov RE, Cooney DF (1964) Takayasu's arteritis and rheumatoid arthritis. Arch Intern Med 114:594–600

71. Feiner HD, Nagesh Raghavendra B, Phelps R, Rooney L (1984) Inflammatory abdominal aortic aneurysms: report of six cases. Hum Pathol 15:454–459

72. Felson B, Akers PV, Hall GS et al. (1977) Mycotic tuberculous aneurysm of the thoracic aorta. JAMA 237:1104–1108

73. Fiessinger JN, Camilleri JP, Chousterman M et al. (1978) Maladie de Horton et maladie de Takayasu: critères anatomopathologiques. Nouv Press Med 7:639–642

74. Fiessinger JN, Housset E (1985) Thromboangiitis obliterans (maladie de Buerger). In: Kahn MF, Peltier AP (eds) Maladies systémiques. Flammarion, Paris, pp 526–531

75. Fiessinger JN, Tawfik-Taher S, Capron L et al. (1982) Maladie de Takayasu: critères diagnostiques. Nouv Press Med 11:583–586

76. Fontaine R, Bouissou H, Batzenschlager A et al. (1968) La thromboangiose. Ses aspects anatomopathologiques. In: Congrès de collège Français de Pathologies vasculaires. L'expansion scientifique, Paris, pp 57–76

77. Forestier J, Certonciny A (1953) Pseudopolyarthrite rhizométique. Rev Rhum Mal Osteoartic 20:854–863

78. Friedman CJ, Tegtmeyer CT (1979) Crohn's disease associated with Takayasu's arteritis. Dig Dis Sci 24:954–958

79. Fujiwara H. Hamashima Y (1978) Pathology of the heart in Kawasaki disease. Pediatrics 61:100–107

80. Gajaraj A, Victor S (1981) Tuberculous aortoarteritis. Clin Radiol 32:461–466

81. Gallacher P, Jones K (1982) Immunohistochemical findings in cranial arteritis. Arthritis Rheum 25:75–79

82. Gamble CN, Wiesner KB, Shapiro RF, Boyer WJ (1979) The immune complex pathogenesis of glomerulonephritis and pulmonary vasculitis in Behcet's disease. Am J Med 66:1031–1039

83. Ghozlan R, Amor B, Delbarre F (1977) Association d'une maladie de Takayasu et d'une spondylartherite ankylosante. Un nouveau cas. Ann Med Interne 128:959–963

84. Gilmour VR (1941) Giant cell chronic arteritis. J Path Bact 53:263–277

85. Goldstone J, Malone JM, Moore WS (1978) Inflammatory aneurysms of the abdominal aorta. Surgery 83:425–430

86. Gore I, Burrows S (1958) Reconsideration of the pathogenesis of Buerger's Disease. Am J Clin Pathol 29:319–330

87. Granato JE, Abben RP, May WS (1980) Familial association of giant cell arteritis. A case report and brief review. Arch Intern Med 141:115–117

88. Gray JA, Welsby PD (1984) Kawasaki disease: a report of 26 patients. J Infect 9:17–21

89. Gulati SM, Singh KS, Thusoo TK et al. (1979) Immunological studies in thromboangitis obliterans (Buerger's disease). J Surg Res 27:287–293

90. Gyototu Y, Kakiuchi T, Nonaka Y et al. (1981) Immune complexes in Takayasu's arteritis. Clin Exp Immunol 45:246–252

91. Hainer JW, Hamilton GW (1969) Aortic abnormalities in relapsing polychondritis. Report of a case with dissecting aortic aneurysm. N Engl J Med 280:1166–1168

92. Hall S, Barr W, Lie JT et al. (1985) Takayasu arteritis. A study of 32 North American patients. Medicine 64:89–99

93. Hamilton CR, Shellex WM, Tumulty PA (1971) Giant cell arteritis: including temporal arteritis and polymyalgia rheumatica. Medicine 50:1–27

94. Hamrin B, Jonsson N, Hellsten J (1968) Polymyalgia arteritica. Further clinical and histopathological studies with a report of six autopsy cases. Ann Rheum Dis 27:397–405

95. Hamrin B, Jonsson H, Landberg T (1964) Arteritis in polymyalgia rheumatica. Lancet I:397–401

96. Hamrin B, Jonsson N, Landberg T (1965) Involvement of large vessels in polymyalgia arteritica. Lancet I:1193–1196

97. Hara M, Bransford RM (1963) Aneurysm of the sub-clavian artery associated with contiguous pulmonary tuberculosis. J Thorac Cardiovasc Surg 46:256–264

98. Harrison CV (1948) Giant cell or temporal arteritis: a review. J Clin Pathol 1:197–211

99. Haynes BF, Kaiser-Kupfer MI, Mason P, Fauci AS (1980) Cogan syndrome: studies in thirteen patients, long-term follow-up, and a review of the literature. Medicine 59:426–441

100. Hazleman B, Goldstone A, Voak D (1977) Association of polymyalgia rheumatica and giant cell arteritis with HLA-B8. Br Med J ii:989–991

101. Hazleman BL, McLennan ICM, Esiri MM (1975) Lymphocyte proliferation to artery antigen as a positive diagnostic test in polymyalgia rheumatica. Ann Rheum Dis 34:122–127

102. Heggtveit HA (1964) Syphilitic aortitis. A clinico-pathologic autopsy study of 100 cases, 1950 to 1960. Circulation 29:346–355

103. Heggtveit HA (1983) Nonatherosclerotic disease of the aorta. In: Silver MD (ed) Cardiovascular pathology, vol 2. Churchill Livingstone, New York, pp 707–737

104. Heggtveit HA (1983) Rheumatic-rheumatoid panaortitis. In: Silver MD (ed) Cardiovascular pathology, vol 2. Churchill Livingstone, New York, pp 713–715

105. Heggtveit HA, Henniga GR, Morrione TG (1963) Panaortitis. Am J Pathol 42:151–172

106. Henochowicz SI, Lindsay J, Furlong MJ et al. (1986) Multiple saccular aortic aneurysms in non specific aortitis. Am J Cardiol 37:377–378

107. Heptinstall RH, Porter KA, Barkley H (1934) Giant cell (temporal) arteritis. J Pathol 67:507–519

108. Hidano A, Watanabe K (1981) Pyoderma gangrenosum et cardio-vasculopathies en particulier artérite de Takayasu. Ann Dermatol Venereol 108:13–21

109. Hills EA (1967) Behcet's syndrome with aortic aneurysms. Br Med J iv:152–154

110. Hirai M, Shionoya S (1979) Arterial obstruction of the upper limb in Buerger's disease: its incidence and primary lesion. Br J Surg 66:124–128

111. Hiraishi S, Yashiro K, Oguni K et al. (1981) Clinical course of cardiovascular involvement in the muco-cutaneous lymph node syndrome. Am J Cardiol 47:323–330

112. Horsch AK, Brechemier D, Robert L et al. (1978) Anti-Elastin-Antickörper bei der Thromboangiitis obliterans (Morbus von Winiwarter-Buerger). Verh Dtsch Ges Inn Med 40:1758–1761

113. Horton BT, Magath TB, Brown GE (1932) An undescribed form of arteritis of the temporal vessels. Mayo Clin Proc 7:700–701

114. Hughes RAC, Berry CL, Seifert M, Lessof MH (1972) Relapsing polychondritis: three cases with a clinico-pathological study and literature review. QJ Med 41:363–380

115. Hunder GG, Ward LE, Burbank MK (1967) Giant cell arteritis producing an aortic arch syndrome. Ann Intern Med 66:578–582

116. Hunder GG, Sheps SG (1967) Intermittent claudication and polymyalgia rheumatica (association with panarteritis). Arch Intern Med 119:638–640

117. Huskisson EC, Dieppe PA, Balme HW (1977) Complicated polymyalgia. Br Med J ii:1459

118. Huston KA, Hunder GG, Lie JT et al. (1978) Temporal arteritis: a 25 years epidemiologie, clinical and pathologic study. Ann Intern Med 88:162–167

119. Hutchinson J (1890) Diseases of the arteries. Arch Surg 1890 1:325–326

120. Igarashi M, Gilmartin RC, Gerald B et al. (1984) Cerebral arteritis and bacterial meningitis. Arch Neurol 41:531–535

121. Imakita M, Sasaki Y, Misugi K et al. (1984) Kawasaki disease complicated with mitral insufficiency: autopsy findings with special reference to valvular lesions. Acta Pathol Jpn 34:605–616

122. Inada K (1976) Aortitis syndrome. In: Shiokawa Y (ed) Vascular lesions of collagen disease and related conditions. University Park Press, Baltimore, pp 143–148

123. Ishikawa K (1978) Natural history and classification of occlusive thromboaortopathy (Takayasu's disease). Circulation 57:27–35

124. Ito I, Saito Y, Nonaka Y (1975) Immunological aspect of aortic syndrome Jpn Circ J 39:459–462

125. Juergens JL (1980) Thromboangiitis obliterans (Buerger's disease). In: Jurgens JL, Spittell JA, Fairbairn JF (eds) Peripheral vascular diseases. Saunders, Philadelphia

126. Kato H, Ichinose E, Yoshioka F et al. (1982) Fate of coronary aneurysms in Kawasaki disease: serial coronary angiography and long-term follow up study. Am J Cardiol 49:1758–1766

127. Kato H, Koike S, Yamamoto M et al. (1975) Coronary aneurysms in infants and young children with acute febrile mucocutaneous lymph node syndrome. J Pediatr 86:892–898

128. Kawai C, Ishikawa K, Kato M et al. (1978) Pulmonary pulseless disease: pulmonary involvement in so-called Takayasu's disease. Chest 73:651–657

129. Kawasaki T, Kosaki F, Okawa S et al. (1974) A new infantile acute febrile mucocutaneous lymph node syndrome (MLNS) prevailing in Japan. Pediatrics 54:271–276

130. Kemp A (1977) Monozygotic twins with temporal arteritis and ophthalmic arteritis. Acta Ophthalmol (Copenh) 55:183–190

131. Kinare SG (1970) Aortitis in early life in India and its association with tuberculosis. J Pathol 100:69–76

132. Kinare SG, Kher YR, Rao G et al. (1976) Pattern of occlusive peripheral vascular disease in India. Angiology 27:165–180

133. Kjeldsen K, Mozes M (1969) Buerger's disease in Israel: investigation on carboxyhemoglobin and serum cholesterol levels after smoking. Acta Clin Scand 135:495–498

134. Klein RG, Campbell RJ, Hunder GG et al. (1976) Skip lesions in temporal arteritis. Mayo Clin Proc 51:504–510

135. Klein RG, Hunder GG, Stanson AW et al. (1975) Large artery involvement in giant cell (temporal) arteritis. Ann Intern Med 83:806–812

136. Knorring J Von, Somer T (1974) Malignancy in association with polymyalgia rheumatica and temporal arteritis. Scand J Rheum 3:129–135

137. Kummer A, Widmer LK, Da Silva A et al. (1977) Thromboangiitis obliterans, zhum Morbus Winiwarter-Buerger. Vasa 6:384–391

138. Kurozumi T, Tanaka K (1978) Buerger's disease: histology and pathogenesis. Vasc Surg 12:63–68

139. Lakhanpal S, Tani K, Lieg T et al. (1985) Pathologic features of Behcet's syndrome: a review of Japanese autopsy registry data. Hum Pathol 16:790–795

140. Lande A, Berkmen R (1976) Aortitis: pathologic, clinical and arteriographic review. Radiol Clin North Am 14:219–240

141. Lande A, La Porta A (1976) Takayasu arteritis. An

arteriographic pathological correlation. Arch Pathol Lab Med, 100:437–440

142. Lande A, Rossi P (1975) The value of total aortography in the diagnosis of Takayasu's arteritis. Radiology 114:287–297

143. Landing BH, Larson EJ (1977) Are infantile periarteritis nodosa with coronary artery involvement and fatal mucocutaneous lymph node syndrome the same? Comparison of 20 patients from North America with patients from Hawai and Japan. Pediatrics 59:651–662

144. Lehner T, Batchelor JR, Challacombe SJ, Kennedy L (1979) An immunogenetic basis for the tissue involvement in Behcet's syndrome. Immunology 37:895–900

145. Liang GC, Simkin PA, Hunder GG et al. (1974) Familial aggregation of polymyalgia rheumatica and giant cell arteritis. Arthritis Rheum 17:19–24

146. Liang GC, Simkin PA, Mannik M (1974) Immunoglobulins in temporal arteritis. An immunofluorescent study. Ann Intern Med 81:19–24

147. Lie JT, Brown AL, Carter ET (1970) Spectrum of aging changes in temporal arteries. Arch Pathol 90:278–285

148. Lindenberg R, Spatz H (1939) Liber die Thromboendoarteritis obliterans der Hirngelfasse: Cerebrae form der von Winiwarter Buergensehen Krankheit. Virchows Arch [A] 305:531–535

149. Lindsay I, Scholtz CL (1978) Recent presentation of neurological tuberculosis. J R Soc Med 71:223–227

150. Liozon F, Catanzano G (1982) L'artérite temporale de Horton. Etude anatomopathologique en microscopie électronique. A propos de 123 biopsies temporales. Rev Med Interne 3:295–301

151. Little AG, Zarins CK (1982) Abdominal aortic aneurysm and Behcet's disease. Surgery 91:359–362

152. Lowenstein MB, Bridgeford PH, Vasey FB (1983) Increased frequency of HLA-DR3 and DR4 in polymyalgia rheumatica giant cell arteritis. Arthritis Rheum 26:925–927

153. Lupi-Herrera E, Sanchez-Torres G, Horwitz S (1975) Pulmonary artery involvement in Takayasu's arteritis. Chest 67:69–74

154. Lupi-Herrera E, Sanchez-Torres G, Marcushamer J et al. (1977) Takayasu's arteritis; clinical study of 107 cases. Am Heart J 93:94–103

155. Maeda S, Murao S, Sugiyama T et al. (1983) Generalized sarcoidosis with "sarcoid aortitis". Acta Pathol Jpn 33:183–188

156. Malhotra KK, Sharma RK, Pradhakar S èt al. (1983) Aortoarteritis as a major cause of renovascular in the young. Indian J Med Res 77:487–494

157. Malmvall BE, Bengstsson BA, Kauser B et al. (1976) Serum levels of immunoglobulin and complement in giant cell arteritis. JAMA 236:1876–1878

158. Manigand G, Taillandier J, Benhamou A et al. (1982) Dissection aortique et artérite de Horton. J Mal Vasc. 7:217–220

159. McKenzie AH (1969) The polymyalgia rheumatica syndrome. Geriatrics 24:158–166

160. McKusick VA, Harris WS, Ottesen OE et al. (1962) Buerger's disease: a distinct clinical and pathologic entity. JAMA 181:93–100

161. McLoughlin GA, Helsby CR, Evans CC et al. (1976) Association of HLA-A9 and HLA-B5 with Buerger's disease. Br Med J ii:1165–1166

162. McNamara MF, Finnegan MO, Bakshi KR (1985) Abdominal aortic aneurysms infected by Escherichia coli. Surgery 98:87–91

163. McPherson JR, Juergens JL, Gilford RW JR (1963) Thromboangiitis obliterans and arteriosclerosis obliterans: clinical and prognostic differences. Ann Intern Med 59:288–296

164. Meadows SP (1966) Temporal or giant cell arteritis. Proc R Soc Med 32:329

165. Melish ME (1981) Kawasaki syndrome: a new infectious disease? J Infect Dis 143:317–324

166. Mendelowitz DS, Ramstedt R, Yao JST, Bergan JJ (1979) Abdominal aortic salmonellosis. Surgery 85:514–519

167. Mitchinson MJ (1970) The pathology of idiopathic retroperitoneal fibrosis. J Clin Pathol 23:681–689

168. Mitchinson MJ (1972) Aortic disease in idiopathic retroperitoneal and mediastinal fibrosis. J Clin Pathol 25:287–293

169. Mitchinson MJ (1984) Chronic periaortitis and periarteritis. Histopathology 8:589–600

170. Mitchinson MJ, Wight DGD, Arno J, Milstein BB (1984) Chronic coronary periarteritis in two patients with chronic periaortitis. J Clin Pathol 37:32–36

171. Moore JW, Reardon MJ, Cooley DA et al. (1985) Severe Takayasu's arteritis of the pulmonary arteries: report of a case with successful surgical treatment. J Am Col Cardiol 5:369–373

172. Morgan SH, Asherson RA, Hughes GRV (1984) Distal Aortitis complicating Reiter's syndrome. Br Heart J 52:115–116

173. Morooka S (1983) Experimental aortitis. Aortic lesions induced by a serine protease. Jpn Heart J 24:615–622

174. Morris PB, Imber MJ, Heinsimer JA et al. (1986) Rheumatoid arthritis and coronary arteritis. Am J Cardiol 57:689–690

175. Mozes M, Cahansky G, Doitsch W et al. (1970) The association of atherosclerosis and Buerger's disease: a clinical and radiological study. J Cardiovasc Surg 11:52–59

176. Myerowitz RL, Friedman R, Grossman WL (1971) Mycotic "mycotic aneurysm" of the aorta due to Aspergillus fumigatus. Am J Clin Pathol 55:241–246

177. Nakanishi T, Takao A, Nakazawa M (1985) Mucocutaneous lymph node syndrome: clinical, hemodynamic and angiographic features of coronary obstructive disease. Am J Cardiol 55:662–668

178. Nakao K, Ikeda M, Kimatas S et al. (1967) Takayasu's arteritis. Clinical report of 7 cases. Circulation 35:1141–1155

179. Namano F, Isohisa J, Kishi U et al. (1978) Takayasu's disease in twin sisters possible genetic factors. Circulation 58:173–177

180. Namano F, Isohisa I, Maezawa H et al. (1979) HLA antigens in Takayasu's disease. Am Heart J 98:153

181. Nasu T (1963) Pathology of pulseless disease. Angiology 11:225–242

182. Neumann MA (1952) Tuberculous lesions of the circulatory system: reported two cases. Am J Pathol 28:919–930

183. Numano F, Maezawa H, Sawada S et al. (1981) Circulating immune complexes in Takayasu's disease. Lack of evidence for a causative role. Arch Intern Med 141:162

184. O'Brien JP, Argyle JC (1981) The role of actinically provoked systemic elastolysis in polymyalgic vascular disease. Am J Dermatopathol 3:273–286

185. O'Duffy JD, Carney JA, Deodhar S (1971) Behcet's disease: report of 10 cases, 3 with new manifestations Ann Intern Med 75:561–570

186. Ogilvie AL, James PD, Toghill PJ (1981) Hepatic artery involvement in polymyalgia arteritica. J Clin Pathol 34:769–772

187. Oliveira DBG, Foale RA, Bensaid J (1984) Coronary aneurysms and Kawasaki's disease in an adult. Br Heart J 51:91–93
188. O'Leary M, Nollet DJ, Blomberg DJ (1977) Rupture of a tuberculous pseudo-aneurysm of the innominate artery into the trachea and esophagus: report of a case and review of the literature. Hum Pathol 8:458–467
189. Onouchi Z, Shimazu S, Kiyosawa N et al. (1982) Aneurysms of the coronary arteries in Kawasaki disease. Circulation 66:6–13
190. Östberg G (1973) On arteritis with special reference to polymyalgia arteritica. Acta Pathol Microbiol Scand [A] 237:1–59
191. Owyang C, Miller LJ, Lie JT et al. (1979) Takayasu's arteritis in Crohn's disease. Gastroenterology 76:825–828
192. Papaioannou CC, Hunde GG, McDuffie FC (1979) Cellular immunity in polymyalgia rheumatica and giant cell arteritis: lack of response to muscle or artery homogenates Arthritis Rheum 22:740–745
193. Papaioannou CC, Gupta RC, Hunder GG et al. (1980) Circulating complexes in giant cell arteritis and polymyalgia rheumatica. Arthritis Rheum 23:1021–1025
194. Papajiannis SP, Spina M, Cotte L (1970) Sequential degradation and phagocytosis of heterologous elastin. Arch Pathol 89:434–439
195. Park JR, Hazleman BL (1978) Immunological and histological study of temporal arteries. Ann Rheum Dis 37:238–243
196. Parker F, Healey LA, Wilske KR et al. (1975) Light and electron microscopic studies on human temporal arteries with special reference to alterations related to senescence, atherosclerosis and giant cell arteritis. Am J Pathol 79:57–69
197. Patel S, Johnston KW (1977) Classification and management of mycotic aneurysms. Surg Gynecol Obstet 144:691–694
198. Paulley JW (1980) Coronary ischaemia and occlusion in giant cell (temporal) arteritis. Acta Med Scand 208:257–263
199. Paulus HE, Pearson CM, Pitts W Jr (1972) Aortic insufficiency in five patients with Reiter's syndrome: a detailed clinical and pathologic study. Am J Med 53:464–472
200. Pearson CM, Kline HM, Newcomer VD (1960) Relapsing polychondritis. N Engl J Med 263:51–58
201. Philippe JM, Lavallard JF, Veyssier P (1978) Atteinte hépatique au cours de la maladie de Horton. Deux cas. Nouv Press Med 7:1118–1119
202. Plouvier B, Francois M, Wattre P et al. (1978) Etude en immunoflourescence directe sur coupes d'artères temporales: intérêt et limites. Nouv Press Med 7:1719–1721
203. Pounder DJ (1985) Coronary artery aneurysms presenting as sudden death 14 years after Kawasaki disease in infancy. Arch Pathol Lab Med 109:874–876
204. Reid JVO (1957) Dilatation of the aorta due to granulomatous giant cell arteritis. Br Heart J 19:206
205. Reinecke RD, Kuwabara T (1969) Temporal arteritis. I. Smooth muscle cell involvement. Arch Ophthalmol 82:446–453
206. Restrepo C, Tejeda C, Correa P (1969) Non syphilitic aortitis. Arch Pathol 87:1–12
207. Reza MJ, Demanes DJ (1978) Behcet's disease: a case with hemoptysis, pseudotumor cerebri, and arteritis. J Rheumatol 5:320–326
208. Rob CG, Eastcot HHG (1955) Aortic aneurysm due to tuberculous lymphadenitis. Br Med J 1:378–379
209. Roberts WC, Wibin EA (1966) Idiopathic panaortitis supra-aortic arteritis granulomatous myocarditis and pericarditis. Am J Med 41:453–461
210. Robinowitz M, Virmani R, Mc Allister HA Jr (1980) Rheumatoid heart disease: a clinical and morphologic analysis of thirty-four autopsy patients. Lab Invest 42:145–147 (abstract)
211. Rodnan GP, Benedek TG, Shaver JA. Fennell RH Jr (1964) Reiter's syndrome and aortic insufficiency. JAMA 189:889–894
212. Rose AG, Dent DM (1981) Inflammatory variant of abdominal atherosclerotic aneurysm. Arch Pathol Lab Med 105:409–413
213. Rose AG, Halper J, Factor SM (1984) Pulmonary arteriopathy in Takayasu's disease. Arch Pathol Lab Med 108:644–648
214. Rose AG, Sinclair-Smith CC (1980) Takayasu's arteritis: a study of 16 autopsy cases. Arch Pathol Lab Med 104:231–237
215. Rosen N, Gaton E (1972) Takayasu's arteritis of coronary arteries. Arch Pathol 94:225–229
216. Rosen N, Sommer I, Knobel B (1985) Intestinal Buerger's disease. Arch Pathol Med Lab 109:962–963
217. Rosenthal M, Muller W, Albert ED et al. (1975) HLA antigens in polymyalgia rheumatica. N Engl J Med 292:595–596
218. Roth LM, Kissane JM (1964) Panaortitis and aortitic valvulitis in progressive systemic sclerosis (scleroderma). Am J Clin Pathol 41:287–296
219. Russo JV, Esterly JR (1967) Retroperitoneal and periarterial fibrosis. Arch Pathol 83:396–398
220. Sasaguchi Y, Kato H (1982) Regression of aneurysms in Kawasaki disease a pathological study. J Pediatr 100:225–231
221. Sauerbruch T, Stuhlinger B, Kaess H (1973) Immunohistogische untersuchungen bei einem fall von riesenzellarteritis (arteritis temporalis). Dtsch Med Wochenschr, 98:283–284
222. Savory WS (1956) Case of young women in whom the main arteries of both helper extremities and of the left side of the neck were throughout completely obliterated. Med Clin Trans (Lond) 39:205–211
223. Schimizu K, Sano K (1951) Pulseless disease. J Neuropathol Clin Neurol 1:37–47
224. Schrire V, Asherson RA (1964) Arteritis of the aorta and its major branches. Q J Med 33:439–463
225. Schwartz CJ, Mitchell JRR (1962) Cellular infiltration of the human arterial adventitia associated with atheromatous plaques. Circulation 26:73–78
226. Self J, Hammarsten JF, Lyne B, Peterson DA (1967) Relapsing polychondritis. Arch Intern Med 120:109–112
227. Sen PK, Kinare SG, Kelkar MD et al. (1973) Non specific aorto-arteritis. A monograph based on a study of 101 cases. Tata McGraw-Hill, New Delhi
228. Serra RM, Engle JE, Jones RE, Schoolwerth AC (1980) Perianeurysmal retroperitoneal fibrosis. Am J Med 68:149–153
229. Sheehan JP (1983) Bacteroides aortitis and aneurysm formation following arteriography. J Infect 7:153–155
230. Shimizu T, Ehrlich GE, Inaba G, Hayashi K (1979) Behcet disease (Behcet syndrome). Semin Arthritis Rheum 8:223–260
231. Shionoya S, Ban I, Nakata Y et al. (1974) Diagnosis, pathology and treatment of Buerger's disease. Surgery 75:695–700
232. Shionoya J, Ban I, Nakata Y et al. (1976) Vascular reconstruction in Buerger's disease. Br J Surg 63:841–846

233. Silbergleit A, Arbulu A, Defever BA, Nedwicki EG (1965) Tuberculous aortitis – surgical resection of ruptured abdominal false aneurysm. JAMA 193:333–335

234. Smith KR (1969) Electron microscopy of giant cell (temporal) arteritis. J Neurol Neurosurg Psychiatr 32:348–353

235. Smolen JS, Yougchaiyd U, Weidinger P et al. (1978) Autoimmunological aspects of thromboangeitis obliterans (Buerger's disease). Clin Immunol Immunopathol 11:168–177

236. Sommers SC, Higgins TE, Kimelblatt BJ (1984) Chronic aortitis following furosemide therapy. Arch Pathol Lab Med 108:293–294

237. Stanson AW, Klein RG, Hunder GG (1976) Extracranial angiographic findings in giant cell (temporal) arteritis. AJR 127:957–963

238. Sutherland JC, Mackham RV, Mardiney MR (1974) Subclinical immune complexes in the glomeruli of kidneys postmortem. Am J Med 57:536–541

239. Swinson DR, Goodwill CJ, Talbot IC (1976) Giant cell arteritis presenting as subclavian artery occlusion. A report of two cases. Post Grad Med J 52:525–529

240. Takahashi T, Sakave M, Nagasawa S et al. (1979) A case of aortitis syndrome with focal glomerulonephritis. Jpn J Nephrol 21:165–174

241. Takayasu M (1908) Case with unusual changes in the central vessels in the retina. Acta Soc Ophthalmol Jpn 12:554–555

242. Tanaka N, Sekimoto K, Naoe E (1976) Kawasaki disease. Relationship with infantile periarteritis nodosa. Arch Pathol Lab Med 100:81–86

243. Teoh PC, Ton LKA, Chia BL et al. (1978) Non specific aorto-arteritis in Singapore with special reference to hypertension. Am Heart J 95:683–690

244. Thomas RD, Croft DN (1974) Thyrotoxicosis and giant cell arteritis. Br Med J i:408

245. Thompson JR (1966) Vascular changes in sarcoidosis. Chest 50:357–361

246. Thompson JR, Simmons CR, Smith LL (1971) Polymyalgia arteritica with bilateral subclavian artery occlusive disease. A case report. Radiology 101:595–596

247. Tucker CR, Fowles RE, Calin A, Popp RL (1982) Aortitis in ankylosing spondylitis: early detection of aortic root abnormalities with two dimensional echocardiography. Am J Cardiol 49:680–686

248. Ueda H, Saito Y, Ito I et al. (1967) Immunological studies of aortitis syndrome. Jpn Heart J 8:4–18

249. Ueda H, Sugiura M, Ito I et al. (1967) Aortic insuffiency associated with aortitis syndrome. Jpn Heart J 8:107–120

250. Van Den Oord JJ, De Wolf Peeters C, Pacchetti F et al. (1984) Cellular composition of hypersensitivity type granuloma. Hum Pathol 15:559–565

251. Vinijchaikul K (1967) Primary arteritis of the aorta and its main branches (Takayasu's arteriopathy). A clinicopathologic autopsy study of 8 cases. Am J Med 43:15–27

252. Virmani R, McAllister HA Jr (1986) Pathology of the aorta and major arteries. In: Lande A, Berkmen YM, McAllister HA (eds) Aortitis: clinical, pathologic and radiologic aspects. Raven Press, New York pp 7–53

253. Volini FI, Olfield C Jr, Thompson JR, Kent G (1962) Tuberculosis of the aorta. JAMA 181:78–83

254. Walker DI, Bloor K, Williams G (1972) Inflammatory aneurysms of the abdominal aorta. Br J Surg 59:609–614

255. Warlow R, Vernea S, Mijc A (1984) Angioim-munoblastic lymphadenopathy associated with large-muscle arteritis. Hum Pathol 15:486–490

256. Wessler S, Ming SC, Guerewich V et al. (1960) A critical evaluation of thromboangiitis obliterans. The case against Buerger's disease. N Engl J Med 262:1149–1160

257. Wiest JW, Traverso LW, Dainko EA et al. (1980) Atrophic coordination of the abdominal aorta. Ann Surg 191:224–227

258. Wiggelinkhuizen J. Cremin BJ (1978) Takayasu arteritis and renovascular hypertension in childhood. Pediatrics 62:209–217

259. Wilkes KR, Healey LA (1967) Polymyalgia rheumatica. A manifestation of systemic giant cell arteritis. Ann Intern Med 66:77–86

260. Wilkins RF, Arnett FC, Bitter T (1983) Reiter's syndrome: evaluation of preliminary criteria for definite disease. Bull Rheum Dis 32:31–34

261. Wilkinson MS, Russell RWR (1972) Arteries of the head and neck in giant cell arteritis. Arch Neurol 27:378–391

262. Wilson SE, van Wagenen P, Passaro E Jr (1978) Arterial infection. Curr Probl Surg 40:6–89

263. Wilson SE, Gordon HE, van Wagenen PB (1978) Salmonella arteritis. A precursor of aortic rupture and pseudoaneurysm formation. Arch Surg 113:1163–1166

264. Wilson WA, Morgan OS, Bain B et al. (1979) Takayasu's arteritis: association with Still's disease in an adult. Arthritis Rheum 22:684–688

265. Winocour PH, Williams GR, Boyd JF, Kennedy DH (1983) Septic arteritis complicating salmonellosis. Br Med J 287:972–973

266. Yutani C, Go S, Kamiya T et al. (1981) Cardiac biopsy of Kawasaki disease. Arch Pathol Lab Med 105:470–473

267. Zak FG, Strauss L, Saphra I (1958) Rupture of diseased large arteries in the course of enterobacterial (salmonella) infections. N Engl J Med 258:824–828

268. Zilko P, Currey HLF, Vernon-Roberts B (1977) Polymyalgia rheumatica (PMP) and giant cell arteritis: lack of response of peripheral blood lymphocytes to arterial wall homogenate. Ann Rheum Dis 36:286–287

269. Zillervelo GE, Ferrer P, Garcia OL et al. (1978) Takayasu's arteritis associated with glomerulonephritis. Am J Dis Child 132:1009–1013

270. Zulch KJ (1969) The cerebral form of Von Winiwarter-Buerger's disease. Does it exist? Angiology 20:61–69

271. Zvaifler NJ, Weintraub AM (1963) Aortitis and aortic insufficiency in the chronic rheumatic disorders: a reappraisal. Arthritis Rheum 6:241–245

Principles of Diagnosis and Therapeutic Approach

J. N. Fiessinger and J. M. Cormier

Takayasu's Disease

First described because of the involvement of the aortic arch, Takayasu's disease has been considered subsequently as a non-specific inflammatory involvement of the aorta and its branches [275,333,338,349]. The histopathological pattern enables arterial biopsy to play an important role in the diagnostic approach [291,292,306,339] (see above).

Principles of Diagnosis

Clinical Approach

Takayasu's disease is clinically characterized by two phases: the early pre-occlusive phase which is accompanied by constitutional symptoms, and the late occlusive phase, exhibiting visceral and peripheral consequences of stenotic and ectatic arterial lesions of the aorta and its branches.

The Early Pre-occlusive Phase. Constitutional symptoms often occur in adolescents [306, 333, 346]. They include fever, weight loss, myalgias, arthralgias and pleural–pericardial effusions, suggesting infections or rheumatic or connective tissue disease rather than arteriopathy. Spontaneous or palpation-induced pains along the affected arteries are sometimes relevant. However, the diagnosis is rarely suspected during this phase, and these symptoms are retrospectively related to arterial disease when the arteriopathy is obvious.

The Occlusive Phase. The consequences of the arterial lesions usually suggest the diagnosis of Takayasu's disease [294,306,333,349]. Hypertension, frequently of renovascular origin, is a common clinical presentation. In the upper limbs, the involvement of axillary and subclavian arteries may be revealed by unequal blood pressure measurements in the two arms, the absence of a humoral pulse, subclavian bruit, acquired unilateral Raynaud's phenomenon or exertional ischaemia. Intermittent claudication of the lower limbs is infrequent and usually not severe. The involvement of the cervicoencephalic arteries remains asymptomatic for a long period of time. If both vertebral and carotid arteries are involved cerebral circulatory insufficiency may occur, revealed by retinal ischaemia, dizziness and diplopia. There is a relationship between the lowering of systemic retinal pressure and the severity of the carotid and vertebral involvement [316]. The retinal involvement has been classified in four stages: stage 1, dilatation of retinal vessels; stage 2, microaneurysms; stage 3, arteriovenous anastomoses; stage 4, ocular complications [314].

In young patients an incidental finding of an aneurysm favours an inflammatory or dysplastic arteriopathy. Angina or aortic insufficiency may indicate Takayasu's disease [272,355]. However, they exceptionally occur in isolation from arterial involvement in other sites. In children sudden cardiac insufficiency may be the initial clinical presentation [333].

Finally, Takayasu's disease is characterized by the following findings [292,294]: prominence in women; onset of the disease before 40 years; pre-occclusive phase with rheumatic or systemic symptoms; previous history of tuberculous or streptococcal infection; and segmental arterial involvement often in several sites.

Complementary Exploration

Vascular imaging often allows the diagnosis of Takayasu's disease, ruling out atherosclerosis.

Ultrasound scanning associated with Doppler yields a satisfactory first assessment of the extent of arterial lesions, allowing the choice of later angiographic techniques. It also provides qualitative data about arterial lesions, showing arterial wall thickening without endoluminal plaque [294].

Angiography is essential in the diagnostic approach [327,328]: in order to explore all the involved sites, the whole aorta and its branches must be visualized. Digitalized angiography makes the angiographic exploration easier. Intravenous computerized angiography [324] and an ultrasound scan and Doppler are a sensible approach in out-patients. However, computerized angiography following intra-arterial injection is often necessary before elaborate surgical treatment [287,344]. The angiographic exploration includes (in the same procedure) the examination of the aortic arch and its cervical branches, the whole aorta, renal and digestive branches, iliac arteries and pulmonary arteries. (Fig. 28.29).

Takayasu's disease is strongly suggested by some angiographic patterns such as: (a) the association of ectatic and stenotic aortic lesions; (b) long subdiaphragmatic aortic coarctation [326,328] and (c) involvement of the post-vertebral subclavian artery extending to the carotid and the axillary artery [294]. Moreover the involvement of several arterial sites is very characteristic. Thus the association of aortic and large branch involvement with pulmonary artery lesions is almost pathognomonic for Takayasu's disease [294].

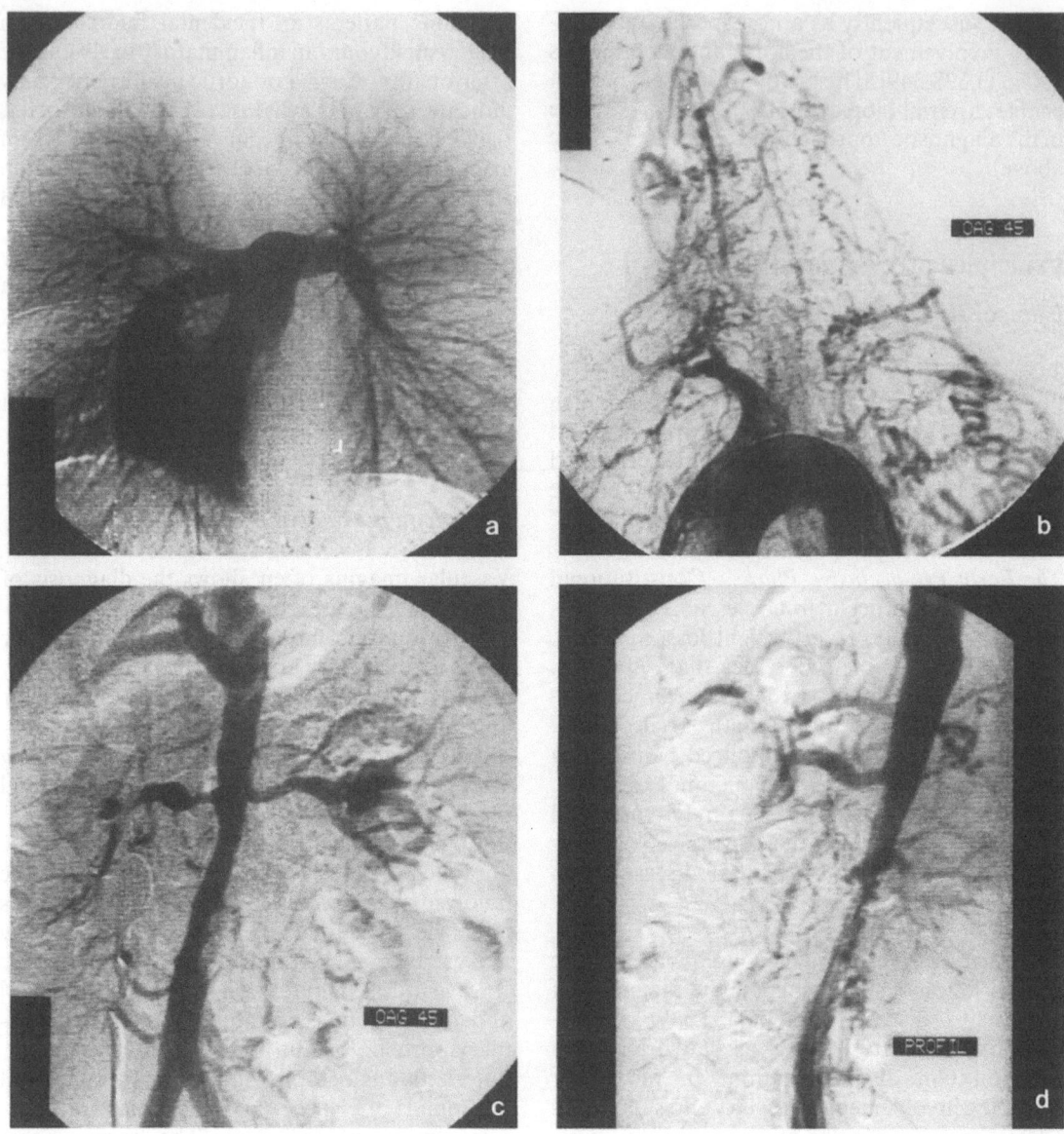

Fig. 28.29 a–e.

Laboratory Findings

Laboratory findings play a minor role in the diagnosis of Takayasu's disease. The inflammatory syndrome is not specific and is not present in half the cases. Anti-aorta antibodies [338,349] are inconstant and have no diagnostic value. There is no specific biological marker in Takayasu's disease.

Pathological Examination

In most cases clinical and angiographic patterns are highly suggestive of Takayasu's disease.

However, this is not true in some instances when arterial involvement is isolated, e.g. segmental subclavian artery lesions. In these cases a surgical arterial biopsy is indicated for diagnostic purposes [292].

Therapeutic Approach

The treatment of Takayasu's disease depends on assumptions about physiopathological hypotheses and the risks of evolving arterial lesions.

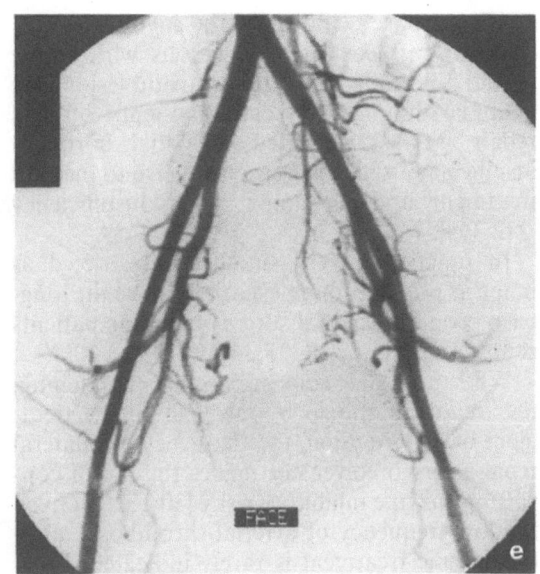

Fig. 28.29 a–e. Exploration of Takayasu's disease using computerized angiography (courtesy Prof. J. C. Gaux, Vascular Radiology, Hôpital Broussais, Paris, France) Methods: Through a peripheral vein, a 5 F catheter is placed in the right atrium where 60 cc of Hexabrix is injected. Two series of pictures are taken of the pulmonary arteries. Through the femoral artery a 5 F catheter is placed in the aorta and five series of pictures are taken of the aortic arch (front and lateral views), the abdominal aorta (front and anteroposterior views) and the iliac arteries (total injection of 200 cc of 50% diluted Hexabrix). **a** Absence of pulmonary artery lesion. **b** Proximal occlusion of the left carotid and subclavian arteries below the right innominate artery. The vertebral arteries are occluded but are injected through retrograde extracranial circulation. **c–d** Stenosis of the abdominal aorta involving the renal arteries. Occlusion of the superior mesenteric artery. **e** Absence of iliac artery lesion

Physiopathological Treatment

This treatment is based on immunosuppressive therapy and on the treatment of possibly associated conditions. Systemic symptoms, an inflammatory syndrome and some animal models of the disease [317] support a dysimmune mechanism in Takayasu's disease. This leads to the use of corticosteroids. In some cases, previously absent pulses recover following corticosteroid use [296,298]. A prospective study of 20 patients with evolving Takayasu's disease showed improvement of the prognosis with prednisone (1 mg/kg) followed by cyclophosphamide (2 mg/kg) if corticosteroids failed [345]. Nevertheless, given the uncertainty of the dysimmune hypothesis, the inconstant improvement with corticosteroids [333,339,349] and their side-effects (especially on blood pressure), the use of corticosteroids should be restricted to patients without contraindications in those cases with progressive disease and the inflammatory syndrome. Tuberculosis may be related to Takayasu's disease [294,333,349]. Anti-tuberculous therapy is therefore indicated in cases of Takayasu's disease, and is widely recommended if corticosteroids are used. However, improvement of Takayasu's disease following corticosteroid therapy has never been demonstrated. Inflammatory rheumatic disease and streptococcal infection may be associated with aortoarteritis. However, the efficiency of their specific treatment is unknown.

Symptomatic Treatment

The treatment of arterial lesions or their consequences may be medical or surgical. The indications are based on the natural history and risks of Takayasu's disease [314].

The surgical procedure must take into acccount the perivascular fibrosis which makes endarterectomy impossible, and the frequency of aortic involvement which hampers straight arterial reimplantation. Thus vascular prostheses are widely indicated. Surgical treatment should be performed outside the inflammatory and evolving phase of Takayasu's disease. Large or evolving aortic aneurysms should be treated with prosthetic replacement which is rarely indicated in aortic occlusions and stenoses. Renal artery lesions are usually operated on in cases of severe hypertension which are non-responsive to anti-hypertensive drugs. Lagneau and Michel [325] reported 12 of 18 patients recovering after a maximum 7 years follow-up. In some cases, renal artery lesions were treated during a surgical procedure for aortic aneurysm. Surgery for protection of renal function is indicated only in patients with severe proximal stenosis of the renal artery and a single functioning kidney. Besides abdominal aortic prosthetic replacement, surgical treatment of digestive artery lesions is only indicated if mesenteric ischaemia is evident. Surgical treatment of aortic arch branch lesions is not clearly defined [316], except for surgical biopsy of the subclavian artery for diagnostic purposes. However, the involvement of the four cervicoencephalic vessels with severe retinal ischaemia, and patent carotid lesions are an indication for surgical treatment. Surgical repair is also indicated in ostial coronary insufficiency which is non-responsive to medical treatment and in major aortic insufficiency [272,355].

In practice surgery should be restricted as much as possible, more especially since the long-term prognosis of that surgery in young patients is not clearly known [325,349].

When arterial lesions are not life-threatening, the treatment should be medical. In the treatment of hypertension, the frequency of bilateral renal artery involvement makes the use of converting enzyme inhibitors risky [304,325]. Given the low frequency of arterial thrombosis, antithrombotic treatment is rarely indicated. Pregnancy does not change the evolution of Takayasu's disease but increases the risk of decompensation of hypertension [315].

Temporal Arteritis

Temporal arteritis is a well-defined disease within the spectrum of arteritides. Its incidence is evaluated at 2.4 per 100 000 in Minnesota [313] but is as high as 9.3 per 100 000 persons and 28.6 per 100 000 persons older than 50 years in a Swedish study [282]. Its diagnosis depends on well-characterized lesions of the temporal artery [276,301,308], but recent studies emphasized atypical clinical presentations in which temporal involvement is absent or minor [299,309, 329,343].

Clinical Diagnosis

Temporal arteritis is usually associated with temporal symptoms and polymyalgia rheumatica [301,308]. These signs can be isolated or variously associated, yielding multiple or atypical clinical presentations.

Temporal symptoms are the usual presenting signs [282,301,313] and may be diagnostic. Frontal headaches are often unilateral with a burning sensation. They are increased by pressure (touch, glasses, hat). The temporal artery appears tender, nodular, painful, pulseless. Jaw claudication is present in nearly one-half of the

cases and is characteristic of the disease [301,309].

Polymyalgia rheumatica is the most frequent manifestation [282,301]. Pains are prominent in the limb girdles and often start abruptly. They are symmetrical and muscular rather than articular. They appear as morning stiffness, hampering standing up rather than the fine movements in dressing [308]. Physical examination finds mild pains of shoulders and hips, while muscular force testing, electromyogram and muscular enzymes are normal.

Ocular involvement is the main complication of temporal arteritis. It can be an early and solitary sign [318]. Visual loss is sudden and irreversible. Without treatment, it often becomes bilateral: in one-third of cases contralateral involvement occurs within the 24 h [318]. Visual loss is related to ischaemia of the optical nerve. Rarely the retinal artery is involved. Transient symptoms, such as amaurosis or diplopia, may precede blindness [282,318].

Involvement of other arteries occurs in 10% of the cases [322]. The subclavian artery is the most frequent site, often bilaterally [322,345]. It can be manifest by Raynaud's phenomenon or claudication of the upper extremity. Aortic involvement may be complicated by aortic dissection, rupture of which is a cause of death during the course of the disease [303,322]. Lower extremities are less frequently involved and the lesions are located above the knees. Involvement of the carotids, vertebral [359], coronary and mesenteric arteries has been described [354].

Atypical manifestations are present in 30%–40% of the histologically proved cases [309,318]: these include fever [299], weight loss, inflammatory anaemia and respiratory signs [329,343]. In such cases the differential diagnosis is neoplasia especially as ESR and alkaline phosphatases may be elevated [309].

Other Investigations

Determination of the erythrocyte sedimentation rate (ESR) is the first step in the diagnosis. The ESR is exceptionally normal [282,308,318,319] but usually is above 50 mm in 1 h. The elevation is not related to the duration of the disease nor to the presence of giant cells in biopsies [318].

Angiography is useful when large arteries are involved. It shows signs suggesting an inflammatory arterial disease [345,354] such as long, smooth stenoses, arterial obliterations and the absence of atherosclerosis. On the subclavian

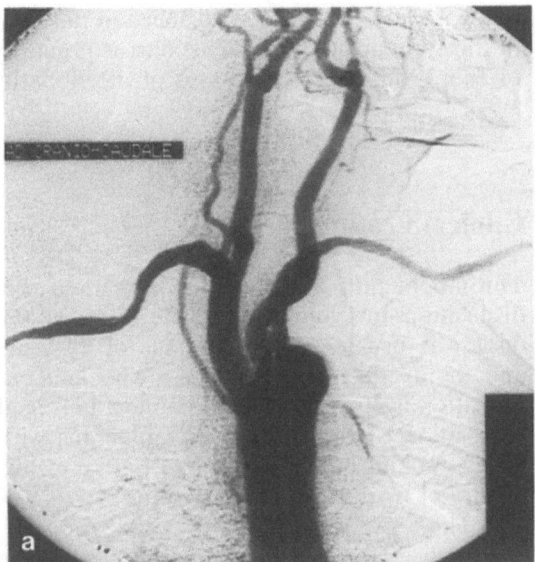
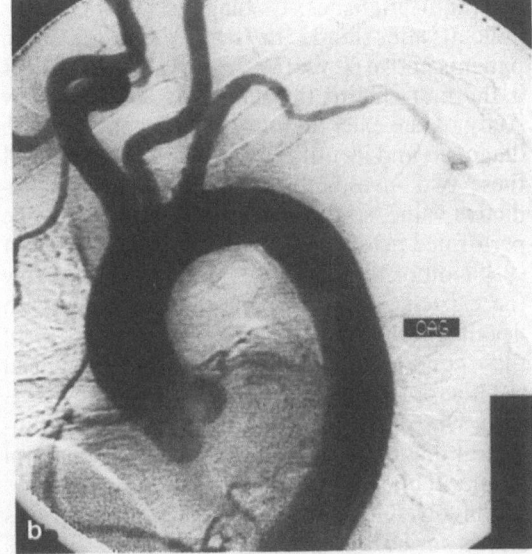
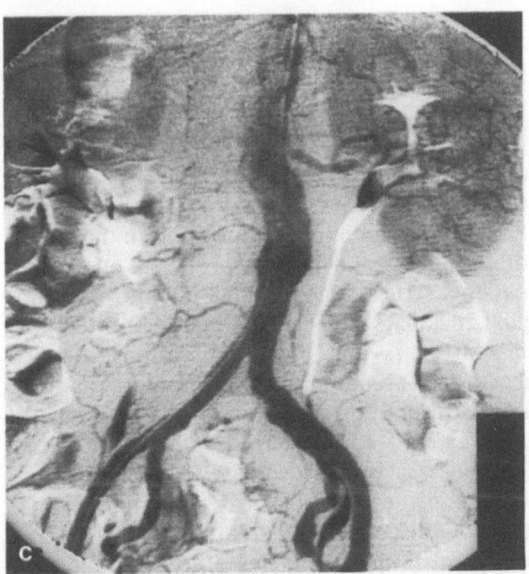

Fig. 28.30 a–c. Digitalized angiogram in temporal arteritis presenting with ischaemia of both upper extremities. (Courtesy Prof. J. C. Gaux. Vascular Radiology, Hôpital Broussais, Paris, France.) **a** Long and smooth stenoses of both subclavian arteries. The carotids are normal. **b** Occlusion of the left axillary artery. **c** Atheromatous lesions of the abdominal aorta and iliac arteries

arteries, lesions are located below the origin of the vertebral arteries and extend along the axillary arteries (Fig. 28.30). Selective angiography of the external carotid shows stenoses on its branches [300,337], but this pattern is not specific for temporal arteritis so that this aggressive exploration is no longer indicated.

Doppler flow studies can investigate internal and external carotids and temporal and ophthalmic arteries [278,331,358]. There is good correlation between perturbations of Doppler flow and temporal biopsy data. However, a normal Doppler flow study can be associated with typical lesions at biopsy.

HLA-DR4 antigen is significantly prevalent in temporal arteritis [277]. This supports the hypothesis that an immunological disorder is implicated in the mechanism of the disease, but it has no diagnostic value. Nor has an elevated factor VIII antigen [290].

Temporal biopsy is the key test for diagnosis [330,341]. As arterial involvement is focal, a long biopsy specimen is recommended [321]. Lesions are bilateral in 90% of the cases [321] allowing the temporal biopsy to be unilateral. If it is nega-

tive in a highly suggestive clinical background, a second biopsy should be taken from the opposite artery [330]. A retrospective study including 132 patients [274] shows a significant increase of negative results if the biopsy is performed after the corticosteroid therapy is started. Moreover this study suggests that the histological lesions respond quickly to this treatment.

Strategy of Exploration

Temporal arteritis is a generalized disease exhibiting various clinical patterns and involving several arterial sites. Its diagnosis is sometimes difficult and should be readily considered in patients above 60 years of age. An elevated ESR is the first clue to the diagnosis. Doppler flow study of the external carotid may contribute to diagnosis and identify the biopsy site. Neither of these two investigations has an absolute predictive value, so that temporal biopsy has to be performed readily in cases when there is the least suspicion of this diagnosis. All the studies show the existence of true temporal arteritides with negative temporal biopsies [282,301,313]. In these conditions some authors proposed clinical criteria for the diagnosis [284,341] allowing treatment when biopsy is negative. However, pathological diagnosis should be the rule. If bilateral temporal biopsies are negative and large arteries are involved, the need for a definite diagnosis allows surgical biopsies of large vessels such as subclavian arteries [291,345].

Treatment

Temporal arteritis needs initial high doses of corticosteroids. Treatment must be started rapidly to prevent ocular involvement [282,318]. However, it does not seem to shorten the course of the disease [280]. The prognosis in treated patients series is the same as in control groups [280,282,303,313,318]. The treatment often lasts several years and the complications of corticosteroids are important [297]. Temporal arteritis is usually treated with initial doses of 1 mg/kg/d of corticosteroids for 6 to 8 weeks. The doses are then decreased to 5 mg/d every 15 d, under control of ESR. For maintenance therapy, discontinuous treatment can be given every 2 d [281]. In the cases necessitating prolonged high doses or presenting side-effects of corticosteroids dapsone [284], immunosuppressive and antimalarial agents have been proposed to decrease corticosteroid doses [276].

Buerger's Disease

The pathological pattern of Buerger's disease, an inflammatory thrombo-occlusive vascular disease involving small- and medium-sized arteries and veins [285,323,334,348,353], is of limited diagnostic value. The anatomical lesion, as well as clinical, biological or arteriographic data, are non-specific [293,295,302,320,360] and the diagnostic approach to the patient consists primarily of ruling out the other causes of arteriopathy [307].

Clinical Features

Thromboangiitis obliterans has a marked predisposition for young men. The onset of the disease is usually around the age of 20 years and the average age of patients who undergo operation is 40 years. There is almost universal association of thromboangiitis obliterans with heavy smoking but it is not rare to find a previous history of frostbite or trauma to the extremities [310,340].

Superficial thrombophlebitis may occur as the first clinical manifestation. It most commonly involves the small veins and usually develops as "phlebitis saltans" [283] independently of possible varicose veins. The venous lesions are distinctly segmental, migratory and recurring, and can occur anywhere in the arm or leg. They may be diagnosed as an erythema nodosum until the biopsy has shown the venous nature of the subcutaneous inflammatory nodule. Thrombophlebitis of deep veins occurs infrequently.

Arterial involvement of the lower limits is distal (infrapopliteal). The clinical manifestation may be a sense of numbness of the foot after exercise, caused for example by intermittent claudication of the arch of the foot [311]. Most frequently painful trophic changes develop early in the course of the disease [310]. Confirmation of the diagnosis may be difficult. Distal pulsations may be present [340] and are more often abolished asymmetrically, raising the question of the significance of an isolated impaired distal pulsation [289]. The presence of functional manifestations of impaired arterial circulation, the bilaterality of the changes and the ischaemic nature of the trophic changes of the foot (pulpar atrophy, shrinkage of the heel and delayed return of colour to the skin) are good evidence for the diagnosis of arteriopathy.

An associated finding of involvement of the upper extremities is of distinctive value for the diagnosis. Almost constant in arteriopathy [335], it presents clinically as Raynaud's phenomenon, often asymmetrically associated with a positive Allen test and manifestations of permanent ischaemia. The presence of digital gangrene or ulceration in a young man favours the diagnosis of Buerger's disease [355]. Other arterial loca-

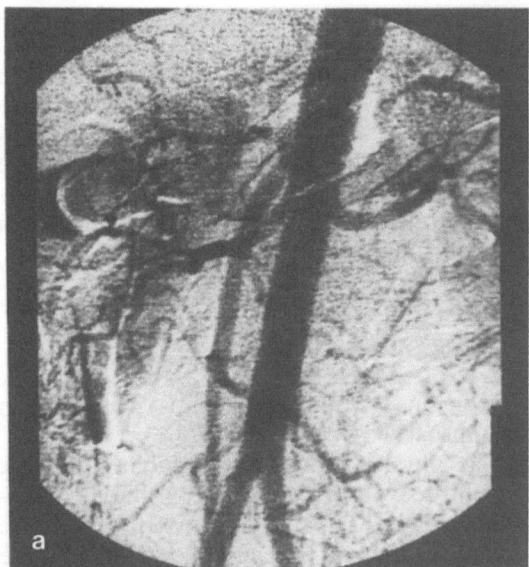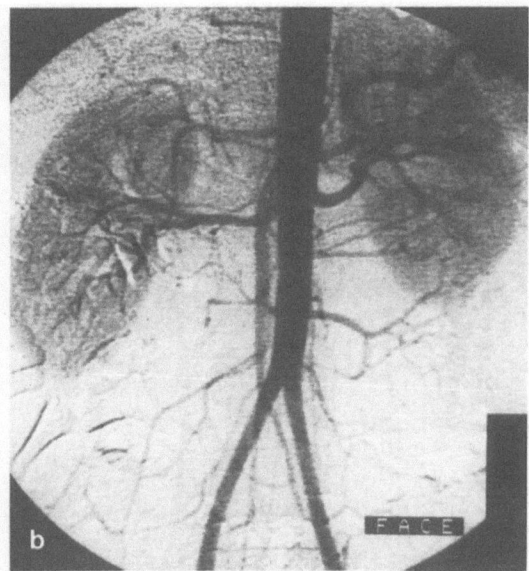

Fig. 28.31. Digitalized angiogram of the aorta and the iliac arteries (front and lateral views). The aortic bifurcation, as well as the renal, coeliac and superior mesenteric arteries are not significantly altered in their proximal segments. (Courtesy Prof. J. C. Gaux, Hôpital Broussais, Paris, France)

lizations, including notably involvement of mesenteric vessels [282,342,345], can be observed but are very infrequently the initial manifestation of the disease. The diagnosis is mainly supported by recognizing peripheral arterial and venous involvement.

Special Investigations

Doppler study is of limited interest. Distal pressures are normal or slightly decreased and ultrasound imaging confirms the absence of atherosclerotic plaques. Measurement of blood pressure in the digital arteries shows a gradient between ankle and toe [286]. Transcutaneous measurement of the partial pressure of oxygen allows assessment of the severity of ischaemia [357].

Investigations are needed to check the absence of heart lesions which are a source for arterial embolism [273,293]. Anamnesis, physical examination, electrocardiography and ultrasound imaging are required to complete the diagnostic approach.

The absence of arterial calcification on roentgenograms is classically of diagnostic value [285,336]. Actually vascular opacification requires to be systematically performed. Aortography is needed to eliminate a possible arterial source of distal embolism, i.e. aneurysms [293],

so correct visualization of the arteries of the lower extremities is required, particularly the popliteal artery and the arteries of the leg down to the plantar arteries. The arteriographic appearances are non-specific (Figs. 28.31 and 28.32) [340,343], i.e. there is an absence of ectasic lesions and atherosclerotic stenoses with abrupt changes from normal to involved arterial segments in one or several arteries and a corkscrew aspect of fine collateral vessels. In young patients, digitalized arteriography with aortic catheter placement is of great value in evaluation of the arterial condition.

Usual blood tests yield no information concerning diagnosis and treatment of Buerger's disease [293]. In the absence of extended trophic changes, there is no inflammatory syndrome and the erythrocyte sedimentation test remains normal. There is neither diabetes nor hyperlipidaemia. The value of tests for hypercoagulability [288] has not been confirmed. No anti-nuclear antibodies can be found; complement and immunoglobulin levels can be increased [283,305] but are often in the normal range [352]. Cell-mediated sensitivity to types I and III collagens has been found recently to be significantly increased [273,283,305,352]. Such abnormalities of cellular immunity can be associated with significant levels of anti-collagen antibodies and seem to be linked to the activity of the disease. They are not observed in patients

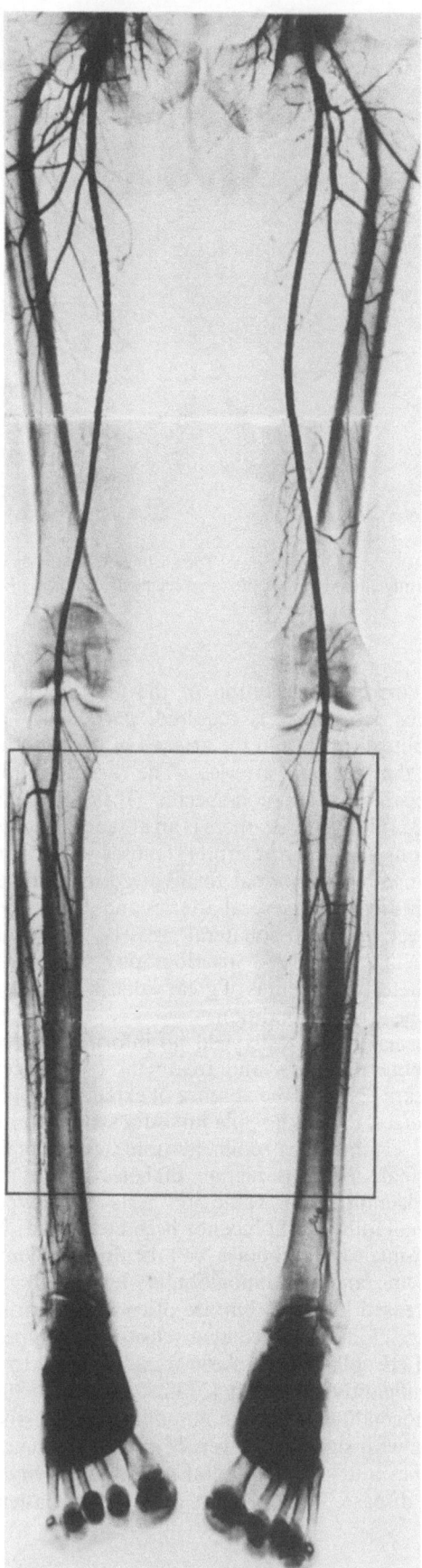

with atherosclerosis, but their specificity and their putative role in other kinds of arteritides have not been established [353]. The increased prevalence of HLA-A9 and HLA-B5 antigens and the absence of HLA-B12 [335] may be considered as predisposing factors but are of little practical help in the diagnosis. Arterial biopsy is usually out of the question in the patient with thromboangiitis obliterans.

Strategy of Diagnosis

The diagnosis of thromboangiitis obliterans can be approached in two stages.

The physician must first establish that the patient does not have diabetes mellitus, hyperlipidaemia or haematological disorders. Cardiac investigation must have established the absence of heart disease as a source of arterial embolism and the clinical and biological examination must have ruled out a suspicion of collagenous disease (the differential diagnosis of Behcet's disease can be difficult). Physical examination and various investigations, notably aortography, do not

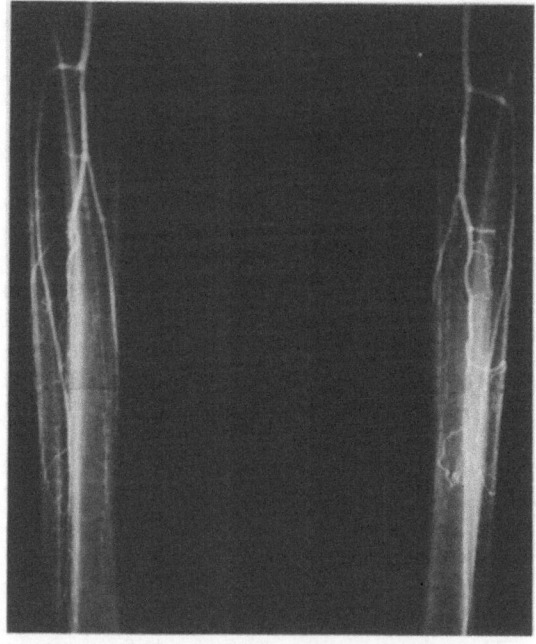

Fig. 28.32. Bifemoral arteriography. The iliac, femoral and popliteal arteries appear smooth and of even calibre. Bilateral obliteration of the three arterial axes of the leg; there is an extensive network of collateral vessels with corkscrew appearance of fine vessels. (Courtesy Prof. J. C. Gaux, Hôpital Broussais, Paris, France)

show any evidence of atherosclerotic disease, thoracic outlet or popliteal entrapment syndromes.

The clinical diagnosis can be suspected in a patient who has isolated occlusive disease of infrapopliteal arteries. Associated findings of either a migratory superficial thrombophlebitis or involvement of the upper extremities (Raynaud's phenomenon, digital ischaemia) allows a tentative diagnosis. Finally, the diagnosis can be made with reasonable certainty if all three manifestations are present. It is possible that further immunological investigations may lead to a biological definition of the entity, considered to date as a syndrome [307]. A good prognosis appears to be one of the distinctive features [336].

Principles of Treatment

Various therapeutic approaches are based on the pathogenic hypothesis. Bilateral sympathectomy is no longer indicated in the treatment of Buerger's disease. Tobacco abstinence is essential for an improved prognosis [310,334]; only complete abstinence may control the progress of the disease and, if the patient smokes again, the disease tend to progress despite treatment. Cold, humidity, trauma and vasoconstrictor agents have a deleterious effect and should be avoided [310]. Use of immunosuppressive agents has been suggested on the basis of an immunological cause. Bollinger et al. [283] suggested high doses of acetylsalicylic acid (3 g/d), replaced by corticosteroid therapy (prednisone 30 mg/d) or azathioprine (100 mg/d) if necessary. No controlled trial has clearly established the value of this therapeutic strategy.

In most cases the treatment of ischaemia is medical. Distal arterial bypass surgery is rarely recommended in such patients [351] with a poor long-term prognosis, but it may allow healing of trophic changes. In order to evaluate the feasibility of surgery, peripheral arteriography is required before amputation is performed. Sympathectomy can be indicated in association with amputation or in order to facilitate healing of the trophic changes [310,340]. Antibiotic therapy is needed to prevent secondary infection. Analgesic procedures are needed to relieve pain and allow the patient to maintain bed rest. They can reduce oedema that tends to increase the ischaemia. Raising the head of the patient's bed can help to reduce rest pain and oedema. Local treatment is essential and requires a close partnership between patients and physicians.

Occasionally, surgical debridement of gangrenous tissue will allow the formation of granulation tissue at the base of the lesion and the avoidance or limitation of the extension of amputation. Massive perfusions, anti-coagulant therapy, vasodilating drugs and calcium inhibitors can be used at the time of progression of the disease.

References

272. Acar J, Leurent B, Slama M et al. (1983) Insuffisance aortique et maladie de Takayasu. Ann Med Interne 134:606–613
273. Adar R, Papa MZ, Halpern Z et al. (1983) Cellular sensitivity to collagen in thromboangiitis obliterans. N Engl J Med 308:1114–1116
274. Allison MC, Gallagher PJ (1984) Temporal artery biopsy and corticosteroid treatment. Ann Rheum Dis 43:416–417
275. Ask-Upmark E, Fajers CM (1956) Further observations of Takayasu's syndrome. Acta Med Scand 155:275–291
276. Barrier J (1985) Maladie de Horton. In: Kahn MF, Peltier AP (eds). Maladies systémiques, 2nd edn. Flammarion, Paris, pp 517–1, 517–5
277. Barrier J, Bignon JD, Soulillou JP, Grolleau J (1981) Increased prevalence of HLA-DR4 in giant-cell arteritis. N Engl J Med 305:104–105
278. Barrier J, Potel G, Renault-Hovasse H et al. (1982) The use of doppler flow studies in the diagnosis of giant cell arteritis. Selection of temporal artery biopsy site is facilitated. JAMA 248:2158–2159
279. Beevers DG, Harpur JE, Turk KAD (1973) Giant cell arteritis. The need for prolonged treatment. J Chronic Dis 26:571–584
280. Bengtsson BA, Malmvali BE (1981) Prognosis of giant cell arteritis including temporal arteritis and polymyalgia rheumatica. A follow up study on ninety patients treated with corticosteroids. Acta Med Scand 209:337–345
281. Bengtsson BA, Malmvall BE (1981) An alternate-day corticosteroid regimen in maintenance therapy of giant cell arteritis. Acta Med Scand 209:347–350
282. Bengtsson BA, Malmvall BE (1981) The epidemiology of giant cell arteritis including temporal arteritis and polymyalgia rheumatica. Incidences of different clinical presentations and eye complications. Arthritis Rheum 24:899–904
283. Bollinger A, Hollmann B, Schneider E, Fontana A (1979) Thromboangiitis obliterans: Diagnose und Therapie im licht neuer immunologischer Befunde. Schweiz Med Wochenschr 109:537–543
284. Bonnetblanc JM, Vidal E, Hessel L et al. (1981) Traitement adjuvant dans la maladie de Horton. Nouv Presse Med 10:2120–2121
285. Buerger L (1908) Thrombo-angiitis obliterans: a study of the vascular lesions leading to presenile spontaneous gangrene. Am J Med Sci 136:567–580
286. Carter SA (1973) The relationship of distal systolic pressures to healing of skin lesions in limbs with arterial occlusive disease with special reference to diabetes mel-

litus. Scand J Clin Lab Invest (Suppl 128) 31:239–243

287. Chaufour J, Melki JP, Cormier JM (1985) Maladie de Takayasu: angiographie numérisée veineuse ou artériographie numérisér. Presse Med 14:981

288. Craven JL, Cotton RC (1967) Haematological differences between thromboangiitis obliterans and atherosclerosis. Br J Surg 54:862–867

289. Criqui MH, Fronek A, Klauber MR et al. (1985) The sensitivity, specificity and predictive value or traditional clinical evaluation of peripheral arterial disease: results from noninvasive testing in a defined population. Circulation 71:516–522

290. Federici AB, Fox RI, Espinoza LR, Zimmerman TS (1984) Evaluation of von Willebrand factor is independent of erythrocyte sedimentation rate and persists after glucocorticoid treatment in giant cell arteritis. Arthritis Rheum 27:1046–1049

291. Fiessinger JN, Camilleri JP, Chousterman M et al. (1978) Maladie de Horton et maladie de Takayasu: critères anatomopathologiques. Deux observations. Nouv Press Med 7:639–642

292. Fiessinger JN, Camilleri JP, Cormier JM, Housset E (1983) La maladie de Takayasu: le diagnostic. Ann Med Interne 134:441–443

293. Fiessinger JN, Housset E (1985) Thromboangéite oblitérante (maladie de Buerger) In: Kahn MF, Peltier AP (eds) Maladies systémiques. 2nd edn. Flammarion, Paris, pp 526–531

294. Fiessinger JN, Tawfik-Taher S, Capron L et al. (1982) Maladie de Takayasu. Critères diagnostiques. Nouv Presse Med 11:583–586

295. Fontaine R, Bouissou H, Batzenschlager A et al. (1968) La thromboangiose ses aspects anatomopathologiques. Premier congrès du collège Français de pathologie vasculaire. L'Expansion Scientifique, Paris, pp 57–76

296. Fraga A, Mintz G, Valle L, Flores-Izquierdo G (1972) Takayasu's arteritis: frequency of systemic manifestations (study of 22 patients) and favorable response to maintenance steroid therapy with adrenocorticosteroids (12 patients). Arthritis Rheum 15:617–624

297. Frampton MW (1981) Temporal arteritis. Seven variations. NY State J Med 81:1179–1182

298. Gardner JD, Lee KR, Abdou NI (1984) Takayasu's arteritis: reversal of pulse deficits after early treatment with corticosteroids. J Rheumatol 11:92–93

299. Ghose MK, Shensa S, Lerner PI (1976) Arteritis of the aged (giant cell arteritis) and fever of unexplained origin. Am J Med 60:429–436

300. Godeau P, Bletry O, Garin JL et al. (1979) Maladie de Horton. Problèmes d'actualité. Sem Hôp Paris. 55:347–353

301. Goodman VW (1979) Temporal arteritis. Am J Med 67:839–852

302. Gore I, Burrows S (1958) Reconsideration of the pathogenesis of Buerger's disease. Am J Clin Pathol 29:319–330

303. Graham E, Holland A, Avery A, Ross Russel RW (1981) Prognosis in giant-cell arteritis. Br Med J. 282:269–271

304. Grossman E, Morag B, Nussinovitch N et al. (1984) Clinical use of captopril in Takayasu's disease. Arch Intern Med 144:95–96

305. Gulati SM, Singh KS, Thusoo TK, Saha K (1979) Immunological studies in thromboangiitis obliterans (Buerger's disease). J Surg Res 27:287–293

306. Hall S, Barr W, Lie JT et al. (1985) Takayasu arteritis. A study of 32 North American patients. Medicine 64:89–99

307. Herman BE (1975) Buerger's syndrome. Angiology 26:713–716

308. Healey LA, Wilske KR (1977) Manifestations of giant cell arteritis. Med Clin North Am 61:261–270

309. Healey LA, Wilske KR (1980) Presentation of occult giant cell arteritis. Arthritis Rheum 23:641–643

310. Hill GL (1974) A rational basis for management of patients with the Buerger syndrome. Br J Surg 61:476–481

311. Hirai M, Shionoya S (1978) Intermittent claudication in the foot and Buerger's disease. Br J Surg 65:210–213

312. Hirai M, Shionoya S (1979) Arterial obstruction of the upper limb in Buerger's disease: its incidence and primary lesion. Br J Surg 66:124–128

313. Huston KA, Hunder GG, Lie JT et al. (1978) Temporal arteritis. A 25-year epidemiologic, clinical and pathologic study. Ann Intern Med 88:162–167

314. Ishikawa K (1981) Survival and morbidity after diagnosis of occlusive thrombo-aortopathy (Takayasu's disease). Am J Cardiol 47:1026–1032

315. Ishikawa K, Matsuura S (1982) Occlusive thromboaortopathy (Takayasu's disease) and pregnancy. Clinical course and management of 33 pregnancies and deliveries. Am J Cardiol 50:1293–1300

316. Ishikawa K, Uyama M, Asayama K (1983) Occlusive thromboaortopathy (Takayasu's disease): cervical stenoses, retinal arterial pressure, retinal microaneurysms and prognosis. Stroke 14:730–735

317. Ito I, Saito Y, Nonaka Y (1975) Immunological aspects of aortitis syndrome. Jpn Circ J 39:459–462

318. Jonasson F, Cullen JF, Elton RA (1979) Temporal arteritis. A 14-year epidemiological, clinical and prognostic study. Scott Med J 24:111–117

319. Kansu T, Corbett JJ, Savino P, Schatz NJ (1977) Giant cell arteritis with normal sedimentation rate. Arch Neurol 34:624–625

320. Kinare SG, Kher YR, Rao G, Sen PK (1976) Pattern of occlusive peripheral vascular disease in India. Angiology 27:165–180

321. Klein RG, Campbell RJ, Hunder GG, Carney JA (1976) Skip lesions in temporal arteritis. Mayo Clin Proc 51:504–510

322. Klein RG, Hunder GG, Stanson AW, Sheps SG (1975) Large artery involvement in giant cell (temporal) arteritis. Ann Intern Med 83:806–812

323. Kurozumi T, Tanaka K (1978) Buerger's disease: histology and pathogenesis. Vasc Surg 12:63–68

324. Lacombe P, Frija G, Dubourg O et al. (1986) Place de l'angiographie numérisée intraveineuse dans la maladie de Takayasu. Arch Mal Cœur 79:273–280

325. Lagneau P, Michel JB (1985) Renovascular hypertension and Takayasu's disease. J Urol 134:876–879

326. Lande A, Berkmen YM (1976) Aortitis. Pathologic, clinical and arteriographic review. Radiol Clin North Am 14:219–240

327. Lande A, Gross A (1972) Total aortography in the diagnosis of Takayasu's arteritis. Am J Roentgenol 116:165–178

328. Lande A, Rossi P (1975) The value of total aortography in the diagnosis of Takayasu's arteritis. Radiology 114:287–297

329. Larson TS, Hall S, Hepper NGG, Hunder GG (1984) Respiratory tract symptoms as a clue to giant cell arteritis. Ann Intern Med, 101: 594–597

330. Liozon F, Catanzano G (1982) L'artérite temporale de Horton. Etude anatomopathologique en microscopie optique. A propos de 123 biopsies temporales. Rev Med Interne 3:295–301

331. Liozon F, Dany F, Catanzano R (1983) Exploration des artères temporales par effet doppler dans la maladie de Horton. Presse Med 12:2113–2114

332. Lippman HI (1952) Cerebrovascular thrombosis in patients with Buerger's disease. Circulation 5:680–682

333. Lupi-Herrera E, Sanchez-Torres G, Marcushamer J et al. (1977) Takayasu's arteritis. Clinical study of 107 cases. Am Heart J 93:94–103

334. McKusick VA, Harris WS, Ottesen OE et al. (1962) Buerger's disease: a distinct clinical and pathologic entity. JAMA 181:93–100

335. McLoughlin GA, Helsby CR, Evans CC, Chapman DM (1976) Association of HLA-A9 and HLA-B5 with Buerger's disease. Br Med J 2:1165–1166

336. McPherson JR, Juergens JL, Gifford RW (1963) Thromboangiitis obliterans and arteriosclerosis obliterans. Clinical and prognostic differences. Ann Intern Med 59:288–296

337. Morris HH (1977) Angiographic findings in giant cell arteritis. South Med J 70:1014–1016

338. Nakao K, Ikeda M, Kimata SI et al. (1967) Takayasu's arteritis. Clinical report of eighty-four cases and immunological studies of seven cases. Circulation 35:1141–1155

339. Nasu T (1963) Pathology of pulseless disease. A systematic study and critical review of twenty-one autopsy cases reported in Japan. Angiology 14:225–242

340. Nielubowicz J, Rosnowski A, Pruszynski B et al. (1980) Natural history of Buerger's disease. J Cardiovasc Surg 21:529–540

341. Parker F, Healey LA, Wilske KR, Odland GF (1975) Light and electron microscopic studies on human temporal arteries with special reference to alterations related to senescence, atherosclerosis and giant cell arteritis. Am J Pathol 79:57–69

342. Pellat J, Perret J, Pasquier B et al. (1976) Etude anatomoclinique et angiographique d'une observation de thromboangiose disséminée a manifestations cérébrales prédominantes. Rev Neurol 132:517–535

343. Quilichini R, Chaffanjon P, Miro I, Aubert L (1981) Maladie de Horton: forme tussigène. Nouv Presse Med 10:2831

344. Riche MC, Chaufour J, Menault JY (1985) Les artériopathies non athéromateuses. In: Melki JP, Hovasse D, Riche MC et al. (eds) L'angiographie numérisée. Techniques, indications. Vigot, Paris pp 101–107

345. Rivers SP, Baur GM, Inahara T, Porker JM (1982) Arm ischemia secondary to giant cell arteritis. Am J Surg 143:554–558

346. Roberts WC, MacGregor RR, DeBlanc HJ (1969) The prepulseless phase of pulseless disease or pulseless disease with pulses. A newly recognized cause of cardiac disease, monoclonal gammopathy and "fever of unknown origin". Am J Med 46:313–324

347. Rosen N, Sommer I, Knobel B (1985) Intestinal Buerger's disease. Arch Pathol Med Lab 109:962–963

348. Schatz IJ, Fine G, Eyler WR (1966) Thromboangiitis obliterans. Br Heart J 28:84–91

349. Sen PK (1973) Non-specific aorto-arteritis. A monograph based on a study of 101 cases. Tata McGraw-Hill, New Delhi

350. Shelhamer JH, Volkman DJ, Parrillo JE et al. (1985) Takayasu's arteritis and its therapy. Ann Intern Med 103:121–126

351. Shionoya S, Ban I, Nakata Y et al. (1976) Vascular reconstruction in Buerger's disease. Br J Surg 63:841–846

352. Smolen JS, Youngchaiyud U, Weidinger P et al. (1978) Autoimmunological aspects of thromboangiitis obliterans (Buerger's disease). Clin Immunol Immunopathol 11:168–177

353. Spittell JA (1983) Thromboangiitis obliterans. An autoimmune disorder? N Engl J Med 308:1157–1158

354. Stanson AW, Klein RG, Hunder GG (1976) Extracranial angiographic findings in giant cell (temporal) arteritis. AJR 127:957–963

355. Thomas D, Dubourg O, Bletry O et al. (1984) L'atteinte coronarienne dans la maladie de Takayasu. A propos de 3 cas dont 2 opérés et revue de la littérature. Arch Mal Cœur 77:386–396

356. Cayssairat M, Fiessinger JN, Housset E (1977) Les nécroses digitales du membre supérieur. 86 cas. Nouv Presse Med 6:931–934

357. Vayssairat M, Mathieu JF, Priollet P et al. (1984) Mesure transcutanée de la pression partielle d'oxygène. Une nouvelle méthode d'exploration fonctionnelle en pathologie vasculaire. Presse Med 13:1683–1686

358. Vinckier L, Hatron PY, Coget J, Devulder B (1984) Le Dopler artériel dans la maladie de Horton. Etude prospective de 24 observations. Rev Med Intern 5:291–297

359. Vuja O (1973) Giant cell angiitis with localizations in the large arteries of the head (vertebral and internal carotid). Eur Neurol 10:197–204

360. Wessler S, Ming SC, Gurewich V, Freiman DG (1960) A critical evaluation of thromboangitis obliterans. The case against Buerger's disease. N Engl J Med 262:1149–1160

Section VI

Arteriopathies of Diverse Origin

Chapter 29

Fibromuscular Arterial Dysplasias

Pathological Aspects and Current Concepts

J.-P. Camilleri

Fibromuscular dysplasia (FMD) or arterial fibrodysplasia is a clinico-pathological entity whose clinical importance has increased since the development of angiography. Moreover, the tremendous progress achieved in the understanding of renovascular hypertension as the leading cause of surgically curable hypertension, has focused attention on this non-atherosclerotic stenotic angiopathy. Since the reports of Hunt et al. [31], MacCormack et al. [46], Harrison et al. [23], Stewart et al. [83], Harrison and MacCormack [24], Abe [1] and Stanley et al. [82], it is well-known that two diseases can be responsible for renovascular hypertension: atherosclerosis and FMD, characterized by segmental non-atherosclerotic fibrous stenoses. FMD affects 2% of patients with hypertension [89] and is responsible for 20%–40% of cases of renovascular hypertension. It is probable that the incidence in the general population of this arteriopathy is higher than previously reported. This is supported by the angiographic demonstration of FMD of the renal arteries in normotensive individuals [89] and the presence of FMD in 1% of 819 consecutive postmortem examinations [27].

Although FMD was first described in the renal artery, its occurrence in extrarenal arteries has been highlighted by Palubinska et al. [61, 62]. Since the mid-1960s multiple arterial involvement outside the renal arteries has been reported, suggesting that the arteriopathy is a systemic multifocal disease. In a review of the literature, Mettinger [53] collected about 1100 cases in 1982 involving, in decreasing order of frequency, the renal artery (58%), the cervicoencephalic (35%), intra-abdominal (3%) and iliac (2.5%) arteries and other sites (1.5%).

FMD of the Renal Artery

The first renal FMD associated with hypertension was reported in 1983 by Leadbetter and Burkland [40] in a 5-year-old boy and since then reports of this entity have appeared in large series. In 1971 two teams of investigators from the Mayo Clinic and the Cleveland Clinic met in Rochester and formulated a classification [23]. Since then most of the authors have agreed with this international formulation based on the arterial layer mainly involved in the lesions, although some modifications have been proposed [8, 81].

The lesions usually affect the distal part of the renal artery and the primary branches. Involvement can be unilateral, more commonly in the right renal artery, but is bilateral in more than 50% of cases.

Background Histopathology

Medial FMD with Mural Aneurysms (Diffuse Medial Fibrodysplasia)

This is the most common form, representing nearly 70% of all dysplastic lesions. It occurs

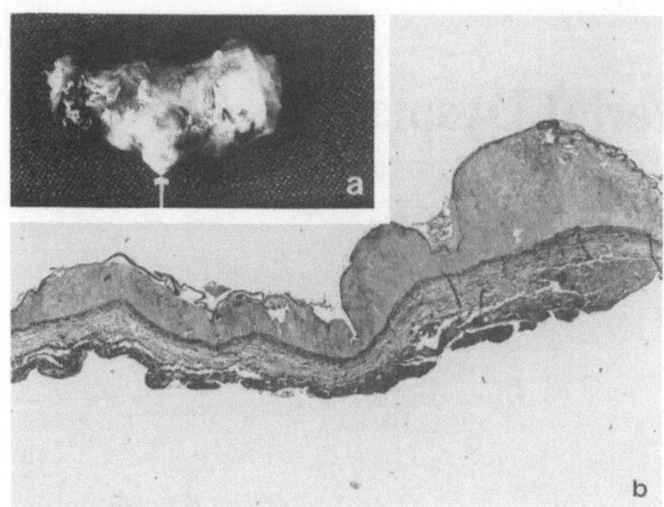

Fig. 29.1a,b. Medial fibromuscular dysplasia of the renal artery. **a** Gross appearance of resected renal artery showing aneurysmal dilatation and semilunar ridges (*arrow*). **b** Longitudinal section showing focal thickenings alternating with medial thin areas. Masson's trichrome stain, × 33

mainly in women between the ages of 25 and 40 years [82]: the mean age of affected women (40 years) exceeds that of men (32.2 years). In contrast this prevalence in females is not evident in the paediatric age group.

Diffuse medial FMD occurs primarily in the distal two-thirds of the renal artery and often extends into the segmental branches. Gross pathological examination shows mural aneurysms and thickened semilunar or helical ridges forming stenotic areas (Fig. 29.1a). On microscopic examination of longitudinal sections, fibromuscular thickenings alternate with aneurysms formed by complete loss of medial cells and

a deficient internal elastic lamina (Fig. 29.1b). Focal intimal thickening can be found, particularly in dilated areas. Sometimes part of the media appears to be detached from the vessel wall and protrudes into the vascular lumen (Fig. 29.2). The smooth-muscle cell component is rarefied and replaced by loose collagen (Fig. 29.3). The remaining smooth-muscle cells show a disorganized pattern and degenerative changes with cytoplasmic vacuolization. The adventitia can be markedly thickened by fibrosis, sometimes associated with hyperplasia of longitudinal smooth-muscle fibrosis, sometimes associated with hyperplasia of longitudinal smooth-muscle

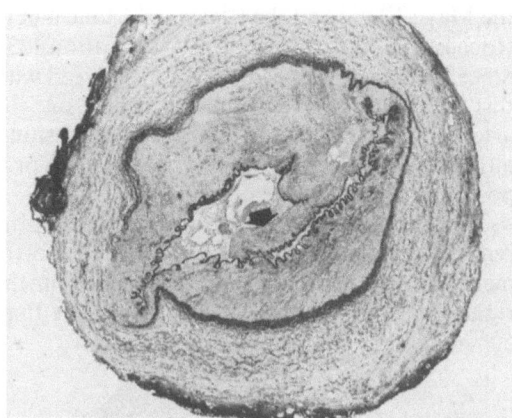

Fig. 29.2. Medial FMD. Cross-section showing mural aneurysms resulting from medial thinnings. Part of the medial coat seems likely to be detached from the vessel wall, protruding into the irregular lumen (*arrow*). The adventitia appears thickened. Orcein stain, × 35

Fig. 29.3. Medial FMD. Detail of the medial coat on longitudinal section, showing loss and disorganization of smooth-muscle cells, which are dispersed within loose connective tissue. Masson's trichrome stain, × 150

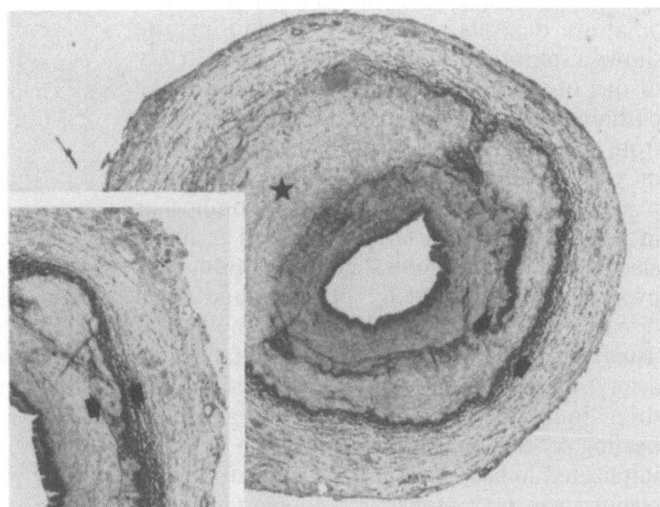

Fig. 29.4. Perimedial FMD. Foci of fibrous tissue appear to develop through the media–adventitia interface (*arrows*) and progressively replace the outer part of the media (*star*). Orcein stain, × 37. *Inset.* Early nodular lesion dissociating the external elastic lamina (*arrows*). Orcein stain, × 38

bundles. The angiographic picture is characterized by the well-known "string of beads" appearance, the beads being larger than the uninvolved artery.

Perimedial (Subadventitial) FMD

This is a form of medial involvement [24] which is characterized by clinical, angiographic and histopathological features [45], and represents about 10%–20% of all medial FMD of the renal artery. It seems likely that this form affects mainly women 10 to 15 years younger than the diffuse form described above. Histopathologically, dense fibrous tissue replaces the outer part of the media resulting in a layer sometimes thicker than the remaining inner media. (Fig. 29.4). In early lesions, nodular fibrous foci penetrate through the media–adventitia interface, dissociating internal elastic lamina and merging into a peripheral collar of dense fibrous tissue which envelops the renal artery over a variable length. Usually the fibrosis includes remaining smooth-muscle cells. The adventitia appears normal or thickened, sometimes showing hyperplastic smooth-muscle foci. Newly formed vessels sprouting from the adventitia and penetrating the perimedial fibrosis can be found [26]. The angiographic picture is characterized by focal tubular stenoses, associated with post-stenotic dilatation and collateral circulation. "Beads" are small and do not exceed the size of the normal uninvolved artery.

Some authors [82] have suggested that the medial and perimedial forms are two aspects of the same process of medial disorganization.

Gradations and associations of both types can be demonstrated (using serial sections) in the same lesion or in two different locations in the same patient. The perimedial (peripheral) form would represent an early lesion, which might progress either to rapidly progressive stenosis and increased blood pressure, or to a more stable form involving the entire media and true macroaneurysms. The younger age of patients with the perimedial form supports this hypothesis. The apparently increasing incidence of perimedial (subadventitial) lesions in recent series may be due to the development of angiography since the mid-1960s.

Little is known of the spontaneous behaviour of this disease. Angiographic evidence of the progression of existing lesions or of the development of new lesions has been reported [18,37,38,50] and must be taken into account in the therapeutic approach to the disease.

Finally, diffuse and peripheral medial forms as defined above are usually characterized as a multifocal vascular disease. The association of lesions affecting both the renal and carotid arteries is found in about 10% of patients with FMD. Intracranial aneurysms are found in 8%–16% of cases and subarachnoid bleeding represents a major risk in these patients. The meaning of such aneurysms will be discussed further.

Intimal FMD

This represents 1% to 5% of all cases of FMD [24,82] and occurs primarily in children and young adults, affecting the two sexes equally. It consists of a circumferential, moderately cellular intimal thickening with no calcium or atheromatous deposits. The angiographic picture shows a smooth focal stenosis of the renal artery or one of the major branches. The task for the pathologist is to distinguish dysplastic lesions from secondary endarteritis which may be found in the vicinity of fibrolipidic atherosclerotic plaques with associated organizing thrombus, or in association with arterial trauma. The fibroelastic endarteritis found in association with hypertensive arteriopathy or end-stage renal disease in a patient on chronic haemodialysis usually involves more distal branches of the renal artery, mainly the interlobular arteries. In children intimal thickening may represent the healing process of arteritis, and rubella has been implicated in such instances [84]. Renal arterial stenosis can be responsible for hypertension in children and young adults presenting with Von Recklinghausen disease.

Medial Hyperplasia

This is a rare variety of FMD occurring in young adults. Its incidence varies in reports from 5% to 15% [24] or may be as low as 1% [82]. The findings are those of medial smooth-muscle hyperplasia associated with a certain degree of structural disorganization and marked luminal stenosis, without intimal or adventitial changes. The arteriographic picture shows smooth tubular stenoses on the main branches of the renal artery. The natural history includes an association with spontaneous dissection.

Periarterial or Adventitial Forms of FMD

These are rare in most series (less than 1% in Harrison and MacCormack's series). They are characterized by a dense adventitial fibrosis including thick-walled vasa vasora. These cases have to be distinguished from FMD of the medial type associated with secondary adventitial thickening and longitudinal smooth-muscle bundle hyperplasia. One also needs to investigate multiple sections in order to detect inflammatory cellular infiltration which might suggest a fibrous form of Takayasu's disease (see Chap. 28). The presence of aortic lesions might support such a diagnosis of inflammatory arteriopathy. When the patients are operated on, aortic wall biopsy is indicated when possible. Finally, periarterial lesions associated with retroperitoneal fibrosis have been reported. A peculiar form characterized by accumulation of elastic fibres within adventitia in the renal artery has been described

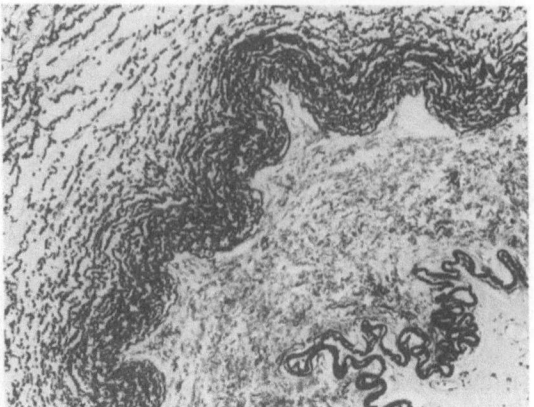

Fig. 29.5. Periarterial fibrodysplasia. Hyperplasia of external elastic lamina. Orcein stain, × 80

as a perimedial type by Stanley et al. [82] (Fig. 29.5).

Complications

Luminal narrowing, resulting in ischaemia of the territory supplied, macroaneurysms and medial dissection are all potential complications of FMD.

The macroaneurysms result from mural dilatations with rupture of the internal elastic lamina. They are usually associated with diffuse medial FMD but can be found in association with the perimedial form. The frequency of FMD with macroaneurysms has been evaluated at 9% and the risk of rupture at 6%. The wall of the aneurysms is made up of a thickened intima supported by external elastic lamina and adventitia.

Intramedial or subadventitial dissection is associated with FMD in 5%–10% of cases. This accident can be seen in diffuse medial forms, for example, in perimedial forms. It can increase the luminal stenosis and sometimes progresses to complete obliteration (Fig. 29.6). The weakness of the media in this disease may facilitate this complication. The role of dilated newly formed vessels penetrating the fragile medial layer has been evoked as one of the factors at work, particularly in the perimedial forms [26]. A FMD can allow a post-traumatic dissection, for example after arterial catheterization. Most often it is a chronic subadventitial dissection, either completely obliterated by organizing thrombus, or forming a new channel bordered with loose connective tissue showing numerous

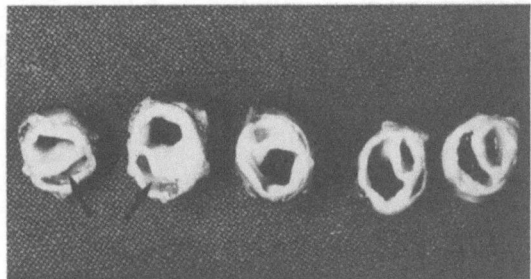

Fig. 29.6. Gross appearance of renal artery showing perimedial dissection.

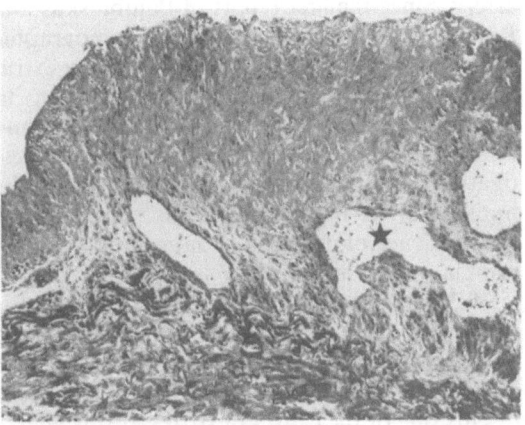

Fig. 29.8. Chronic dissection with abrupt interruption of the medial coat which is replaced by an organizing fibrous tissue including new vessels (*stars*). Masson's trichrome stain, × 135

haemosiderin-containing macrophages. Some authors have outlined the risk of rapidly progressive hypertension at the time of medial dissection [54]. Small foci of unexpected chronic dissection can be found in resected arterial specimens (Figs. 29.7 and 29.8). The association with arteriovenous fistulae has been reported in rare cases [32].

FMD of the Renal Artery in Children

In paediatric patients renal artery stenoses are mainly of a dysplastic nature [75]. Some of them are associated with stigmata of neurofibromatosis, idiopathic hypercalcaemia or one of the inheritant diseases of connective tissue. The others present as idiopathic FMD [68,15]. Among the 25 children of the Stanley series [82] there was no predilection for side or sex affected. Histopathologically, intimal lesions and medial

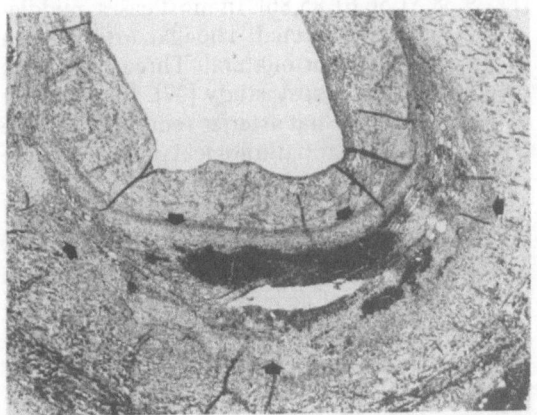

Fig. 29.7. Small focus of chronic dissection (*arrows*) found at the histological examination of a resected renal artery with diffuse medial FMD. Masson's trichrome stain, × 40

hyperplasia are the main findings in children. Familial occurrence of FMD is discussed below.

FMD of the Renal Artery of Transplanted Kidney

Stenoses of the renal artery have been implicated as a factor in systemic hypertension in about 10% of patients following renal transplantation [36,39,41,58,69]. Intimal thickening is usually found either at the site of anastomoses or at a distance from it, sometimes extending into the intrarenal branches of the vessels. The change has to be distinguished from kinking or atherosclerosis of the hypogastric artery of the recipient and from stenoses due to technical errors involving the anastomotic site. It is usually associated with post-stenotic dilatation [60]. Rarely, stenosis of the renal artery may be the natural evolution of an unexpected FMD of the transplant renal artery [58].

Several mechanisms have been suggested: (a) trauma, (b) healing of injured endothelium and (c) an immunological process [59].

FMD of the Cervicoencephalic Arteries

The cervicoencephalic arteries are the most frequent site of involvement in FMD after the renal artery. Since the first report of Palubinskas and Ripley in 1964 [61], several large angiographic series have estimated the incidence of the extracranial cerebrovascular FMD at 0.23% to 0.70%. The vascular abnormality is often found at the time of a systematic angiographic investigation in patients with vascular disease and the number of asymptomatic forms suggests that the incidence of dysplastic lesions in the general population is higher than is usually reported [5,30,33,49,52,53,60]. The predilection for women is evident, but the FMD seems to be diagnosed earlier in men. FMD of the cerebral arteries has been found in children and represents one of the causes of stroke at this age.

In 75% of patients, the internal carotid artery is involved in the segment adjacent to the second vertebra. Involvement of either the proximal part or the intracranial portion is less frequent. FMD has also been described in other sites, such as the common and the external carotid arteries, the vertebro-basilar system and even the intracranial arteries [70].

Histopathologically, the various types of lesions have been found to be identical to those described in the renal arteries. Most of the histologically verified FMD of the cervical arteries have been shown to be of the diffuse medial type with mural aneurysms. Only 5% were perimedial forms. Spontaneous dissection has been detected more often since the development of angiography and the increasing number of histological examinations made of resected specimens [78]. Complete occlusion of the affected artery is rare [30,49]. Thrombosis has been reported in patients on oestrogen or progesterone therapy. A risk of microembolization from platelet aggregates formed at the site of dysplastic lesions has been reported [51].

Intracranial aneurysms can be found in 20% to 35% of cases. Mettinger and Ericson detected 25 intracranial aneurysms on angiograms in 19 of a series of 37 patients [52]. If one takes into account all the reported cases, it is evident that the incidence of intracranial aneurysms is higher in patients with FMD when compared with 0.5%–1.5% reported incidence in the general population. The majority of aneurysms appear to be located in the internal carotid and the middle cerebral arteries, often on the same side as the most affected cervical arteries. Their macroscopic appearance is usually that of the classical "berry" aneurysms. Systemic hypertension has been suggested as one of the causative factors of such aneurysms. Nevertheless, it appears that most patients are not hypertensive [52] and that hypertension merely plays a role in increasing the risk of bleeding in affected patients.

Other sites of FMD

The FMD of the splanchnic arteries may involve the coeliac artery [9,27,38,49,71,82], the superior and inferior mesenteric arteries [7,49,61,81,82], the hepatic artery [63,66,82] and the cystic or gastric arteries [28]. In Stanley's series [82], these locations were found in ten of the 152 adult patients, most often associated with renal artery FMD. Patients with abdominal pain remain infrequent [7,71]. In most cases, intra-abdominal arterial stenoses are discovered in asymptomatic patients at the time of angiographic exploration for systemic hypertension. Medial dissection has been reported, particularly in FMD of the hepatic artery [66].

Splenic artery involvement has been noted with variable frequency. Stanley and Fry [81] reported eight cases of FMD in a series of 60 aneurysms of the splenic artery. Similar lesions were noted in splenic arteries systematically studied in patients operated on for spleno-renal anastomosis with splenectomy [44]. Nevertheless it seems difficult to distinguish medial fibrosis secondary to portal hypertension from true fibromuscular dysplasia in such patients.

The peripheral arteries are rarely affected. Since the report of Wylie [88], about 20 FMD of the iliac arteries have been noted in the literature [11,29,38,51,56,61,85,86]. In most cases, middle-aged women are affected. The iliac artery lesion may be unilateral or bilateral. Three cases were detected in an autopsy study [27]. Involvement of axillary or brachial arteries remains very rare [16,28,77]. The histopathological pattern is of the diffuse medial type.

Four cases of aortic FMD have been collected by Hata and Hosoda [25] and another case was recently reported by Letsch et al. [42]. It is an uncommon type of acquired aortic coarctation, a lesion which is more often due to other causes such as atherosclerosis, organization of incomplete post-traumatic rupture with medial dissection (as observed after deceleration), Takayasu's aortoarteritis or compression by a periaortic lesion.

Pathophysiology

Several comments on pathogenesis and localization may be based on clinico-pathological observations.

FMD is a Systemic Arterial Disease

Hormonal influences are suggested by the predilection for middle-aged women [64]. Such an hypothesis is supported by the following arguments: (a) the possible effects of pregnancy on connective tissue [2,10,48,87]; (b) the possible role of oral oestrogen or progesterone treatment in the genesis and evolution of the disease [3,22,34]; and (c) the effect of oestrogen on collagen synthesis in primary fibroblast and smooth-muscle cell cultures [72,73]. Nevertheless no demonstration of a direct role for hormonal factors is available, and the vascular complications which have been reported in association with longstanding anti-ovulant medications were characterized by intimal lesions and thrombosis involving arteries of smaller diameter than FMD.

Evidence for the developmental nature of the vascular disease is supported by (a) the frequency of asymptomatic forms; (b) the multifocal localization of the disease in the same patient; (c) the association of FMD with intracranial aneurysms; (d) the association of renovascular FMD with renal or renal artery abnormalities in 15% of cases [53]; (e) the occurrence of FMD in children and the newborn [68]; and (f) the occurrence of FMD in first degree relatives of those affected [1,20,21,47,55,65,67,82]. Recently, in pedigrees of 20 families of patients with FMD the authors have found from one to 11 relatives with clinical symptoms compatible with FMD.

It has been suggested that FMD is a dominant trait with reduced penetrance in males. Gladstein et al. [17] estimated the penetrance to between 0.44 and 0.52. Patients without a family history may have relatives with asymptomatic disease or they may represent a new mutation. In a study from the Karolinska Hospital [52], one-third of the patients with FMD of the cervical arteries had a family history of stroke, migraine or systemic hypertension. In this study, the prevalence in women and the young age of patients were useful criteria for suspecting FMD. Although true estimation of the risk would be based on histologically verified diagnoses, it is likely that such pedigree studies favour a congenital com-

ponent which might predispose to the disease [76].

The Predilection for FMD to Occur in Elective Sites Suggests a Role for Local Factors

The vulnerability of the renal and carotid arteries to FMD could be attributed to haemodynamic factors, these two territories receiving 20% and 14% of the cardiac output, respectively.

Stretching of the arterial wall has been invoked as a causative factor by Stanley et al. [82]. The increase in strain might result in fibrosis of the media either directly by modifying the rate of smooth-muscle cell biosynthesis of matrix proteins, or indirectly by mural ischaemia. Such an hypothesis might explain the predilection for renovascular FMD to occur in the right renal artery, associated findings of renal ptosis [35,38], the frequent occurrence of FMD in the internal carotid artery (stretching movements of the head on the upper cervical vertebrae [82]) and the involvement of the superior mesenteric or the coeliac arteries which support the weight of splanchnic viscera. It has been shown experimentally that repeated stretching of smooth-muscle cells may increase collagen biosynthesis in culture [43]. Such findings may be relevant to the medial lesions observed in the splenic artery wall in patients with portal hypertension [44]. Thus unusual haemodynamic and physical stresses might play a role in the genesis of dysplastic changes.

Associated findings of conditions predisposing to the development of generalized arterial spasm in patients with FMD give support to Bercquist's hypothesis of a spastic origin of FMD [4]. The vasoconstrictive side-effects of prolonged used of ergot preparations have been reported by Fievez et al. [13,14] in association with perimedial FMD of the renal artery; angiographically, the patient presented with multiple stenoses of femoral, radial and carotid arteries and the authors highlighted the histological similarity between FMD changes and the segmental arterial lesions of ergotamine intoxication [13]. The adventitial smooth-muscle cell hyperplasia could favour the role of a prolonged arterial spasm. Moreover the lesion of the coronary artery reported in patients with Prinzmetal variant angina [19] would give some support to the spastic hypothesis. It may thus be wise to search for a toxic factor in cases of FMD. Finally a parallelism has been noted on arteriographic films between FMD and the reversible picture observed in patients with phaeochromocytoma, and fibrodysplastic lesions have been found in association with phaeochromocytoma [12].

The sparsity of vasa vasora nourishing the renal, extracranial, carotid and external iliac arteries has been suggested as a possible factor in mural ischaemia [82]. This concept is supported by the involvement of the external part of the media in focal forms. The perimedial type of FMD has been considered to be an early form of the disease. Such findings are in agreement with experimental data, particularly obliteration of vasa vasora using a thrombin–gelatin mixture which produces peripheral fibrosis of the media in the dog [57,79].

FMD Affects Mainly the Media of Middle-Sized Muscular Arteries

If one excludes the intimal lesions and the true medial hyperplasia which affect the young, it is likely that the vast majority of arterial FMD can be defined as medial lesions, affecting middle-aged women, involving medium-sized muscular arteries and characterized pathologically by segmental disorganization of the vascular wall, marked inequalities in thickness, loss of smooth-muscle cells, fibrosis and disruption of the elastic laminae. Lipid deposits and inflammatory cell infiltrates are absent. Thrombosis is rare. Stenoses and saccular aneurysms are the usual vascular expressions of FMD.

Such a pathological pattern focuses attention on the medial smooth-muscle cell, a multi-potential cell controlling the organization of the media. In the study of six cases of perimedial dysplasia of the renal artery, Hata et al. [26] could not identify fibroblasts in the abnormal media; according to these authors, the cellular component in the fibrous areas seems to be chiefly smooth-muscle cells exhibiting various degrees of hyperfunction (Fig. 29.9). Sottiurai et al. [80] described transformation of smooth-muscle cells into myofibroblasts as the basic mechanism at the cellular level. A relationship between insufficient vessel wall nourishment and myofibroblastic transformation has been evoked [80]. These findings have been confirmed by Bragin and Cherkason [6]. In our material, disorganization and vacuolization of the medial muscle cells appear prominent, together with abundant extracellular matrix (Fig. 29.9).

Whatever the primary factor, it is likely that FMD is a multifactorial disease and may be

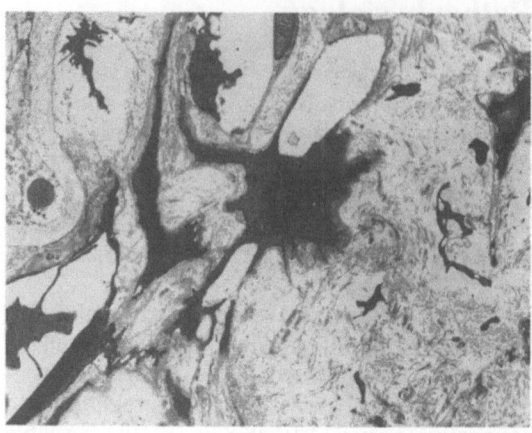

Fig. 29.9. Electron micrograph of the media of a diffuse medial FMD. In the media of the renal artery, smooth-muscle cells exhibit irregular shapes and cytoplasmic vacuoliation. Smooth-muscle cells have lost their concentric architecture and are embedded in an abundant fibrous extracellular matrix. Internal elastic lamina (→). Uranyl lead, × 10 500

considered as a genetically pre-conditioned abnormal myofibroblastic proliferative response of the media to hormonal, mechanical, ischaemic or spastic factors.

References

1. Abe K (1973) Renovascular hypertension. Diagnosis and surgical indication. Jpn J Clin Med 31:858–872
2. Albert EN, Bhussry BR (1967) The effects of multiple pregnancies and age on the elastic tissue of uterine arteries in the guinea pig. Am J Anat 121:259–270
3. Altuhuler JH, McLaughlin RA, Neubuerge KT (1968) Neurologic catastrophe related to oral contraceptive. Arch Neurol 19:264–273
4. Bercquist E, Erikson U, Uzfewdahl HR (1971) Stationary wave on segmental vaso-constriction. Acta Radiol (Stockh) 11:497–505
5. Boudin G, Guillard A, Romion A (1974) Dysplasies fibromusculaires des artères carotides et vertébrales. Ann Med Interne 12:863–875
6. Bragin MA, Cherkasov AP (1979) Morphogenesis of fibromuscular dysplasia of the renal artery. An ultrastructural study. Arkh Patol 41:46–52
7. Claiborne TS (1970) Fibromuscular hyperplasia: report of a case with involvement of multiple arteries. Am J Med 49:103–105
8. Croker DW (1968) Renal artery stenosis. In: Somniers SC (ed) Pathology annual. Appleton-Century-Crafts, Connecticut, pp 187–212
9. Crummy AB (1966) The sphenic artery arising from the superior mesenteric artery in a case of stenosis of the coeliac axis. Vasc Dis 3:266–268
10. Danforth DN, Buckingham JC (1964) Connective tissue mechanisms and their relation to pregnancy. Obstet Gynecol Surv 19:715–732

11. Daskalakis MK (1978) Fibromuscular hyperplasia of external iliac arteries. West J Med 128:345–347

12. De Mendoca WC, Espat PA (1981) Pheochromocytoma associated with arterial fibromuscular dysplasia. Am J Clin Pathol 75:749–754

13. Fievez M, Philippart F, Hustin J (1975) Ergotism. Anatomo-clinical study of a case. Angiology 26:491–498

14. Fievez M, Koerperich G, Dulieu J (1975) Dysplasie fibromusculaire artérielle et ergotisme. Am Anat Pathol 20:357–366

15. Fry WJ, Ernst CB, Stanley JC et al. (1973) Renovascular hypertension in the pediatric patient. Arch Surg 107:696–698

16. Garrett HE, Hodosh S, De Bakey ME (1967) Fibromuscular hyperplasia of the left axillary artery. Arch Surg 94:737–738

17. Gladstein K, Rushton AR, Kidd KK (1980) Penetrance estimates and recurrence risks for fibromuscular dysplasia. Clin Genet 17:115–116

18. Goncharenko V, Gerlock AJ, Shaff MI et al. (1981) Progression of renal artery fibromuscular dysplasia in 42 patients as seen on angiography. Radiology 139:45–51

19. Guermonprez JL, Gueret P, Camilleri JP et al. (1972) Angor de Prinzmetal. Etude histogique coronaire de prélèvements per-opératoires. A propos de deux cas. Arch Mal Coeur 10:301

20. Halpern HH, Sanford HS, Viamonte M (1965) Renal artery abnormalities in three hypertensive sisters. JAMA 194:124–125

21. Hansen J, Holter C, Thorberg JV (1965) Hypertension in two sisters caused by so-called fibromuscular hyperplasia of the renal arteries. Acta Med Scand 178:461–474

22. Hardy-Godon S, Fredy D, Chodkiewicz JP et al. (1979) Aspects angiographiques des accidents vasculaires cérébraux sous oestroprogestatifs. J Neuroradiol 6:239–254

23. Harrison EG, Hunt JC, Bernatz PE (1967) Morphology of fibromuscular dysplasia of the renal artery in renovascular hypertension. Am J Med 43:97–112

24. Harrison EG, McCormack LJ (1971) Pathologic classification of renal artery disease in renovascular hypertension. Mayo Clin Proc 46:161–167

25. Hata J, Hosoda Y (1976) Tubular stenosis of aorta with aortic fibromuscular dysplasia. Arch Pathol Lab Med 100:652–655

26. Hata JI, Hosoda Y (1979) Perimedial fibrodysplasia of the renal arteries. Arch Pathol Lab Med 103:229–223

27. Heffelfinger MJ, Holley KE, Harrison EG Jr et al. (1970) Arterial fibromuscular dysplasia studied at autopsy. Am J Clin Pathol 54:274

28. Hill LD Antonius JI (1965) Arterial dysplasia. An important surgical lesion Arch Surg 90:585–595

29. Horne TW (1975) Fibromuscular hyperplasia of the iliac arteries. Aus NZ J Surg 45:415–417

30. House OW, Baker HL, Sandok BA et al. (1971) Cephalic arterial fibromuscular dysplasia. Neuroradiology 101:605–611

31. Hunt JC, Harrison EG Jr, Kinlaid OW et al. (1962) Idiopathic and fibromuscular stenoses of the renal arteries associated with hypertension. Mayo Clin Proc 37:181–216

32. Imray TJ, Cohen AJ, Hahn L (1984) Renal ateriovenous fistula associated with fibromuscular dysplasia. Urology 23:378–380

33. Losue A, Kier EI, Ostrow D (1972) Fibromuscular dysplasia involving the intracranial vessels. J Neurosurg 37: 749–752

34. Irey NS, Norris HJ (173) Intimal vascular lesions associated with female steroids. Arch Pathol 96:227–234

35. Kaufman JJ, Maxwell MH (1963) Upright aortography in the study of nephroptosis stenotic lesion of the renal artery and hypertension. Surgery 53:736–742

36. Kaufman HM, Sampson D, Fox PS et al. (1977) Prevention of transplant renal artery stenosis. Surgery 81:161–167

37. Kelly TF, Morris GC (1982) Arterial fibromuscular disease. Am J Surg 143:232–236

38. Kincaid OW, Davis GD, Hallerman FJ et al. (1968) Fibromuscular dysplasia of the renal artery, arteriographic features, classification and observation on natural history of the disease. Am J Radiol 104:271–282

39. Lacombe M (1975) Arterial stenosis complicating renal allotransplantation in man (a study of 38 cases). Am Surg 181:283–288

40. Leadbetter WF, Burkland CE (1988) Hypertension in unilateral renal disease. J Urol 39:611–626

41. Lee HM, Madge GE, Mendez-Pilon G et al. (1978) Surgical complications in renal transplant recipient. Surg Clin North Am 58:285

42. Letsch R, Kantartzis M, Sommer T (1980) Arterial fibromuscular dysplasia: report of a case with involvement of the aorta and review of the literature. Thorac Cardiovasc Surg 28:206

43. Leung DY, Glagov S, Mathews L (1976) Cyclic stretching stimulates synthesis of matrix components by arterial smooth muscle cell in vitro. Science 191:475

44. L'Hermine C, Houcke M, Quandalle P et al. (1982) Dysplasie fibromusculaire de l'artère splénique et hypertension portale. Larc Medical 2:117–124

45. MacCormack LJ, Noto TJ, Meaney FF (1967) Subadventitial fibroplasia of the renal artery: a disease of young women. Am Heart J 73:602–606

46. MacCormack LJ, Pourasse EE, Meaney TF et al. (1966) A pathologic–arteriographic correlation of renal arterial disease. Am Heart J 72:188–198

47. Major P, Genest J, Cartier P et al. (1977) Hereditary fibromuscular dysplasia in the renovascular hypertension. Ann Intern Med 86:583

48. Manalo-Estrella P, Barker AE (1967) Histopathologic findings in human aortic media associated with pregnancy: a study of 16 cases. Arch Pathol 83:336–341

49. Manelfe C, Clariesse J, Fredy D et al. (1974) Dysplasie fibromusculaire des artères cervicocéphaliques. J Neuroradiol 1:149–231

50. Meaney TF, Dustan HP, MacCormack LJ (1968) Natural history of renal arterial disease. Radiology 91:881–887

51. Mehigan JT, Stoney RJ (1977) Arterial microemboli and fibromuscular dysplasia of the external iliac arteries. Surgery 81:484–486

52. Mettinger KL, Ericson K (1982) Fibromuscular dysplasia and the brain. I. Observation on angiographic, clinical and genetic characteristics. Stroke 13:46—52

53. Mettinger KL (1982) Fibromuscular dysplasia and the brain. II. Current concept of the disease. Stroke 13:53–58

54. Michel JB, Camilleri JP, Lagneau P et al. (1980) Fibrodysplasies de l'artère rénale. Nouv Presse Med 9:697–700

55. Morimoto S, Kurodo M, Uchioa K (1976) Occurence of renovascular hypertension in two sisters. Nephron 17:314–320

56. Nafaji H (1966) Fibromuscular hyperplasia of the external iliac arteries. Arch Surg 92:394–396

57. Nakata Y (1967) An experimental study on the vascular lesions caused by obstruction of the vasa vasorum. Jpn Circ J 31:275–287

58. Nghiem DD, Schulak JA, Bonsib JM et al. (1984) Fibromuscular dysplasia: an unusual cause of hypertension in the transplant recipient. Transplant Proc 16:555–558

59. O'Connel TX, Mowbray JF (1973) Arterial intimal thickening produced by alloantibody and xenoantibody. Transplantation 15:262

60. Osborne AG, Anderson RE (1977) Angiographic spectrum of cervical and intracranial fibromuscular dysplasia. Stroke 8:617–626

61. Palubinskas AJ, Ripley HR (1964) Fibromuscular hyperplasia in extrarenal arteries. Radiology 82:451–455

62. Palubinskas AJ, Perloff D, Neuton TH (1966) Fibromuscular hyperplasia: an arterial dysplasia of increasing clinical importance. AJR 98:907–913

63. Patchefsky AS, Paplanus SH (1967) Fibromuscular hyperplasia and dissecting aneurysm of the hepatic artery. Arch Pathol 83:141–144

64. Paulson WG, Boesel CP, Evans WE (1978) Fibromuscular dysplasia. Arch Neurol 35:287–290

65. Petit H, Bouchez B, Destee A et al. (1982) Forme familiale de dysplasie fibromusculaire des artères carotides internes. J Neuroradiol 9:15–22

66. Pinkerton JA, Wood WG, Fowler D (1976) Fibrodysplasia with dissection of aneurysms of the hepatic artery. Surgery 79:721–723

67. Plagnol P, Gillet JM, Cambuzat JM et al. (1975) Hypertension rénovasculaire familiale. J Radiol 56:173–174

68. Price RA, Vawter GF (1972) Arterial fibromuscular dysplasia in infancy and childhood. Arch Pathol Lab Med 93:419–426

69. Ricotta JJ, Schaff HV, Williams GM et al. (1978) Renal artery stenosis following transplantation. Etiology diagnosis, and prevention. Surgery 83:595

70. Rinaldi I, Harris WO, Kopp JE et al. (1976) Intracranial fibromuscular dysplasia: report of two cases, one with autopsy verification. Stroke 7:511–516

71. Ripley HR, Levin SM (1966) Abdominal angina associated with fibromuscular hyperplasia of the iliac and superior mesenteric arteries. Angiology 17:297–310

72. Ross R, Klebanoff SJ (1967) Fine structural changes in uterine smooth muscle and fibroblasts in response to estrogen. J Cell Biol 32:155–167

73. Ross R, Klebanoff SJ (1971) The smooth muscle cell. In vivo synthesis of connective tissue protein. J Cell Biol 50:172

74. Rothfield NJH Fibromuscular arterial disease. Experimental studies. Ann Radiol 14:294–295

75. Royer P, Habib R, Mathieu H et al. (1973) Hypertension artérielle de l'enfant. Nephrologie pédiatrique. Flammarion, Paris, pp 88–294

76. Rushton AR (1980) The genetics of fibromuscular dysplasia. Arch Intern Med 140:233–236

77. Saha SP, Goff RD, Stephenson SE (1975) Arm ischemia due to fibromuscular hyperplasia of the axillary artery. South Med J 68:645–646

78. Sato S, Hata JI (1982) Fibromuscular dysplasia. Its occurrence with a dissecting aneurysm of the internal carotid artery. Arch Pathol Lab Med 106:332–335

79. Sottiurai V, Fry W, Stanley JC (1978) Ultrastructural characteristics of experimental arterial fibroplasia induced by vasa-vasorum occlusion. J Surg Res 24:169–177

80. Sottiurai VS, Fry WJ, Stanley JC (1978) Ultrastructure of medial smooth muscle and myofibroblasts in human arterial dysplasia. Arch Surg 113:1280–1288

81. Stanley JC, Fry WJ (1974) Pathogenesis and clinical significance of splenic artery aneurysms. Surgery 76:898–909

82. Stanley JC, Gewertz BL, Bove EL et al. (1975) Arterial fibrodysplasia. Histopathologic character and current etiologic concepts. Arch Surg 110:561–566

83. Stewart BH, Dustan HP, Kiser WS et al. (1970) Correlation of angiography and material history in evaluation of patients with renovascular hypertension. J Urol 104:231–238

84. Stewart DR, Price RA, Nebesar R et al. (1973) Progressive peripheral fibromuscular hyperplasia in an infant: a possible manifestation of the rubella syndrome. Surgery 73:374–380

85. Twigg HL, Palmisano PJ (1965) Fibromuscular hyperplasia of the iliac artery a case report AJR 95:418–423

86. Walter JF, Stanley JC, Mehigan JT et al. (1978) External iliac artery fibrodysplasia. AJR 131:125–128

87. Wexler BC (1970) Vascular degenerative changes in the uterine arteries and veins of multiparous rats. Am J Obstet Gynecol 107:6–16

88. Wylie EJ, Binkley FM, Palubinskas AJ (1966) Extrarenal fibromuscular hyperplasia. Am J Surg 112:149–155

89. Youngberg JP, Sheps SG, Strong CG (1977) Fibromuscular disease of the renal arteries. Med Clin North Am 61:623–641

Principles of Diagnosis and Therapeutic Approaches

Internal Carotid Artery

C. F. Degos

Developmental abnormalities of the cervicoencephalic arteries are not uncommon if hypoplasia, agenesis and arterial aneurysms, particularly intracranial aneurysms, are included. However, FMD is the most frequently recognized entity in clinical practice [90,101,111,116, 120].

FMD of the internal carotid artery [115] affects primarily white women (80% of cases) [112,122]. The mean age of patients is about 50 years [112]. The vascular abnormality is bilateral in nearly half of the cases and associated involvement of other sites within the cervicoencephalic territory needs to be sought for systematically, particularly FMD of the vertebral artery (22% of cases) which is sometimes revealed by vertebrovertebral arteriovenous fistula [91,93,102,118] and intracranial aneurysms (30% of cases) [117]. The carotid artery is the site most frequently involved after the renal artery and involvement of both arteries is observed in more than 50% of cases. FMD of other systemic arteries (splanchnic and peripheral arteries) are far more rarely found.

Clinical Characterization

FMD of the carotid artery can be diagnosed either in patients with symptoms of cerebral ischaemia (from 40% to 60% of cases in various reports) [95,122], or at the time of subarachnoid aneurysmal bleeding (15% of cases) [122]. It seems likely to be more difficult to ascertain whether the FMD is responsible for symptoms such as Horner's syndrome, headaches (which have been reported in 25% of cases), epilepsy or paresis of cranial nerves [105,108]. FMD appears to be a neglected diagnosis in most of these cases, and it is often only at reinvestigation that the angiopathy is recognized.

Symptoms of cerebral ischaemia [95] have no distinctive features when compared with other aetiologies, such as atherosclerosis. Transient ischaemic attacks or various syndromes corresponding to the territories supplied by the affected vessel have been reported. Monocular blindness occurs rarely [122]. Careful auscultation of the arteries of the neck will detect a bruit in about 25% of cases.

The diagnosis of the vascular lesion is often made by ultrasound [97]. Doppler flow studies can detect stenoses and echotomography examination visualizes them. These methods of vascular imaging do not replace angiography, because false negative results may be due to the higher localization of the stenosis, and the morphological analysis of the lesions is not always of good quality. However, ultrasound examination is useful for detection of other sites of involvement and in follow-up studies.

Subarachnoid bleeding following rupture of intracranial aneurysms is not a major problem as far as the diagnosis is concerned because angiography is systematically carried out under such conditions. The risk would be overlooking the carotid artery abnormality in the neck or involvement of other sites. However, arteriographic examination may be contraindicated in aged and/or hypertensive women (renal artery involvement neglected), or patients with severe atherosclerosis. Under such circumstances ultrasound examination is very useful.

Angiographic Characteristics

Conventional or digital intravenous arteriography [114] is the basic method used to make the diagnosis of FMD. The typical appearance resembles a "string of beads" pattern [109] with multiple constrictions and aneurysmal dilatations of the lumen in the cervical portion of the internal carotid artery along a variable length (Fig. 29.10). Rarely the lesion extends to the carotid siphon. This appearance is usually easily recognized. "Stationary waves" [126] can be ruled out because the arterial calibre is spared, the stenoses appear regular and ultrasound examination gives normal results.

Focal tubular stenoses and atypical semi-circumferential lesions with diverticulum-like outpouchings have been described [114].

Therapeutic Approach

Asymptomatic FMD of the carotid artery, and lesions revealed by non-characteristic symptoms, need only survey [94] and possibly anti-platelet aggregation treatment.

Surgical procedures are not indicated and are not beneficial for FMD recognized after completed stroke. A careful search for other sites of localization in such patients is recommended.

The FMD revealed by transient ischaemic attacks may require surgical resection of the affected arterial segment with or without venous grafting. In cases of general or local contraindications for surgery, medical treatment would consist of anti-platelet aggregation agents rather than anti-coagulant therapy, if co-existing intracranial aneurysm has not been ruled out. Some authors [96,99,125] have suggested progressive arterial dilatation in patients with extensive arterial involvement; nevertheless the risk of arterial dissection remains. Hormonal oral contraception must be avoided in women with FMD [100].

Renal Artery

J.-B. Michel, P. Lagneau and J. Ménard

Fibromuscular dysplasias (FMD) of renal arteries are the second major cause of renovascular hypertension. Stenosis of the renal artery leads to a decrease of the perfusion pressure in the kidney distal to the dysplasia. The kidney reacts by activating the renin–angiotensin system which has general and local consequences. The general consequence is an increase in blood pressure fol-

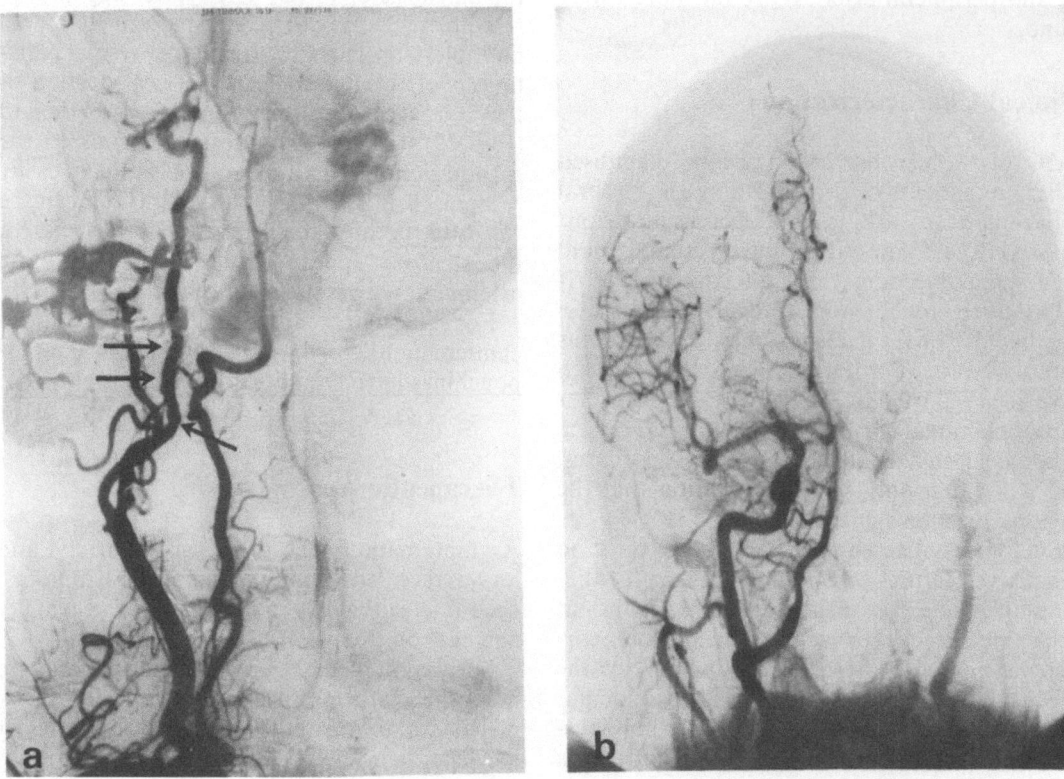

Fig. 29.10 a,b. Fibromuscular dysplasia of the right carotid artery: "string of beads" pattern (*arrows*). Intracranial arterial aneurysm of the first segment of the sylvian artery from the front intracranial view. **a** Lateral cervical view. **b** Front intracranial view.

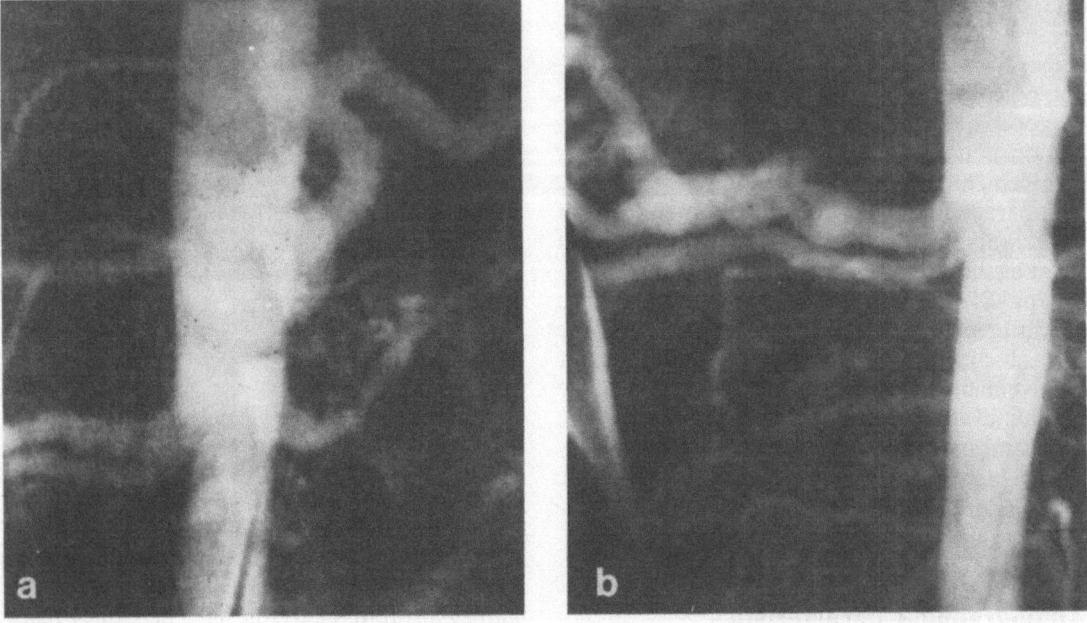

Fig. 29.11a, b. Example of lesional evolutivity of a medial FMD of renal artery. **a** 1974: severe hypertension, bilateral lesions predominant on the left side, left nephrectomy. **b** 1984: reappearance of severe hypertension related to the evolution of FMD on the right side, already present in 1974. Right renal revascularization.

lowing vasoconstriction induced by angiotensin II, and the local consequence is the maintenance of a perfusion pressure in the glomeruli and of the glomerular filtration rate. The clinical problems of fibrodysplasia of the renal artery are essentially related to hypertension. In fact, it is rare that an asymptomatic lesion of fibrodysplasia is discovered during screening for another localization. The evolution of the lesions of fibrodysplasia of the renal artery is slow but constant in all reports: 16% progression of the disease for Meaney et al. [110], 33% for Schreiber et al. [121] (Fig. 29.11). FMD may infrequently lead to the occlusion of the renal artery and to destruction of the kidney: two renal infarcts among 42 fibrodysplasias were reported by Goncharenko et al. [103]. Non-iatrogenic renal insufficiency secondary to dysplasia is very rare.

The diagnosis and therapeutic methods described for atheromatous renal artery stenosis (Chap. 21) also adapt to renovascular hypertension related to FMD [127]. We shall describe, from our own experience [104], what the differences are between FMD-induced hypertension and that induced by renal artery atheromatous stenosis.

Strategy of Diagnosis

Positive Diagnosis

Hypertension associated with FMD of the renal artery is characterized by the youth of the patients (34 ± 12 years in fibrodysplasia, versus 51 ± 8.6 years in atheromatous patients, $P < 0.001$) (Fig. 29.12) and by the predominance of female patients: 76% of patients with FMD are women, whereas only 14% of patients with atheromatous disease are women [104,110]. A history of gravidic hypertension should always be sought for. Hypertension secondary to fibrodysplasia of the renal artery is pure renovascular hypertension. Thus, for an identical high level of diastolic blood pressure secondary to the increase of peripheral resistances, systolic blood pressure will often be lower in FMD than in atheroma (Fig. 29.13). Generally, hypertension is less severe in FMD than in renovascular hypertension of atheromatous origin, necessitating the use of fewer types of drugs in FMD (2.3 ± 1) than in atheroma (3.2 ± 0.6, $P < 0.001$). In FMD there are fewer significant visceral consequences and plasma creatinine is less often elevated (12% of creatinine $> 115 \mu$mol/l in FMD versus 38% in atheroma). In contrast, cardiac consequences

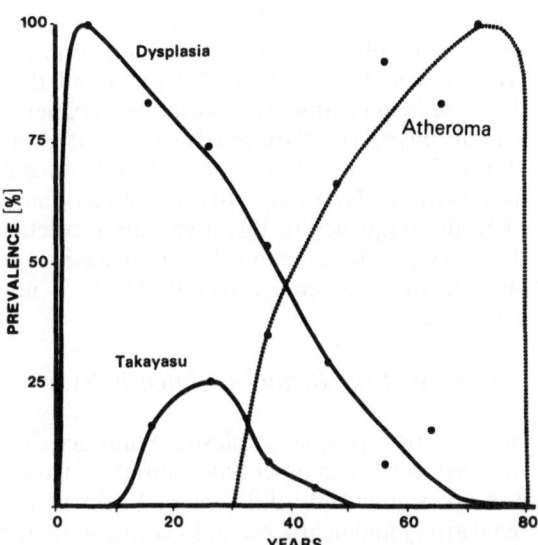

Fig. 29.12. Relative frequency of the different aetiologies in renovascular hypertension.

and particularly cardiac hypertrophy seem to be similar in fibrodysplasia and in atheroma, in our series.

Other Localizations

Associated disease of cervical arteries and other visceral branches of the aorta is rare. Localization in the iliac and splanchnic arteries should be sought on aortography. Detection of cervical localization depends on the clinical search for a

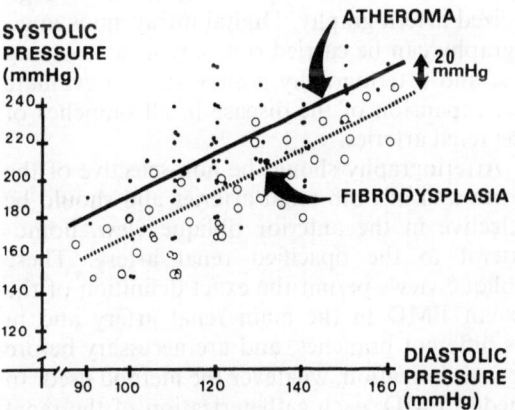

Fig. 29.13. Covariance between systolic and diastolic blood pressure in two populations of patients with renovascular hypertension: atheroma and FMD. The two regression lines were parallel, did not differ in their slopes but differed in their origin intercept, i.e. for the same diastolic blood pressure, systolic blood pressure was more elevated in atheroma than in FMD in relation to the diffusion of atheroma processes to large arteries.

bruit in the carotids and on Doppler ultrasonic and echographic examination of the cervical arteries. We should always keep in mind that the most frequent disease association in patients with fibrodysplasia of the renal artery is the intracranial aneurysm with risks of cerebro-meningial haemorrhage. No prospective study has quantified the frequency of this association exactly, thus, systematic screening for its detection in all patients with renal artery dysplasia is not axiomatic.

Screening of the Renin–Angiotensin System

In our series, peripheral plasma renin activity, sampled in the recumbent and upright posture, was not significantly different in FMD of the renal artery and in atheroma. In contrast, in our experience, the renal vein renin ratio was less frequently significant in fibrodysplasia than in atheroma. This result can be explained by the population of patients who are predominantly female, or by the existence of minor contralateral lesions. In this situation, measurement of renal vein renin content must be done with great technical care. In particular, simultaneous sampling in both renal veins, sampling in the ostium and in the hilus of the vein and acute stimulation of renin release by intravenous injection of converting enzyme inhibitors should be performed.

Definition of the Lesions

Complete definition of the lesions of the disease should be obtained by conventional or digitalized arteriography. Digital intravenous angiography can be carried out only as a screening test and arteriography is necessary to evaluate the expansion of the disease in all branches of the renal arteries.

Arteriography should be non-selective of the aorta and of both renal arteries and should be selective in the anterior oblique view, homolateral to the opacified renal artery. These oblique views permit the exact definition of the extent FMD in the main renal artery and in its different branches, and are necessary before revascularization, whatever the method used. In medial FMD, each catheterization of the renal artery includes a risk of dissection, particularly when the leader is pushed distally in the trunk of the renal artery. FMDs of the renal artery visualized by arteriography were characterized by their frequency on the right side (57%), by the frequency of bilateral forms (25%) and also by the spread of the disease in renal arterial branches. FMD may induce either a focal or segmental stenosis in the intimal and perimedial forms, or an image of stenosis associated with microaneurysm in medial forms. It is in medial forms, the most frequent, that oblique views are the most necessary, the risk of iatrogenic dissection by catheterization the most significant and the existence of true aneurysm the most frequent.

Therapeutic Approach

The proposed therapy in fibrodysplastic stenosis of renal arteries depends predominantly on hypertension. Revascularization of the kidney is a possibility depending on the severity of hypertension and of the resistance to classical medical treatment. In these patients in particular, sensitivity of the blood pressure to treatment with converting enzyme inhibitors should help evaluate the impairment of renal function in the kidney distal to the stenosis.

Renal vascularization may be done by surgery or by endoluminal dilatation. The results obtained by endoluminal dilatation in FMD of the renal artery seem to be better than in atheromatous stenosis [124]. Conversely the iatrogenic risk of dissection, induced by the leader of the catheter, is particularly significant in the medial forms with intramedial microaneurysms. The ruptured internal elastic lamina permits a haemorrhage into the external part of the media. This risks described with selective arteriography of renal artery fibrodysplasia are particularly true for endoluminal dilatation.

Revascularization of the kidney by surgical methods may be complex due to the extension of the disease into the branches of the renal artery, necessitating a bifurcated or trifurcated graft, or ex vivo surgical repair.

Whatever the method used, revascularization of the kidney gives good results in fibrodysplasia. Fifty-four per cent of patients were cured of their hypertension by revascularization, 32% were clearly improved. These results are better than in hypertension due to atheromatous stenosis (24% cured and 50% improved). It is probably due to the absence of associated large artery disease.

References

90. Abdul-Rahman AM, Abu-Salih, Brun A et al. (1978) Fibromuscular dysplasia of the cervico-cephalic arter-

ies. Surg Neurol, 9:217–222

91. Bahar S, Chiras J (1984) Spontaneous vertebro-vertebral arteriovenous fistula associated with fibromuscular dysplasia. Neuroradiology 26:45–49

92. Bellot J, Gherardi R, Poirier J et al. (1985) Fibromuscular dysplasia of cervico-cephalic arteries with multiple dissections and carotido-cavernous fistula. A pathological study. Stroke 16:225–261

93. Bonduelle M, Ruscalleda J, Zalzal P (1973) Dysplasie fibromusculaire avec fistule arterioveineuse de l'artère vertébrale extracranienne. Rev Neurol 128:204–206

94. Boudin G, Guillard A, Romion A (1974) Dysplasies fibromusculaires des artères carotides et vertébrales. Ann Med Interne 12:863–875

95. Corrin LS, Sandok BA, Houser DW (1981) Cerebral ischemic events in patients with carotid artery fibromuscular dysplasia Arch Neurol 38:616–618

96. Dublin AB, Baltaxe HA (1983) Percutaneous transluminal carotid angioplasty in fibromuscular dysplasia. J Neurosurg 59:162–165

97. Edell SL, Huang P (1981) Sonographic demonstrations of fibromuscular hyperplasia of the cervical internal carotid artery. Stroke 12:518–520

98. Garcia-Merino JA, Gutierrez JA (1983) Double lumen dissecting aneurysm of the internal carotid artery in fibromuscular dysplasia. Case report. Stroke 14:815–818

99. Garrido E, Montoya J (1981) Transluminal dilatation of internal carotid artery in fibromuscular dysplasia. A preliminary report. Surg Neurol 16:469–471

100. Gautier JC (1977) Accidents neurologiques des contraceptifs oraux. Arch Suisses Neurol Neurochir Psychiatr 120:335–338

101. Gautier JC, Awada A (1983) Les dysplasies carotidiennes Ann Med Interne 133:465–469

102. Geraud J, Manelfe C, Caussanel JP et al. (1973) Fistule artérioveineuse spontanée de l'artère vertébrale. Rôle éventuel de la dysplasie fibro-musculaire dans sa pathogénie. Rev Neurol 128:206–213

103. Goncharenko V, Gerlock AJ, Shaff MI et al. (1981) Progression of renal artery fibromuscular dysplasia in 42 patients as seen on angiography. Radiology 139:45–51

104. Harrison EG, McCormack LJ (1971) Pathologic classification of renal arterial disease in renovascular hypertension. Mayo Clin Proc 46:161–167

105. Havelius U, Hindfelt B (1982) Carotid fibromuscular dysplasia and paresis of lower cranial nerves (Collet-Sicard Syndrome). J Neurosurg 56:850–853

106. Klaus E, Nekula J (1978) Stenosis (vascular dysplasia) of the internal carotid artery in a child with von Recklinghausen disease. Neuroradiology 15:57–58.

107. Lagneau P, Michel JB (1981) Arterial reconstructive surgery for renovascular hypertension. Arch Surg 116:999–1002

108. Lederman RJ, Salanga V (1976) Fibromuscular dysplasia of the internal carotid artery. A cause of Raeder's paratrigeminal syndrome. Neurology 26:353.

109. Manelfe C, Clarisse J, Fredy D et al. (1974) Fibromuscular dysplasia of the cervico-cephalic arteries. Report of 70 cases. Neuroradiology 1:149–321

110. Meaney TF, Dustan HP, McCormack LJ (1968) Natural history of renal arterial disease. Radiology 9:877—887

111. Mettinger KL (1982) Fibromuscular dysplasia and the brain. II. Current concept of the disease. Stroke 13:53–58

112. Mettinger KL, Ericson K (1982) Fibromuscular dysplasia and the brain. I. Observations on angiographic clinical and genetic characteristics. Stroke 13:46–52

113. Montfort JC, Degos JD, Eizenbaum JF (1981) Aspect artériographique de dysplasie fibromusculaire révélé par une dissection de la carotide. Ann Med Interne 132:333–336

114. Osborn AG (1976) The angiographic spectrum of cervical and intracranial fibromuscular dysplasia. Neuroradiology, 12:1–45

115. Palubinskas AJ, Ripley HR (1964) Fibromuscular hyperplasia in extra-renal arteries. Radiology 82:451–455

116. Pollock M, Jackson BM (1971) Fibromuscular dysplasia of the carotid arteries. Neurology 21:1226–1230

117. Rebollo M. Quintana F. (1983) Giant aneurysm of the intra-cavernous artery and bilateral carotid fibromuscular dysplasia. J Neurol Neurosurg Psychiatry 46:284–285

118. Reddy SV, Karnes WE, Earnest. Spontaneous extracranial vertebral arteriovenous fistula with fibromuscular dysplasia. J Neurosurg 54:399–402

119. Ringel SP, Harrison SH, Norenberg HD et al. (1977) Fibromuscular dysplasia: multiple spontaneous dissecting aneurysm of the major cervical arteries. Ann Neurol 1:301–304

120. Sandok BA (1983) Fibromuscular dysplasia of the internal carotid artery. Neurol Clin 1:17–26

121. Schreiber MJ, Pohl, MA, Novick AC (1984) The natural history of atherosclerotic and fibrous renal artery disease. Urol Clin North Am 11:383–392

122. So EL, Toole, JF, Dolol P et al (1981) Cephalic fibromuscular dysplasia in 32 patients. Arch Neurol 3:619–622

123. So EL, Toole JF, Moody DM (1979) Cerebral embolism form septal fibromuscular dysplasia of the common carotid artery. Ann Neurol 6:75–78

124. Sos TA, Pickering PG, Saddekni S et al. (1984) The current role of renal angioplasty in the treatment of renovascular hypertension. Urol Clin North Am 11:503–513

125. Starr DS, Lawrie GM, Morris GC. (1981) Fibromuscular disease of carotid arteries: long term results of graduated internal dilatation. Stroke 12:196–199

126. Theander G (1960) Arteriographic demonstration of stationary arterial waves. Acta Radiol 53:417–425

127. Vaughan ED (1985) Renovascular hypertension: Kidney Int 27:811–827

128. Vernant JC, Kieffer E, Buisson G (1981) La dysplasie fibromusculaire des artères à destinée cérébrale. A propos de 9 cas observés et opérés en Martinique. Med Int 16:55–63

129. Wells RP, Smith RR (1982) Fibromuscular dysplasia of the internal carotid artery: a long term follow up. Neurosurg 10:39–43

Chapter 30

Cardiovascular Alterations in Inherited Connective Tissue Disorders

M. Depairon and C. M. Lapière

Introduction

Inherited connective tissue diseases are numerous and varied. The different syndromes are subdivided into better defined ailments based on genetic transmission, symptomatology or the molecular alteration causing the disease. The morbidity of these illnesses depends on the type of supporting structure altered, but risk to life often depends on cardiac and peripheral vascular changes.

Cardiovascular pathology is part of a set of manifestations involving functional disturbance of other tissues such as skin, tendons, bones and joints. Other organs can also be altered through defects in their supporting tissue. Vascular diseases found in the heritable connective tissue disorders have in common only those symptoms with which they are associated. Based on this observation, we would like to structure our description by first starting with a separate section on the vascular pathology that can be observed in each disease, before examining the criteria by which clinical diagnosis can be established and considering the pathogenesis at the molecular level.

For each section, the diseases will be described in the order that we proposed in 1976 [72], starting with congenital alterations of the fibrous proteins, i.e. collagen, elastin and other glycoproteins involved in the structure of the connective tissue. We will then examine those diseases caused by defective degradation of various types of proteoglycons and glycos-

aminoglycans, glycoproteins or cell membrane constituents. In a third class are certain inherited disorders, where metabolites affect the composition and functional properties of the connective tissue. In the last group are those heritable diseases which combine cardiovascular manifestations with alterations in specific connective tissues (for review see Pyeritz [109]).

Pathological changes occur in all parts of the cardiovascular tree.

In the heart, fibrous elements, essentially made up of the structural proteins collagen and elastin, form the support for contractile muscles, valves and the base of the large blood vessels. In peripheral blood vessels, these same elements of the connective tissue are found in proportions which vary depending on the type of vessel (see Chaps. 2 and 3). They support the smooth muscles and the endothelium to modulate blood flow, arterial pressure and vascular wall permeability. It is difficult to classify the alterations systematically as a function of a particular type of genetic disease. To help the clinician, we have gathered the cardiac and vascular disease frequencies into Tables 30.1–30.5. The rarity of some of these connective tissue diseases explains reports in the literature of forms presenting with cardiovascular symptomatology only exceptionally, shown in the table with the designation (+). In contrast to this, cardiovascular pathology may be a cardinal symptom of the disease, which is then shown as (+ + +). The intermediate frequency (+ +) indicates an incidence higher than that observed in a normal population. We have flagged with (?) those diseases where more documentation is needed.

Table 30.1. Cardiac abnormalities

Frequency	Myocardial infiltration	Mitral valve prolapses	Valve thickening	Cor pulmonale	Pericardial constriction
+		ED I ED II Cutis laxa (AD) OI III			
+ +		ED III ED VI OI Marfanoïde hypermobility	OI II Scheie Hunter's (mild form) Morquio's	Hurler's Hurler–Scheie Hunter's Maroteaux–Lamy	
+ + +	Hurler's Hurler–Scheie Maroteaux–Lamy I-cell disease Fucosidosis Mannosidosis Lipidoses	Marfan's	Hurler's Hurler–Scheie Hunter's Maroteaux–Lamy I-cell disease Pseudo-Hurler dystrophy Mannosidosis Alkaptonuria	Cutis laxa (AR)	Dwarfism with pericardial constriction (Mulibrey nanism)
?	Sialidoses	ED VIII ED X	Sanfilippo MPS VII Aspartylglycosaminuria		

See text for explanation of +, + +, + + + and ?.
ED, Ehlers–Danlos syndrome; OI, Osteogenesis imperfecta; PXE, pseudoxanthoma elasticum; AD, Autosomal dominant form; AR, Autosomal recessive form.

Table 30.2. Aortic root and pulmonary artery abnormalities

Frequency	Aortic dilatation	Aortic stenosis	Pulmonary artery stenosis
+	Congenital contractural arachnodactyly ED I ED II Cutis laxa (AD)	OI I	
+ +	OI I		
+ + +	Marfan's		Cutis laxa (AR)
?	ED X OI IV		

Key as in Table 30.1.

Fibrous Protein Diseases

Marfan's Syndrome

This disease is transmitted as an autosomal dominant trait, with an incidence of 5 in 100 000. The main alterations are cardiovascular, skeletal and ocular. Significant manifestations in two of the three principal classes of symptoms are required to support the diagnosis [112].

The cardiac abnormality is essentially localized on the valves, especially the mitral valve, in the form of prolapse. The tricuspid valve is altered in the same way, but to a lesser degree. Alterations of aortic and pulmonary valves are also observed [83,101,110]. Anatomical examination of the mitral valve shows redundant leaflets with thickened edges [115].

The most frequent large vessel abnormality is dilation of the proximal aorta. It is progressive and starts with a symmetric enlargement of the sinuses of Valsalva. Onset is in childhood, and

Table 30.3. Peripheral artery abnormalities

Frequency	Rupture	Calcification	Occlusion	Atherosclerosis	Thromboembolism	Tortuosity	High blood pressure
+							
+ +		OI II		Alkaptonuria			Hurler's PXE (AR) I PXE (AD) I Alkaptonuria
+ + +	ED IV	PXE (AR) I PXE (AR) I Arterial calcification in infancy	Hurler's Hurler– Scheie Hunter's Maroteaux– Lamy I-cell disease		Homocystinuria	Menkes arterial tortuosity	
?	ED VI						

Key as in Table 30.1

Table 30.4. Vein abnormalities

Frequency	Varicosities	Thromboembolism
+		
+ +	ED II ED IX	
+ + +		Homocystinuria

Table 30.5. Capillary abnormalities

Frequency	Angiokeratoma
+	
+ +	
+ + +	Fucosidosis Fabry's disease

systematic searches should be made by echo-cardiographic screening [103] in patients and their close relatives [18]. Lethal complications are aortic regurgitation following sinus of Valsalva dilatation and aortic dissection. This starts in the ascending aorta and spreads through the descending aorta or the coronary arteries. Non-invasive tests such as echocardiography are not reliable to establish or exclude the existence of a dissection. Digitized arteriography is required for confirmation [36]. However, echocardiography has been found to be the best technique for exploring patients with Marfan phenotypes [51].

The wall of the ascending aorta is thinner than normal. The main modifications are localized in the media where smooth-muscle cells are sparse and disorganized by the accumulation of baso-philic material. Elastic fibres are partially fragmented (Fig. 30.1). In fact, the so-called cystic necrosis, a descriptive term, in no way allows Marfan lesions to be distinguished from those of other aortic dilatations [52] (see Chap. 32). Skin biopsy can show extensive rarefaction of the elastic fibres (Fig. 30.2).

Dilatation can also be observed, to a lesser degree, in the pulmonary artery.

Musculo-skeletal manifestations are practically constant. Arachnodactyly, the most frequent, is characterized by abnormally long fingers and can be demonstrated by the thumb sign, that is the overflowing of the thumb beyond the hypothenar eminence with the hand closed, and the wrist sign, where the distal phalanges of the first and fifth digit of one hand overlap when encircling the opposite wrist. Dolichostenomely (long, thin arms and legs) is frequent and is accompanied by a size out of proportion with the upper segment. Height is often elevated, and found in the 95th percentile of the patients age group. Frequent thoracic deformities are pectus excavatus or pectus carinatum. The palate may be deformed and highly arched. The existence of flat feet as well as scoliosis is found in close to 50% of patients. Skin striae on the back and inguinal hernia are signs of alterations in the dermis and facias. Hyperextensibility of the joints is often less pronounced than in the Ehlers–Danlos syndrome. This has resulted in a poorly justified classification of some patients as having marfanoid hyperextensibility syndrome.

Ocular manifestations exist in over 60% of patients. Upward lens subluxation is found at

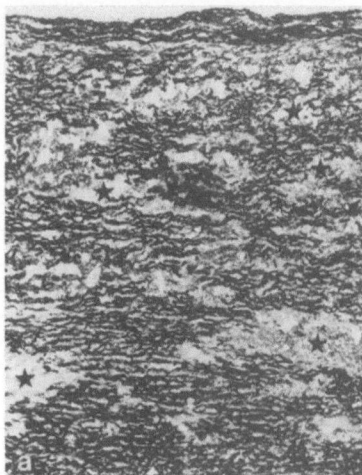

Fig. 30.1a,b. Marfan's syndrome. a Aortic wall showing mucoid within the altered media (*arrows*); elastic fibres are fragmented. Orcein stain, × 75. b Severe changes characterized by extensive destruction of the elastic framework of the media, which is replaced by loose connective tissue, sparing fragmented remnant elastic fibres in the subadventitial area (*arrow*). Orcein stain, × 54

the first examination and seems to exist in utero. Traumatic glaucoma by luxation of the lens in the anterior chamber of the eye has been reported. Myopia is frequent (30%), and the cornea is abnormally flat.

Pulmonary alterations, pneumothorax and bullous emphysema with a reduction of lung capacity and increased residual volume can aggravate pulmonary functional problems due to kyphoscoliosis and thoracic cage deformations.

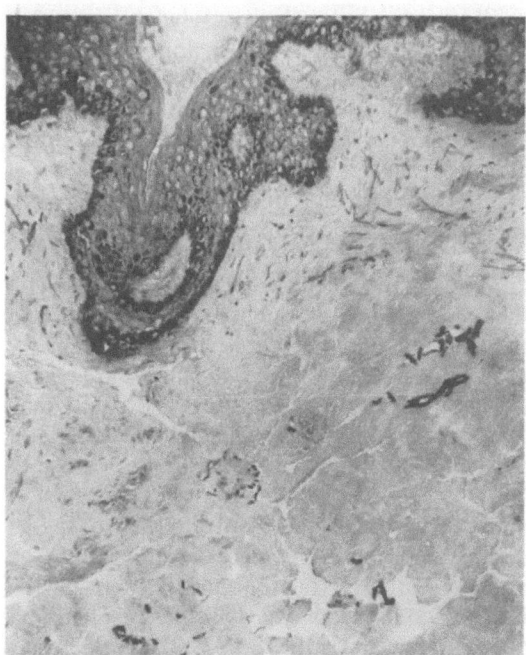

Fig. 30.2. Marfan's syndrome. Skin biopsy specimen showing extensive rarefaction of elastic fibres in the dermis. Orcein stain, × 180 (see normal aspect in Chap. 4)

Different biochemical anomalies involving fibrous proteins or their associative capacities demonstrate the heterogeneity of Marfan's syndrome. Around 50% of patients show increased urinary secretion of hydroxyproline and a lowered proline/hydroxyproline ratio (95), suggesting increased collagen catabolism. A deficiency in synthesizing the cross-links to stabilize collagen and elastin has been noted in some patients [16]; elongation of one of the collagen polypeptides was observed as well [55]. Fibroblasts in certain patients demonstrate, when cultured, abnormally high hyaluronic acid synthesis [71], suggesting interference between an excess of glycosaminoglycans and polymerization of the fibrous protein.

In the marfanoid hypermobility syndrome, without prejudice towards its analogy with Marfan's syndrome, some cardiovascular alterations have been observed, including mitral [48] or aortic [47] regurgitation, mitral valve prolapse [130,3], aortic dilation [27] and coarctation [32].

A syndrome resembling Marfan's disease was described by Beals and Hecht in 1971 (8) under the name "congenital contractural arachnodactyly". Skeletal modifications similar to those in Marfan's syndrome exist, but cardiovascular and occular manifestations are absent. The principal skeletal alteration is arachnodactyly, accompanied by contractures that improve with age. Ears are small and deformed.

Ehlers–Danlos Syndrome (EDS)

The Ehlers-Danlos syndrome is currently described in ten different forms according to clinical manifestations, genetic transmission and, for some forms, a defined molecular disorder. The cardinal signs of the Ehlers–Danlos syndrome, found to different degrees and in different associations in the various forms of EDS, are cutaneous, articular and vascular. Alterations of other connective tissues are also observed (for a review of types I to VIII see [84,21]).

The skin is hyperelastic without laxity. Under traction, it stretches abnormally but returns quickly to its initial position. Teguments are velvety to touch. Cutaneous wounds gape due to retraction of their lips. Healing proceeds normally, but forms thin (cigarette paper) scars. At pressure points, patients can develop molluscoid tumours, probably of traumatic origin. Hypermobility may be evident in small and/or large joints. In the hand, this is characterized by hyperextensibility of fingers in dorsal flexion, a rubbery hand sensation during hand shaking and the ability to project the thumb beyond the hypothenar eminence. Hyperextensibility in the elbow joint is manifested by extension beyond 180°, with greatly increased rotation. Genu recurvatum is frequent. Dislocations are possible. Scoliosis is frequent and varied. In particular forms, rheological abnormalities of other mesenchymal tissues are found such as foetal membrane fragility, diaphragmatic or inguinal hernias, diverticulum, pneumothorax, intestinal ruptures, ocular fragility as well as periodontopathy and so on. Associated with cardiac and peripheral vascular lesions, these different signs with their various degrees of gravity allow the classification on a clinical basis of the different forms of Ehlers–Danlos syndrome.

Genetic transmission is an important consideration, both for the classification as a particular form of the disease and for the determination, within a given syndrome, of the gravity of the case. Genetic investigations, however, are positive in only two-thirds of the observations, suggesting frequent mutations.

Ehlers–Danlos syndrome type I (EDS I) or gravis form is autosomal and dominant. All cutaneous manifestations are present at maximum intensity, with the exception of bruising. Articular hyperlaxity is extensive in both small and large joints, allowing these patients to perform unusual contortions. Besides the valvular and peripheral vascular alterations, connective tissue fragility is also responsible for premature foetal membrane rupture, hernias, diverticula in different locations and articular dislocations. The underlying biochemical defect is unknown. Within connective tissues, including the skin, the bundles of collagen fibres are thin and rarefied [102]. Abnormal polymers can be seen by electron microscopy [134].

EDS II or *mitis* form is similar in its genetic transmission and clinical signs to EDS I, except that the different symptoms are less intense. The two forms, I and II, represent around 75% of patients with suspected Ehlers–Danlos syndrome. No biochemical anomaly has been found, and the morphological modifications are similar to those observed in the preceding form.

Few studies describe cardiovascular manifestations in EDS types I and II, and no certitude exists that their frequency is higher than in normal populations.

Some cardiovascular problems have been reported including atrial and ventricular septal defects, dextrocardia and a bicuspid tricuspid valve [118,11,85,4,75]. The most frequent anomaly remains mitral valve prolapse associated with thoracic pain, arrhythmia or minor regurgitation [17]. Large vessel abnormalities, such as aortic coarctation and symmetric sinus of Valsalva dilatation associated with aortic regurgitation, are also rare [132,121,75]. Aortic dissection is seldom found [75,84]. Stenosis of the peripheral pulmonary artery and a bifid pulmonary artery may occur [75]. Venous abnormalities in the form of varicose veins in the lower limbs are frequent in EDS type II [7,11,17].

EDS III or benign hypermobility is also autosomal dominant in inheritance, and represents around 10% of known forms. Cutaneous symptomatology is minimal. The hyperlaxity of large joints is major, while that of small articulations is variable. The biochemical anomaly is unknown. Morphological modifications are minimal.

Mitral valve prolapse is rare [84,23,75]. Atrial septum defects [75], conduction abnormalities and ectopic beats [23] can be associated.

EDS IV or arterial form is autosomal, either dominant or recessive. This type, especially the recessive form, is obviously the most serious because of the major vascular involvement. Although the skin is not hyperextensible, it is abnormally thin and transparent. The major cutaneous symptom is the presence of ecchymoses. Articular modifications are minimal. Other connective tissues are fragile, explaining the intestinal and uterine ruptures, and pneumothorax.

In both forms, there is a reduction of type III collagen due to a diminution of synthesis in the recessive form, or intracellular retention in the dominant form. A type III collagen synthesis deficiency has also been reported in patients presenting with rupture of a cerebral aneurysm without evident clinical EDS III manifestations [106,93].

The arterial form is characterized by the fragility of vascular walls and can result in ruptures, either spontaneous or following minor trauma [82]. This is in turn responsible for haemorrhages or arteriovenous fistula formation [7,77,117]. Clinical exploration of these patients must be done by digital angiography, as arteriography may lead to haemorrhage at the puncture site [77,31].

Morphological abnormalities are described in this form. Arterial alterations are shown by thinning of the vessel wall, especially the media, and fragmentation of the internal elastic lamina [70]. Saccular aneurysms are rare. In the dermis, skin biopsy specimens can exhibit changes in the collagenous framework which appears loosened, with hypertrophy of the elastic fibres, especially in the upper dermis around arterioles; macrophages and deposits of haemosiderin can be associated and are due to previous microhaemorrhages (Fig. 30.3).

EDS V is one of the few X-linked transmitted forms. Cutaneous and articular symptomatology is similar to EDS II. It is uncommon. The molecular defect is unknown. Cardiac manifestations including mitral valve prolapse with regurgitation have been reported in the rare cases studied [37].

EDS VI or ocular form is also infrequent. The genetic transmission is autosomal recessive. All symptoms are found with an intermediate severity. Ocular globe fragility is very striking, and accompanied by keratocone, microcornea and retinal detachment. Scoliosis is frequent. There can be a certain degree of arachnodactyly.

The disease results from lysine hydroxylase inactivity [103] and the formation of abnormal cross-links due to the absence of hydroxylysine.

Cardiac manifestations consist of mitral valve prolapse [103] and mitral regurgitation [126]. Large-vessel abnormalities are represented by aortic rupture [109]. The few descriptions of this form in the literature involved young subjects whose clinical evolution maust be followed. They must be considered as high-risk patients and be frequently re-examined until more information is available.

EDS VII or arthrochalasis multiplex is an uncommon form transmitted in autosomal dominant or recessive form. Cutaneous alterations are minimal while articular alterations are major, height is small and scoliosis frequent.

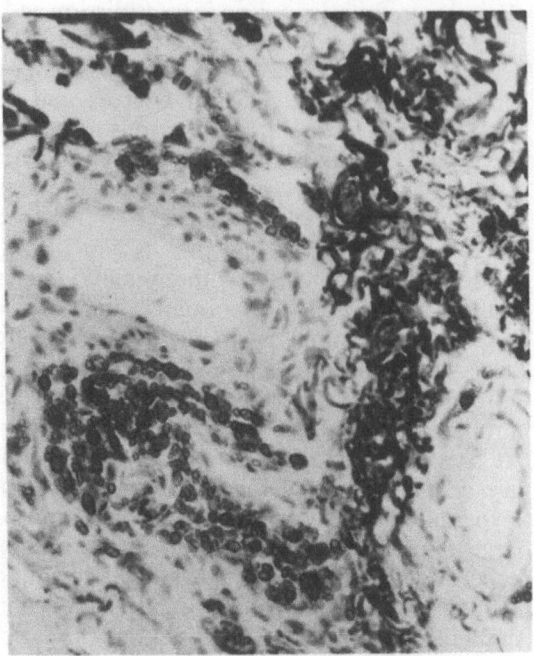

Fig. 30.3. Type IV Ehlers–Danlos syndrome. Elastic fibres appear thickened, densely packed around small vessels of the reticular dermis and associated with haemosiderin deposits. Orcein stain, ×450 (Courtesy Prof. Chomette, Pitié Salpétrière, Paris)

No cardiovascular manifestations have been reported.

In the dominant form, the substitution of an amino acid at the cleavage site of the polypeptidic extension has been found in one patient [127]. Accumulation the precursor of type I collagen in the recessive form seems to be related to pro-collagen peptidase inactivity [76].

EDS VIII or periodontal form is also rare and transmitted in dominant autosomal mode. Cutaneous symptomatology is quite similar to that of EDS II. Articular alterations predominate in small joints. The distinctive characteristic is a major periodontitis of early onset, resulting in precocious toothlessness if no care is taken. The biochemical anomaly is unknown [59].

The only cardiac manifestation reported was mitral valve prolapse [59].

EDS IX is rare, with an X-linked transmission mode. Cutaneous symptomatology is not manifest, with the exception of a certain degree of hyperelasticity. Small joints are hypermobile. This form is characterized by occipital exotosis and bladder diverticula [78]. Only severe abdominal and lower limb varicose veins have been described [109].

A defect in the metabolism of copper, a lysine oxidase co-factor, seems to be responsible for the reduction in activity of this enzyme [22,69]. Collagen and elastin cross-linking is reduced.

EDS X or fibronectin linked form is apparently autosomal recessive. Cutaneous manifestations are of intermediate intensity as in EDS II, while hyperlaxity is mainly present in small joints. The characteristic is biochemical and related to platelet dysfunction, corrected by fibronectin [5].

Mitral valve prolapse, auriculo-ventricular conduction problems and moderate aortic dilatation were reported in the same patient [5].

Cutis Laxa

The phenotype of these patients is characterized by abnormally loose, but not hyperelastic, skin with deep folds giving an appearance of premature senility. In addition, the pulmonary parenchyma is greatly altered with emphysema and bronchiectasis [12].

The autosomal dominant form is rare, principally characterized by cutaneous lesions. The autosomal recessive form reveals itself very soon after birth, by cutaneous alterations, severe pulmonary manifestations and urinary and intestinal diverticula. Joints may be hypermobile and dislocations are present. Psychomotor retardation can exist. An acquired form is beyond the scope of this chapter.

This disease is the result of disorganization of the elastic fibres by a mechanism that is, as yet, unknown.

In the autosomal recessive form, cardiac manifestations in the form of heart failure are often present from birth. Electrocardiograms often display right ventricular hypertrophy associated with conduction delay [137]. A systolic murmur is generally observed at the pulmonary level. Evidence of cardiomegaly can be present in clinical or radiographic examinations. These symptoms should be related to the development of lung alterations, severe in these patients, whether associated with pulmonary artery stenosis or not [137]. Large-vessel disorders are pulmonary artery stenosis or dilatation of the ascending aorta [135,137].

Microscopically, large-vessel alterations are similar to those found in Marfan's syndrome and are accompanied by a reduction in the number of elastic fibres (Figs. 30.4 and 30.5) [109].

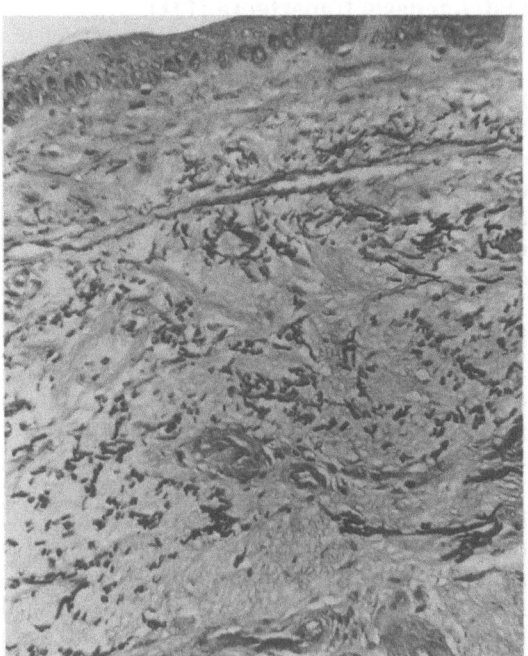

Fig. 30.4. Cutis laxa. Skin biopsy showing fragmentation of the elastic fibres. Orcein stain, × 180

In the autosomal dominant form, various cardiac manifestations have been reported including right ventricular conduction delay [118], ventricular septal defects [12] and mitral valve prolapse with moderate dilation of the sinus of Valsalva [20]. Peripheral artery disorders are marked by dilated and tortuous carotid vessels [53].

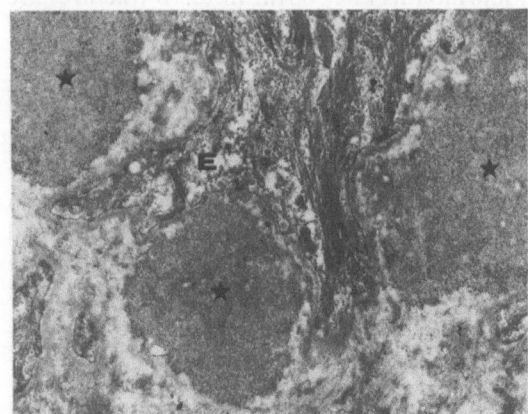

Fig. 30.5. Cutis laxa. Electron micrograph. In the upper dermis elastic fibres are replaced by microfilamentous material (*Stars*) *E*, epidermal cells. Uranyl lead, × 50 000

Osteogenesis Imperfecta (OI)

The symptoms characteristic of OI are fractures from minor trauma, accompanied by deafness, blue sclerae and dentinogenesis imperfecta (opalescent teeth). The different clinical forms are classified into four types, depending on the age of onset, genetic transmission and the existence of one or another of the cardinal symptoms of determined intensity [119,120].

OI is the syndrome in which is found the largest number of alterations in the genes for type I collagen, including deletions and duplication of exons, as well as other mutations [107]. These genetic alterations extend the heterogeneity of each clinical class.

OI I is an autosomal dominant form where fractures of varying intensity, number and age of onset are found, accompanied or not by blue sclerae, sometimes by deafness, as well as inconstant dental anomalies. Already on a clinical basis, this class is heterogeneous. Fractures are very rarely present at birth, they increase in number until puberty, and decrease thereafter.

Cardiac manifestations are the result of the malfunction of mitral and aortic valves, with or without regurgitation [19,54]. Aortic root dilatation is a large-vessel abnormality found with average frequency in these patients [29,84,38,57]. it is difficult to know whether this dilatation is characteristic of this disease, as in Marfan's syndrome, or the consequence of valve incontinence [111].

OI II is the lethal perinatal form. It is autosomal recessive and characterized by multiple fractures in utero, congenital deformations and incomplete skull bone formation. Sclerae are blue and the infant usually dies in utero or within a few months after birth. In utero fractures are often multiple, as opposed to single fractures in the other forms.

The descriptions of cardiovascular anomalies in this disease are based on pathological studies, due to the early deaths of affected children as a result of fractures, haemorrhages or respiratory infections. Cardiac manifestations result from alterations in the valve leaflets [49,113] and arterial manifestations from wall calcification [44,84,125]. The pathological changes observed are imprecise, calcification frequently being encountered. Other non-specific lesions are observed.

OI III is autosomal recessive. Fractures can occur in utero and persist throughout the rest of life. This results in progressive bone deformation. Dentinogenesis is imperfect in one-half of the patients. This is the most commonly encountered form of OI. The rare cardiac manifestations result from mitral valve prolapse, which are usually asymptomatic.

OI IV is the most uncommon and least severe form. Transmission is autosomal dominant. Fractures are infrequent and occur mostly during childhood. Other signs are absent, except dentinogenesis which may be imperfect. The only cardiac abnormality described in the literature is aortic regurgitation with dilatation of the aortic root [109].

Pseudoxanthoma Elasticum (PXE)

The characteristic clinical feature of PXE is the presence of yellowish vascular lesions of the skin at the armpit, the inner side of the elbow or the lateral side of the neck though they may sometimes be more widespread, depending on the clinical form. Ocular lesions are present as angioid streaks of the fundus that can lead to blindness. This syndrome also includes digestive haemorrhages and vascular manifestations. The different clinical forms are classified as proposed by Pope [105]. The pathogenesis of the lesions is unclear, beyond the morphological alterations of elastic fibres, that are fragmented and calcified.

PXE I is autosomal recessive. This is the most frequent form, where the cutaneous manifestations and ocular alterations are most marked. The ocular prognosis and vascular manifestations are responsible for the severity of the disease.

Primary cardiac manifestations are due to alteration and calcification of elastic fibres in the endocardium and valves. This is rarely of clinical significance [24,61,84,14,89]. It can be secondary to altered coronary artery or hypertension and result in ischaemia and myocardial infarction. The most frequently encountered lesions are mitral valve prolapse, cardiomegaly, cardiac failure, variable blocks, angina, myocardial infarction and valve dysfunction leading to sudden death [25].

Large-vessel manifestations are rare. The aorta is relatively spared, with the exception of reports of two cases of dilatation [84] and one case of stenosis [90]. Muscular and elastic arteries suffer from wall fragility and constriction of the lumen. Clinical manifestations are intermittent lower limb claudication, angina, abdominal claudication, cerebrovascular accidents and renovascular hypertension. The fragility is responsible, at all ages, for gastrointestinal,

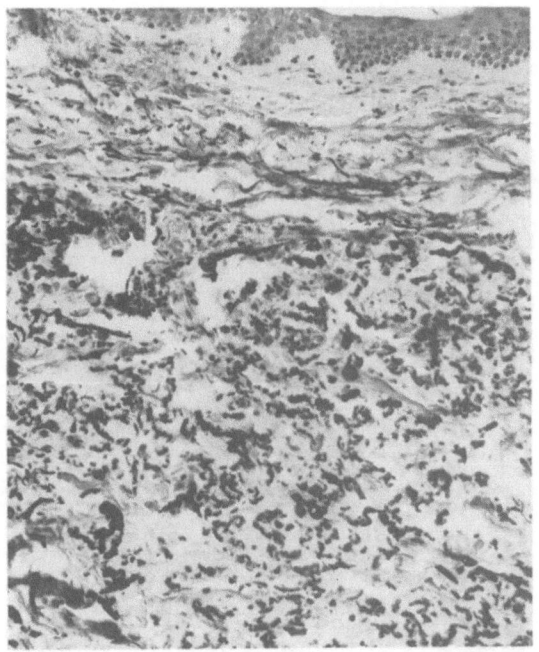

Fig. 30.6. Pseudoxanthoma elasticum (PXE). Skin biopsy. Elastic fibres appear numerous, fragmented and coiled. Orcein stain, × 125 (Courtesy Dr. Lessano-Leibovitch, Hôpital Tarnier, Paris)

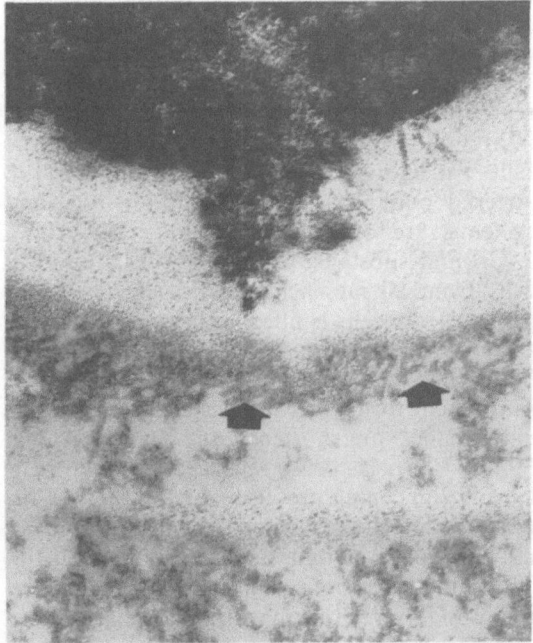

Fig. 30.7. Pseudoxanthoma elasticum (PXE). Electron micrograph. Dermal elastic fibres exhibit massive electron dense calcifications of their core of elastin, whereas peripheral microfilamentous component is normal (*arrows*). Note that collagen fibres are ultrastructurally normal. Phosphotungstic acid–uranyl lead counterstaining, × 116 000

cerebral, retinal, renal and uterine haemorrhages [42,136,50,84].

Morphological modifications of elastic fibres are most characteristic in the skin. They consist of elastic fibre fragmentation and deposition of calcium salts and polyanions (Figs. 30.6 and 30.7) [81]. In large arteries, calcification occurs in punctuate, or occasionally circular, subendothelial deposits. In muscular arteries, the internal elastic lamina is fragmented and calcified, muscle cell density is reduced, while the wall is thickened by deposition of fibroelastic material. An identical fibroelastic material deposit is observed in the endocardium [89].

PXE II is a rare autosomal recessive disease. Only cutaneous symptomatology is present. No cardiac abnormalities have been reported [104,90,2].

PXE III is autosomal dominant, and differs from type I only by its mode of genetic transmission. Cardiovascular alterations are similar to type I.

PXE IV is autosomal dominant and characterized by less intense cutaneous and ocular symptomatology. No cardiac abnormalities have been reported.

Storage Diseases

Clinical symptomatology of ground substance diseases will be briefly mentioned to differentiate those in which peripheral vascular anomalies are observed. Most are accompanied by cardiac alterations. They are classified as a function of the molecular alteration, depending on the nature of the incomplete degradation products which cause the storage disease.

Polysaccharidoses

Polysaccharidoses result from the incomplete degradation of glycosaminoglycans. These lysosomal diseases are now better defined on the basis of the missing enzyme activity responsible for the defect [88,86].

Hurler's Syndrome

Hurler's syndrome is autosomal recessive and depends on an alpha-L-iduronidase deficiency. It is characterized by dwarfism, coarse facies, progressive skeletal alterations, mental deterio-

522 Diseases of the Arterial Wall

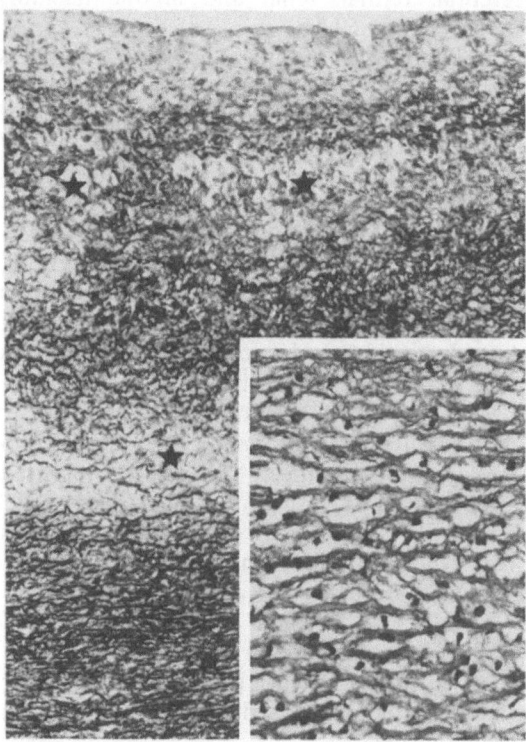

Fig. 30.8. Hurler's syndrome. Elastic artery showing areas of clear cells within the media (*stars*). H & E, ×108. *Inset.* Detail of the cells displaying finely vacuolized cytoplasm. H & E, ×200 (Courtesy Prof. Nezelof, Hôpital Necker, Paris)

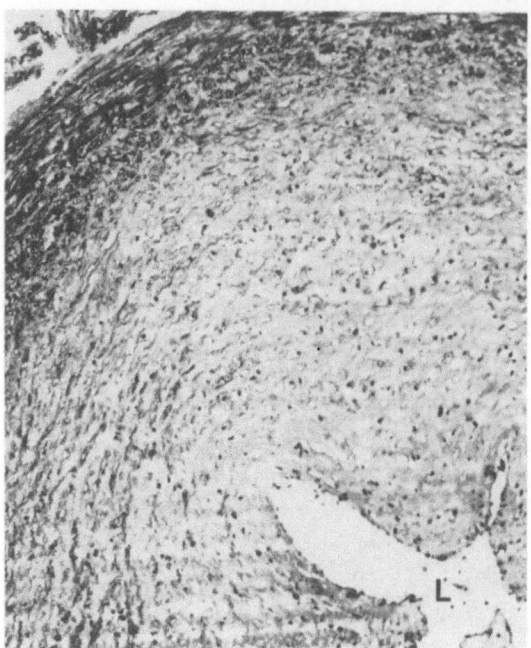

Fig. 30.10. Hurler's syndrome. Involvement of the coronary artery. Marked intimal thickening with reduction of the lumen (*L*). H & E, ×110

ration, hepatosplenomegaly, corneal clouding and retinal degeneration. Normally, death occurs before age 10 years.

Cardiac manifestations are aortic and mitral regurgitation [13], angina [28] and myocardial infarction [40] resulting in congestive heart failure in more than half the patients [66,114]. Arterial manifestations, both pulmonary and systemic, are found. Hypertension is frequent.

Ground substance accumulation is found in cells from all mesenchymal tissues. The cytoplasm of the cells is filled with vacuoles, giving them a ghost- or gargoyle-like appearance (Fig. 30.8). This change appears in varying degrees in different forms of polysaccharidosis. Cardiac valve leaflets are shortened and thickened (Fig. 30.9). Fibroelastosis, myocarditis and pericardial thickening are also observed [67]. Coronary arteries are sometimes narrowed by a circular non-atheromatous fibrosis (Fig. 30.10). The same alterations, but to different degrees, are found in other forms of polysaccharidosis.

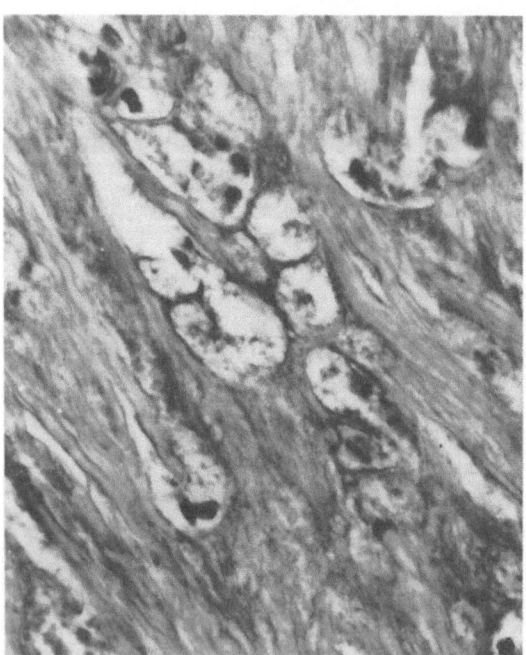

Fig. 30.9. Hurler's syndrome. Involvement of a valvular cusp showing clusters of vacuolized cells partially positive with alcian blue stain, within collagenous tissue. Alcian blue stain, ×560

Scheie Syndrome

The Scheie syndrome is autosomal recessive and also depends on an alpha-L-iduronidase deficiency, different, however, from the preceding one. It resembles Hurler's syndrome, but is less pronounced. Size is normal, and psychomotor development is almost unaltered. The cornea is cloudy and hepatosplenomegaly is of medium severity. Arthropathies can develop prematurely. Aortic regurgitation often results in decreased life expectancy [84] due to heart failure [34].

Hurler–Scheie Syndrome

The Hurler–Scheie syndrome is autosomal recessive and is the result of a combination of the two preceding alterations. It associates a Hurler phenotype with normal intelligence, hepatomegaly, some facial swelling, corneal clouding and multiple exostosis. The vital prognosis is less compromised than in Hurler's syndrome.

Cardiac manifestations are systolic and diastolic murmurs, without a predilection for the aortic valve. Generally, signs of cardiac decompensation bring patients for cardiological evaluation. Echocardiography may show valvular thickening and hypertrophy of both ventricles [128,66].

Hunter's Syndrome

Hunter's syndrome is X-linked and depends on an iduronate sulphatase deficiency. Clinical symptomatology resembles that of Hurler's syndrome, but is less severe, and mortality is delayed. Psychomotor development is less altered, and a minor form seems to exist as well. A recessive form has been described.

Those patients with the severe form develop cardiovascular anomalies similar to those found in Hurler's syndrome. In its moderate form, the cardiovascular anomalies are similar, but appear at a slightly more advanced age. A predilection for pulmonary hypertension is observed [84]. This may be indirect in origin and dependent on right ventricular insufficiency. In the recessive autosomal form, no cardiac anomaly has been reported.

Sanfilippo Syndrome

The Sanfilippo syndrome groups four types of diseases, similar from a clinical point of view,

related to Hunter's syndrome, but with less marked hepatomegaly, corneal clouding, skeletal modifications and facial changes. Mental retardation can be severe. The four subgroups depend on different enzymatic alterations. Cardiovascular abnormalities are rare. One case of severe initial regurgitation has been reported [58].

Morquio's Syndrome

Morquio's syndrome is autosomal recessive and dependent on a galactosamine-6-sulphatase deficiency. Patients are small and present with thoracic deformations, pigeon-chest, articular hyperlaxity, corneal clouding, progressive deafness and dental alterations. Prognosis is conditioned by spinal deformities which can provoke neurological or pulmonary complications. The syndrome can exist in a mild form. The rare cardiovascular abnormalities noted have been aortic regurgitations [84] not functional murmurs.

Maroteaux–Lamy Syndrome

The Maroteaux–Lamy syndrome is autosomal recessive and depends on an N-acetylgalactosamine-4-sulphatase deficiency. It is characterized by multiple dysotoses. As well as the classic form, intermediate and mild forms exist.

Cardiovascular anomalies are frequent, similar to those found in Hunter's syndrome and dominated by cardiac valve dysfunction [84]. In the intermediate form, aortic stenosis with regurgitation can evolve towards failure [122,139].

Type VII and VIII Syndromes

Patients with type VII and VIII syndromes excrete large amounts of heparan- and keratansulphates or dermatan-sulphate. Clinical aspects are represented by one or another of the signs described in the preceding syndromes. These syndromes are rare.

Mucolipidoses

The absence of a specific hydrolase [94] can result in the accumulation of degradation products from different types of glycoprotein and lipoprotein. Recognized more recently, through progress in biochemical analysis, most of these diseases are named after the incomplete degradation product which accumulates in lysosomes, except for the following two whose identification is different.

I-Cell Disease

I-cell disease is autosomal recessive and is severe, due to visceral and neurological alterations which appear in childhood and result in early death. Mental retardation, hernias, hip dislocations and thoracic deformations accompany facial and bony alterations similar to those described in the preceding syndromes.

Cardiovascular manifestations such as valve thickening, myocardial infiltration [80] and cardiomegaly play an important part in the clinical evolution [123]. Large-vessel manifestations depend on subintimal proliferation in the aortic wall [80].

Pseudo-Hurler Polydystrophy

Pseudo-Hurler polydystrophy is also a complex autosomal recessive disease, less severe than Hurler's syndrome, presenting analogous symptomatology, but with a later onset. Pseudo-Hurler polydystrophy displays cardiovascular manifestations, usually of the same type as in the preceding form, notably with stenosis and mitral and aortic valve regurgitation [84,65].

Glycoproteinoses

Glycoproteinoses are classified according to the enzyme alteration responsible for the disease [9].

Sialidoses

The sialidoses are autosomal recessive diseases. Symptomatology is similar to that for Hurler's syndrome although onset is later and death less precocious. In one type where myoclonus and cherry red spots are found, symptomatology is less marked and there is an absence of mental modifications. An even less severe partial deficiency has also been described.

Cardiomegaly is the main cardiac manifestation [64], and can be accompanied by pericarditis and a nephrotic syndrome [6], or by mitral regurgitation [63]. No cardiovascular anomalies have been described in the myoclonus–cherry red spot syndromes, nor in partial neuraminidase deficiencies.

Aspartyl Glycosaminuria

Aspartyl glycosaminuria is probably autosomal recessive. It is characterized by progressive facial coarsening, reduced skin elasticity, bone fragility, hoarseness and lenticular opacities and is accompanied by progressive mental retardation.

Cardiac manifestations such as cardiomegaly, mitral regurgitation and aortic valve thickening [62] are present. Arterial anomalies are distinguished by tortuous cerebral arteries of small diameter, with consequent slow blood flow [98].

Fucosidoses

Fucosidoses are autosomal recessive and are characterized by psychomotor retardation from 2 years of age onwards, and signs of storage in skin, liver and nervous system. In some patients, evolution is much less severe and they may present with angiokeratoma similar to that found in Fabry's disease.

Patients show cardiac manifestations such as cardiomegaly, extrasystoles and incomplete right bundle branch block [41]. However, parenchymal infiltration of the lungs and endothelium by incomplete degradation products often leads to death from neurological alterations or infectious complications before cardiovascular problems become clinically significant. No cardiovascular manifestations are noted in the mild form accompanied by angiokeratoma [68]. Cardiac muscle cells are deformed by cytoplasmic vacuoles containing clear or lamellar bodies [41]. Skin lesions similar to those in Fabry's disease can be found [68].

Mannosidoses

These are also autosomal recessive diseases and clinical characteristics include a coarse face with a flat nose and prognathism, gingival hyperplasia, posterior cataract, hepatomegaly and multiple dysotoses accompanying psychomotor retardation and deafness. Clinical severity is variable. No cardiovascular abnormalities have been reported.

Lipidoses

Lipidoses rarely affect the connective tissue of the cardiovascular system, except in severe forms

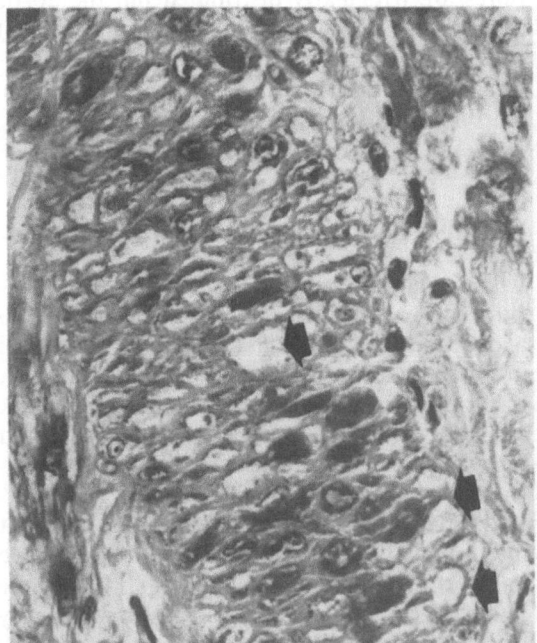

Fig. 30.11. Fabry's disease. Muscular artery showing intimal thickening associated with intense vacuolization of the cytoplasm of the medial smooth-muscle cells (*arrows*). Masson's trichrome, × 750

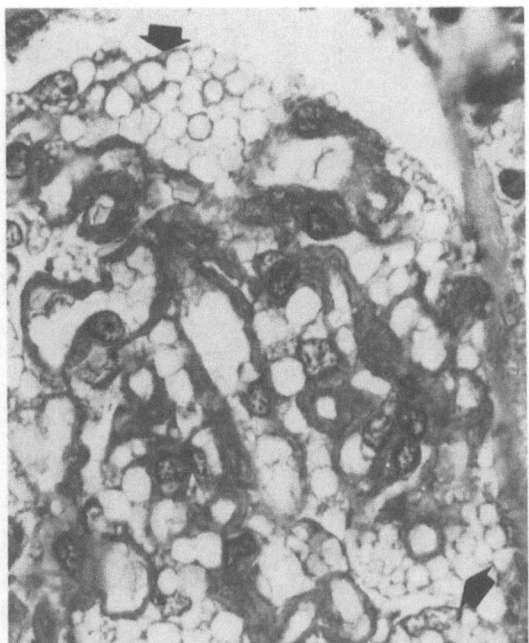

Fig. 30.12. Fabry's disease. The cytoplasm of glomerular epithelial cells appears markedly vacuolized (foam cells). Masson trichrome, × 850

where cardiac manifestations result from infiltration of the valves [84,87].

GM1 Gangliosidosis

GM1 gangliosidosis, rapidly lethal, is autosomal recessive. The phenotype is similar to that of Hurler's syndrome. A late onset form with less marked manifestations is, however, fatal before age 10 years. Manifestations are mainly neurological.

Fabry's Disease

Fabry's disease, with its characteristic angiokeratoma (angiokeratoma corporis diffusum) consists of an inborn X-linked error of glycosphingolipid metabolism, resulting from the deficient activity of the enzyme ceramide trihexosidase, a specific alpha galactosidase [66a]. It is a rare condition showing a combination of skin lesions and involvement of cutane organs, particularly the heart, blood vessels and kidneys [20a, 43a, 118a]. The lesions result from accumulation of glycolipid ceramide trihexoside in the cells which appear vacuolated (foam cells). Changes in the blood vessels concern primarily

the smooth-muscle cells (Fig. 30.11). Involvement of the media is associated with marked thickening of the intima and reduction of the lumen. In the kidney, there is foamy vacuolation of the glomerular cells, particularly the epithelial cells (Fig. 30.12). usually proteinuria and hypertension precede renal failure. The cardiovascular changes can be responsible for cardiomegaly and congestive cardiac failure. Electron microscopy shows characteristic dense laminated lysosomal inclusions (Fig. 30.13).

Inborn Metabolic Diseases with Secondary Effects on Fibrous Proteins

There are three diseases of this type. The first two result in the production of metabolites which react with specific sites on fibrous proteins and interfere with their association as polymers. The third disease results from altered copper metabolism affecting lysyl oxidase function, an enzyme required to insure the stability of fibrous protein polymers.

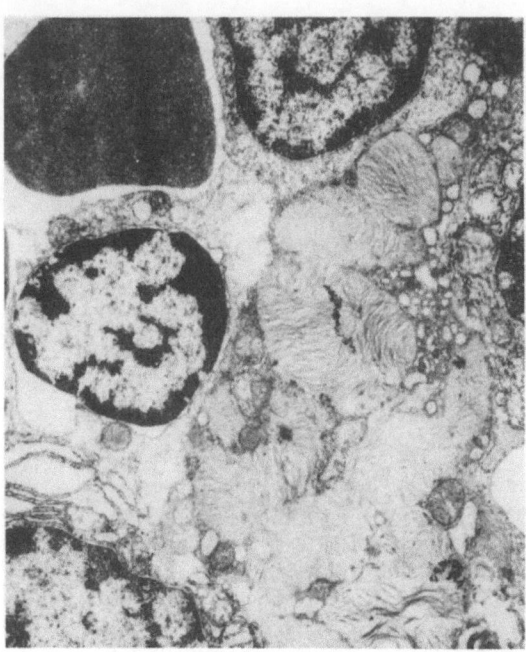

Fig. 30.13. Fabry's disease. Electron micrograph showing intracellular accumulation of lysosomal laminated membrane inclusions. Uranyl lead, × 20 000

Homocystinuria

Homocystinuria is autosomal recessive and presents a phenotype reassembling that found in Marfan's syndrome except that there is progressive early onset mental retardation, osteoporosis, malar flush and characteristic vascular lesions [91].

Congenital cardiopathic, valve lesions and primary myocardial manifestations are more frequent than in normal populations. In particular, ischaemic heart disease is relatively frequent in young subjects. Major complications are thrombosis and lower limb arterial insufficiency, pulmonary embolism, cerebrovascular accidents and thrombophlebitis [116]. Less frequently observed are cerebral [45,84], renal or mesental venous thromboses with hypertension or renal infarction [45,116].

Some vascular signs can be observed on the skin, such as livedo reticularis, malar flush and dilatation of superficial thoracic veins, often as a result of thrombosis of the inferior vena cava.

Microscopic alterations in arteries are characterized by thinning of the media, fewer smooth-muscle cells and an increased intercellular space. The internal elastic lamina may be irregular and fragmented. In the aorta, elastic fibres are irregularly distributed but the aortic alterations seen in Marfan's syndrome are not present. The intima is generally the site of fibro-elastic proliferation which deforms the lumen. Thrombi are frequent [2].

Alkaptonuria

Alkaptonuria is autosomal recessive. The deposition of homogentisic acid (ochronosis) due to a deficiency of its own oxidase is manifested in all connective tissues, especially in cartilage, resulting in degenerative arthropathy and a dark staining of sclerae. Urine is brown coloured when exposed to air. The clinical manifestations worsen with age [124].

Cardiac manifestations in alkaptonuria result in chronic valve disease. Stenosis of the aortic valve is the most frequent [10,38,142]. These anomalies are often associated with coronary arterial disease which may lead to myocardial infarction [26,46,84]. Moderate arterial hypertension is often observed in these patients [38,92]. It cannot be ruled out that this contributes to the development of arteriosclerosis as well as cardiac decompensation. The characteristic lesion is a deposit of granular black pigment in fibroblasts or in neighbouring collagen fibres. The myocardium is often less pigmented than other connective tissues. Arteries are sclerotic and deposits in the intima are often pigmented and calcified [84].

Menkes Syndrome

The Menkes syndrome (kinky hair disease) has an X-linked inheritance pattern and is manifested from birth by thin, curly, red hair, rapidly accompanied by growth retardation and cardiovascular and neurological manifestations [33]. Copper transport defects are complex and different from those in EDS IX [100].

Its clinical presentation is essentially dominated by the severity of the neurological problems. Arteriography has, however, shown dilatation and elongation of the cerebral arteries [1,138,39,33]. The femoral artery is also involved [138], as are the visceral arteries [108]. Many arteries present macroscopic and microscopic alterations; they may be diffusely dilated, with thin and irregular walls but without intimal proliferation. The internal elastic lamina is irregular and fragmented [96].

Various Hereditary Disorders Affecting Connective Tissue and the Vascular Network

Osteochondrodysplasia. Osteochondrodysplasia often results in rapid death. Postmortem studies show cardiovascular manifestations that are probably secondary to pulmonary disease or thoracic malformations [30,79].

Larsen's Syndrome. Larsen's syndrome [73] is characterized by a prominent forehead and hypertelorism with a depressed nasal bridge. Dislocations of large joints are frequent. Hands and feet are deformed. There are inconsistent cardiovascular anomalies [56]. The syndrome may present with cardiac manifestations such as a large ventricular septal defect [56] or with simultaneous heart and large-vessel manifestations [131].

Infantile Arterial Calcification. Infantile arterial calcification is an autosomal recessive. Infants seem normal during their first months, before manifesting clinical symptoms dependent on the localization of the vascular calcification [140].

A diffuse, calcifying arteriopathy strikes the coronary arteries, while often sparing cerebral vessels (Fig. 30.14) [129].

Arterial Tortuosity

Arterial tortuosity in its isolated form involves principally the large- and medium-calibre muscular arteries and is sometimes accompanied by dilatation of the aortic root with regurgitation [43] or stenosis of the pulmonary artery [15]. This type of vascular alteration has also been reported associated with the Ehlers–Danlos syndrome [74] and cutis laxa [135].

Dwarfism

Dwarfism with pericardial constriction is characterized by intrauterine and post-natal growth retardation. It is accompanied by muscular hypotonia and dilatation of cerebral ventricles, as well as retinopathy [133] and a thickening of the pericardium with pericardial constriction [133]. Microscopic examination shows a thickened pericardium, often adhering to the epicardium. Calcification may be present [133].

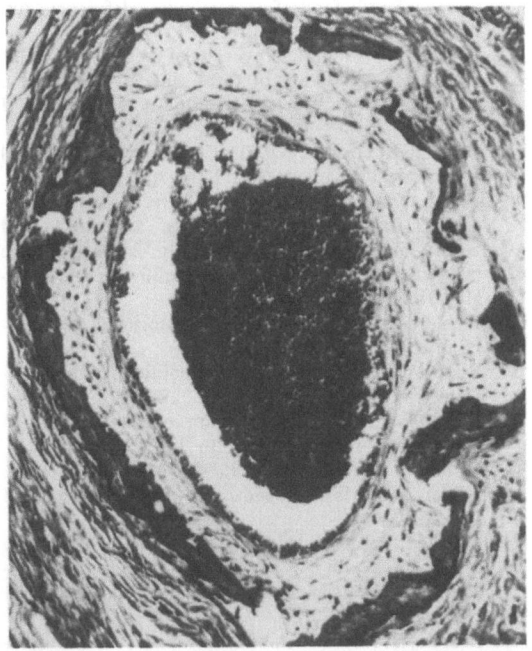

Fig. 30.14. Infantile arterial calcification. Massive calcium deposits in the media of the coronary artery, associated with intimal thickening. H & E, ×85

Winchester Syndrome

The Winchester syndrome is characterized by joint stiffness occurring in infancy, contractures and kyphoscoliosis. Precocious osteoporosis is observed, as well as delayed eruption of the teeth and atlanto-axial subluxation [60]. Cardiovascular alterations are inconsistent.

References

1. Aguilar MJ, Chadwick DL, Okuyama K et al. (1966) Kinky hair disease. I. Clinical and pathological features. J Neuropathol Exp Neurol 25:507–522
2. Almgren B, Eriksson I, Hemmingsson A et al. (1978) Abdominal aortic aneurysm in homocystinuria. Acta Chir Scand 144:545–548
3. Anderson RE, Koch S, Camerini-Otero RD (1984) Cardiovascular findings in congenital contractural arachnodactyly: report of an affected kindred. Am J Med Genet 18:265–271
4. Antani J, Srinivas HV (1973) Ehlers–Danlos syndrome and cardiovascular abnormalities. Chest 63:214-217
5. Arneson MA, Hammerschmidt DE, Furcht LT et al. (1980) A new form of Ehlers–Danlos syndrome: fibronectin corrects defective platelet function. JAMA 244:144–147
6. Aylsworth AS, Thomas GH, Hood JL (1979) The

severe infantile form of neuraminidase deficiency. Am J Hum Genet 31:68A (abstract)

7. Barabas AP (1967) Heterogeneity of the Ehlers–Danlos syndrome: description of three clinical types and a hypothesis to explain the basic defect(s). Br Med J ii:612–613

8. Beals RK, Hecht F (1971) Congenital contractural arachnodactyly. J Bone Joint Surg [Am] 53:987–993

9. Beaudet AL (1983) Disorders of glycoprotein degradation: mannosidosis, fucosidosis, sialfydosis and aspartylglucosaminuria. In: Stanbury JB et al. (eds) The metabolic basis of inherited diseases, 5th edn. McGraw-Hill, New York, pp 788–802

10. Beddard AP (1910) Ochronosis associated with carboluria. QJ Med 3:329–337

11. Beighton P (1969) Cardiac abnormalities in the Ehlers–Danlos syndrome. Br Heart J 31:227–232

12. Beighton P (1972) The dominant and recessive forms of cutis laxa. J Med Genet 9:216–221

13. Berenson GS, Geer JC (1963) Heart disease in the Hurler and Marfan syndromes. Arch Intern Med 111:58–69

14. Bete JM, Banas JS, Moran J et al. (1975) Coronary artery disease in an 18-year-old girl with pseudoxanthoma elasticum: successful surgical therapy. Am J Cardiol 36:515–520

15. Beuren AJ, Hort W, Kalbfeisch H et al. (1969) Dysplasia of the systemic and pulmonary arterial system with tortuosity and lengthening of the arteries: a new entity, diagnosed during life, and leading to coronary death in early childhood. Circulation 39:109–115

16. Boucek RJ, Noble NL, Gunja-Smith Z et al. (1981) The Marfan syndrome: a deficiency in chemically stable collagen cross-links. N Engl J Med 305:988–991

17. Brandt KD, Sumner RD, Ryan TJ et al. (1975) Herniation of mitral leaflets in the Ehlers–Danlos syndrome. Am J Cardiol 36:524–528

18. Bruno L, Tredici S, Mangiavacchi M et al. (1984) Cardiac, skeletal, and ocular abnormalities in patients with Marfan's syndrome and in their relatives; comparison with the cardiac abnormalities in patients with kyphoscoliosis. Br Heart J 51:220–230

19. Browell JN, Drake EH (1966) Aortic valve lesions associated with osteogenesis imperfecta. Henry Ford Hosp Med Bull 14:245–247

20. Brown FR, Holbrook KA, Byers PH et al. (1982) Cutis laxa. Johns Hopkins Med J 150:148–153

20a. Burkholder PM, Ware RA, Reese VG (1980) Clinicopathologic, enzymatic and genetic features in a case of Fabry's disease. Arch Pathol Lab Med 104:17–25

21. Byers PH, Barsh GS, Holbrook KA (1981) Molecular mechanisms of connective tissue abnormalities in the Ehlers–Danlos syndrome. Coll Relat Res 1:475–489

22. Byers PH, Siegel RC, Holbrook KA et al. (1980) X-linked cutis laxa. N Engl J Med 303:61–65

23. Cabeen WR, Reza MJ, Kovick RB et al. (1977) Mitral valve prolapse and conduction defects in Ehlers–Danlos syndrome. Arch Intern Med 137:1127–1231

24. Carlborg U, Ejrup B, Gronblad E et al. (1959) Vascular studies in pseudoxanthoma elasticum and angioid streaks. Acta Med Scand 350:1–84

25. Coffman JD, Sommers SC (1959) Familial pseudoxanthoma elasticum and valvular heart disease. Circulation 19:242–250

26. Coodley EL, Greco AJ (1950) Clinical aspects of ochronosis. Am J Med 8:816–822

27. Cotton DJ, Brandt KD (1976) Cardiovascular abnor-

malities in the marfanoid hypermobility syndrome. Arthritis Rheum 19:763–768

28. Craig WS (1954) Gargoylism in a twin brother and sister. Arch Dis Child 29:293–296

29. Criscitiello MG, Ronan JA, Besterman EMM et al. (1965) Cardiovascular abnormalities in osteogenesis imperfecta. Circulation 31:255–261

30. Curran JP, Sigmon BA, Opitz JM (1974) Lethal forms of chondrodysplastic dwarfism. Pediatrics 53: 76–85

31. Dalton ML, Bricker DL, Nannini L (1974) Spontaneous rupture of the subclavian artery and innominate vein. Arch Surg 190:552–554

32. Daneshwar A, Tavakoli D, Nazarian J (1979) Marfanoid hypermobility syndrome associated with coarctation of the aorta. Br Heart J 41:621–623

33. Danks DM (1983) Hereditary disorders of copper metabolism in Wilson's disease and Menke's disease. In: Stanbury JB et al. (eds) The metabolic basis of inherited disease 5th edn. McGraw-Hill, New York, pp 1251–1268

34. Dekaban AS, Herman MM, Constantopoulos G et al. (1976) Mucopolysaccharidosis type V (Scheie syndrome). Arch Pathol Lab Med 100:237–245

35. De Los Arco E, Urquia M, Torrano E et al. (1973) Osteogenesis imperfecta e insufficiencia aortica. Rev. Clin Espano 133:531–536

36. Detrano R, Moodie DS, Gill CC et al. (1985) Intravenous digital subtraction aortography in the preoperative and postoperative evaluation of Marfan's aortic disease. Chest 88:249–253

37. DiFerrante N, Leachman RD, Angelini P et al. (1975) Ehlers–Danlos type V (X-linked form): a lysyl oxidase deficiency. Birth Defects 11:31–37

38. Dijkstra PF, Van Vugt AC (1977) Alkaptonurie. Ned Tijdschr Geneeskd 121:2069–2073

39. Dobrescu O, Larbrisseau A, Dubie LJ et al. (1980) Trichopoliodystrophy or Menkes disease. Can Med Assoc J 123:490–497

40. Dorfman A, Matalon R (1972) The mucopolysaccharidoses. In: Stanbury JB et al. (eds) The metabolic basis of inherited diseases, 3rd edn. McGraw-Hill, New York, pp 1218–1272

41. Durand P, Borrone C, Della Cella C (1969) Fucosidosis. J Pediatr 75:665–674

42. Eddy DD, Farber EM (1962) Pseudoxanthoma elasticum – internal manifestations – a report of cases and a statistical review of the literature. Arch Dermatol 86:729–740

43. Ertugrul A (1967) Diffuse tortuosity and lengthening of the arteries. Circulation 36:400–407

43a. Ferrans VJ, Hibbs RG, Burda CD (1969) The heart in Farbry's disease, a histochemical and electromicroscopic study. Am J Cardiol, 24:95–110

44. Follis RH (1952) Osteogenesis imperfecta congenita: a connective tissue diathesis. J Pediatr 41:713–721

45. Gibson JB, Carson NAJ, Neill DW (1964) Pathological findings in homocystinuria. J Clin Pathol 17:427–437

46. Goldston MJ, Steel JM, Dobriner K (1952) Alcaptonuria and ochronosis. Am J Med 13:432–452

47. Goodman RM, Baba N, Wooley CF (1969) Observations on the heart in a case of combined Ehlers–Danlos and Marfan syndrome. Am J Cardiol 24:734–742

48. Goodman RM, Wooley CF, Frazier RL (1965) Ehlers–Danlos syndrome occuring together with the Marfan syndrome. N Engl J Med 273:514–519

49. Gordon CA (1928) Osteogenesis imperfecta congenita. Am J Obstet Gynecol 16:214–218

50. Grison D, Eisenmann B, Baehrel B et al. (1976) Les complications vasculaires de l'élastorrhexie systématisée. Chirurgie 102:440–448

51. Gruber MA, Graham TP, Engel E et al. (1978) Marfan syndrome with contractural arachnodactyly and severe mitral regurgitation in a premature infant. J Paediatr 93:80–82

52. Haleston PS, Leonard JC (1979) Dissecting aortic aneurysms. II. Histopathology of the aorta. QJ Med 48:63–76

53. Hayden JG, Talner NS, Klaus SN (1968) Cutis laxa associated with pulmonary artery stenosis. J Pediatr 72:506–509

54. Heckman BA, Steinberg I (1968) Congenital heart disease (mitral regurgitation) in osteogenesis imperfecta. AJR 103:601–607

55. Henke E, Leader M, Tajima S et al. (1985) A 38 base pair insertion in the pro alpha 2 (I) collagen gene of a patient with Marfan syndome. J Cell Biochem 27:169–174

56. Henriksson P, Ivarsson S, Theander G (1977) The Larsen syndrome and glial proliferation in the brain. Acta Pediatr Scand 66:653–657

57. Hepper RL, Babitt HI, Bianchine JW et al. (1973) Aortic regurgitation and aneurysm of sinus of valsalva associated with osteogenesis imperfecta. Am J Cardiol 31:754–657

58. Herd JK, Robinson H (1973) Type III mucopolysaccharidosis: report of a case with several mitral valve involvement. J Pediatr 82:101–104

59. Hollister DW (1982) Clinical features of Ehlers–Danlos syndrome, types VIII and IX. In: Glimcher MJ, Bornstein P (eds) AAOS symposium on heritable disorders of connective tissue. Mosby, St Louis

60. Hollister DW, Rimoin DL, Lachman RS et al. (1974) The Winchester syndrome A nonlysosomal connective tissue disease. J. Pediatr 84:701–709

61. Huang S, Kuman G, Steele HD et al. (1967) Cardiac involvement in pseudoxanthoma elasticum. Am Heart J 74:680–686

62. Isenberg JN, Sharp HL (1975) Aspartylglucosaminuria: psychomotor retardation masquerading as a mucopolysaccharidosis. J. Pediatr 86:713–717

63. Kelly TE, Bartoshesky L, Harris HJ et al. (1981) Mucolipidosis I (acid neuraminidase deficiency): three cases and delineation of the variability of the phenotype. Am J Dis Child 135:703–708

64. Kelly TE, Graetz G (1977) Isolated acid neuraminidase deficiency: a distinct lysosomal storage disease. Am J Med Genet 1:31–46

65. Kelly TE, Thomas GH, Taylor HA et al. (1975) Mucolipidosis III (pseudo-Hurler polydystrophy): clinical and laboratory studies of 12 patients. Johns Hopkins Med J (1978) 137:156–175

66. Kohn G, Bach G, Lasch E et al. (1978) A new phenotypic variant of alpha L-iduronidase deficiency. Monogr Hum Genet 19:7–10

66a. Kolodny EH (1976) Current concepts in genetica-lysosomal storage disease N Engl J Med 294:1217–1220

67. Korvetz LJ, Schiebler GL (1972) Cardiovascular manifestations of the genetic mucopolysaccharidoses. Birth Defects 8:192–196

68. Koucef BG, Beratis NG, Strauss L et al. (1976) Fucosidosis type 2. Pediatrics 57:205–213

69. Kuivaniemi H, Peltonen L, Palotie A et al. (1982) Abnormal copper metabolism and deficient lysyl oxidase activity in heritable connective tissue disorder.

J Clin Invest 69:730–733

70. Krane SM, Trelstad RL (1979) Case records of the Massachusetts General Hospital. N Engl J Med 300:129–135

71. Lamsberg SJ, Dorfman A (1973) Synthesis and degradation of hyaluronic acid in the cultured fibroblasts of Marfan's disease. J Clin Invest 52:2428–2433

72. Lapiere CM, Nusgens B (1976) Collagen pathology at the molecular level. In: Ramachandran GN, Reddi AH (eds) Biochemistry of collagen. Plenum Press, New York pp 377–447

73. Larsen LT, Schottstaedt ER, Bost FC (1950) Multiple congenital dislocations associated with characteristic facial abnormality. J Pediatr 37:574–581

74. Lees MH, Menashe VD, Sunderland CO et al. (1969) Ehlers–Danlos syndrome associated with multiple pulmonary artery stenoses and tortuous systemic arteries. J Pediatr 75:1031–1036

75. Leier CV, Call TD, Fulkerson PK et al. (1980) The spectrum of cardiac defects in the Ehlers–Danlos syndrome, types I and III. Ann Intern Med 98:171–178

76. Lichtenstein JR, Martin GR, Kohn LD et al. (1973) Defect in conversion of procollagen to collagen in a form of Ehlers–Danlos syndrome. Science 182:298–300

77. Lynch HT, Larsen AL, Wilson R et al. (1965) Ehlers–Danlos syndrome and "congenital" arteriovenous fistulae. A clinicopathologic study of a family. JAMA 194:163–166

78. MacFarlane JD, Hollister DW, Weaver DD et al. (1980) A new Ehlers–Danlos syndrome with skeletal dysplasia. Am J Hum Genet 32:118 (abstract)

79. Maroteaux P, Stanescu V, Stanescu R (1976) The lethal chondrodysplasias. Clin Ortho 114:31–45

80. Martin JJ, Leroy JG, Farriaux JP et al. (1975) I-cell disease (Mucolipidosis II). Acta Neuropathol 33:285–305

81. Martinez-Hernandez A, Huffer WE et al. (1978) Resolution and repair of elastic tissue calcification in pseudoxanthoma elasticum. Arch Pathol Lab Med 102:303–305

82. McFarland W, Fuller DE (1964) Mortality in Ehlers–Danlos syndrome due to spontaneous rupture of large arteries. N Engl J Med 271:1309–1310

83. McKusick VA (1955) The cardiovascular aspects of Marfan's syndrome: a heritable disorder of connective tissue. Circulation 11:321–341

84. McKusick VA (1972) Heritable disorders of connective tissue. Mosby, St Louis

85. McKusick VA (1972) Association of congenital bicuspid aortic valve and Erdheim's cystic medial necrosis. Lancet I:1026–1027

86. McKusick VA, Neufeld E (1983) The mucopolysaccharide storage disorders. In: Stanbury JB et al. (eds) The metabolic basis of inherited disease, 5th edn. McGraw-Hill, New York, pp 751–777

87. McKusick VA, Neufeld EF, Kelly TE (1978) The mucopolysaccharide storage diseases. In: Stanbury JB, Wyngaarden JB, Fredrickson DS (eds) The metabolic basis of inherited disease, 4th edn. McGraw-Hill, New York pp 1282–1307

88. McKusick VA, Pyeritz RE (1980) Genetic heterogeneity and allelic variation in the mucopolysaccharidoses. Johns Hopkins Med J 146:71–72

89. Mendelsohn G, Bulkley BH, Hutchins GM (1978) Cardiovascular manifestations of pseudoxanthoma elasticum. Arch Pathol Lab Med 102:298–302

90. Messis CP, Budzilovich GN (1970) Pseudoxanthoma elasticum: Report of an autopsied case with cerebral

involvement. Neurology 20:703–709

91. Mudd SH, Levy HL (1983) Disorders of trans-
sulfuration. In: Stanbury JB et al. (eds) The metabolic
basis of inherited disease, 5th edn. McGraw-Hill, New
York, pp 522–559

92. Muhlen-Hoffmeister E, Reinwein D (1979) Endokrine
Ausfalle bei Alkcaptonurie? Med Klin 74:877–882

93. Neildwyer G, Bartlett JR, Nicholss AC et al. (1983)
Collagen deficiency and ruptured cerebral aneurysms.
A clinical and biochemical study. J Neurosurg 59:16

94. Neufeld EF, McKusick VA (1983) Disorders of lysosomal enzyme synthesis and localization: I-cell disease
and pseudo-Hurler polydystrophy. In: Stanbury JB et
al. (eds) The metabolic basis of inherited diseases, 5th
edn. McGraw-Hill, New York, pp 778–787

95. Nusgens B, Lapiere CM (1973) The relationship
between proline and hydroxyproline urinary excretion
in human as an index of collagen catabolism. Clin Chim
Acta 48:203–211

96. Oakes BW, Danks DM, Campbell PE (1976) Human
copper deficiency: ultrastructural studies of the aorta
and skin in a child with Menke's syndrome. Exp Mol
Pathol 25:83–98

97. Painvin GA, Weisel RD, David TE et al. (1980) Surgical
treatment of annuloaortic ectasia. Can J Surg 23:445–
449

98. Palo J, Mattsson K (1970) Eleven new cases of aspartylglucosaminuria. J Ment Defic Res 14:168–173

99. Papaioannou AC, Matsaniotis N, Cantez T et al. (1970)
Marfan syndrome. Onset and development of cardiovascular lesions in Marfan syndrome. Angiology
21:580–590

100. Peltonen L, Kuivaniemi H, Palotie A et al. (1983) Alter-
ations in copper and collagen metabolism in the
Menkes syndrome and a new subtype of the Ehlers–
Danos syndrome. Biochemistry 22:6156–6163

101. Phornphutkul C, Rosenthal A, Nadas AS (1973)
Cardiac manifestations of Marfan syndrome in infancy
and childhood. Circulation 48:587–596

102. Pierard GE, Pierard-Fanchimont C, Lapiere CM (1983)
Histopathological aid at the diagnosis of the Ehlers–
Danlos syndrome, gravis and mitis types. Int J Derm-
atol 22:300–304

103. Pinnell SR, Krane SM, Kenzora JE et al. (1972) A
heritable disorder of connective tissue: hydroxylysine-
deficient collagen disease. N Engl J Med 286:1013–1020

104. Pope FM (1974) Autosomal dominant pseudoxanthoma elasticum. J Med Genet, 11:152–157

105. Pope FM (1975) Historical evidence for the genetic
heterogeneity of pseudoxanthoma elasticum. Br J
Dermatol 92:493–509

106. Pope FM, Nicholls AC, Narcisi P et al. (1981) Some
patients with cerebral aneurysms are deficient in type
III collagen. Lancet I:973–975

107. Prockop DJ, Kivirikko KI (1984) Heritable diseases of
collagen. N Engl J Med 311:376–386

108. Procopis P, Camakaris J, Danks DM (1981) A mild
form of Menkes steely hair syndrome. J Pediatr 98:97–
99

109. Pyeritz RE (1983) Cardiovascular manifestation of
heritable disorders of connective tissue. In: Steinberg
AG et al. (eds) Genetics of cardiovscular diseases. Saunders, Philadelphia, pp 191–302 (Progress in medical
genetics V)

110. Pyeritz RE, Brinker JA, Varghese PJ (1979) Clinical
and echocardiographic correlates in 127 young Marfan
patients. Clin Res 27:196 (abstract)

111. Pyeritz RE, Levin LS (1981) Aortic root dilatation

and valvular dysfunction in osteogenesis imperfecta.
Circulation 64:IV–311 (abstract)

112. Pyeritz RE, McKusick VA (1979) The Marfan syndrome: diagnosis and management. N Engl J Med
300:772–777

113. Remigio PA, Grinvaslky HT (1970) Osteogenesis
imperfecta congenita: association with conspicuous
extraskeletal connective tissue dysplasia. Am J Dis
Child 119:524–528

114. Renteria VG, Ferrans VJ (1976) Intracellular collagen
fibrils in cardiac valves of patients with the Hurler
syndrome. Lab Invest 34:263–272

115. Roberts WC (1979) Congenital cardiovascular abnor-
malities usually "silent" until adulthood. In: Roberts
WC (ed) Congenital heart disease in adults. FA Davis
Co, Philadelphia

116. Schimke RN, McKusick VA, Tuang T et al. (1965)
Homocystinuria: studies of 20 families with 38 affected
members. JAMA 193:711–719

117. Schoolman A, Kepes JJ (1967). Bilateral spontaneous
carotid-cavenous fistulae in Ehlers–Danlos syndrome.
J Neurosurg 26:82–86

118. Sestak K (1962) Ehlers–Danlos syndrome and cutis
laxa: an account of families in the Oxford area. Ann
Hum Genet 25:313–321

118a. Schatzki PF, Kipreos B, Payne J (1979) Fabry's
disease. Primary diagnosis by electron microscopy. Am
J Surg Pathol 3:11–214

119. Sillence DO, Romoin DL, Danks DM (1979) Clinical
variability in osteogenesis imperfecta – variable ex-
pressivity or genetic heterogeneity. Birth Defects 5B:
113–119

120. Sillence DO, Senn A, Danks DM (1979) Genetic heterogeneity in osteogenesis imperfecta. J Med Genet
16:101–116

121. Simon AP, Stein PD (1978) Aortic insufficiency in
Ehlers–Danlos syndrome. Angiology 25:290–296

122. Spranger JW, Koch F, McKusick VA et al.
(1970) Mucopolysaccharidosis VI (Maroteaux-Lamy's
disease). Helv Paediatr Acta 25:337–362

123. Spritz RA, Doughty RA, Spackman TJ et al. (1978)
Neonatal presentation of I-cell disease. J Pediatr
93:954–958

124. Srsen S (1979) Alkaptonuria. Johns Hopkins Med J
145:217–226

125. Stefani M (1977) Lesioni aortiche in un caso di osteogenesis imperfecta. Boll Soc Ital Cardiol 22:867–869

126. Steinmann BU, Gitzelmann UR, Vogel A et al. (1975)
Ehlers–Danlos syndrome in two siblings with deficient
lysyl hydroxylase activity in cultured skin fibroblasts
but with only mild hydroxylysine deficit in skin. Helv
Paediatr Acta 30:255–274

127. Steinmann B, Tuderman L, Peltonen L et al. (1980)
Evidence for a structural mutation of procollagen type
I in a patient with the Ehlers–Danlos syndrome type
VII. J Biol Chem 155:8887–8893

128. Stevenson RE, Cain JR, Riopel DA et al. (1981) Early
death in Hurler-Scheie disease (MPS IH/IS compound).
Proc Greenwood Genet Cen 1:79–83

129. Stryker WA (1946) Arterial calcification in infancy with
special reference to the coronary arteries. Am J Pathol
22: 1007–1031

130. Sulica LO, Siggers DC (1975) The marfanoid hypermobility syndrome. Birth Defects 11:335–358

131. Swensson RE, Linnebur AC, Paster SB (1975) Striking
aortic root dilatation in a patient with the Larsen syndrome. J Pediatr 86:914–915

132. Tucker DH, Miller DE, Jacoby WJ (1963) Ehlers–

Danlos syndrome with a sinus of valsalva aneurysm and aortic insufficiency simulating rheumatic heart disease. Am J Med 35:715–720

133. Tuuteri L, Perheentupa J, Rapola J (1974) The cardiopathy of Mulibrey nanism, a new inherited syndrome. Chest 65:628–631

134. Vogel A, Holbrook KA, Steinmann B et al. (1979) Abnormal collagen fibril structure in the gravis form (type I) of Ehlers–Danlos syndrome. Lab Invest 40:201–206

135. Wagstaff LA, Firth JC, Levin SE (1970) Vascular abnormalities in congenital generalized elastolysis (cutis laxa): report of a case. S Afr Med J 44:1125–1127

136. Wahlquist ML, Fox RM, Beech AM et al. (1977) Peripheral vascular disease as a mode of presentation of pseudoxanthoma elasticum. Aust NZ J Med 7:523–525

137. Weir EK, Joffe HS, Blaufuss AH et al. (1977) Cardiovascular abnormalities in cutis laxa. Eur J Cardiol 5:255–261

138. Wesenberg RL, Gwinn JL, Barnes GR (1969) Radiological findings in the kinky-hair syndrome. Radiology 92:500–506

139. Wilson CS, Mankin HT, Pluth JR (1980) Aortic stenosis and mucopolysaccharidosis. Ann Intern Med 92:496–498

140. Witzleben CL (1970) Idiopathic infantile arterial calcification – a misnomer ? Am J Cardiol 26:305–309

141. Wong FL, Friedman S, Yakovac W (1961) Cardiac complications of Marfan's syndrome in a child. Am J Dis Child 107:404–409

142. Wyre WJ Jr (1979) Alkaptonuria with extensive ochronosis. Arch Dermatol 115:461–463

Chapter 31

Aneurysms and Arterial Dissections

M. Fievez

Aneurysms are permanent, irreversible, localized dilatations of arteries. They differ from arterial ectasias which are regular, non-expansive and mostly generalized dilatations of vessels. A true aneurysm is formed by dilatation of the vessel, and its wall is composed of recognizable components of the arterial wall. In contrast, false aneurysms result from rupture of the vessel wall and are in fact organized perivascular haematomas in which no elastic and/or muscle tissue remnants can be identified.

Arterial dissection (dissecting haematoma, dissecting aneurysm) is caused by blood dissecting between the layers of the arterial wall, and arteriovenous aneurysms (cirsoid aneurysms, racemose haemangiomas) result from varying degrees of arteriovenous shunting (see Chap. 36).

A saccular aneurysm is an outpouching resulting from a localized thinning of the vessel wall; it communicates with the parent vessel by a relatively narrow channel. Fusiform aneurysms involve the whole circumference of the arterial wall. Whatever the aetiology, the aneurysmal wall is composed of dense, hyalinized fibrous tissue of variable thickness. Calcification is found on x-ray or in resected specimens. The adventitia includes many vessels and perivascular infiltration of mononuclear cells is seen. In saccular aneurysms, organized thrombus may lead to obliteration and apparent healing.

Aortic Aneurysms

The frequency of aortic aneurysms at postmortem examination has varied in various series

from 0.5%–3% in early studies, to 4.5%–6.6% in more recent work [7,33]. The increased frequency of aneurysms overall in the general population predominantly results from the increasing number of aneurysms of the abdominal aorta [14,18,25,30]. Asymptomatic small aneurysms of less than 5 cm in diameter were recognized ten times more often in the 1970s than in the 1950s and the incidence of aneurysms measuring 5–7 cm or more than 7 cm has increased 2.4 and 2.6 times respectively over the same period [62]. This is in contrast with the decreasing number of aneurysms of the thoracic aorta [14]. These changes may have resulted from the decline of syphilis, the increasing life span of the population and recent improvements in diagnosis. However, other factors may be at work, such as alcoholism and increased use of tobacco. Auerbach [7] has noted that aortic aneurysms are about eight times more frequent in cigarette smokers (20 to 40 cigarettes a day) than in non-smokers; in ex-smokers and cigar smokers, the ratios are 6:1 and 5:1 respectively when compared with non-smokers.

Aortic aneurysms are six to nine times more frequent in men than in women and usually develop over the age of 60. They are rare under the age of 50. However, aortic aneurysms have been reported in paediatric patients with coarctation and in young adults (for example, aneurysm of the sinus of Valsalva [9]). Aortic aneurysms may be single or multiple, and are often associated with aneurysms at other locations (Figs. 31.1 and 31.2). Among the 1510 patients in Crawford's series [20], 191 patients (12.6%) had multiple aortic aneurysms either simultaneously or successively. Aneurysms of the descending segment of the thoracic aorta were associated with lesions in the abdominal

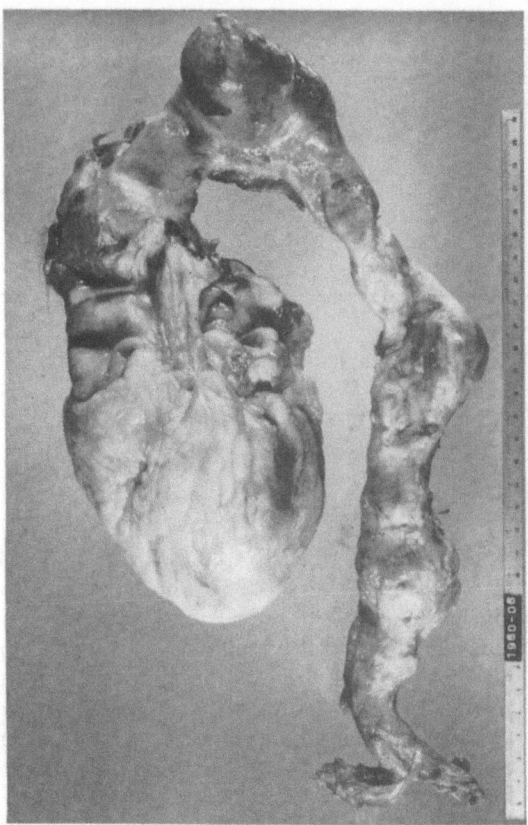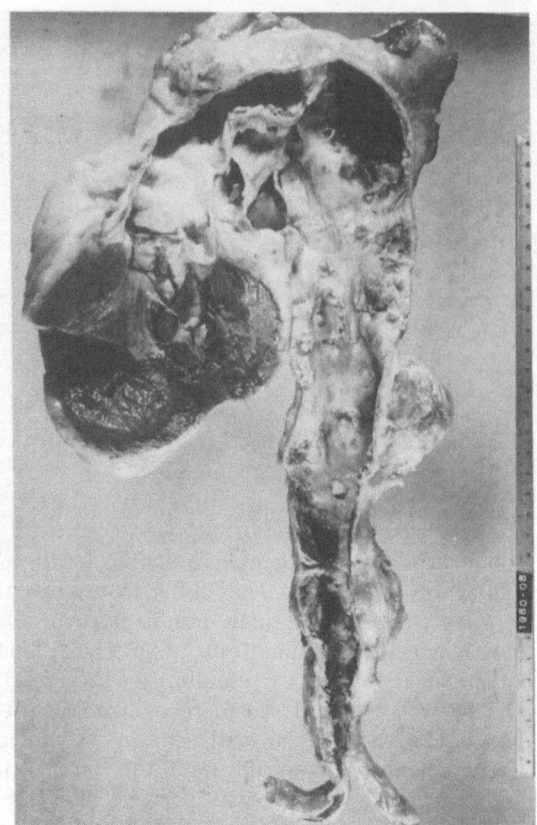

Fig. 31.1. Multiple aneurysms of the saccular and fusiform types in the aorta.

Fig. 31.2. Multiple aneurysms. Same case as Fig. 31.1. Calcified and ulcerated atherosclerosis with intra-aneurysmal thromboses.

segment below the renal artery in 44% of cases. In decreasing order of frequency aneurysms are located in subdiaphragmatic, supradiaphragmatic and rarely the sub-combined and supradiaphragmatic segments. Their diameters range from 1 to 15 cm (mean 5.2 ± 0.25 cm) [7].

The natural history of aneurysms is characterized by expansion, resulting from haemodynamic factors operating in accordance with Laplace's law, and the progressive weakening of the vessel wall. The mean increase in diameter of aortic aneurysms has been evaluated at 0.4 cm per year using repeat echotomography. Nevertheless the growth rate, which appears to be independent of the age of the patient and other factors such as diastolic and systolic blood pressures, may be rapid, with enlargement by 0.5 cm within a brief period of 3 months. In other cases the growths may be slow or arrested for long periods of time [12]. Such expansion may be conditioned by two factors. The first is concerned with the geometry of aneurysms which usually acquire a fusiform shape. The configuration is better approximated to by a sphere than a cyl-

inder of the same diameter. Secondly, increased collagen in the wall may limit expansion in the early stages. Localized disruption of the aneurysmal wall with progressive excavation and dilatation of the underlying adventitia and media may lead to the formation of intramedial or perianeurysmal haematomas (secondary false aneurysms) (Fig. 31.3). Small breaks in the wall may allow lipoproteins to enter the surrounding tissues, where they may produce the inflammatory aneurysms reported by Feiner et al. [32] (see Chap. 28).

Expansion of an aortic aneurysm can lead to the compression of surrounding organs and favour thrombosis and embolization, which may secondarily increase ischaemia in organs such as the kidneys or colon [30]. Eventual rupture seems inevitable and this risk increases with the diameter of the aneurysm. Rupture can occur into the pericardial cavity, the pleural cavity, the lung or the oesophagus for intrathoracic aneurysms, and into the retroperitoneum, the peritoneal cavity, the duodenum (Fig 31.4) or the urinary system for intra-abdominal aneurysms. Rupture into

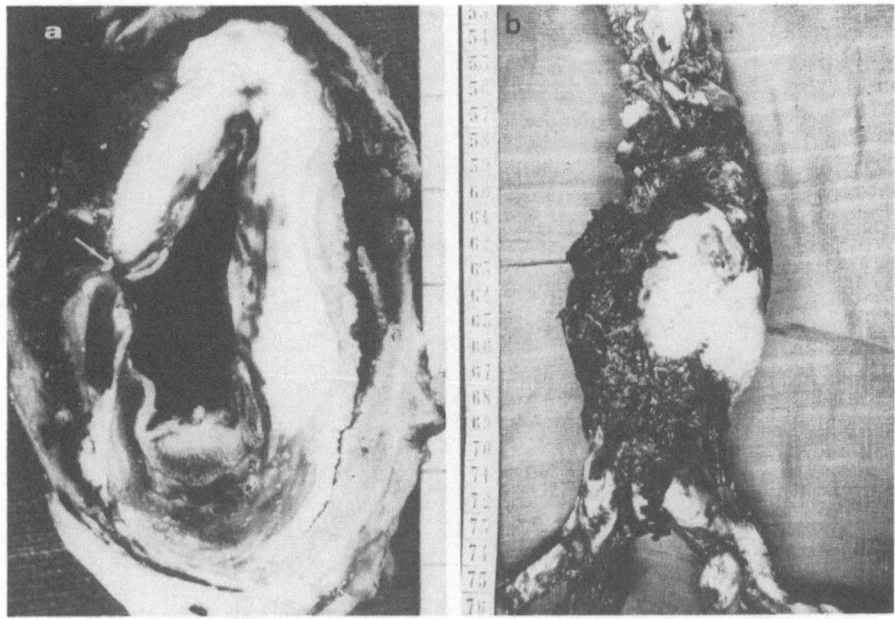

Fig. 31.3. a Section of an atherosclerotic aneurysm of the abdominal aorta, showing thrombi of various stages and fissuration (*arrow*) at the orifice of an intramedial haematoma. **b** Exterior aspect of the periaortic haematoma.

the vena cava produces an arteriovenous fistula.

The commonest cause of aortic aneurysms is currently atherosclerosis. In 1974, according to a

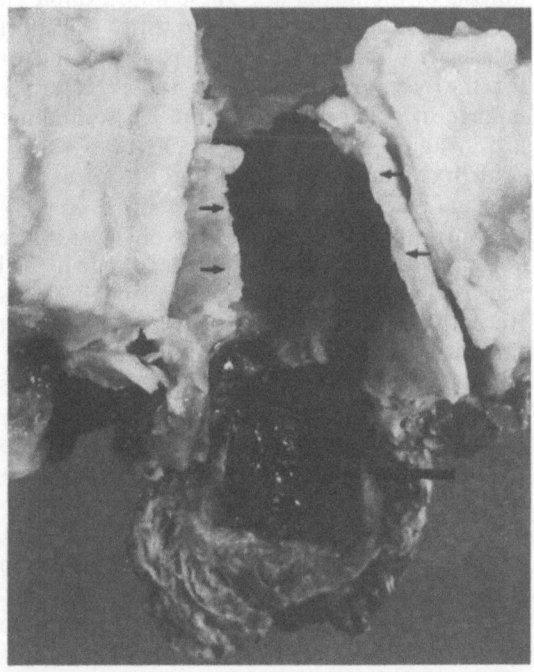

Fig. 31.4. Posterior view of a ruptured abdominal aneurysm (*thick arrow*). The duodenal wall is marked by *thin arrows*.

study in the National Health Service, aneurysms were responsible for 0.6% of deaths in women, and 1.2% of deaths in men over 65 years old. Three-quarters of all aneurysms [41], in particular aneurysms of the abdominal aorta [18] are found to be atherosclerotic; 69% of aneurysms of the descending aorta [54] and 29% of aneurysms of the thoracic aorta [13] are atherosclerotic. Aneurysms of the abdominal aorta can be the source of distal embolization and of both sizeable clots and microscopic debris. Histological examination may show typical atheromatous emboli at various stages of organization (Fig. 31.5). Multiple organs include kidney, pancreas, digestive tract and spleen can be involved.

The direct role of atherosclerosis in the formation of aneurysms is uncertain. Aneurysm formation may be one manifestation of arteriosclerosis resulting from gradual weakening of the medial coat due to various factors such as ageing, repeated injury and repair during life, or impaired nutrition [73]. Recent experiments performed on isolated arteries have demonstrated the respective roles of elastin and collagen in the development of aneurysms. Treatment with purified elastase causes the vessel to dilate but not to rupture. In contrast, all the vessels treated with purified collagenase ruptured, emphasizing the primordial role of collagen in maintaining wall integrity [24]. Associ-

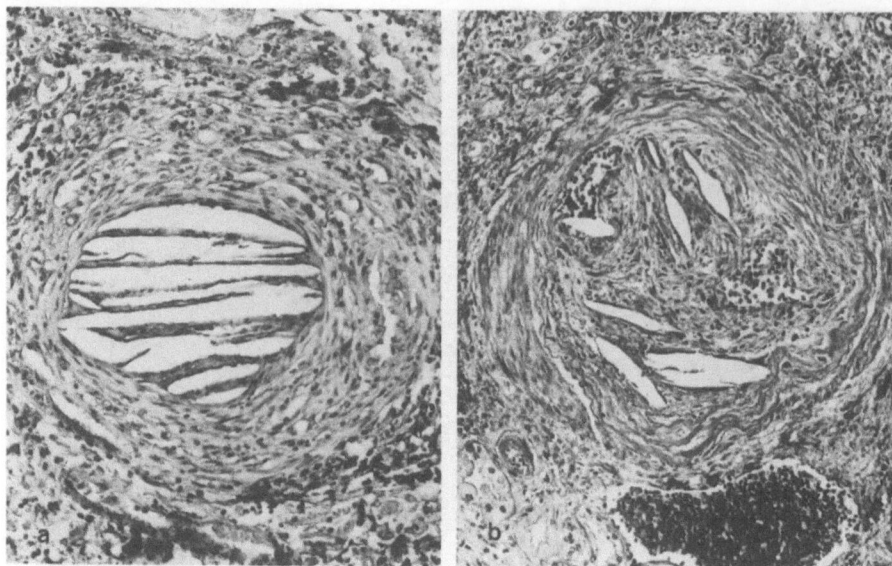

Fig. 31.5a,b. Atheromatous embolus in small muscle artery of the lower limb. **a** Recent stage with marked cellular infiltration. **b** Aged embolus showing obliterative fibrous reaction circumscribed by crystalline clefts. Haematoxylin–eosin–safran, × 150

ated findings of diffuse arteriomegaly in patients with aneurysms give support to this hypothesis [56]. Arteriomegaly is characterized by lengthening and dilatation of the vessel resulting in tortuosity and irregular enlargement in angiograms [36]. According to some statistical studies [81], 57%–66% of aneurysms are associated with arteriomegaly [81]. Tilson and Dang [83] consider arteriomegaly to be present when the aorta exceeds 25 mm diameter and using this definition showed that the diameter of the aorta above the renal arteries is 45% greater in patients with aneurysms than in patients presenting with atherosclerosis but without aneurysms.

Congenital predisposition may be a factor for the following reasons: (a) there are familial forms [19]; (b) the marked prevalence in males has been suggested to be due to a hypothetical mutation of the X chromosome; and (c) pedigree analysis of 16 families, including 41 patients with aneurysms, has suggested a possible sex-linked dominant transmitted trait. Further, arteriomegaly and aneurysms can be associated with inherited metabolic diseases (see Chap. 30). Such association has been reported in Menkes syndrome (kinky hair syndrome), a sex-linked recessively transmitted copper deficiency state related to impaired absorption of copper from the gastrointestinal tract. Significant tissue copper deficiency is found in patients with aortic aneurysms when compared with a control group com-

prising patients with severe atherosclerosis but without aneurysms [82]. Finally, only males are involved in the vascular disease of Blotchy mice, who have spontaneous aortic aneurysms characterized by early elastic fragmentation and fibrosis of the media [2]. These mice exhibit an X chromosome allele associated with a transport deficiency of copper, which acts as a co-factor for the lysyloxidase necessary to effect cross-linkage of collagen and elastin. These findings are in agreement with the decrease of scleroprotein content of abdominal aortic aneurysms in man [79].

Three-quarters of aneurysms of the thoracic aorta are associated with so-called cystic medial necrosis, a term coined by Erdheim in 1931. Recently attention has been drawn to the fact that the lesion is neither truly necrotic nor cystic, and the term "medial mucoid degeneration" has been proposed. The lesion is characterized by focal accumulation of mucoid homogeneous material within the media; this material appears metachromatic using toluidine blue and exhibits weakly PAS-positive mucopolysaccharides. These mucoid foci may be large, resulting in pools of various sizes. The finding is not specific. It may be associated with major changes in elastic tissue (marked fragmentation and/or loss of elastic fibre in the aortic media) (Fig. 31.6) and it can be found in relatively young patients. Some of the aneurysms associated with this

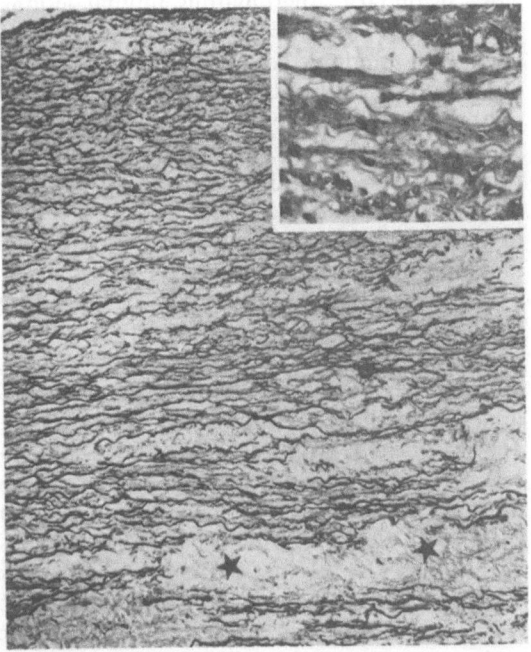

Fig. 31.6. Aneurysm of the ascending aorta. Fragmentation of elastic framework around small foci of cystic change, mainly within the outer part of the vascular wall. Orcein, × 70 *Inset*. Detail of the foci of cystic change. Haematoxylin-eosin-safron, × 145

finding are compatible with Marfan's syndrome and the remaining cases are considered as "formes frustes" of the disease (see Chap. 30). They are usually located in the ascending aorta.

Dilatation of the aortic ring and myomatous degeneration of the aortic valve leaflets can produce aortic regurgitation before aneurysmal dilatation is evident. The evolution may be either progressive towards cardiac failure, or dramatically complicated by acute aortic dissection or rupture. Nevertheless, dissection is rare in patients with fusiform aortic root aneurysms. Intimal and medial tears in the wall of such aneurysms are frequent but cannot be considered to be true dissections.

Aneurysms of the ascending aorta without extensive changes in elastic tissue are found in older patients. Various degrees of cystic medial change, elastic fragmentation of medial fibrosis can be found in the resected specimens. Several factors have been considered in the pathogenesis of these aneurysms, including ageing and systemic hypertension. They may all act as factors of injury and repair of the arterial wall (see below). According to Klima et al. [51] both cystic medial changes and elastic fragmentation appear inversely correlated with the severity of athero-

sclerosis. They suggested that these changes could reflect "tissue insufficiency" in early life leading to aortic dilatation.

Although declining in incidence, evidence of syphilis is still found in 4% [13] or 6% [8,30] of patients with aneurysms of the thoracic aorta, and in 9% of patients with aneurysms of the ascending aorta. Syphilitic aneurysms may reach 20 cm in diameter and may be responsible for compression syndromes of the neighbouring organs, erosion of ribs or vertebral bodies and rupture with fatal haemorrhage. Aortic insufficiency and coronary ostial stenosis are the two other major complications of syphilitic aortitis. Histopathological criteria are discussed elsewhere (see Chap. 28).

Besides those seen in syphilitic aortitis, inflammatory aortic aneurysms are rarely observed. Giant cell arteritis (temporal arteritis, Horton's disease) may involve the aorta in 10%–15% of cases, and produce aneurysms of the ascending aorta with aortic regurgitation [4]. Aneurysms may result from Takayasu's disease (see Chap. Behcet's disease mainly occurs in small-calibre vessels, i.e. vessels having a diameter of less than 1 mn. However involvement of medium-sized arteries and veins has been described. Little [55] has reported aortic aneurysms in four patients, suggesting an arterial lesion via the vasa vasorum. Kawasaki disease in children [17] may also be associated with thoracic aortic aneurysms.

Congenital aneurysms have been described in the newborn [46], in children [77] and in young adults [42]. They are often multiple and characterized by lack of elastic tissue, showing diffuse mural fibrosis with few remaining muscle cells. They may be related to Ehlers-Danlos disease (see Chap. 30). Aneurysms of the sinus of Valsalva [9], commonly associated with a supracristal ventricular defect or various vascular abnormalities, are rare congenital conditions. Aortic aneurysms can be found in patients with congenital aortic stenosis, patent ductus arteriosus or congenital coarctation; spontaneous dissection and rupture may cause death in these conditions.

Extra-aortic Aneurysms

Except for aneurysms of the cerebral arteries extra-aortic isolated aneurysms are uncommon but can occur in any artery.

McReady et al. [60] in 1983, collecting recent cases from the files of the Mayo Clinic, reported 50 patients with 71 aneurysms of the iliac artery (17 cases with multiple aneurysms). They represented 0.9% of the aneurysms which had been diagnosed over a period of 12 years. Almost always atherosclerotic in nature, those aneurysms occurred principally in the common iliac artery, mostly in men aged from 41 to 92 years (mean age 69.7 years). They measured from 2 to 20 cm in diameter (mean 4.7 cm), and had a growth rate of about 0.4 cm per year. The authors evaluated the risk of rupture at 14%, a low level when compared with other studies (50%–70%). Rare cases of associated syphilis or pregnancy reported in the previous literature were quoted by the authors. Finally iliac aneurysms have been described in arterial fibromuscular dysplasia (see Chap. 29).

Kasulke [50] collected 14 cases of isolated aneurysms of the internal iliac artery from the English literature. Another series of six cases (two false and four true aneurysms) was published recently [66]. Such unilateral aneurysms, located deeply in the pelvis, are often voluminous, leading to compression of neighbouring organs such as the gut or the urinary conduits [70]. These complications are associated with a high mortality rate.

Aneurysms of the splenic artery are the third most frequent type after aneurysms of the abdominal aorta and iliac artery. Deterling in 1971 [23] found 665 splenic aneurysms out of 1424 aneurysms of visceral arteries (47%). They are twice as frequent as aneurysms of the renal artery. Aneurysms of the splenic artery are often singular, saccular and located in the distal third of the artery. In contrast to arterial aneurysms in general, they are three times more frequent in women then in men. In most series the female patients are younger than the males. There is a predilection for multiparous women [85] (mean age 58 years). With an average diameter of 2 cm, splenic aneurysms are often asymptomatic but can present with rupture. The frequency of rupture in the general population is 40%, rising to 80% in pregnant women. It usually occurs in the last trimester of pregnancy and is characterized by a low survival rate for the patient, and always in death for the foetus [61]. Atherosclerosis is the cause of more than 50% of the cases, but 10%–30% of splenic artery aneurysms are believed to be congenital; their wall exhibits an abrupt defect of the internal elastic lamina and is mainly composed of endothelial cells and fibrous tissues [76]. These aneurysms often disclose secondary atheromatous intimal lesions. The absence of atherosclerosis in other sites, associated findings of pregnancy or fibromuscular dysplasia of the renal artery in 4%–9% of cases [78,85] and the predilection for the bifurcations of the artery, support the view of a primary deficiency of the connective tissue of the arterial wall. Similar changes have been found in the aortas of pregnant women [57] and of animals treated with oestrogens [72].

Aneurysms of the *popliteal artery* are the most frequent of the peripheral aneurysms. They occur almost exclusively in men, aged from 60 to 80 years; very rare cases are observed in patients under 40 years. They are mostly atheromatous and associated with severe occlusive atherosclerosis of coronary and cerebral arteries in 40% of cases; 50%–68% of them are bilateral and or associated with aortic, femoral or iliac aneurysms [29,86,91]. Because of their great size (3 to 9 cm), popliteal artery aneurysms may lead to complications such as distal thromboembolism and compression of satellite veins and nerves (notably the posterior tibial nerve). Rupture remains rare.

The frequency of aneurysms of the renal artery varies from 0.1% in autopsy material [1] to 0.97% in angiograms [80], and up to 9.7% in postmortem angiograms [75]. These discrepancies might result partly from the small size of most renal artery aneurysms. This is particularly true for those associated with systemic polyarteritis or arterial fibromuscular dysplasia. Except in these conditions, renal artery aneurysms measure around 0.3 to 4 cm in diameter (mean 1.5 cm) [47] and involve women as well as men in their 60s; they are rarely mutliple or bilateral, and are associated with systemic hypertension in two-thirds of cases. Aneurysms of the renal artery are generally saccular, atherosclerotic and calcified. However, atheromatous changes may be secondary to congenital deficiency of the vessel wall, as suggested by their usual location at bifurcation sites, their occurrence in children and the absence of generalized atherosclerosis. Moreover mural calcification resulting from mineralization of prothrombin polypeptide fragments at the interface between the luminal thrombus and the vessel wall, is not part of the atherosclerotic process. Fusiform aneurysms of the renal artery are infrequent and associated with arterial fibromuscular dysplasia (see Chap. 29); medial dissection and post-traumatic false aneurysms are discussed below. Rupture, probably favoured by pregnancy, remains very rare [47].

Aneurysms of the hepatic arteries are rare (200 cases collected in 1970) [39]. They are most often discovered at systematic autopsy examination and sometimes revealed by abdominal pain, icterus [53] or intestinal bleeding. These aneurysms are saccular and three times more frequent in men than in women. Some of them are of congenital origin; however, in young adults, they may result from bacterial infection secondary to cholecystitis or endocarditis, while in aged persons they appear to be mostly atherosclerotic.

Aneurysms of the gastric arteries are uncommon; twenty-four cases were gathered in 1965. They are atherosclerotic, usually post-traumatic and occur mainly in middle-aged men; they are located in the posterior wall of the stomach. Rupture may produce bleeding into the gastric cavity or the peritoneal cavity [6].

Aneurysms of the mesenteric arteries and their branches, such as the jejunal artery, have been reported very rarely (one case in 12000 autopsies) [59]. They are of small calibre (a few mm up to 1.5 cm in diameter). They may result from either atherosclerosis or bacterial infection secondary to endocarditis [3]. Rare cases of congenital aneurysms have been observed in association with Ehlers–Danlos syndrome, or isolated and resulting from a defeat of the media at the site of bifurcations. Associated findings of rheumatoid arteritis or necrotizing arteritis have been reported [59].

Coronary artery aneurysms can be recognized in 1%–2% of patients presenting with cardiopathy. Unfortunately the diagnosis is still rarely made during life by coronary arteriography (five cases in a series of 1500 consecutive arteriographies) [65]. Congenital coronary artery aneurysms, unique or multiple, affect predominantly the left circumflex artery; they can form fistulas in cardiac cavities or the coronary sinus. Some are observed in children [5]. They are characterized by fibrosis and the accumulation of mucopolysaccharides within the intima. Acquired coronary aneurysms are usually atherosclerotic, involving predominantly the right coronary artery [65]. They may be infectious [58] or inflammatory, associated with polyarteritis nodosa [36], systemic lupus erythematosus, scleroderma or the mucocutaneous lymph node syndrome (MLNS or Kawasaki syndrome). In the latter, coronary arteriogram may demonstrate diffuse irregular aneurysms, sometimes associated with aneurysms in other sites. Death may result from thrombosis, heart failure or rupture. Some lesions apparently regress spontaneously [38].

Most aneurysms of the cerebral arteries are saccular aneurysms ("berry" aneurysms); they are rarely found in children and are usually recognized in women and men over 40 years old [35]. Recently a predilection for women has been found [89]. Berry aneurysms measure several millimetres in diameter, and are multiple in 20% of cases [15]. They involve predominantly the apical angle of the bifurcation of the internal carotid artery or the origin of the posterior communicating artery, the fork formed by the anterior communicating artery and one of the anterior cerebral arteries, the branchings of the middle cerebral arteries and the bifurcation of the basilar artery. The behaviour of these aneurysms is unpredictable. Consecutive arteriograms have demonstrated that some of them remain stable, while others increase in volume. Those that rupture and may cause death usually exceed 0.7 cm in diameter [31,64]. This complication is thought to be the leading cause of non-traumatic subarachnoid bleeding, and can result in arterial spasm or homo- or contralateral cerebral infarction. Saccular aneurysms are considered to be due to medial gaps in the apical angle of the arterial fork, a site bearing direct marked haemodynamic stresses (Fig. 31.7).

Atherosclerotic aneurysms of cerebral arteries are less common, fusiform and located in the

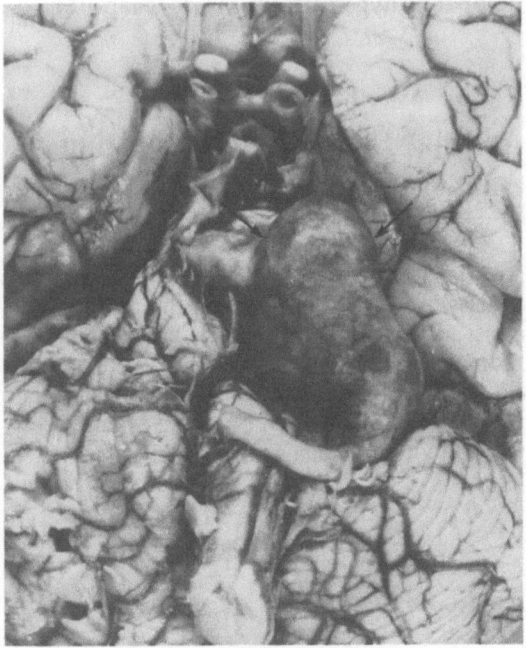

Fig. 31.7. Large intracerebral aneurysm (*arrows*).

internal carotid and basilar arteries. Infectious aneurysms have been found in the branches of the middle cerebral artery.

Charcot–Bouchard intracerebral microaneurysms may be found in up to 50% of hypertensive patients in carefully performed autopsy studies and older patients have a greater incidence of such aneurysms. It is largely accepted that massive intracerebral haemorrhage results from their rupture.

Aneurysms of the main pulmonary artery and its branches are considered to be one of the rarest lesions of these vessels. About 20 cases were published between 1942 to 1972 [11]. Isolated aneurysms can be found in elastic branches and in peripheral muscular branches. They are almost always associated with acquired pulmonary hypertension and may be multiple. Heart failure and rupture of the aneurysm are the most common causes of death.

Most pulmonary aneurysms are mycotic resulting from lung abcess, sometimes tuberculous in origin (aneurysms of Rasmussen), or following bacterial endocarditis, notably in young patients. Anomalies of the heart or great vessels, patent ductus arteriosus or interventricular septal defect may be associated. Some pulmonary artery aneurysms are congenital, associated with recurrent peripheral endophlebitis and pulmonary embolism (Hughes–Stovens syndrome), or with diffuse lengthening and tortuosity of systemic and pulmonary arteries (idiopathic hypercalcaemia) (see Chap. 36).

Arterial Dissections/Dissecting Aneurysms

Arterial dissections or dissecting aneurysms consist of haematomas within the arterial wall, classically resulting from intimal tears and longitudinal cleavage of the media by a dissecting column of blood. Such lesions mostly occur in the aorta, but have been observed almost everywhere in the systemic and pulmonary vasculature.

Aortic Dissections

Unlike atherosclerotic changes, dissecting aneurysms occur most commonly in the thoracic aorta.

Since the De Bakey series [22], three types of aortic dissection have been recognized:

1. Type 1 where a primary tear is located in the ascending aorta just above the aortic valves and dissection involves the aortic arch to the origin of the left subclavian artery.
2. Type 2 where dissection involves only the ascending aorta.
3. Type 3 where the origin of the dissection is located just beyond the left subclavian artery and the dissection extends in the distal aorta for a variable distance.

Types 1 and 3 are more frequent than type 2. Other types can be found, including the retrograde type 3 where the cleavage extends proximally to involve the ascending aorta and the rare dissections exclusively located in the abdominal aorta.

Aortic dissections are far less common than aneurysms. They were observed in 0.28% of 175 405 postmortem examination in one series [45] and in 15% of 22 792 postmortem examinations by Parkhurst and Decker [69]. In 1960 only 1416 cases were collected by Erb and Tullis [27]; according to these authors they represented 12%–15% of all types of aortic aneurysms. The condition affects patients between 40 and 60 years of age (mean age 50.4 years) [51], but can

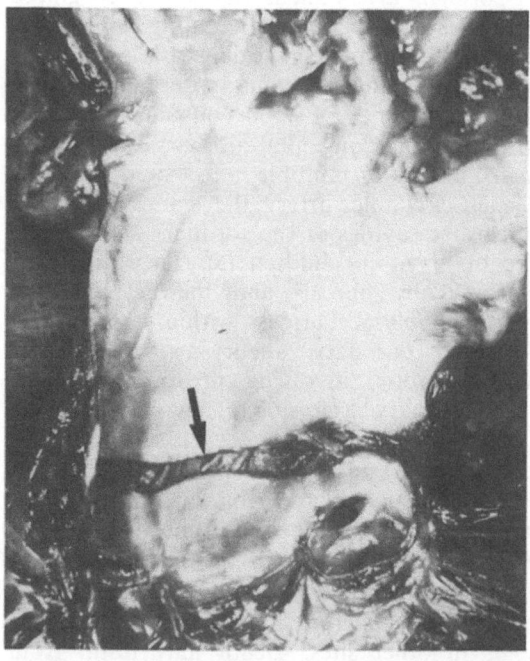

Fig. 31.8. Transverse linear rupture of the intima (*arrow*) in the ascending aorta. Type I acute aortic dissection.

occur at any age [48]. Whatever the age of onset, aortic dissections are six times less frequent in women than in men; under the age of 40 both sexes appear to be equally affected.

Macroscopically the site of origin of the aneurysms consists of a transverse intimal tear, most commonly located in the ascending aorta (61%) (Figs. 31.8 and 31.9), less commonly in the aortic isthmus and in the descending aorta (16%) [68]. In 15% of the 204 cases reported by Wilson [90] and in 22 cases out of the Gore series [37], no intimal tear was found. The most common cause of interruption of the longitudinal dissection appears to be extensive atherosclerotic plaque [71]. Usually the outer wall of the false channel is thin, comprising only the external elastic lamina and a few medical elastic fibres, and is likely to rupture [71]. A more distal or re-entry intimal tear may be present in the abdominal aorta or iliac arteries, resulting in the so-called double aorta. Dissection usually extends within the media involving the circumference partly or completely. It may affect major branch vessels (the brachiocephalic trunk, coronary arteries, visceral arteries), where thrombosis can occur (Fig. 31.10). External rupture is most often found in the ascending aorta, very rarely in the abdominal aorta. Rupture occurs into the pericardium, the pleural cavity, the mediastinum and

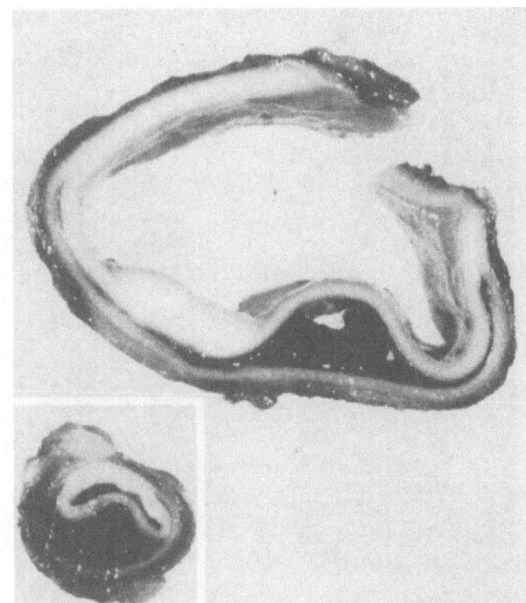

Fig. 31.10. Cross-section of the dissecting aneurysm of a cerebral artery as an extension of the dissection of the aorta.

rarely into the lung parenchyma. Retrograde dissection in the ascending aorta can extend into the aortic sinus and the media of a coronary artery, usually the right coronary artery. Dissection of the aortic may lead to the development of aortic regurgitation [71]. The extension of blood into the free wall of the adjacent atria and the septum may result in complete atrioventricular dissociation. Re-entry into the aortic lumen may result in a double aorta and favour survival. Such cases have tended to be more frequent since recent progress in surgical and medical (anti-hypertensive) treatments. Apparent healing may occur and the lining of the false channel is secondarily endothelialized with eventual atheromatous changes (Figs. 31.11 and 31.12). Occasionally the channel is filled by thrombi and may show multiple saccular aneurysms. In general the chance of healing depends on the distance of the intimal tear from the aortic valve. Occasionally limited healed changes can be found at the moment of an acute dissection.

Histopathologically the changes are consistent with the belief that structural weakening of the aortic media predisposes to dissection. Medial cystic necrosis is frequently cited as a specific hallmark associated with arterial dissection, although this lesion is clearly not the only factor at work. Similar changes can be observed in the apparently normal aorta [73,74] particularly in aged and/or hypertensive patients. In a recent

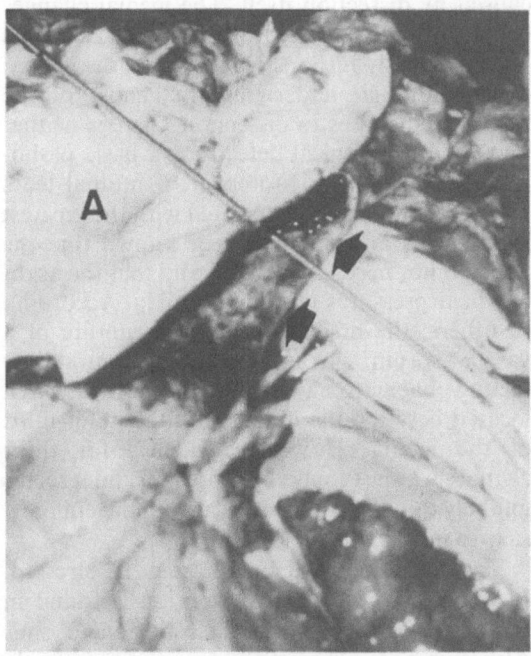

Fig. 31.9. Aortic dissection in a patient with Marfan's syndrome. The stylet is placed within the internal orifice of the rupture.

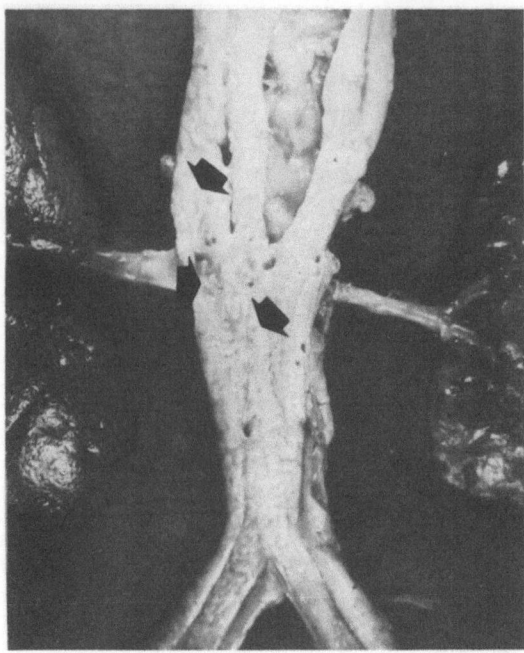

Fig. 31.11. Gross aspect of chronic dissection of the abdominal aorta. The opened false channel exhibits secondary atheromatous changes (*arrows*).

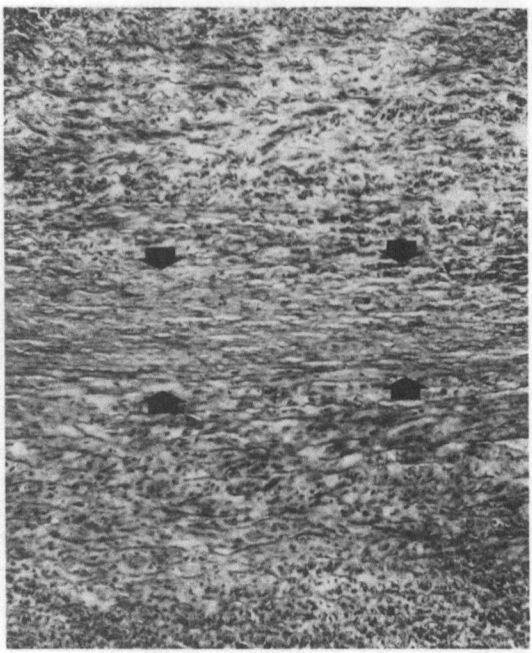

Fig. 31.13. Laminar necrosis of the media in contact with aortic dissection. Masson's stain, × 70

study [73,74] of 204 dissecting aneurysms [90], Wilson reported only 20 cases of medial cystic necrosis; on the other hand, the authors described scarring fibrotic changes in 36 cases and "laminar" necrosis in 63 cases (Fig 31.13). These lesions, often found in sites adjacent to but not continuous with the plane of the section, can be shown in normal ageing aorta of the

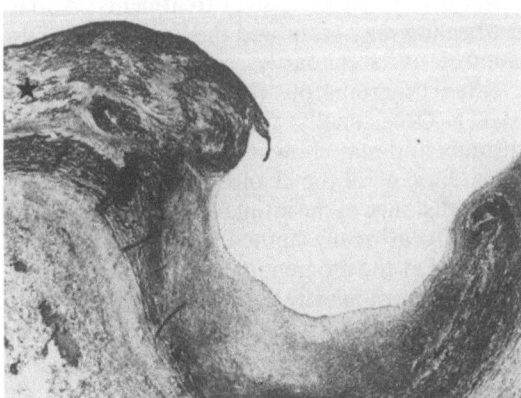

Fig. 31.12. Chronic dissection of the aorta. Changes in remaining media are marked (*stars*). The intima and the media are interrupted at the origin of the false channel, the external wall of which appears covered by fibrous thickening. Orcein, × 11

control group; they may simply reflect previous medial traumas or ischaemic injury, perhaps caused by dissection itself. The medial changes could not be considered to be specific causative factors directly responsible for aortic dissection. Thus, although underlying weakening of the media predisposes to dissection, there is neither a specific histological defect nor a basic pathogenetic mechanism. Moreover, if intimal tears usually occur in sites associated with the greatest shear stress and pressure, it is known that the tears do not necessarily occur and that the aorta can bear pressures up to 600 mmHg. According to others, the first step might be rupture of a vasa vasorum, giving rise to intramural dissecting haematoma within the critical area located between the middle and the outer third of the media. However, evidence for these changes was not found by Gore [37] and it seems unlikely that the vasa vasora have sufficient pressure to initiate dissection [90].

Several aetiological factors may be stressed. Evidence of systemic hypertension is found in two-thirds of patients with dissecting aneurysms. Although aortic dissection does not commonly occur in malignant hypertension, its incidence increases in patients treated with hypotensive drugs, suggesting a role for pressure fluctuations in promoting dissection. Aortic dissection is observed in one-third of patients with Marfan's syndrome (see Chap. 30) and in certain congenital cardiovascular abnormalities, particularly coarctation, aortic stenosis and non-stenotic bicuspid aortic value [51]; such bicuspid aortic values can be observed in 1%–2% of the general population, and in 9%–24% of patients with aortic dissection. Familial cases were described in 1957 [48] and dissection has been reported to be associated with pregnancy. This susceptibility may be related to structural changes in the aortic media as described by Man-allo-Estrella [57]. The association of dissection with tuberculosis, syphilis, hypothyroidism, hyperparathyroidism, ergotism or lupus ery-thematosus [92] has been reported rarely. Arterial dissection is uncommon in Takayasu's disease but represents one of the complications observed in great vessel involvement of giant cell temporal arteritis.

Under experimental circumstances, several factors inducing increased weakness of the aortic media can lead to dissection. These include lathyrism, feeding beta-aminopropionitrile or similar chemical compounds, a diethylstil-boestrol-low protein diet and copper deficiency in growing animals.

Extra-aortic Dissecting Aneurysms

Isolated dissecting aneurysms of visceral or peripheral arteries are rare; in 1972 Guthrie and McLean [40] collected 130 cases recorded after the first report of dissecting aneurysm of the iliac artery by Turner in 1829 [88]. In order of decreasing frequency such aneurysms have been reported in the renal artery, the coronary arteries and the carotid and mesenteric arteries. These aneurysms, sometimes multiple, affect men during the fifth decade and middle-aged women (mean age 40 years); the ages range from 11 years (carotid artery) to 87 years (mesenteric artery); the lesions may measure from several millimetres to several centimetres in length and may involve the circumference partly or completely.

Microscopically, dissection of the vascular wall develops between the external elastic lamina and the media in the muscular arteries and within the outer part of the media in elastic arteries such as the pulmonary, innominate, subclavian or vertebral arteries [34]. Nonetheless anatomical changes, aetiological factors and the various mechanisms at work do not differ in peripheral dissecting aneurysms when compared with aortic aneurysms.

The following additional findings are of interest. In 16 cases of aneurysms of the renal artery collected in 1966 [26] only two exhibited intimal tears; the author suggested that ischaemia of the vascular wall secondary to the atherosclerotic process might play a role in the degenerative changes of the intima. Dissecting aneurysms of the coronary arteries seem to be more frequent in women, notably during the post-partum period; cases have been reported as complications of contraceptive therapy [43]. In dissecting aneurysms of intracranial arteries, the plane of cleavage occurs between the elastic lamina and the media. Dissecting aneurysms of the hepatic artery [16,39,44] are associated with fibro-muscular dysplasia of the hepatic artery and/or the renal artery.

References

1. Abeshouse BS (1951) Aneurysm of renal artery: report of 2 cases and review of literature. Urol Cut Rev 55:451–463
2. Andrews EJ, White WJ, Bullock LP (1975) Spontaneous aneurysms in blotchy mice. Am J Pathol 78:199–207
3. Anon. (1975) Case report 19. N Engl J Med 292:1068–1073
4. Anon. (1979) Case report 21–1979. N Eng J Med 300:1204–1209
5. Anon. (1980) Case 35. Congenital aneurysm of left circumflex coronary artery. N Engl J Med 303:571–577
6. Anon. (1981) Case 25. False aneurysm of branch of splenic artery after repair of hiatus hernia. N Engl J Med 304:1533–1538
7. Auerbach O, Garfinkel L (1980) Atherosclerosis and aneurysms of aorta in relation to smoking habits and age. Chest 78:805–809
8. Baskervilla PA, Blakeney CG, Young AE et al, (1983) The diagnosis and treatment of periaortic fibrosis ("inflammatory aneurysms"). Br J Surg 70:381–385
9. Batsenschlager A, Fievez M, Million P (1964) Anévrysme du sinus de Valsava. Ann Anat Pathol 9:231–238
10. Bergan JJ (1984) A personal view of abdominal aortic aneurysms. Br J Surg 71:297–301
11. Bernheim J, Griffel B (1972) A propos d'un cas d'anévrysme disséquant de l'artère pulmonaire. Ann Anat Pathol 17:83–90
12. Bernstein E, Dilley RB, Goldberger LE et al. (1976) Growth rates of small abdominal aortic aneurysms. Surgery 80:765–772
13. Bickerstaff LK, Pairolero PC, Hollier LH et al. (1982) Thoracic aortic aneurysms: a population-based study. Surgery 92:1103–1108
14. Blau SA, Kerstein MD, Deterling RA: Abdominal aortic aneurysms. In: Kerstein MD, Moulder PV, Webb WR (eds) Aneurysms. Williams and Wilkins, Baltimore, pp 127–196
15. Burger PC, Vogel ST (1978) Cerebro-vascular disease. Am J Pathol 92:257–314

16. Callicott JH, Hoke HF (1968) Dissecting aneurysm of the common hepatic artery. Arch Pathol Lab Med 85:681–685

17. Canter CE, Bower RJ, Strauss AW (1981) Atypical Kawasaki disease with aortic aneurysm. Pediatrics 68:885–888

18. Cinqualbre J, Kieny R (1979) Les anévrysmes athéromateux de l'aorte abdominale. Rev Prat 29:765

19. Clifton MA (1977) Familial aortic aneurysms. Br J Surg 64:765–766

20. Crawford ES, Cohen ES (1982) Aortic aneurysms: a multifocal disease. Arch Surg 117:1393–1400

21. Darling CL (1976) Growth rates of small abdominal aortic aneurysms (discussion) Surgery 80:771

22. De Bakey ME, Walter SH, Colley DA et al. (1965) Dissecting aneurysms of aorta. J Thorac Surg 49:130–149

23. Deterling RA (1971) Aneurysms of the visceral arteries. J Cardiovasc Surg 12:209–322

24. Dobrin PB, Baker WH, Gley WC (1984) Elastolytic and collagenolytic studies of arteries. Arch Surg 119:405–409

25. Elveback LR, Connolly DC, Kurland LT (1981) Coronary heart disease in residents of Rochester Minnesotta. Mayo Clin Proc 56:665–672

26. Englund GW (1966) Primary dissecting aneurysm of the renal artery. Report of a case and review of the literature. Am J Clin Pathol 45:472–479

27. Erb BD, Tullis IF (1960) Dissecting aneurym of the aorta. The clinical features of thirty autopsied. Circulation 22:315 (cited by Hudson)

28. Ertugrul A (1967) Diffuse tortuosity and lengthening of the arteries. Circulation 36:400–407

29. Evans WE, Conley JE, Bernhard V (1971) Popliteal aneurysms. Surgery 70:762–767

30. Fairbairn JF, Juergens JL, Spittel JA (1980) Peripheral vascular diseases. Saunders, Philadelphia

31. Ferguson GG, Peerless SJ, Drake CG (1981) Natural history of intracranial aneurysms. N Engl J Med 305:99

32. Feiner HD, Raghavendra BN, Phelps R et al (1984) Inflammatory abdominal aortic aneurysms. Hum Pathol 15:454–459

33. Fomon JJ, Kurzweg FT, Broadway RK (1967) Aneurysms of the aorta: a review. Ann Surg 165:557–563

34. Foord AG, Lewis RD (1959) Primary dissecting aneurysms of peripheral and pulmonary arteries. Arch Pathol Lab Med 68:561–577

35. Freytag E (1966) Fatal rupture of intracranial aneurysms. Arch Pathol Lab Med 81:418–424

36. Glantz S, Bittner SJ, Berman MA et al. (1976) Regression of coronary artery aneurysms in infantile polyarteritis nodosa. N Engl J Med 294:939–941

37. Gore I, Hirst AE (1973) Dissecting aneurysm of the aorta: clinical–pathologic correlations. In: Edwards SE (ed) Cardiovascular clinics 5. FA Davis, Philadelphia, pp 239–260

38. Gorin R, Sorin M, Meyer A et al. (1978) Le syndrome cutanéo-muqueux et ganglionnaire de Kawasaki. Semin Hop Paris 54:442–447

39. Guerrero EC (1970) Primary dissecting aneurysm of the hepatic artery. Arch Pathol Lab Med 89:569–573

40. Guthrie W, McLean H (1972) Dissecting aneurysms of arteries other than the aorta. J Pathol 108:219–232

41. Halbert B, Willems RK (1962) Aneurysms of the aorta: an analysis of 249 necropsies. Arch Pathol Lab Med 74:163–168

42. Haynes CD (1982) Multiple congenital aneurysms associated with spontaneous vascular rupture. Surgery 92:910–912

43. Heefner WA (1973) Dissecting hematoma of the coronary artery: a possible complication of oral contraceptive therapy. JAMA 223:550–551

44. Hill DE, Lobell M, Edwards JK (1974) Primary dissecting aneurysm of the hepatic artery. Arch Intern MEd 133:471–474

45. Hirst AE, Johns VJ, Kime DW (1958) Dissecting aneurysms of the aorta: review of 505 cases. Medicine 37:217–279

46. Howorth MB Jr (1967) Aneurysm of abdominal aorta in the newborn infant. N Engl J Med 276:1133

47. Hubert JP, Pairolero PC, Kazmier FJ (1980) Solitary renal artery aneurysm. Surgery 88:557–565

48. Hudson REB (1965) Cardiovascular pathology. Edward Arnold, London

49. Janney C (1983) Chest pain, collapse and death in late pregnancy. Clinicopathologic conference. Am J Med 75:694–696

50. Kasulke RJ, Clifford A, Nichols K et al (1982) Isolated atherosclerotic aneurysms of the internal iliac artery. Arch Surg 117:73–77

51. Klima T, Spjut HJ, Coelho A et al. (1983) The morphology of ascending aortic aneurysms. Hum Pathol 14:810–817

52. Layman TE, Wang Y (1968) Idiopathic cystic medianecrosis and aneurysmal dilatation of the ascending aorta. Med Clin North Am 52:1145–1156

53. Lewis R, Kung H, Connon JJ (1982) Biliary obstruction secondary to hepatic artery aneurysm: cholangiographic appearance and diagnostic consideration. Gastroenterology 82:1446–1451

54. Liddicoat JE, Bekassy SM, Rubio PA et al. (1979) Circulation 1975 (case report 21). N Engl J Med 300:1204–1209

55. Little AG, Zarins CH (1982) Abdominal aortic aneurysm and Behcet's disease. Surgery 91:359–362

56. Loire R, Desscotes J (1984) Les anevrysmes atheromateaux et par mediadystrophie. J Mal Vasc 9:232–234

57. Manallo-Estrella P, Barker E (1967) Histopathologic findings in human aortic media associated with pregnancy. Arch Pathol Lab Med 83:336–341

58. McGee MB, Khan MY (1980) Ruptured mycotic aneurysm of a coronary artery: a fatal complication of salmonella infection. Arch Intern Med 140:1097–1098

59. McNamara R, Griska LB (1980) Superior mesenteric artery branch aneurysm. Surgery 88:626–630

60. McReady RA, Pairolero PC, Gilmore J et al. (1983) Isolated iliac artery aneurysms. Surgery 93:688–693

61. Mehrotra D, di Benedetto R, Theriot E et al. (1983) Spontaneous rupture of splenic artery aneurysm: sixth instance of both maternal and fetal survival. Obstet Gynecol 62:665–666

62. Melton LJ, Bickerstaff LK, Hollier LH et al. (1984) Changing incidence of abdominal aortic aneurysms: a population-based study. Am J Epidemiol 120:379–386

63. Mitchinson MJ (1972) Aortic disease in idiopathic retroperitoneal and mediastinal fibrosis. J Clin Pathol 25:287–293

64. Ojemann RG (1981) Management of the unruptured intracranial aneurysms. N Engl J Med 304:724–726

65. Oliveros RA, Falsetti HL, Carroll RS et al. (1974) Atherosclerotic coronary artery aneurysm. Arch Intern Med 134:1072–1076

66. Perdue GD, Mittenthal MJ, Smith RB, et al. (1983) Aneurysms of the internal iliac artery. Surgery 93:243–246

67. Pereira M, Filho AA, Lastoria S et al. (1981) Inflammatory aneurysm of the abdominal aorta with coronary arteritis. Arch Pathol Lab Med 105:678–679

68. Pomerance A, Yacoub MH, Gula G (1977) The surgical pathology of thoracic aortic aneurysms. Histopathology 1:257–276

69. Parkhurst GF, Decker JP (1955) Bacterial aortitis and mycotic aneurysms of the aorta. Am J Pathol 31:821–836

70. Rennick JM, Link DP, Palmer JM (1975) Spontaneous rupture of an iliac artery aneurysms into a ureter: a case report and review of the literature. J Urol 116:111–113

71. Roberts WC (1981) Aortic dissection: anatomy, consequences and causes. Am Heart J 101:195–204

72. Ross R, Klebanoff SJ (1971) The smooth muscle cell. In vivo synthesis of connective tissue proteins. J Cell Biol 50:172

73. Schlatmann TJM, Becker AE (1977) Histologic changes in the normal aging aorta: implications for dissecting aortic aneurysms. Am J Cardiol 39:13–20

74. Schlatmann TJM, Becker AE (1977) Pathogenesis of dissecting aneursym of aorta. Comparative histopathologic study of significance of medial changes. Am J Cardiol 39:21–26

75. Schwartz CJ, White TA (1965) Aneurysm of the renal artery. J Pathol 89: 349–356

76. Sheps SG, Spittel JA, Fairbairn JF et al. (1958) Aneurysms of the splenic artery with special reference to bland aneurysms. Proc Mayo Clin 33:381–390

77. Short DW (1978) Multiple congenital aneurysms in childhood: report of a case. Br J Surg 65:509–512

78. Stanley JC, Fry WJ (1974) Pathogenesis and clinical significance of splenic artery aneurysms. Surgery 76:898–909

79. Sumner DS, Hokanson DE, Strandness DE Jr (1970) Stress–strain characteristics and collagen–elastin content of abdominal aortic aneurysms. Surg Gynecol Obstet 130:459–466

80. Tham G, Ekelund L, Herrlin K et al. (1983) Renal artery aneurysm. Ann Surg 197:348–352

81. Thomas L (1981) Arteriomegaly. Arch Surg 116:1030–1032

82. Tilson D (1982) Decreased hepatic copper levels: a possible chemical marker for the pathogenesis of aortic aneurysms in man. Arch Surg 117:1212–121?

83. Tilson D, Dang C (1981) Generalized arteriomegaly: a possible predisposition to the formation of abdominal aortic aneurysms. Arch Surg 116:1030–1032

84. Tilson D, Seashore MR (1984) Human genetics of the abdominal aortic aneurysm. Surg Gynecol Obstet 158:129–132

85. Trastek VF, Pairolero PC, Joyce JW et al. (1982) Splenic artery aneurysm. Surgery 91:694–699

86. Vermillion BD, Kimmins SA, Pace WG et al. (1981) A review of 147 popliteal aneurysms with long term follow-up. Surgery 90:1009–1014

87. Walker DI, Bloor K, Williams G et al. (1972) Inflammatory aneurysms of the abdominal aorta. Br J Surg 59:609–614

88. Watson AJ (1956) Dissecting aneurysm of arteries other than the aorta. J Pathol 72:439–449

89. Wiebers DO, Whisnant JP, O'Fallon WM (1981) The natural history of unruptured intracranial aneurysms. N Engl J Med 304:696–698

90. Wilson SK, Hutchins GM (1982) Aortic dissecting aneurysms: causative factors in 204 subjects. Arch Pathol Lab Med 106:175–180

91. Wychulis AR, Spittel JA, Wallace RB (1970) Popliteal aneurysms. Surgery 68:942–952

92. Youngson G, Engeset J, Hussey J et al. (1983) Dissecting aneurysm of infrarenal abdominal aorta. Surgery 94:521–523

Chapter 32

Traumatic Arterial Disease

P. N. Vuong and P. Lagneau

Most arterial injuries are due to direct trauma. Advances in the surgical repair of these injuries are due primarily to surgical experience acquired during war [19,43,71]. The destructiveness of weapons used in recent military conflicts, in particular in the war in Vietnam [39,67] accounts for the wide variety of lesions encountered. In hospital practice [6,10,22,42,48,60] certain invasive diagnostic procedures, used for vascular imaging, [29,31,47] or even to repair damaged arteries [11,16,52,53,64] may be responsible for iatrogenic lesions. The pathologist often must examine specimens of arterial resection collected during the surgical repair of an artery. His task thus involves not only confirming the traumatic nature of the lesion but also detecting whether or not the artery is the site of occult lesions, a finding of considerable interest in forensic medicine [64,65].

Direct Trauma to Arteries

Evidence of traumatic injury may be obvious. Occasionally, these lesions are occult, and manifest only when complications arise, but they remain similar whatever the circumstances of onset may be. There is wide discrepancy between the gross appearances of a circumscribed lesion and the histological changes identified in the arterial wall. This fact underlines the importance of excision of the whole damaged portion of the vessel and the value of a systematic histological study of every surgical specimen. Lesions are of four basic types: rupture of the arterial intima, rupture of the intima and the media, complete rupture of all the three layers, rupture of the adventitia alone or of the external elastic lamella. Traumatic injury to arteries is rarely an isolated occurrence. Frequently, concomitant lesions of other vessels or structures are present.

Arterial Lesions

Rupture of the Intima

The tunica intima or inner layer is the most fragile layer of the artery [90]. Careful examination to detect a possible rupture of the intima must be undertaken systematically whenever an arterial specimen excised following trauma is studied.

The outward appearance of the surgical specimen is normal. A thrombus may occlude the lumen of the vessel. Upon dissection, removal of the thrombus at its base exposes a rupture of the intima. This break may be either longitudinal, parallel to the main axis of the artery, or circumferential, causing a ring defect [24]. In this instance, as a result of the circulatory shock wave, the distal edge of the wound becomes detached from the underlying parietal tissue. It thus forms an intimal leaflet [18,33] (Fig. 32.1), which floats freely within the lumen of the artery. When two or more circumferential lesions are present, detachment may be complete giving rise to an intimal cast [56,57,61,66].

Histologically, rupture of intima is characterized by laceration of the endothelium

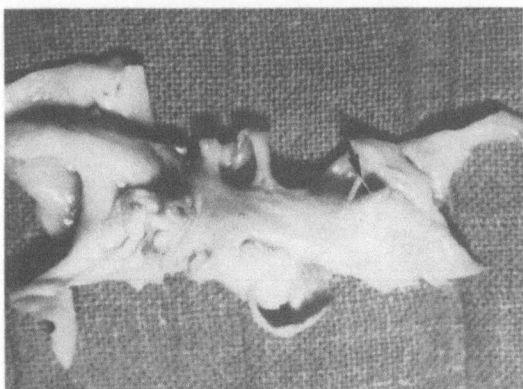

Fig. 32.1. Aorta. Old intimal rupture with formation of an open-work fence leaflet.

together with rupture of the internal elastic lamina, thereby exposing the underlying tissue (Fig. 32.2). Polymorphonuclear cells and monohistiocytes infiltrate the edges of the rupture site. The wound crater is filled by a mural thrombus. The intimal leaflet remains supported by the internal elastic lamella which is drawn back and inward. The underlying line of cleavage is the site of haematoma formation.

The outcome of a rupture of the intima depends on the extent of the intimal damage and the size of the artery. In large arteries, punctate or linear-shaped rupture may heal without consequence. Organization of a mural thrombus leads to formation of a fibrous plaque which may subsequently become re-endothelialized [8,35, 49]. In small arteries, or if the intimal lesion is large with an intimal leaflet, the mural thrombus may become extensive and completely occlude the artery.

Rupture of the Intima and Media

Rupture of the tunica intima may occur with partial or complete rupture of the underlying tunica media. This rupture is frequently complicated by parietal dissection with arterial occlusion (Fig. 32.3). Due to the effect of the systolic shock wave, blood lifts up the upper edge of the intimo-medial lesion, separates the media and expands in the subadventitial space. The haematoma just formed occludes the lumen of the artery and raises up the adventitia.

Histologically, the internal elastic lamina is interrupted, thereby exposing the media, which is dissociated by haemorrhage. Inflammatory cells infiltrate the edges of the intimo-medial rupture. Myocytes which border the medial rupture are necrotic and lose their staining ability. The parietal defect is filled in by thrombus.

Intimo-medial rupture always causes occlusion of the artery. Healing of the intimo-medial wound leads to the formation of scar tissue composed of newly formed vessels and haemosiderin-loaded macrophages. This tissue replaces the damaged intima and media. When a circumferential rupture occurs, organization of an occlusive haematoma transforms the arterial segment to a sclerotic cord. Serial histological sections with appropriate staining of elastic

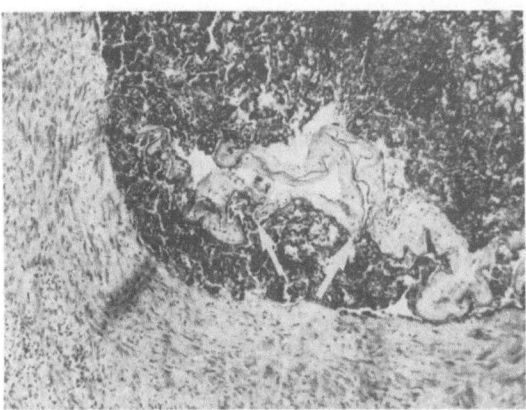

Fig. 32.2. Humeral artery. Recent intimal rupture with intimal leaflet detachment. Masson's trichrome, × 95

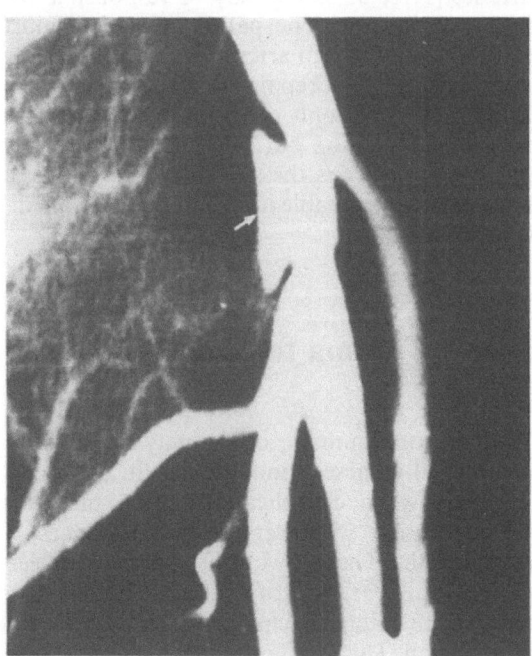

Fig. 32.3. Iliac artery. Angiographic appearance of a rupture of the intima and media.

fibres demonstrate the external elastic lamella which remains intact as well as the remains of the internal elastic lamina at the stump of this sclerotic cord.

Complete Rupture of the Three Layers of the Artery

This lesion consists of complete sectioning of the artery and results in haemorrhage, the importance of which depends on the extent of the wound and the size and location of the artery affected. If the lesion is transverse in a small-size artery, the upper stump may retract. The lumen thus becomes occluded by a thrombus which results in temporary haemostasis. Complete rupture of the three arterial layers rarely goes unnoticed during surgery (Fig. 32.4). Healing of these lesions is similar to that observed in intimal or intimo-medial lesions [32,69].

Rupture of the Adventitia Alone or of the External Elastic Lamella

This lesion is due to abrupt stretching of the artery at the point of impact. It may go unnoticed and only become manifest when complications develop.

Examination of histological preparations shows there is a dissociation of or even the dis-

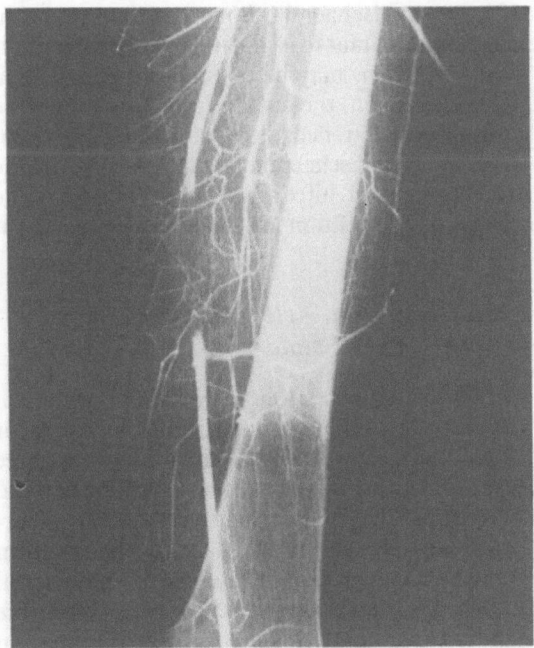

Fig. 32.4. Deep femoral artery. Angiographic aspect of a complete rupture of the three arterial layers.

appearance of the external elastic lamella with sclerotic scarring of the adventitia. A few newly formed vessels are apparent, together with haemosiderin pigments.

This rupture may constitute an area of parietal weakness, the starting point for formation of a hernia of the media into the adventitia [2]. Most often, its healing leads to formation of a fibrous callus, a true ring of sclerotic tissue strangling the artery. Differential diagnosis may be made to rule out sclerotic lesions occurring during the course of certain inflammatory arteriopathies with primarily media and adventitia involvement, in particular, in the course of Takayasu's disease.

Concomitant Lesions

Traumatic injuries to arteries rarely occur alone. They often are associated with lesions in adjacent tissues [41]: veins, nerves, bones, muscles.

Veins

Venous lesions may consist of simple compression of a vein or rupture, leading to occlusion which thereby impedes the return flow of blood. Whenever possible, surgical repair of veins should be performed at the same time as repair of an artery.

Nerves

Nerves may be damaged by either partial or complete sectioning, or by blunt trauma with subsequent blocking of nervous impulses. Nerve lesions imply a poor prognosis in cases of vascular lesions because of the muscular sequelae which result. Generally, neurological disorders are treated later on, after surgical repair of arteries.

Bones

Traumatic injuries to bones may be associated with arterial lesions. The damaged bones may also injure the artery. Bone and arterial injuries must be corrected sequentially by two different surgical teams, during the same surgical operation, with bone lesions being treated first.

Muscles

Muscles may be lacerated or sectioned. Muscle lesions have an adverse effect on collateral circulation. Traumatic occlusion of an artery, in particular, creates a real internal tourniquet. When this tourniquet is removed during surgery to repair an injured artery, an abrupt discharge of toxic substances into the circulation may occur with resultant shock and acute renal tubular necrosis. For this reason, extensive debridement of injured tissue is essential and traumatic injuries of the arteries examined at a late stage require aponeurotomy to prevent extension of the initial muscular injury.

Internal Organs

Lesions of internal organs are frequently observed when traumatic injury of the aorta or its collateral branches occurs, for example open wounds or blunt trauma to intrathoracic organs (heart, lungs) or to intra-abdominal organs (liver, spleen, small intestine, colon). When present, these injuries carry a much poorer prognosis.

Results of Arterial Lesions

Spasm

Spasm of an artery is defined as a contracture of the arterial wall with significant narrowing of the lumen which may ultimately result in complete occlusion. It is a purely functional process. The three layers of the artery remain intact. However, it is always difficult to eliminate an organic lesion, especially when the intima is affected.

Thrombosis

A thrombus is always the result of an alteration of the endothelial lining. Once initiated, a thrombus increases in size until it occludes the lumen of an artery. It may extend to include the collateral circulation. Examination of the stage of organization of a thrombus together with the history makes it possible to identify when the arterial lesion occurred.

Microemboli

Microemboli are either thrombotic or atheromatous. The former arise from non-occlusive mural thrombi, the latter form atheromatous plaques, shaken loose by the trauma [3,58].

Macroemboli

Macroemboli may be produced by bullet wounds of arteries. They are of different types and contain tissue debris or projectile fragments [21].

Haemorrhage

Haemorrhage may occur internally in case of rupture of the intima or the intima and media, resulting in formation of an intraparietal haematoma. External haemorrhage will occur when there is a complete rupture of the three arterial layers. The severity of the haemorrhage depends on the size of the artery. In certain anatomical regions spontaneous haemostasis may occur [14]. A periarterial haematoma confined by the adjacent anatomical structures rapidly becomes taut and ensures haemostasis.

Ischaemia

Ischaemia is the result of thrombosis and of haemorrhage. Its onset may be abrupt or gradual. The vulnerability of different tissues to ischaemia varies. Nerves tolerate ischaemia poorly and persistent neurological disturbance after surgical repair of arteries translates into severe and irreversible ischaemic lesions. Muscle requires much more oxygen than connective tissues and necrosis of muscle is complete after 12 h of total ischaemia. Bone, cartilage and skin are more resistant to ischaemia. For skin, reversibility of lesions may be possible even after 48 h. For this reason, the cutaneous status cannot be considered to reflect accurately the degree of severity of the ischaemia. Ischaemia favours the development of infection, compromising tissue repair by separation of sutures with resultant haemorrhage.

False Aneurysm, Arterial Dissection, Arteriovenous Fistula

1. False aneurysm is always the result of arterial rupture, either partial (of the intima or of intima–media) or total (involving the three layers of the artery). Organization of the parietal haematoma leads to formation of a false aneurysm [88]. Grossly, it is a saccular or fusiform nodule which communicates with the arterial lumen, of variable size and pulsatile while the lumen is still patent. Histological examination reveals the presence of the remains of sectioned arterial coats within the organized haematoma. If there is an intimal or intimo-medial injury, the

internal elastic lamina and media are found to be lacerated while the external elastic lamella and the adventitia are unaffected. When there is complete rupture of an artery, these structures are no longer distinguishable. The haematoma is transformed into a more or less dense block of sclerotic tissue encircling newly formed vessels mixed with haemosiderin pigment-laden macrophages.

2. A dissecting haematoma is formed when extension of a parietal haematoma beyond the area of a parietal rupture into the subadventitial space or directly into the media occurs. On histological sections, the dissecting artery has two lumina [15]. One is the true lumen which is partially collapsed, the other is a channel in the shape of an incurving slit, moulded against the arterial lumen. This channel corresponds to the line of cleavage caused by the haematoma (Fig. 32.5). Histological examination defines the characteristics of these two lumina. The true lumen is delineated by the intima, supported by the internal elastic lamina. The media which forms a semicircle bordering the new channel contains sclerotic tissue. The channel is lined with granulation tissue which comes from the organization of the dissecting haematoma. In the second stage, this tissue is replaced by sclerotic scar tissue. The whole formation is circumscribed by the external elastic lamella and adventitia (Fig. 32.6). In some cases the luminal aspect of the channel is invested by endothelial cells which may be identified by certain biological markers such as factor VIII, blood group antigens and so on. The presence of an orifice allowing the channel to communicate with the lumen of the artery produces a true dissection, when the artery presents a double-barrel aspect on

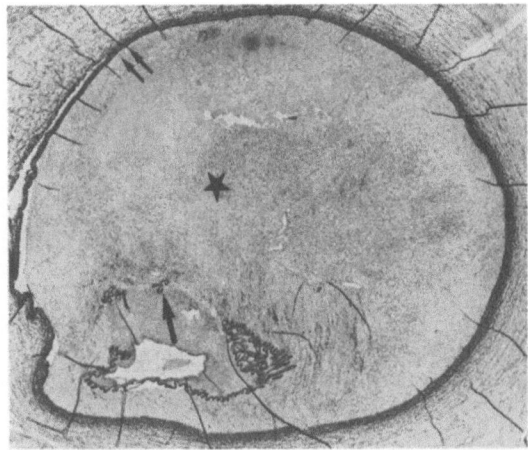

Fig. 32.6. Renal artery. Histological aspect of a parietal dissection secondary to a rupture of the intima and media. The internal elastic lamina is disrupted (→). The whole formation is circumscribed by normal external elastic lamella (⇉). Orcein, × 31

cross-section. There are many intermediate lesions possible between a dissecting haematoma and true dissection.

3. An arteriovenous fistula may occur in anatomical regions where an artery and its satellite vein follow the same approximate path (Fig. 32.7). This lesion occurs following a simultaneously occurring transparietal injury to two blood vessels, or the rupture of a false aneurysm into its satellite vein. The length of time needed for the development of an arteriovenous fistula is variable, ranging from 2 to 16 h or up to 8 d. In certain instances, a longer period is necessary which may extend from 3 to 9 years [37,38, 40,73,82].

Macroscopically, the vein shows a fusiform dilatation, extending over both sides of the fistula orifice. The lumen of the artery is dilated above this lesion site and is stenotic below it. These changes are due to the diversion of arterial blood into the venous circulation [77].

Histologically, only sections perpendicular to the main axis of the vessels permit the identification of the fistula orifice. This orifice corresponds to the junction of the arterial rupture and the adjacent satellite vein. Depending on the age of the fistula, the communication may be surrounded by sclerotic scar tissue of varying density. This tissue is the site of newly formed blood vessels and haemosiderin pigment. Staining of elastic fibres reveals the remains of intimal and external elastic lamella, as well as of the adventitia.

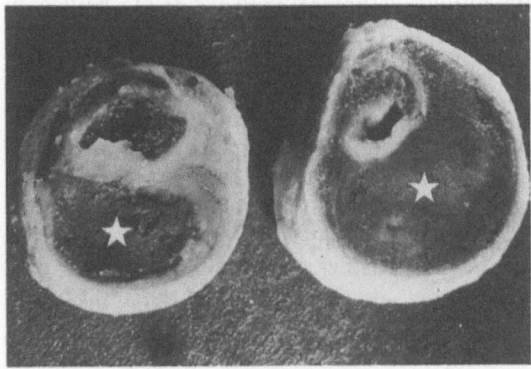

Fig. 32.5. Superficial femoral artery. Gross aspect of a subadventitial parietal dissection. The line of cleavage is occupied by a haematoma. The vascular lumen is collapsed.

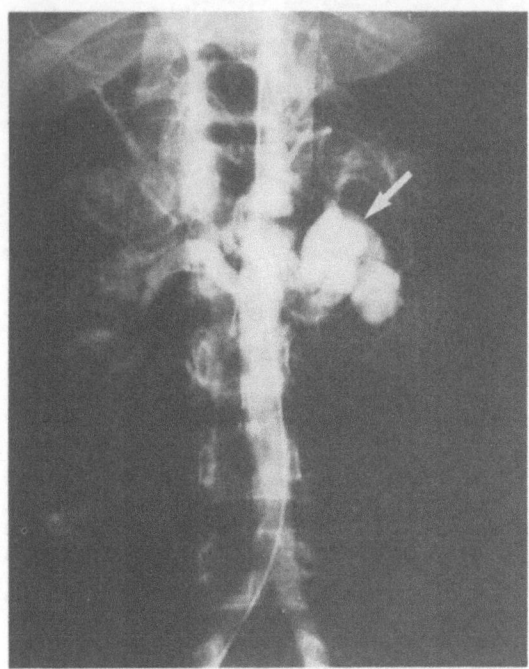

Fig. 32.7. Renal arteriovenous fistula secondary to bullet wounds. Angiographic aspect.

During the clinical course of the fistula, changes arise in the affected artery and vein. On the arterial side, the intima of the vessel above the lesion becomes thickened by development of sclerotic endarteritis. The segment below the lesion conserves its original structure. On the venous side, the endothelial lining becomes detached, beginning around 1 h after the process starts. The internal area of the media is infiltrated by polymorphonuclear and mononuclear inflammatory cells. Four weeks after the initial injury, the endovein exhibits thickened areas in the form of small cushion-like plaques. The media is the site of focal hypertrophy of myocytes. In the second stage of this process, increased transmural pressure weakens the wall of the vein, which is replaced by sclerotic tissue. Gradually, the medial layer of the artery is transformed into a sclerotic wall, encompassing a few islets of hypertrophied myocytes. Elastic fibres predominate and are formed on the luminal side of the vessel. In certain instances, the production of elastic fibres forms a true internal elastic lamina.

Exclusion of the fistula by surgical repair of the artery with or without ligation of the vein makes it possible to avoid later haemorrhages, cardiovascular complications and the consequences of increased venous pressure.

Traumatic Mechanisms and Agents

Open Trauma

Open trauma or arterial wounding may cause complete rupture of the three layers of the artery. Frequently this rupture develops from without. There is a punctate wound in the case of iatrogenic injuries following arterial puncture (e.g. catheterization using Seldinger's technique). It may be a partial lesion if caused by cutting tools or pointed instruments (bullet or grenade fragment, knife, swords, bull's horn, bony sequestrum, etc). The lesion is total when a complete cross-section of the artery occurs (by means of a sword, knife or accidental amputation of a limb extremity). In these circumstances, the degree of periarterial tissue destruction determines the prognosis.

Closed Trauma

Closed trauma or blunt injury to arteries is the cause of a parietal rupture of the wall of the artery, leaving the adventitia intact. Several mechanisms are involved.

Direct Blow. Trauma occurs directly to the artery which is either crushed or compressed. The severity of the lesion depends not only on the force of the blow but also on the location of the artery, in particular when superficial blood vessels are compressed against a bony surface (superficial temporal artery, popliteal artery, subclavian artery, etc). The type of trauma varies (a blow by an iron bar or metal girder, a fall from a motorcycle, a kick by a horse, a fist fight or slap, a shot from a hockey puck, a blow caused by a ball in ball games) [87].

Crush Injury. Crush injury causes rupture of the intimo-medial layers of the artery with stenosis. Effective ligation is produced by concentric crush injury of the artery. In closed traumatic injuries of the thorax, the force of the blow is transmitted to the column of blood contained in the aorta, thereby causing a crush injury to the aorta. In cases of strangulation or "balanced" hanging, compression involves the hyoid bone and thyroid cartilage and the internal carotid arteries are crushed against these two structures, just below the area where they bifurcate. Imprinting on these structures is revealed at dissection [55].

Deceleration, Crashes, Falls and Flexion. These mechanisms are a major cause of closed traumatic injuries to the aorta.

A sudden, violent deceleration produces a resultant bidirectional force with shearing of the artery at the junction between fixed and relatively mobile segments of the vessel. In the thorax, the aortic arch is held in place on one side by the supra-aortic trunk and on the other side by the left pulmonary pedicle. The thoracic aorta is secured to the vertebral column by the intercostal arteries and the posterior diaphragmatic arch. During deceleration, the aortic arch is drawn downwards by the heart. The descending aorta, securely held to the vertebral column, is pushed up and forward. There is a resultant shearing force with rupture of the aorta at this level. Sixty per cent of deaths due to deceleration accidents are from severe damage to the aorta. The localization of the lesion varies with the type of injury [75,83,91]: ascending aorta (23%), descending aorta outside the isthmus (12.7%), aortic arch (5%) and abdominal aorta (4.7%).

In an automobile crash, passengers who are not wearing seat belts are abruptly thrown forward and upward on impact. The mechanisms of the traumatic action exerted on the aortic arch are then slightly different. In addition to the two forces of deceleration, there is a third force of torsion due to rotation of the heart leftward. The resultant of these three forces ruptures the aortic arch at the point of insertion of the arterial ligament, in front of the origin of the left subclavian artery. The car driver receives the impact of the steering wheel against his chest which accentuates torsion of the heart. A second rupture of the ascending aorta may occur just above the aortic valve (Fig. 32.8). Additionally, there is blunt trauma to the heart with possible rupture of one of the chambers [79].

During steep vertical falls (e.g. paratroopers landing feet first or the fall of a lift cage when the cable snaps, etc), decelerating forces act on a fixed point of the body, thereby causing rupture of the aortic arch with concomitant fracture of the liver or the base of the skull [20].

Wearing a two-point seat belt system causes abrupt flexion of automobile accident victims so that the aortic segment above the area of flexion is carried downward and backward, and the inferior segment upward and forward. The zone of flexion thus undergoes shearing forces with consequent rupture of the aorta.

In airplane crashes [84], rupture of the aorta may occur in the thoracic or abdominal portions of this vessel or at its point of bifurcation. This injury frequently occurs in combination with rib fractures, spinal cord injuries, blunt trauma to the heart, lungs, liver or spleen and with lacer-

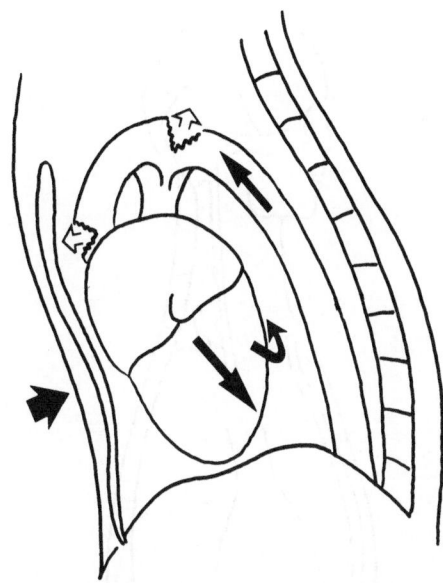

Fig. 32.8. Mechanism of rupture of the thoracic aorta and of the aortic arch in car driver, victim of an automobile crash.

ations of intercostal and abdominal muscles. A particular aspect of these injuries is that passengers who were seated in the same area frequently have similar injuries.

Stretch and Tear Injuries. Injuries to pediculate arteries frequently are observed in cases of blunt trauma to the abdomen. Initial force abruptly displaces the internal organs, while the aorta remains relatively fixed. This motion causes abrupt stretching of the arteries with propagation of the systolic shock wave due to increased intra-abdominal pressure [13,30]. As a result, an incomplete rupture of the intimal and medial layers of the arteries occurs, leaving the adventitia intact. The site of the rupture is frequently at the ostium or in the para-ostial portion of the vessel [18,51,62,74].

Injuries to the upper part of the internal carotid arteries occur in some victims of automobile crashes with fronto-facial impact. In effect, when an accident happens, the three-point seat belt exerts thoracic compression on the wearer and decreases the antero-posterior diameter (Fig. 32.9) [68]. Deceleration tilts the heart and its contents downward and backward. This motion accentuates the hyperflexion of the aortic arch and the supra-aortic trunk is stretched downward. This stretching is tolerated by the body when the cervical vertebrae also undergo hyperflexion. Fronto-facial impact causes an abrupt arrest of cervical flexion. Moreover,

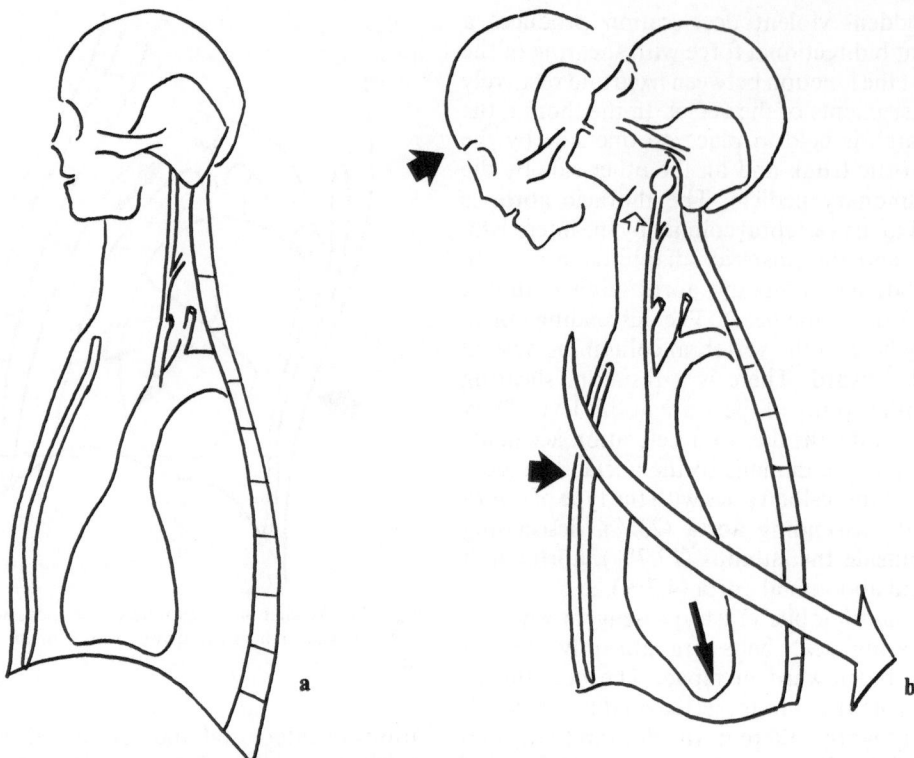

Fig. 32.9a, b. Mechanism of rupture of the internal carotid arteries occurring in the victim of an automobile crash wearing a three-point seat belt and having fronto-facial impact.

deceleration as well as thoracic compression increase the pressure of the circulatory shock wave, which propagates upward. This circulatory wave is transmitted to the carotid vessels, which are already stretched in a longitudinal fashion, and extends up to the orifices of the carotid vessels at the base of the skull. At this level the wave of blood strikes inextensible bony structures, and ampullary dilatation occurs in the carotid arteries, with "bursting" of the intima or intimal and medial layers of the vessel.

Bullet or projectile wounds in arteries have a differing morphology according to the speed of penetration of the bullet. There are two types of bullets [2]: those with a slow speed of penetration (300 m/s) and those with a high speed of penetration (600–900 m/s). Arterial lesions are caused by the phenomenon of temporary cavitation, caused by a sudden release of energy when the bullet explodes in the body tissue. The amount of kinetic energy released is directly proportional to the weight and speed of penetration of the bullet.

Temporary cavitation causes stretching or even tearing of the artery which may contain one or more ruptures. There is a correlation between the number of ruptures produced and the speed of penetration of the bullet. The severity of these ruptures in the artery depends on the proximity of the vessels to the centre of the temporary cavity. A bullet having a low speed of penetration may push back the artery as it penetrates body tissues. There is then a rupture of the adventitia alone, and of the external elastic lamella.

Arterial Injuries Due to Biological Effects of Certain Physical Agents

Arterial Injuries Caused by Electromagnetic Radiation

Within the spectrum of electromagnetic waves, x-rays (wavelength 1 nm) have harmful biological effects on the walls of arteries. There is no evidence that prolonged exposure to ultraviolet rays (wavelength 290–320 nm), even in second degree burn, causes arterial injury. The same is true for microwaves (wavelength 1 mm) and for ultrasound which may release heat energy. On the contrary, the thermal effects of laser light have been applied to treat vascular disease conditions, especially to destroy certain occluding lesions in arteries.

Radiation-Induced Lesions

Radiation-induced lesions responsible for clinical manifestations appear to be exceptional cases, but the true incidence of these injuries is probably underestimated. The patient's age at the time of exposure to radiation does not seem to affect the onset of lesions. The duration of time between exposure and development of post-radiation arteries varies from 3 to 36 years, with a mean of 15 years, and further vascular complications may occur much later. No correlation has been established between the severity of the injuries, the localization of the anatomical area irradiated and the radiation dose received.

All arteries are vulnerable [7]: the aorta, pulmonary, coronary, renal, mesenteric, iliac and femoral arteries have all been affected. Macroscopically, these lesions vary in their severity from stenosis with thrombosis or even rupture of the wall with false aneurysm to microcalcification. Histological examination reveals changes in all the three layers of the artery. Endothelial cells are vacuolated, destroyed and detached from the intima. This damage predisposes to thrombosis which occludes the lumen of the small-size arteries located in the vicinity of the irradiated areas [34]. The intima is then thickened by sclerotic tissue and there is a build-up of foam cells. In some instances, foam cell infiltration is noted in the media (Fig. 32.10). The entire process results in the formation of genuine fibrolipid plaques, narrowing the lumen of the artery [76]. The sclerosis is dense and hyalinized in the subintimal space. Later on, changes which occur result in the formation either of sclerotic tissue affecting mainly the intima and adventitia after 5 to 10 years of development or to atheromatous plaques which undergo calcification to a certain extent after 10 to 25 years. These lesions differ little from those caused by atherosclerosis. Some investigators do not rule out the part played by atheroma-inducing factors such as hypercholesterolaemia [36,78], in initiating this disease process. Arterial calcifications in a previously irradiated arm or leg have been reported [47]. One must be careful in interpreting post-radiation rupture of arteries.

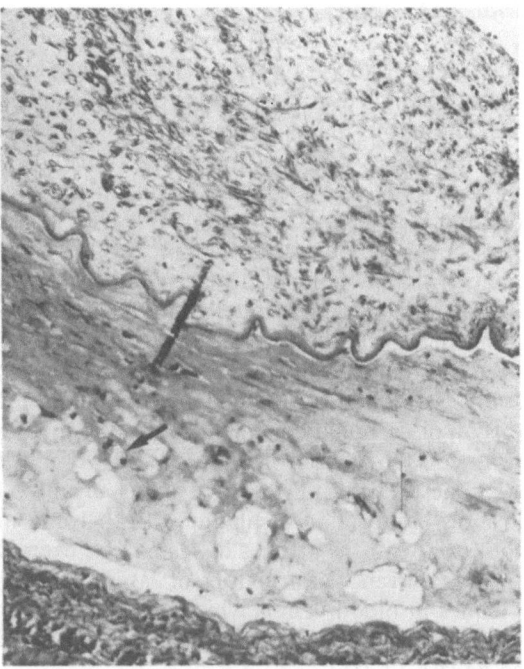

Fig. 32.10. Muscular artery. Histological changes in radiation-induced arteritis. Foam cells (→) are noted in the media. Masson's trichrome, × 190

It may be the result of necrosis of tumour tissue, infiltration of the wall of the artery or, in certain cases, iatrogenic disorders arising after surgery [9,23,25,27,70].

The diagnosis of post-radiation arteritis may be envisaged in the following situations: (a) the presence of arterial and periarterial sclerosis without concomitant atheroma in the irradiated area in a patient who does not have a history of inflammatory arteriopathy, smoking, diabetes mellitus or hypercholesterolaemia; (b) the presence of sclerotic lesions affecting the intima and media of arteries with concomitant atheromatous plaques in the irradiated area while adjacent arteries exhibit normal morphology; (c) a selective preponderance of atheromatous lesions in those areas which show post-radiation sequelae, for example cutaneous or intestinal lesions [59].

From the therapeutic standpoint, when post-radiation arteritis is symptomatic, the surgical treatment proposed consists of an arterial bypass, since endarteriectomy is contraindicated theoretically. Periarterial sclerosis may hinder dissection; sclerotic changes occurring in the wall of the artery are a cause of separation of the anastomosis and lesions in the soft tissue and of the skin may impede the healing process.

Injuries Caused by Laser

The thermal action of laser light has been successfully applied in many areas of medicine. In the continuous mode, thermal activity produces major changes. In effect, the energy transported by laser light is absorbed by living tissue and dissipated as heat. The target tissue then is the site of coagulative necrosis and the result is complete "volatilization" of the tissue targeted, producing a crater whose edges contain coagulated tissue debris and burnt tissue [86]. Research has been conducted with different types of lasers: solid lasers (ruby, YAG-neodyme) and gas lasers (argon or carbon dioxide). Several experimental protocols have been carried out in vivo, for example clearing of occluded areas in animal arteries, destruction of plaques of atheroma in the open with the arteries dissected and excised from cadavers, destruction of atheromatous plaques in blood vessels which are occluded or in arteries left in place in amputated limbs and so on [1,4,12,28].

When a laser is fired at a fluid, it is necessary to increase its total energy. This is true for several reasons: part of the energy is absorbed by water and there is loss of energy due to cooling. Importantly, the size of the target area to be destroyed determines the direction of the firing. Firing perpendicular to the artery wall causes complete rupture of all three layers of the artery. This rupture is bordered by coagulative necrosis with tissue debris on the surface. When the laser is fired coaxially, tissue destruction is limited. Depending on the power of the laser, the intima alone is the site of coagulative necrosis, producing a genuine intimectomy. The internal elastic lamina is preserved. The myocytes present in the internal portion of the media exhibit coagulative necrosis.

Present experimental data confirm the possibility of producing limited destruction of an occluding lesions in an artery, but at the price of possible perforation of the vessel. Additional research is underway aimed at employing the photochemical effects of laser light. Use of a dye to focus these effects on atheromatous lesions (such as haematoporphyrin derivatives) may prove useful [81].

Effects of Vibration

Vibrating instruments may cause osteo-articular, musculo-aponeurotic, neurological or vaso-motor disturbances in subjects who use these instruments on a regular basis [26,50,89].

Development of Raynaud's syndrome in workers who use jack-hammers, drills, polishing tools or chain-saws has been known to occur for a considerable time. The mean duration of time between the start of regular use of the tool and onset of a medical disorder is 7.85 ± 5.16 years [85]; in most cases, the disease becomes manifest when an impairment of function occurs, for example decrease in warmth of the hand with reduced blood flow, decrease in the number of subungual capillary loops per millimetre as seen by capillaroscopy or arterial spasm with the appearance of "amputation" of digital, ulnar and radial arteries seen on arteriography.

Histological study reveals the presence of thickening of the intima in small-size arteries with narrowing of the lumen and medial sclerosis.

The pathogenesis of this disease process is debatable. Abnormal spasm occurring in the digital arteries has been suggested in all studies. Some investigators believe that vibrations induce changes in the contractility of myocytes with medial hyperplasia, while the capillaries remain intact. These changes are accompanied by development of thrombosis and digital arteritis and later trophic disturbances. Exacerbation of this process by noise and cold has been suggested by other authors.

Effects of Cold

Cold may soothe, worsen or induce certain vaso-motor syndromes of varying aetiology (erythro-thermalgia, acrocyanosis).

Chilblains

Chilblains are caused by moderately cold temperature with concomitant dampness. This lesion occurs in early winter, affects women and children mainly and tends to recur each year. Chilblains are manifest by shiny red spots (erythema pernio) limited to the dorsal aspect of the hands, feet or knees, the nose and the ears. The spots become purplish, with burning and itching. The skin, which is cold when touched, may develop cracks and become infected. Most often, this process involves hypersensitivity to cold, but may be due to underlying organic diseases such as cryoglobulinaemia, cold agglutinins etc. The only effective treatment is adequate protection from cold.

Frostbite

Frostbite is a local phenomenon produced in certain subjects when exposed to cold. The lesion is characterized by intense vasoconstriction with numbness, pain and later absence of pain. Vasodilatation with reactional hyperthermia then produces acute pain with redness. The next stage involves cutaneous oedema with blistering and superficial and/or deep tissue necrosis which may require amputation. Sequelae are many and include cutaneous disorders, excessive sweating, peripheral occlusive arteriopathy and Raynaud's phenomenon.

Many studies have been undertaken to determine the influence of different factors in the pathogenesis of frostbite, for example the direct effect of cold, arterial spasms, thrombosis, anoxia and so on. The role of these factors varies but the direct effect of cold appears to play a fundamental role. According to some investigators, alternating arterial spasm and vasodilatation with an increase in blood flow in the capillaries are responsible for lesions in myocytes located in arterioles, with subsequent opening of arteriovenous anastomoses.

Prolonged exposure to less intense cold does not cause any specific arterial lesions. In fishmongers [54], exposure to wide differences in temperature causes vascular disturbances similar to Raynaud's phenomenon.

Electrocution

Blood-filled arteries are good electrical conductors. When an electric current crosses body tissues, electrical energy is converted into heat energy [44,45]. The heat thereby produced causes coagulative necrosis with burning of tissue.

In most cases, there is a limited area of skin burn with adjacent muscle necrosis whose extent is difficult to evaluate. Experimental research has shown that nerves and arteries are damaged when the heat dissipated reaches 45°C [72].

In man, since large-sized arteries have a large blood flow, there are no notable changes. The circulation causes the heat to dissipate and thus prevents coagulative necrosis from occurring. In contrast, small-sized arteries show necrosis of their endothelium with formation of thromboses which cause necrosis of adjacent tissues. Fasci-

otomy is the only way to verify if tissues and adjacent muscles are still viable [80].

Injuries Caused by High Pressure

There are frequent household accidents to do-it-yourselfers who use pressurized fluid sprayers [5]. These accidents frequently involve fingers or hands. The high pressure jet of paint causes tissue necrosis by crush injury at the point of impact [17]. These areas of necrosis contain wide paths of modified polymorphonuclear cells. Small-sized arteries and veins may present rupture of their walls with formation of thromboses. These lesions cause muscle necrosis due to ischaemia. The presence of a foreign body (paint) facilitates onset of infection and may necessitate amputation in severe cases [46,63].

References

1. Abela GS, Normann S, Cohen D et al. (1982) Effects of carbon dioxide, Nd-YAG and argon laser radiation on coronary atheromatous plaques. Am J Cardiol 50:1199–1205
2. Amato JJ, Rich NM, Billy LJ et al. (1971) High-velocity arterial injury. A study of the mechanism of injury. J Trauma 11:412–416
3. Banis JC, Rich N, Wheelan TJ (1977) Ischemia of the upper extremity due to noncardiac emboli. Am J Surg 134:131–139
4. Becquemin JP, Benhaim-Sigaux N, Melliere D (1986) Effets du laser sur les plaques d'athérome: aspects actuels et perspectives d'avenir. In: Meyet Ph, Gilbert JC (eds) Les médicaments de l'athérosclérose. Masson, Paris, pp 188–198
5. Beguin JM, Poilvache G, Van Meerbeeck J et al. (1985) Lésions de la main par injection sous haute pression. Apport de l'anesthérie loco-régionale. Ann Chir Main 4:37–42
6. Bergqvist D, Erikson U, Grevsten S (1972) False aneurysm in the deep femoral artery as a complication of osteosynthesis of intertrochanteric fracture. Acta Chir Scand 138:630–632
7. Bigot JM, Mathieu D, Reizine D (1983) Les artériopathies radiques. Ann Med Interne (Paris) 134:411–415
8. Bjorkerud S, Bondjers G (1973) Arterial reparation and atherosclerosis after mechanical injury. Part 5. Tissue response after induction of a large superficial transverse injury. Atherosclerosis 18:235–255
9. Bole PV, Hintz G, Chander P et al. (1975) Bilateral carotid aneurysms secondary to radiation therapy. Ann Surg 181: 888–892
10. Bole PV, Purdy RT, Munda RT et al. (1976) Civilian arterial injuries. Ann Surg 183:13–23
11. Boontje ABH (1978) Iatrogenic arterial injuries. J Cardiovasc Surg 19:335–340
12. Choy SDJ, Stertzer SH, Rotterdam HZ et al. (1982)

Transluminal laser catheter angioplasty. Am J Cardiol 50:1206–1208

13. Clark DE, Georgitis JW, Ray FS (1981) Renal arterial injuries caused by blunt trauma. Surgery 90:87–96

14. Cormier JM, Florent J (1967) Les hématomes rétro-péritonéaux par blessures des vaisseaux iliaques au cours des fractures du bassin. Mem Acad Chir (Paris) 93:458–462

15. Cormier JM, Moulonguet A, Diebold J et al. (1972) Hématome disséquant du tronc coeliaque et des artères rénales et mésentérique supérieure. Nouv Presse Med 1:113–116

16. Cormier JM, Sautot J, Frileux Cl (1977) Artères–veines–lymphatiques. In: Patel J, Léger L, Patel JC (eds) Nouveau traité de techniques chirurgicales, vol 4. Masson, Paris

17. Craig EV (1984) A new high-pressure injection injury of the hand. J Hand Surg 9A:240–242

18. David D, Blumenberg RM (1970) Subintimal aortic dissection with occlusion after blunt abdominal trauma. Arch Surg 100:302–304

19. De Bakey ME, Simeone FA (1946) Battle injuries of the arteries in World War II: an analysis of 2,471 cases. Ann Surg 123:534–579

20. De Haven H (1942) Mechanical analysis of survival in fall from heights of fifty to one hundred and fifty feet. War Medicine 2:586–596

21. Derobert L (1937) Les projectiles migrateurs intra-vasculaires. A propos de leur mode de pénétration. Strasbourg Méd 25:455–457

22. Drapanas T, Hewitt RL, Weichert RF et al. (1970) Civilian vascular injuries: a critical appraisal of three decades of management. Ann Surg 172:351–360

23. Elerding SC, Fernandez RN, Grotta JC et al. (1981) Carotid artery disease following external cervical irradiation. Ann Surg 194:609–615

24. Elliot JA (1956) Acute arterial occlusion: an unusual case. Surgery 39: 825–826

25. Fajardo LF, Lee A (1975) Rupture of major vessels after radiation. Cancer 36:904–913

26. Fiessinger JN, Sautarel E, Chousterman M et al. (1974) Le syndrome des vibrations Arch Mal Prof Med Trav Sécur Soc 35:901–905

27. Fonkalsrud EW, Sanchez M, Zerubavel R et al. (1977) Serial changes in arterial structure following radiation therapy. Surg Gynecol Obstet 145:395–400

28. Geschwind H, Vieilledent C, Boussignac G et al. (1985) Traitement des artériopathies des membres inférieurs par laser. Angéiologie (Paris) 37:51–55

29. Greenfield LJ (1984) Complications in surgery and trauma. Lippincott, Philadelphia

30. Guerriero WC, Carlton CE, Scott R et al. (1971) Renal pedicle injuries. J Trauma 11:53–62

31. Haimovici H (1982) Vascular emergencies. Appleton-Century-Crofts, New York

32. Hall R (1971) Vascular injuries resulting from arterial puncture or catheterization. Br J Surg 58:513–516

33. Hare RR, Gaspar MR (1971) The intimal flap. Arch Surg 102:552–555

34. Hasleton PS, Carr N, Schofield PF (1985) Vascular changes in radiation bowel disease. Histopathology 9:517–534

35. Haudenschild CC, Schwartz SM (1979) Endothelial regeneration. II. Restitution of endothelial continuity. Lab Invest 41:407–418

36. Hayward RH (1972) Arteriosclerosis induced by radiation. Surg Clin North Am 52:359–366

37. Hegarty MM, Angorn IB, Gollogly J et al. (1975) Trau-matic arteriovenous fistulas. Injury 7:20–28

38. Hewitt RL, Collins DJ (1969) Acute arteriovenous fistulas in war injuries. Ann Surg 169:447–449

39. Hewitt RL, Collins DJ, Hamit HF (1969) Arterial injuries at a surgical hospital in Vietnam. Arch Surg 98:313–316

40. Hewitt RL, Smith AD, Drapanas T (1973) Acute trau-matic arteriovenous fistulas. J Trauma 13:901–906

41. Holin JM (1973) Aspects anatomo-pathologiques et thérapeutiques des traumatismes artériels: a propos de 90 observations. Thèse Doctorat en Médecine, Université d'Amiens

42. Holscher EC (1968) Vascular and visceral injuries during lumbar-disc surgery. J Bone Joint Surg [Am] 50:383–393

43. Hughes CW (1954) Acute vascular trauma in Korean war casualties: an analysis of 180 cases. Surg Gynecol Obstet 99:91–100

44. Hunt JL, McManus WF, Haney WP et al. (1974) Vascular lesions in acute electric injuries. J Trauma 14:461–473

45. Hunt JL, Mason AD, Masterson TS et al. (1976) The pathophysiology of acute electric injuries. J Trauma 16:335–340

46. Jouglard JP, Bureau H, Tramier H et al. (1970) A propos de quatre cas de corps étrangers de la main injectés sous pression. Marseille Chir 4:411–414

47. Juergens JL, Pluth JR (1980) Trauma and peripheral vascular disease. In: Juerger JL, Spitell JA, Fairbairn II JF (eds) Peripheral vascular disease. Saunders, Philadelphia, pp 607–627

48. Kelly GL, Eiseman B (1975) Civilian vascular injuries. J Trauma 15:507–514

49. Krupski W, Thal ER, Gewertz BL et al. (1979) Endothelial response to venous injury. Arch Surg 114:1240–1248

50. Laroche PG (1976) Traumatic vasospastic disease in chain-saw operators. Can Med Assoc J 115:1217–1220

51. Leandoer JL, Tremann JA, Oishi RH et al. (1972) Bilateral renal artery thrombosis following blunt trauma: report of two cases. J Trauma 12:166–169

52. Liddicoat JE, Bekassy SM, Daniell MB et al. (1975) Inadvertent femoral artery "stripping": surgical management. Surgery 77:318–320

53. Litchford B, Okies JE, Sugimura S et al. (1976) Acute aortic dissection from cross-clamp injury. J Thorac Cardiovasc Surg 72:709–713

54. Mackiewicz Z, Piskorz A (1977) Raynaud's phenomenon following long-term repeated action of great differences of temperature. J Cardiovasc Surg 18:151–154

55. Martin E (1930) Mécanisme de la compression des carotides dans la pendaison. Ann Med Leg 10:565–567

56. Matthews GB, Buzzi A (1965) Acute obstruction of the external iliac artery due to circumferential intimal fracture. J Cardiovasc Surg 6:422–425

57. McGough EC, Helfrich LR, Hughes RK (1972) Traumatic intimal prolapse of the common carotid artery. Am J Surg 123:724–725

58. Mellière D, Bourgeois A, Becquemin JP et al. (1981) Les micro-emboles des membres supérieurs d'origine artérielle. Chirurgie 107:428–432

59. Melliere D, Becquemin JP, Moehne M et al. (1983) Vraies et fausses artérites radiques. J Mal Vasc 8:321–327

60. Moore CH, Wolma FJ, Brown RW et al. (1971) Vascular trauma. A review of 250 cases. Am J Surg 122:576–578

61. Moore TC (1958) Acute arterial obstruction due to traumatic circumferential intimal fracture. Ann Surg 148:111–114

62. Morton JR, Crawford ES (1972) Bilateral traumatic renal artery thrombosis. Ann Surg 176:62–67

63. Nahigian SH (1966) Airless spray gun. A new hand hazard. JAMA 195:688–691

64. Natali J, Benhamou AC (1979) Iatrogenic vascular injuries. A review of 125 cases (excluding angiographic injuries). J Cardiovasc Surg 20:169–176

65. Orcel L, Chomette G (1978) Anatomie pathologique vasculaire. In: Delarue J, Laumonier R (eds) Traité d'anatomie pathologique. Flammarion, Paris, pp 154–197b

66. Ouchi H, Ohara I, Kijma M (1965) Intraluminal protrusion of completely disrupted intima. An unusual form of acute arterial injury. Surgery 57:220–224

67. Rich NM, Gaugh JH, Hughes CW (1969) Acute arterial injuries in Vietnam: 1,000 cases. J Trauma 10:359–368

68. Richaud J, Lazorthes Y (1980) Pathogénie des lésions traumatiques fermées de la carotide interne du cou. Déduction thérapeutique à propos de 17 cas. J Traumatology 3:161–170

69. Rimpau A (1957) Etude macroscopique et histologique des orifices de ponction après artériographie cérébrale. Ann Anat Pathol (Paris) 2:227–288

70. Rotman M, Seidenberg B, Rubin I et al. (1969) Aortic arch syndrome secondary to radiation in childhood. Arch Int Med 124:87–90

71. Saade B, Merhej S (1978) Traumatisme artériel. Expérience de la guerre au Liban. A propos de 83 cas. Ann Chir 32:345–348

72. Sances A, Myklebust JB, Larson SJ et al. (1981) Experimental electrical injury studies. J Trauma 21:589–597

73. Schenk WG, Martin JW, Leslie MB et al. (1960) The regional hemodynamics of chronic experimental arteriovenous fistulas. Surg Gynecol Obstet 110:44–50

74. Sethi GK, Scott SM, Takaro T (1975) False aneurysm of the abdominal aorta due to blunt trauma. Ann Surg 182:33–36

75. Sevitt S (1968) Fatal road accidents. Injuries complications and causes of death in 250 subjects. Br J Surg 55:481–505

76. Sheehan JF (1944) Foam cell plaques in the intima of irradiated small arteries. Arch Pathol Lab Med 37:297–308

77. Shumacker HB (1970) Aneurysm development and degenerative changes in dilated artery proximal to arteriovenous fistula. Surg Gynecol Obstet 130:636–640

78. Silverberg GD, Britt RH, Don Goffinet R (1978) Radiation-induced carotid artery disease. Cancer 39:130–137

79. Slatis P (1962) Injuries in fatal traffic accidents. An analysis of 349 medico-legal autopsies. Acta Chir Scand [Suppl] 297: 40 pp

80. Solem L, Fischer RP, Strate RG (1977) The natural history of electrical injury. J Trauma 17:487–491

81. Spears JR, Serur J, Shropshire D et al. (1983) Fluorescence of experimental atheromatous plaques with hematoporphyrin derivatives. J Clin Invest 71:395–399

82. Stehbens WE (1968) Blood vessel changes in chronic experimental arteriovenous fistulas. Surg Gynecol Obstet 127:327–338

83. Tabbara W, Proteau J, Dumont G et al. (1962) A propos de la rupture traumatique de l'aorte. Etude anatomo-clinique de 7 nouveaux cas. Ann Med Leg 42:390–404

84. Teare D (1951) Post-mortem examinations on air-crash victims. Br Med J ii:707–708

85. Vayssairat MP, Patri B, Guilmot JL et al. (1982) La capillaroscopie dans la maladie des vibrations. Nouv Presse Med 11:3111–3115

86. Vuong NP, Baviera E, Houissa-Vuong S et al. (1984) Conisation du col utérin au laser CO2. Une étude anatomo-pathologique des altérations tissulaires et du processus de réparation à propos de huit observations. Arch Anat Cytol Pathol 32:26–29

87. Vuong NP, Ecourrou J, Desoutter P et al. (1986) Faux anévrysme post-traumatique de l'artère temporale superficielle. A propos d'une observation avec revue de la littérature médicale. J Mal Vasc 11:375–378

88. Vuong NP, Sebban E, Sobiak S et al. (1986) Pseudo-anévrysme traumatique de l'artère temporale superficielle ou complication vasculaire d'une gifle: a propos d'une observation avec revue de la littérature médicale. Sem Hôp Paris 62:2697–2703

89. Welsh C (1980) The effect of vibration on digital blood flow. Br J Surg 67:708–710

90. Wood NE, Stutzman FL (1963) Intimal separation in various arterial injuries. Angiology 14:265–267

91. Zhender MA (1960) The symptomatology and course of aortic rupture in closed thoracic injury based on 12 cases. Thorac Surg 8:1–46

Chapter 33

Perioperative Arterial Changes

P. N. Vuong and J. M. Cormier

Any reconstructive procedure in the repair of an artery has the potential to cause immediate or later complications, which may be related to the surgical technique itself and/or the type of prosthetic material employed [21,26,72,100,113,135, 136,143,147].

In this chapter, we will discuss pathological disorders related to surgical procedures, pathological disorders arising from the implantation of a foreign body within the wall of the artery and disorders due to changes which occur at the blood–arterial wall interface.

Disorders of Operated Arteries Related to Surgical Procedures

A surgical operation is itself a "controlled" traumatic injury, having the potential for producing complications [30,55,99,100,108,127,199]. Preoperative assessment of the patient must thus take into consideration the dual risk of the natural outcome of untreated arteriopathy and of the trauma associated with surgical repair of a vessel [57,101].

Arterial Exposure

An artery is surrounded by a sheath of connective-adipose tissue or peri-adventitia. This sheath separates the artery from neighbouring tissue and from the organs adjacent to it. It contains nutrient vessels or vasa vasora, lymphatic vessels and nerve fibres which form true nervous plexuses whose branches penetrate the wall of the artery. There is no clear demarcation between the tunica adventitia and the peri-adventitia [111,124].

Exposure of the artery requires the separation of the vasculo-nervous bundle after incision of the peri-adventitia. The arterial segment is thus uncovered and temporarily disconnected from its nutrient vessels and nervous supply. This denervation is the basis for the theoretical explanation of longstanding vasomotor disturbances causing dilatation and tortuosity of arteries [193]. In practice, possible adverse ischaemic and denervation effects on the exposed artery have not been demonstrated. Experimentally, regeneration of nerve fibres in the wall may occur over a period of time, depending on the type of blood vessels involved and extent of injury.

The ease with which a healthy artery can be exposed contrasts with the delicate and even dangerous procedure needed to expose a diseased artery presenting with inflammation of the peri-adventitia. The difficulty is even greater in case of repeat operations where the formation of post-operative sclerotic tissue welds the artery to adjacent structures, in particular to veins.

Traumatically induced complications are varied. They may arise as the result of difficult dissection and include in particular extensive haemorrhage in the peri-adventitia and wounding of the exposed axis of the artery or at an arterial bifurcation. The accidental sectioning of a collateral vessel enveloped in the peri-adventitia which has been thickened by sclerosis may occur, as may traumatic rupture of the intima or intimal and medial layers by excessive traction on a segment of exposed artery. This happens in particular in re-operations on thromboendarterectomized arteries.

Arterial Clamping

This procedure temporarily cuts off the flow of blood. Blood stoppage may be partial in the case of lateral clamping (using a curved metal clamp). Total occlusion occurs when transverse clamping is applied (compression with a finger, loop-shaped compression, metal clamp, a balloon catheter inserted into the artery, etc). Complications caused by clamping may be due either to the cessation of flow of blood or to traumatic injury caused by the clamping itself.

Clamping causes haemodynamic changes both above and below the site of intervention. In the thoracic aorta, blockage of the circulation produces increased arterial pressure above the clamp, increasing the residual volume which may lead to acute left ventricular failure. In an area where the collateral circulation is insufficient, prolonged clamping of a nutrient artery causes hypotension in the area below the clamp with resultant tissue anoxia. The result may include acute thrombosis of a stenotic artery, a venous thrombosis, ischaemic tissue lesions leading to metabolic disturbances (acidosis, renal tubular necrosis, hyperkalaemia) and in case of unchecked ischaemia, gangrene or Volkman's syndrome may result [57,199,135].

Traumatic injury to the artery due to clamping depends not only on the anatomical area involved but also on the technique employed. Selection of the site where the clamp is to be placed is of primary importance. If placed on an atheromatous or already altered artery thickened by medial sclerosis, a clamp may crush or detach a plaque or calcified area. These lesions are the starting point for arterial dissection beneath the plaque, with mural thrombosis and embolism ocurring below [164]. Excessive tightening of a loop tourniquet may cause actual strangulation of the artery with partial mural rupture affecting the intima and media of the artery. When a loop is applied twice, the artery is both strangled and pivoted on its axis, and the line of fracture is thus a spiral shape.

A metal clamp which is too tightly fastened may crush the artery and cause various types of damage to its wall, ranging from rupture of the intima or the intima and media, to necrosis of all three layers of the artery wall. Forceps-type clamps with fine points on each blade impress multiple pricks or punctate injuries on the arterial wall. Depending on the length of these points, each little wound formed causes a partial rupture of the adventitia and media, or a complete rupture of all three arterial layers [15,81,103,129,152]. Damage to the wall of the artery thus depends on the closeness of these wounds to each other in the area squeezed by the clamp blades. Overinflation of a balloon catheter used to clamp a vessel from inside the lumen causes abrasion of the endothelial lining of the vessel with "compression" of intimal lesions. This mechanism is similar to the one used in transluminal angioplasty.

Arterial Ligation

During surgery, haemostasis of small arteries is obtained by means of a clip or by electric cautery. Such a procedure causes coagulative necrosis of the wall of the blood vessel. Coagulation causes retraction of all three arterial layers, thereby reducing the lumen of the artery. Necrosis of the vessel wall also causes formation of a mural thrombus, leading to complete occlusion of the lumen. The entire process causes a temporary "haemostatic plug" to form in a first stage. In the second stage this plug becomes permanent, after organization of the thrombus.

Simple ligation consists of passing a suture thread underneath the artery and tying a knot around the area to be ligated. In middle-sized arteries, simple ligation must be doubled using a single stitch suture, reinforcing the ligation and preventing its unfastening. In large arteries, ligation must be ensured by using simple ligatures placed on each side of the area to be ligated, thus effectively separating the area from the circulation. In the following step, the wall of the arterial segment thus cut off is incised and the blood drained. Whatever technique is used, ligation exerts concentric compression on the arterial wall, thereby causing rupture of the intimal and medial layers with diminution of the lumen and formation of a mural thrombus which extends on both sides of the ligated area. Once organized, this thrombus definitively occludes the ligated areas.

Two types of complications may occur. Firstly, haemorrhage may result from loosening of the band of the ligature which may slip if it is

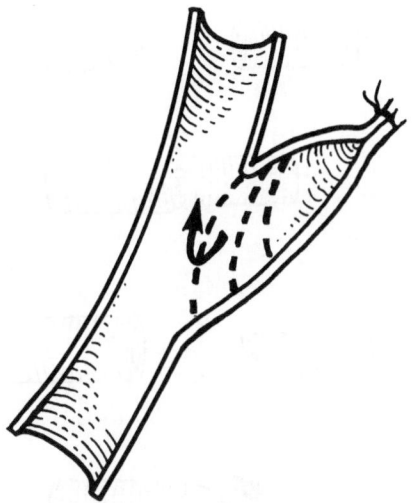

Fig. 33.1. Arterial ligation. In an arterial bifurcation, a too long stump of a ligated collateral vessel may be a starting point for extensive mural thrombus which continues in the main arterial trunk (→).

not secured in large arteries, or when the arterial segment to be ligated has been incompletely exposed and separated from neighbouring tissue. Secondly, an extensive mural thrombosis may occur in an arterial bifurcation after ligation of a collateral vessel which has a long stump. In effect, the stump of an artery forms a dead end whose lumen is occluded by a mural thrombus in the ligated area. When there is an insufficient blood flow from above, this thrombus becomes extensive and continues into the lumen of the second branch of bifurcation or into the main arterial trunk. This process leads to gradual occlusion of the entire lumen of the arterial bifurcation or is the starting point for an embolus (Fig. 33.1).

Arteriotomy

An arteriotomy is a controlled incision into an artery. It may be linear, i.e. parallel to the main axis of the artery, or be transverse resulting in a complete section. It may also represent the removal of a piece of artery wall, producing a "window". This procedure, which is easy to perform on a healthy artery, becomes more delicate when the vessel is diseased.

Complications depend primarily on the location of the arteriotomy. This procedure may initiate the start of a parietal dissection, either intimal or intramedial, with occlusion of the

lumen of the artery, formation of a proximal mural thrombus and embolism. These complications occur in certain disease conditions, for example in atheromatous arteries whose linings contain partially necrotic or calcified lesions or in arteries containing degenerative lesions (diseases involving the media, Marfan's syndrome, fibro-dysplasia, etc.).

Arterial Suture

This procedure aims at joining the edges of an arteriotomy in such a way that the repaired area has an anatomical structure which is almost normal. During suturing, the needle must pass through all three layers of the arterial wall. Each stitch causes a microscopic parietal wound of the edges of the arteriotomy. The tightening of each stitch allows the edges of the arteriotomy to line up with each other exactly. Histologically, each hole caused by the needle produces a very limited detachment of the vascular endothelium, thereby exposing the tissue below the intima. When clamping ends, this area of detachment facilitates the adhesion of a layer of fibrin and platelets. This phenomenon persists for 24 to 48 h as has been demonstrated by certain studies with radio-active isotopes using platelets tagged with indium-III [132]. Endothelial repair follows with endothelial cells sliding from the periphery of the cut. The mechanism which permits healing of the edges of the arteriotomy is similar to the one which occurs in every parietal wound.

When the sutures are tied too tightly, the suture thread shears the edges of the arteriotomy, causing rupture of the intimal and medial layers of the wall, resulting in formation of either a parietal haematoma or a localized thrombus. Complete rupture caused by the suture thread may cause loosening of the suture with a later haemorrhage. The outcome of the sutured area depends on the type of suture used which determines the haemodynamic profile of the repaired area. In single-stitch sutures, stitches spaced too far apart may cause gaping of the edges of the arteriotomy, a factor which may cause later haemorrhage or a false aneurysm. An overcast-type suture, composed of either simple stitches or mattress stitches, may cause a defect in lining up the faces of the wound; an overcast stitch which is too tight may create a stenosis and stitches too close together increases the formation of scar fibrosis, making the wall of the artery too stiff and giving rise to progressive stenosis [120,176].

Arterial Anastomosis

An anastomosis aims at restoring the continuity of the vessel while respecting the anatomical configuration of the artery and the mechanical proprieties of the arterial wall. There are three types of anastomosis: end-to-end anastomosis, end-to-side anastomosis and side-to-side anastomosis.

End-to-End Anastomosis

The follow-up period for an end-to-end anastomosis of two blood vessels of the same size is comparable to that for the suture of an artery. When different size vessels are involved, end-to-end anastomosis may result in formation of a functional stenosis if the ends of these vessels do not fit each other, or if they are not exactly congruent. After drawing the vascular ends towards each other, the diameter of the anastomosis is smaller than that of the larger size vessel. The latter forms a post-stenotic dilatation compared with the anastomosis (Fig. 33.2a). This functional stenosis facilitates the creation of turbulent blood flow which may cause thrombosis. There are several methods which can be used to make the ends of the vessels to be anastomotized fit each other. The basic principle involves enlarging the end of the small size vessel by creating two slits, thus allowing it to fit the larger vessel. It is also possible to make an oblique incision or a wide cut. In this case, the incision must involve both ends of the arteries. A single incision in one artery causes an angulation or a folding just before the anastomosis after it is sutured. This defect in the configuration of the artery causes the later development of fibro-blastic endarteritis (Fig. 33.2b–c).

End-to-Side Anastomosis

End-to-side anastomosis may be necessary either to transpose an artery (e.g. reimplantation of the subclavian artery in the adjacent internal carotid) or to anastomose a bypass. The purpose of an end-to-side anastomosis is either to preserve collateral vessels in the excised segment, or to permit a lengthy anastomosis, eliminating the risk of stenosis. Blood flow is not interfered with since on each arterial circumference, at the level of the anastomosis, there are only two stitches. In contrast, a similar anastomosis for an end-to-side implanted bypass creates an anastomotic chamber, a source of turbulence which may later give rise to an aneurysm or a false

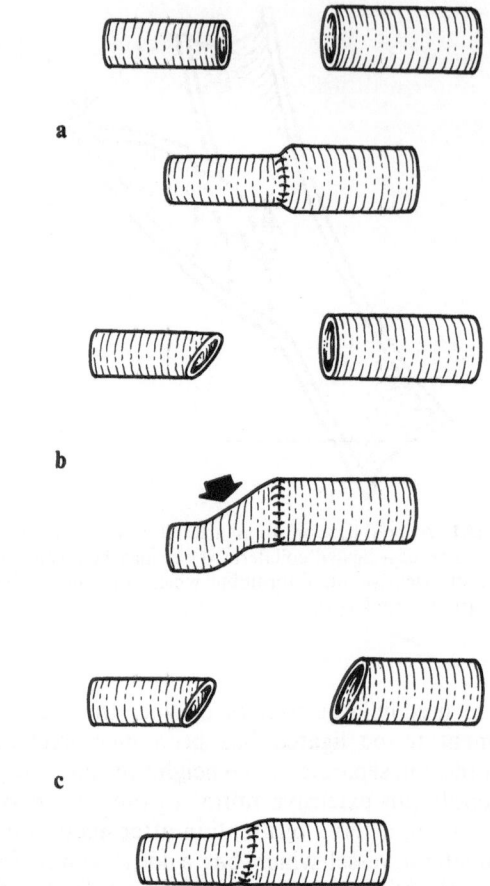

Fig. 33.2a–c. End-to-end anastomosis. **a** In direct anastomosis of vascular ends of different size, the larger vessel forms a functional post-stenotic dilatation, facilitating creation of turbulent blood flow. **b** A single oblique incision in one artery causes an angulation or a folding just before the anastomosis. **c** Complications are prevented when the incisions are performed in both ends of the arteries.

aneurysm. In addition, depending on the outline of the section of the anastomosis, haemodynamic complications may occur. When the cut is perpendicular, the suture may cause a stenosis of the anastomosis. A bevelled incision permits the connection of the ends at a more or less acute angle compared to the side branch. If the bevel is too long, the end may present an angulation or fold just in front of the anastomosis after suturing. Generally, the greater the angle of connection, the greater the turbulence in the blood flow. This results in a decreased blood flow below the anastomosis. A 45° connecting angle produces blood flow close to normal. Correct tension at the end also helps to regularize the haemodynamics of the anastomosis. Too great a tension at this end deforms and flattens the side branch. In contrast, insufficient tension causes redundancy and produces a type of "dolicho-collateral" vessel. All of these defects in arterial configuration cause turbulent blood flow, a thrombosis-inducing factor, or detachment of the intima in the anastomosis.

Side-to-Side Anastomosis

This is possible when both arteries are parallel. Complications which may develop involve the sutures.

Intimal Hyperplasia

Intimal hyperplasia is an intermediate-term cause of failure of an arterial anastomosis [16,70,121,130,157,183]. At the interconnection of vessels or in a bypass, intimal hyperplasia is frequently observed near the distal or lower anastomosis. The time of onset varies, ranging from 3 to 12 months after surgery for certain femoro-popliteal bypass procedures using tet-rafluoroethylene material [76]. Macroscopi-cally, there is a whitish bulging of the intima, formed around the anastomosis, including the suture thread. The distribution of this swelling varies. In an end-to-end anastomosis, it is circum-ferential; in an end-to-end anastomosis it is focal. Histologically, there is a plaque of fibroblastic tissue on the intima, which has developed from the principal collateral vessel.

Pathology of the Operated Artery Due to Inclusion of a Foreign Body in the Wall of the Artery

This condition primarily involves reactions between the wall of the artery and suture threads inserted through it in sutures or anastomoses.

The ideal suture thread must be pliable, elastic, resistant and biologically inert. It must be free from capillarity, i.e. it must not act as a wick, allowing blood to ooze out of the artery. However, the ideal suture thread does not exist and as in all tissues, it is a foreign body in the artery wall. It induces an inflammatory reaction leading to the formation of giant-cell granulomata and organization of fibrous scar tissue [4]. Factors which determine the severity of the different stages of the inflammatory process have not yet been etablished. However, the intensity

of the reaction to a foreign body varies according to the types of suture thread employed.

Resorbent thread (catgut) is broken down and granulation tissue progresses to form sclerotic scar tissue, the final stage of the process of inflammation. When non-resorbent thread is employed (linen, silk, nylon, tergal covered thread, stiff cotton thread, etc.), inflammation persists for a long time. Foreign body gran-ulomata are developed with braided nylon but are scarce with single-strand nylon, terga or covered thread. With the latter material, this reaction is quickly replaced by the formation of sclerotic bands of tissue.

The onset of a foreign body reaction changes the tensile strength of the suture thread con-siderably. Thus, with ordinary catgut, a resorb-able suture thread, the tensile strength is halved within 2 d; with chromic catgut this occurs in around 10 d. This decrease in the tensile strength of the thread may be the starting point for loos-ening of the suture.

Whatever the type of suture material employed, the final stage in the foreign body reaction is the formation of sclerotic scar tissue. This makes the sutured area stiffer than the rest of the artery wall. Sclerotic scar tissue causes parietal retraction or even stenosis, if the stitches are close together.

Pathological Changes in the Operated Artery Due to Interactions Between the Blood and the Artery Wall

Thromboendarterectomy

Techniques

The object of this operation is to remove the internal portion of an atheromatous artery to restore a "normal" lumen. Unblocking the artery may be performed in two planes: either in the external part of the media, or more externally, with the remaining wall of the artery containing only adventitia and external elastic lamina lined by a few muscle fibres from the media (Fig. 33.3).

There are several techniques for endar-terectomy: a "turnover" procedure, after an oblique or transverse sectioning of the artery; or

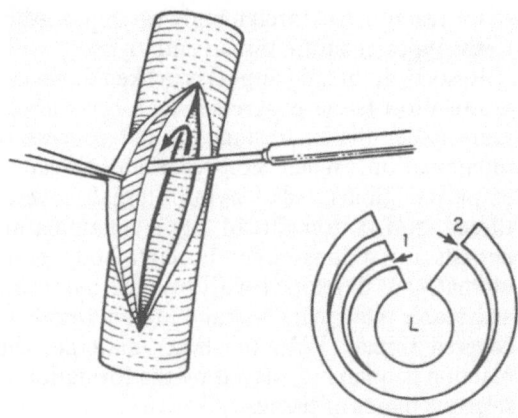

Fig. 33.3. Thromboendarterectomy. Unblocking the artery may be performed either in the external part of the media (→1) or more externally in the subadventitial space (→2). *L* lumen. (→), cleavage plane.

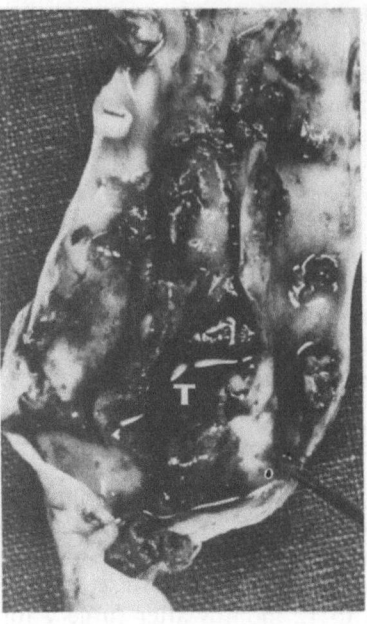

Fig. 33.4. Core of thromboendarterectomy. The intima is thickened by lesions of atheroma with a thrombosis on the surface.

unblocking by a continuous longitudinal arteriotomy or by discontinuous arteriotomy. Initial separation of the cleavage planes is accomplished using a pliable spatula. The "atheromatous" cylinder is detached from the external portion of the artery either by direct visualization, or by means of a stripping instrument or by injection of pressurized water or gas.

Whatever the techniques used, the boundaries of the unblocked area must be clear and gently sloped below to prevent detachment of the intima from the arterial segment situated beyond the endarterectomized area.

Results

The core of the thromboendarterectomy is cylindrically shaped and its face towards the lumen is irregular. The intima is thickened by lesions of atheroma. Some of these are ulcerated with a luminal thrombosis on the surface. Others are calcified. The external face, which corresponds to the plane of cleavage, is haemorrhagic (Fig. 33.4).

The endarterectomized face of the lumen is smooth. After detachment in the external plane of the media is performed, the newly formed artery often has a larger size than before the operation (Fig. 33.5).

Healing of the endoarterectomized area is poorly understood [11]. An experimental study using indium-III-tagged platelets in dogs undergoing carotid endarterectomy shows aggregation of platelets in the surgical wound immediately after the replacement of the newly formed artery [132]. This effect is maximal 1 h after and appears

to stabilize towards 24 h. These experimental findings are corroborated by ultrastructural study. In the deeper layer of the artery, there is a proliferation of myocytes from the remaining portion of the arterial media. This process tends

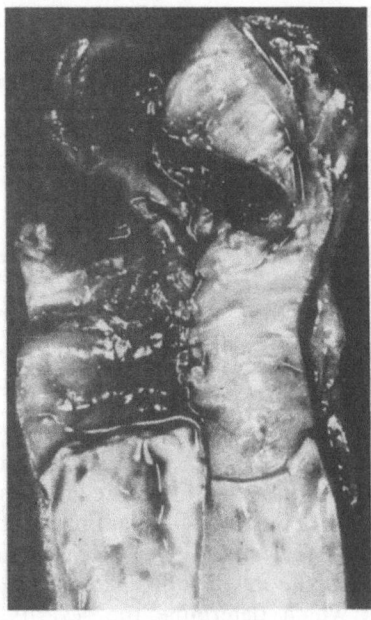

Fig. 33.5. Carotid arteries. Gross aspect of the endarterectomized area.

to include the coating of fibrin on the surface. The ends of the neo-artery are lined with endothelial cells which come from the endothelium of the wound edges by sliding over. In man, study shows that wound healing is obtained towards the 8th post-operative day.

Complications

Early Complications. The thromboendarterectomized area is characterized by a haemorrhagic surface which may give rise to distal emboli. The distal projection of the intima may become detached by a flow of blood. The resulting dissecting haematoma formed thus produces a stenosis. The detached intima may turn over in the lumen of the artery and beome a factor in the formation of an acute thrombosis. An arterial embolus may precede acute occlusion due to a thrombus formed; surgical suturing of the projection in the intima avoids these complications.
Late Complications. After thromboendarterectomy, the repaired artery no longer possesses the characteristics of a normal artery. Fibrous organization may give rise to complications according to the portion of the artery repaired. In the aorta and the common iliac arteries following thromboendarterectomy [79], there is a dilatation of the axis of the artery giving the appearance of an aneurysmal polydystrophy, usually developing after 10 to 12 years (Fig. 33.6). In the external iliac artery, fibrous organization from the superficial femoral artery results in gradual stenosis, evident chiefly after 5 or 6 years. In the internal carotid artery, an early stenosis due to hyperplasia of the intima begins after 3 months,

and progressively worsens up to 12 or 18 months. It occurs in about 10% of operated arteries [51,184]. Whatever the anatomical localization of the thromboendarterectomy, the repeated occurrence of the resultant atheroma is a constant finding, as is calcification of the unobstructed area; this evolution is manifest after 5 years and gradually becomes more apparent; in the ulcerated parietal areas, mural thrombi may cause distal emboli which damage the vascular bed below.

Arterial Embolectomy

This procedure consists of incising the artery to remove an embolus and then closing the arteriotomy. It may be accomplished directly under visual control or by a retrograde technique using either clamping forceps as employed in the common bile duct operation or Fogarty's balloon catheter.

The time of embolectomy determines the adherence of the blood clot to the intima. When the operation is carried out soon after formation of the embolus, removal of the clot is easy. If the operation is delayed, adherence of the clot to the wall of the artery is complicated by a rapidly extending thrombosis below [126].

Embolectomy is always associated with a linear wound in the intima of the artery. Insertion of Fogarty's catheter to perform thrombectomy from below causes sloughing of the endothelial lining [84]. In small-size arteries, these lesions may cause repeated thrombosis.

Other complications are due to faulty surgical technique: excessive "sweeping" with Fogarty's catheter may cause tearing of the intima or even of the intima and media which may result in perforation of the artery wall or even arteriovenous fistula [89,153]. With atheromatous arteries, a balloon catheter may cause limited or extensive parietal dissection if it goes in the wrong way. Withdrawal of the catheter may bring out an atheromatous plaque. Any trauma of the artery may thus be complicated by thrombosis [85,109,146,158]. Rupture of the Fogarty's balloon catheter with distant migration is an exceptional occurrence.

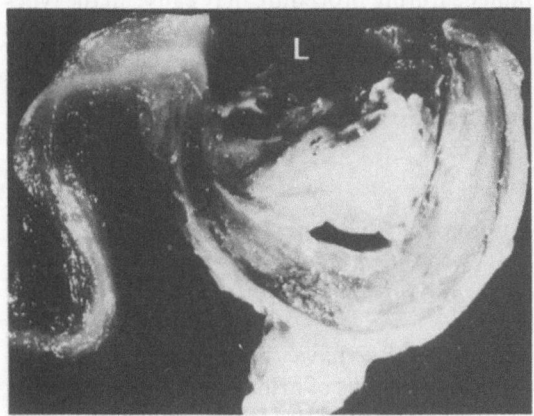

Fig. 33.6. Endarterectomized iliac artery. Gross aspect of an aneurysmal polydystrophy. *L*, lumen.

Transluminal Angioplasty

This procedure was first used by Dotter [74] to treat arteriopathy of the lower extremities, and was quickly applied to other arteries including iliac, femoro-popliteal, renal and coronary arteries. The basic principle involved is to dilate the stenotic artery by inflating a balloon catheter under high pressure which already has been inserted into the artery [2,6,7,23,114,149,174, 200].

The object of this procedure is to restore an adequate degree of patency to the artery, that is to eliminate any significant stenosis (for the lower limbs, both at rest and upon effort). The morphological changes procured by dilatation have been the subject of much controversy. The initial mechanism of this process involves compression of the plaque of atheroma or its impaction into the wall of the artery. In effect, experiments with rabbits show that, in the dilated segment of the artery, there is a fraying of the endothelial cells with rupture of the intima and the media. The entire area is crowned by a fibrinous thrombocytic mural thrombus. If angiography is performed immediately after dilatation of the artery, the presence of parietal dissection with cracks and displacement of the plaques of atheroma is confirmed with diffusion of the radioactive contrast medium into the adjacent artery wall below. Echotomography carried out after dilataton occurs corroborates these findings.

Endoluminal angioplasty causes trauma to any artery. Complications resulting from this procedure are linked to the effects of the dilatation itself and to manoeuvring of the catheter [5,22,39,47,60,165,170].

Cracking of the atheromatous lesions corresponds to an arterial wound. In places, the intima is broken as well as the internal elastic lamina. This arterial wound is accompanied by formation of a thrombosis. Medial injury associated with intimal hyperplasia can be one of the factors of arterial restenosis [8,33,82].

Dislodging of plaques of atheroma which thus have become fragile may cause an atheromatous or fibrinous embolism in distal arteries. Ischaemia often regresses spontaneously. This complication occurs in about 3% of patients undergoing angioplastic dilatation for arteriopathy of the lower extremities.

In the dilated portion of the artery, over-inflation of a balloon catheter may cause rupture of the intima and media of the artery with formation of a haematoma. Organization of this rupture site leads to development of a false aneurysm with mural thrombosis. Complete rupture of the wall of the artery may lead to haemorrhage, requiring an emergency procedure to obtain haemostasis.

Finally, all the complications secondary to catheterization are seen at the site of arterial puncture: haematoma, parietal dissection, detachment of atheromatous plaques, cholesterol emboli, infectious complications, and when the catheter is being introduced, parietal dissection, insertion in the wrong direction or even perforation may occur.

Patch-Graft Angioplasty

One of the possible risks connected with closure of a longitudinal arteriotomy in a small- or medium-sized artery is stenosis, which is a risk factor for the development of a thrombosis. A patch-graft angioplasty is then necessary. This procedure involves covering the arteriotomy with a patch of tissue. This patch of tissue is attached to the edges of the arteriotomy by suture.

The type of patch used varies. For example, it may be the wall of an endarterectomized artery or an artery autograft, a venous autograft or a synthetic prosthesis. After the patch is put into place, it is attached to the wall of the artery in the same way as a graft or prosthesis. It is difficult to locate this area angiographically after surgery; in resected portions of arterial segments containing a patch, only the suture stitches enable the pathologist to locate its position [192].

Possible complications are due either to faulty surgical technique or to changes in the patch itself. From a technical standpoint, every patch may be effectively a defect in the configuration of the artery. A patch which is too large can cause turbulence, leading to formation of a mural thrombus and aneurysmal dilatation. A faulty suturing procedure may cause all the types of complications described above. Complications which occur as a result of changes in the patch are those which occur with any graft or prosthesis.

Bypass by Graft or Prosthesis

General Considerations

Venous grafts have been used to replace damaged portions of arteries from the early 20th century [57,77,101,110,159]. Introduction of synthetic materials as a substitute for natural tissue by Voorhes, Jaretzki and Blackmore led to a considerable development of surgery to implant arterial bypasses [57].

Biological grafts must be differentiated from arterial prostheses [57,101,194]. Biological grafts consist of tissues (arteries or veins) which come either from the patient himself (autograft), or from a subject of the same species (xenograft). Prostheses are made of different types of synthetic material which are relatively well tolerated by the body [137,163,188,196].

Generally speaking, the graft material for the ideal prosthesis must have the following properties : (a) it must be relatively easy to obtain in various shapes and sizes; (b) it must be easy to handle and suture; (c) there must be perfect biocompatibility with the host organism allowing excellent healing to occur; (d) there must be a similar degree of vascular compliance to the replaced vessel, with permeability completely assured; (e) it must be mechanically resistant allowing it to last a long period, or be a definitive replacement for the remainder of the patient's life; (f) it must be resistant to infection; and (g) it must be free of any toxic or oncogenic effects.

Despite the many improvements in the techniques for collection of tissue specimens or their preparation, the graft materials and prostheses currently in use are hardly perfect. Implantation of a graft or a prosthesis leads to series of similar types of reactions in adaptation to the site and resistance to new haemodynamic constraints [73,118,192]. Each of these phenomena may involve complications whose severity or incidence may vary with the type of material used.

Outline of the Clinical Course and General Complications Affecting a Bypass

Healing of Sutures and Lines of Anastomosis. In the recipient artery the suture thread causes a foreign body reaction. In the bypass, it behaves as a wick and thus allows leakage of plasma and permits the penetration of inflammatory cells into the wall of the bypass. An inflammatory granuloma forms around it. Sclerotic organization of the granuloma thus modifies the performance of the bypass.

There is never any real continuity between the recipient artery and the bypass at the line of anastomosis. Lining up of the edges of the line of anastomosis is always accompanied by a projection. Immediately after it is implanted, the external side of the line of the bypass anastomosis becomes surrounded by loose connective tissue. A few rare connective septa infiltrate the line of anastomosis, and reach the luminal side in the form of a fibrotic wedge. Sclerotic covering of the external side of the line of anastomosis continues and provides a certain strength to this structure which still remains as a zone of weakness in the bypass.

Adaptation of the Bypass to the Neighbouring Tissue and the Recipient Artery. In the period following the implantation of a bypass, an inflammatory reaction gives rise to sclerotic scar tissue which develops in the connective tissue which forms the bed of the graft or of the prosthesis (Fig. 33.7). This tissue connects the bypass to the surrounding tissue [40,145].

Changes which occur at the blood–graft or blood–prosthesis interface vary according to location; different changes occur in the intermediate portion and at the lines of anastomosis [9,95,107]. As soon as the bypass is completed, the blood–bypass interface is carpeted by a thin layer of fibrin, containing platelets, monocytes, and polymorphonuclear leucotyes. Between the end of the 1st week and the 3rd week following the bypass operation, a layer of fibrin becomes organized into a pseudo-intima. At the line of anastomosis, this formation is partially re-endothelialized for a few millimetres. This finding has been confirmed by ultrastructural studies and immunomorphological techniques using endo-

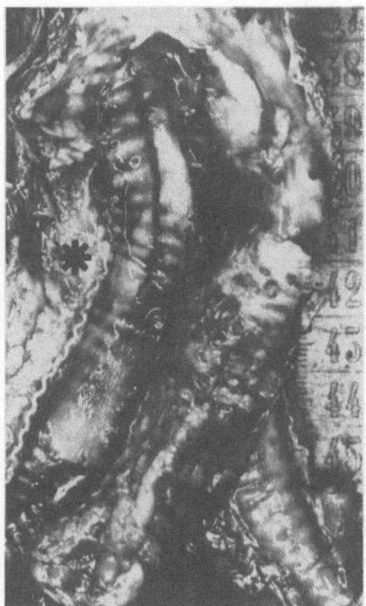

Fig. 33.7. Aorto-bifemoral bypass. Gross aspect of external capsule developed around a synthetic prosthesis (*).

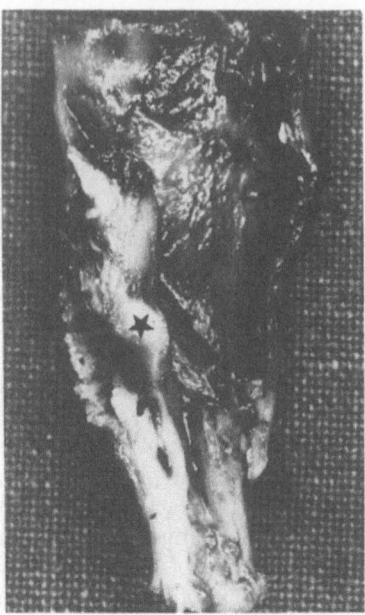

Fig. 33.8. Femoro-popliteal bypass. Gross aspect of the intimal leaflet coming from the recipient artery (★) and sliding over the femoral PTFE prosthesis.

thelial cell markers such as factor VIII and major blood group antigens. Endothelial cells come from the endothelium of the recipient artery by sliding over. This process approximates to repair of the endothelium in case of wounds of the intima. Development of the endothelium distant from the line of anastomosis remains controversial since, beyond the line of anastomosis,

Fig. 33.9. Femoro-popliteal bypass. The fibroblastic leaflet develops from the intima of the recipient artery. Its formation causes stenosis of the bypass. Masson's trichrome, × 30

the endothelium does not appear to form continuously. In the months which follow, a fibroblastic leaflet of tissue develops, thickening the pseudo-intima. This leaflet comes from the intima of the recipient artery by sliding over. Its formation, similar to hyperplasia of the intima in arterial anastomoses, causes stenosis of the bypass [18] (Fig. 33.8, 33.9).

General Complications of Bypasses. Each step in the course of the bypass may involve early- or late-onset complications [195].

Thrombosis occurs preferentially at the line of anastomosis and may extend to the arterial side. There are many causes, for example initial technical reasons (a defect in the geometrical configuration, poor alignment of the prosthesis, localized dissection of the atherosclerotic arterial edge when the suture thread is passed), detachment of the intimal plaque covering the pseudo-intima at the anastomosis, false aneurysm or a combination of intimal hyperplasia from defective anastomotic compliance and progression of the atherosclerotic process [54].

Thrombosis is initially localized, but when the stenosis reaches a certain point it produces acute occlusion of the vessel with extensive thrombosis [20,142]. Certain types of material may favour onset of thrombosis along the path of the bypass.

Rupture of the line of anastomosis occurs as a result of partial or total separation of the sutures. This rupture is facilitated by several factors: exceptionally the suture thread breaks, or more frequently rupture of the edge of the artery which has been sheared by the suture thread occurs. Several possible factors may cause this rupture of the edge of the artery including per- or post-operative infection, inflammatory reaction with foreign body granuloma production, inadequate alignment of the edges of the anastomosis, weakening of the wall of the artery by thrombo-endarterectomy, or post-operative periprosthetic lymphocoele which has led to poor healing. This rupture causes an externalized haemorrhage with haematoma formation; this haemotoma extends to the external part of the anastomosis, separating the bypass from the adjacent tissue. Organization of this haematoma produces a false aneurysm [3,177,185] (Fig. 33.10).

Changes in shape or size of the artery occur after localized or wide breakdown of the bypass. The two most frequent causes are structural alterations due to a lack of tolerance of the graft material with foreign body reaction, and mechanical fatigue occurring after the haemodynamic

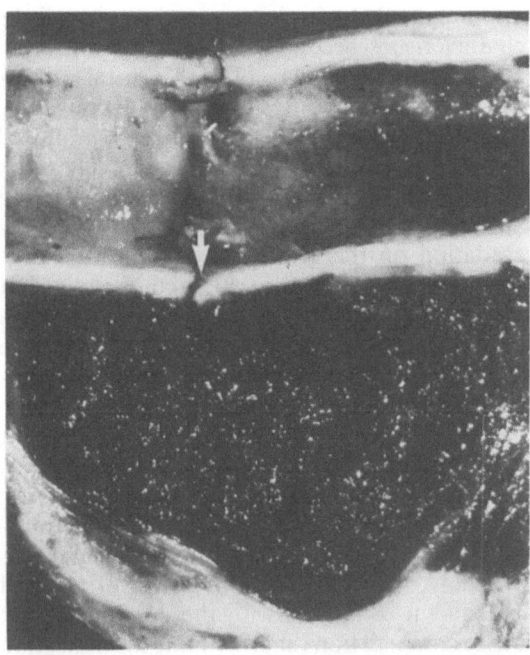

Fig. 33.10. Vascular bypass with end-to-end anastomosis of two PTFE prostheses. Organization of a haematoma secondary to rupture of the line of anastomosis producing a false aneurysm.

constraints created by the new anatomical site (excessive tension, repeated movements) [14,166]. Bending of the bypass or aneurysm are the adverse effects which develop in these circumstances.

Aneurysm of the recipient artery occurs as a post-stenotic dilatation of the vessel by turbulent blood flow as it crosses the anastomosis. It occurs preferentially in end-to-side anatomoses, and it is the very basis of any procedure to create an anastomotic cavity.

Infections may occur either immediately or long after the operation. Early-onset infections or infection developing up to 4 or 5 years after surgery are secondary to inadequate aseptic surgical technique or the spreading of adjacent infection to the wall of the artery (necrosis of a surgically exposed area, infected haematoma or lymphocoele). Infection occurring after a long period (10 to 15 years later) usually occurs as the result of bacteria reaching the anastomosis during a phase of bacteraemia or septicaemia or more rarely may be due to adjacent infections. These are serious complications resulting in rupture or haemorrhage and may cause septic shock or embolism [36,56,75,87,94,112,128,141, 182].

A fistula creates a connection between the bypass and adjacent hollow organs (duodenum, ileum, ureter). It may occur as a result of erosion in the organ due to the bypass, causing infection. It is most often situated at the line of anastomosis. Mechanical factors may be the cause, as in previously cited cases. However, this complication usually results from infection of the suture line, giving rise to a septic false aneurysm which bursts into the adjacent organ [37,186].

Lipid deposits and calcifications may overload the pseudo-intima, producing true plaques of atheroma in the long term; these lesions are identical to those observed in the recipient artery [150].

Pathology of Biological Grafts

Biological grafts may be arterial or venous. Biological grafts can be classified according to their origin as xenografts, homografts or autografts.

Xenografts. Fresh xenograft procedures are no longer used [35]. Currently, certain investigators use xenografts previously "treated" to make them biologically inert [59,155,156]. In most cases, the grafts are bovine carotid arteries taken from adult cattle [62,116]. The peri-arterial tissue is peeled away and the xenografts are immersed in a proteolytic enzyme solution (1% ficin at 37°C for 3 h), washed in a citrate phosphate buffer at a pH 5, rinsed with 1% sodium chloride for 18 h and then treated with a 1.3% dialdehyde solution at pH 8.8. The sodium chloride inactivates the enzymatic residues adsorbed by the heterografts. After washing, ligation of collateral vessels and verification of water-tightness, the xenografts are preserved in a mixture of 4% ethyl alcohol and 1% propylene oxide at room temperature. Storage is possible for up to 3 years. The speciments are thereby transformed into a tubular segment of connective tissue, devoid of any antigenic action. Sawyer et al. [160] have proposed substitution of dialdehyde by glutaraldehyde according to the Carpentier et al. [44] technique for xenografts of cardiac valves. This treatment improves clinical results [155,156].

Bovine arterial xenografts are used to fashion arteriovenous fistulas, now the preferred route for the administration of chemotherapeutic agents or for repeated blood transfusions in haemophilic patients. After implantation, the xenograft becomes rapidly surrounded by dense connective tissue. In man, the graft tissue is still recognizable 2 years later.

Rosenberg et al. [154,155,156] have obtained good results for iliac artery bypasses in human subjects over a 13-year period. In the femoro-popliteal bypasses, the percentage of unobstructed grafts is about 41% after 3 years. It is then unchanged for the next 3 years. This result is clearly improved (73%) with heterografts treated with glutaraldehyde. These statistics are optimistic and the long-term course is peppered with possible complications: stenosis chiefly at the line of anastomosis, true aneurysm (3%–6% of implanted xenografts) and recurrences of atheroma. These complications limit the indications for xenografts [154]. To eliminate some of these disadvantages, some investigators have proposed adding Dacron to strengthen the graft.

Homografts.   The homografts may be arterial or venous.

The arterial homograft is a segment of artery collected from a cadaver and preserved. The ideal time to collect the specimen is about 12 h after death. Strict aseptic technique must be used and the excised artery must be free of disease. Some disease conditions should preclude the use of this method (septicaemia, cancer, certain virus infections). There are several techniques which can be used to preserve the specimen including fixation in either absolute alcohol or a 4% solution of formaldehyde at pH 5.6; the specimen must then be immersed in a sterile, isotonic nutrient medium (Hank's solution, Collin's solution) at 4°C. Freezing then chills the specimen to −72°C rapidly. It is then kept in cold storage at −30°C. The graft is warmed to room temperature when it is to be implanted. The procedure of lyophylization combines freezing (for 24 h) with drying in a vacuum (for 4 d). Once implanted, arterial homografts behave exactly the same regardless of the preservation technique used. Most frequently the endothelium and the intima disappear before implantation. Myocytes and fibroblasts in the media disappear after a few days. In the first weeks thereafter, the media continues to be denuded of its component cells. Elastic lamellae break into pieces. One month after, sclerotic tissue surrounds the adventitia [90,180]. Eighteen months post-operatively, the homograft appears as a fibrous hyalin tube whose softness and elasticity have been reduced. Long- or short-term complications arise and include thrombosis, rupture and thinning of the wall of the artery with ectasia and atheroma.

Venous homografts collected from cadavers are no longer used. However, a homograft of saphenous vein excised during a stripping oper-

ation and subsequently prepared retains all the characteristics necessary for the creation of an arteriovenous fistula or for distal revascularization procedures employed in salvaging an extremity. The venous homograft employed may also be a human umbilical vein treated with glutaraldehyde which binds to collagen and amino groups, conferring resistance to biodegradation on the graft. After sterilization, the specimen is preserved in 50% ethyl alcohol. It may be supported on a Dacron mesh and when implanted, a foreign body reaction develops in contact with the mesh, and at the periphery sclerotic tissue covers this reinforcing coat. The configuration of the vein remains recognizable; it is hyalinized in places and cracked in some areas. Long-term complications include thromboses, recurrence of atheromatous lesions and transverse cracks. Development of these cracks is probably due to greater elongation of the wall of the vein compared with the coat of dacron. These complications sharply limit the indication for venous homografts [179].

Autografts   There are two types of autografts: arterial and venous. An autograft is not subject to the rejection phenomenon.

An arterial autograft is a segment of artery transplanted to a new site within the patient. The splenic, hypogastric, external iliac, superficial femoral and internal mammary arteries may be excised for this purpose. The value of an arterial homograft is due to the fact that a healthy arterial segment can be used which heals normally and conserves all its physiological and mechanical qualities. It is, then, the best possible material available for a bypass, since it is not subject to any changes over time other than those observed during "ageing" of vessels in a given subject. However, this procedure may only be applied to young patients with arteries free from atherosclerosis. This problem may be overcome by using a thromboendarterectomized segment of artery collected by eversion. This in situ autograft is subject to all of the disadvantages observed with a thromboendarterectomized artery, namely stenosis and aneurysmal dilatation.

A venous autograft is a material of exceptional quality when employed as a patch angioplasty, a short bypass (coronary bypass, mesenteric bypass, carotid–subclavian bypass) or a long bypass (carotid–humeral bypass, femoro-popliteal or femoro-peroneal bypass). Generally the graft used is the internal saphenous vein excised surgically; less frequently the external saphenous vein is used. More rarely the cephalic vein is used and in exceptional cases the basilar veins whose texture is much thinner. Their use involves the risk of aneurysmal dilatation. Excision of the internal jugular vein for grafting is no longer carried out. The vein chosen must be free from dystrophic changes (varicose dilatation or endophlebitis). For some authors, these changes do not constitute a factor determining the patency of the bypass in the long run. For ex situ bypasses, the vein must be sufficiently large. The specimen is collected using spread incisions after ligation of all collateral vessels and gentle dilatation with the patient's heparinized blood or with Collin's solution, the medium in which the vein is immersed before being surgically implanted [25,27,32,45,61,71,98,102,105,123, 125,168].

The venous autograft used as a bypass may be employed in three different ways:

1. As an inverted venous graft. This graft has not undergone valvular destruction. There are two pitfalls: the vein is disconnected from its tissue bed, and for long bypasses it is inversely adapted since the smaller end of the vein is anastomosed to the larger end of the artery and vice versa.

2. As a non-inverted in situ venous graft. Here the vein remains in its tissue bed, the sizes of the vessels correspond and haemodynamic considerations are maintained, with the vein calibre decreasing in size and running parallel to that of the arterial axis. However, this procedure requires destruction of the valves, a cause of trauma which can lead to operative or post-operative complications.

3. As a non-inverted ex situ venous graft which may be placed in a distant site (e.g. on internal saphenous vein placed in the upper limb, in the iliac trunk, or in the ilio-popliteal or ilio-peroneal segments). This graft has additional disadvantages compared with the non-inverted in situ graft, in that it is disconnected from the tissue bed.

A venous autograft is a pliable material, easy to suture [1,2,13,31,34,38,48,50,88,96,131,172, 175]. Immediately after its implantation, the endothelial lining is detached from the endovein. The denuded subendothelial tissue and the inner portion of the media are infiltrated by inflammatory elements. The whole area is covered by a fibrinous coating. From 4 weeks on, the endovein becomes thicker, with proliferation of myocytes which are often arranged parallel to the blood

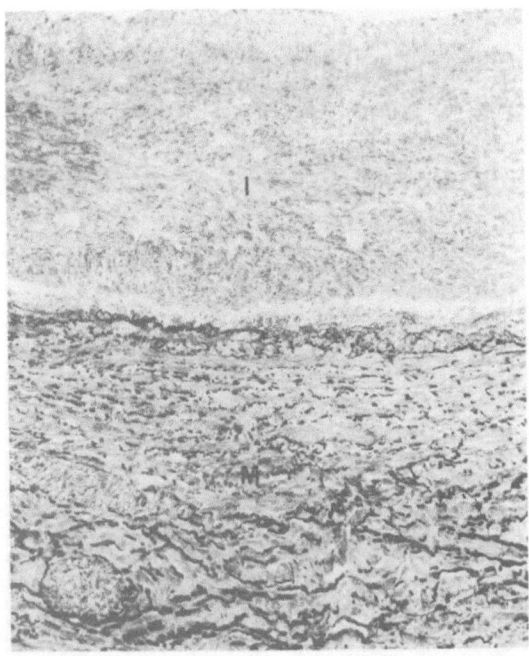

Fig. 33.11. Venous autograft. Histological aspect of a venous autograft with fibroblastic thickening of the endovein. *I*, intima. Orcein, × 120

flow (Fig. 33.11). Maximum thickening of the endovein occurs after 12 months and is combined with a newly formed internal elastic lamella, which is often interrupted in places, without reference to a real arterial internal elastic lamella (Fig. 33.12). In the media, relative ischaemia of the autograft when it is collected causes a certain loss of cellular elements which are replaced by sclerotic tissue; however, the configuration of the autograft remains unchanged. The graft remains patent if the surgical technique has been correctly performed [86,115,191].

Several types of complications are possible. The deterioration of the vascular bed below the graft due to progressive atherosclerosis accounts for the formation of thromboses, but it should be emphasized that the venous autograft is the only bypass which maintains a patent vessel when blood flow diminishes due to peripheral vessel loss.

Stenosis of the venous autograft may affect either the anastomosis or the graft itself [171,181]. Stenosis of the anastomosis (superior or inferior) may of course be due to the atheromatous process (common femoral, deep femoral), but it may be due to an uncorrected technical error or hyperplasia of the intima. These three mechanisms are frequently combined [12,90,98]. There are several different

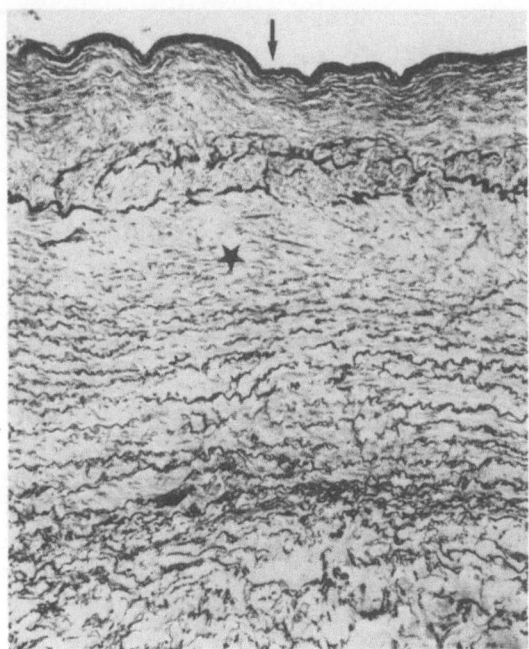

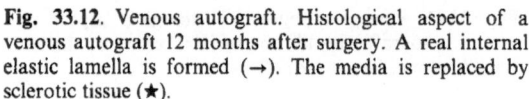

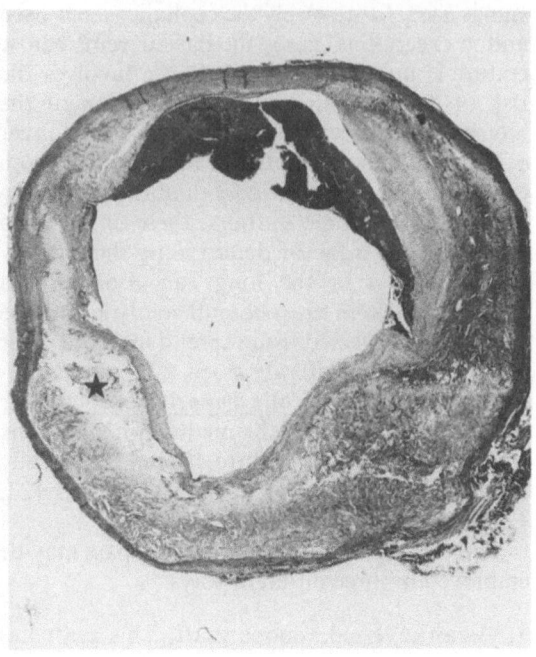

Fig. 33.12. Venous autograft. Histological aspect of a venous autograft 12 months after surgery. A real internal elastic lamella is formed (→). The media is replaced by sclerotic tissue (★).

Fig. 33.13. Venous autograft. Deposits of lipids in the endovein in some areas may give rise to true atheromatous lesions. Masson's trichrome, × 13

causes of stenosis of the graft beyond the lines of the anastomosis. Faulty operating technique (causing parietal ischaemia by denuding the wall of the vein, dilatation due to too great a pressure in the saphenous vein with an inappropriate vascular bed, or peri-arterial fibrosis following organization of a haematoma) are factors implicated in the pathogeneis of tubular stenosis of the graft segment. Segmental tubular stenosis [28,29,131] may be due either to an atheromatous overloading, or to a process of endarteritis developing from the long dead end of a collateral vessel or to organization of the parietal thrombus which was formed at the time of surgical repair. Localized diaphragm-like stenosis caused by decreasing vessel calibre may be due to an uncorrected faulty technique in the ligation of a collateral vessel or to stenosis of the valves. Two hypotheses have been suggested to explain this. Firstly, a thrombus may develop in the valve convexity and its organization could cause valvular fibrosis. Secondly, the valves could create turbulent blood flow, a factor causing intimal hyperplasia and stenosis. In contrast, stenosis or bending are never encountered at the site of flexure. This is another important benefit of the venous graft when compared with other procedures for bypasses.

Thromboses occur as the result of untreated stenosis. When stenoses attain a certain critical point, perhaps from worsening of atheromatous lesions above the bypass site, they facilitate the onset of this complication.

Ectasias develop in various anatomical locations and with different degrees of severity. Ectasia from proximal anastomotic implantation is due to a dilation of a stump of a collateral vessel tied too far from the graft. Single fusiform ectasia occurs following a dilatation situated at a valve, or at a varicose area. A multistaged ectasia, a genuine aneurysmal polydystrophy, is due to thinning of the wall of the autograft following loss of fibroblastic activity of the media of the vein with time.

Deposition of lipids in the endovein may also occur creating true atheromatous lesions with calcifications identical to those observed in the adjacent artery [17,28,29,68,69,80,92] (Fig. 33.13).

Pathology of Arterial Prostheses

The concept of strengthening the wall of a blood vessel goes back to the early 20th century. Carrel, Tuffier and Guthrie inserted glass tubes or tubes made of silver, vitallium or inert metals into arteries. These attempts usually led to complications (thrombosis, haemorrhage) and resulted in failure.

The development of the plastics industry facilitated the development of bypass techniques. From 1952, synthetic prostheses made of the following materials were used in experiments: Vinyon N (acrylonitrile/vinyl chloride), Nylon (polyprolactain), Orlon (polycrylonitride), Ivalon, Teflon (polytetrafluoroethylene) or Dacron (polyethylene terephthalate). Some of these early prosthetic materials are no longer used. Currently, the most widely used prostheses are made of Teflon or Dacron. Several types of prostheses with specific characteristics may be distinguished according to their texture [57,101,197]:

1. Crape-type prostheses (Wesolowski's prostheses) [194] are soft and possess good cell permeability. These are gradually being abandoned because they lead to the formation of peri-prosthetic haematoma around muscle fibres.

2. Knitted prostheses (Debakey's ultra-light prostheses), have good cell permeability; they require careful pre-coagulation and thus may not be employed if anti-coagulant drugs are administered.

3. Woven prostheses are stiffer and rapidly become waterproof since cell permeability is low; they are used preferentially in patients receiving anti-coagulant drugs or in those who have disorders of haemostasis.

4. Velvet prostheses are covered by a Dacron velvet material which resists thromboses and cellular deposits.

5. Expansive reinforced porous prostheses appear to possess durable resistance and can last for over 20 years [197]. PET (polyethylene terepthalate) prosthese and PTFE (polytetrafluoroethylene) prostheses are of this sort.

Other types of prostheses which are still currently experimental have been used. They are intended for small vessels and include polyurethane grafts and graft materials containing carbon fibres which provide a negative charge to the luminal side of the prosthesis [160].

Some biological prostheses contain a biological structure with a synthetic material used as a method of reinforcement, for example human umbilical vein treated with glutaraldehyde and reinforced by a Dacron coating [63,64,65,66,67,138]. Others require in vivo manufacture. This consists of implanting a synthetic framework into the recipient of the prosthesis which ensures fibrous proliferation during a "maturation period" of 5 or 6 weeks. The conduit thus formed can be employed as a bypass. The Spark-mandril is an example [173].

Apart from the general complications arising from the vascular bypasses already described, the pathology of arterial prostheses is essentially dominated by the problem of the resistance of the material used since the installation of a prosthesis is usually a definitive measure [14,19,24]. The many types of material available make it difficult to select the best, and it must be remembered that the host response has an impact on the structure of the material employed. The stability of the prosthesis over a long period of time is the determining factor in the reliability of the prosthesis.

Synthetic Prostheses (Nylon, Orlon, Teflon and Dacron). As soon as a prosthesis is implanted, an inflammatory reaction develops around the site. Inflammation followed by the organization of sclerotic tissue forms the "external capsule". As time passes, this connective tissue penetrates the external part of the prosthesis in the form of fibrous septa which fix the prosthesis to the peri-arterial tissue. This penetration of connective tissue increases with time, infiltrates the mesh of the prosthesis and can reach the luminal side [144,151,161,178] (Fig. 33.14). Sauvage and others [162] consider it to be one of the ideal qualities of a prosthesis, allowing it to be integrated into the organism. These characteristics contitute the basis of the "latticework" concept [40,166]. However, insertion of the prosthesis is always accompanied by a foreign body reaction which develops around the prosthetic meshes (Fig. 33.15). After a few years, the external part of the prosthesis breaks down. The prosthesis, which is altered to a certain extent, becomes a rigid conduit surrounded by a sclerotic magma. These changes are known to occur with knitted, woven and velvet types of synthetic prostheses and decrease the strength or elasticity of the wall of the prosthesis. A comparative study of different materials was performed by Harrison in experiments undertaken since 1958 [77]. After being implanted for 100 d, the tensile wall strength decreases to 81% for Nylon, 6.9% for Orlon, 10.1% for Dacron and increases to 3.2%

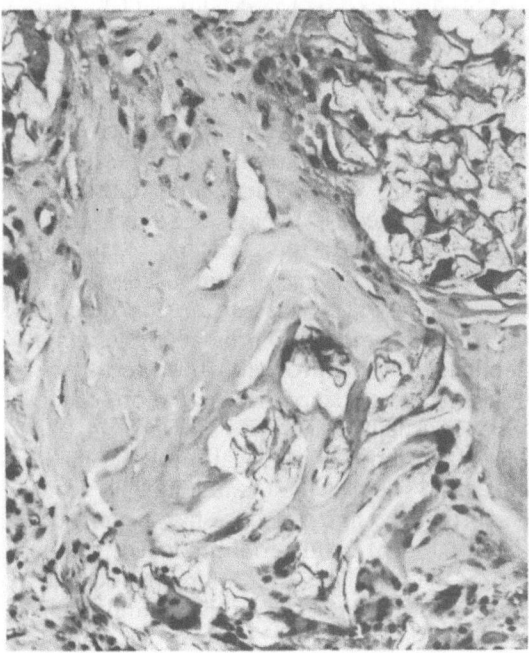

Fig. 33.14. Polyethylene terephtalate (Dacron) prosthesis. Histological aspect of the prosthetic mesh infiltrated by sclerotic tissue. Haematoxylin–eosin–safran, × 120

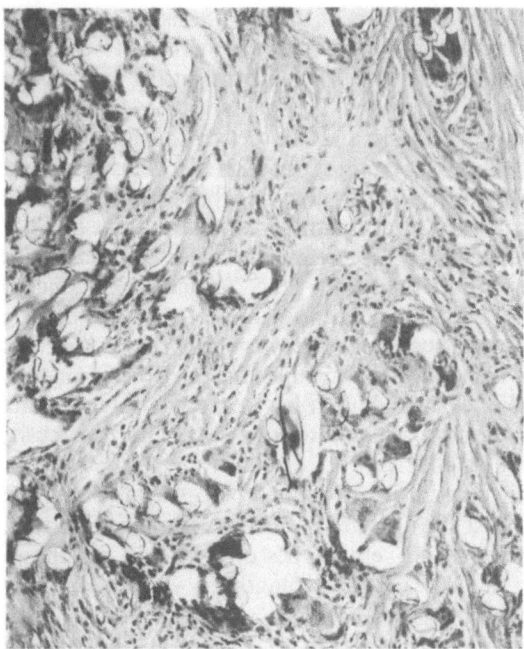

Fig. 33.15. Polyethylene terephtalate (Dacron) prosthesis. Histological aspect of the foreign body reaction developed around the prosthetic meshes. Haematoxylin–eosin–safran, × 210

for Teflon. These findings have been confirmed in studies in man [78]. Once implanted, the prostheses dilate by 15%–20%, so that small-calibre prostheses are of practical interest. With Dacron prostheses, the tensile strength appears to decrease immediately after implantation, then remains stable thereafter. Teflon prostheses have a tensile wall strength which is unchanged for years.

The texture of the prosthesis also contributes to its long-term breakdown [97,134,180]. Thus, with Dacron prostheses (polyesters), woven compact types which are rigid when implanted appear satisfactory [14] since the constraints of mechanical breakdown are less pronounced [52]. Knitted prostheses, which may or may not have a velvety surface, are more pliable and thus easy to implant after prior coagulation, but they fray at the extremities. In addition, with the action of the systolic shock-wave the knots stretch and this produces permanent and gradual distension of the prosthesis [106,117,118,119,122,148,150, 198].

A particular complication of synthetic prostheses is the development of stenoses along their path. For Dacron prostheses which cross the knee joint flexure, stenosis may occur either by detachment of the neo-intima or of the internal capsule, with adjacent haematoma formation, or by alloplastization of the prosthesis. When this stenosis reaches a certain critical stage, it gives rise to occlusion by acute thrombosis [19,150].

Tetrafluoroethylene Prostheses (TFEP). This material is used in a porous, expansive form. The prostheses are composed of 15% pure TFEP and 85% air by volume. Manufacturing procedures allow us to modify the sizes of knots, the shape and the length of the prosthetic fibres at will [58,83,91,195]. The first TFEP bypasses in the carotid and femoral arteries were implanted in dogs around 1973 [43,133]. They have been used in man for arteriovenous fistulas since 1976 [139,167]. The first series of femoral-popliteal bypasses performed in man was reported by Veith in 1978 [189]. Half of these bypasses exhibited a fusiform dilatation which occurred in the following months. This complication was the result of a defect in circumferential tensile strength [140]. This led the manufacturer to reinforce the TFEP prosthesis with a bandage wound around the prosthesis.

A reinforced TFEP prosthesis does not require previous clotting before it is implanted. Once implanted, connective tissue penetration stops at

Fig. 33.16. PTFE prosthesis. Histological aspect of the external capsule (→) organized around the reinforcing sheath (★) Haematoxylin–eosin–safran, × 180

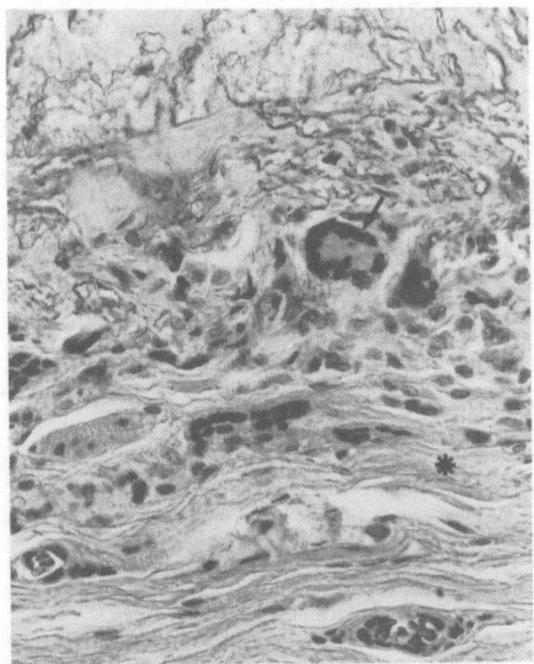

Fig. 33.17. PTFE prosthesis. Histological aspect of the external capsule (*). Foreign body reaction developed just along the reinforcing sheath. Haematoxylin–eosin–safran, × 475

the reinforcing sheath (Fig. 33.16) which seems to limit the foreign body reaction (Fig. 33.17). In the interstitial spaces of the prosthesis, there is a plasma insudation and a few blood cells are present. The organization of the neo-intima has been the focus of much research and results are contradictory [49,53,104]. Campbell's study with dogs demonstrates connective tissue penetration with neovascularization, producing a true neo-intima. The same results have been obtained in man [42,43]. A study of biopsies of permeable TFEP prostheses implanted for a period longer than 60 months, undertaken by Camilleri et al. [41], seems to refute these results. In contast, it shows a lack of connective tissue penetration, the absence of neovascularization and that endothelial proliferation is limited to the lines of anastomosis. The neo-intima is basically made of fibrin, platelets, blood cells and cell debris. Collagen and elastin fibres are never observed even in prostheses implanted for a long time (Fig. 33.18 a–d).

Apart from general complications [47,76], postural occlusion of TFEP prostheses crossing the crural arch may occur. An excessive inflammatory reaction caused by the external reinforcing bandage of the prosthesis leads to formation of blocks of fibrous tissue around the prosthesis. When the thigh is flexed for a long period, the block of periprosthetic tissue leads to kinking of the prosthesis with consequent thrombosis [10,93,187].

Bioprostheses. Bioprostheses of human umbilical vein have been discussed in the paragraph concerning homografts.

The Spark-mandril concept dates to the beginning of 1960s [173]. The patient who is to receive a bypass has a mandril implanted on which a loose polyester fabric is inserted. This structure ensures fibrous penetration for 5 to 6 weeks. The mandril is then removed and the conduit formed is used as a prosthesis. Experimental trials have been conducted with dogs, using various materials (steel rods with metal thread, tubes of nylon, a silaxane elastomer mandril and a polyester tube). The Spark's model has reached the stage of clinical evaluation and results obtained so far show that this bypass prosthesis is a failure. Examination of prostheses resected after implantation both in man and in animals shows the absence of elastin fibres and myocytes in newly formed tissue during maturation. Stretching of the polyester fabric causes ectasia and thrombosis.

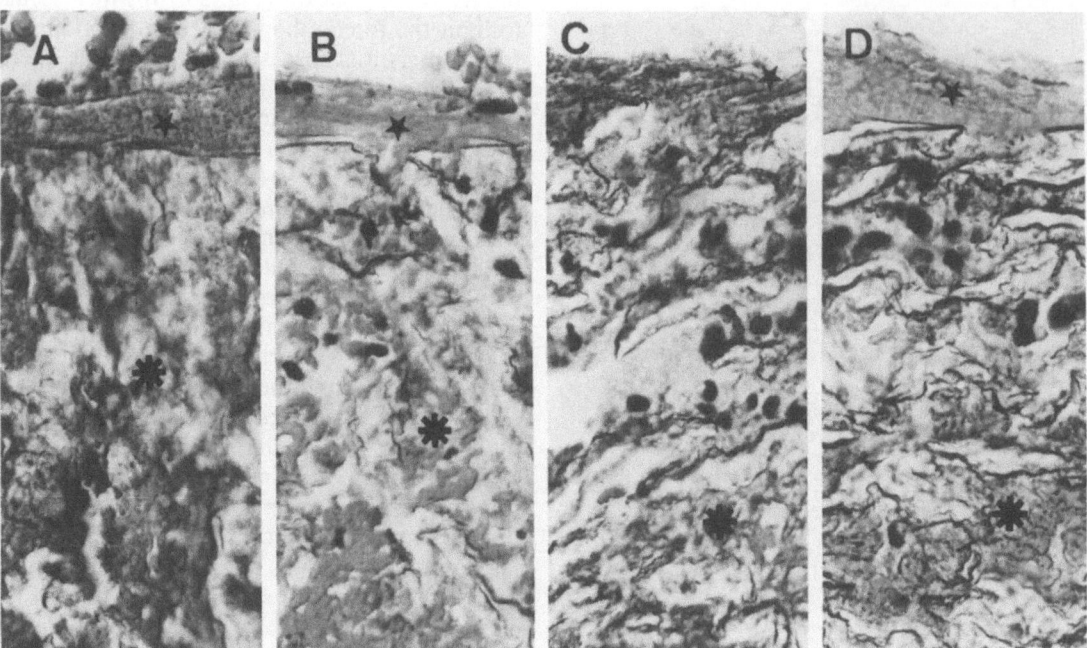

Fig. 33.18a–d. Histologic aspect of the "neo-intima" of a patent PTFE prosthesis at 24 h (**a**), 6 weeks (**b**), 10 weeks (**c**) and 5 years (**d**) after. At 24 h (**a**), the neo-intima (★) is composed mainly of fibrin and some blood cells. In the interstitial space, there is a plasma insudation (*). Polymorphonuclear cells and mononuclear cells are still noted in the neo-intima and in the interstitial space 6 weeks (**b**) and 10 weeks (**c**) after. Endothelial cells are never observed even 5 years (**d**) after. Haematoxylin–eosin–safran, × 475

References

1. Abbott WM, Wieland S, Austen WG (1974) Structural changes during preparation of autogenous venous grafts. Surgery 76:1031–1040
2. Abele J (1980) Balloon catheters and transluminal dilatation. AJR 135:901–906
3. Agrifoglio G, Constantini S, Zanetta M et al. (1979) Infections and anastomotic false aneurysms in reconstructive vascular surgery. J Cardiovasc Surg 20:25–32
4. Alexander JW, Kaplan JZ, Altemeir WA (1967) Role of suture materials in the development of wound infection. Ann Surg 165:192–199
5. Alpert J, Bhaktan EK, Gielchinsky I et al. (1976) Vascular complications of intra-aortic balloon pumping. Arch Surg 111:1190–1195
6. Alpert J, Ring EJ, Freiman DB et al. (1980) Treatment of stenosis of the iliac artery by balloon catheter dilatation. Surg Gynecol Obstet 150:481–485
7. Athanasoulis CA (1980) Percutaneous transluminal angioplasty: general principles AJR 135:893–900
8. Austin GE, Ratliff NB, Hollman J et al. (1985) Intimal proliferation of smooth muscle cells as an explanation for recurrent coronary stenosis after percutaneous transluminal coronary angioplasty. J Am Coll Cardiol 6:369–375
9. Baier RE (1978) Physical chemistry of the vascular interface: composition, texture and adhesive quality. In: Sawyer PN, Kaplitt MJ (eds) Vascular grafts. Appleton Century Crofts, New York, pp 76–107

10. Baker WH, Hadcock MM, Littooy FN (1980) Management of polytetrafluoroethylene graft occlusion. Ach Surg 115:508–513
11. Baker WH, Hayes AC, Mahler D et al. (1983) Durability of carotid endarterectomy. Surgery 94:112–115
12. Barboriak JJ, Pintar K, Van Horn DL et al. (1978) Pathologic findings in the aortocoronary vein grafts. A scanning electron microscope study. Atherosclerosis 29:69–80
13. Batayias GE, Barboriak JJ, Korns ME et al. (1977) The spectrum of pathologic changes in aortocoronary saphenous vein grafts. Circulation [Suppl] 56:18–22
14. Batt M, King M, Guidoin R et al. (1984) Fatigue mécanique d'une prosthèse artérielle. Press Med 13:1997–2000
15. Becker RM, Wexler J, Frater RWM (1981) False aneurysm of aorta secondary to partial occlusion clamp injury. Diagnosis by nuclear flow study. Chest 80:331–333
16. Becquemin JP, Deleuze Ph, Leandri J et al. (1984) Influence du type d'anastomose artérielle sur le développement des hyperplasies endothéliales. J Mal Vasc 9:90–94
17. Beebe HG, Clark WF, De Wesse JA (1970) Atherosclerotic change occurring in autogenous venous arterial graft. Arch Surg 101:85–88
18. Ben-Shachar G, Nicoloff DM, Edwards JE (1981) Separation of neointima from Dacron graft causing obstruction: case following Fontan procedure for tricuspid atresia. J Thorac Cardiovasc Surg 82:268–271
19. Berger K, Sauvage LR (1981) Late fiber deterioration in Dacron arterial grafts. Ann Surg 193:477–491

20. Bernhard VM, Rory LI, Towne JB (1977) The reoperation of choice for aorto-femoral graft occlusion. Surgery 82:867–874

21. Bernhard VM, Towne JB (1985) Complications in vascular surgery. Grune and Stratton, Orlando, Florida

22. Biddle TL, Stewart S, Stuart ID (1976) Dissection of the aorta complicating intra-aortic balloon counterpulsation. Am Heart J 92:781–784

23. Block PC, Baughman KL, Pasternak RC et al. (1980) Transluminal angioplasty: Correlation of morphologic and angiographic findings in an experimental model. Circulation 61:778–785

24. Blumenberg RM, Gelfand ML (1977) Failure of knitted Dacron as an arterial prosthesis. Surgery 81:493–496

25. Bonchek LI (1980) Prevention of endothelial damage during preparation of saphenous veins for bypass grafting. J Thorac Cardiovasc Surg 79:911–915

26. Boontje AbH (1978) Iatrogenic arterial injuries. J Cardiovasc Surg 19:335–340

27. Bosher LP (1983) Fresh and preserved homologous vein as a vascular conduit. In: Wright CB, Hobson II RW, Hiratzka LF et al. (eds) Vascular grafting. Clinical applications and techniques. John Wright-PSG, Boston, pp 128–132

28. Bouchet A, Loire R (1972) Transformation athéromateuse précoce d'une auto-greffe veineuse. CML 48:529–532

29. Bouchet A, Loire R, Le Bihan J (1979) La maladie des auto-greffons veineux: à propos d'un cas traité par homogreffe veineuse et suivi pendant cinq ans. Chirurgie 105:565–574

30. Bove EL, Fry WJ, Gross WS et al. (1979) Hypotension and hypertension as consequences of baroreceptor dysfunction following carotid endarterectomy. Surgery 85:633–637

31. Breyer RH, Spray TL, Kastl DG et al. (1976) Histologic changes in saphenous vein aorta-coronary bypass grafts. J Thorac Cardiovasc Surg 72:916–924

32. Brook WH (1975) A historical review of the histology of patent autogenous vein grafts and vein patches. J Cardiovasc Surg 16:43–52

33. Bruneval P, Guermonprez JL, Perrier P et al. (1986) Coronary artery restenosis following transluminal coronary angioplasty. Arch Pathol Lab Med 110:1186–1187

34. Buchbinder D, Singh JK, Karmody AM et al. (1981) Comparison of patency rate and structural changes of in situ and reversed vein arterial bypass. J Surg Res 30:213–222

35. Bulkley BH, Roberts WC (1975) Heterografts as aortocoronary bypass conduits in human beings. Am J Cardiol 36:823–828

36. Bunt TJ (1983) Synthetic vascular graft infections. I. Graft infections. Surgery 93:733–746

37. Bunt TJ (1983) Synthetic vascular graft infections II. Graft-enteric erosions and graft-enteric fistulas. Surgery 94:1–9

38. Buxton B, Lambert RP, Pitt TTE (1980) The significance of vein wall thickness and diameter in relation to the patency of femoropopliteal saphenous vein bypass grafts. Surgery 87:425–431

39. Byrnes G, McGowen WAL (1975) The injury potential of Fogarty balloon catheters. J Cardiovasc Surg 16:590–593

40. Callow AD (1985) Problems with graft incorporation. In: Bernhard VM, Towne JB (eds) Complications in vascular surgery. Grune and Stratton, Orlando, Florida pp 527–536

41. Camilleri JP, Vuong NP, Bruneval P et al. (1985) Surface healing and histologic maturation of patent polytetrafluoroethylene grafts implanted in patients for up to 60 months. Arch Pathol Lab Med 109:833–837

42. Campbell CD, Brooks DH, Webster MW et al. (1979) Expanded microporous polytetrafluoroethylene as a vascular substitute. A two year follow-up. Surgery 85:177–183

43. Campbell CD, Goldfarb D, Roe R (1975) A small arterial substitute: expanded microporous polytetrafluoroethylene patency versus porosity. Ann Surg 182:138–143

44. Carpentier A, Deloche A, Relland J et al. (1974) Six year follow-up of glutaraldehyde-preserved heterografts. J Thorac Cardiovasc Surg 68:771–782

45. Carrel A, Guthrie CC (1906) Uniterminal and biterminal venous transplantations. Surg Gynecol Obstet 2:266–286

46. Carson SN, Hunter G, French S et al. (1980) Occurrence of occlusive intimal changes in an expanded polytetrafluoroethylene graft. J Cardiovasc Surg 21:503–508

47. Charlesworth PM, Brewster DC, Darling RC (1984) Renal artery injury from a Fogarty balloon catheter. J Vasc Surg 1:573–576

48. Cheanvechai C, Effler DB, Hooper JR et al. (1975) The structural study of the saphenous vein. Ann Thorac Surg 20:636–645

49. Chignier E, Guidollet J, Heynen Y et al. (1983) Macromolecular, histological, ultrastructural and immunocytochemical characteristics of the neointima developed within PTFE vascular grafts: experimental study in dogs. J Biomed Mater Res 17:623–636

50. Choux R, Payan MJ, Juhan-Vague I et al. (1982) Etude morphologique de différents types de greffons veineux. Essai de corrélation avec l'activité fibrinolytique pariétale. Ann Pathol 2:293–300

51. Clagett GP, Rich NM, McDonald PT et al. (1983) Etiologic factors for recurrent carotid artery stenosis. Surgery 93:313–318

52. Clagett GP, Salander JM, Eddleman WL (1983) Dilatation of knitted Dacron aortic prostheses and anastomotic false aneurysms: etiologic considerations. Surgery 93:9–16

53. Clowes AW, Gown AM, Hanson SR et al. (1985) Mechanisms of arterial graft failure. I. Role of cellular proliferation in early healing of PTFE prostheses. Am J Pathol 118:43–54

54. Collins GJ Jr (1983) Thombosis. In: Wright CB, Hobson II RW et al. (eds) Vascular grafting. Clinical applications and techniques. John Wright-PSG, Boston, pp 245–273

55. Collins GJ, Rich NM, Andersen CA et al. (1978) Stroke associated with carotid endarterectomy. Am J Surg 135:221–225

56. Conn JF, Hardy JD, Chavez CM et al. (1970) Infected arterial grafts. Ann Surg 171:704–712

57. Cormier JM, Sautot J, Frileux Cl (1977) Artères-veines-lymphatiques. In: Patel J, Léger L, Patel JC (eds) Nouveau traité de techniques chirurgicales, vol 4. Masson, Paris

58. Courbier R, Jausseran JM, Reggi M et al. (1979) Les prothèses artérielles en polytétrafluoréthylène. Premiers résultats cliniques. J Mal Vasc 4:151–154

59. Cutler BS, Thompson JE, Patman RD et al. (1974) The modified bovine arterial graft: a clinical study. Surgery 76:963–973

60. Dainko E (1972) Complication of the use of the Fogarty balloon catheter. Arch Surg 105:79–82

61. Dale WA (1966) Autogenous vein grafts for femoro-popliteal arterial repair. Surg Gynecol Obstet 123:1282–1288

62. Dale WA, Lewis MR (1976) Further experiences with bovine arterial grafts. Surgery 80:711–721

63. Dardik H (1984) Biologic behavior of glutaraldehyde-stabilized human umbilical cord vein grafts. In: Haimovici H (ed) Vascular surgery. Appleton Century Crofts, Connecticut, pp 125–134

64. Dardik H, Baier RW, Meenaghan M et al. (1982) Morphologic and biophysical assessment of long-term umbilical cord vein implants used as vascular conduits. Surg Gynecol Obstet 154:17–26

65. Dardik H, Ibrahim IM, Baier R et al. (1976) Human umbilical cord: a new source for vascular prosthesis. JAMA 236:2859–2862

66. Dardik H, Ibrahim IM, Dardik I (1975) Modified and unmodified umbilical vein allografts and xenograft employed as arterial substitutes: a morphologic assessment. Surg Forum 26:286–287

67. Dardik H, Ibrahim IM, Jarrah M et al. (1980) Three-year experience with glutaraldehyde-stabilized umbilical vein for limb salvage. Br J Surg 67:229–232

68. Davidson E, De Palma RG (1972) Atherosclerotic aneurysm occurring in an autogenous vein graft. Am J Surg 124:112–114

69. De La Rocha AG, Peixoto RS, Baird RJ (1973) Atherosclerosis and aneurysm formation in a saphenous vein graft. Br J Surg 601:72–73

70. De Weese JA (1985) Anastomotic neointimal fibrous hyperplasia. In: Bernhard VM, Towne JB (eds) Complications in vascular surgery. Grune and Stratton, Orlando, Florida, pp 157–170

71. De Weese JA, Rob CG (1977) Autogenous venous grafts ten years later. Surgery 82:775–784

72. De Weese MS, Fry WJ (1962) Small-bowel erosion following aortic resection. JAMA 179:882–886

73. Diebold J, Cormier JM (1970) Devenir évolutif des greffes et prothèses: exposé introductif. Journ Angéiol Lang Fr; Sandoz, Rueil-Malmaison 4:469–479

74. Dotter CT, Judkins MP (1964) Transluminal treatment of arteriosclerotic obstruction. Circulation 30:654–670

75. Douglas BP, Bulkley BH, Hutchins GM (1979) Infected saphenous vein coronary artery bypass graft with mycotic aneurysm. Fatal dehiscence of the proximal anastomosis. Chest 75:76–77

76. Echave V, Koornick AR, Haimov M et al. (1979) Intimal hyperplasia as a complication of the use of the polytetrafluoroethylene graft for femoral-popliteal bypass. Surgery 86:791–798

77. Edwards WS (1978) Arterial grafts: past, present and future. Arch Surg 113:1225–1233

78. Edwards WS, Snyder R, Botzko K et al. (1985) Healing characteristics, dilatation and durability of synthetic grafts. In: Bernhard VM, Towne JB (eds) Complications in vascular surgery. Grune and Stratton, Orlando, Florida, pp 537–544

79. Ehrenfeld WK, Hays RJ (1972) False aneurysm after carotid endarterectomy. Arch Surg 104:288–291

80. Ejrup B, Hiertonn T, Moberg A (1961) Atheromatous changes in autogenous venous grafts. Acta Chir Scand 121:211–218

81. Elliott DP, Roe BB (1965) Aortic dissection during cardiopulmonary bypass. J Thorac Cardiovasc Surg 50:357–361

82. Essed CE, Brand MVD, Becker AE (1983) Transluminal coronary angioplasty and early restenosis. Br Heart J 49:393–396

83. Florian A, Cohn LH, Dammin GJ et al. (1976) Small-vessel replacement with goretex (expanded poly-tetrafluoroethylene). Adv Surg 111:267–270

84. Fogarty TV, Cranely JJ, Krause RJ et al. (1963) A method of extraction of arterial emboli and thrombi. Surg Gynecol Obstet 116:241–244

85. Foster JH, Carter JW, Graham C et al. (1970) Arterial injuries secondary to the use of the Fogarty catheter. Ann Surg 171:971–978

86. Fournier Cl, Paraiso N, Nguyen Van Tuyen G et al. (1978) Etude critique de la maladie des greffons saphènes aorto-coronaires. Arch Mal Coeur 71:1239–1247

87. Fry WJ, Lindenauer SM (1967) Infection complicating the use of plastic arterial implants. Arch Surg 94:600–609

88. Fush JCA, Mitchner JS, Hagen P (1978) Postoperative changes in autologous vein grafts. Ann Surg 188:1–15

89. Gaspard DJ, Gaspar MR (1972) Arteriovenous fistula after Fogarty catheter thrombectomy. Arch Surg 105:90–92

90. Gauthier-Villars P, Oudot J (1953) Etude anatomique des greffes artérielles. Sem Hôp Paris 29:127–131

91. Gazzaniga AB, Lamberti JJ, Siewers RE et al. (1976) Arterial prosthesis of microporous expanded poly-tetrafluoroethylene for construction of aorta-pulmonary shunts. J Thorac Cardiovasc Surg 72:357–363

92. Geiringer E (1949) Venous atheroma. Arch Pathol 48:410–420

93. Gigou F, Bouillon P, Fichelle JM et al. (1984) Pontages axillo-fémoraux PTFE. Résultats précoces et secondaires, 104 patients. J Mal Vasc 9:83–89

94. Goldstone J, Moore WS (1974) Infection in vascular prostheses. Clinical manifestations and surgical management. Am J Surg 128:225–233

95. Graham LM, Vinter DW, Ford JW et al. (1980) Endothelial cell seeding of prosthetic vascular grafts. Arch Surg 115:929–933

96. Green RM, Thomas M, Luka N et al. (1979) Comparison of rapid-healing prosthetic arterial grafts and autogenous veins. Arch Surg 114:944–947

97. Guidoin R, Gosselin C, Domurado D et al. (1977) Dacron as arterial prosthetic material: nature, properties, brands, fate and perspectives. Biomater Med Devices Artif Organs 5:177–203

98. Gundry SR, Jones M, Ishihara T (1980) Optimal preparation techniques for human saphenous vein grafts. Surgery 88:785–794

99. Haimovici H (1979) Muscular, renal and metabolic complications of acute arterial occlusions: myonephropathic-metabolic syndrome. Surgery 85:461–468

100. Haimovici H (1982) Technical pitfalls in vascular surgery. In: Haimovici H (ed) Vascular emergencies. Appleton Century Crofts, New-York, pp 611–618

101. Haimovici H (1984) Vascular surgery. Appleton Century Crofts, Connecticut

102. Hall KV, Rostad H (1978) In situ vein bypass in the treatment of femoro-popliteal atherosclerotic disease: a ten year study. Am J Surg 136:158–161

103. Hall R (1971) Vascular injuries resulting from arterial puncture or catheterization. Br J Surg 58:513–516

104. Hanel KC, McCabe C, Abbott WM et al. (1982) Current PTFE grafts: a biochemical, scanning electron, and light microscopic evaluation. Ann Surg 195:456–463

105. Harrison LH (1976) Historical aspects in the development of venous autografts. Ann Surg 183:101–106

106. Hayward RH, White RR (1971) Aneurysm in a woven Teflon graft. Angiology 22:188–190

107. Herring MB (1983) Endothelial seeding of blood flow surfaces. In: Wright CB, Hobson II RW, Hiratzka LF et al. (eds) Vascular grafting. Clinical applications and techniques. John Wright-PSG, Boston, pp 275–314

108. Hertzer N, Feldman BJ, Beven EG et al. (1980) A prospective study of the incidence of injury to cranial nerves during carotid endarterectomy. Surg Gynecol Obstet 151:781–784

109. Holm J, Schersten T (1974) Subintimal dissection secondary to the use of the Fogarty catheter. J Cardiovasc Surg 15:684–686

110. Hufnagel CA (1983) History of vascular grafting. In: Wright CB, Hobson II RW, Hiratzka LF et al. (eds) Vascular grafting. Clinical applications and techniques. John Wright-PSG, Boston pp 1–12

111. Husni EA (1967) The edema of arterial reconstruction. Circulation [Suppl 1] 35:169–174

112. Jamieson GG, De Weese JA, Rob CG (1975) Infected arterial grafts. Ann Surg 181:850–852

113. Johnson WC, Nabseth DC (1974) Visceral infarction following aortic surgery. Ann Surg 180:312–318

114. Johnston KW, Colapinto RF, Baird RJ (1982) Transluminal dilatation. Arch Surg 117:1604–1609

115. Kern WH, Wells WJ, Meyer BW (1981) The pathology of surgically excised aorto-coronary saphenous vein bypass grafts. Am J Surg Pathol 5:491–496

116. Keshishian JM, Smyth NP, Adkins PC et al. (1970) Clinical experience with the modified bovine arterial heterograft. Ann Surg 172:690–702

117. Kim GE, Imparato AM, Nathan I et al. (1979) Dilatation of synthetic grafts and junctional aneurysms. Arch Surg 114:1296–1303

118. King M, Guidoin R, Gosselin C et al. (1982) Protocole d'étude d'une greffe artérielle en polyester après exérèse. Les standards sont-ils utiles ou désagréables? RBM 4:26–33

119. Kinley CE, Paasche PE, MacDonald AS et al. (1974) Stress at vascular anastomosis in relation to host artery: synthetic graft diameter. Surgery 75:28–30

120. Klein SR, Miranda R, Nelson R et al. (1982) Effect of suture technique on arterial anastomotic compliance. Arch Surg 117:45–47

121. Knox WG (1976) Peripheral vascular anastomotic aneurysms: a fifteen year experience. Ann Surg 183:120–123

122. Komoto Y, Kawakami S, Uchida H (1978) Prosthetic aneurysm in an axillo-femoral Dacron bypass graft. Vasc Surg 12:274–279

123. Kulin J (1949) Le traitement de l'artérite oblitérante par la greffe veineuse. Arch Mal Coeur 42:371–372

124. Kwaan JHM, Bernstein JM, Connolly JE (1979) Management of lymph fistula in the groin after arterial reconstruction. Arch Surg 114:1416–1418

125. Leather RP, Shah DM, Karmondy AM (1981) Infrapopliteal arterial bypass for limb salvage: increased patency and utilization of the saphenous vein used "in situ". Surgery 90:1000–1008

126. Levin BH, Giordano JM (1982) Delayed arterial embolectomy. Surg Gynecol Obstet 155:549–551

127. Liebman PR, Barnes RW (1984) Complications of vascular surgery and trauma. In: Greenfields LJ (ed) Complications in surgery and trauma. Lippincott, Philadelphia, pp 353–369

128. Liekweg WG, Greenfield LJ (1977) Vascular prosthetic infections: collected experience and results of treatment. Surgery 81:335–342

129. Litchford B, Okies JE, Sugimura S et al. (1976) Acute aortic dissection from cross-clamp injury. J Thorac Cardiovasc Surg 72:709–713

130. Lo Gerfo FW, Quist WC, Nowak MD et al. (1983) Downstream anastomotic hyperplasia. Ann Surg 197:479–483

131. Loire R, Tahib A (1982) Les remaniements structuraux histologiques des greffons veineux aorto-coronariens. Ann Chir Thorac Cardiovasc 36:604–606

132. Lusby RJ, Ferrell LD, Englestad BL et al. (1983) Vessel wall and indium-111 labelled platelet response to carotid endarterectomy. Surgery 93:424–432

133. Matsumoto H, Hasegawa T, Fuse K et al. (1973) A new vascular prosthesis for a small caliber artery. Surgery 74:519–523

134. May J, Stephen M (1978) Multiple aneurysms in Dacron velour graft. Arch Surg 113:320–321

135. McCombs PR, Roberts B (1979) Acute renal failure following resection of abdominal aortic aneurysm. Surg Gynecol Obstet 148:175–178

136. Meinke AH, Estes NC, Ernst CB (1979) Chylous ascites following abdominal aortic aneurysmectomy. Ann Surg 190:631–633

137. Mellière D, Lasry G, Lange F (1979) Résultats et indications des prothèses en chirurgie artérielle. J Chir 116:285–296

138. Mindich B, Silverman M, Elguezabel A et al. (1977) Human umbilical cord vein for vascular replacement: preliminary report and observations. Surgery 81:152–160

139. Mohaideen AH, Avram MM, Mainzer RA (1976) Polytetrafluoroethylene grafts for arterio-venous fistulae. NY State J Med 76:2152–2155

140. Mohr L, Smith L (1980) Polytetrafluoroethylene graft aneurysms. Arch Surg 115:1467–1470

141. Moore WS, Rosson CT, Hall AD et al. (1969) Transient bacteremia: A cause of infection in prosthetic vascular grafts. Am J Surg 117:342–343

142. Najafi H, Dye WS, Javid H et al. (1975) Late thrombosis affecting one limb aortic bifurcation graft. Arch Surg 110:409–412

143. Natali J, Benhamou AC (1979) Iatrogenic vascular injuries. A review of 125 cases (excluding angiographic injuries). J Cardiovasc Surg 20:169–176

144. Natali J, Orcel L, Delcour M (1964) Contribution à l'étude anatomo-pathologique des prothèses artérielles au Dacron. Arch Anat Path 12:A183–A187

145. Newton WT, Stokes JM, Butcher HR (1959) Changes in the elasticity of arterial substitutes following implantation. Surgery 46:579–588

146. Ochlert WH (1972) A complication of the Fogarty arterial embolectomy catheter. Am Heart J 84:484–486

147. Orcel L, Chomette G (1978) Anatomie pathologique vasculaire. In: Delarue J, Laumonier R (eds) Traité d'anatomie pathologique. Flammarion, Paris, pp 154–197b

148. Paasche PE, Kinley CE, Donald FG et al. (1973) Consideration of suture line stresses in the selection of synthetic grafts for implantation. J Biochem 6:253–258

149. Pickering TG, Sos TA, Laragh JH (1984) Role of balloon dilatation in the treatment of renovascular hypertension. Am J Med 77:61–66

150. Pinkerton JA Jr (1979) Erosion of Dacron graft by atherosclerotic plaque. J Cardiovasc Surg 20:385–387

151. Reichle FA, Stewart GJ, Essa N (1973) A transmission and scanning electron microscopic study of luminal surfaces in Dacron and autogenous vein bypass in man and dog. Surgery 74:945–960

152. Rimpau A (1957) Etude macroscopique et histologique

des orifices de ponction après artériographie cérébrale. Ann Anat Pathol (Paris) 2:227–288

153. Rob C, Battle S (1971) Arteriovenous fistula following the use of the Fogarty balloon catheter. Arch Surg 102:144–145

154. Rosenberg N (1983) The modified bovine arterial graft. In: Wright CB, Hobson II RW, Hiratzka LF et al. (eds) Vascular grafting. Clinical applications and techniques. John Wright-PSG, Boston, pp 148–152

155. Rosenberg N, Lord GH, Henderson J et al. (1970) Collagen arterial graft of bovine origin: seven year observations in the dogs. Surgery 67:951–956

156. Rosenberg N, Martinez A, Sawyer PN et al. (1966) Tanned collagen arterial prosthesis of bovine carotid origin in man. Ann Surg 164:247–256

157. Satiani B, Kazmers M, Evans WE (1980) Anastomotic arterial aneurysms: a continuing challenge. Ann Surg 192:674–682

158. Sautot J, Poulat R (1960) A propos des embolectomies du carrefour aortique. Lyon Chir 56:927–930

159. Sawyer PN, Kaplitt MJ (1978) Vascular grafts. Appleton Century Crofts, New-York

160. Sawyer PN, Kirschenbaum DM, Boyle RE et al. (1983) A six-year clinical experience with negatively charged glutaraldehyde tanned (NCGT) grafts. In: Wright CB, Hobson II RW, Hiratzka LF et al. (eds) Vascular grafting. Clinical applications and techniques. John Wright-PSG, Boston, pp 331–343

161. Sauvage LR, Berger K, Beilin LB et al. (1975) Presence of endothelium in the axillary-femoral graft of knitted Dacron with an external velour surface. Ann Surg 182:749–853

162. Sauvage LR, Berger K, Wood SJ et al. (1974) Interspecies healing of porous arterial prostheses. Observations 1960–1974. Arch Surg 109:698–705

163. Sauvage LR, Walker MW, Berger K et al. (1979) Current arterial prostheses. Arch Surg 114:687–693

164. Schechter DC (1979) Atheromatous embolization to lower limbs. NY State J Med 79:1180–1186

165. Schubart PJ, Porter JM (1985) Arterial complications associated with the use of balloon catheters. In: Bernhard VM, Towne JB (eds) Complications in vascular surgery. Grune and Stratton, Orlando, Florida, pp 87–109

166. Seifert KB, Ablo DR Jr, Knowlton H et al. (1979) Effect to elasticity of prosthetic wall on patency of small diameter arterial prostheses. Surg Forum 30:206–208

167. Shack RB, Neblett WW, Richie RE et al. (1977) Expanded polytetrafluoroethylene as dialysis access grafts: serial study of histology and fibrinolytic activity. Am Surg 43: 817–825

168. Shah DM, Buchbinder D (1981) Modified technique to produce valvular incompetence in in situ saphenous vein arterial bypass. Arch Surg 116:356–358

169. Silver MD, Wilson GJ (1983) Pathology of cardiovascular prostheses including coronary artery bypass and other vascular grafts. In: Silver MD (ed) Cardiovascular pathology, vol 2. Churchill Livingstone, Edinburgh pp 1255–1296

170. Simonetti G, Rossi P, Passariello R et al. (1983) Iliac artery rupture: a complication of transluminal angioplasty. AJR 140:989–990

171. Sladen JG, Gilmour JG (1981) Vein graft stenosis. Characteristics and effect of treatment. Am J Surg 141:549–553

172. Smith SH, Geer JC (1983) Morphology of saphenous vein-coronary artery bypass grafts. Arch Pathol Lab Med 107:13–18

173. Sparks CH (1972) Silicone mandril method of femoropopliteal artery bypass. Am J Surg 124:244–248

174. Spence RK, Freiman DB, Gatenby R (1981) Long-term results of transluminal angioplasty of the iliac and femoral arteries. Arch Surg 116:1377–1386

175. Spray TL, Roberts WC (1977) Changes in saphenous veins used as aorto-coronary bypass grafts. Am Heart J 94:500–516

176. Starr DS, Weatherford SC, Lawrie GM et al. (1979) Suture material as a factor in occurrence of anastomotic false aneurysms: an analysis of 26 cases. Arch Surg 114:412–415

177. Stoney RJ, Albo FJ, Wylie EJ (1965) False aneurysms occurring after arterial grafting operations. Am J Surg 110:153–161

178. Stratton JR, Thiele BL, Ritchie JL (1983) Natural history of platelet deposition on Dacron aortic bifurcation grafts in the first year after implantation. Am J Cardiol 52:371–374

179. Streinchenberger R, Brault A, Vieville C et al. (1970) Place des homogreffes veineuses fraiches dans la chirurgie artérielle restauratrice des membres inférieurs (à propos de 18 cas). CML 46:2279–2288

180. Szilagyi DE, McDonald FT, Smith RF et al. (1957) Biologic fate of human arterial homografts. Arch Surg 75:506–529

181. Szilagyi DE, Smith RF, Elliott JP Jr et al. (1965) Long-term behavior of Dacron arterial substitute: clinical roentgenologic and histologic correlations. Ann Surg 162:453–477

182. Szilagyi DE, Smith RF, Elliott JP Jr et al. (1972) Infection in arterial reconstruction with synthetic grafts. Ann Surg 176:321–333

183. Szilagyi DE, Smith FR, Elliot JP et al. (1975) Anastomotic aneurysms after vascular reconstruction: Problems of incidence, etiology and treatment. Surgery 178:800–816

184. Thomas M, Otis SM, Rush M et al. (1984) Recurrent carotid artery stenosis following endarterectomy. Ann Surg 200:74–79

185. Thompson BW, Read RC, Campbell GS (1979) Uninfected false aneurysms after arterial reconstruction with prosthetic grafts. Am J Surg 138:921–923

186. Tingaud R, Serise JM, Quancard X (1982) Hémorragie digestive par érosion séromusculeuse de la paroi d'une anse grêle au niveau d'une prothèse aortique mise en place cinq ans auparavant. J Mal Vasc 7:179–182

187. Veith FJ, Gupta S, Daly V (1980) Management of early late thrombosis of expanded polytetrafluoroethylene (PTFE) femoropopliteal bypass grafts: favorable prognosis with appropriate reoperation. Surgery 87:581–587

188. Veith FJ, Gupta SK, Samson RH et al (1981) Progress in limb salvage by reconstructive arterial surgery combined with improved adjunctive procedures. Ann Surg 194:386–401

189. Veith FJ, Moss CM, Fell SC et al. (1978) Expanded polytetrafluoroethylene grafts in reconstructive arterial surgery: preliminary report of the first 110 consecutive cases for limb salvage. JAMA 240: 1867–1869

190. Veith FJ, Moss CM, Fell SC et al. (1978) Comparison of expanded polytetrafluoroethylene and autologous saphenous vein grafts in high risk arterial reconstructions for limb salvage. Surg Gynecol Obstet 147:749–752

191. Vlodaver Z, Edwards JE (1973) Pathologic analysis in fatal cases following saphenous vein coronary arterial bypass. Chest 64: 555–563

192. Vuong PN, Baviera E, Houissa-Vuong S (1986) Protocole d'étude en anatomie et cytologie pathologiques. Du prélèvement à la technique morphologique. Vigot, Paris
193. Warembourg H, Niquet G, Merlen JF (1963) Les dolicho-méga-artères. Lille Med 8:661–668
194. Wesolowski SA, Fries CC, Martinez A et al. (1968) Arterial prosthetic material. Ann NY Acad Sci 146:325–344
195. Wright CB, Ehrenhaft JL (1983) Vascular grafts and technical failures. In: Wright CB, Hobson II RW, Hiratzka LF et al. (eds) Vascular grafting. Clinical applications and techniques. John Wright-PSG, Boston, pp 237–244
196. Wright CB, Hiratzka LF (1983) Vascular standards development. In: Wright CB, Hobson II RW, Hiratzka LF. (eds) Vascular grafting. Clinical applications and techniques. John Wright-PSG, Boston, pp 365–370
197. Wright CB, Hobson II RW, Hiratka LF et al. (1983) Vascular grafting. Clinical applications and techniques. John Wright-PSG, Boston
198. Yashar JJ, Richman MH, Dyckman J et al. (1978) Failure of Dacron prostheses caused by structural defect. Surgery 84:659–663
199. Youkey JR, Clagett GP, Rich NM et al. (1983) Vascular trauma secondary to diagnostic and therapeutic procedures: 1974 through 1982. A comparative review. Am J Surg 146:788–791
200. Zarins CK, Lu C, McDonnell AE et al. (1980) Limb salvage by percutaneous transluminal recanalization of the occluded superficial femoral artery. Surgery 87:701–708

Section VII

Pathology of the Pulmonary Arteries

Chapter 34

Pulmonary Vasculitides

R. P. Michel

Introduction

In this chapter, discussion will revolve around the vasculitides which specifically affect the lung, or in which there is major lung involvement. Since several of them also have a granulomatous component, the angiitides and granulomatoses may be considered to be one group. Their diagnosis is a problem for the clinician in a patient presenting with multiple nodules or infiltrates in the lung, with or without systemic involvement and for the pathologist who must make a diagnosis on a histological section of the lung. Close collaboration between these two physicians is required to arrive at the specific diagnosis upon which therapy depends so heavily.

The naming and classification of these disorders, which primarily affect the lung, have been difficult for several reasons [57]. Firstly, the anatomical distribution, the types of vessels involved and the morphology of the lesions all overlap to varying degrees. Secondly, the terminology is confusing, since different names may be used for one condition and conversely, one term may encompass more than one entity.

A review of the history of these disorders, although done in other publications [8, 18,20,21,34], helps to explain some of the problems of nomenclature and classification.

In 1866, Kussmaul and Maier [47] reported the first case of periarteritis nodosa. Thirty years later, McBride [56] published clinical photographs of a case of rapid destruction of the nose and face. Until the 1930s, two streams of publications followed: the first, after Kussmaul and Maier, grouped all cases of vasculitis with peri-

arteritis nodosa; the second, after McBride, described patients with "lethal midline granuloma". A partial link between these two streams was provided by Klinger [41] and by Wegener [83,84], who respectively described the clinical aspects and precisely characterized the entity now known as Wegener's granulomatosis. Still, several authors considered lethal midline granuloma as a separate entity: Stewart [73] in 1933 described ten patients with "progressive lethal granulomatous ulceration of the nose", which he believed was a chronic inflammatory process. Walton in 1959 [80] and Eichel et al. [22] in 1966 recognized three major causes of destructive nasal lesions, other than infections: cancer, Wegener's granulomatosis and lethal midline granuloma (Stewart's syndrome). Eichel et al. [22] described 33 patients with nasal lymphoreticular lesions of which nine had a polymorphic microscopic picture and survived longer: for these patients, they coined the term "polymorphic reticulosis", which corresponds to Stewart's syndrome.

In 1951, Churg and Strauss [11] described 13 patients with a new entity, previously lumped with periarteritis nodosa, characterized by prominent pulmonary manifestations including asthma, fever and eosinophilia as well as vascular and extravascular inflammatory (including granulomatous) lesions. They called the disease, which now bears their names, allergic granulomatosis. Also in the 1950s, Fienberg [29,30,31] reported two cases of necrotizing granulomatosis and angiitis of the lungs with systemic involvement and a third with localized pulmonary disease, for which he coined the term "pathergic granulomatosis", today synonymous with Wegener's granulomatosis [57].

Liebow [51] regrouped in 1973 the pulmonary vasculitides and granulomatoses into five morphological entities: classical Wegener's granulomatosis, limited Wegener's granulomatosis, lymphomatoid granulomatosis, necrotizing sarcoid granulomatosis and bronchocentric granulomatosis. He emphasized that they were not necessarily variants of the same disease, since their aetiology and pathogenesis were unknown. This terminology remains the most widely used; it encompasses entities previously described and although advances in our knowledge of the pathogenesis of these disorders has resulted in their partial regrouping with other diseases [8], it is still valuable.

Five years after the classic papers of Liebow, Saldana et al. [67] correlated the histology with the natural history and response to therapy in 62 patients with pulmonary angiitis and granulomatosis: the authors distinguished a "lymphocyte-depleted" angiitis and granulomatosis (Wegener's granulomatosis), a "malignant lymphoproliferative" form (lymphomatoid granulomatosis) and described a new entity, benign lymphocytic angiitis and granulomatosis. The status of the latter remains unclear today: some believe it may represent a form of lymphomatoid granulomatosis with a more favourable histology and outcome [8,71,86]. DeRemee et al. [20,21] consolidated the older literature with the concepts of Liebow and grouped lymphomatoid granulomatosis with polymorphic reticulosis, emphasizing their distinction from Wegener's granulomatosis and allergic granulomatosis.

From this brief survey, one can make a few generalizations. First, the allergic granulomatosis of Churg and Strauss is, with its prominent pulmonary involvement, distinct from periarteritis nodosa. The latter, with several of its variants, is considered a systemic disorder [57]; the lung is only involved rarely insofar as its systemic blood supply (the bronchial arteries) is concerned. Second, Wegener's granulomatosis remains a distinct entity; whether the localized form is truly separable from the classic form, however, is still unclear (vide infra). Third, the midline granuloma syndrome is not a distinct disease but rather a symptom complex of upper respiratory involvement which has multiple causes [8,14,15]. Fourth, the possibility of lymphomatoid granulomatosis or polymorphic reticulosis must be entertained in the differential diagnosis of the pulmonary angiitides and granulomatosis and of the lethal midline granuloma syndrome, even if it is probably not inflammatory but rather a lymphoproliferative (prelymphomatous or frankly lymphomatous) disorder [8,57,71]. Fifth, necrotizing sarcoid granulomatosis is an accepted entity although its relationship to classic sarcoidosis remains unresolved. Sixth, bronchocentric granulomatosis, originally described and grouped by Liebow [51] with the vasculitides, has been separated from them because of its bronchocentric pattern.

One can construct a working classification of these diseases (Table 34.1) consistent with modern concepts [16,37,38,49,51,57,71]. The granulomatoses due to infections and to the inhaled or injected agents are included in the list because they must be considered first in the differential diagnosis. A detailed account of these may be found elsewhere [1,6,55,62,64,69,71, 74,75,76,79]. In addition, several connective tissue disorders may produce lesions resembling the pulmonary vasculitic disorders [16,24,28, 38,49,59]. This chapter will be devoted, however, to the idiopathic disorders in which a vasculitis figures prominently. Emphasis will be placed on the clinicopathological features and the differential diagnosis.

Table 34.1. Classification of the necrotizing angiitides and/or granulomatoses in the lung

A. Infectious
  1. Bacterial (e.g. mycobacterial)
  2. Fungal
  3. Parasitic

B. Inhaled or injected substances
  1. Occupational diseases (e.g. talc, beryllium)
  2. Drugs (e.g. drug-abuser's lung)
  3. Aspiration (chronic form, or lipoid)
  4. Extrinsic allergic alveolitis

C. Connective tissue disorders

D. Idiopathic disorders
  1. Wegener's granulomatosis
  2. Allergic granulomatosis and angiitis (Churg–Strauss syndrome)
  3. Sarcoidosis and necrotizing sarcoid granulomatosis
  4. Lymphomatoid granulomatosis or polymorphic reticulosis
  5. Bronchocentric granulomatosis
  6. Pulmonary hyalinizing granuloma

Wegener's Granulomatosis (WG)

After the description of the syndrome by Wegener [83,84], Godman and Churg [34] in 1954 described its pathology. They observed the following features: (a) necrotizing granulomatous lesions of the upper air passages or lower respiratory tract; (b) generalized focal necrotizing vasculitis of both arteries and veins; and (c) glomerulonephritis. Fauci et al. [27] consider the diagnosis of WG definitive when a patient has "clinical evidence of disease in at least two of the following three areas: upper airways, lung, kidney", and advocate biopsy proof of the disease in at least one and preferably two of these sites.

The question of the existence of a distinct group of patients with a limited form of WG has already been raised. Indeed, Fienberg [29] described one such patient in 1953. Carrington and Liebow [4] published a report of 16 cases in 1966: "limited" implied the absence of glomerulonephritis, since renal involvement in the form of angiitis and necrotic granulomas was seen in five patients. Furthermore skin and intestinal tract disease were present, and the authors suggested steroids as effective therapy. Cassan et al. [5] reported four patients with limited WG and no renal involvement at all, but could not substantiate the response to steroids. Later, Liebow [51], in his classic article, reviewed 84 cases of WG without glomerulonephritis and found (as had Carrington and Liebow [4]), that they fared better than when glomerulonephritis was present. Renal involvement, however, in the form of an angiitis and granulomatous lesion, was seen in two-thirds of the fatal cases. The extent of organ involvement in the definition of limited WG varies, since for Liebow, upper respiratory tract disease was absent, while in previous reports [4,5], it could occur.

To unify the nomenclature of WG, the group from the Mayo Clinic [18] proposed a classification based on anatomical sites of involvement with upper airway described as E, lung as L and kidney as K. In the 50 patients classified according to this scheme, all combinations were seen except K alone, since the necrotizing glomerulonephritis of WG is not specific and involvement elsewhere is required to make the diagnosis. Patients with E involvement only have the type of "lethal midline granuloma" due to WG, whereas those with L or EL have the limited form of WG (if absence of K means only absence of necrotizing glomerulonephritis), and those with EK, LK and ELK have classic WG.

The clinical features have been reported in several series of patients [3,18,27,32,53,88], and are described in Chap. 27. Most patients present with upper respiratory tract signs and have pulmonary infiltrates. Renal manifestations are unusual initially, and tend to follow extrarenal involvement. In the upper airway, the nose, the

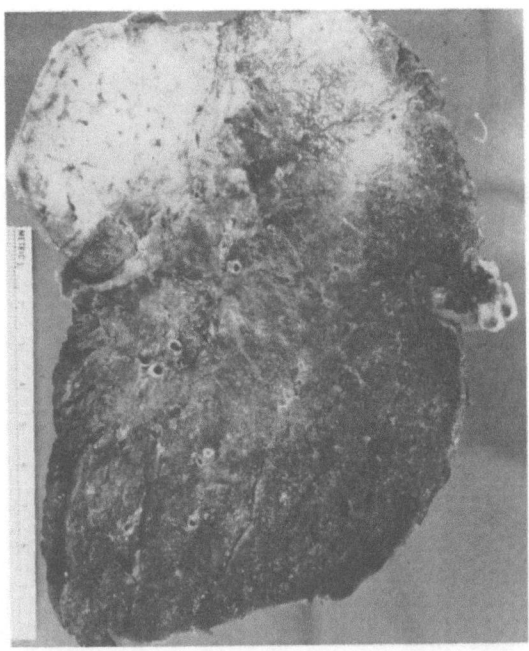

Fig. 34.1. Slice of a right upper lobectomy lung specimen from a patient with Wegener's granulomatosis. The single lesion is at the top left, poorly defined, with central necrosis, and gradually merges with the surrounding parenchyma.

ears and sinuses may be affected. Fauci et al. [27] emphasize that while WG can erode through the walls of the sinuses, it does not penetrate through the skin of the face or nose, nor does it perforate the palate, phenomena which may complicate the malignant forms of lethal midline granuloma including polymorphic reticulosis.

The lung fields on chest radiographs show multiple, usually bilateral, nodular densities with frequent cavitation, and according to some authors [51], a propensity for lower lobe involvement. Fleeting infiltrates or single masses are less common. Focal areas of atelectasis and pleural effusion may also occur.

Macroscopically, the lungs show one or more lesions, frequently cavitating, with irregular margins infiltrating the surrounding parenchyma (Fig. 34.1). By light microscopy, the cardinal features are an extensive destruction of the parenchyma with granulomatous inflammation and a necrotizing vasculitis. The necrosis has a characteristic irregular "geographic" pattern, often surrounded by palisaded histiocytes, with giant cells, and a mixed inflammatory infiltrate containing lymphocytes, plasma cells and fibroblasts (Fig. 34.2). Eosinophils and discrete sarcoid-like granulomas are rarely prominent. The necrosis may be caseous or infarct-like, with

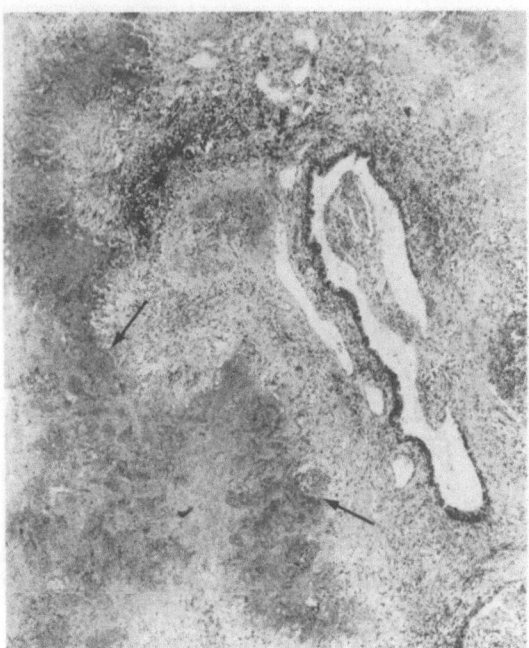

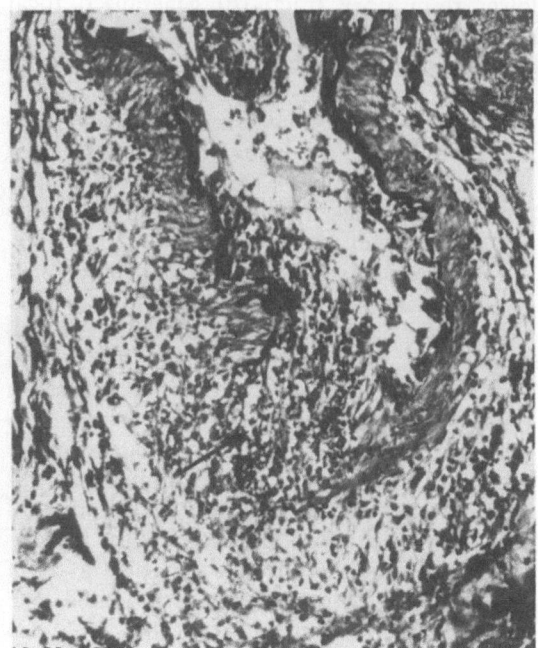

Fig. 34.2. Low power photomicrograph of an area of the lung of a patient with Wegener's granulomatosis. The zone of "geographic" necrosis on the left borders a bronchovascular bundle in which the artery is partly involved by inflammation. The airway also shows an ulceration. H & E, × 50

Fig. 34.4. Large muscular artery in Wegener's granulomatosis. Elastic laminae are totally disrupted by mononuclear inflammatory cells at the lower left corner of the vessel. Elastic van Gieson, × 120

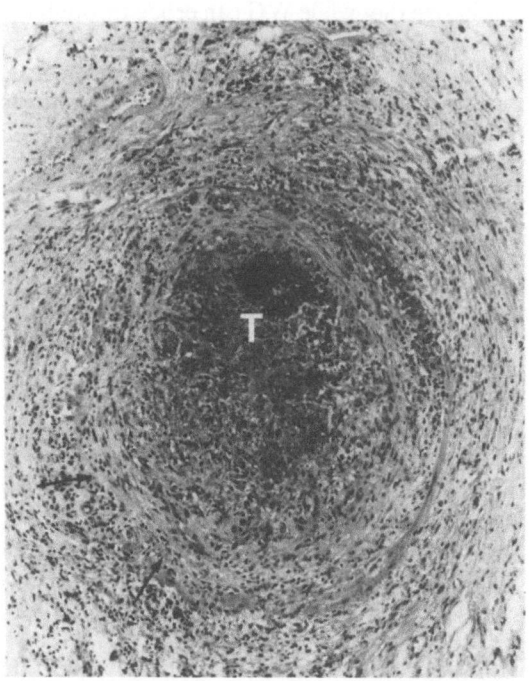

Fig. 34.3. Large muscular artery, away from any necrosis, in Wegener's granulomatosis. Note the extensive transmural inflammation, the disruption of the elastic laminae and the luminal thrombosis (*T*). H & E × 120

preservation of parenchymal outlines, or sometimes purulent. The lesions are angiocentric (Figs. 34.2–34.4), affecting both arteries and veins, and in the less severely diseased areas, airways are preserved. Katzenstein and Askin [38] underline two features of the necrotizing angiitis in WG. First, the inflammation must be transmural and second, there must be necrosis of either the media or the elastic laminae. Elastic stains help to delineate the disrupted elastic laminae and to locate damaged vessels within necrotic areas (Fig. 34.4). One useful feature which separates WG from the infectious granulomatoses is to find the necrotizing vasculitis at a distance from the necrosis (Fig. 34.3), although this may be difficult, particularly if only a small piece of tissue is available.

In a recent autopsy study of 22 patients, Yoshikawa and Watanabe [89] divided the pulmonary lesions of WG into (a) a fulminant type, characterized by an acute alveolitis with neutrophils, macrophages and fibrinous exudates, and early granuloma formation; (b) a granulomatous type, the most characteristic, with central necrosis, epithelioid and giant cells, chronic inflammation and fibroblastic proliferation; (c) a fibrous scar type with nodules of connective tissue, and vessels with sclerosed walls; and (d) a mixed type, combining the alveolitis and vasculitis, with the granulomatous and ciccatricial areas. These authors correlated the fulminant pulmonary lesions with systemic involvement, including systemic vasculitis and glomerulonephritis.

In the upper respiratory tract, the characteristic feature is a granulomatous vasculitis. Unfortunately, it is not always present and there may only be non-specific acute and chronic inflammation with necrosis [27]. This occurred in 50% of the biopsies reported by Crissman et al. [15] although in their final analysis, after multiple biopsies, five of seven patients with WG had the diagnostic features. Part of the reason for the non-specificity of the inflammation may be the superficial nature of the biopsy and the bacterial superinfection (frequently with *Staphylococcus aureus*).

The characteristic renal lesion in WG is a focal and segmental necrotizing glomerulonephritis. In addition, granulomata and a necrotizing vasculitis may be observed. Even if the renal glomerular lesions are not pathognomonic, their demonstration in a patient with pulmonary angiitis and granulomatosis virtually eliminates all other diagnostic possibilities (see Chap. 27).

Other aspects of WG, including therapy and prognosis, and aetiology and pathogenesis, are covered in Chap. 27 and in several references [16,19,33,36,38,49,57,68,72,77].

Allergic Granulomatosis and Angiitis (Churg–Strauss Syndrome)

Even if some authors, including Churg himself, have used the term "allergic granulomatosis" alone [10,25,42], most authors prefer the original complete designation of "allergic granulomatosis and angiitis" proposed in 1951 by Churg and Strauss [11]. This is justified since there is an acute angiitis or at least a residue of vascular disease in nearly all cases: Churg and Strauss [11] found it in 12 of their 13 cases, and Chumbley et al. [7] in all of their 30 cases. Although not originally emphasized by Churg and Strauss, a diagnostic triad for allergic granulomatosis and angiitis has been suggested [42] based on their description. It includes (a) asthma, (b) tissue and blood eosinophilia and (c) granulomatous vasculitis and extravascular granulomas.

The largest series reported to date by Chumbley et al. [7] includes 30 patients, 21 males and 9 females. In smaller series, a female preponderance or an equal sex distribution have been observed [11,42,48]. The age range is wide with a mean of about 45 years. Nearly all patients have asthma and a peripheral eosinophilia. Seventy per cent of the cases of Chumbley et al. [7] had allergic rhinitis. Fever and weight loss are also common. Laboratory abnormalities include anaemia, leucocytosis and an elevated erythrocyte sedimentation rate.

Chest abnormalities by roentgenography are variable, and were seen in eight of the 30 (27%) patients reported by Chumbley et al. [7] and in 93% of cases reviewed by Leavitt and Fauci [49] (Table 34.2). Appearances range from the more frequent patchy or diffuse pneumonic infiltrates to less common nodular masses; cavitation is not a feature of the Churg–Strauss syndrome, unlike WG [38,49].

Table 34.2. Allergic granulomatosis and angiitis: frequency of organ system involvement (%)

Organ	Chumbley et al. [7]	Leavitt and Fauci [49]
Lung	100	96
Asthma or wheezing	100	82
Radiographical abnormality	27	93
Skin	67	67
Peripheral nerves	63	63
Gastrointestinal tract	17	42
Heart	NS	38
Genitourinary tract	20	38
Joints	20	21

NS = not specified.

Aside from the lung, other organ systems are affected, as illustrated in Table 34.2, with variable frequency. For example, the heart was often involved in the original series of Churg and Strauss [11] with myocardial infarction due to coronary artery vasculitis, and heart failure due to myocardial disease. In the series of Chumbley et al. [7], in contrast, no mention is made of cardiac involvement although five patients died from some form of heart disease. In addition, gastrointestinal, cutaneous and peripheral nervous system involvement is quite frequent, while renal disease is unusual, although glomerulonephritis has been demonstrated at autopsy [34].

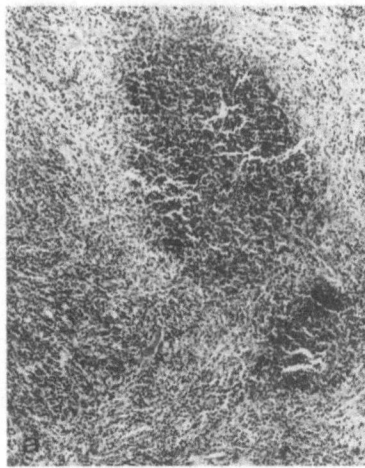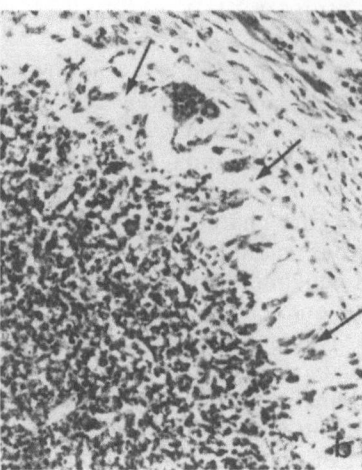

Fig. 34.5. a Low power photomicrograph of an area of a lung from a patient with allergic granulomatosis and angiitis showing coalescing zones of necrosis with many eosinophils. H & E, × 75. RC10–11. b Higher power showing palisaded histiocytes and giant cells bordering the central eosinophilic necrosis. H & E, × 190

Macroscopically, the disease resembles periarteritis nodosa in many organs except in the lung. Nodular swellings are scattered along the small arteries of the heart, liver, spleen, kidneys, gastrointestinal tract and other sites [11]. Secondary haemorrhage, infarcts and scars complicate the vascular lesions. In the lung, zones of consolidation or nodularities, and vascular lesions (wall thickening or thrombi) are seen.

By light microscopy, the hallmarks of the disease are a necrotizing vasculitis of arteries and veins and necrotizing extravascular granulomas, superimposed in the lung on a picture of chronic eosinophilic pneumonia. Chumbley et al. [7] found necrotizing vasculitis in all of their 30 cases and extravascular granulomas in 22. The necrosis is fibrinoid and contains numerous eosinophils and Charcot–Leyden crystals. As in WG, it is surrounded by palisaded histiocytes and giant cells, with some lymphocytes and plasma cells (Fig. 34.5).

For a discussion of the therapy and prognosis, and of the aetiology and pathogenesis, the reader should consult Chap. 27 and other references [7,38,42,46,49].

Necrotizing Sarcoid Granulomatosis (NSG)

Liebow [51] first wrote about this entity in 1973, based on 11 patients, and characterized its features as follows: (a) a chest radiographic picture of multiple bilateral nodules or infiltrates with rare hilar adenopathy; (b) an absence of clinical extrapulmonary involvement; (c) a histological picture of sarcoid granulomata, a granulomatous vasculitis and varying degrees of necrosis; and (d) a benign course compared with the other diseases in the necrotizing angiitis and granulomatosis group. Today, according to some authors, NSG is a variant of sarcoidosis corresponding to so-called nodular sarcoid [9], although others still consider it distinct from classical sarcoidosis [38,43,49].

Clinically, patients present with lower respiratory tract symptoms (cough, chest pain, dyspnoea) and systemic symptoms (fever, night sweats, malaise). About 20% of patients are asymptomatic. Table 34.3 summarizes the clinical and chest radiographic features from three series. Features which distinguish NSG from classical sarcoidosis are the paucity of extrapulmonary disease [70], the lower incidence of hilar adenopathy and the finding of cavitation on chest radiographs.

The histology, well described by Liebow [51], shows a destructive vasculitis and areas of necrosis on a background of non-caseating granulomas. Three types of vasculitis, which is frequently transmural, occur: (a) a destructive or occlusive vasculitis caused by well-formed granulomata, involving both arteries and veins, but particularly the latter; (b) a "giant cell arteritis" picture in which the granulomas are less well formed and more loosely arranged; and (c) a diffuse lymphoplasmacytic vasculitis, without granulomas. The vasculitis, however, is not specific for NSG since it has been observed in classical sarcoidosis. Indeed, Rosen et al. [66] described a granulomatous angiitis in 69% of

Table 34.3. Necrotizing sarcoid granulomatosis: clinical and radiographical data

Parameter	Liebow [51]	Churg et al. [9]	Koss et al. [43]
No. patients	11	12 + 9	13
Sex ratio, M : F	1 : 1	1 : 5	1 : 3.3
Age range (mean) (years)	27–55 (NS)	23–68 (50)	24–75 (48)
% asymptomatic	18	25	15
Extrapulmonary disease (%)	0	14	15
Chest radiographical findings (%)			
Diffuse bilateral lesions	82	67	50
Cavitation	NS	0	15
Hilar adenopathy	9	50	8

NS = not specified.

open lung biopsies from 128 patients; they also noted the more prominent venous involvement.

Therefore, the principal feature which separates NSG from classical sarcoidosis is the necrosis. It must be extensive since small areas of fibrinoid necrosis not infrequently occupy the centre of otherwise non-caseating granulomas. Furthermore, in the series presented in Table 34.3, the extent of necrosis varies markedly, suggesting a significant overlap exists between classical sarcoidosis and NSG.

The therapy of NSG is conservative: corticosteroids alone are sufficient, if required at all. The only death in the series of Churg et al. [9] occurred from infection in a patient treated with cyclophosphamide, which underscores the importance of separating NSG from the other necrotizing angiitides and granulomatoses.

The aetiology and pathogenesis, like those of most of the other disorders in this group, are presumed to have an immune origin. A flu-like prodrome in some patients with NSG suggests an infectious inciting agent. Koss et al. [43] provided some support for the idea of a hypersensitivity-type pneumonitis to a fungus when they found marked staining by immunofluorescence in one of their 13 patients, using anti-sera to *Aspergillus fumigatus*.

Lymphomatoid Granulomatosis (LYG)

As noted above, the term LYG was introduced in 1972 by Liebow et al. [52] with a description of 40 patients; the number expanded to 72 one

year later with the publication of the J. Burns Amberson Lecture [51]. The principal characteristics of LYG, according to Liebow, were as follows. First, the disease was more prevalent in males, mainly in "early middle age", and the patients presented with respiratory symptoms and a chest radiographic picture of nodular lower lobe masses. Second, associated nervous system, cutaneous and renal lesions (the latter without glomerulonephritis) were frequent. Third, the histology of the lung showed an angiocentric polymorphic infiltration with atypical lymphoreticular cells, and apparently combined features of a malignant lymphoproliferative disorder and of a vasculitic process. Fourth, the prognosis was poor, with progression to a malignant lymphoma in 10%–15% of patients. Subsequently, case reports and series of patients with LYG appeared in the literature (Table 34.4). It also became evident that the entity was not entirely new, but rather focused on the pulmonary aspects of a disease which in the upper respiratory tract was known as polymorphic reticulosis.

The clinical and radiological features are summarized in Table 34.4. Most patients reported under the title of LYG present with lower respiratory and systemic symptoms, although over 20% of those in the series of Katzenstein et al. [39] presented with neurological complaints. Just under 3% of the patients in this series were asymptomatic. In the series of DeRemee et al. [20], reported under the name of polymorphic reticulosis, 20% of patients had only the upper airways involved and 50% had the upper airways plus another system involved. In most series, however, the lung is always involved, with, as reported by Liebow, prominent nervous system

Table 34.4. Lymphomatoid granulomatosis: clinical and radiographical data

Parameter	Liebow [51,52]	Saldana et al. [67]	DeRemee et al. [20]	Katzenstein et al. [39]	Fauci et al. [26]	Koss et al. [44]
No. patients	74	24	40	152*	15	42
Sex ratio, M:F	2:1	1:1	3:1	1.7:1	6.5:1	3:1
Age range (mean)	8–70	18–70 (48)	15–80 (45)	7–85 (48)	16–65 (34)	20–69 (47)
Frequency of involvement (%)**						
Lung	100	100	38	100	100	100
Nervous system	33	33	5	30	33	29
Skin	45	46	25	39	20	33
Kidney	45	NS	8	32	100	NS
URT	NS	13	70	NS	NS	NS
% lymphoma	13	4	10	12	47	7
% survival**	30	35 (1 year) 22 (5 year)	53	37	47	62

*Includes cases of Liebow [51,52]; **Variably defined; NS = not specified; URT = upper respiratory tract.

and cutaneous disease. Renal involvement is also frequent, but unlike in WG, glomerulonephritis is absent. The gastrointestinal tract is rarely affected [63].

Radiographically, most patients have bilateral lower lobe nodular lesions which may be mis-

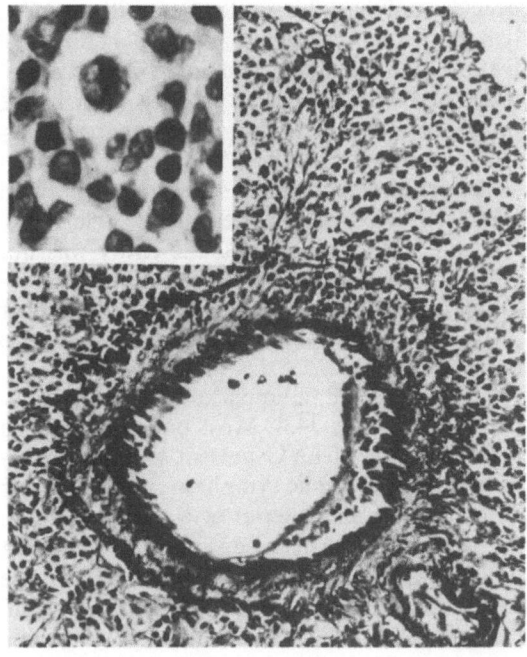

Fig. 34.6. Artery in the lung of a patient with lymphomatoid granulomatosis markedly infiltrated by atypical "lymphoreticular" cells. Note the disrupted internal elastic lamina. Elastic van Gieson, × 180. *Inset* shows detail of the polymorphic infiltrate with an atypical cell. H & E, × 550

taken for metastases. Wechsler et al. [82] compared chest radiographs in WG and in LYG and found that "the frequency and distribution of nodular masses, cavitation, and migratory lesions were similar in the two disorders, but reticulonodular infiltrates occurred only in LYG".

Liebow et al. and other authors have described the pathological features in detail [20,38, 39,51,52]. Macroscopically, the lesions resemble those of WG, and are poorly defined masses of variable size, frequently necrotic. By light microscopy, the preserved areas show a polymorphic "lymphoreticular" infiltrate composed of histiocytes, plasma cells and lymphocytes, with varying degrees of atypia (Figs. 34.6 and 34.7): many of the atypical cells have features of immunoblasts of the B or T cell series [13,58]. The former may assume a plasmacytoid appearance. This polymorphic infiltrate displays a marked propensity for vascular involvement, with transmural infiltration of both arteries and veins, and frequent narrowing or complete occlusion of the lumen. The necrosis, which may be extensive and resembles tumour necrosis, is presumably due to the obliteration of the vessels. Despite the appellation of the disease, granulomas with epithelioid and multinucleated giant cells are not characteristic of LYG.

The differentiation of LYG from malignant lymphoma may be difficult, due to the nature of the disease (vide infra). Most authors, however, believe that a polymorphic infiltrate characterizes LYG, while a monomorphic infiltrate,

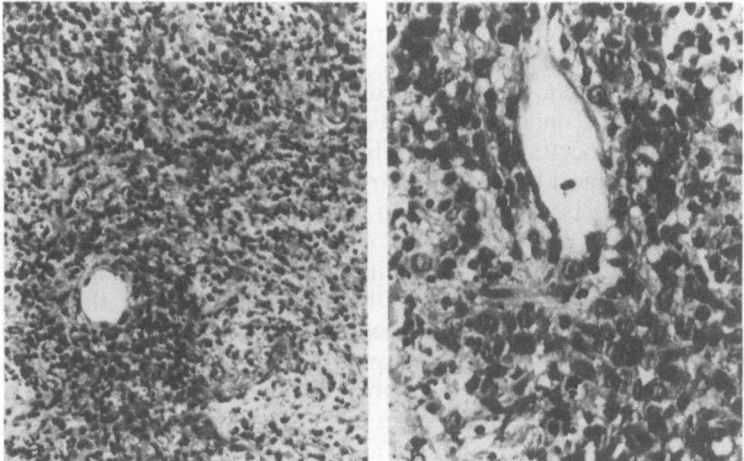

Fig. 34.7. a Medium power photomicrograph of islands of atypical cells, centred around vessels, in a sea of necrosis, from a patient with LYG. H & E, × 230. **b** Higher power of same, showing greater atypia of the cells (immunoblasts) than in Fig. 34.6. H & E, × 490

with sheets of clearly malignant cells, is required to diagnose a malignant lymphoma [12,13,38]. Vascular invasion per se may not be a valid criterion for differentiating between them, since it is also present in a high percentage of primary or secondary malignant lymphomas [12,35]. A more promising criterion may be the demonstration of a monoclonal population of lymphoid cells with immunofluorescence or immunoperoxidase, which, however, is useful today only in B cell lymphomas.

The survival rates are shown in Table 34.5, although sometimes without knowledge of the length of time over which the patients were followed up. In the series of Katzenstein et al. [39], 63.5% of patients followed up died, with a median survival of 14 months; of the 50 survivors, three-quarters were alive without

Table 34.5. Differential diagnosis of the principal pulmonary angiitides and granulomatoses: clinical features

	Wegener's granulomatosis	Allergic granulomatosis and angiitis	Necrotizing sarcoid granulomatosis	Lymphomatoid granulomatosis	Bronchocentric granulomatosis
Age (years)	~45	~45	~48	~45	~40
M:F ratio	1.5–2:1	~1:1	~1:3	2–3:1	1:1
Systems involved	Lung, kidney, URT	Lung, skin, nerves, GI, heart	Lung mainly	Lung, NS, skin, URT, kidneys	Lung
Asthma	No	Yes	No	No	Yes in ~50%
Eosinophilia	No	Yes	No	No	Yes in ~50%
Commonest chest radiological patterns	Bilateral lower lobe masses, cavitation	Infiltrate or masses	Multiple bilteral nodules or infiltrate	Bilateral lower lobe masses	Unilateral upper lobe infiltrate, atelect, or masses
Therapy	Cyclophosphamide, steroids	Steroids and cytotoxic immunosuppressant drugs	None or steroids	Cytotoxic, immunosuppressant drugs, radiotherapy	Steroids, resection
Prognosis	90%–95% remission rates	90% 1-year survival	Very good	30%–50% survival	Very good

NB: see also references [37,38,67]. GI, gastrointestinal tract; NS, nervous system; URT, upper respiratory tract.

disease, one-quarter alive with disease. The authors looked for prognostic indicators and found that clinically, unilateral lesions were favourable indicators, while neurological manifestations were unfavourable indicators; histologically, histiocytes were favourable predictors, increased atypical lymphoreticular cells (immunoblasts) adverse predictors. In their recent report, Koss et al. [44] could not confirm these indicators, although their definitions for the clinical parameters differed from those of Katzenstein et al. and their series was significantly smaller. As shown in Table 34.4, progression to frank lymphoma occurs in a variable percentage of instances, from 4% in the series of Saldana et al. [67], to nearly 50% in that of Fauci et al. [26].

DeRemee et al. [20] reported that localized upper airway LYG responded well to radiation therapy. In general, however, with systemic spread, and as indicated by the survival figures in Table 34.4, response to therapy, usually corticosteroids and cytotoxic agents, is poor. Fauci et al. [26] advocated treatment with cyclophosphamide and prednisone: of 13 patients, seven had remissions of about 5 years. Nevertheless, their overall mortality rates and incidence of malignant lymphomas were about 50%. Bernstein et al. [2] recently reported one patient treated with bone marrow transplantation.

The high incidence of transformation of LYG to an overt lymphoma prompts the question of the relationship between them and indeed of whether LYG may not be a covert or early lymphoma. Liebow [51] in his initial description classified LYG with the other necrotizing "angiitides and granulomatoses". The "angiitis" component, however, hardly seems inflammatory, and the "granulomatous" component is essentially never seen. Therefore, several possibilities exist. First, LYG is allied to the inflammatory angiitides and granulomatoses, such as WG and the Churg–Strauss syndrome. All authors agree, on the basis of the light microscopy findings including the cellular atypia and the absence of glomerulonephritis, that this is not the case. The second possibility is that LYG is a "pre-lymphoma" analogous to angio-immunoblastic lymphadenopathy and some cases of lymphoid interstitial pneumonitis and Sjögren's syndrome [12,13]. Indeed, some patients with these syndromes eventually develop LYG [85], and some with "benign lymphocytic angiitis and granulomatosis" [67] may also fit into this category. These entities may evolve into overt malignant lymphomas, with or without extension to lymph nodes. The third possibility, that LYG is really a

malignant lymphoma, albeit disguised, is gaining popularity. Support for this comes from (a) the presence of the atypical cells; (b) the fact that a phenomenon of "dilution" and obscuring of the atypical cells by secondary parenchymal inflammatory changes frequently occurs in the lung, and that as a result of sampling problems, acceptable foci of lymphoma are difficult to find [13]; and (c) that LYG may be a T cell lymphoma, for example a T immunoblastic lymphoma, in which the malignant infiltrate is polymorphic and in which skin and nervous system involvement is the main extrapulmonary symptom of the disease [13,61]. Furthermore, as mentioned earlier, vascular invasion by lymphomatous cells in the lung is seen in a significant percentage of instances.

In summary, as stated by Colby and Carrington [12], it appears "that most cases of LYG are lymphoreticular neoplasms". Some cases, however, may fit better into a pre-lymphoma or angio-immunoblastic lymphadenopathy category. The distinct clinical picture and polymorphous infiltrates seen by light microscopy are probably reason enough to keep the lesion separate, until larger numbers of cases have been studied with immunological techniques, in particular to ascertain how many are T cell lymphomas.

Knowledge of the aetiology and pathogenesis of LYG will probably evolve parallel to that of the malignant lymphomas. The occurrence of LYG in patients with altered or compromised immunity [54,78] suggests this. The hypothesis of Liebow [51] that LYG is initiated by a viral agent which alters pulmonary tissue antigens and results in an immune reaction with lymphoreticular proliferation may eventually be borne out by the facts.

Bronchocentric Granulomatosis (BCG)

BCG, although first described by Liebow in 1973 [51] with the angiitides and granulomatoses, differs from them in several ways: first, the inflammation, as its name suggests, begins and is centred around bronchi and bronchioles, and vascular involvement is secondary; second, extrapulmonary involvement is essentially absent; and third, the aetiology and pathogenesis have been more clearly defined. Several small- or medium-sized series of patients with BCG have been reported (totalling 61 patients) and the key features from them will be summarized [40,45,50,51].

The age range of patients with BCG is wide, although the mean age is close to 40 years [40,45]. The overall male : female ratio approximates to 1 : 1. In the largest series, Katzenstein et al. [40] divide the patients into two groups: those with asthma (one-quarter to one-half of the total) and those without. The patients with asthma constitute a uniform group: the male : female ratio is higher (about 2 : 1) and the mean age lower (22 years) than in those without asthma. A peripheral eosinophilia occurs in a high percentage (90% in the series of Katzenstein et al.), and correlates with an eosinophilic infiltrate on histology. In contrast, the group without asthma is more heterogeneous [38,40]: the patients are older, have a more equal sex distribution, no atopy and more variable signs and symptoms; eosinophilia is unusual.

The roentgenographic findings in BCG are variable, although unilateral lesions (in 50%–70% of instances) in the upper lobes were frequent. Katzenstein et al. [40] noted more pneumonic infiltrates or atelectasis (95%), while in the series of Robinson et al. [65], mass lesions were more frequent (60%). Cavitation may be seen on occasion and hilar adenopathy is most unusual. The radiographic pattern appears similar whether asthma is present or not, although Katzenstein et al. [40] mention that asthmatics exhibit twice as much sublobar atelectasis or consolidation as non-asthmatics. In general, the differentiation, with roentgenography, of BCG from a number of entities, including other granulomatoses, infections and neoplasms, is difficult in individual cases. A diagnosis therefore must be made by lung biopsy or resection.

The macroscopic pathology of BCG indeed differs from the other angiitides and granulomatoses since the airways are affected primarily: their lumina are filled with inspissated mucus or caseous material, and their walls are thickened. By light microscopy, the bronchioles and smaller bronchi show focal or diffuse mucosal ulceration with numerous necrotic palisaded granulomata in the wall, and necrotic or mucoid material in the lumen. Large numbers of eosinophils with eosinophilic debris and Charcot–Leyden crystals may be seen; these are more frequent in the asthmatic patients. The extensive necrosis may obliterate the airways, then only identifiable by their proximity to arteries. Elastic stains may aid in outlining their rem-

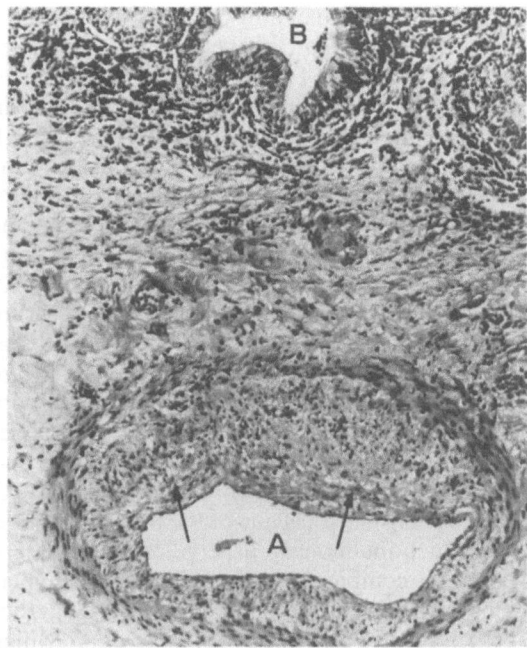

Fig. 34.8. Detail of a bronchovascular bundle in bronchocentric granulomatosis, showing the submucosal inflammation in the wall of the small bronchus (*left*) and the lymphocytic infiltrate and intimal thickening of the artery adjacent to the airway. The opposite side of the artery is almost completely spared. H & E, × 130

nants. In and around the walls of the airways, a mixed lymphoplasmacytic cellular infiltrate surrounds the necrotizing granulomatous inflammation. In the more proximal bronchi, mucoid impaction, with a few non-caseating granulomata, and submucosal infiltration with lymphocytes and plasma cells replaces the necrotizing inflammatory process.

The vascular involvement in BCG is limited to those arteries adjacent to involved airways, and then only on the side of the airway (Fig. 34.8), with infiltration by lymphocytes and plasma cells sometimes disrupting the elastic laminae, and an associated eccentric intimal thickening. The latter resembles the vascular changes in interstitial fibrosis (see Chap. 35). The severity of the vascular lesions varies, however, and may lead to confusion with WG. Features which help to separate the two include, in BCG, the absence of venous lesions and of vascular damage away from the necrosis. Secondary parenchymal changes may, particularly with time, complicate the airway disease: they consist of chronic inflammation, bronchiolitis obliterans and areas of obstructive or organizing pneumonitis.

The histological finding of fungal hyphae, particularly *Aspergillus*, in the luminal mucoid or necrotic contents of airways has helped to elucidate the aetiology and pathogenesis of BCG in about half of the patients (those with asthma). Katzenstein et al. [40] identified hyphae in nine of ten asthmatics and Koss et al. [45] in one of five patients with masses of eosinophils in the necrosis seen by light microscopy. This finding, coupled with reports of positive sputum cultures for *Aspergillus* (usually *A. fumigatus*) or of serum precipitins to *Aspergillus*, and occasionally to *Candida*, suggests that many cases of BCG are closely related to the group of hypersensitivity reactions to fungi, which also encompasses allergic bronchopulmonary asperigillosis and mucoid impaction of the bronchi. Indeed, BCG may be considered as one of the tissue responses of bronchopulmonary aspergillosis, which appears to combine type I and type III hypersensitivity reactions. The aetiology and pathogenesis of BCG without asthma, or in which eosinophils are absent on tissue sections, is less clear, since hyphae are identified only in a small percentage of cases. Myers and Katzenstein [60] in a recent report, underline the importance of searching for an infectious aetiology in BCG, particularly in non-asthmatics: they identified mycobacteria, histoplasma and blastomyces by histology or culture in four otherwise typical cases of BCG. In these, the organisms were invasive and directly responsible for the tissue response, rather than an immune phenomenon. In another case, Den Hertog et al. [17] observed remnants of a pulmonary echinococcal infection on lung sections and surmised that the BCG was either a direct reaction to the foreign material or an allergic response to echinococcal antigens.

Most patients with BCG are treated with steroids although several are treated only by resection of the lesion. Cytotoxic agents are usually not advocated. Wiedemann et al. [87], however, used cyclophosphamide to treat a scleritis in one patient with BCG; whether in this instance the two diseases were related is unclear. Katzenstein et al. [40] and Warren et al. [81] emphasize the importance of separating WG from BCG (in which there may be concurrent but unrelated renal disease), since their therapies differ so widely.

Pulmonary Hyalinizing Granuloma

This disorder is briefly mentioned here since its histology may show a granulomatous or a mild vasculitic component [38]. Engleman et al. [23] reported their observations in 20 patients.

Clinically, the patients are younger than those with the other diseases considered thus far. They present with mild respiratory or systemic symptoms; about 25% are asymptomatic. The chest radiographs show multiple, usually bilateral and well-defined, nodules which may be mistaken for metastases. These nodules gradually enlarge over a period of years.

By light microscopy, the nodules are sharply circumscribed and composed predominantly of

Table 34.6. Differential diagnosis of the principal pulmonary angiitides and granulomatoses: histological features

	Wegener's granulo- matosis	Allergic granulomatosis and angiitis	Necrotizing sarcoid granulo- matosis	Lymphomatoid granulo- matosis	Bronchocentric granulo- matosis
Topography	Angiocentric	Angiocentric	Angiocentric	Angiocentric	Bronchocentric
Angiitis	Arteries, veins	Arteries, veins	Veins and esp. arteries	Infiltrative arteries, veins	Arteries adjacent to airways
Granulomata	Mainly nectrotizing or caseous	Necrotizing with eosinophils	Many non-caseating and necrosis	Absent	Caseating in bronchioles
Necrosis	Coagulative "geographic"	Fibrinoid and eosinophilic	Caseating and fibrinoid	Coagulative or "tumour"	In airways with eosinophilia
Cellular infiltrate	Scant, benign N, L, P. H	Benign, many eosinophils and also L, P, H	L, P Benign	Atypical lymphoreticular	Benign mixed, with eosinophils
Glomerulonephritis	No	Occasional	No	No	No

NB: see also references [37,38,67]. H, histiocytes; L, lymphocytes; N, neutrophils; P, plasma cells.

dense hyalinized connective tissue in a lamellar arrangement. Chronic inflammatory cells surround the fibrotic areas and focally infiltrate the walls of blood vessels. Necrotic zones and foreign body giant cells are also sometimes seen.

The differential diagnoses includes WG, rheumatoid nodules, sarcoidosis, amyloidosis, plasma cell granuloma and the infections already considered. These are discussed in detail elsewhere [23,38], and in general, the histology of pulmonary hyalinizing granuloma is sufficiently characteristic to make a diagnosis. The pathogenesis of this disorder may be related to that of sclerosing mediastinitis and of retroperitoneal fibrosis, which complicated the course of five patients in the series of Engleman et al. [23].

Conclusion

The problems presented at the beginning of the chapter remain, and will not be solved until a clearer understanding of the aetiology and pathogenesis of the disorders described here has emerged. Every effort should be made to find an offending infectious agent, and to outline more precisely the putative immune mechanisms. Until this happens, pathologists and clinicians will be confronted with patients suffering from these diseases and will have to differentiate between them. The clinical and histological features of the principal so-called immune or idiopathic disorders are summarized in Tables 34.5 and 34.6.

References

1. Abraham JL (1978) Recent advances in pneumoconiosis: the pathologist's role in etiologic diagnosis. In: Thurlbeck WM, Abell MR (eds) The lung, structure, function and disease. Williams and Wilkins, Baltimore, pp 96–137
2. Bernstein ML, Reece ER, de Chadarevian J-P et al. (1986) Bone marrow transplantation in lymphomatoid granulomatosis: report of a case. Cancer 58:969–972
3. Brandwein S, Esdaile J, Danoff D et al. (1983) Wegener's granulomatosis. Clinical features and outcome in 13 patients. Arch Intern Med 143:476–479
4. Carrington CB, Liebow AA (1966) Limited forms of angiitis and granulomatosis of Wegener's type. Am J Med 41:497–527
5. Cassan SM, Coles DT, Harrison EG Jr (1970) The concept of limited forms of Wegener's granulomatosis. Am J Med 49:366–379
6. Chen W, Monnat RJ Jr, Chen M et al. (1978) Aluminium induced pulmonary granulomatosis. Hum Pathol 9:705–711
7. Chumbley LC, Harrison EG Jr, DeRemee RA (1977) Allergic granulomatosis and angiitis (Churg–Strauss syndrome). Report and analysis of 30 cases. Mayo Clin Proc 52:477–484
8. Churg A (1983) Pulmonary angiitis and granulomatosis revisited. Hum Pathol 14:868–883
9. Churg A, Carrington CB, Gupta R (1979) Necrotizing sarcoid granulomatosis. Chest 76:406–413
10. Churg J (1963) Allergic granulomatosis and granulomatous-vascular syndromes. Ann Allergy 21:619–628
11. Churg J, Strauss L (1951) Allergic granulomatosis, allergic angiitis, and periarteritis nodosa. Am J Pathol 27:277–301
12. Colby TV, Carrington CB (1982) Pulmonary lymphomas simulating lymphomatoid granulomatosis. Am J Surg Pathol 6:19–32
13. Colby TV, Carrington CB (1983) Lymphoreticular tumors and infiltrates of the lung. Pathol Annu 18:27–70
14. Crissman JD (1979) Midline malignant reticulosis and lymphomatoid granulomatosis. A case report. Arch Pathol Lab Med 103:561–564
15. Crissman JD, Weiss MA, Gluckman J (1982) Midline granuloma syndrome. A clinicopathologic study of 13 patients. Am J Surg Pathol 6:335–346
16. Cupps TR, Fauci AS (1981) The vasculitides, Saunders, Philadelphia
17. Den Hertog RW, Wagenaar S, Westermann CJJ (1982) Bronchocentric granulomatosis and pulmonary echinococcosis. Am Rev Respir Dis 126:344–347
18. DeRemee RA, McDonald TJ, Harrison EG Jr et al. (1976) Wegener's granulomatosis. Anatomic correlates, a proposed classification. Mayo Clin Proc 51:777–781
19. DeRemee RA, McDonald TJ, Weiland LH (1985) Wegener's granulomatosis: observations on treatment with antimicrobial agents. Mayo Clin Proc 60:27–32
20. DeRemee RA, Weiland LH, McDonald TJ (1978) Polymorphic reticulosis, lymphomatoid granulomatosis. Two diseases or one? Mayo Clin Proc 53:634–640
21. DeRemee RA, Weiland LH, McDonald TJ (1980) Respiratory vasculitis. Mayo Clin Proc 55:492–498
22. Eichel BS, Harrison EG Jr, Devine KD et al. (1966) Primary lymphoma of the nose including a relationship to lethal midline granuloma. Am J Surg 112:597–605
23. Engleman P, Liebow AA, Gmelich J et al. (1977) Pulmonary hyalinizing granuloma. Am Rev Respir Dis 115:997–1008
24. Eraut D, Evans J, Caplin M (1978) Pulmonary necrobiotic nodules without rheumatoid arthritis. Br J Dis Chest 72:301–306
25. Fauci AS (1976) Pulmonary vasculitis. In: Kirkpatrick CH, Reynolds HY (eds) Immunologic and infectious reactions in the lung. Marcel Dekker, New York, pp 243–259
26. Fauci AS, Haynes BF, Costa J et al. (1982) Lymphomatoid granulomatosis. Prospective clinical and therapeutic experience over 10 years. N Engl J Med 306:68–74
27. Fauci AS, Haynes BF, Katz P et al. (1983) Wegener's granulomatosis: prospective clinical and therapeutic experience with 85 patients for 21 years. Ann Intern Med 98:76–85
28. Fayemi AO (1976) Pulmonary vascular disease in systemic lupus erythematosus. Am J Clin Pathol 65:284–290
29. Fienberg R (1953) Necrotizing granulomatosis and angi-

itis of the lungs with massive splenic necrosis and focal thrombotic granulomatous glomerulonephritis. Am J Clin Pathol 23:413–428

30. Fienberg R (1953) Necrotizing granulomatosis and angiitis of the lungs and its relationship to chronic pneumonitis of the cholesterol type. Am J Pathol 29:913–931

31. Fienberg R (1955) Pathergic granulomatosis. Am J Med 19:829–831 (editorial)

32. Fienberg R (1981) The protracted superficial phenomenon in pathergic (Wegener's) granulomatosis. Hum Pathol 12:458–467

33. Gephardt GN, Ahmad M, Tubbs RR (1983) Pulmonary vasculitis (Wegener's granulomatosis). Immunohistochemical study of T and B cell markers. Am J Med 74:700–704

34. Godman GC, Churg J (1954) Wegener's granulomatosis. Pathology and review of the literature. Arch Pathol Lab Med 58:533–553

35. Herbert A, Wright DH, Isaacson PG et al. (1984) Primary malignant lymphoma of the lung: histopathologic and immunologic evaluation of nine cases. Hum Pathol 15:415–422

36. Hui AN, Ehresmann GR, Quismorio FP et al. (1981) Wegener's granulomatosis. Electron microscopic and immunofluorescent studies. Chest 80:753–756

37. Katzenstein A (1980) The histologic spectrum and differential diagnosis of necrotizing granulomatous inflammation in the lung. In: Fenoglio CM, Wolff M (eds) Progress in surgical pathology. Masson, New York, pp 41–70

38. Katzenstein A, Askin FB (1982) Surgical pathology of non-neoplastic lung disease. Saunders, Philadelphia

39. Katzenstein AA, Carrington CB, Liebow AA (1979) Lymphomatoid granulomatosis. A clinicopathologic study of 152 cases. Cancer 43:360–373

40. Katzenstein A, Liebow AA, Friedman PJ (1975) Bronchocentric granulomatosis, mucoid impaction and hypersensitivity reactions to fungi. Am Rev Respir Dis 111:497–537

41. Klinger H (1931) Grenzformen der Periarteritis nodosa. Frankfurt Z Pathol 42:455–480

42. Koss MN, Antonovych T, Hochholzer L (1981) Allergic granulomatosis (Churg–Strauss syndrome). Pulmonary and renal morphologic findings. Am J Surg Pathol 5:21–28

43. Koss MN, Hochholzer L, Feigin DS et al. (1980) Necrotizing sarcoid-like granulomatosis: clinical, pathologic, and immunopathologic findings. Hum Pathol (Suppl) 11:510–519

44. Koss MN, Hochholzer L, Langloss JM et al. (1986) Lymphomatoid granulomatosis: a clinicopathologic study of 42 patients. Pathology 18:283–288

45. Koss MN, Robinson RG, Hochholzer L (1981) Bronchocentric granulomatosis. Hum Pathol 12:632–638

46. Kus J, Bergin C, Miller R et al. (1985) Lymphocyte subpopulations in allergic granulomatosis and angiitis (Churg–Strauss syndrome). Chest 87:826–827

47. Kussmaul A, Maier R (1866) Ueber eine bisher nicht beschriebene eigenthumliche Arterienerkrankung (Periarteritis nodosa), die mit Morbus Brightii und rapid fortschreitender allgemeiner Muskellahmung einhergeht. Dtsch Arch Klin Med 1:484–517

48. Lanham JC, Elkon KB, Pusey CD et al. (1984) Systemic vasculitis with asthma and eosinophilia: a clinical approach to the Churg–Strauss syndrome. Medicine 63:65–81

49. Leavitt RY, Fauci AS (1986) Pulmonary vasculitis. Am Rev Respir Dis 134:149–166

50. Lee JH, Joihovsky T, Yan K (1982) Bronchocentric granulomatosis: review of 14 patients. Thorax 37:779 (abstract)

51. Liebow AA (1973) Pulmonary angiitis and granulomatosis. The J Burns Amberson Lecture. Am Rev Respir Dis 108:1–18

52. Liebow AA, Carrington CRB, Friedman PJ (1972) Lymphomatoid granulomatosis. Hum Pathol 3:457–558

53. Littlejohn GO, Ryan PJ, Holdsworth SR (1985) Wegener's granulomatosis: clinical features and outcome in seventeen patients. Aust NZ J Med 15:241–245

54. Louie S, Daoust PR, Schwartz RS (1980) Immunodeficiency and the pathogenesis of non-Hodgkin's lymphoma. Semin Oncol 7:267–284

55. Mark EJ (1984) Lung biopsy interpretation. Williams and Wilkins, Baltimore

56. McBride P (1897) Photographs of a case of rapid destruction of the nose and face. J Laryngol Otol 12:64–66

57. McCluskey RT, Fienberg R (1983) Vasculitis in primary vasculitides, granulomatoses, and connective tissue diseases. Hum Pathol 14:305–315

58. Michel RP, Case BW, Moinuddin M (1979) Immunoblastic lymphosarcoma: a light, immunofluorescence and electron microscopic study. Cancer 43:224–236

59. Miller LR, Greenberg SD, McLarty JW (1985) Lupus lung. Chest 88:265–269

60. Myers JL, Katzenstein A (1986) Granulomatous infection mimicking bronchocentric granulomatosis. Am J Surg Pathol 10:317–322

61. Nichols PW, Koss M, Levine AM et al. (1982) Lymphomatoid granulomatosis: a T-cell disorder? Am J Med 72:467–471

62. Pare JA, Fraser RG, Hogg JC et al. (1979) Pulmonary "mainline" granulomatosis: talcosis of intravenous methadone abuse. Medicine 58:229–239

63. Rattinger MD, Thaddeus LD, Christian CD et al. (1983) Gastrointestinal involvement in lymphomatoid granulomatosis. Report of a case and review of the literature. Cancer 51:694–700

64. Reyes CN, Wenzel FJ, Lawton BR et al. (1982) The pulmonary pathology of farmer's lung disease. Chest 81:142–146

65. Robinson RG, Wehunt WD, Tsou E et al. (1982) Bronchocentric granulomatosis: roentgenographic manifestations. Am Rev Respir Dis 125:751–756

66. Rosen Y, Moon S, Huang CT et al. (1977) Granulomatous pulmonary angiitis in sarcoidosis. Arch Pathol Lab Med 101:170–174

67. Saldana MJ, Patchefsky AS, Israel HI et al. (1977) Pulmonary angiitis and granulomatosis. The relationship between histological features, organ involvement, and response to treatment. Hum Pathol 8:391–409

68. Shasby DM, Schwarz MI, Forstot JZ et al. (1982) Pulmonary immune complex deposition in Wegener's granulomatosis. Chest 81:338–340

69. Siegel H (1972) Human pulmonary pathology associated with narcotics and other addictive drugs. Hum Pathol 3:55–66

70. Singh N, Cole S, Krause PJ et al. (1981) Necrotizing sarcoid granulomatosis with extrapulmonary involvement. Clinical, pathologic, ultrastructural, and immunologic features. Am Rev Respir Dis 124:189–192

71. Spencer H (1985) Pathology of the lung, Pergamon, Oxford

72. Steinman TI, Jaffe B, Monaco AP et al. (1980) Recurrence of Wegener's granulomatosis after kidney transplantation: successful reinduction of remission with cyclophosphamide. Am J Med 64:458–460

73. Stewart JP (1933) Progressive lethal granulomatous ulceration of the nose. J Laryngol Otol 48:657–674
74. Tomashefski JR Jr, Hirsch CS (1980) The pulmonary vascular lesions of intravenous drug abuse. Hum Pathol 11:133–145
75. Ulbright TM, Katzenstein A (1980) Solitary necrotizing granulomas of the lung. Differentiating features and etiology. Am J Surg Pathol 4:13–28
76. Vallyathan NV, Green FH, Craighead JE (1980) Recent advances in the study of mineral pneumoconiosis. Pathol Annu 15:77–104
77. Van der Woude FJ, Rasmussen N, Lobatto S et al. (1985) Autoantibodies against neutrophils and monocytes: tool for diagnosis and marker of disease activity in Wegener's granulomatosis. Lancet 1:425–429
78. Veltri RW, Raich PC, McClung JE et al. (1982) Lymphomatoid granulomatosis and Epstein-Barr virus. Cancer 50:1513–1517
79. Voith MA, Lichtenfeld KM, Schimpff SC et al. (1979) Systemic complications of MER immunotherapy of cancer. Pulmonary granulomatosis and rash. Cancer 43:500–504
80. Walton EW (1959) Non-healing granulomata of the nose. J Laryngol Otol 73:242–260
81. Warren J, Pitchenik AE, Saldana MJ (1985) Bronchocentric granulomatosis with glomerulonephritis. Chest 87:832–834
82. Wechsler RJ, Steiner RM, Israel HL et al. (1984) Chest radiograph in lymphomatoid granulomatosis: comparison with Wegener's granulomatosis. AJR 142:79–83
83. Wegener F (1936) Uber generalisierte, septische Gefasserkrankungen. Verh D Ges Pathol 29:202–209
84. Wegener F (1939) Uber eine eigenartige rhinogene Granulomatose mit besonderer Beiteiligung des Arteriensystems und der Nieren. Beitr Pathol Anat 102:36–68
85. Weisbrot IM (1976) Lymphomatoid granulomatosis of the lung, associated with a long history of benign lymphoepithelial lesions of the salivary glands and lymphoid interstitial pneumonitis. Report of a case. Am J Clin Pathol 66:792–801
86. Weiss MA, Rolfes DB, Alvira MA et al. (1984) Benign lymphocytic angiitis and granulomatosis: a case report with evidence of an autoimmune etiology. Am J Clin Pathol 81:110–116
87. Wiedemann HP, Bensinger RE, Hudson LD (1982) Bronchocentric granulomatosis with eye involvement. Am Rev Respir Dis 126:347–350
88. Wolff SM, Fauci AS, Horn RG et al. (1974) Wegener's granulomatosis. Ann Intern Med 81:513–525
89. Yoshikawa Y, Watanabe T (1986) Pulmonary lesions in Wegener's granulomatosis: a clinicopathologic study of 22 autopsy cases. Hum Pathol 17:401–410

Chapter 35

Pulmonary Hypertension

R. P. Michel

The pulmonary vascular bed, situated in series with the systemic circulation, differs from it in physiology, morphology and in its response to pathological stimuli. Many of these differences are explained by the simple fact that the lung vasculature is designed for oxygen uptake and the systemic circulation for oxygen delivery. In order that the lung can function without becoming oedematous, its circulation has a high flow rate, a low pressure and a low resistance. Indeed, the pulmonary circulation is nearly always maximally dilated, and the muscularity of its arteries is considerably lower than that of systemic arteries. In the lung, blood flow is regulated by vasoconstriction rather than vasodilatation as in the systemic circulation. The two vascular beds respond to stimuli such as hypoxia and histamine in a diametrically opposed fashion: constriction occurs on the pulmonary side, dilatation on the systemic.

From the pathological viewpoint, the mechanisms and disease entities which result in hypertension in the lung differ markedly from those causing systemic hypertension, and there is essentially no overlap between them. Indeed, a variety of diseases produce hypertension in the lung, each with its own characteristic morphological features which are not encountered in the systemic vascular bed. The most obvious example is the plexiform lesion which is never seen in systemic hypertension. In addition, therapy and ,prognosis vary with the cause of hypertension, making it imperative to arrive at a precise diagnosis which is based, in a significant number of instances, on morphological changes. In systemic hypertension, the histological lesions are much more stereotyped.

This chapter will cover the pulmonary circulation in health and disease, and will look at particularly those diseases causing pulmonary hypertension. Emphasis will be placed on morphology and morphometry and their relationships to physiological changes, i.e. on pathophysiology as well as pathogenesis. Therapy will be dealt within in a cursory manner only. Several of the references are to recent reviews or books which are themselves the source of numerous individual articles.

The Normal Pulmonary Artery and Circulation

Normal Macroscopic Morphology and Morphometry

The circulation of the lung extends from the pulmonary valve to the orifices of the veins in the left atrium. The pulmonary trunk divides into two main extrapulmonary arteries beneath the aortic arch. The right, slightly larger than the left, runs horizontally to the right and divides into two main branches: the larger lower one supplies the middle and lower lobes, and the smaller upper branch follows the epi-arterial bronchus to the upper lobe. The left main pulmonary artery courses horizontally, crosses the descending aorta, lying above the left main bronchus until it gives off a branch to the upper lobe, after which it runs downward, behind and lateral to the bronchus. The arteries then divide into lobar branches in a manner analogous to the airways. The intrapulmonary arteries, which follow the airways, share a connective tissue sheath with them. More distally, they separate from the airways and gradually lose their sheath altogether to become entirely surrounded by alveoli. In addition, there are supernumerary arteries that do not follow the airways (vide infra).

The human pulmonary venous system drains all of the pulmonary arterial blood flow and much of the bronchial arterial flow [30,132]. The veins lie away from the airways, in interlobular septa; each lobule is drained by a single vein [46]. The smaller intrapulmonary veins gradually merge with one another to form a single lobar vein. The superior veins drain both upper lobes and the right middle lobe, while the inferior veins drain the lower lobes. All four veins then enter the upper portion of the left atrium.

Although the pulmonary circulation carries the main blood supply to the lungs, it is important to remember the bronchial circulation which conveys oxygenated blood to the airways, pleura, parts of the interstitium and to the large hilar vessels and nerves via the vasa vasorum. The features of this "private" circulation to the lungs are reviewed elsewhere [17,30,46,155] and will only be considered here briefly. Most bronchial arteries arise from the thoracic aorta or intercostal arteries and enter the lung at the hilus. They form an annulus around each main bronchus from which bronchial and pleural arteries emerge. The former follow the bronchi distally approximately to the level of the respiratory bronchioles. Bronchial venous drainage is bidirectional, and it is estimated that the central one-third drains to the right side of the heart while the peripheral two-thirds reach the left atrium by anastomosing with the pulmonary veins [17].

The morphometry of the entire pulmonary vasculature has been described for human, canine [95] and more recently feline [166] lungs. In man, data were obtained by Horsfield, Cumming and their colleagues [18,60,61,130] from resin casts of postmortem lungs. Branches of the arterial and venous trees are classified by Strahler orders, with the most distal branches starting as the first order. The vascular tree can be described knowing three ratios, listed as the branching, diameter and length ratios. The branching ratio is defined as the ratio of the number of branches in the daughter order to those of the parent order; the diameter and the length ratios are defined respectively as the ratios of diameter and length in the parent order to those of the daughter order. Using this system, the arterial vasculature has 17 orders, a branching ratio of 3.12, a diameter ratio of 1.63 and a length ratio of 1.60. For the venous system, which has 15 orders, the respective ratios are 3.31, 1.67 and 1.56. The smaller number of orders on the venous side is explained by the four pulmonary veins which enter the atrium [61]. With these data, several parameters (cross-sectional area, blood volume and linear velocity) can be derived to predict the distribution of pulmonary vascular resistance (vide infra).

Normal Microscopic Morphology and Morphometry

Nomenclature

Because of the important structural differences between the pulmonary and systemic vasculatures, controversy exists over the nomenclature of the classes of pulmonary blood vessels. Brenner [10] in 1935 laid down criteria for a system of classification, where he defined elastic pulmonary arteries as those over $1000\,\mu m$ in external diameter, and muscular pulmonary arteries as those between 100 and $1000\,\mu m$ with a distinct media bordered by an internal and an external elastic lamina; arterioles were defined as having a single elastic lamina and no muscular media, being usually under $100\,\mu m$. Harris and Heath [46] subscribe to these definitions. However, as stated by Brenner himself [10], the term arteriole has "no satisfactory definition". Wagenvoort and Wagenvoort [155] thought an arteriole should be "an artery with a complete media in its beginning and none in its remaining course". Because of the difficulties in definition, Reid and her colleagues [28,114] define the classes of vessels according to the structure of their walls. Elastic arteries have an internal and an external elastic lamina and more than seven additional laminae between them. Muscular arteries have less than four laminae between the internal and external ones and transitional arteries are intermediate between elastic and muscular arteries. Partially muscular arteries are surrounded by only a crescent of smooth muscle and non-muscular arteries have no muscle at all. The nomenclature of the pulmonary veins poses similar problems, and because of the more haphazard arrangement of the elastic tissue in their walls, they are best classifed as either muscular, partially muscular or non-muscular [56,87,91].

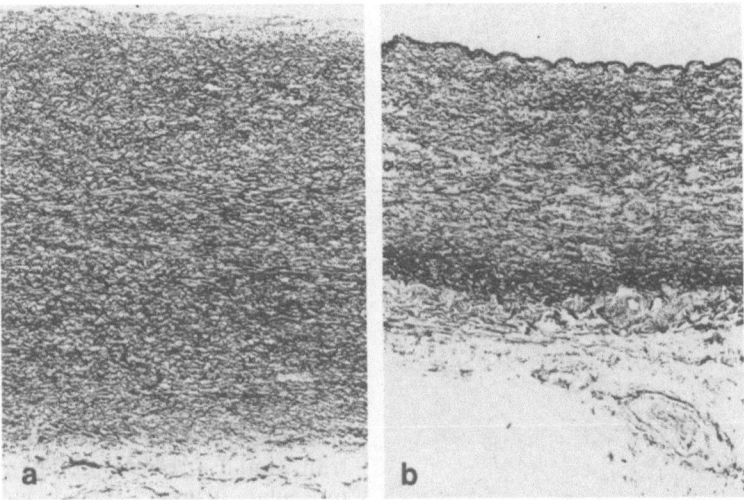

Fig. 35.1. a Light photomicrograph of normal aortic wall from a 28-year-old male showing entire medial thickness with regular, parallel, wavy elastic laminae and a small amount of intima. Elastic van Gieson, × 80. **b** Pulmonary artery trunk showing thinner media with sparser and fragmented elastic fibrils. Elastic van Gieson, × 80

Light Microscopy of Larger Arteries and Veins

The pulmonary trunk and main pulmonary arteries in the adult have a wall thickness ranging

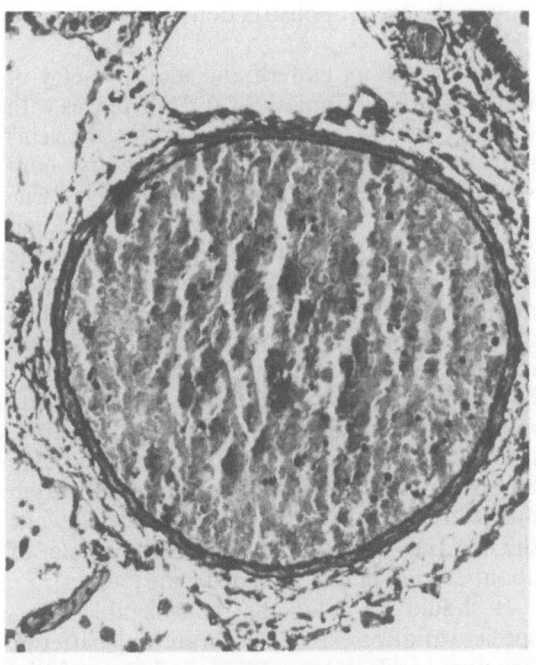

Fig. 35.2. a Normal muscular artery with thin media bordered by distinct dark internal and external elastic laminae. Elastic van Gieson, × 300

from 40% to 75% of aortic thickness (Fig. 35.1). In all elastic arteries of young adults, whether extrapulmonary or intrapulmonary, the intima is thin, and is composed of endothelium overlying a basement membrane. Between the inner and outer elastic laminae are numerous alternating layers of smooth muscle and elastic laminae. The latter are fragmented in the extrapulmonary arteries whereas in the aorta they are in parallel (Fig. 35.1). In the intrapulmonary arteries, the elastic laminae are well preserved. The generally sparse adventitia is composed of fibrous tissue, vasa vasora and nerves. The muscular arteries also have a thin intima and scanty adventitia, between which are sandwiched an internal and an external elastic lamina bordering a media composed predominantly of smooth-muscle cells (Fig. 35.2). The media of muscular pulmonary arteries is much thinner than that of systemic arteries of comparable size. Larger intrapulmonary arteries may be recognized, therefore, by their structure and by their close relationship to airways.

Muscular veins may be distinguished from their arterial counterpart in man by their location at the periphery of the lobule away from the airways, their thinner media and thicker fibrous adventitia and by the absence of a distinct external elastic lamina (Fig. 35.3); their internal elastic lamina, as in the arteries, is a distinct structure. In the larger veins, the elastic fibrils are disorganized, precluding a designation of "elastic vein". The largest veins, near their inser-

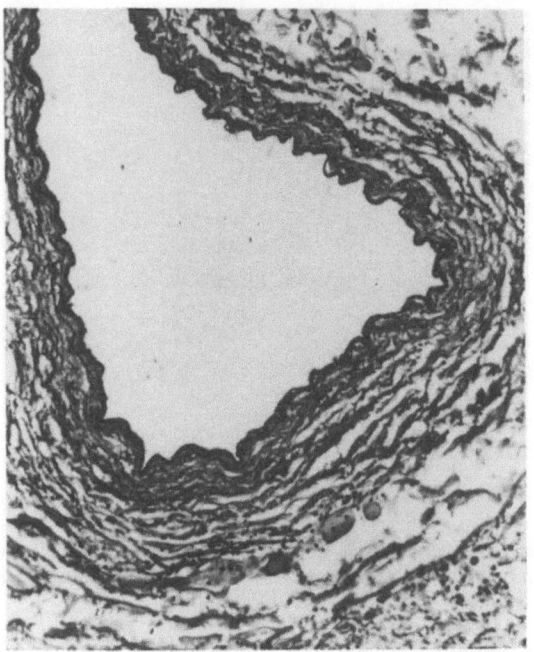

Fig. 35.3. a Part of muscular vein wall with well-defined crenated internal elastic lamina, poorly defined media with sparse elastic fibrils blending imperceptibly with thicker adventitia. Elastic van Gieson, × 200

tion into the left atrium, are covered by a layer of cardiac muscle in continuity with the atrium.

Branching Pattern and Morphometry of the Human Vasculature

Brenner's descriptions [10] of the microscopic appearance of human pulmonary vasculature are based on lungs fixed by inflation without prior vascular injection. To identify arteries and veins separately, however, and to achieve a uniform degree of vascular distension for morphometry, it is useful to inject gelatin and barium mixtures into the arteries and veins prior to fixation [28,56,58,91,114,115]. With these methods, readily applicable to the study of pulmonary hypertensive states, the pulmonary arteries can be described in quantitative terms. First, the degree of muscularity of a vessel can be expressed numerically in relation to its external diameter (ED): this is called the per cent medial muscle thickness (%MT) which is defined as the ratio of the sum of the medial thickness of each wall relative to the external diameter $(2 \times MT/ED) \times 100$. The %MT varies with age, being 1.5%–5% in adult pulmonary arteries [114]. Second, the structure of vessels can be

related to diameter: in the adult, most arteries over $1000\,\mu m$ are elastic, while most arteries between 1000 and $150\,\mu m$ are muscular. Overlap occurs however and the change from one class to another in arteries of the same size is not identical along all pathways [115]. For example, of arteries with a diameter of $100\,\mu m$, 45% are muscular, 45% partially muscular and 10% non-muscular. Third, the concentration of arteries and veins can be expressed as an alveolar/arterial ratio. And fourth, the arterial structure can be related to airway generations: since the arteries follow both the development and course of the airways, it is useful to describe the class of artery with the type of airway. In general, elastic and transitional arteries accompany bronchi, while muscular arteries extend more peripherally, accompanying bronchioles, respiratory bronchioles and alveolar ducts, and even reach the level of alveoli. More distally, the percentage of non-muscular and partially muscular arteries increases.

Arteries which accompany airways are termed "conventional", and are outnumbered in a ratio of 2.5–4:1 by "supernumerary" arteries which do not travel with the airways. Supernumerary arteries are usually muscular, arise in an irregular manner at right angles to elastic arteries and are more frequent in the intra-acinar regions. They supply alveoli directly and provide collateral flow in the event of obstruction of another vessel [114].

The branching pattern and morphometry of veins show both similarities and differences with those of arteries [58]. Like the arteries, the veins may be conventional or supernumerary: conventional veins run in their own connective tissue sheath along an axial pathway (seen on angiograms) and are equivalent in number to the airways and conventional arteries. Supernumerary veins outnumber the conventional by about 3–4:1 and are more frequent at the periphery of the lung. Compared with arteries, veins have a higher density per unit alveolar area and more supernumerary branches. Veins also have a smaller %MT (mean 1.5%) than arteries [56]. Smooth muscle is first seen in veins of 40–60 μm diameter (therefore partially muscular), while the first fully muscular veins have a diameter of about 200 μm in the distended state [56,155].

It should be noted that considerable interspecies variation exists in the branching patterns, structure and degree of muscularity of the normal lung vasculature, which must be taken into account when studying animal models of human disease [86,87,91,94,117,118].

Morphology of the Microvasculature

Even if a detailed discussion of the microcirculation is impossible here, a brief review of relevant facts is warranted. Although definitions of the pulmonary microcirculation (or microvasculature) vary, it is reasonable to define it as including capillaries and partially muscular and non-muscular arteries and veins, all of which are intra-acinar and may participate in gas exchange and fluid filtration in the lung [63,92,116]. Capillaries are easily recognized morphologically and defined on the basis of size (15 μm diameter). Their walls consist of endothelial cells resting on a basement membrane, and of pericytes that are present in smaller numbers than in systemic capillaries [11,160]. Small intra-acinar arteries and veins (muscular, partially muscular and non-muscular), however, may be difficult to distinguish from each other by light and electron microscopy, since the usual differentiating criteria (location within the lobule, per cent medial muscle thickness and amount of adventitia) are not readily applied. It may therefore be necessary to resort to methods such as vascular injection with gelatin–barium mixtures, serial sections from a known artery or vein or in vivo visualization of blood flow [28,88,91,114,225].

For electron microscopy, Meyrick and Reid [87] advocate perfusion of fixative through the arteries, with the veins tied off at the hilum, so that the displaced erythrocytes label the veins. Partially muscular arteries and veins are very similar at the light microscopic level: in them, the elastic laminae are simplified and only a crescent of smooth muscle is seen [114]. Non-muscular vessels, naturally, have no smooth-muscle cells. Instead, in both partially muscular and non-muscular vessels, pericytes and "intermediate" (between smooth muscle and pericytic) cells are interposed between the endothelium and sparse adventitia [57,87,88,115,160]. Both of these cell types may differentiate into smooth-muscle cells under abnormal conditions, including hypoxia, and therefore have also been termed "precursor smooth-muscle cells".

One ultrastructural feature which may help to separate very small arteries from veins is the presence, in the veins, of numerous myoendothelial junctions, which have been demonstrated in dogs and cats [91,119]. These junctions are also seen in other species, including rats and humans, and are formed between endothelial cells on one hand and smooth-muscle cells, intermediate cells or pericytes on the other

through small fenestrations in the internal elastic lamina [91,116,119,152]. Although the role of these junctions is unresolved, it is conceivable that they play a role in pulmonary vascular reactivity since they are a potential link between the endothelial cells, exposed to and synthesizing various mediators, and the smooth-muscle cells or their precursors (vide infra). The significance of the morphological variations in myoendothelial junctions (veins versus arteries) remains an enigma.

Similar regional differences in endothelial cell structure have been noted [87,160]. For example, the endothelium of arteries and veins is thicker than that of capillaries, and Weibel–Palade bodies are more numerous in hilar than in peripheral vessels. In addition, arterial endothelial cells are elongated with their long axis parallel to the direction of blood flow while venous endothelial cells are polygonal. Intercellular junctions also vary regionally, as they do in the systemic circulation: by freeze-fracture, arteries have been shown to have tight junctions that contain the most numerous rows of particles and the most numerous gap junctions, which suggests that they are the least permeable and have the highest degree of coupling between endothelial cells. The venous microvascular segment, with its simpler tight junctions, may be the most permeable [127].

Growth and Development

Pre-natal Development

As reviewed by several authors [11,46,58,132], in the *embryonic period*, the main extrapulmonary arteries develop from the sixth pair of aortic arches. Early on, in the 4–5-mm stage (28–32 d), the primitive laryngotracheal bud derives its blood supply from the dorsal aorta through paired segmental arteries which end in a capillary plexus. The right and left sixth arches appear at the 5-mm stage as projections from the ventral aspect of the fourth arch, and from each of them, dorsal and ventral anlages form: the ventral anlages remain in connection with the pulmonary trunk when the truncus arteriosus divides, and become the main right and left pulmonary arteries. These grow into the lung buds, to join with their plexus of vessels, which loses its original connections with the segmental dorsal aortic branches. The dorsal part of the left sixth aortic arch becomes the ductus arteriosus while that of the right disappears. By the end of the 7th week (18-mm stage), this adult-type blood supply to the lungs is established.

The embryonic development of the pulmonary venous drainage is intimately related to that of the rest of the foregut, which at the 6–7-mm stage drains into the cardinal veins and associated visceral veins, and hepatic sinusoids [46]. Later, a single vein forms from the left atrium and divides into four branches which grow towards the lung bud and connect with the splanchnic plexus. With time, the common pulmonary venous trunk is absorbed into the left atrium, leaving the four tributaries to enter it separately. Some of these early connections are maintained in the adult as anastomoses between the portal and pulmonary venous systems.

In the late embryonal and early foetal periods, during the glandular (or pseudoglandular) stage of lung development (5 to 16 weeks) the pre-acinar branching pattern of both conventional and supernumerary arteries is established along with that of the airways [49,58,107]. After 16 weeks, pre-acinar arteries no longer multiply, and only increase in size. The main and large elastic pulmonary arteries have the same structure as the aorta, with numerous elastic laminae, and by 19 weeks, elastic arteries extend to the same level as in the adult. Mature smooth-muscle cells appear in arterial walls at about 23 weeks gestation. Within the acinus, however, arterial branching continues in sequence with the development of respiratory bronchioles and later of saccules, resulting in an increase in number of arteries per unit area with foetal age. During the entire foetal period, the peripheral size limits of muscular, partially muscular and non-muscular arteries are in the same range as in the adult. Their position with respect to airways differs, however, and the change from muscular to partially muscular to non-muscular is more proximal in the foetus. At birth, there is no muscle in arterial walls distal to the terminal bronchiole. For the muscularized foetal vessels, the %MT is about double that in the adult, being particularly high in the small vessels.

In summary, at birth the main pulmonary artery has a structure identical to that of the aorta, and pre-acinar vessels are fully developed. Muscular vessels, all pre-acinar, increase in size from just below 100 μm diameter; partially muscular vessels are in the size range of 35–175 μm; non-muscular below 125 μm.

The foetal pulmonary venous development lags behind arterial development since the adult branching pattern of pre-acinar veins is not reached until 20 weeks gestation [58]. However, the intra-acinar veins continue to multiply like the arteries. Smooth muscle appears significantly later in veins: there are only minimal amounts of it by 28 weeks, and a full coat is only seen at or near birth.

Post-natal Development

At birth when the lungs fill with air, pulmonary arterial pressure (Ppa) and pulmonary vascular resistance (PVR) fall dramatically: most of the decline occurs within the first 2 to 3 d after birth and it is complete by 2 weeks [123]. Important anatomical changes occur in the vasculature, particularly on the arterial side. The extra-pulmonary arteries thin during the first few months to adult levels. By 6 months, their elastic laminae, although still parallel, have fragmented, and by the end of the 2nd year they resemble those of the adult main pulmonary artery. The larger intrapulmonary pre-acinar arteries dilate and their media atrophies, which produces an increase in their diameter over the first 4–10 months of life, after which time adult levels of %MT are attained. The walls of acinar vessels under 200 μm thin much more rapidly by dilatation, which correlates with the fall of the pulmonary artery pressure to adult levels within a few days after birth. Part of the thinning in these vessels also occurs because muscle development lags behind the increase in external diameter. For example, in an 11-year-old child, only 40% of acinar arteries of 200 μm are muscular, while in the adult, 100% are muscular. Between 4 months and 3 years particularly, and even up to 8 years, there is an important increase in the number of acinar arteries which accompanies the development of new small airways and alveoli. During adolescence, the arteries no longer multiply, but the proximal pulmonary arteries increase in size; also muscle extends to the periphery, so that by 19 years of age, muscle is found in arteries related to alveolar walls. The veins also increase in size and number, but unlike the arteries, the distribution of muscle in their walls changes little, so that the overall %MT of veins remains low.

Descriptions of the normal development of the lung are available for several species including rats, pigs and lambs, and provide a basis for the study of various models of pulmonary hypertension and of the pharmacology of the pulmonary vasculature.

Age-Related Changes

With age, structural changes occur in both arteries and veins. In the pulmonary trunk, elastic tissue diminishes and collagen and calcium content rise. These changes correlate with a decrease in the extensibility and an increase in capacity of the main pulmonary trunk [156]. The extrapulmonary arteries show intimal fibrosis and patches of atheroma although considerably lower in frequency and severity than in the aorta. Therefore, while atherosclerosis is a pathological process, most individuals over 40–50 years of age have at least some traces of it, usually near ramifications of secondary and tertiary pulmonary arterial branches [10]. Pulmonary atherosclerosis increases not only with age but with cigarette smoking, with aortic athero-sclerosis and with pulmonary hypertension from any cause [46,98,155]. Although the %MT in muscular arteries does not appear to vary significantly with age [156], intimal fibrosis occasionally presents in young individuals and increases sharply in older age groups. It has been suggested that the intimal thickening is related to in situ thrombosis. Similar, albeit less severe, changes occur in veins, in which a characteristic almost acellular hyaline intimal fibrosis may develop.

Normal Physiology

In addition to its obvious and crucial role in oxygenating blood and removing carbon dioxide for the systemic circulation, the lung circulation functions, directly or indirectly, in several other ways. First, it has a metabolic role [4,26, 30,64,113,124,125] in (a) the activation of angiotensin I to angiotensin II; (b) the inactivation of compounds such as bradykinin, serotonin, norepinephrine and prostaglandins of the E and F series; and (c) the synthesis and release of a number of biologically active substances including amines (e.g. histamine), polypeptides (e.g. bombesin, vasoactive intestinal polypeptide, adrenocorticotropic hormone and substance P), as well as complement, arachidonate metabolites (e.g. PGD_2, PGE_2, $PGF_{2\alpha}$, prostacyclin, thromboxanes and leukotrienes) and platelet-activating factor. In addition, as in the systemic circulation, evidence exists for an endothelium-derived relaxing factor which activates smooth-muscle relaxation [30,33]. Second, the lung circulation filters the systemic venous drainage, and has defence functions which are carried out by both non-specific (e.g. by blood phagocytes) and specific (by immunological) mechanisms. Third, it supplies much of the pulmonary parenchyma

itself with O_2 and nutrients, the bronchial circulation providing the balance [17,30]. Fourth, it is a reservoir for left ventricular output and finally, it has a role in fluid absorption: the latter is particularly imporant after birth, but may also operate in several forms of lung oedema.

Let us now consider briefly the haemodynamics of the pulmonary circulation. Since in post-natal life, after expansion of the lungs and closure of the ductus and of the foramen ovale, the pulmonary circulation accommodates 98%–99% of the total measured cardiac output, it comes as no surprise that it is a low pressure, high flow and low resistance circuit [30, 39]. Unlike the systemic vascular bed in which both vasodilation and constriction control blood flow, the latter, in the lung, is controlled by vasoconstriction. Indeed, under normal conditions, pulmonary vasculature is near maximally dilated to permit gas exchange. Prostacyclins may play a role in maintaining this normal low pulmonary vascular tone [24].

Pulmonary haemodynamics are usually assessed in humans by measuring pulmonary arterial pressure (Ppa), pulmonary arterial wedge pressure (Ppw) and cardiac output (CO). In most instances, these variables are readily obtained by the introduction of flow-directed balloon catheters (Swan–Ganz) into the pulmonary artery; the thermistor at their tip serves to measure CO by thermodilution. From these measured variables, total pulmonary vascular resistance (PVR) is calculated from the equation PVR = (Ppa − Ppw)/CO. The normal Ppa is about 22/10 mmHg with a mean of 15 mmHg, and the normal Ppw, which closely approximates large pulmonary venous and left atrial pressures, averages 8–10 mmHg. Cardiac output has a mean value of 5–8 l/min at rest in upright man. With exercise, it can rise three- to fourfold, which results in only a small elevation in Ppa. This is explained by a decrease in calculated total PVR due to recruitment of previously unopened vessels (particularly in the upper zones of the lung) and to distension of vessels already perfused. The PVR, in normal resting adults, is approximately 1 mmHg/l/min or 80 dynes-/s/cm^{-5}, a value about ten times lower than in the systemic circulation. The total PVR depends on a number of factors apart from flow which are dictated by the Poiseuille equation: PVR is proportional to the length of the vessels and to the viscosity of blood (which may be altered in polycythaemia or in the hyperviscosity syndrome); it is inversely proportional to the fourth power of the radius, which implies that small decreases in vascular radius result in large rises in PVR.

The interpretation of the measured total PVR, however, is not without problems. One difficulty is that a rise in PVR does not necessarily indicate vasomotion, and vice versa. Indeed, as emphasized recently [80], when flow (or cardiac output) to one lung increases, Ppa goes up in a linear fashion, assuming vasoconstriction does not occur, which means true resistance has not changed. When PVR is calculated using the above equation, however, it appears to fall. To circumvent this problem, Ppa and flow should be measured at several points to obtain a pressure–flow curve with a slope and an intercept [37,80,96]: the latter is called the critical pressure, the zero-flow pressure or the Starling resistor pressure, and the former the upstream resistance, meaning upstream from the locus of the critical pressure. Thus Ppa may rise because of an increase in critical closing pressure (for example with an elevation of airway pressure or with hypoxia), or because of a rise in the slope of the pressure–flow curve (for example with vasoconstrictors such as serotonin and histamine). Similarly, disease states may affect the pressure–flow curve either by raising its critical pressure or its slope, although the relationship between these measured changes and morphology needs to be studied further.

A second problem with measurements of total PVR is that it does not provide information on the distribution of this resistance between arterial, venous and microvascular segments. In the systemic circulation, under normal circumstances, most of the resistance is in pre-capillary arterioles. In the lung, however, the distribution of PVR, which has important functional implications both in pulmonary hypertension and as a determining factor in fluid filtration [30,39,93], was not known for a long time. Recent exciting experimental techniques, however, such as arterial and venous occlusion [24,42] and direct micropuncture of small subpleural pulmonary vessels [7], have provided new, although somewhat contradictory, data on the distribution of PVR. With arterial and venous occlusion, the total PVR can be partitioned into relatively indistensible arterial and venous segments separated by a compliant middle (? microvascular) segment [42]. Each segment responds specifically to physiological, pharmacological and pathological stimuli: serotonin increases the arterial segment pressure drop, histamine the venous drop, and elevation of airway pressure or hypoxia raise middle

segment resistance [42,43]; creation of systemic-to-pulmonary shunts raises arterial segment resistance [94].

In general, stimuli which raise the arterial pressure drop also increase the slope of the pressure–flow curve [94,96], while the intercept increases with middle segment resistance. With the other recently introduced technique, direct micropuncture of small subpleural pulmonary arteries and veins [7,103], most of the resistance appears to lie in the capillaries, data that conflict with those obtained with arterial and venous occlusion. In addition, these data must be reconciled with predictions on the distribution of PVR obtained from the morphometric data of Cumming et al. [18,60,130], which suggest that most of the resistance is in small arteries. It is clear that there are advantages and disadvantages to each of these techniques: for example, arterial and venous occlusion does not provide anatomical information on the size of vessels in each segment; micropuncture studies are based on the assumption that subpleural vessels accurately represent the vasculature of the rest of the lung. Zhuang et al. [166] tried to reconcile some of these contradictory results and, based on different models of the lung vasculature of cats, concluded that about 29%, 22% and 49% of the total arteriovenous pressure drop (Ppa − Ppw) was in the arterial, capillary and venous segments respectively.

In remains true that these new techniques are useful in defining, with greater precision the sites of action of hypoxia and of drugs and with morphology, in defining the sites of disease. The arterial occlusion method has recently been applied to intact dogs and to humans by observing the pressure profile after rapid inflation of the balloon of the Swan–Ganz catheter [15,59]. Using this procedure in man, about two-thirds of the resistance lies on the arterial side and one-third on the venous side of the circulation, although some variability occurs. This technique may help to identify sites of resistance in pulmonary hypertensive states and to measure precisely capillary hydrostatic pressure, so

Table 35.1. Mechanisms of pulmonary arterial hypertension

Increased pulmonary blood flow (hyperkinetic)
Increased pulmonary venous pressure (passive transmission)
Vasoconstriction (e.g. by hypoxia, humoral mediators or pharmacological agents)
Organic vascular lesions (intraluminal, mural, extramural)
Destruction of vessels
Increased blood viscosity

important in determining fluid filtration in the lung.

Pulmonary Hypertension and the Pulmonary Artery

General Features and Morphological Grading

Definitions and Classification

Pulmonary hypertension (PHT) exists when the mean Ppa, measured at catheterization, exceeds 20 to 25 mmHg or when systolic Ppa exceeds 30 mmHg [38,47,155]. These values must be obtained at rest since significant variations are observed with exercise and in association with other conditions [46]. A concise classification of pulmonary hypertension is not easy. One based on mechanisms such as those listed in Table 35.1 is certainly possible, and was one of the first proposed. However, it is confusing since in some instances, mechanisms are unknown (e.g. primary PHT), while in others more than one mechanism may be operating, (e.g. in the parenchymal lung diseases). Moreover, once the Ppa starts to rise, secondary organic lesions develop, in the form of medial hypertrophy and muscularization of peripheral arteries, which further raise Ppa. A classification based on purely clinical criteria is incomplete since for example "clinical" primary pulmonary hypertension traditionally encompasses three pathologically distinct entities including plexogenic pulmonary arteriopathy, recurrent pulmonary thromboembolism and veno-occlusive disease [27,155]; more recently, capillary haemangiomatosis was added to this list. From the viewpoint of the pathology, some overlap also occurs, since plexiform lesions are the hallmark of several aetiologically different forms of PHT. Despite these shortcomings, most authorities agree that a classification combining aetiology and pathology, such as the one presented in Table 35.2, is quite suitable. This classification will be followed here. Emphasis will be placed on mechanisms of disease, on morphological features which aid in the diagnosis and on their physiological implications, rather than on clinical and therapeutic details.

The classification of PHT and the understanding of the mechanisms involved in its

Table 35.2. Classification of pulmonary arterial hypertension

1. Congenital cardiac left-to-right shunts
     Pre-tricuspid
     Post-tricuspid
2. Persistent pulmonary hypertension of the newborn (PPHN)
3. Primary plexogenic pulmonary arteriopathy (PPPA)
     Primary pulmonary hypertension
     Drug-induced and dietary pulmonary hypertension
     Pulmonary hypertension due to hepatic disease
4. Embolic pulmonary hypertension
5. Pulmonary veno-occlusive disease (VOD)
6. Capillary haemangiomatosis of the lungs
7. Pulmonary hypertension secondary to pulmonary venous hypertension
8. Hypoxic pulmonary hypertension
9. Pulmonary hypertension in parenchymal lung diseases

genesis have been aided by the examination of lung tissue, both at autopsy and by open lung biopsy. The latter, which has enjoyed a recent surge in popularity, championed by Wagenvoort [149,155] and others [69,107,108], is indicated to: (a) arrive at a precise diagnosis in "clinically unexplained pulmonary hypertension"; (b) assess the progression of arterial lesions in congenital cardiac shunts or the regression of lesions after banding of the pulmonary artery; (c) ascertain, on occasion, the nature of a cardiac anomaly or acquired cardiac disease; and (d) determine the contribution of parenchymal disease to pulmonary hypertension. Therefore biopsy is done to make a diagnosis, which affects therapy and prognosis. In most forms of PHT, the changes are diffuse, so that biopsy of any part of the lung is representative [108,149,155]. The only exception is the lingula, which is not infrequently the site of non-specific fibrosis and vascular changes, and this should be avoided whenever possible [48,104].

Morphological Grading

Although the morphology of the response of the pulmonary vasculature to increased pressure or flow has been known for over 35 years, it was only in 1958 that Heath and Edwards [50] proposed a histological grading of pulmonary hypertension to describe changes due primarily to congenital cardiac defects and to idiopathic pulmonary hypertension. The six grades, presented in Table 35.3, may be divided into two groups: the first includes grades 1 to 3, in which the changes are at least potentially reversible; the second includes grades 4 to 6, considered irreversible. While good correlation between this

Table 35.3. Histological structural grading of pulmonary hypertension*

Grade 1
Increased medial thickness in muscular pulmonary arteries
Muscularization of small pulmonary arteries
Grade 2
Medial hypertrophy with cellular intimal proliferation in pulmonary arteries less than 300 μm diameter
Grade 3
Grade 2 plus acellular intimal fibrosis (may be concentric, onion-skin pattern)
Intimal proliferation in arteries greater than 300 μm
Grade 4
Grade 3 plus multifocal dilatation lesions including plexiform lesions, vein-like branches of hypertrophied muscular arteries and angiomatoid lesions
Grade 5
Grade 4 plus pulmonary haemosiderosis
Grade 6
Grade 5 plus necrotizing arteritis, with neutrophils and thrombosis

---

* Modified from references [46] and [50].

grading system and the levels of Ppa and PVR has been demonstrated in several studies, advances in our understanding of disease mechanisms and in therapy, particularly of congenital heart disease, have resulted in re-evaluation and refinement of the grading system. Indeed several criticisms have been levelled at the Heath and Edwards grading system [138,155]. For example, although individual grades 1 to 3 appear to correlate with the level of PHT and prognosis, grades 4 to 6 do not. To circumvent this problem, Wagenvoort [148,155] suggested that grades 4 to 6 be replaced with a single grade, or with the descriptive term "plexogenic pulmonary arteriopathy", in accordance with the World Health Organization [47]. This view is supported by the notion that necrotizing arteritis (grade 6) may precede the plexiform lesions and indeed play a role in their formation [155]. The second criticism of the grading of Heath and Edwards is that it does not apply to all forms of PHT, since plexiform lesions are only found in certain entities. Furthermore, even for the lower grades, morphological patterns vary with the aetiology. In addition, the grading applies only to arteries, and excludes venous, microcirculatory and other parenchymal lesions. Finally, it is qualitative, not quantitative and important variations may exist within a given grade particularly for the degree of medial hypertrophy.

Recently, Rabinovitch et al. [108] analysed the early vascular changes in congenital heart disease using quantitative methods and devised the following grading system: grade A is an abnormal extension of muscle into small arteries, grade B

an increase in the %MT of the arterial medial mucle and grade C a reduction in the number of small peripheral arteries, judged by a rise in the alveolar/arterial ratio. These three morphological grades were correlated with elevated pulmonary blood flows, high Ppa and high PVR respectively. This grading scheme can be assessed in uninjected lung biopsies, although in them the normal %MT in arteries of 50–100 μm diameter is about 2.5 times higher than in injected lungs, being 20% up to 4 months of age, and 10% thereafter. Nevertheless, this grading system has some problems since it is not applicable to all forms of PHT. Moreover, the finding of a decrease in the concentration of arteries (grade C) in human lungs of patients with congenital cardiac defects [137], in normal rats [97] and in two animal models of PHT [71] has been questioned. This decrease may be the result of functional closure of vessels rather than of true obliteration [63].

In summary, while the Heath and Edwards grading system remains valuable for the assessment of patients with congenital cardiac defects and with primary pulmonary hypertension, it has its limitations and should be supplemented, whenever possible, with quantitative data and with descriptions of the veins, of the microvasculature and of the rest of the parenchyma.

Pulmonary Hypertension in Congenital Cardiac Left to Right Shunts

Classification and Pathogenesis

Congenital cardiac left to right shunts frequently give rise to pulmonary hypertension and are usually divided [46] into (a) pre-tricuspid shunts, caused most frequently by atrial septal defects (septum secundum or sinus venosus), less commonly by anomalous pulmonary venous drainage into the right atrium or its tributary vessels, and (b) post-tricuspid shunts, which are usually caused by large ventricular septal defects, single ventricle, persistent truncus arteriosus, transposition of the great vessels or widely patent ductus arteriosus.

Important differences exist between these two types of shunts. Patients with pre-tricuspid shunts do not develop PHT until later in life since little shunting occurs across the two low-pressure chambers at birth. Therefore the postnatal development of the pulmonary arteries proceeds nearly normally and the lung vasculature can accommodate the flow from the shunt, which increases with time, at a low pressure.

In contrast, patients with significant post-tricuspid shunts develop PHT early in life since blood flows and frequently pressures are high right from birth. Indeed, the distension produced by the markedly increased flow, often four to six times normal [38], raises Ppa and constricts vessels whose walls are still thick. The mechanism of the vasoconstriction may be a myogenic reflex elicited by the vascular distension (the Bayliss effect). The importance of vasoconstriction has been emphasized by several authors, based on its partial reversibility with vasodilators in humans, on morphological assessment and on hyper-reactivity to vasoconstrictors in experimental models of systemic to pulmonary shunts [46,94,146,155]. The vasoconstriction further raises Ppa and sets up a vicious circle in which the morphological vascular changes also participate. Vasoconstriction due to the hypoxia of high altitude may aggravate the vascular changes in congenital cardiac shunts. Alt and Shikes [2] found six infants under 1 year of age with grade 4 lesions among 280 who died of congenital cardiac shunts at high altitude. The pulmonary vascular lesions may result from (a) failure of the foetal musculature to develop normally; (b) abnormal muscularization due to work hypertrophy and vasoconstriction; and (c) possibly the release of factors mitogenic for smooth-muscle cells [109]. While blood flow across the shunt undoubtedly contributes to the pathogenesis of the morphological lesions, the importance of the high Ppa cannot be minimized. Indeed, Wagenvoort et al. [153] showed, in both atrial and ventricular septal defects, that the severity of the anatomical lesions correlated with Ppa, and not with blood flow.

Morphology and Morphometry

The spectrum of vascular changes in congenital left-to-right cardiac shunts evident using light microscopy will be described for patients of all ages, followed by a summary of recent quantitative data from children under 2 years of age. The changes in the elastic pulmonary arteries are well described by Harris and Heath [46] and Wagenvoort and Wagenvoort[155]. The main pulmonary artery is dilated compared with the aorta, and shows initimal thickening and more pronounced atherosclerosis than normal, which may be complicated by aneurysm formation. When PHT is present from birth, the main pulmonary artery has the same thickness as the aorta and shows persistence of the foetal pattern, with regular and parallel elastic laminae interspersed between bundles of hypertrophied muscle. The media also contains more mucopolysaccharide ground substance. The pulmonary trunk of patients with pre-tricuspid shunts usually shows an adult pattern.

The muscular, partially muscular and nonmuscular pulmonary arteries show the spectrum of changes outlined above in the Heath and Edwards grading system. The initial and most frequent abnormalities are medial hypertrophy, to which both hyperplasia and hypertrophy of smooth-muscle cells contribute, and peripheral muscularization of small arteries. In addition, longitudinal muscle bundles may appear in the intima, media and adventitia, although this feature is more prominent in hypoxic PHT. Intimal changes are observed beginning in grade 2 of the Heath and Edwards classification with cellular proliferation, progressing in grade 3 to a more fibrotic lesion, both of which may occlude the arterial lumen; when severe, these assume a characteristic concentric laminar or onion skin appearance (Fig. 35.7). Smaller arteries are affected initially, and larger arteries are involved later when the PHT becomes severe. In grades 4 to 6, there are multifocal complex dilatation lesions, a term which includes plexiform lesions, vein-like branches of hypertrophied muscular arteries and angiomatoid lesions [46].

Plexiform lesions, the hallmark of plexogenic pulmonary arteriopathy, consist of a hypertrophied artery, often with concentric laminar intimal thickening, leading into a dilated segment which contains a plexus of thin-walled channels with areas of intraluminal thrombosis and a characteristic complex proliferation of intimal "fibrous tissue" and endothelial cells. Distally, there is frequently a zone of dilated vascular channels. Some overlap occurs between the appearance of the plexiform lesion, particularly in its earlier stages, and that of the other "dilatation" lesions. For example, the vein-like branches of hypertrophied arteries display only the wide vascular channels emerging from an artery with a markedly thickened wall; in certain instances, these may cluster together to form an "angiomatoid lesion". Plexiform lesions are believed to be located at the border between a high pressure vessel and alveolar wall arteries whose pressure is normal [48].

By electron microscopy, it can be shown that most of the cells in the intima of concentrically thickened arteries and in the proliferating part of plexiform lesions are not endothelial cells, but myofibroblasts which can contract and which can produce collagen, proteoglycans and elastin [52,131]. The myofibroblasts are often in a close relationship with fibrin and Wagenvoort and Wagenvoort [155] believe that plexiform lesions arise from the complex organization of in situ thromboses, which are secondary to a necrotizing arteritis (Fig. 35.8). The latter is therefore believed to be a precursor of the plexiform lesion rather than the last stage of the spectrum of plexogenic arteriopathy. The origin of plexiform lesions may not be entirely settled in view of the recent report by Shure et al. [129] who produced them in dogs by chronic ligation of the pulmonary artery; the lesions developed between 3 and 11 weeks after ligation and were related to the extent of bronchopulmonary anastomoses.

Although overshadowed by the arterial abnormalities, subtle alterations in venous morphology have been reported, particularly in patients with pre-tricuspid shunts having associated left heart disease [108,150,155]. These include medial hypertrophy, intimal fibrosis and increased muscularization of non-muscular veins. In a morphometric study, Samuelson et al. [126] found a correlation between the %MT of veins and pulmonary venous pressure. In pure post-tricuspid shunts, the veins are nearly always normal.

Much information can be derived by qualitative assessment of the lung vasculature, but more subtle alterations are detected with quantitative methods. This is particularly important since grades 4 to 6 of Heath and Edwards are distinctly unusual under the age of 2 years, when corrective surgery is performed today. Rabinovitch et al. [108] evaluated lung biopsies in 95 patients by quantitative histology: 47 had ventricular septal defects, 33 had transposition of the great vessels and 15 had defects of the atrioventricular canal. The results showed a good correlation between morphological grades A to C (described above) and physiological parameters: 21 patients had abnormal extension of muscle (grade A) and high pulmonary blood flows, but normal Ppa; 74 patients had increased medial muscle thickness (grade B), which correlated with an elevation of Ppa; 15 of the 95 patients also demonstrated a reduction in small peripheral arteries. In all but one of these, PVR was moderately to markedly elevated. This scheme has been further refined by dividing grade B into an early phase with mild medial hypertrophy, correlating with a mild increase in Ppa, and a late phase, in which severe medial

hypertrophy and decreased artery size correlated with a moderate to severe increase in Ppa and a small rise in PVR. Ultrastructural analysis [88] confirms the increased muscular thickness and muscularization of normally non-muscular arteries. The mechanism for this, which also occurs in hypoxic PHT and experimental crotalaria-induced PHT, is hypertrophy of pericytes and intermediate cells which then divide and differentiate into mature smooth-muscle cells.

Relationship Between Morphology, Haemodynamics and Prognosis

In general, haemodynamics correlate well with morphology. Significant variability, however, may exist between patients and be due to (a) the type and severity of cardiac shunt: patients with atrial septal defects behave differently from those with ventricular defects. PHT is particularly severe in patients with complete transposition; (b) the presence of modifying factors such as left heart disease, polycythaemia or hypoxia; (c) variability between individuals, partly related to genetic influences; and (d) variability in haemodynamic measurements obtained at cardiac catheterization, since these are frequently measured only once and may vary with sedation and other factors.

Despite these limitations, certain patterns valuable in predicting outcome have emerged. Earlier studies, reviewed in detail by Harris and Heath [46], show good correlation between the grade of hypertensive pulmonary vascular disease and Ppa, pulmonary blood flow and PVR. In particular, patients with grades 4 to 6 of Heath and Edwards have irreversible disease with high Ppa, low blood flow and high PVR. Surgical correction of the shunt is contraindicated, since it increases the load on an already overburdened right ventricle. In many of these patients, PVR is so high that the shunt is reversed (termed Eisenmenger's syndrome). At the other end of the scale, patients with grades 1 and 2 lesions are nearly always operable, showing a decrease in Ppa and PVR after surgery. A grey zone occurs in grade 3 lesions which requires closer scrutiny. It appears that as long as the intimal lesions are cellular, they are at least partially reversible and surgery is warranted. When dense fibrosis supervenes, however, reversibility is in serious doubt.

The good correlation between the morphometric grading of Rabinovitch et al. [108] and haemodynamics has already been described. To assess reversibility, this same group of investigators [109] examined the outcome of 74 patients after repair of congenital (predominantly post-tricuspid) cardiac defects taking into account morphometry (grades A to C), the Heath and Edwards grades and age. In 67 of these patients, haemodynamic measurements were repeated 1 year after corrective surgery: Ppa and PVR were normal in (a) all infants operated on prior to 9 months of age, (b) those with grade A or B (mild) regardless of the Heath and Edwards grade and (c) patients with grade B (severe) and grade 1 of Heath and Edwards. However, the Ppa and PVR were increased in half of the paients with grade B (severe) and Heath and Edwards grade 2, or with grade C, and increased in all patients over 2 years with grade C and/or grade 3 of Heath and Edwards. Similar findings were recently reported in a biopsy-based study by Wagenvoort [150] who concluded that medial hypertrophy and intimal lesions due to cellular proliferation, longitudinal smooth muscle in arteries, and venous lesions were reversible; in contrast, severe concentric laminar intimal fibrosis, plexiform lesions, fibrinoid necrosis of arterial walls and probably numerous dilatation lesions were considered irreversible.

In summary, reversibility can be predicted largely by knowing measured and calculated haemodynamic variables such as blood flow, Ppa and PVR, by a complete assessment of vascular morphology and morphometry with light microscopy and by taking into account the age of the patient. The latter factor has also been demonstrated to be important in experimental models of congenital cardiac shunts [118]. Additional functional predictions can be made by the response to oxygen and hypoxia, to vasodilators and to exercise.

Primary Plexogenic Pulmonary Arteriopathy

This term encompasses the clinicopathological entity of "classic primary pulmonary hypertension". Before a discussion of the entity itself however, mention should be made of semantic problems which arise because the term "primary pulmonary hypertension" has different meanings for the clinician and for the pathologist. For the clinician, primary or unexplained PHT includes the entities, apparently unrelated, listed in Table 35.4 [27,31,46,47,69,155]. While they may sometimes be distinguishable from each other by clinical criteria, it is frequently necessary

Table 35.4. Entities in "clinically unexplained pulmonary hypertension"

Primary plexogenic pulmonary arteriopathy
Chronic thromboembolic pulmonary hypertension
Pulmonary veno-occlusive disease
Pulmonary capillary haemangiomatosis

to resort to lung biopsy to make a specific diagnosis. An estimate of the relative frequency of the first three entities listed in Table 35.4 is obtained by combining data from the three largest series of clinically unexplained PRT [8,147a,154]: out of a total of 246 patients, about 60% had primary plexogenic pulmonary arteriopathy, 35% had chronic thromboembolism and 5% veno-occlusive disease. Capillary haemangiomatosis, the fourth entity, is extremely rare. The histological features used to distinguish these diseases are discussed below under their respective headings. From the viewpoint of the pathologist, plexiform lesions, which are the most characteristic feature of primary plexogenic pulmonary arteriopathy, are also found in other diseases (Table 35.5). In practice however, the forms of PHT with plexiform lesions are readily separated with clinical information.

Table 35.5. Plexogenic pulmonary arteriopathy: associated conditions

Congenital heart disease
Primary plexogenic pulmonary arteriopathy
Drug-induced and dietary pulmonary hypertension
Pulmonary hypertension in liver diseases
Pulmonary hypertension due to schistosomiasis

In this section, discussion will centre on primary plexogenic pulmonary arteriopathy (PPPA) (Fig. 35.4). In addition, related entities such as drug-induced and dietary PHT, and PHT in liver disease will be mentioned since they share morphological features with PPPA and may shed light on its pathogenesis. Also, persistent pulmonary hypertension of the newborn will be discussed briefly, with the understanding that several authors [34,35,54,102] classify it separately from PPPA.

Epidemiology, Clinical Features and Outcome

PPPA, apparently known since 1891 when it was described by Romberg (cited in [46]), is a rare disease causing less than 1% of all cases of pulmonary hypertension. Wagenvoort in 1970 [154]

described 110 patients with PPPA out of 156 with clinically unexplained PHT. This probably represents the single largest series reported. At that time, he found 602 cases in the literature and more recently stated there were about 1000 recorded cases [155]. While the age of onset may range from a few days after birth to the 6th decade, most patients are in their 20s or 30s. The female to male ratio is 4 or 5:1, except before adolescence when it is 1:1.

The clinical features of PPPA are well described elsewhere [31,38,46,145]. Briefly, most patients present with dyspnoea on exertion and fatigue. Syncope and chest pain may occur, while haemoptysis, cough, cyanosis and hoarseness are rarer features. Physical examination discloses findings compatible with pulmonary hypertension and right ventricular hypertrophy or failure, and this is confirmed by electrocardiography. The chest roentgenograph shows enlargement of the main pulmonary artery and its large tributaries, with peripheral oligaemia. The diagnosis is confirmed by cardiac catheterization: the NIH registry [161] accepts a mean Ppa greater than 25 mmHg at rest or 30 mmHg with exercise as being diagnostic. In most patients, mean Ppa varies between 50 and 100 mmHg and rises markedly with exercise. Furthermore, the cardiac index, for a given level of Ppa, tends to be much lower than in the chronic lung diseases [111], a feature which may explain the symptoms of PPPA and the substantial incidence of sudden death after catheterization, angiography and surgery [145].

Patients with PPPA do not fare well, with survival averaging 2–3 years, although considerable variability exists and some patients live for many years following diagnosis. While it has been said that the lower the Ppa the better the prognosis [31], the correlation between survival and Ppa is generally poor [145]. Rich and Levy [120] found that stroke volume and right atrial pressure were the best predictors of survival. Therapy of PPPA has been disappointing and vasodilators or heart and lung transplantation are the only modalities which offer any hope of prolonging life today [31,73,106]. Recently, Barst [5] found a close correlation between the pulmonary vasodilator response to prostacyclin and to nifedipine, and suggested that the acute response to the former predicts a chronic response to the latter. The use of dilators is based on the premise that vasoconstriction or a hyperreactive pulmonary vasculature are important in the pathogenesis of PPPA. The absence of a response, however, does not necessarily mean

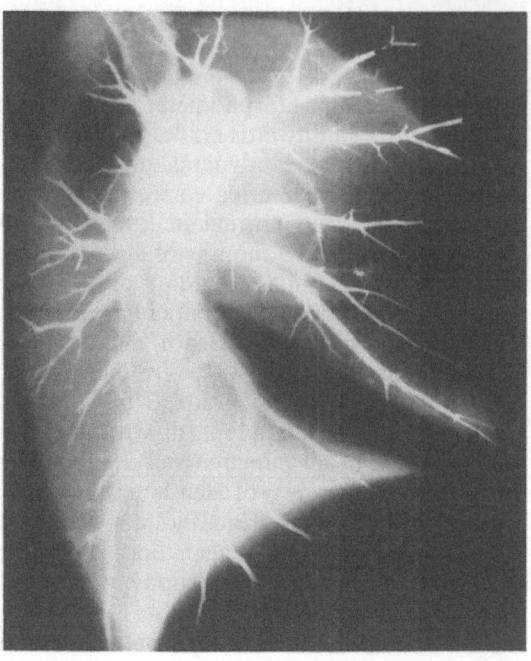

Fig. 35.4. Postmortem arteriogram of right lung from a 23-year-old patient who died with primary plexogenic pulmonary arteriopathy. Arteries were injected at a pressure of 100 mmHg with a coloured gelatin and barium mixture. Note only the large arteries are filled.

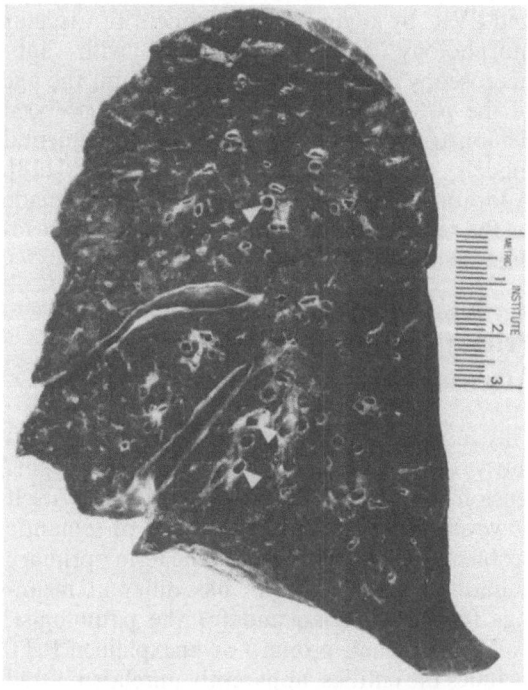

Fig. 35.5. Lung slice from 7-year-old female who died with primary pulmonary hypertension. Note prominent vessels on background of normal-appearing parenchyma.

Fig. 35.6. Large intraparenchymal elastic artery from patient with primary plexogenic pulmonary arteriopathy. Note mildly thickened media (*bottom*) and overlying intima with severe atherosclerosis. *M*, media; *I*, intima, *A*, atherosclerosis. H & E, × 50

vasoconstriction does not occur, since it can be explained by the irreversible morphological lesions.

Pathology

Macroscopically, the main pulmonary trunk and its immediate tributaries are dilated and thicker than normal and the peripheral vasculature is diminished (Fig. 35.5): this is particularly evident on angiograms done during life or post-mortem (Fig. 35.4). By light microscopy, the pattern of the elastic laminae can be seen to be, with rare

exceptions, of the adult type, consistent with the acquired nature of the disorder. Atherosclerosis is usually prominent (Fig. 35.6) and may be complicated by saccular aneurysms with occasional rupture.

The most striking findings are in the muscular arteries which show the entire spectrum of changes found in the lungs of children with congenital cardiac shunts. These include marked medial hypertrophy and peripheral muscularization of small arteries, intimal thickening, often of the concentric laminar form (Fig. 35.7), and frequent plexiform lesions (Fig. 35.9) and necrotizing arteritis (Fig. 35.8). In the two series of Wagenvoort [147a,154], plexiform lesions were present in 70% of cases, necrotizing arteritis in 30% of cases: in general, both are found in larger numbers than in association with congenital cardiac shunts. While some authors require plexiform lesions to diagnose PPPA [8], others accept the characteristic severe concentric laminar arterial thickening [69,147a,154] in the absence of significant lesions in the veins and in the rest of the lung parenchyma. The main differential diagnosis of PPPA, from the morphological viewpoint, is recurrent pulmonary thromboembolism. In the latter however, the intimal thickening is patchy and eccentric, not concentric, and the characteristic intravascular fibrous septae are seen. In addition, there is much less medial hypertrophy, and plexiform lesions and necrotizing arteritis are absent. On occasion, thrombosis may be superimposed on typical

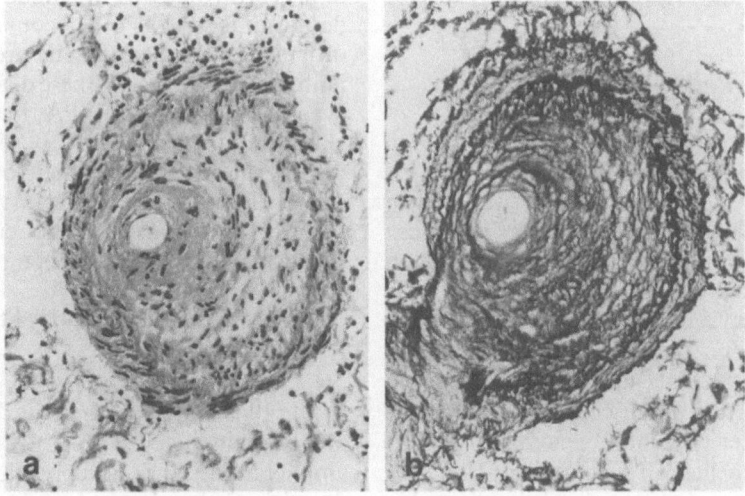

Fig. 35.7. Muscular artery from patient with PPPA. Most of the thickening is due to intimal proliferation of myofibroblasts. H & E, × 300. **b** Same artery, stained with elastic van Gieson, with prominent concentric laminar (onion-skin) intimal thickening.

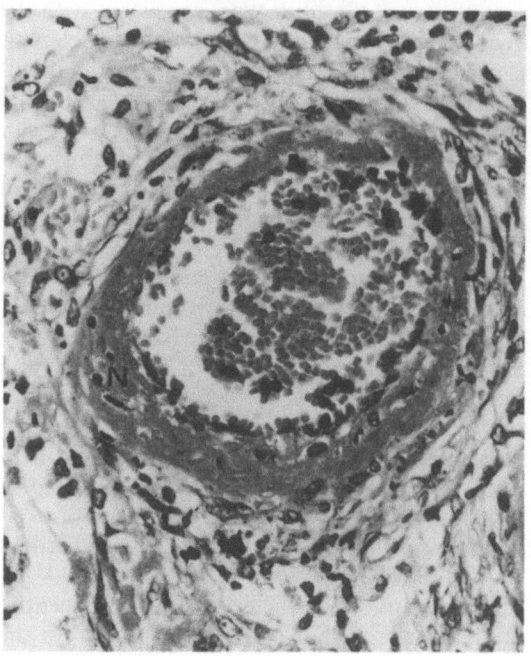

Fig. 35.8. Necrotizing arteritis involving wall of muscular artery (grade 6 of Heath and Edwards) in PPPA. H & E, ×450

plexiform lesions and give rise to confusion in very small biopsies. In general, however, most authors agree that the two entities are readily separated by histology. In a quantitative histological study, Anderson et al. [3] reported a reduction of small arteries associated with ghost arteries in PPPA, features absent in early thromboembolic PHT.

Aetiology and Pathogenesis

Although the cause of PPPA is unknown, several theories have been proposed [46,62,73,145,155, 161]. An early theory, that PPPA was due to chronic silent thromboembolism, has now been discarded on morphological grounds, although it is still possible that the coagulation cascade and platelets play a role in the in situ thrombosis which occurs in the plexiform lesions. In addition, some patients may exhibit a reduced fibrinolytic activity which could be primary or "... be a marker of a more generalized abnormality in endothelial cell function" [161].

Vasoconstriction appears clearly implicated in the pathogenesis of PPPA: indeed, several patients also have Raynaud's phenomenon, migraine or angina, believed to be vasospastic in

origin, and they may respond to vasodilators. In fact, the term "vasoconstrictive PHT" has been used to describe PPPA. However, vasoconstriction occurs under other circumstances (including PHT in congenital cardiac defects, hypoxia, etc.) and in any case, the cause of the vasoconstriction remains to be determined. Endothelial cells may be implicated since they show pronounced swelling [83] and perhaps, in these patients, are unable to extract or metabolize vasoconstrictors, or produce fewer endogenous vasodilators such as prostacyclin. Furthermore, platelets adhering to damaged endothelium may elaborate growth factors which contribute to the medial smooth-muscle and intimal cellular proliferation [73,145,161].

Genetic factors have been advanced in the aetiology of PPPA since there are well-documented familial occurrences, some associated with Raynaud's phenomenon. The mode of inheritance, however, is controversial. Some experimental support for genetic involvement exists since pulmonary hypertension is genetically transmissible in cattle. The mechanism of the genetic contribution to pulmonary vascular hyper-creativity remains unknown.

Similarly, the female preponderance and the association with pregnancy need to be explained, particularly since pregnancy normally results in a reduction in Ppa. Cows that have congenitally reactive pulmonary vessels develop PHT during pregnancy and the same phenomenon may be occurring in PPPA [161].

Autoimmune factors have been proposed because of the association of PPPA with systemic lupus erythematosus, Raynaud's disease, scleroderma and several other connective tissue disorders. The frequent female preponderance and familial occurrences of these disorders links them even more closely to PPPA, but fails to provide clues about mechanisms. It seems that in this disorder, as in several other forms of PHT, the endothelium may hold the answer.

Let us now consider other entities, related to PPPA, which may shed light on its pathogenesis.

Drug-Induced and Dietary Pulmonary Hypertension

In late 1965 and 1966, the drug Aminorex fumarate, an anorexigen with a structure resembling amphetamine and norepinephrine, was introduced in Switzerland, Austria and the Federal Republic of Germany. In 1968 Gurtner et al. [40,41] reported the association of pulmonary hypertension with Aminorex and the drug was

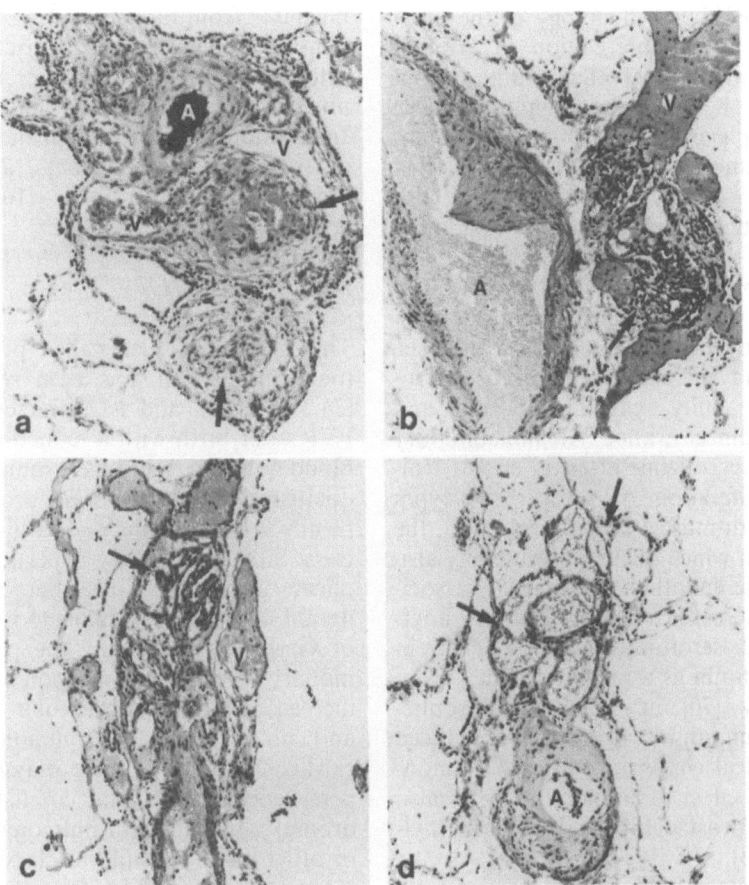

Fig. 35.9 a–d. Plexiform and dilatation lesions from a patient with PPPA. **a** Plexiform lesion after injection of pigmented barium–gelatin mixture (cf. Fig. 35.4), present only in proximal artery (*A*) which shows medial and intimal hyperplasia. Distally, plexiform proliferations (*arrows*) and dilated vein-like branches (*V*) are seen. H & E, × 180. **b** Plexiform lesion with parent artery (*A*) leading to the complex plexus (*arrow*) and dilated branches (*V*). H & E, × 140. **c** Less complex plexiform lesion with angiomatoid features (*arrow*) and vein-like branches (*V*). H & E, × 150. **d** Dilated angiomatoid lesion (*arrows*) distal to artery (*A*) with thick intima. H & E, × 140

withdrawn. Between then and 1970, about 600 cases of PPPA occurred, representing a 20-fold increase in incidence; mortality rates varied between 12% and 20% after 2 years. Although only 1%–2% of individuals who took Aminorex developed PPPA, the risk rose by a factor of 50, and correlated with the amount ingested. The amount injested did not, however, correlate with haemodynamic measurements of Ppa or PVR in affected individuals, nor with the severity of lesions observed by light microscopy. These, studied in detail by Widgren [163] in 37 patients, were in all respects identical to those observed in PPPA unrelated to Aminorex; plexiform lesions were observed in 13 cases and arteritis in three.

Duplication of the chronic pulmonary hypertension with its characteristic morphological lesions has been attempted in several species of

animals but has not been successful, although acute doses of Aminorex did raise Ppa and PVR under some conditions. Nevertheless, the epidemiological data clearly point to a cause–effect relationship. The low incidence and differences in species susceptibility emphasize the variability of pulmonary vasoconstriction and suggest other factors (e.g. genetic) are involved. As in PPPA, the role of the endothelium has been invoked [40].

The experimental model that resembles PPPA and drug-induced PHT the most (albeit imperfectly) uses rats fed or injected with the alkaloid monocrotaline, the agent present in *Crotalaria spectabilis* seeds [46,89,136,155]. Like Aminorex, its effect is species-specific and there is a delay of several weeks between its administration and the haemodynamic and morphological effects [68].

In established PHT, the histology of the lungs shows peripheral muscularization of arteries, and medial and initimal thickening; venous lesions, although less prominent, are also present, while plexiform and dilatation lesions are absent. Haemodynamic studies indicate that not only the morphological changes but also vasoconstriction are delayed, since Ppa takes about 2 weeks to rise [84]. Observation of the events which precede these full-blown abnormalities suggests that endothelial damage, possibly mediated by oxygen radicals or platelets, occurs first and increases pulmonary microvascular permeability, causing oedema and reduced serotonin clearance. An inflammatory reaction with necrotizing arteritis ensues, followed by proliferation of several cell types including smooth-muscle cells. In addition, the vasoconstriction which occurs, presumably also secondary to the endothelial damage, is associated with an increased responsiveness to angiotensin II and serotonin [55]. Evidence in support of serotonin as a mediator comes from a recent study [70] in which p-chlorophenylalanine, which inhibits its synthesis, reduced the morphological changes produced by monocrotaline. The effect of monocrotaline is also inhibited by prostacylin and by methylprednisolone [63].

A form of dietary PHT in man was recently described in Spain after ingestion of a toxic cooking oil [29]. Its main effect is diffuse lung damage, and PHT develops in a small percentage of cases. The female preponderance, associated Raynaud's phenomenon, eosinophilia, elevated IgE levels and thrombocytopenia suggest a complex interplay of factors is involved in the genesis of this most unusual form of PHT.

Pulmonary Hypertension and Liver Disease

Since 1951, when the co-existence of portal and pulmonary hypertension was first resported, controversy has existed as to whether this association is coincidental or real, and possible mechanisms have been discussed [79]. It appears that 1% or less of patients with severe liver disease, usually cirrhosis or portal hypertension, may develop PHT with plexiform lesions [46,155]. Mechanisms which were proposed initially to account for this form of PPPA include increased pulmonary blood flow, attributed to the portal–pulmonary venous anastomoses, and thromboembolism. The first is indeed possible, but the second unlikely in view of the morphology. The preferred hypothesis today is that an unknown

mediator from the gut bypasses metabolic inactivation in the liver and constricts the pulmonary arteries. Fishman [30] has described this theory under the heading of the "gut–liver–lung axis". In addition, autoimmune phenomena have been described in these patients, as in PPPA not associated with liver disease [100].

Persistent Pulmonary Hypertension of the Newborn (PPHN)

This syndrome, also called persistence of the foetal circulation, has been reviewed recently [34,35,54,102], and is characterized by a high PVR after birth with a reduction in pulmonary blood flow and right to left shunting through the ductus arteriosus or foramen ovale. The shunt results in hypoxaemia and cardiorespiratory distress. Since pulmonary hypertension may complicate a variety of disorders in the newborn period, it seems reasonable to follow the dictum of Gersony [35]: "Only the patient with pulmonary vascular constriction (with or without increased pulmonary vascular smooth muscle) and no parenchymal pulmonary disease or cardiac lesion should be diagnosed as having persistence of the fetal circulation (PFC syndrome)". Infants with pulmonary parenchymal or other disease should be classified as having a secondary form of PHT of the newborn. The relationship between PPHN and PPPA remains unclear [48], although most authors subscribe to the notion that while there may be a small yet undefined overlap, they represent separate entities. By light microscopy, lungs of patients with "idiopathic" PPHN show prominent peripheral muscularization of acinar arteries, a feature normally absent at birth [34,102]. The media of preacinar vessels is usually normal or slightly thickened; occasionally marked hypertrophy is seen. In addition, there is a pronounced proliferation of adventitial tissue. The significance of this finding, which has also been observed in experimental animals with chronic hypoxic PHT, is unclear [57,86,134].

The combination of the morphological changes and their timing suggests that the pulmonary vascular bed has been remodelled in utero. At least three explanations for the changes have been proposed. The first states that intrauterine hypoxaemia coupled with a hyper-reactivevasculature or a decreased availability of dilators is responsible. The second stems from the observation that in occasional cases, maternal administration of an inhibitor of prostaglandin synthesis such as indomethacin or aspirin constricts the ductus arteriosus in utero and results in PHT and increased medial muscle thickness. The idea therefore is that cyclo-oxygenase inhibition could either reduce the synthesis of dilator prostanoids or increase the synthesis of leukotrienes by shunting arachidonate metabolism to the lipoxygenase pathway [112]. In support of this notion, Meyrick et al. [85] recently produced a rise in Ppa and PVR in sheep after repeated administration of indomethacin, with morphological evidence of vasoconstriction and of increased sequestration of granulocytes at the periphery of the lung. A third explanation for PPHN may be meconium aspiration. Indeed, recently, 11 infants who died with meconium aspiration were studied and all but one showed changes by light microscopy identical to those of PPHN [34]. Whether common aetiological and pathogenetic mechanisms link these two entities is unknown.

Pulmonary Hypertension Due to Thromboembolic Disease

Since the advent of the autopsy many centuries ago, it has been recognized that thrombi may be found in the pulmonary arteries. Despite significant advances in our understanding of pulmonary thromboembolism (PTE) in the late 19th and the 20th centuries, this cause of PHT still eludes clinical diagnosis today. In this section, pulmonary embolism secondary to thromboemboli only will be discussed, since the other forms due to air, fat, tumour, amniotic fluid, schistosomiasis and foreign body emboli are less frequent and are reviewed elsewhere [36,132]. Emphasis will be placed on pathophysiology, pathology and outcome, while diagnosis and therapy will not be covered in detail. Sharma et al. [128] provide a useful list of key references.

The incidence of PTE at autopsy ranges from under 5% to over 65%. The reasons for this wide range include: (a) the diligence applied to the search; (b) the minimum size of vessel considered embolized before the diagnosis is made; and (c) the nature of the population studied. Laissue et al. [72] recently summarized the data of over 30 publications, with a total of nearly 370 000 patients: the incidence of PTE ranged from 0.8% to 69%, with a mean of 20%; of these, one-third suffered massive pulmonary embolism (variably defined). A reasonable figure for the postmortem incidence of PTE probably lies between 25% and 50%. Using postmortem and angiographic data, Dalen and Alpert [19] estimated that PTE

accounts for 200 000 deaths in the USA each year. In half the cases PTE is the sole cause of death, in the other half it is a major contributing factor, making it the third most frequent cause of death. These authors also estimated the incidence of symptomatic PTE at 630 000 per annum, of which 11% die within 1 hour. Of the 89% who live beyond 1 hour, the diagnosis is made in 29% of which 92% survive with therapy. The diagnosis therefore is missed in 71% and about one-third of these die, usually of recurrent embolism.

Pulmonary hypertension is believed to occur in about 80% of patients with PTE [82], and the main mechanism responsible (vide infra) is occlusion of the main pulmonary artery or its branches by emboli. These arise from thrombi usually in the large veins of the thigh, leg and pelvis, less commonly in the right side of the heart or superior vena cava. Patients at risk have heart disease, malignancy or trauma, or are recovering from surgery particularly on the extremities, back and abdomen. In these, the delicate balance between the coagulation system, platelets and the fibrinolytic system may be altered [36].

Pathology

Most of the descriptions of PTE come from autopsy studies, or less frequently from angiographical studies. Pulmonary thromboembolism may be divided into acute and chronic forms although the distinction between them is not always sharp, since up to 80% of recent pulmonary emboli are accompanied by organized thromboemboli [155]. The most massive form of acute pulmonary embolism, the "saddle" embolus, occludes the trunk of the main pulmonary artery at its bifurcation, and usually lies loosely within the pulmonary artery with minimal attachment to its wall. Other forms of "massive" PTE may occlude the left or right main pulmonary arteries, lobar arteries or even smaller vessels. Definitions of massive and submassive PTE, however, vary. According to the urokinase pulmonary embolism trial [144], massive pulmonary embolism is defined as filling defects in two or more lobar arteries, or an equivalent amount in smaller vessels, so that at least 40% of the vasculature is occluded. Submassive PTE is defined as a filling defect in at least one segmental artery, with the sum of defects being less than for massive embolism. Based on angiographical data, 30%–60% of pulmonary thromboemboli are massive [36,144]. In a recent autopsy study, Bergqvist and Lindblad [6] described 1274 surgical patients with PTE: the emboli were massive and fatal in 27%, contributed to death (occluding at least one lobar artery) in 28% and were an incidental finding in 45%.

Pulmonary emboli preferentially lodge in the lower lobes and in the right lung; this distribution, probably related to blood flow patterns [36,164], was confirmed in the urokinase pulmonary embolism trial [144], although upper lobe involvement was still surprisingly frequent. Emboli are commonly multiple and of different ages when seen at autopsy, which confirms their recurrent nature. At autopsy, fresh thrombi are characteristically laminated and cause the vessel to bulge from the surface of the lung.

When the time between embolism and death increases, the emboli may either lyse by fibrinolysis, migrate or extend distally or become firmly adherent to the vessel wall and organized. The sequence of the organization is difficult to ascertain in man due to the recurrent nature and multiplicity of PTE. Sequence has, however, been determined for some experimental animals [164]. Macroscopic examination of 10- to 14-day-old emboli in dogs at varying times after embolization, in the absence of any therapy, reveals the following sequence. By day 7, no change is seen; however by day 14, fine strands anchor the embolus to the artery. At 6 weeks, the embolus is reduced to delicate threads and has almost completely disappeared by 10 weeks. Light microscopy at 2 days shows a denuded arterial endothelium and neutrophils at the periphery of the thrombus. The endothelium begins to cover the thrombus by 7 days, a process which is complete by 14 days. Macrophages (often haemosiderin-laden) and capillaries invade the embolus. By 48 days, all embolic fragments are vascularized.

This slow process of organization apparently contradicts other data obtained from dogs by Dalen and co-workers [19] who found, by angiography, significant relief of obstruction within hours of embolization with radio-opaque autologous clots also 10–14 d old. This was attributed however to distal migration of the clots since their size had not changed. The effectiveness of in vivo fibrinolysis in removing emboli appears to depend on the age of the thrombus: Moser et al. [101] showed that the volume of freshly formed thrombi (30 min) was already reduced by 50% 3 h after embolization.

In man, the urokinase pulmonary embolism trial [144] and other studies [19] indicate that minimal resolution on angiogram occurs 1 d after embolism, while complete resolution, which on occasion occurs as early as 14 d, usually takes several weeks. The late vestiges of acute emboli resemble those found in chronic embolism (vide infra) and include eccentric intimal thickening of elastic arteries, and bands and webs composed of dense fibrous tissue which span the arterial lumen of elastic and smaller arteries.

Chronic recurrent thromboembolic PHT may occur either in patients with known PTE, or may present insidiously as clinical primary PHT. Indeed, Wagenvoort and Wagenvoort [155] and Edwards and Edwards [27] exclude from "clinically unexplained PHT" patients with known emboli or in situ thrombosis.

The pathology of chronic recurrent PTE has been well described [46,69,82,149,155] and differs from that of PPPA and veno-occlusive disease (VOD). The hallmark is the presence of obstructing emboli of various ages and in different stages of organization and recanalization, located predominantly in muscular arteries although larger elastic arteries are also affected (Fig. 35.10). Well-healed lesions show eccentric intimal fibrosis and webs which may divide the arterial lumen into a colander-like arrangement (Fig. 35.10). Non-obstructed arteries show only mild medial hypertrophy, and there is no significant peripheral muscularization [3,155]. The concentric laminar intimal thickening and plexiform lesions of PPPA, and the venous lesions of VOD are absent. As stated by Wagenvoort [154], "as long as one has sufficient material at one's disposal, cases of chronic thromboembolism or thrombosis are clearly identifiable as such by histologic study". The pulmonary trunk and main arteries in chronic embolic PHT show some medial hypertrophy and atherosclerotic plaques, and their elastic laminae show an adult pattern, as expected in this acquired condition.

Pathogenesis of Pulmonary Hypertension in Thromboembolism

A modern account of the mechanisms of the hypertension which accompanies emboli is provided by Messer [82]. Most reports agree that a decrease in cross-sectional area by the emboli accounts for most of the increased PVR in thromboembolic PHT. While it is possible to acutely occlude, for example with a balloon, up to 50% of the lung vasculature without a pronounced elevation in Ppa, a smaller degree of embolic occlusion may raise Ppa significantly. Indeed, in their classic study of 20 patients who

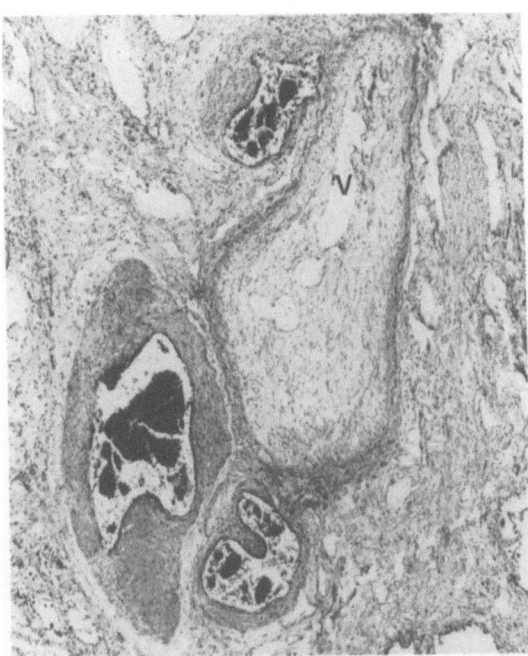

Fig. 35.10. Venous lesions from an 18-year-old male with veno-occlusive disease. Low-power view of eccentrically thickened veins in interlobular septum surrounding another vein (*V*) nearly completely obstructed by loose fibrous tissue. H & E, × 45

were *free* of pre-existing cardiopulmonary disease, McIntyre and Sarahara [81] found (a) mean Ppa consistently increased when the obstruction, demonstrated by angiography, exceeded 30%; (b) a good correlation between the extent of obstruction and mean Ppa; and (c) that the highest Ppa was 40 mmHg, suggesting that this is the maximum load which can be sustained by a previously normal right ventricle, above which mortality is high. The discrepancies between this study and other data which relate extent of obstruction to elevation of Ppa may be explained by additional emboli not visible by angiography, and by vasoconstriction (vide infra).

In a companion study, McIntyre and Sasahara [81] studied the response to PTE in patients *with* pre-existing cardiopulmonary disease. Mean percentage obstruction in this group was lower (23%) than in the group free of prior disease, and no patient had more than 50% obstruction. Mean Ppa however was higher (40 mmHg), and unrelated to the degree of obstruction. In this group of patients, the hypertrophied right ventricle could therefore sustain a much greater load. The importance of pre-existing cardio-

pulmonary disease was also emphasized by the findings of the urokinase pulmonary embolism trial [144]: indeed, patients with right ventricular hypertrophy could generate higher Ppa, but developed circulatory failure out of proportion to the extent of embolization.

Whether mechanical obstruction can account for all of the increase in PVR with embolism is controversial. Superimposed vasoconstriction by such agents as serotonin, histamine, catecholamines, prostaglandins, thromboxanes (released from platelets) or by bradykinin and vasoactive fibrinopeptides (from breakdown of fibrin) has been suggested [36,82,155]. In addition, there may be a component of hypoxic vasoconstriction or of activation of neurogenic reflexes by acute distension of the main pulmonary artery due to the acute rise in Ppa [39,65]. Evidence in support of vasoactive mechanisms in PTE includes some reversal of PHT with dilators, the transient nature of the vasoconstriction or bronchoconstriction and the discrepancies between the percentage estimated obstruction and symptoms or degree of rise of Ppa. However, while the release of several mediators certainly occurs in pulmonary microembolism [74], there is doubt that it plays a role in more than the occasional patient with macroembolism. Presumably however, some patients may throw off microemboli from larger emboli which could constrict the airways and vessels.

Effects and Pathophysiology of Pulmonary Thromboembolism

Pulmonary emboli can have several consequences [21,82,121,164]. In some patients they may be asymptomatic. Other patients will develop acute dyspnoea with hypoxaemia. The mechanisms for this are still unclear and may be multifactorial: Manier et al. [76] attributed the hypoxaemia to a fall in mixed venous PO_2, although shunting, ventilation–perfusion abnormalities and oxygen diffusion impairment also contributed. A third group of patients will develop haemodynamic abnormalities, in the form of acute or chronic cor pulmonale. Acute cor pulmonale often follows massive PTE, with systemic hypotension occurring in about one-quarter of cases [121]. A fourth consequence is pulmonary infarction, manifested by pleuritic chest pain, a friction rub and haemoptysis, which follows 10%–50% of PTE [20,143,164]. Infarcts occur mostly in the lower lobes and by light microscopy show a spectrum of changes ranging from haemorrhage to frank necrosis [45,143]. Finally, chronic cor pulmonale characterized by right ventricular hypertrophy, with or without dilatation and failure, may occur, usually after recurrent episodes of PTE. In a recent study, abnormal oxygenation was used to predict the presence or subsequent development of PHT in patients with chronic dyspnoea after PTE [75].

Pulmonary Veno-Occlusive Disease and Capillary Haemangiomatosis

Pulmonary veno-occlusive disease (VOD) is a rare entity of unknown aetiology which accounts for about 5% of cases of clinically unexplained PHT [27]. It tends to occur in children and young adults but has been reported at both age extremes [99,135]. There is no sex predilection. The onset is frequently insidious and many patients have a history of a short preceding febrile illness. The diagnosis may be suggested by the characteristic clinical and haemodynamic triad of PHT, pulmonary oedema and congestion with Kerley B lines on chest roentgenograph, and normal Ppw. Frequently however, lung biopsy must be performed to make the diagnosis which not uncommonly is made only at autopsy.

Morphological findings by light microscopy are characterized by an extensive, although sometimes focal, narrowing or obliteration of the small, medium and large pulmonary veins and sometimes of bronchial veins (Fig. 35.11). The obliteration results from pronounced intimal proliferation, which is nearly always paucicellular and oedematous. On occasion, in older lesions, the intima may form eccentric nodules of dense collagen and elastic tissue, or give a colander-like appearance to the veins [69, 139,155]. Moragas et al. [99] reported one case of VOD in a newborn, and described extensive eccentric myxoid thickening of small veins with dense fibrosis in some areas. In the large veins, the lesions were more sparse and always myxoid. The authors suggested that the disease started in the smallest veins and secondarily involved the larger ones. Medial thickening is not a prominent finding [155], and is probably secondary to the elevated vascular pressures. Sometimes, an important phlebitis is seen (Fig. 35.11d) [9,13,147].

Since microvascular pressures are higher than normal, patchy parenchymal changes develop

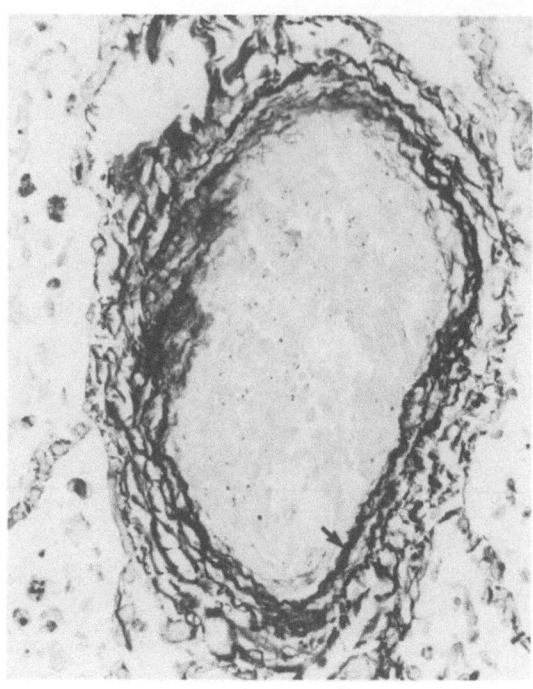

Fig. 35.11. Vein from a 54-year-old patient with pulmonary hypertension secondary to left ventricular failure due to mitral stenosis and regurgitation. Note thickened media with "arterialization". H & E, × 290. **b** Same vein showing elastic laminae (*arrows*) resembling those of an artery. Elastic van Gieson, × 290

due to capillary congestion and lung oedema: the interlobular septa and the connective tissue surrounding bronchovascular bundles are widened by oedema fluid and distended lymphatics. The alveolar spaces contain haemosiderin-laden macrophages, and type II alveolar epithelial cells proliferate in their walls. Interstitial pneumonitis with lymphocytes, plasma cells and macrophages, and fibrosis have been described [155]. VOD also results in arterial changes, usually thought to be secondary to the venous changes. Marked arterial thickening and plexiform lesions are never seen. In a recent review, however, Wagenvoort et al. [159] found significant arterial narrowing or obliteration in about half of 26 patients, with recent arterial and venous thrombi in seven patients. The same group also reported one case with both arterial and venous lesions, which they termed pulmonary vascular occlusive disease [105]. Davies and Reid [23], in a quantitative study of VOD in siblings, described medial hypertrophy of pre-acinar and acinar arteries, extension of smooth muscle into pre-capillary vessels and a reduction in the number of veins.

Kay et al. [67] examined one case with electron microscopy and described, in the occluded veins, an intact endothelium, a haphazard proliferation of smooth-muscle cells, and collagen fibrils. They also found electron-dense deposits in the basement membrane which they attributed to disintegrating erythrocytes rather than immune complexes.

It is widely believed, based on the morphology which strongly suggests a thrombotic process with organization and recanalization, that thrombosis is the primary event, even in young infants [23,46,69,155]. Descriptions of acute thrombi in VOD, however, are very rare except in the report of Wagenvoort et al. [159]. Moragas et al. [99] and Stoler et al. [135], who could find no thrombi, only intimal proliferation, suggested that endothelial injury was the main event. These two theories are certainly not incompatible since thrombosis frequently follows endothelial damage. The mechanisms of endothelial damage are unclear however and no definite aetiology for the disease has emerged.

Of the possible aetiologies, environmental and toxic factors are probably not important and it does not appear that coagulation abnormalities can account for the thrombosis [147]. Furthermore there is no indication that herbal medicines such as bush tea, which may produce veno-occlusive disease of the liver in humans, cause pulmonary VOD [159]. Familial and genetic factors are involved, but the mechanisms remain unclear. Greater attention has been paid to infection as a possible aetiological factor in VOD. Viral infections, in particular, have been implicated by the pyrexial flu-like syndrome which may precede the onset of the disease [147]. McDonnell et al. [78] described a 47-year-old male with VOD who displayed strong morphological evidence of a viral infection. Toxoplasmosis has also been suggested but appears much less likely. An immune aetiology was suggested by Corrin et al. [16] who found immune complexes in one case; no others have been reported however and Kay et al. [67] could not demonstrate any in their patient. Recent publications have described VOD after hypertrophic cardiomyopathies, the sudden infant death syndrome, renal transplantation and chemotherapy for malignant disease [12,69]. These multiple associations emphasize the heterogeneous nature of VOD, and the notion that it is a syndrome rather than a single entity must seriously be entertained [147].

In the differential diagnosis, common lesions such as left heart disease (e.g. due to mitral stenosis), sclerosing mediastinitis, occlusion or stenosis of the large extrapulmonary veins, congenital venous anomalies and the other causes of clinical primary PHT must be ruled out. While the morphological changes on biopsy are characteristic, a normal Ppw may be helpful in making a *clinical* diagnosis of VOD [67,110]. This normal Ppw has been explained [13,46,110] by the predominant stenosis of small veins: as long as there is a static column of blood from the vessel in which the catheter is wedged to a point distal to the stenosis, a normal pressure will be measured. If much larger veins are occluded however, Ppw may be elevated. Sometimes when the Swan–Ganz catheter is wedged, a normal Ppw is not recorded immediately; when it is flushed with saline, however, the Ppw rises disproportionately then falls very slowly to a normal level. This may be explained by the trapping or slow run-off of saline between the catheter tip and the partially obstructed veins downstream. Another factor which determines the measured Ppw is the size of catheter used for the measurement and the size of the vessel: indeed, if a smaller vessel is occluded, the measured Ppw will be higher since venous resistance accounts for a significant portion of total PVR [42,93,167].

Therapy of VOD has been disappointing and although steroids, azathioprine, anti-inflammatory, anti-platelet and anti-coagulant drugs have been advocated, the prognosis remains uniformly dismal [14,110].

Pulmonary Capillary Haemangiomatosis

First described in 1978 by Wagenvoort and his colleagues [151], this entity must be added to the list of diagnoses to be considered in clinical primary PHT. There are apparently only four cases recorded in the world literature [69]. It is characterized by an aggressive proliferation of vessels, mainly capillary, sinusoidal or venous, which infiltrate many areas of the parenchyma, particularly walls of vessels and airways, and the pleura. The infiltration of the veins occludes their lumina and produces oedema with bleeding, interstitial fibrosis and haemosiderin, which lead to the clinical manifestations of lung haemorrhage and progressive respiratory insufficiency. The nature of the lesion is unknown: it may be a neoplasm or a hamartoma. The prominent veno-occlusion in certain cases of capillary haemangiomatosis suggests some overlap with VOD.

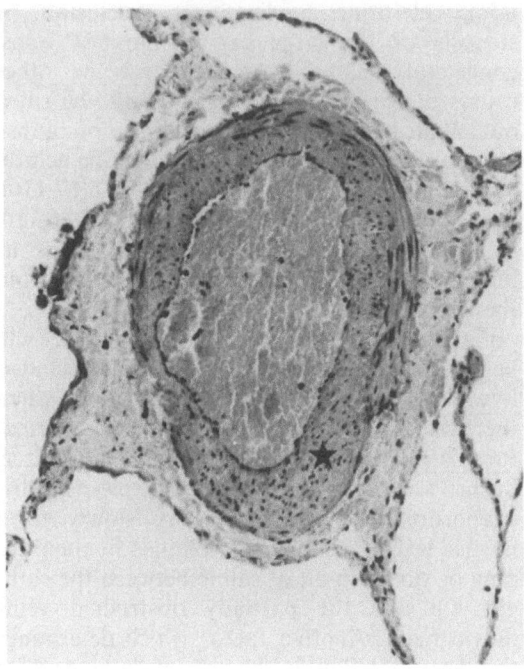

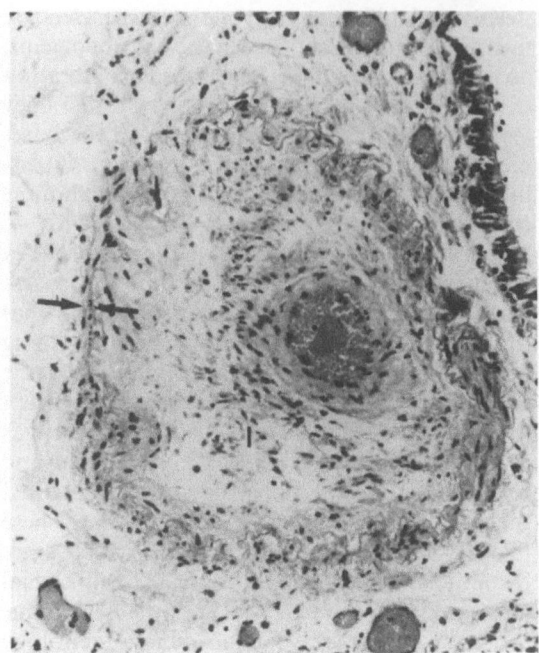

Fig. 35.12. Cross-section of muscular artery from patient with severe chronic obstructive lung disease showing characteristic longitudinal smooth muscle in intima. H & E, × 200

Fig. 35.13. Muscular pulmonary artery in a patient with idiopathic lung fibrosis. Note attenuated media with close apposition of internal and external elastic laminae and marked intimal thickening with sparse longitudinal smooth-muscle cells. H & E, × 180

Pulmonary Hypertension Secondary to Pulmonary Venous Hypertension

Most cases of PHT are secondary to acquired disease in the left heart, including left ventricular failure due to marked systemic hypertension, myocardial infarction and aortic or mitral valvular disease. Unusual causes include stenosis of the large pulmonary veins, congenital cardiac lesions such as cor triatriatum and mediastinal fibrosis. Mitral stenosis, with or without regurgitation, is the best studied for several reasons [25]: (a) the haemodynamics are well documented since these patients are operable; (b) biopsies are readily obtained at surgery; and (c) the disease is slowly progressive leaving ample time for lesions to develop. The morphology of the lung in PHT secondary to pulmonary venous hypertension is distinctive, and the veins, arteries and parenchyma are affected. The veins show medial hypertrophy due to increased numbers and size of smooth-muscle cells and to fibrosis and elastosis. When medial hypertrophy is associated with condensation of elastic fibres into distinct internal and external elastic laminae, the process of "arterialization" whereby the

veins come to resemble arteries, is said to be occurring (Fig. 35.12). Venous intimal fibrosis may also be prominent, although it must be separated from age-related changes (Fig. 35.13).

The abnormalities in the arteries resemble those in several other forms of PHT [157]. The elastic arteries show medial thickening with fibrosis and increased mucopolysaccharides, intimal fibrosis and atherosclerosis. Muscular arteries show mild medial hypertrophy. Peripheral muscularization of non-muscular arteries, medial and intimal fibrosis and longitudinal smooth-muscle fibres are additional features. The intimal fibrosis is characterized by eccentric patches or cushion-like thickenings which differ from the active cellular proliferation and onion skin arrangement seen in congenital cardiac shunts, and from the recanalization with septa of thromboembolic PHT. Rarely, there is a "fibrinoid vasculosis" in which fibrin is deposited in the vessel wall.

Changes in the pulmonary parenchyma, related to recurrent episodes of congestion, pulmonary oedema and microhaemorrhage, include prominent dilated lymphatics, large collections of haemosiderin-laden macrophages within alveolar spaces and sometimes encrustation of vascular walls with iron. Interstitial fibrosis, frequently mild, may in some instances be prominent and present a differential diagnostic problem with primary lung diseases such as pulmonary fibrosis. The characteristic firm, brown appearance of the lungs in longstanding mitral stenosis due to the fibrosis and haemosiderosis is termed brown induration. Other lesions include microlithiasis and foci of ossification, which may result from organization of a fibrinous exudate.

Two interesting aspects of the haemodynamics of pulmonary venous hypertension deserve comment. The first is the correlation between the vascular lesions and haemodynamics. In general, it is not good, compared for example with PHT due to congenital cardiac shunts. If the lung vessels behaved as an entirely passive system, elevation of the left atrial or pulmonary venous pressure should produce an equal rise in Ppa. Instead, particularly in patients with longstanding mitral stenosis, Ppa rises disproportionately to Ppw, and PVR also increases. This usually occurs when the Ppw exceeds 25 mmHg, but may be seen with lower values [25,46]. The reasons for this include (a) a decrease in pulmonary blood flow, which raises PVR; (b) the morphological changes in the arteries and veins; and (c) vasoconstriction, since Ppa generally falls rapidly after mitral valve surgery. It is likely that all three mechanisms act in concert to raise Ppa and PVR.

While it is generally believed the morphological lesions do not contribute to the outcome of patients after mitral valve surgery, a recent study by Tryka et al. [142] suggests this may not always be so. They found that three out of 16 patients with persistently elevated Ppa who died after mitral valve surgery had vascular disease severe enough to account for their death.

The second interesting haemodynamic feature is the change in blood flow patterns seen on chest roentgenographs when Ppa and Ppw are raised. Under normal circumstances, blood flow is greatest in the basal regions of the lung [39]. When Ppw and Ppa are increased acutely in normal individuals, blood flow becomes evenly distributed between apical and basal vessels, presumably due to the recruitment of upper lung vessels. In patients with chronic elevation of Ppa and Ppw, for example with mitral stenosis, a clear redistribution occurs and apical vessels are more prominent than basal vessels. This is explained by the hydrostatic pressure gradient which at the base (a) increases vascular reactivity by the Bayliss effect and (b) results in more severe

arterial and venous lesions. These two effects further interact since increased medial smooth muscle may raise vascular reactivity.

Hypoxic Pulmonary Hypertension

Over the 40 years which have elapsed since the first description by Von Euler and Liljestrand of hypoxic vasoconstriction in cats, it has become clear that this response, which may be acute or chronic, occurs in "virtually all animal species that have lungs" [39]. The function of hypoxic vasoconstriction in the lung is to divert blood flow away from regions hypoxic due to localized diseases such as pneumonia and atelectasis, thereby preserving the ventilation/perfusion ratio and arterial blood oxygenation. Under certain conditions, however, when hypoxia is generalized rather than localized, pulmonary hypertension supervenes without improvement in arterial oxygenation. Indeed, it has been shown that the larger the hypoxic region, the less effective the diversion of blood flow [39]. This maladaptation occurs in some humans and animals who reside at high altitudes, in patients hypoxic because of hypoventilation due to obesity (Pickwickian syndrome) or musculoskeletal disorders, or because of pulmonary parenchymal disorders such as the obstructive and restrictive lung diseases. This section will describe the haemodynamics of acute hypoxic vasoconstriction, and the effects of chronic hypoxia on experimental animals and on humans without pulmonary parenchymal disease.

Haemodynamics and Mechanisms of Hypoxic Vasoconstriction

Within seconds of exposure to hypoxia, there is a rapid and widespread vasoconstriction, with an increase in Ppa and PVR, which peaks within minutes. The relationship between Ppa and arterial PO_2 is a curve in the form of a hyperbola. Under most circumstances, the response is potentiated by acidosis and by an increase in arterial PCO_2, the latter acting probably by raising hydrogen ion concentration. The magnitude of the hypoxic response varies, particularly after chronic exposure, due to differences (a) among species (weak response in dogs, sheep and hamsters, brisk response in cattle, pigs and rats); (b) in degree and duration of the hypoxia; and (c) in ventilatory and haematopoietic responses. In addition, hereditary factors are important since there is a genetically transmissible susceptibility or resistance to PHT in cattle: the greater amount of smooth muscle in arteries of susceptible cattle may increase the pulmonary response to hypoxia. In humans, the variability of Ppa at high altitude among individuals is probably due to differences in (a) alveolar ventilation in response to a given altitude and (b) degree of vasoconstriction to a given hypoxic stimulus.

The components of the hypoxic vasoconstrictor response include an oxygen sensor, a transduction process and an effector [39]. The exact nature and site of the oxygen sensor is unknown, but it appears to be closely related to small airways, alveoli or arteries. Since airway and alveolar hypoxia are better stimuli to vasoconstriction than pulmonary artery PO_2, the sensor may be located on the gas-exchanging (alveolar or airway) side. Molecular sensing mechanisms may involve both ATP-generating and non-ATP-generating O_2-dependent biochemical reactions, although details of these are far from clear.

The transduction process of hypoxic vasoconstriction, despite years of intense research, remains an enigma. It is clear, however, that the hypoxic pressor response is intrinsic to the lung, independent of systemic neural or humoral input, since it is readily elicited in isolated perfused lungs. Beyond this there is much speculation, and three principal mechanisms have been proposed. The first suggests that one or more humoral mediators are released from the lung parenchyma or endothelial cells: catecholamines, histamine, bradykinin and more recently prostaglandins and related arachidonic acid metabolites have been proposed, but none conclusively implicated. According to a second theory, hypoxia acts directly on smooth-muscle cells, which therefore act both as sensor and as effector. While this mechanism, which involves calcium-dependent reactions, remains plausible, definite proof for it is also lacking. The third theory proposes that the release or activity of a pulmonary vasodilator, for example a prostaglandin, is blunted by hypoxia; this mechanism, supported by the fact that cyclo-oxygenase inhibitors potentiate the hypoxic response, still remains to be proven.

The effector in hypoxic vasoconstriction is the smooth-muscle cell. Much of the evidence, both physiological and morphological, indicates that the site of response is in the small arteries [1,24,39,66,103,133]: physiological data are derived from studies on the distribution of vascular resistance using the occlusion technique and the micropuncture technique. Morphological support for the notion of a predominant arterial effect comes from direct radiological or histological observations, or from results of studies of exposure to chronic hypoxia. In addition, some evidence exists for venous or alveolar vessel constriction in hypoxia, in which case the smooth-muscle cells in veins or the contractile interstitial cells in alveolar walls would be the effector cells.

Pathology of Hypoxic Pulmonary Hypertension

Most morphological studies have investigated the effects of chronic hypoxia on the vasculature (vide infra). Efforts have been made, however, to examine acute vasoconstriction usually evaluated with physiological techniques only, by morphological methods. The advantage of morphology is that it can be used to identify precisely the vascular segment which is constricting. The effects of vasoconstriction have been studied in rats after hypoxia, and after fulvine, one of the *Crotalaria* alkaloids [152,156]. By light microscopy, it can be seen that when a vessel constricts its wall thickens, its lumen narrows and its internal elastic lamina becomes crenated. At the ultrastructural level, cytoplasmic excrescences of smooth-muscle cells indent the internal elastic lamina and protrude through fenestrations in it. Unfortunately, fixation may still produce a baseline vasoconstriction, which is difficult to separate from true vasoconstriction. To circumvent this problem, Davies et al. [22] studied rat lung slices incubated with norepinephrine with quantitative histology and showed constriction above baseline levels in muscular arteries under $200\,\mu m$ diameter. Nonmuscular arteries and veins also constricted, probably due to contraction of pericytes and intermediate cells. Other methods to assess hypoxic vasoconstriction have included measurement of changes in arterial diameters of frozen cat lungs [66] with hypoxia, and use of radiological evidence of arterial constriction [1].

The morphology of chronic hypoxic pulmonary vascular disease can be considered from two viewpoints: (a) by examining the lungs of patients who live at high altitude or suffer from PHT due to the hypoventilation syndromes; and (b) by looking at the sequential changes that occur in experimental animals after exposure to hypobaric hypoxi. We will begin by briefly reviewing the latter [32,57,63,86,90]. When rats are exposed to an ambient barometric pressure of 380 mmHg, Ppa rises significantly by 3 d and doubles itself by 10 d. Morphologically, the earliest changes are also seen at 3 d and are well developed by 10–14 d. There is a striking neomuscularization of partially muscular and nonmuscular arteries: pericytes and intermediate cells, located inside the single elastic lamina, first hypertrophy due to an increase in the amount of actin and myosin in their cytoplasm, then divide and differentiate into mature smooth-muscle cells with myofilaments and dense bodies. A new "internal" elastic lamina is produced later. In the larger arteries, all three layers double or triple in thickness, due to an initial increase in cell size (smooth-muscle cells, endothelial cells and fibroblasts). followed by a small rise in cell number (particularly fibroblasts and endothelial cells) and an increase in oedema fluid and extracellular collagen and elastin. In the smaller arteries, there is an increase in the thickness and also slightly later (day 7) in the number of endothelial cells.

When animals are returned to room air, there is recovery of some but not all of the features described above [32]. The peripheral muscularization of the smallest vessels disappears, while the medial hypertrophy and other changes may not return to baseline. These experimental findings have implications for humans in whom return to sea level from high altitude may not be accompanied by a full return to normal of the vasculature [90].

In man, the purest form of hypoxic PHT occurs in high-altitude residents, and most of the descriptions are derived from them [46,53,155]. In infants born at high altitude, since Ppa remains elevated until about 5 years of age, the pulmonary trunk shows an aortic pattern which, depending on the altitude, will remain for 3 to 10 years, and be followed by a "persistent pattern" in which there is a high content of elastic tissue as in the aorta, but in which the fibres are more fragmented. The normal adult configuration appears much later, usually after the age of 50 [46,53]. The intraparenchymal arteries show characteristic features which taken together strongly suggest hypoxic PHT [155]. These include peripheral muscularization of small arteries, development of longitudinal smooth-muscle fibres in an otherwise normal intima and a modest medial hypertrophy.

Similar although much milder changes have been reported in veins. In six of 12 adults resident at altitudes over 3000 m, Wagenvoort and Wagenvoort [158] found medial hypertrophy

and longitudinal smooth-muscle cells in the intima.

Mention should be made of Monge's disease which occurs in a small number of high-altitude residents, particularly males [53]. The mean age of onset is 40 and patients present with neuropsychiatric disturbances, easy fatigability and decreased exercise tolerance. They exhibit cyanosis due to a markedly elevated haematocrit and diminished oxygen saturation. The primary abnormalities appear to be alveolar hypoventilation and alveolar hypoxia, which cause pulmonary vasoconstriction and PHT, to which the polycythaemia and resultant hyperviscosity also contribute. The morphology of Monge's disease has not been studied in detail but is consistent with hypoxic PHT, although emboli (probably related to the high haematocrit) may also be found.

A high mountain disease of cattle, Brisket disease, is also characterized by severe PHT but, in the words of Heath and Williams [53], "should not be regarded as a bovine form of Monge's disease", since arterial oxygen saturation is not markedly impaired. Rather it is postulated that the arteries of certain animals have more muscle and are hyper-reactive to hypoxia [39].

Pulmonary Hypertension in Parenchymal Lung Disease

The chronic respiratory diseases affect nearly 50 million people in the USA and account for over 80 000 deaths each year. Mortality in these disorders correlates best with the presence and amount of PHT [122] and most patients indeed die of cor pulmonale. The latter term was defined by the World Health Organization as "hypertrophy of the right ventricle resulting from diseases affecting the function and/or structure of the lung, except when these pulmonary alterations are the result of diseases that primarily affect the left side of the heart or of congenital heart disease" [165]. While it has been suggested that the term be dropped [155], it remains ingrained in the medical literature and used widely. Several of the heterogeneous causes of cor pulmonale have already been discussed, including diseases directly affecting the pulmonary vasculature (thromboembolism, PPPA) and hypoxic PHT due to high altitude and to the hypoventilation syndromes. Emphasis will be placed here on PHT that results from emphysema, chronic bronchitis and small airway disease, and that which results from pulmonary interstitial fibrosis.

Pulmonary Hypertension in Chronic Bronchitis, Emphysema and Small Airway Disease

A significant proportion of patients with this group of diseases will develop PHT and pulmonary vascular alterations, some of which can be explained by hypoxia. Indeed, many of the histological lesions resemble those found in hypoxic PHT, and include peripheral muscularization of non-muscular and partially muscular arteries, the development of longitudinal fasciculi of smooth muscle, with some increase in the %MT of the media. It has been suggested that the longitudinally oriented smooth muscle cells are a reaction of the vessel to stretching by the increased lung volume or emphysematous spaces; this suggestion may be questioned however, since the same features are seen in pure hypoxic PHT. An additional feature, absent in pure hypoxic PHT, is pronounced intimal fibrosis, attributable to associated airways inflammation, previous pneumonia or less commonly to organized thromboemboli. Hale et al. [44] studied the muscular pulmonary arteries of 25 long-term cigarette smokers and found increased medial smooth muscle and intimal thickening which correlated with the severity of small airway disease and of emphysema.

It has been known for many years that the correlation between the amount of pure emphysema assessed with light microscopic morphometry and the degree of right ventricular hypertrophy is poor, suggesting that the destruction of the capillary bed which parallels the destruction of alveolar walls does not account for PHT and cor pulmonale in emphysema [46]. The close resemblance of the vascular lesions in chronic bronchitis and emphysema with those of hypoxic PHT suggests that hypoxia (with hypercarbia and acidosis) is important in the genesis of the PHT. Indeed, patients with the poorest gas exchange (bronchitics with centri-acinar emphysema and small airway disease) are also those with the greatest degree of PHT and right ventricular hypertrophy. In addition, exposure to hypoxic gas mixtures raises Ppa further. In contrast, the reversibility of the elevated Ppa with supplemented oxygen is variable, being slight initially, but more significant over 4 to 8 weeks. This is not unexpected since the organic vascular lesions are not immediately reversible, as reviewed in the experimental studies above. It remains true that while much of the PHT with chronic bronchitis and emphysema is due to hypoxaemia, the modest improvement with O_2 suggests other factors are operating. Several have been proposed: (a) the increased airway resistance, gas trapping and prolonged expiration, which raise vascular transmural pressures and PVR; (b) polycythaemia, which raises viscosity; (c) the increased blood volume; (d) superimposed thromboembolism in occasional patients; and (e) left ventricular failure, which occurs in a variable percentage of these patients, may aggravate the cor pulmonale [77]. In summary, the mechanisms of PHT in the chronic obstructive lung diseases are multiple, with hypoxia playing a prominent role.

PHT in Pulmonary Fibrosis and Related Disorders

Interstitial pulmonary fibrosis and "honeycomb" lung are the final result of a wide variety of disease entities or insults to the lung, some of immune or occupational origin, others related to drugs and toxins, and many idiopathic. The adult respiratory distress syndrome encountered in patients with multiple trauma and sepsis may also, in its late stages, produce extensive lung fibrosis. Patients with these disorders may, depending on the extent of involvement and on the degree of hypoxia, develop PHT and cor pulmonale. Indeed, hypoxia plays a role in elevating Ppa, as shown by Weitzenblum et al. [162] who reported a good correlation between pulmonary haemodynamics and arterial PO_2 in 65 patients with interstitial lung disease.

However, in contrast to the situation seen in high-altitude residents or in patients with chronic obstructive lung disease, patients with pulmonary fibrosis have more severe anatomical parenchymal and vascular changes to account for the PHT. In a recent series, of 22 patients who died with the adult respiratory distress syndrome reported by Tomashefski et al. [140], pre-mortem Ppa was 35 mmHg in an early group (less than 9 d between intubation and death) and 42 mmHg in intermediate and late groups (10 to 20 or more days). The authors noted an early decrease in arterial luminal diameter, probably due to endothelial damage and swelling and late vascular remodelling, with an increase in %MT and prominent fibrocellular intimal obliteration of arteries and veins. In addition, in patients with this syndrome, thromboemboli and infective vasculitis are frequent findings which may be responsible for some of the rise in Ppa.

The morphology of the vessels in pulmonary fibrosis has been described by several authors [46,51,155]. It is useful, as advocated by Heath and Smith [51], to consider the vessels (a) in fibrotic areas, (b) bordering fibrotic areas and (c) outside fibrotic areas. In the fibrosis, which may be patchy, both arteries and veins are markedly abnormal and show severe intimal fibrosis and fibroelastosis, followed by progressive infiltration, compression and destruction by fibrous tissue; the elastic laminae are distorted. Thrombosis with recanalization also contributes to the vascular obstruction. These changes undoubtedly contribute to the ventilation–perfusion abnormalities present in the interstitial fibrotic lung disorders. In addition, dilated, thin-walled vessels are described which may represent pulmonary to systemic and pulmonary venoarterial anastomoses, contributing to the shunting.

For the arteries adjacent to the fibrosis, there is prominent development of longitudinal smooth-muscle cells in the intima, which is gradually replaced by fibrous tissue. The appearance of the small arteries is said to differ from that in chronic hypoxia: in fibrosis, there is a single thick elastic lamina with longitudinal muscle and fibrosis in its inner aspect, while in chronic hypoxia, there is a distinct media with circular muscle, bounded by the original outer elastic lamina and a new internal lamina. As in chronic bronchitis and emphysema, stretching of the arteries has been invoked to explain the longitudinal muscle. The vessels outside the fibrotic areas display the lesions seen in chronic hypoxia, and result from the abnormalities in gas exchange due to disease in the affected areas.

Different changes occur in newborns who die with the special form of pulmonary fibrosis, bronchopulmonary dysplasia, which follows the infant respiratory distress syndrome: indeed Tomashefski et al. [141] found that the %MT of most pre-acinar and intra-acinar arteries was reduced compared with normal *foetuses* of the same age, while there were more intra-acinar arteries. Therefore in this setting, both an attempt at normal adaptation to extrauterine life and a pathological remodelling are taking place.

Acknowledgements. The research of the author is supported by the Medical Research Council of Canada. The author also thanks Drs T. Hakim, M. Ostrowski, P. Paré and E. Wood for reading over the manuscript, and Mrs S. Totilo for secretarial assistance.

References

1. Allison DJ, Stanbrook HS (1980) A radiological and physiologic investigation into hypoxic pulmonary vasoconstriction in the dog. Invest Radiol 15:178–190
2. Alt B, Shikes RH (1983) Pulmonary hypertension in congenital heart disease; irreversible vascular changes in young infants. Pediatr Pathol 1:423–434
3. Anderson EG, Simon G, Reid L (1973) Primary and thrombo-embolic pulmonary hypertension: a quantitative pathological study. J Pathol 110:273–293
4. Bakhle YS, Ferreire SH (1985) Lung metabolism of eicosanoids: prostaglandins, prostacyclins, thromboxane, and leukotrienes. In: Fishman AP, Fisher AB (eds) Handbook of physiology: section 3, the respiratory system, vol I. Circulation and nonrespiratory functions. American Physiological Society, Bethesda, pp 365–386
5. Barst RJ (1986) Pharmacologically induced pulmonary vasodilatation in children and young adults with primary pulmonary hypertension. Chest 89:497–503
6. Bergqvist D, Lindblad B (1985) A 30-year survey of pulmonary embolism verified at autopsy: an analysis of 1274 patients. Br J Surg 72:105–108
7. Bhattacharya J, Nanjo S, Staub NC (1982) Factors affecting lung microvascular pressure. Ann NY Acad Sci 384:107–114
8. Bjornsson J, Edwards WD (1985) Primary pulmonary hypertension: a histopathologic study of 80 cases. Mayo Clin Proc 60:16–25
9. Braun A, Greenberg SD, Malik S et al. (1973) Pulmonary veno-occlusive disease associated with pulmonary phlebitis. Arch Pathol Lab Med 95:67–70
10. Brenner O (1935) Pathology of the vessels of the pulmonary circulation, part 1. Arch Intern Med 56:211–237
11. Burri PH Development and growth of the human lung. In: Fishman AP, Fisher AB (eds) Handbook of Physiology: section 3, the respiratory system, vol I. Circulation and nonrespiratory functions. American Physiological Society, Bethesda, pp 1—46
12. Canny GJ, Arbus GS, Wilson GJ et al. (1985) Fatal pulmonary hypertension following renal transplantation. Br J Dis Chest 79:191–195
13. Carrington CB, Liebow AA (1970) Pulmonary veno-occlusive disease. Hum Pathol, 1:322–324
14. Chawla SK, Kittle CF, Faber LP et al. (1976) Pulmonary veno-occlusive disease. Ann Thorac Surg 22:249–253
15. Cope DK, Allison RC, Parmentier JL et al. (1986) Measurement of effective pulmonary capillary pressure using the pressure profile after pulmonary artery occlusion. Crit Care Med 14:16–12
16. Corrin B, Spencer H, Turner-Warwick M et al. (1974) Pulmonary veno-occlusion – an immune complex disease? Virchows Arch [A] 364:81–91
17. Cudkowicz L (1979) Bronchial arterial circulation in man. Normal anatomy and responses to disease. In: Moser KM (ed) Pulmonary vascular diseases. Marcel Dekker, New York, pp 111–232
18. Cumming G, Harding LK, Horsfield K et al. (1970) Morphological aspects of the pulmonary circulation and of the airways. In: Fluid dynamics of blood circulation and respiratory flow. AGARD Conf Proc 65:23–0–23–6
19. Dalen JE, Alpert JS (1975) Natural history of pul-

monary embolism. Prog Cardiovasc Dis 17:259–270

20. Dalen JE, Haffajee CI, Alpert JS et al. (1977) Pulmonary embolism, pulmonary hemorrhage and pulmonary infarction. N Engl J Med 296:1431–1435

21. Dantzker DR, Bower JS (1982) Alterations in gas exchange following pulmonary thromboembolism. Chest 81:495–501

22. Davies P, Maddalo F, Reid L (1984) The response of microvessels in rat lung explants to incubation with norepinephrine. Exp Lung Res 7:93–100

23. Davies P, Reid L (1982) Pulmonary veno-occlusive disease in siblings: case reports and morphometric study. Hum Pathol 13:911–915

24. Dawson CA (1984) Role of pulmonary vasomotion in physiology of the lung. Physiol Rev 64:544–616

25. Dexter L (1979) Pulmonary vascular disease in acquired heart disease. In: Moser KM (ed) Pulmonary vascular diseases. Marcel Dekker, New York, pp 427–487

26. Dey RD, Said Si (1985) Lung peptides and the pulmonary circulation. In: Said SI (ed) The pulmonary circulation and acute lung injury. Futura, Mount Kisco, New York, pp 101–122

27. Edwards WD, Edwards JE (1977) Clinical primary pulmonary hypertension. Three pathological types. Circulation 56:884–888

28. Elliott FM, Reid L (1965) Some new facts about the pulmonary artery and its branching pattern. Clin Radiol 16:193–198

29. Fernandez-Segoviano P, Esteban A, Martinez-Cabruja R (1983) Pulmonary vascular lesions in the toxic oil syndrome in Spain. Thorax 38:724–729

30. Fishman AP (1965) Pulmonary circulation. In Fishman AP, Fisher AB (eds) Handbook of physiology. Section 3, the respiratory system, vol I. Circulation and nonrespiratory functions. American Physiological Society, Bethesda, pp 93–165

31. Fishman AP, Pietra GG (1980) Primary pulmonary hypertension. Ann Rev Med 31:421–431

32. Fried R, Reid LM (1984) Early recovery from hypoxic pulmonary hypertension: a structural and functional study. J Appl Physiol 57:1247–1253

33. Furchgott RF, Martin W (1985) Interactions of endothelial cells and smooth muscle cells of arteries. Chest 88:210S–213S

34. Geggel RL, Reid LM (1984) The structural basis of PPHN. Clin Perinatol 11:525–549

35. Gersony WM (1984) Neonatal pulmonary hypertension: pathophysiology, classification, and etiology. Clin Perinatol 11:517–524

36. Goldhaber SZ (1985) Pulmonary embolism and deep vein thrombosis. Saunders, Philadelphia

37. Grant BJB, Dantzker DR (1984) Pathophysiology of the pulmonary circulation. In: Rubin LJ (ed) Pulmonary heart disease. Martinus Nijhoff, Boston, pp 11–64

38. Grossman W, Alpert JS, Braunwald E: (1984) Pulmonary hypertension. In: Braunwald E (ed) Heart disease. A textbook of cardiovascular medicine. Saunders, Philadelphia, pp 823–848

39. Grover RF, Wagner WW, McMurtry IF et al. (1983) Pulmonary circulation. In Shepherd JT, Abboud FM (eds) Handbook of physiology. Section 2, the cardiovascular system, vol III. Peripheral circulation and organ blood flow, part I. American Physiological Society, Bethesda, pp 103–136

40. Gurtner HP (1985) Chronische pulmonale Hypertonie vaskularen Ursprungs, plexogene pulmonale Arteriopathie und der Appetitzugler Aminorex: Nachlese zu einer Epidemie. Schweiz Med Wochenschr 115:782–789, 818–827

41. Gurtner HP, Gertsch M, Salzmann C et al. (1968) Haufen sich die primar vaskularen Formen des chronischen Cor pulmonale? Schweiz Med Wochenschr, 98:1579–1589, 1695–1707

42. Hakin TS, Michel RP, Chang HK (1982) Partitioning of pulmonary vascular resistance in dogs by arterial and venous occlusion. J Appl Physiol 52:710–715

43. Hakim TS, Michel RP, Minami H et al. (1983) Site of pulmonary hypoxic vasoconstriction studied with arterial and venous occlusion. J Appl Physiol 54:1298–1302

44. Hale KA, Niewoehner DE, Cosio MG (1980) Morphologic changes in the muscular pulmonary arteries: relationship to cigarette smoking, airway disease, and emphysema. Am Rev Respir Dis 122:273–278

45. Hampton AO. Castleman B (1940) Correlation of postmortem chest teleroentgenograms with autopsy findings. With special reference to pulmonary embolism and infarction. AJR 43:305–326

46. Harris P, Heath D (1977) The human pulmonary circulation. Churchill Livingstone, Edinburgh

47. Hatano S, Strasser T (1975) Primary pulmonary hypertension. Report on a WHO meeting. World Health Organization, Geneva

48. Haworth SG (1983) Primary and secondary pulmonary hypertension in childhood: a clinicopathological reappraisal. Curr Top Pathol 73:91–52

49. Haworth SG, Hislop AA (1981) Adaptation of the pulmonary circulation to extra-uterine life in the pig and its relevance to the human infant. Cardiovasc Rev 15:108–119

50. Heath D, Edwards JE (1958) The pathology of hypertensive pulmonary vascular disease. A description of six grades of structural changes in the pulmonary arteries with special reference to congenital cardiac defects. Circulation, 18:533–547

51. Heath D, Smith P (1979) Pulmonary vascular disease secondary to lung disease. In: Moser KM (ed) Pulmonary vascular diseases. Marcel Dekker, New York, pp 387–426

52. Heath D, Smith P (1983) Electron microscopy of hypertensive pulmonary vascular disease. Br J Dis Chest 77:1–13

53. Heath D, Williams Dr (1981) Man at high altitude. The pathophysiology of acclimatization and adaptation. Churchill Livingstone, Edinburgh

54. Heymann MA, Hoffman JIE (1984) Persistent pulmonary hypertension syndromes in the newborn. In: Weir EK, Reeves JT (eds) Pulmonary hypertension. Futura, Mount Kisco, New York, pp 45–71

55. Hilliker KS, Roth RA (1985) Increased vascular responsiveness in lungs of rats with pulmonary hypertension induced by monocrotaline pyrrole. Am Rev Respir Dis 131:46–50

56. Hislop A, Reid L (1973) Fetal and childhood development of the intrapulmonary veins in man – branching pattern and structure. Thorax 28:313–319

57. Hislop A, Reid L (1976) New findings in pulmonary arteries of rats with hypoxia-induced pulmonary hypertension. Br J Exp Pathol 57:542–554

58. Hislop A, Reid L (1981) Growth and development of the respiratory system: anatomical development. In: Davies JA, Dobbing J (eds) Scientific foundations of paediatrics. William Heinemann, London, pp 390–432

59. Holloway H, Perry M, Downey J et al. (1983) Estimation of effective pulmonary capillary pressure in

intact lungs. J Appl Physiol 54:846–851

60. Horsfield K (1978) Morphometry of the small pulmonary arteries in man. Circ Res 42:593–597

61. Horsfield K, Gordon WI (1981) Morphometry of pulmonary veins in man. Lung 159:211–218

62. Hughes JD, Rubin LJ (1986) Primary pulmonary hypertension. An analysis of 28 cases and a review of the literature. Medicine 65:56–72

63. Jones R, Langleben D, Reid LM (1985) Patterns of remodelling of the pulmonary circulation in acute and subacute lung injury. In: Said SI (ed) The pulmonary circulation and acute lung injury. Futura, Mount Kisco, New York, pp 137–188

64. Junod AF (1985) 5-hydroxytryptamine and other amines in the lung. In: Fishman AP, Fisher AB (eds) Handbook of physiology: section 3, the respiratory system, vol I. Circulation and nonrespiratory functions. American Physiological Society, Bethesda, pp 337–349

65. Juratsch CE, Grover RF, Rose CE et al. (1985) Reversal of reflex pulmonary vasoconstriction induced by main pulmonary arterial distension. J Appl Physiol 58:1107–1114

66. Kato M, Staub NC (1966) Response of small pulmonary arteries to unilobar hypoxia and hypercapnia. Circ Res 19:426–440

67. Kay JM, De Sa DJ, Mancer JFK (1983) Ultrastructure of lung in pulmonary veno-occlusive disease. Hum Pathol 14:451–456

68. Kay JM, Harris P, Heath D (1967) Pulmonary hypertension produced in rats by ingestion of *Crotalaria spectabilis* seeds. Thorax 22:176–179

69. Kay JM, Heath D (1985) Pathologic study of unexplained pulmonary hypertension. Semin Respir Med 7:180–192

70. Kay JM, Keane PM, Suyama KL (1985) Pulmonary hypertension induced in rats by monocrotaline and chronic hypoxia is reduced by p-chlorophenylalanine. Respiration 47:48–56

71. Kay JM, Suyama KL, Keane PM (1982) Failure to show decrease in small pulmonary blood vessels in rats with experimental pulmonary hypertension. Thorax 37:927–930

72. Laissue JA, Gebbers JO, Musy JP (1984) Embolie pulmonaire: épidémiologie et pathologie. Schweiz Med Wochenschr 114:1711–1720

73. Lockhart A (1983) Actualité de l'artériopathie pulmonaire plexogénique de cause inconnue. Bull Eur Physiopathol Respir 19:521–529

74. Malik AB (1983) Pulmonary miroembolism. Physiol Rev 63:1114–1207

75. Mammosser M, Albert RK, Schrijen F et al. (1986) Decreased arterial oxygenation predicts pulmonary hypertension in patients with persistent dyspnea after pulmonary emboli. Am Rev Respir Dis 133:A129 (abstract)

76. Manier G, Castaing Y, Guenard H (1985) Determinants of hypoxemia during the acute phase of pulmonary embolism in humans. Am Rev Respir Dis 132:332–338

77. Matthay RA, Berger HJ (1981) Cardiovascular performance is chronic obstructive pulmonary diseases. Med Clin North Am 65:489–524

78. McDonnell PJ, Summer WR, Hutchins GM (1981) Pulmonary veno-occlusive disease. Morphological changes suggesting a viral cause. JAMA 246:667–671

79. McDonnell PJ, Toye PA, Hutchins GM (1983) Primary pulmonary hypertension and cirrhosis: are they related? Am Rev Respir Dis 127:437–441

80. McGregor M, Sniderman A (1985) On pulmonary vascular resistance: the need for more precise definition. Am J Cardiol 55:217–221

81. McIntyre KM, Sasahara AA (1974) Hemodynamic and ventricular responses to pulmonary embolism. Prog Cardiovasc Dis 17:175–190

82. Messer JV (1984) Thromboembolic pulmonary hypertension. In: Weir EK, Reeves JT (eds.) Pulmonary hypertension. Futura, Mount Kisco, New York, pp 169–249

83. Meyrick B, Clarke SW, Symons C et al. (1974) Primary pulmonary hypertension. A case report including electron microscopic study. Br J Dis Chest 68:11–20

84. Meyrick B, Gamble W, Reid L (1980) Development of Crotalaria pulmonary hypertension: hemodynamic and structural study. Am J Physiol 239:H692–H702

85. Meyrick B, Niedermeyer ME, Ogletree ML et al. (1985) Pulmonary hypertension and increased vasoreactivity caused by repeated indomethacin in sheep. J Appl Physiol 59:443–452

86. Meyrick B, Reid L (1978) The effect of continued hypoxia on rat pulmonary arterial circulation. An ultrastructural study. Lab Invest 38:188–200

87. Meyrick B, Reid L (1979) Ultrastructural features of the distended pulmonary arteries of the normal rat. Anat Rec 193:71–98

88. Meyrick B, Reid L (1979) Ultrastructural findings in lung biopsy material from children with congenital heart defects. Am J Pathol 101:527–542

89. Meyrick BO, Reid LM (1982) Crotalaria-induced pulmonary hypertension. Uptake of ^{3}H-thymidine by the cells of the pulmonary circulation and alveolar walls. Am J Pathol 106:84–94

90. Meyrick B, Reid L (1983) Pulmonary hypertension. Anatomic and physiologic correlates. Clin Chest Med 4:199–217

91. Michel RP (1982) Arteries and veins of the normal dog lung: qualitative and quantitative structural differences. Am J Anat 164:227–241

92. Michel RP (1985) Lung microvascular permeability to dextran in alpha-naphthylthiourea-induced edema. Sites of filtration, patterns of accumulation, and effects of fixation. Am J Pathol 119:474–484

93. Michel RP, Hakim TS, Chang HK (1984) Pulmonary arterial and venous pressures measured with small catheters in dogs. J Appl Physiol 57:309–314

94. Michel RP, Hakim TS, Hanson RE et al. (1985) Distribution of lung vascular resistance after chronic systemic-to-pulmonary shunts. Am J Physiol 249:H1106–H1113

95. Miller WS (1893) The structure of the lung. J Morphol 8:165–188

96. Mitzner W (1983) Resistance of the pulmonary circulation. Clin Chest Med 4:127–137

97. Mooi W, Wagenwoort CA (1983) Decreased numbers of pulmonary blood vessels: reality or artifact? J Pathol 141:441–447

98. Moore GW, Smith RRL, Hutchins GM (1982) Pulmonary artery atherosclerosis. Correlation with systemic atherosclerosis and hypertensive pulmonary vascular disease. Arch Pathol Lab Med 106:378–380

99. Moragas A, Huguet P, Toran N et al. (1983) Morphogenesis of pulmonary veno-occlusive disease in a newborn. Image analysis study. Pathol Res Pract 176:176–184

100. Morrison EB, Gaffney FA, Eigenbrodt EH et al. (1980) Severe pulmonary hypertension associated with macronodular (postnecrotic) cirrhosis and auto-

immune phenomena. Am J Med 69:513–519

101. Moser KM, Guisan M, Bartimmo EE (1973) Resolution rates of experimental venous thromboemboli. In: Moser KM, Stein M (eds) Pulmonary thromboembolism. Year Book Medical Publishers, Chicago, pp 104–113

102. Murphy JD, Rabinovitch M, Goldstein JD et al. (1981) The structural basis of persistent pulmonary hypertension of the newborn infant. J Pediatr 98:962–967

103. Nagasaka Y, Bhattacharya J, Nanjo S et al. (1984) Micropuncture measurement of lung microvascular pressure profile during hypoxia in cats. Circ Res 54:90–95

104. Newman SL, Michel RP, Wang NS (1985) Lingular lung biopsy: is it representative? Am Rev Respir Dis 132:1084–1086

105. Paakko P, Sitinen S, Remes M et al. (1985) A case of pulmonary vascular occlusive disease: comparison of post-mortem radiography and histology. Histopathology 9:253–262

106. Palevsky HI, Fishman AP (1985) Vasodilator therapy for primary pulmonary hypertension. Ann Rev Med 36:563–578

107. Rabinovitch M (1985) Morphology of the developing pulmonary bed: pharmacologic implications. Pediatr Pharmacol 5:31–48

108. Rabinovitch M, Haworth SG, Vance Z et al. (1980) Early pulmonary vascular changes in congenital heart disease studied in biopsy tissue. Hum Pathol 11:499–509

109. Rabinovitch M, Keane JF, Norwood WI et al. (1984) Vascular structure in lung tissue obtained at biopsy correlated with pulmonary hemodynamic findings after repair of congenital cardiac defects. Circulation 4:655–667

110. Rambihar VS, Fallen EL, Cairns JA (1979) Pulmonary veno-occlusive disease: antemortem diagnosis from roentgenographic and hemodynamic findings. Can Med Assoc J 120:1519–1522

111. Reeves JT, Groves BM (1984) Approach to the patient with pulmonary hypertension. In: Weir EK, Reeves JT (eds) Pulmonary hypertension. Futura, Mount Kisco, New York, pp 1–44

112. Reeves JT, Herget J (1984) Experimental models of pulmonary hypertension. In: Weir EK, Reeves JT (1984) Pulmonary hypertension. Futura, Mount Kisco, New York, pp 361–391

113. Reeves JT, Stenmark KR, Voelkel NF (1985) Possible role of leukotrienes in the pathogenesis of pulmonary hypertensive disorders. In: Said SI (ed) The pulmonary circulation and acute lung injury. Futura, Mount Kisco, New York, pp 337–356

114. Reid L (1968) Structural and functional reappraisal of the pulmonary artery system. In: The scientific basis of medicine annual reviews. Athlone Press, London, pp 289–307

115. Reid LM (1979) The pulmonary circulation: remodeling in growth and disease. Am Rev Respir Dis 119:531–546

116. Reid L, Meyrick B (1982) Microcirculation: definition and organization of the tissue level. Ann NY Acad Sci 384:3–20

117. Rendas A, Lennox S, Reid L (1979) Aorta-pulmonary shunts in growing pigs. Functional and structural assessment of the changes in the pulmonary circulation. J Thorac Cardiovasc Surg 77:109–118

118. Rendas A, Reid L (1983) Pulmonary vasculature of piglets after correction of aorta-pulmonary shunts. J Thorac Cardiovasc Surg 85:911–916

119. Rhodin JAG (1978) Microscopic anatomy of the pulmonary vascular bed in the cat lung. Microvasc Res 15:169–193

120. Rich S, Levy PS (1984) Characteristics of surviving and nonsurving patients with primary pulmonary hypertension. Am J Med 76:573–578

121. Rosenow EC, Osmundson PJ, Brown ML (1981) Pulmonary embolism. Mayo Clin Proc 56:161–178

122. Rubin LJ (1984) Pulmonary hypertension secondary to lung disease. In: Weir EK, Reeves JT (eds) Pulmonary hypertension. Futura, Mount Kisco, New York, pp 291–320

123. Rudolph AM (1970) The changes in the circulation after birth. Their importance in congenital heart disease. Circulation, 41:343–359

124. Ryan US (1985) Processing of angiotensin and other peptides by the lungs. In: Fishman AP, Fisher AB (eds) Handbook of physiology: section 3, the respiratory system, vol I. Circulation and nonrespiratory functions. American Physiological Society, Bethesda, pp 351–364

125. Ryan US (1986) Metabolic activity of pulmonary endothelium: modulations of structure and function. Ann Rev Physiol 48:263–277

126. Samuelson A, Becker AE, Wagenvoort CA (1970) A morphometric study of pulmonary veins in normal infant and infants with congenital heart disease. Arch Pathol Lab Med 90:112–116

127. Schneeberger EE (1982) Structure of intercellular junctions in different segments of the intrapulmonary vasculature. Ann NY Acad Sci 384:54–63

128. Sharma GVRK, Schoolman M, Cella G et al. (1983) Pulmonary embolism, parts I and II. Circulation 67:245–247, 474–477

129. Shure D, Dockweiler DW, Lammers RJ et al. (1986) Post-obstructive pulmonary arteriopathy is related to the extent of bronchopulmonary anastamotic formation in an animal model. Am Rev Respir Dis 133:A161 (abstract)

130. Singhal S, Henderson R, Horsfield K et al. (1973) Morphometry of the human pulmonary arterial tree. Circ Res 33:190–197

131. Smith P, Heath D (1979) Electron microscopy of the plexiform lesion. Thorax 34:177–186

132. Spencer H (1985) Pathology of the lung, Pergamon, Oxford

133. Staub NC (1985) Site of hypoxic pulmonary vasoconstriction. Chest 88:240S–244S

134. Stenmark KR, Fasules J, Tucker A et al. (1986) Severe pulmonary hypertension and arterial adventitial changes in newborn calves at 4300 m. Am Rev Respir Dis 133:A227 (abstract)

135. Stoler MH, Anderson VM, Stuard ID (1982) A case of pulmonary veno-occlusive disease in infancy. Arch Pathol Lab Med 106:645–647

136. Sugita T, Hyers TM, Dauber IM et al. (1983) Lung vessel leak precedes right ventricular hypertrophy in monocrotaline-treated rats. J Appl Physiol 54:371–374

137. Takahashi T, Wagenvoort CA (1983) Density of muscularized arteries in the lung. Its role in congenital heart disease and its clinical significance. Arch Pathol Lab Med 107:23–28

138. Taylor WE (1984) Pathology of pulmonary heart disease. In: Rubin LJ (ed) Pulmonary heart disease. Martinus Nijhoff, Boston, pp. 65–105

139. Thadani U, Burrow C, Whitaker W et al. (1975) Pulmonary veno-occlusive disease. QJ Med 173:133–159

140. Tomashefski JF Jr, Davies P, Boggis C et al. (1983) The

pulmonary vascular lesions of the adult respiratory distress syndrome. Am J Pathol 112:112–126

141. Tomashefski JF Jr, Oppermann HC, Vawter GF et al. (1984) Bronchopulmonary dysplasia: a morphometric study with emphasis on the pulmonary vasculature. Pediatr Pathol 2:469–487

142. Tryka AF, Godleski JJ, Schoen FJ et al. (1985) Pulmonary vascular disease and hypertension after valve surgery for mitral stenosis. Hum Pathol 16:65–71

143. Tsao MS, Schraufnagel D, Wang NS (1982) Pathogenesis of pulmonary infarction. Am J Med 72:599–606

144. Urokinase Pulmonary Embolism Trial (1973) A national cooperative study. Circulation (Suppl II) 47:1–108

145. Voelkel N, Reeves JT (1979) Primary pulmonary hypertension. In: Moser KM (ed). Pulmonary vascular diseases. Marcel Dekker, New York, pp 573–628

146. Wagenvoort CA (1960) Vasoconstriction and medial hypertrophy in pulmonary hypertension. Circulation 22:535–546

147. Wagenvoort CA (1976) Pulmonary veno-occlusive disease. Entity or syndrome? Chest 69:82–86

147a. Wagenvoort CA (1980) Lung biopsy specimens in the evaluation of pulmonary vascular disease. Chest 77:614–625

148. Wagenvoort CA (1981) Grading of pulmonary vascular lesions – a reappraisal. Histopathology 5:595–598

149. Wagenvoort CA (1984) Lung biopsies and pulmonary vascular disease. In: Weir EK, Reeves JT (eds) Pulmonary hypertension. Futura, Mount Kisco, New York, pp 393–427

150. Wagenvoort CA (1985) Open lung biopsies in congenital heart disease for evaluation of pulmonary vascular disease. Predictive value with regard to corrective operability. Histopathology 9:417–436

151. Wagenvoort CA, Beetstra A, Spijker J (1978) Capillary hemangiomatosis of the lungs. Histopathology 2:401–406

152. Wagenvoort CA, Dingemans KP (1985) Pulmonary vascular smooth muscle and its interaction with endothelium. Morphologic considerations. Chest 88:200S–202S

153. Wagenvoort CA, Nauta J, Van Der Schaar PJ et al. (1967) Effect of flow and pressure on pulmonary vessels. A semiquantitative study based on lung biopsies. Circulation 35:1028–1037

154. Wagenvoort CA, Wagenvoort N (1970) Primary pulmonary hypertension. A pathologic study of the lung vessels in 156 clinically diagnosed cases. Circulation 42:1163–1184

155. Wagenvoort CA, Wagenvoort N (1977) Pathology of pulmonary hypertension. John Wiley, New York

156. Wagenvoort CA, Wagenvoort N (1979) Pulmonary vascular bed. Normal anatomy and responses to disease. In: Moser KM (ed) Pulmonary vascular diseases. Marcel Dekker, New York, pp 1–109

157. Wagenvoort CA, Wagenvoort N (1982) Smooth muscle content of pulmonary arterial media in pulmonary venous hypertension compared with other forms of pulmonary hypertension. Chest 81:581–585

158. Wagenvoort CA, Wagenvoort N (1982) Pulmonary veins in high-altitude residents: a morphometric study. Thorax 37:931–935

159. Wagenvoort CA, Wagenvoort N, Takahashi T (1985) Pulmonary veno-occlusive disease: involvement of pulmonary arteries and review of the literature. Hum Pathol 16:1033–1041

160. Weibel ER (1985) Lung cell biology. In: Fishman AP, Fisher AB (eds) Handbook of physiology: section 3, the respiratory system, I. Circulation and non-respiratory functions. American Physiological Society, Bethesda, pp 47–91

161. Weir EK (1984) Diagnosis and management of primary pulmonary hypertension. In: Weir EK, Reeves JT (1984) Pulmonary hypertension. Futura, Mount Kisco, New York, pp 115–168

162. Weitzenblum E, Ehrhart M, Rasaholinjanahary J et al. (1983) Pulmonary hemodynamics in interstitial pulmonary fibrosis and other interstitial pulmonary diseases. Respiration, 44:118–127

163. Widgren S (1977) Pulmonary hypertension related to aminorex intake. Histologic, ultrastructural, and morphometric studies of 37 cases in Switzerland. Curr Top Pathol 64:1–64

164. Wolfe WG, Sabiston DC Jr (1980) Pulmonary embolism. Saunders. Philadelphia

165. World Health Organization (1963) Chronic cor pulmonale. Report of an expert committee. Circulation 27:594–615

166. Zhuang FY, Fung YC, Yen RT (1983) Analysis of blood flow in cat's lung with detailed anatomical and elasticity data. J Appl Physiol 55:1341–1348

167. Zidulka A, Hakim TS (1985) Wedge pressure in large vs. small pulmonary arteries to detect pulmonary venoconstriction. J Appl Physiol 59:1329–1332

Section VIII

Angiodysplasias, Tumours and Tumour-like Lesions of the Blood Vessels

Chapter 36

Arteriovenous Malformations and Arterial Dysplasias

C. L. Berry

Introduction

The vascular system is uniform in man only in terms of its major components; there is considerable variability peripherally, variability which is individual-specific. It is worth remembering that it was suggested that prints of the pattern of veins on the dorsum of the hand might be used in identification before fingerprinting was adopted.

This variability, however, extends only to well-defined limits and it is rare for a group of individuals to have a vascular abnormality of pattern

Table 36.1. Clinical syndromes with arteriovenous malformations

---

Klippel–Trenaunay–Weber Syndrome
    Increased bone growth, limb hypertrophy, varicose veins, cutaneous haemangiomatous naevi

Parkes–Weber Syndrome
    Limb hypertrophy associated with a cutaneous capillary haemangioma and multiple arteriovenous fistulas

Hippel–Lindau
    Angiomas of the retinal and central nervous system. Cystic change in pancreas, kidney and CNS and tumours of the kidney and CNS

Sturge–Weber Syndrome
    Vascular hamartoma of face, ipsilateral vascular malformation of meninges often with calcification

Maffucci's Syndrome
    Multiple visceral and cutaneous angiomata with dyschondroplasia

Rendu–Osler–Weber Disease
    Inherited as Mendelian dominant. Telangiectasia in nose and gut. Pulmonary arteriovenous fistulas may occur

---

producing clinical symptoms. Aortic arch anomalies are uncommon and peripheral variation with symptomatology very rare indeed (see Appleberg [1] and Dent et al. [12] for examples). It should be remembered that arteriovenous communications normally occur in the palms, soles, eyelids, nose and at the tip of the tongue in man, and in large numbers in the heel. However, the clinical syndromes characterized by the development of arteriovenous communications (Table 36.1) are probably abnormalities of development rather than exaggerations of normal phenomena.

Arteriovenous Malformations

There is an extraordinary range of arteriovenous lesions, from cutaneous lesions of no significance, to lesions where shunting affects cardiovascular function and where malformations form massive ulcerating tumours which may be directly life threatening.

There is no reliable estimate of the frequency of these lesions in general due to differences in ascertainment and classification and to very varied reporting. However, from several series [9,38,59] it appears that around two-thirds of clinically evident cases occur in the lower limbs.

Arteriovenous malformations have been classified in terms of the developmental stage at which they may be presumed to have arisen. The relevant text is that of Woolard [61] but the principles discussed in Chap. 3 should also be borne in mind. Woolard described three stages of vascular development (in the limb of the pig). The first is the capillary network stage, which is followed by the development of larger trunks, fusion of vessels and early media formation and then by the third stage in which the major stem vessels of the limb develop (Fig. 36.1).

The pattern of development makes it clear that anomalies could be considered to have different forms depending on the part of the system from which they develop, or the stage of development at which they form. However, many complex anomalies contain microvascular nets, arterioles, small veins and larger vessels, although rarely large elastic arteries with a well-developed medial structure. Their form depends more on the size of the principle feeder vessel, which is probably determined by anatomical considerations, i.e. the site of establishment of venous/arterial communication. To suggest that capillary malformations are produced in the network stage and arteriovenous fistulas in the trunk-forming stage suggests a rigidity not generally evident in this system. For this reason the author does not support the classification of Szilagyi et al. [56] although, in agreement with Dean [10], considers the complex collection of "descriptive phases and synonyms" a poor alternative.

Thus the terms cirsoid aneurysm, racemose aneurysm, arteriovenous haemangioma, erectile tumour aneurysm by anastomosis, naevus angiectoid angiodysplasia with arteriovenous shunt etc. add little to our knowledge or understanding and the global "arteriovenous malformation" is to be preferred.

Evidence for major failure of the large artery pattern may exist, however. As an example, there are several case reports of direct communication between the anterior and posterior tibial arteries [9,36,38]. It is doubtful whether these communications are actually commoner in the lower limb than elsewhere; what is clear is that at this site these anomalies are more often investigated angiographically.

The Relationship Between Haemangiomas and Vascular Malformations

Vascular tumours are described elsewhere in this volume but certain differences between normal vessel growth, growth of vascular anomalies and haemangiomata are worth emphasizing here.

The pattern of changes in the formation of new vessels in an established microcirculation is widely accepted [8]. The normal cell–cell and cell–basement membrane junctional attachments loosen, presumably to facilitate cell movement. Locally active proteases interfere with the structure of the basement membrane and modify the local connective tissue in a way which permits the freed endothelial cells to protrude from the vessel. They form solid sprouts and move towards the angiogenic stimulus with cell division occurring in a region a number of cells away from the tip. However, this cell division does not occur for some hours after movement

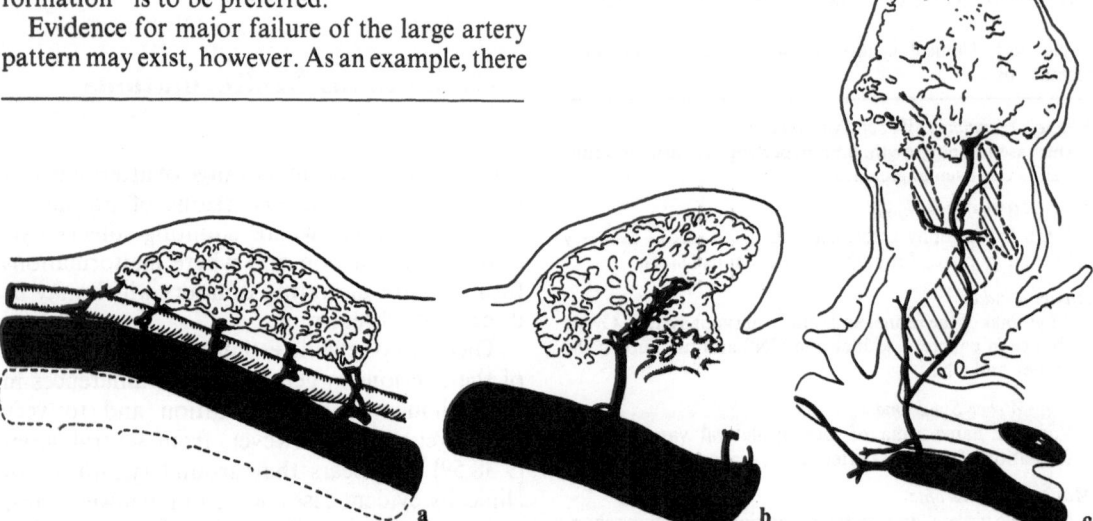

Fig. 36.1. a The earliest stage of lymphovasculature development. The capillary net is supplied by segmental artery branches and drains to an underlying vein. **b** As the limb develops the distinction between arteries and veins is becoming more obvious. **c** With the onset of segmentation in the limb, the arterial and venous pattern is distinguished.

It is assumed that the arteriovenous malformations form as stage C is reached but there is little evidence for this.

begins. Lumen formation occurs probably within endothelial cells, with intracellular lumina subsequently linking to make a complete vessel, as described in cultured capillary endothelium by Folkman and Haundenschild [16].

This ordered sequence is clearly controlled and is apparently mimicked in haemangiomas. However, Mulliken, Zetter and Folkman [46] reported important differences in the behaviour of endothelium from haemangiomas and from vascular malformations in tissue culture. Collagenase-digested tissue from the lesions was sieved, producing a mixture of capillary segments, pericytes, smooth-muscle cells and fibroblasts. Non-endothelial cells were "weeded" out of the gelatin cultures at intervals and the resultant endothelial cells were grown on gelatin and in plasma clots.

Capillary-like structures could be formed in vitro from endothelial cells isolated from haemangiomas in their proliferative phase but not from endothelial cells isolated from the vascular malformations studied. This result correlates well with clinical observations on the two types of lesion. Haemangiomas often grow rapidly in early life, regressing at 5 or 6 years of age, a change accompanied by decreased labelling of endothelial cells and gradual fibro-fatty replacement [45]. Vascular malformations however, only grow "commensurately with the child" or in response to haemodynamic load.

In previous work, Glowacki and Mulliken [19] had investigated the role of mast cells in these two groups of lesions. Mast cells have been thought to produce angiogenic factors, a finding suspected since the experiments of Azizkhan et al. [2] showed that mast-cell-conditioned media stimulated the migration of microvascular endothelial cells in vitro.

Fraser and Simpson [17] have located mast cells in close apposition to vessels involved in angiogenesis in studies of tumour angiogenic factor on chick chorioallantoic membrane. They point out that histamine, 5-hydroxytryptamine (5-HT) and heparin, all found in mast cells, stimulate endothelial cell growth in vivo.

In addition, there are many reports of the high numbers of mast cells found in vascular tumours (see Riley [51] for historical review) but no systemic survey of malformations or haemangiomas has been made. Glowacki and Mulliken [19] found a large number of mast cells in haemangiomas, a result which, considered with their labelling data, suggests that in these neoplasms mast cells play an important role in angiogenesis, in particular, Glowacki and Mulliken suggest, by

inducing directional elongation of new capillary sprouts by sustained release of heparin.

Whether these processes are disturbed in vascular dysgenesis has not been studied.

Pathophysiology

In limbs with an arteriovenous fistula there are changes in the circulation with temperature, and oxygenation changes and trophic changes in the skin [25,26,32]. Takiff and his colleagues [57] have recently described the effects on wound healing and blood flow of an experimentally induced groin arteriovenous fistula. They examined the healing strength of two wounds, one 3 cm distal to the groin incision and one 5 cm proximal to the foot, using a tensiometer at 7 d to measure wound strength after removal of the sutures. Wound healing occurred more rapidly in the proximal incision on the treated side, but was less effective on the treated side. These changes can be correlated with changes in flow, as these authors showed using radioactive ^{133}Xe studies.

Well-defined changes occur around arteriovenous fistulas, including decreased distal arterial perfusion, collateral formation adjacent to the fistula and venous stasis and hypertension distally. Local collaterals presumably explain the "enhanced" healing proximally in this experiment and the arterial and venous changes found will have adverse effects on the distal wound. In general the magnitude of the distal hypoperfusion depends on the size of the vessels. Lough et al. [37] found that with large fistulas in the dog a pressure drop of 94 mmHg (to 42 mmHg average pressure) was found. The venous pressure was 35 mmHg, providing a very small arteriovenous gradient. This reduces distal flow and promotes arterial shunting. Venous stasis may be accompanied by ulceration and venous diameter may increase by up to 80%.

Arteriovenous Malformations in the CNS

Warkany and Lemine [60] classify the clinical consequences of these lesions into two groups, which may be briefly considered to be:

1. Anomalies resulting from the cerebrovascular abnormality, e.g. shunting, heart failure, hydrocephalus, convulsions and so on.

2. Associated congenital malformations. These include congenital heart disease, unilateral renal hyperplasia, supernumerary digits, cervical spondylosis and systemic angiomatosis.

Arteriovenous Malformations of the Spinal Cord

Clinical symptomatology with severe difficulties for the patient may occur with arteriovenous malformations of the spinal cord. These appear to be commonest in the lumbosacral region and usually give rise to symptoms in the 3rd or 4th decades. Although a spinal bruit is mentioned in classical descriptions it is very seldom found clinically.

They occur in three forms as classified by Doppman et al. [13]:

1. A single coiled vessel along the posterior surface of the cord – the commonest type.
2. A glomus type – a plexus of coiled vessels fed by a single artery.
3. A "juvenile" type with multiple feeder vessels supplying an often extensive and bulky lesion.

With the first two types bleeding is a major problem but the third will often present with the signs of a space-occupying lesion or with cord infarction as a consequence of thrombosis.

Cobb Syndrome

In 1976 Kaplan et al. [31] reported a family in which spinal cord damage due to an intraspinal arteriovenous malformation was associated with cutaneous vascular anomalies in the related dermatome in the proband. Vascular malformations in the skin were also found in the mother and in a total of three generations in the family.

Where a cutaneous haemangioma is found in the same dermatome as a spinal vascular anomaly, the association is sometimes referred to as Cobb syndrome but such cases are extremely rare, there being fewer than 20 in the literature (see Kissel and Dureux [33] for review).

Aneurysm of the Great Vein of Galen

Carroll and Jakoby [7] described a number of cases of congenital arteriovenous malformations occurring in the central nervous system and pre-

senting with congestive heart failure. Eight of the 12 cases they reported had abnormalities of the great vein of Galen as did four of eight cases with similar presentation described by Holden et al. [22].

The vein of Galen is formed posterior to the pineal gland by the joining of the two internal cerebral veins. It runs backwards to empty into the straight sinus at the junction of the falx cerebri and the tentorium cerebelli. Arteriovenous fistulas affecting the system are presumed to be remnants of the branches of the middle cerebral artery which communicate directly with veins draining into the vein of Galen in 18–25 cm embryos, but the aneurysm may also be fed by one or both posterior cerebral arteries or one of their branches. Less commonly there are communications with posterior branches of the middle cerebral artery or with anomalous branches of the carotid and/or basilar system.

The entire system of venous sinuses tends to be distended in these cases. Periventricular haemorrhage and infarction are common due to high pressure from the tumour-like vascular mass. Calcification or thrombosis may occur [35,55].

Some Specific Entities

Clinical Points

The hypertrophy which occurs in limbs where arteriovenous malformations are present is sometimes considered to be due to increased blood flow (Holman 1968) [24]. However, the mechanism by which this may work is unclear and there are reports of shortening of the affected limb [9]. In addition, studies with isotope-labelled microemboli have failed to show arteriovenous communications in legs affected by gigantism [48] and occasionally a limb affected may not be that in which the arteriovenous malformation exists.

Sturge–Weber Syndrome

Classically, the syndrome consists of naevus formation in the distribution of the sensory branches of the fifth cranial nerve and of choroidal angiomas and other ocular abnormalities, including glaucoma (Fig. 36.2). The meninges appear deeply congested due to the great density of small veins, and calcification occurs in the outer layers of the cerebral cortex. This calcification is mainly perivascular and is

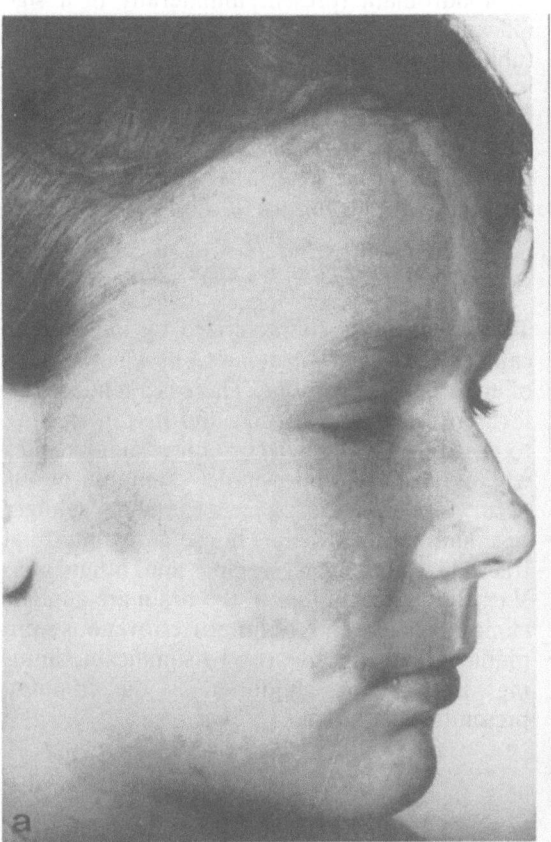

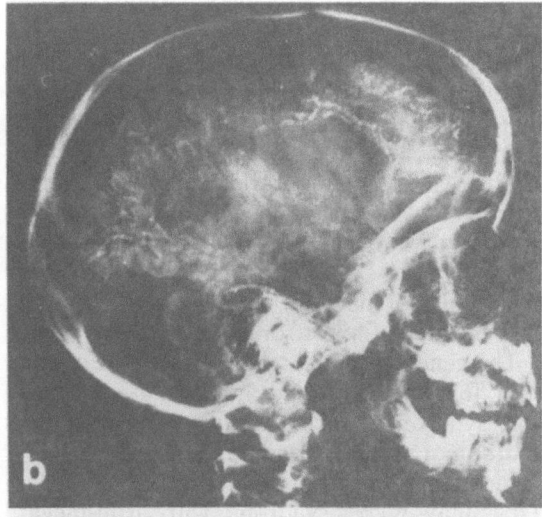

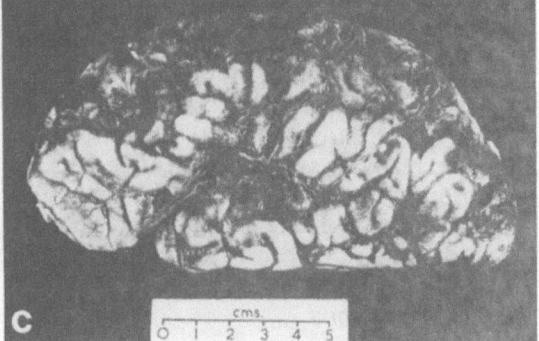

Fig. 36.2. a Facial haemangioma in a case of Sturge–Weber syndrome. **b** X-ray appearance showing diffuse calcification of the vasculature. **c** Cerebral angiomatosis.

accompanied by iron deposition. It is presumably acquired as it is not present in the 1st year of life. Symptoms develop after 3 or 4 months, usually in the form of grand mal seizures.

The disease is apparently sporadic, no genetic pattern being evident.

Von Hippel–Lindau Angiomatosis

In this syndrome of capillary angiomas in the skin and viscera there are associated retinal angiomas and cysts in the cerebellum, pancreas, kidneys, liver and lung. The cerebellar tumours are haemangioblastomas and similar lesions may be found in the brain stem or spinal cord. In familial cases transmission is by an autosomal dominant gene.

Symptoms are usually cerebellar dysfunction and raised intracranial pressure, but these do not appear until adolescence and retinal lesions are not usually seen until after 20 years of age.

There is a greatly increased risk of renal tubular carcinoma in these patients and its not infrequent bilaterality has encouraged the use of transplantation in therapy [49].

In a case of von Hippel–Lindau syndrome Ishmail et al. [27] examined cerebellar, retinal and spinal haemangioblastomas by immunocytochemistry and electron microscopy. The tumours were found to be positive for neuron-specific enolase and variably positive for somatostatin, pancreatic polypeptide and bombesin. These findings were considered in regard to the uncertain histogenesis of these tumours and led the authors to propose a neuro-electodermal origin for them, an origin supported by the finding of membrane-bound electron-dense granules as reported by Ishwar et al. [28] in these tumours.

Heterotopia and micropolygyria have been reported and hemi-megalencephaly can occur, often without associated hemi-hypertrophy.

Klippel–Trenaunay–Weber Syndrome

In this syndrome there is typically congenital hypertrophy of one or, less commonly, more than one limb. Rarely hypertrophy develops soon after birth.

There are numerous varied cutaneous haemangiomata which may be cavernous or capillary together with phlebectasia and varicosities. The lower part of the body is most often involved but the haemangiomata are not always co-terminal with the area of hypertrophy.

The genetics are unknown, although Zonana et al. [62] have described a dominantly inheritable syndrome which resembles this with the addition of multiple lipomas.

In a recent study of eight cases of Klippel–Trenaunay–Weber syndrome with hemi-megalencephaly and hemi-hypertrophy, Matsubara et al. [40] have shown that the hemi-hypertrophy produced is accompanied by a normal form and normal DNA content of the neurons and glial cells in the affected part of the brain. Hemimegalencephaly is associated with an increase in the absolute number of nerve and glial cells, and the authors suggested an effect mediated by mesenchyme.

Maffucci Syndrome

The major anomalies in this syndrome are skeletal. Bowing of the long bones with asymmetrical retardation of growth develops at any time from birth to adolescence, and is associated with

enchondromata (present unilaterally in a significant number of cases) in the hands, feet and tubular long bones (Fig. 36.3a).

The haemangiomas are usually associated with the enchondromata in the surrounding subcutaneous fat and develop in the 1st year of life (Fig. 36.3b). Phlebectasia is common with the formation of phleboliths.

Rendu–Osler–Weber Disease

This condition is characterized by multiple or capillary-like vessels, having walls which consist of endothelial cells alone. There is a tendency to form arteriovenous fistulas and it is interesting to note that the telangiectases are found at sites where normal arteriovenous communications occur (tongue, lips, ears, face, conjunctiva, finger tips, nail beds) as well as in the gastrointestinal tract, bladder, lungs, vagina and other sites. Vascular abnormalities in the brain are a major clinical problem and the lung arteriovenous communications may give rise to significant shunting. Epistaxis in childhood is the common presenting symptom.

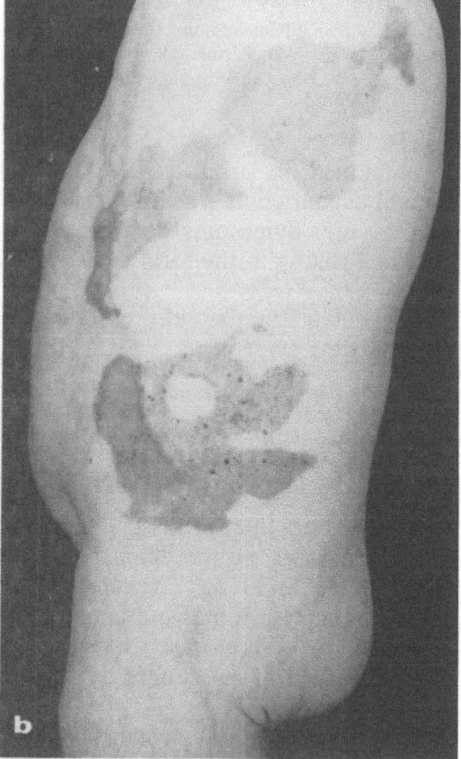

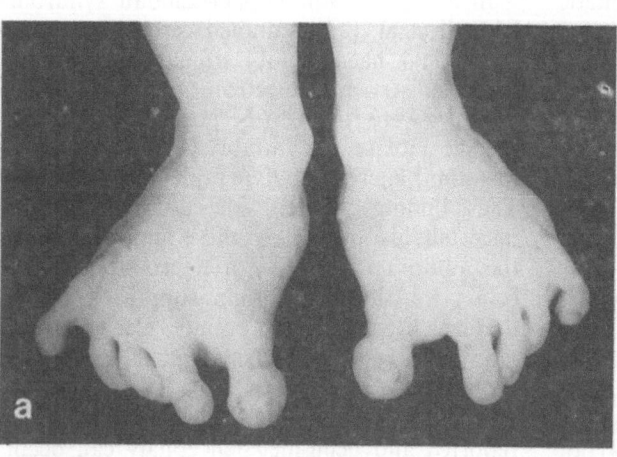

Fig. 36.3. a Deformity of the foot due to osteochondromas. **b** Trunk haemangiomas from the same case.

The condition is apparently inherited as an autosomal dominant gene.

Giant Haemangioma with Thrombocytopenia

These lesions are usually present at birth but purpura and thrombocytopenia develop later [21]. The haemangioma may be in any part of the body, may penetrate muscles and bones and may be multiple. Radiotherapy to the tumour may bring prompt restoration of the platelet count. Most large or multiple haemangiomata are not accompanied by thrombocytopenia.

Neonatal Haemangiomatosis

Holden and Alexander [23] reviewed a total of eight cases of diffuse neonatal haemangiomatosis where many lesions were found in the central nervous system, muscle, thymus, cervix uteri, iris, skin, liver, lung, kidneys, bowel and other tissues. Thrombocytopenia may be an associated finding (Fig. 36.4).

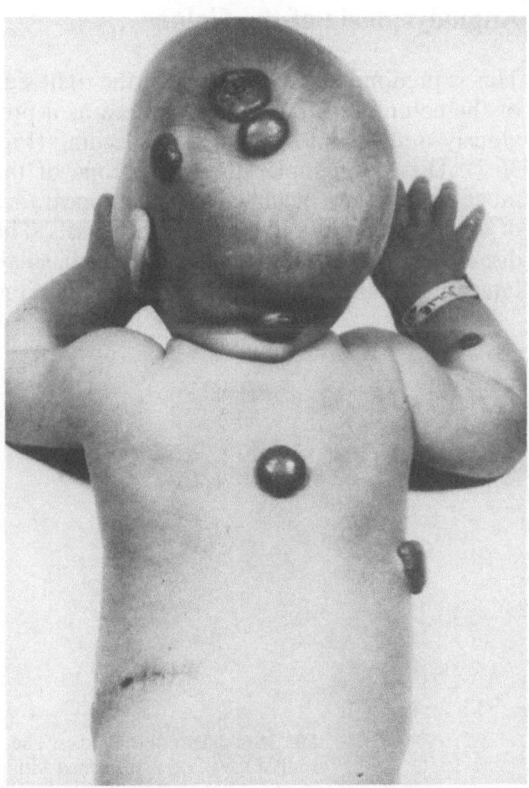

Fig. 36.4. Numerous fleshy haemangiomas in a child with visceral lesions and thrombocytopenia. A scar of a splenectomy is seen.

Arterial Dysplasias

Idiopathic Arterial Calcification in Infancy

This is a rare condition characterized by calcification mainly localized to the internal elastic lamina and with associated intimal thrombosis due to luminal narrowing. Almost all vessels except those of the central nervous system may be affected and in some instances the spread of calcification into the media may result in x-ray changes detectable in life. Myocardial ischaemia is the commonest cause of death but apparent recovery with loss of calcium from arteries has been recorded.

There is no associated inflammatory change although some giant cells may be present. For a review see Moran [44].

Similar changes can occur as a result of metastatic calcification.

Supravalvular Aortic Stenosis

This anomaly occurs in three distinct clinical settings. It may rarely occur as an apparently isolated congenital defect in otherwise normal patients [11]. More commonly the defect occurs as part of a syndrome of aortic obstruction with multiple peripheral pulmonary artery stenoses having a pattern of inheritance suggesting it is transmitted by an autosomal recessive gene [30]. Finally, supravalvular stenosis occurs as a manifestation of the syndrome of "hypervitaminosis D" or vitamin D hypersensitivity in which increased intestinal absorption of calcium occurs. In this syndrome it is accompanied by "elfin" facies (retrousse nose with a flat nasal bridge, heavy orbital ridges, wide mouth, see Fig. 36.5), psychomotor and growth retardation, craniosynostosis, strabismus, dental anomalies, pulmonary and peripheral vascular stenoses and precocious puberty in females (see Friedman [18] for review).

It is not clear that these three types of supravalvular change are pathogenically related although the presence of stenotic vascular lesions in two forms is interesting. Recent review of a large group of children with idiopathic hypercalcaemia [39] reveals a wider symptom complex with feeding difficulties a major feature. Most patients do not have vascular disease.

Fig. 36.5. Two children showing the typical facies of the idiopathic hypercalcaemia syndrome.

Macaroni Arteries

McDonald et al. [41] described a family with multiple peripheral pulmonary artery and supravalvular aortic stenosis. Most elastic arteries were involved and the vessels were thick walled with abundant elastic tissue. The syndrome is apparently a very rare one since there are no other reports.

Neurofibromatosis

Individuals with neurofibromatosis not infrequently have associated haemangiomas, and the neurofibromatous tissue itself often contains large dilated thin-walled vessels and frankly angiomatous areas.

There have been many detailed classifications of the vascular lesions found in association with neurofibromatosis, but they are probably all best considered to be examples of a general mesodermal dysplasia. Their main feature is an anomalous proliferation of smooth muscle [20].

The changes are found most often in the aorta or its abdominal branches but may occur in the coronary vessels or the carotid or vertebral arteries. The changes resemble fibromuscular dysplasia with nodular cellular aggregates in the media but may show largely intimal changes (Fig. 36.6). Reviews are given by Salyer and Salyer [53] and Mikuz et al. [42].

Tissue resembling glomus arteriovenous shunts may be found in association with neurofibromas.

Angiodysplasia of the Colon

This is predominantly a disease of the right side of the colon, and is being recognized as a previously underestimated cause of bleeding (Fig. 36.7). Dissection under the microscope of the mucosa shows a typical "coral reef" appearance of dilated mucosal and submucosal vessels. The disease has been described in the stomach, small intestine and the left side of the colon and in the

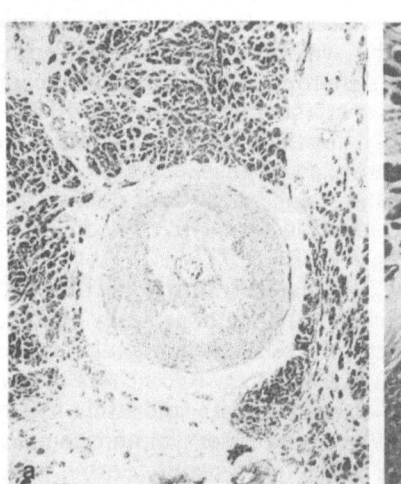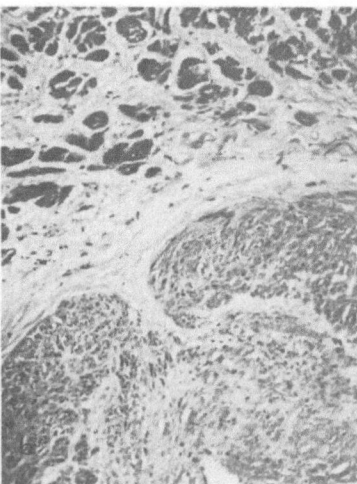

Fig. 36.6. Neurofibromatosis. The wall of the vessel is thickened with an abnormal media in which smooth-muscle cells are occasionally seen with some production of new elastic tissue. **a** Myocardium. H & E, × 80. **b** Myocardium. H & E, × 240

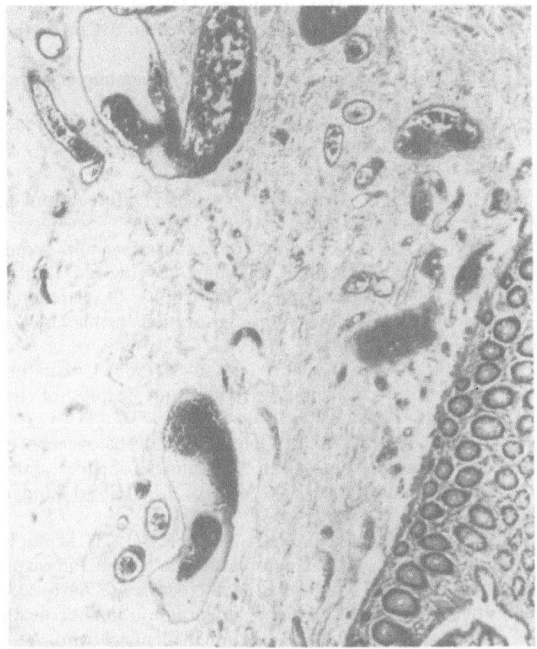

Fig. 36.7. Dilated submucosal vessels from a case of angiodysplasia in a 65-year-old woman presenting with melaena.

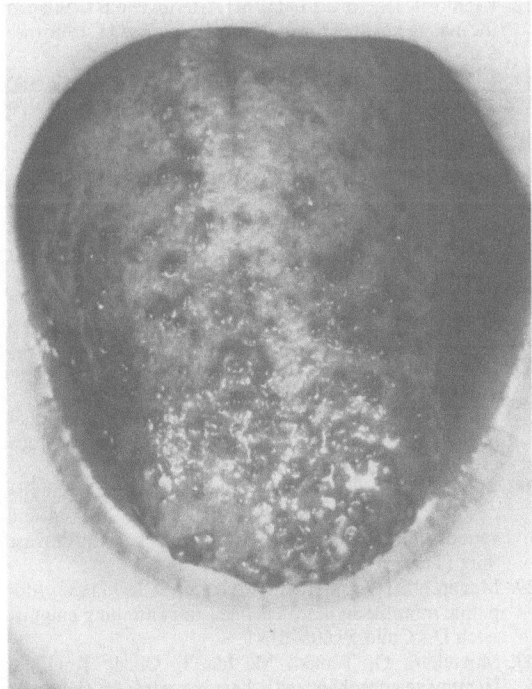

Fig. 36.8. Capillary haemangiomas on the tongue in a child with evidence of gastrointestinal bleeding.

latter two sites it often occurs in younger patients [43] where congenital abnormality is more likely to be a possible cause. However, in angiographic studies arteriovenous communications are not found [3] and careful vascular studies suggest that the lesions are degenerative in nature [6].

Other Gastrointestinal Vascular Anomalies

In general these anomalies present in childhood with severe anaemia and persistent ill health. They are of three main types:

1. Cavernous haemangiomas, usually circumscribed but occasionally diffuse.
2. Capillary haemangiomas (Fig. 36.8).
3. Gastrointestinal haemangiomas associated with cutaneous lesions.

All types tend to be multiplied and although usually confined to the submucosa they may involve the whole thickness of the wall and cause intestinal obstruction. A useful review is that of Nader and Margolin [47].

Haemangiomas of the small intestine may form part of the Kippel–Trenaunay syndrome, when they are accompanied by thrombocytopenia.

A distinct entity which may present as massive haemorrhage in the adult is the so-called cirsoid aneurysm of the stomach. In these cases there is usually a dilated and tortuous artery in the submucosa but no abnormal arteriovenous communication which the name suggests. A good description and brief review is found in Richter [50].

Intracranial Aneurysms

A comment on the aetiology of berry aneurysms has been made in Chap. 3 which deals with information that suggests that there may be a genetic factor determining the nature and frequency of the aneurysms. Familial aggregates of intracranial aneurysms have been observed repeatedly [4,53,58] and the lesions have been observed in identical twins [15,29]. Lagos, in a review in 1977 [34], said that 39 families with more than one affected member had been reported to that date. More recently Evans et al. [14] described a family with apparent autosomal dominant inheritance.

Ehlers–Danlos syndrome may occur with multiple intracranial aneurysms [52]. The association of intracranial aneurysms with polycystic kidneys was first described in detail by Bigelow [5].

References

1. Appleberg M (1975) Congenital atresia of the external iliac artery. S Afr Med J 49:1885–1886
2. Azizkhan RG, Azizkhan JC, Zetter BR, Folkman J (1980) Mast cells heparin stimulates migration of capillary endoethelial cells in-vitro. J Exp Med 150:931–944
3. Baer JW, Ryan S (1976) Analysis of cecal vasculature in the search for vascular malformations. AJR 126:394–405
4. Bannerman RM, Ingall GB, Graff CJ (1970) The familial occurrence of intracranial aneurisms. Neurology 20:283–292
5. Bigelow NH (1953) The association of polycystic kidneys with intracranial aneurisms and other related disorders. Am J Med Sci 225:485–494
6. Boley SJ, Sammartano R, Adams A, Di Biase A, Kleinhaus S, Spagregen S (1977) On the nature and aetiology of vascular ectasias of the colon. Gastroenterology 72:650–660
7. Carroll CP, Jakoby RK (1966) Neonatal congestive heart failure as the presenting symptom of cerebral arterio-venous malformations. J Neurosurg 25:159–163
8. Cliff WJ (1963) Observations on healing tissue: a combined light and electron microscopic investigation. Philos Trans R Soc London [Biol] 246:3205–325
9. Cotton LT, Sykes BJ (1969) The treatment of diffuse congenital arteriovenous fistulae of the leg. Proc R Soc Med 62:245–247
10. Dean RH (1983) Congenital arteriovenous malformations. In: Dean RH, O'Neill JA Jr (eds) Vascular disorders of childhood. Lea and Febiger, Philadelphia
11. Denie JJ, Verheugt (1958) Supravalvular aortic stenosis. Circulation 18:902
12. Dent DM, Fataar S, Rose AG (1981) Ainhum and angiodysplasia. Lancet II:396–398
13. Doppman JL, Di Chiro G, Ommaya AK (1971) Selective arteriography of the spinal cord. Warren Green, St Louis
14. Evans TW, Venning MC, Strang FA (1981) Dominant inheritance of intracranial berry aneurysm. Br Med J 283:824–825
15. Fairburn B (1973) "Twin" intracranial aneurysms causing sub-arachnoid haemorrhage in identical twins. Br Med J i:210–211
16. Folkman J, Haudenschild CC (1980) Angiogenesis in-vitro. Nature 288:551–556
17. Fraser RA, Simpson JG (1983) Role of mast cells in experimental tumour angiogenesis. In: Nugent J, O'Connor M (eds) Development of the vascular system. Pitman, London, pp 120–131 (Ciba Foundation Symposium 100)
18. Friedman WF (1968) Vitamin D and the supravalvular aortic stenosis syndrome. In: Wollam DHM (ed) Advances in teratology 3 Logos Press, London, pp 85–96
19. Glowacki J, Mulliken JB (1982) Mast cells in hemangiomas and vascular malformations. Pediatrics 70:48–51
20. Green JF, Fitzwater JE, Burgess J (1974) Arterial lesions associated with neurofibromatosis. Am J Clin Pathol 62:481
21. Hill GJ II, Longino LA (1962) Giant haemangioma with thrombocytopenia. Surgery 114:304
22. Holden AM, Flyer DC, Shillito J, Jr, Nadas AS (1972) Congestive heart failure from intracranial arteriovenous fistula in infancy. Pediatrics 49:30–39
23. Holden KR, Alexander F (1970) Diffuse neonatal hemangiomatosis. Pediatrics 46:411–421
24. Holman E (1968) Abnormal arteriovenous communications, 2nd edn. CC Thomas, Springfield, Illinois
25. Horton BT (1932) Hemihypertrophy of extremities associated with congenital ateriovenous fistula. JAMA 98:373–379
26. Ingebrigsten R, Krog J, Lerand S (1963) Circulation distal to experimental arteriovenous fistulas of the extremities. Acta Chir Scand 125:3089–317
27. Ishmail SM, Jasani B, Cole G (1985) Histogenesis of haemangioblastomas: an immunocytochemical and ultrastructural study in a case of von Hippel Lindau syndrome. J Clin Pathol 38:417–421
28. Ishwar S, Taniguchi RM, Vogel FS (1971) Multiple supratentorial haemangioblastomas. Case study and ultrastructural characteristics. J Neurosurg 35:396–405
29. Jokl E, Wolffe JB (1954) Sudden, non-traumatic death associated with physical exertion in identical twins. Acta Genet Med Gemme Ilol (Roma) 3:245
30. Kahler RL, Braunwald E, Plauth WH Jr, Morrow AG (1966) Familial occurrence of atrial septal defect with AV conduction abnormalities, of supravalvular aortic and pulmonic stenosis, and of ventricular septal defect. Am J Med 40:384
31. Kaplan P, Hollandberg RD, Clarke Fraser F (1976) A spinal arteriovenous malformation with heriditary cutaneous haemangiomas. Am J Dis Child 130:1329–1331
32. Kinmorth JB, Negus D (1974) Arteriovenous fistulae in the management of lower limb discrepancy. J Cardiovasc Surg 15:447–453
33. Kissel P, Dureux JB (1972) Cobb syndrome: cutaneomeningeospinal angiomatosis. In: Vinken P, Bruyn GW (eds) Handbook of clinical neurology, vol 14. North Holland, Amsterdam. pp 429–445
34. Lagos JC (1977) Congenital aneurysms and arteriovenous malformations. In: Vinken PJ, Bruyn GW (eds) Handbook of clinical neurology, Vol 31. Congenital malformations of the brain and skull, part II. North Holland, Amsterdam, pp 137–209
35. Lazar ML (1974) Vein of Galen aneurysm: successful excision of a completely thrombosed aneurysm in an infant. Surg Neurol 2:22–24
36. Lewis D de W (1930) Congenital arteriovenous fistulae. Lancet II:621
37. Lough FC, Giordano JM, Hobson RW (1976) Regional hemodynamics of large and small femoral arteriovenous fistulas in dogs. Surgery 79:346–349
38. Malan E, Puglionisi A (1965) Congenital angiodysplasias of the extremities. II. Arterial, arterial and venous, and haemolymphatic dysplasias. J Cardiovasc Surg (Torino) 6:255–345
39. Martin NDT, Snodgrass GJAI, Cohen RD (1984) Idopathic infantile hypercalcaemia – a continuing enigma. Arch Dis Child 59:605–613
40. Matsubara O, Tanaka M, Ida T, Okeda R (1983) Hemimegalencephaly with hemihypertrophy (Klippel–Trenaunay–Weber syndrome). Virchows Arch [A] 400:155–162

41. McDonald AH, Gerlus LM, Somerville J (1969) Familial arteriopathy with associated pulmonary and systemic arterial stenoses. Br Heart J 31:375

42. Mikuz G, Weiser G, Propst A (1975) Vascular neurofibromatosis. Pathol Microbiol 43:195

43. Miller KD, Tutton RH, Bell KA, Simon BK (1979) Angiodysplasia of the colon. Diagn Radiol 132:309–313

44. Moran JJ (1975) Idiopathic arterial calcification of infancy: a clinicopathologic study. In: Sommers SC (ed) Cardiovascular pathology decennial, 1966–1975. Appleton Century Crofts, New York, p 47

45. Mulliken JB, Glowacki J (1982) Haemangiomas and vascular malformations in infants and children: a classification based on endothelial characteristics. Plast Reconstr Surg 49:412–420

46. Mulliken JB, Zetter BR, Folkman J (1982) In-vitro characteristics of endothelium from hemangiomas and vascular malformations. Surgery 92:348–353

47. Nader PR, Margolin F (1966) Haemangioma causing gastrointestinal bleeding. Am J Dis Child 11:215

48. Partsch H, Lafferer O, Mostbeck A (1974) Zur Diagnostik von arteriovenousen Fistein bei Angiodysplasien der Extremitaten. Vasa 4:288–294

49. Petersen GJ, Codd JE, Cuddihee RE, Newton WT (1977) Renal transplantation in Lindau–Von Hippel disease. Arch Surg 112:841–842

50. Richter RM (1975) Massive gastric hemorrhage from sub-mucosal arterial malformation. Am J Gastroentrol 64:324–326

51. Riley JF (1959) The mast cells. Livingstone, Edinburgh

52. Rubenstein MK, Cohen NH (1964) Ehlers–Danlos syndrome associated with multiple intracranial aneurysms. Neurology 14:125–132

53. Sakai N, Sakata K, Yamada H (1974) Familial occurrence of intracranial aneurysms. Surg Neurol 2:25–29

54. Salyer WR, Salyer DC (1974) The vascular lesions of neurofibromatosis. Angiology 25:510–519

55. Siqueria EB, Murray KJ (1972) Calcified aneurysms of the vein of Galen: report of a presumed case and review of the literature. Neurochirugica 15:106–112

56. Szilagyi DE, Elliot JP, De Russo FJ, Smith RF (1965) Peripheral congenital arteriovenous fistulas. Surgery 57:61

57. Takiff H, Owens ML, Williams RA (1984) Effect of arteriovenous fistula on wound healing and skin blood flow. Br J Exp Pathol 65:677–681

58. Thierry A, Ballivet J, Dumas R (1972) Les cas famileaux d'aneurisms intracranienne. Neurochirurgie 18:267–276

59. Tice DA, Clauss RA, Keirle AM, Reed GE (1963) Congenital fistulae of the extremities. Arch Surg 86:460

60. Warany J, Lemine RJ (1984) Arteriovenous malformations of the brain: A teratologic challenge. Teratology 29:333–353

61. Woolard HH (1922) The development of the principal arterial stems in the forelimbs of the pig. Contrib Embryol Carnegie Inst (Washington) 14:139

62. Zonana J, Rimion RL, Davis DC (1976) Macrocephaly with multiple lipomas and hemangiomas. J Pediatr 89:600

Chapter 37

Tumours and Tumour-like Lesions

G. Chomette and J. Diebold

Tumour-like Lesions

Pseudohaemangiomas

It is often difficult to make the distinction between benign vascular tumours (haemangioma) and vascular tumour-like lesions (reactive vascular hyperplasia). Two granulation tissue-type pseudohaemangiomas, pyogenic granuloma and granuloma gravidarum, will be considered first.

Pyogenic Granuloma

This lesion is a friable, polypoid vascular excrescence that is reddish-black in colour, sessile or pedunculated, and located in the integument or in mucosal membranes. It is thought to result from reactive vascular hyperplasia in a zone of inflammation. Existence of this lesion underlines the fact that adult endothelium retains growth potential. Pyogenic granulomata are most often found at digital extremities, but are also seen on the face, lips and tongue. They generally appear after local trauma, and may grow quickly to a size of several centimetres. There is no sex or age predilection for the development of these lesions.

Histologically, pyogenic granulomata consist of numerous capillary knots arranged in lobules supplied by an arteriole of larger calibre [128]. Some of the capillaries may be fine and are often impermeable, whereas others are distinct with a mature endothelium. Overlying purulent exudates and ulcerations may be present. On occasion, certain of these "botryomycomas" may re-epithelialize. The capillary knots lie within a fibromyxoid connective tissue matrix which harbours inflammatory cells (lymphocytes, plasmocytes and histiocytes). As these lesions age, this matrix becomes sclerotic and impinges upon the capillary lumina.

Diagnosis is generally straightforward, although it is important to rule out certain malignancies such as Kaposi's sarcoma.

Pyogenic granulomata have a characteristically benign evolution even though they seldom regress spontaneously. There are generally no recurrences after complete excision, but multiple satellite nodules may appear if ablation is incomplete [180].

Granuloma Gravidarum

This reactive lesion affects the buccal mucosa of susceptible individuals under certain hormonal influences, notably that of oestrogen [119].

Associated risk factors include local trauma and poor dental hygiene. Such angiogranulomata generally appear during the 3rd month of pregnancy and may grow considerably thereafter. After delivery, complete regression is the rule, but small nodules may persist that can flare up during subsequent pregnancies.

Gingival involvement is either localized or diffuse. Discrete reddish, ulcerated and friable lesions may exist between two teeth, or alternatively diffuse hypertrophy of the gums may be evident. The latter is often associated with ulceration, infection and haemorrhage.

Histologically, these lesions resemble pyogenic granulomata. Hyperplastic capillaries are seen, generally embedded in inflammatory, oedematous connective tissue. Occasionally the capillary network consists of a few large vessels also embedded in a fibro-oedematous matrix.

Surgery is generally not required for this form of gingival hyperplasia since the condition regresses spontaneously after delivery.

Subadventitial Cystic Degeneration of the Popliteal Artery (Cystic Disease of the Popliteal Artery, Adventitial Mucoid Cyst)

This lesion, frequent only in young men, was first described by Ejrup and Hiertonn in 1954 [58]. A review of published cases by Flanigan et al. in 1979 enumerated a total of 115 cases [65].

The hallmark of this lesion is the abrupt onset of lower limb ischaemia accentuated by bending of the knee in a young person. There is an associated highly characteristic angiographic "hourglass" appearance. It may occasionally be difficult to distinguish this lesion from a dissecting aneurysm or from extrinsic compression. Surgery is the treatment of choice, and should involve either cyst evacuation (leading to the risk of recurrence) or resection and grafting.

Histologically, one or more pseudocysts are present within the external media and the adventitia of the popliteal artery (Fig. 37.1). These cavities are either empty or they contain acid muco-substances that stain intensely with alcian blue at acidic pH. The hyaluronic acid content is probably considerable given the digestion observed with hyaluronidase treatment. PAS staining is weak. Macrophages and lymphocytes are often seen. Territories of myxoid degeneration are sometimes found adjacent to pseudocysts. Popliteal pseudocysts are unilateral.

Rarely, a similar process can involve iliac or femoral arteries.

It is has been suggested that pseudocysts form during embryological development by incorporation of mucous-secreting cells into the arterial wall, analogous to the formation of synovial cysts [89]. In fact, pseudocysts are often adjacent to joints and in some cases have been reported to communicate with joint cavities [163]. Other authors suggest that a primary defect in endothelial development is at play [48], and still others invoke repeated trauma as the probable cause [82,86,89]. The latter hypothesis was recently favoured by Mark et al. [115] in a case of nail patella syndrome associated with knee malformation and hypertrophy of the femoral condyles.

Haemangiomas

Haemangiomas may be difficult to distinguish from those lesions previously described, from hamartomas (arteriovenous malformations) and from vascular tumours [6, 41 and Chap. 36].

Haemangiomas are the most frequent soft tissue tumours, mainly affecting children and adolescents. In women, capillary or cavernous haemangiomas may vary in size during pregnancy and menopause. Endothelial cells are thus sensitive to hormonal influences, particularly those of oestrogen [60].

Several types of haemangioma may be identified either by their morphology or their localization.

Capillary Haemangiomas

These lesions are found primarily in the skin of the head and neck, on the lips, in the mouth and the pharynx and rarely in the nose or the digestive tract. There is no danger of recurrence after complete surgical excision.

Capillary angiomas are variable in size and are either reddish, superficial and spreading, or blue, nodular and somewhat lobulated. These tumours infiltrate underlying adipose tissue and may invade muscular aponeuroses. They may be rather amorphous or sometimes loosely encapsulated by fibrous tissue. Brownish discolouration may be present in areas of old haemorrhage. Surface ulceration and inflammation may render the lesion superficially similar to a botryomycoma.

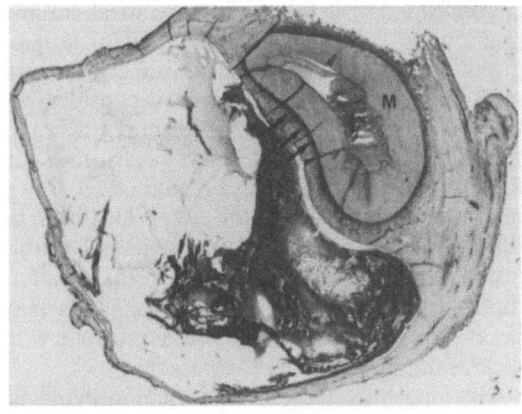

Fig. 37.1. Subadventitial cystic degeneration of the popliteal artery. Voluminous cystic area displacing the popliteal artery which appears to be compressed (*arrow*). *M*, media of the popliteal artery. Haematoxylin–eosin–safran, × 14

Histologically, a pulsating, blood-filled mass of capillaries is seen. The endothelium is generally flattened but occasionally turgid, and endothelial cell nuclei are uniform in appearance and somewhat vesicular. There is no stratification or mitosis. The capillaries are separated by a fine connective tissue reticulum. This reticulum may be more or less prominent giving rise to predominantly vascular lesions or to "older" angiomas with few capillaries and a dense connective tissue (largely collagen and fat) capsule. Lobular segmentation does not occur as it does in pyogenic granuloma. The capillary network is supplied by large vascular trunks and drained by venules. Inflammatory cells may surround capillary haemangiomas, and haemosiderin may be present following haemorrhage in advanced lesions (sclerosing angiomas).

These lesions are highly differentiated tumours with marked alkaline phosphatase activity and normal factor VIII production [25]. The endothelial cells visualized by electron microscopy (Fig. 37.2) contain Palade bodies, endocytotic vesicles and microfilaments and are linked by desmosomes.

A basement membrane covers the anti-luminal endothelial cell pole and encompasses adjacent adult pericytes. These cells are elongated and contain endocytotic vesicles and abundant filaments.

Specific Subtypes

Capillary haemangiomas may be further categorized into juvenile and senile forms. Juvenile haemangiomas, also called childhood haemangioendotheliomas, represent an immature form of capillary haemangioma. They most often involve the cervico-facial region, especially the parotid. They may be multiple and are generally noticed shortly after birth. They are impressive in size and are most prominent at 6 months of age. Spontaneous regression occurs subsequently, leaving a more or less pigmented scar. Most of these lesions are dark red but deeper haemangiomas may be lightly coloured. All are supplied by a single arteriole [60]. After the formation of pigmented scars, aesthetic surgery may be envisaged.

Senile haemangiomas or Morgan spots are characteristically seen in adults and occasionally in adolescents. These lesions are tiny and are found over the trunk and the extremities. They represent ectatic thin-walled capillaries infiltrating the superficial dermis under a thinned epidermis.

Cavernous Haemangiomas

Cavernous haemangiomas may co-exist with capillary haemangiomas and represent a more evolved form of the latter. They are also found in the head and neck region (lips, tongue, scalp). However, cavernous haemangiomas are deep-seated and often barely visible (e.g. macroglossia may be a manifestation of haemangioma of the tongue). Some are situated in the digestive tract: fairly large lesions may be present in the stomach or colon and provide a locus for ulceration and bleeding. The genitourinary tract may also be involved (kidney, bladder, uterus). Frequently, small (1 or 2 cm), multiple haemangiomas may be present under the splenic or the hepatic capsules. Deeper lesions, invisible to the naked eye, may also be found within these organs often constituting masses of considerable size. In the mesentery or the retroperitoneum, angiomas with a supporting adipose component (angiolipomas) may be seen. Bony haemangiomas give rise to osteolytic lesions found primarily in the vertebral column, long bones and facial bones, notably the mandible. Such bony haemangiomas may bleed and be responsible for calamitous complications such as tooth loss.

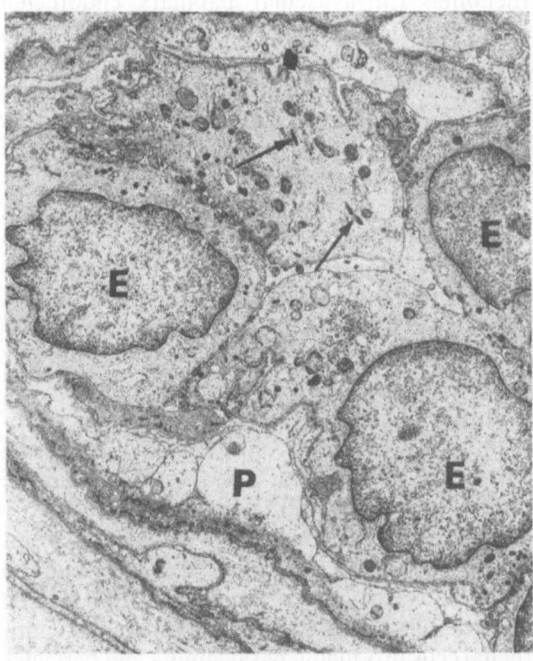

Fig. 37.2. Electron micrograph of a capillary haemangioma showing turgescent endothelial cells (*E*) and pericytes (*P*). Weibel–Palade bodies are present (*arrow*). Uranyl lead, ×5000

Macroscopically, cavernous haemangiomas when superficial may resemble haematomas. Visceral haemangiomas are reddish-black and also ill-defined. They may be encrusted with calcification or show ossification that appears as curvilinear opacities or nodular phleboliths on radiography, the latter being the more specifically diagnostic.

Histologically, angiomas of this nature are large, filled with blood and lined by a flattened endothelium. They comprise multiple anastomotic cavities that are separated by fibrous tissue, and occasionally entrap lymphocytes and adipocytes. In the more differentiated haemangiomas, smooth-muscle cells may be seen.

Cavernous haemangiomas may cause compression (cervical angiomas), haemorrhage (angiomas of the neuroaxis) and thrombocytopenic purpura [Kasabach–Merrit syndrome, 178]. The latter condition occurs in childhood, particularly in individuals with multiple, large cavernous haemangiomas; rarely other vascular tumours are implicated (capillary angioma, angiomatoses, angiosarcoma).

The angioma appears to trigger a consumptive coagulopathy resulting in cutaneous and visceral purpuric haemorrhage. This process is associated with an abrupt increase in the size of the angioma. Finally, the presence of multiple hepatic angiomas in the child may precipitate heart failure [22].

Venous Haemangioma

This is a rare tumour [60] found principally in the retroperitoneum, the mesentery and in skeletal muscle of the extremities. The haemangioma is deep-seated and often difficult to identify. Calcifications may mark the location of the angiomaon plain radiographs. Angiography alone is insufficient for opacification, and venography is required. Macroscopically, venous haemangiomas are clusters of dilated veins that are thick-walled due to an accumulation of fibrous and muscular tissue. In addition to calcification, thrombosis may occur within these angiomas.

Specific Types

The circulatory consequences of haemangiomas and indications for therapy are discussed elsewhere.

The most frequent type of soft tissue angioma is the intramuscular angioma [3]. It is charac-teristically latent, ill-delimited and often multiple. Lower limb involvement is classical, and both radiography (because of calcification) and angiography may be used to visualize these lesions. The most common malformation is the capillary haemangioma which may at times have an alarming appearance (proliferating endothelium with mitotic figures). Cavernous angiomas in this location are embedded in adipose tissue (angiolipoma).

Synovial haemangiomas (villonodular tenosynovitis) are rare, affecting tendon sheaths and joint cavities. The knee joint is almost always the site of involvement, and multiple episodes of haemarthrosis are characteristic. The angioma may be amorphous or pedunculated, is generally of the cavernous type and lies within a connective tissue matrix containing inflammatory cells and haemosiderin. It is unclear whether these lesions represent hamartomas or whether they arise after trauma. Individuals with synovial haemangiomas may have other angiomas elsewhere.

Intranodal haemangiomas (nodal angiomatosis) [63] are composed of small, well-developed and differentiated capillaries, and are found in the subcapsular zone of lymph nodes replacing the marginal sinus. Diagnostic confusion may arise with Kaposi's sarcoma.

Intravascular vegetant haemangioendotheliomas (intravascular papillary endothelial hyperplasia were first identified by Masson [117] and more recently reviewed by Amerigo and Beazy [5]. These tiny reddish-blue nodules are found within certain angiomatous lesions (laryngeal polyps, haemorrhoids, urethral caruncles). These nodules are distinct [106], generally superficial, firm and small in size (2–3 cm). They are located near cephalic veins, and on the fingers, thorax and abdomen. Rarely of traumatic origin, these lesions may arise in pre-existing vascular malformations (pyogenic granuloma, haemangioma). Histologically, multiple papillary fronds are seen within a dilated vessel. Endothelial cells are strewn along the fibro-hyaline axis of these fronds (Fig. 37.3), and the vestiges of a thrombus may be present on the underlying vessel wall. Veins are preferentially involved and lesions may sometimes erupt through the vessel wall into the extra-parietal space. Ultrastructural studies confirm a benign and vascular origin. Well-differentiated endothelial cells producing basal lamina and pericytes are seen. The lesion is thought to represent an exaggerated endothelial cell reaction to stasis, and thrombus formation, and is not a true angioma.

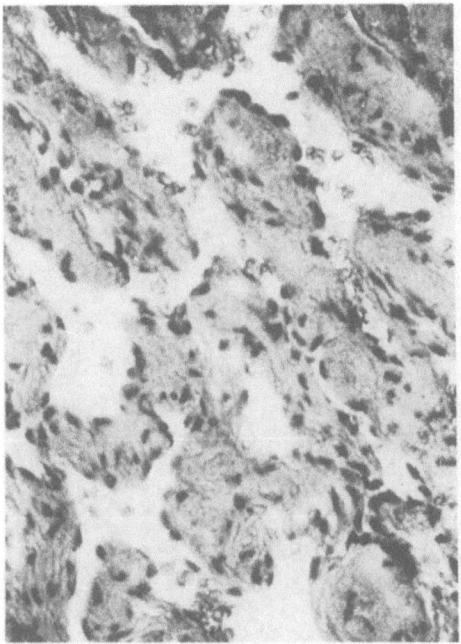

Fig. 37.3. Intravascular papillary endothelial hyperplasia. Fibrohyaline papillary formations are covered by swollen endothelial cells. Haematoxylin–eosin–safran × 220

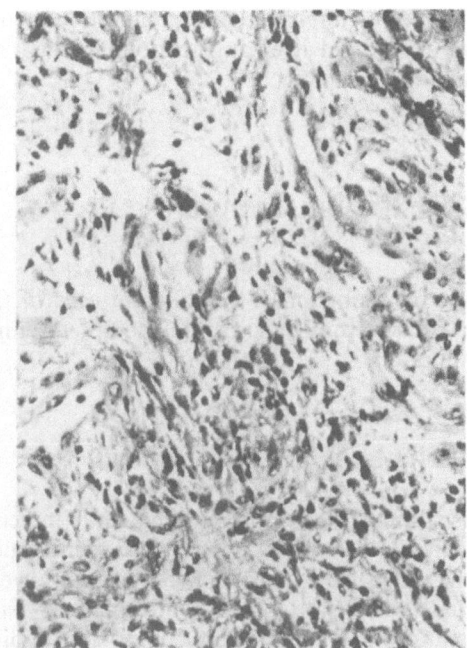

Fig. 37.4. Angiolymphoid hyperplasia showing numerous mature and immature vessels of various diameter associated with lymphoid aggregates. Haematoxylin–eosin–safran, × 220

Epithelioid haemangioma (Kimura's disease) is also referred to as angiolymphoid hyperplasia with eosinophilia [332], and is related to the condition called Kimura's disease in the Japanese literature.

Generally seen in adults, particularly women, this type of angioma is found on facial skin (forehead, temples, nose), and rarely on the skin of the limbs. Some cases of deep, visceral involvement have been reported [105]. The lesions are nodular, friable and covered by an excoriated crust. They may become confluent and can be confused with tumours or with erythroderma. Systemic eosinophilia (10%–20%) is present, but there is no associated alteration in general wellbeing and no lymphadenopathy. Epithelioid haemangiomas are usually superficial, but may invade the skin and spread into underlying aponeuroses or muscle. They are composed of a cluster of capillaries that often surround a larger, muscular-walled vessel. The endothelial cells lining these capillaries are of variable maturity (Fig. 37.4). In some cases, well-differentiated cells with acidophilic cytoplasm and occasionally vacuoles are seen. Still other capillaries are very immature and are lined by a cuboidal pseudoglandular endothelium. Solid cellular cords may also be observed made up of one or two rows of chromophilic cuboidal cells in continuity with more mature capillaries. Marked inflammation accompanies these capillary clusters. Lymphocytes, sometimes in nodular configurations with germinal centres, leucocytes and especially eosinophils, mastocytes and plasmocytes are present. An abundant collagenous matrix may also be observed.

Histoenzymatic studies reveal cells that are rich in acid phosphatase activity and in oxidative enzymes [55]. Alkaline phosphatase activity is low. Factor VIII can be identified in the endothelium by immunopathological techniques. Electron microscopy [55,100] confirms that well-differentiated, typical endothelial cells are present (pinocytosis, Palade bodies, basement membrane) as are pericytes. Unusual features include the rarity of desmosomes and the prominence of free ribosomes, mitochondria and microfilaments.

This lesion is benign and responds to steroids. Its origin remains controversial. Enzymatic studies indicate a possible relationship to histiocytoid angiomas [151], whereas the surrounding inflammation and eosinophilia suggest a possible immunological lesion. It may sometimes be difficult to distinguish these proliferative lesions from epithelioid haemangioendotheliomas and from angiosarcomas.

Other Benign Tumours

Apart from angiomas, the different cells of the vascular wall may be susceptible to other benign tumours.

Miscellaneous Parietal Tumours

These tumours develop more frequently in veins than in arteries, and in low-pressure circuits such as the pulmonary bed. They are generally well-circumscribed fibromas, chondromas or leiomyomas which may be difficult to distinguish from leimyosarcomas.

Glomus Tumours and Glomangiomatosis

These tumours were first described by Masson in 1924. They appear as angiomas composed of distinctive vessels, very specialized and richly supplied by nerves, belonging to the neuro-myoarterial glomus (arteriovenous anastomosis of Hoyer).

These tumours, usually found alone and always of small size (1.5 cm maximum diameter), are extremely painful, either spontaneously or in paroxysms produced on contact or with change in temperature. They are cutaneous, sometimes having an angiomatous aspect, or deep, lying in the subdermis; in the latter case are they only detectable by the pain provoked. In half of the cases they are found in the ungual bed, but may also be found on the forearm, the upper arm, the thigh and the leg. Extracutaneous localizations can also be observed; for example, the kneecap, bone, thoracic wall, stomach, eyelid, nose, rectum, cervix, vagina, lip and mesentery can be affected [60].

These tumours are rare and have an incidence of only 1.6% in a series of 500 tumours of the soft tissues [162]. They occur slightly more frequently in women (sex ratio 3:1) [162] occurring mainly in the adult. A notion of local trauma is sometimes found in the previous history.

The glomus tumours are made up of multiple ramifications of arterioles in vascular segments

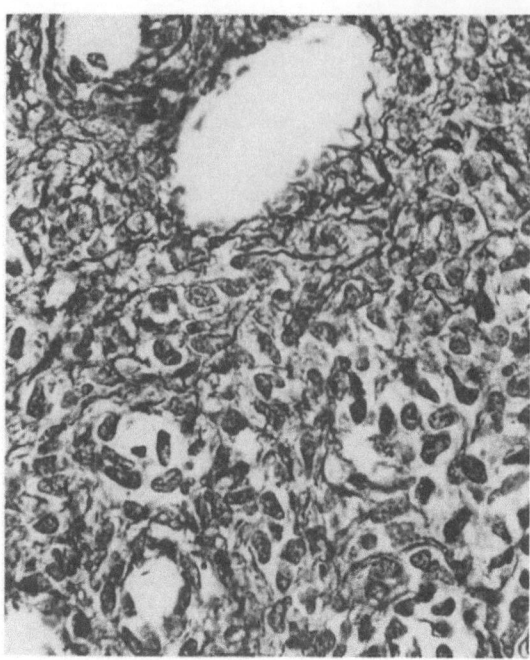

Fig. 37.5. Cutaneous glomus tumour. Cords of "epithelioid" cells associated with collagen fibres are arranged around vascular segments of various size. Haematoxylin–eosin–safran, × 560

[60,118,168]. Some have a lumen which may be clearly visible, lined by endothelial cells; more often it is collapsed. Several layers of large cells with eosinophilic or amphophilic cytoplasm, called "epithelioid" cytoplasm, are distributed around the lumen; they possess a very chromophilic central round nucleus with a big nucleolus (Fig. 37.5). Incomplete partitions of collagen fibres cross the cords of epithelioid cells. Other intermediary vascular segments between the artery and the preceding vessels show typical smooth-muscle cells arranged in several layers under the endothelium bordering the lumen. At the periphery of this layer, the smooth-muscle cells progressively give way to epithelioid cells with all the elements intermediate between these two cell types. All these vessels together form masses of varying density separated by loose connective tissue or connective tissue rich in collagen fibres. On the edge of the tumour there are venules with irregular lumina and with a wall becoming more and more attenuated but made up of epithelioid cells which drain into the normal venules. The presence of these vessels is useful for differentiating a glomus tumour from a paraganglioma. Between the vascular segments, numerous myelinized nerve fibres, and often non-myelinized fibres, run inside connective tissue rich in collagen fibres and containing many mastocytes.

On electron microscopy, the most remarkable aspect of the glomus cell is the presence of intermediary filaments and bundles of microfilaments of 80 Å diameter filling the cytoplasm and resembling actin [60,129]. These filaments are orientated and possess typical dense bodies. They sometimes terminate in the form of dense attachment plates on the cytoplasmic membrane. The cell is surrounded by a basal membrane and has several peripheral cytoplasmic extensions which are less developed than in the pericytes. These ultrastructural aspects are compatible with a smooth muscle origin [60,129].

Using immunohistochemical methods [129] the cells of glomus tumours appear to contain only one type of intermediary filament: vimentin. They contain no desmin. On the other hand, myosin is present in their cytoplasm and laminin can be detected in the membranes surrounding the cells. The glomus cells do not express any of the markers of endothelial cells (factor VIII-associated antigen, lectin Ulex Europeaus I). These data suggest a resemblance to smooth-muscle cells, in particular vascular smooth-muscle cells, since the latter also contain desmin [129]. The glomus cells present ATPase activity on their membrane [113].

There are several histopathological forms of glomus cell:

1. The epithelioid or compact form is composed of groups of light cells with an epithelioid appearance separated by connective tissue septae or nerve filaments grouped in fascicles. The vascular lumen is poorly visible and collapsed, and the endothelial coating is difficult to recognize.

2. The neuromatous form is characterized by the presence of numerous nerves arranged as thick trunks and plexuses isolating the vascular segments.

3. The angiomatous or telangiectatic form is characterized by wide, gaping lumens and by abundant collagenous sclerosis separating the vascular segments and dissociating the epithelioid cells. This aspect of glomangioma predominates in the multiple forms.

4. The form with hyaline sclerosis and sometimes mucoid degeneration appears to be made up of lobules separated by areas of sclerosis and sometimes appears pseudo-cylindromatous. Each lobule consists of an arteriole surrounded by nerve filaments, vascular segments with epithelioid walls and peripheral draining venules.

5. The transitional form with an angiomyoma or glomangiomyoma is made up of blood vessels of the arterial type with smooth thick muscle walls and of venous lakes of large size. Groups of epithelioid cells appear in places. However, there are no filaments and no nerve trunks. This absence of components of the nervous system which are usually so characteristic of the glomus tumours, is the principle characteristic of the glomangiomyomas which represent less than 10% of all the glomus tumours [60].

The only treatment is complete surgical excision. Glomus tumours always have a benign evolution.

Multiple Glomus Tumours or Glomangiomatosis

This rare disease is usually transmitted by autosomal dominant inheritance. [9,60,146]. It seems to be more frequent in women and develops during childhood [9,102,146]. Blue-black nodular lesions develop which are painless and these lesions can attain a diameter of 4 to 6 cm [44]. They may resemble those of the "Blue rubber bleb naevus" syndrome [16,155]. They involve mainly the limbs, and can be localized in one region, spread to several regions or even be diffuse and disseminated [44]. Sometimes they are associated with other malformations (vascular, bone) or with a Recklinghausen neurofibromatosis. They themselves probably represent vascular malformations and not a tumour [44,146]. Their evolution is always benign. Relapses are due to lesions undetected at the first clinical examination. Arteriography is thus indicated for their detection [44].

The lesions present as rounded or oval masses, having an incomplete whitish capsule from 0.5 to 1 cm in thickness. The cross-section reveals an angiomatous aspect with cavities of varying size containing blood clots more or less organized. Calcified zones are associated [44] (Fig. 37.6).

The histological appearance of glomangiomas is typical [44,60,127,146]. The numerous dilated cavities are bordered by a flattened epithelium. They contain blood, and thrombi at different stages of organization. Between these cavities, connective tissue with varying amounts of collagen is seen, sometimes with associated thick sheets of hyaline sclerosis. Areas of typical glomus cells and numerous sinuous vessels are also observed in glomangiomas. The angio-

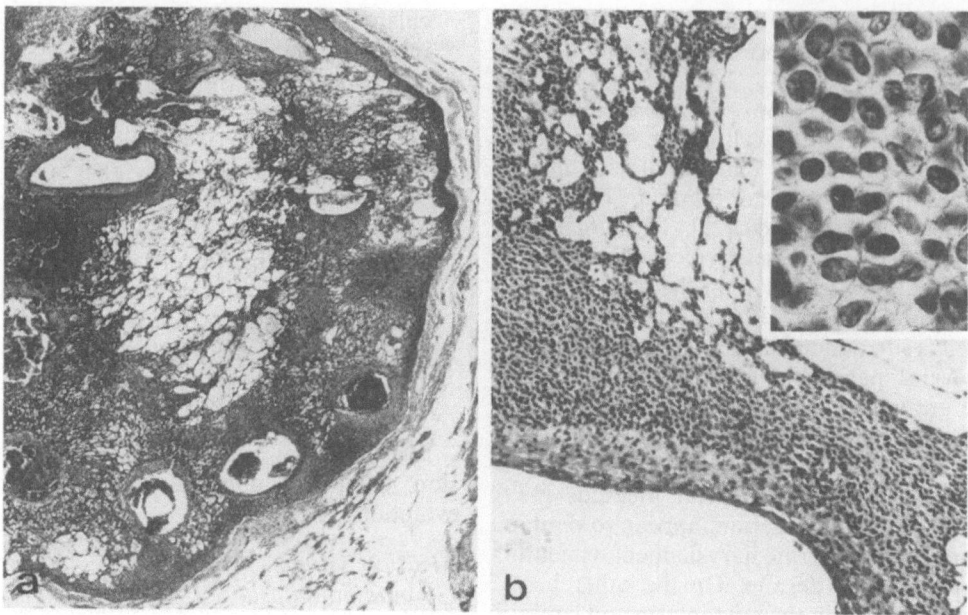

Fig. 37.6. Glomangiomatosis. **a** Vessels of various size, sometimes occluded by organized thrombi, are surrounded by areas of glomus cells. Haematoxylin–eosin–safran, × 24. **b** Area of glomus cells are in close contact with a vessel of arterial type. Haematoxylin–eosin–safran, × 120. *Inset.* Detail of the glomus cells showing very chromophilic central nuclei with prominent nucleoli. Haematoxylin–eosin–safran, × 106

matous cavities and certain vessels are surrounded by one to six layers of glomus cells. Under the endothelium, other vessels possess doubled smooth-muscle fibres towards the exterior of the glomus cells. There is no elastic tissue. Some included vessels have the appearance of normal veinules or arteries. Finally, the presence of a great number of mastocytes must be mentioned.

Differential diagnosis mainly includes the haemangiomas. However, the presence of glomus cells in the walls of the cavities of certain vessels should confirm the diagnosis. Glomangiomatosis may also be confused with the "blue rubber bleb naevus" syndrome. This syndrome is characterized by the formation of multiple single spongy haemangiomas, having no glomus cells, in the skin or certain internal organs, particularly the gastrointestinal tract [16,146,155], which could be at the origin of the haemorrhage. Treatment is exclusively surgical.

Vascular Leiomyoma

Vascular leiomyoma (angiomyoma) occurs preferentially at the extremities and usually develops within the deep dermis or subcutaneous areas [168]. It presents as a solitary nodule, very rarely exceeding 1 cm in diameter, which slowly enlarges. Pain is a prominent feature in about half of the cases. Exceptionally, multiple nodules have been described.

The tumours consist of circumscribed nodules of smooth-muscle cells, arranged in a circumferential fashion around small patent vascular lumina (Fig. 37.7). Obvious elastic laminae are lacking within these thick-walled vessels. Nerve fibres have been demonstrated in some cases and may account for the pain. Surgical excision is adequate therapy.

Tumours of the Juxtaglomerular Apparatus

(J.-P. Camilleri)

The tumours of the juxtaglomerular apparatus, of which about 20 cases have been reported in 15 years, represent the most well-known form of primary hyper-reninism. They are distinguished from other renal or extrarenal tumours by their

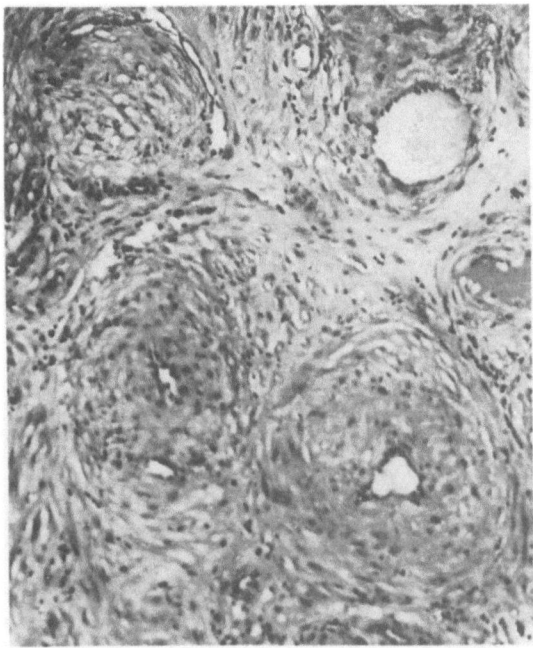

Fig. 37.7. Angioleiomyoma of the dermis made of circumscribed nodules of smooth-muscle cells arranged around small lumina. Haematoxylin–eosin–safran, × 120

synthesis and secretion of renin (for review see J. Menard et al. [123]).

The first clinical description was by Robertson et al. [150]. The tumours were discovered in young subjects in association with severe high blood pressure and extensive hypokalaemia. The diagnosis is based on: (a) the existence of a high plasma level of renin, whether this is measured in terms of activity or concentration or by direct radioimmunological assay [79]; (b) the absence of other causes of hyper-reninism (stenosis of the renal artery, malignant high blood pressure, segmentary renal hypoplasia, sodium depletion, Bartter's syndrome) and; (c) the visual detection of the tumour based on the presence of an avascular zone on arteriography. Surgical excision of the tumour cures the hypertension and the hypokalaemia.

They are restricted tumours, cortical and most often polar, measuring from a few millimetres to 4 cm and are seen to be yellowish-white when cut. Their histological structure, first described by Kihara et al. [99], consists of polygonal cells involved in endocrine frameworks, often associated with numerous mastocytes (Fig. 37.8). The

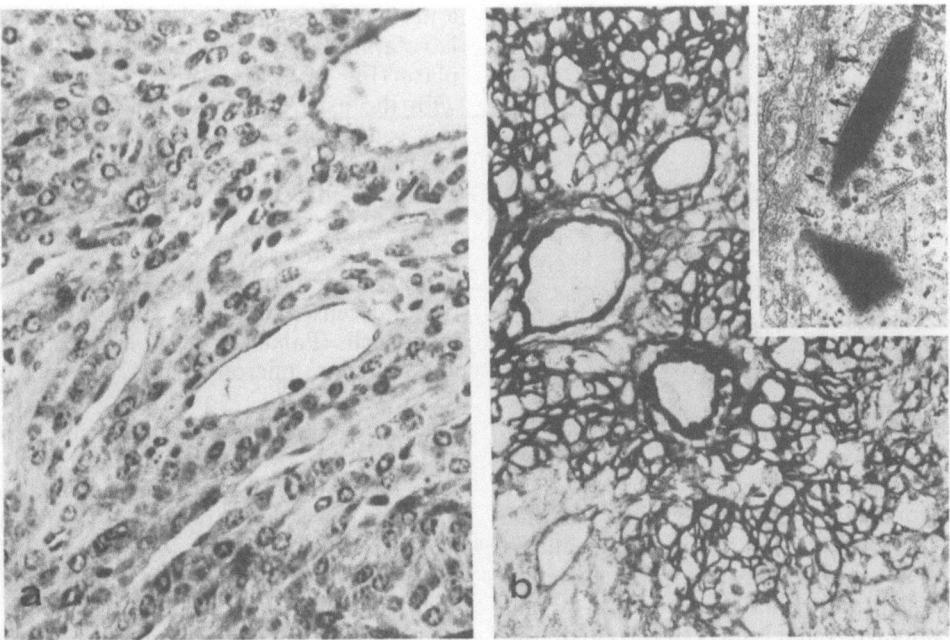

Fig. 37.8. Tumours of the juxtaglomerular apparatus. **a** Histological pattern showing cords of polygonal cells associated with rich capillary network. Masson's stain, × 410. **b** Silver impregnation (according to Marianozzi) showing centrifuging maturation between the perivascular and the endocrine-like areas whose argyrophilic framework is rarefied and dispersed. × 410. *Inset.* Electron micrograph of cristalloid protogranules in the Golgi apparatus. Bundles of microfilaments can be noted (*arrows*). Uranyl lead, × 21 000

richness of the capillary network contrasts with the avascular character of the arteriographical image. The arterioles and small muscular arteries often show medial hyperplastic lesions in the form of onion bulbs and/or deposits of hyalin or fibrin. These vascular anomalies are probably evidence of local activation of the renin–angiotensin system [27]. Endocrine maturation centrifuging of the tumour cells from the vascular walls seems possible, thus illustrating the phenomenon of phenotypic transformation already described for ischaemic kidney [29]. Using electron microscopy the tumour cells can be seen to possess both myofibrils and two types of secretory granules (Fig. 37.8). The presence of acid phosphatase activity suggests that certain of these granules are of lysosomal origin [27].

Immunochemical studies using specific antihuman renin antibodies [28] confirm using both photon and electron microscopy the presence of renin in the granules [27,110]. These data are confirmed by strong renin activity in the middle of tumour cells put into culture.

The histogenesis of these tumours is questioned. The presence of nerve filaments and elements of the distal tube suggest a hamartomatous origin [12,27]. No malignant tumour of the juxtaglomerular apparatus has yet been described.

Tumours with Malignant Potential

Certain vascular tumours cannot be classified as benign or malignant on histological grounds alone. These include epithelioid haemangioendotheliomas, haemangiopericytomas and chemodectomas or paragangliomas.

Epithelioid Haemangioendothelioma

This rare tumour was originally referred to as a haemangioma [141], and has a growth potential intermediate between that of an angioma and an angiosarcoma [177]. It may be superficial or deep (e.g. retroperitoneal), and may involve organs such as bone, liver and lung [93,116]. Often, this lesion arises in association with medium- and large-sized veins.

The tumour is whitish, soft and gelatinous, and may resemble a myxoma. Histologically, thin trabeculae of a homogeneous cell type are

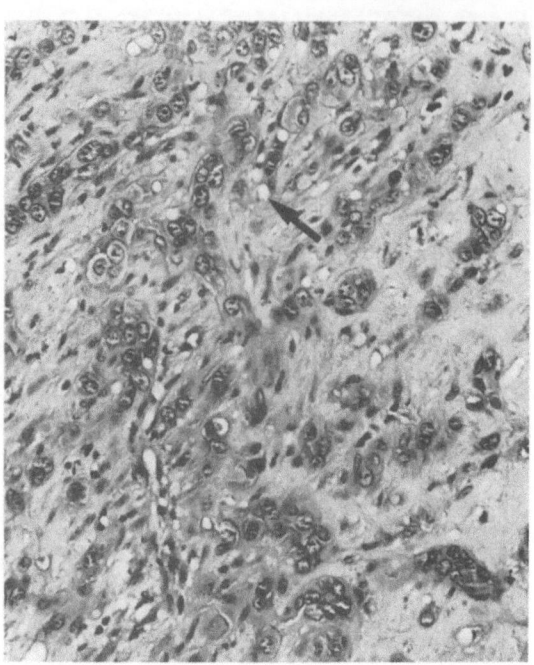

Fig. 37.9. Epithelioid haemangioendothelioma, showing branched trabecular structures made of vacuolated eosinophilic cells (*arrow*). Haematoxylin–eosin–safran, × 200

observed. Rarely, these trabeculae may assume a more elaborate, branched structure. The cells have a markedly vacuolated, eosinophilic cytoplasm (Fig. 37.9). The vacuoles cannot be stained with the usual mucin-staining dyes, indicating that some degree of intracellular angiogenesis is occurring. Whatever the cellular conformation, a fine reticular sheath surrounds the individual cells that are in turn embedded in a acid mucopolysaccharide-rich matrix.

Ultrastructural studies confirm that the tumoral cells have the usual features of endothelial cells (Palade bodies, pinocytotic vesicles) and abundant microfilaments. The cells contain factor VIII.

Diagnosis is often very difficult, and confusion may arise with other tumours particularly of the lung (bronchiolo-alveolar epithelioma) and of the liver (cholangiocarcinoma). It may also be difficult to distinguish an epithelioid haemangioendothelioma from an epithelioid haemangioma. However, the latter generally has a well-defined structure and a prominent inflammatory component. A small percentage of haemangioendotheliomas (10%–20%) may become malignant, and thus easily confused with angiosarcomas. Malignant haemangioendotheliomas can recur locally and may also metastasize.

Haemangiopericytoma

After the identification of the pericyte by Zimmerman in 1923 [181], Stout and Murray recognized the haemangiopericytoma as a distinct entity in 1942 [169]. This tumour affects both sexes at any age, but occurs especially in childhood [95]. Although ubiquitous [59], the most frequent sites of involvement are the skin of the upper limbs, distal extremities, the neck, the scalp and the face [121]. The retroperitoneum is the second most frequent localization [168]. Other sites include the nasal fossa [43,71], the mouth [37] and the pharynx, and rarely, the digestive tract [138], mediastinum and uterus [76,165]. Exceptional cases have been reported of eye, skeletal [97], renal, cerebral and meningeal involvements (angioblastic meningioma, [18]). This neoplasm is painless, in contrast to glomus tumours. Its growth rate is variable (months to several years). Particularly large tumours in the pelvis and retroperitoneum may be associated with hypoglycaemic episodes [143] that resolve after surgery. The angiographic picture is characteristic [81,172]. The injection of contrast material is followed almost immediately by diffuse capillary opacification and during the venous phase of the study, dilated veins are seen at the periphery of the tumour.

Macroscopically, the tumour is either white or brown, and generally 3–5 cm in size although larger tumours have been identified. The lesion is well delimited and finely encapsulated. Dilated veins may surround the lesion and significant intraoperative haemorrhage may occur. Haemorrhagic and necrotic foci may also be present within the tumour, particularly if it is malignant.

Histologically, cells of the pericytic type surround a rich capillary network. These cells may be round, ovoid or even fusiform. Their nuclei are sometimes vesicular, and sometimes densely chromophilic, regular and displaying few mitoses. The pericytes may aggregate into confluent sheets or may be otherwise distributed around vessels (Fig. 37.10a,b). Occasionally they form palisades or assume a storiform or whorled appearance. The underlying capillaries may be distinct with large lumina, or tiny and difficult to identify within the enveloping pericytic mass. These vessels are lined by a flattened endothelium easily distinguished from adjacent pericytes. Between the capillary network and the sheets of tumoral cells, a fine reticulum can be identified with appropriate stains. This reticulum outlines even the smallest capillaries, accentuating the marked degree of tumour vascularization (Fig. 37.10b). The pericytes are also individually outlined by this reticulum and concentric aggregates of pericytes are sometimes highlighted by this fine web. In addition to this argyrophilic network, collagen may be distributed in a similar fashion within the tumour.

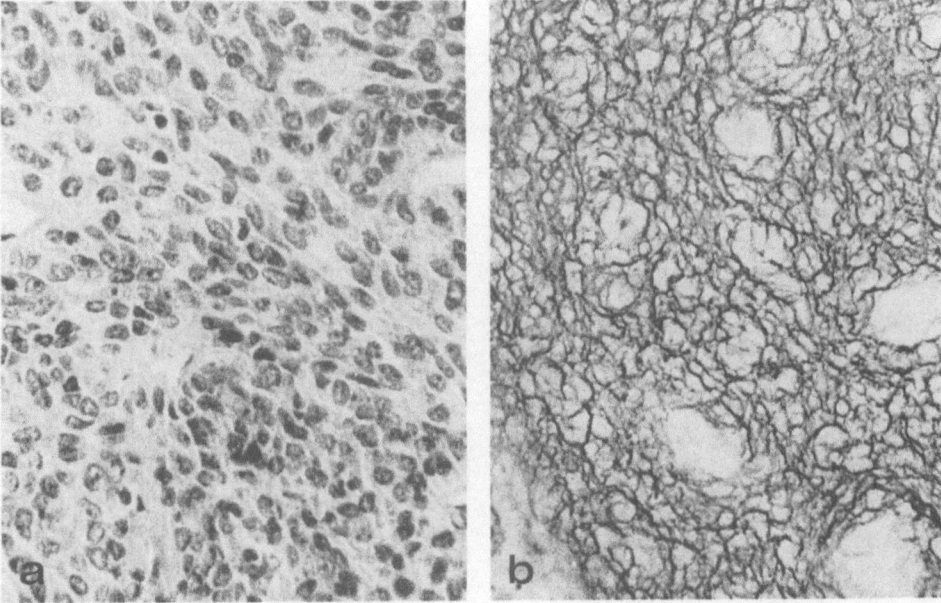

Fig. 37.10. Haemangiopericytoma. a Oval or fusiform cells, sometimes arranged around capillaries. Haematoxylin–eosin–safran, × 350 b Rich reticulinic network outlining cells and vessels. Silver impregnation, × 260

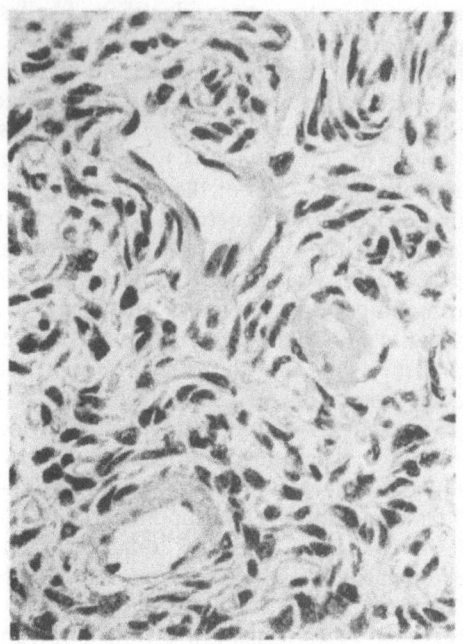

Fig. 37.11. Haemangiopericytoma. Abundant collagen can be seen within the tumour. H & E, × 260

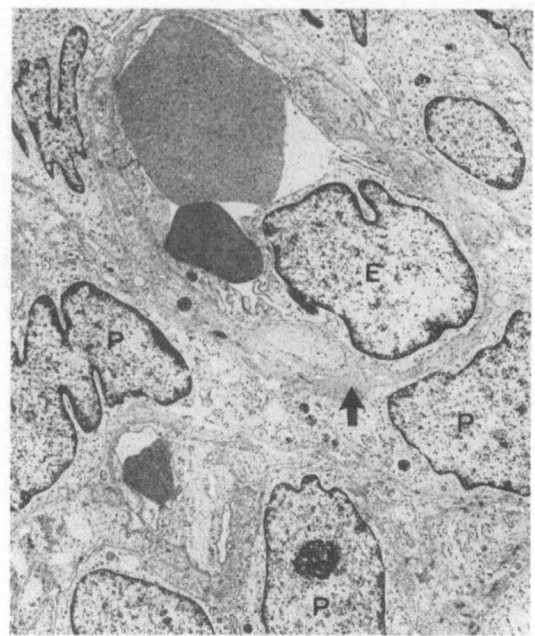

Fig. 37.13. Electron micrograph of haemangiopericytoma. Endothelial cell (*E*) separated from tumoral pericytes (*P*) by a continuous basement membrane (*arrow*). Uranyl lead, × 3150

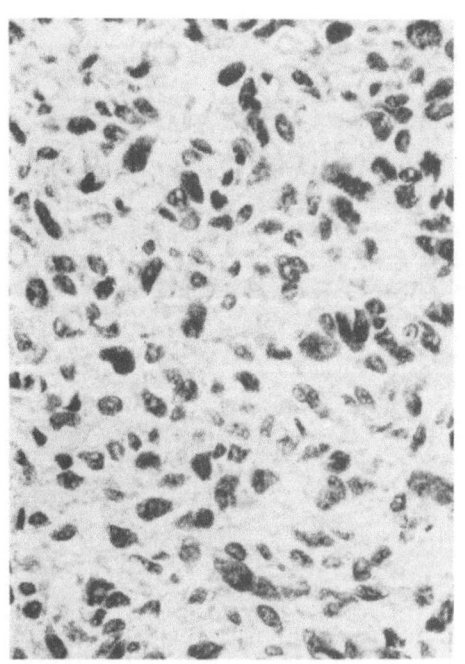

Fig. 37.12. Malignant haemangiopericytoma. Marked cytological abnormalities and mitotic figures. H & E, × 260

Myxoid and even chondroid zones may be evident within haemangiopericytomas. An interstitial infiltrate composed of lymphocytes, mastocytes and scattered xanthomatous cells is usually present. In contrast to glomus tumours, no nerve radicles are seen.

The heterogeneous appearance of haemangiopericytomas often leads to diagnostic confusion. Some of these lesions are correctly identified only because of the reticulum that outlines the capillaries and highlights the pericytic proliferation within these tumours. In some cases, the presence of a thick collagen sheath separating capillaries from pericytes serves as a diagnostic aid (Fig. 37.11). Malignant transformation is associated with marked cytological anomalies, numerous mitotic figures, haemorrhage and necrosis (Fig. 37.12). In such a tumour, the cells are polymorphic and often orientated perpendicular to the vascular lumina (perithelial implantation).

Ultrastructural studies [8,15,81,135] confirm the existence of pericytic proliferation (Fig. 37.13). The pericytes are sometimes globular, but more often elongated with numerous arborescent processes. They contain speckled, heterochromatic nuclei. Their cytoplasm is poor in ribosomes and mitochondria, but endocytotic vesicles adjacent to the plasma membrane are prominent. The cells contain peripheral microfilaments. A thick basement membrane with many infoldings surrounds the pericytes. Only some cells are connected by desmosomes. The pericytes are clearly separated from endothelial cells by a continuous basement membrane. In malignant tumours, certain morphological variations are possible: large nuclei with prominent nucleoli, mitotic figures, poor cytoplasmic differentiation (many ribosomes) and discontinuities in the basement membrane. A population of very immature cells thought to be of endothelial cell origin [40,134] may also be present within malignant tumours. The cells are arranged in trabeculae and are identified by their irregular heterochromatic nuclei. It may be difficult to distinguish such cells from tumour pericytes.

The vessels within haemangiopericytomas are seen to be normal by electron microscopy. A well-differentiated endothelium is characteristic with cells containing filaments and endocytotic vesicles. The cells are linked by desmosomes and lie upon a distinct basement membrane. Abundant interstitial collagen is noted within the most highly differentiated tumours, particularly around capillary lumina.

Electron microscopy and immunopathological studies may be necessary to correctly distinguish haemangiopericytomas from other tumours. In the skin a haemangiopericytoma may resemble a glomus tumour. The latter, however, has a more organoid structure and contains nerve fibrils. Electron microscopy can distinguish between the two [131]. Palisades within a haemangiopericytoma may evoke the erroneous diagnosis of Schwannoma and confusion is possible with malignant fibrous histiocytomas and synovial sarcomas. Distinguishing features are the generalized polymorphism of the former, and the presence of a "biphasic" pseudoepithelial population in the latter. In the skeleton, it may be difficult to distinguish haemangiopericytomas from mesenchymal chondrosarcomas which may entrap haemangiopericytic foci. Adding to the difficulty is the fact that cartilage may be present within haemangiopericytomas.

Neoplasms composed of fusiform cells and a prominent vascular network, such as liposarcomas and mesotheliomas, may also pose a diagnostic problem.

Electron microscopy is indispensable in some cases in order to distinguish certain malignant and occasionally benign haemangiopericytomas from angiosarcomas, as immature, angioblastic structures are sometimes visible within haemangiopericytomas. However, the pericytic nature of the proliferation can be demonstrated conclusively by electron microscopy in doubtful cases [149].

Childhood Haemangiopericytoma

This lesion is almost always subcutaneous and evident at birth or before 15 months of age. The tumour is composed of a central mass and multiple satellite nodules. Histologically, sheets of very immature cells are present. These cells are sometimes necrotic and they obliterate the underlying vessels. Their appearance is often atypical and numerous mitotic figures are observed. An increased proportion of immature endothelial cells can also be identified by electron microscopy [57]. Despite this disturbing appearance, these tumours generally have a favourable prognosis. Recurrences and metastases are rare.

Prognosis and Evolution

Several factors influence the prognosis in haemangiopericytomas; these include:

1. Age: congenital lesions are generally benign.
2. Location: nasal and uterine tumours have a better prognosis than those of the mediastinum or of the lower limbs.
3. Histological appearance [15,37,135]: the prognosis is poor for immature haemangiopericytomas harbouring necrotic or haemorrhagic foci, indistinct vascular lumina and large ovoid pericytes with irregular nuclei, as well as discontinuous basement membranes. This appearance contrasts with the relatively benign nature of haemangiopericytoma which contains distinct fusiform pericytes separated from well-defined patent capillaries by a basement membrane and by a collagenous matrix.
4. Mitotic index: prognosis is guarded if more than four mitoses per ten microscopic fields are identified.

After initial treatment, recurrences are common in up to 40%–50% of cases. These recurrences may be late, and they generally herald the appearance of metastases. In 10%–30% of cases, metastases are evident 5 or more years after diagnosis. They are most often located in the lungs and in bone.

Histogenesis

Ultrastructural studies and tissue cultures [127] have established that pericytic proliferation is the origin of this rare tumour. It is unclear whether tumour pericytes arise from an autonomous cell layer or from an immature fibroblastic or muscular endothelial cell. It is to be emphasized that there is no relationship between haemangiopericytomas and glomus tumours.

Chemodectomas/Paragangliomas

These tumours arise from neural crest cells within the paraganglionic chain. Only perivascular chemodectomas will be discussed, excluding those within the adrenal gland. This system [74] includes branchiomeric paragangliomas (jugulo-tympanic, carotid, supraclavicular, laryngeal, coronary, aortopulmonary and orbital), intravagal paragangliomas and aorticosympathetic paragangliomas (intrathoracic, bifurcation of the aorta and iliofemoral).

Paragangliomas of the Carotid Body

These tumours develop in a chemoreceptor sensitive to arterial oxygen level and blood pH, and because of this they are often called chemodectomas. They represent approximately 60% of cervicocephalic paragangliomas [107]. They are more frequent in subjects living at altitude (Peru, Mexico), and usually occur in adults between 40 and 60 years of age, in both sexes equally [60]. They are painless, cervical, usually unilateral or sub-maxillary at the angle of the jaw and can cause paralysis of vocal cords or the Claude Bernard Horner syndrome. Selective arteriography shows the extent of the tumour. They are bilateral in 2%–5% of cases [60,107] and sometimes occur in a family context (autosomal dominant). They are often bilateral in this case. Chemodectomas sometimes are associated with other paragangliomas [60].

The evolution is usually benign. Clinical effects are linked to the close connection with the carotid artery. Nevertheless, malignant chemodectomas do exist, and are often revealed by lymph-node metastases, or even mediastino-pulmonary and vertebral dissemination. These forms, whose malignancy is testified by the metastases, have a very long evolution [26].

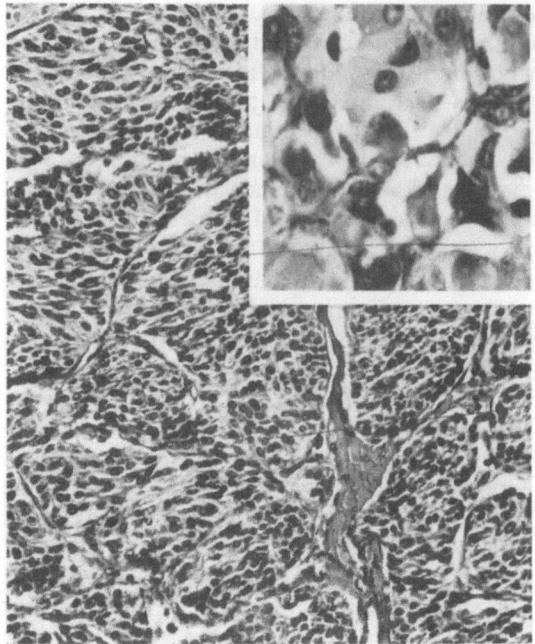

Fig. 37.14. Chemodectoma. Cell cords separated by connective tissue septa containing vessels. Haematoxylin–eosin–safran, × 180. *Inset.* The tumour is made of small nests, showing abundant granular cytoplasm with a central nucleus and a small nucleolus. Haematoxylin–eosin–safran, × 120

The benign form [60,74,118] consists of red-brown lobulated tumours, sometimes having haemorrhagic areas, with a diameter varying from 2 to 5 cm [60,118]. They adhere very strongly to the vascular wall creating surgical problems and often requiring ligature of the vessel. They are composed of cell masses of unequal size, separated by connective tissue septa which contain the vessels (Fig. 37.14). These partitions, thickened by numerous collagen fibres, cut the lobules into small groups of cells. There are numerous mastocytes in these partitions, accompanied by lymphocytic and plasmocytic infiltrates. Each lobule is composed of rounded nests of cells (Zellballen), of variable size, separated by slender walls which are followed by the capillaries thus forming an endocrine pattern (Fig. 37.14). The rounded, polygonal or globular tumour cells with abundant granular cytoplasm can be eosinophilic or amphophilic. The cytoplasm may contain numerous vacuoles or appear lacey. More rarely, the cells may assume a fusiform appearance. The central nucleus is rounded with fine chromatin and a small nucleolus. Mitoses are rare.

When the tumour cells are retracted due to poor fixation, the "Zellballen" appearance is lost. The cells are arranged in cords of variable thickness, producing a pseudo-glandular pseudo-rosette appearance [60].

A thin, incomplete capsule surrounds the tumour. At the periphery the nerve filaments are thinned or included. The tumour lobules extend into the adventitia of the carotid. Extension into the media is exceptional.

According to the predominating cellular or vascular component, adenomatous or angiomatous forms can be produced.

The chromaffin reaction is usually negative [118], but this technique does not allow very small amounts of catecholamines to be detected [60]. However, in either fresh or frozen tissue, it is possible to demonstrate the presence of catecholamines by the formol-induced fluorescence technique. On paraffin section, a modified Grimelius reaction may be used, to show the presence of silver-stained grains, but this is subject to interpretation difficulties.

Ultrastructural study [47,60] demonstrates that the tumour cells, some light, some dark, contain a variable quantity of ribosomes and mitochondria. Fundamental is the demonstration in some cells of small grains (100 to 200 nm in diameter) with an electron-dense centre of Golgian origin [47,60]. These represent the stockpiling of catecholamines. The subtentacular cells surrounding the "Zellballen" of normal carotid corpuscles are lacking in the paragangliomas [60].

Malignant change can be suspected when the lobulated aspect is absent, when deformed extensive rows exist with cells piled up in several layers and with thick conjunctival walls containing several capillaries between the rows. Sometimes foci of necrosis at the centre of the cellular masses are associated with this more or less complete deletion of the endocrine structure. The tumour cells can present a great inequality of both size and form. In particular, in certain areas, they take on a fusiform appearance, producing bundles simulating a sarcoma. A more striking anisocytosis, the presence of monstrous nuclei and evidently atypical forms may all also draw the attention, together with numerous mitoses, which are sometimes abnormal. However, often nothing of the architectural organization nor of the tumour cell appearance allows the benign form to be distinguished from the malignant forms. The aggressiveness of the tumour is indicated by an infiltration at the periphery of the tumour with inclusion of large nerves, and perineural neoplastic lymphatic and vascular invasion. Sometimes, lymph-node involvement is the only evidence of malignancy [26]. It is important to emphasize the absence of any precise clinical, biological or histopathological criteria for malignancy. Only the occurrence of metastasis allows the unequivocal diagnosis of malignancy [26,47,60].

The only treatment is complete surgical excision, preferably conserving the carotid. The resection of this blood vessel with reconstitution by grafting is sometimes necessary. Radiotherapy, used on inoperable tumours or metastases, has proved to be ineffective [60].

The Other Paragangliomas

The jugulo-tympanic paragangliomas spread to the base of skull and to the temporal and the middle ear. The histological appearance is identical to that described for the carotid body. Very vascular pseudo-angiomatous forms are frequent; they are rarely malignant, their clinical significance being linked to their spread towards the base of the skull. Because of this they are often inoperable. Radiotherapy is also risky because it may lead to necrosis of the mandible [60].

Vagal paragangliomas and mediastinal paragangliomas also resemble those of the carotid body.

The extra-suprarenal retroperitoneal paragangliomas appear in younger patients (between 30 and 45 years of age). They may secrete norepinephrine (leading to high blood pressure). Angiography reveals and localizes the tumour; certain tumours are associated with a phaeochromocytoma. Malignant cases, detected by metastases, have been described: the histological aspect is sometimes that of a carotid paraganglioma, sometimes that of a suprarenal phaeochromocytoma. The only effective treatment is complete surgical excision. However, the tumours often have aggressive behaviour and metastasis occurs in 20%–42% of cases (lymph node, bone, liver, lung). Radiotherapy and chemotherapy seem to be of little help [60].

Paragangliomas at other sites (nasopharynx, larynx, eye socket) are very rare.

Malignant Tumours

Angiosarcoma

This neoplasm, also called malignant haemangioendothelioma, affects both sexes equally and can occur at any age. Rarely, it may arise at the site of a pre-existing angioma [73] or of a neurofibroma in von Recklinghausen's disease [33]. These cases may in fact be related to irradiation, as angiosarcomas are known to arise following radiation therapy for a variety of tumours (endometrial, ovarian, Hodgkin's disease [35]. When Thorotrast was widely used for angiography, hepatic angiosarcomas occurred as a recognized complication. Other chemical substances such as arsenical insecticides used by vintners, and polyvinyl chloride used to make synthetic rubber and plastic [148], have been implicated in the genesis of angiosarcomas. Treatment with anabolic steroids or oestrogens has also been incriminated [61, 87].

Angiosarcomas are most frequently located on the skin [13], in particular the uncovered portions of the head and neck [62], suggesting a possible relationship with sun exposure. In the mouth, these tumours infiltrate widely and are generally ulcerated [13]. Angiosarcomas of the breast occur [56,91] and give rise to a rapid increase in breast size and reddish discolouration of the overlying skin, but no nipple retraction or adenopathy. The prognosis of such tumours is generally unfavourable, but if the cells are relatively well differentiated, a better prognosis can be given [52]. Angiosarcomas can arise in the skeletal musculature and in the viscera, notably

the liver [10]. Hepatic angiosarcomas (apart from those induced by Thorotrast) often arise in patients with cirrhosis, suggesting that the endothelial and hepatocytic regeneration characteristic of this condition may have a triggering role. In children, angiosarcomas have a benign appearance resembling angiomas, and have a relatively favourable prognosis.

The tumours have also been identified in the spleen, the colon, the thyroid gland [101] and in the skeleton [166], generally causing multicentric, osteolytic lesions. Although rare, constrictive pericardial angiosarcomas may develop and in fact represent the most common form of cardiac sarcoma.

Macroscopically, angiosarcomas of the skin or of mucous membranes are reddish-black, occasionally with ulcerated plaques, surrounded by congested blood vessels. In histological sections, these confluent, haemorrhagic tumours may resemble a haematoma, and may separate deep planes from fascia and from aponeuroses. In organs such as the liver, the neoplasm consists of reddish nodular clusters, or alternatively of more scattered foci.

Relatively well-differentiated vascular structures may be present within angiosarcomas (Fig. 37.15a). These vessels are distinct and lined by a swollen endothelium resembling that seen in angiomas. However, some vascular lumina may

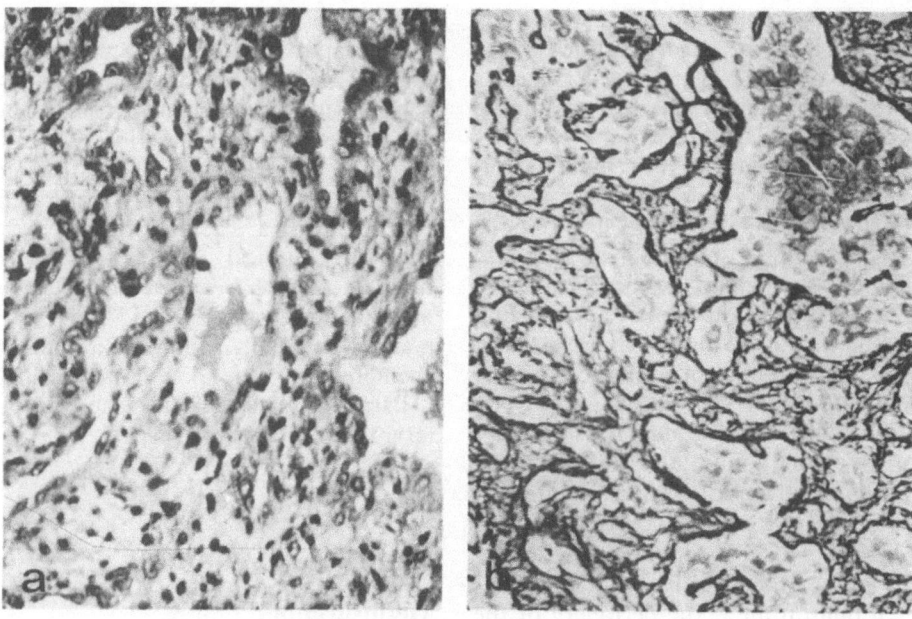

Fig. 37.15. Well-differentiated malignant haemangioendothelioma. **a** Patent vascular lumina lined by swollen tumoral endothelial cells. Haematoxylin–eosin–safran, ×260. **b** Reticular network outlining poorly differentiated vascular profiles with polystratified endothelial cells. Silver impregnation, ×120

be lined by a jumble of polystratified endothelial cells, or even occluded by highly characteristic papillary projections. In some poorly differentiated areas, barely patent passages are created by the juxtaposition of swollen tumour cells. At the extremities of these cords, multinucleated giant cells with enormous nuclei and vacuolated cytoplasm (intracellular lumen formation) may be seen. Even more immature forms are possible. Aggregates of immature cells may give rise to pseudo-epithelial lobules. Sheets of polymorphic polygonal or fusiform cells may obliterate recognizable vascular structures leaving only a few microcavities lined by cuboidal cells.

Angiosarcomatous cells are polymorphic (cuboidal, polygonal, fusiform or flattened) and have numerous atypical features including large, hyperchromatic nuclei and frequent mitoses. The interstitial matrix may be prominent in the better differentiated tumours, or reduced to a fine reticular network in highly malignant cases. The identification of this reticulum is important as it outlines the tumour cells individually and thus confirms that the sarcomatous proliferation is intraluminal (Fig. 37.15b).

Histoenzymatic studies reveal alkaline phosphatase and ATPase activity in periluminal or pericavitary zones [36]. Other features point to the endothelial origin of these tumours, notably the presence of factor VIII in some cases [77,132] and the presence of certain blood group antigens [20].

By electron microscopy, the angiosarcomatous cells are either scattered with no particular organization, or arranged in layers around capillary lumina (Fig. 37.16). Often two or three malignant cells only surround a red blood cell, and in some sections red blood cells appear trapped within cytoplasmic projections of the malignant endothelial cells [36]. All the cells contain large, irregular nuclei and prominent nucleoli. Apart from a few dark, elongated cells containing peripheral microfilaments and resembling pericytes, the majority of cells are of endothelial origin [152]. Pinocytotic vesicles, intracellular junctions and often Weibel–Palade bodies are seen. Some cells are immature and contain few intracellular organelles.

Angiosarcomas are surrounded by a finely granular reticulum that is layered and often infolded.

Depending on the degree of cellular differentiation, angiosarcomas may be difficult to distinguish from some other tumours. Differentiated examples containing well-structered capillaries must be distinguished from

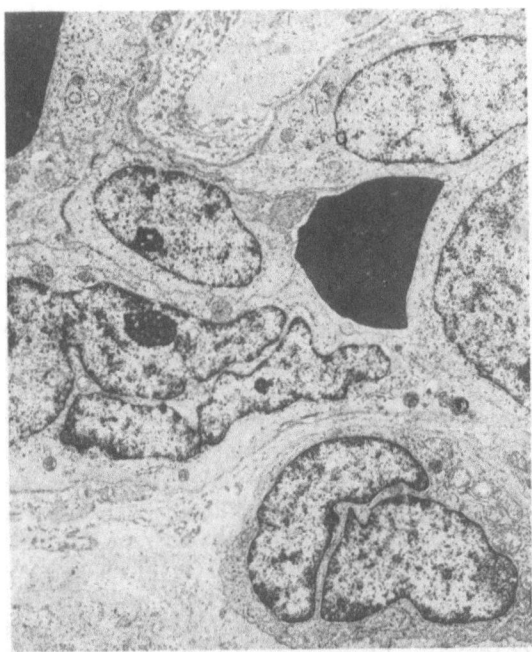

Fig. 37.16. Electron micrograph of tumoral endothelial cells in close contact with a vascular cleft (erythrocyte). Uranyl lead, × 5000

benign angiomas. Angiosarcomas of this type generally have a swollen, stratified endothelium, and areas of increased cellular density as well as intravascular vegetations. A differentiated but cellular angiosarcoma may be confused with an intravascular papillary angioma. Angiosarcomas, however, display greater cellular heterogeneity and stratification. They also contain more rudimentary structures such as vascular cords, and may extend into the extravascular space. In contrast to angiomas, vegetative angiosarcomas rarely have a hyaline connective tissue axis. Malignant synovial tumours and papillary epitheliomas may sometimes be difficult to distinguish from papillary forms of angiosarcoma. Less differentiated tumours with a lobular organization must be distinguished from poorly differentiated glandular carcinomas. It is also difficult to distinguish between angiosarcomas and epithelioid haemangioendotheliomas with malignant potential.

Massive angiosarcomas composed of fusiform cells may resemble fibrosarcomas or malignant schwannomas. In this case, a search for cavities within the lesion is required.

An angiosarcoma containing giant cells may mimic a malignant teratoma of the chorioepithelioma type.

Finally, angiosarcomas often resemble other tumours with a prominent vascular component. These include endocrine epitheliomas (hepatomas, renal cell tumours), certain "angioblastic" osteosarcomas and malignant melanomas, as well as the necrotic connective tissue sarcomas referred to as perithelial tumours.

Evolution

Cutaneous and mammary angiosarcomas are highly malignant and generally lead to death in less than 2 years. The prognosis is tempered for hepatic and osseous angiosarcomas as well as for those tumours found in childhood [96] and for well-differentiated tumours. Metastatic lymphadenopathy is rare. In contrast, metastases are frequently found in liver, lung and bone. Primary or metastatic angiosarcomas in the liver may become haematopoietic [10]. Intravascular coagulation may occur within angiosarcomas [4,176] resulting in haemolysis and thrombocytopenia, and tumour rupture may be responsible for alarming haemorrhage [1,170]. Angiosarcomas may be multicentric, particularly within the liver, spleen and bone. This may pose a therapeutic problem as the treatment for tumours with multiple foci may differ from that for metastases.

Specific Types

Post-mastectomy Angiosarcoma. This tumour has been considered to be lymphangiosarcoma [64,164] and analogous to tumours arising in cases of congenital lymphoedema [34,179] or following trauma [124]. However, the angioendothelial proliferation in cases of angiosarcoma has not been clearly localized to lymphatic vessels.

This neoplasm arises 5–10 years following mastectomy with lymph node resection and/or radiotherapy for cancer of the breast. Multiple reddish-brown, cutaneous and subcutaneous nodules appear covering the upper arm and axilla. These nodules may become confluent and ulcerated, and may extend to the chest wall.

Histologically, these tumours are highly undifferentiated [14]. The nodules are fusiform, lobulated in a pseudo-epithelial fashion and sometimes arranged around blood vessels. Rarely, but more characteristically, intravascular papillae are seen. The tumour is clearly different from late metastatic breast cancer, as ultrastructural studies [164] and the detection of

factor VIII [30] have established the endothelial cell origin of this tumour [114,120].

Prognosis is poor and death from pulmonary metastases generally follows within 1 year. The pathogenesis of these tumours is controversial. It is possible that vascular stasis associated in some cases with irradiation may promote endothelial proliferation as in congenital elephantiasis [159]. Other authors invoke the absence of immunological surveillance in the upper limb following local lymphoid ablation as a contributing factor.

Rare Form of Haemangiosarcoma/ Malignant Intravascular Papillary Angioendothelioma

This rare tumour of the skin or subcutaneous tissue arises in childhood, is relatively quiescent and was first described by Dabska in 1969 [46]. The lesion either spreads diffusely under the skin, or is represented by an intradermal nodule that can penetrate more deeply if left untreated. The tumour is composed of vascular anastomoses lined by a proliferative endothelium forming papillary masses. The latter are composed of round, chromophilic, lymphocytoid cells. Some of these vegetations are centred on small hyaline corpuscles. Prognosis is relatively favourable even in cases of regional lymph node involvement. This tumour is considered by some [142] to be of borderline malignancy.

Systemic Angioendotheliomatosis. This condition is rare and represents malignant transformation of endothelial cells in multiple sites [80,160]. Clinically, multiple erythematous nodules and cutaneous plaques are seen, resembling erythema nodosum. The abnormal cutaneous and subcutaneous vessels are filled with mononuclear cells in continuity with the endothelium. These neoplastic cells can be found in some cases in visceral vessels and in the nervous system. Electron microscopy and immunohistochemistry [101] may be required to distinguish this tumour from a lymphoma.

Angioendotheliosarcoma

This exceptional tumour develops from vascular endothelium, particularly that of the inferior vena cava and distal veins, as well as of the pulmonary artery [38]. Streams of tumour cells obstruct the vascular lumen and may secondarily spread to adjacent extravascular tissue. Histologically, this tumour is a typical extravascular haemangiosarcoma.

Leiomyosarcoma

These rare tumours affect veins more frequently than arteries. Low-pressure circuits such as the pulmonary bed are preferentially involved. The inferior vena cava [98] harbours these tumours more often than does the superior vena cava [1]. Leiomyosarcomas affect women more often than men, and in 60% of cases affect the inferior vena cava leading eventually to obstruction and to the Budd–Chiari syndrome [31]. These tumours may also arise from the musculature of small vessels in extra-smooth muscle sites (soft tissue, striated muscle, bone, etc.).

Macroscopically, these tumours rarely protrude into the vascular lumen. Instead, they form discrete parietal nodules that may become voluminous and directed away from the obstructed vein towards adjacent viscera (Fig. 37.17a). The tumours are composed of fusiform cells arranged in fascicles or palisades (Fig. 37.17b) and surrounded by a prominent reticulum. The cells are of muscular origin as evidenced by their acidophilic cytoplasm, glycogen content and by the presence of occasional epithelioid or plasmodial cells. In equivocal cases, electron microscopy can be used to identify myofibrils, endocytotic vesicles and pericellular basement membranes [39].

Two difficulties arise in the diagnosis of leiomyosarcoma. Tumours with little cellular atypia must not be confused with leiomyomas. Secondly, it is important to combine histological information with surgical observations to define the point of origin of leiomyosarcomas.

Prognosis is poor, especially in highly mitotic tumours (more than four mitotic figures per field at a 25-fold magnification). Recurrences and regional extension are common after excisional surgery. Significant visceral and retroperitoneal involvement is also common. Massive thrombosis occurs in the distal portions of veins obstructed by tumour. Endoluminal proliferation may extend to involve the right heart chambers [49,88]. Finally, multiple arterial pulmonary emboli and metastases generally supervene.

Glomangiosarcoma

This type of tumour is exceptional. The diagnosis of malignancy is made on the presence of tumour zones resembling a fibromyosarcoma or a leiomyosarcoma. Nevertheless, none of these tumours has given rise to metastatic dissemination after complete surgical excision [60].

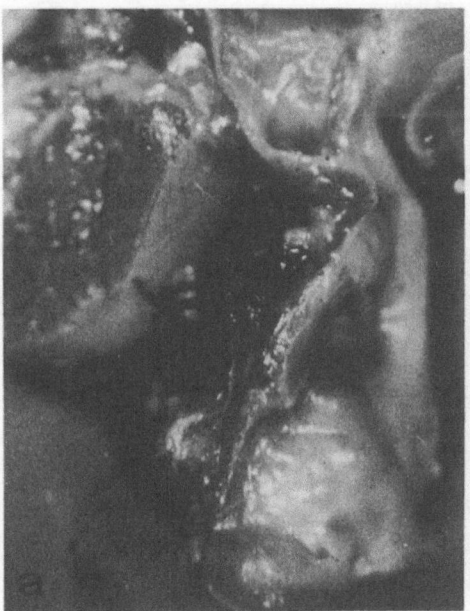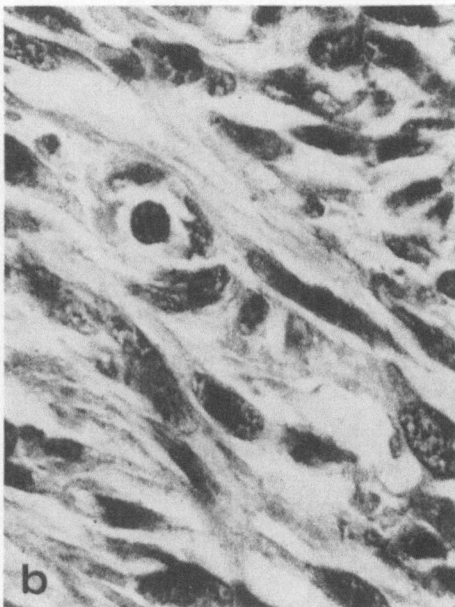

Fig. 37.17. Leiomyosarcoma of the inferior vena cava. **a** Gross view of the parietal infiltration leading to stenosis of the lumen. **b** Histopathological aspect showing marked cytological abnormalities. Haematoxylin–eosin–safran, × 450

Kaposi's Sarcoma (KS)

This vascular proliferation was first described in 1872 by Moricz Kaposi under the name of "multiple idiopathic pigmented sarcoma of the skin" and afterwards under that of "multiple idiopathic haemorrhagic sarcoma".

Incidence and Geographical Distribution

In Europe, classical KS is mainly observed in Mediterranean countries (especially Italy) and the Balkans. Ashkenaze Jews have a particularly high frequency, especially the men [136,153]. In the USA, KS is still rarely found, its incidence being from 0.02% to 0.06%. However, this frequency has increased since the appearance of epidemic KS linked to the acquired immuno-deficiency syndrome (AIDS).

In Africa, especially tropical Africa, the number of cases is much greater than in the rest of the world. KS is endemic. The highest frequency is observed in western Zaire, where it accounts for 90% of all malignant tumours.

The frequency decreases from Zaire towards western or southern African countries. In South Africa, KS represents 1.1% of malignant tumours found in the black population. This frequency varies greatly, sometimes within the same country. Men are particularly affected [92].

The lesions take on three patterns in the classical form (Fig. 37.18): initially as erythematous spots having an angiomatous appearance, later as infiltrated patches and sometimes as nodules. These three patterns tend to follow one another with time.

The histological appearance includes a proliferation of blood vessels and spindle-shaped cells.

The initial lesions, macular and pseudo angiomatous, represent the proliferation of thin-walled capillaries which have flattened endothelial cells without atypical or monstrous nuclei (Fig. 37.19). The lumina of the dilated capillaries present bizarre forms, sometimes consisting of simple fissures. Small foci of extravasated blood and deposits of haemosiderin can already be observed in the connective tissue situated

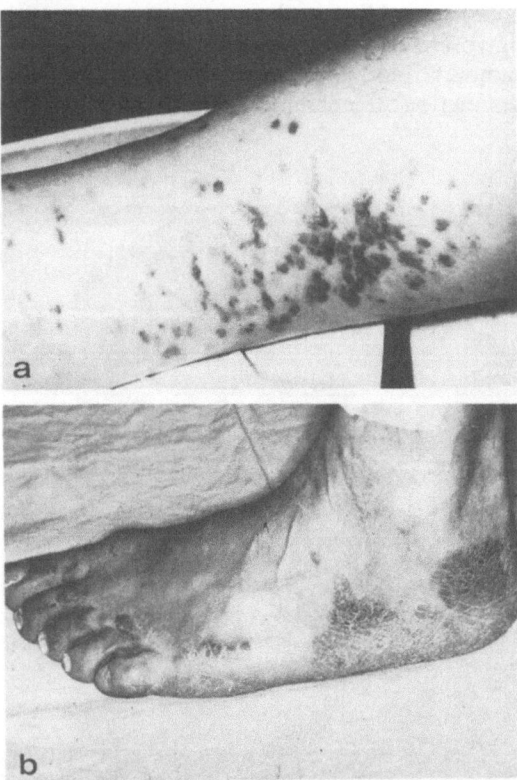

Fig. 37.18. Classical Kaposi's sarcoma. **a** Cutaneous violine papulomacular lesions. **b** Infiltrated patches. (Courtesy Dr Lessana-Leibovitch, Hôpital Tarnier, Paris, France)

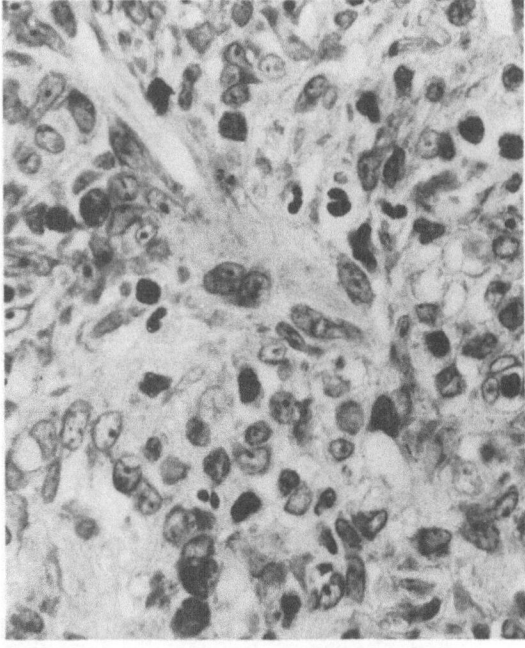

Fig. 37.19. Initial lesion of Kaposi's sarcoma within a lymph node. Proliferation of fusiform cells surrounding capillary profiles or directly lining vascular clefts. Lymphocytes and plasmocytes are present. Giemsa stain, × 760

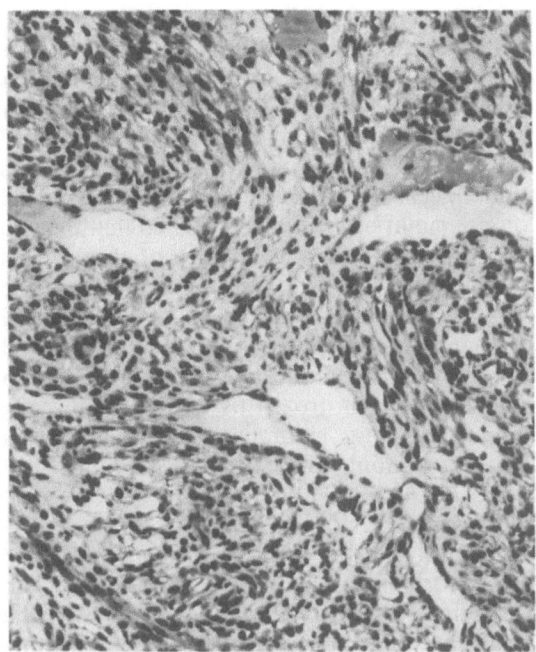

Fig. 37.20. Classical Kaposi's sarcoma. Cutaneous nodules showing bundles of fusiform cells directly lining vascular lumina containing blood. Numerous vessels of various size. Haematoxylin–eosin–safran, × 180

between the vessels. This connective tissue is often very rich in fusiform cells with a varying quantity of reactive cells, especially lymphocytes, plasmocytes and histiocytes. Such appearances are often difficult to distinguish from a fleshy bud or a capillary haemangioma.

The evolved tumour lesions appear as nodular groups of fusiform cells with anisocytosis and sometimes atypical nuclei (Fig. 37.20). These cells are grouped in bundles which are often crescentic, and vascular fissures appear in places between these cells. These fissures contain red blood cells (Fig. 37.21a,b,). Zones of necrosis, haemorrhagic foci and haemosiderin deposits (either intracellular or in phagocytic histiocytes) can be seen. Differentiation from other vascular tumours can be difficult, especially in the most anaplastic forms.

In the intermediary forms, the two previously described forms are juxtaposed or associated in variable ways. Columns of cells with voluminous nuclei creep between the collagen fibres, giving an appearance of infiltration with dissection of the connective tissue. Finally, between the cells or in the cytoplasm of the fusiform cells or phagocytic histiocytes, it is possible to recognize

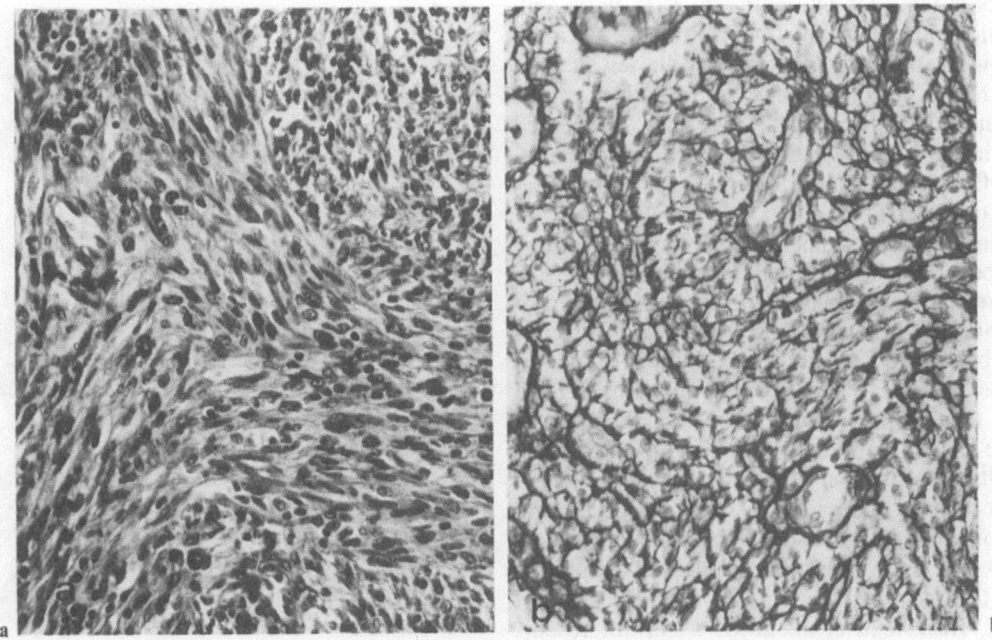

Fig. 37.21. Kaposi's sarcoma in lymph node. **a** Proliferation of fusiform cells looking at fibroblastic sarcoma. Vascular clefts are rare but clearly visible among the fusiform cells. Haematoxylin–eosin–safran, × 300. **b** Silver impregnation shows incomplete fibres between the cells and outlines numerous small vascular profiles, some of them displaying organized vascular wall. Gordon Sweet, × 300

eosinophilic hyalin globules which resemble Russell bodies. The presence of such inclusions is considered by some authors to be a factor in the differential diagnosis, especially from that of a fleshy bud [53].

Ultrastructural studies have not demonstrated the origin of the proliferating fusiform cells convincingly [2,24,84,108,122,167,175]. They appear as stretched-out cells having no special ultrastructural characteristics. Most of the publications in the literature point out the absence of micropinocytosis and Weibel–Palade bodies. Only a few intracytoplasmic filaments exist, and no myofilaments are seen at all. At the surface of the cells, no basal membrane can be seen, or only a few small fragments. The pseudo-angiomatous dilated capillaries have an altered structure: the endothelial surface is intermittent; the endothelial cells have ony a few junctional structures; and the basal membrane is extremely fragmented and disappears, partially or completely, in places. Pericytes may be lacking or when present are in a reduced number, with cytoplasmic expansion. In places, capillary rupture with tissue haemorrhage can be seen. Finally, vascular structures with virtually no lumen can be seen, which seem to correspond to the columns of cells dissecting the connective tissue. They are made up of large concentric cells, sometimes separated by collagen fibres, creating kinds of cylinders. Undifferentiated cells with voluminous nuclei occupy their centre. Between the vascular structures and the capillaries, several different types of cells can be recognized mixed in with the fusiform cells. These vary as reported in the literature, and consist of fibroblasts, myofibroblasts, smooth-muscle cells, pericytes and/or histiocytes.

Immunohistochemical studies have also furnished discordant results. The endothelial cells of the blood vessels contain an antigen linked to factor VIII which can be detected by immunolabelling on fixed tissue sections and on frozen tissue sections. Several groups of authors have thus been able to demonstrate the presence of these factors in the endothelial cells of the dilated capillaries and in varying percentages of the fusiform cells in KS cases [19,66,78,130,133]. The tumour cells and the endothelial cells also express HLA-DR. However, other studies have reported negative results [2,17,25,126,161]. Faced with these, Beckstead et al. noted a certain resemblance between the behaviour of the tumour cells in KS and that of lymphatic endothelial cells, thus confirming an origin already proposed by Dorfman. Ulex Europeaus I lectin is another

marker of endothelial cells. The same disagreement for this marker is also found in the literature: an absence of reactivity of the tumour cells in some cases [126], and focal positivity for others [140]. Several hypotheses may be evoked to explain these conflicting results: the KS tumour cells, even though they are of endothelial origin, might be so undifferentiated that they do not contain, or contain very little of, the proteins revealed by these techniques. Moreover, the small amount of factor VIII-related antigen and the low number of cells containing it could explain the negative results obtained in certain studies. Work on frozen tissue with amplifying immunolabelling methods using avitin–biotin complexes should allow the position to be clarified.

KS can be localized on a surface, predominantly on the skin or on the digestive or respiratory mucosa. It may also be generalized, spreading to numerous cutaneous areas or the gastrointestinal tract. Finally, it can be disseminated, affecting several internal organs.

The forms with multiple lymph-node involvement should be considered separately. In the severe forms, lymph-node localization may be massive and thus easy to identify (Fig. 37.21).

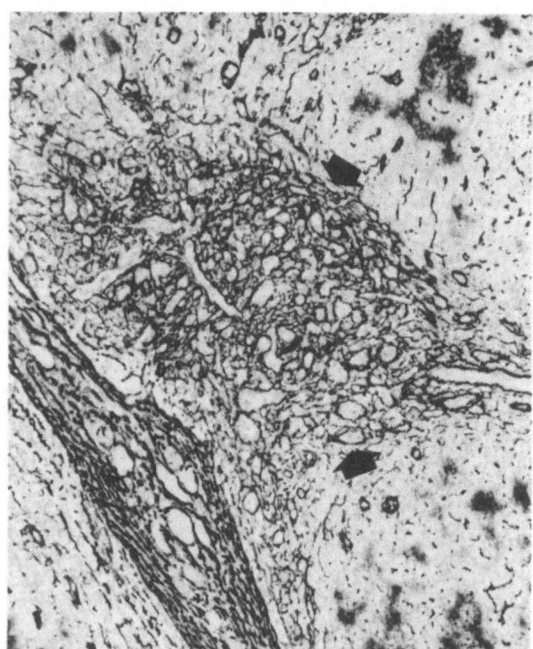

Fig. 37.22. Kaposi's sarcoma. Early lesion in lymph node in a patient with AIDS. Silver impregnation outlines the vascular proliferation and highlights the small tumoral focus which appears rich in fibres and small vessels. Gordon Sweet, × 120

However, it may be in the form of minor pseudo-angiomatous foci in the capsule, the vessel-carrying conjunctival framework or the ganglial parenchyma (Fig. 37.22). The lymph node sometimes presents a follicular hyperplasia with plasmocytosis. More rarely, the lymphoid hyperplasia produces the appearance of an angio-follicular hyperplasia of Castleman's disease. Such a picture associated with a KS necessitates the search for other risk factors of AIDS [50]. Sometimes, the lymph node may resemble an angioimmunoblastic lymphadenopathy [50,174].

A retrospective study on KS shows the high frequency (37% of cases) of association with another malignant proliferation [156], usually Hodgkin's disease or non-Hodgkin malignant lymphomas [72,173]. The incidence of these lymphomas is 20 times greater in patients with KS. Thus, in the series already cited [156], 58% of the patients having a KS were also affected by malignant lymphoma. These findings, initially demonstrated in classical KS and then in endemic forms, are also apparent in the epidemic form linked to AIDS.

KS is expressed in different clinical–pathological forms, which are often related to a particular geographical distribution and/or special aetiological circumstances.

Classical European KS

This occurs mainly in men between 50 and 70 years of age (the sex ratio is 3:1), especially if they, or their family, originate from south or east Europe (Italy, Russia, Poland). The affection begins and predominates on the lower limbs in the form of brownish or reddish irregular spots. They evolve, building up to wheals then to confluent nodules in patches, with ulceration. The evolution occurs by successive increases (sometimes regression occurs), and can take from 10 to 15 years. Lymph-node involvement is exceptional, as is involvement of internal organs, in spite of what has been discovered on autopsy in certain subjects.

A generalized form with ganglial visceral lesions is sometimes observed [156].

Endemic African KS

This form is essentially observed in equatorial Africa in the same zones as Burkitt's lymphoma [92], and involves mainly Africans. It is much rarer in Europeans and Asians living in these regions.

African KS can present in a clinical form similar to that of the classical European form: it predominates in males (sex ratio 17:1) but appears in younger subjects (between 25 and 44 years).

Another African form affects children between 1 and 10 years of age, with a male predominance (3:1). This infantile form is rare, and evolves rapidly to death. It is expressed by polyadenopathy due to tumour localization, with few or no cutaneous lesions.

KS in Immunodepressed Subjects

Since 1970, KS has been described in another group of patients. These are subjects undergoing prolonged immunosuppression therapy for kidney transplant or chemotherapy for various malignant diseases [85,94,144,171].

Sarcomatous lesions appear after a delay varying from 3 months to 4 years [90]. In certain cases, the lesions are predominantly in cutaneous areas, but more often diverse visceral involvement is present including pulmonary changes (Fig. 37.23). Evolution is often rapid with pluri-visceral dissemination. Nevertheless, several cases of regression of the Kaposi lesions have been reported after modification or discontinuation of the immunosuppressor [67].

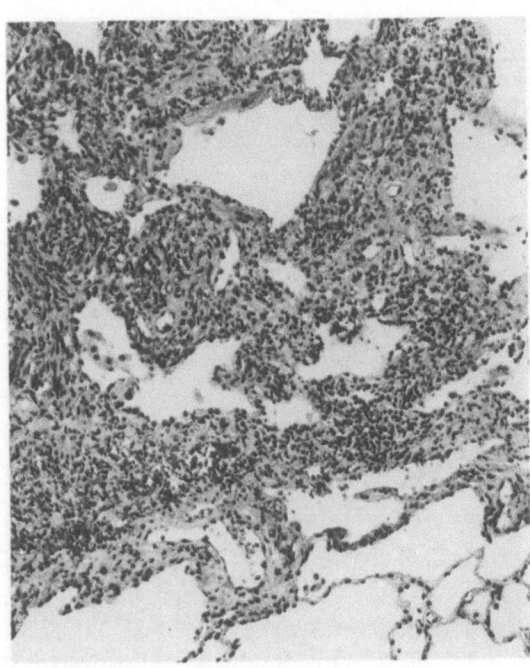

Fig. 37.23. Kaposi's sarcoma in immunodepressed patient. Haematoxylin–eosin–safran, × 180

AIDS-Associated Epidemic KS

More recently, a fourth group of patients presenting with KS was described. Indeed, the observation of a real "epidemic" of KS occurring in a population of young homosexuals (average age 35 years) in large urban centres in the USA, then in Europe, was the origin of the description of a "new" disease, the acquired immunodeficiency syndrome or AIDS.

Approximately one-third of the subjects with AIDS develop KS. In nearly 94% of the cases, the patients are homosexual, only 22% of drug abusers developing a KS [72].

The KS lesions in AIDS-related complex (ARC) or AIDS are a little unusual, being expressed by the sudden appearance of one or several cutaneous lesions, or by mucosal lesions in the mouth. The cutaneous lesions are predominantly facial, but do occur commonly on the thorax and upper limbs. They are reddish or pink, flat or slightly nodular and of variable size (a few millimetres to several centimetres) (Fig. 37.24). They can occur in groups or, in contrast, can be dispersed, and may resemble the cutaneous lesions of purpura, pytiriasis or secondary syphilis. Clinical diagnosis can be made even in the absence of cutaneous manifestations. The histological appearance is the same as that of the other types of KS, but very often the first small foci can be very difficult to distinguish

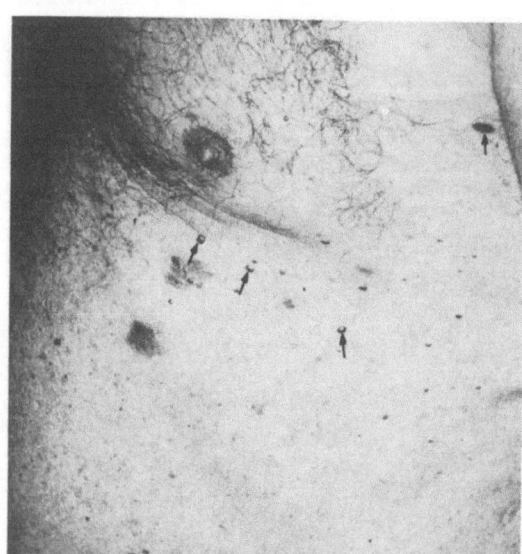

Fig. 37.24. Kaposi's syndrome in patient with AIDS. Small flat or slightly nodular lesions on the thorax. (Courtesy Dr Lessana-Leibovitch)

from those lesions due to a simple inflammatory reaction.

Aetiopathogenesis

A viral origin has been proposed for KS, involving an immune defect and perhaps a genetic predisposition: evidence includes a pluricentric origin; endemic evolution with foci of infected subjects in Africa; and incidence of the classical form in certain populations.

The virus incriminated is mainly cytomegalovirus (CMV). In the classical endemic African form and the epidemic form linked to AIDS, the patients have a significantly elevated level of anti-CMV antibody in their blood [72]. Similarly, in the three types of KS, the studies on tumour biopsies and tumour cell cultures using DNA–DNA recombination methods, in situ cyto-hybridation and anti-complement immunofluorescence tests have allowed the detection of either early CMV antigens, early antigens and RNA sequences of CMV or early antigens and RNA and DNA sequences of CMV in the tumour cells.

CMV can be isolated from the urine, the sperm and hepatic and pulmonary tissue in the epidemic forms; it has also been isolated from tumour cells in culture in the endemic forms, but more rarely [72]. Exceptionally, herpes-type viral particles have been recognized in tumour cells [72,175].

No formal proof of the direct oncogenic power of CMV in the classical and endemic forms has been provided as yet. Nevertheless, very recently, the transfection of DNA from a KS which appeared in a subject with AIDS to NIH 3T3 cells, allowed clones to be obtained which induced the appearance of haemorrhagic vascular tumours resembling KS when they were injected into nude mice. The tumour cells of this type of KS thus contain a transforming gene [112].

The possible existence of a genetic predisposition [67] is suggested by the observation of an HLA haplotype frequently occurring in subjects suffering from classical and epidemic KS, HLA-DR5 [72]. Moreover, it is known that this antigen is very frequent in the black African population.

The working hypothesis proposed at present to explain the development of KS involves the following stages [67,72]:

1. Non-tumoural polyclonal stimulation and proliferation [45,67] of endothelial cells by the

angiogenic substances liberated by stimulated lymphocytes (lymphokines) and/or activated histiomonocytes (monokines) during repeated or chronic inflammatory reactions.

2. Adjuvant role of CMV during persistent infection with a massive production of viral particles and/or multiple reactivation. HLA-DR could intervene as a receptor for the CMV virus on the endothelial cells thus favouring infection, reinfection, latent infection and reactivation.

3. Increase in the synthesis of CMV in endothelial cells, often with accumulation of incomplete particles.

4. Possible amplification of viral and perhaps cellular oncogenes in these cells.

5. Transformation of endothelial cells having normal phenotype to tumour cells with a modified phenotype.

6. Monoclonal proliferation of the transformed cells, favoured by a defect in immunosurveillance (immunodepression, iatrogene, AIDS).

7. Possible secretion by the KS cells of a growth factor for endothelial cells which diffuses through the blood system thus favouring the appearance of multiple foci [67].

Treatment

This depends on the extent of the disease. In the slowly developing localized forms, radiotherapy or intratumoral injection of vinblastine have been proposed. In the localized but more aggressive forms, intravenous or intra-arterial injection of vinblastine may be used in association with the preceding treatment. In the disseminated forms, polychemotherapy is usually administered using doxorubicin, bleomycin and vinblastine. Immunomodulating treatment has also been proposed (bovine thymus extracts, thymopoietin, anti-T suppressor monoclonal antibodies, α-interferon, interleukin 2) for the KS appearing during the course of AIDS [103].

References

1. Abratt RP, Williams M, Raff M et al. (1983) Angiosarcoma of the superior vena cava. Cancer 52:740–743
2. Akhtar M, Bunuan H, Ashraf M et al. (1984) Kaposi's sarcoma in renal transplant recipients: ultrastructural and immunoperoxidase study of 4 cases. Cancer 53:258–266
3. Allen PW, Enzinger FM (1972) Hemangioma of skeletal muscle. An analysis of 89 cases. Cancer 29:8–22
4. Alpert LI, Benisch B (1970) Hemangioendothelioma of the liver associated with microangiopathic hemolytic anemia. Am J Med 48:624–628
5. Amerigo J, Berry CL (1980) Intravascular papillary endothelial hyperplasia in the skin and subcutaneous tissue. Virchows Arch [A] 387:81–90
6. Ashley DJB (1978) Evan's histological appearances of tumours. Churchill Livingstone, Edinburgh, pp 66–72
7. Ayella RJ (1970) Hemangiopericytoma. A case report with arteriographic findings. Radiology 97:611–612
8. Backwinkel KD, Diddams JA (1970) Hemangiopericytoma. Report of a case and comprehensive review of the literature. Cancer 25:896–906
9. Bailey OT (1935) The cutaneous glomus and its tumours: glomangioma. Am J Pathol 11:915–935
10. Baker HDC, Paget GE, Davidson J (1956) Hemangioendothelioma (Kupffer cell sarcoma) of the liver. J Pathol Bact 72:173–182
11. Baldet P, Mimran A (1977) Tumeur bénigne du rein à sécrétion de rénine dite de "l'appareil juxtaglomérulaire". Etude optique et ultrastructurale. Ann Anat Pathol (Paris) 22:21–40
12. Barajas L, Bennett CM, Connor G et al. (1977) Structure of a juxtaglomerular cell tumour: the presence of a neural component. A light and electron microscopic study. Lab Invest 37:357–368
13. Bardwil JM, Mocega EE, Butler JI et al. (1968) Angiosarcoma of the head and neck region. Am J Surg 116:548–560
14. Barnett WO, Hardy JD, Hendrix JH (1969) Lymphangiosarcoma following post-mastectomy lymphedema. Ann Surg 169:960–968
15. Battifora H (1973) Hemangiopericytoma. Ultrastructural study of five cases. Cancer 331:1418–1432
16. Bean WB (1958) Vascular spiders and related lesions. CC Thomas, Springfield, Illinois, pp 178–185
17. Beckstead JH, Wood GS, Fletcher V (1985) Evidence for the origin of Kaposi's sarcoma from lymphatic endothelium. Am J Pathol 119:294–300
18. Begg CF, Garret R (1954) Hemangiopericytoma occurring in the meninges. Cancer 7:602–606
19. Bendelac A, Kanitakis J, Chouvet B et al. (1985) Sarcome de Kaposi: étude immunohistochimique comparative et intérêt histogénétique des marqueurs endothéliaux. Ann Pathol 5:45–52
20. Berry CL, Amerigo J (1980) Blood group antigens in vascular tumours. Virchows Arch [A] 388:167–174
21. Bizard T, Melliere D, Lange F (1979) Kystes adventiciels de la fémorale commune. Ann Chir 33:36–39
22. Blumenfeld TA, Fleming ID, Johnson WW (1969) Juvenile hemangioendothelioma of the liver: report of a case and review of the literature. Cancer 24:853–857
23. Blumenfeld W, Egbert BM, Sagebiel RW (1985) Differential diagnosis of Kaposi's sarcoma. Arch Pathol Lab Med 109:123–127
24. Braun-Falco O, Schmoeckel C; Hubner G (1976) Zur Histognese des Sarcoma Idiopathicum Multiplex Haemorrhagicum (Morbus Kaposi). Virchows Arch [A] 369:215–227
25. Burgdorf WHC, Mukai K, Rosai J (1981) Immunohistochemical identification of factor VIII-related antigen in endothelial cells of cutaneous lesions of alleged vascular nature. Am J Clin Pathol 75:167–171
26. Camilleri JP, Diebold J, Delarue J (1971) Les tumeurs malignes du corpuscule carotidien. Arch Anat Pathol 3:237–246

27. Camilleri JP, Hinglais N, Bruneval P et al. (1984) Renin storage and cell differentiation in juxtaglomarular cell tumours. An immunohistochemical and ultrastructural study of three cases. Hum Pathol 15:1069–1079
28. Camilleri JP, Phat VN, Bariety J et al. (1980) Use of specific antiserum for renin detection in human kidney. J Histochem Cytochem 28:1343–1346
29. Cantin M, Araujo-Nascimento MF, Benchimol S et al. (1977) Metaplasia of smooth muscle cells into juxtaglomarular cells in the apparatus, arteries and arterioles of the ischemic (endocrine) kidney. Am J Pathol 87:581–590
30. Capo V, Ozzello L, Fenoglio CM et al. (1985) Angiosarcoma arising in edematous extremities: immunostaining for factor VIII-related antigen and ultrastructural features. Hum Pathol 16:144–150
31. Cardell BS, McGill DAF, Williams R (1971) Leiomyosarcoma of inferior vena cava producing Budd-Chiari syndrome. J Pathol 104:283–286
32. Castro C, Winkelmann RK (1974) Angiolymphoid hyperplasia with eosinophilia in the skin. Cancer 34:1696–1705
33. Chaudhuri B, Ronan SG, Manaligod JR (1980) Angiosarcoma arising in a plexiform neurofibroma. A case report. Cancer 46:605–610
34. Chen KTK, Gilbert EF (1979) Angiosarcoma complicating generalized lynphangiectasia. Arch Pathol Lab Med 103:86–88
35. Chen KTK, Hoffman KD, Hendricks EJ (1979) Angiosarcoma following therapeutic irradiation. Cancer 44:2044–2048
36. Chomette G, Auriol M (1984) Angiosarcome primitif du coeur. Etude histoenzymologique et ultrastructurale d'une observation. Arch Anat Cytol Pathol 32:311–315
37. Chomette G, Auriol M, Bertrand JC et al. (1983) Les hémangiopéricytomes de la cavité buccale. Etude anatomaclinique et ultrastructurale. Ann Med Interne 134:7–14
38. Chomette G, Auriol M, Delsol M (1978) Sarcome primitif de l'artère pulmonaire à type d'hémangioendothéliosarcome. Ann Anat Pathol (Paris) 23:161–170
39. Chomette G, Auriol M, Princ G et al. (1984) Le léiomyosarcome osseux. Etude ultrastructurale et histoenzymologique. A propos d'un cas de siège mandibulaire. Arch Anat Cytol Pathol 32:20–25
40. Chomette G, Auriol M, Tereau Y (1978) Hémangiopéricytome malin. A propos de l'analyse ultrastructurale et histoenzymologique d'une observation. Ann Anat Pathol (Paris) 23:41–52
41. Chomette G, Auriol M, Tranbaloc P et al. (1983) Aspects anatomopathologiques des angiodysplasies artérielles. Ann Med Interne 134:444–450
42. Chomette G, Pinaudeau Y, Brocheriou C et al. (1967) A propos d'une observation d'hémangioendothéliosarcome mammaire avec métastases multiples. Arch Anat Pathol 15:304–307
43. Compagno J, Hyans J (1976) Hemangiopericytoma-like intranasal tumours. A clinico-pathological study of 23 cases. Am J Clin Pathol 66:672–683
44. Cormier M, Diebold J, Camilleri JP et al (1970) Glomangiomatosis. Anatomoclinical study of a case. J Cardiovasc Surg 13:158–163
45. Costa J, Rabson AS (1983) Generalized Kaposi's sarcoma is not a neoplasm. Lancet I:58
46. Dabska M (1969) Malignant endovascular papillary angioendothelioma of skin in childhood. Cancer 24:503–510

47. Darrouzet V, Rivel J, Deminiere C et al (1982) Chemodectome carotidien malin avec incidence familiale et métastases ganglionnaires. Etude optique et ultrastructurale d'un cas et revue de la littérature. Ann Pathol 2:163–167
48. De Laurentin DA, Wolferth CC, Wolf FM et al. (1973) Mucin adventitial cysts of the popliteal artery in an 11-year-old girl. Surgery 74:456–459
49. Deutsh V, Fraenkel O, Frand U et al. (1968) Leiomyosarcoma of the inferior vena cava propagating into the right atrium. Br Heart J 30:571–574
50. Diebold J, Marche CI, Audouin J et al. (1985) Lymph node modification in patients with the acquired immunodeficiency syndrome (AIDS) or with AIDS related complex (ARC) Pathol Res Pract 180:590–611
51. Diebold J, Tulliez M, Audouin J et al. (1980) Sarcome de Kaposi viscéral avec lymphadénopathie angioimmunoblastique et syndrome sec latent. Bull Cancer (Paris) 67:131–138
52. Donnel RM, Rosen PP, Lieberman PH et al. (1981) Angiosarcoma and other vascular tumours of the breast: pathologic analysis as a guide to prognosis. Am J Surg Pathol 5:629–637
53. Dorfman RF (1962) Kaposi's sarcoma: the contribution of enzyme histochemistry to the identification of cell types. Acta Un Int Cancer 18:464–476
54. Dorfman RF (1985) Cutaneous and lymphadenopathic Kaposi's sarcoma in Africa and the USA with observations on persistent lymphadenopathy in homosexual men at risk from the acquired immunodeficiency syndrome. Front Radiat Ther Oncol 9:105–116
55. Eady RAJ, Wilson JE (1977) Pseudopyogenic granuloma, enzyme histochemical and ultrastructural study. Hum Pathol 8:653–668
56. Edwards AT, Kellen HS (1968) Hemangiosarcoma of breast. J Pathol Bact 95:457–465
57. Eimoto T (1977) Ultrastructure of an infantile hemangiopericytoma. Cancer 40:2161–2170
58. Ejrup B, Hiertonn T (1954) Intermittent claudication: three cases treated by free vein graft. Acta Chir Scand 108:217–230
59. Enzinger FM, Smith BH (1976) Hemangiopericytoma. An analysis of 106 cases. Hum Pathol 7:61–82
60. Enzinger FM, Weiss SW (1983) Soft tissue tumours. Mosby, St Louis
61. Falk H, Thomas LB, Popper H et al. (1979) Hepatic angiosarcoma associated with androgenic anabolic steroids. Lancet II:1120–1123
62. Farr HW, Carandang CH, Huvos AG (1970) Malignant vascular tumours of the head and neck. Am J Surg 120:501–504
63. Fayemi AO, Toker (1975) Nodal angiomatosis. Arch Pathol 99:170–172
64. Fisher JH (1965) Post-mastectomy lymphangiosarcoma in the lymphedematous arm. A review of four cases. Can J Surg 8:350–362
65. Flanigan DP, Burnham SJ, Goodreau JJ et al. (1979) Summary of cases of adventitial cystic disease of the popliteal artery. Ann Surg 180:165–175
66. Flotte TJ, Hatcher VA, Friedman-Kien AE (1984) Factor VIII-related antigen in Kaposi's sarcoma in young homosexual men. Arch Dermatol 120:180–182
67. Friedman-Kien AE (1984) Kaposi's sarcoma: an opportunistic neoplasm. J Invest Dermatol 82:446–448
68. Friedman-Kien AE, Laubenstein LJ (1984) AIDS: the epidemic of Kaposi's sarcoma and opportunistic infections. Masson New York

69. Friedman-Kien AE, Laubenstein LJ, Rubinstein P et al. (1982) Disseminated Kaposi's sarcoma in homosexual men. Ann Intern Med 6:693–700

70. Galen FX, Devaux C, Corvol MT et al. (1984) Renin biosynthesis by human tumoral juxtaglomerular cells. Evidence for a renin precursor. J Clin Invest 73:1144–1155

71. Gill BS, Mehra YN (1968) Haemangiopericytoma in nasal cavity. J Laryngol Otol 82:839–844

72. Giraldo G, Beth E, Buonaguro FM (1984) Kaposi's sarcoma: a natural model of interrelationship between viruses, immunologic responses, genetics and oncogenesis. Antibiot Chemother 32:1–11

73. Girard C, Johnson WC, Graham JH (1970) Cutaneous angiosarcoma. Cancer 26:868–883

74. Glenner GG, Grimley PM (1984) Tumours of the extra-adrenal paraganglion system (including chemoreceptors). In: Atlas of tumour pathology, 2nd series, fasc 9. Armed Forces Institute of Pathology, Washington

75. Gottlieb GJ, Ackerman AB (1982) Kaposi's sarcoma, an extensively disseminated form in young homosexual men. Hum Pathol 13:882–892

76. Greene RR, Gerbie A (1954) Hemangiopericytoma of the uterus. Obstet Gynecol 3:150–159

77. Guarda LA, Ordonez NG, Smith JL et al. (1982) Immunoperoxydase localization of factor VIII in angiosarcoma. Arch Pathol Lab Med 106:515–516

78. Guarda LG, Silva EG, Ordonez NG et al. (1981) Factor VIII in Kaposi's sarcoma. Am J Clin Pathol 76:197–200

79. Guyenne TT, Galen FX, Devaux C et al. (1980) Direct radioimmunoassay of human renin. Comparison with renin activity in plasma and amniotic fluid. Hypertension 2:465–470

80. Haber H, Harris-Jones J, Wells A (1974) Intravascular endothelioma (endothelioma in situ, systemic endotheliomatosis). J Clin Pathol 17:608–612

81. Hahn MJ, Dawson R, Esterly JA et al. (1973) Hemangiopericytoma. An ultrastructural study. Cancer 31:255–261

82. Hansen JPH (1966) Cystic mucoid degeneration of the popliteal artery. Acta Chir Scand 131:171–177

83. Harris HR (1971) Angiosarcoma of the heart. J Clin Pathol 24:520–523

84. Harrison AC, Kahn LB (1978) Myogenic cells in Kaposi's sarcoma: an ultrastructural study. J Pathol 124:157–160

85. Harwood AR (1984) Kaposi's sarcoma in renal transplant patients. In: Friedman-Kien AE, Laubenstein LJ (eds) AIDS: the epidemic of Kaposi's sarcoma and opportunistic infections. Masson, New York

86. Hiertonn T, Lindberg K, Rob C (1967) Cystic degeneration of the popliteal artery. Br Med J iii:411–415

87. Hoch-Ligeti C (1978) Angiosarcoma of liver associated with diethylstilbesterol. JAMA 240:1510–1511

88. Hoffbrand AV, LLoyd-Thomas HF (1964) Leiomyosarcoma of the inferior vena cava leading to obstruction of the tricuspid valve. Br Heart J 26:709–712

89. Holmes JG (1960) Cystic adventitial degeneration of the popliteal artery. JAMA 173:126–129

90. Hood AF, Farmer ER, Weiss RA (1982) Kaposi's sarcoma. Johns Hopkins Med J 151:222–230

91. Horne WI, Percival WL (1975) Hemangiosarcoma of the breast. Can J Surg 18:81–84

92. Hutt MSR (1984) Classical and endemic form of Kaposi's sarcoma. A review of antibiotics and chemotherapy. Karger, Basel, pp 12–17

93. Ishak K, Sesterhenn I, Goodman MZ et al. (1984) Epithelioid hemangioendothelioma of the liver. A clinicopathological and follow-up study of 32 cases. Hum Pathol 15:839–852

94. Kapadia SB, Krause JR (1977) Kaposi's sarcoma after long term alkylating agent therapy for multiple myeloma. South Med J 70:1011–1013

95. Kauffman SL, Stout AP (1960) Hemangiopericytoma in children. Cancer 13:695–710

96. Kauffman SL, Stout AP (1961) Malignant hemangioendothelioma in infants and children. Cancer 14:1186–1196

97. Kennedy JC, Fisher JH (1960) Hemangiopericytoma: its orthopaedic manifestations. J Bone Joint Surg [Br] 42:80–92

98. Kieffer E, Berrod JL, Chomette G (1985) Primary tumours of the inferior vena cava. Surgery of the vein. Grune and Stratton, New York, pp 423–443

99. Kihara L, Kitamura S, Hoshino T et al. (1968) A hitherto unreported vascular tumour of the kidney: a proposal of "juxtaglomarular cell tumour". Acta Pathol Jpn 18:197–206

100. Kindblom LG, Fassina AS (1981) Angiolymphoid hyperplasia with eosinophilia of the skin: light microscopic and ultrastructural study of 4 cases. Acta Pathol Microbiol Scand [A] 89:271–283

101. Kitagawa M, Matsubara O, Song SY et al. (1985) Neoplastic angioendotheliosis. Immunohistochemical and electron microscopic findings in three cases. Cancer 56:1134–1143

102. Kohout E, Stout AP (1961) The glomus tumour in children. Cancer 14:555–566

103. Krown SE, Real FX, Krim M et al. (1985) Interferons and other biological response modifies in the treatment of Kaposi's sarcoma. Front Radiat Therap Oncol 9:138–149

104. Kuhn C, Rosai J (1969) Tumours arising from pericytes. Ultrastructure and organ culture of a case. Arch Pathol 88:653–663

105. Kuo TT, Hsueh S, Su IJ et al. (1985) Histiocytoid hemangioma of the heart with peripheral eosinophilia. Cancer 55:2854–2861

106. Kuo T, Sayers P, Rosai J (1976) Masson's vegetant intravascular hemangioendothelioma: a lesion often mistaken for angiosarcoma. Cancer 38:1227–1236

107. Lack EE, Cubilla AL, Woodruff JM et al. (1977) Paraganglions of the head and neck region. A clinical study of 69 patients. Cancer 39:397–409

108. Leu HJ, Odermatt B (1985) Multicentric angiosarcoma (Kaposi's sarcoma). Light and electron microscopic and immunological findings of idiopathic cases in Europe and Africa and of cases associated with AIDS. Virchows Arch [A] 408:29–41

109. Lewis GJJ, Douglas DM, Reid W et al. (1967) Cystic adventitial disease of the popliteal artery. Br Med J iii:411–415

110. Lindop GBM, Stewart JA, Downie TT (1983) The immunocytochemical demonstration of renin in a juxtaglomarular cell tumour by light and electron microscopy. Histopathology 7:421–431

111. Llombart-Bosch A, Peydro-Olaya A, Pellin A (1982) Ultrastructure of vascular neoplasms: a transmission and scanning electron miroscopical study based upon 42 cases. Pathol Res Pract 74:1–4

112. Lo SC, Liotta LA (1985) Vascular tumours produced by NIH/3T3 cells transfected with human AIDS Kaposi's sarcoma DNA. Am J Pathol 118:7–14

113. Machinami R, Adachi S (1984) Histochemical evidence

of adenosine triphosphatase activity in glomus tumour. Report of a case. Pathol Res Pract 79:1–6

114. Mackenzie DR (1971) Lymphangiosarcoma arising in chronic congenital and idiopathic lymphedema. J Clin Pathol 24:524–529

115. Mark TM, Rywlin AM, Unger H (1983) Cystic adventitial degeneration of the popliteal artery. Its occurrence in a patient with the Nail-Patella Syndrome. Arch Pathol Lab Med 107:186–188

116. Maruyama N, Kumagai Y, Ishida Y et al. (1985) Epithelioid haemangioendothelioma of the bone tissue. Virchows Arch [A] 407:159–166

117. Masson P (1923) Hemangioendotheliome végétant intravasculaire. Bull Soc Anat (Paris) 93:517–522

118. Masson P (1956) Tumeurs humaines. Histologie, diagnostic et techniques. Maloine, Paris

119. McCarthy PL, Shklar G (1980) Diseases of the oral mucosa, 2nd edn. Lea and Febiger, Philadelphia, pp 423–443

120. McConnel EM, Harris HR (1966) Angiosarcoma in postmastectomy lymphedema. Br J Surg 53:572–577

121. McCormack LJ, Gallivan WF (1954) Hemangiopericytoma. Cancer 7:595–601

122. McNutt NS, Fletcher V, Conant MA (1983) Early lesions of Kaposi's sarcoma in homosexual men: an ultrastructural comparison with other vascular proliferations in skin. Am J Pathol 111:62–77

123. Menard J, Soubrier F, Bariety J et al. (1983). Primary reninism. In: Genest J, Kuchel O, Hamet P, Cantin M (eds) Hypertension, 2nd edn. McGraw Hill, New York, pp 1034–1040

124. Merrick T, Erlandson RA, Hajdu SI (1971) Lymphangiosarcoma of a congenitally lymphedematous extremity. Arch Pathol 91:365–371

125. Mesmin F, Gomes H, Behar C et al. (1979) Leucémie lymphoblastique et sarcome de Kaposi. Pediatr Radiol 8:185–187

126. Miettinen M, Holthofer H, Lehto V et al. Ulex europeaus I Lectin as a marker for tumours derived from endothelial cells. Am J Clin Pathol 79:32–36

127. Mignot J, Simard CI (1961) Tumeurs glomiques multiples. Arch Anat Pathol 9:236–240

128. Mills SE, Cooper PH, Fechner RE (1980) Lobular capillary hemangioma: the underlying lesion of pyogenic granuloma. Am J Surg Pathol 4:471–475

129. Mittinen M, Lehto VP, Virtanen I (1983) Glomus tumour cells: evaluation of smooth muscle and endothelial cell properties. Virchows Arch [B] 43:139–150

130. Modlin RL, Hofman FM, Kempf RA et al. (1983) Kaposi's sarcoma in homosexual men: an immunohistochemical study. J Am Acad Dermatol 8:620–627

131. Murad TM, Von Haarm E, Murthy MSN (1968) Ultrastructure of a hemangiopericytoma and a glomus tumour. Cancer 22:1239–1249

132. Nadji M, Gonzalez MS, Castro A et al (1980) Factor VIII-related antigen: an endothelial cell marker. Lab Invest 42:139 (abstract)

133. Nadji M, Morales AR, Ziegler-Weisman J et al. (1981) Kaposi's sarcoma. Immunohistochemical evidence for an endothelial origin. Arch Pathol Lab Med 105:274–275

134. Newland RC, Maxwell LE, Constance TJ (1978) Malignant hemangiopericytoma. Case report and ultrastructural study. Pathology 10:277–283

135. Nunnery EW, Kahn LB, Reddick RL et al. (1981) Hemangiopericytoma. A light microscopic and ultrastructural study. Cancer 47: 906–914

136. Oettle AG (1963) Geographical and racial differences in the frequency of Kaposi's sarcoma as evidence of environmental or genetic causes. In: Ackermann M (ed) Symposium on Kaposi's sarcoma Karger, Basel, pp 17–50 (Unio Internationalis contra cancrum vol 13)

137. Ognibene FP, Steis RG, Macher AM et al. (1985) Kaposi's sarcoma causing pulmonary infiltrates and respiratory failure in the acquired immunodeficiency syndrome. Ann Intern Med 102:471–475

138. Olsen EG, Wellwood JM (1970) Hemangiopericytoma of the small intestine. Br J Surg 57: 66–69

139. Orcel L, Chomette G (1978) Anatomie pathologique vasculaire. Flammarion, Paris

140. Ordonez NG, Batskis JG (1984) Comparison of Ulex europeaus I Lectin and factor VIII-related antigen in vascular lesions. Arch Pathol Lab Med 108:129–132

141. Orsos G (1934) Gefassprossgechwulst (hemangioma). Beitr Anat Pathol 93:121–142

142. Patterson K, Chandra RS (1985) Malignant endovascular papillary angioendothelioma. Cutaneous borderline tumour. Arch pathol Lab Med 109:671–673

143. Paullada JJ, Lisci-Gramilla A, Gonzales-Angulo A et al. (1968) Hemangiopericytoma associated with hypoglycemia. Metabolic and electron microscopic studies of a case. Am J Med 44:990–999

144. Penn I (1979) Kaposi's sarcoma in organ transplant recipient. Transplantation, 27:8–11

145. Penn I (1983) Kaposi's sarcoma in immunosuppressed patients. J Clin Lab Immunol 12:1–10

146. Pepper MC, Laubenheimer R, Cripps DJ (1977) Multiple glomus tumours. J Cutan Pathol 4:244–257

147. Pfaltz M, Hedinger C, Saremaslani P et al. (1983) Malignant hemangioendothelioma of the thyroid and factor VIII-related antigen. Virchows Arch [A] 401:177–184

148. Popper H, Thomas LB, Telles NC et al. (1978) Development of hepatic angiosarcoma in man induced by vinyl-chloride, thorotrast and arsenic. Am J Pathol 92:349–369

149. Ramsey HJ (1966) Fine structure of hemangiopericytoma and hemangioendothelioma. Cancer 19:2005–2018

150. Robertson PW, Klidjian A, Harding LK et al. (1967) Hypertension due to a renin-secreting renal tumour. Am J Med 43:963,976

151. Rosai J, Gold J, Landy R (1979) The histiocytoid hemangiomas: unifying concept embracing several previously described entities of skin, soft tissue, large vessels, bone and heart. Hum Pathol 10:707—760

152. Rosai J, Sumner HW, Kostianovsky M et al. (1976) Angiosarcoma of the skin. A clinico-pathologic and inner structural study. Hum Pathol 7:83–109

153. Rothman S (1963) Remarks on sex, age, and racial distribution of Kaposi's sarcoma and possible pathogenic mechanisms. In: Ackermann, M (ed): Symposium on Kaposi's sarcoma. Karger, Basel pp 13–16 (Unio Internationalis contra cancrum vol 13)

154. Rutgers JL, Wieczorek R, Bonetti F et al. (1986) The expression of endothelial cell surface antigens by AIDS associated Kaposi's sarcoma: evidence for a vascular endothelial cell origin. Am J Pathol 122(3):493–499

155. Sablet M, Mascaro JM (1967) Tumeurs glomiques multiples et blue rubber bleb naevus. Ann Dermatol Venereol 94:35–46

156. Safai BJ, Good RA (1981) Kaposi's sarcoma. In: Safai BJ, Good RA (eds) Comprehensive immunology. New York (Immunodermatology vol 7)

157. Salm R (1963) The nature of the so-called post-mastectomy lymphangiosarcoma. J Pathol 85:445–456

158. Saltz RK, Kurtz RC, Lightdale CJ et al. (1984) Kaposi's sarcoma: gastrointestinal involvement correlation with skin findings and immunologic function. Plenum, New York (Digestive diseases and sciences. New series, vol 29, no 9)

159. Schruber H, Barry FM, Russell WC et al. (1979) Stewart's Treves syndrome. A lethal complication of post-mastectomy lymphedema and regional immune deficiency. Arch Surg 114:82–91

160. Scott PWB, Silvers DN, Helwig EB (1975) Proliferating angioendotheliomatosis. Arch Pathol 99:323–326

161. Sehested M, You-Jensen K (1981) Factor VIII-related antigen as an endothelial cell marker in benign and malignant disease. Virchows Arch [A] 391:217–225

162. Shugart RR, Soule EH, Johnson EW (1963) Glomus tumours. Surg Gynecol Obstet 117:334–340

163. Shute K, Rothnic NG (1973) The aetiology of cystic arterial disease. Br J Surg 60:397–400

164. Silverberg SG, Kay S, Koss LG (1971) Post-mastectomy lymphangiosarcoma: ultrastructural observations. Cancer 27:100–108

165. Silverberg SG, Willson MA, Board JA (1971) Hemangiopericytoma of the uterus. An ultrastructural study. Am J Obstet Gynecol 110:397–404

166. Steiner GC, Dorfman HD (1972) Ultrastructural study of hemangioendothelial sarcoma of bones. Cancer 29:122–135

167. Sterry W, Steigleder GK, Bodeux E (1979) Kaposi's sarcoma: venous capillary hemangioblastoma. A histochemical and ultrastructural study. Arch Dermatol Res 266:253–267

168. Stout AP, Lattes R (1967) Tumors of the soft tissues. In: Atlas of tumour pathology, 2nd series, fasc 7. Armed Forces Institute of Pathology, Washington

169. Stout AP, Murray MR (1942) Hemangiopericytoma. A vascular tumour featuring Zimmerman's pericyte. Ann Surg 116:26–33

170. Strate SM, Rutledge JG, Weinberg AG (1984) Delayed development of angiosarcoma in multinodular infantile hepatic hemangioendothelioma. Arch Pathol Lab Med 108:943–944

171. Stribling J, Weitzner S, Smith GV (1978) Kaposi's sarcoma in renal allograft recipient. Cancer 42:442–446

172. Sutton D, Pratt AE (1967) Angiography of hemangiopericytoma. Clin Radiol 18:324–328

173. Ulbright TM, Santa-Cruz DJ (1981) Kaposi's sarcoma: relationship with hematologic lymphoid and thymic neoplasia. Cancer 47:963–973

174. Varsano S, Manor Y, Steiner Z et al (1984) Kaposi's sarcoma and angioimmunoblastic lymphadenopathy. Cancer 54:1582–1585

175. Walter P, Philippe E, Khalil Th (1984) Le sarcoma de Kaposi. Un néoplasme vasculaire présumé d'origine virale. Caractères histologiques et ultrastructuraux. Ann Pathol 4:19–23

176. Weinblatt ME, Kahn E, Kochen J (1984) Hemangioendothelioma with intravascular coagulation and ischemic colitis. Cancer 54:2300–2304

177. Weiss SW, Enzinger FN (1982) Epithelioid hemangioendothelioma. A vascular tumour often confused with carcinoma. Cancer 50:970–981

178. Wind MS, Pillari G (1979) Deep soft tissue hemangioma of infancy: Kasabach-Meritt syndrome. NY State J Med 79:373–379

179. Woodward AH, Ivins JC, Soule EH (1972) Lymphangiosarcoma arising in chronic lymphedematous extremities. Cancer 30:562–572

180. Zaynoune ST, Juljulian HH, Kurban AK (1974) Pyogenic granuloma with multiple satellites. Arch Dermatol 9:689–691

181. Zimmermann KW (1923) Der feinere Bau der Blutcapillaren. Anat Embryol (Berl) 68(2):29–109

Chapter 38

Diagnostic and Therapeutic Principles in Angiodysplasia

J. M. Cormier

Haemodynamic criteria are essential for a useful classification of angiodysplasia, for only such an approach can provide the necessary clinical, prognostic and therapeutic information about vascular malformations. Pathological data give only indirect insight into the circulatory conditions which characterize the vascular malformations [1,3,5].

Vascular malformations can affect vessels in one or several tissues and may be focal and segmental, widespread, circumscribed or diffuse. Other abnormalities are often associated including arteriovenous communications, arterial aneurysms, agenesis, venous ectasia and persistence of vestigial veins which accounts for anomalous venous drainage patterns. All these topographical features have distinct therapeutic implications: the more widespread and diffuse the malformations, the smaller the chance of a "cure".

Haemodynamic Categories

Four clinical and haemodynamic categories of vascular malformation may be defined.

Sequestrated Vascular Malformations

These lesions do not communicate with the normal vascular tree, and are either soft or firm

depending upon their content. Regional haemodynamic alterations are not transmitted to sequestrated vascular malformations.

Haemodynamically Venous Vascular Malformations

These lesions communicate with the venous system and vary in size in response to pressure changes generated by coughing, orthostatism or elevation. Different venous territories may be involved.

Varices may be widespread or well localized. A typical feature is a flat angioma of the lateral surface of the leg or of the thigh. The angioma drains into the external marginal vein which retains vestigial sural, crural or gluteal portions.

Haemangiomas may be superficial or deep. These lesions will slowly increase in size if their drainage route is compressed. Occasional painful episodes signal the occurrence of thrombosis (the so-called "cavernous" angioma).

Low-Flow Arteriovenous Malformations

Skin temperature in the vicinity of these lesions is elevated.

The associated varices or angiomas of subcutaneous or deep soft tissue empty in response to pressure but refill rapidly. On elevation, these lesions decrease in size only modestly. If their drainage route is compressed, however, they will readily become very tense.

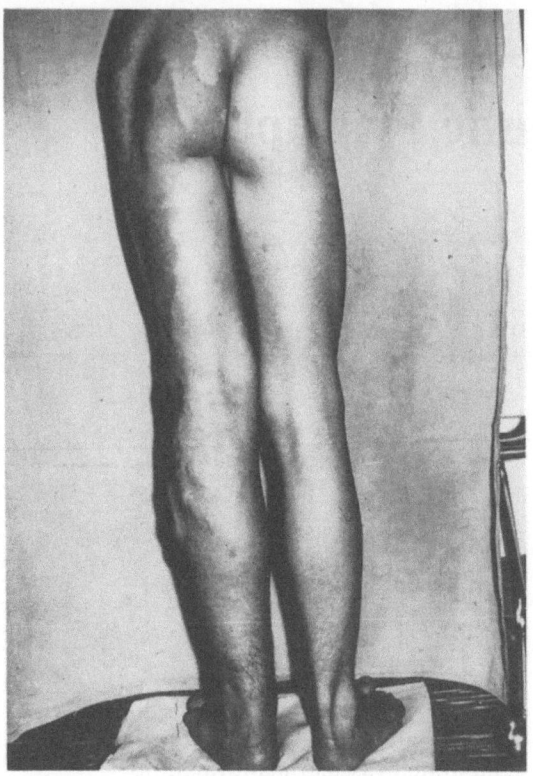

Fig. 38.1. Klippel–Trenaunay syndrome with persistence of the external vestigial vein.

High-Flow Arteriovenous Malformations

These lesions have all the features of arterio-venous fistulas. They are generally pulsatile varices of soft-tissue angiomas associated with cutaneous hyperaemia, bruits and thrills. Postural and compressive manoeuvres have little effect upon the size of these lesions. However, afferent arterial occlusion will immediately abolish or decrease the importance of arterio-venous fistulas.

Various types of vascular malformations may be associated with the Klippel–Trenaunay syndrome, which presents with cutaneous angioma, unilateral varices and hypertrophic bony lengthening of the limb (Fig. 38.1) [6].

Ancillary Investigations

The clinical diagnosis of angiodysplasia is corroborated and completed with a series of ancillary investigations.

Plain radiography may reveal phleboliths or evidence of intraosseous arteriovenous fistulas or angiomas (sometimes associated with neuro-fibromatosis).

Ultrasound examination should be performed prior to the more invasive methods [4]. Doppler echography permits the categorization of angiodysplasias as follows:

1. Sequestrated vascular malformations.
2. Venous malformations with a typical ultra-sound picture in response to postural changes or to the valsalva manoeuvre and with associated superficial or deep venous insufficiency.
3. Arteriovenous malformations with diastolic flow in the afferent vein. Doppler echography allows precise determination of the number and site of fistulas using compressive manoeuvres, the assessment of the significance of shunts and the identification of the arterial sources of complex fistulas.

Echotomography permits visualization of the structure, location, extent, nature (venous lakes, cavernous aspect) and content (phleboliths, thrombus) of soft-tissue angiomas. Afferent and efferent vessels can be identified. Anomalous veins, arterial and/or venous dilatation and areas of agenesis or hypoplasia can also be detected.

Arteriography is useful only for arteriovenous fistulas. Digital non-selective arteriography is sufficient to confirm the presence of an arteriovenous communication. Afferent vessels and the structure of the malformation are readily appreciated by injection of contrast material into the venous system during the arterial phase and by venous compression (this prevents the passage of contrast material from the arterial into the venous system) (Fig. 38.2). The successive compression of different radicles is the equivalent of selective angiography and permits the identification of major and minor afferent vessels and of the territory which they supply.

Phlebography is performed for three reasons:

1. Under free-flow conditions to evaluate the deep venous system.
2. To follow a vestigial crural vein to determine its mode of drainage (e.g. superficial femoral vein, deep femoral vein, common femoral vein, ischial or gluteal veins).
3. At the inferior pole of a malformation or in a superficial venous pocket using upstream and downstream compression to better appreciate the structure of the angiodysplastic lesion.

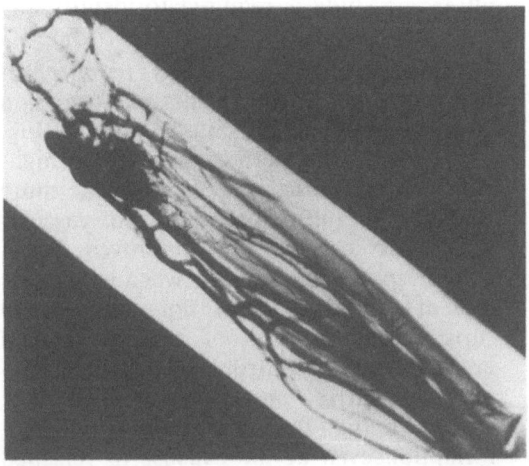

Fig. 38.2. Arteriography with venous compression revealing high-flow congenital arteriovenous communications in the forearm.

Computerized axial tomography performed with upstream and downstream compression is used to appreciate the depth of tissue invasion, particularly in the case of bone (Fig. 38.3).

The vascular lesions identified by these techniques may be associated with other abnormalities of neural crest structures, either at the ectodermal or at the mesodermal level. These anomalies are most frequently vascular (lymphoedema, haemolymphangioma), but other possibilities include syndactyly, supernumerary digits or bony malformations (agenesis, hypoplasia, genu varum or valgum, coxaplana or valga, etc.).

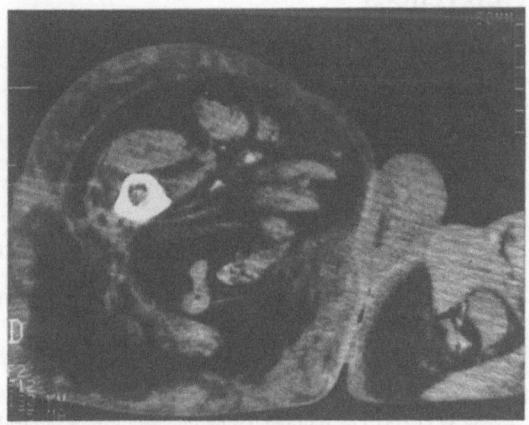

Fig. 38.3. Computerized axial tomography showing invasion of the thigh by haemodynamically venous angiodysplasia.

Associated abnormalities may be located at a distance from the site of angiodysplasia. In addition to other vascular malformations of the pulmonary, retinal or cerebral circulations urogenital malformations or phacomatosis (particularly neurofibromatosis) may be found. It is thus important to conduct a general search for associated lesions. A pelvic scan with injection of contrast medium may be required to detect extension of lower limb haemangiomas to the gluteal, pelvic or ischiorectal regions and to investigate the possibility of urogenital or rectal malformations necessitating cystoscopy or proctoscopy.

Evolution

The evolution of vascular malformations is episodic and related to trauma or to pregnancy. However, the fate of most of these lesions varies according to the haemodynamic subtype.

Sequestrated malformations may increase in volume or become painful due to thrombosis, but are otherwise relatively benign. Venous malformations may be associated with a variety of complications related to elevated venous pressure. These include stasis changes, limb hypertrophy and local thrombosis. When there is muscular involvement, repeated thromboses may eventually cause muscular atrophy or Volkman's contractures. Invasion of neural tissue is generally painful and joint involvement may lead to haemarthroses and stiffening. Low-flow arteriovenous malformations may remain quiescent for many years, but retain the potential for development into major arteriovenous shunts.

High-flow arteriovenous fisculas are associated with all of the complications mentioned for venous malformations. More importantly, they carry a risk of haemorrhage and eventual amputation if bleeding cannot be controlled.

Finally, in the newborn and the young child, local intravascular coagulation can occur causing defibrination and thrombocytopenia.

Treatment

Therapy is adapted to the haemodynamic subtype of the vascular malformation.

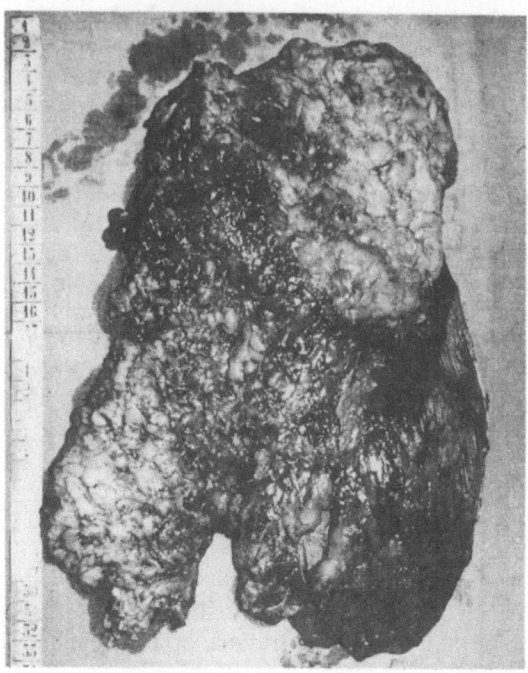

Fig. 38.4. Macroscopic view of an "en bloc" excision of a high-flow arteriovenous malformation within the quadriceps muscle.

Sequestrated haemangiomas are simply excised.

A number of different solutions are possible in the case of venous malformations:

1. Superficial or deep venous malformations may be removed "en bloc" with underlying muscle of little or no functional importance. In the case of malformations that cause major muscle dysfunction, excision followed by tendon transplantation is a possible solution.
2. Painful but voluminous malformations not amenable to complete excision may be treated with partial excision or with sclerosing agents to induce secondary thrombosis.
3. Major insufficiency of a lower limb vestigial vein is treated either by excision if this is possible, or by balloon occlusion of the afferent hypogastric or gluteal veins. Sclerosing agents may also be employed in some cases.

Arteriovenous malformations are the most problematic:

1. Arteriovenous communications involving major vessels are treated by direct ablation. Care must be taken to preserve flow in the normal parent vessels, and intra-operative

ultrasonic study is required to identify any accessory fistulas [2].

2. Complex angiodysplasia may require "en bloc" excision of the vascular malformation and adjacent "healthy" muscle or bone (Fig. 38.4). This technique is employed for high-flow malformations, and has become more widely applicable as procedures for vascular repair after excision have improved. When angiodysplasia is diffuse or when lesions are present in critical areas not amenable to surgery, embolization of afferent radicles under ultrasonic control is the most efficient, albeit palliative, approach [7]. This may convert a high-flow into a low-flow malformation even at the expense of repeated embolizations. Ligation of arterial afferent vessels or other bypass procedures are contraindicated since they invite the formation of collateral vessels which may worsen the haemodynamic picture. In extreme cases, inexorable progression and haemorrhage may necessitate the amputation of a limb.

Surgical intervention and therapeutic angiography should be undertaken with caution only for evolving lesions, since any intervention carries a risk of haemodynamic deterioration.

Regardless of the type of lesion, external compression is indispensable to prevent venous stasis. In children, epiphyseal growth plate ablation at a time dictated by growth charts may be advisable to avoid limb asymmetry.

References

1. André JM (1973) Les dysplasies vasculaires systématisées. L'Expansion Scientifique Francaise, Paris
2. Cormier JM, Laurian Cl, Franceschi Cl, Luizy F (1981) Traitement chirurgical des fistules artério-veineuses des membres sous controle ultrasonographique. Chirurgie 107:424–427
3. Fairbairn JF, Bernatz PE (1980) Arteriovenous fistulas. In: Juergens JL, Spittell JA, Fairbairn JF (eds) Peripheral vascular diseases. Saunders, Philadelphia, p 441
4. Franceschi Cl, Franco G, Luizy F, Tanitte M (1986) Précis d'échotomographie vasculaire. Vigot, Paris
5. Malan E (1974) Vascular malformations. Carlo Erba Foundation, Milan 1974
6. Phillips GN, Gordon DH, Martin EC, Haller JO, Casarella W (1978) The Klippel-Trenaunay syndrome: clinical and radiological aspects. Radiology 128:429–434
7. Riche MC, Melki JP, Reizine D, Merland JJ (1985) L'angiographie thérapeutique. In: Melki JP, Hovasse D, Riche MC, Chaufour J, Merland JJ (eds). L'angiographie numerisée. Techniques – indications. Vigot, Paris

Subject Index